Edited by
Julio Alvarez-Builla,
Juan Jose Vaquero,
and José Barluenga

Modern Heterocyclic Chemistry

Related Titles

Majumdar, K. C., Chattopadhyay, S. K. (eds.)

Heterocycles in Natural Product Synthesis

2011

ISBN: 978-3-527-32706-5

Eicher, T., Hauptmann, S., Speicher, A.

The Chemistry of Heterocycles

Structure, Reactions, Synthesis, and Applications

Third Edition

2011

ISBN: 978-3-527-32868-0

Yudin, A. K. (ed.)

Catalyzed Carbon-Heteroatom Bond Formation

2010

ISBN: 978-3-527-32428-6

Ma, S. (ed.)

Handbook of Cyclization Reactions

2009

ISBN: 978-3-527-32088-2

L. Ackermann (Ed.)

Modern Arylation Methods

2009

ISBN: 978-3-527-31937-4

H. Yamamoto, K. Ishihara (Eds.)

Acid Catalysis in Modern Organic Chemistry

2 Volumes

2008

ISBN. 978-3-527-31724-0

J. Royer (Ed.)

Asymmetric Synthesis of Nitrogen Heterocycles

2009

ISBN: 978-3-527-32036-9

Edited by
Julio Alvarez-Builla, Juan Jose Vaquero, and José Barluenga

Modern Heterocyclic Chemistry

Volume 1

WILEY-VCH Verlag GmbH & Co. KGaA

The Editors

Prof. Dr. Julio Alvarez-Builla
Universidad de Alcalá
Facultad de Farmacia
Dpto. de Química Organíca
Campus Universitario s.n.
Alcalá de Henares
28871 Madrid
Spain

Dr. Juan Jose Vaquero
Universidad de Alcalá
Dpto. de Química Organíca
Ctra. Madrid-Barcelona km 33
Alcalá de Henares
28871 Madrid
Spain

Prof. Dr. José Barluenga
Universidad de Oviedo
Instituto Universitario de
Química Organometálica "Enrique Moles"
33071 Oviedo
Spain

Library of Congress Card No.: applied for

British Library Cataloguing-in-Publication Data
A catalogue record for this book is available from the British Library.

Bibliographic information published by the Deutsche Nationalbibliothek
The Deutsche Nationalbibliothek lists this publication in the Deutsche Nationalbibliografie; detailed bibliographic data are available on the Internet at http://dnb.d-nb.de.

Typesetting Thomson Digital, Noida, India
Printing and Binding betz-druck GmbH, Darmstadt
Cover Design Schulz Grafik-Design, Fußgönheim

Printed in the Federal Republic of Germany
Printed on acid-free paper

Print ISBN: 978-3-527-33201-4
oBook ISBN: 978-3-527-63406-4

Contents

Volume 2

Volume 3

List of Contributors

Ramón Alajarín
Universidad de Alcalá
Departamento de Química Orgánica
Alcalá de Henares
28871 Madrid
Spain

Benito Alcaide
Universidad Complutense de Madrid
Facultad de Química
Departamento de Química Orgánica I
28040 Madrid
Spain

Pedro Almendros
Instituto de Química Orgánica General
(CSIC)
Juan de la Cierva, 3
28006 Madrid
Spain

Julio Alvárez-Builla
Universidad de Alcalá
Facultad de Farmacia
Departamento de Química Orgánica
Alcalá de Henares
28871 Madrid
Spain

Venkataramanarao G. Anand
Regional Research Laboratory (CSIR)
Chemical Sciences and Technology
Division
Photosciences and Photonics Section
Trivandrum 695 019
India

José Barluenga
Universidad de Oviedo
Instituto Universitario de Química
Organometálica "Enrique Moles"
Julián Clavería 8
33006 Oviedo
Spain

Miguel Bayod
Asturpharma S.A.
Peña Brava 23
Polígono Industrial Silvota
33192 Llanera, Asturias
Spain

Jan Bergman
Karolinska Institute
Department of Biosciences and
Nutrition
Unit of Organic Chemistry
Novum Research Park
141 57 Huddinge
Sweden

Ulhas Bhatt
Albany Molecular Research, Inc.
Albany, NY 12212
USA

Carolina Burgos
Universidad de Alcalá
Departamento de Química Orgánica
Alcalá de Henares
28871 Madrid
Spain

María-Paz Cabal
Universidad de Oviedo
Instituto Universitario de Química
Organometálica "Enrique Moles"
Julián Clavería 8
33006 Oviedo
Spain

Luis Castedo
Universidad de Santiago de Compostela
Facultad de Química
Departamento de Química Orgánica
15782 Santiago de Compostela
Spain

Tavarekere K. Chandrashekar
Regional Research Laboratory (CSIR)
Chemical Sciences and Technology
Division
Photosciences and Photonics Section
Trivandrum 695 019
India

and

Indian Institute of Technology
Department of Chemistry
Kanpur 208 016
India

Ugo Chiacchio
Università di Catania
Dipartimento di Scienze Chimiche
Viale Andrea Doria 6
95125 Catania
Italy

Stefano Cicchi
Università degli Studi di Firenze
Dipartimento di Chimica "Ugo Schiff"
via della Lastruccia 13
50019 Sesto Fiorentino-Firenze
Italy

Ana M. Cuadro
Universidad de Alcala
Departamento de Química Orgánica
Alcalá de Henares
28871 Madrid
Spain

José Elguero
University of Aveiro
Instituto de Química Médica (CSIC)
Department of Chemistry
Juan de la Cierva, 3
28006 Madrid
Spain

José L. García-lvarez
Universidad de Oviedo
Instituto Universitario de Química
Organometálica "Enrique Moles"
Departamento de Química Orgánica e
Inorgánica
Unidad asociada al CSIC
Julian Claveria 8
33006 Oviedo
Spain

Cristina Gómez de la Oliva
Instituto de Química Médica (CSIC)
Juan de la Cierva, 3
28006 Madrid
Spain

Concepción González-Bello
Universidad de Santiago de Compostela
Facultad de Ciencias
Departamento de Química Orgánica
27002 Lugo
Spain

Pilar Goya Laza
Instituto de Química Médica (CSIC)
Juan de la Cierva, 3
28006 Madrid
Spain

Bernardo Herradón
Instituto de Química Orgánica (CSIC)
Juan de la Cierva, 3
28006 Madrid
Spain

Xue-Long Hou
The Chinese Academy of Sciences
Shanghai Institute of Organic Chemistry
Shanghai-Hong Kong Joint Laboratory in Chemical Synthesis and State Key Laboratory of Organometallic Chemistry
354 Feng Lin Road
Shanghai 200032
China

Hui Huang
The Chinese Academy of Sciences
Shanghai Institute of Organic Chemistry
Shanghai-Hong Kong Joint Laboratory in Chemical Synthesis
354 Feng Lin Road
Shanghai 200032
China

Tomasz Janosik
Karolinska Institute
Department of Biosciences and Nutrition
Unit of Organic Chemistry
Novum Research Park
141 57 Huddinge
Sweden

Amparo Luna
Universidad Complutense de Madrid
Facultad de Química
Departamento de Química Orgánica I
28040 Madrid
Spain

Fabrizio Machetti
Instituto Chimica dei Composti Organometallica del CNR c/o
Dipartimento di Chimica Organica "Ugo Schiff"
Via Madonna del Piano 10
50019 Sesto Fiorentino (Firenze)
Italy

François Mathey
Nanyang Technical University
School of Physical and Mathematical Sciences
Division of Chemistry and Biological Chemistry
21 Nanyang Link
637371 Singapore
Singapore

S. Shaun Murphree
Allegheny College
Department of Chemistry
520 N, Main Street
Meadville, PA 16335
USA

Carmen Ochoa de Ocariz
Instituto de Química Médica (CSIC)
Juan de la Cierva, 3
28006 Madrid
Spain

Teresa M.V.D. Pinho e Melo
Universidade de Coimbra
Departamento de Química
3004-535 Coimbra
Portugal

Julia Revuelta
Instituto de Quimica Organica General (CSIC)
Grupo de Quimica Organica Biologica
C/Juan de la Cierva, 3
28006 Madrid
Spain

Sylvie Robin
Université de Paris-Sud
ICMMO
Laboratoire ed Synthèse Organique et Méthodologie
Université de Paris-Sud
91405 Orsay
France

Giovanni Romeo
Università di Messina
Dipartimento Farmaco-Chimico
Via SS Annunziata
98168 Messina
Italy

Gérard Rousseau
Université de Paris-Sud
ICMMO
Laboratoire ed Synthèse Organique et Méthodologie
Université de Paris-Sud
91405 Orsay
France

Javier Santamaría
Universidad de Oviedo
Instituto Universitario de Química Organometálica "Enrique Moles"
Departamento de Química Orgánica e Inorgánica
Unidad asociada al CSIC
Julian Claveria 8
33006 Oviedo
Spain

Artur M.S. Silva
University of Aveiro
Department of Chemistry
3810-193 Aveiro
Portugal

Alagar Srinivasan
Regional Research Laboratory (CSIR)
Chemical Sciences and Technology Division
Photosciences and Photonics Section
Trivandrum 695 019
India

Augusto C. Tomé
University of Aveiro
Department of Chemistry
3810-193 Aveiro
Portugal

Carlos Valdés
Universidad de Oviedo
Instituto Universitario de Química Organometálica "Enrique Moles"
Julián Clavería 8
33006 Oviedo
Spain

Juan J. Vaquero
Universidad de Alcala
Departamento de Química Orgánica
Alcalá de Henares
28871 Madrid
Spain

José M. Villalgordo
Villalpharma S.L.
Polígono Industrial Oeste
C/Paraguay, Parcela 7/5-A, Módulo A-1
30169 Murcia
Spain

David J. Wilkins
Key Organics Ltd.
Highfield Industrial State
Camelford
Cornwall PL32 9QZ
UK

Henry N.C. Wong
The Chinese University of Hong Kong
Institute of Chinese Medicine, and
Central Laboratory of the Institute of
Molecular Technology for Drug
Discovery and Synthesis
Department of Chemistry
Center of Novel Functional Molecules
Shatin, New Territories
Hong Kong SAR, China

and

The Chinese Academy of Sciences
Shanghai Institute of Organic
Chemistry
Shanghai-Hong Kong Joint Laboratory
in Chemical Synthesis
354 Feng Lin Road
Shanghai 200032
China

Jie Wu
Fudan University
Department of Chemistry
220 Handan Road
Shanghai 200433
China

Larry Yet
299 Georgetown Ct
Albany, NY 12203
USA

Kap-Sun Yeung
Bristol-Myers Squibb Pharmaceutical
Research Institute
5 Research Parkway
P.O. Box 5100
Wallingford, CT 06492
USA

1
Heterocyclic Compounds: An Introduction

Julio Alvárez-Builla and José Barluenga

1.1
Heterocyclic Compounds: An Introduction

The *IUPAC Gold Book* describes *heterocyclic compounds* as:

"Cyclic compounds having as ring members atoms of at least two different elements, e.g. quinoline, 1,2-thiazole, bicyclo[3.3.1]tetrasiloxane" [1].

Usually they are indicated as counterparts of *carbocyclic compounds,* which have only ring atoms from the same element. Another classical reference book, the *Encyclopaedia Britannica*, describes a heterocyclic compound, also called a *heterocycle,* as:

"Any of a class of organic compounds whose molecules contain one or more rings of atoms with at least one atom (the heteroatom) being an element other than carbon, most frequently oxygen, nitrogen, or sulfur" [2].

Although heterocyclic compounds may be inorganic, most contain within the ring structure at least one atom of carbon, and one or more elements such as sulfur, oxygen, or nitrogen [3]. Since non-carbons are usually considered to have replaced carbon atoms, they are called heteroatoms. The structures may consist of either aromatic or non-aromatic rings.

Heterocyclic chemistry is the branch of chemistry dealing with the synthesis, properties, and applications of heterocycles.

Heterocyclic derivatives, seen as a group, can be divided into two broad areas: aromatic and non-aromatic. In Figure 1.1, five-membered rings are shown in the first row, and the derivative **1** corresponds to the aromatic derivative, furan, while tetrahydrofuran (**2**), dihydrofuran-2-one (**3**), and dihydrofuran-2,5-dione (**4**) are not aromatic, and their reactivity would be not unlike that expected of an ether, an ester, or

Modern Heterocyclic Chemistry, First Edition.
Edited by Julio Alvarez-Builla, Juan Jose Vaquero, and José Barluenga.

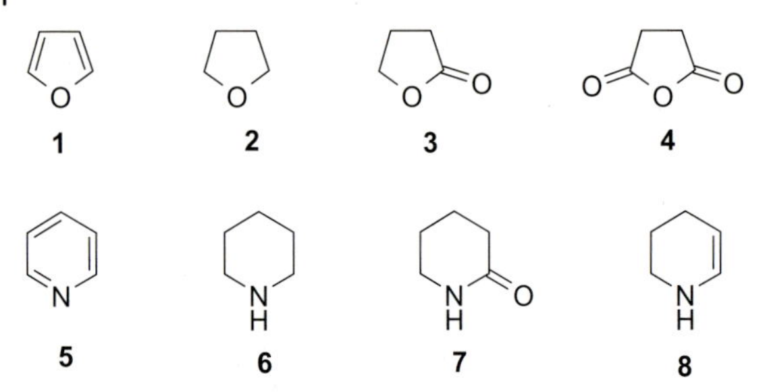

Figure 1.1 Examples of heterocyclic compounds.

a carboxylic anhydride, respectively. The second row shows six-membered rings, initially in an aromatic form as pyridine (**5**), while piperidine (**6**), piperidin-2-one (**7**), and 1,2,3,4-tetrahydropyridine (**8**) are not aromatic; their reactivity would not be very different from that expected of an amine, amide, or enamine, respectively. In general, the reactivity of aromatic heterocycles, which is a combination of that expected from an aromatic system combined with the influence of the heteroatoms involved, is usually more complex, while the reactivity of the non-aromatic systems is not too different from the usual non-cyclic derivatives. Thus, most books on heterocyclic chemistry are mainly devoted to the reactivity of aromatic compounds.

Tables 1.1–1.4 indicate models of the heterocyclic derivatives described in these volumes. Table 1.1 shows simple heterocyclic systems of three or four members. In this case, the literature examples are mainly non-aromatic, as indicated in the table, and the expected reactivity is always related to the ring strain present in all of them, which produces a release of energy when they are opened to give aliphatic products.

Table 1.1 Main three- and four-membered heterocycles.

Ring size	**Heteroatom**			
	N	**O**	**S**	**Other**
3	Aziridine	Oxirane	Thiirane	
	Diaziridine	Dioxirane		Oxaziridine
4	Azetidine	Oxetane	Thietane	Seletane Phosphetane

Table 1.2 Main five-membered heterocycles.

Ring size	Heteroatom				
	N	Benzo	O	Benzo	S
5	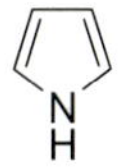 Pyrrole	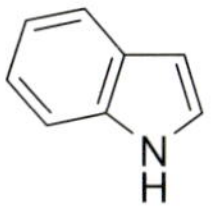 Indole	 Furan	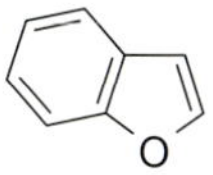 Benzofuran	 Thiophene
5	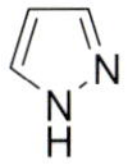 Pyrazole		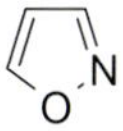 Isoxazole		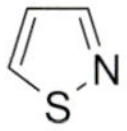 Isothiazole
	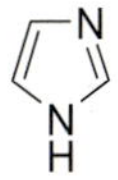 Imidazole		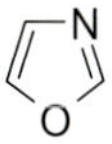 Oxazole		 Thiazole
	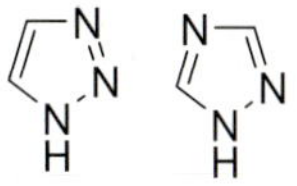 Triazoles		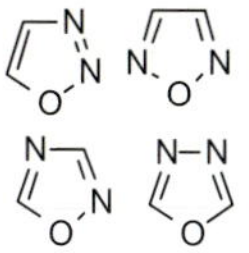 Oxadiazoles		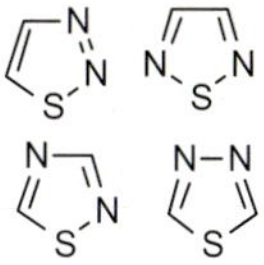 Thiadazoles
	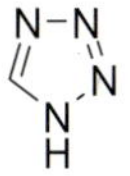 Tetrazole				

Table 1.2 indicates five-membered heterocyclic systems, such as pyrrole, furan, their benzo derivatives, and thiophene, and a set of heterocycles with more than one heteroatom, as 1,2-azoles, 1,3-azoles, triazoles, oxa- and thiadiazoles, and tetrazole.

Table 1.3 shows six-membered rings, namely, pyridine, its benzo derivatives quinoline and isoquinoline, the pyrilium cation, and, as in Table 1.2, other common heterocycles with more than one heteroatom, such as diazines, triazines, and tetrazines.

Finally, Table 1.4 shows the simplest seven-membered ring, that is, azepine and its benzo derivative, as well as examples of the nitrogen bridgehead bicyclic systems, pyrrolizine, indolizines, and quinolizinium cation.

Other additional chapters have been included with special systems relevant from different points of view: 2-azetidinones or β-lactams, benzodiazepines, and two general chapters on new materials based on heterocyclic systems and solid phase and combinatorial chemistry related to heterocyclic derivatives.

Table 1.3 Main six-membered heterocycles.

Ring size	N	Benzo	O
6	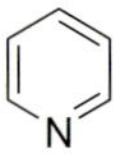 Pyridine	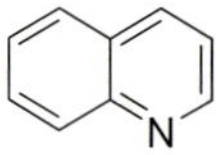 Quinoline 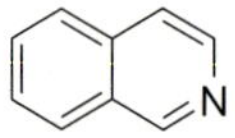 Isoquinoline	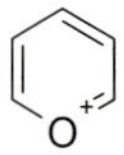 Pyrilium
	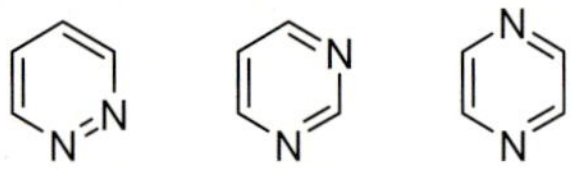 Diazines Pyridazine Pyrimidine Pyrazine		
	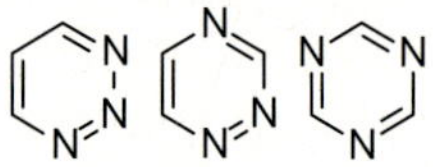 Triazines		
	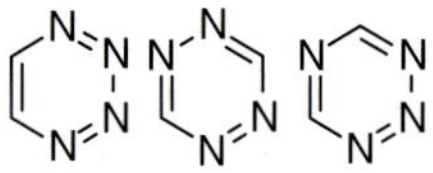 Tetrazines		

Table 1.4 Other simple heterocycles.

Ring size	Heteroatom	
	N	Benzo
7	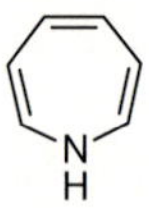 Azepine	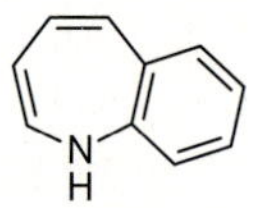 Benzoazepine
5-5, 5-6, 6-6	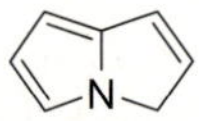 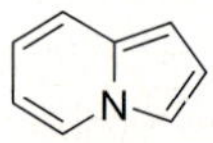 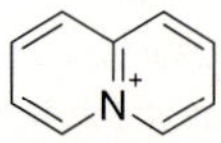 Pyrroline Indolizine Quinolizinium	

1.2
Structure and Reactivity of Aromatic Five-Membered Systems

As is indicated in most handbooks of heterocyclic chemistry [3, 4], a pictorial valence bond resonance description is used in most chapters, as a simple way to rationalize the reactivity of the most important aromatic heterocycles. Two examples are described in detail as representative of most of the aromatic rings considered: pyrrole as a model of the π-excessive rings, and pyridine as a model of the π-deficient ones.

Pyrrole has a structure that is isoelectronic with the cyclopentadienyl anion, but is electrically neutral, having a nitrogen atom with a pair of electrons, which is part of the aromatic sextet, and its resonance hybrid can be represented as a combination of main forms I–V (Scheme 1.1), one without charge, and the others with charge separation. As expected, not all forms contribute equally to the structure of the pyrrole, with the order of importance being I > III, IV > II, V, that is, the major contribution is produced by the non-charged form, and, of the charged ones, those in which the nitrogen is using its lone pair of electrons. As a combination of all forms, structure **9** indicates how the heteroatom bears a partial positive charge, while the carbon positions show an increase in electronic density, compared with the typical aromatic system, benzene. Thus, a π-excessive system such as pyrrole would be easily attacked by electrophiles and not by nucleophiles.

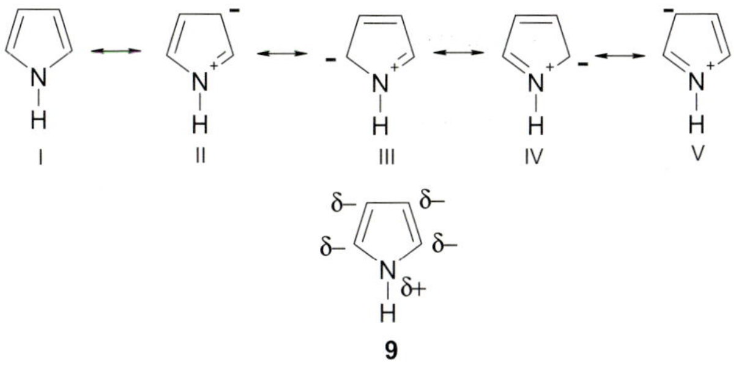

Scheme 1.1 Resonance hybrids of pyrrole.

Scheme 1.2 indicates how the attack of an electrophile usually proceeds. The major isomer **13** is formed through intermediates **10–11–12**, of which the intermediate **10** contributes most to the stabilization of the intermediate. Alternatively, a minor isomer **16** is produced through the less stable intermediates **14** and **15**.

Alternatively, Scheme 1.3 shows the attack of a nucleophile on pyrrole. Intermediate **17** is not stabilized, and the lone pair of electrons on the heteroatom does not contribute to the progress of the process. The only process that usually can be detected is deprotonation of the N–H bond to generate the pyrrolate (**18**), which can be used to make a bond with a suitable electrophile (i.e., an alkyl halide) to produce the N-substituted pyrrole **19**.

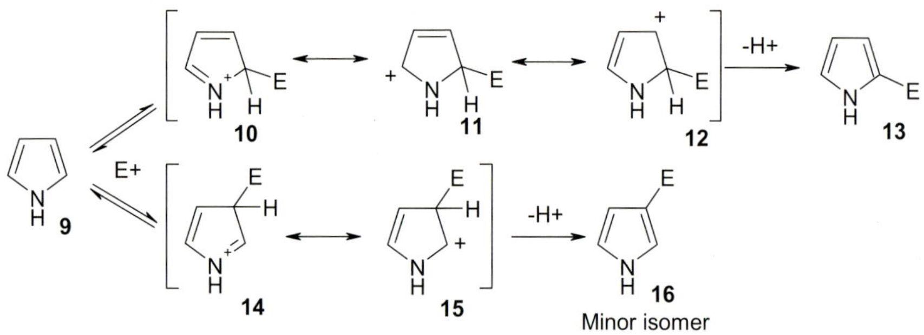

Scheme 1.2 Electrophilic attack on pyrrole.

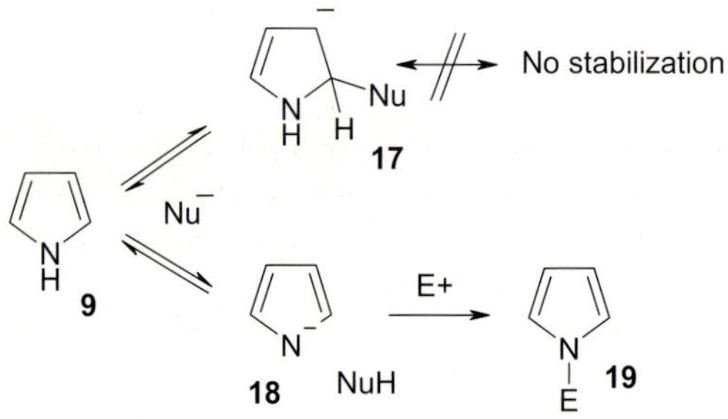

Scheme 1.3 Attack on pyrrole by nucleophiles.

This behavior can be extended with small differences to other π-excessive heterocycles, with the limit due to the existence or not of a N–H bond at position 1. In the case of rings like thiazole or isoxazole, the lack of the acidic bond makes the process **9–18–19** impossible. Attack by radicals or complex organometallic reagents are more complex and are discussed in every chapter.

1.3
Structure and Reactivity of Aromatic Six-Membered Systems

The structure of pyridine is analogous to that of benzene, with one of the carbons replaced by a nitrogen atom. This produces alterations in the geometry, which is no longer perfectly hexagonal, due to the shorter CN bonds; the existence of an unshared pair of electrons, not related with the aromatic sextet, gives the pyridine basic character, along with a permanent dipole in the ring, due to the electronegative character of the heteroatom compared with carbon.

Scheme 1.4 indicates the main canonical forms (I–V) that contribute to the resonance hybrid of the structure of pyridine. Obviously, not all of them contribute equally – the two Kekulé forms I and II, which are not charged, are the more stable

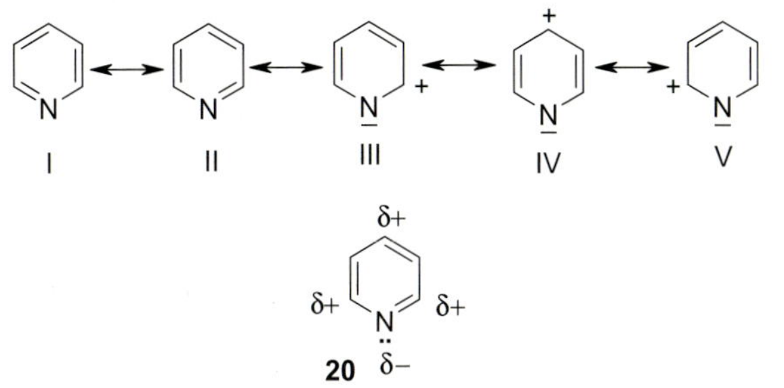

Scheme 1.4 Resonance hybrids of pyridine.

forms, followed by those in which nitrogen is negatively charged. Other forms can be envisaged, but their contributions can be neglected. Thus, the combination of the main forms can be represented as structure **20**, in which the nitrogen bears a partial negative charge, and positions 2, 4, and 6 are electron deficient; usually, positions 2 and 6 are the most deficient due to the inductive effect produced by the heteroatom. Positions 3 and 5 can be considered neutral, comparable to benzene carbons. Thus, the more characteristic reactivity of the pyridine ring would be against nucleophiles, which would attack the more electron-deficient positions.

As expected from the structure of pyridine, Scheme 1.5 describes the attack of a nucleophile on the system. The main process goes through intermediates **21–22–23** to produce the major isomer **24**, substituted at position 2. Alternatively, attack can also occur at position 4, through intermediates **25–26–27**, yielding the minor isomer **28**.

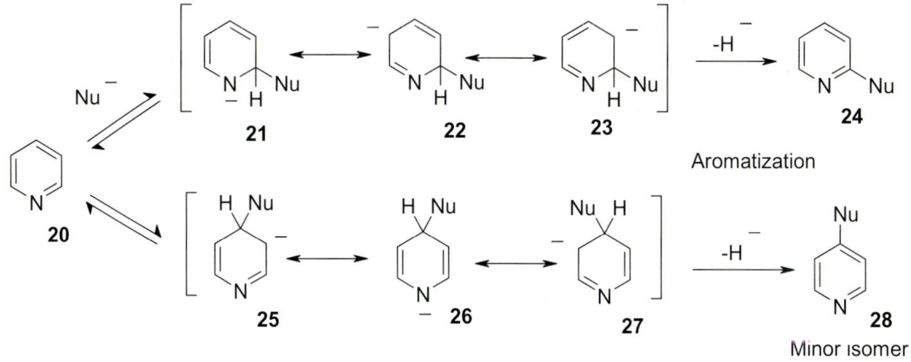

Scheme 1.5 Nucleophilic attack on pyridine.

Scheme 1.6 describes an electrophilic attack on pyridine. The initial attack of the electrophile usually takes place on the pyridine nitrogen. When the attacking species can produce a stable bond, the product should be the pyridinium salt **33**, but when this product is not stable enough the process goes through intermediates **29–30–31**, that is, by attacking the neutral carbons, to produce the 3-substituted derivative **32**. As a general view, the reactivity of pyridine can be taken as a model for other π-deficient systems, and can be easily extended to diazines, triazines, or pyrilium derivatives. Other processes, like radical attack or reaction with complex organometallic reagents are described in every chapter.

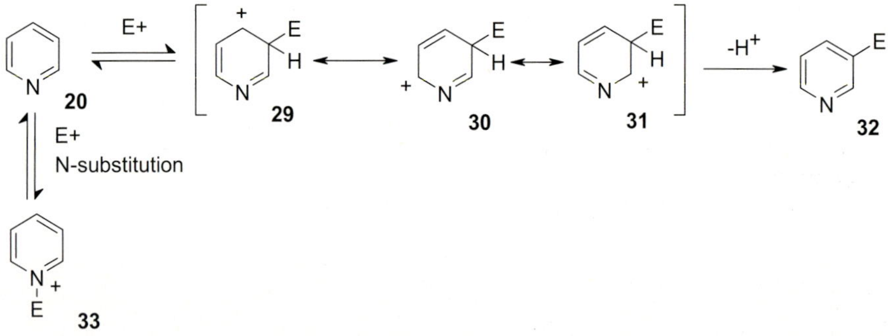

E+ -H+ N-substitution 20 29 30 31 32 33

Scheme 1.6 Electrophilic attack on pyridine.

1.4 Basic Literature on Heterocyclic Compounds

To introduce the recent literature in heterocyclic chemistry, it is necessary to indicate, among the textbooks available [3–6], two of them: one [3] from Eicher and Hauptmann with a highly structured organization, which is simple and efficient and can be used as the basis of a heterocyclic course. The other [4], from Joule and Mills, combines the condensed format with extensive information about the basic heterocycles considered. As reference books, it is necessary to cite the collection *Comprehensive Heterocyclic Chemistry* from Katritzky and colleagues [7–9]; this is associated with the *Handbook of Heterocyclic Chemistry* [10], which is regularly updated with the *Comprehensive* edition. Heterocyclic series are also of great interest, becoming readable collections that allow an update of the literature in the field. *Progress in Heterocyclic Chemistry* [11] describes mostly the advances in every relevant field of heterocyclic chemistry in a yearly volume. The series of monographs *Advances in Heterocyclic Chemistry* [12], which consists of 101 volumes to date, covers in depth very different topics in the field.

Other recent monographs are of interest in various topics on the field, a good guide called *Name Reactions in Heterocyclic Chemistry* has been given by Li [13] and the monograph *Aromaticity in Heterocyclic Compounds* [14] is also a good basic help for heterocyclic chemists, as is the *Synthesis of Heterocycles via Multicomponent Reactions* [15]. Other recent monographs have centered on synthetic techniques such as palladium chemistry [16], chemistry of heterocyclic carbenes [17–19], or synthesis with microwaves [20]. In addition, a recent monograph on general heterocyclic chemistry emphasizes the importance of heterocyclic compounds in the field of medicinal chemistry and natural products [21].

References

1 IUPAC (2009) IUPAC Compendium of Chemical Terminology - the Gold Book heterocyclic compounds: http://goldbook.iupac.org/H02798.html (accessed on).

2 Encyclopædia Britannica, Encyclopædia Britannica Online. The official website: heterocyclic compound http://www.britannica.com (accessed on).

3 Eicher, T. and Hauptmann, S. (2003) *The Chemistry of Heterocycles: Structure, Reactions, Syntheses, and Applications*, 2nd edn, Wiley-VCH Verlag GmbH, Weinheim.

4 Joule, J.A. and Mills, K. (2000) *Heterocyclic Chemistry*, 4th edn, Blackwell, Oxford.

5 Sainsbury, M. (2002) *Heterocyclic Chemistry*, Royal Society of Chemistry, Cambridge.

6 Nylund, K., Johansson, P., Puterova, Z., and Krutosikova, A. (2010) *Heterocyclic Compounds: Synthesis, Properties and Applications*, Nova Science Publishers, Hauppauge, New York.

7 Katritzky, A.R. and Rees, C.W. (eds) (1984) *Comprehensive Heterocyclic Chemistry I*, Pergamon Press, Oxford.

8 Katritzky, A.R., Rees, C.W. and Scriven, F.V. (eds) (1996) *Comprehensive Heterocyclic Chemistry II*, Pergamon Press, Oxford.

9 Katritzky, A.R., Ramsden, C.A., Scriven, E.F.V., and Taylor, R.J.K. (2008) *Comprehensive Heterocyclic Chemistry III*, Elsevier, Amsterdam.

10 Katritzky, A.R., Ramsden, C.A., Joule, J.A., and Zhdankin, V.V. (2010) *Handbook of Heterocyclic Chemistry*, 3rd edn, Elsevier, Amsterdam.

11 Gribble, G.W. and Joule, J. (eds) (2009) *Progress in Heterocyclic Chemistry*, Elsevier, Amsterdam.

12 Katritzky, A.R. (ed.) (2010) *Advances in Heterocyclic Chemistry*, Elsevier, Amsterdam.

13 Li, J.J. (2004) *Name Reactions in Heterocyclic Chemistry*, John Wiley & Sons, Inc., Hoboken, New Jersey.

14 Krygowski, T.M. and Cyranski, M.K. (2009) *Aromaticity in Heterocyclic Compounds (Topics in Heterocyclic Chemistry)*, Springer, Heidelberg.

15 Orru, R.V.A., Ruijter, E., and Maes, B.U.W. (2010) *Synthesis of Heterocycles via Multicomponent Reactions I (Topics in Heterocyclic Chemistry)*, Springer, Heidelberg.

16 Li, J.J. and Gribble, G.W. (2006) *Palladium in Heterocyclic Chemistry: A Guide for the Synthetic Chemist*, 2nd edn, Pergamon, Amsterdam.

17 Nolan, S.P. (2006) *N-Heterocyclic Carbenes in Synthesis*, Wiley-VCH Verlag GmbH, Weinheim.

18 Kühl, O. (2010) *Functionalised N-Heterocyclic Carbene Complexes*, John Wiley & Sons, Ltd., Chichester.

19 McGuinness, D. (2009) *Heterocyclic Carbene Complexes: Reaction Chemistry and Catalytic Applications*, Lambert Academic Publishing, Koeln.

20 Van der Eycken, E. and Kappe, C.O. (2006) *Microwave-Assisted Synthesis of Heterocycles (Topics in Heterocyclic Chemistry)*, Springer, Heidelberg, http://www.amazon.com/Comprehensive-Heterocyclic-Chemistry-III-15-/dp/0080449913/ref=sr_1_26?s=books&ie=UTF8&qid=1280185250&sr=1-26.

21 Quin, L.D. and Tyrell, J. (2010) *Fundamentals of Heterocyclic Chemistry: Importance in Nature and in the Synthesis of Pharmaceuticals*, John Wiley & Sons, Inc., Hoboken, New Jersey.

2
Three-Membered Heterocycles. Structure and Reactivity

S. Shaun Murphree

2.1
Aziridines

The field of aziridine chemistry is brimming with activity and, as a consequence, it has been the subject of multiple recent reviews [1–5]. Of particular note are the efficient and engaging article by Sweeney [6], a brief overview of properties and chemistry [7] and a newer monograph with a more encyclopedic sweep [8]. The present work aims not to be comprehensive, but rather to provide a general landscape and to capture the spirit of best practices available to the synthetic chemist, with an emphasis on preparative utility.

2.1.1
Properties of Aziridines

The smallest and most functionally spartan of the nitrogen heterocycles, aziridine (**1**), is an isolable liquid at room temperature, but prone to polymerization and other thermal degradation pathways because of its inherent ring strain [9]. It is weakly basic, with a pK_a of 7.98 [7], and the basicity trends of variously substituted aziridines have been the subject of a recent theoretical study [10]. From the standpoint of molecular geometry, aziridine describes an almost equilateral triangle, with a C–N–C bond angle of 60.58° (Figure 2.1) [11]. The N-inversion energy is relatively high, at almost 17 kcal mol^{-1}; however, conjugating substituents decrease the barrier significantly [12]. The ^{1}H-NMR signals of the methylene protons are centered at 1.4 ppm, while the carbons resonate at about 27 ppm. Coupling constants range from 3.8 Hz for trans vicinal ^{1}H-^{1}H coupling, to 6.3 Hz for cis vicinal ^{1}H-^{1}H coupling, and 168.1 Hz for ^{1}H-^{13}C coupling [13].

Aziridine moieties are imbedded in structurally diverse natural products from various sources (Figure 2.2), and the reader is directed to Lowden's excellent recent treatise on this topic [14]. Isolated from *Streptomyces griseofuscus* [15, 16], azinomycin A (**2**) and azinomycin B (**3**, also known as carzinophilin) are among the most intensely studied members of this class [17–19]. As potent antitumor agents, their

Modern Heterocyclic Chemistry, First Edition.
Edited by Julio Alvarez-Builla, Juan Jose Vaquero, and José Barluenga.

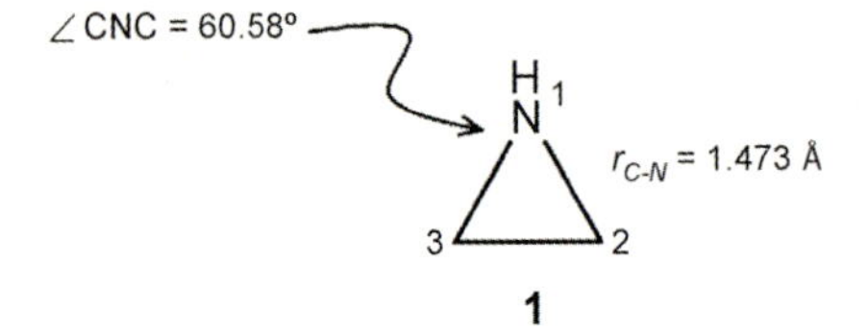

Figure 2.1 Geometry of aziridine.

biological activity springs from an ability to form interstrand purine base crosslinks in duplex DNA [20–23]. Central to this behavior is the aziridine ring bound up in the 1-aza-bicyclo[3.1.0]hexane system, a structural feature also found in ficellomycin (**4**), an antibacterial isolated from *Streptomyces ficellus* [24]. A similar 3,6-diaza-bicyclo [3.1.0]hexane system is at the functional heart of mitomycin C (**5**), a notable representative of the mitosanes extracted from *Streptomyces verticillatus* [25] and the target of many synthetic studies [26]. For decades, mitomycin C has found place in the arsenal of clinically relevant antibiotic and anti-tumor drugs, and the mitomycins have inspired studies into many promising non-natural analogs [27].

Also equipped with a 2,3-dialkylaziridine residue is the protease inhibitor miraziridine A (**6**), which is isolated from the marine sponge *Theonella* aff. *mirabilis* [28] and which exhibits a linear peptide structure vaguely similar to madurastatin A1 (**7**), a compound demonstrated in a culture of a pathogenic *Actinomadura madurae IFM 0745* strain, which shows activity against *Micrococcus luteus* [29]. An even more exposed aziridine ring is seen in the azicemicins A (**8**) and B (**9**), antibacterials isolated from *Amycolatopsis* sp. *Mj126-NF4* [30, 31]. These naturally occurring aziridine alkaloids have also inspired a genre of semi-synthetic and synthetic analogs of medicinal interest [32].

2.1.2 Synthesis of Aziridines

Generally speaking, there are three major synthetic routes to the aziridines (Scheme 2.1): (a) the addition of monovalent nitrogen species to alkenes; (b) the addition of divalent carbon centers to imines; and (c) the N-alkylative ring closure of amines equipped with β-leaving groups . A fourth route (pathway d) is less frequently encountered, but is nevertheless included here because of its potential synthetic utility. Enantioselective protocols in the first two categories have been the subject of a review [33], and aziridine synthesis as a whole has been more generally summarized by Sweeney [34].

2.1.2.1 Aziridination of Alkenes

Analogous to epoxidation, in which olefins react with electrophilic oxygen reagents, the aziridination of alkenes involves the addition of nitrenes (or nitrenoids) to a β-bond. Common nitrene precursors include [*N*-(*p*-toluenesulfonyl)imino]phenyliodinane (PhI=NTs), *N*-chloro-*p*-toluenesulfonamide sodium salt (Chloramine-T), and

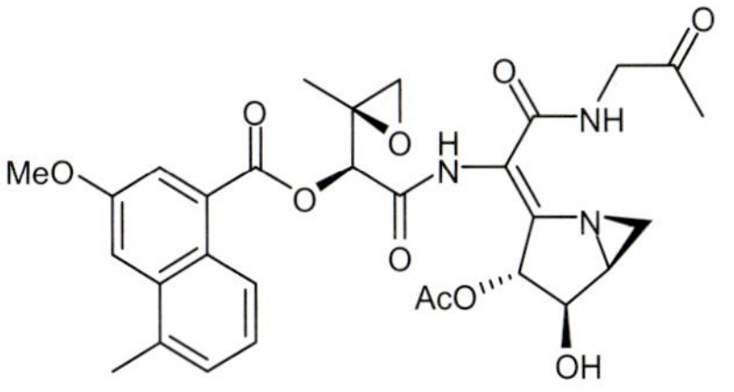

2, Azinomycin A

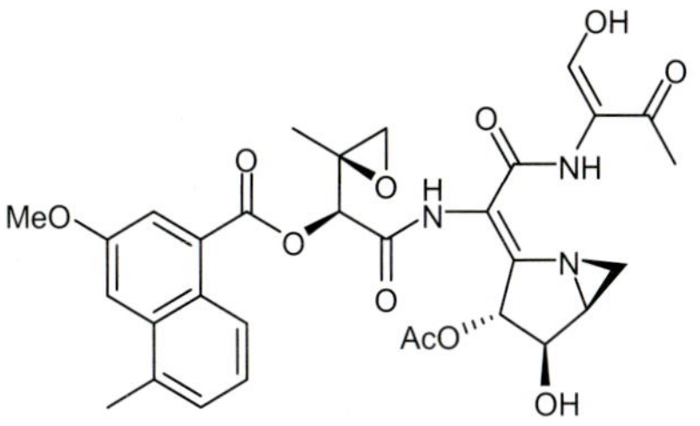

3, Azinomycin B

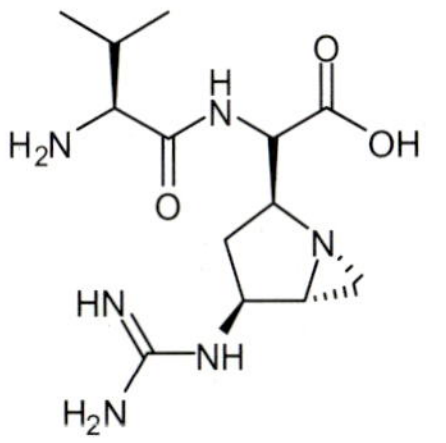

4, Ficellomycin

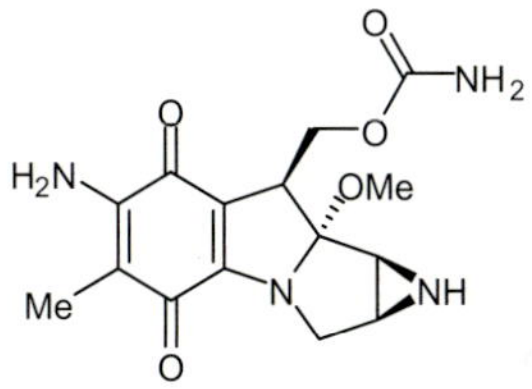

5, Mitomycin C

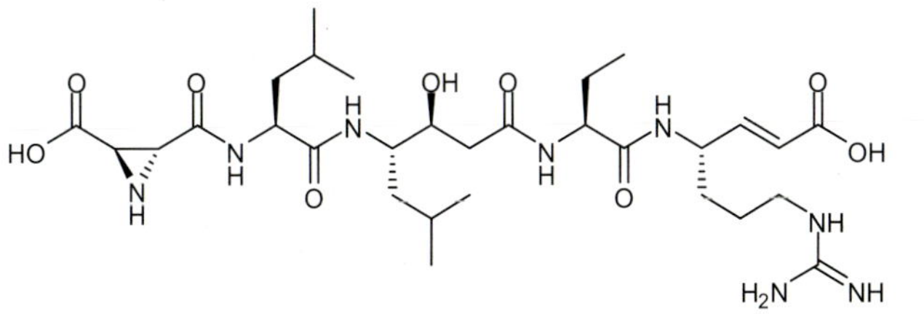

6, Miraziridine A

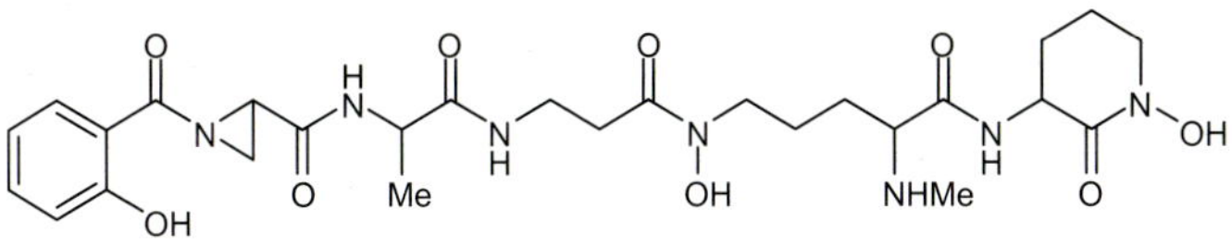

7, Madurastatin A1

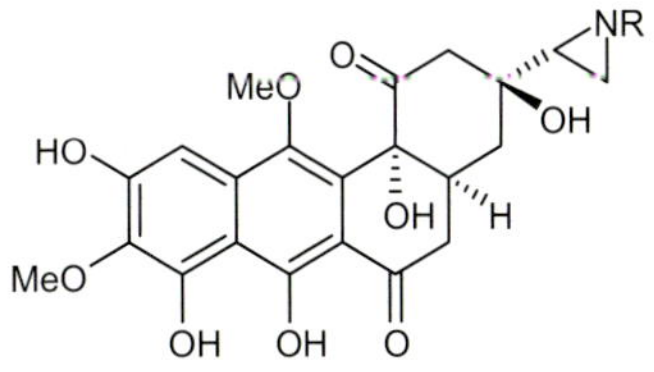

8, Aicemicin A, R=H
9, Aicemicin B, R=Me

Figure 2.2 Some naturally occurring aziridines.

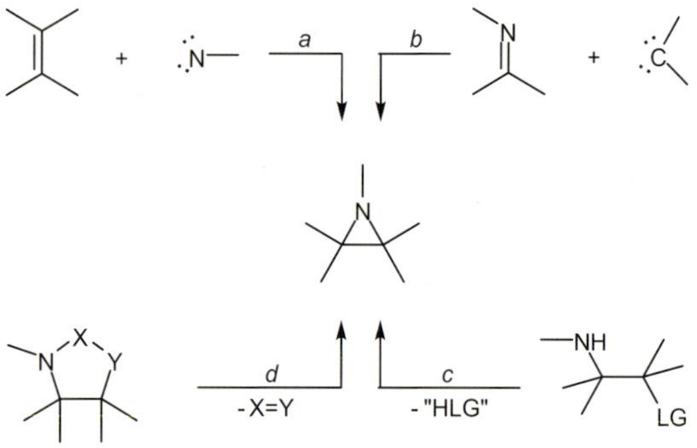

a
b
d
c
-X=Y
-"HLG"
NH
LG

Scheme 2.1 General synthetic routes to aziridines.

its N-bromo analogue (Bromamine-T), although use of the more esoteric *N*-iodo-*N*-potassio-*p*-toluenesulfonamide (TsN·KI) has also been reported [35]. These precursors can be activated using various transition-metal catalysts (Figure 2.3).

Also mirroring epoxidation methodology, a common model reaction is the aziridination of styrene (Scheme 2.2). Some illustrative examples are summarized in Table 2.1. For example, the copper(I) complex of a fluorinated tris(pyrazolyl)borate (or homoscorpionate) ligand forms an adduct with ethylene (**12a**), which catalyzes the aziridination of styrene with great efficiency using PhI=NTs as a nitrene precursor [36]. The more readily available Chloramine-T can be used effectively in the presence of methyl homoscorpionate complex **12b**, even with equimolar charges of olefin and nitrene precursor [37]. This catalyst motif has been incorporated into a heterogeneous system [38]. Other interesting copper(I) catalysts include those derived from pyridyl-1,5-diazacyclooctanes (e.g., **13**) [39] and bispidones (e.g., **14**) [40]. A particularly intriguing protocol using copper(I)iodide under aqueous phase-transfer conditions (entry 5) has also been reported [41].

Examples of copper(II) catalysts include the 1,4,7-triazacyclononane complex **15**, which requires a rather large excess of olefin [42], copper(II) acetylacetonate immobilized in ionic liquids (entry 7), which can be recycled many times without loss of activity [43], and Cu^{2+} exchanged zeolite Y (CuHY) in acetonitrile (entry 8), which allows for respectable conversion using almost equimolar alkene : nitrene ratios [44]. Other transition metals can be used to advantage, as well. For example, the fluorinated iron(III) porphyrin catalyst **16a** [45], although certainly dearer to synthesize, exhibits marked advantages over its older manganese-based cousin **16b** [46]; and even iron(II) triflate is effective in promoting high-yielding aziridination reactions [47] (Table 2.1, entry 11).

In the realm of precious metals, a novel and structurally interesting disilver(I) complex (**17**) has been shown to function as a competent catalyst in aziridination, a process that may involve high-valent silver intermediates [48]. Some polymer-supported ruthenium porphyrin catalysts have been employed for this transformation; however, conversions tend to be low [51, 52].

The use of a metal catalyst can be circumvented in some cases. In one particularly convenient example, a nitrene precursor is generated *in situ* from *p*-toluenesulfo-

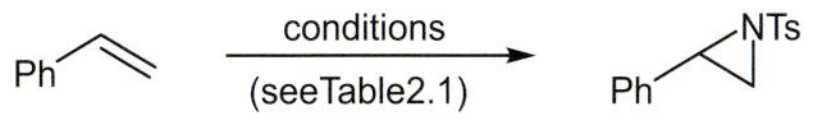

Ph/= —conditions (seeTable2.1)→ Ph–aziridine(NTs)

10 → 11

Scheme 2.2 Aziridination of styrene.

namide using iodobenzene diacetate. The aziridination is facilitated by substoichiometric quantities of iodine [49] (Table 2.1, entry 13). A conceptually related protocol is carried out using t-butylhypoiodite, prepared *in situ* from t-butylhypochlorite and sodium iodide [50] (Table 2.1, entry 14).

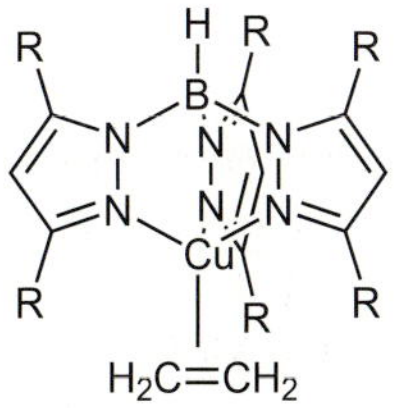

12a, R = CF_3
12b, R = Me

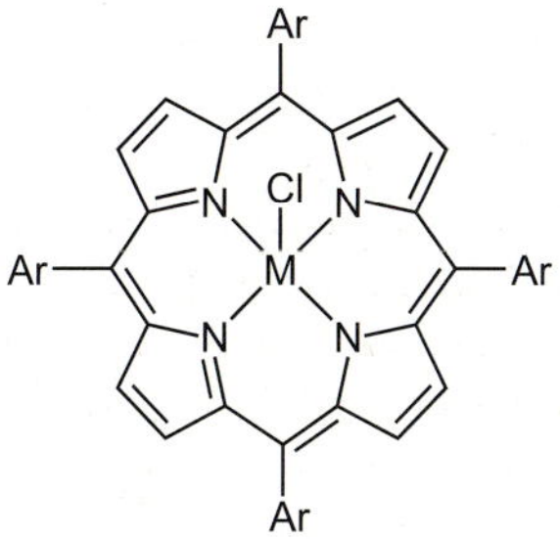

16a, Ar = C_6F_5; M = Fe^{III}
16b, Ar = C_6H_5; M = Mn^{III}

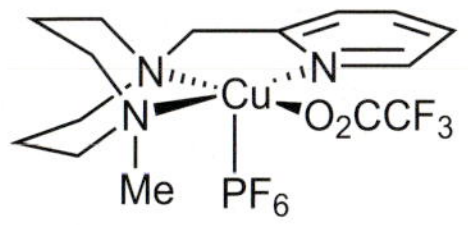

13

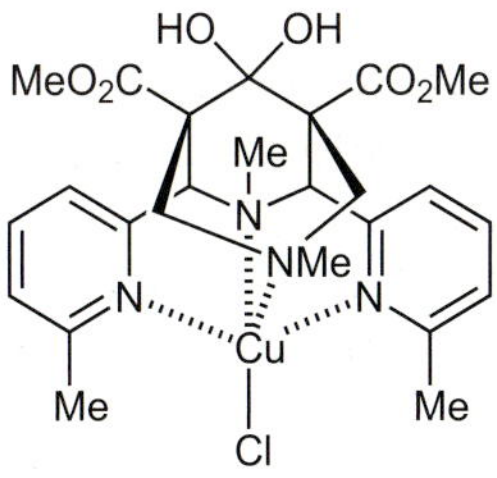

14

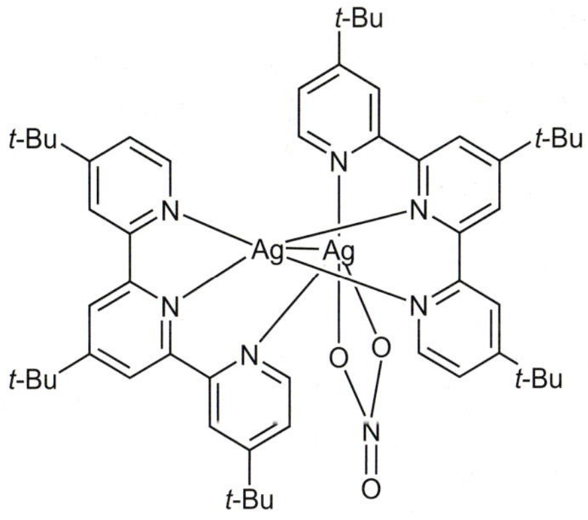

17

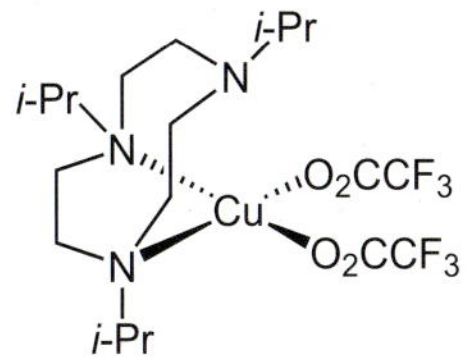

15

Figure 2.3 Representative catalysts for racemic aziridination.

Table 2.1 Reaction conditions for styrene aziridination.

Entry	Catalyst (mol.%)[a]	Nitrene source	Styrene : nitrene	Solvent	Time (h)	Yield (%)[a]	Reference
Copper(I) catalysts							
1	**12a** (5)	PhI=NTs	1 : 1.5	CH_3CN	16	99	[32]
2	**12b** (5)	Chloramine-T	1 : 1	CH_3CN	n.r.	84	[33]
3	**13** (5)	PhI=NTs	3.8 : 1	CH_3CN	1.5	99	[34]
4	**14** (3.5)	PhI=NTs	2 : 1	CH_3CN	7	80	[35]
5	CuI(10)	Chloramine-T	2 : 1	H_2O[b]	3	91	[36]
Copper(II) catalysts							
6	**15** (5)	PhI=NTs	20 : 1	CH_3CN	16	96	[37]
7	$Cu(acac)_2$ (8)[c]	PhI=NTs	5 : 1	CH_3CN	1	95	[38]
8	CuHY	PhI=NTs	1 : 1.5	CH_3CN	3	86	[39]
Other metal catalysts							
9	**16a** (5)	Bromamine-T	5 : 1	CH_3CN	12	80	[40]
10	**16b** (5)	PhI=NTs	100 : 1	CH_2Cl_2	n.r.	80	[41]
11	$Fe(OTf)_2$ (5)	PhI=NTs	7 : 1	CH_3CN	3	82	[47]
12	**17** (2)	PhI=NTs	5 : 1	CH_3CN	6	91	[42]
No metal catalyst							
13	none	$TsNH_2/PhI(OAc)_2/I_2/t$-BuOK	1 : 3	DCE	2	88	[49]
14	none	$TsNH_2/t$-BuOCl/NaI	2 : 1	CH_3CN	5	95	[50]

a) Based on nitrene source.
b) Bu_4NBr used as PTC.
c) Immobilized in $bmimBF_4$.

Progress continues to be made in the asymmetric aziridination of olefins using the same general approach (Scheme 2.3), but with chiral catalyst systems (Table 2.2). For example, impressive enantioselectivity has been reported for copper-exchanged zeolite Y (CuHY) modified with the chiral bis(oxazoline) **18a** (Figure 2.4) using [*N*-(*p*-nitrosulfonyl)imino]phenyliodinane (PhI=NNs) as the nitrene precursor [53]. Inferior results are obtained when [*N*-(*p*-toluenesulfonyl)imino]phenyliodinane (PhI=NTs) is used [54]. A one-pot homogeneous variant using bis(oxazoline) **18b** and commercially available iodobenzene diacetate has also been reported [55]. Evidence suggests that the ultimate stereochemical outcome may be affected by a secondary reaction between the aziridines formed and other components in the reaction mixture [56].

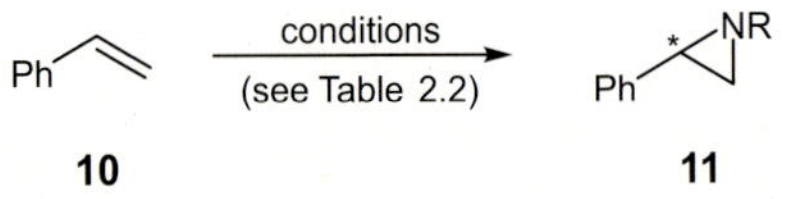

Ph **10** — conditions (see Table 2.2) → Ph **11** (NR, *)

Scheme 2.3 Asymmetric aziridination of styrene.

Table 2.2 Reaction conditions for asymmetric styrene aziridination.

Entry	Catalyst (mol.%)[a]	Nitrene source	Styrene : nitrene	Solvent	Time (h)	Yield (%)[a]	ee (%)	Reference
1	CuHY + **18** (7%)	PhI=NNs	1 : 1.3	CH_3CN	16	82	91	[45]
2	**19** (5)	PhI=NTs	1 : 5	CH_2Cl_2	n.r.	76	94	[49]
3	**20**(0.1)	TsN_3	1 : 1	CH_2Cl_2	24	78	85	[51]

a) based on nitrene source.

Another major avenue for enantioselective aziridination is offered through the use of (salen)manganese(III) complexes, such as the Katsuki catalyst (**19**) [57]. Evidence from the Jacobsen group suggests that the high enantiofacial selectivity observed for aryl alkenes may derive from well-defined bidentate aromatic interactions between substrate and catalyst [58]. Analogous ruthenium-based catalysts (e.g., **20**) allow for the use of tosyl azide as a nitrene precursor and are effective even at extremely low catalyst loadings [59]. Metalloporphyrin catalysts continue to show some promise for asymmetric aziridination, although enantioselectivities remain modest [60].

Some very convenient methodology has developed around bromine-catalyzed aziridination reactions using Chloramine-T as the source of electrophilic nitrogen

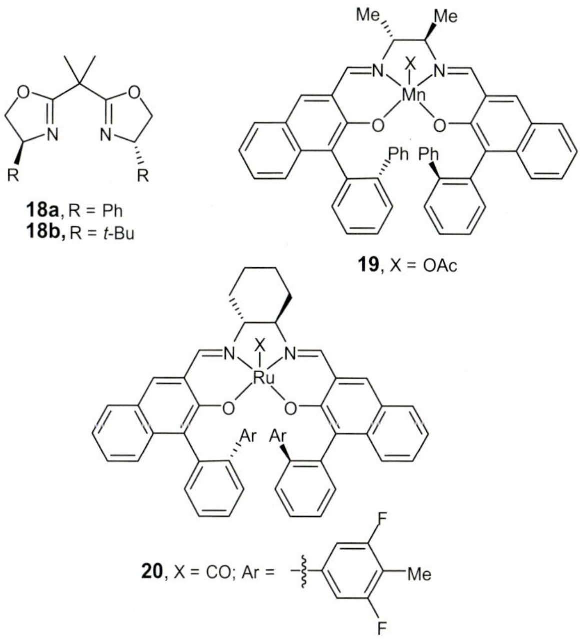

18a, R = Ph
18b, R = *t*-Bu

19, X = OAc

20, X = CO; Ar =

Figure 2.4 Chiral catalysts for asymmetric alkene aziridination.

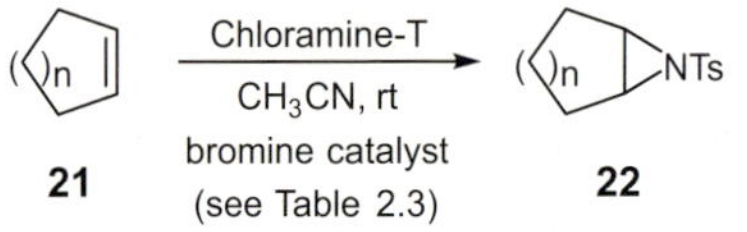

Scheme 2.4 Bromine-catalyzed aziridination.

(Scheme 2.4). For example, when cyclohexene is treated with Chloramine-T trihydrate in the presence of substoichiometric quantities of hydrogen peroxide and hydrobromic acid, the corresponding bicyclic aziridine is produced in good yield (Table 2.3, entry 1). The process involves the *in situ* generation of hypobromous acid, which in turn gives rise to bromonium intermediates [61]. *N*-Bromosuccinimide is also a competent catalyst in this regard (entry 2) [62]. Both of these protocols might be seen as modifications to an earlier report by Sharpless [63], which describes the use of phenyltrimethylammonium bromide (PTAB) as both bromine source and phase-transfer catalyst (entry 3).

The Sharpless protocol is in some ways complementary to prior art. For example, methylcyclohexadiene oxide (**23**) can be aziridinated in good yield using Chloramine-T trihydrate and catalytic amounts of PTAB (Scheme 2.5), a conversion that failed using PhI=NTs and $Cu(acac)_2$. Interestingly, only the trans aziridino epoxide (i.e., **24**) is observed, presumably due to preferential formation of the *cis*-epoxy bromonium intermediate **25**, which has been calculated to lie about 2.4 kcal mol^{-1} lower in energy than the corresponding *trans* isomer [64].

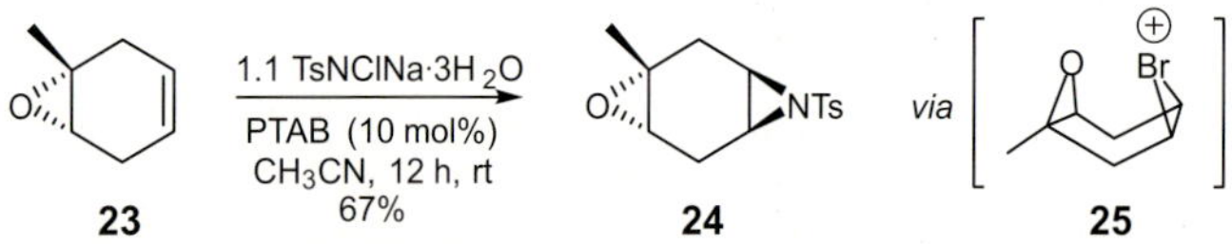

Scheme 2.5 Aziridination of epoxyalkenes.

Some progress has been made in using sulfonamides as starting materials for aziridination. Thus, the pyridinesulfonamide **26** is converted into a nitrene precursor (i.e., **28**) *in situ* using commercially available iodosobenzene diacetate as an oxidant, providing aziridine **27** in very good yield (Scheme 2.6). Another notable aspect of this

Table 2.3 Reaction conditions for bromine-catalyzed aziridination.

Entry	*n*	Catalyst (loading mol.%)	Chloramine-T type	Olefin : nitrene	Time (h)	Product	Yield (%)	Reference
1	2	H_2O_2/HBr (20)	trihydrate	1 : 1.3	5	**22b**	75	[53]
2	2	NBS (20)	anhydrous	1 : 1.0	3	**22b**	82	[54]
3	1	PTAB (30)	anhydrous	1 : 1.1	12	**22a**	86	[55]

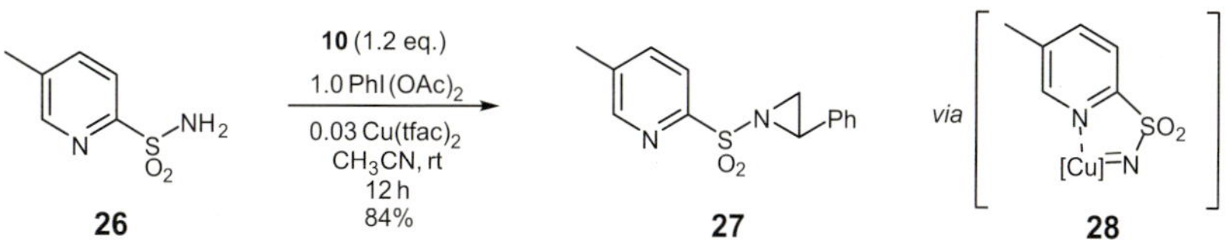

Scheme 2.6 Chelating sulfonamides.

system is that it obviates the need for external ligands and bases, since the pyridyl nitrogen provides intermolecular chelation. The free aziridine can be accessed by deprotection using magnesium in methanol [65]. A copper-catalyzed aziridination of tosylamides using iodine has also been reported [66].

DuBois and Guthikonda [67] have developed a similar rhodium-based strategy for the aziridination of sulfonamides, which they have applied to various unfunctionalized alkenes. With ω-butenyl sulfonamide **29a**, an intramolecular process can ensue to provide the bicyclic aziridine in good yield (Scheme 2.7). These and other investigations have shown the process to be stereospecific (Table 2.4, entry 2), whereby alkene geometry is preserved in the product [68]. Moreover, existing chiral centers can impose diastereoselectivity, as shown by the intramolecular aziridination of alkenyl sulfonamide **29c**, which proceeds with a 10 : 1 syn : anti ratio [69].

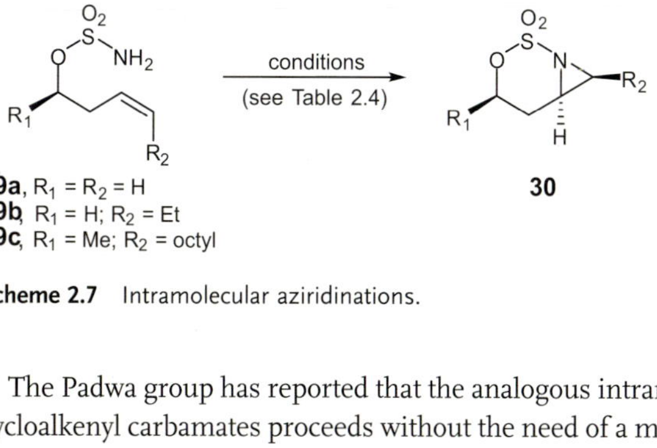

Scheme 2.7 Intramolecular aziridinations.

The Padwa group has reported that the analogous intramolecular aziridination of cycloalkenyl carbamates proceeds without the need of a metal catalyst (Scheme 2.8). Thus, cyclohexenyl carbamate **31a** underwent clean conversion into the tricyclic heterocycle **32a** upon treatment with 2 equivalents of iodosobenzene [70] (Table 2.5,

Table 2.4 Reaction conditions for intramolecular aziridinations.

Entry	Substrate	Catalyst (loading mol.%)	Oxidant	Solvent	Yield (%)	Reference
1	**29a**	$Rh_2(tfacam)_4$ (1)	$PhI(OAc)_2$	Benzene	84	[59]
2	**29b**	$Cu(CH_3CN)_4PF_6$ (10)	PhI=O	CH_3CN	80	[60]
3	**29c**	$Rh_2(Ooct)_4$ (2)	$PhI(OAc)_2$	CH_2Cl_2	84	[61]

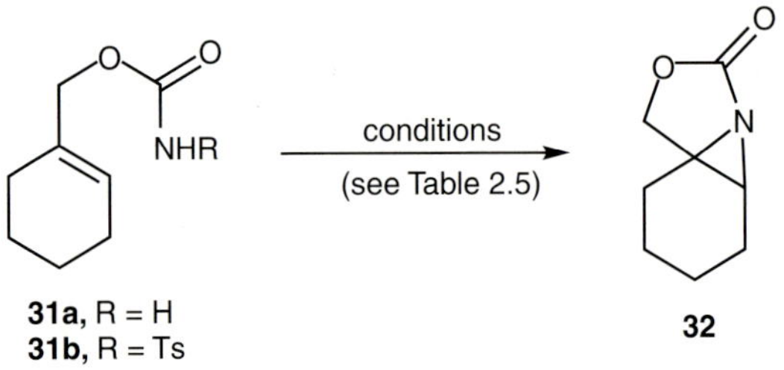

Scheme 2.8 Intramolecular aziridination of carbamates.

Table 2.5 Reaction conditions for the intramolecular aziridination of carbamates.

Entry	Substrate	Conditions	Yield (%)	Reference
1	**31a**	PhIO (2.0 eq), CH_2Cl_2, 40°C	75	[71]
2	**31b**	K_2CO_3 (7 eq), $Rh_2(OAc)_4$ (5 mol %), acetone, 25°C	79	[72]

entry 1). A similar rhodium-catalyzed variant has been reported for N-tosyloxycarbamates [71] (Table 2.5, entry 2).

Electron-deficient olefins often require different conditions for efficient aziridination than their unactivated counterparts. Along these lines, while certainly not limited to electron-poor alkenes, N-aminophthalimide (**34**) acts as a versatile nitrogen donor for aziridinations under various oxidizing conditions. The classical protocol involves the mild but environmentally questionable reagent lead tetraacetate [72], under which conditions the active aziridinating agent is believed to be an N-acetoxy species rather than a nitrene [73]. Meanwhile, other innovative methodologies have evolved. For example, using the conventional oxidant of iodosylbenzene diacetate (Table 2.6, entry 1), chalcone (**33**) is aziridinated in excellent yield (Scheme 2.9) [74].

Table 2.6 Reaction conditions for phthalimide aziridinations.

Entry	Eq 34	Oxidant (loading where appropriate)	Additive	Solvent	Time (h)	Yield (%)	Reference
1	1.4	$PhI(OAc)_2$ (1.5 eq)	K_2CO_3	CH_2Cl_2	12	93	[65]
2	1.4	*p*-MeOPhI/mCPBA (1.4 eq)	K_2CO_3	CH_2Cl_2	12	94	[66]
3	1.3	+ 1.80 V (vs. Ag wire)	Et_3NHOAc	CH_3CN	4	83	[67]

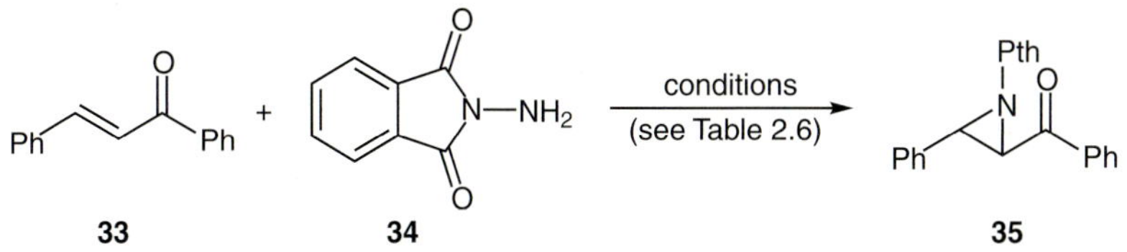

33 + 34 → conditions (see Table 2.6) → 35

Scheme 2.9 N-Aminophthalimide as nitrogen donor.

These conditions have also been used to advantage for the aziridination of allylic alcohols [75].

The hypervalent iodine reagent can also be generated *in situ* by combining equimolar amounts of *p*-iodoanisole and *m*-chloroperbenzoic acid (entry 2) with no negative impact on yield [76]. An even more atom-economical approach can be realized using electrochemical conditions (entry 3), a stereospecific process that has been described as a *click* preparation of aziridines [77], and which may proceed via a nitrene intermediate [78–80]. A similarly efficient oxidation has been reported using superoxide ion [81].

Addition of the chiral camphor-derived ligand **36** (Scheme 2.10) can result in an enantioselective process. Thus, the unsaturated oxazolidinone imide **37** is converted into the corresponding aziridine (**38**) in good yield and impressive enantiomeric excess. By comparison, (+)-tartaric acid gave only 42% ee. The choice of solvent is important, as migration to THF results in no loss of yield but an almost total disappearance of enantioselectivity [82].

Chiral bis(oxazoline) (BOX) ligands allow for a tunable aziridination of chalcones (Scheme 2.11) merely by changing the connecting backbone moiety (Figure 2.5). Thus, use of the cyclohexyl catalyst **41** provides the 2*R*,3*S* product (Table 2.7, entry 1) with very good enantioselectivity [83], while the anthracene derivative **42** yields the

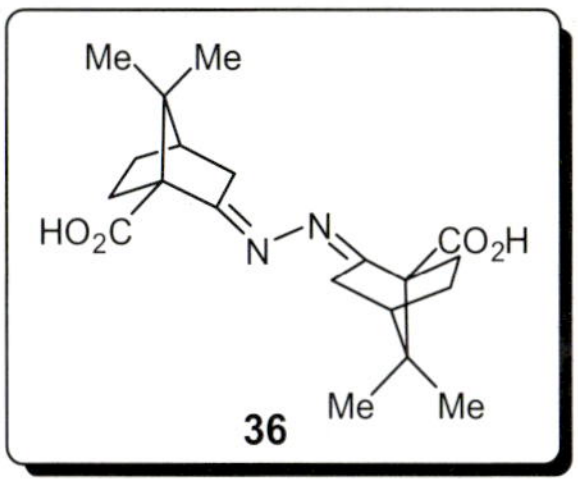

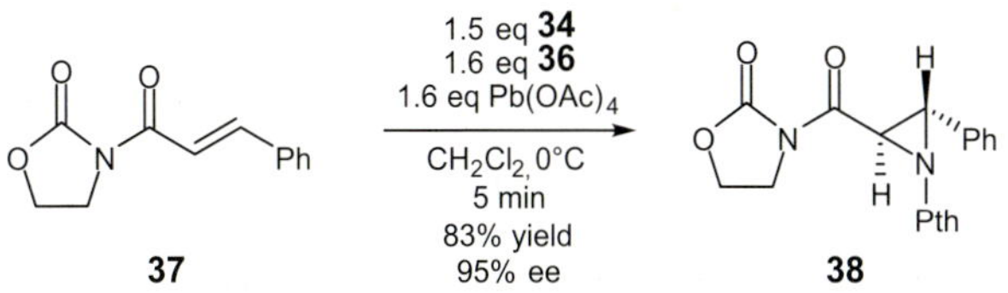

36

37 → 1.5 eq **34**, 1.6 eq **36**, 1.6 eq $Pb(OAc)_4$; CH_2Cl_2, 0°C, 5 min; 83% yield, 95% ee → 38

Scheme 2.10 Asymmetric N-aminophthalimide-mediated aziridination.

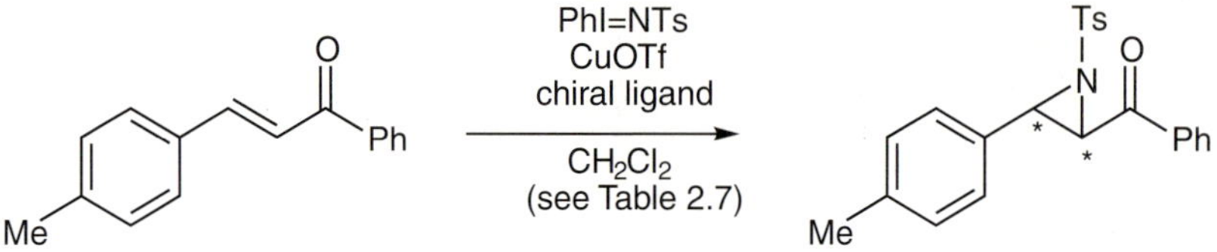

Scheme 2.11 Tunable BOX-mediated asymmetric aziridination.

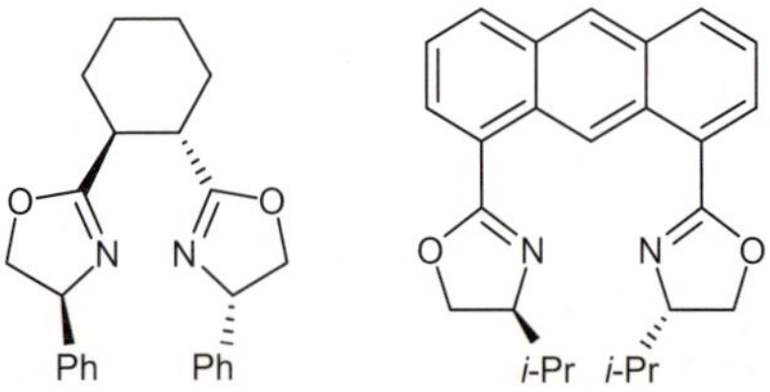

Figure 2.5 cHBOX and AnBOX ligands.

other antipode with even better yield and enantioselectivity (entry 2). The origin of this interesting crossover has been rationalized on the basis of a more crowded steric environment in the latter case [84]. The scope of the organocatalytic asymmetric aziridination of enones has been expanded to substrates other than chalcones by using a hydroquinine-derived catalyst and N-protected hydroxylamine tosylates as nitrogen donors [85].

Cinnamate esters can be aziridinated using axially dissymmetric binaphthyldiimine copper(I) catalysts, such as those derived from salen-type ligand **43** (Scheme 2.12). Thus, *trans-t*-butyl cinnamate (**44**) was aziridinated stereospecifically and enantioselectively to provide the product in excellent yield [86]. A later report from another set of investigators using essentially identical conditions gave the same high enantioselectivity but significantly lower yield [87].

Table 2.7 Reaction conditions for tunable BOX-mediated asymmetric aziridination.

Entry	PhI=NTs (eq)[a]	CuOTf (eq)[a]	Ligand (loading mol.%)[a]	Time (h)	Yield (%)[b]	Configuration	ee (%)	Reference
1	0.67	0.03	**41** (4)	5	62	2*R*,3*S*	94	[71]
2	0.67	0.03	**42** (4)	5	86	2*S*,3*R*	98	[72]

a) Based on olefin.
b) Based on PhI=NTs.

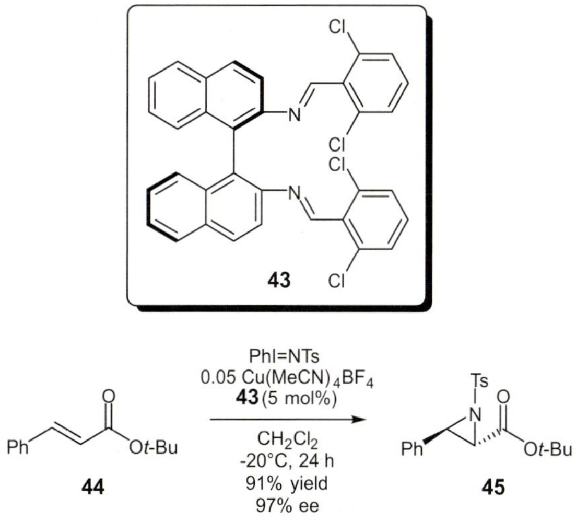

Scheme 2.12 Asymmetric aziridination of cinnamate esters.

Diactivated alkenes can be converted into aziridines by a somewhat different set of conditions. For example, in the presence of calcium oxide, ethyl nosyloxycarbamate (**46**) functions as a nitrogen source that engages Knoevenagel adducts (e.g., **47**) in aziridination (Scheme 2.13), presumably via nitrene intermediates [88]. Modest diastereoselectivities have been achieved by incorporating a menthol-derived chiral auxiliary into the carbamate reagent. Unfunctionalized alkenes give the products of nitrene C–H insertions under the same conditions [89].

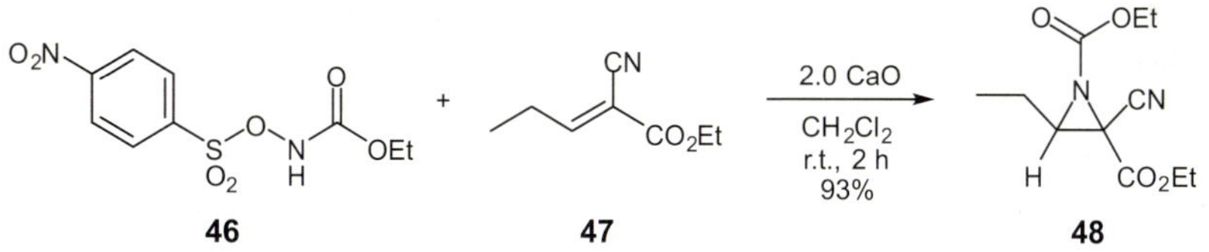

Scheme 2.13 Aziridination of diactivated alkenes.

2.1.2.2 Aziridination of Imines

Another powerful approach to the aziridine moiety is through the aziridination of imines using carbene- or ylide-type species [90–92]. One very popular carbene precursor is the commercially available ethyl diazoacetate (**49**). This species can be induced to react with imines to form aziridines (Scheme 2.14) under various conditions, including indium trichloride in methylene chloride [93], lanthanide triflates in protic media [94], boron trifluoride in ether [95], a cyclopentadienyl iron(II) dicarbonyl complex [96], a bis(cyclooctadienyl) iridium(II) chloride complex [97] and

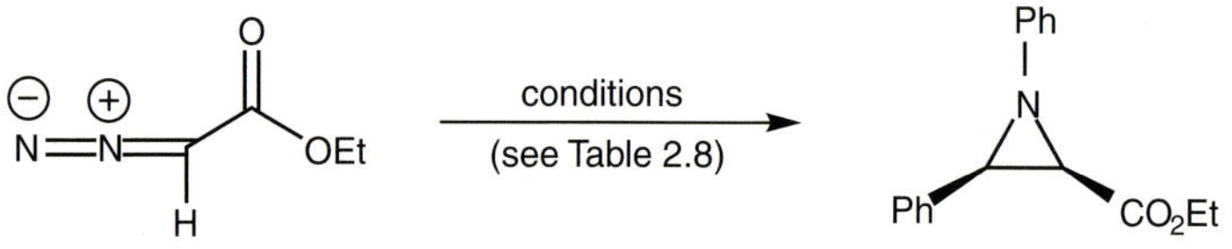

Scheme 2.14 Racemic aziridination using ethyl diazoacetate.

Table 2.8 Reaction conditions for racemic aziridinations with ethyl diazoacetate.

Entry	Aldimine components	Catalyst (loading mol.%)	Solvent	Time (h)	Yield (%)	Reference
1	PhCH=NPh	$SnCl_4$ (1)	CH_2Cl_2	n.r.	>90	[84]
2	PhCH=NPh	None	$bmimPF_6$	5	93	[85]
3	PhCH=O + $PhNH_2$	$LiClO_4$ (10)	CH_3CN	4.5	89	[86]

copper(II) triflate in methylene chloride or tetrahydrofuran [98]. These reactions are often very high-yielding, as demonstrated by the tin(IV) chloride mediated reaction of ethyl diazoacetate with *N*-phenylbenzaldimine (Table 2.8, entry 1), which proceeds with a $Z:E$ selectivity of 15 : 1 [99]. Interestingly, when the same reaction is run using an ionic liquid as solvent (entry 2), the reaction does not require the addition of a catalyst. In this case, the solvent itself is presumed to fulfill the role of Lewis acid [100]. An operationally attractive procedure has been reported in which the aldimine is formed *in situ* using lithium perchlorate as a catalyst (entry 3), providing exclusively the *cis* isomer [101].

Acive methylene compounds serve as useful carbenoid precursors (in the form of phenyliodonium ylides) by treatment with iodobenzene diacetate and a catalytic amount of base under very mild conditions. Thus, tosylaldimines are converted into the corresponding aziridines in a one-pot procedure without the need for a metal catalyst [102].

Some noteworthy asymmetric protocols have been developed using diazoesters, particularly those employing vaulted biaryl ligands [103]. For example, a catalyst derived from *(S)*-VAPOL (**51**) and BH_3·THF complex promotes an enantioselective aziridination reaction between ethyl diazoacetate and *N*-diphenylmethyl *p*-bromobenzaldimine (**52**) with very good yields, almost exclusive *cis* selectivity and excellent enantiomeric excess (Scheme 2.15). Curiously, impure commercial samples of BH_3·THF give the best results [104]. Recent crystallographic evidence in an analogous system suggests an unexpected boroxinate species as the active catalyst, the formation of which should be facilitated by adventitious water in the commercial boron reagent [105].

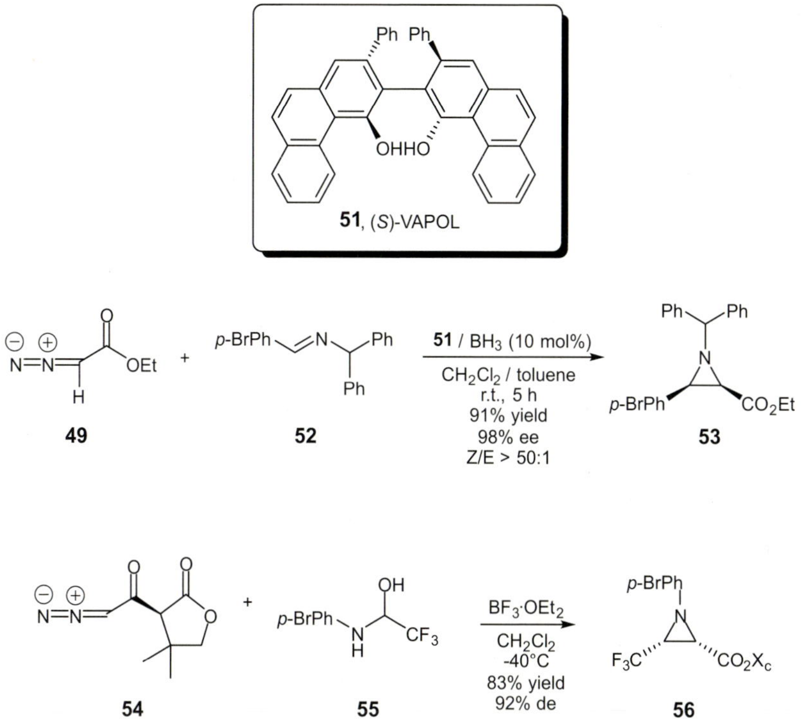

51, (S)-VAPOL

49 + 52 → 53 (51 / BH_3 (10 mol%), CH_2Cl_2 / toluene, r.t., 5 h, 91% yield, 98% ee, Z/E > 50:1)

54 + 55 → 56 ($BF_3 \cdot OEt_2$, CH_2Cl_2, -40°C, 83% yield, 92% de)

Scheme 2.15 Chiral aziridination using diazoesters.

An alternative approach to asymmetric induction is to attach a chiral auxiliary to the diazoester itself, as exemplified by the (*R*)-pantolactone derived ester **54**. Using boron trifluoride as a Lewis acid catalyst, the trifluoromethyl hemiaminal **55** is converted into the corresponding imine and subsequently aziridinated to give products (i.e., **56**) in high yield and diastereomeric excess [106].

Other carbon donors for the aziridination of aldimines can be drawn from the ranks of ylide haloanions. For example, the chloroethylthiazole derivative **57** (Scheme 2.16) can be deprotonated with *n*-butyllithium to give an anion that reacts with *N*-phenylbenzaldimine (**58**) to provide the corresponding *cis*-aziridinyl thiazole **59** in good yield [107]. In an asymmetric variant of this protocol, the anion derived from chloromethyl benzothiazole **60** adds to the chiral aldimine **61** in a highly diastereoselective fashion [108]. In the same vein, Davis has reported the aza-Darzens reaction between the lithium anion of diethyl iodomethylphosphonate (**63**) and chiral non-racemic arylsulfinyl imines (e.g., **64**), which proceeds in good yield and excellent diastereomeric excess [109, 110]. An analogous reaction occurs between *t*-butylsulfinyl imines and the ylide derived from trimethylsulfonium iodide [111] or ylides derived from substituted allyltetrahydrothiophenium salts, which provide access to chiral non-racemic vinyl aziridines [112].

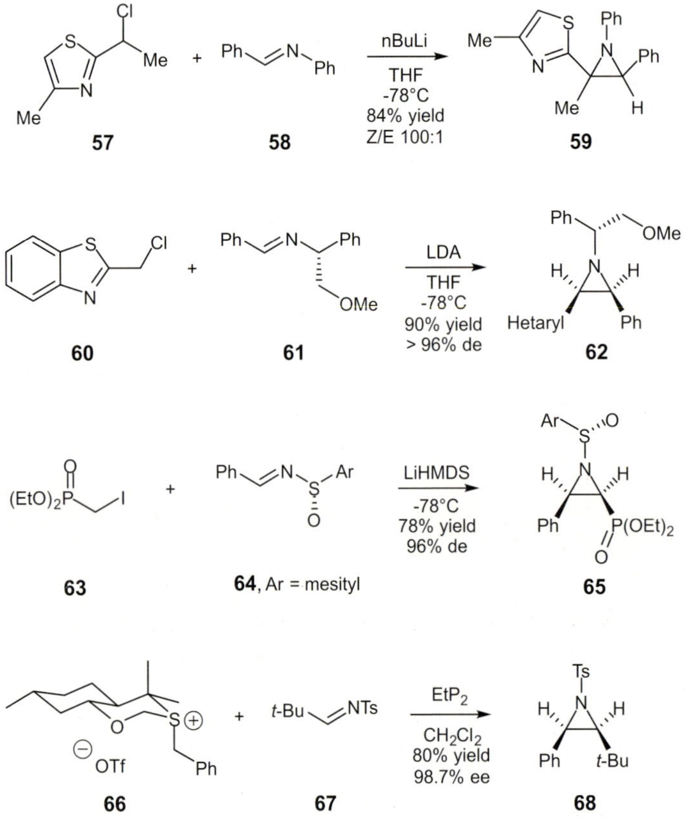

Scheme 2.16 The ylide/haloanion approach to aziridines.

The chiral sulfur ylide approach has been developed less for aziridine chemistry than for epoxides; nevertheless, some very useful strategies have been pioneered, largely by Dai and Aggarwal [113, 114], and new methodologies continue to appear on the scene. For example, the chiral *S*-benzyl sulfonium triflate **66** can be deprotonated with the commercially available phosphazene base Et_2P to give an ylide that aziridinates *N*-tosyl pivaldimine (**67**) in good yield and excellent enantioselectivity [115]. Aggarwal and Vasse have applied a catalytic version of the sulfur ylide methodology to the synthesis of the taxol side chain [116] which uses a novel camphor-derived ylide scaffold [117]. The *cis* : *trans* ratios in these reactions are very substrate-dependent. A recent computational study of this reaction manifold provides a useful theoretical framework in good agreement with observed experimental results [118]. The catalytic strategy has also been adapted to novel arsonium ylides [119].

The substrate dependent nature of *cis* : *trans* ratios is not limited to sulfur ylide chemistry. A particularly striking example is seen in the reaction of the anion from camphorsultam bromoacetamide **69** with various *o*-substituted benzaldimines **70** (Scheme 2.17). Both the phenyl and the nitrophenyl derivatives give exclusively *cis*-

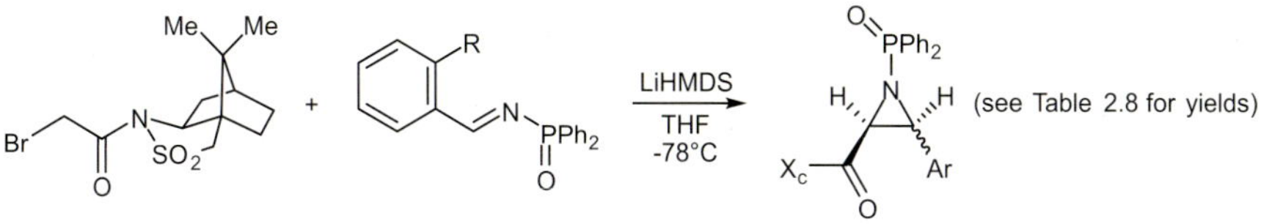

Scheme 2.17 Camphorsultam as chiral auxiliary in aziridination.

Table 2.9 Yield data for camphorsultam-mediated aziridination.

Entry	R	Yield (%)	Cis : trans ratio
1	–H	71	100 : 0
2	$-NO_2$	72	100 : 0
3	–Me	87	50 : 50
4	–OMe	65	0 : 100

aziridines (Table 2.9, entries 1 & 2), whereas the *o*-anisole imine provides solely the *trans* isomer (entry 4), and the *o*-tolyl analog yields an equimolar mixture of diastereomers (entry 3). The mechanistic underpinnings for this selectivity are not well understood, but are believed to derive from a complex mixture of polar, steric and chelation effects [120, 121].

2.1.2.3 **Ring Closure of Amines**

The synthesis of aziridines via the ring closure of 2-aminoalcohols has been known for three-quarters of a century [122], and the method has been adapted for such activating agents as triphenylphosphine dibromide [123], diphosphorus tetraiodide [124], the Mitsunobu reagent [125] and molecular sieves [126, 127]. This time-honored approach is, however, still in currency. For example, the chiral nonracemic aminoalcohol **72** (Scheme 2.18) derived from alanine [128] undergoes double tosylation and ring closure in one pot to provide the corresponding *N*-tosylaziridine (**73**) in excellent yield [129]. Similarly, the *N*-(*t*-butoxycarbonyl)-*N*-(2-nitrobenzenesulfonyl) aminoalcohol **74**, derived from the ring-opening of an epoxide, undergoes a sequence of deprotection, *O*-mesylation and carbonate-mediated ring closure to give aziridine **76**, proceeding with inversion of configuration and no loss in optical activity [130].

This concept of ring closure is, of course, open to any good leaving group. For example, an interesting protocol has been reported in which an *N*-2-bromoalkylimine (e.g., **78**), prepared by the addition of the 2-bromoalkylamine hydrobromide to an aldehyde (e.g., **77**), suffers nucleophilic attack by methoxide to give an incipient imide anion that engages in immediate 3-*exo-tet* ring closure [131]. Cyclization of a 2-bromoalkylamine is also at the heart of an aziridination of simple amines, such as ammonia, which engages in a tandem series of conjugate addition and ring closure

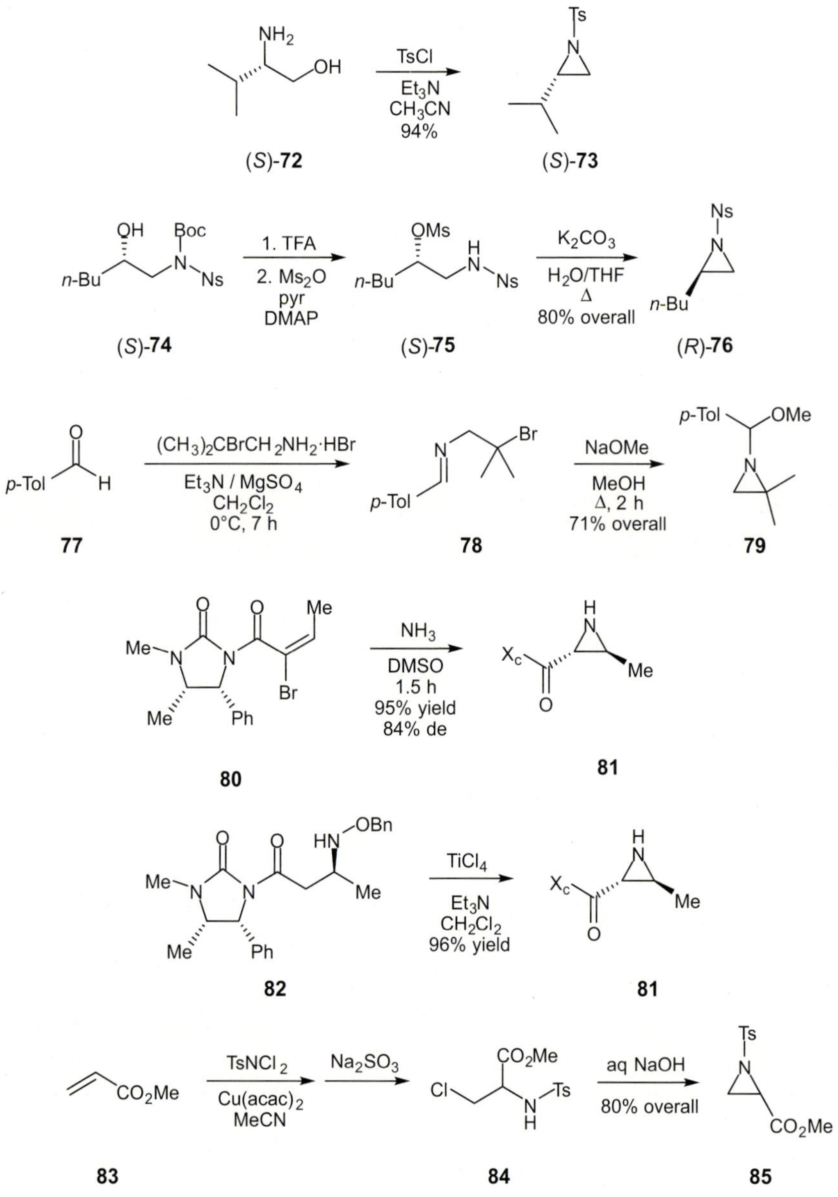

Scheme 2.18 Aziridines from ring-closing protocols.

reactions with the chiral α bromoamide **80** to provide aziridine **81** with good diastereoselectivity [132]. This methodology has also been adapted to a solid-phase protocol using Wang resin derivatives [133].

The same chiral auxiliary can be used to advantage in the synthesis of chiral β-amino acid precursors (e.g., **82**), which can be converted into the corresponding amino acids by zinc-copper reduction of the N–O bond and lithium peroxide hydrolysis of the amide linkage [134]. However, in the presence of a suitable Lewis acid, **82** is converted into aziridine **81** with remarkable efficiency [135]. Aziridinyl esters (e.g., **85**) are conveniently prepared from methyl acrylate (**83**) or its derivatives by the copper-catalyzed addition of *N,N*-dichlorotosyl sulfonamide and subsequent ring closure [136]. Alternatively, these products can be accessed by treating α,β-dibromoesters with simple primary amines [137].

2.1.2.4 **Ring Contraction of Other Heterocycles**

Aziridines can be accessed through the extrusion of elements from larger nitrogenous heterocycles, most notably triazolines. Thus, for example, azanorbornene **86** undergoes 1,3-dipolar cycloaddition with phenyl azide to yield a 2 : 3 mixture of regioisomeric tricyclic triazolines in quantitative combined yield (Scheme 2.19). Photolysis of these compounds in a quartz reaction vessel using a medium-pressure mercury lamp led to efficient formation of the fused tricyclic aziridine **88** [138].

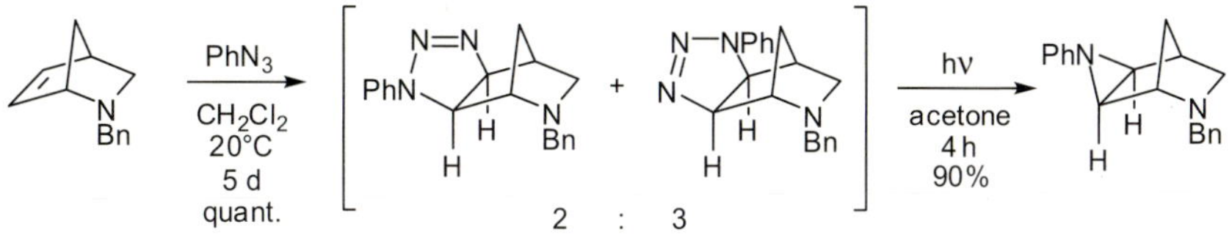

PhN_3, CH_2Cl_2, 20°C, 5 d, quant.; hν, acetone, 4 h, 90%; **86**, **87a**, **87b** (2 : 3), **88**

Scheme 2.19 Azide addition and ring contraction.

The rather sluggish initial cycloaddition is accelerated significantly by tethering the dipole to the dipolarophile, as seen with the ω-alkenylaryl azide **89a** (Scheme 2.20), which undergoes complete cycloaddition within 3 h in refluxing toluene [139]. Even with very highly functionalized substrates (Table 2.10), the cycloaddition step proceeds with excellent yield and complete diastereoselectivity, the latter presumably the result of a preferred chair-like reactive conformer in which the TMS group adopts a pseudo-equatorial attitude [140]. Subsequent irradiation with a Hanovia lamp afforded the corresponding aziridines **91** in fair yield.

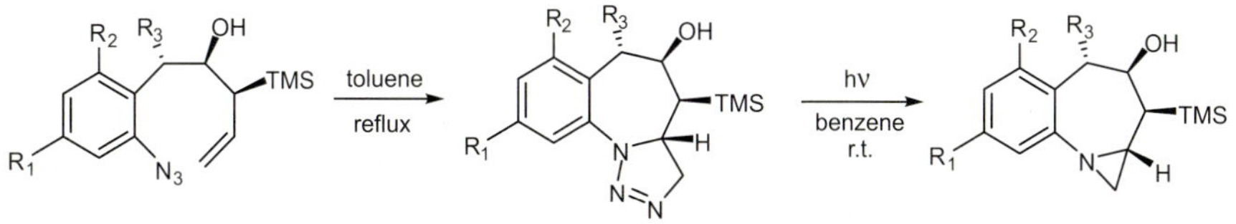

toluene, reflux; hν, benzene, r.t.; **89**, **90**, **91**

Scheme 2.20 Intramolecular azide addition and ring contraction.

Table 2.10 Yield data for intramolecular azide addition and ring contraction.

Entry	Substrate	R_1	R_2	R_3	Yield **90** (%)	Yield **91** (%)	Reference
1	**89a**	–H	–H	–H	90	77	[118]
2	**89b**	$-CH_2OBn$	–OBn	$-CH_2OBn$	99	68	[119]

Molteni and Del Buttero [141] have reported the direct aziridination of ethyl acrylate (**93**, Scheme 2.21) using an alkyl azide supported on poly(ethylene glycol) monomethyl ether (**92**), which they suggest proceeds through the intermediacy of a triazole. In their studies of triflic acid mediated aziridination of electron-deficient alkenes, however, Johnston and coworkers [142] propose an intriguing concerted mechanism involving a multicentered transition state (i.e., **96**).

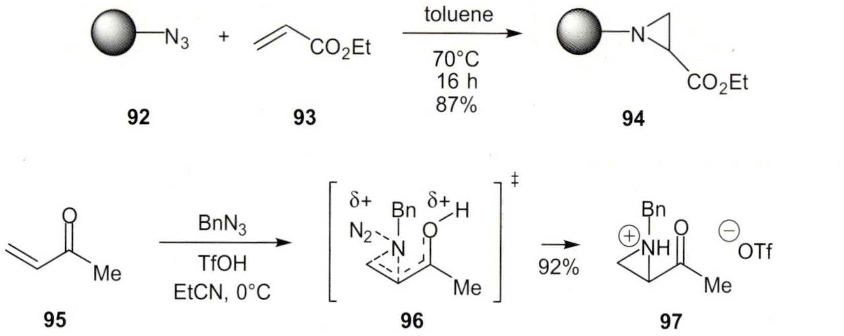

Scheme 2.21 Direct aziridination with alkyl azides.

2.1.3 Reactivity of Aziridines

The reactions of aziridines are legion, but most fit into a few broad categories: (i) nucleophilic ring opening, (ii) N-substitution, (iii) aziridinyl anion chemistry and (iv) ring expansion to larger heterocyclic species . Although no longer the most recent, a review of the synthetic applications of chiral aziridines by McCoull and Davis [4] still offers an excellent overview of the diverse field of aziridine chemistry. A more recent (but more specialized) chapter on aziridinecarboxylate esters is equally worthwhile [143].

2.1.3.1 Nucleophilic Ring Opening

Perhaps the most common of aziridine reactions is the ring opening by nucleophiles. This topic has been reviewed recently and fairly comprehensively by Hu [144], and an interesting computational study on CN vs. CC bond cleavage has been published [145]. Some illustrative synthetic examples are given here.

Ring opening can be used as a means of carbon–carbon bond formation when carbon-based nucleophiles are employed [146]. For example, in the presence of catalytic amounts of cuprous iodide, phenyllithium attacks the less substituted site of the *N*-tosyl aziridine **98** (Scheme 2.22) to give the corresponding protected amine [147]. The analogous addition of Grignards onto *meso*-aziridines (e.g., **22b**)

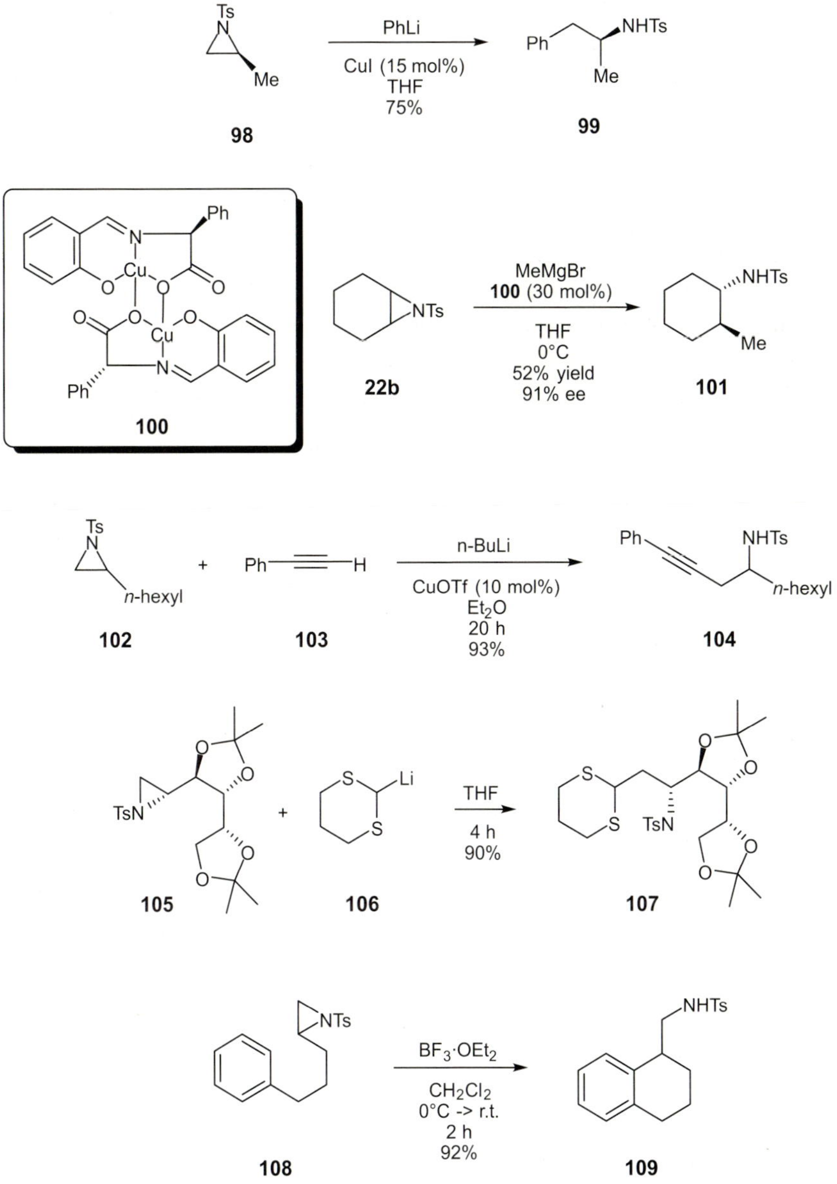

Scheme 2.22 Ring opening of aziridines with carbon nucleophiles.

can exhibit impressive enantioselectivity under the influence of the Schiff base/amino acid copper(II) complex dimer **100**, although relatively high catalyst loadings must be used [148]. Acetylide anions are also competent nucleophiles when a copper(I) catalyst is used, as demonstrated by the high-yielding conversion of the 2-alkylaziridine **102** into the homopropargylamine **104**, resulting from attack at the less hindered carbon [149]. Wu and Zhu have used the diastereoselective addition of an aryl Grignard onto a chiral aziridine as a key step in their total synthesis of (–)-renieramycin M and G and (–)-Jorumycin [150].

This exclusive regioselectivity was used to advantage in an approach to 1-deoxymannojirimycin analogs from deoxyglucitol-derived aziridine **105**, which engages in a very well-behaved reaction with 1,3-dithiane anion **106** [151]. Similar regiochemical outcomes are observed for cyanide addition using stoichiometric trimethylsilyl cyanide (TMSCN) and tetrabutylammonium fluoride (TBAF) in catalytic amounts [152], as well as an enantioselective protocol used to alkylate active methine compounds with unsymmetrical aziridines under mild basic conditions using a cinchona derived phase-transfer catalyst [153].

The regioselectivity of ring opening can be reversed under more cationic conditions, however, as illustrated by the Lewis acid-promoted intramolecular ring-opening reaction of phenypropylaziridine **108**, in which the product outcome is doubly supported by the stability of the Friedel-Crafts-like transition state and geometric considerations [154]. The same regiochemistry is observed in the iron(III)-catalyzed attack of electron-rich arenes onto unsymmetrical C-aryl aziridines [155].

Heteroatomic nucleophiles are equally useful in unlocking the synthetic utility of the aziridine ring. For example, **22b** reacts smoothly with aniline to provide the corresponding diamine **110** (Scheme 2.23) under various mild conditions (Table 2.11), including bismuth trichloride in acetonitrile [156], indium tribromide in methylene chloride [157], and lithium perchlorate in acetonitrile [120, 158]. Aqueous conditions have been developed using a cyclodextrin catalyst [159], and silica gel allows for a completely solventless system [160].

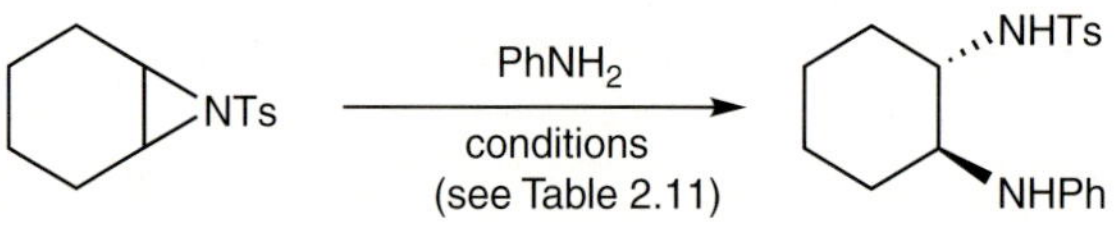

NTs
PhNH$_2$
conditions
(see Table 2.11)
NHTs
NHPh
22b
110

Scheme 2.23 Ring opening of aziridines with aniline.

Table 2.11 Yield data for ring opening of aziridines with aniline.

Entry	Reagent/catalyst	Solvent	Time (h)	Yield (%)	Reference
1	$BiCl_3$	CH_3CN	1.5	96	[130]
2	$InBr_3$	CH_2Cl_2	5.5	90	[131]
3	$LiClO_4$	CH_3CN	5.5	90	[132]
4	β-Cyclodextrin	$H_2O/Me_2C{=}O$	24	89	[133]
5	Silica gel	none	1	91	[134]

The asymmetric ring-opening of meso aziridines such as **22b** is a useful approach for accessing variously substituted chiral amines, and it has been the subject of some very good reviews [144, 146, 161]. One recently reported protocol involves the use of a titanium binolate catalyst, which can achieve ees of 99% [162].

Still other conditions have been worked out for alkylamines, as exemplified by addition of benzylamine (Scheme 2.24). In what may be one of the biggest catalyst sleepers of the century, the commercially available tris(pentafluorophenyl) borane [163] has escaped the niche of specialized polymerization catalysis and now finds application in various other useful synthetic transformations, including the ring opening of aziridines with simple amines (Table 2.12, entry 1). Interestingly, the active catalytic species involves a water-borane complex [164]. Other novel protocols for this reaction include ceric ammonium nitrate (CAN) in acetonitrile [165] and tributyl phosphite in an organic/aqueous medium [166]. Microwave conditions have been developed using resin-bound alkylamines in a protocol suitable for parallel synthesis [167].

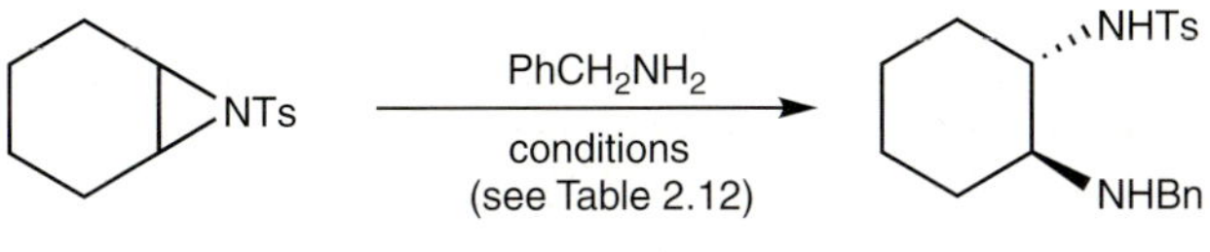

Scheme 2.24 Ring opening of aziridines with benzylamine.

Azide is also a popular nitrogen-based nucleophile by virtue of its relatively low basicity, high nucleophilicity and ability to undergo subsequent reduction to the primary amine (i.e., a surrogate for ammonia). To achieve virtually neutral conditions, trimethylsilyl azide can be used as the source of azide (Table 2.13, entry 1), along with catalytic amounts of tributylammonium fluoride (TBAF) to liberate the azide *in situ* [152]. Of course, sodium azide itself can be used as a reactant, with the addition being promoted efficiently by cerium trichloride heptahydrate [168], lithium perchlorate [169] or Oxone [170] (Scheme 2.25). Meso aziridines can be enantioselectively desymmetrized using a dimeric salen yttrium catalyst in near quantitative yield and excellent enantioselectivity [161].

Synthetically useful β-aminoalcohols and aminoethers can be obtained by using oxygen-centered nucleophiles in the ring-opening reaction, and several mild and

Table 2.12 Yield data for ring opening of aziridines with benzylamine.

Entry	Reagent/catalyst	Solvent	Time (h)	Yield (%)	Reference
1	$B(C_6F_5)_3$	CH_3CN	12	99	[136]
2	CAN	CH_3CN	3	93	[137]
3	PBu_3	H_2O/CH_3CN	12	99	[138]

Table 2.13 Yield data for ring opening of aziridines with azide.

Entry	Azide source	Additive	Solvent	Time (h)	Yield (%)	Reference
1	$TMSN_3$	TBAF (5 mol.%)	THF	6	97	[128]
2	NaN_3	$CeCl_3 \cdot 7H_2O$	CH_3CN	6	90	[139]
3	NaN_3	$LiClO_4$	CH_3CN	6	90	[140]
4	NaN_3	Oxone	H_2O/CH_3CN	3	89	[141]

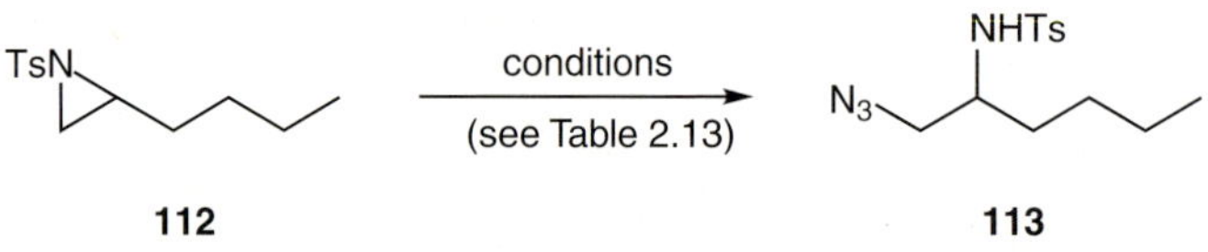

TsN (aziridine) **112** → conditions (see Table 2.13) → N_3, NHTs **113**

Scheme 2.25 Ring opening of aziridines with azide.

efficient methods have been reported to effect this transformation. For example, the alcoholysis of phenylaziridine **11** (Scheme 2.26) is promoted by ceric ammonium nitrate (Table 2.14, entry 1), which is effective for both alcohols [171] and water [165]; boron trifluoride etherate (entry 2) or tin(II) triflate [172]; montmorillonite KSF clay (entry 3), which serves as a solid-supported mild acid catalyst [174]; phosphomolybdic acid (PMA) on silica gel (entry 4), which gives excellent yields with various aziridines and alcohols [178]; and copper(II) triflate (entry 5) [175]. The water-tolerant bismuth(III) triflate is particularly well suited to catalyze hydrolysis reactions (entry 6) [176]. Note that in all cases the oxygen is attached to the benzylic position. With branched alkylaziridines (e.g., cyclohexylaziridine) the regioselectivity is reversed. Ring opening can also occur under basic conditions in the absence

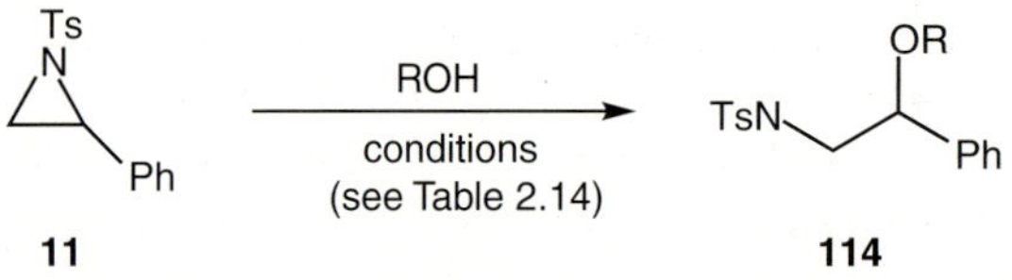

Ts, N, Ph **11** → ROH, conditions (see Table 2.14) → TsN, OR, Ph **114**

Scheme 2.26 Ring opening of aziridines with alcohols.

Table 2.14 Yield data for ring opening of aziridines with oxygen-centered nucleophiles.

Entry	*R*	Catalyst	Solvent	Yield (%)	Reference
1	Me	CAN	MeOH	90	[142]
2	Me	$BF_3 \cdot OEt_2$	MeOH	99	[143]
3	Et	KSF	CH_2Cl_2	86	[144]
4	*t*-Bu	PMA/SiO_2	MeCN	94	[145]
5	CH_2CH_2Cl	$Cu(OTf)_2$	$HOCH_2CH_2Cl$	87	[175]
6	H	$Bi(OTf)_3$	$MeCN/H_2O$	88	[176]

of a Lewis acid catalyst, in which case nucleophilic attack occurs at the less hindered position [177].

It was discovered that the regiochemistry of the hydrolysis of α-aminoaziridine **115** (Scheme 2.27) could be controlled by manipulation of the reaction conditions. Thus, use of a protic acid (such as *p*-toluenesulfonic acid) led to attack at the terminal carbon, presumably via the highly reactive aziridinium salt, yielding the hydroxydiamine **116**. In contrast, if a classic Lewis acid was employed (e.g., boron trifluoride etherate) the opposite regiochemistry predominated, providing the secondary alcohol **117**. The stereochemical outcome for the latter product was rationalized on the basis of a double inversion from anchimeric assistance by the dibenzylamino substituent [179].

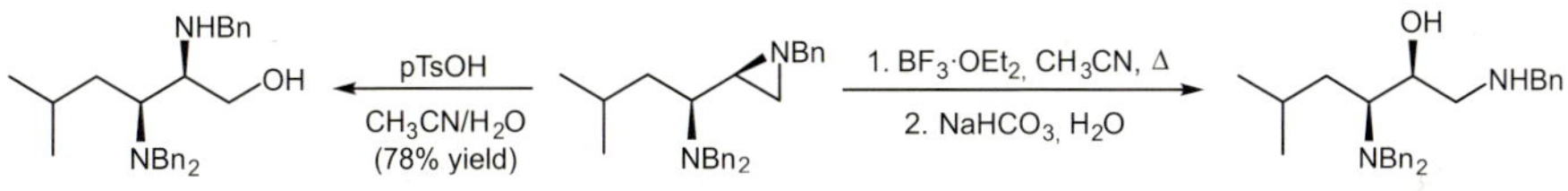

Scheme 2.27 Tunable hydrolysis conditions.

The scope is by no means limited to carbon, nitrogen and oxygen-centered nucleophiles. The nucleophilic palette embraces sulfur-based species, such as thiocyanates and thiols, the addition of which is promoted by a heterogeneous recyclable sulfated zirconia catalyst [180] and poly(ethylene glycol) [181]. The regiochemistry tends to follow the conventional course, with benzylic attack dominating for arylaziridines and terminal attack for alkylaziridines. Thiols can also engage in the enantioselective ring-opening of aziridines under the catalysis of a VAPOL phosphoric acid derivative [182].

The aziridine ring can also be cleaved by phenylselenide [183], reductively cleaved under transfer hydrogenation conditions [184], or opened with halides under fairly straightforward conditions, such as tetrabutylammonium fluoride in DMF [185], aqueous HCl in acetone [186], magnesium bromide in ether [187], and indium triiodide in acetonitrile [188].

2.1.3.2 *N*-Elaboration Reactions

Aziridines bearing no N-substituent can be elaborated in various ways. For example, cyclohexene imine (**118**) engages in conjugate addition onto methyl acrylate under solvent-free conditions (Scheme 2.28, route a), although two equivalents of the Michael acceptor is needed [164]. The same aziridine undergoes smooth palladium-catalyzed allylation with prenyl acetate, whereby the electrophile is captured at the more substituted terminus (route b), although this regiochemistry is substrate-dependent [189]. N-Arylation is also possible using a palladium/BINAP protocol (route c), in which aryl bromides and arylboronic acids act as suitable reaction partners [190].

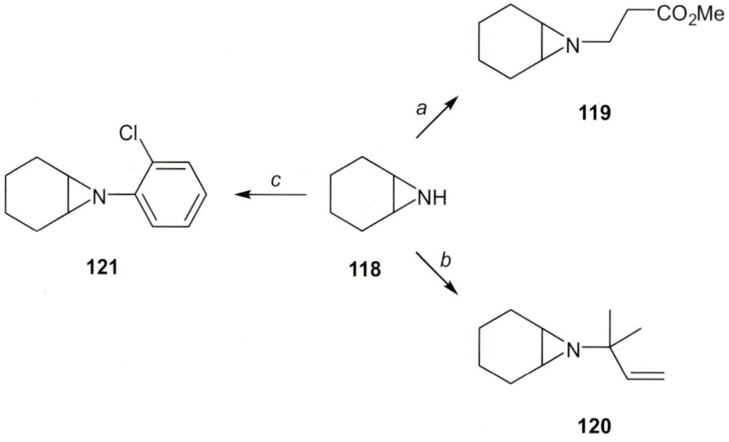

Scheme 2.28 N-Elaboration of cyclohexene imine (Table 2.15).

Table 2.15 Conditions for N-elaboration of cyclohexene imine.

Entry	Route	Electrophile	Catalytic system	Solvent	Yield (%)	Reference
1	a	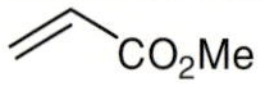 CO_2Me	None	Neat	89	[136]
2	c	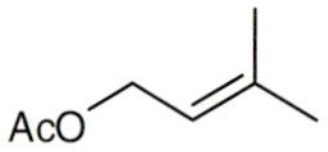 AcO	$[Pd(\eta^3\text{-}C_3H_5)Cl]_2$, BINAP	THF	89	[153]
3	b	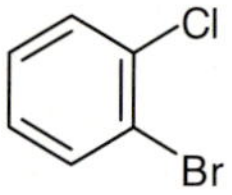 Cl, Br	$Pd(dba)_3$, BINAP, *t*-BuONa	Toluene	95	[154]

Many other protocols exist in which unprotected aziridines function as nucleophiles, including alkylations using epoxides [191] or alkyl bromides [192] as electrophiles. N-Acylation can be effected with acyl chlorides [193] or a combination of carboxylic acid and dicyclohexylcarbodiimide (DCC) [3, 194] (Scheme 2.29).

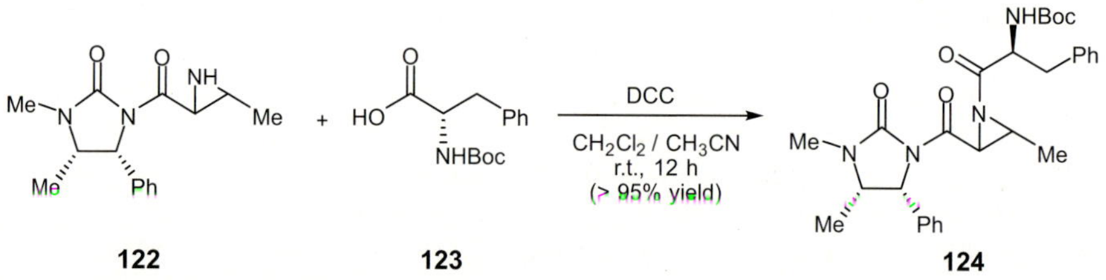

Scheme 2.29 N-Acylation using DCC.

2.1.3.3 Azirdinyl Anion Chemistry

Aziridinyl anion chemistry is of growing interest, as evidenced by its inclusion in a symposium-in-print [195]. Hodgson *et al.* have also authored a noteworthy review dedicated to this topic [196], to which the reader is directed, and computational studies on the physical properties of aziridinyl anions are emerging [197].

Protected aziridines can be deprotonated, and the resulting carbanions add to a range of electrophiles. For example, the *t*-butylsulfonyl (Bus) protected pentylaziridine **125** (Scheme 2.30) is deprotonated at the less substituted carbon (i.e., more stable anion) upon treatment with an excess of lithium 2,2,6,6-tetramethylpiperidide (LTMP) at low temperature (−78 °C), and the lithiate adds smoothly to 3-pentanone to give the aziridinyl alcohol **126** in good yield [198]. The aziridinyl anion can also exhibit carbene-like character, so that when alkenes are tethered to the substrate (e.g., **127**)

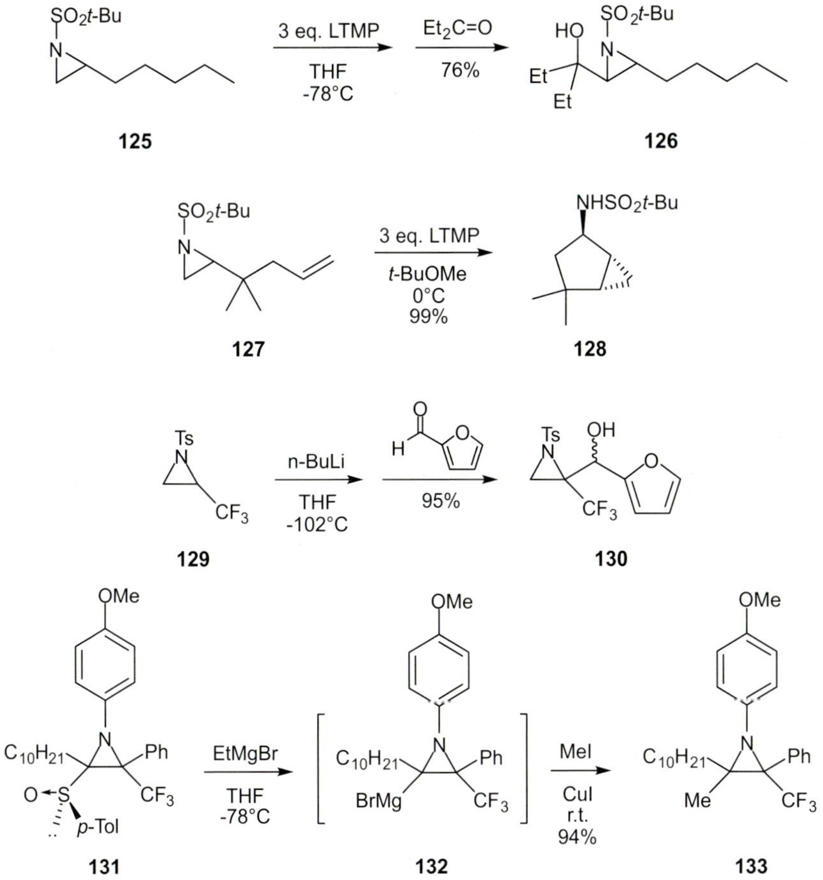

125: SO_2t-Bu, N; 3 eq. LTMP, THF, -78°C; $Et_2C{=}O$, 76%; 126: HO, SO_2t-Bu, N, Et, Et

127: SO_2t-Bu, N; 3 eq. LTMP, *t*-BuOMe, 0°C, 99%; 128: $NHSO_2t$-Bu

129: Ts, N, CF_3; n-BuLi, THF, -102°C; H, O, O, 95%; 130: Ts, N, OH, CF_3, O

131: OMe, N, $C_{10}H_{21}$, Ph, O, S, CF_3, *p*-Tol; EtMgBr, THF, -78°C; 132: OMe, N, $C_{10}H_{21}$, Ph, BrMg, CF_3; MeI, CuI, r.t., 94%; 133: OMe, N, $C_{10}H_{21}$, Ph, Me, CF_3

Scheme 2.30 Reactions of aziridinyl anions.

and the anion is formed at somewhat higher temperatures, a remarkably high-yielding intramolecular cyclopropanation ensues [199].

When an electron-withdrawing group is attached to the aziridine ring, it directs deprotonation and stabilizes the resulting anion. Thus, aziridine **129** is lithiated adjacent to the trifluoromethyl group, after which treatment with 2-furaldehyde provides the aziridinyl alcohol **130** in excellent yield [200]. Satoh and coworkers have developed a very useful protocol involving sulfinylaziridines (e.g., **131**), in which the sulfinyl group (having served as a chiral auxiliary in the prior aziridination reaction) is quantitatively removed in the presence of ethyl Grignard to give the corresponding aziridinylmagnesium bromide (i.e., **132**). This anion can then be cross-coupled with alkyl iodides in the presence of cuprous iodide to give alkylated products (i.e., **133**) with net retention of configuration [201, 202].

2.1.3.4 Ring Expansions

Aziridines can serve as handy templates for access to other heterocyclic subunits, and quite a few interesting methodologies have been reported. In the case of vinyl aziridines, near quantitative rearrangement to 3-pyrroline derivatives is promoted by copper(II) hexafluoroacetylacetonate in toluene at elevated temperature [203]. Another general approach to ring expansion is through cycloaddition strategies. For example, in the presence of zinc chloride, *p*-methoxyphenyl protected aziridine **134** (Scheme 2.31) engages in a formal [4 + 2] cycloaddition with norbornene (**135**) to provide the tetracyclic piperidine **136**. The mechanism proceeds through a Mannich reaction with the initially formed ylide and subsequent intramolecular Friedel–Crafts alkylation [204]. Highly substituted pyrroles can be accessed by reacting aziridines with electron-deficient allenes, a process that involves a formal [3 + 2] cycloaddition [205]. Pyrroles are also formed by the platinum(II)-catalyzed electrophilic iodocyclization of propargylic aziridines [206].

The ylide itself can be trapped with olefins under thermal conditions [207], as demonstrated by the intramolecular 1,3-dipolar cycloaddition of the terminal aziridine **137** to give the bicyclic pyrrolidine **138** [208]. Harrity and coworkers [209] have developed a [3 + 3] cycloaddition strategy for access to the piperidine ring system, in which a palladium-trimethylenemethane complex (derived from acetoxy trimethylsilyl methylene compounds such as **139**) adds to an aziridine (e.g., **98**) to form methylenepiperidines such as **140** in generally good yields. The substituent on nitrogen greatly impacts the efficiency of this reaction, with *p*-tosyl and *p*-methoxyphenyl-sulfonyl moieties providing the best results.

The next broad class of transformations could be characterized as carbonyl insertion reactions (Scheme 2.32), although it encompasses the insertion of carbon monoxide and carbon dioxide through various modes. Thus, in the presence of sodium iodide, Boc anhydride serves as a carbon dioxide surrogate in the conversion of the (+) pseudoephedrine-derived aziridine **141** into oxazolidinone **143**, in which the stereochemistry of the ring substituents is preserved [210, 211]. Hydroxymethylaziridines (e.g., **144**) can be bridged with phosgene into labile fused bicyclic oxazolidinone structures (i.e., **145**), which undergo *in situ* nucleophilic attack by chloride to give chloromethyloxazolidinones (e.g., **146**) in excellent

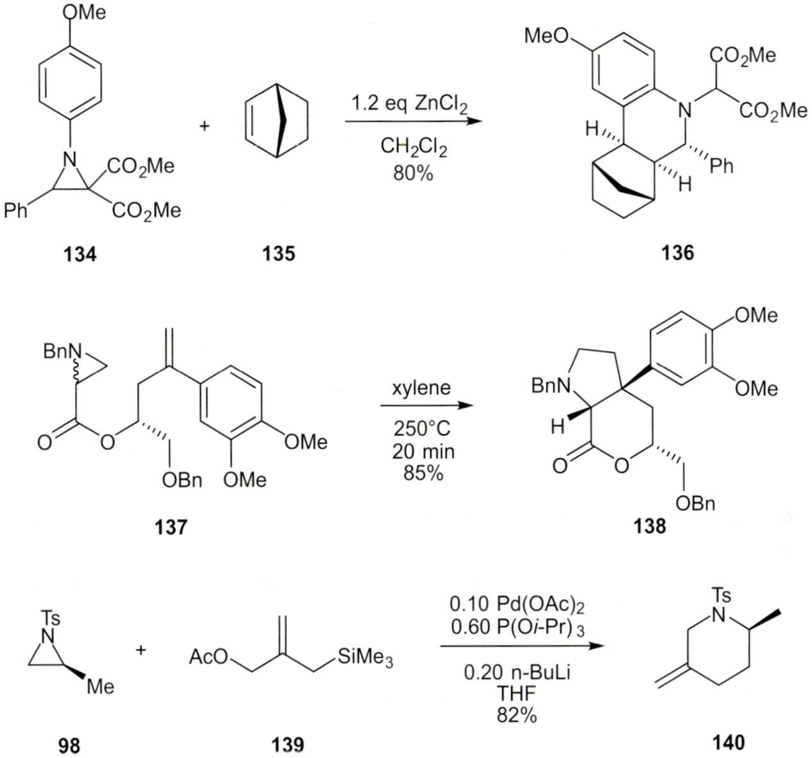

Scheme 2.31 Cycloaddition reactions of aziridines.

yield [212]. Also, gaseous carbon dioxide itself can be trapped by aziridines at atmospheric pressure using lithium bromide as a catalyst, as shown by the conversion of arylaziridine **147** into the corresponding oxazolidinone **148**. The mechanism proceeds through a series of bromide-induced ring opening, carboxylation of the resulting amide anion, and alkylative ring closure [213]. This transformation has also been shown to proceed under the catalysis of a (salen) chromium(III)/DMAP complex [214].

A novel rhodium-complexed dendrimer (**149**) is effective in promoting the carbonylative ring expansion of aziridines to β-lactams. The near-quantitative yield exhibited by *N*-*t*-butyl-2-phenylaziridine (**150**) is not unusual, although relatively high pressures [approx. 27 atm (400 psi)] and elevated temperatures (90 °C) are required. Nevertheless, the catalyst appears to be easily isolated and recycled without loss of activity [215].

Nitrogen and sulfur can also be introduced via ring expansion strategies. Thus, styrene imine derivative **11** engages in [3 + 2] cycloaddition with acetonitrile in the presence of boron trifluoride etherate at room temperature to give the corresponding imidazoline (**152**) in good yield (Scheme 2.33) [216]. The iminothiazolidine nucle-

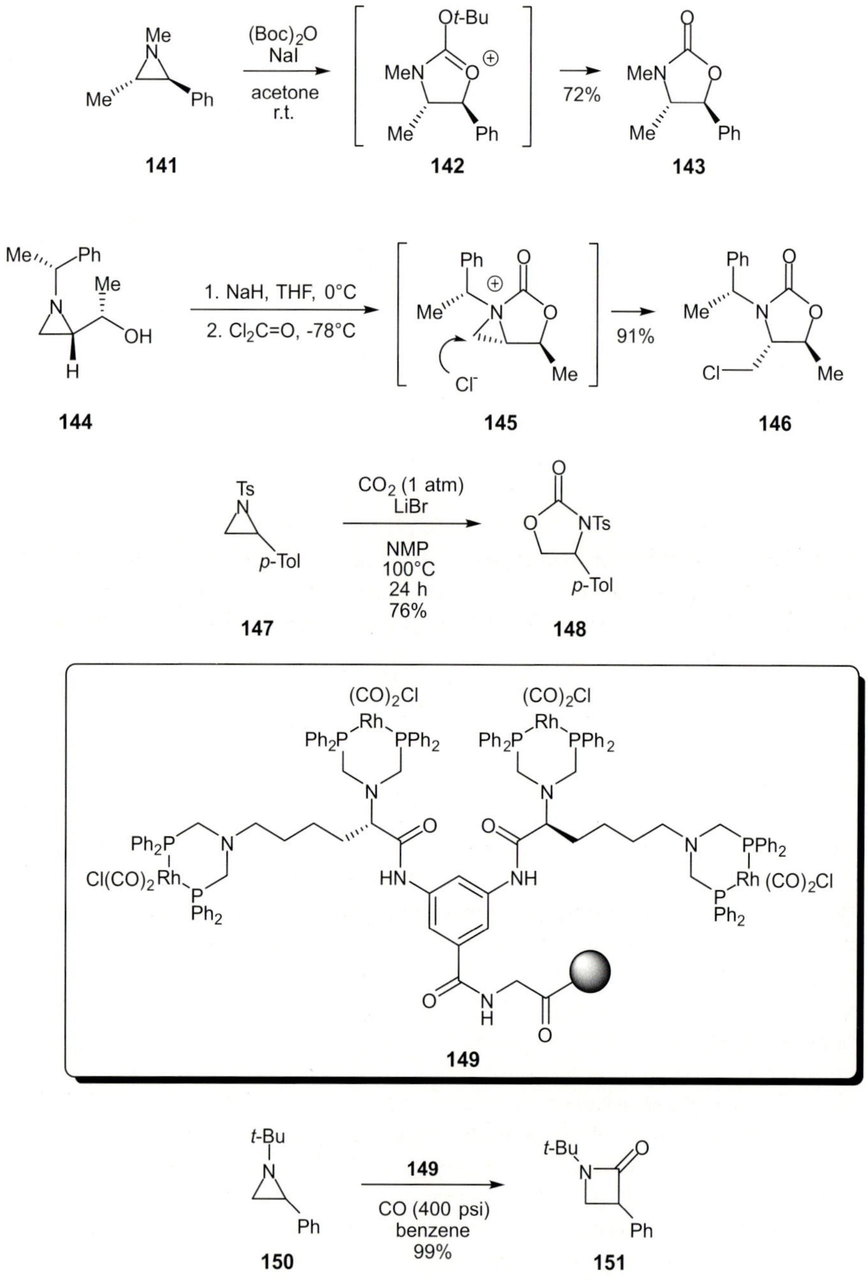

Scheme 2.32 Carbonyl insertions into aziridines.

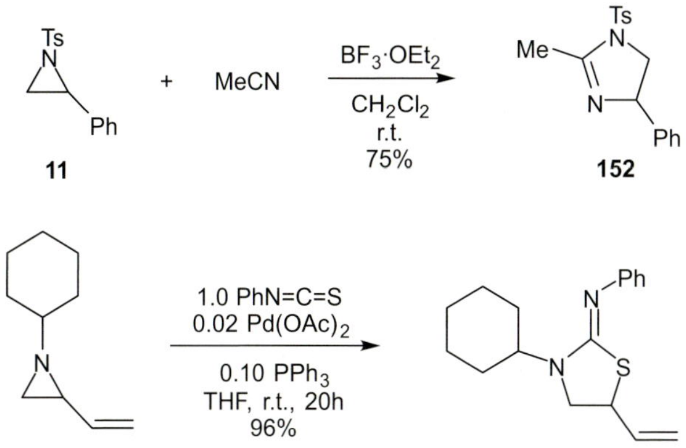

Scheme 2.33 Insertion of nitrogen and sulfur into aziridines.

us [217] can be derived from aziridines at room temperature using a palladium(II) acetate/triphenylphosphine catalyst system. Thus, phenylisothiocyanate is inserted into the C–N bond of vinylaziridine **153** to give the ring-expanded product **154** in excellent yield [218].

2.2 2*H*-Azirines

Although the level of activity does not match that of the aziridines, research into the chemistry of azirines is healthy and on the increase. Several reviews have appeared in recent years. The reader is directed particularly to two excellent treatises by Palacios and coworkers [219, 220] as well as an earlier work by Rai and Hassner [1].

2.2.1 Properties of Azirines

Also referred to as azacyclopropene (and confoundingly as 1-azirine), 2*H*-azirine is the unsaturated cousin of the aziridine ring, which might well be described metaphorically as a cyclic imine. The alternative 1*H*-tautomer lies some 35 kcal mol^{-1} higher in energy, in part because it constitutes an antiaromatic ring system [221–223]. It is a weaker base than aziridine, which is analogous to the trend in corresponding open-chain amines and imines [221]. Geometrically, according to calculation at the B3LYP/6-311 + G(3df,2p) level of theory, the triangular ring is somewhat scalene, with the C–N single bond being the longest side. Consequently, the sp^2 carbon vertex is the widest at almost 70° (Figure 2.6) [224]. The ^{1}H-NMR signals of the methylene protons resonate at 1.26 ppm, while the olefinic proton is shifted significantly downfield to 14.4 ppm. The C3 carbon appears at 164 ppm [225].

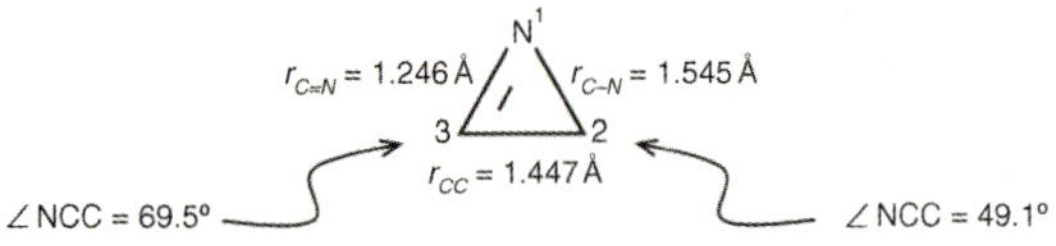

$r_{C=N}$ = 1.246 Å, r_{C-N} = 1.545 Å, r_{CC} = 1.447 Å, ∠NCC = 69.5°, ∠NCC = 49.1°

Figure 2.6 Geometry of 2*H*-azirine.

Remarkably, the azirine moiety is found in some naturally occurring compounds. For example, azirinomycin (**155**, Figure 2.7), isolated from *Streptomyces aureus* [226], exhibits antibiotic properties [227]. Two more biologically active azirines, dysidazirine (**156**) and antazirine (**157**), were identified in extracts of the marine sponge *Dysidea fragilis* [228]. While both enantiomers are found in nature, only the *(R)*-isomers of each are associated with desirable cytotoxic activity, the other antipodes being inactive [229].

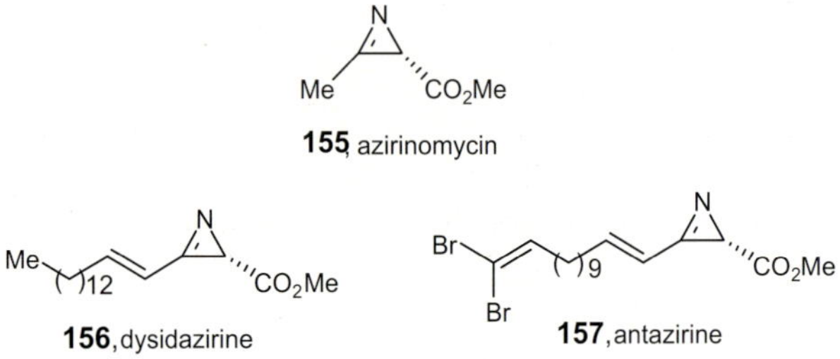

155 azirinomycin

156, dysidazirine

157, antazirine

Figure 2.7 Some naturally occurring 2*H*-azirines.

2.2.2 Synthesis of Azirines

2.2.2.1 Neber Route

This synthetic methodology has a very interesting history. The Neber rearrangement, first reported in 1926 [230], is the base-mediated conversion of oxime derivatives (originally sulfonates) into α-aminoketones (Scheme 2.34). Ultimately, it represents a useful protocol for the α-amination of ketones, which has recently been used to advantage in the total synthesis of dragmacidin F [231, 232]. While most applications still involve this complete transformation, it became clear early on that the mechanism involved an azirine intermediate that could be isolated under the right conditions [233].

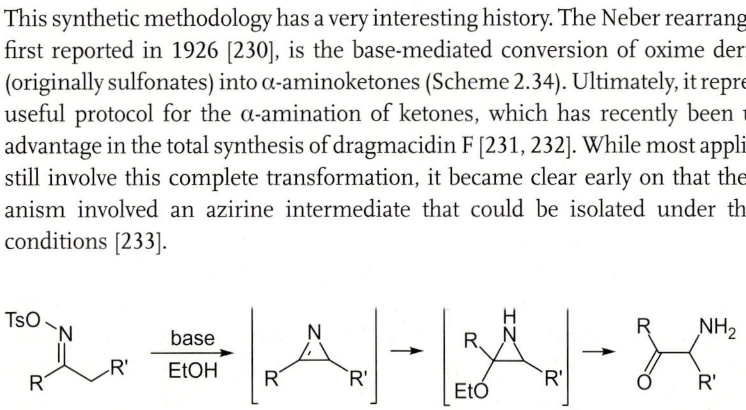

base, EtOH

Scheme 2.34 Neber rearrangement.

Generally speaking, there are two regiochemical possibilities for α-amination proceeding, in turn, through two regioisomeric azirines. House and Berkowitz [234] demonstrated that the regiochemistry of the Neber rearrangement is driven not by oxime geometry (as with the Beckmann rearrangement), but rather by the relative acidities of the α-protons. In other words, the initial tosylate displacement is effected by the more readily formed carbanion. Thus, in Neber precursors with electron-withdrawing groups, the active methylene almost always ends up as the saturated carbon of the azirine. For example, treatment of phosphonatoketoxime **158** (Scheme 2.35) with triethylamine at room temperature afforded in smooth fashion azirine **159** as the exclusive regioisomer [235]. Equally satisfactory results could be obtained using a solid-supported triethylamine analog (i.e., **160**) [236].

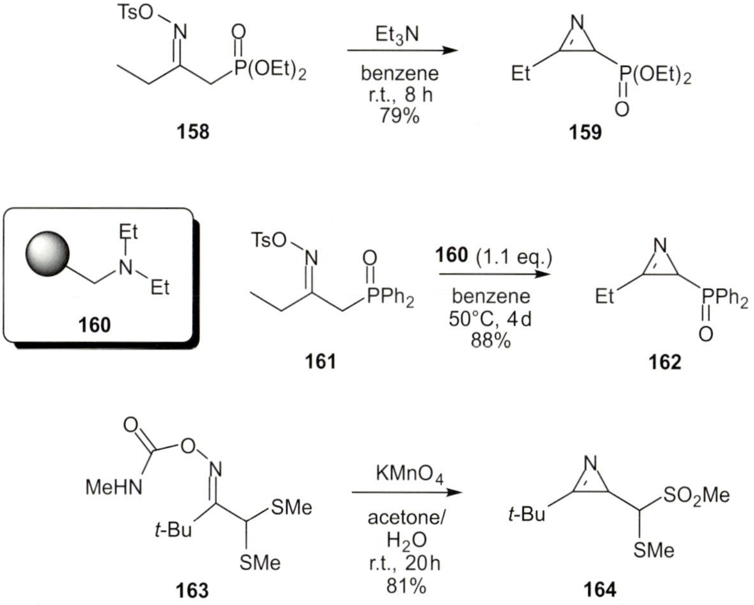

Scheme 2.35 Azirines from oximes with activating groups.

An unusual set of oxidative conditions has been reported for the thioacetal oxime carbamate **163**, which suffers oxidation of only one sulfide linkage in the presence of potassium permanganate and then cyclizes to azirine **164**, presumably under the influence of hydroxide liberated during the oxidation [237].

The method is tolerant to other leaving groups on the imino nitrogen. One frequently encountered class of compounds in this regard incorporates the quaternary hydrazonium moiety [238]. For example, the hydrazonium salt **165** (Scheme 2.36) derived from propiophenone is converted into the corresponding azirine **166** in fair yield upon treatment with sodium hydride in DMSO [239], and the similar unsaturated system **167** gives an excellent yield of azirine **168** under identical

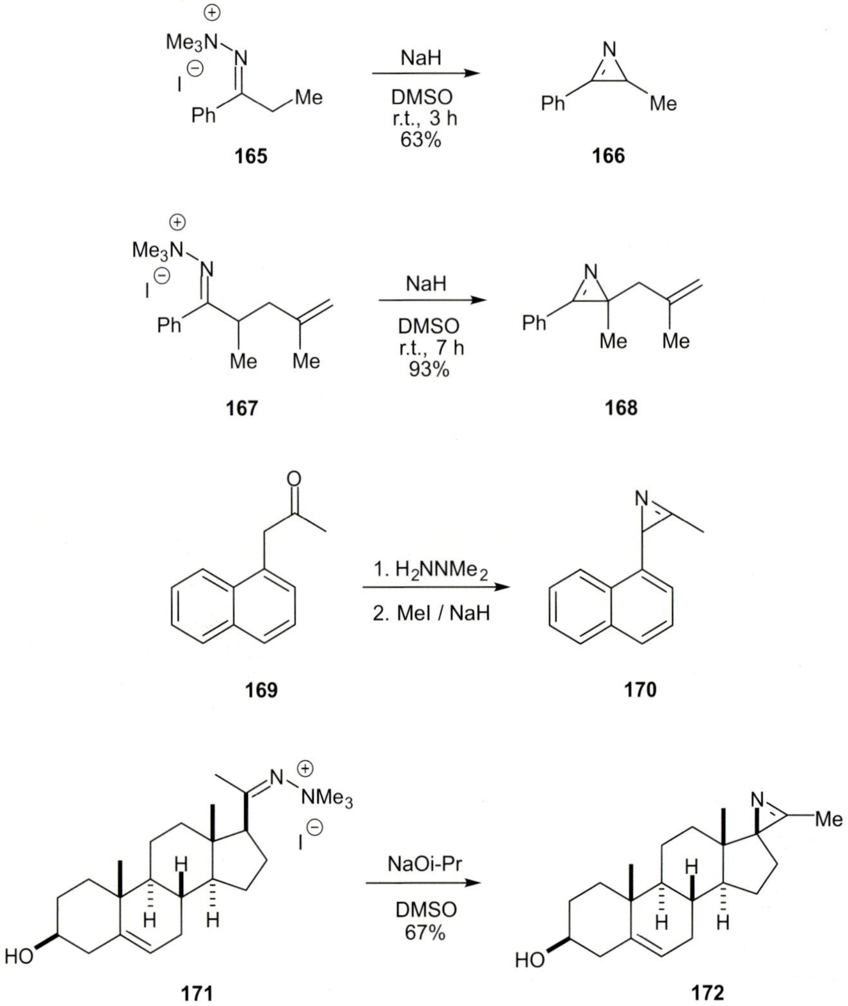

Scheme 2.36 Azirines from quaternary hydrazonium salts.

conditions [240]. In these cases, the regiochemistry is unambiguous, as there is only one set of α-protons. However, the process can be selective even when two regioisomers are possible. Thus, the naphthylacetone derivative **169** provides azirine **170** through sequential hydrazone formation, exhaustive methylation and deprotonation, whereby the product formation proceeds via the benzylic anion [241]. Even more subtle is the reaction course of the pregnenolone hydrazonium salt **171**, which gives a reasonably good yield of azirine **172**, corresponding to ring closure via the more substituted aza-enolate species. The latter case provides a striking example of the stability of azirines, as the authors report storage in ethanol for two weeks at room temperature without decomposition [242, 243].

There appears to be no universally applicable approach to chiral nonracemic azirines; however, strides are being made. In one example, the chiral amido oxime derivative **173** (Scheme 2.37) is reported to undergo conversion into azirine **174** with exclusive diastereoselectivity [244]. It is also interesting to note here the ability to isolate the azirine from the type of protic media usually encountered in the Neber rearrangement, which appears promising for adapting such protocols to azirine synthesis. The use of chiral bases has met with marginal success. In one case, modest enantioselectivities (up to 70% ee) were achieved using a chiral phase-transfer agent, which is assumed to form a tight ion pair with hydroxide and thus impose an asymmetric environment around the proton-transfer transition state [245]. A chiral base can also be used with good results, as demonstrated by the preparation of azirine **177** mediated by quinidine (**175**). Best results are obtained using stoichiometric quantities of organic base. However, a 20 mol.% loading could be used in the presence of potassium carbonate with only a slight decrease in selectivity [246]. This strategy has also been applied to the synthesis of azirines derived from phosphine oxides (e.g., **179**) [247].

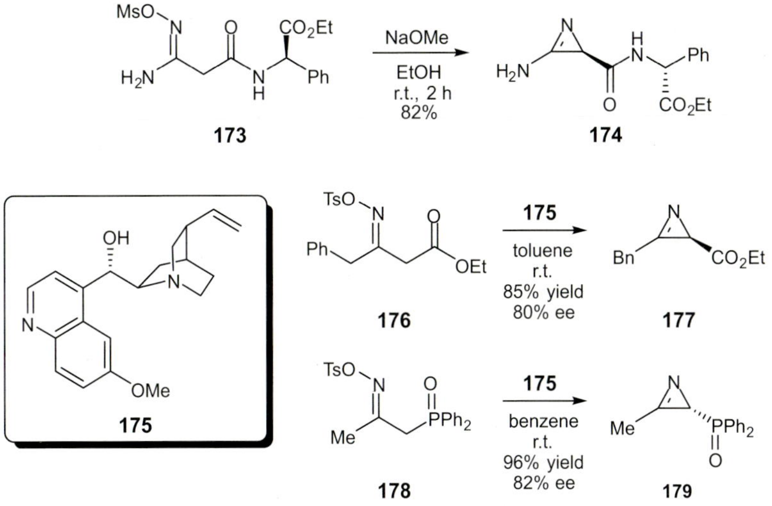

Scheme 2.37 Asymmetric induction in azirine formation.

2.2.2.2 From Vinyl Azides

Smolinsky reported in 1962 that the vapor-phase thermolysis of α-aziridinylstyrene (**180**, Scheme 2.38) produced substantial quantities of phenylazirine **181** [248, 249]. Subsequently, the method was adapted to be more amenable to the preparative scale, as exemplified by the synthesis of the trifluoromethylphenylazirine **184**, which was carried out on a 10-gram scale [250]. The vinyl azide in this case was synthesized from the corresponding alkene (i.e., **182**) through sequential bromination, azide displace-

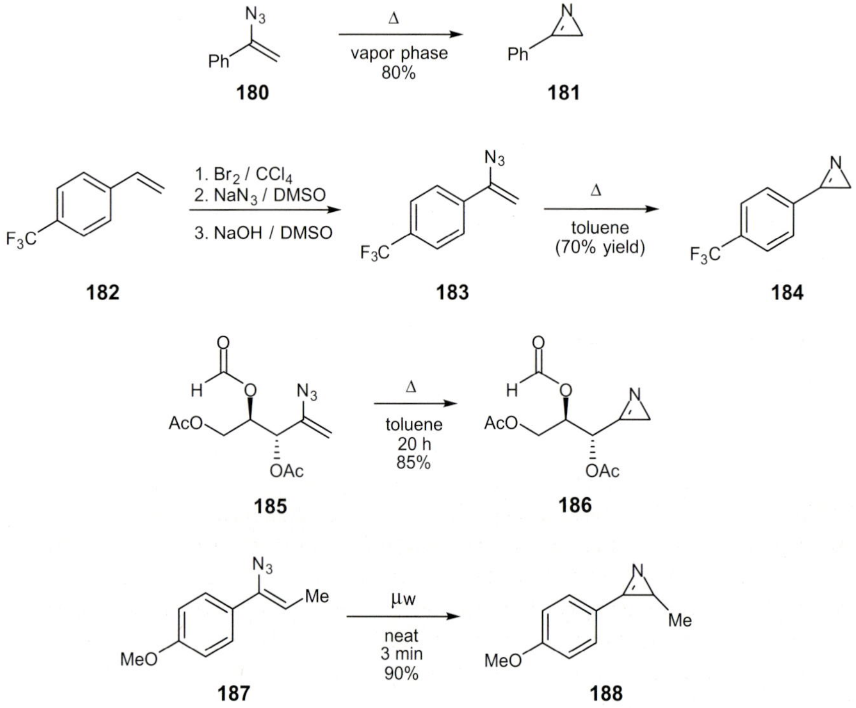

Scheme 2.38 Thermal decomposition of vinyl azides to azirines.

ment and dehydrobromination, a protocol used by Gilchrist and coworkers to access azirinyl esters [251, 252].

More recently, Somfai and coworkers [253] reported a much improved yield of azirine **184** (>95%) by carrying out the thermolysis in methylene chloride at 150 °C for 20 min in a sealed tube. Even at atmospheric pressure the process can be quite high yielding and remarkably tolerant to other functionality. Thus, heating the hexopyranose-derived vinyl azide **185** in toluene for 20 h resulted in the smooth formation of azirine **186** [254]. Perhaps the most convenient route is the neat thermolysis of aryl vinyl azides (e.g., **187**) in an open container under microwave irradiation, which proceeds in very good to excellent yields within a matter of minutes [255].

The course of the thermolytic reaction is substrate-dependent, and Hassner has generalized that azirines are usually formed when the substituent geminal to the azide is an aryl, alkyl, heteroatomic or ester group, whereas unsubstituted or keto-substituted substrates tend to form nitriles and other heterocycles upon thermolysis [256].

The same transformation can also be mediated by photolysis. Thus, irradiation of β-phenylethyl vinyl azide **189** (Scheme 2.39) induces loss of nitrogen and concomitant formation of azirine **190** in excellent yield [257]. The obvious advantage of this approach is that the decomposition can be carried out at low temperature, thus

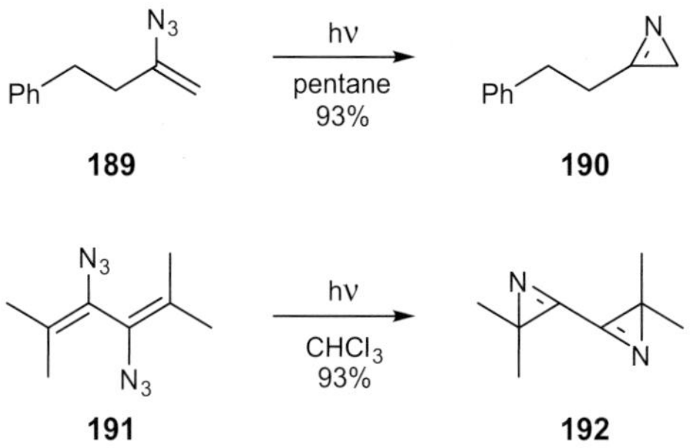

Scheme 2.39 Photolytic decomposition of vinyl azides to azirines.

affording access to more strained products, including fused bicyclic azirines and bis-azirines such as **192**. In the latter case, the loss of nitrogen occurs stepwise, and the intermediate vinyl azirine can be isolated [258]. One bottleneck for both the thermolytic and photolytic methodology is access to the requisite vinyl azide substrates [259]. One promising recent contribution to solving this problem is the copper (I)/L-proline mediated coupling of sodium azide with (*Z*)-1-iodo-1-alkenes [260] which, in turn, can be accessed stereoselectively via the Wittig reaction of aldehydes with iodomethylenetriphenylphosphorane [261].

The mechanism of this reaction has been the matter of some debate. Careful kinetic studies in solution phase point towards a concerted mechanism involving a multi-centered transition state in which the nascent sp^2 azirinyl carbon exhibits partial positive character (Figure 2.8). This is supported by Hammett studies in which electron-donating substituents on the aryl ring accelerate the decomposition. Thus, in a solvent of 1-butanol at 80 °C, the conversion of the *p*-methoxy derivative (R=4-OMe) is about 50% faster than that of the phenyl substrate (R=H) and almost three times faster than the *m*-nitro analog (R=3-NO_2) [262].

For α-aminoalkenyl azides, then, decomposition to the corresponding aminoazirine is usually spontaneous and rapid even at room temperature. In fact, many protocols simply form the vinyl azide *in situ* from other precursors. For example, keteneiminium salts (**194**, Scheme 2.40), available from the treatment of amides

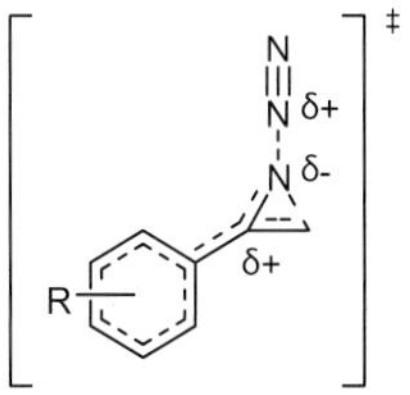

Figure 2.8 Proposed transition state for thermolysis of aryl vinyl azides.

(e.g., **193**) with phosgene, add sodium azide to form transient α-aminovinyl azides (e.g., **195**) that subsequently lose nitrogen at room temperature to provide azirine products (e.g., **196**) [263]. The use of phosgene can be avoided with diphenyl phosphorazidate (DPPA), an aziridinating agent that reacts with amidate anions to give the corresponding azirines directly, as shown by the conversion of *N*-methyl-*N*-phenyl amide **197** into the aminoazirine **199** in 94% yield [264]. This methodology is tolerant of various functionalities [265], but some substrates must first be converted into the thioamide using Lawesson's reagent [266, 267]. Other suitable precursors include α-chloroenamines (e.g., **200**), which undergo sequential azide displacement and rapid thermal conversion into the azirine (e.g., **202**) under very mild conditions [268].

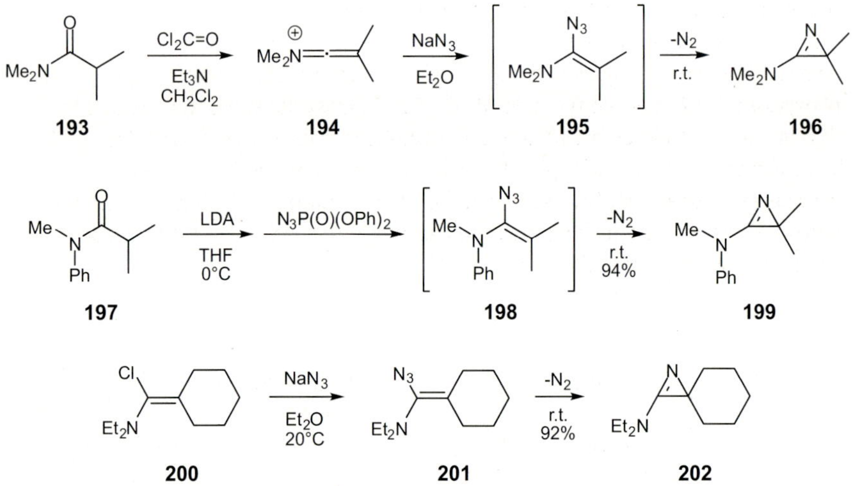

Me_2N **193** —($Cl_2C{=}O$, Et_3N, CH_2Cl_2)→ Me_2N **194** —(NaN_3, Et_2O)→ [N_3, Me_2N **195**] —($-N_2$, r.t.)→ Me_2N, N **196**

Me, N, Ph **197** —(LDA, THF, 0°C)→ —($N_3P(O)(OPh)_2$)→ [N_3, Me, N, Ph **198**] —($-N_2$, r.t., 94%)→ Me, N, Ph, N **199**

Cl, Et_2N **200** —(NaN_3, Et_2O, 20°C)→ N_3, Et_2N **201** —($-N_2$, r.t., 92%)→ N, Et_2N **202**

Scheme 2.40 Synthesis of azirines with electron-donating groups.

There are also high-yielding methods for preparing azirines appended with electron-withdrawing substituents. Thus, allenic phosphonates (**204**, Scheme 2.41), which can be accessed from propargyl alcohol derivatives (**203**), are themselves efficient precursors for phosphonylmethyl vinyl azides (**205**). Irradiation of the latter results in extremely high-yielding photolysis to the azirine (**206**) [269]. As an additional entry under the rubric of electron-withdrawing substituents, bromoazirinyl ester **210** is formed in almost quantitative yield from the azidoacrylate derivative **209** under thermal conditions in refluxing hexane [270]. Analogous iodo azirinyl esters can be prepared through similar methodology [271].

2.2.2.3 From Other Heterocycles

Synthetically useful yields of azirines can be obtained from the thermolysis of isoxazoles (Scheme 2.42), as demonstrated by the thermal rearrangement of ami-

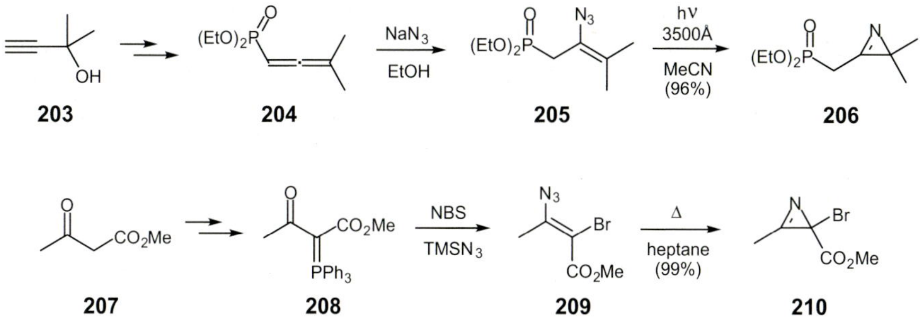

Scheme 2.41 Synthesis of azirines with electron-withdrawing groups.

noisoxazole **211** to azirinyl carboxamide **212** at high temperature under argon. Neat thermolysis is the preferred mode, as the use of solvent was found to give lower yields [272]. Alternatively, iron dichloride promotes the analogous rearrangement of alkoxyisoxazole **213** at room temperature. The 5-alkoxy substituent is crucial, as 5-alkyl or 5-aryl substituents lead to enaminoketones under the same conditions [273]. Benzisoxazole derivatives (e.g., **215**) exhibit a regiochemically distinct mode of thermal rearrangement. Instead of bonding to the internal α-carbon, the nitrogen instead forms the azirine ring with the adjacent carbon of the substituent, so that aromaticity is preserved [274].

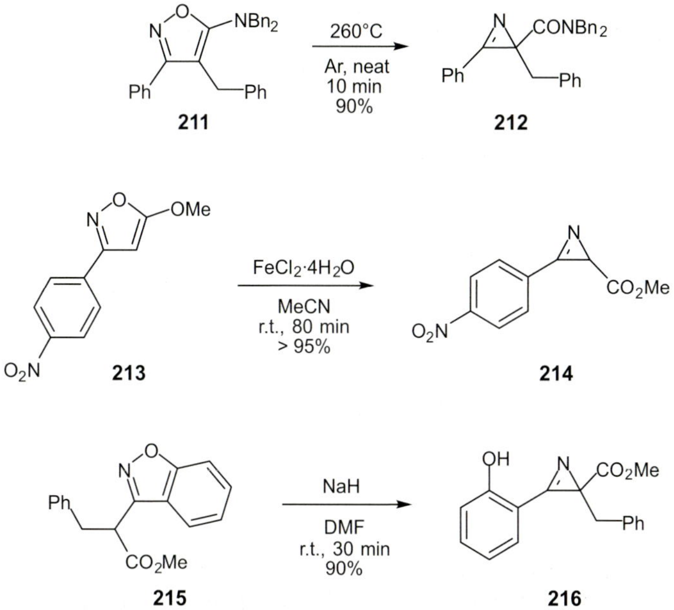

Scheme 2.42 Synthesis of azirines from isoxazoles.

In a similar vein, oxazaphosphole **217** (Scheme 2.43) underwent pyrolysis at 400 °C under high vacuum to produce azirine **218**, which was trapped at low temperature and characterized by IR spectroscopy. Pyrolysis at higher temperatures (700 °C) gave nitrile ylides instead [275]. A somewhat more synthetically friendly procedure was reported for oxazaphospholine **220** (available in four steps from α-bromoketoxime **219**), which decomposed at 120 °C and atmospheric pressure to provide methylazirine **221** as a distillate in good yield [276].

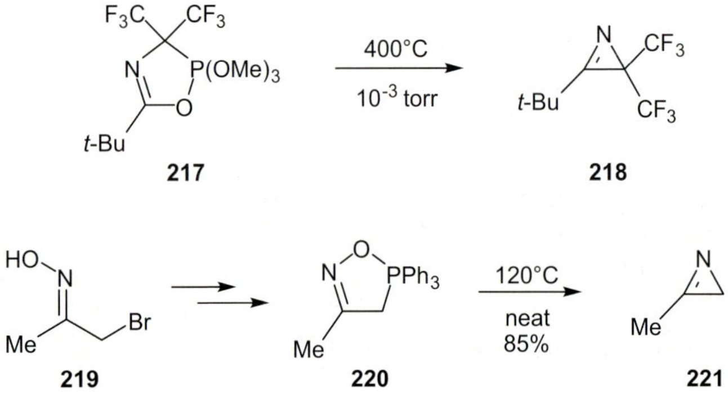

Scheme 2.43 Synthesis of azirines from dihydrooxazaphosphole derivatives.

Finally, azirines can be synthesized from suitably equipped aziridine precursors. For example, aziridinyl ester **222** (Scheme 2.44) was smoothly oxidized using *t*-butyl hypochlorite to give the N-chloro derivative **223** in excellent yield. Subsequent DBU-mediated elimination to the azirine **224**, however, was considerably less efficient [277]. A convenient workaround to this problem was found by using Swern conditions, which effects the same transformation in 86% overall yield [278]. Davis and coworkers have reported an interesting eliminative pathway by treating *N*-sulfinyl or *N*-sulfonylaziridines (e.g., **225**) with lithium diisopropylamide (LDA) at low temperature [279, 280]. Methyleneaziridines can easily be isomerized to azirines, since the latter lie about 9 kcal mol^{-1} lower in energy [281]. Thus, *N*-silyl-methyleneaziridine **227** quantitatively isomerizes to azirine **228** via an alumina-mediated Brooks rearrangement at room temperature [282].

2.2.3
Reactivity of Azirines

2.2.3.1 Addition of Nucleophiles

Inasmuch as azirines incorporate an unsaturated nitrogen center within a three-membered ring, it would be reasonable to think of their expected chemical behavior as that of an activated imine. Consequently, nucleophilic addition to the sp^2 carbon can be a synthetically useful reaction for azirines. For example, azirinyl phosphonate

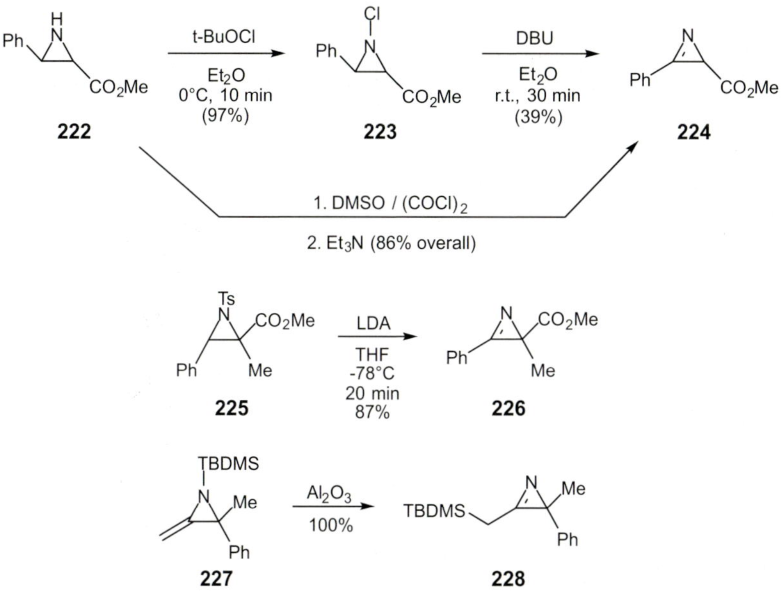

Scheme 2.44 Synthesis of azirines from aziridines.

229 (Scheme 2.45) suffers high-yielding nucleophilic allylation in the presence of excess allyl Grignard [283]. Analogous results can be obtained in excellent yield using allylindium reagents generated *in situ* from the corresponding iodide and indium metal, as shown in the conversion of azirine **231** into allylaziridine **232**. The stereochemistry of the latter reaction depends upon the substituent at the saturated carbon: chelating groups (i.e., keto or hydroxy moieties) led to *cis*-delivery of the nucleophile, whereas alkyl and aryl groups resulted in *trans*-addition [284]. When the azirine is sufficiently activated, even weak nucleophiles can engage in addition to the ring. Thus, exposure of azirinyl phosphonate **233** to acryloyl chloride results in initial N-acylation followed by subsequent rapid addition of chloride to form chloroaziridine **234** [285].

Aziridines can also be accessed by the stereoselective reduction of azirines with hydride reagents. For example, sodium borohydride reacts smoothly in ethanolic medium with azirine **229** to yield the *cis*-aziridine **235**, resulting from the trans-addition of hydride [286]. Alternatively, an asymmetric transfer hydrogenation protocol using the chiral aminoalcohol **236** and a ruthenium catalyst can be used to generate enantiomerically enriched aziridines from achiral azirine precursors. Thus, tolylazirine **237** was converted into *(S)*-tolylaziridine **238** in good yield and promising enantiomeric excess. The mechanism follows the route elucidated in analogous reactions of ketones, in which hydride addition and proton transfer occur simultaneously [287].

A very clever protocol has been developed by Heimgartner and coworkers, which takes advantage not only of the propensity of the azirine to add nucleophiles but also

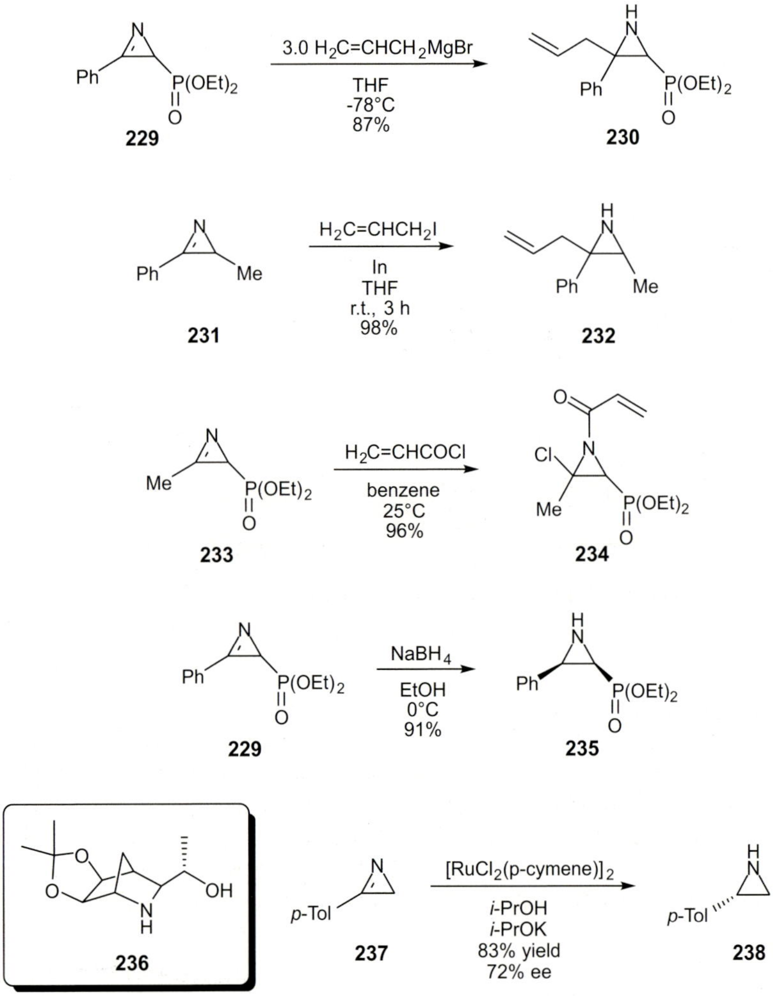

Scheme 2.45 Reaction of azirines with nucleophiles.

of the residual ring strain in the resulting aziridine ring. Thus, aminoazirines such as **240** (Scheme 2.46) engage carboxylates in nucleophilic addition to form transient aziridine intermediates that spontaneously rearrange to give amino acid derivatives (e.g., **241**) [288, 289]. This strategy, dubbed the "azirine/oxazolone method," allows quite a bit of flexibility in introducing structural variation into peptide backbones, as exemplified by the spiroazirines **243** [290] and **245** [266]. Quite a few synthons have been developed within this manifold, including those for 2-methylaspartate [291], α-methylglutamate [267], as well as for dipeptides [292] such as 2-aminoisobutyric acid/4-hydroxyproline [293] and other 2,2-disubstituted glycines [294, 295]. The methodology has been showcased in the synthesis of Trichovirin I derivatives [296] and has been further expanded into the realm of solid-phase synthesis [297, 298].

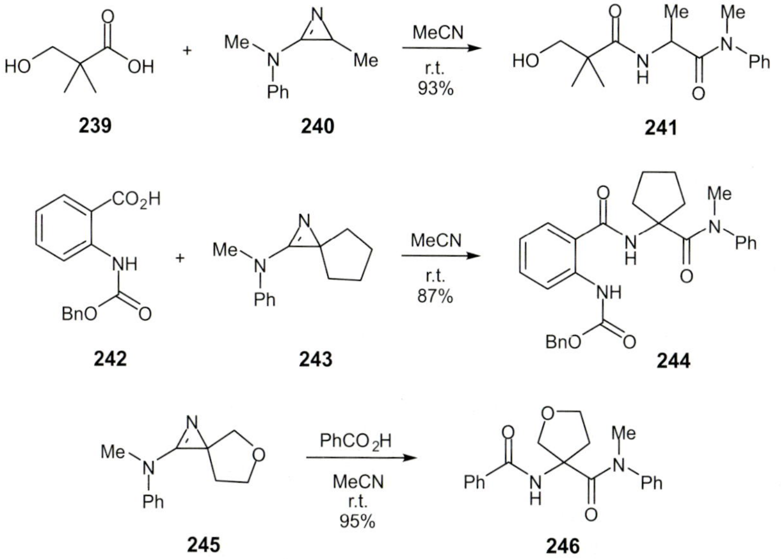

Scheme 2.46 Addition of carboxylic acids to aminoazirines.

Nucleophilic substitution can actually be carried out at the saturated carbon of certain uniquely functionalized azirines. For example, the benzotriazolyl azirine **247** (Scheme 2.47) takes part in an intermolecular S_N2 reaction with thiophenolate to give the aziridinyl sulfide **248** in fair yield. Carbon-centered nucleophiles, such as Grignard reagents, can also be used [299]. Bromoazirine derivative **250** reacts with *o*-phenylenediamine (**249**) in both modes (nucleophilic addition and S_N2 displacement) to give the disubstituted quinoxaline **251** in very good yield [300].

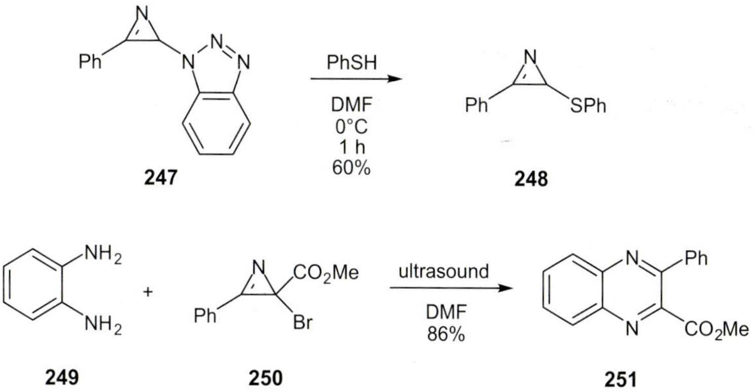

Scheme 2.47 S_N2-Type additions to azirines.

2.2.3.2 Cycloadditions

Perhaps some of the most fascinating chemistry associated with the azirines springs from their propensity to act as 2π donors in cycloaddition reactions. For example, Diels–Alder reactions using azirines as dienophiles [301] can provide synthetically valuable fused azabicyclo[4.1.0]heptane ring systems. In some cases, the reaction proceeds under simple thermal conditions and in very high yield, as illustrated by the quantitative reaction of azirinyl ester **252** (Scheme 2.48) with furan to give the Diels–Alder adduct **253** [302]. The method is amenable to other electron-rich dienes, such as bis(siloxy)diene **254** [303], as well as other dienophiles, such as aziridinyl carboxamides [304]. The reaction rate can be dramatically accelerated and the scope extended to alkyl and aryl aziridines (e.g., **258**) by the use of Lewis acids [305]. A particularly high-yielding result was obtained in the magnesium bromide-mediated Diels–Alder reaction between trimethylsiloxycyclohexadiene (**260**) and chiral non-racemic azirinyl ester **261**, which proceeded with almost exclusive endo- and regioselectivity (but apparently low facial selectivity) [306, 307].

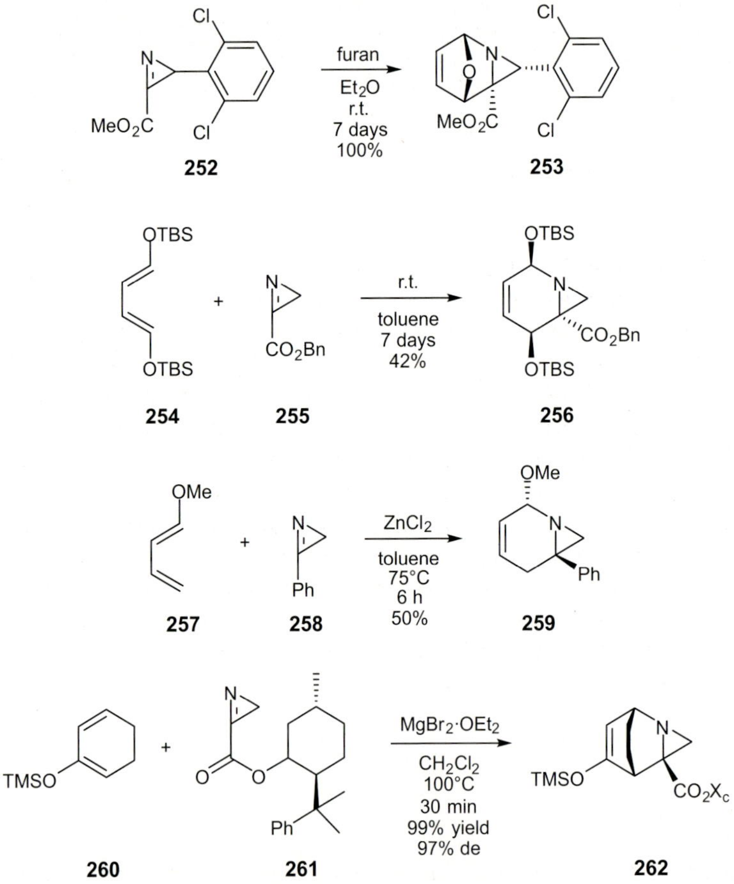

Cl, N, MeO_2C, Cl, **252**; furan, Et_2O, r.t., 7 days, 100%; Cl, N, O, MeO_2C, Cl, **253**

OTBS, OTBS, **254**; +; N, CO_2Bn, **255**; r.t., toluene, 7 days, 42%; OTBS, N, CO_2Bn, OTBS, **256**

OMe, **257**; +; N, Ph, **258**; $ZnCl_2$, toluene, 75°C, 6 h, 50%; OMe, N, Ph, **259**

TMSO, **260**; +; N, O, O, Ph, **261**; $MgBr_2{\cdot}OEt_2$, CH_2Cl_2, 100°C, 30 min, 99% yield, 97% de; TMSO, N, CO_2X_c, **262**

Scheme 2.48 Diels–Alder reactions with azirines.

The field is not limited to [4 + 2] methodology. For example, arylazirine **264** and fulvene **263** engage in a formal [6 + 3] cycloaddition catalyzed by yttrium triflate and mediated by adventitious water, providing a novel entry into the [2]pyridine system (Scheme 2.49) [308]. The *p*-nitrophenylazirine **268** was also shown to be an effective 1,3-dipolarophile in the presence of azomethine ylide **267** (generated by the thermolysis of bicyclic oxazolidinone **266**), forming a fused tricyclic adduct (**269**) that was parlayed into an approach to 1-azacephams [309].

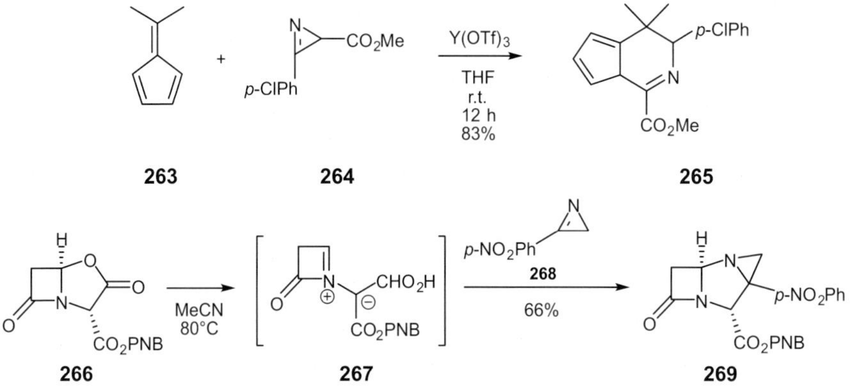

Scheme 2.49 Other cycloadditions with azirines.

2.2.3.3 **Rearrangements into other Heterocycles**

Owing to their strain and functionality, azirines are prone to molecular rearrangement. For example, under neutral thermolytic conditions the alkyl aziridinylphosphonate **270** (Scheme 2.50) is converted almost quantitatively into the pyrazine **271**. The mechanism is believed to involve the dimerization of unstable nitrile ylides generated from the initial thermal ring-opening of the azirine [310]. Furthermore, Padwa and Stengel have reported some fascinating azirine rearrangements promoted by Grubb's catalyst. Thus, the azirinyl aldehyde **272** is transformed cleanly into the isoxazole **273** at room temperature. Similarly, vinyl azirine **274** undergoes thermal rearrangement to form exclusively the 2,5-disubstituted pyrrole **276**, which is complementary to photolytic rearrangement, a process providing only the 2,3-product **275** [311, 312]. Finally, Taber has reported on the high-yielding synthesis of indoles (e.g., **278**) from arylazirines **277**, which are themselves synthesized from the oximes of aryl ketones [313].

2.3 Oxiranes

It may well be said that the field of oxirane chemistry has escaped the confines of a brief comprehensive overview. Indeed, an encyclopedic summary of just the last

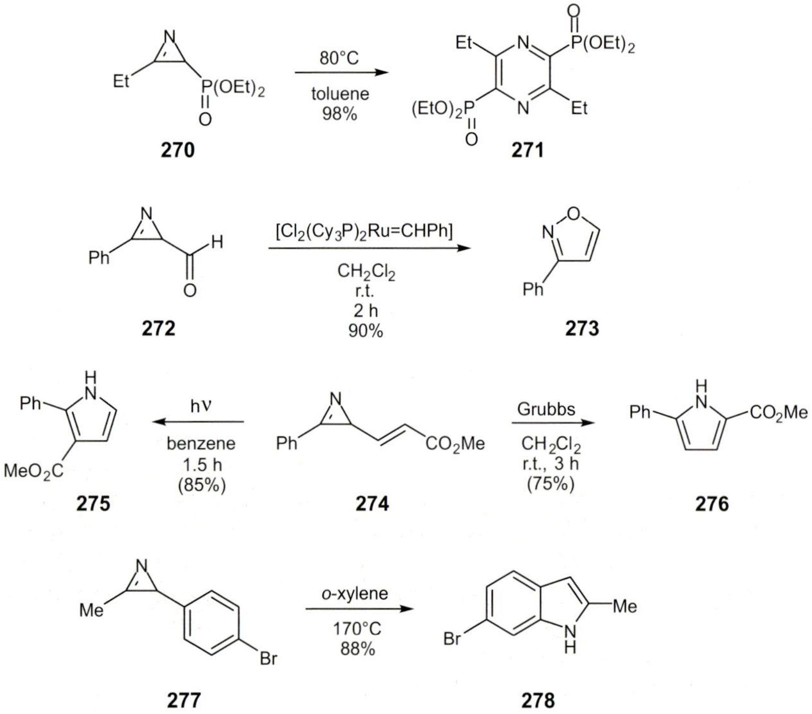

Scheme 2.50 Some rearrangements of azirines.

year's activity would occupy volumes. Several significant reviews have appeared recently [8], many of which pertain to specific areas within epoxide chemistry and will therefore be cited in the appropriate context. The present chapter seeks to assemble a sampling of the diverse applications for organic synthesis.

2.3.1 Properties of Oxiranes

Oxiranes (also known as epoxides and oxacyclopropanes) owe much of their utility, whether as synthetic intermediates or biologically active compounds, to ring strain. For example, oxirane itself (bp 10.5 °C) is associated with some 27 kcal mol^{-1} of strain enthalpy. Like most of its congeners, this molecule exhibits the very useful balance of being stable enough for isolation, transportation and so on, while still harboring a remarkable propensity for reaction. Geometrically, oxirane describes an almost equilateral triangle, with a slightly relaxed bond angle at the oxygen center (Figure 2.9). The ^{1}H-NMR signals of the methylene protons resonate at 2.54 ppm and the carbon atoms appear at 39.7 ppm. The increased s-character of the C–H bond leads to an unusually high ^{13}C-H coupling of 176 Hz [7].

The oxirane ring is found in a host of naturally occurring compounds of biological relevance. Occasionally, the structures can be quite simple, with the oxirane being the

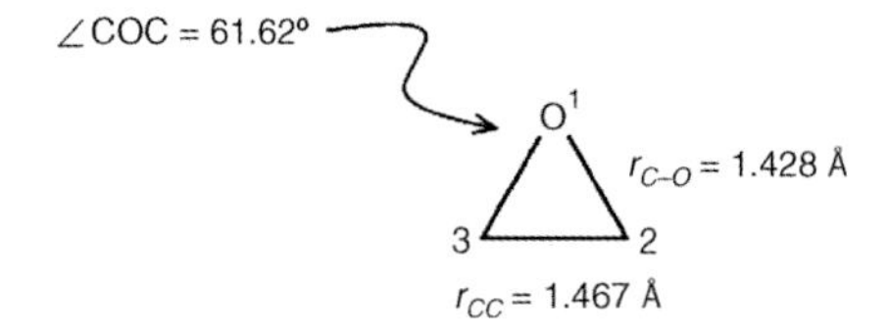

Figure 2.9 Geometry of oxirane.

dominant functional group. Such is the case for the newly discovered stilbene oxide derivative **279** (Figure 2.10), which was isolated from the larval *G. mellonella* infected by the nematode-bacterium complex *H. megidis* 90/*P. luminescens* C9. This compound displays broad antimicrobial activity against many troublesome pathogens, including the drug-resistant strain *Staphylococcus aureus* RN4220 [314].

More commonly, however, the epoxide ring is imbedded within a molecule carrying elaborate functional embellishments. A good example is altromycin B (**280**), a secondary metabolite of soil *Streptomyces* which exhibits both antibiotic and anticancer activity [315]. The mode of action has been traced to the inhibition of DNA synthesis caused by the alkylation of guanine through epoxide ring-opening [316]. Another natural product sporting the oxirane and anthraquinone moieties is dynemicin A (**281**), isolated from *Micromonospora chersina*, which is cytotoxic at concentrations approaching the parts per trillion range. Here the epoxide ring functions as the trigger for a mouse trap: nucleophilic ring opening of the epoxide results in a conformational relaxation that allows for cyclization of the enediyne array, forming the very reactive arene diradical [317].

Ambuic acid (**282**), isolated from the rain forest fungi *Pestalotiopsis* spp. and *Monochaetia* spp., has shown activity against several plant pathogenic fungi [318]. This polyketide-derived natural product is one representative from a broad range of cyclohexane epoxides found in nature, a class that has been the subject of a quite recent and fairly comprehensive review [319]. The structural and biological diversity even within this classification is impressive. Fumagillin (**283**), a cyclohexane spiroepoxide produced by *Aspergillus fumigatus*, is an angiogenic inhibitor that binds to methionine aminopeptidase. Consequently, it is a promising antineoplastic agent and may even inhibit atherosclerosis [320, 321].

Polyketide-derived oxiranyl natural products also extend into macrocyclic species. Amphidinolide B1 (**284**) is one such cytotoxic 26-membered macrolide, isolated from the marine dinoflagellate *Amphidinium* sp. Y-5 [322–324]. Similar cytotoxic activity exhibited by the 16 membered macrolide epothilone A (**285**), a metabolite of the cellulose-degrading myxobacterium *Sorangium cellulosum* (Myxococcales), has made it an interesting anti-cancer candidate and subsequently the subject of active investigation from both a synthetic organic and chemical biology standpoint [325, 326]. Also of marine origin is the bromotyrosine-derived dimeric spiroisoxazoline (+)-calafianin (**286**), which is found in methylene chloride extracts of the Mexican sponge *Aplysina gerardogreeni* n. sp (Aplysinidae). Members in this class of compounds have demonstrated activity against *Mycobacterium tuberculosis* [327, 328].

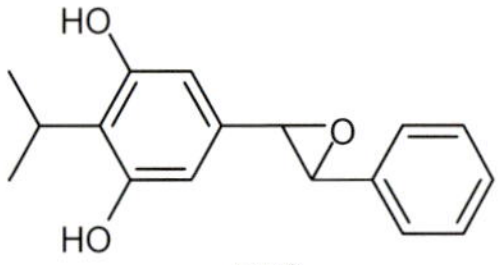

279

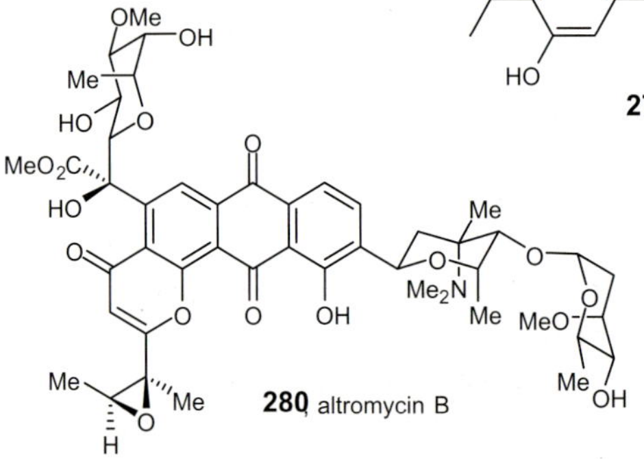

280, altromycin B

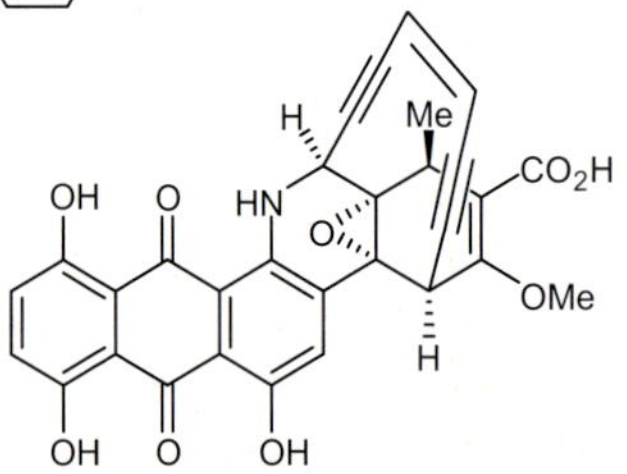

281, dynemicin A

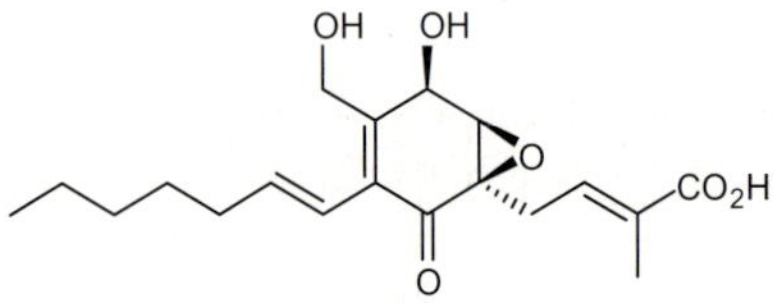

282, ambuic acid

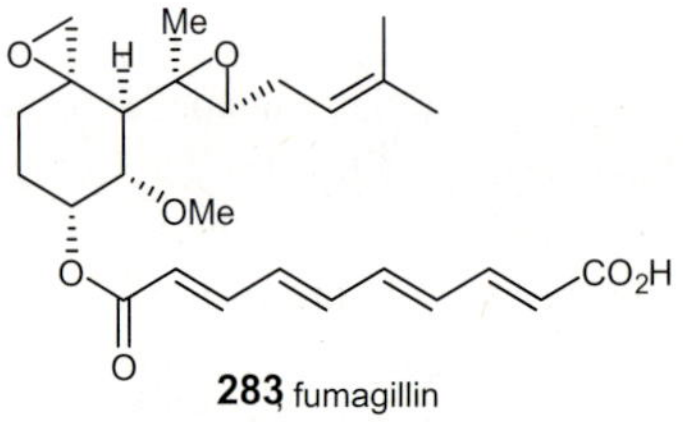

283, fumagillin

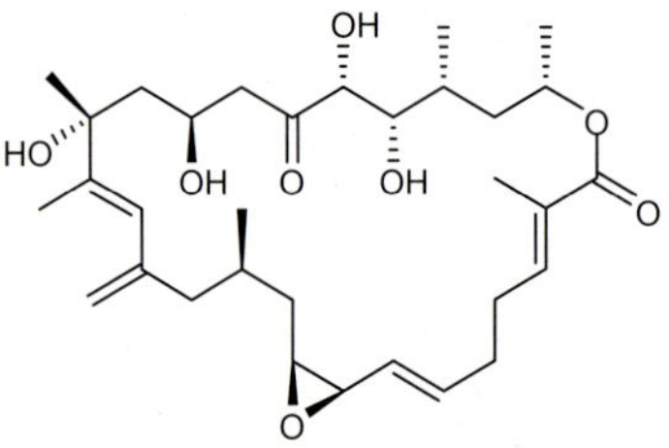

284, amphidinolide B1

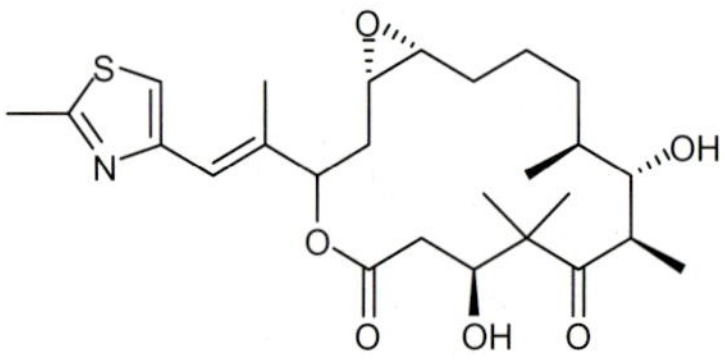

285, epothilone A

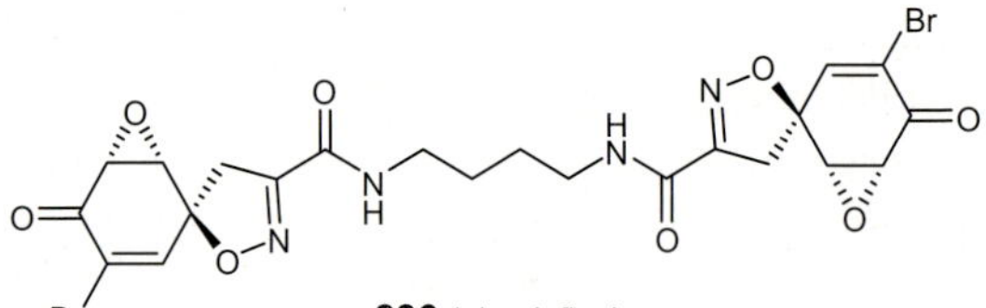

286, (+)-calafianin

Figure 2.10 Some naturally occurring oxiranes.

2.3.2 Synthesis of Oxiranes

The synthesis of oxiranes has been the subject of several recent reviews, encompassing biosynthetic routes [329] and asymmetric methodologies [330], including those proceeding through homo- and heterogeneous catalysis [331] and epoxidations under the influence of chiral auxiliaries [332]. The preparation of vinyl epoxides has also been reviewed recently [333], as have as other topical works mentioned in the appropriate context below.

2.3.2.1 Using Dioxiranes

Epoxides can be prepared by the action of dioxiranes on alkenes under mild conditions and low temperature [334]. For unstable epoxides, such as those derived from enol ethers, dioxirane epoxidation is the method of choice. Several reports have exploited this reaction to obtain the previously unisolable acyloxo, alkoxo and silyloxooxiranes. An example of this approach involves the preparation of epoxide **288** from enol lactone **287** (Scheme 2.51) using one of the simplest dioxirane

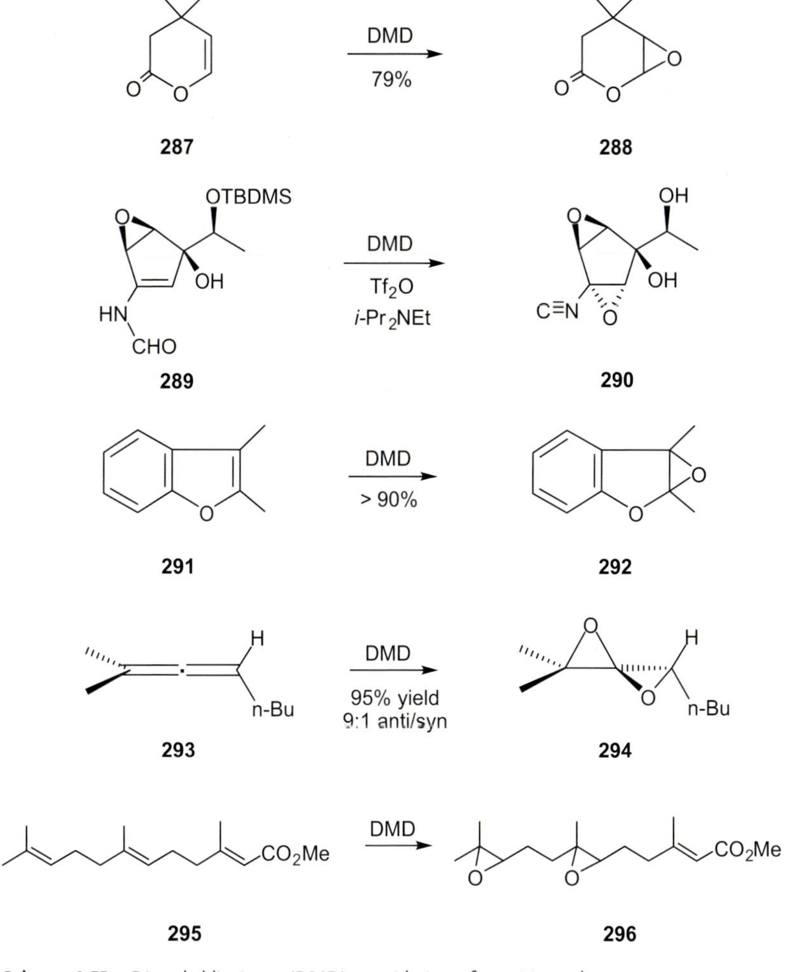

Scheme 2.51 Dimethyldioxirane (DMD) epoxidation of sensitive substrates.

derivatives, dimethyldioxirane (DMD). Similarly, DMD epoxidation of silyl enol ethers afforded the corresponding epoxides in excellent yields. These substrates readily undergo an acid-catalyzed rearrangement to α-trimethylsiloxy carbonyl derivatives [335, 336]. Danishefsky has described the use of DMD for the direct epoxidation of glycals. The 1,2-anhydrosugars produced were employed in the stereospecific construction of β-linked oligosaccharides [337].

Epoxy isonitriles cannot be prepared by the direct epoxidation of vinyl isonitriles. However, the DMD oxidation of a series of vinyl formamides was found to produce epoxy formamides in good yield. These compounds were readily converted into the epoxy isonitriles using triflic anhydride and Hünig's base. A total synthesis of isonitrin B (**290**) was carried out using this methodology [338]. Adam and coworkers reported the first benzofuran epoxide synthesis (i.e., **292**) using DMD in their study on the mutagenesis of benzofuran dioxetanes [339]. The same oxidant has been used to advantage in accessing other labile products, including 1,2-dialkoxyoxiranes [340] and flavone epoxides [341].

Crandall and coworkers have studied the DMD-mediated epoxidation of various allenes **293**. The corresponding 1,4-dioxiaspiro[2.2]pentanes **294** were produced in good yields, the mono- and di-substituted allenes giving anti diastereomers with good stereoselectivity. These spirodiepoxides then underwent nucleophilic cleavage under buffered conditions (Bu_4NOAc/HOAc) to give highly functionalized α,α′-dihydroxyketone derivatives [342]. Messeguer and coworkers have employed dimethyldioxirane to prepare the diepoxide **296**, a proposed metabolic intermediate of juvenile hormone III [343].

Dioxiranes can also be employed in a catalytic fashion. Practically unrivaled in efficiency and ease of use, DMD itself can be generated *in situ* from acetone in an appropriate buffer. Thus, the dropwise addition of an aqueous solution of Oxone to a stirred mixture of *cis*-carveol (**297**, Scheme 2.52), sodium bicarbonate, and acetone at 0 °C led to the selective formation of epoxide **298** in 92% yield [344]. This general methodology has been expanded to include a wide variety of oxygen carriers. For example, in the area of natural products, Marples and coworkers have used dioxiranes generated *in situ* from a range of ketones to effect 5,6-epoxidation of cholesterol or its acetate in high yield [345]. Bortonlini *et al.* have used sodium dehydrocholate as an organomediator for the sodium perborate (SPB) mediated preparation of acid-sensitive epoxides, in which the intermediacy of a steroidal dioxirane has been suggested [346].

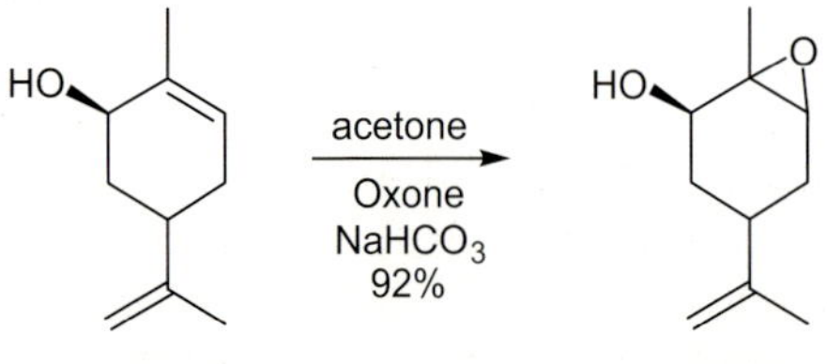

Scheme 2.52 Catalytic epoxidation using DMD.

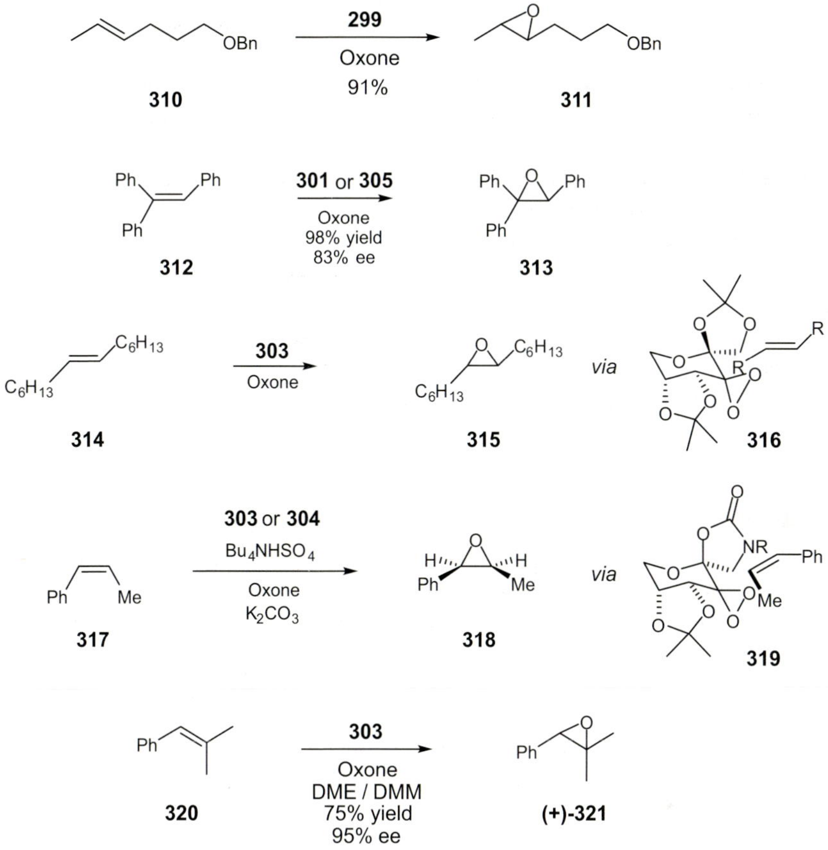

Scheme 2.53 Epoxidation of alkenes using catalytic dioxiranes.

Denmark has developed a practical phase-transfer protocol for the catalytic epoxidation of alkenes, which uses Oxone as a terminal oxidant. The olefins studied (e.g., **310**, Scheme 2.53) were epoxidized in 83–96% yield. Of the many reaction parameters examined in this biphasic system, the most influential were found to be the reaction pH, the lipophilicity of the phase-transfer catalyst and the counterion present. In general, optimal conditions feature 10 mol.% of the catalyst 1-dodecyl-1-methyl-4-oxopiperidinium triflate (**299**, Figure 2.11) and a pH 7.5–8.0 aqueous methylene chloride biphasic solvent system [347], although these systems are tolerant to neutral and basic pH ranges [348]. The dioxirane derived from the bis(ammonium) ketone **300** also exhibits remarkable reactivity, stability, and water solubility, and has been used to advantage on various substrates [349]. The activation barrier for these reactions is drastically reduced by hydrogen bonding in the transition state, whether by solvent or intramolecularity [350].

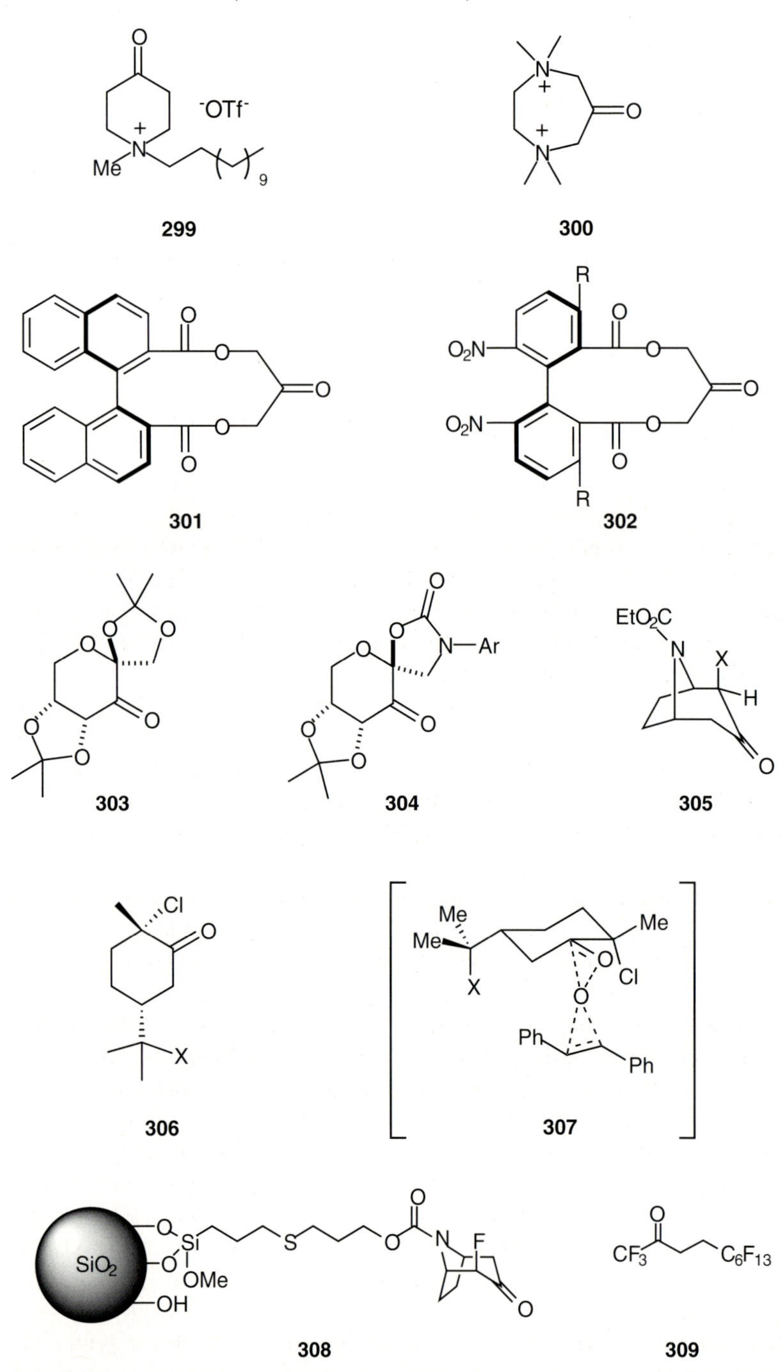

Figure 2.11 Ketone precursors for dioxirane oxidations.

The latest twist to this development is the use of chiral ketones in catalytic amounts for the induction of asymmetry during the epoxidation [351]. However, the resultant dioxiranes have two reacting sites, a fact that makes the prediction and execution of asymmetric induction problematic. Yang and coworkers addressed this problem by designing the C2 symmetric, eleven-membered ring ketone **301** as a chiral dioxirane precursor [352]. In this case, both active sites are characterized by the same chiral environment. In fact, a catalytic amount of ketone **301**, in the presence of Oxone as a terminal oxidant, was capable of epoxidizing *trans*-disubstituted and trisubstituted alkenes in excellent yield and modest enantiomeric excess (e.g., **312** → **313**, Scheme 2.53). Electronic and steric embellishments on the ketones (e.g., **302**) can lead to mild enhancements [353]. The main conduit of asymmetric induction is assumed to occur via the "steric sensors" on the aromatic ring. Curiously, enantioselectivity increases with the size of the substituent only to a certain point, then begins to decrease again.

Shi and coworkers have successfully developed ketalized D-fructose derivatives (e.g., **303**) as chiral dioxirane precursors [354]. Excellent ees were obtained, even for the recalcitrant *trans*-disubstituted alkenes (e.g., **314** → **315**, 81% yield, 90% ee) [355] and trisubstituted alkenes (e.g., **320** → **321**, 75% yield, 95% ee) [356]. However, this system appears to be highly substrate dependent, owing to a complex interaction of steric and electronic factors [357–359]. There are several important features of the key dioxirane intermediate. First, the stereogenic centers are in close proximity to the ultimate reactive site of the dioxirane; second, the carbonyl group is flanked by a fused ring on one side and a quaternary center on the other, preventing epimerization; and, finally, only one face of approach is available, since the other is sterically blocked. As for the actual transitions state, the results are consistent with a spiro configuration (**316**) that is directed by steric interactions. The protocol has been optimized so that the chiral ketone **303** can be used in catalytic quantities with Oxone as the stoichiometric oxidant. The key to preserving the lifetime of the chiral auxiliary is pH control during the reaction; the optimum range was found to be 10.5 or above, which is conveniently maintained with potassium carbonate [360, 361].

The related oxazolidinone ketone catalyst **304**, prepared in six steps from D-glucose [362], has the advantage of exhibiting high ees for both *cis*- and terminal olefins [363]. Interestingly, for olefins with aromatic substituents, it appears that the transition state shows a preference for positioning the π-system proximal to the oxazolidinone moiety (as in **319**), so that aromatic groups can be efficiently differentiated during the epoxidation. In studies involving the epoxidation of *cis*-methylstyrene (**317**), the electronic character of the oxazolidino *N*-aryl group was found to influence the outcome of the reaction, presumably by modulating the interaction between the catalyst and the aromatic substituent of the substrate [364]. Similarly, increasing the steric demand adjacent to the amide carbonyl can improve selectivity [365].

Other catalysts also exhibit a combination of these factors. For example, the tropinone-derived chiral ketone **305** owes its enantioselectivity to the structurally rigid and compact asymmetric ring structure. Incorporation of an electron-withdrawing fluoro substituent at the α-position enhanced the catalytic reactivity. Enantiomeric excesses tend to be modest, but occasionally are quite good, as seen in the epoxidation of phenylstilbene (**312**), which takes place in quantitative yield and with 83% ee [366]. The impact of these electronic effects is intriguing, and can be quite dramatic. A particularly noteworthy example is the non-conjugated electronic interactions that result from a remote substituent in the carvone-derived catalyst **306**. In the epoxidation of *trans*-stilbene, the series of small substituents (F, Cl, OH, OEt and H) give ees of 42–87%, exhibiting a very high correlation with the respective Hammett ϱ values. These results have been rationalized on the basis of stabilization of the favored transition state (**307**) by field effects [367]. Some interesting immobilized dioxirane precursors have also been reported, such as the novel heterogeneous ketone **308** [368] and the fluoroketone **309**, designed for use in fluorous media [369].

2.3.2.2 **Using other Oxidants without Metal Catalysts**

The epoxidation of alkenes without metal catalysis represents a large and diverse group of preparative methods for oxiranes, and here the various systems can be characterized largely by two key components: the oxygen carrier (or catalyst) and the terminal (stoichiometric) oxidant. In this regard, *m*-chloroperbenzoic acid (mCPBA) is a tried and true reagent [370], and has been adapted to the large-scale practical synthesis of epoxides [371]. Buffered mCPBA systems are useful for epoxidations in which the alkenes and/or resultant epoxides are acid-sensitive. For example, 2,6-di-*tert*-butylpyridine was shown to give superior results in the case of certain allyl acetals (e.g., **322**, Scheme 2.54) [372]. Bicarbonate [373] and phosphate [374] buffers are also frequently encountered in this context.

Multifunctional alkenes offer some interesting possibilities. For example, enone **324** undergoes ketone-directed epoxidation when treated with mCPBA to give exclusively the syn epoxyketone **326**. As for the mechanism, hydrogen bonding effects were discounted on the basis of solvent insensitivity. Intramolecular attack by some oxidized form of the ketone moiety could be operative, although ^{18}O labeling studies have ruled out a dioxirane intermediate as the active epoxidizing species. Thus, the observed stereoselectivity was rationalized on the basis of intramolecular epoxidation by an α-hydroxy peroxide (i.e., **325**) or possibly by a carbonyl oxide intermediate [375].

Whereas aminoalkenes cannot be converted into epoxides by usual methods (competing N-oxidation), protonation by an arenesulfonic acid and subsequent treatment with mCPBA allows for chemoselective epoxidation. Furthermore, when properly disposed, the pendant ammonium functionality can serve as a potent directing group for the oxidant. Thus, under these conditions cyclohexenylamine **327** affords exclusively the syn epoxide **328** [376].

In the case of dual functionality, the two sites may interact in either a constructive or destructive fashion. This was illustrated by a set of stereoselective epoxidations on a series of allylic carbamates that were appended with a carbomethoxy group, a hydroxymethyl group, or an acetoxymethyl group. In all cases, *threo* epoxides were favored (syn to the carbamate) upon treatment with mCPBA, which reflects the strong directing power of the carbamate group. However, the magnitude of the syn : anti ratio was dependent upon the type and configuration of the other

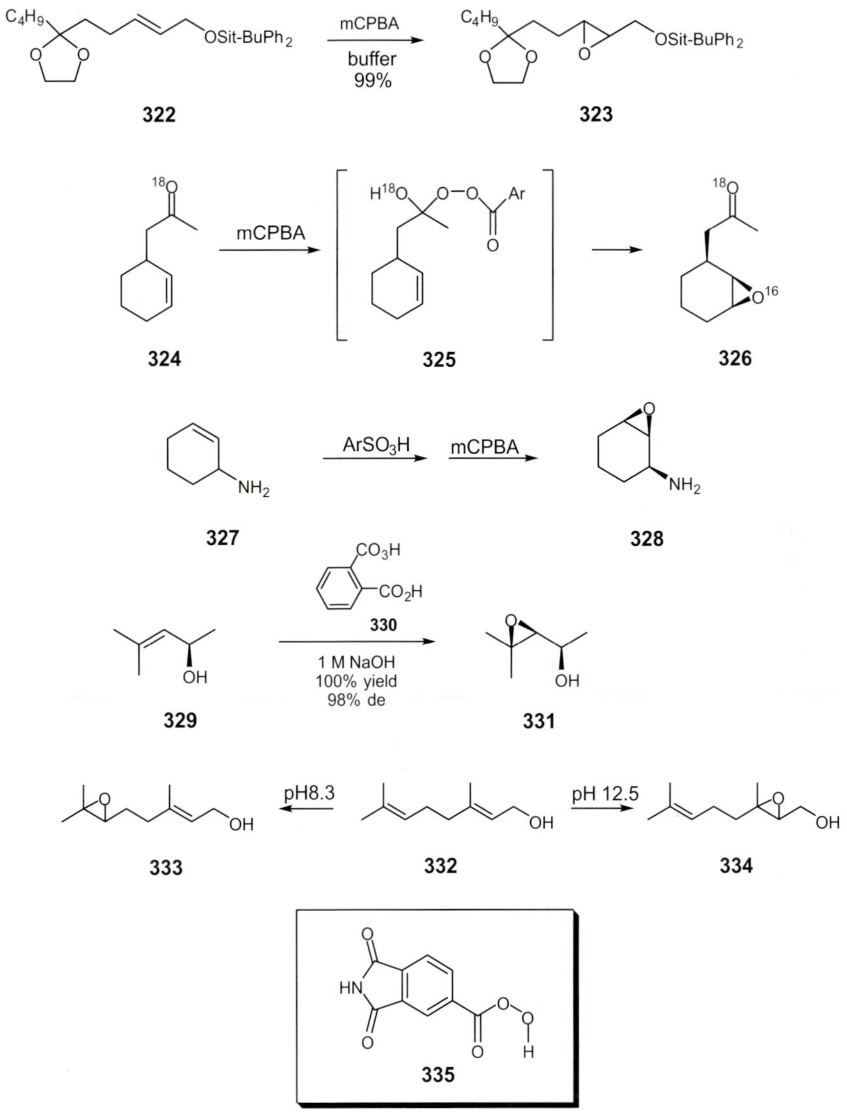

Scheme 2.54 Epoxidation of alkenes using peracids.

functionality. A relatively low ratio was observed when the two groups compete for face selectivity, whereas "cooperative coordination" leads to higher selectivity. From the magnitude of the perturbation, the following order of directing ability was proposed: carbamate > methyl ester > homoallylic alcohol = acetate [377].

The reaction of allyl alcohol with peroxyformic acid has been examined extensively using molecular modeling calculations [378]. Prompted by the observation that peracid epoxidations can be far more selective in basic aqueous medium than in

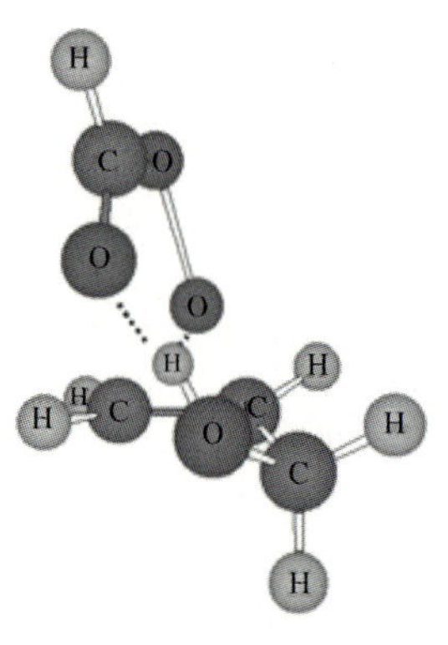

Figure 2.12 TS for epoxidation of allylic alcohol with performate.

organic solvent [379], Washington and Houk studied transition structures from the epoxidation of allylic alcohol with performate ion at the B3LYP/6-31 + G(d,p) level of theory in a CPCM continuum model for water [380]. Their findings indicate a preferential hydrogen bonding of performate with the substrate hydroxyl group rather than with water, leading to a directed epoxidation via the transition state depicted in Figure 2.12. This reaction is similar to the corresponding cyclopropanation of allylic alcohols in the presence of aqueous sodium hydroxide. To test this model experimentally, the chiral allylic alcohol **329** was treated with monoperoxyphthalic acid (MPPA, **330**) in 1 M NaOH to give the syn epoxy alcohol **332** in quantitative yield with 98% syn/anti selectivity.

An interesting regiochemical anomaly has been reported by Fringuelli and coworkers [381]. In the epoxidation of geraniol (**332**) by MPPA, the reaction could be directed to either double bond by simple modification to the experimental conditions. In the presence of cetyl(trimethyl)ammonium hydroxide (CTAOH) at pH 12.5, 2,3-epoxygeraniol (**334**) is formed exclusively; however, at pH 8.3 in the absence of CTAOH, the formation of 6,7-epoxygeraniol (**333**) is favored. The magnesium salt of this reagent, magnesium bis(monoperoxyphthalate) hexahydrate (MMPP), is touted as a less expensive and more stable surrogate for mCPBA [382].

With an eye towards industrial applications, Johnstone and coworkers have developed 5-hydroperoxycarbonylphthalimide (**335**) as a new reagent for epoxidation. An easily prepared, shock-stable, crystalline solid, this peroxy acid was designed to exhibit all the desirable properties of more hazardous or expensive reagents (i.e., ease of work-up, low acidity in reaction medium, etc.). Yields, using various substrates, are excellent [383].

Hydrogen peroxide is another frequently used terminal oxidant for epoxidations, and its use with manganese [384] and palladium [385] catalysts has been the subject of recent reviews. Garcia-Bosch and coworkers have demonstrated that metal-catalyzed disproportionation of hydrogen peroxide in some catalytic systems can be suppressed by using a large excess of acetic acid as an additive, presumably facilitating the formation of peracetic acid [386]. Alternatively, Ti(salan) compounds have been applied effectively to hydrogen peroxide epoxidation systems [387, 388], as well as boron trifluoride in the absence of metal co-catalysts [389].

Another relatively simple metal-free system for the epoxidation of tri- and *cis*-disubstituted olefins (e.g., **336** → **337**, Scheme 2.55) is formamide-hydrogen peroxide in an aqueous medium. This reagent has the advantage of being pH-independent, which makes it attractive for biochemically mediated transformations. No reaction was observed in the case of *trans*-disubstituted and terminal olefins. With bifunctional alkenes, the more reactive double bond is selectively epoxidized [390].

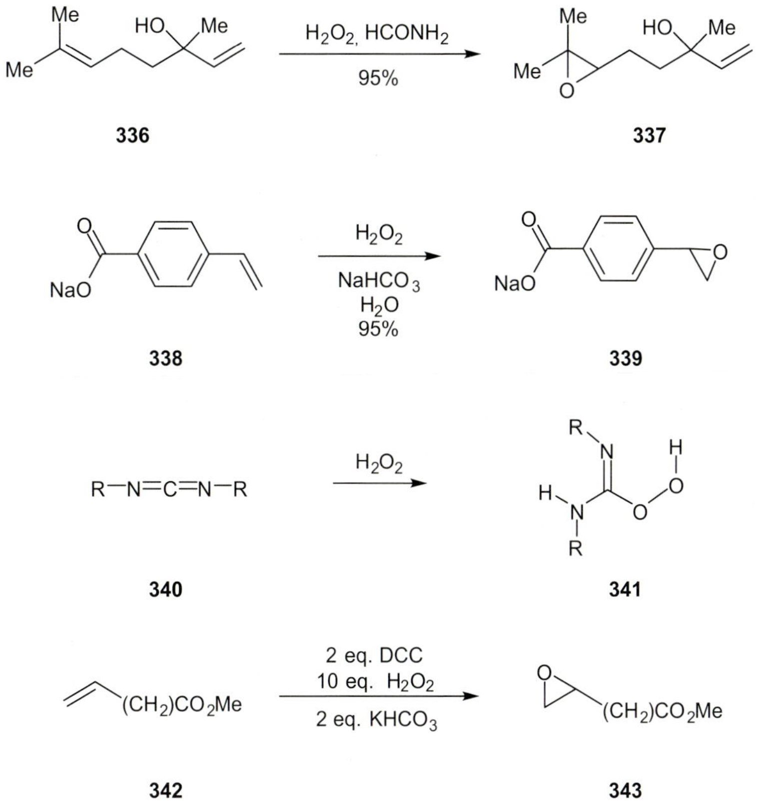

Scheme 2.55 Epoxidation of alkenes using hydrogen peroxide.

Water-soluble alkenes can be epoxidized in remarkably high yields using bicarbonate-activated hydrogen peroxide (BAP). Thus, epoxide **339** was obtained in >95% yield from sodium *p*-vinylbenzoate (**338**). Diol formation is a competing side reaction with some substrates [391].

The epoxidation of olefins with hydrogen peroxide can also be promoted by the addition of carbodiimides, presumably by the initial formation of a peroxyisourea species (**341**) [392]. Majetich and Hicks have reported on the epoxidation of isolated olefins (e.g., **342**) using a combination of 30% aqueous hydrogen peroxide, a carbodiimide (e.g., DCC) and a mildly acidic or basic catalyst. This method works best in hydroxylic solvents and not at all in polar aprotic media. The type and ratios of reagents are substrate dependent, and steric demand about the alkene generally results in decreased yields [393]. The methodology can be adapted to asymmetric epoxidation by using an aspartate-containing tripeptide [394].

Olefins containing free hydroxyl groups or carboxylic acid moieties can be oxidized rapidly and efficiently at room temperature using an easily prepared acetonitrile complex of hypofluorous acid ($HOF{\cdot}CH_3CN$). The reagent does not induce formation of peroxides with free hydroxy groups, and aromatic rings do not interfere with the reaction. Thus, oleic acid (**344**, Scheme 2.56) was epoxidized in 10 min and in 90% yield [395].

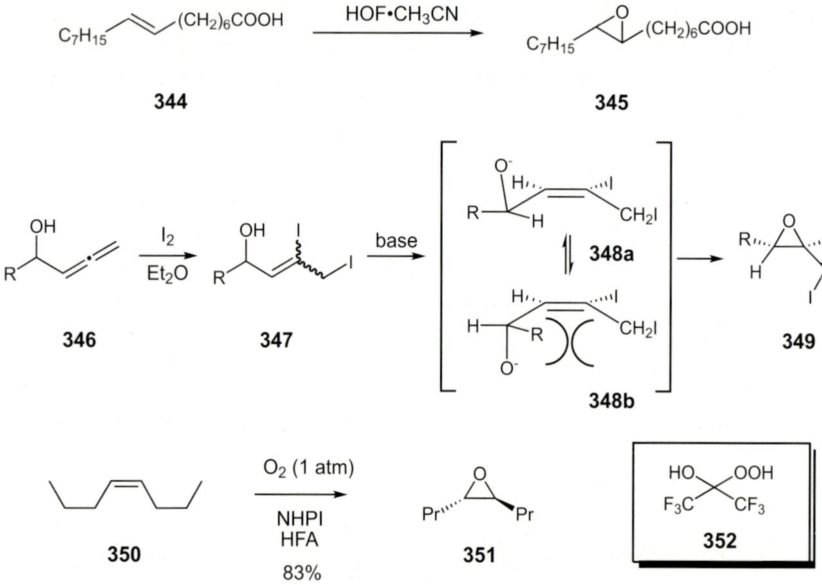

Scheme 2.56 Epoxidation of alkenes using other non-metal oxidizing agents.

Allenic alcohols **346** are converted in the presence of iodine into a mixture of *(Z)*- and *(E)*-diiodides (**347**), which, upon subsequent treatment with base, form the *trans*-iodovinyl epoxides **349** with a diastereomeric excess of >99%. This high degree of selectivity is rationalized on the basis of steric interactions between the R group and the iodine atom in the transition state leading to epoxide formation (i.e., **348a** vs. **348b**) [396].

One of the most attractive oxidants for this chemistry is dioxygen, both from an environmental and cost standpoint. In this vein, a metal-free epoxidation protocol was reported that proceeded via the *in situ* generation of hydrogen peroxide from O_2 through a complex series of steps involving *N*-hydroxyphthalimide (NHPI). Once formed, the H_2O_2 is activated by addition onto the somewhat esoteric solvent, trifluoroacetone, to give 2-hydroperoxy-hexafluoropropan-2-ol (**352**) as the oxygen transfer reagent. The reaction appears to be general and yields are very good. Regardless of the starting configuration of the alkene, a strong preference for the formation of *trans*-epoxides was observed (e.g., **350** → **351**) [397].

2.3.2.3 Metal-Catalyzed Epoxidation of Alkenes

The topic of metal-based catalysis for alkene epoxidation is expansive, and the reader is directed to two excellent reviews of this chemistry by Adolfsson [398] and Adolfsson and Balan [399], as well as an outstanding treatise on mechanism and kinetics by Oyama [400] and an overview of asymmetric methods by Matsumoto and Katsuki [401]. The development of (salen)metal complexes (Figure 2.13) for the epoxidation of alkenes has been nothing less than revolutionary, providing access to epoxides from simple olefins in much the same way Sharpless chemistry paved the way for the epoxidation of functionalized alkenes. An extensive review specifically on the chromium- and magnesium-salen catalyzed epoxidation of alkenes has recently appeared [402].

In the absence of other functionality, typical peroxyacid epoxidation of dienes favors reaction on the more substituted double bond. However, the (salen)manganese complex **353** promotes epoxidation of the less substituted double bond (e.g., **362** → **363**, Scheme 2.57), thus providing a complement to conventional methods. This protocol is also useful for substrates that polymerize under peracid conditions [403]. Allylic alcohols are equally suitable substrates, as demonstrated by the epoxidation of **364** using the vanadyl salen oxo-transfer catalyst **354** in supercritical carbon dioxide with *tert*-butyl hydroperoxide as a terminal oxidant [404].

The lion's share of research activity in this area has centered around asymmetric epoxidation using chiral salen catalysts. Thus, the chiral (salen)Mn(III) complex **356**, which is readily available on a large scale in high yield from commercially available starting materials [405, 406], is the centerpiece of Jacobsen's enantioselective synthesis of the taxol side chain **369** [407], in which the epoxy ester **367** is prepared from *cis*-ethyl cinnamate with very high enantiomeric excess (Scheme 2.58). Typically under these conditions, *cis*-double bonds are converted stereospecifically into *cis*-epoxides. However, in the case of conjugated dienes, *trans*-epoxides are the major products. The crossover has been ascribed to a step-wise oxygen transfer mechanism involving an intermediate radical that undergoes bond rotation to give the observed products. This anomalous behavior has been leveraged in a method for the regioselective epoxidation of *cis,trans*-dienes to give *trans,trans*-diene monoepoxides (e.g., **370** → **371**) [408].

Cyclic and acyclic trisubstituted alkenes are also epoxidized with high enantioselectivity under the catalysis of **356**, as exemplified by the conversion of phenylstilbene **312** into the (*S*)-epoxide **313** with 92% enantiomeric excess. The sense of enantioselection is opposite that of disubstituted alkenes, and a global mechanistic model has been developed to rationalize the stereochemical outcomes of this class of

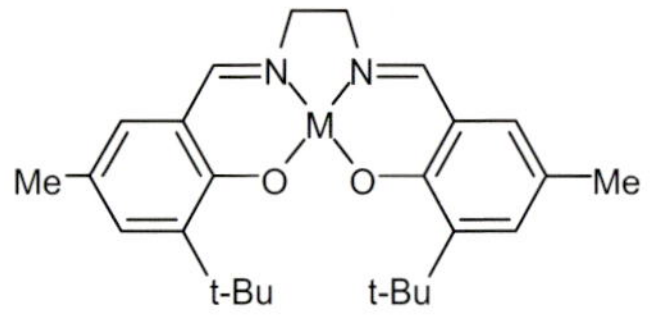

353, M = Mn-Cl
354, R = M = V(O)

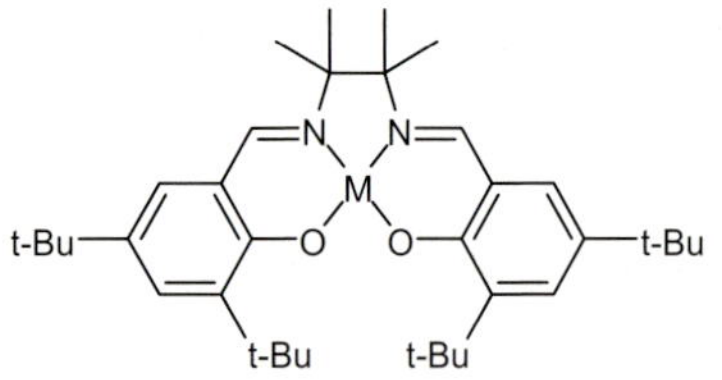

355, M = Mn-Cl

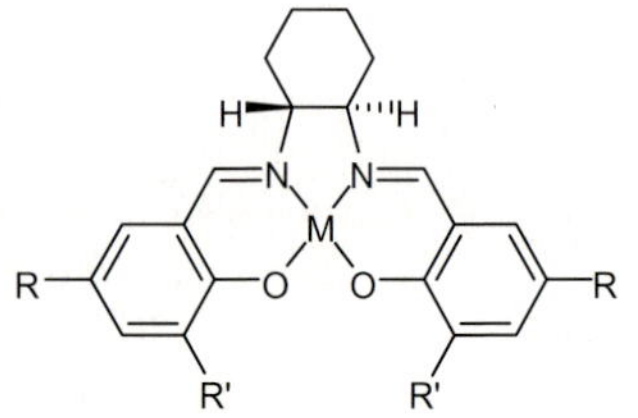

356, R = R' = t-Bu; M = Mn-Cl
357, R = (i-Pr)$_3$SiO-;R' = t-Bu; M = Mn-Cl
358, R = R' = t-Bu; M = Cr(O)-Cl
359, R = R' = Cl; M = Cr(O)-PF$_6$

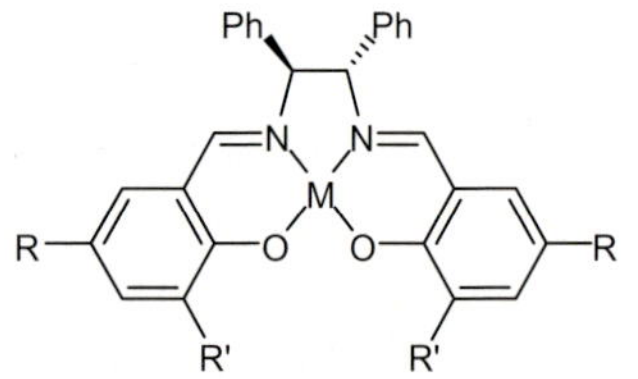

360, R = (i-Pr)$_3$SiO-;R' = t-Bu; M = Mn-Cl

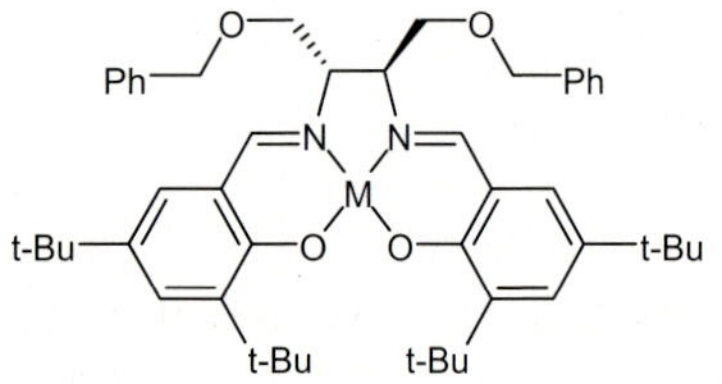

361, M = Cr-OTf

Figure 2.13 Salen metal catalysts for alkene epoxidation.

epoxidations [409]. Furthermore, a wide range of terminal oxidants have been employed with this catalyst, including hydrogen peroxide [410], dimethyldioxirane [411], periodates [403] and the organic-soluble oxidant tetrabutylammonium monopersulfate [412].

Terminal olefins typically give relatively low enantiomeric purities, which might be due to poor enantiofacial selectivity during the oxygen addition or facile rotation of a

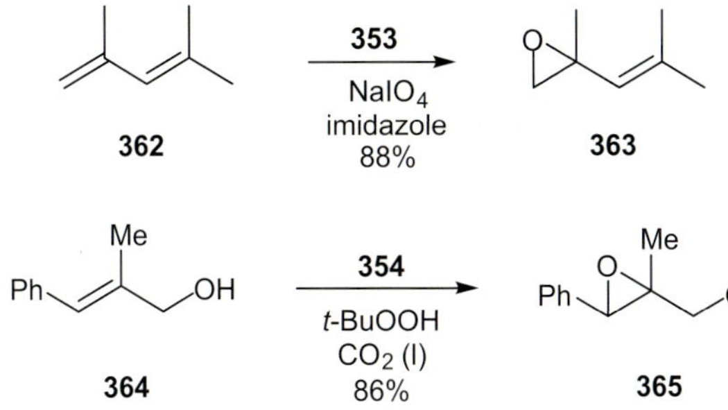

Scheme 2.57 Alkene epoxidations using achiral salen metal catalysts.

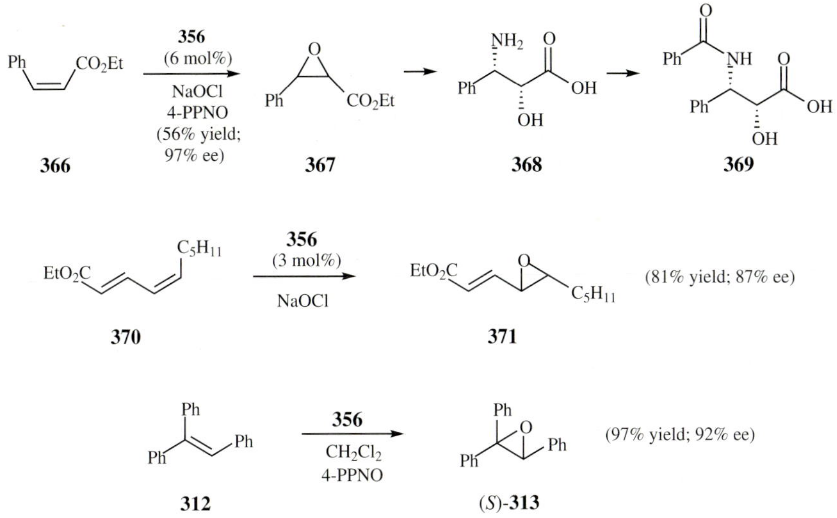

Scheme 2.58 Alkene epoxidations using chiral salen metal catalysts.

radical intermediate unencumbered by α-substitution. Either of these impediments should be positively impacted by decreasing thermal energy, and this has been borne out by experimental evidence [413]. Katsuki has reported increased ees by adding NaCl to the aqueous hypochlorite system, thus allowing for sub-zero (−18 °C) reaction temperatures [414]. The asymmetric induction may be further enhanced by modification of the catalyst. Replacement of the *tert*-butyl group with the triisopropylsiloxy substituent affords a catalyst (i.e., **357**) that is not only sterically more defined but also electronically attenuated, and thus is milder and more selective [415]. The diphenyl variant **360** was equally effective in epoxidizing a series of recalcitrant olefins, such as vinylbenzoic acid **372** (Scheme 2.59) in the presence of *N*-methylmorpholine *N*-oxide (NMO) as an additive and mCPBA as a terminal oxidant in methylene chloride at −78 °C [416]. These conditions were also advantageous for the epoxidation

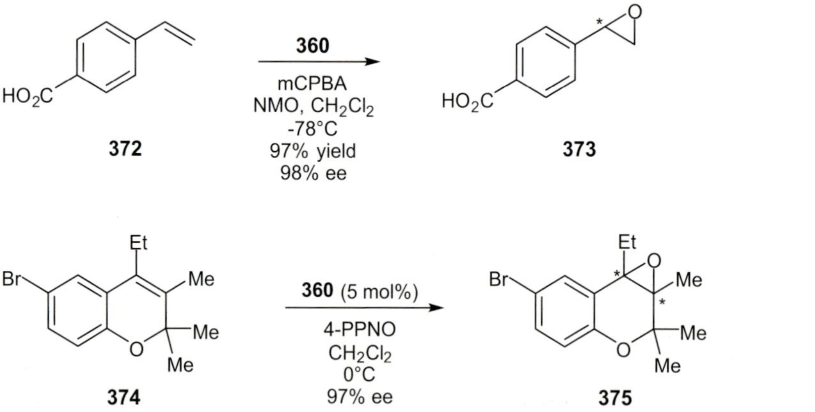

HO_2C **372** → **360**, mCPBA, NMO, CH_2Cl_2, -78°C, 97% yield, 98% ee → HO_2C **373**

374 (Br, Et, Me, Me, Me, O) → **360** (5 mol%), 4-PPNO, CH_2Cl_2, 0°C, 97% ee → **375** (Br, Et, O, Me, Me, Me)

Scheme 2.59 More alkene epoxidations using chiral salen metal catalysts.

of tetrasubstituted alkenes, as demonstrated by the conversion of chromene derivative **374** into the corresponding epoxide with 97% ee [417].

Changing the metal center from manganese to chromium can have surprising results. For example, one aggravating phenomenon associated with the (salen)Mn complexes is that the epoxidation of *trans*-olefins proceeds typically with low ees. However, the analogous chromium complexes (e.g., **359**) catalyze such epoxidations with greater selectivity than for the corresponding *cis*-olefins under the same conditions. Here the mechanism is presumed to involve an electrophilic process, which is supported by the fact that only electron-rich alkenes are effectively epoxidized. In the case of *trans*-β-methylstyrene (**376**, Scheme 2.60), enantioselectivities

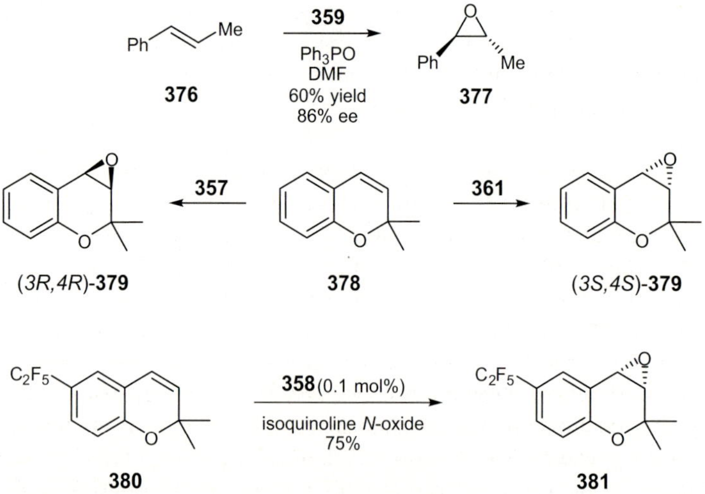

Ph–CH=CH–Me **376** → **359**, Ph_3PO, DMF, 60% yield, 86% ee → **377** (Ph, O, Me)

(3*R*,4*R*)-**379** ← **357** — **378** — **361** → (3*S*,4*S*)-**379**

C_2F_5 **380** → **358** (0.1 mol%), isoquinoline *N*-oxide, 75% → C_2F_5 **381**

Scheme 2.60 Salen(Cr)-catalyzed alkene epoxidation.

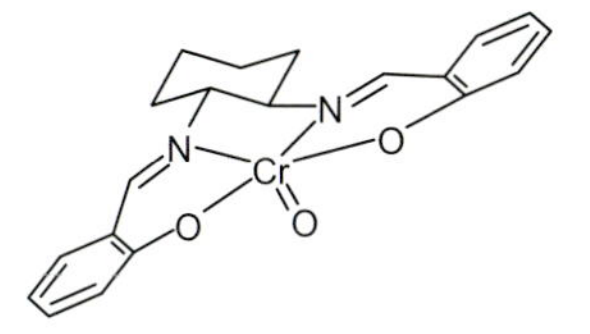

Figure 2.14 Stepped conformation of a salen(Cr) catalyst.

of up to 86% are achieved [418]. Gilheany and coworkers [419] have observed high (>90%) enantioselectivities for *trans*-alkenes using stoichiometric chromium complexes, albeit with modest chemical yields, presumably resulting from the formation of a μ-oxo Cr(IV) dimer *in situ*. A systematic study of the effect of aromatic substituents on enantioselectivities [420] is consistent with an oblique approach of the substrate to a nonplanar (stepped) oxidized catalyst (Figure 2.14).

An interesting reversal of chiral induction in chromium(III)-salen complexes using a tartaric derived alicyclic diamine moiety (i.e., **361**) has been observed by Mosset, Saalfrank and coworkers [421]. Thus, epoxidation of the chromene **378** using catalyst **361** and an oxidant consisting of mCPBA/NMO afforded epoxide **379** in the (3*S*,4*S*) configuration, whereas a classical Jacobsen catalyst (**357**) provided the corresponding (3*R*,4*R*) enantiomer. This approach has been applied to the chiral epoxidation of chromene **380** using the readily available chromium salen catalyst **358** in a synthesis of the novel potassium channel activator BRL55834 [422].

The Katsuki group have focused their attention on (salen)Mn(III) catalysts of a slightly different configuration (e.g., **382–386**, Figure 2.15), which are characterized as having chiral residues at the aromatic 3,3′-positions. These catalysts have been used to advantage in the epoxidation of conjugated *cis*-olefins [423], including chromenes [424, 425], benzocycloheptenes [426], dihydronaphthalenes [427] and enynes [428]. The proposed mechanism involves a flanking attack by the substrate, which is steered by both steric interactions (e.g., the cyclohexyl residue) as well as π–π repulsive forces. Generally speaking, the enantiofacial selection of *cis*-olefins in these catalyst systems appears to be influenced mainly by the chirality on the ethylenediamine bridge, whereas *trans*-olefin epoxidation seems to be directed more by the C3 and C3′ substituents [429].

More subtle arguments have been invoked to rationalize the dichotomous behavior of the so-called "second-generation" Mn-salen catalysts **384** and **385** towards unfunctionalized and nucleophilic olefins. For example, higher yields and ees are obtained with the (*R*,*S*)-complex (**384**) for the epoxidation of indene (**387**, Scheme 2.61). However, *N*-toluenesulfonyl-1,2,3,4-tetrahydropyridine (**389**) gave better results using the (*R*,*R*)-diastereomer (**385**). An analysis of the transition-state enthalpy and entropy terms indicates that the selectivity in the former reaction is enthalpy driven, while the latter result reflects a combination of enthalpy and entropy factors [430].

Other structural modifications to salen catalysts can confer operational advantages. For example, hydrogen peroxide is attractive as a terminal oxidant due to its low cost and ready availability, but epoxidations using this reagent tend to be less enantioselective and more prone to radical-induced side reactions. In some cases, these

382

383

384 (*R*,*S*)
385 (*R*,*R*)

386

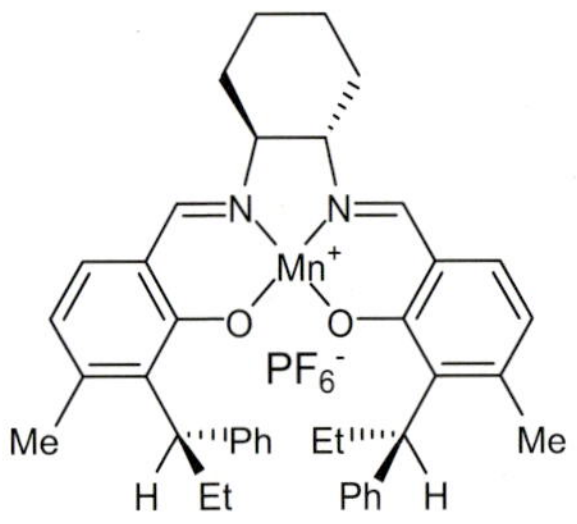

Figure 2.15 Salen catalysts with chirality at the 3-position.

387 388

389 390

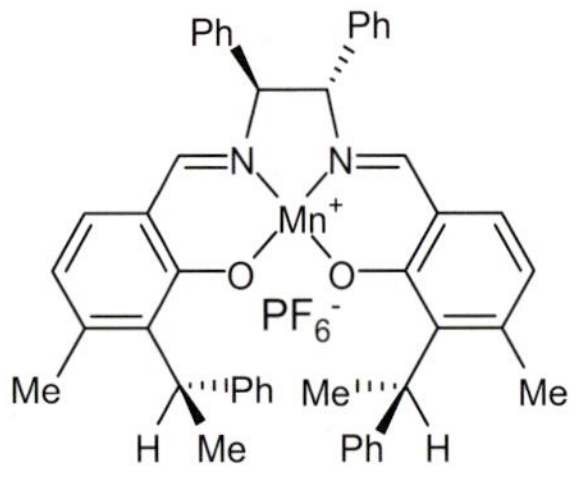

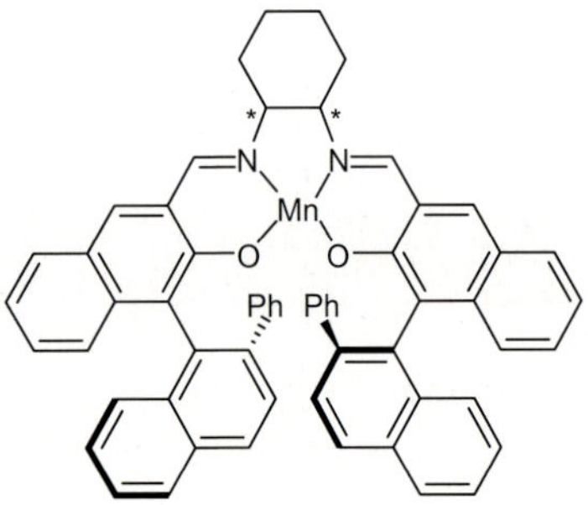

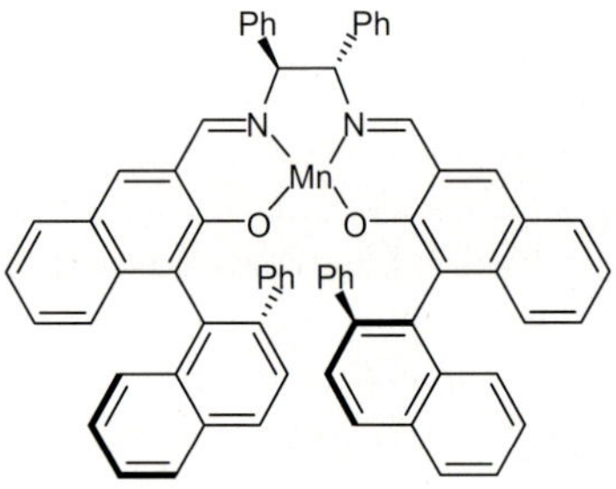

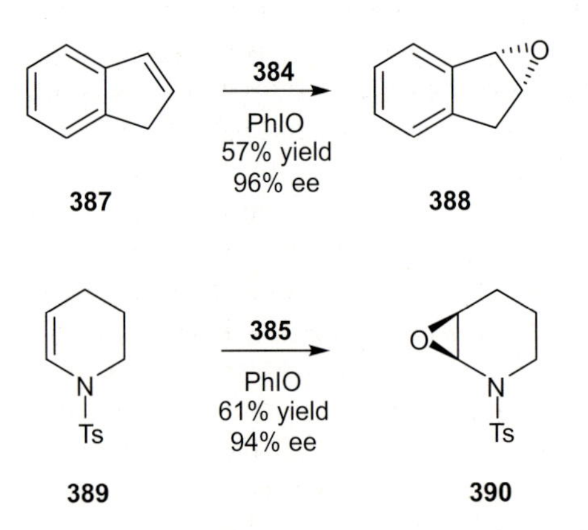

Scheme 2.61 Salen(Mn) catalyzed alkene epoxidation.

disadvantages have been circumvented by using bioinspired models [431]. For example, by tethering an imidazole moiety to a chiral salen-type Mn(III) catalyst, an axial ligand is provided that imitates a peroxidase coordination sphere while still taking advantage of the asymmetric active site of the chiral (salen)Mn(III) species.

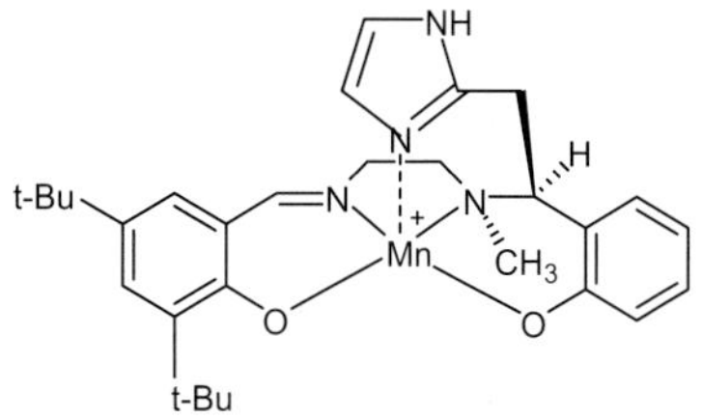

391

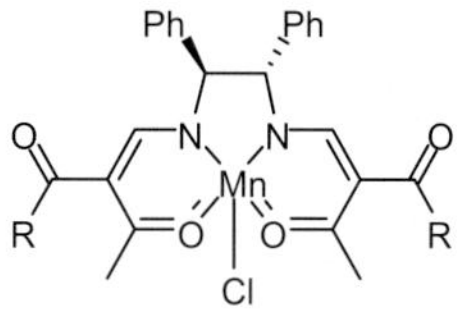

392

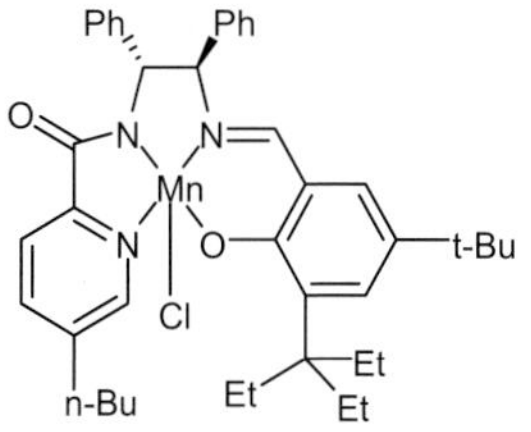

393

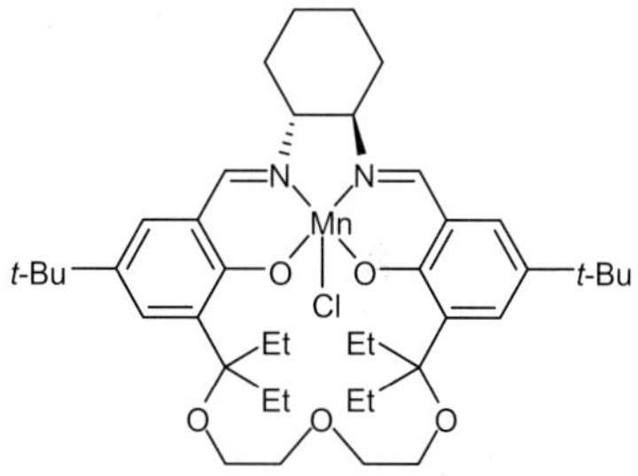

394

Figure 2.16 Modified salens and salen analogs.

The resulting catalyst (**391**, Figure 2.16) can be used at 10 mol.% loadings with dilute hydrogen peroxide as an oxidant [432]. In a similar vein, the chiral dicarbonyliminato manganese(III) complex **392** is effective using molecular oxygen as the terminal oxidant [433], and the manganese-picolinamide-salicylidene complex **393** exhibits excellent turnover using sodium hypochlorite [434]. The recently disclosed macrocyclic analogue **394** gives very promising results indeed. At a 5 mol.% loading with two equivalents of sodium hypochlorite as a terminal oxidant, chromene **378** is epoxidized in quantitative yield and 93% ee [435].

No small amount of effort has been directed towards the development of immobilized catalysts, both for ease of catalyst recovery and for application to solid-phase combinatorial synthesis. Toward this end, the catalytic moiety has been tethered to a solid support via either the ethylenediamine portion [436] or the salicylaldehyde subunit [437] to give immobilized catalysts of type **395** and **396**, respectively (Figure 2.17). These are the first gel-type resins to give results rivaling solution-phase counterparts. The backbone of Jacobsen's catalyst has also been immobilized on silica gel by radical grafting (e.g., **397**) [438] and it has even been prepared in polymeric form (i.e., **398**) [439]. Other approaches include the use of perfluoroalkyl-substituted catalysts (e.g., **399**) in a fluorous biphasic system [440] and the conventional Jacobsen's catalyst **356** in a medium of the air- and moisture-stable ionic liquid [bmim][PF_6] [441].

A somewhat different approach to catalyst separation has been devised by engineering the chiral salen catalyst to have built-in phase-transfer capability, as exemplified by the Mn(III) complex **400** [442]. Thus, enantioselective epoxidation of

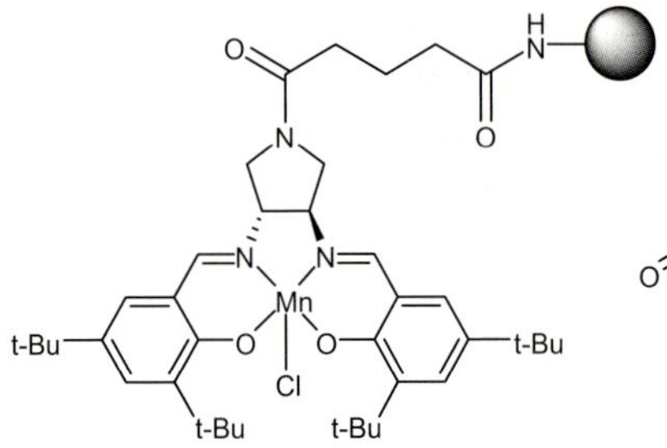

395 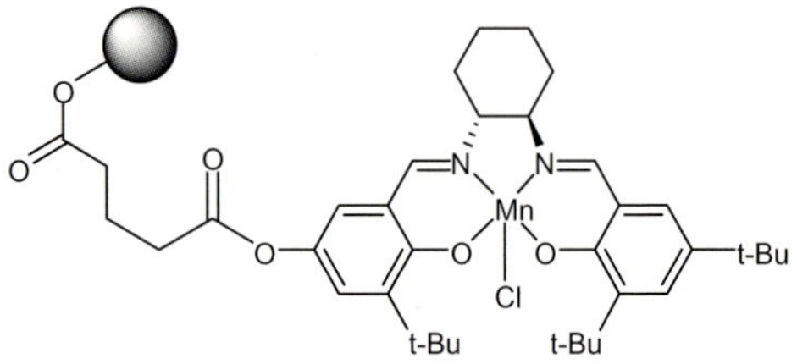 **396**

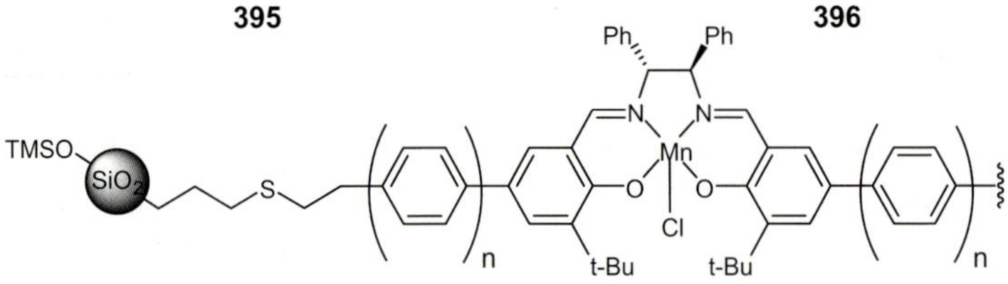

397

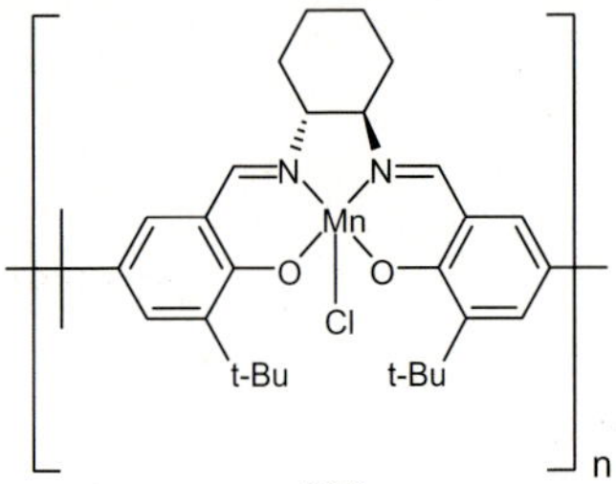

398 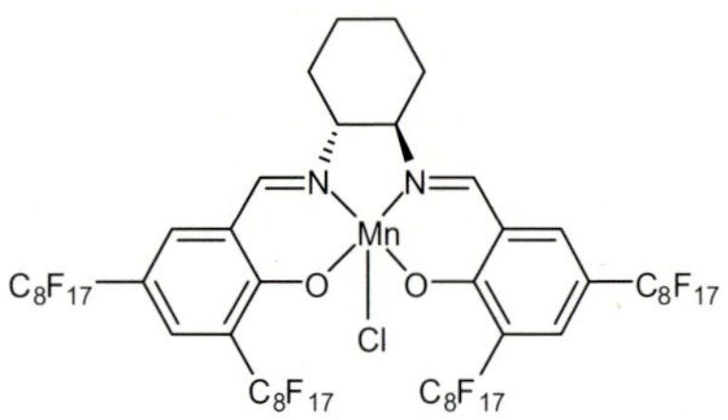 **399**

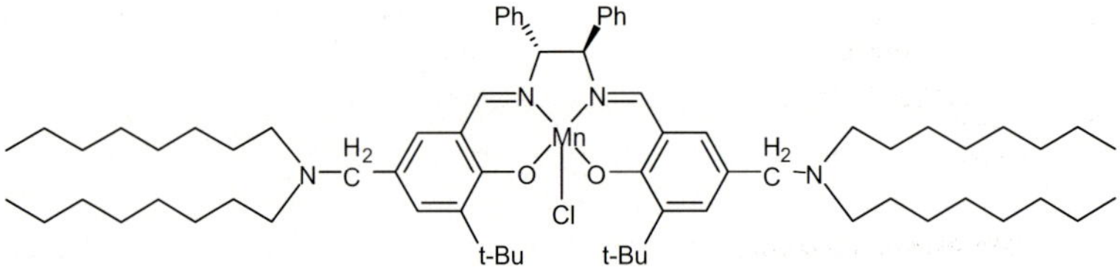

400

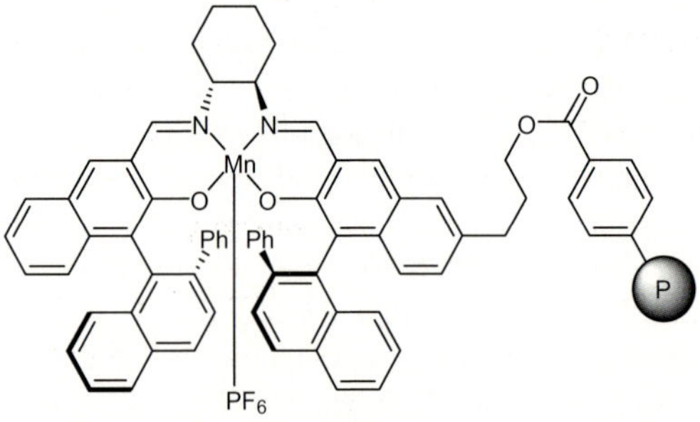

401

Figure 2.17 Salens designed for biphasic systems.

chromene derivatives (e.g., **402**) in the presence of 2 mol.% catalyst **400** and pyridine *N*-oxide (PNO) under phase-transfer conditions (methylene chloride and aqueous sodium hypochlorite) proceeded in excellent yield and very good ees (Scheme 2.62). The catalyst loading could be reduced to about 0.4% with only marginal loss of efficiency. Finally, Smith and Liu [443] immobilized a Katsuki-type salen ligand by an ester linkage to Merrifield's resin to produce catalyst **401**. In a test epoxidation of 1,2-dihydronaphthalene (**404**) using sodium hypochlorite as an oxidant and 4-phenylpyridine *N*-oxide (4-PPNO) as an activator, the immobilized (salen)Mn complex sustained high enantioselectivity (>90%), even after being recycled six times.

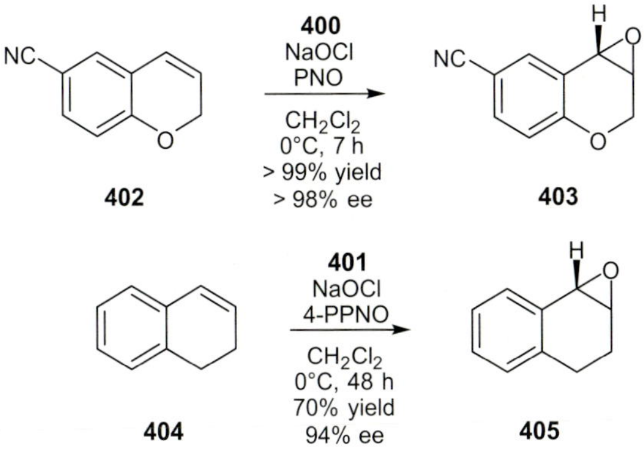

402 → 403: **400**, NaOCl, PNO, CH_2Cl_2, 0°C, 7 h, > 99% yield, > 98% ee

404 → 405: **401**, NaOCl, 4-PPNO, CH_2Cl_2, 0°C, 48 h, 70% yield, 94% ee

Scheme 2.62 Epoxidation with immobilized metal salen catalysts.

No discussion of metal-catalyzed epoxidation could be complete without addressing Sharpless chemistry. Now an imbedded part of the synthetic organic canon, this topic has been very nicely summarized in a recent review article [444]. Two common catalysts in this regard are $VO(acac)_2$ and $Ti(O\textit{i}\text{-}Pr)_4$, and although their first applications were reported decades ago, the methodology is still very much in currency. Thus, the vanadium-mediated diastereoselective epoxidation of allylic alcohols, a key step in the synthesis of *Cecropia* juvenile hormone (i.e., **406** → **407**, Scheme 2.63) described in 1974 [445], was employed in much the same form in 2006 in the epoxidation of the highly functionalized allylic alcohol **408**, providing a key intermediate for an approach to the quartromicins [446]. Likewise, the classic tartrate/titanium-mediated asymmetric epoxidation of allylic alcohols (i.e., **410** → **411**), the scope of which was described in detail by Sharpless in 1987 [447], is an almost indispensable tool for the synthetic chemist, as evidenced by its application in the construction of the bicyclic ether core of (+)-sorangicin A (i.e., **412** → **413**) reported by Crimmins and Haley [448].

These protocols continue to provide springboards for further development and innovation. For example, Hussain and Walsh [449] have developed a Sharpless-inspired tandem alkylation/epoxidation of prochiral enones to provide chiral non-racemic hydroxyepoxides. Yamamoto and coworkers [450] have developed a chiral

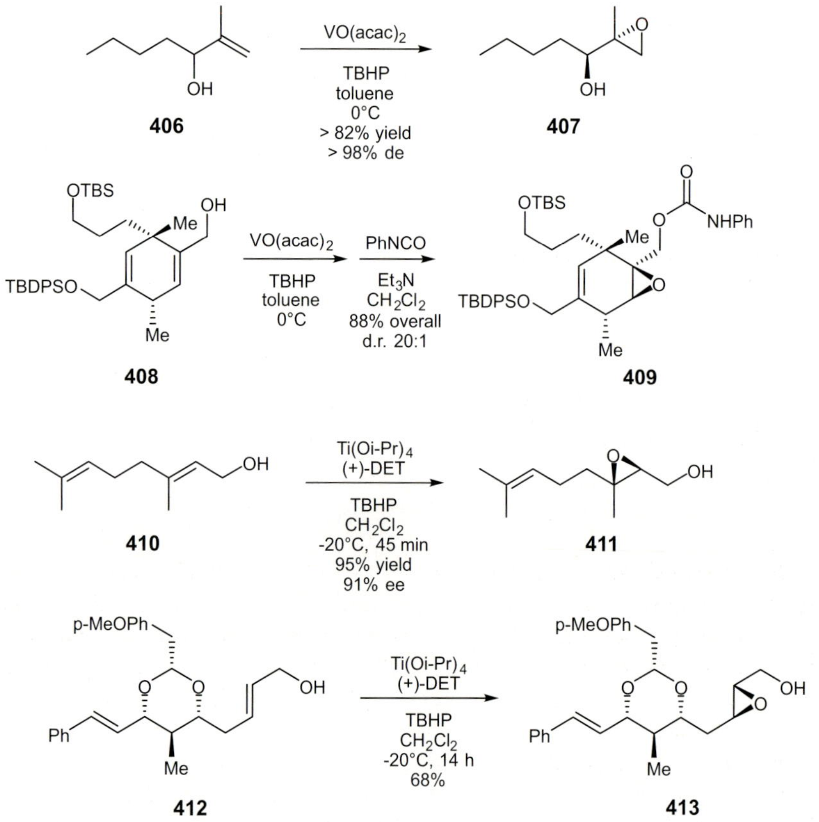

Scheme 2.63 Epoxidation under Sharpless conditions.

hydroxamic acid (**414**) derived from binaphthol, which serves as a coordinative chiral auxiliary when combined with $VO(acac)_2$ or $VO(i\text{-}PrO)_3$ in the epoxidation of allylic alcohols. In this protocol, triphenylmethyl hydroperoxide (TrOOH) provides markedly increased enantiomeric excess, compared to the more traditional *t*-butyl hydroperoxide. Thus, the epoxidation of *(E)*-2,3-diphenyl-2-propenol with 7.5 mol.% $VO(i\text{-}PrO)_3$ and 15 mol.% of **414** in toluene (−20 °C; 24 h) provided the (*2S,3S*) epoxide **416** in 83% ee. Malkov and coworkers were able to carry out the same transformation in 92% yield and 94% ee using a cyclohexylamine-derived sulfonamide hydroxamic acid catalyst [451].

An alternative organic peroxide source is also at the heart of another modified catalytic Sharpless epoxidation of allylic alcohols, in which the tertiary furyl peroxide **417** serves as the terminal oxidant in the presence of L-diisopropyl tartrate (L-DIPT). Thus, *trans*-2-methyl-3-phenylprop-2-en-1-ol (**364**) was converted into epoxide **365** in 87% yield and with 97% ee using a catalyst loading of 20 mol.% [452] (Scheme 2.64).

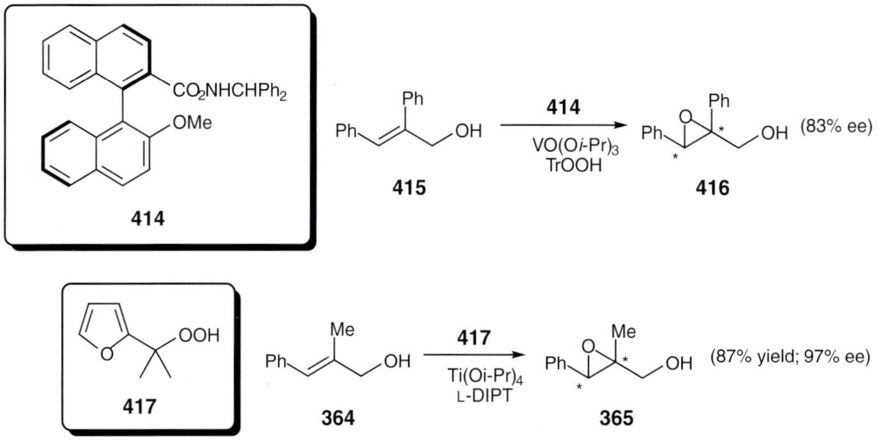

Scheme 2.64 Modified Sharpless conditions.

Biological models have inspired the adaptation of metalloporphyrin complexes for synthetic processes. For example, chloroperoxidase (CPO) catalyzes the epoxidation of many simple *cis*-alkenes with high enantioselectivity, although bulkier substrates tend to lead to low conversion and terminal alkenes alkylate the catalyst [453]. A more generally applicable analog can be found in the iron(III) tetrakis(pentafluorophenyl) porphyrin **418a** (Scheme 2.65), for which considerable mechanistic studies have been carried out [454, 455]. Cyclooctene (**419**) is smoothly epoxidized by **418a** using hydrogen peroxide as the terminal oxidant. Many interesting modifications have been made to the porphyrin template, most notably for the purposes of asymmetric induction, which has been the subject of a recent review [456]. Thus, the binaphthyl strapped iron-porphyrin catalyst **424** promotes the enantioselective epoxidation of styrene (**10**) with iodosylbenzene to give (*R*)-styrene oxide (**421**) in excellent yield and enantiomeric excess [457]. Effective non-heme iron catalysts using pentadentate bispidine ligands have also been studied [458].

Some interesting ruthenium porphyrins have also been reported, including a dioxoruthenium(VI) species [459] and the ruthenium(II) catalyst **418b**, which functions as a photosensitizer capable of effecting the selective epoxidation of alkenes (e.g., **422**) using water as an oxygen source, although the method suffers from the limitation of requiring hexachloroplatinate as an electron acceptor [460]. Metalloporphyrins of all stripes have been appended to solid supports, and the reader is directed to a recent and effective review on the topic for further information [461].

Polyoxometallates (POMs) have been in the research crosshairs lately, as evidenced by a recent review [462]; this interest stems in some portion from their ruggedness and environmental acceptability. As an example, the sandwich-type POM $[WZnMn^{II}_2(ZnW_9O_{34})_2]^{12-}$ catalyzes the selective epoxidation of chiral allylic alcohols with aqueous hydrogen peroxide under mild conditions. Thus, 4-methylpent-3-en-2-ol **425** is converted into the *threo* epoxide **426** in 88% yield and 84% de (Scheme 2.66). The diastereoselectivity is highly sensitive to the substitution about

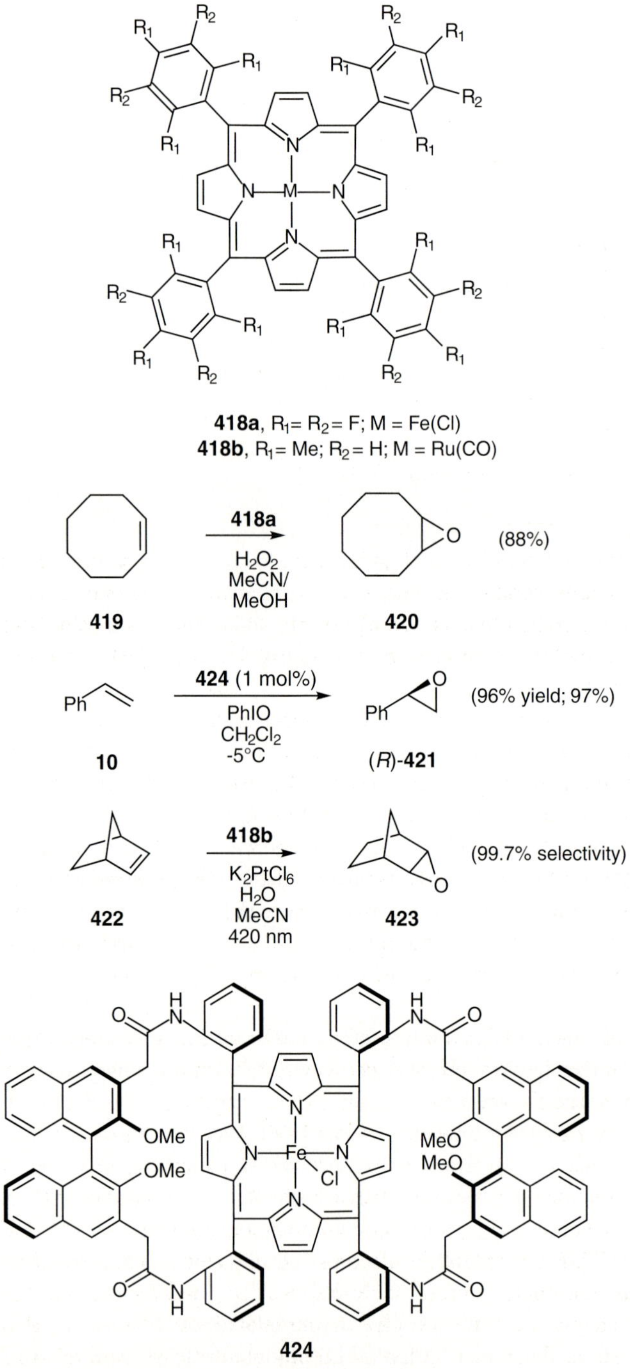

418a, $R_1 = R_2 = F$; M = Fe(Cl)
418b, R_1 = Me; R_2 = H; M = Ru(CO)

419 → **420** (88%); **418a**, H_2O_2, MeCN/MeOH

10 → (*R*)-**421** (96% yield; 97%); **424** (1 mol%), PhIO, CH_2Cl_2, -5°C

422 → **423** (99.7% selectivity); **418b**, K_2PtCl_6, H_2O, MeCN, 420 nm

424

Scheme 2.65 Epoxidations using metalloporphyrins.

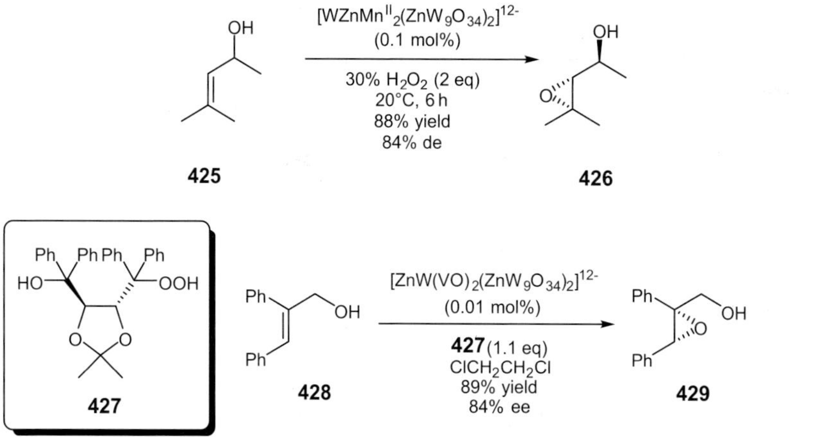

Scheme 2.66 Epoxidations using polyoxometallates (POMs).

the double bond, with *trans*-pent-3-en-2-ol giving a 1 : 1 mixture of *threo* and *erythro* epoxides under the same conditions. The key intermediate is believed to be a tungsten peroxo complex rather than an oxo-Mn species [463]. Self-assembled POM catalysts can also be immobilized within layered double hydroxides (LDH) using ion-exchange techniques [464].

A polyoxometallate is also at the heart of an enantioselective epoxidation of allylic alcohols using C-2 symmetric chiral hydroperoxide **427** derived from 1,1,4,4-tetraphenyl-2,3-*O*-isopropylidene-D-threitol (TADDOL). Thus, in the presence of the oxovanadium(IV) sandwich-type POM $[ZnW(VO)_2(ZnW_9O_{34})_2]^{12-}$ and stoichiometric amounts of hydroperoxide **427**, the stilbenemethanol derivative **428** is converted into the (2*R*) epoxide **429** in 89% yield and 83% ee. The proposed catalytic cycle invokes a vanadium(V) template derived from the POM, substrate and hydroperoxide – a hypothesis supported by the lack of enantioselectivity with unfunctionalized alkenes. The catalytic turnover is remarkably high at about 40 000 TON [465].

Under the rubric of other metal catalysts, methyltrioxorhenium (MTO) represents a fascinating entry. Unlike titanium catalysts (see above), MTO appears not to engage allylic alcohols in tight metal-alcoholate binding, although hydrogen bonding with the substrate can play a role in nonpolar solvents [466]. However, in polar protic media alkenes proximal to a hydroxy group no longer command preferential epoxidation. Thus, treatment of geraniol (**410**) with MTO affords epoxide **430** as the major product (Scheme 2.67), providing a complementary alternative to conventional methods [467]. For simple allylic alcohols (e.g., **431**) formation of the *threo* epoxide (e.g., **432**) predominates – presumably the result of 1,3-allylic strain in the hydrogen bonded catalyst–substrate complex. Unfunctionalized alkenes are also efficiently epoxidized, as illustrated by the practically quantitative conversion of vinylcyclohexane (**433**) into epoxide **434** [468]. Compatible oxygen donors include hydrogen peroxide and its urea adduct (UHP) [469]; perfluoroalkanol solvents tend to

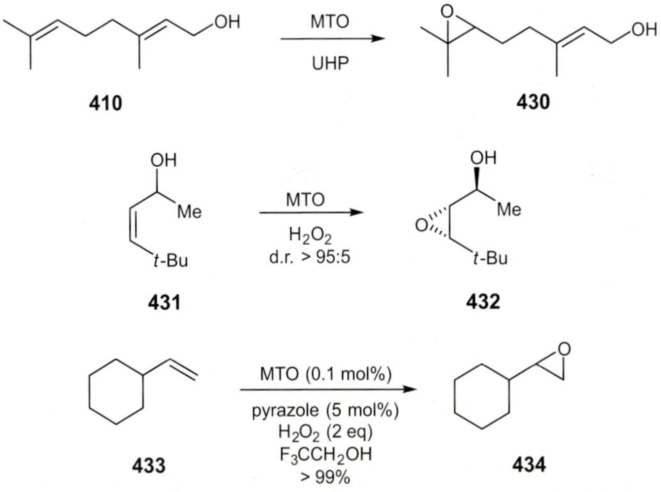

Scheme 2.67 Epoxidations using methyltrioxorhenium (MTO).

give superior results [470]. A polymer-supported version of MTO has also been disclosed [471].

Other noteworthy biphasic and supported metal catalyst systems [472] include hydrotalcite with hydrogen peroxide [473], cobalt-modified hydrotalcite with molecular oxygen [474], manganese dicarboxylate coordination polymers with hydrogen peroxide [475], a binuclear manganese carboxamide array with dioxygen [476], and aqueous sodium tungstate under phase-transfer conditions [477].

A fascinating "triphase" catalyst for epoxidation of allylic alcohols has been prepared from the combination of phosphotungstic acid and an amphiphilic poly-(*N*-isopropylacrylamide)-derived polymer, which yields a macroporous complex (**435**, Scheme 2.68). Thus, treatment of allylic alcohol **436** with 0.003 mol.% catalyst **435** and 2 equivalents of hydrogen peroxide in aqueous medium furnished the corresponding epoxide **437** in 96% yield. The catalyst exhibits a very high turnover rate (35 000), is easily recoverable by filtration and is reusable without loss in efficacy [478]. Also in the category of organic–inorganic hybrids, titanium-silsesquioxane catalysts have been prepared by the complexation of titanium to incompletely condensed silsesquioxanes [479].

Lipophilic alkenes such as **404** can be epoxidized in a triphasic system using ionic liquids, with epoxidation components being provided via an aqueous phase and the reaction products (e.g., **405**) extracted into a pentane layer [480]. Simple alkenes are also converted into epoxides in high efficiency using a recyclable immobilized molybdenum catalyst (**438**) prepared by the reaction of aminated polystyrene and molybdenum hexacarbonyl. For example, cyclohexene (**439**) is quantitatively epoxidized within 5 h using 1 mol.% of catalyst **438** with *t*-butyl hydroperoxide (TBHP) as an oxygen donor [481].

Finally, a combination of wet copper(II) sulfate and potassium permanganate in *t*-butanol (Parish conditions) represents a simple and inexpensive reagent for the

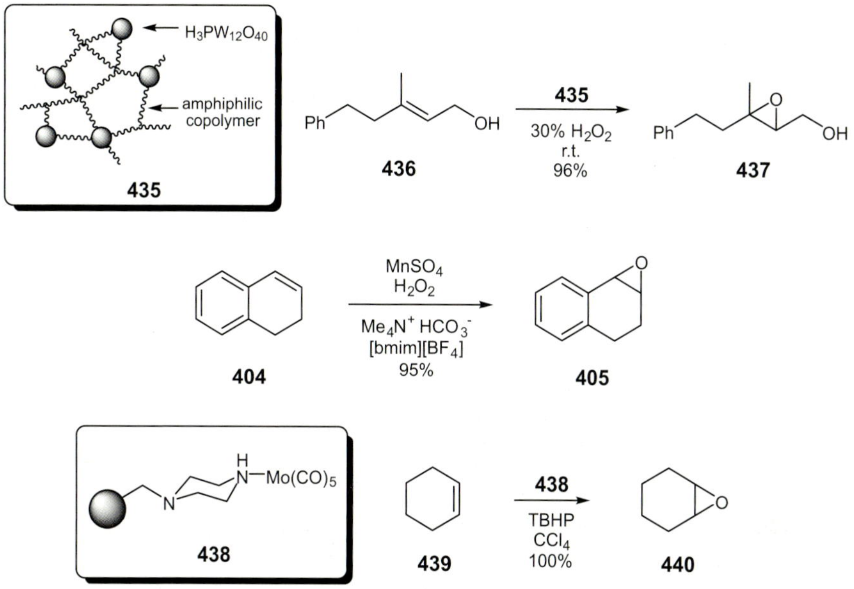

Scheme 2.68 Other immobilized metal epoxidation catalysts.

epoxidation of cyclic alkenes. Thus, stigmasteryl acetate was selectively converted into the 5,6-epoxide in near quantitative yield. The mechanism is believed to involve a series of electron transfers mediated by copper(II) permanganate [482].

2.3.2.4 Epoxidation of Electron-Deficient Alkenes

Different conditions usually apply for the epoxidation of electron-poor olefins [483], most of which capitalize on the susceptibility of the substrate toward nucleophilic attack. More recent innovations in this regard include the use of basic hydrogen peroxide in an ionic liquid/aqueous biphasic system [484] or aqueous hydrogen peroxide in the presence of natural phosphate [485] or hydrotalcite with ultrasound [486]. Non-aqueous systems commonly employ TBHP, and this oxidant can be effectively catalyzed by a non-nucleophilic base, such as the guanidine derivative 1,5,7-triazabicyclo[4.4.0]dec-5-ene (TBD) [487], or even by potassium fluoride adsorbed onto alumina [488]. The methodology is not limited to conventional nucleophilic chemistry: electron-deficient alkenes can also be epoxidized under electrochemical conditions using a silver(III)oxo bis(2,2′-bipyridine) catalyst [489] as well as with more electrophilic oxidizing agents, such as iodosylbenzene [490].

As with epoxidation protocols in general, vigorous activity has surrounded the asymmetric synthesis of epoxides from electron-deficient alkenes, and many chiral catalysts and auxiliaries have been developed for this purpose (Figure 2.18). A common test reaction for comparing yields and enantioselectivities is the epoxidation of chalcone (**453**, Scheme 2.69); Table 2.16 summarizes some illustrative contribu-

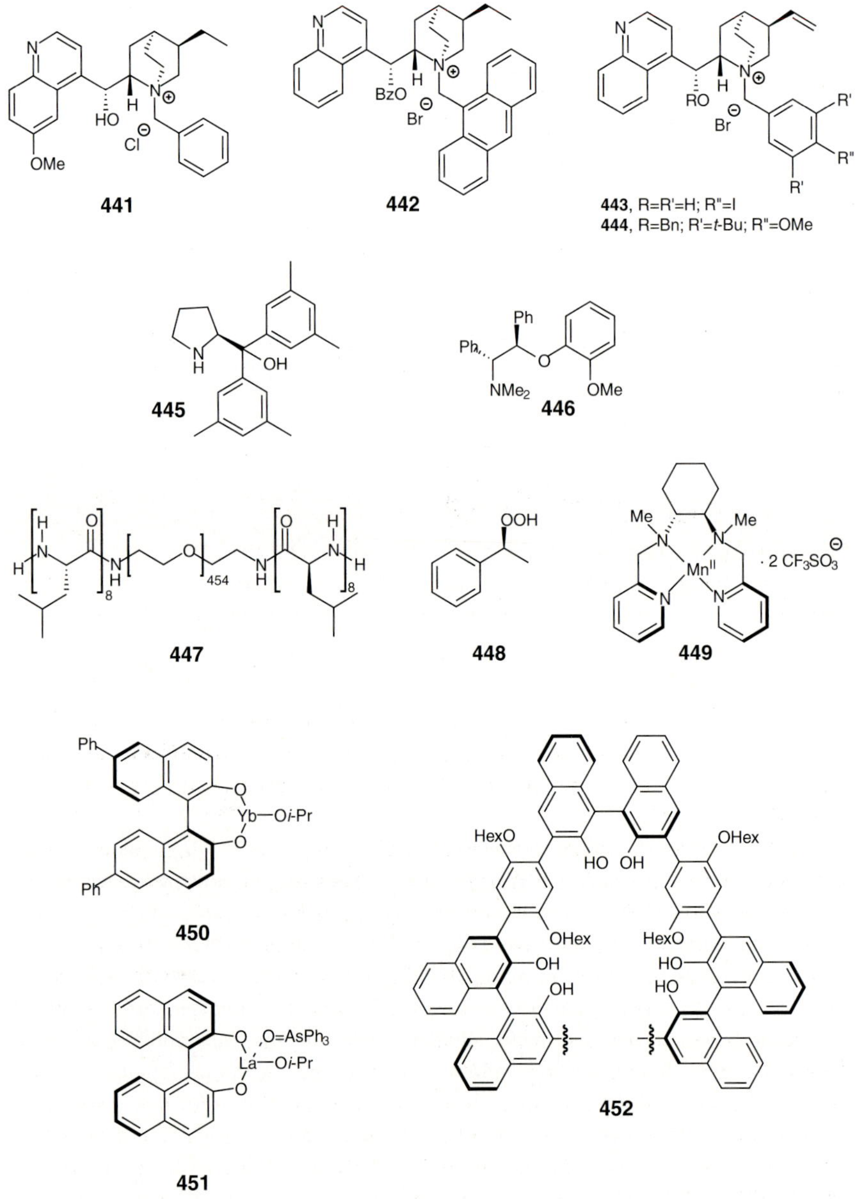
441
442
443, R=R'=H; R"=I
444, R=Bn; R'=t-Bu; R"=OMe
445
446
447
448
449
450
451
452

Figure 2.18 Chiral catalysts and auxiliaries for electron-deficient alkenes.

tions to this methodology. For example, *Cinchona*-derived phase-transfer catalysts (e.g., **441–444**) are effective in aqueous organic systems (entries 1–3) [491–493]. As a further demonstration of chiral pool inspired catalysts, proline-derived aminoalcohols (e.g., **445**) promote asymmetric epoxidation of enones through non-covalent catalysis [494, 495]. Oligopeptides also show promise as chiral auxiliaries [496], as

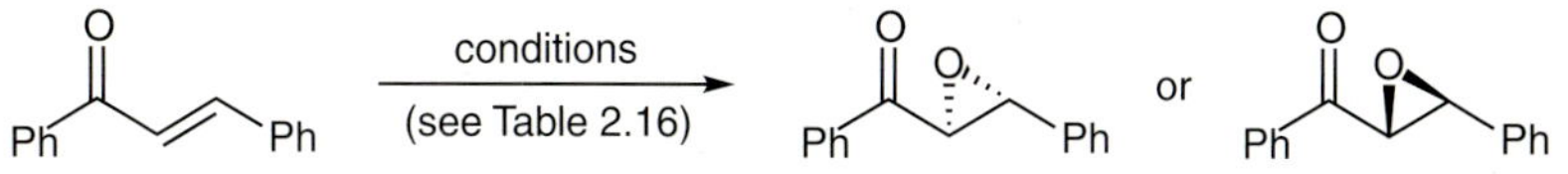

conditions (see Table 2.16)

453 **(+)-454** or **(-)-454**

Scheme 2.69 Asymmetric epoxidation of chalcone.

Table 2.16 Yield data for the asymmetric epoxidation of chalcone.

Entry	Catalyst	Oxidant	Solvent	Yield (%)	ee (%)	Conf.	Ref
1	**442**	NaOCl	Toluene	98	86	(+)	[424]
2	**443**	H_2O_2/LiOH	Bu_2O	97	84	(+)	[425]
3	**444**	NaOCl	Toluene	91	60	(+)	[426]
4	**445**	TBHP	Hexane	90	91	(−)	[494]
5	**447**	Urea-H_2O_2		>99	94	(−)	[428]
6	**450**	CHP	THF	91	97	(−)	[429]
7	**451**	TBHP	THF	95	97	(+)	[430]
8	**452**/Et_2Zn	TBHP	Et_2O	95	74	(−)	[432]

illustrated by the novel poly(ethylene glycol)-supported oligo(L-leucine) catalyst **447** used to carry out the Juliá–Colonna epoxidation of chalcone through a continuously operated "chemzyme" membrane reactor with urea-hydrogen peroxide as the terminal oxidant [497].

The chiral ytterbium complex formed from Yb(*i*-PrO)$_3$ and 6,6′-diphenyl-BINOL (**450**) catalyzed the epoxidation in 91% yield and 97% ee using cumene hydroperoxide as the oxygen source [498]. A similar outcome, but a much more rapid conversion, is achieved using the lanthanoid-BINOL-triphenylarsine complex **451**, which provides complete epoxidation in three minutes [499]. This protocol was used as a key step in the synthesis of (+)-decursin from commercially available esculetin [500]. Solid-supported catalysts have also been reported, including the complex formed by treating binaphthyl polymers (e.g., **452**) with diethyl zinc [501].

The tridentate aminodiether ligand **446** has been used with lithium cumene hydroperoxide to give fair to good enantioselectivities in the epoxidation of certain enones (e.g., **456**, Scheme 2.70), presumably through a tetracoordinate lithium complex [502]. The chiral peroxide **448** was effective in the epoxidation of 2-methylene-1-tetralone derivatives, such as **458** [503], as was the immobilized synthetic peptide poly-L-leucine (*i*-PLL) in the presence of urea-peroxide and DBU in a solvent of isopropyl acetate [504]. As is the case with unfunctionalized alkenes, electron-deficient olefins are also subject to asymmetric epoxidation using chiral dioxirane reagents [505]. Interestingly, the chiral cationic manganese bis(pyridyl) catalyst **449** provided for very little asymmetric induction, although it was nevertheless quite efficient in promoting the epoxidation of enones such as cyclohexenone

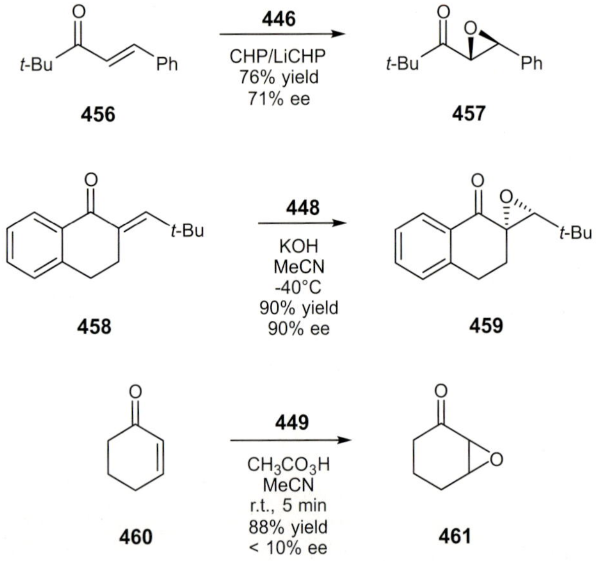

Scheme 2.70 Other epoxidations with chiral catalysts and reagents.

(**460**). When both electron-rich and electron-poor olefins are affixed to the same substrate, the former are preferentially oxidized [506].

In the realm of on-board chiral auxiliaries, proline-derived cinnamides (e.g., **462**, Scheme 2.71) were epoxidized with lithium *t*-butyl hydroperoxide with excellent diastereoselectivity, although in moderate yield [507]. Use of a 2,2-dimethyloxazolidine chiral auxiliary (e.g., **464**) led to superior yields but somewhat attenuated diastereomeric excess [508]. The nucleophilic epoxidation of γ-hydroxy-vinyl sulfoxide derivatives (e.g., **466**) proceeds both in high yield and exclusive diastereoselectivity. The latter outcome has been rationalized by invoking a geometrically constrained chair-like transition state [509].

2.3.2.5 Epoxidation of Carbonyl Compounds

Just as aziridines can be prepared by the addition of carbenes (or carbene equivalents) across imines, so too can epoxides be synthesized from carbonyl compounds, particularly aldehydes. A recent review has brilliantly captured the synthetic utility of this approach [90]. One prototypical example is the conversion of 4-chlorobenzaldehyde (**468**) into the corresponding styrene oxide (**469**) by treatment with diiodomethane and methyllithium at 0 °C. The mechanism is believed to proceed through a sequence of lithium–halogen exchange, carbonyl addition and rapid ring closure of the intermediate iodoalkoxide [510].

The Corey–Chaykovsky synthesis [511], by now a standard method, is nevertheless still the subject of current innovation [512]. This reaction, like most other methylenations, relies upon an intermediate sulfur ylide to serve as a methylene transfer reagent. Recent reports have shown that dry mixtures of trimethylsulfonium iodide

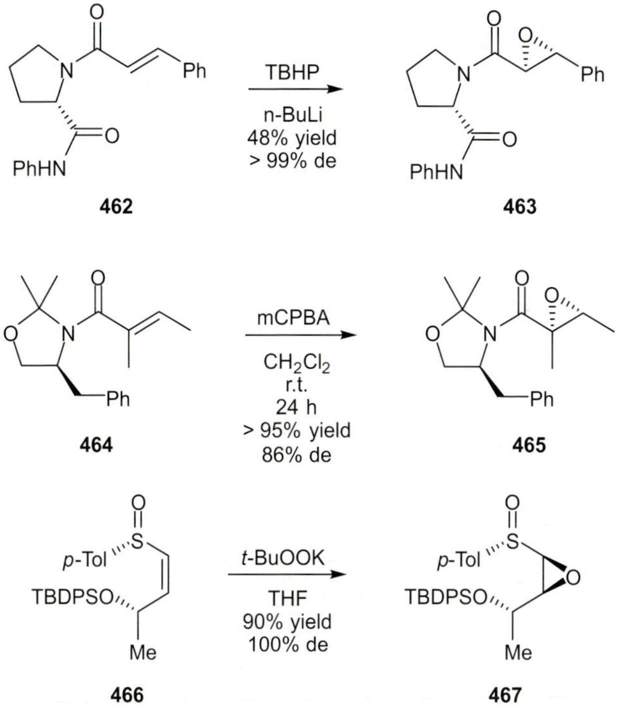

Scheme 2.71 Diastereoselective epoxidations with chiral auxiliaries.

and sodium hydride form a shelf-stable source of "instant methylide" upon treatment with carbonyl compounds in a polar aprotic solvent. Thus, combination of 4-methylacetophenone (**470**, Scheme 2.72) with the "instant methylide" reagent in dimethyl sulfoxide (DMSO) resulted in the high-yielding formation of the corresponding epoxide **471** [513]. The Corey–Chaykovsky epoxidation has also been adapted for use in an ionic liquid medium, such as (bmim)PF_6 [514].

Some methylides are conveniently available through a novel thermal decarboxylation of carboxymethylsulfonium betaines. Thus, treatment of the sulfonium bromide **472** with silver oxide affords the corresponding betaine **473**, which exhibits a half-life of 5 h in chloroform at room temperature, but can be stored for months neat at $<0\,^{\circ}C$. At elevated temperatures, however, a rapid decarboxylation provides the methylide **474**, which reacts with 2,6-dichlorobenzaldehyde (**475**) to give the epoxide **476**. Unsurprisingly, electron-deficient aldehydes give higher yields, with benzaldehyde itself failing to provide any epoxide at all, presumably due to competing thermal decomposition of the ylide **474** [515].

Synthetically useful alkynyl epoxides can be accessed through the treatment of aldehydes with propargyl ylides in the presence of trialkylgallium bases, which lead to (Z)-stereoselectivity [516]. These products are also available through a non-ylide route by treating carbonyls with 1-bromoalkynes and t-butoxide. In this interesting cascade

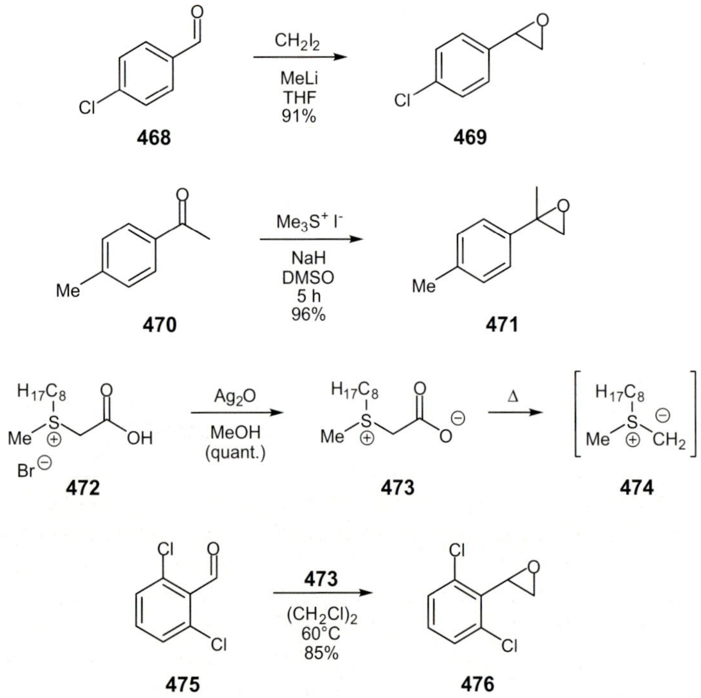

Scheme 2.72 Epoxidation of carbonyls with methylene equivalents.

reaction, the 1-bromoalkyne is believed to function as both electrophilic halogen source and acetylide equivalent [517].

Asymmetric variants of this protocol have been reported using chiral organic sulfides (Figure 2.19) that are converted into the corresponding ylides, usually either stoichiometrically in a separate step or catalytically *in situ*. The conversion of

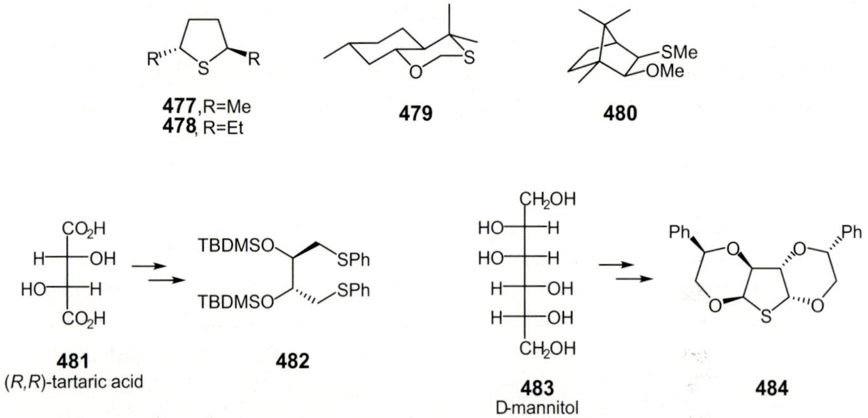

Figure 2.19 Some chiral sulfur ylide precursors.

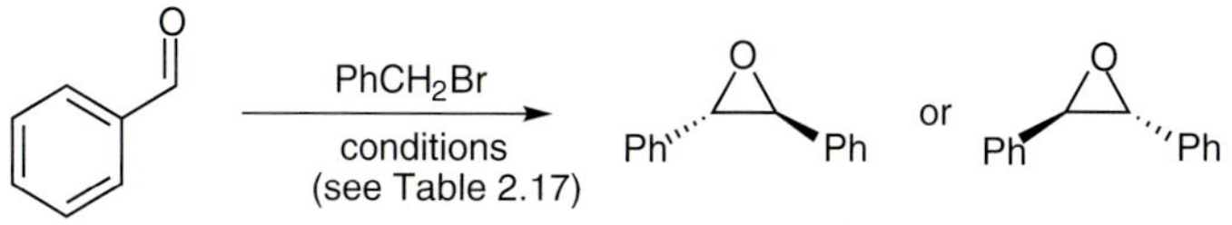

PhCH$_2$Br
conditions
(see Table 2.17)
or

485 (*S*,*S*)-**486** (*R*,*R*)-**486**

Scheme 2.73 Sulfur-mediated asymmetric epoxidation of benzaldehyde.

benzaldehyde (**485**) into *trans*-stilbene oxide (**486**) is a convenient test system for such procedures (Scheme 2.73). The 2,5-dimethylthiolane (**477**) was used in a one-pot stoichiometric variant of this reaction (Table 2.17, entry 1), providing the (*S*,*S*)-epoxide in excellent yield and good enantioselectivity [518]; even higher optical purity was obtained simply by using the diethyl analog **478** [519].

One challenge for this general approach is ready access to chiral auxiliaries in optically pure form. The ideal precursors, therefore, should be available in relatively few steps and on large scale from the chiral pool. The bis-sulfide **482**, obtained from (*R*,*R*)-tartaric acid (**481**), arguably satisfies these criteria. However, the level of asymmetric induction (Table 2.17, entry 3) falls somewhat short of synthetic utility, most likely because of its conformational flexibility [520]. However, the more rigid tricyclic sulfide **484**, derived from D-mannitol (**483**), provides excellent enantioselectivity (albeit with moderate yields) while operating in a catalytic one-pot environment [521].

A highly enantioselective synthesis of glycidic amides has been reported using stoichiometric amounts of the chiral sulfide **480**, which is available in three steps and in high yield from camphor. Thus, optically pure sulfonium bromide **488** was prepared by treatment of sulfide **480** with bromoamide **487** (Scheme 2.74). Further exposure to benzaldehyde under basic conditions affords glycidic amide **489** in excellent yield and optical purity [522].

One other major route from carbonyls to epoxides involves the addition of metal-stabilized carbenoids to carbonyls. For example, the donor–acceptor rhodium carbenoids derived from aryldiazoacetates **490** add across the carbonyl moiety of α,β-unsaturated aldehydes, such as *trans*-crotonaldehyde (**491**), to give vinyl epoxides

Table 2.17 Yield data for the sulfur-mediated epoxidation of benzaldehyde.

Entry	Precursor	Base	Solvent	Yield (%)	de (%)	ee (%)	Conf.	Reference
1	**477**	KOH	*t*-BuOH/H$_2$O	92	88	84	(*S*,*S*)	[446]
2	**478**	KOH	*t*-BuOH/H$_2$O	97	88	93	(*S*,*S*)	[447]
3	**482**	NaOH	*t*-BuOH/H$_2$O	22	48	68	(*R*,*R*)	[448]
4	**484**	NaOH	MeCN/H$_2$O	42	82	94	(*R*,*R*)	[449]

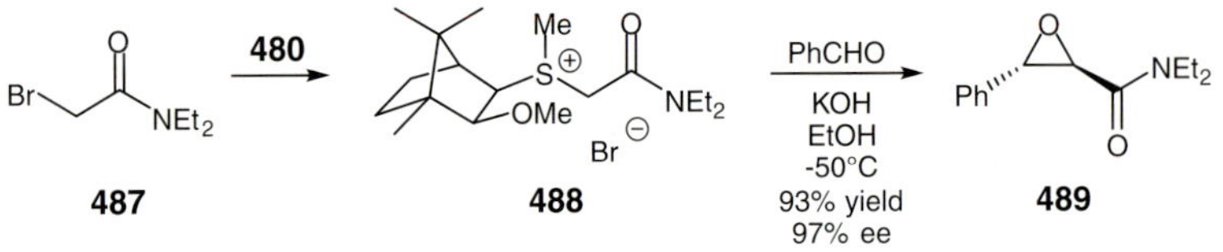

Scheme 2.74 Enantioselective synthesis of glycidic amides.

(e.g., **492**) in good to excellent yield (Scheme 2.75) [523]. Modification of these conditions to include triphenylarsine leads to the formation of intermediate arsonium ylides that function as the active carbon-transfer reagents, resulting in excellent *trans*-selectivities [524]. Asymmetric protocols are also available. Thus, treatment of benzaldehyde (**485**) with a slight excess of the tosyl hydrazone sodium salt **497** in the presence of catalytic amounts of rhodium acetate and 20 mol.% of the camphor-derived sulfide **496** furnishes 1,2-diarylepoxide **498** with 89% ee [525]. However, when the substrate is a heteroaromatic, *n*-alkyl aliphatic, α,β-unsaturated or acetylenic aldehyde, stoichiometric quantities of the preformed chiral sulfur ylide must be used [526].

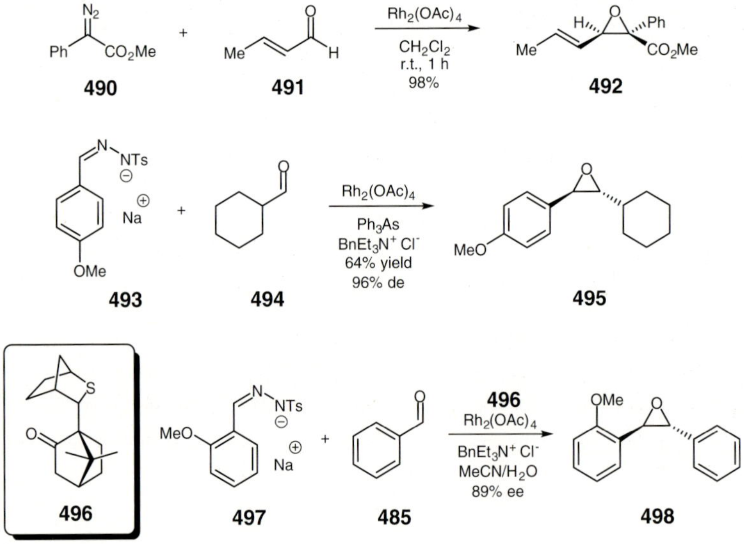

Scheme 2.75 Rhodium-catalyzed epoxidations of carbonyls.

2.3.2.6 Ring-Closing Reactions

One of the oldest techniques for preparing epoxides is the base-promoted ring closure of halohydrins, used by Wurtz to synthesize ethylene oxide from β-chloroethanol as early as 1859 [527]. The same procedure is still routinely used with various modifications. For example, in their approach to the taxol side chain antipodes, Stewart and coworkers treated the optically pure chlorohydrin **499** (Scheme 2.76) with potassium carbonate in DMF to obtain the *cis*-epoxide **500** in almost quantitative yield [528]. A sequential process was used as an early-stage key step in the total synthesis of ovalicin, in which the doubly protected cyclic triol **502** undergoes mesylation, deprotection and ring closure to give the spirocyclic epoxide **503** [529]. A similar procedure assembled the epoxide ring late in the total synthesis of the neocarzinostatin chromophore aglycone [530]. The development of biocatalytic methods for preparing chiral non-racemic chlorohydrins from α-chloroketones has opened a potentially useful pathway towards chiral epoxides by this route [531].

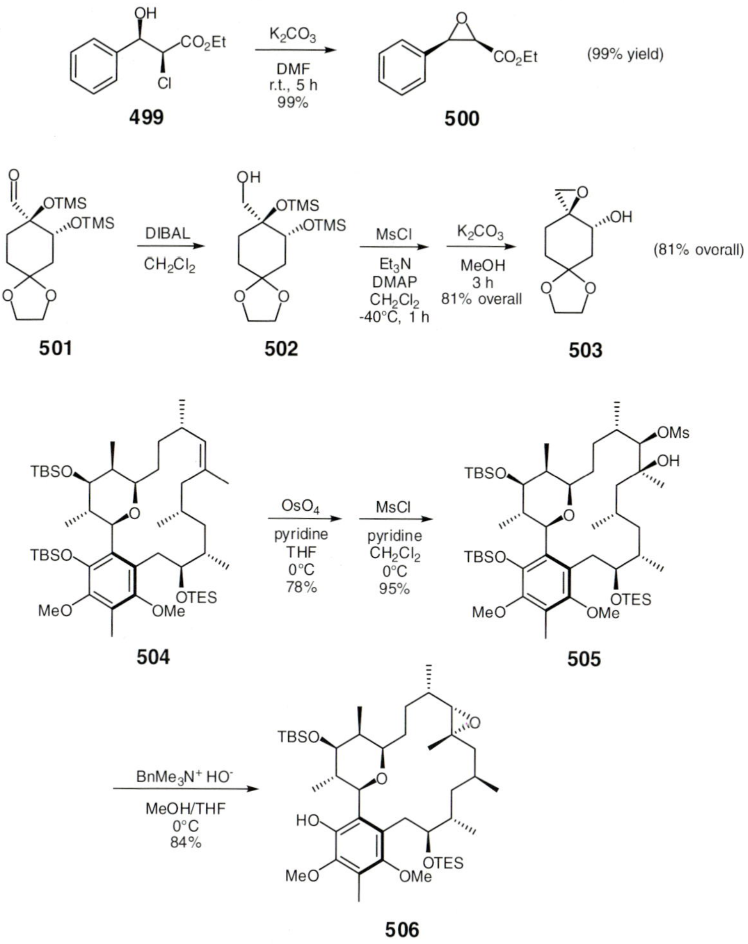

Scheme 2.76 Epoxides from ring-closing reactions.

Ring closure methodology can be employed as a nice complement to other concerted oxygenations. For example, in their total synthesis of (−)-kendomycin, Smith and coworkers carried out a *cis*-dihydroxylation on the macrocyclic alkene **504** to give the corresponding diol. Mesylation of the secondary alcohol afforded the hydroxy mesylate **505**, which suffered ring-closure with inversion under phase-transfer conditions. In this way, a *cis*-alkene was efficiently and controllably converted into the *trans*-epoxide [338].

2.3.3 Reactivity of Oxiranes

2.3.3.1 Nucleophilic Ring Opening

Cleavage of the epoxide ring by various nucleophiles is one of the most frequently encountered behaviors of this system, both biologically and synthetically. In the latter realm, the extreme versatility of this simple reaction lends it considerable preparative power. The nucleophilic palette runs the gamut, and protocols are being developed continually to direct the nucleophilic ring opening in an enantioselective fashion [532].

Unadorned carbon nucleophiles may be used, as exemplified by the one-pot conversion of alkenyl epoxide **507** (Scheme 2.77) to the homologous silyl ether **508** using a system of trimethylaluminium and silyl triflate. The methyl group is delivered via backside attack on the less substituted terminus of the epoxide, and the alkoxide so formed is silylated *in situ* [533]. An ethyl group can be appended in like fashion using triethylaluminium catalyzed by triphenylphosphine [534]. Similar ring openings also can be carried out using indoles with lithium perchlorate [535] or ruthenium trichloride [536], lithium enolates [537], vinyl magnesium bromide [538], aluminum ester enolates [539], and silyl enol ethers catalyzed by titanium tetrachloride [540]. Vinyl epoxides undergo ring opening at the allylic position using diethylzinc catalyzed by trifluoroacetic acid [541] and alkyllithium reagents catalyzed by boron trifluoride [542].

Alkynyl anions react smoothly with epoxides, as well. For example, Kumar and Naida used this strategy to stitch together the functionalized lithium acetylide derivative **509** and epoxide **510** in their total synthesis of microcarpalide [543]. Other conditions for this reaction include the use of alkynyllithiums with catalytic trimethylaluminium [544] and lithium TMS-acetylide in dimethyl sulfoxide [545]. The addition of vinyl anions per se is not a general reaction; however, there are some interesting vinyl anion equivalents such as trimethylsulfonium iodide, which converts the tetrahydrofuranyl epoxide **512** into the corresponding allyl alcohol (**513**) [546]. Confoundingly, the counterion can play an enormous role in the overall reaction yields; in many cases, the trifluoromethylsulfonyl anion can give superior results [547].

The cyanide anion is a common carbon nucleophile that is also capable of epoxide ring opening. For example, treatment of the terminal epoxide **514** with trimethylsilyl cyanide (TMSCN) in the presence of lithium perchlorate resulted in the delivery of cyanide to the less substituted position in excellent yield [548]. The binaphthyl derived

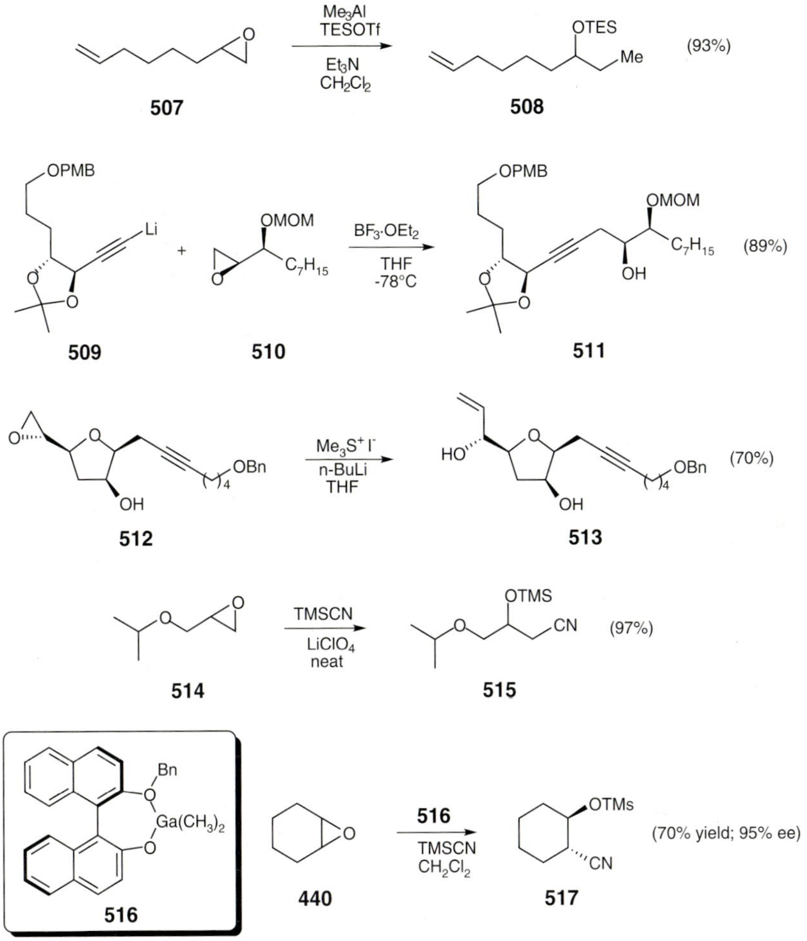

Scheme 2.77 Epoxide ring opening with carbon nucleophiles.

gallium catalyst **516** represents an asymmetric variant of this process, promoting the addition of cyanide to *meso* epoxides (e.g., **440**) in good yield and excellent enantioselectivity [549].

In the category of nitrogen-based nucleophiles, simple amines add smoothly to epoxides in predictable ways. For example, benzylamine attacks the less hindered carbon of the epoxyether **518** (Scheme 2.78) under the influence of lithium bis-trifluoromethanesulfonimide to give the aminoalcohol **519** in 95% yield [550]. When the epoxide ring bears an aromatic substituent, the regiochemistry is often reversed, as shown by the ring-opening of styrene oxide (**421**) with aniline in the presence of zinc chloride [551]. Amines can also be added using calcium triflate in acetonitrile [552], and in water with erbium(III) triflate as a catalyst [553], with no catalyst under conventional conditions [554], or with the assistance of ultrasound [555].

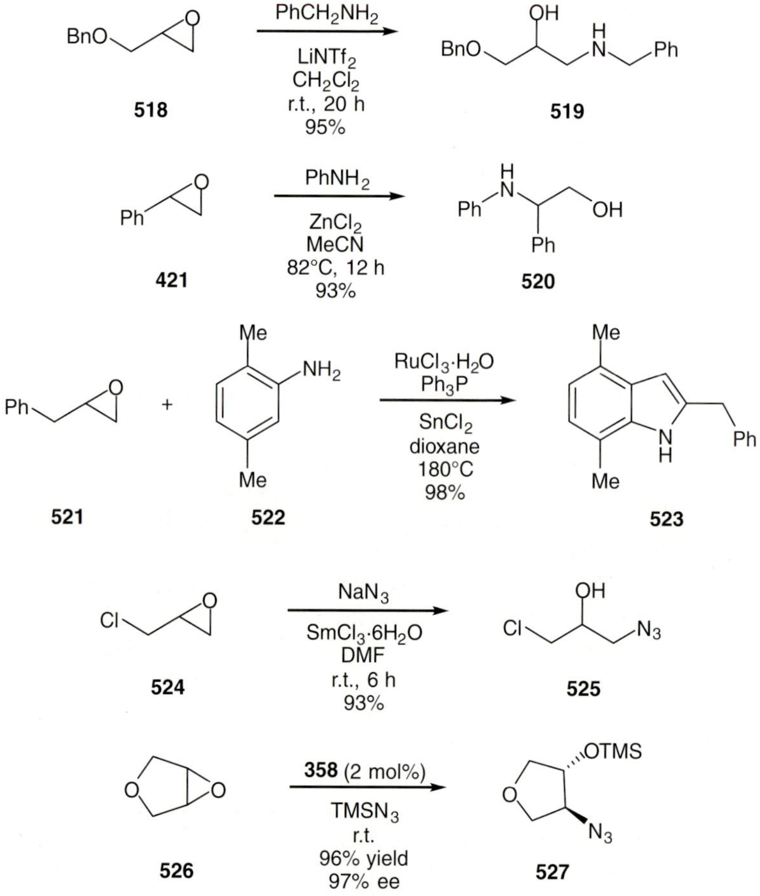

Scheme 2.78 Epoxide ring opening with nitrogen nucleophiles.

The addition of aromatic amines can be catalyzed by stannic or cupric triflate [556, 557]; β-cyclodextrin [558]; zirconium(IV) chloride [559] or bismuth trichloride [560, 561] in acetonitrile; and zinc oxide [562], ytterbium(III) nitrate [563], or a mesoporous silica immobilized cobalt complex [564] under solvent-free conditions. A ruthenium catalyst in the presence of tin chloride also results in an S_N1-type substitution behavior with aniline derivatives (e.g., **522**), but further provides for subsequent cyclization of the intermediate amino alcohol, thus representing an interesting synthesis of 2-substituted indoles (e.g., **523**) [565]. Certain meso-epoxides can be desymmetrized with aromatic amines under catalytic conditions, for example, using a proline-based N,N'-dioxide-indium tris(triflate) complex [566].

Azide represents a simple, versatile and selective nitrogen nucleophile. In the presence of catalytic quantities of samarium(III) chloride in a medium of *N,N*-dimethylformamide (DMF), sodium azide attacks the terminal epoxide carbon of epichlorohydrin (**524**) to give the highly functionalized three-carbon fragment **525** [567]. Azide addition can also be promoted by the use of lithium tetrafluoroborate in *t*-butanol [568] and lithium perchlorate in propionitrile [569]. Attack at the more substituted position is favored by some conditions, including diethylaluminium azide [570] and sodium azide in the presence of ceric ammonium nitrate [571]. Like cyanide, azide can be used to desymmetrize meso-epoxides, as shown by the asymmetric azidolysis of the bicyclic epoxide **526** using TMS-azide and catalytic amounts (2 mol.%) of the salen-chromium complex **358** [572, 573].

Oxygen-centered nucleophiles are also of synthetic importance in this regard, and the reactions of alicyclic epoxy compounds with these nucleophilic species are the subject of a review [574]. Arguably the most readily available oxygen-centered nucleophile for epoxide ring opening is water, but the course of the hydrolysis reaction is dependent upon the reaction environment and structural features of the substrate [575]. Conditions can be exceedingly mild, as shown by the high-yielding hydrolysis of spiroepoxide **528** (Scheme 2.79) using tetrabutylammonium bisulfate [576]; catalytic bismuth triflate in wet acetonitrile can also be used to advantage [176]. In fact, many epoxides can be hydrolyzed simply by heating in water without any catalyst at all [577]. The conversion of epoxides into diols in this manner is the basis for an enormously important method for preparing optically pure epoxides from racemic mixtures through hydrolytic kinetic resolution (HKR) [578]. Chiral nonracemic diols are also available from the hydrolytic desymmetrization of meso-epoxides (e.g., **440**) using an oligomeric Jacobsen-type catalyst (**530**) [579].

Alcohols can be added with equal efficiency. Thus, phenoxymethyl epoxide **532** suffers nucleophilic attack by methanol in the presence of catalytic amounts of potassium dodecatungstocobaltate to provide hydroxyether **533** in quantitative yield [580]. As with other nucleophiles, when an aromatic group is attached to the epoxide ring, attack often predominates at the benzylic position. For example, treatment of stilbene oxide (**421**) with methanol under the catalysis of $TiO(TFA)_2$ yields the primary alcohol **534**. The electrophilic center is cleanly inverted in the process [581]. This addition can also be carried out using ferric perchlorate [579], molybdenum(VI) dichloride dioxide [582], hydrazine sulfate [583], amberlyst-15 resin [584], copper(II) tetrafluoroborate [585], and aminopropylsilica gel (APSG) supported iodine [586]. The hydroperoxide anion functions as a competent nucleophile under the catalysis of silica-supported antimony trichloride [587].

The scope of oxygen nucleophiles extends to carboxylic acids, as illustrated by the titanium-catalyzed addition of pivalic acid to the terminal epoxide **535** [588]. Chromium(III) acetate is a useful catalyst for such additions of carboxylic acids in industrial processes [589]. Finally, nitrates can also be coaxed into serving as oxygen-centered nucleophiles by reagents such as bismuth(III) nitrate [590] and tetranitromethane [591].

Other oxygen-containing heterocycles with varying degrees of structural complexity are conveniently prepared by the intramolecular ring-opening of epoxides [592]. An illustrative example is found with the titanium-promoted cyclization of the highly oxygenated bicyclic epoxide **537** to give the spiroketal **538** with retention of configuration [593]. Similar intramolecular processes have been catalyzed by caesium

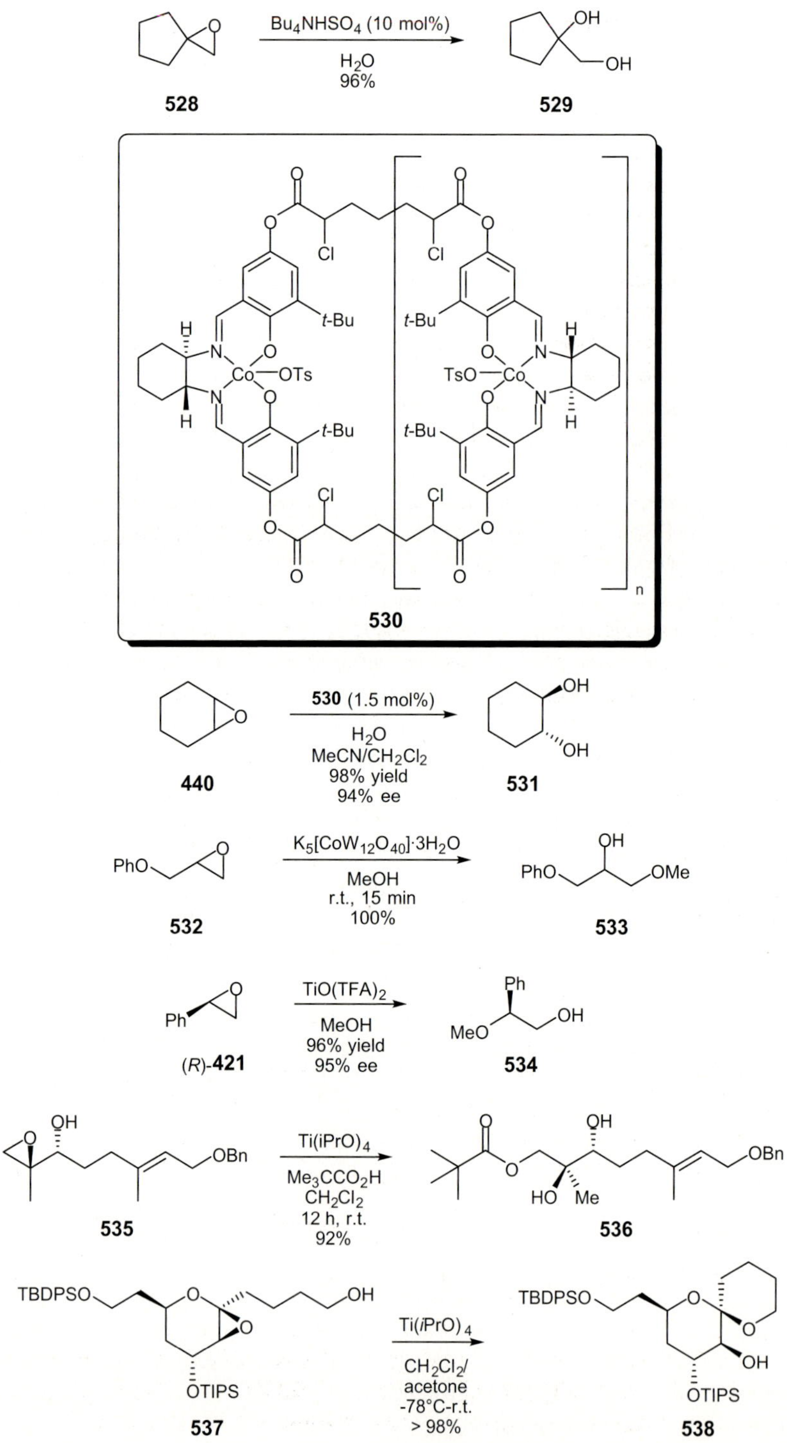

528 → 529: Bu_4NHSO_4 (10 mol%), H_2O, 96%

530

440 → 531: **530** (1.5 mol%), H_2O, $MeCN/CH_2Cl_2$, 98% yield, 94% ee

532 → 533: $K_5[CoW_{12}O_{40}]\cdot 3H_2O$, MeOH, r.t., 15 min, 100%

(*R*)-**421** → 534: $TiO(TFA)_2$, MeOH, 96% yield, 95% ee

535 → 536: $Ti(iPrO)_4$, Me_3CCO_2H, CH_2Cl_2, 12 h, r.t., 92%

537 → 538: $Ti(iPrO)_4$, CH_2Cl_2/acetone, -78°C-r.t., > 98%

Scheme 2.79 Epoxide ring opening with oxygen nucleophiles.

carbonate [594] and *p*-toluenesulfonic acid [595]. In one such intramolecular epoxide ring opening – a key step in the diastereoselective synthesis of α-tocopherol – anhydrous hydrochloric acid in ether/acetonitrile was chosen to promote the disfavored 6-*endo-tet* ring closure mode [596].

Among sulfur-centered nucleophiles, thiocyanate is frequently encountered. For example, the epoxy ester **540** (Scheme 2.80) is smoothly converted into the thiocyanato adduct **541** using stoichiometric ammonium thiocyanate and a quaternized amino functionalized cross-linked polyacrylamide (**539**) as a solid–liquid phase-transfer catalyst in acetonitrile [597]. Hydroxysulfones can be prepared "on water" by the action of sodium benzenesulfinate on epoxides with no added catalyst [598]. Thiophenol is another useful species that adds under very mild conditions, as shown by the ring-opening of the cyclohexadiene oxide derivative **542** catalyzed by ytterbium triflate in toluene, whereby the thiophenol moiety attacks the allylic site of the epoxide ring [599]. In another example, unfunctionalized epoxides (e.g., **544**) can be transformed into allylic alcohols **547** through an initial epoxide ring-opening with thiophenol in hexafluoroisopropanol (HFIP) and *in situ* oxidation to the sulfoxide (**546**), followed by pyrolysis in the presence of potassium carbonate [600]. Thiols can also be added using tributylphosphite [166], lithium perchlorate [601, 602], montmorillonite K-10 clay under solvent-free microwave conditions [603], in water [604], and in ionic liquids without additional catalysts [605]. The addition of Rongalite® allows for the use of disulfide thiol precursors [606], and the entantioselective cleavage of meso-epoxides with thiophenol can be achieved using a heterobimetallic Ti-Ga-salen catalyst [607].

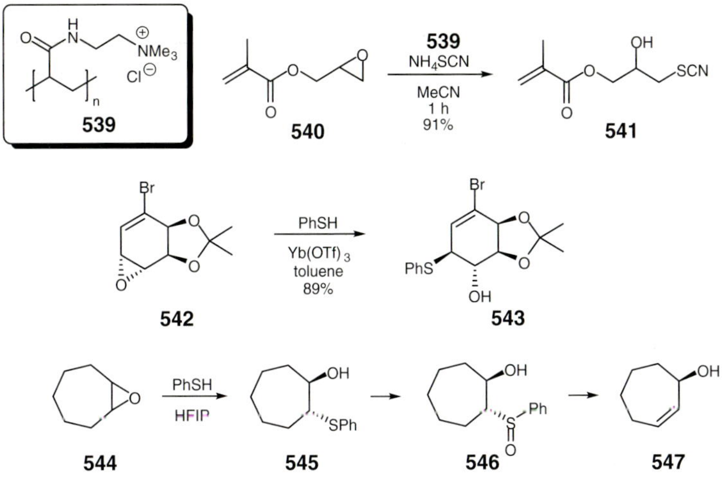

Scheme 2.80 Epoxide ring opening with sulfur nucleophiles.

Finally, halides are interesting nucleophiles inasmuch as they preserve the electrophilic character of the center they substitute. As a representative example of these additions, chloride can be introduced using bis-chlorodibutyltin oxide in chloroethanol, as shown in the conversion of phenylmethyl epoxide **521** (Scheme 2.81) into chlorohydrin **548** [608]. Similarly, the terminal epoxide **549** is converted almost quantitatively into the bromohydrin **550** with lithium bromide in the presence of silica gel in methylene chloride [609]. Bromide-mediated ring-opening can also be executed using a combination of *N*-bromosuccinimide, triphenylphosphine and dimethylformamide [610]. Most nucleophilic of all, iodide readily attacks styrene oxide (**421**) to give the iodohydrin **551**, and the use of β-cyclodextrin directs the addition to the less-substituted carbon, presumably due to steric hindrance caused by guest–host complexation [611]. In the realm of asymmetric synthesis cyclic meso-epoxides can be desymmetrized by chloride attack using silicon tetrachloride catalyzed by PINDOX [612].

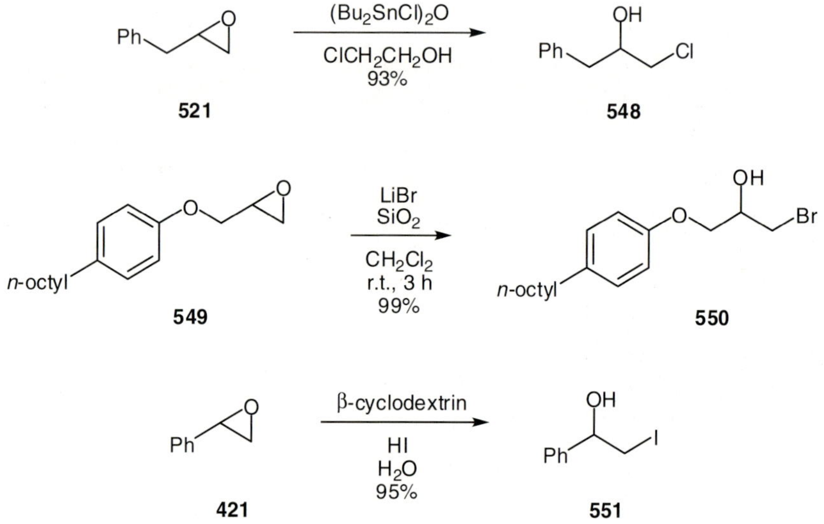

521 → **548**: $(Bu_2SnCl)_2O$, $ClCH_2CH_2OH$, 93%

549 → **550**: LiBr, SiO_2, CH_2Cl_2, r.t., 3 h, 99%

421 → **551**: β-cyclodextrin, HI, H_2O, 95%

Scheme 2.81 Epoxide ring opening with halide nucleophiles.

2.3.3.2 Rearrangements

Epoxides can be isomerized to allylic alcohols using hindered bases. For example, α-pinene oxide **552** (Scheme 2.82) undergoes eliminative ring opening upon treatment with lithium diethylamide. The transformation proceeds in higher yields in the presence of lithium *t*-butoxide, which is believed to disrupt aggregation of the anion [613]. Similarly, the optically pure bicyclic epoxide **554** is converted into the methylenecyclohexenol derivative **555** in excellent yield using diethylaluminium 2,2,6,6-tetramethylpiperidide (DATMP) in toluene [614]. Milder bases can be used when activating groups are nearby. Thus, the hydroxyepoxide **556** is smoothly oxidized to the corresponding ketone under Dess-Martin conditions, making the α-protons acidic enough to remove with sodium hydroxide, leading to the enone

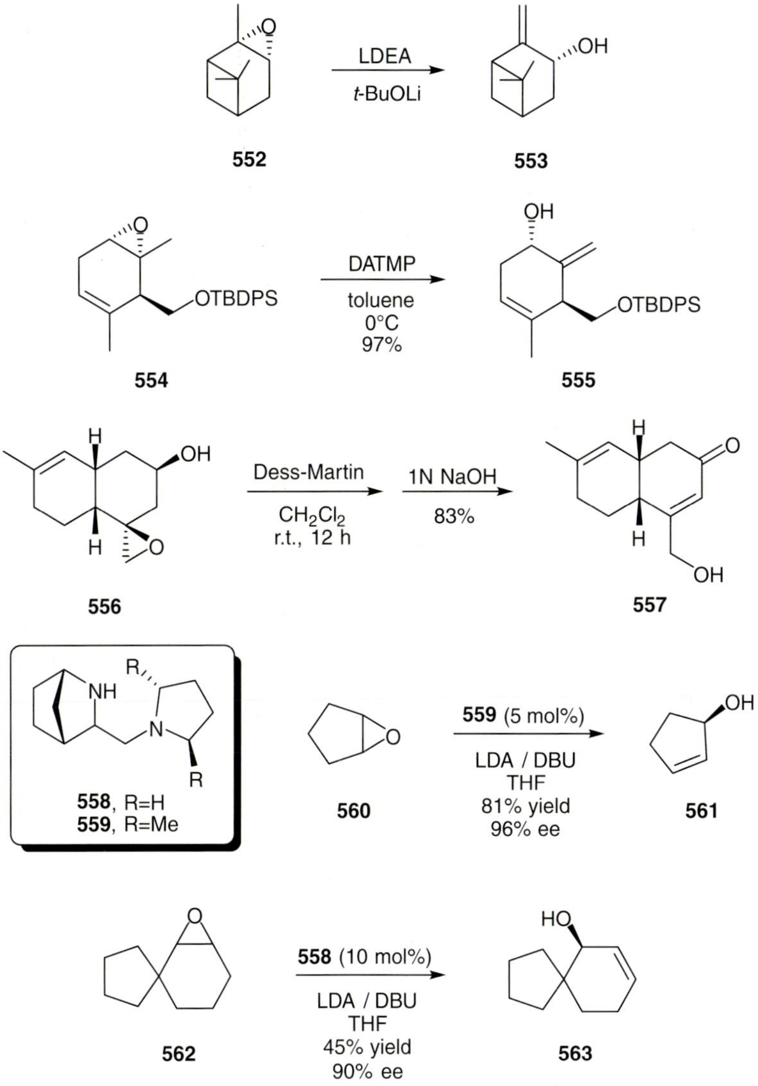

Scheme 2.82 Base-catalyzed rearrangement of epoxides to allylic alcohols.

product **557** [615]. Finally, the isomerization can occur under essentially neutral conditions using titania-supported gold nanoparticles [616].

The enantioselective isomerization of meso epoxides to allylic alcohols continues to be a promising route for the preparation of these materials in high optical purity. In an extension of their ongoing work in this area with lithium amide bases [617], Andersson and coworkers have designed the optically active (1*S*,3*R*,4*R*)-[*N*-(*trans*-2,5-dimethyl)pyrrolidinyl]-methyl-2-azabicyclo[2.2.1]heptane (**559**), which exhibits superior chiral induction in catalytic quantities using lithium diisopropylamide

(LDA) as the stoichiometric base. Thus, the challenging substrate cyclopentene oxide (**560**) was cleanly isomerized to the chiral cyclopentenol **561** in 81% yield and with 96% ee – a significant improvement over the 49% ee obtained with higher loadings of the earlier generation catalyst **558** [618]. Diamines derived from *(R)*-phenylglycine have also given promising results [619]. The chiral amide approach has also been applied to the catalytic kinetic resolution of racemic epoxides. For example, exposure of the tricyclic epoxide **562** to 10 mol.% **558** and stoichiometric LDA at 0 °C led to the recovery of the chiral spiro[4.5]decenol **563** with 90% ee and in 45% isolated yield, compared to the theoretical 50% maximum [620]. Chiral nonracemic aminoepoxides are isomerized stereoselectively to aminoalcohols using the superbasic mixture of *n*-butyllithium, diisopropylamide and potassium *t*-butoxide (LIDAKOR) [621].

In addition to β-elimination, the epoxide moiety also undergoes rearrangement to a carbonyl group, and this reactivity can be quite synthetically useful. The course of the rearrangement is highly dependent upon the nature of the substrate. Generally, the regiochemistry is driven by two factors: (i) the stability of the nascent carbocation generated from ring opening and (ii) the migratory aptitude of the adjacent substituents. For example, the simple monoalkyl-substituted epoxide **564** (Scheme 2.83) undergoes regioselective rearrangement in the presence of iron(III)tetraphenylporphyrin to give the corresponding aldehyde (**565**) via a 1,2-hydride shift onto an incipient secondary cationic center [622], a process also promoted by sodium periodate under ambient conditions [623]. Terminal epoxides react with tetraallyltin in the presence of bismuth(III) triflate to give homoallylic alcohols **568**. The reaction involves an initial 1,2-shift to form aldehyde **567**, which is then attacked by the allyl tin species [624]. A similar but operationally more straightforward protocol is available by combining allyl bromide with indium metal, followed by the addition of epoxide [625].

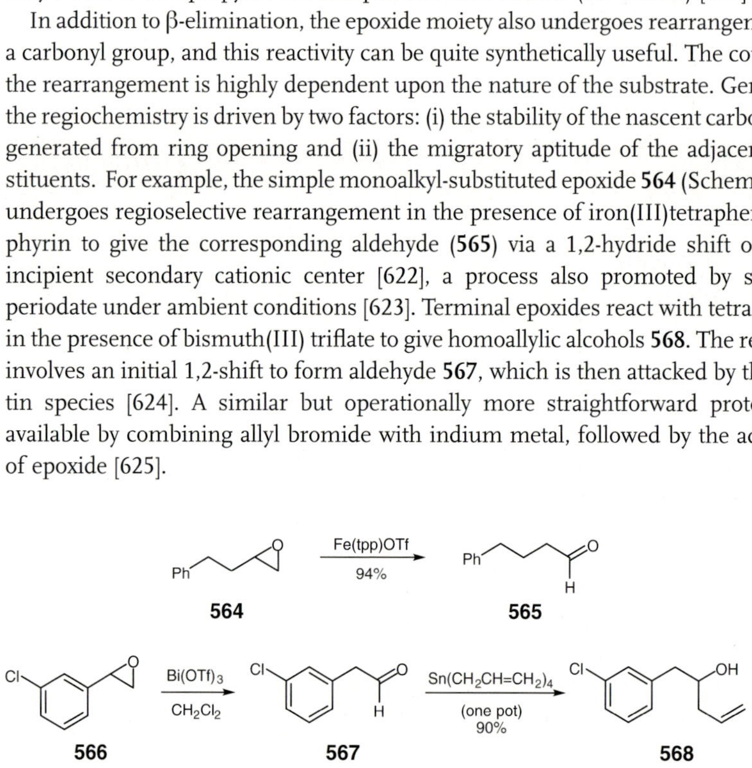

Ph 564 — Fe(tpp)OTf, 94% → Ph 565 (CHO)

566 — $Bi(OTf)_3$, CH_2Cl_2 → 567 — $Sn(CH_2CH{=}CH_2)_4$ (one pot) 90% → 568

Scheme 2.83 Rearrangement of terminal epoxides to aldehydes.

When *trans*-stilbene oxide (**569**, Scheme 2.84) is treated with bismuth triflate, aldehyde **570** is formed through a process of benzylic cation formation and subsequent phenyl migration [626]. However, the structurally very similar epoxide **571** provides a ketone upon treatment with Lewis acid, which reflects a more facile hydride shift to the cationic center [627]. Certain alkoxymethyl groups can also easily migrate, as seen in the rearrangements of epoxides **573** [628] and **575** [629]. In the latter example, the siloxy migration was an unwanted (albeit efficient) side reaction of

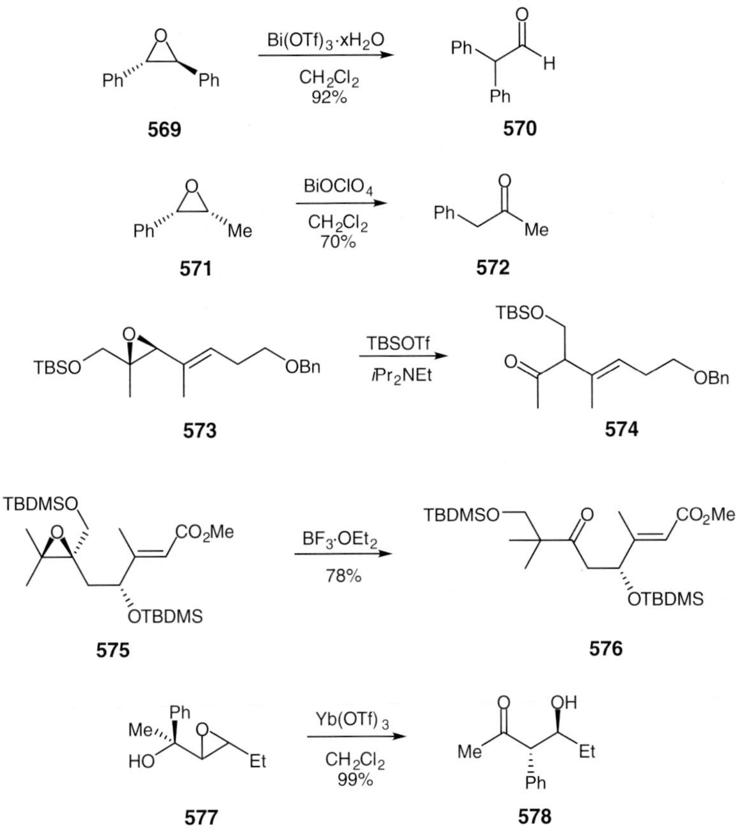

Scheme 2.84 Rearrangement of internal epoxides to aldehydes ketones.

a desired cationic ring closure; changing to a trimethylsilylethoxymethyl (SEM) protecting group resolved this problem. An analogous rearrangement can occur in epoxyalcohols of type **577**, which cleanly produced the hydroxyketone **578** in almost quantitative yield after 3 h upon exposure to 20 mol.% ytterbium triflate in methylene chloride [630].

This process can be used to advantage to access cyclic ketones, as well. For example, the cyclopentene oxide derivative **579** (Scheme 2.85) opens up to the more stable benzylic carbocation (i.e., **580**), which then provides the cyclopentanone derivative **581** via 1,2-methyl migration in 93% yield [631]. An analogous mechanistic step begins the organoaluminium-promoted cyclization of olefinic epoxides (e.g., **583**), whereby the initially formed aldehyde (**584**) undergoes a highly stereoselective Lewis acid-catalyzed intramolecular ene reaction to give the methylenedecalone **585** in the presence of methylaluminium bis(4-bromo-2,6-di-*t*-butylphenoxide) (MABR). This strategy is proposed as a route for the stereoselective synthesis of various terpenes [632].

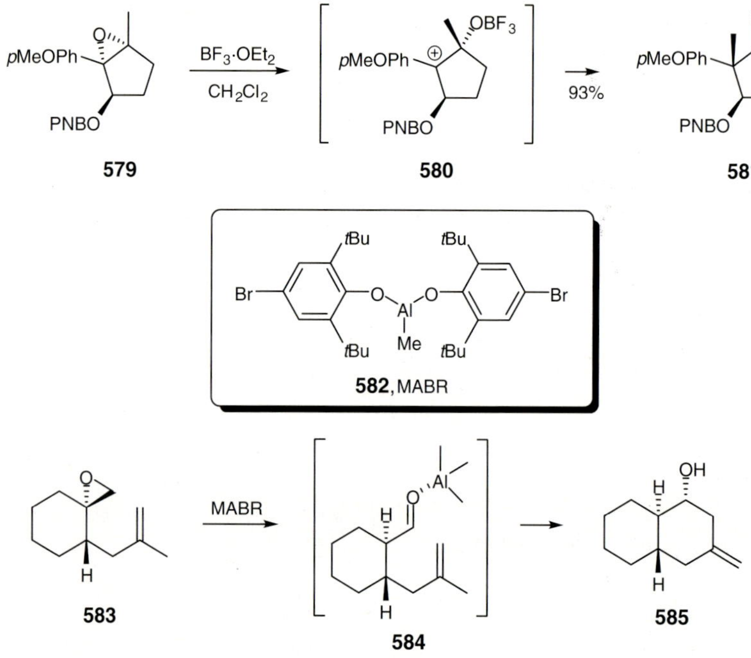

Scheme 2.85 Rearrangement involving cyclic structures.

The rearrangement sometimes occurs with concomitant ring contraction, as seen in the conversion of cyclohexene oxide derivative **586** (Scheme 2.86) into the cyclopentanealdehyde product **587** [633]. In a similar vein, Kita and coworkers [634] have used a novel acid-promoted rearrangement of cyclic α,β-epoxy acylates (e.g., **588**) for the stereoselective synthesis of spirocyclanes (e.g., **589**), a technique which is also found in the total synthesis of (–)-pseudolaric acid B by Trost *et al.* [635], and

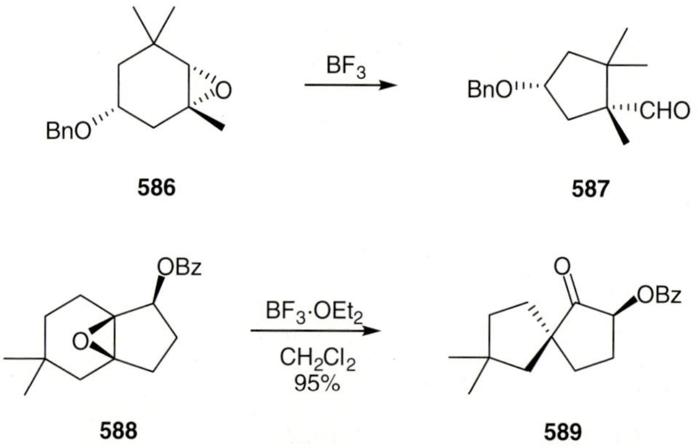

Scheme 2.86 Rearrangement with ring contraction.

which promises broad application to the preparation of optically active compounds of this type.

Ring expansions are also possible. For example, treatment of the cyclopentylepoxide **590** (Scheme 2.87) with boron trifluoride etherate induces a cascade reaction involving desilylation of the alcohol and rearrangement to the cyclohexanone derivative **591** [636]. Similarly, cyclohexyl epoxides (e.g., **592**) expand to form tropinones (**593**) [637]. Finally, although the process proceeds through a mechanis-

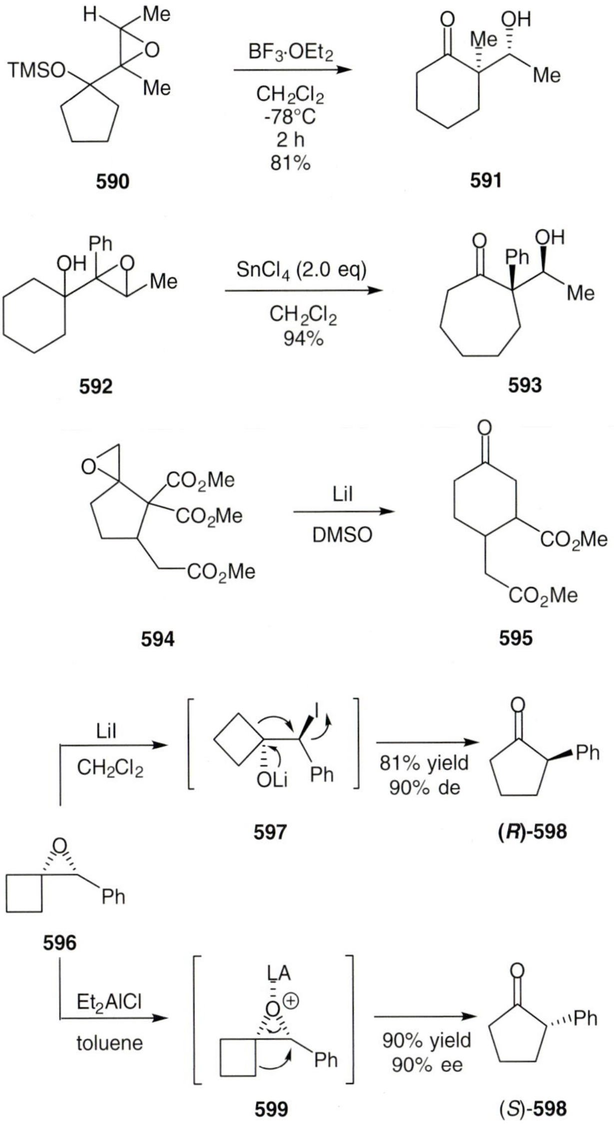

Scheme 2.87 Rearrangement with ring expansion.

tically distinct pathway, the iodide-mediated rearrangement of the spiroepoxide **594** produces the ring-expanded cyclohexanone **595** [638]. This was used in a very clever way to access both antipodes of phenylcyclopentanone (**598**) from spiroepoxide **596**. The anionic iodide pathway led to smooth conversion into the *(R)*-isomer, while Lewis acid catalysis provided equally high optical purity of the *(S)*-enantiomer [639].

2.3.3.3 Radical Chemistry

Li has authored an excellent overview of the radical reactions of epoxides, to which the reader is directed [640]. While this field encompasses some fascinating chemistry, yields and selectivities tend to be rather widely distributed. Exceptions to this generalization are found in specifically functionalized epoxides. For example, the vinyl epoxide **600** (Scheme 2.88) suffers radical addition of tributyltin radical to give an α-epoxy radical, which immediately opens to the allylic alkoxy radical **601**. This species engages in radical ring closure onto the pendant alkene to give, after hydrolysis, the bicyclic alcohol **604** in 89% yield [641, 642]. Samarium iodide promotes similar reactivity in ketoepoxides such as **605**. Thus, single electron transfer from SmI_2 to the carbonyl moiety gives the typical radical anion, which then isomerizes to the allylic alkoxy radical **607**. Trapping of the samarium stabilized enolate **608** with phenylpropionaldehyde gives the dihydroxyketone **609** in excellent yield [643].

Titanocene-mediated radical cyclization of epoxides has been reviewed very recently [644], and this area is rapidly expanding in synthetic utility. Here a titanocene reagent such as Cp_2TiCl engages the epoxide ring itself in single electron transfer, typically cleaving the weakest C–O bond and forming the more substituted carbon-centered radical, which can take part in further reactivity. For example, the epoxynitrile **610** is smoothly converted into the hydroxymethylcyclohexanone derivative **613** through a cascade of radical ring opening and subsequent cyclization onto the nitrile [645] (the intermolecular version of this methodology is promoted by titanocene [646]). Similarly, the propargylic epoxide **614** undergoes ring cleavage to give the secondary radical (**615**), which proceeds through 6-*exo-dig* radical cyclization to provide the methylenetetrahydropyran derivative **617** in high yield [647].

2.3.3.4 Reduction and Deoxygenation

Epoxides can undergo reductive ring opening using various reagents [648, 649]. One extremely mild protocol involves the use of bis(cyclopentadienyl)titanium(III) chloride. Significantly, the regioselectivity of the epoxide cleavage is often quite high, being determined by the stability of a radical intermediate, and sometimes opposite to what is expected for a classical S_N2 epoxide ring opening. For example, treatment of spiroepoxide **618** (Scheme 2.89) with Cp_2TiCl leads to an intermediate carbon radical that can be trapped by a H-atom donor (in this case cyclohexadiene) to give the secondary alcohol **620**. By comparison, a "classical" reductive ring opening with lithium triethylborohydride gives only the tertiary alcohol **619** [650]. Iyer [651] and Dragovich [652] have independently reported the regiospecific ring opening of epoxides by way of a palladium-catalyzed transfer hydrogenolysis using ammonium formate as the hydrogen source. Under these conditions, hydride attacks at the less

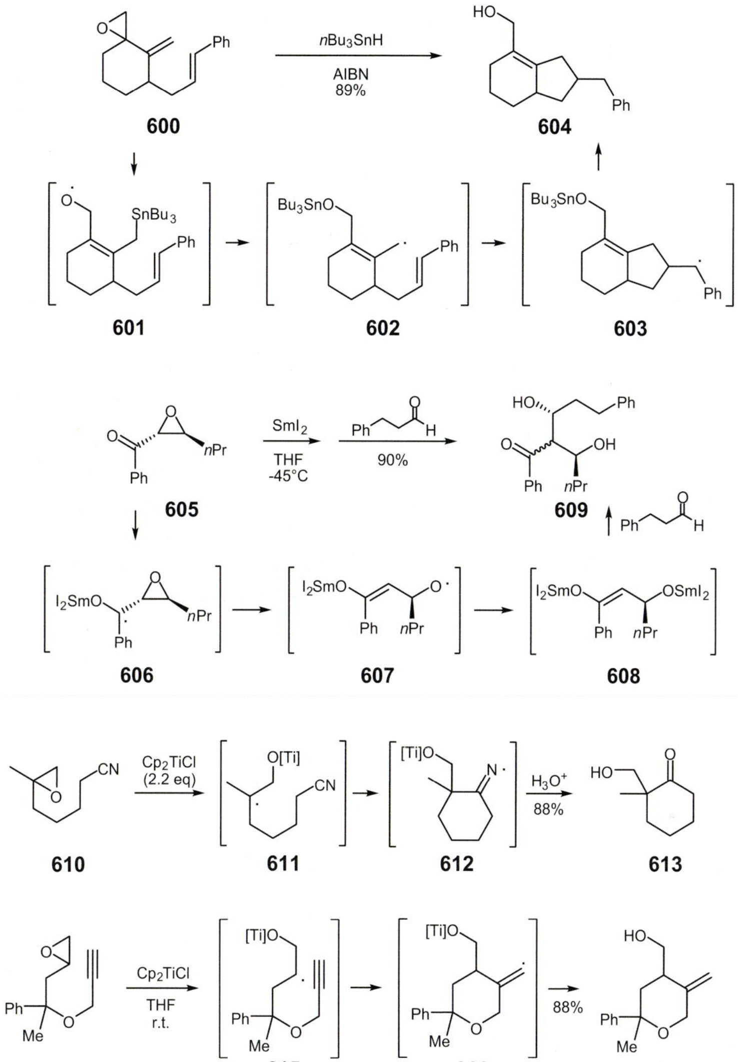

Scheme 2.88 Some radical reactions of functionalized epoxides.

hindered carbon atom (e.g., **621** → **622**), except in the case of aryl-substituted epoxides, where ring opening occurs exclusively at the benzylic position (e.g., **421** → **623**). Lithium aluminium hydride has been used for the regio- and diastereoselective reductive ring opening of chiral nonracemic epoxides, such as **624** [653] (a key step featured in the enantiospecific synthesis of (+)-hernandulcin [654]), and racemic epoxides can be reduced to an enantiomerically enriched mixture of alcohols by treatment with zirconium tetrachloride–sodium borohydride in the presence of L-proline as a chiral auxiliary [655].

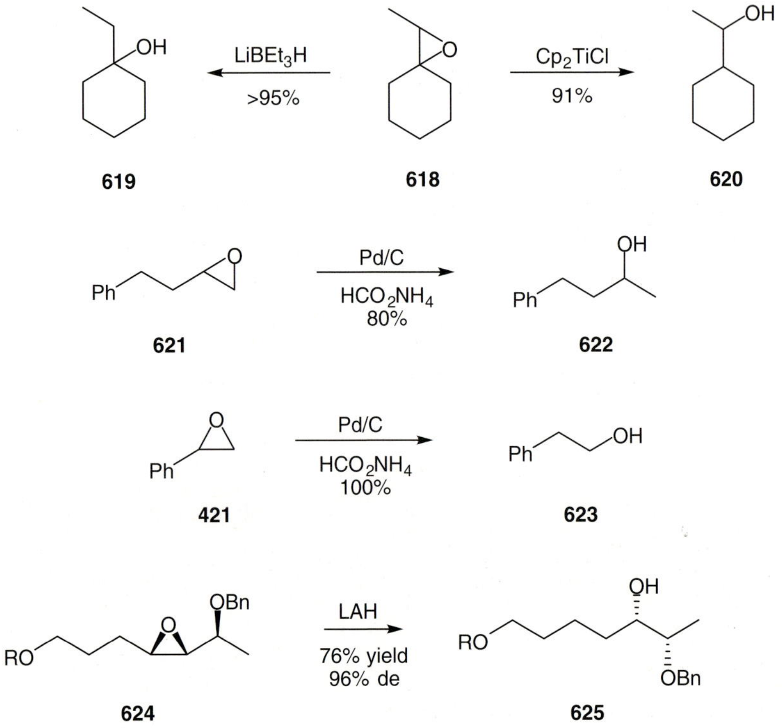

Scheme 2.89 Reductive ring opening of epoxides.

The net reversal of the epoxidation reaction, namely the eliminative deoxygenation of epoxides, has been carried out in various ways. For example, tungsten reagents react with epoxides to form tungsten(IV)-oxo complexes that ultimately lead to the corresponding olefins with predominant retention of configuration [656]. Glycidyl acetates **626** (Scheme 2.90) undergo deoxygenation and concomitant deacetylation upon treatment with lithium telluride, which is conveniently prepared *in situ* from tellurium metal and lithium triethylborohydride [657]. Another extremely mild epoxide deoxygenation protocol involves the use of bis(cyclopentadienyl)titanium (III) chloride, which promotes homolytic cleavage of the epoxide C–O bond. The mildness of this reagent is showcased in the deoxygenation of epoxide **628**, which gives the highly sensitive methoxydihydrofuran derivative **629** in 66% yield [650].

Electrophilic halogen reagents are also useful in this regard. Thus, the system of iodine and triphenylphosphine in dimethylformamide effected the quantitative deoxygenation of the allyloxymethyl epoxide **630** [658a]. In addition, a novel deoxygenation protocol has been reported for the conversion of epoxyketones into the corresponding enones using thiourea dioxide as a reducing agent under phase transfer conditions [658b]. Essentially neutral conditions are obtained using molybdenum hexacarbonyl in refluxing benzene. The mechanism proceeds through initial loss of carbon monoxide followed by a complexation of the molybdenum center with

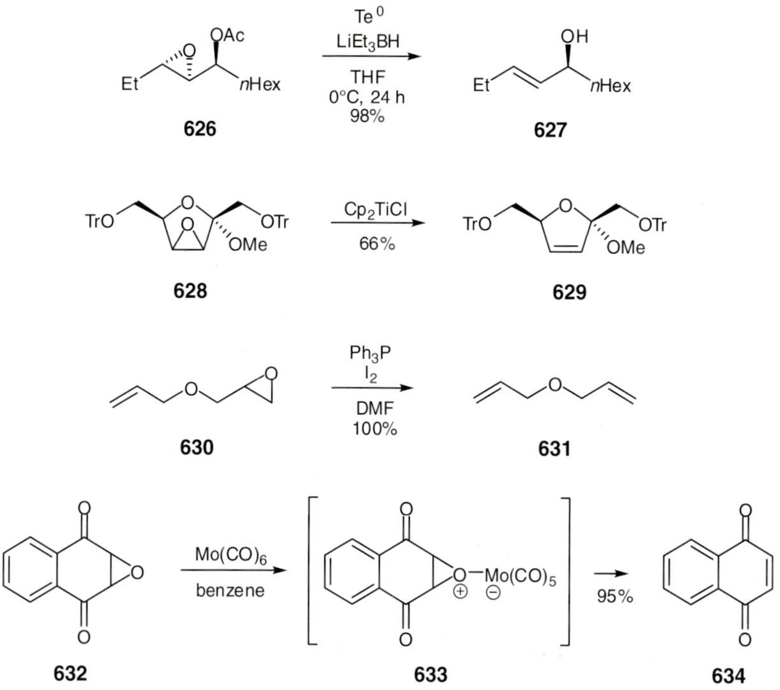

Scheme 2.90 Reductive deoxygenation of epoxides.

the epoxide oxygen to provide an activated species (i.e., **633**) that collapses to form the alkene (i.e., **634**) [659]. The low-valent titanium catalyst Cp_2TiCl, readily available by the *in situ* reduction of Cp_2TiCl_2 with activated zinc, has also been used for this type of deoxygenation [660, 661].

2.3.3.5 **Oxiranyl Anions**

Epoxides can be deprotonated on the ring, and the anions thus formed undergo interesting and synthetically useful chemistry, much of which has been summarized in recent review articles [196, 662, 663]. When electron-withdrawing groups are present, these stabilized oxiranyl anions engage in smooth S_N2 reaction with various electrophiles. Thus, the sulfonyl epoxide **635** (Scheme 2.91) is deprotonated using *n*-butyllithium in HMPA/THF at low temperature and treated with triflate **636** to give the highly functionalized adduct **637** in 81% yield. This protocol was used for the construction of the ABCDEF-ring systems of yessotoxin and adriatoxin [664]. Similar sulfonyl oxirane strategies have been used in other synthetic applications [665–668]. The oxazolinyl group also provides a useful stabilizing moiety for such alkylations (e.g., **638** → **639**) [669].

In certain cases, even non-stabilized oxiranyl anions can be coaxed into well-behaved conversions. Hodgson and coworkers [670] have reported on a convenient method of deprotonating terminal oxiranes with lithium 2,2,6,6-tetramethylpiperidide (LTMP), followed by trapping of the anion with silyl-based electrophiles, to

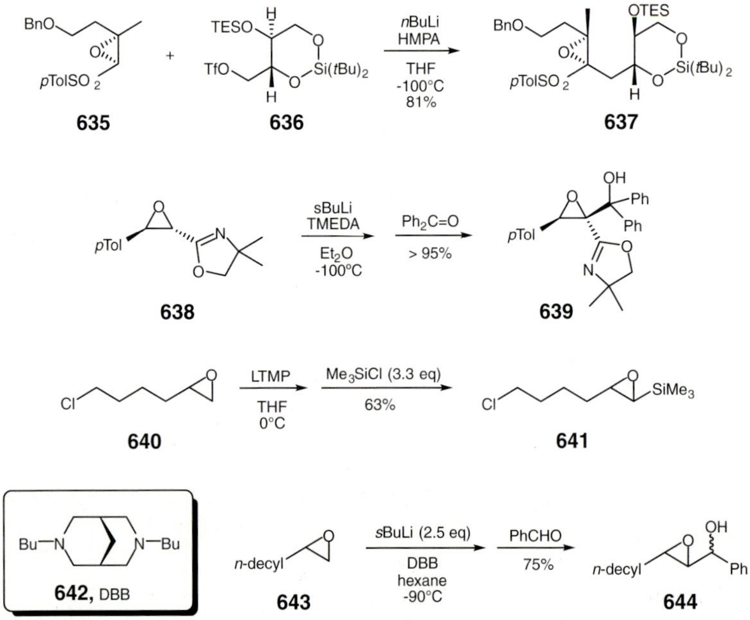

Scheme 2.91 Reaction of oxiranyl anions with electrophiles.

provide α,β-epoxysilanes in good yield. For example, chloro-epoxide **640** underwent clean conversion into epoxysilane **641** at 0 °C. This approach improves upon an earlier method, which employed sparteine derivatives at very low temperature (−90 °C) [671]. The initial proton–lithium exchange can be facilitated by diamines, such as dibutylbispidine (DBB, **642**). Thus, dodecene oxide **643** was deprotonated with *sec*-butyllithium and then treated with benzaldehyde to give the epoxyalcohol **644** in 75% yield [672, 673].

Often, though, when no stabilizing group is present, oxiranyl anions tend to exhibit significant carbenoid behavior, as indicated by resonance structure **646** (Scheme 2.92). Indeed, α-alkyloxyepoxide **647** can be regioselectively deprotonated (presumably under chelation control) to form an oxiranyl anion that undergoes α-eliminative ring opening and alkyl insertion to give cyclic allylic alcohols **648** in good to excellent yield. The carbenoid nature of the intermediates was supported by the isolation of the tricyclic alcohol **650**, the product of intramolecular trapping by an olefin [674]. Other examples of such cyclopropanation reactions include the allylic epoxide **651**, which is deprotonated by phenyllithium to give a highly strained tricyclic intermediate (**652**) that hydrolyzes to provide spirocyclic ketone **653** [675]. Lithium tetramethylpiperidide (LMTP) is also effective in the deprotonation, as shown in the conversion of alkenyl epoxide **654** into the fused bicyclic epoxide **655** [676].

As is the case with other carbene species, the carbene-like oxiranyl anions can engage in C−H insertion reactions. Hodgson and Lee [677] have devised a clever method for accessing enantiopure bicyclic alcohols from meso-epoxides by such a reaction. For example, treatment of cyclononene oxide (**656**) with isopropyllithium in the presence of an excess of (−)-sparteine leads to an enantioselective α-deprotona-

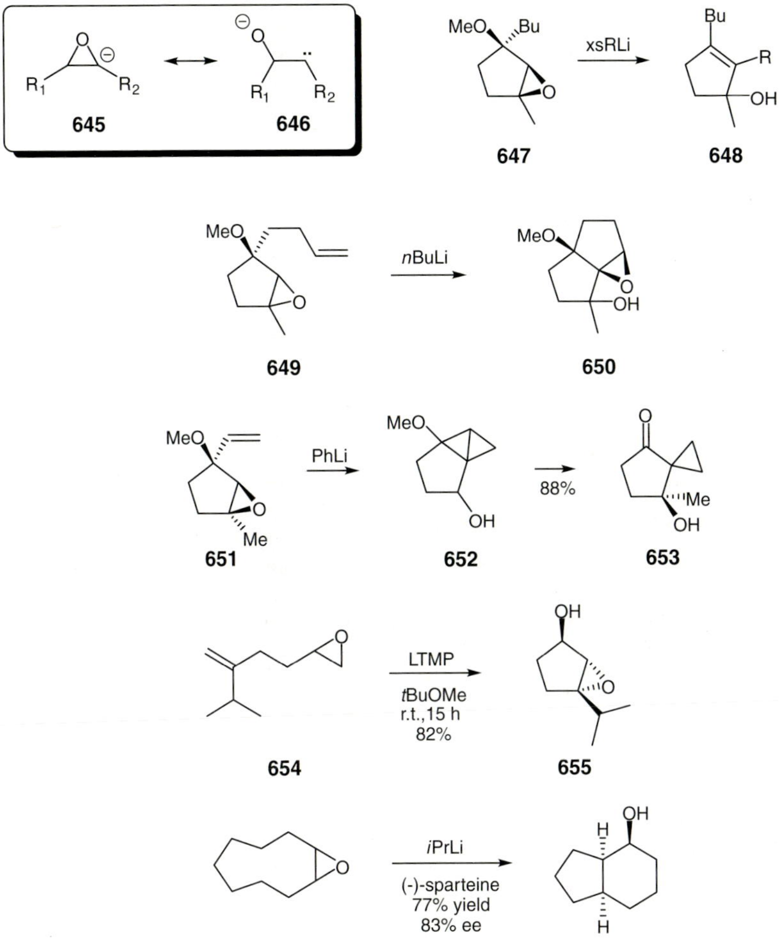

Scheme 2.92 Carbenoid behavior of oxiranyl anions.

tion, followed by intramolecular C–H insertion, to give bicyclononanol **657** in 77% yield and 83% enantiomeric excess.

2.4 Thiiranes

2.4.1 Properties of Thiiranes

Thiiranes [678] (also known as episulfides and thiacyclopropanes) can be thought of as row 3 epoxide analogs. For example, thiirane itself (bp 55 °C) exhibits a ring strain enthalpy of about 20 kcal mol^{-1}, which is about 7 kcal mol^{-1} less than oxirane. This

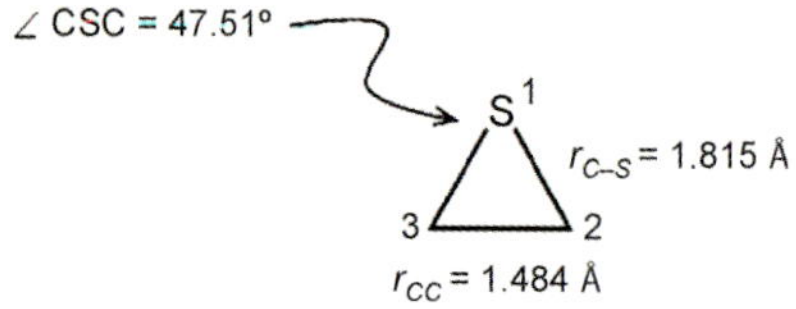

Figure 2.20 Geometry of thiirane.

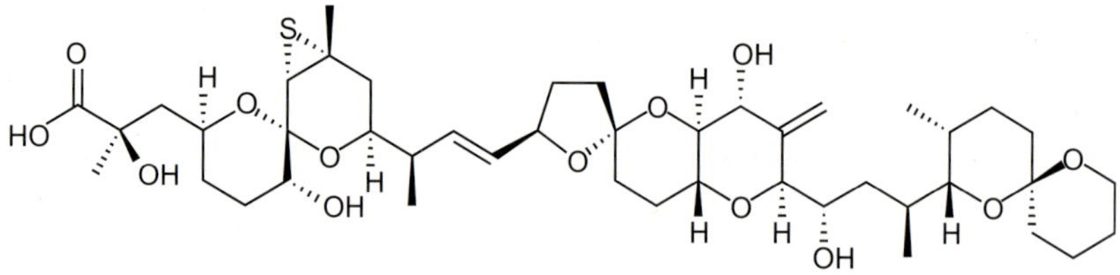

660, acanthifolicin

Figure 2.21 Natural occurrence of thiiranes.

increased stability can be attributed to more flexible geometry about the sulfur center, which is manifested in the extremely acute C–S–C bond angle of less than 48° (Figure 2.20). The ^{1}H-NMR signals of the methylene protons resonate at 2.27 ppm and the carbon atoms appear at 18.1 ppm. The proton NMR shows vicinal coupling of about 6 Hz and geminal coupling of less than 1 Hz [6].

Thiiranes are produced in nature. For example, 3,4-epithiobutanenitrile, also known as 4ETN, is found in many cruciferous vegetables and may function as a weak biological alkylating agent [679], and humulene-4,5-episulfide is one of the components in the essential oil of hops [680]. Acanthifolicin (**660**, Figure 2.21) is an antibiotic polyether carboxylic acid isolated from the extracts of the marine sponge *Pandaros acanthifolium* [681], the activity of which is linked to protein phosphatase inhibition [682]. Synthetic episulfides have also been designed as mechanism-based matrix metalloproteinase inhibitors [683].

2.4.2 Synthesis of Thiiranes

2.4.2.1 From Epoxides

Preparatively, the broadest and most useful technique for obtaining episulfides is to launch from an existing epoxide, essentially exchanging an oxygen atom for a sulfur. One common reagent used to effect this transformation is thiourea, which is preferred for its relative stability and ease of handling, and quite a few experimental conditions have been developed for its use. For example, cyclohexene oxide (**440**, Scheme 2.93) is cleanly converted into the corresponding episulfide using thiourea at elevated temperatures in the absence of solvent (Table 2.18, entry 1) [684]. The reaction also proceeds in methylene chloride under the catalysis of silica gel [685], in

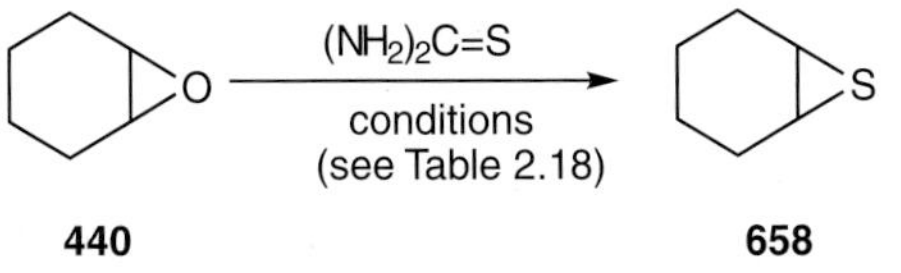

(NH2)2C=S
conditions
(see Table 2.18)

440 **658**

Scheme 2.93 Conversion of epoxides into episulfides using thiourea.

Table 2.18 Yield data for the conversion of epoxides into episulfides using thiourea.

Entry	Activator	Solvent	Temp (°C)	Time	Yield (%)	Ref
1	None	Neat	120	25 min	92	[577]
2	SiO_2	CH_2Cl_2	r.t.	30 min	92	[578]
3	SiO_2-$AlCl_3$	MeCN	45	1.3 h	89	[579]
4	β-Cyclodextrin	H_2O	r.t.	6 h	82	[580]

acetonitrile with silica-supported aluminium chloride [686] and in water using β-cyclodextrin as a catalyst [687].

Notably, this sulfurization proceeds with inversion of configuration, and the degree of stereospecificity depends upon the reaction conditions. Thus, when (*R*)-styrene oxide (**421**, Scheme 2.94) is treated with thiourea without solvent, very little optical purity is lost during the reaction. However, in methylene chloride at 0 °C the enantiomeric excess drops to 70% [685].

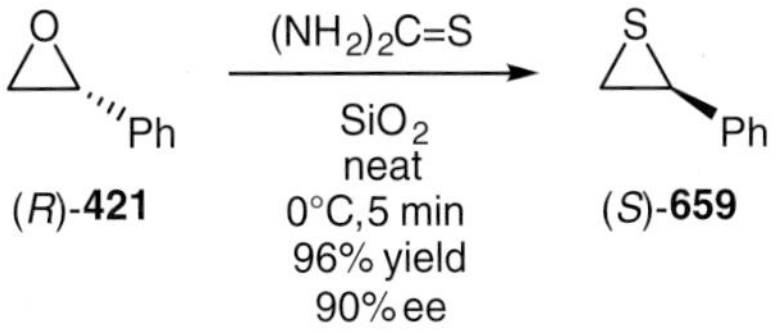

(NH2)2C=S
SiO2
neat
0°C,5 min
96% yield
90% ee

(*R*)-**421** (*S*)-**659**

Scheme 2.94 Enantiospecific preparation of episulfides from epoxides.

Ammonium thiocyanate is another convenient sulfur donor for these processes, and a very practical method has been described using cyanuric chloride as a co-reagent. Thus, when styrene oxide (**421**, Scheme 2.95) is treated with stoichiometric ammonium thiocyanate and cyanuric chloride in the absence of solvent, the episulfide **659** is produced in almost quantitative yield within 15 min. The mechanism involves nucleophilic ring opening of the epoxide to give an alkoxide intermediate (**660**); reaction with cyanuric chloride activates the oxygen towards displacement by sulfur, yielding a cyanatoepisulfenium species (**662**), which undergoes hydrolysis and subsequent loss of carbon dioxide to provide the observed

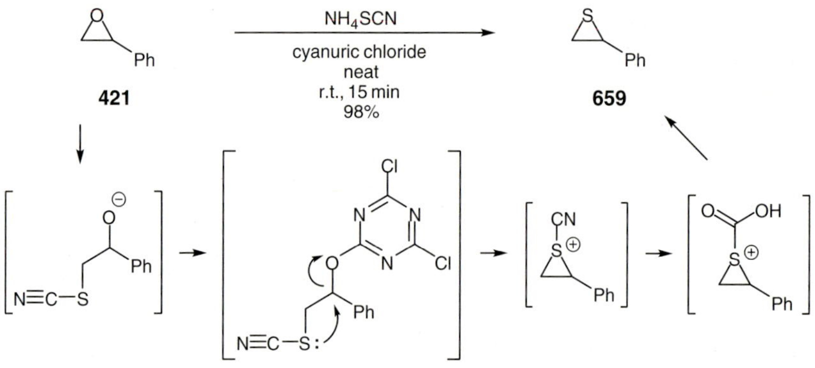

Scheme 2.95 Activation of epoxides using cyanuric chloride.

product [688]. Magnesium hydrogen sulfate has also proven to be an effective activating agent [689].

The addition of thiocyanate is also catalyzed by poly(allylamine) (PAA) under slightly basic aqueous conditions. For example, phenyloxymethyl epoxide **532** (Scheme 2.96) is converted into the corresponding episulfide in excellent yield under these conditions (Table 2.19) [690]. Very good yields have been reported using potassium thiocyanate in a biphasic medium of ionic liquid and water with no additional catalyst [579]. Finally, elemental sulfur can be employed as the donor using diethylphosphite, ammonium acetate and alumina in the absence of solvent under microwave irradiation. The mechanism involves a thiophosphate species [691].

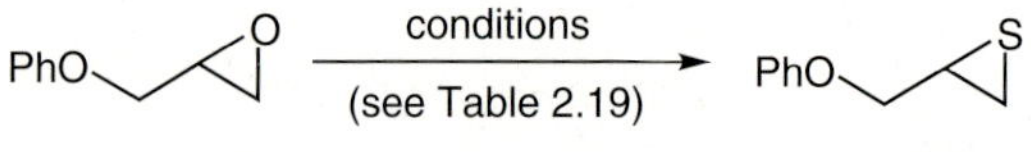

Scheme 2.96 Epoxide–episulfide conversion using other sulfur sources.

Table 2.19 Yield data for epoxide–episulfide conversions.

Entry	S source	Additives	Solvent	Temp (°C)	Time	Yield (%)	Reference
1	NH_4SCN	PAA/NaOH	H_2O	45	70 min	93	[583]
2	KSCN	None	[bmim] PF_6-H_2O	45	1.3 h	89	[579]
3	S/HP(O)(OEt)$_2$	NH_4OAc/Al_2O_3	neat	μw	2 min	68	[584]

2.4.2.2 **From Alkenes**

If episulfides were strictly analogous to their cousin epoxides, their preparation would spring predominantly from the direct sulfurization of alkenes. This is, however, not the case. The few direct methods have been nicely summarized in a recent review [692]. Nevertheless, there are some protocols that merit special attention. For example, the sultene **665** (Scheme 2.97), an isolable cycloadduct from fluorinethione-*S*-oxide and *trans*-cyclooctene, has been shown to exhibit promising sulfur-transfer capabilities. Thus, treatment of norbornene (**666**) with sultene **665** in the presence of catalytic quantities of trifluoroacetic acid furnishes *exo*-episulfide **667** in 83% yield [693]. The dithiophosphate molybdenum complex **668** actually catalyzes the transfer of sulfur from one episulfide to another alkene. This allows the use of a more readily available substrate (e.g., **659**) as a sulfur donor, as shown in the episulfidation of *trans*-cyclononene **669** [694]. Thiatriazole **671** also functions as a stoichiometric sulfur transfer reagent for a wide range of alkene substrates (e.g., **672** → **673**) [695]. Dinitrogen sulfide has been implicated as the active sulfur-transfer species [696].

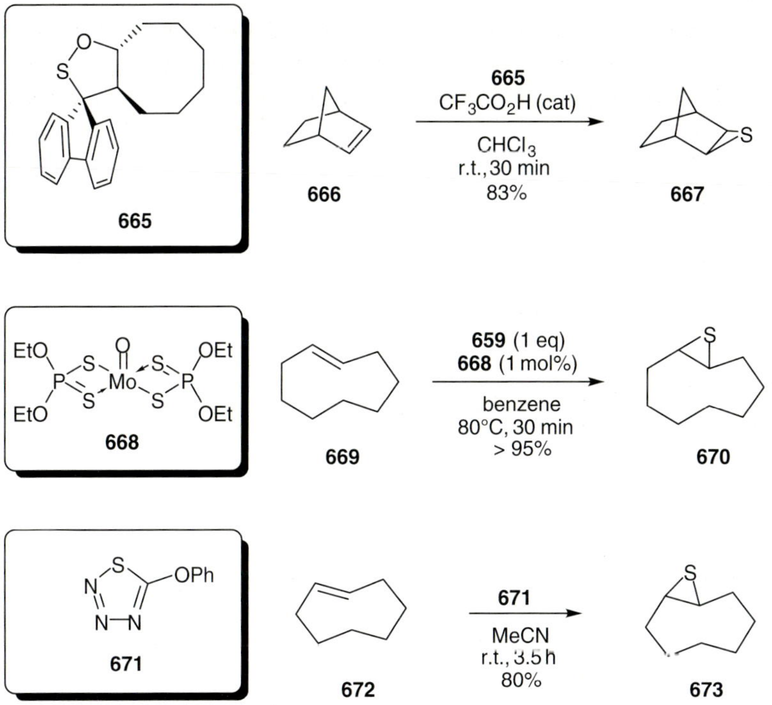

665; 666; 665, CF_3CO_2H (cat), $CHCl_3$, r.t., 30 min, 83%; 667

EtO, S, O, S, OEt, P, Mo, P, S, S, EtO, OEt; 668; 669; 659 (1 eq), 668 (1 mol%), benzene, 80°C, 30 min, > 95%; S; 670

S, OPh, N, N—N; 671; 672; 671, MeCN, r.t., 3.5 h, 80%; S; 673

Scheme 2.97 Thiiranes from alkenes.

2.4.2.3 **From Haloketones**

One very intriguing method that has received surprisingly little attention is the conversion of α-haloketones into episulfides using the commercially available

O,*O*-diethyl hydrogen phosphodithioate as a sulfur donor under microwave conditions. The procedure appears to be fairly general and high yielding. Thus, bromoacetophenone **674** (Scheme 2.98) provides excellent yield of the corresponding episulfide (**659**) within 3 min. The mechanism is thought to involve a sequence of nucleophilic displacement, phosphorus transfer and cyclization (inset Scheme 2.98). For substrates that yield 1,2-disubstituted episulfides, diastereoselectivity is very high, with trans : cis ratios generally being greater than 10 : 1, as shown in the preparation of diethylepisulfide **677** [697].

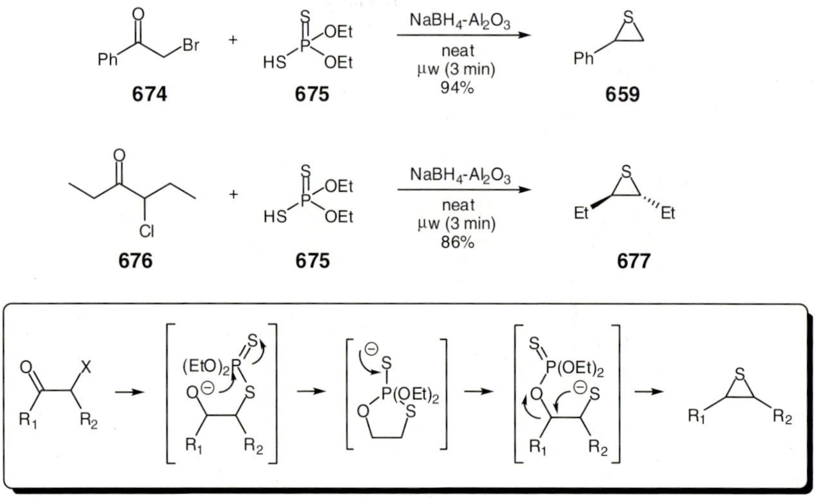

Scheme 2.98 Thiiranes from haloketones.

2.4.3 Reactivity of Thiiranes

2.4.3.1 Nucleophilic Ring Opening

Like the epoxides, episulfides are prone to ring-opening reactions induced by nucleophiles, whereby the C–S bond is cleaved. However, sulfur stabilizes a negative charge better than oxygen and therefore functions as a more active leaving group. The sulfide so-formed is also more nucleophilic than the analogous alkoxide. Furthermore, sulfur supports radical centers more readily than oxygen. Taken together, these behaviors contribute to the fact that the ring opening of episulfides is usually attended by some degree of uncontrolled polymerization and other yield-reducing processes. However, many well-behaved conversions are known, and those outlined below are meant to provide a general impression of synthetic possibilities.

Acetate is a frequently employed nucleophile. For example, in their protocol for preparing sulfur-containing disaccharides, Santoyo-González and coworkers [698] heated a mixture of episulfide **678** (Scheme 2.99), sodium acetate and acetic anhydride in acetic acid to obtain diacetate **679** in very good yield. Similarly, the thiiranyl acetal **680** undergoes ring opening in the presence of silver(I) acetate and

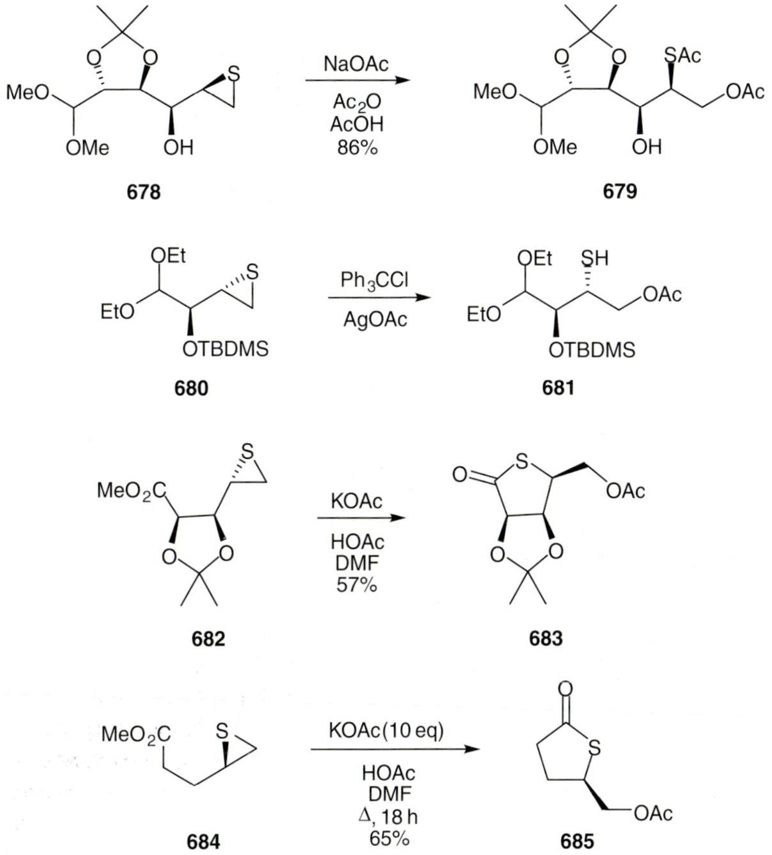

Scheme 2.99 Ring opening of thiiranes with acetate.

triphenylmethyl chloride to provide the acetoxythiol **681**, a key intermediate in the asymmetric total synthesis of thietanose [699]. These ring openings can also trigger subsequent cyclizations from the sulfur center. Thus, the thiiranyl acetonide ester **682** suffers attack by acetate to give an intermediate sulfide, which engages in intramolecular attack of the ester carbonyl to yield the thiolactone **683** [700]. A similar strategy was used to access the acetoxymethyl thiobutyrolactone derivative **685** [701].

Amines are also competent nucleophiles in these reactions. For example, the benzyloxymethylthiirane (**686**, Scheme 2.100) is attacked by dibenzylamine at the less substituted position, providing aminothiol **687** in good yield [702]. In like fashion, spiroepisulfide **688** undergoes aminolysis by 4-hydroxypiperidine (**689**) in solventless thermal conditions to give excellent yield of the β-aminoethanethiol **690** [703]. A twist on this protocol has been reported for the synthesis of taurine derivatives from episulfides. Thus, dibenzylthiirane **691** was treated with ammonia in the presence of silver nitrate to give a silver-chelated adduct that was sequentially treated with hydrogen sulfide (generated *in situ* from sodium sulfide and hydrochloric acid), sodium hydroxide and performic acid (generated *in situ* from formic

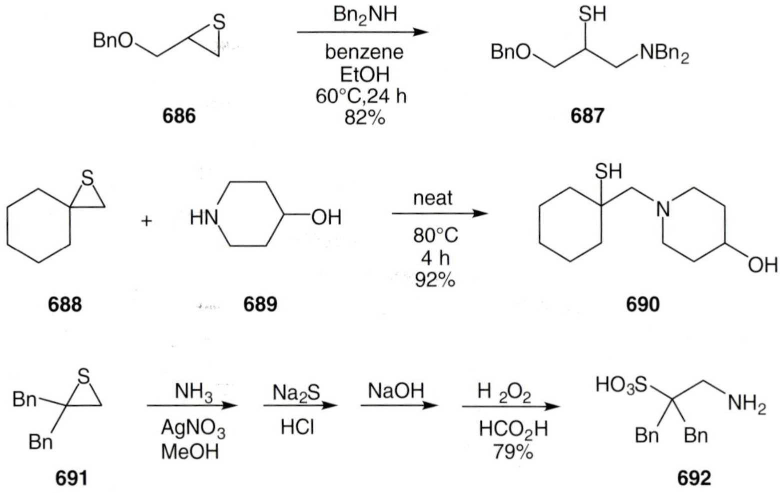

Scheme 2.100 Ring opening of thiiranes with amines.

acid and hydrogen peroxide), ultimately generating the 1,1-disubstituted taurine **692** in good overall yield [704].

High-yielding processes using other nucleophiles have also been reported. For example, treatment of the terminal episulfide **693** (Scheme 2.101) with lithium aluminium hydride in ether resulted in hydride-mediated reductive ring opening to give 2-hexanethiol (**694**) in good yield. The internal thiirane **695** underwent nucleophilic ring opening with allyl Grignard, forming the hydroxythiol **696**. The regiochemistry of the attack can be rationalized by chelation control in this case [705]. Even halides can engage in this process when appropriate activating agents are employed. Thus, thiiranyl acetal **697** was converted into the disulfide **698** using methanesulfenyl bromide in a medium of tetramethylurea (TMU) and methylene chloride through a process of electrophilic S-activation followed by nucleophilic ring-opening by bromide [706]. A similar process is promoted by acyl chlorides, as shown in the quantitative conversion of terminal episulfide **699** into the chlorothioester **700** upon treatment with acetyl chloride and catalytic tetrabutylammonium bromide (TBAB) [707].

2.4.3.2 **Desulfurization**

Thiiranes can be converted into alkenes through a process of desulfurization, often in very high yields. For example, methyltrioxorhenium (MTO) catalyzes the stereospecific removal of sulfur by triphenylphosphine, as shown in the quantitative conversion of alkenyl sulfide **701** (Scheme 2.102) to 1,5-hexadiene (**702**). Performance is enhanced when the catalyst is pretreated with hydrogen sulfide; a Re^{V} species has been implicated as the catalytically relevant species [708]. An extremely efficient copper-mediated desulfurization was used as the key step in the synthesis of C_2-symmetric dibenzosuberane (DBS) helicene **704**, which is of interest as a potential chiroptical switch [709].

Alkyllithium reagents, such as *n*-butyllithium and phenyllithium, have been known for some time to function as desulfurization agents, as exemplified by the

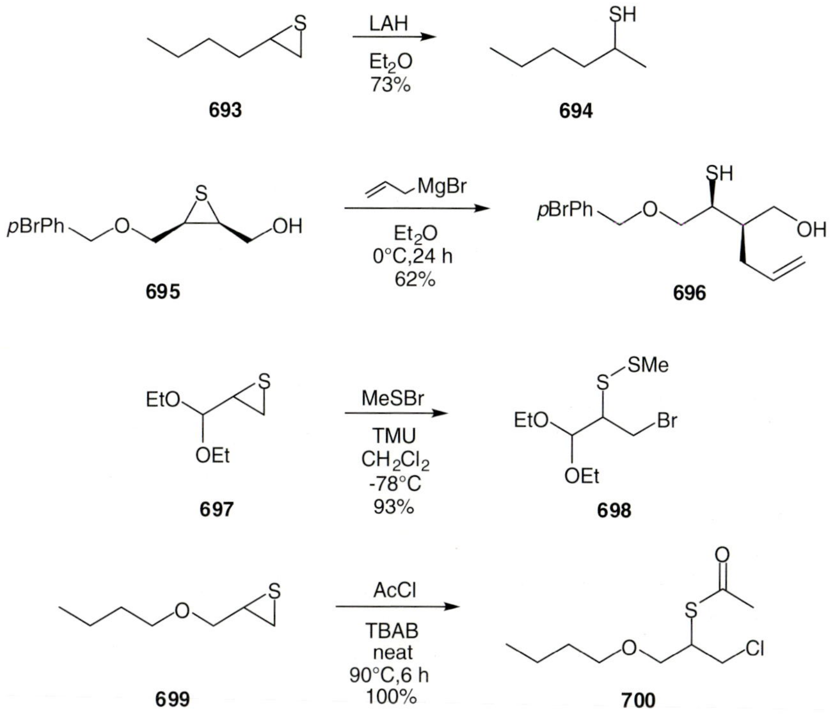

Scheme 2.101 Ring opening of thiiranes with other nucleophiles.

butyllithium-mediated conversion of cyclohexene oxide **658** into the parent alkene **439**. The mechanism involves initial ring-opening by attack on sulfur to give an α-thio anion (i.e., **705**) that rapidly undergoes elimination of butyl sulfide to form the observed product [710]. Tributyltin hydride also effects a reductive desulfurization in the presence of a radical initiator, such as triethylborane at low temperature or AIBN at elevated temperatures. Under these conditions, the thiiranyl alcohol **706** is converted into the corresponding allylic alcohol (**707**) in excellent yield and without the need for protecting group chemistry [711]. Calculations have shown that triethylphosphite should function as a desulfurizing agent through a concerted process, thus promising high stereospecificity [712].

2.5 Diaziridines

2.5.1 Properties of Diaziridines

Diaziridines can be thought of as the smallest cyclic hydrazine derivatives. At 25.5 $kcal\ mol^{-1}$, the calculated ring strain for diaziridine itself is slightly less than that of oxirane; and just as the open-chain analog, hydrazine, is more stable toward

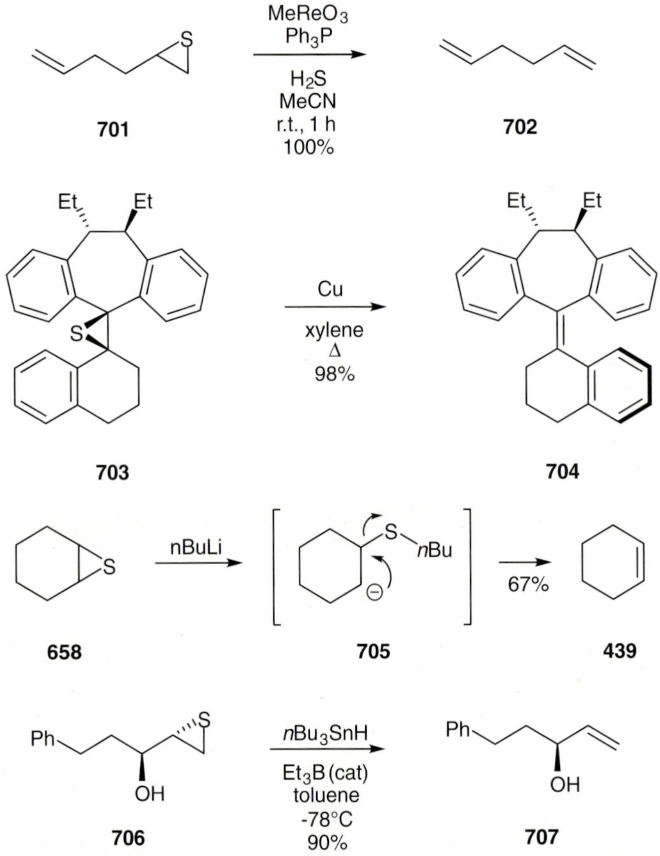

Scheme 2.102 Desulfurization of thiiranes.

interheteroatom cleavage than hydrogen peroxide so too is diaziridine less prone to undergo N–N ring opening than dioxiranes are to suffer O–O scission. Calculations have also indicated a proton affinity for diaziridine similar to that of ammonia. Geometrically, the diaziridine ring describes an almost equilateral triangle (Figure 2.22), with the N–C–N bond angle opening about 2° wider than the other two; consequently, the N–N bond is slightly longer than the C–N bond [713]. The ^{1}HNMR signals of the methylene protons resonate at about 2.2 ppm [714].

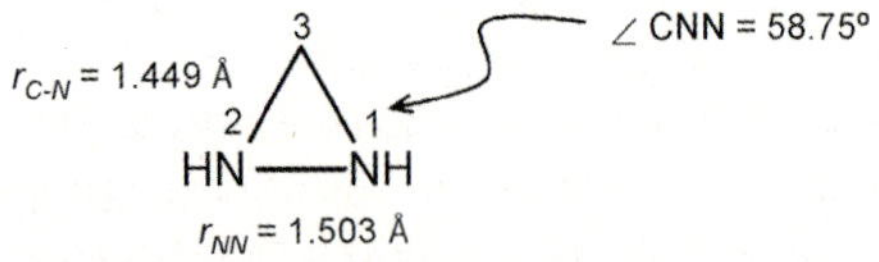

Figure 2.22 Geometry of diaziridine.

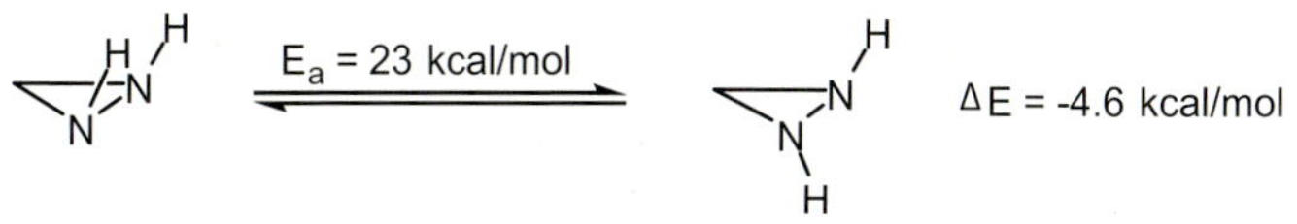

E_a = 23 kcal/mol

ΔE = -4.6 kcal/mol

Figure 2.23 Cis–trans isomerism in diaziridine.

From a physical organic perspective, diaziridine is a fascinating species. As there is about a 23 kcal mol^{-1} barrier to inversion about the nitrogen center (Figure 2.23), diaziridines exist as isolable *cis*- and *trans*-isomers, the latter lying about 5 kcal mol^{-1} lower in energy. For *N,N*-dialkyl derivatives, this barrier is even higher. For the di-*t*-butyl variant, it has been calculated at 30 kcal mol^{-1} [715]. As a result, *trans-N,N*-disubstituted diaziridines have been recognized as novel C_2-symmetric compounds, and efforts have been made to elaborate methods for their resolution [716, 717]. Diaziridines have also been investigated as possible high-energy materials (although not particularly promising in this regard) [713], and there is at least one report of diaziridine derivatives exhibiting psychotropic activity, particularly with respect to monoamine oxidase inhibition and antidepressant behavior [718].

2.5.2 Synthesis of Diaziridines

The synthesis of monocyclic diaziridines and their fused derivatives is the subject of a recent review [719]. Some of the more common synthetic routes are outlined below.

2.5.2.1 Oxidative Methods using Hypohalites

One of the more common routes for accessing diaziridines is the oxidative ring closure of aminals, which are usually formed *in situ* from the respective amines and carbonyl compounds. Thus, *N,N*-dibutyldiaziridine (**710**, Scheme 2.103) is prepared in good yield by combining *n*-butylamine and formaldehyde in the presence of aqueous sodium hypochlorite [720]. The method can be applied to diamine substrates, such as propylenediamine (**711**) to construct fused bicyclic derivatives (e.g., **712**) in excellent yield. Furthermore, the diamines can be condensed onto substituted aldehydes (e.g., **714**) and ketones (e.g., **716**) with equal efficiency. The pH of the reaction medium can have an impact on yields and product distributions [721].

The aminals can be formed from other precursors, as demonstrated by the aminoalkylimine **718** (Scheme 2.104), which produces the fused tricyclic diaziridine **719** in the presence of aqueous sodium hypochlorite [722]. The benzimidamide **720** serves as a precursor for diazirines in similar oxidizing environments; however, an intermediate chlorodiaziridine (**721**) has been identified in the reaction mixture, and it is stable enough to isolate [723]. Similarly, diaziridinimines such as **723** can be prepared in good yield by subjecting tosyl guanidine **722** to sequential treatment by *t*-butyl hypochlorite and *t*-butoxide [724].

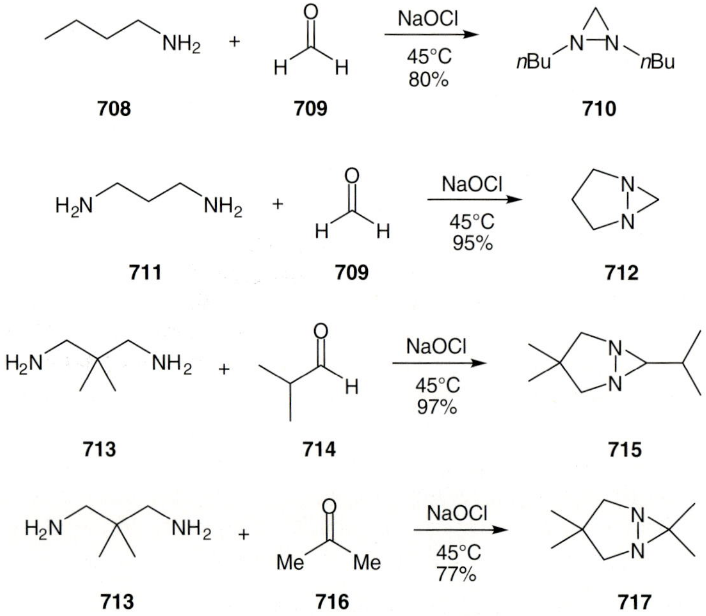

Scheme 2.103 Synthesis of diaziridines from carbonyls and amines.

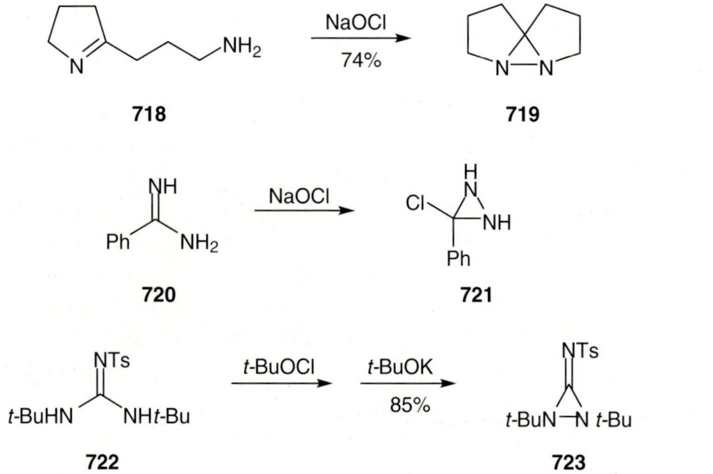

Scheme 2.104 Synthesis of diaziridines from imines, imidamides, and guanidines.

2.5.2.2 **Via Hydroxylamine Derivatives**

Another alternative for diaziridine synthesis is presented in the use of O-sulfonated hydroxylamine derivatives, which offers the advantage of a much less oxidizing environment. As an example, the cyclic imino ester **724** (Scheme 2.105) is converted into diaziridine **725** upon treatment with hydroxylamine *O*-sulfonic acid [725], and

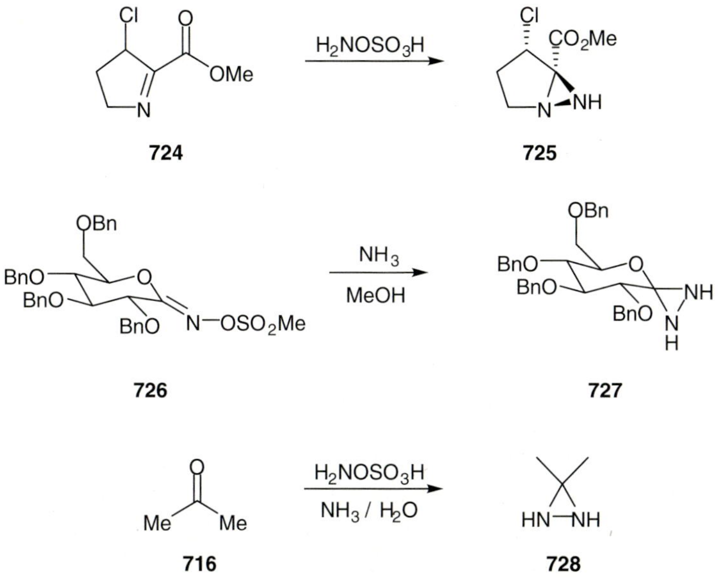

Scheme 2.105 Synthesis of diaziridines from hydroxylamine derivatives.

the same method is effective in preparing the glycosylidene-derived diaziridine **727**, which serves as a precursor for the corresponding diazirine and thus a glycosylidene carbene [726]. An operationally more straightforward variant of the methodology can be used for preparing various 2,2-disubstituted derivatives. Thus, a ketone such as acetone (**716**) is treated with hydroxylamine *O*-sulfonic acid in the presence of aqueous ammonia to give the desired diaziridine (**728**) in one pot [727].

2.5.2.3 **Other Methods**

Although the generality of the methods has not been firmly established, two protocols merit special attention. The first is the direct electrochemical synthesis of diaziridines from amines and aldehydes in a bicarbonate buffered solution, which is directly analogous to methods described in Section 2.5.2.1 but obviating the need for chemical oxidants. Thus, propylenediamine (**711**, Scheme 2.106) and formaldehyde

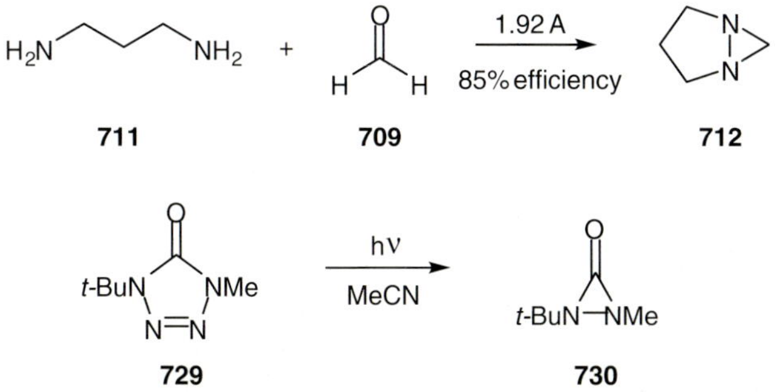

Scheme 2.106 Synthesis of diaziridines via electrical and photochemical means.

cleanly provided bicyclic diaziridine **712** under these electrochemical conditions. The current efficiency reached 85% under optimum conditions [728]. The second method is the photolysis of tetrazolone derivatives (e.g., **729**) to give diaziridinones (e.g., **730**), which could be obtained in high purity [729].

2.5.3
Reactivity of Diaziridines

2.5.3.1 Diaziridines

Like their open-chain analogs, diaziridines are nucleophilic species that engage in well-behaved reaction with various electrophiles. For example, 3,3-dimethyldiaziridine (**728**, Scheme 2.107) is smoothly N-acylated with 3-phenyoxybenzyl chloroformate (**731**) in the presence of triethylamine in a medium of methylene chloride to give the carbamate **732** in 63% yield. The remaining nitrogen was further functionalized to produce the diazapyrethroid analog **733**, and the diaziridine moiety proved to be remarkably stable to the experimental conditions (particularly zinc in acetic acid) [727]. Diaziridines also engage in conjugate addition onto traditional Michael

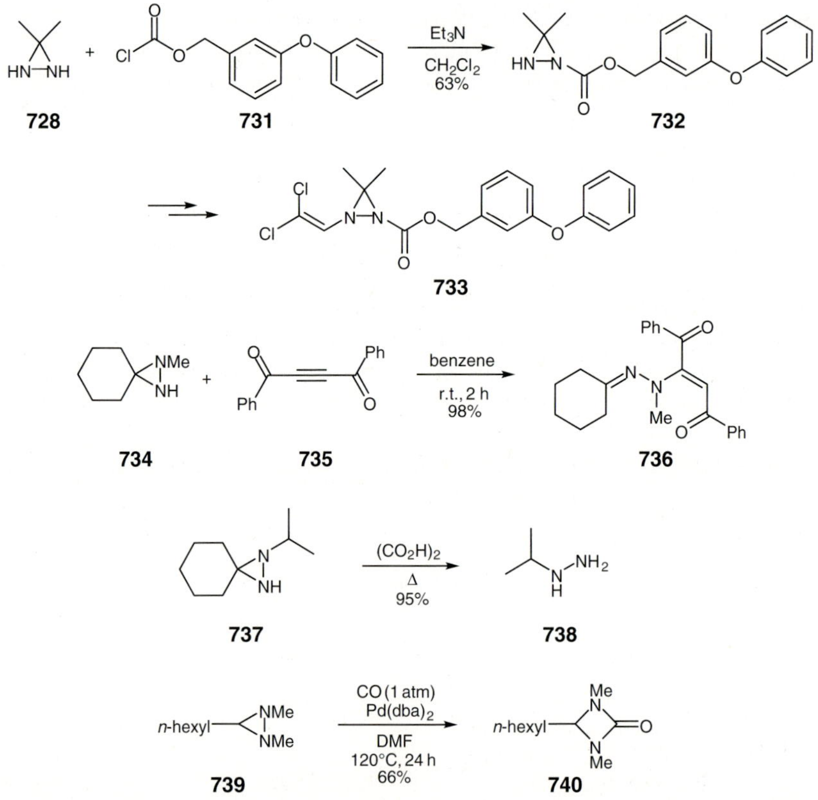

Scheme 2.107 Some reactions of diaziridines.

acceptors, such as DMAD [730] and dibenzoylacetylene [731]. The intermediate diaziridinium species formed by the initial addition suffers ring-opening through scission of the C–N bond to give a hydrazone derivative (e.g., **736**). In fact, under acidic conditions diaziridines spontaneously decompose to form the component hydrazines and carbonyl compounds, as shown by the recovery of *N*-isopropylhydrazine (**738**) in 95% yield upon exposure of the precursor diaziridine **737** to oxalic acid [732, 733]. The cyclic N–N bond is not totally inert, however. For example, palladium catalyzes the insertion of carbon monoxide into the N–N bond, converting diaziridines (e.g., **739**) into the corresponding azalactam derivatives (e.g., **740**) [734].

2.5.3.2 Diaziridinones and Diaziridinimines

Diaziridinones also exhibit some interesting chemistry. For example, they too have been reported to engage in carbonyl insertion reactions in the presence of a nickel catalyst to give diazetidine-2,4-diones (e.g., **742**, Scheme 2.108) in reasonable yields [735]. Usually, however, their reactivity involves the nucleophilic attack of the carbonyl carbon, as illustrated by the reaction of *N,N*-di-*t*-butyldiaziridinone (**741**)

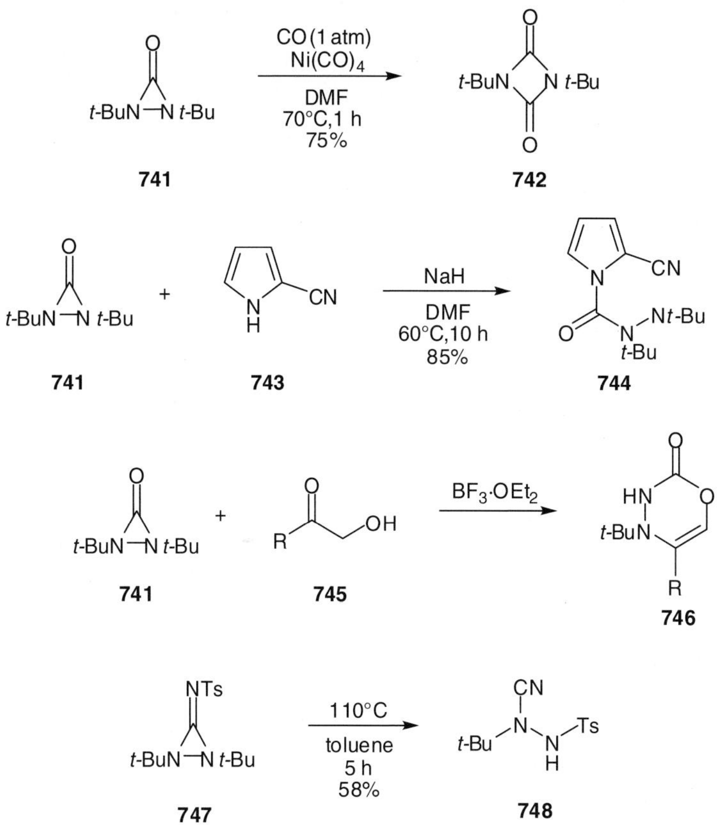

Scheme 2.108 Some reactions of diaziridinones and diaziridinimines.

with the sodium salt of cyanopyrrole **743**, which engages in nucleophilic addition to the carbonyl, followed by ring-opening to provide the acylhydrazine **744** in 85% yield [736]. A similar addition event is the first step of many in the Lewis acid-catalyzed conversions of diaziridines into oxadiazinones (e.g., **746**) by α-hydroxy ketones (e.g., **745**) [737]. Finally, the analogous diaziridinimes (e.g., **747**) are thermally unstable, rearranging to *N*-cyanohydrazine derivatives (e.g., **748**) at elevated temperatures [724].

2.6 3*H*-Diazirines

2.6.1 Properties of Diazirines

With a literature age of less than 50 years old [738, 739], diazirines are relative youngsters among their companion three-membered heterocycles, and an interesting lot they are. Although this ring system might intuitively appear quite unstable, the strain energy has been calculated at a remarkably low 21 kcal mol^{-1} [740]. This, combined with very low basicity, is largely responsible for the observed stability (even *in vivo*) of the diazirine ring at room temperature. The double bond character between the two nitrogens results in a short N=N bond distance (1.23 Å), which, in turn, significantly compresses the N–C–N bond angle (Figure 2.24) [741]. Likewise, the diaziridine moiety is planar and symmetrical. Since molecular nitrogen can be rapidly extruded under thermal or photochemical conditions, neat diazirines should be afforded the respect in handling that all potentially explosive compounds deserve.

2.6.2 Synthesis of Diazirines

The lion's share of protocols for the preparation of diazirines proceed through a diaziridine intermediate. For example, 4-aziadamant-1-amine (**750**, Scheme 2.109) was synthesized by the chromium(VI) mediated oxidation of the corresponding diaziridine (**749**) [742]. However, diaziridine precursors can serve as suitable substrates for various one-pot procedures. Thus, the camphor-derived iminium salt **751** was converted into the sterically hindered chiral diaziridine 2-azicamphane **752** by

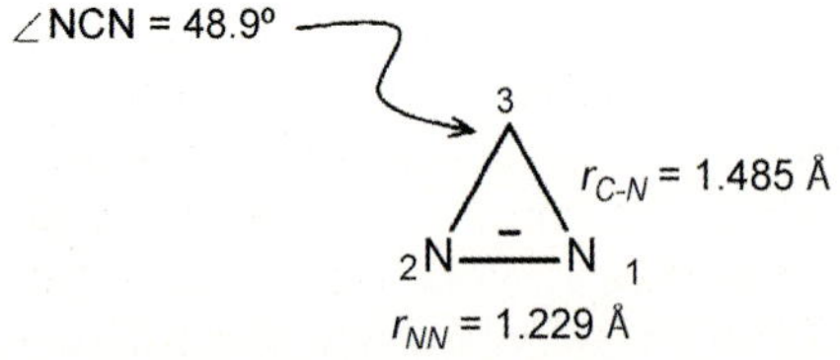

∠NCN = 48.9°
3
$r_{C\text{-}N}$ = 1.485 Å
2N—N 1
r_{NN} = 1.229 Å

Figure 2.24 Geometry of 3*H*-diazirine.

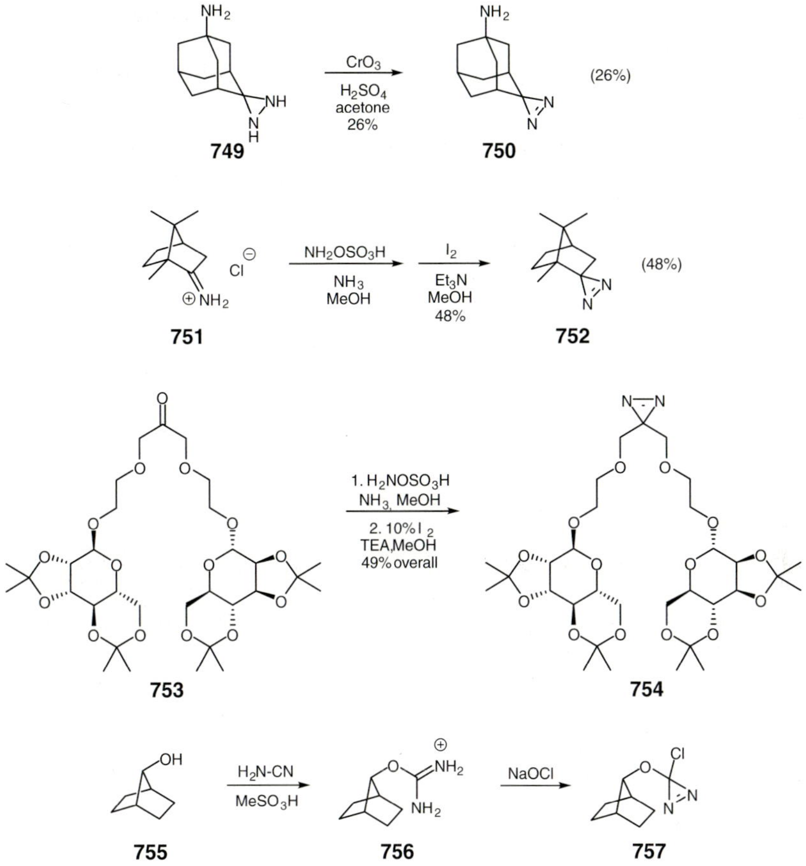

Scheme 2.109 Synthesis of diazirines.

treatment with hydroxylamine O-sulfonic acid in methanolic ammonia, followed by oxidation with iodine in the presence of triethylamine [743]. Ketones can also be used to advantage in this regard, as illustrated by the construction of the diaziridinyl cluster mannoside **754** from ketone **753** [744]. Alkoxy substituted diazirines can be prepared from alcohols. Thus, norboranol **755** was treated with cyanamide and methanesulfonic acid to afford an intermediate isouronium salt (**756**) that was oxidized with hypochlorite to give the diazirine **757** [745]. Diazirines have also been prepared by the partial hydrolysis of nitriles, followed by hypochlorite oxidation [746].

Since they have desirable end-application properties, 3-trifluoromethyldiazirines are of particular interest, and these are prepared in similar fashion. For example, the diazirinyl antiotensin II analogue **759** (Scheme 2.110) was produced in very good yield from the silver(I) oxide mediated oxidation of diaziridine **758** [747]. An equally efficient diazirine synthesis was achieved launching from the O-tosyl oxime **760**, which was converted into the diaziridine by treatment with liquid ammonia. After the fluoride-mediated desilylation of the alcohol functionality, PDC oxidation afforded

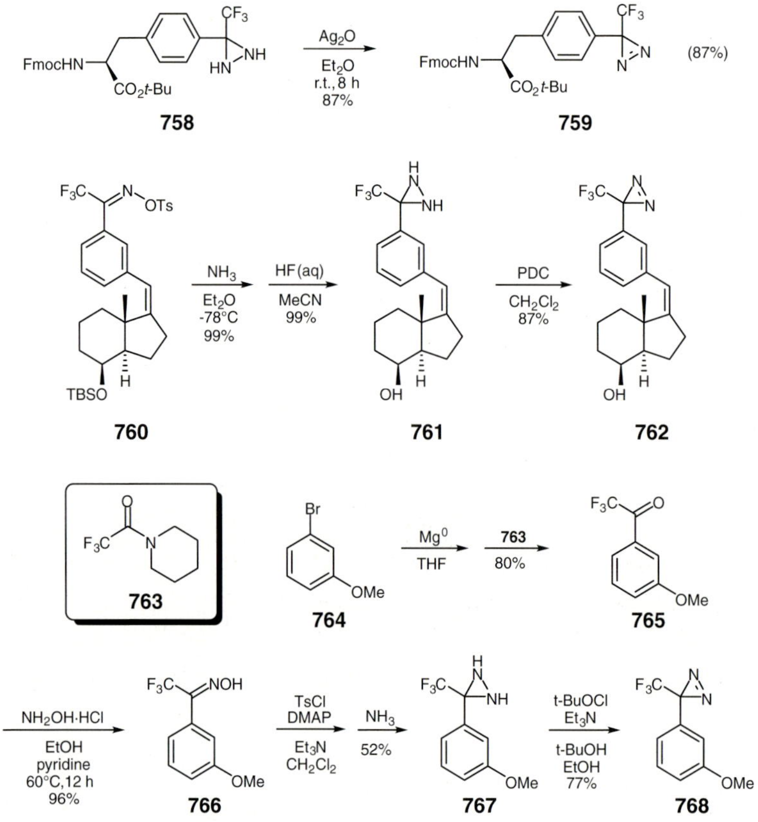

Scheme 2.110 Synthesis of trifluoromethyldiazirines.

the desired diazirine **762** in remarkably good overall yield. The target was used as the centerpiece for the construction of a vitamin D analog, a testament to the robustness of the diazirinyl moiety toward various experimental conditions [748].

Ultimately, the trifluoromethyl group is usually introduced by way of an activated trifluoroacetic acid derivative. For example, the aryl Grignard derived from *m*-bromoanisole (**764**) smoothly added to *N*-trifluoroacetylpiperidide (**763**) to give the trifluoroacetophenone derivative **765**, which was converted into the corresponding oxime (**764**) in excellent yield using conventional methods. Subsequent *O*-tosylation and treatment with ammonia afforded the diaziridine **767**, which was oxidized to the diazirine **768** using *t*-butyl hypochlorite [749].

2.6.3 Reactivity of Diazirines

Diazirines with leaving groups at the 3-position can undergo substitution reactions without affecting the azo moiety [750]. As one example, the bromodiazirine **769**

(Scheme 2.111) is converted into the fluoro analog **770** upon treatment with anhydrous tetrabutylammonium fluoride (TBAF) under solvent-free conditions [751]. It would be fair to state, however, that the spotlight shines on diazirines for their propensity to extrude molecular nitrogen to form reactive carbene intermediates [752, 753]. Thus, the 3-acyldiazirine **771** suffers loss of nitrogen under thermal conditions in the presence of alkenes to yield products of carbene–olefin cycloaddition [754].

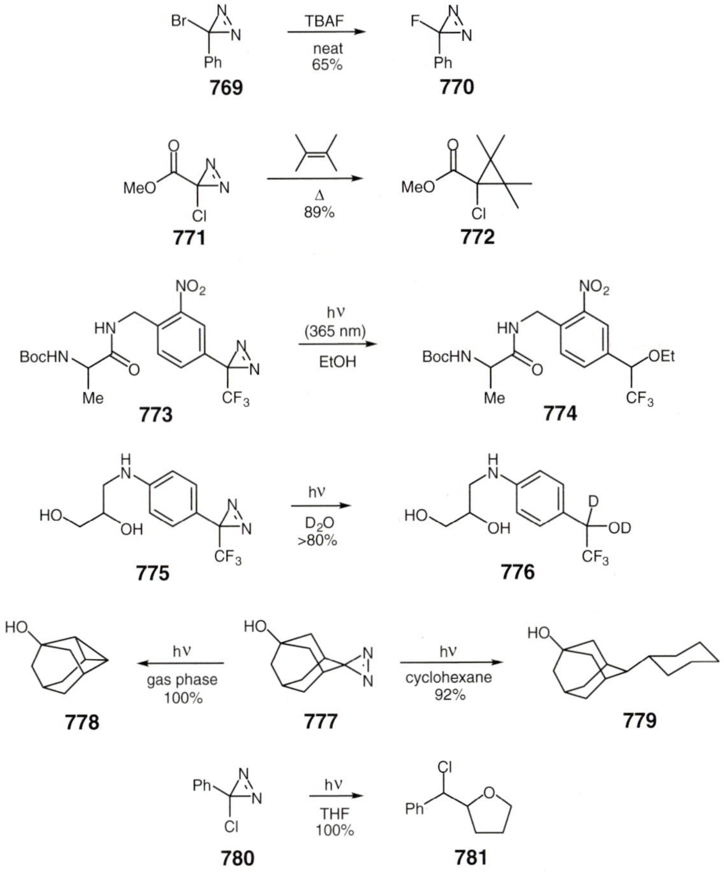

Scheme 2.111 Some reactions of diazirines.

Photolysis of diazirines also generates carbenes, and this procedure is more significant in terms of application [755]. Once the carbene intermediates are formed, the reactivity is much the same as for carbenes generated by any other method. For example, when the aryl diazirine **773** is photolyzed in ethanol, the major product (**774**) results from the insertion of carbene in the O–H bond of the solvent [756]. Similar behavior is observed in aqueous media (i.e., **775** → **776**) [757]. Photolysis of

2-azi-5-hydroxyadamantane (**777**) in the gas phase results in the quantitative formation of the intramolecular 1,3 C−H insertion product **778**. When the same substrate is photolyzed in an organic solvent, then the intermolecular process dominates, as illustrated by the cyclohexane adduct **779** [758]. This behavior is fairly general – the carbene derived from arylchlorodiazirine **780** also quantitatively traps THF when generated in that solvent [759].

To summarize, diazirines can be synthesized with a high degree of regioselectivity, launching from various readily available functional groups. The diazirines themselves have a very long half-life at room temperature and under a broad range of conditions. Reaction can be triggered by photolysis in a manner that engages few, if any, other functionalities. It can be assumed that carbene generation is practically quantitative, and these carbenes react quickly with a wide variety of substrates. This particular confluence of behaviors has earned these substrates an important niche, namely within the realm of photoaffinity labels (PAL), an application that has been nicely summarized in a recent review [760].

Interestingly, for the same reasons, diazirines have been investigated as materials surface modifiers. In their study of the underlying chemistry of these processes, Hayes and coworkers photolyzed the fluorenone-modified diazirine **782** in the presence of acetic acid, *n*-butylamine, and *N*-butylbutanamide as models of the functional group environment encountered in a medium of nylon 6,6 (and, incidentally, proteins *in vivo*). In all cases, they observed extremely efficient (>95%) insertion reactions, leading to the products **783**, **784**, and **785**, deriving from O−H insertion, amine N−H insertion, and amide N−H insertion, respectively [761] (Scheme 2.112).

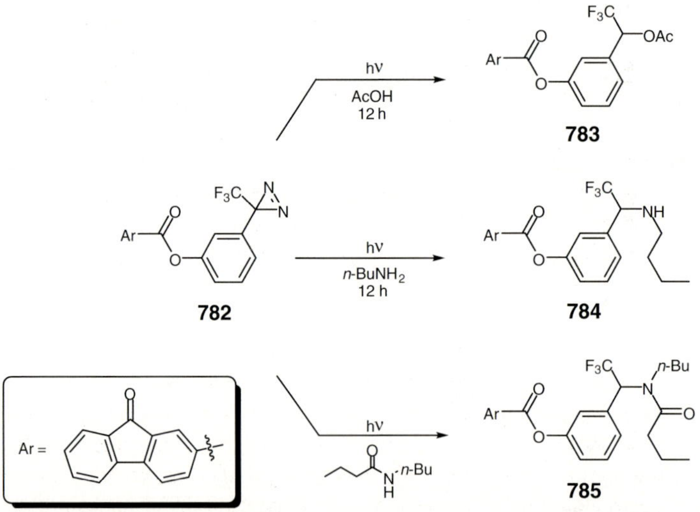

Scheme 2.112 Insertion reactions of a photoaffinity label (PAL).

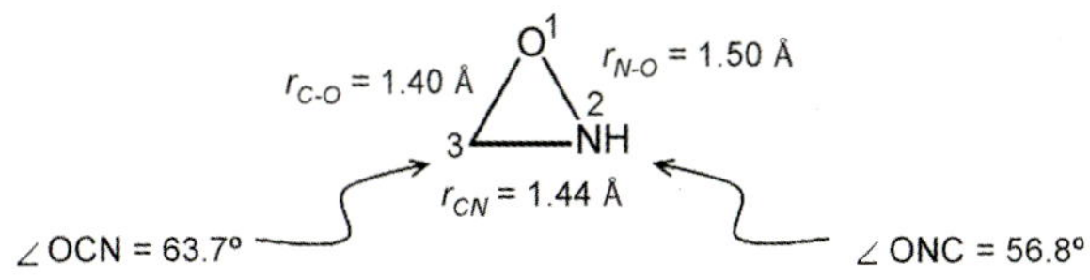

$r_{C\text{-}O}$ = 1.40 Å
$r_{N\text{-}O}$ = 1.50 Å
r_{CN} = 1.44 Å
∠OCN = 63.7°
∠ONC = 56.8°

Figure 2.25 Geometry of oxaziridine.

2.7 Oxaziridines

2.7.1 Properties of Oxaziridines

The oxaziridine moiety incorporates an oxygen, nitrogen and carbon within a cyclic three-atom array (Figure 2.25). The conventional strain energy (CSE) for the parent compound has been calculated at 27.6 kcal mol^{-1}, lying between those of cyclopropane and cyclobutane [762]. Each bond in the ring is unique, with lengths ranging from 1.40 Å (C–O) to 1.50 Å (N–O). It follows that the bond angles are also non-identical, with the N–C–O angle being the widest [763]. These physical characteristics are responsible in part for the fascinating reaction diversity exhibited by this class of compounds. Aside from their synthetic utility, oxaziridines have excited interest due to their potential as antifungal [764], antibiotic [765] and antitumor agents [766, 767].

2.7.2 Synthesis of Oxaziridines

There are reports of oxaziridine preparation from nitrones [768], and a calculational study has been carried out regarding this isomerization [763]. Oxaziridines are also the products of oxidative amination of ketones [769]. However, these methods have not found wide application in synthetic methodology. Instead, almost all preparative protocols launch from the oxidation of imines. Nevertheless, there is considerable diversity even within this one general category.

The *N*-alkoxysulfonyl oxaziridine **787** (Scheme 2.113) was efficiently prepared from the precursor imine **786** using the fairly traditional oxidant *m*-chloroperbenzoic acid (mCPBA) [770]. The sulfinylimine **788** was converted into sulfonyloxaziridine **789** in short order and in excellent yield using mCPBA in first acidic and then basic medium [771]. A trichloroacetonitrile–hydrogen peroxide system was found to convert imines such as **790** into oxaziridines in very high yield and exclusive *(E)*-stereochemistry under essentially neutral conditions [772]. Several other near-neutral systems are also available. For example, pyridylimine **792** was oxidized using a buffered monophasic Oxone system [773], and phase-transfer conditions were applied to the mild and quantitative conversion of arylaldimine **794** into the corresponding oxaziridine (**795**) using tetrabutylammonium Oxone in acetonitrile [774]. The combination of storage-stable urea-hydrogen peroxide adduct (UHP) and maleic anhydride in methanolic solution was also effective in the high-yielding

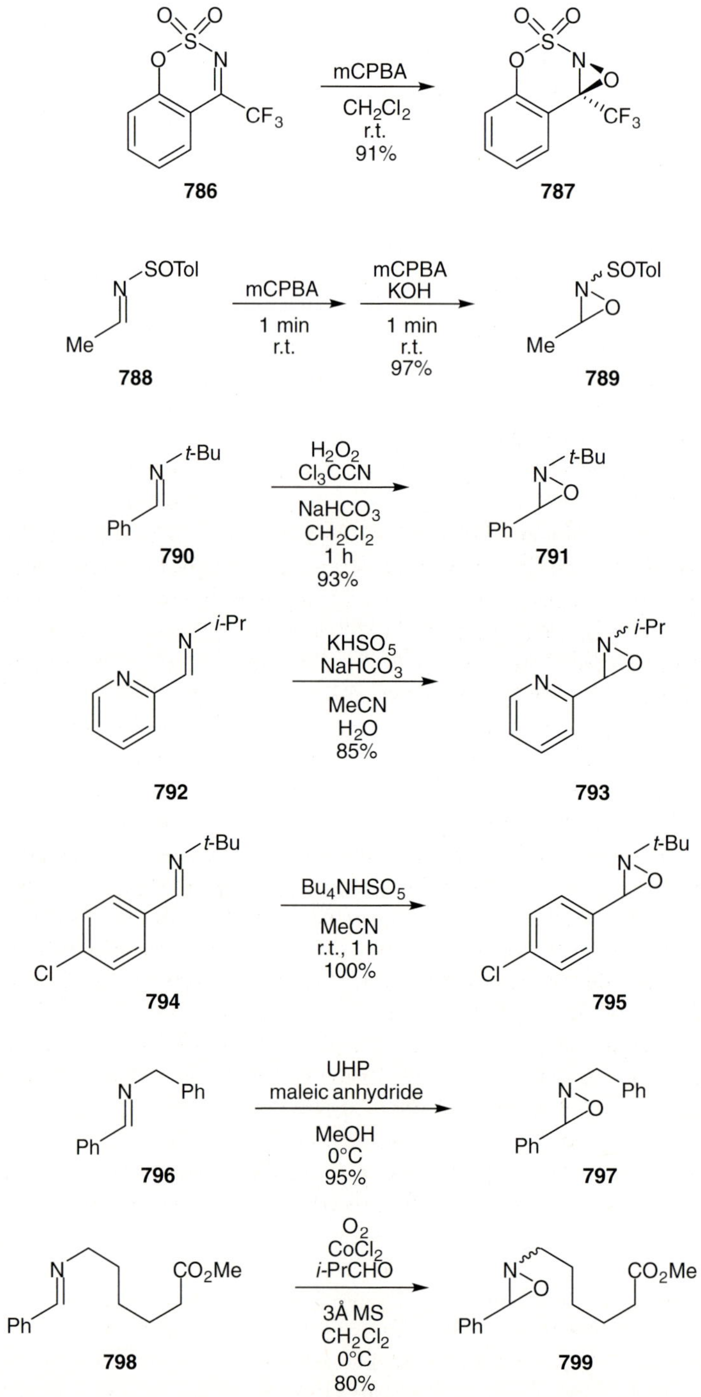

Scheme 2.113 Synthesis of oxaziridines from imines.

oxaziridination of *N*-benzylimine **796** [775]. Even molecular oxygen can be used as the terminal oxidant, using a cobalt catalyst (e.g., **798** → **799**), although the catalyst must be preformed for optimum yields [776].

2.7.3 Reactivity of Oxaziridines

Oxaziridines have quickly become very important synthetic reagents [777–780]. Furthermore, the related oxaziridium ions, which bear a quaternary nitrogen center, are significant in their own right, both as stoichiometric reagents [781] and putative intermediates in catalytic cycles [782–784]. However, the present chapter is limited to neutral oxaziridines.

2.7.3.1 Nitrogen Transfer Reactions

An interesting aspect of oxaziridine chemistry is that either heteroatom can be transferred to other compounds, depending upon the nature of the substrate and the substituents on the oxaziridine [785]. Common oxaziridines used in nitrogen transfer reactions include the spirocyclic derivative of cyclohexanone (**800**, Figure 2.26), as well as those prepared from electron-deficient carbonyls, such as trichloroacetaldehyde (i.e., **801**) and diethyl oxomalonate (i.e., **802**). An overview of the oxaziridine-mediated electrophilic amination of organic compounds can be found in a review from the not-too-distant past [786].

Aziridination has been reported using oxaziridines as nitrogen donors, although this reaction is not general. For example, certain styrene derivatives (**803**, Scheme 2.114) are aziridinated in moderate yield under thermal conditions in the presence of oxaziridine **800** [787], but many other substrates give complex mixtures. The structural influence of this variable behavior has been studied computationally [788]. A more common outcome is the amination of active methylene groups. Thus, treatment of barbituric acid (**805**) with oxaziridine **800** in a medium of dilute sodium hydroxide led to the production of the 5-aminobarbituric acid (**806**) in 78% yield [787]. Similarly, deprotonation of phenylacetonitrile **807** with lithium hexamethyldisylazide (LiHMDS), followed by treatment with oxaziridine **802**, provided the Boc-protected benzylamine **808** in 46% yield [785]. Oxaziridine **802** also mediates the stereospecific conversion of allylic sulfide **809** into the allylic *N*-Boc-sulfimide **810**, a process that involves a [2,3]-sigmatropic rearrangement [789].

Nitrogen can also be transferred to amines. For example, phenethylamine (**811**, Scheme 2.115) is aminated by oxaziridine **802** under very mild conditions, providing the Boc-protected hydrazine derivative **812** in good yield. These compounds were

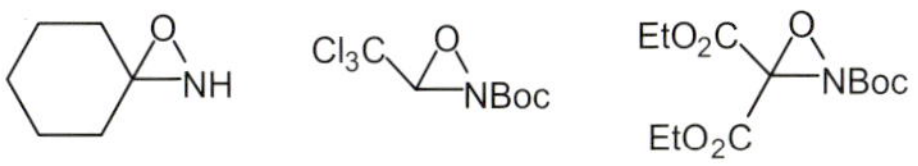

NH | Cl_3C NBoc | EtO_2C NBoc EtO_2C

800 **801** **802**

Figure 2.26 Some common oxaziridines used for nitrogen transfer.

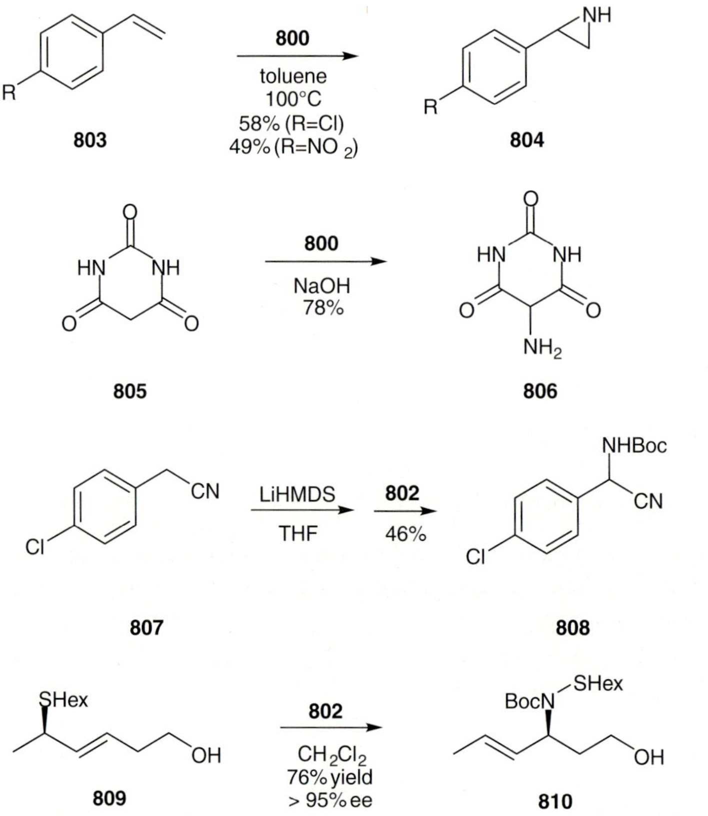

Scheme 2.114 Nitrogen transfer to carbon.

then converted into pyrazoles in a one-pot reaction [790]. Hannachi and co-workers [791] have applied this methodology to synthesize orthogonally diprotected L-hydrazino acids (e.g., **814**), which are useful in the biological and structural studies of pseudopeptides containing the N–N–C–C=O fragment.

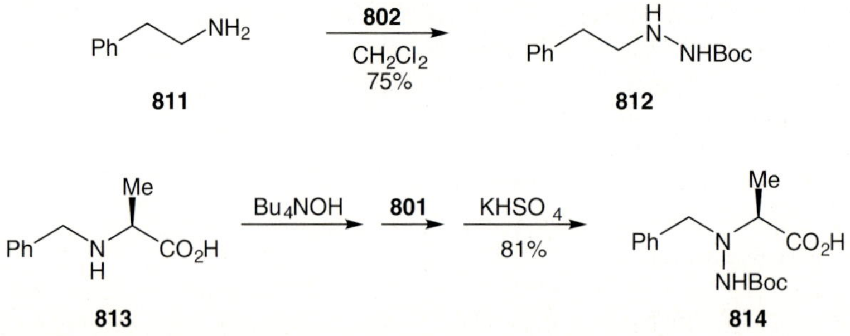

Scheme 2.115 Nitrogen transfer to amines.

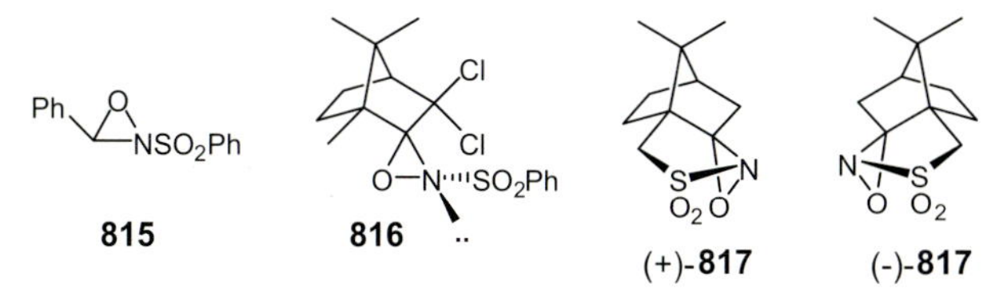

Figure 2.27 Some common oxaziridines used for oxygen transfers.

2.7.3.2 **Oxygen Transfer Reactions**

Some very exciting and synthetically useful methodology is available from the oxaziridine-mediated oxygen transfer onto organic compounds, and in many regards this technology is complementary to other methods. Oxygen-donating oxaziridines are generally of the 3-alkyl or 3-aryl variety (Figure 2.27), although there are examples of a given oxaziridine exhibiting both O-donating and N-donating behavior under different experimental conditions.

In isolated cases, oxaziridines can engage in epoxidation reactions, both in an intramolecular sense [792] and as a bimolecular event, as illustrated by the high-yielding conversion of 1,1-diphenylethene (**818**, Scheme 2.116) into the corresponding epoxide (**819**) [793]. However, this reaction is not general, and it is usually carried out more conveniently using conventional epoxidation conditions. In contrast, oxaziridines are superb reagents for selective α-hydroxylation reactions. For example, in their synthetic approach towards the microbial immunosuppressive agent FR901483, Weinreb and coworkers [794] employed the readily accessible oxaziridine **815** for the diastereoselective α-hydroxylation of the bicyclic lactam **820**. Asymmetric hydroxylations are also possible using chiral oxaziridines, such as the camphor derived reagents **816** and **817**. As an illustrative example, the sodium enolate of phenylacetophenone (**822**) was treated with (+)-(10-camphorsulfonyl) oxaziradine (**817**) to provide the α-hydroxyketone **823** in 80% yield and 94% ee. This reagent was also useful in the diastereoselective preparation of 13^2-hydroxylated chlorophylls (e.g., **825**) [795].

Oxaziridines occupy another synthetic niche in the realm of sulfur chemistry, as they engage in some very specific and selective sulfur oxidation reactions. For example, oxaziridine **815** oxidized the lithium salt of 4-fluorothiophenol (**826**) selectively to the sulfinate salt, which could be trapped with active electrophiles such as methyl iodide (Scheme 2.117). This particular strategy was used to access ^{11}C-labeled methyl sulfones [796]. Of particular synthetic utility is the oxaziridine-mediated asymmetric oxidation of prochiral sulfides to chiral nonracemic sulfoxides, which are receiving increasing attention in their own right. Thus, 4-methylthiotoluene (**829**) is converted into *(S)*-methyl tolyl sulfoxide (**830**) in excellent yield and enantiomeric excess under essentially neutral conditions [797].

2.7.3.3 **Rearrangements**

Some interesting metal-mediated rearrangements of oxaziridines have been reported, although none appear to be widely general in their scope. For example, the copper(I) catalyst $[Cu(CH_3CN)_4]PF_6$ induces the conversion of oxaziridine **831**

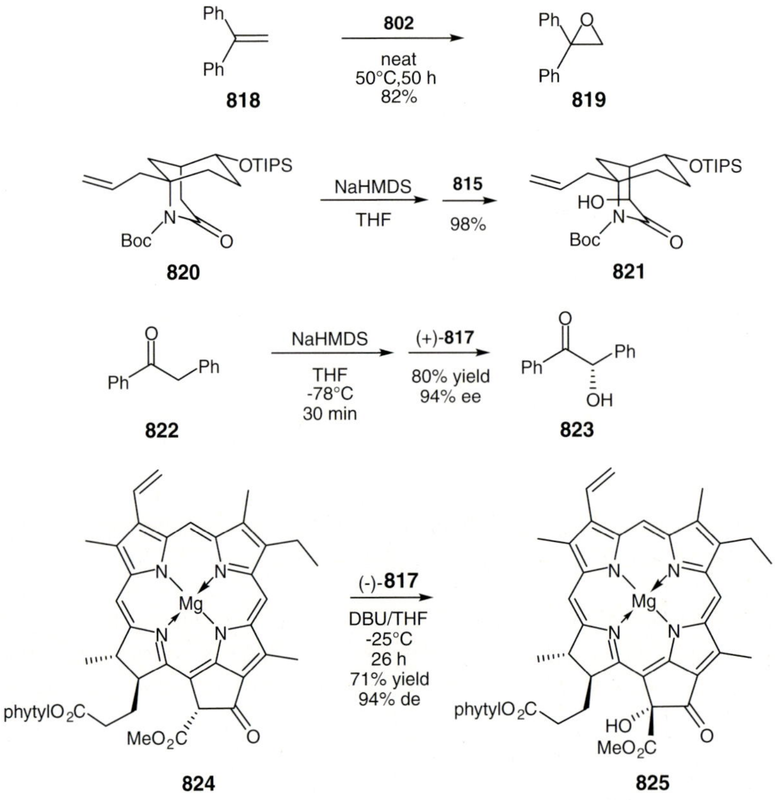

Scheme 2.116 Oxygen transfer to carbon.

into the dihydro-2*H*-pyrrole **832** through a fascinating mechanistic sequence involving initial N–O bond cleavage by single electron transfer, followed by radical cyclization, phenyl migration and loss of acetaldehyde [798]. When the 3-substituent is secondary (or presumably tertiary), the intermediate radical can collapse with

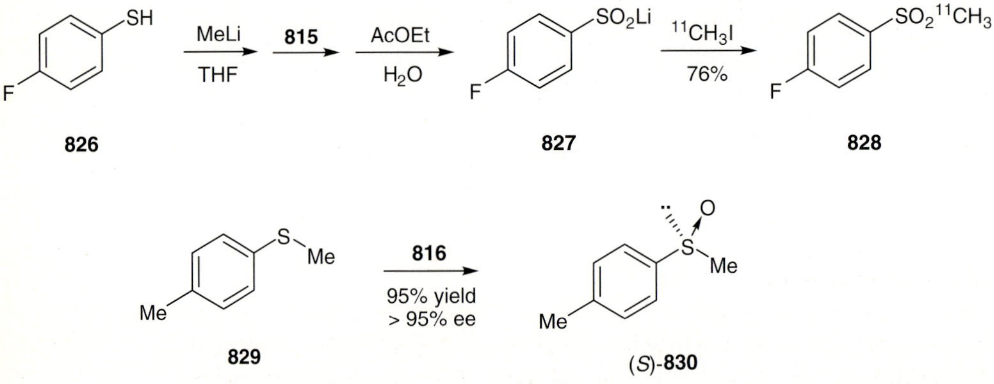

Scheme 2.117 Oxygen transfer to sulfur.

concomitant C–C bond cleavage to give amide derivatives. Thus, the 3-isopropyl oxaziridine **833** yields amide **834** under the same conditions [799]. Under the influence of Lewis acids, the oxaziridine ring can be opened by nucleophiles. For example, the 3-methoxyphenyl-oxaziridine **835** suffers nucleophilic attack at the ring carbon by *O*-benzylhydroxylamine (Scheme 2.118, inset) in the presence of aluminium trichloride, ultimately leading to the liberation of the oxaziridinyl N–O fragment, which is subsequently *O*-protected by the tri(isopropyl)silyl group *in situ* [800].

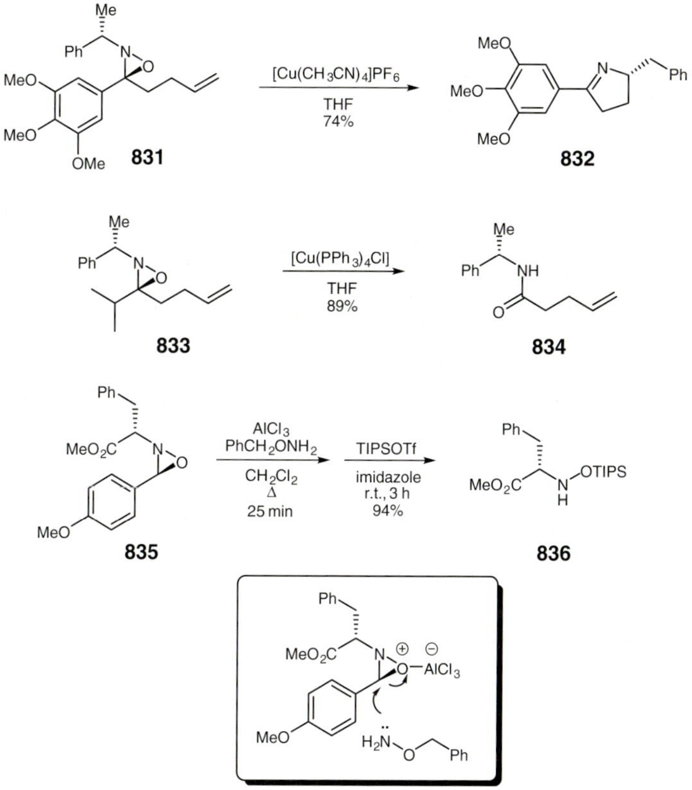

Me
Ph
N
O
MeO
MeO
OMe **831**
$[Cu(CH_3CN)_4]PF_6$
THF
74%
MeO
MeO
MeO
N
Ph
832

Me
Ph
N
O
833
$[Cu(PPh_3)_4Cl]$
THF
89%
Me
Ph
NH
O
834

Ph
MeO_2C
N
O
MeO
835
$AlCl_3$
$PhCH_2ONH_2$
CH_2Cl_2
Δ
25 min
TIPSOTf
imidazole
r.t., 3 h
94%
Ph
MeO_2C
N
H
OTIPS
836

Ph
MeO_2C
N ⊕ ⊖
O–$AlCl_3$
MeO
H_2N
O
Ph

Scheme 2.118 Metal-mediated rearrangements of oxaziridines.

2.8 Dioxiranes

2.8.1 Properties of Dioxiranes

Dioxiranes are cyclic peroxides, and in many respects their chemical behavior can be described as activated peroxy species, or what Greer has dubbed “unusual peroxides” [801]. The strain energy of the parent compound (Figure 2.28) has been

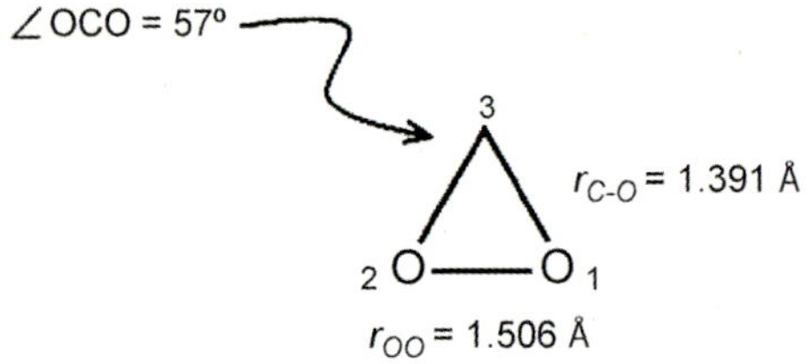

∠OCO = 57°
3
r_{C-O} = 1.391 Å
2 O—O 1
r_{OO} = 1.506 Å

Figure 2.28 Geometry of dioxirane.

calculated at 19.6 kcal mol^{-1}. However, this figure drops by almost half to 10.6 kcal mol^{-1} in the case of 3,3-dimethyldioxirane (DMD) [802, 803]. Geometrically, the unsubstituted dioxirane ring describes a mildly distorted triangle [804].

2.8.2 Synthesis of Dioxiranes

A survey of the preparative synthetic methods for dioxiranes is a little peculiar, because the field is dominated by virtually a single technique. There are certainly isolated alternative conditions that lead to the formation of the dioxirane system, such as the photolytic oxidation of diaryl diazoalkanes (e.g., **837** → **838**, Scheme 2.119) [805]; however, to date they have not been widely adopted by the synthetic laboratory.

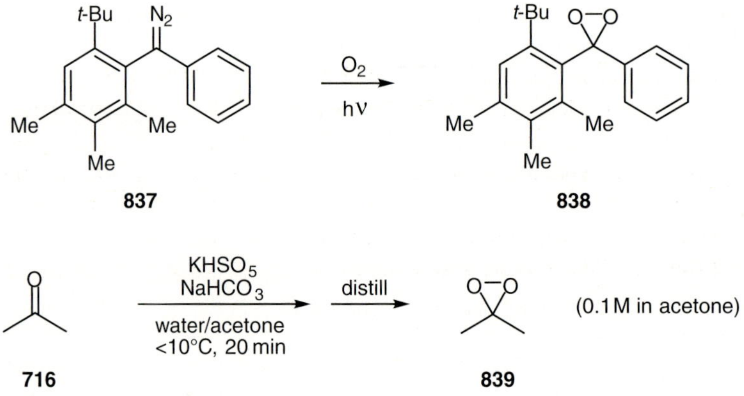

t-Bu N_2 Me Me Me **837** $\xrightarrow[h\nu]{O_2}$ t-Bu O–O Me Me Me **838**

716 $\xrightarrow[\text{water/acetone, <10°C, 20 min}]{KHSO_5,\ NaHCO_3}$ distill → **839** (0.1 M in acetone)

Scheme 2.119 Preparation of dioxiranes.

In fact, most procedures involving dioxiranes as reactive intermediates are designed to generate these species *in situ*, usually as part of a catalytic cycle, and usually by oxidizing a ketone precursor. This not only brings the obvious benefit of requiring less than stoichiometric amounts of the ketone precursor (some of which are quite dear), but it also obviates the need to isolate, store and continually titrate mixtures of unstable dioxiranes. Nevertheless, it is sometimes convenient, or even necessary, to work with "isolated" dioxirane reagents. Toward this end, dilute solutions of dimethyldioxirane (DMD) are available by treating acetone with Oxone in an aqueous bicarbonate buffer at low temperature, after which the dioxirane is distilled at reduced pressure (Scheme 2.119). The distillate so-obtained contains about a 0.1 M solution of DMD in acetone (towards which the dioxirane is stable [806]), and this is usually titrated before use [807, 808]. The dilution factor can be inconvenient [809], although some practical modifications have been disclosed [810, 811], including phase-transfer conditions, which can be applied even to large-scale preparations [812].

Generally, almost all synthetically relevant dioxiranes are prepared by this method; however, the reader is directed to Murray's excellent review of dioxiranes for more detailed information and an historical perspective on the development of dioxirane preparations [813].

2.8.3 Reactivity of Dioxiranes

2.8.3.1 Epoxidation of Alkenes

Arguably the most high-profile member of the dioxiranes reactive portfolio, the epoxidation of alkenes is treated in Section 2.3.2.1. The reader is also directed to an outstanding comprehensive review by Adam, Saha-Möller and Zhao [814], as well as a more recent overview by Srivastava [815].

2.8.3.2 Hydroxylation of Alkanes

Another fascinating (and remarkably underutilized) reaction mediated by dioxiranes is the hydroxylation of alkanes, a process that can be quite clean and high-yielding. For example, treatment of 2,3-dimethylbutane (**840**, Scheme 2.120) with 3-methyl-3-trifluoromethyldioxirane (**841**) (TFD) at low temperature results in rapid and almost quantitative conversion to tertiary alcohol **842** [816]. In general, tertiary sites are preferred to secondary, a phenomenon underscored by the oxidation of adamantane (**843**) with excess TFD, in which only tertiary sites are affected [817]. In a selectivity study, Curci and coworkers [818] found that these bridgehead sites were also preferred to cyclopropane methylene positions, as demonstrated by the smooth conversion of the spiro cyclopropyladamantane **846** into the monohydroxy derivative **847**.

Secondary sites are not immune to oxidation. In fact, careful kinetic studies have shown that the reactivity exhibits an almost perfect Hammett correlation to the electron density of the C–H sigma bond [819], which strongly suggests a concerted insertion mechanism. Computational studies also support this idea, although a diradical mechanism cannot be ruled out [804]. In any event, lability towards oxidation is enhanced by adjacent heteroatoms. For example, ethyl *t*-butyl ether (**848**) suffers practically quantitative oxidative degradation to *t*-butanol (**850**) and acetaldehyde (**851**) via an initially formed hemiacetal intermediate (**849**) [820], which explains the previously reported observation that DMD solutions lose titer in the presence of adventitious ether contaminants [821]. Through a similar mechanism, dioxiranes convert alcohols into ketones, as illustrated by the oxidation of epoxyal-

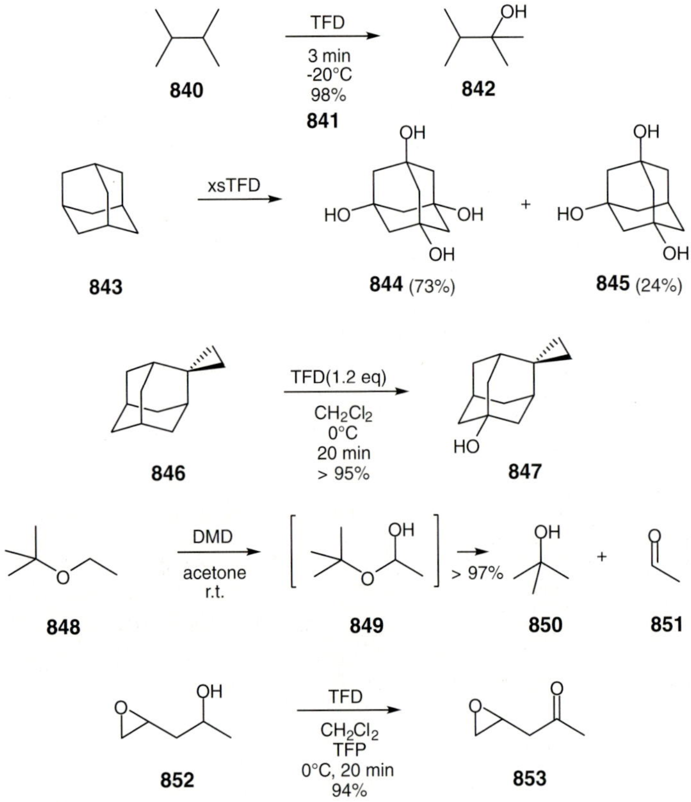

Scheme 2.120 Dioxirane-mediated hydroxylation.

cohol **852** to the corresponding ketone (**853**) in excellent yield upon treatment with TFD in methylene chloride and trifluoropropanone (TFP) [120].

Interestingly, this oxidation can be carried out in an intramolecular fashion – a sort of remote functionalization. For example, when the trifluoromethyl ketone **854** (Scheme 2.121) is treated with Oxone in buffered medium, it forms a dioxirane that oxidizes the δ-methylene position. The hydroxyketone so formed subsequently cyclizes to the stable hemiacetal **855** [822]. In this intramolecular manifestation, geometric factors override electronic preferences. Thus, the dioxirane generated from ketone **856** also engages the secondary δ-position over the adjacent tertiary position, ultimately forming the bicyclic hemiacetal **857** [823].

2.8.3.3 **Oxidation of Sulfur**

Dioxiranes have been applied to the selective oxidation of sulfides to sulfoxides [824]. For example, the thiochromanone **858** (Scheme 2.122) is converted into the corresponding sulfoxide (**859**) upon treatment with a slight excess of DMD at near-0 °C temperatures. For best selectivity, the reaction was halted at partial conversion (about 75%); longer reaction times and higher loadings of DMD led to excellent yields of

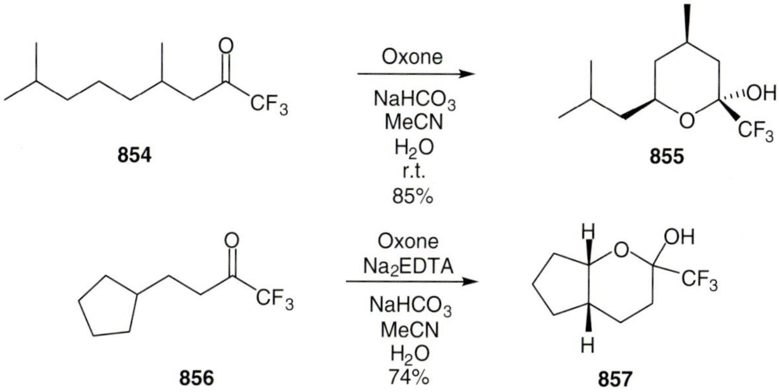

Scheme 2.121 Intramolecular dioxirane-mediated hydroxylation.

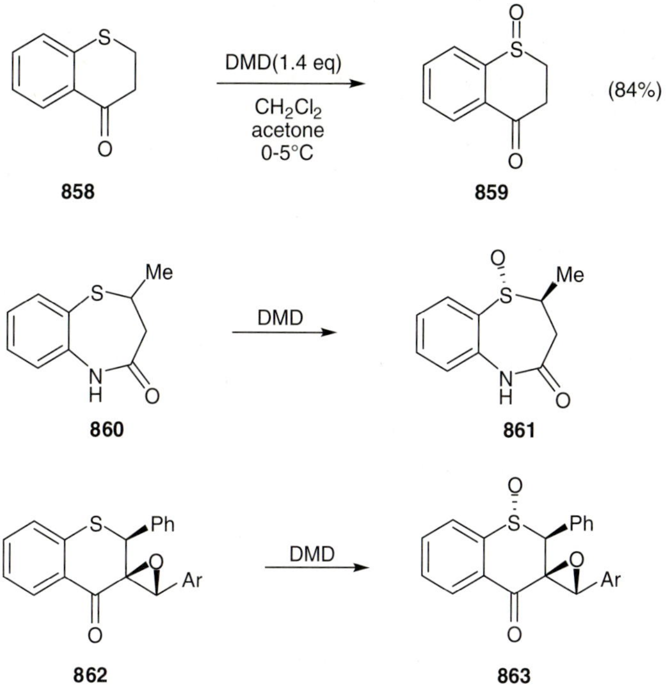

Scheme 2.122 Dioxirane-mediated sulfoxidation.

sulfone derivatives [825]. These oxidations can be quite diastereoselective. Thus, the DMD oxidation of 2,3-dihydro-1,5-benzothiazepinone **860** provided an overwhelming majority of the *trans*-sulfoxide **861** [826], and the epoxy thiochromanone **862** provided the *trans/cis*-sulfoxide **863** exclusively [827]. The same diastereomeric bias was observed even with the opposite epoxide stereochemistry, which seems to point toward a pivotal role of the α-substituent. Enantioselective sulfoxidation using chiral ketone precursors appears promising [828], although results are highly variable and substrate-dependent.

References

1 Rai, K.M.L. and Hassner, A. (2000) in *Advances in Strained and Interesting Organic Molecules* (ed. B. Halton), JAI, Stamford, pp. 187–257.

2 Padwa, A. and Murphree, S.S. (2005) *Arkivoc*, 6–33.

3 Cardillo, G., Gentilucci, L., and Tolomelli, A. (2003) *Aldrichimica Acta*, **36**, 39–50.

4 McCoull, W. and Davis, F.A. (2000) *Synthesis*, 1347–1365.

5 Singh, G.S., D'hooghe, M., and De Kimpe, N. (2007) *Chemical Reviews*, **5**, 2080–2135.

6 Sweeney, J.B. (2002) *Chemical Society Reviews*, **31**, 247–258.

7 Eicher, T., Hauptmann, S., and Speicher, A. (2003) *Three-membered Heterocycles, from The Chemistry of Heterocycles: Structure, Reactions, Syntheses, and Applications*, Wiley-VCH Verlag GmbH, Weinheim, p. 556.

8 Yudin, A.K. (2006) *Aziridines and Epoxides in Organic Synthesis*, Wiley-VCH Verlag GmbH, Weinheim, p. 492.

9 Barić, D. and Maksić, Z.B. (2005) *Theoretica Chimica Acta*, **114**, 222–228.

10 Hadjadj-Aoul, R., Bouyacoub, A., Krallafa, A., and Volatron, F. (2008) *THEOCHEM*, **849**, 8–16.

11 Ebrahimi, A., Deyhimi, F., and Roohi, H. (2001) *Journal of Molecular Structure: THEOCHEM*, **535**, 247–256.

12 Park, G., Kim, S., and Kang, H. (2005) *Bulletin of the Korean Chemical. Society*, **26**, 1339.

13 Mortimer, F.S. (1961) *Journal of Molecular Spectroscopy*, 199–205.

14 Lowden, P.A.S. (2006) Aziridine natural products–discovery, biological activity and biosynthesis, in *Aziridines and Epoxides in Organic Synthesis* (ed. A.K. Yudin), Wiley-VCH Verlag GmbH, Weinheim, p. 399.

15 Nagaoka, K., Matsumoto, M., and Oono, J. (1986) *Journal of Antibiotics*, **39**, 1527–1532.

16 Yokoi, K., Nagaoka, K., and Nakashima, T. (1986) *Chemical and Pharmaceutical Bulletin*, **34**, 4554–4561.

17 Alcaro, S., Ortuso, F., and Coleman, R.S. (2005) *Journal of Chemical Information and Modeling*, **45**, 602–609.

18 Coleman, R.S., Li, J., and Navarro, A. (2001) *Angewandte Chemie, International Edition*, **41**, 1736–1739.

19 Biosynthetic studies: Corre, C. and Lowden, P.A.S. (2004) *Chemical Communications*, 990–991.

20 Hodgkinson, T.J. and Shipman, M. (2001) *Tetrahedron*, **57**, 4467–4488.

21 Fujiwara, T., Saito, I., and Sugiyama, H. (1999) *Tetrahedron Letters*, **40**, 315–318.

22 Armstrong, R.W., Salvati, M.E., and Nguyen, M. (1992) *Journal of the American Chemical Society*, **114**, 3144–3145.

23 Lown, J.W. and Majumdar, K.C. (1977) *Canadian Journal of Biochemistry*, **55**, 630–635.

24 Argoudelis, A.D., Reusser, F., and Whaley, H.A. (1976) *Journal of Antibiotics*, **29**, 1001–1006.

25 Kasai, M. and Kono, M. (1992) *Synlett*, 778–790.

26 Papaioannou, N., Evans, C.A., Blank, J.T., and Miller, S.J. (2001) *Organic Letters*, **3**, 2879–2882.

27 Yoshimoto, M., Miyazawa, H., and Nakao, H. (1979) *Journal of Medicinal Chemistry*, **22**, 491–496.

28 Schaschke, N. (2004) *Bioorganic & Medicinal Chemistry Letters*, **14**, 855–857.

29 Harada, K., Tomita, K., Fujii, K., *et al.* (2004) *Journal of Antibiotics*, **57**, 125–135.

30 Tsuchida, T., Iinuma, H., Kinoshita, N., *et al.* (1993) *Journal of Antibiotics*, **46**, 1772–1774.

31 Tsuchida, T., Iinuma, H., Kinoshita, N., *et al.* (1995) *Journal of Antibiotics*, **48**, 217–221.

32 Ismail Fyaz, M.D., Levitsky, D.O., and Dembitsky, V.M. (2009) *European Journal of Medical Chemistry*, **44**, 3373–3387.

33 Muller, P. and Fruit, C. (2003) *Chemical Reviews*, **103**, 2905–2920.

34 Sweeney, J.B. (2006) Synthesis of aziridines, in *Aziridines and Epoxides in Organic Synthesis* (ed. A.K. Yudin), Wiley-VCH Verlag GmbH, Weinheim, p. 117.

35 Jain, S.L. and Sain, B. (2003) *Tetrahedron Letters*, **44**, 575–577.

36 Rasika Dias, H.V., Lu, H., Kim, H., *et al.* (2002) *Organometallics*, **21**, 1466–1473.

37 Mairena, M.A., Diaz-Requejo, M.M., Belderrain, T.R., *et al.* (2004) *Organometallics*, **23**, 253–256.

38 Diaz-Requejo, M. and Perez, P.J. (2001) *Journal of Organometallic Chemistry*, **617–618**, 110–118.

39 Halfen, J.A., Fox, D.C., Mehn, M.P., and Que, L. (2001) *Journal of Inorganic Chemistry*, **40**, 5060–5061.

40 Comba, P., Merz, M., and Pritzkow, H. (2003) *European Journal of Inorganic Chemistry*, 1711–1718.

41 Wu, H., Xu, L., Xia, C., *et al.* (2005) *Catalysis Communications*, 221–223.

42 Halfen, J.A., Hallman, J.K., Schultz, J.A., and Emerson, J.P. (1999) *Organometallics*, **18**, 5435–5437.

43 Lakshmi Kantam, M., Neeraja, V., Kavita, B., and Haritha, Y. (2004) *Synlett*, 525–527.

44 Gullick, J., Taylor, S., Kerton, O., *et al.* (2001) *Catalysis Letters*, **75**, 151–154.

45 Vyas, R., Gao, G., Harden, J.D., and Zhang, X.P. (2004) *Organic Letters*, **6**, 1907–1910.

46 Evans, D.A., Faul, M.M., and Bilodeau, M.T. (1994) *Journal of the American Chemical Society*, **116**, 2742–2753.

47 Nakanishi, M., Salit, A., and Bolm, C. (2008) *Advanced Synthesis and Catalysis*, **350**, 1835–1840.

48 Cui, Y. and He, C. (2003) *Journal of the American Chemical Society*, **125**, 16202–16203.

49 Fan, R., Pu, D., Gan, J., and Wang, B. (2008) *Tetrahedron Letters*, **49**, 4925–4928.

50 Minakata, S. (2009) *Accounts of Chemical Research*, **42**, 1172–1182.

51 Zhang, J. and Che, C. (2002) *Organic Letters*, **4**, 1911–1914.

52 Zhou, Z., Zhao, Y., Yue, Y., *et al.* (2005) *Arkivoc*, 130–136.

53 Gullick, J., Taylor, S., McMorn, P., *et al.* (2002) *Journal of Molecular Catalysis A: Chemical*, **182**, 571–575.

54 Taylor, S., Gullick, J., McMorn, P., *et al.* (2001) *Journal of the Chemical Society, Perkin Transactions 2*, 1714–1723.

55 Kwong, H., Liu, D., Chan, K., *et al.* (2004) *Tetrahedron Letters*, **45**, 3965–3968.

56 Gullick, J., Taylor, S., Ryan, D., *et al.* (2003) *Chemical Communications*, 2808–2809.

57 Nishikori, H. and Katsuki, T. (1996) *Tetrahedron Letters*, **37**, 9245–9248.

58 Quan, R.W., Li, Z., and Jacobsen, E.N. (1996) *Journal of the American Chemical Society*, **118**, 8156–8157.

59 Omura, K., Uchida, T., Irie, R., and Katsuki, T. (2004) *Chemical Communications*, 2060–2061.

60 Liang, J., Huang, J., Yu, X., *et al.* (2002) *Chemistry – A European Journal*, 1563–1572.

61 Jain, S.L., Sharma, V.B., and Sain, B. (2004) *Tetrahedron Letters*, **45**, 8731–8732.

62 Thakur, V.V. and Sudalai, A. (2003) *Tetrahedron Letters*, **44**, 989–992.

63 Jeong, J.U., Tao, B., Sagasser, I., *et al.* (1998) *Journal of the American Chemical Society*, **120**, 6844–6845.

64 Sureshkumar, D., Maity, S., and Chandrasekaran, S. (2006) *The Journal of Organic Chemistry*, **71**, 1653–1657.

65 Han, H., Bae, I., Eun, J.Y., *et al.* (2004) *Organic Letters*, **6**, 4109–4112.

66 Jain, S.L., Sharma, V.B., and Sain, B. (2005) *Synthetic Communications*, 9–13.

67 Guthikonda, K. and Du Bois, J. (2002) *Journal of the American Chemical Society*, **124**, 13672–13673.

68 Duran, F., Leman, L., Ghini, A., *et al.* (2002) *Organic Letters*, **4**, 2481–2483.
69 Wehn, P.M., Lee, J., and Du Bois, J. (2003) *Organic Letters*, **5**, 4823–4826.
70 Padwa, A., Flick, A.C., Leverett, C.A., and Stengel, T. (2004) *The Journal of Organic Chemistry*, **69**, 6377–6386.
71 Lebel, H., Leogane, O., Huard, K., and Lectard, S. (2006) *Pure and Applied Chemistry*, **78**, 363.
72 Jones, D.W. and Thornton-Pett, M. (1995) *Journal of the Chemical Society, Perkin Transactions 1*, 809–815.
73 Atkinson, R.S., Grimshire, M.J., and Kelly, B.J. (1989) *Tetrahedron*, **45**, 2875–2886.
74 Li, J., Liang, J., Chan, P.W.H., and Che, C. (2004) *Tetrahedron Letters*, **45**, 2685–2688.
75 Zhang, E., Tu, Y., Fan, C., Zhao, X., Jiang, Y., and Zhang, S. (2008) *Organic Letters*, **10**, 4943–4946.
76 Li, J., Chan, P.W.H., and Che, C. (2005) *Organic Letters*, **7**, 5801–5804.
77 Gil, M.V., Arevalo, M.J., and Lopez, O. (2007) *Synthesis*, 1589–1620.
78 Siu, T. and Yudin, A.K. (2002) *Journal of the American Chemical Society*, **124**, 530–531.
79 Caiazzo, A., Dalili, S., Picard, C., *et al.* (2004) *Pure and Applied Chemistry*, **76**, 603–613.
80 Watson, I.D.G., Yu, L., and Yudin, A.K. (2006) *Accounts of Chemical Research*, **39**, 194–206.
81 Singh, S. and Singh, K.N. (2005) *Synthetic Communications*, 2597–2602.
82 Yang, K. and Chen, K. (2002) *Organic Letters*, **4**, 1107–1109.
83 Ma, L., Du, D., and Xu, J. (2005) *The Journal of Organic Chemistry*, **70**, 10155–10158.
84 Xu, J., Ma, L., and Jiao, P. (2004) *Chemical Communications*, 1616–1617.
85 Pesciaioli, F., De Vincentiis, F., Galzerano, P., Bencivenni, G., Bartoli, G., Mazzanti, A., and Melchiozze, P. (2008) *Angewandte Chemie International Edition*, **47**, 8703–8706.
86 Shi, M., Wang, C., and Chan, A.S.C. (2001) *Tetrahedron: Asymmetry*, **12**, 3105–3111.
87 Suga, H., Kakehi, A., Ito, S., *et al.* (2003) *Bulletin of the Chemical Society of Japan*, **76**, 189–199.
88 Fioravanti, S., Morreale, A., Pellacani, L., and Tardella, P.A. (2004) *Synlett*, 1083–1085.
89 Fioravanti, S., Morreale, A., Pellacani, L., and Tardella, P.A. (2003) *Tetrahedron Letters*, **44**, 3031–3034.
90 Aggarwal, V.K., Badine, D.M., and Moorthie, V.A. (2006) Asymmetric synthesis of epoxides and aziridines from aldehydes and imines, in *Aziridines and Epoxides in Organic Synthesis* (ed. A.K. Yudin), Wiley-VCH Verlag GmbH, Weinheim, p. 1.
91 Sweeney, J. (2009) *European Journal of Organic Chemistry*, 4911–4919.
92 Sun, X. and Tang, Y. (2008) *Accounts of Chemical Research*, **41**, 937–948.
93 Sengupta, S. and Mondal, S. (2000) *Tetrahedron Letters*, **41**, 6245–6248.
94 Xie, W., Fang, J., Li, J., and Wang, P.G. (1999) *Tetrahedron*, **55**, 12929–12938.
95 Casarrubios, L., Perez, J.A., Brookhart, M., and Templeton, J.L. (1996) *The Journal of Organic Chemistry*, **61**, 8358–8359.
96 Mayer, M.F., Wang, Q., and Hossain, M.M. (2001) *Journal of Organometallic Chemistry*, **630**, 78–83.
97 Ishii, Y. and Sakaguchi, S. (2004) *Bulletin of the Chemical Society of Japan*, **77**, 909–920.
98 Rasmussen, K.G. and Jørgensen, K.A. (1995) *Journal of the Chemical Society, Chemical Communications*, 1401–1402.
99 Rasmussen, K.G., Juhl, K., Hazell, R.G., and Jørgensen, K.A. (1998) *Journal of the Chemical Society, Perkin Transactions 2*, 1347–1350.
100 Sun, W., Xia, C., and Wang, H. (2003) *Tetrahedron Letters*, **44**, 2409–2411.
101 Yadav, J.S., Reddy, B.V.S., Shesha Rao, M., and Reddy, P.N. (2003) *Tetrahedron Letters*, **43**, 5275–5278.
102 Fan, R. and Ye, Y. (2008) *Advanced Synthesis and Catalysis*, **350**, 1526–1530.
103 Zhang, Y., Desai, A., Li, Z., Hu, G., Ding, Z., and Wulff, W.D. (2008) *Chemistry—A European Journal*, **14**, 3785–3803.

104 Antilla, J.C. and Wulff, W.D. (2000) *Angewandte Chemie, International Edition*, **40**, 4518–4521.

105 Hu, G., Huang, L., Huang, R.H., and Wulff, W.D. (2009) *Journal of the American Chemical Society*, **131**, 15615–15617.

106 Akiyama, T., Ogi, S., and Fuchibe, K. (2003) *Tetrahedron Letters*, **44**, 4011–4013.

107 Bona, F., De Vitis, L., Florio, S., *et al.* (2003) *Tetrahedron*, **59**, 1381–1387.

108 De Vitis, L., Florio, S., Granito, C., *et al.* (2004) *Tetrahedron*, **60**, 1175–1182.

109 Davis, F.A., Ramachandar, T., and Wu, Y. (2003) *The Journal of Organic Chemistry*, **68**, 6894–6898.

110 Davis, F.A., Wu, Y., Yan, H., *et al.* (2003) *The Journal of Organic Chemistry*, **68**, 2410–2419.

111 Morton, D., Pearson, D., Field, R.A., and Stockman, R.A. (2006) *Chemical Communications*, 1833–1835.

112 Chigboh, K., Morton, D., Nadin, A., and Stockman, R.A. (2008) *Tetrahedron Letters*, **49**, 4768–4770.

113 Dai, L., Hou, X., and Zhou, Y. (1999) *Pure and Applied Chemistry*, **71**, 369–376.

114 Li, A., Dai, L., and Aggarwal, V.K. (1997) *Chemical Reviews*, **97**, 2341–2372.

115 Solladie-Cavallo, A., Roje, M., Welter, R., and Sunjic, V. (2004) *The Journal of Organic Chemistry*, **69**, 1409–1412.

116 Aggarwal, V.K. and Vasse, J. (2003) *Organic Letters*, **5**, 3987–3990.

117 Aggarwal, V.K., Alonso, E., Hynd, G., *et al.* (2001) *Angewandte Chemie, International Edition*, **41**, 1430–1433.

118 Robiette, R. (2006) *The Journal of Organic Chemistry*, **71**, 2726–2734.

119 Zhu, S., Liao, Y., and Zhu, S. (2005) *Synlett*, 1429–1432.

120 Song, L., Servajean, V., and Thierry, J. (2006) *Tetrahedron*, **62**, 3509–3516.

121 Sweeney, J.B., Cantrill, A.A., McLaren, A.B., and Thobhani, S. (2006) *Tetrahedron*, **62**, 3681–3693.

122 Wenker, H. (1935) *Journal of the American Chemical Society*, **57**, 2328–2328.

123 Okada, I., Ichimura, K., and Sudo, R. (1970) *Bulletin of the Chemical Society of Japan*, **43**, 1185–1189.

124 Suzuki, H. and Tani, H. (1984) *Chemistry Letters*, 2129–2130.

125 Pfister, J.R. (1984) *Synthesis*, 969–970.

126 Olson, K.D. and Kaiser, S.W. (1989) Patent WO 8906229.

127 Olson, K.D. and Kaiser, S.W. (1989) Patent WO 8905797.

128 Bieber, L.W. and De Araujo, M.C.F. (2002) *Molecules*, 902–906.

129 Ye, W., Leow, D., Goh, S.L.M., *et al.* (2006) *Tetrahedron Letters*, **47**, 1007–1010.

130 Kim, S.K. and Jacobsen, E.N. (2004) *Angewandte Chemie, International Edition*, **44**, 3952–3954.

131 D'hooghe, M., Hofkens, A., and De Kimpe, N. (2003) *Tetrahedron Letters*, **44**, 1137–1139.

132 Cardillo, G., Gentilucci, L., Tomasini, C., and Castejon-Bordas, M.P.V. (1996) *Tetrahedron: Asymmetry*, **7**, 755–762.

133 Olsen, C.A., Franzyk, H., and Jaroszewski, J.W. (2007) *European Journal of Organic Chemistry*, 1717–1724.

134 Amoroso, R., Cardillo, G., Sabatino, P., *et al.* (1993) *The Journal of Organic Chemistry*, **58**, 5615–5619.

135 Cardillo, G., Casolari, S., Gentilucci, L., and Tomasini, C. (1996) *Angewandte Chemie, International Edition in English*, **36**, 1848–1849.

136 Nadir, U.K. and Singh, A. (2004) *Synthetic Communications*, 1337–1347.

137 Taylor, A.M., and Schreiber, S.L. (2009) *Tetrahedron Letters*, **50**, 3230–3233.

138 Malpass, J.R., Belkacemi, D., Griffith, G.A., and Robertson, M.D. (2002) *Arkivoc*, 164–174.

139 Ducray, R., Cramer, N., and Ciufolini, M.A. (2001) *Tetrahedron Letters*, **42**, 9175–9178.

140 Ciufolini, M.A. (2005) *Farmaco (Societa Chimica Italiana: 1989)*, 627–641.

141 Molteni, G. and Del Buttero, P. (2005) *Tetrahedron*, **61**, 4983–4987.

142 Mahoney, J.M., Smith, C.R., and Johnston, J.N. (2005) *Journal of the American Chemical Society*, **127**, 1354–1355.

143 Zhou, P., Chen, B., and Davis, F.A. (2006) Asymmetric syntheses with aziridinecarboxylate and aziridinephosphonate building blocks, in *Aziridines and Epoxides in Organic*

Synthesis (ed. A.K. Yudin), Wiley-VCH Verlag GmbH, Weinheim, p. 73.

144 Hu, X.E. (2004) *Tetrahedron*, **60**, 2701–2743.

145 Paasche, A., Arnone, M., Fink, R.F., Schirmeister, T., and Engels, B. (2009) *Journal of Organic Chemistry*, **74**, 5244–5249.

146 Pineschi, M. (2006) *European Journal of Organic Chemistry*, 4979–4988.

147 Nenajdenko, V.G., Karpov, A.S., and Balenkova, E.S. (2001) *Tetrahedron Asymmetry*, **12**, 2517–2527.

148 Muller, P. and Nury, P. (2001) *Helvetica Chimica Acta*, **84**, 662–677.

149 Ding, C., Dai, L., and Hou, X. (2004) *Synlett*, 2218–2220.

150 Wu, Y. and Zhu, J. (2009) *Organic Letters*, **11**, 5558–5561.

151 Mao, H., Joly, G.J., Peeters, K., *et al.* (2001) *Tetrahedron*, **57**, 6955–6967.

152 Wu, J., Hou, X., and Dai, L. (2000) *The Journal of Organic Chemistry*, **65**, 1344–1348.

153 Moss, T.A., Fenwick, D.R., and Dixon, D.J. (2008) *Journal of the American Chemical Society*, **130**, 10076–10077.

154 Bergmeier, S.C., Katz, S.J., Huang, J., *et al.* (2004) *Tetrahedron Letters*, **45**, 5011–5014.

155 Wang, Z., Sun, X., and Wu, J. (2008) *Tetrahedron*, **64**, 5013–5018.

156 Swamy, N.R. and Venkateswarlu, Y. (2003) *Synthetic Communications*, 547–554.

157 Yadav, J.S., Reddy, B.V.S., Vishweshwar Rao, K., *et al.* (2002) *Synthesis*, 1061–1064.

158 Yadav, J.S., Reddy, B.V.S., Jyothirmai, B., and Murty, M.S.R. (2002) *Synlett*, 53–56.

159 Reddy, M.A., Reddy, L.R., Bhanumathi, N., and Rao, K.R. (2001) *Chemistry Letters*, 246–247.

160 Anand, R.V., Pandey, G., and Singh, V.K. (2002) *Tetrahedron Letters*, **43**, 3975–3976.

161 Schneider, C. (2009) *Angewandte Chemie International Edition*, **48**, 2082–2084.

162 Peruncheralathan, S., Teller, H., and Schneider, C. (2009) *Angewandte Chemie International Edition*, **48**, 4849–4852, S4849/1-S4849/55.

163 Piers, W.E. and Chivers, T. (1997) *Chemical Society Reviews*, **97**, 345–354.

164 Watson, I.D.G. and Yudin, A.K. (2003) *The Journal of Organic Chemistry*, **68**, 5160–5167.

165 Chakraborty, T.K., Ghosh, A., and Raju, T.V. (2003) *Chemistry Letters*, 82–83.

166 Fan, R. and Hou, X. (2003) *The Journal of Organic Chemistry*, **68**, 726–730.

167 Crestey, F., Witt, M., Frydenvang, K., Staerk, D., Jaroszewski, J.W., and Franzyk, H. (2008) *Journal of Organic Chemistry* **73**, 3566–3569.

168 Sabitha, G., Babu, R.S., Rajkumar, M., and Yadav, J.S. (2002) *Organic Letters*, **4**, 343–345.

169 Yadav, J.S., Reddy, B.V.S., Parimala, G., and Reddy, P.V. (2002) *Synthesis*, 2383–2386.

170 Sabitha, G., Babu, R.S., Reddy, M.S.K., and Yadav, J.S. (2002) *Synthesis*, 2254–2258.

171 Chandrasekhar, S., Narsihmulu, C., and Sultana, S.S. (2002) *Tetrahedron Letters*, **43**, 7361–7363.

172 Prasad, B.A., Sekar, G., and Singh, V.K. (2000) *Tetrahedron Letters*, **41**, 4677–4679.

173 Wang, S., Zhu, Y., Wang, Y., and Lu, P. (2009) *Organic Letters*, **11**, 2615–2618.

174 Yadav, J.S., Reddy, B.V.S., Balanarsaiah, E., and Raghavendra, S. (2002) *Tetrahedron Letters*, **43**, 5105–5107.

175 Ghorai, M.K., Shukla, D., and Das, K. (2009) *Journal of Organic Chemistry*, **74**, 7013–7022.

176 Venkat Narsaiah, A., Reddy, B. V. S., Premalatha, K., Reddy, S.S., and Yadav, J.S. (2009) *Catalysis Letters*, **131**, 480–484.

177 Wang, L., Liu, Q., Wang, D., Li, X., Han, X., Xiao, W., and Zhou, Y. (2009) *Organic Letters*, **11**, 1119–1122.

178 Kishore Kumar, G.D. and Baskaran, S. (2004) *Synlett*, 1719–1722.

179 Concellon, J.M. and Riego, E. (2003) *The Journal of Organic Chemistry*, **68**, 6407–6410.

180 Das, B., Ramu, R., Ravikanth, B., and Reddy, K.R. (2006) *Tetrahedron Letters*, **47**, 779–782.

181 Kamal, A., Reddy, D.R. and Rajendar (2006) *Tetrahedron Letters*, **47**, 2261–2264.

182 Larson, S.E., Baso, J.C., Li, G., and Antilla, J.C. (2009) *Organic Letters*, **11**, 5186–5189.

183 Perez-Bautista, J.A., Sosa-Rivadeneyra, M., Quintero, L., Hoepfl, H.,

Tejeda-Dominguez, F.A., and Sartillo-Piscil, F. (2009) *Tetrahedron Letters*, **50**, 5572–5574.

184 Ammetto, I., Gasperi, T., Loreto, M.A., Migliorini, A., Palmarelli, F., and Tardella, P.A. (2009) *European Journal of Organic Chemistry*, 6189–6197.

185 Dureault, A., Tranchepain, I., and Depezay, J.C. (1989) *The Journal of Organic Chemistry*, **54**, 5324–5330.

186 Gnecco, D., Orea, F.L., Galindo, A., *et al.* (2000) *Molecules*, 998–1003.

187 Righi, G. and Catullo, S. (2004) *Synthetic Communications*, 85–97.

188 Yadav, J.S., Subba Reddy, B.V., and Mahesh Kumar, G. (2001) *Synlett*, 1417–1418.

189 Watson, I.D.G., Styler, S.A., and Yudin, A.K. (2004) *Journal of the American Chemical Society*, **126**, 5086–5087.

190 Sasaki, M., Dalili, S., and Yudin, A.K. (2003) *The Journal of Organic Chemistry*, **68**, 2045–2047.

191 Kim, H.Y., Talukdar, A., and Cushman, M. (2006) *Organic Letters*, **8**, 1085–1087.

192 Ahman, J. and Somfai, P. (1994) *Synthetic Communications*, 1121–1127.

193 Sommerdijk, N.A.J.M., Buynsters, P.J.J.A., Akdemir, H., *et al.* (1997) *The Journal of Organic Chemistry*, **62**, 4955–4960.

194 Cardillo, G., Gentilucci, L., and Tolomelli, A. (1999) *Chemical Communications*, 167–168.

195 For graphical abstracts, see: *Tetrahedron* (2003), **59**, 9687–9691.

196 Hodgson, D., Bray, C., and Humphreys, P. (2006) *Synlett*, 0001–0022.

197 Capriati, V., Florio, S., Luisi, R., Musio, B., Alkorta, I., Blanco, F., and Elguero, J. (2008) *Structural Chemistry*, **19**, 785–792.

198 Hodgson, D.M., Humphreys, P.G., and Ward, J.G. (2005) *Organic Letters*, **7**, 1153–1156.

199 Hodgson, D.M., Humphreys, P.G., and Ward, J.G. (2006) *Organic Letters*, **8**, 995–998.

200 Yamauchi, Y., Kawate, T., Itahashi, H., *et al.* (2003) *Tetrahedron Letters*, **44**, 6319–6322.

201 Satoh, T. and Fukuda, Y. (2003) *Tetrahedron*, **59**, 9803–9810.

202 Satoh, T., Ozawa, M., Takano, K., Chyouma, T., and Okawa, A. (2000) *Tetrahedron*, **56**, 4415–4425.

203 Brichacek, M., Lee, D., and Njardarson, J.T. (2008) *Organic Letters*, **10**, 5023–5026.

204 Pohlhaus, P.D., Bowman, R.K., and Johnson, J.S. (2004) *Journal of the American Chemical Society*, **126**, 2294–2295.

205 Ribeiro Laia, F.M. and Pinho e Melo, T. M. V. D. (2009) *Tetrahedron Letters*, **50**, 6180–6182.

206 Yoshida, M., Al-Amin, M., and Shishido, K. (2009) *Tetrahedron Letters*, **50**, 6268–6270.

207 For an early report of intermolecular cycloaddition see: Heine, H.W. and Peavy, R. (1965) *Tetrahedron Letters*, **6**, 3123–3126.

208 Coldham, I. and Hufton, R. (2005) *Chemical Reviews*, **105**, 2765–2809.

209 Hedley, S.J., Moran, W.J., Price, D.A., and Harrity, J.P.A. (2003) *The Journal of Organic Chemistry*, **68**, 4286–4292.

210 Testa, L., Akssira, M., Zaballos-Garcia, E., *et al.* (2003) *Tetrahedron*, **59**, 677–683.

211 Testa, M.L., Hajji, C., Zaballos-Garcia, E., *et al.* (2001) *Tetrahedron: Asymmetry*, **12**, 1369–1372.

212 Park, C.S., Kim, M.S., Sim, T.B., *et al.* (2003) *The Journal of Organic Chemistry*, **68**, 43–49.

213 Sudo, A., Morioka, Y., Sanda, F., and Endo, T. (2004) *Tetrahedron Letters*, **45**, 1363–1365.

214 Miller, A.W. and Nguyen, S.T. (2004) *Organic Letters*, **6**, 2301–2304.

215 Lu, S. and Alper, H. (2004) *The Journal of Organic Chemistry*, **69**, 3558–3561.

216 Prasad, B.A.B., Pandey, G., and Singh, V.K. (2004) *Tetrahedron Letters*, **45**, 1137–1141.

217 D'hooghe, M. and De Kimpe, N. (2006) *Tetrahedron*, **62**, 513–535.

218 Butler, D.C.D., Inman, G.A., and Alper, H. (2000) *The Journal of Organic Chemistry*, **65**, 5887–5890.

219 Palacios, F., de Retana, A.M.O., Marigorta, E.M., and de los Santos, J.M. (2001) *European Journal of Organic Chemistry*, 2401–2414.

220 Palacios, F., de Retana, A.M.O., de Marigorta, E.M., and de los Santos, J.M. (2002) *Organic Preparations and Procedures International*, **34**, 219–269.

221 Alcami, M., Mo, O., and Yaniez, M. (1993) *Journal of the American Chemical Society*, **115**, 11074–11083.

222 Mayer, P.M., Taylor, M.S., Wong, M.W., and Radom, L. (1998) *Journal of Physical Chemistry A*, **102**, 7074–7080.

223 Mó, O., de Paz, J.L.G., and Yáñez, M. (1987) *The Journal of Physical Chemistry*, **91**, 6484–6490.

224 Calvo-Losada, S., Quirante, J.J., Suárez, D., and Sordo, T.L. (1998) *Journal of Computational Chemistry*, **19**, 912–922.

225 Guillemin, J., Denis, Jean-Marc, Lasne, M., and Ripoll, J.-L. (1988) *Tetrahedron*, **44**, 4447–4455.

226 Miller, T.W., Tristram, E.W., and Wolf, F.J. (1971) *Journal of Antibiotics*, **24**, 48–50.

227 Stapley, E.O., Hendlin, D., Jackson, M., *et al.* (1971) *Journal of Antibiotics*, **24**, 42–47.

228 Salomon, C.E., Williams, D.H., and Faulkner, D.J. (1995) *Journal of Natural Products*, **58**, 1463–1466.

229 Molinski, T.F. and Ireland, C.M. (1988) *The Journal of Organic Chemistry*, **53**, 2103–2105.

230 Neber, P.W. and Friedolsheim, A.V. (1926) *Justus Liebig's Annalen der Chemie*, **449**, 109–134.

231 Garg, N.K., Caspi, D.D., and Stoltz, B.M. (2005) *Journal of the American Chemical Society*, **127**, 5970–5978.

232 Garg, N.K., Caspi, D.D., and Stoltz, B.M. (2004) *Journal of the American Chemical Society*, **126**, 9552–9553.

233 O'Brien, C. (1964) *Chemical Reviews*, **64**, 81–89.

234 House, H.O. and Berkowitz, W.F. (1963) *The Journal of Organic Chemistry*, **28**, 2271–2276.

235 Palacios, F., de Retana, O.A.M., and Gil, J.I. (2000) *Tetrahedron Letters*, **41**, 5363–5366.

236 Palacios, F., Aparicio, D., de Retana, O.A.M., *et al.* (2002) *The Journal of Organic Chemistry*, **67**, 7283–7288.

237 Corkins, H.G., Storace, L., and Osgood, E. (1980) *The Journal of Organic Chemistry*, **45**, 3156–3159.

238 Smith, P.A.S. and Most, J.E.E. (1957) *The Journal of Organic Chemistry*, **22**, 358–362.

239 Nair, V. (1968) *The Journal of Organic Chemistry*, **33**, 2121–2123.

240 Padwa, A. and Carlsen, P.H.J. (1977) *Journal of the American Chemical Society*, **99**, 1514–1523.

241 Barcus, R.L., Wright, B.B., Platz, M.S., and Scaiano, J.C. (1983) *Tetrahedron Letters*, **14**, 3955–3958.

242 Morrow, D.F., Butler, M.E., and Huang, E.C.Y. (1965) *The Journal of Organic Chemistry*, **30**, 579–587.

243 Morrow, D.F. and Butler, M.E. (1964) *Journal of Heterocyclic Chemistry*, **1**, 53–54.

244 Piskunova, I.P., Eremeev, A.V., Mishnev, A.F., and Vosekalna, I.A. (1993) *Tetrahedron*, **49**, 4671–4676.

245 Ooi, T., Takahashi, M., Doda, K., and Maruoka, K. (2002) *Journal of the American Chemical Society*, **124**, 7640–7641.

246 Verstappen, M.M.H., Ariaans, G.J.A., and Zwanenburg, B. (1996) *Journal of the American Chemical Society*, **118**, 8491–8492.

247 Palacios, F., de Retana, A.M.O., Gil, J.I., and Ezpeleta, J.M. (2000) *The Journal of Organic Chemistry*, **65**, 3213–3217.

248 Smolinsky, G. (1962) *The Journal of Organic Chemistry*, **27**, 3557–3559.

249 Smolinsky, G. (1961) *Journal of the American Chemical Society*, **83**, 4483–4484.

250 Hortmann, A.G., Robertson, D.A., and Gillard, B.K. (1972) *The Journal of Organic Chemistry*, **37**, 322–324.

251 Alves, M.J. and Gilchrist, T.L. (1998) *Tetrahedron Letters*, **39**, 7579–7582.

252 Gilchrist, T.L. and Mendonca, R. (2000) *Synlett*, 1843–1845.

253 Timen, A.S., Risberg, E., and Somfai, P. (2003) *Tetrahedron Letters*, **44**, 5339–5341.

254 Alonso-Cruz, C.R., Kennedy, A.R., Rodriguez, M.S., and Suarez, E. (2003) *Organic Letters*, **5**, 3729–3732.

255 Singh, P.N.D., Carter, C.L., and Gudmundsdottir, A.D. (2003) *Tetrahedron Letters*, **44**, 6763–6765.

256 Hassner, A., Wiegand, N.H., and Gottlieb, H.E. (1986) *The Journal of Organic Chemistry*, **51**, 3176–3180.

257 Hassner, A. and Fowler, F.W. (1968) *Journal of the American Chemical Society*, **90**, 2869–2875.

258 Banert, K. (1985) *Tetrahedron Letters*, **26**, 5261–5264.

259 Hassner, A. (1971) *Accounts of Chemical Research*, **4**, 9–16.

260 Zhu, W. and Ma, D. (2004) *Chemical Communications*, 888–889.

261 Stork, G. and Zhao, K. (1989) *Tetrahedron Letters*, **30**, 2173–2174.

262 Jordan, D. (1989) *The Journal of Organic Chemistry*, **54**, 3584–3587.

263 Heimgartner, H. (1991) *Angewandte Chemie, International Edition in English*, **31**, 238–264.

264 Villalgordo, J.M., Enderli, A., Linden, A., and Heimgartner, H. (1995) *Helvetica Chimica Acta*, **78**, 1983–1998.

265 Mekhael, M.K.G. and Heimgartner, H. (2003) *Helvetica Chimica Acta*, **86**, 2805–2813.

266 Stamm, S., Linden, A., and Heimgartner, H. (2003) *Helvetica Chimica Acta*, **86**, 1371–1396.

267 Hilty, F.M., Brun, K.A., and Heimgartner, H. (2004) *Helvetica Chimica Acta*, **87**, 2539–2548.

268 Rens, M. and Ghosez, L. (1970) *Tetrahedron Letters*, **11**, 3765–3768.

269 Abramovitch, R.A., Konieczny, M., Pennington, W., *et al.* (1990) *Journal of the Chemical Society, Chemical Communications*, 296–270.

270 Pinho e Melo, T.M.V.D., Lopes, C.S.J., Cardoso, A.L., and Gonsalves, A.M.d.R. (2001) *Tetrahedron*, **57**, 6203–6208.

271 Pinho e Melo, T.M.V.D., Lopes, C.S.J., and Gonsalves, A.M.d.R. (2000) *Tetrahedron Letters*, **41**, 7217–7220.

272 Lipshutz, B.H. and Reuter, D.C. (1988) *Tetrahedron Letters*, **29**, 6067–6070.

273 Auricchio, S., Bini, A., Pastormerlo, E., and Truscello, A.M. (1997) *Tetrahedron*, **53**, 10911–10920.

274 Ueda, S., Naruto, S., Yoshida, T., *et al.* (1988) *Journal of the Chemical Society, Perkin Transactions 1*, 1013–1021.

275 Wentrup, C., Fischer, S., Berstermann, H., *et al.* (1986) *Angewandte Chemie, International Edition in English*, **26**, 85–86.

276 Hassner, A. and Alexanian, V. (1979) *The Journal of Organic Chemistry*, **44**, 3861–3864.

277 Legters, J., Thijs, L., and Zwanenburg, B. (1992) *Recueil des Travaux Chimiques des Pays-Bas*, **111**, 75–78.

278 Gentilucci, L., Grijzen, Y., Thijs, L., and Zwanenburg, B. (1995) *Tetrahedron Letters*, **36**, 4665–4668.

279 Davis, F.A., Reddy, G.V., and Liu, H. (1995) *Journal of the American Chemical Society*, **117**, 3651–3652.

280 Davis, F.A., Liu, H., Liang, C., *et al.* (1999) *The Journal of Organic Chemistry*, **64**, 8929–8935.

281 Goumans, T.P.M., Ehlers, A.W., Lammertsma, K., and Wuerthwein, E. (2003) *European Journal of Organic Chemistry*, 2941–2946.

282 Belloir, P.F., Laurent, A., Mison, P., *et al.* (1985) *Tetrahedron Letters*, **36**, 2637–2640.

283 Palacios, F., de Retana, A.M.O., and Alonso, J.M. (2005) *The Journal of Organic Chemistry*, **70**, 8895–8901.

284 Hirashita, T., Toumatsu, S., Imagawa, Y., *et al.* (2006) *Tetrahedron Letters*, **47**, 1613–1616.

285 Palacios, F., de Retana, A.M.O., Gil, J.I., and Alonso, J.M. (2004) *Tetrahedron*, **60**, 8937–8947.

286 Palacios, F., Aparicio, D., de Retana, A.M.O., *et al.* (2003) *Tetrahedron: Asymmetry*, **14**, 689–700.

287 Roth, P., Andersson, P.G., and Somfai, P. (2002) *Chemical Communications*, 1752–1753.

288 Koch, K.N., Linden, A., and Heimgartner, H. (2001) *Tetrahedron*, **57**, 2311–2326.

289 Koch, K.N., Linden, A., and Heimgartner, H. (2000) *Helvetica Chimica Acta*, **83**, 233–257.

290 Philipova, I., Linden, A., and Heimgartner, H. (2005) *Helvetica Chimica Acta*, **88**, 1711–1733.

291 Brun, K.A. and Heimgartner, H. (2005) *Helvetica Chimica Acta*, **88**, 2951–2959.

292 Breitenmoser, R.A. and Heimgartner, H. (2002) *Helvetica Chimica Acta*, **85**, 885–912.

293 Breitenmoser, R.A., Hirt, T.R., Luykx, R.T.N., and Heimgartner, H. (2001) *Helvetica Chimica Acta*, **84**, 972–979.

294 Brun, K.A., Linden, A., and Heimgartner, H. (2001) *Helvetica Chimica Acta*, **84**, 1756–1777.

295 Brun, K.A., Linden, A., and Heimgartner, H. (2002) *Helvetica Chimica Acta*, **85**, 3422–3443.

296 Luykx, R.T.N., Linden, A., and Heimgartner, H. (2003) *Helvetica Chimica Acta*, **86**, 4093–4111.

297 Stamm, S., Linden, A., and Heimgartner, H. (2006) *Helvetica Chimica Acta*, **89**, 1–15.

298 Stamm, S. and Heimgartner, H. (2004) *European Journal of Organic Chemistry*, 3820–3827.

299 Katritzky, A.R., Wang, M., Wilkerson, C.R., and Yang, H. (2003) *The Journal of Organic Chemistry*, **68**, 9105–9108.

300 Pinho e Melo, T.M.V.D., Lopes, C.S.J., Gonsalves, A.M.d.R., *et al.* (2002) *The Journal of Organic Chemistry*, **67**, 66–71.

301 Gilchrist, T.L. (2001) *Aldrichimica Acta*, 51–55.

302 Alves, M.J. and Gilchrist, T.L. (1998) *Journal of the Chemical Society, Perkin Transactions 1*, 299–304.

303 Alves, M.J., Azoia, N.G., Bickley, J.F., *et al.* (2001) *Journal of the Chemical Society, Perkin Transactions 1*, 2969–2976.

304 Gilchrist, T.L. and Mendonca, R. (2000) *Arkivoc*, 769–778.

305 Ray, C.A., Risberg, E., and Somfai, P. (2001) *Tetrahedron Letters*, **42**, 9289–9291.

306 Timen, A.S. and Somfai, P. (2003) *The Journal of Organic Chemistry*, **68**, 9958–9963.

307 Timen, A.S., Fischer, A., and Somfai, P. (2003) *Chemical Communications*, 1150–1151.

308 Hong, B., Gupta, A.K., Wu, M., and Liao, J. (2004) *Tetrahedron Letters*, **45**, 1663–1666.

309 Brown, D., Brown, G.A., Andrews, M., *et al.* (2002) *Journal of the Chemical Society, Perkin Transactions 1*, 2014–2021.

310 Palacios, F., de Retana, A.M.O., Gil, J.I., and de Munain, R.L. (2002) *Organic Letters*, **4**, 2405–2408.

311 Padwa, A. and Stengel, T. (2005) *Arkivoc*, 21–32.

312 Padwa, A. and Stengel, T. (2004) *Tetrahedron Letters*, **45**, 5991–5993.

313 Taber, D.F. and Tian, W. (2006) *Journal of the American Chemical Society*, **128**, 1058–1059.

314 Hu, K., Li, J., Li, B., *et al.* (2006) *Bioorganic and Medicinal Chemistry*, **14**, 4677–4681.

315 Fei, Z. and McDonald, F.E. (2005) *Organic Letters*, **7**, 3617–3620.

316 Sun, D., Hansen, M., and Hurley, L. (1995) *Journal of the American Chemical Society*, **117**, 2430–2440.

317 Danishefsky, S.J. and Shair, M.D. (1996) *The Journal of Organic Chemistry*, **61**, 16–44.

318 Mehta, G. and Pan, S.C. (2005) *Tetrahedron Letters*, **46**, 3045–3048.

319 Marco-Contelles, J., Molina, M.T., and Anjum, S. (2004) *Chemical Reviews*, **104**, 2857–2899.

320 Taber, D.F. and Christos, T.E. (1999) *Journal of the American Chemical Society*, **121**, 5589–5590.

321 Datta, B., Majumdar, A., Datta, R., and Balusu, R. (2004) *Biochemistry*, **43**, 14821–14831.

322 Mandal, A.K., Schneekloth, J.S., Jr, Kuramochi, K., and Crews, C.M. (2006) *Organic Letters*, **8**, 427–430.

323 Kobayashi, J. and Tsudi, M. (2004) *Natural Product Reports*, **21**, 77–93.

324 Kobayashi, J., Shimbo, K., Kubota, T., and Tsuda, M. (2003) *Pure and Applied Chemistry*, **75**, 337–342.

325 White, J.D., Carter, R.G., and Sundermann, K.F. (1999) *The Journal* *of Organic Chemistry*, **64**, 684–685.

326 Nicolaou, K.C., Roschangar, F., and Vourloumis, D. (1998) *Angewandte Chemie, International Edition*, **38**, 2014–2045.

327 Bardhan, S., Schmitt, D.C., and Porco, J.A., Jr (2006) *Organic Letters*, **8**, 927–930.

328 Encarnacion, R.D., Sandoval, E., Malmstrom, J., and Christophersen, C. (2000) *Journal of Natural Products*, **63**, 874–875.

329 Grüschow, S. and Sherman, D.H. (2006) The biosynthesis of epoxides, in *Aziridines and Epoxides in Organic Synthesis* (ed. A.K. Yudin), Wiley-VCH Verlag GmbH, Weinheim, p. 349.

330 Liu, M. (2001) Epoxidation of Alkenes, in *Rodd's Chemistry of Carbon Compounds*, 2nd edn, Vol. **V** (ed. M. Sainsbury), Elsevier, New York, p. 1.

331 Xia, Q., Ge, H., Ye, C., *et al.* (2005) *Chemical Reviews*, **105**, 1603–1662.

332 Adam, W. and Zhang, A. (2005) *Synlett*, 1047–1072.

333 Olofsson, B. and Somfai, P. (2006) Vinylepoxides in organic synthesis, in *Aziridines and Epoxides in Organic Synthesis* (ed. A.K. Yudin), Wiley-VCH Verlag GmbH, Weinheim, p. 315.

334 Murray, R.W., Jeyaraman, R., and Mohan, L. (1986) *Journal of the American Chemical Society*, **108**, 2470–2472.

335 Adam, W., Hadjiarapoglou, L., and Wang, X. (1989) *Tetrahedron Letters*, **30**, 6497–6500.

336 Chenault, H.K. and Danishefsky, S.J. (1989) *The Journal of Organic Chemistry*, **54**, 4249–4250.

337 Halcomb, R.L. and Danishefsky, S.J. (1989) *Journal of the American Chemical Society*, **111**, 6661–6666.

338 Smith, A.B., Mesaros, E.F., and Meyer, E.A. (2006) *Journal of the American Chemical Society*, **128**, 5292–5299.

339 Adam, W., Hadjiarapoglou, L., Mosandl, T., *et al.* (1991) *Journal of the American Chemical Society*, **113**, 8005–8011.

340 Adam, W., Hadjarapoglou, L., and Wang, X. (1991) *Tetrahedron Letters*, **32**, 1295–1298.

341 Adam, W., Golsch, D., Hadjiarapoglou, L., and Patonay, T. (1991) *Tetrahedron Letters*, **32**, 1041–1044.

342 Crandall, J.K., Batal, D.J., Sebesta, D.P., and Lin, F. (1991) *The Journal of Organic Chemistry*, **56**, 1153–1166.

343 Messeguer, A., Sanchez-Baeza, F., Casas, J., and Hammock, B.D. (1991) *Tetrahedron*, **49**, 1291–1302.

344 Ferraz, H.M.C., Muzzi, R.M., de, O., Vieira, T., and Viertler, H. (2000) *Tetrahedron Letters*, **41**, 5021–5023.

345 Marples, B.A., Muxworthy, J.P., and Baggaley, K.H. (1991) *Tetrahedron Letters*, **32**, 533–536.

346 Bortolini, O., Fantin, G., and Fogagnolo, M. (2009) *Synthesis*, 1123–1126.

347 Denmark, S.E., Forbes, D.C., Hays, D.S., *et al.* (1995) *The Journal of Organic Chemistry*, **60**, 1391–1407.

348 Frohn, M., Wang, Z., and Shi, Y. (1998) *The Journal of Organic Chemistry*, **63**, 6425–6426.

349 Denmark, S.E. and Wu, Z. (1998) *The Journal of Organic Chemistry*, **63**, 2810–2811.

350 Miaskiewicz, K. and Smith, D.A. (1998) *Journal of the American Chemical Society*, **120**, 1872–1875.

351 Shi, Y. (2004) Organocatalytic Oxidation. Ketone-Catalyzed Asymmetric Epoxidaton of Olefins, in *Modern Oxidation Methods*, (ed. J.E. Bäckvall), Wiley-VCH Verlag GmbH, Weinheim, pp. 51.

352 Yang, D., Yip, Y., Tang, M., *et al.* (1996) *Journal of the American Chemical Society*, **118**, 491–492.

353 Yang, D., Wong, M., Yip, Y., *et al.* (1998) *Journal of the American Chemical Society*, **120**, 5943–5952.

354 Tu, Y., Wang, Z., Frohn, M., *et al.* (1998) *The Journal of Organic Chemistry*, **63**, 8475–8485.

355 Tu, Y., Wang, Z., and Shi, Y. (1996) *Journal of the American Chemical Society*, **118**, 9806–9807.

356 Tian, H., She, X., Shu, L., *et al.* (2000) *Journal of the American Chemical Society*, **122**, 11551–11552.

357 Frohn, M., Dalkiewicz, M., Tu, Y., *et al.* (1998) *The Journal of Organic Chemistry*, **120**, 2948–2953.

358 Cao, G., Wang, Z., Tu, Y., and Shi, Y. (1998) *Tetrahedron Letters*, **39**, 4425–4428.

359 Zhu, Y., Tu, Y., Yu, H., and Shi, Y. (1998) *Tetrahedron Letters*, **39**, 7819–7822.

360 Wang, Z., Tu, Y., Frohn, M., and Shi, Y. (1997) *The Journal of Organic Chemistry*, **62**, 2328–2329.

361 Wang, Z., Tu, Y., Frohn, M., *et al.* (1997) *Journal of the American Chemical Society*, **119**, 11224–11235.

362 Shu, L., Shen, Y., Burke, C., *et al.* (2003) *The Journal of Organic Chemistry*, **68**, 4963–4965.

363 Tian, H., She, X., Yu, H., *et al.* (2002) *The Journal of Organic Chemistry*, **67**, 2435–2446.

364 Shu, L., Wang, P., Gan, Y., and Shi, Y. (2003) *Organic Letters*, **5**, 293–296.

365 Wong, O.A., Wang, B., Zhao, M., and Shi, Y. (2009) *Journal of Organic Chemistry*, **74**, 6335–6338.

366 Armstrong, A. (1998) *Chemical Communications*, 621–622.

367 Yang, D., Yip, Y., Chen, J., and Cheung, K. (1998) *Journal of the American Chemical Society*, **120**, 7659–7660.

368 Sartori, G., Armstrong, A., Maggi, R., *et al.* (2003) *The Journal of Organic Chemistry*, **68**, 3232–3237.

369 Legros, J., Crousse, B., Bourdon, J., *et al.* (2001) *Tetrahedron Letters*, **42**, 4463–4466.

370 Hagen, T.J. (2007), The Prilezhaev reaction, in *Name Reactions of Functional Group Transformations*, (eds. J.J. Li and E.J. Corey), Wiley-VCH Verlag GmbH, Weinheim, p. 274.

371 Lowe, J.T., Youngsaye, W., and Panek, J.S. (2006) *The Journal of Organic Chemistry*, **71**, 3639–3642.

372 Svensson, A., Lindstroem, U.M., and Somfai, P. (1996) *Synthetic Communications*, 2875–2880.

373 Gentric, L., Le Goff, X., Ricard, L., and Hanna, I. (2009) *Journal of Organic Chemistry*, **74**, 9337–9344.

374 Wan, X. and Joullie, M.M. (2008) *Journal of the American Chemical Society*, **130**, 17236–17237.

375 Armstrong, A., Barsanti, P.A., Clarke, P.A., and Wood, A. (1994) *Tetrahedron Letters*, **35**, 6155–6158.

376 Asensio, G., Mello, R., Boix-Bernardini, C., *et al.* (1995) *The Journal of Organic Chemistry*, **60**, 3692–3699.

377 Jenmalm, A., Berts, W., Luthman, K., *et al.* (1995) *The Journal of Organic Chemistry*, **60**, 1026–1032.

378 Bach, R.D., Estevez, C.M., Winter, J.E., and Glukhovtsev, M.N. (1998) *Journal of the American Chemical Society*, **120**, 680–685.

379 Ye, D., Fringuelli, F., Piermatti, O., and Pizzo, F. (1997) *The Journal of Organic Chemistry*, **62**, 3748–3750.

380 Washington, I. and Houk, K.N. (2002) *Organic Letters*, **4**, 2661–2664.

381 Fringuelli, F., Germani, R., Pizzo, F., *et al.* (1992) *The Journal of Organic Chemistry*, **57**, 1198–1202.

382 Carvalho, J.F.S., Silva, M.M.C., and Sa e Melo, M.L. (2009) *Tetrahedron*, **65**, 2773–2781.

383 James, A.P., Johnstone, R.A.W., McCarron, M., *et al.* (1998) *Chemical Communications*, 429–430.

384 Brinksma, J., De Boer, J.W., Hage, R., and Feringa, B.L. (2004) Manganese-Based Oxidation with Hydrogen Peroxide, in *Modern Oxidation Methods* (ed. J.E. Bäckvall), Wiley-VCH Verlag GmbH, Weinheim, pp. 295.

385 Muzart, J. (2007) *Journal of Molecular Catalysis A: Chemical*, **276**, 62–72.

386 Garcia-Bosch, I., Ribas, X., and Costas, M. (2009) *Advanced Synthesis and Catalysis*, **351**, 348–352.

387 Matsumoto, K., Sawada, Y., and Katsuki, T. (2008) *Pure and Applied Chemistry*, **80**, 1071–1077.

388 Arends, I.W.C.E. (2006) *Angewandte Chemie International Edition*, **45**, 6250–6252.

389 Terent'ev, A.O., Boyarinova, K.A., and Nikishin, G.I. (2008) *Russian Journal of General Chemistry*, **78**, 592–596.

390 Chen, Y. and Reymond, J. (1995) *Tetrahedron Letters*, **36**, 4015–4018.

391 Yao, H. and Richardson, D.E. (2000) *Journal of the American Chemical Society*, **122**, 3220–3221.

392 Majetich, G., Hicks, R., Sun, G., and McGill, P. (1998) *The Journal of Organic Chemistry*, **63**, 2564–2573.

393 Majetich, G. and Hicks, R. (1996) *Synlett*, 649–651.

394 Berkessel, A. (2008) *Angewandte Chemie International Edition*, **47**, 3677–3679.

395 Rozen, S., Bareket, Y., and Dayan, S. (1996) *Tetrahedron Letters*, **37**, 531–534.

396 Friesen, R.W. and Blouin, M. (1993) *The Journal of Organic Chemistry*, **58**, 1653–1654.

397 Iwahama, T., Sakaguchi, S., and Ishii, Y. (1999) *Chemical Communications*, 727–728.

398 Adolfson, H. (2004) Transition Metal-Catalyzed Epoxidation of Alkenes, in *Modern Oxidation Methods*, (ed. J.E. Bäckvall), Wiley-VCH Verlag GmbH, Weinheim, pp. 21.

399 Adolfsson, H. and Balan, D. (2006) Metal-catalyzed synthesis of epoxides, in *Aziridines and Epoxides in Organic Synthesis* (ed. A.K. Yudin), Wiley-VCH Verlag GmbH, Weinheim, p. 185.

400 Oyama, S.T. (2008) Rates, Kinetics, and Mechanisms of Epoxidation: Homogeneous, Heterogeneous, and Biological Routes, in *Mechanisms in Homogeneous and Heterogeneous Epoxidation Catalysis* (ed. S.T. Oyama), Elsevier, Oxford, p. 3.

401 Matsumoto, K. and Katsuki, T. (2008) Asymmetric epoxidation of non-activated olefins, in *Asymmetric Synthesis: the Essentials* (eds. M. Christmann and S. Bräse), Wiley-VCH Verlag GmbH, Weinheim, p. 123.

402 McGarrigle, E.M. and Gilheany, D.G. (2005) *Chemical Reviews*, **105**, 1563–1602.

403 Pietikainen, P. (1995) *Tetrahedron Letters*, **36**, 319–322.

404 Haas, G.R. and Kolis, J.W. (1998) *Tetrahedron Letters*, **39**, 5923–5926.

405 Larrow, J.F., Jacobsen, E.N., Gao, Y., *et al.* (1994) *The Journal of Organic Chemistry*, **59**, 1939–1942.

406 Cepanec, I., Mikuldas, H., and Vinkovic, V. (2001) *Synthetic Communications*, 2913–2919.

407 Deng, L. and Jacobsen, E.N. (1992) *The Journal of Organic Chemistry*, **57**, 4320–4323.

408 Chang, S., Lee, N.H., and Jacobsen, E.N. (1993) *The Journal of Organic Chemistry*, **58**, 6939–6941.

409 Brandes, B.D. and Jacobsen, E.N. (1994) *The Journal of Organic Chemistry*, **59**, 4378–4380.

410 Pietikainen, P. (1994) *Tetrahedron Letters*, **35**, 941–944.

411 Adam, W., Jeko, J., Levai, A., *et al.* (1995) *Tetrahedron Letters*, **36**, 3669–3672.

412 Pietikainen, P. (1999) *Tetrahedron Letters*, **40**, 1001–1004.

413 Palucki, M., Pospisil, P.J., Zhang, W., and Jacobsen, E.N. (1994) *Journal of the American Chemical Society*, **116**, 9333–9334.

414 Mikame, D., Hamada, T., Irie, R., and Katsuki, T. (1995) *Synlett*, 827–828.

415 Chang, S., Heid, R.M., and Jacobsen, E.N. (1994) *Tetrahedron Letters*, **35**, 669–672.

416 Palucki, M., McCormick, G.J., and Jacobsen, E.N. (1995) *Tetrahedron Letters*, **36**, 5457–5460.

417 Brandes, B.D. and Jacobsen, E.N. (1995) *Tetrahedron Letters*, **36**, 5123–5126.

418 Bousquet, C. and Gilheany, D.G. (1995) *Tetrahedron Letters*, **36**, 7739–7742.

419 Daly, A.M., Renehan, M.F., and Gilheany, D.G. (2001) *Organic Letters*, **3**, 663–666.

420 O'Mahony, C.P., McGarrigle, E.M., Renehan, M.F., *et al.* (2001) *Organic Letters*, **3**, 3435–3438.

421 Scheurer, A., Mosset, P., Spiegel, M., and Saalfrank, R.W. (1999) *Tetrahedron*, **55**, 1063–1078.

422 Bell, D., Davies, M.R., Finney, F.J.L., *et al.* (1996) *Tetrahedron Letters*, **37**, 3895–3898.

423 Sasaki, H., Irie, R., and Katsuki, T. (1994) *Synlett*, 356–358.

424 Irie, R., Hosoya, N., and Katsuki, T. (1994) *Synlett*, 255–256.

425 Hatayama, A., Hosoya, N., Irie, R., *et al.* (1992) *Synlett*, 407–409.

426 Pietikainen, P. (2000) *Tetrahedron*, **56**, 417–424.

427 Sasaki, H., Irie, R., and Katsuki, T. (1993) *Synlett*, 300–302.
428 Hamada, T., Irie, R., and Katsuki, T. (1994) *Synlett*, 479–481.
429 Hosoya, N., Hatayama, A., Irie, R., *et al.* (1994) *Tetrahedron*, **50**, 4311–4322.
430 Nishida, T., Miyafuji, A., Ito, Y.N., and Katsuki, T. (2000) *Tetrahedron Letters*, **41**, 7053–7058.
431 Wu, M., Wang, B., Wang, S., Xia, C., and Sun, W. (2009) *Organic Letters*, **11**, 3622–3625.
432 Schwenkreis, T. and Berkessel, A. (1993) *Tetrahedron Letters*, **34**, 4785–4788.
433 Mukaiyama, T., Yamada, T., Nagata, T., and Imagawa, K. (1993) *Chemistry Letters*, 327–330.
434 Zhao, S., Ortiz, P.R., Keys, B.A., and Davenport, K.G. (1996) *Tetrahedron Letters*, **37**, 2725–2728.
435 Martinez, A., Hemmert, C., Loup, C., Barre, G., and Meunier, B. (2006) *The Journal of Organic Chemistry*, **71**, 1449–1457.
436 Song, C.E., Roh, E.J., Yu, B.M., *et al.* (2000) *Chemical Communications*, 615–616.
437 Reger, T.S. and Janda, K.D. (2000) *Journal of the American Chemical Society*, **122**, 6929–6934.
438 Heckel, A.D.S. (2002) *Helvetica Chimica Acta*, **85**, 913–926.
439 Song, Y., Yao, X., Chen, H., *et al.* (2002) *Journal of the Chemical Society, Perkin Transactions 1*, 870–873.
440 Cavazzini, M., Manfredi, A., Montanari, F., *et al.* (2000) *Chemical Communications*, 2171–2172.
441 Song, C.E. and Roh, E.J. (2000) *Chemical Communications*, 837–838.
442 Kureshy, R.I., Khan, N.H., Abdi, S.H.R., *et al.* (2002) *Tetrahedron Letters*, **43**, 2665–2668.
443 Smith, K. and Liu, C.H. (2002) *Chemical Communications*, 886–887.
444 Ramon, D.J. and Yus, M. (2006) *Chemical Reviews*, **106**, 2126–2208.
445 Tanaka, S., Yamamoto, H., Nozaki, H., *et al.* (1974) *Journal of the American Chemical Society*, **96**, 5254–5255.
446 Trullinger, T.K., Qi, J., and Roush, W.R. (2006) *The Journal of Organic Chemistry*, **71**, 6915–6922.
447 Gao, Y., Klunder, J.M., Hanson, R.M., *et al.* (1987) *Journal of the American Chemical Society*, **109**, 5765–5780.
448 Crimmins, M.T. and Haley, M.W. (2006) *Organic Letters*, **8**, 4223–4225.
449 Hussain, M.M. and Walsh, P.J. (2008) *Accounts of Chemical Research*, **41**, 883–893.
450 Murase, N., Hoshino, Y., Oishi, M., and Yamamoto, H. (1999) *The Journal of Organic Chemistry*, **64**, 338–339.
451 Malkov, A.V., Czemerys, L., and Malyshev, D.A. (2009) *Journal of Organic Chemistry*, **74**, 3350–3355.
452 Lattanzi, A., Iannece, P., and Scettri, A. (2002) *Tetrahedron Letters*, **43**, 5629–5631.
453 Dembitsky, V.M. (2003) *Tetrahedron*, **59**, 4701–4720.
454 Stephenson, N.A. and Bell, A.T. (2005) *Journal of the American Chemical Society*, **127**, 8635–8643.
455 Stephenson, N.A. and Bell, A.T. (2006) *Inorganic Chemistry*, **45**, 5591–5599.
456 Rose, E., Andrioletti, B., Zrig, S., and Quelquejeu-Etheve, M. (2005) *Chemical Society Reviews*, **105**, 573–583.
457 Rose, E., Ren, Q.-Z., and Andrioletti, B. (2004) *Chemistry – A European Journal*, 224–230.
458 Comba, P. and Rajaraman, G. (2008) *Inorganic Chemistry*, **47**, 78–93.
459 Liu, C., Yu, W., Che, C., and Yeung, C. (1999) *The Journal of Organic Chemistry*, **64**, 7365–7374.
460 Funyu, S., Isobe, T., Takagi, S., *et al.* (2003) *Journal of the American Chemical Society*, **125**, 5734–5740.
461 Brule, E. and de Miguel, Y.R. (2006) *Organic and Biomolecular Chemistry*, 599–609.
462 Mizuno, N., Yamaguchi, K., and Kamata, K. (2005) *Coordination Chemistry Reviews*, **249**, 1944–1956.
463 Adam, W., Alsters, P.L., Neumann, R., *et al.* (2003) *The Journal of Organic Chemistry*, **68**, 1721–1728.
464 Liu, P., Wang, C., and Li, C. (2009) *Journal of Catalysis*, **262**, 159–168.
465 Adam, W., Alsters, P.L., Neumann, R., *et al.* (2003) *Organic Letters*, **5**, 725–728.

466 Adam, W., Mitchell, C.M., Paredes, R., *et al.* (1997) *Liebigs Annalen*, 1365–1369.

467 Adam, W., Mitchell, C.M., and Saha-Moller, C.R. (1999) *The Journal of Organic Chemistry*, **64**, 3699–3707.

468 van Vliet, M.C.A., Arends, I.W.C.E., and Sheldon, R.A. (1999) *Chemical Communications*, 821–822.

469 Boehlow, T.R. and Spilling, C.D. (1996) *Tetrahedron Letters*, **37**, 2717–2720.

470 Iskra, J., Bonnet-Delpon, D., and Begue, J. (2002) *Tetrahedron Letters*, **43**, 1001–1003.

471 Saladino, R., Neri, V., Pelliccia, A.R., and Mincione, E. (2003) *Tetrahedron*, **59**, 7403–7408.

472 Lambert, R.M., Williams, F.J., Cropley, R.L., and Palermo, A. (2005) *Journal of Molecular Catalysis A: Chemical*, **228**, 27–33.

473 Yamaguchi, K., Ebitani, K., and Kaneda, K. (1999) *The Journal of Organic Chemistry*, **64**, 2966–2968.

474 Angelescu, E., Ionescu, R., Pavel, O.D., Zavoianu, R., Birjega, R., Luculescu, C.R., Florea, M., and Olar, R. (2009) *Journal of Molecular Catalysis A: Chemical*, **315**, 178–186.

475 Hosseini Monfared, H., Mohajeri, A., Morsali, A., and Janiak, C. (2009) *Monatshefte fur Chemie*, **140**, 1437–1445.

476 Khavasi, H.R., Sasan, K., Pirouzmand, M., and Ebrahimi, S.N. (2009) *Inorganic Chemistry*, **48**, 5593–5595.

477 Sato, K., Aoki, M., Ogawa, M., *et al.* (1997) *Bulletin of the Chemical Society of Japan*, **70**, 905–915.

478 Yamada, Y.M.A., Ichinohe, M., Takahashi, H., and Ikegami, S. (2001) *Organic Letters*, **3**, 1837–1840.

479 Pescarmona, P.P., van der Waal, J.C., Maxwell, I.E., and Maschmeyer, T. (2001) *Angewandte Chemie, International Edition in English*, **41**, 740–743.

480 Tong, K., Wong, K., and Chan, T.H. (2003) *Organic Letters*, **5**, 3423–3425.

481 Grivani, G., Tangestaninejad, S., Habibi, M.H., *et al.* (2006) *Applied Catalysis A: General*, **299**, 131–136.

482 Baqi, Y., Giroux, S., and Corey, E.J. (2009) *Organic Letters*, **11**, 959–961.

483 Diez, D., Nunez, M.G., Anton, A.B., Garcia, P., Moro, R.F., Garrido, N.M., Marcos, I.S., Basabe, P., and Urones, J.G. (2008) *Current Organic Synthesis*, **5**, 186–216.

484 Wang, B., Kang, Y., Yang, L., and Suo, J. (2003) *Journal of Molecular Catalysis A: Chemical*, **203**, 29–36.

485 Fraile, J.M., Garcia, J.I., Mayoral, J.A., *et al.* (2001) *Green Chemistry*, **3**, 271–274.

486 Pillai, U.R., Sahle-Demessie, E., and Varma, R.S. (2003) *Synthetic Communications*, 2017–2027.

487 Genski, T., Macdonald, G., Wei, X., *et al.* (1999) *Synlett*, 795–797.

488 Yadav, V.K. and Kapoor, K.K. (1994) *Tetrahedron Letters*, **35**, 9481–9484.

489 Kandzia, C. and Steckhan, E. (1994) *Tetrahedron Letters*, **35**, 3695–3698.

490 McQuaid, K.M. and Pettus, T.R.R. (2004) *Synlett*, 2403–2405.

491 Lygo, B. and To, D.C.M. (2001) *Tetrahedron Letters*, **42**, 1343–1346.

492 Arai, S., Tsuge, H., Oku, M., *et al.* (2002) *Tetrahedron*, **58**, 1623–1630.

493 Kim, D.Y., Choi, Y.J., Park, H.Y., *et al.* (2003) *Synthetic Communications*, 435–443.

494 Russo, A. and Lattanzi, A. (2009) *Synthesis*, 1551–1556.

495 Russo, A. and Lattanzi, A. (2008) *European Journal of Organic Chemistry*, 2767–2773, S2767/1-S2767/15.

496 Kelly, D.R. and Roberts, S.M. (2006) *Biopolymers – Peptide Science Section*, **84**, 74–89.

497 Tsogoeva, S.B., Woltinger, J., Jost, C., *et al.* (2002) *Synlett*, 707–710.

498 Chen, R., Qian, C., and de Vries, J.G. (2001) *Tetrahedron Letters*, **42**, 6919–6921.

499 Nemoto, T., Ohshima, T., Yamaguchi, K., and Shibasaki, M. (2001) *Journal of the American Chemical Society*, **123**, 2725–2732.

500 Nemoto, T., Ohshima, T., and Shibasaki, M. (2003) *Tetrahedron*, **59**, 6889–6897.

501 Yu, H., Zheng, X., Lin, Z., *et al.* (1999) *The Journal of Organic Chemistry*, **64**, 8149–8155.

502 Tanaka, Y., Nishimura, K., and Tomioka, K. (2003) *Tetrahedron*, **59**, 4549–4556.

503 Adam, W., Rao, P.B., Degen, H., and Saha-Moller, C.R. (2000) *Journal of the*

American Chemical Society, **122**, 5654–5655.
504 Bentley, P.A., Bickley, J.F., Roberts, S.M., and Steiner, A. (2001) *Tetrahedron Letters*, **42**, 3741–3743.
505 Wu, X., She, X., and Shi, Y. (2002) *Journal of the American Chemical Society*, **124**, 8792–8793.
506 Murphy, A., Dubois, G., and Stack, T.D.P. (2003) *Journal of the American Chemical Society*, **125**, 5250–5251.
507 Meth-Cohn, O., Williams, D.J., and Chen, Y. (2000) *Chemical Communications*, 495–496.
508 Adam, W., Pastor, A., Peters, K., and Peters, E. (2000) *Organic Letters*, **2**, 1019–1022.
509 de la Pradilla, F.R., Buergo, M.V., Manzano, P., *et al.* (2003) *The Journal of Organic Chemistry*, **68**, 4797–4805.
510 Concellon, J.M., Cuervo, H., and Fernandez-Fano, R. (2001) *Tetrahedron*, **57**, 8983–8987.
511 Corey, E.J. and Chaykovsky, M. (1962) *Journal of the American Chemical Society*, **84**, 867–868.
512 Brière, J.-F. Metzner, P. (2008) *Synthesis and Use of Chiral Sulfur Ylides in Organosulfur Chemistry in Asymmetric Synthesis* (eds. T. Toru and C. Bolm), Wiley-VCH Verlag GmbH, Weinheim, p. 179.
513 Ciaccio, J.A., Drahus, A.L., Meis, R.M., *et al.* (2003) *Synthetic Communications*, 2135–2143.
514 Chandrasekhar, S., Narasihmulu, C., Jagadeshwar, V., and Venkatram Reddy, K. (2003) *Tetrahedron Letters*, **44**, 3629–3630.
515 Forbes, D.C., Standen, M.C., and Lewis, D.L. (2003) *Organic Letters*, **5**, 2283–2286.
516 Nishimura, Y., Shiraishi, T., and Yamaguchi, M. (2008) *Tetrahedron Letters*, **49**, 3492–3495.
517 Trofimov, A., Chernyak, N., and Gevorgyan, V. (2008) *Journal of the American Chemical Society*, **130**, 13538–13539.
518 Julienne, K., Metzner, P., and Henryon, V. (1999) *Chemical Communications*, 731–736.
519 Zanardi, J., Leriverend, C., Aubert, D., *et al.* (2001) *The Journal of Organic Chemistry*, **66**, 5620–5623.
520 Ishizaki, M. and Hoshino, O. (2002) *Heterocycles*, **57**, 1399–1402.
521 Winn, C.L., Bellenie, B.R., and Goodman, J.M. (2002) *Tetrahedron Letters*, **43**, 5427–5430.
522 Aggarwal, V.K., Charmant, J.P.H., Fuentes, D., *et al.* (2006) *Journal of the American Chemical Society*, **128**, 2105–2114.
523 Davies, H.M.L. and DeMeese, J. (2001) *Tetrahedron Letters*, **42**, 6803–6805.
524 Aggarwal, V.K., Patel, M., and Studley, J. (2002) *Chemical Communications*, 1514–1515.
525 Aggarwal, V.K., Alonso, E., Bae, I., *et al.* (2003) *Journal of the American Chemical Society*, **125**, 10926–10940.
526 Aggarwal, V.K., Bae, I., Lee, H., *et al.* (2003) *Angewandte Chemie, International Edition*, **43**, 3274–3278.
527 Wurtz, A. (1859) *Annalen der Chemie und Pharmacie*, 125–128.
528 Feske, B.D., Kaluzna, I.A., and Stewart, J.D. (2005) *The Journal of Organic Chemistry*, **70**, 9654–9657.
529 Yamaguchi, J., Toyoshima, M., Shoji, M., *et al.* (2006) *Angewandte Chemie, International Edition*, **45**, 789–793.
530 Kobayashi, S., Hori, M., Wang, G.X., and Hirama, M. (2006) *The Journal of Organic Chemistry*, **71**, 636–644.
531 Poessl, T.M., Kosjek, B., Ellmer, B., *et al.* (2005) *Advanced Synthesis & Catalysis*, **347**, 1827–1834.
532 Pastor, I.M. and Yus, M. (2005) *Current Organic Chemistry*, 1–29.
533 Shanmugam, P. and Miyashita, M. (2003) *Organic Letters*, **5**, 3265–3268.
534 Schneider, C. and Brauner, J. (2000) *Tetrahedron Letters*, **41**, 3043–3046.
535 Westermaier, M. and Mayr, H. (2008) *Chemistry–A European Journal*, **14**, 1638–1647.
536 Tabatabaeian, K., Mamaghani, M., Mahmoodi, N.O., and Khorshidi, A. (2008) *Tetrahedron Letters*, **49**, 1450–1454.
537 Posner, G.H., Maxwell, J.P., and Kahraman, M. (2003) *The Journal of Organic Chemistry*, **68**, 3049–3054.

538 Das, B., Laxminarayana, K., Krishnaiah, M., and Kumar, D.N. (2009) *Bioorganic and Medicinal Chemistry Letters*, **19**, 6396–6398.

539 Taylor, S.K. (2000) *Tetrahedron*, **41**, 1149–1163.

540 Lalic, G., Petrovski, Z., Galonic, D., *et al.* (2000) *Tetrahedron Letters*, **41**, 763–766.

541 Xue, S., Li, Y., Han, K., *et al.* (2002) *Organic Letters*, **4**, 905–907.

542 Alexakis, A., Vrancken, E., Mangeney, P., and Chemla, F. (2000) *Journal of the Chemical Society, Perkin Transactions 1*, 3352–3353.

543 Kumar, P. and Naidu, S.V. (2005) *The Journal of Organic Chemistry*, **70**, 4207–4210.

544 Ooi, T., Kagoshima, N., Ichikawa, H., and Maruoka, K. (1999) *Journal of the American Chemical Society*, **121**, 3328–3333.

545 Lee, T.W., Proudfoot, J.R., and Thomson, D.S. (2006) *Bioorganic & Medicinal Chemistry Letters*, **16**, 654–657.

546 Taber, D.F. and Zhang, Z. (2006) *The Journal of Organic Chemistry*, **71**, 926–933.

547 Bode, J.W. and Carreira, E.M. (2001) *The Journal of Organic Chemistry*, **71**, 6410–6424.

548 Mirmashhori, B., Azizi, N., and Saidi, M.R. (2006) *Journal of Molecular Catalysis A: Chemical*, **247**, 159–161.

549 Zhu, C., Yuan, F., Gu, W., and Pan, Y. (2003) *Chemical Communications*, 692–693.

550 Cossy, J., Bellosta, V., Hamoir, C., and Desmurs, J. (2002) *Tetrahedron Letters*, **43**, 7083–7086.

551 Pachon, D.L., Gamez, P., van Brussel, J.J.M., and Reedijk, J. (2003) *Tetrahedron Letters*, **44**, 6025–6027.

552 Cepanec, I., Litvic, M., Mikuldas, H., *et al.* (2003) *Tetrahedron*, **59**, 2435–2439.

553 Procopio, A., Gaspari, M., Nardi, M., Oliverio, M., and Rosati, O. (2008) *Tetrahedron Letters*, **49**, 2289–2293.

554 Azizi, N. and Saidi, M.R. (2005) *Organic Letters*, **7**, 3649–3651.

555 Abaee, M.S., Hamidi, V., and Mojtahedi, M.M. (2008) *Ultrasonics Sonochemistry*, **15**, 823–827.

556 Sekar, G. and Singh, V.K. (1999) *The Journal of Organic Chemistry*, **64**, 287–289.

557 Mancilla, G., Femenia-Rios, M., Macias-Sanchez, A.J., and Collado, I.G. (2008) *Tetrahedron*, **64**, 11732–11737.

558 Reddy, L.R., Reddy, M.A., Bhanumathi, N., and Rama Rao, K. (2000) *Synlett*, 339–340.

559 Swamy, N.R., Goud, T.V., Reddy, S.M., *et al.* (2004) *Synthetic Communications*, 727–734.

560 Ollevier, T. and Lavie-Compin, G. (2002) *Tetrahedron Letters*, **43**, 7891–7893.

561 Swamy, R.N., Kondaji, G., and Nagaiah, K. (2002) *Synthetic Communications*, 2307–2312.

562 Hosseini-Sarvari, M. (2008) *Acta Chimica Slovenica*, **55**, 440–447.

563 Bhanushali, M.J., Nandurkar, N.S., Bhor, M.D., and Bhanage, B.M. (2008) *Tetrahedron Letters*, **49**, 3672–3676.

564 Bordoloi, A., Hwang, Y.K., Hwang, J., and Halligudi, S.B. (2009) *Catalysis Communications*, **10**, 1398–1403.

565 Cho, C.S., Kim, J.H., Choi, H., *et al.* (2003) *Tetrahedron Letters*, **44**, 2975–2977.

566 Gao, B., Wen, Y., Yang, Z., Huang, X., Liu, X., and Feng, X. (2008) *Advanced Synthesis and Catalysis*, **350**, 385–390.

567 Bhaumik, K., Mali, U.W., and Akamanchi, K.G. (2003) *Synthetic Communications*, 1603–1610.

568 Kazemi, F., Kiasat, A.R., and Ebrahimi, S. (2003) *Synthetic Communications*, 999–1004.

569 Chini, M., Crotti, P., and Macchia, F. (1990) *Tetrahedron Letters*, **31**, 5641–5644.

570 Davis, C.E., Bailey, J.L., Lockner, J.W., Coates, R.M. (2003) *The Journal of Organic Chemistry*, **68**, 75–82.

571 Iranpoor, N. and Kazemi, F. (1999) *Synthetic Communications*, 561–566.

572 Schaus, S.E., Larrow, J.F., and Jacobsen, E.N. (1997) *The Journal of Organic Chemistry*, **62**, 4197–4199.

573 Brandes, B.D. and Jacobsen, E.N. (2001) *Synlett, SPEC. ISS*, 1013–1015.

574 Kas'yan, L.I., Kas'yan, A.O., and Okovityi, S.I. (2006) *Russian Journal of Organic Chemistry*, **42**, 307–337.

575 Whalen, D.L. (2005) *Advances in Physical Organic Chemistry*, **40**, 247–298.

576 Fan, R. and Hou, X. (2003) *Organic and Biomolecular Chemistry*, 1565–1567.

577 Wang, Z., Cui, Y., Xu, Z., and Qu, J. (2008) *Journal of Organic Chemistry*, **73**, 2270–2274.

578 Schaus, S.E., Brandes, B.D., Larrow, J.F., *et al.* (2002) *Journal of the American Chemical Society*, **124**, 1307–1315.

579 Salehi, P., Seddighi, B., Irandoost, M., and Behbahani, K.F. (2000) *Synthetic Communications*, 2967–2973.

580 Tangestaninejad, S., Moghadam, M., Mirkhani, V., *et al.* (2006) *Monatshefte für Chemie*, **137**, 235–242.

581 Iranpoor, N. and Zeynizadeh, B. (1999) *Synthetic Communications*, 1017–1024.

582 Jeyakumar, K. and Chand, D.K. (2008) *Synthesis*, 807–819.

583 Leitao, A.J.L., Salvador, J.A.R., Pinto, R.M.A., and Sa e Melo, M.L. (2008) *Tetrahedron Letters*, **49**, 1694–1697.

584 Liu, Y., Liu, Q., and Zhang, Z. (2008) *Journal of Molecular Catalysis A: Chemical*, **296**, 42–46.

585 Barluenga, J., Vazquez-Villa, H., Ballesteros, A., and Gonzalez, J.M. (2002) *Organic Letters*, **4**, 2817–2819.

586 Tamami, B., Iranpoor, N., and Mahdavi, H. (2002) *Synthetic Communications*, 1251–1258.

587 Liu, Y., Zhang, Z., and Li, T. (2008) *Synthesis*, 3314–3318.

588 Nahmany, M. and Melman, A. (2005) *Tetrahedron*, **61**, 7481–7488.

589 Bukowska, A. and Bukowski, W. (2002) *Organic Process Research & Development*, **6**, 234–237.

590 Cavdar, H. and Saracoglu, N. (2008) *European Journal of Organic Chemistry*, 4615–4621.

591 Volkova, Y.A., Ivanova, O.A., Budynina, E.M., Averina, E.B., Kuznetsova, T.S., and Zefirov, N.S. (2008) *Tetrahedron Letters*, **49**, 3935–3938.

592 Salomatina, O.V., Yarovaya, O.I., and Barkhash, V.A. (2005) *Russian Journal of Organic Chemistry*, **41**, 155–185.

593 Moilanen, S.B., Potuzak, J.S., and Tan, D.S. (2006) *Journal of the American Chemical Society*, **128**, 1792–1793.

594 Simpson, G.L., Heffron, T.P., Merino, E., and Jamison, T.F. (2006) *Journal of the American Chemical Society*, **128**, 1056–1057.

595 Furuta, H., Takase, T., Hayashi, H., *et al.* (2003) *Tetrahedron*, **59**, 9767–9777.

596 Chapelat, J., Buss, A., Chougnet, A., and Woggon, W. (2008) *Organic Letters*, **10**, 5123–5126.

597 Tamami, B. and Mahdavi, H. (2002) *Tetrahedron Letters*, **43**, 6225–6228.

598 Narayana Murthy, S., Madhav, B., Prakash Reddy, V., Rama Rao, K., and Nageswar, Y.V.D. (2009) *Tetrahedron Letters*, **50**, 5009–5011.

599 Bellomo, A. and Gonzalez, D. (2006) *Tetrahedron: Asymmetry*, **17**, 474–478.

600 Kesavan, V., Bonnet-Delpon, D., and Begue, J. (2000) *Tetrahedron Letters*, **41**, 2895–2898.

601 Mojtahedi, M.M., Abassi, H., Saeed Abaee, M., and Mohebali, B. (2006) *Monatshefte fur Chemie*, **137**, 455–458.

602 Azizi, N. and Saidi, M.R. (2006) *Catalysis Communications*, 224–227.

603 Mojtahedi, M.M., Ghasemi, M.H., Saeed Abaee, M., and Bolourtchian, M. (2005) *Arkivoc*, 68–73.

604 Mukherjee, C., Maiti, G.H., and Misra, A.K. (2008) *Arkivoc*, 46–55.

605 Chen, J., Wu, H., Jin, C., *et al.* (2006) *Green Chemistry*, **8**, 330–332.

606 Guo, W., Chen, J., Wu, D., Ding, J., Chen, F., and Wu, H. (2009) *Tetrahedron*, **65**, 5240–5243.

607 Sun, J., Yuan, F., Yang, M., Pan, Y., and Zhu, C. (2009) *Tetrahedron Letters*, **50**, 548–551.

608 Salomon, C.J. (2001) *Synlett*, 65–68.

609 Ludwig, J., Bovens, S., Brauch, C., *et al.* (2006) *Journal of Medicinal Chemistry*, **49**, 2611–2620.

610 Iranpoor, N., Firouzabadi, H., Chitsazi, M., and Ali Jafari, A. (2002) *Tetrahedron*, **58**, 7037–7042.

611 Reddy, M.A., Surendra, K., Bhanumathi, N., and Rao, K.R. (2002) *Tetrahedron*, **58**, 6003–6008.

612 Malkov, A.V., Gordon, M.R., Stoncius, S., Hussain, J., and Kocovsky, P. (2009) *Organic Letters*, **11**, 5390–5393.

613 Saravanan, P., DattaGupta, A., Bhuniya, D., and Singh Vinod, K. (1997) *Tetrahedron*, **53**, 1855–1860.

614 Iwasaki, J., Ito, H., Nakamura, M., and Iguchi, K. (2006) *Tetrahedron Letters*, **47**, 1483–1486.

615 Ho, T. and Chein, R. (2006) *Helvetica Chimica Acta*, **89**, 231–239.

616 Raptis, C., Garcia, H., and Stratakis, M. (2009) *Angewandte Chemie International Edition*, **48**, 3133–3136, S3133/1-S3133/28.

617 Bertilsson, S.K. and Andersson, P.G. (2002) *Tetrahedron*, **58**, 4665–4668.

618 Bertilsson, S.K., Sodergren, M.J., and Andersson, P.G. (2002) *The Journal of Organic Chemistry*, **67**, 1567–1573.

619 Bhuniya, D., DattaGupta, A., and Singh, V.K. (1996) *The Journal of Organic Chemistry*, **61**, 6108–6113.

620 Gayet, A., Bertilsson, S., and Andersson, P.G. (2002) *Organic Letters*, **4**, 3777–3779.

621 Mordini, A., Valacchi, M., Pecchi, S., *et al.* (1996) *Tetrahedron Letters*, **37**, 5209–5212.

622 Suda, K., Baba, K., Nakajima, S., and Takanami, T. (1999) *Tetrahedron Letters*, **40**, 7243–7246.

623 Binder, C.M., Dixon, D.D., Almaraz, E., Tius, M.A., and Singaram, B. (2008) *Tetrahedron Letters*, **49**, 2764–2767.

624 Yadav, J.S., Reddy, B.V.S., and Satheesh, G. (2003) *Tetrahedron Letters*, **44**, 6501–6504.

625 Oh, B.K., Cha, J.H., Cho, Y.S., *et al.* (2003) *Tetrahedron Letters*, **44**, 2911–2913.

626 Bhatia, K.A., Eash, K.J., Leonard, N.M., *et al.* (2001) *Tetrahedron Letters*, **42**, 8129–8132.

627 Anderson, A.M., Blazek, J.M., Garg, P., *et al.* (2000) *Tetrahedron Letters*, **41**, 1527–1530.

628 Jung, M.E. and Marquez, R. (1999) *Tetrahedron Letters*, **40**, 3129–3132.

629 Pettersson, L. and Frejd, T. (2001) *Journal of the Chemical Society, Perkin Transactions 1*, 789–800.

630 Bickley, J.F., Hauer, B., Pena, P.C.A., *et al.* (2001) *Journal of the Chemical Society, Perkin Transactions 1*, 1253–1255.

631 Kita, Y., Furukawa, A., Futamura, J., *et al.* (2001) *Tetrahedron*, **57**, 815–825.

632 Murase, N., Maruoka, K., Ooi, T., and Yamamoto, H. (1997) *Bulletin of the Chemical Society of Japan*, **70**, 707–711.

633 Constantino, M.G., Donate, P.M., Frederico, D., *et al.* (2000) *Synthetic Communications*, 3327–3340.

634 Kita, Y., Kitagaki, S., Yoshida, Y., *et al.* (1997) *The Journal of Organic Chemistry*, **62**, 4991–4997.

635 Trost, B.M., Waser, J., and Meyer, A. (2008) *Journal of the American Chemical Society*, **130**, 16424–16434.

636 Baldwin, S.W., Chen, P., Nikolic, N., and Weinseimer, D.C. (2000) *Organic Letters*, **2**, 1193–1196.

637 Marson, C.M., Khan, A., Porter, R.A., and Cobb, A.J.A. (2002) *Tetrahedron Letters*, **43**, 6637–6640.

638 Bouyssi, D., Cavicchioli, M., Large, S., *et al.* (2000) *Synlett*, 749–751.

639 Shen, Y., Wang, B., and Shi, Y. (2006) *Angewandte Chemie, International Edition*, **46** 1429–1432.

640 Li, J.J. (2001) *Tetrahedron*, **57**, 1–24.

641 Kim, S. and Lee, S. (1991) *Tetrahedron Letters*, **32**, 6575–6578.

642 Kim, S., Lee, S., and Koh, J.S. (1991) *Journal of the American Chemical Society*, **113**, 5106–5107.

643 Mukaiyama, T., Arai, H., and Shiina, I. (2000) *Chemistry Letters*, 580–581.

644 Barrero, A.F. Quilez del Moral, Jose F., Sanchez, E.M., and Arteaga, J.F. (2006) *European Journal of Organic Chemistry*, 1627–1641.

645 Fernandez-Mateos, A., Buron, L.M., Clemente, R.R., *et al.* (2004) *Synlett*, 1011–1014.

646 Fernandez-Mateos, A., Madrazo, S.E., Teijon, P.H., and Gonzalez, R.R. (2009) *Journal of Organic Chemistry*, **74**, 3913–3918.

647 Banerjee, B. and Roy, S.C. (2006) *European Journal of Organic Chemistry*, 489–497.

648 Cha, J.S. (2007) *Bulletin of the Korean Chemical Society*, **28**, 2162–2190.

649 Kas'yan, L.I., Kas'yan, A.O., and Golodaeva, E.A. (2008) *Russian Journal of Organic Chemistry*, **44**, 153–183.

650 RajanBabu, T.V. and Nugent, W.A. (1994) *Journal of the American Chemical Society*, **116**, 986–997.

651 Verqhese, J.P., Sudalai, A., and Iyer, S. (1995) *Synthetic Communications*, 2267–2273.

652 Dragovich, P.S., Prins, T.J., and Zhou, R. (1995) *The Journal of Organic Chemistry*, **60**, 4922–4924.

653 Gruttadauria, M., Noto, R., and Riela, S. (1998) *Journal of Heterocyclic Chemistry*, **35**, 865–869.

654 Gatti, F.G. (2008) *Tetrahedron Letters*, **49**, 4997–4998.

655 Santosh Laxmi, Y.R. and Iyengar, D.S. (1997) *Synthetic Communications*, 1731–1736.

656 Atagi, L.M., Over, D.E., McAlister, D.R., and Mayer, J.M. (1991) *Journal of the American Chemical Society*, **113**, 870–874.

657 Dittmer, D.C., Zhang, Y., and Discordia, R.P. (1994) *The Journal of Organic Chemistry*, **59**, 1004–1010.

658a Paryzek, Z. and Wydra, R. (1984) *Tetrahedron Letters*, **25**, 2601–2604.

658b dos Santos, R.B., Brocksom, T.J., and Brocksom, U. (1997) *Tetrahedron Letters*, **38**, 745–748.

659 Patra, A., Bandyopadhyay, M., and Mal, D. (2003) *Tetrahedron Letters*, **44**, 2355–2357.

660 Hardouin, C., Doris, E., Rousseau, B., and Mioskowski, C. (2002) *The Journal of Organic Chemistry*, **67**, 6571–6574.

661 Hardouin, C., Burgaud, L., Valleix, A., and Doris, E. (2003) *Tetrahedron Letters*, **44**, 435–437.

662 Hodgson, D.M., Humphreys, P.G., and Hughes, S.P. (2007) *Pure and Applied Chemistry*, **79**, 269–279.

663 Capriati, V., Florio, S., and Luisi, R. (2008) *Chemical Reviews*, **108**, 1918–1942.

664 Mori, Y., Nogami, K., Hayashi, H., and Noyori, R. (2003) *The Journal of Organic Chemistry*, **68**, 9050–9060.

665 Mori, Y., Yaegashi, K., and Furukawa, H. (1998) *The Journal of Organic Chemistry*, **63**, 6200–6209.

666 Mori, Y., Yaegashi, K., and Furukawa, H. (1996) *Journal of the American Chemical Society*, **118**, 8158–8159.

667 Mori, Y., Yaegashi, K., Iwase, K., *et al.* (1996) *Tetrahedron Letters*, **37**, 2605–2608.

668 Mori, Y., Yaegashi, K., and Furukawa, H. (1997) *Journal of the American Chemical Society*, **119**, 4557–4558.

669 Abbotto, A., Capriati, V., Degennaro, L., *et al.* (2001) *The Journal of Organic Chemistry*, **66**, 3049–3058.

670 Hodgson, D.M., Reynolds, N.J., and Coote, S.J. (2002) *Tetrahedron Letters*, **43**, 7895–7897.

671 Hodgson, D.M. and Norsikian, S.L.M. (2001) *Organic Letters*, **3**, 461–463.

672 Hodgson, D.M., Reynolds, N.J., and Coote, S.J. (2004) *Organic Letters*, **6**, 4187–4189.

673 Hodgson, D.M., Kirton, E.H.M., Miles, S.M., *et al.* (2005) *Organic and Biomolecular Chemistry*, **3**, 1893–1904.

674 Dechoux, L., Doris, E., and Mioskowski, C. (1996) *Chemical Communications*, 549–550.

675 Agami, C., Dechoux, L., Doris, E., and Mioskowski, C. (1997) *Tetrahedron Letters*, **38**, 4071–4074.

676 Hodgson, D.M., Chung, Y.K., and Paris, J. (2004) *Journal of the American Chemical Society*, **126**, 8664–8665.

677 Hodgson, D.M. and Lee, G.P. (1996) *Chemical Communications*, 1015–1016.

678 Sander, M. (1966) *Chemical Reviews*, **66**, 297–339.

679 Uda, Y., Kurata, T., and Arakawa, N. (1986) *Agricultural and Biological Chemistry*, **50**, 2741–2746.

680 Peppard, T.L., Sharpe, F.R., and Elvidge, J.A. (1980) *Journal of the Chemical Society, Perkin Transactions 1*, 311–313.

681 Schmitz, F.J., Prasad, R.S., and Gopichand, Y. (1981) *Journal of the American Chemical Society*, **103**, 2467–2469.

682 Holmes, C.F.B., Luu, H.A., Carrier, F., and Schmitz, F.J. (1990) *FEBS Letters*, **270**, 216–218.

683 Ikejiri, M., Bernardo, M.M., Merouch, S.O., *et al.* (2005) *The Journal of Organic Chemistry*, **70**, 5709–5712.

684 Kiasat, A.R., Kazemi, F., and Jardi, M.F.M. (2004) *Phosphorus, Sulfur and Silicon and the Related Elements*, **179**, 1841–1844.

685 Iranpoor, N., Firouzabadi, H., and Jafari, A.A. (2005) *Phosphorus, Sulfur and Silicon and the Related Elements*, **180**, 1809–1814.

686 Borujeni, K.P. (2005) *Synthetic Communications*, 2575–2579.

687 Surendra, K., Krishnaveni, N.S., and Rao, K.R. (2004) *Tetrahedron Letters*, **45**, 6523–6526.

688 Bandgar, B.P., Joshi, N.S., and Kamble, V.T. (2006) *Tetrahedron Letters*, **47**, 4775–4777.

689 Salehi, P., Khodaei, M.M., Zolfigol, M.A., and Keyvan, A. (2003) *Synthetic Communications*, 3041–3048.

690 Tamami, B. and Kolahdoozan, M. (2004) *Tetrahedron Letters*, **45**, 1535–1537.

691 Kaboudin, B. and Norouzi, H. (2004) *Tetrahedron Letters*, **45**, 1283–1285.

692 Adam, W. and Bargon, R.M. (2004) *Chemical Reviews*, **104**, 251–262.

693 Adam, W. and Weinkotz, S. (1998) *Journal of the American Chemical Society*, **120**, 4861–4862.

694 Adam, W., Bargon, R.M., Schenk, W.A. (2003) *Journal of the American Chemical Society*, **125**, 3871–3876.

695 Adam, W.R.M.B. (2001) *European Journal of Organic Chemistry*, 1959–1962.

696 Yu, Z. and Wu, Y. (2003) *The Journal of Organic Chemistry*, **68**, 6049–6052.

697 Yadav, L.D.S. and Kapoor, R. (2002) *Synthesis*, 2344–2346.

698 Isac-Garcia, J., Calvo-Flores, F.G., Hernandez-Mateo, F., and Santoyo-González, F. (1999) *Chemistry – A European Journal*, 1512–1525.

699 Uenishi, J., Motoyama, M., Kimura, Y., and Yonemitsu, O. (1998) *Heterocycles*, **47**, 439–451.

700 Varela, O. and Zunszain, P.A. (1993) *The Journal of Organic Chemistry*, **58**, 7860–7864.

701 Zunszain, P.A. and Varela, O. (2000) *Tetrahedron Asymmetry*, 765–771.

702 Huang, J., Wang, F., Du, D., and Xu, J. (2005) *Synthesis*, 2122–2128.

703 Dong, Q., Fang, X., Schroeder, J.D., and Garvey, D.S. (1999) *Synthesis*, 1106–1108.

704 Huang, J., Du, D., and Xu, J. (2006) *Synthesis*, 315–319.

705 Branalt, J., Kvarnstrom, I., Svensson, S.C.T., *et al.* (1994) *The Journal of Organic Chemistry*, **59**, 4430–4432.

706 Silvestri, M.G. and Wong, C. (2001) *The Journal of Organic Chemistry*, **66**, 910–914.

707 Kameyama, A., Kiyota, M., and Nishikubo, T. (1994) *Tetrahedron Letters*, **34**, 4571–4574.

708 Jacob, J. and Espenson, J.H. (1999) *Chemical Communications*, 1003–1004.

709 Chen, C. and Chou, Y. (2000) *Journal of the American Chemical Society*, **122**, 7662–7672.

710 Bordwell, F.G., Andersen, H.M., and Pitt, B.M. (1954) *Journal of the American Chemical Society*, **76**, 1082–1085.

711 Uenishi, J. and Kubo, Y. (1994) *Tetrahedron Letters*, **34**, 6697–6700.

712 Kalaiselvan, A. and Venuvanalingam, P. (2006) *Journal of Molecular Structure: THEOCHEM*, **763**, 1–5.

713 Gessner, K.J. and Ball, D.W. (2005) *Journal of Molecular Structure: THEOCHEM*, **730**, 95–103.

714 Mannschreck, A., Radeglia, R., Gründemann, E., and Ohme, R. (1967) *Chemische Berichte*, **100**, 1778–1785.

715 Trapp, O., Schurig, V., and Kostyanovsky, R.G. (2004) *Chemistry – A European Journal*, 951–957.

716 Kostyanovsky, R.G., Malyshev, O.R., Lyssenko, K.A., *et al.* (2004) *Mendeleev Communications*, **14**, 315–318.

717 Mintas, M., Mannschreck, A., and Klasinc, L. (1981) *Tetrahedron*, **37**, 867–871.

718 Kostyanovsky, R.G., Shustov, G.V., and Nabiev, O.G. (1986) *Khimiko*

Farmatsevticheskii Zhurnal, **20**, 671–674.

719 Makhova, N.N., Petukhova, V.Y., and Kuznetsov, V.V. (2008) *Arkivoc*, 128–152.

720 Bronstert, K. (1987) DE 3607993 A1 19870917.

721 Kuznetsov, V.V., Makhova, N.N., and Khmel'nitskii, L.I. (1997) *Russian Chemical Bulletin*, **46**, 1354–1356.

722 Denisenko, S.N., Pasch, E., and Kaupp, G. (1989) *Angewandte Chemie, International Edition in English*, **29**, 1381–1383.

723 Berneth, H.S.H. (1980) *Chemische Berichte*, **113**, 2040–2042.

724 L'abbe, G., Verbruggen, A., Minami, T., and Toppet, S. (1981) *The Journal of Organic Chemistry*, **46**, 4478–4481.

725 Denisenko, S.N. (1998) *Mendeleev Communications*, **8**, 54–56.

726 Karin Briner, A.V. (1989) *Helvetica Chimica Acta*, **72**, 1371–1382.

727 Hwang, K., Yu, C., and Lee, I.Y. (1994) *Bulletin of the Korean Chemical Society*, 523–524.

728 Lyalin, B.V. and Petrosyan, V.A.E. (2002) *Russian Journal of Electrochemistry*, **38**, 1220–1227.

729 Helmut Quast, L.B. (1981) *Chemische Berichte*, **114**, 3253–3272.

730 Carboni, B., Toupet, L., and Carrie, R. (1987) *Tetrahedron*, **43**, 2293–2302.

731 Heine, H.W., Hoye, T.R., Williard, P.G., and Hoye, R.C. (1973) *The Journal of Organic Chemistry*, **38**, 2984–2988.

732 Schmitz, E. (1961) *Angewandte Chemie*, **73**, 220–221.

733 Schmitz, E. (1962) *Chemische Berichte*, **95**, 680–687.

734 Alper, H., Delledonne, D., Kameyama, M., and Roberto, D. (1990) *Organometallics*, **9**, 762–765.

735 Komatsu, M., Tamabuchi, S., Minakata, S., and Ohshiro, Y. (1999) *Heterocycles*, **50**, 67–70.

736 Komatsu, M., Kobayashi, M., Itoh, S., and Ohshiro, Y. (1993) *The Journal of Organic Chemistry*, **58**, 6620–6624.

737 Komatsu, M., Sakai, N., Hakotani, A., *et al.* (2000) **53**, *Heterocycles*, 541–544.

738 Paulsen, S.R. (1960) *Angewandte Chemie*, **72**, 781–782.

739 Schmitz, E. (1964) *Angewandte Chemie, International Edition in English*, **4**, 333–341.

740 Skancke, A. and Liebman, J.F. (1999) *The Journal of Organic Chemistry*, **64**, 6361–6365.

741 Puzzarini, C., Gambi, A., and Cazzoli, G. (2004) *Journal of Molecular Structure*, **695-696**, 203–210.

742 Knoll, W., Bobek, M.M., Giester, G., and Brinker, U.H. (2001) *Tetrahedron Letters*, **42**, 9161–9165.

743 Krois, D. and Brinker, U.H. (2001) *Synthesis*, 379–381.

744 Walter, M., Wiegand, M., and Lindhorst, T.K. (2006) *European Journal of Organic Chemistry*, 719–728.

745 Moss, R.A., Fu, X., and Sauers, R.R. (2005) *Canadian Journal of Chemistry*, **83**, 1228–1236.

746 Graham, W.H. (1965) *Journal of the American Chemical Society*, **87**, 4396–4397.

747 Pillion, D., Deraet, M., Holleran, B.J., and Escher, E. (2006) *Journal of Medicinal Chemistry*, **49**, 2200–2209.

748 Fernández-Gacio, A.M. (2002) *European Journal of Organic Chemistry*, 2529–2534.

749 Hatanaka, Y., Hashimoto, M., Kurihara, H., *et al.* (1994) *The Journal of Organic Chemistry*, **59**, 383–387.

750 Moss, R.A. (2006) *Accounts of Chemical Research*, **39**, 267–272.

751 Cox, D.P., Moss, R.A., and Terpinski, J. (1983) *Journal of the American Chemical Society*, **105**, 6513–6514.

752 Liu, M.T.H., Choe, Y., Kimura, M., *et al.* (2003) *The Journal of Organic Chemistry*, **68**, 7471–7478.

753 Moss, R.A. (1989) *Accounts of Chemical Research*, **22**, 15–21.

754 Martinu, T. and Dailey, W.P. (2004) *The Journal of Organic Chemistry*, **69**, 7359–7362.

755 Arenas, J.F., Lopez-Tocon, I., Otero, J.C., and Soto, J. (2002) *Journal of the American Chemical Society*, **124**, 1728–1735.

756 Mchedlidze, M.T., Sumbatyan, N.V., Bondar', D.A., *et al.* (2003) *Russian Journal of Bioorganic Chemistry*, **29**, 177–184.

757 Hashimoto, M. and Hatanaka, Y. (2006) *Analytical Biochemistry*, **348**, 154–156.

758 Bobek, M.M. and Brinker, U.H. (2000) *Journal of the American Chemical Society*, **122**, 7430–7431.

759 Rosenberg, M.G. and Brinker, U.H. (2003) *The Journal of Organic Chemistry*, **68**, 4819–4832.

760 Blencowe, A. and Hayes, W. (2005) *Soft Matter*, **1**, 178–205.

761 Blencowe, A., Cosstick, K., and Hayes, W. (2006) *New Journal of Chemistry*, **30**, 53–58.

762 Lewis, L.L., Turner, L.L., Salter, E.A., and Magers, D.H. (2002) *Journal of Molecular Structure: THEOCHEM*, **592**, 161–171.

763 Aminova, R.M. and Ermakova, E. (2002) *Chemical Physics Letters*, **359**, 184–190.

764 Peng, L., Chen, C., Gonzalez, C.R., and Balogh-Nair, V. (2002) *International Journal of Molecular Science*, **3**, 1145–1161.

765 Marchand-Brynaert, J., Bounkhala-Khrouz, Z., Vanlierde, H., and Ghosez, L. (1990) *Heterocycles*, **30**, 971–982.

766 Mlochowski, J., Kubicz, B., Kloc, K., *et al.* (1988) *Annalen Der Chemie-Justus Liebig*, 455–464.

767 Said, S.B., Mlochowski, J., and Skarzewski, J. (1990) *Annalen Der Chemie-Justus Liebig*, 461–464.

768 Iesce, M.R., Cermola, F., and Guitto, A. (1997) *Synthesis*, 657–660.

769 Schmitz, E., Ohme, R., and Murawski, D. (1965) *Chemische Berichte*, **98**, 2516–2524.

770 Brodsky, B.H. and Du Bois, J. (2005) *Journal of the American Chemical Society*, **127**, 15391–15393.

771 Ruano, J.L.G., Aleman, J., Fajardo, C., and Parra, A. (2005) *Organic Letters*, **7**, 5493–5496.

772 Kraiem, J., Othman, R.B., and Ben Hassine, B. (2004) *Comptes Rendus Chimie*, 1119–1126.

773 Schoumacker, S., Hamelin, O., Teti, S., *et al.* (2005) *The Journal of Organic Chemistry*, **70**, 301–308.

774 Mohajer, D., Iranpoor, N., and Rezaeifard, A. (2004) *Tetrahedron Letters*, **45**, 631–634.

775 Damavandi, J.A., Karami, B., and Zolfigol, M.A. (2002) *Synlett*, 933–934.

776 Lin, Y. and Miller, M.J. (2001) *The Journal of Organic Chemistry*, **66**, 8282–8285.

777 Colonna, S., Pironti, V., Drabowicz, J., *et al.* (2005) *European Journal of Organic Chemistry*, 1727–1730.

778 Mishra, J.K. (2005) *Synlett*, 543–544.

779 Adam, W., Saha-Moller, C.R., and Ganeshpure, P.A. (2001) *Chemical Reviews*, **101**, 3499–3548.

780 Li, J.J. (2007) in *Name Reactions for Functional Group Transformations*, John Wiley & Sons, Inc., Hoboken, p. 22.

781 Biscoe, M.R. and Breslow, R. (2005) *Journal of the American Chemical Society*, **127**, 10812–10813.

782 Armstrong, A. (2004) *Angewandte Chemie, International Edition*, **44**, 1460–1462.

783 Lacour, J., Monchaud, D., and Marsol, C. (2002) *Tetrahedron Letters*, **43**, 8257–8260.

784 Bohe, L. and Kammoun, M. (2004) *Tetrahedron Letters*, **45**, 747–751.

785 Armstrong, A., Edmonds, I.D., Swarbrick, M.E., and Treweeke, N.R. (2005) *Tetrahedron*, **61**, 8423–8442.

786 Andreae, S. and Schmitz, E. (1991) *Synthesis*, 327–341.

787 Andreae, S. and Schmitz, E. (1991) *Synthesis*, 327–341.

788 Washington, I., Houk, K.N., and Armstrong, A. (2003) *The Journal of Organic Chemistry*, **68**, 6497–6501.

789 Armstrong, A., Challinor, L., Cooke, R.S., *et al.* (2006) *The Journal of Organic Chemistry*, **71**, 4028–4030.

790 Armstrong, A., Jones, L.H., Knight, J.D., and Kelsey, R.D. (2005) *Organic Letters*, **7**, 713–716.

791 Hannachi, J., Vidal, J., Mulatier, J., and Collet, A. (2004) *The Journal of Organic Chemistry*, **69**, 2367–2373.

792 Armstrong, A. and Draffan, A.G. (1999) *Tetrahedron Letters*, **40**, 4453–4456.

793 Armstrong, A., Edmonds, I.D., and Swarbrick, M.E. (2005) *Tetrahedron Letters*, **46**, 2207–2210.

794 Kropf, J.E., Meigh, I.C., Bebbington, M.W.P., and Weinreb, S.M. (2006) *The Journal of Organic Chemistry*, **71**, 2046–2055.

795 Hynninen, P.H., Leppaekases, T.S., and Mesilaakso, M. (2006) *Tetrahedron*, **62**, 3412–3422.

796 Martin, C., Sandrinelli, F., Perrio, C., *et al.* (2006) *The Journal of Organic Chemistry*, **71**, 210–214.

797 Fernandez, I. and Khiar, N. (2003) *Chemical Reviews*, **103**, 3651–3705.

798 Aube, J. (1997) *Chemical Society Reviews*, **26**, 269–277.

799 Usuki, Y., Peng, X., Gulgeze, B., *et al.* (2006) *Arkivoc*, 189–199.

800 Di Gioia, M.L., Leggio, A., Le Pera, A., *et al.* (2005) *The Journal of Organic Chemistry*, **70**, 10494–10501.

801 Greer, A. (2003) *Science*, **302**, 235–236.

802 Bach, R.D. and Dmitrenko, O. (2002) *The Journal of Organic Chemistry*, **67**, 3884–3896.

803 Bach, R.D. and Dmitrenko, O. (2002) *The Journal of Organic Chemistry*, **67**, 2588–2599.

804 Freccero, M., Gandolfi, R., Sarzi-Amade, M., and Rastelli, A. (2003) *The Journal of Organic Chemistry*, **68**, 811–823.

805 Kerstin Schroeder, W.S. (2005) *European Journal of Organic Chemistry*, 496–504.

806 Zeller, K., Kowallik, M., and Schuler, P. (2005) *European Journal of Organic Chemistry*, 5151–5153.

807 Murray, R.W. and Singh, M. (1997) *Organic Syntheses*, 91–97.

808 Murray, R.W. and Sing, M. (1988) *Organic Syntheses Collective Vol. IX*, **9**, 288.

809 Duffy, R.J., Morris, K.A., and Romo, D. (2005) *Journal of the American Chemical Society*, **127**, 16754–16755.

810 Adam, W., Bialas, J., and Hadjiarapoglou, L. (1991) *Chemische Berichte*, **124**, 2377.

811 Broshears, W.C., Esteb, J.J., Richter, J., and Wilson, A.M. (2004) *Journal of Chemical Education*, **81**, 1018–1019.

812 Kachasakul, P., Assabumrungrat, S., Praserthdam, P., and Pancharoen, U. (2003) *Chemical Engineering Journal*, **92**, 131–139.

813 Murray, R.W. (1989) *Chemical Reviews*, **89**, 1187–1201.

814 Adam, W., Saha-Möller, C.R., and Zhao, C. (2002) *Organic Reactions*, **61**, 219–516.

815 Srivastava, V.P. (2008) *Synlett*, 626–627.

816 Curci, R., D'Accolti, L., and Fusco, C. (2006) *Accounts of Chemical Research*, **39**, 1–9.

817 Mello, R., Cassidei, L., Fiorentino, M., Fusco, C., and Curci, R. (1990) *Tetrahedron Letters*, **31**, 3067–3070.

818 D'Accolti, L., Dinoi, A., Fusco, C., *et al.* (2003) *The Journal of Organic Chemistry*, **68**, 7806–7810.

819 Gonzalez-Nunez, M.E., Royo, J., Mello, R., *et al.* (2005) *The Journal of Organic Chemistry*, **70**, 7919–7924.

820 Grabovskiy, S.A., Timerghazin, Q.K., and Kabal'nova, N.N. (2005) *Russian Chemical Bulletin*, 2384–2393.

821 Ferrer, M., Sanchez-Baeza, F., Casas, J., and Messeguer, A. (1994) *Tetrahedron Letters*, **34**, 2981–2984.

822 Wong, M., Chung, N., He, L., and Yang, D. (2003) *Journal of the American Chemical Society*, **125**, 158–162.

823 Wong, M., Chung, N., He, L., *et al.* (2003) *The Journal of Organic Chemistry*, **68**, 6321–6328.

824 Levai, A. (2003) *Archivoc*, 14–30.

825 Patonay, T., Adam, W., Levai, A., *et al.* (2001) *The Journal of Organic Chemistry*, **66**, 2275–2280.

826 Patonay, T., Adam, W., Jeko, J., *et al.* (1999) *Heterocycles*, **51**, 85–94.

827 Levai, A., Patonay, T., Toth, G., Kovacs, J., and Jeko, J. (2002) *Journal of Heterocyclic Chemistry*, **39**, 817–821.

828 Dieva, S.A., Eliseenkova, R.M., Efremov, Y.Y., *et al.* (2006) *Russian Journal of Organic Chemistry*, 12–16.

3
Four-Membered Heterocycles: Structure and Reactivity

Gérard Rousseau and Sylvie Robin

3.1
Azetidines

3.1.1
Introduction

This chapter deals with azetidines, which are four-membered rings containing one nitrogen atom. The particular case of azetidin-2-ones (β-lactams) is examined in Chapter 24. Different important reviews have already been reported on their preparation and reactivity [1–3].

Different natural products of terrestrial or marine origins having an azetidine ring have been isolated. We indicate in this chapter only some representative examples. The simple (*R*) and (*S*)-azetidine-2-carboxylic acids **1** were found in numerous plants [4, 5]. N-Alkyl derivatives such as nicotianamine (**2**) [6] (isolated from a culture of *Streptomyces*) and mugineic acid (**3**) (isolated from a plant) [7] have also been reported.

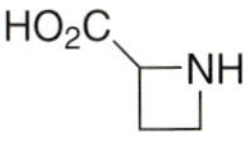

HO_2C NH **1** HO_2C N CO_2H N H H CO_2H NH_2 **2** HO_2C N CO_2H OH N H CO_2H **3**

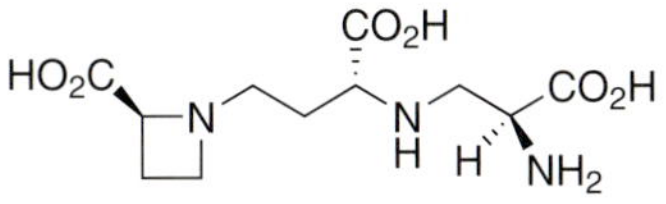

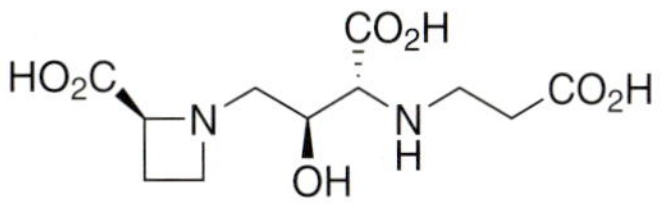

Some C-alkyl derivatives are known, such as penaresidine-B (**4**) isolated from various sponges [8] and *cis*-polyoximic acid (**5**) isolated from a culture of *Streptomyces* [9, 10].

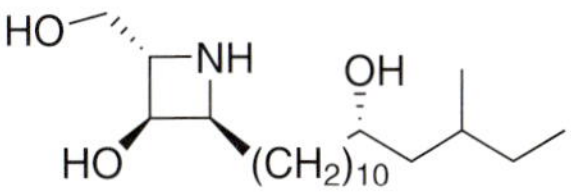

HO NH OH HO $(CH_2)_{10}$ **4** HO_2C NH Me **5**

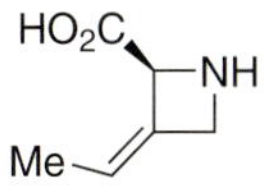

Modern Heterocyclic Chemistry, First Edition.
Edited by Julio Alvarez-Builla, Juan Jose Vaquero, and José Barluenga.

The nucleus azetidine is also present in more complex products, such as polyoxin A (**6**, isolated from culture broth of *Streptomyces cacaoi* var. *asoensis*) [11], gelsemoxonine (**7**) isolated from the creeper *Gelsemium elegans* and vioprolides A ($n = 1$) and B ($n = 2$) **8** [12], obtained by fermentation of a strain of the myxo bacterium *Cystobacter violaceus* [13].

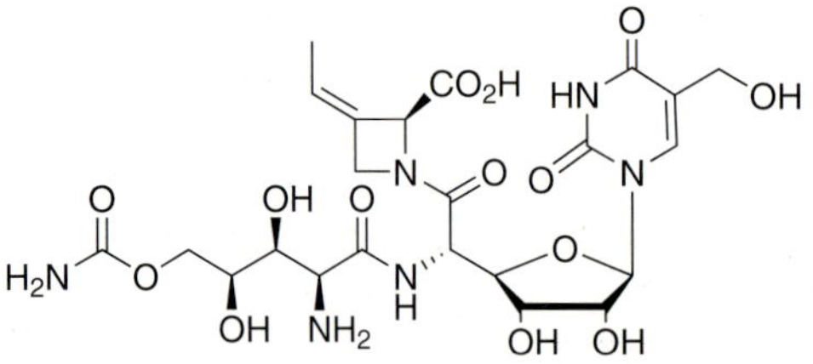

6 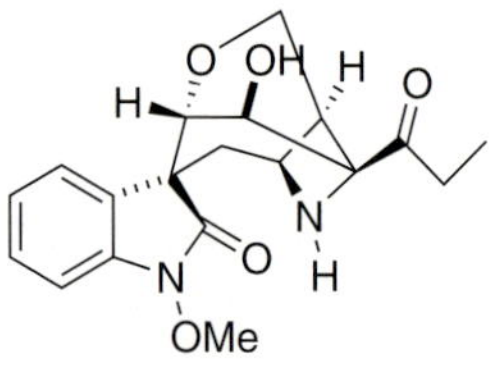 **7**

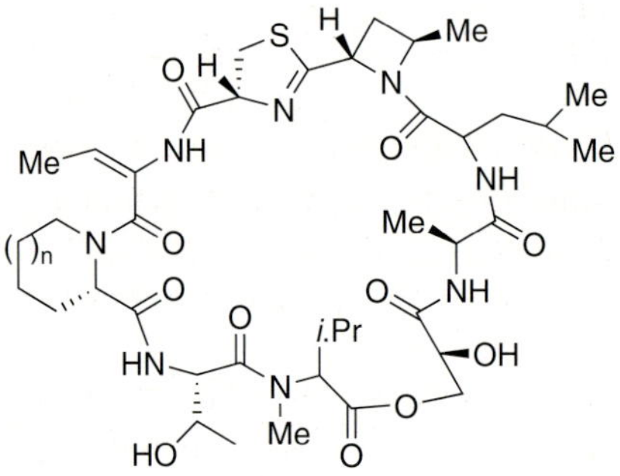

8

Agrochemical applications of azetidine derivatives have been reported [3]. Actions on the central nervous system with various activities (antiepileptic, anticonvulsant, antihypertensive, antidepressant, analgesic, etc.) have been also described [3]. A promising compound appears to be ABT-594 (**9**), synthesized from (+)-epibatidine (**10**), present in frogs [14]. Compound **9** proved to be more powerful than morphine as a painkiller [15].

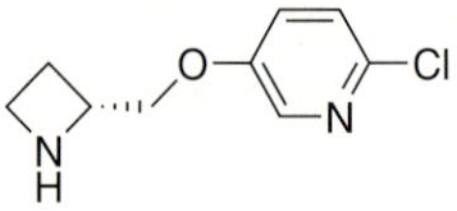

9 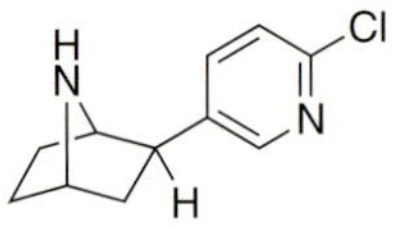 **10** 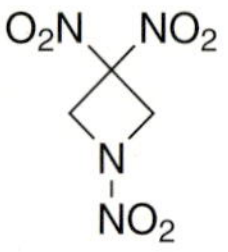 **11**

The preparation of 1,3,3-trinitroazetidine (**11**) has been reported [16]. This compound was evaluated as a substitute for TNT, due to its high thermal stability. However, its cost of preparation and its high volatility limit its utilization as explosive or propellant.

3.1.2 Physicochemical Data

Azetidines appear to be thermally stable. They are unreactive towards numerous reagents and can be prepared without special precautions. They can be analyzed using gas chromatography or liquid chromatography over silica gel, for example. Their reactivities appear closer to these of higher cyclic amines than aziridines. The strain in azetidines explains their difficult formation by cyclization, which is comparable to these of azepines [17]. Electron diffraction and X-ray crystallographic studies have shown the non-planarity of the ring. For azetidine, the angle between the planes formed by the CCC and CNC bonds is 37°, which is similar to that found for cyclobutane. Inversion of the pyramidal nitrogen is very easy for these heterocycles ($\Delta G^{\#} \approx 10\,\text{kcal mol}^{-1}$).

Azetidines have been studied by NMR spectroscopy. In the absence of substituents the chemical shift of the hydrogens fixed on the carbon at the 2-position are in the range 2.5–3.5 ppm and the hydrogens fixed on the carbon at the 3-position 1 ppm higher. In general the magnitude of the coupling constants for vicinal hydrogens is in the range 6–8 Hz for the cis stereoisomers and 2–4 Hz for the trans stereoisomers. With 3-hydroxyazetidine derivatives, the determination of the stereochemistries could be come very difficult. Structural identification was reported to be possible when the ^{1}H NMR spectra were recorded out in the presence of known amounts of Eu(Fod)$_3$ [18]. In ^{13}C NMR, the carbon shifts in 2,4 positions are close to those of carbons in five- and six-membered cyclic amines. Figure 3.1 shows some representative examples of these chemical shifts. The ^{15}N chemical shift of azetidine (from NH_3) was reported to 25.3 ppm. N-Substituted azetidines showed chemical shifts comparable to other nitrogen heterocycles [19].

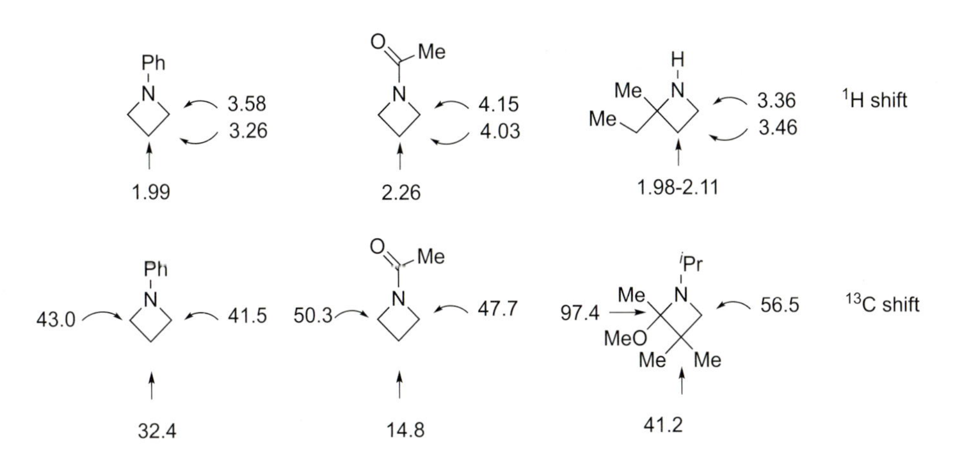

Figure 3.1 Representative examples of chemical shifts.

3.1.3 Synthesis

This chapter reports the main methods that lead to azetidines with acceptable yields. Some less synthetically useful methods can be found in previous reviews. These preparations can be divided into three groups: formation (i) by cyclization, (ii) from other heterocycles and (iii) by [2 + 2] cycloadditions.

3.1.3.1 Cyclization Reactions

3.1.3.1.1 Formation of a C–N Bond

Ring Closure of γ-Amino Derivatives A powerful and simple preparation of azetidines is the intramolecular displacement of a leaving group fixed on a carbon by a γ-amino group. Bromo, iodo, tosyloxy, mesyloxy and triflyloxy were mainly used as leaving groups, and alkyl, aryl and tosyl were used as protecting groups on the nitrogen atom (Scheme 3.1) [3].

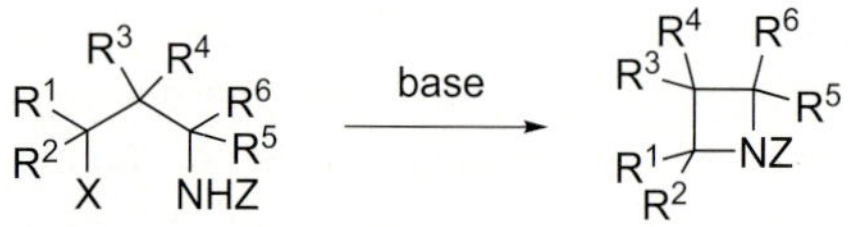

Scheme 3.1

This approach was reported as the key step in a synthesis of penaresidine B (**4**) (Scheme 3.2) [20]. It was subsequently reported that diastereoselective cyclizations of the mesylates lead to similar results (Scheme 3.3) [21].

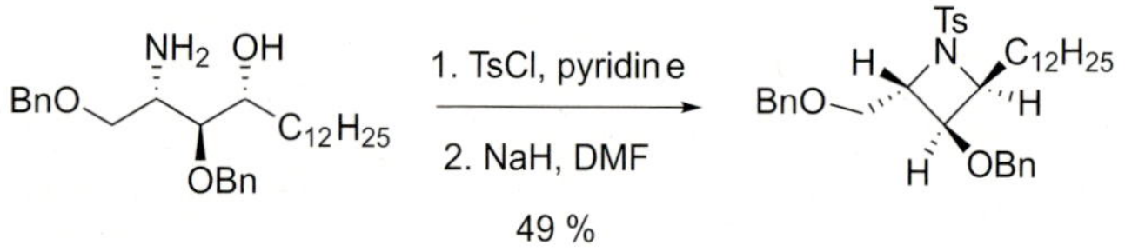

Scheme 3.2

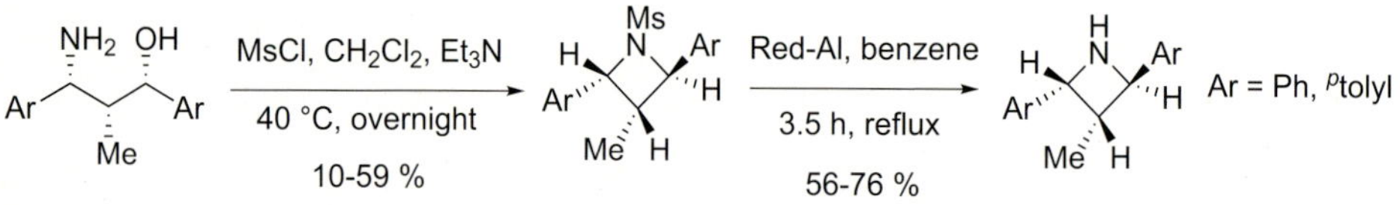

Scheme 3.3

A modification of this method has been investigated. It was shown that utilization of Mitsunobu conditions (X = OH in Scheme 3.1) could be very efficient [22]. This cyclization was applied to the preparation of substituted azetidines, without any epimerization of the chiral centers (Scheme 3.4) [23].

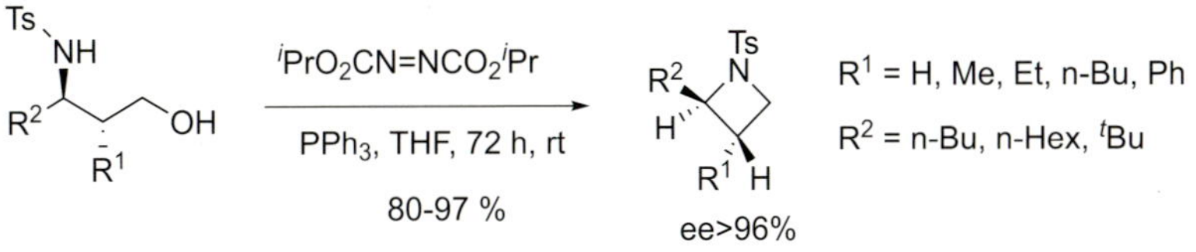

Scheme 3.4

Cyclization of protected 1-azidopropan-1-ols was also reported to lead to the formation of azetidines. This cyclization occurred when the azido group was reduced using Raney-nickel (Scheme 3.5) [24]. This method appears more efficient than when using PPh_3 as reducing agent [25].

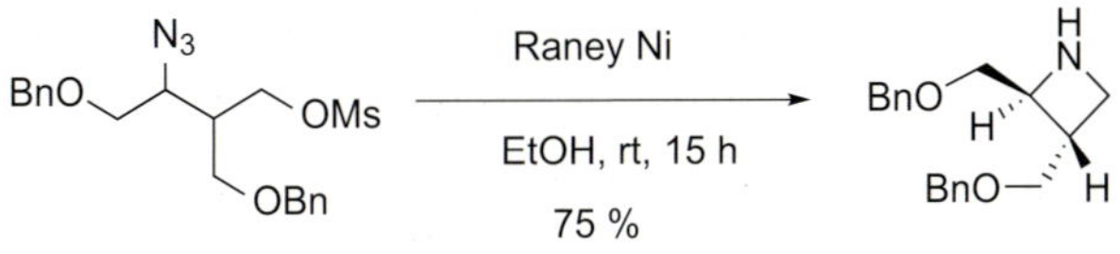

Scheme 3.5

This cyclization was found to be easy with 3-chloroalkylamines obtained by reaction of 1-aminoalkyl chloromethyl ketones with organocerium reagents. The cyclizations occurred during evaporation of the reaction solvents (Scheme 3.6) [26]. These isolable azetidinium salts were subsequently transformed into azetidines by catalytic hydrogenation. Chloromethyl ketones react with ester enolates in similar fashion (Scheme 3.7) [27].

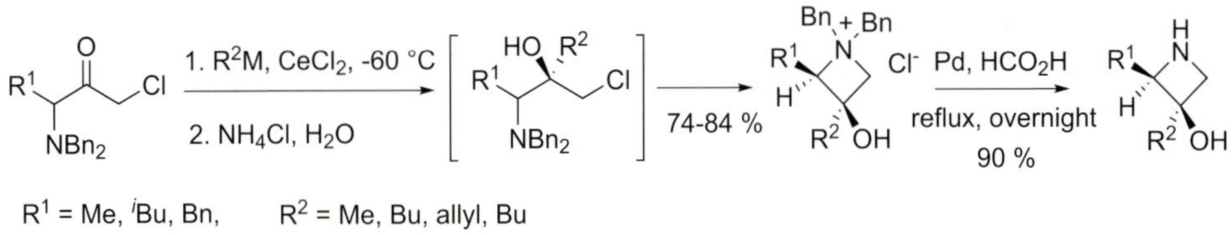

Scheme 3.6

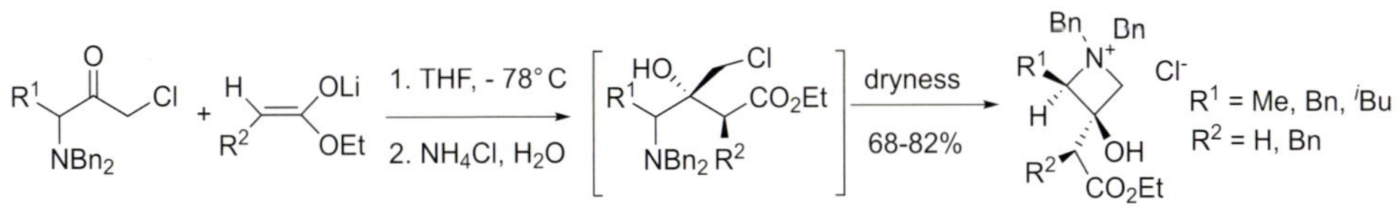

Scheme 3.7

Formation of 3-azetidinones from 1-aminoalkyl chloromethyl ketones was also reported to occur by reaction with tributyltin hydride (Scheme 3.8) [28].

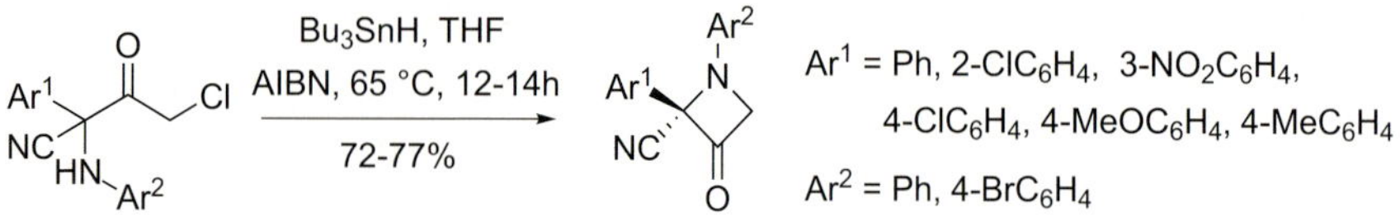

Bu_3SnH, THF
AIBN, 65 °C, 12-14h
72-77%

Ar^1 = Ph, 2-ClC_6H_4, 3-$NO_2C_6H_4$, 4-ClC_6H_4, 4-$MeOC_6H_4$, 4-MeC_6H_4
Ar^2 = Ph, 4-BrC_6H_4

Scheme 3.8

Another interesting preparation of 3-azetidinones involves carbenoid insertion into a N–H bond. 1-Aminoalkyl α-diazomethyl ketones reacts with rhodium acetate to give optically active 2-substituted 3-azetidinones (Scheme 3.9) [29]. These ketones allowed highly diastereoselective additions of nucleophilic reagents or Wittig reactions ($Ph_3P{=}CHCO_2Me$). Utilization of copper acetylacetonate as catalyst has been subsequently reported [30].

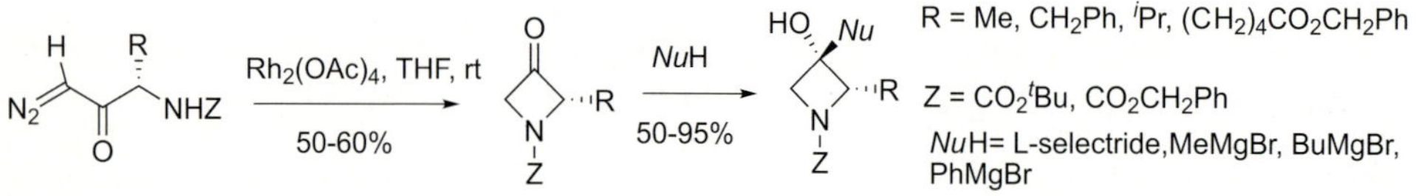

$Rh_2(OAc)_4$, THF, rt
50-60%
*Nu*H
50-95%

R = Me, CH_2Ph, ^{i}Pr, $(CH_2)_4CO_2CH_2Ph$
Z = $CO_2{}^tBu$, CO_2CH_2Ph
*Nu*H= L-selectride, MeMgBr, BuMgBr, PhMgBr

Scheme 3.9

Azetidines can be obtained by $MgBr_2$ catalyzed isomerization of aminomethyloxiranes. A mixture of diastereomers was generally obtained (Scheme 3.10). With the trans epoxide substituted by an alkyl group, only one azetidine was detected [31].

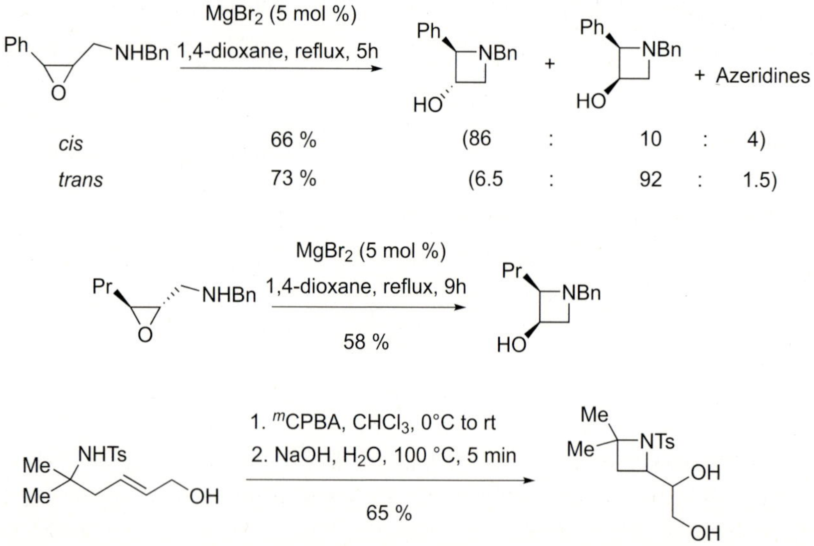

$MgBr_2$ (5 mol %)
1,4-dioxane, reflux, 5h

	Yield			+ Azeridines
cis	66 %	(86	10	4)
trans	73 %	(6.5	92	1.5)

$MgBr_2$ (5 mol %)
1,4-dioxane, reflux, 9h
58 %

1. ^{m}CPBA, $CHCl_3$, 0°C to rt
2. NaOH, H_2O, 100 °C, 5 min
65 %

Scheme 3.10

Utilization of sodium hydroxide to carry out ring opening of oxiranes was reported to be possible, if the epoxide function was β of the amino group (4-*exo* mode cyclization) (Scheme 3.10) [32].

Ring Closure of 1,3-Functionnalized Propane Derivatives with Amines 1,3-Dielectrophiles react with primary amines to afford azetidines (Scheme 3.11).

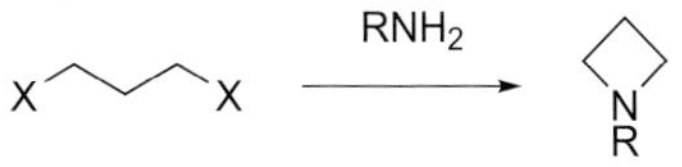

Scheme 3.11

1,3-Dihalopropane derivatives have been used with more or less success, due to the possible formation of side products [3]. However, satisfactory yields were obtained when the halogen atoms were in activated positions (Scheme 3.12) [33].

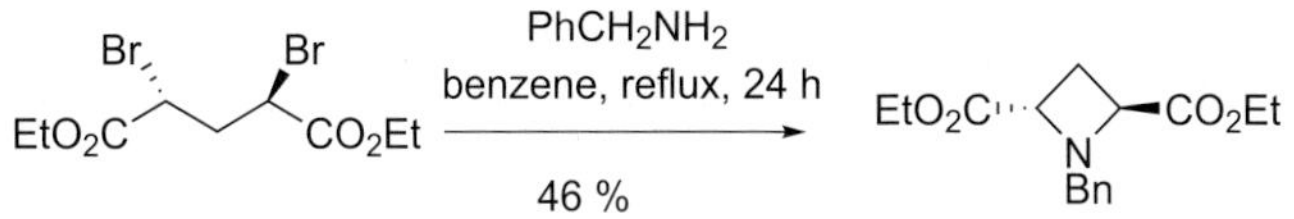

Scheme 3.12

Better yields were observed with 1,3-triflates (Scheme 3.13, reactions carried out a 1 kg scale) [34], 1,3-ditosylates [35] or 1,3-dimesylates (Scheme 3.14) [36].

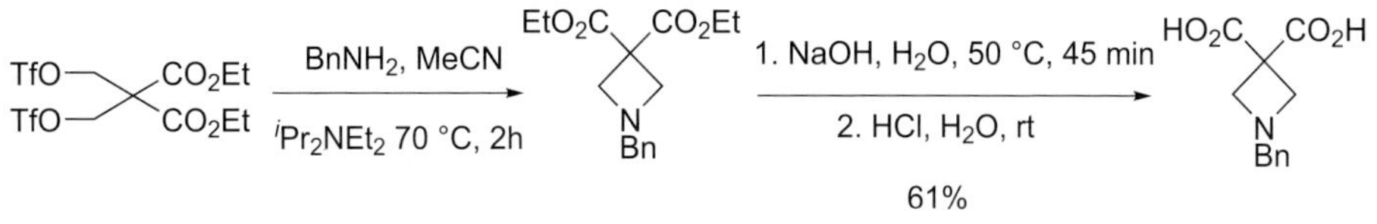

Scheme 3.13

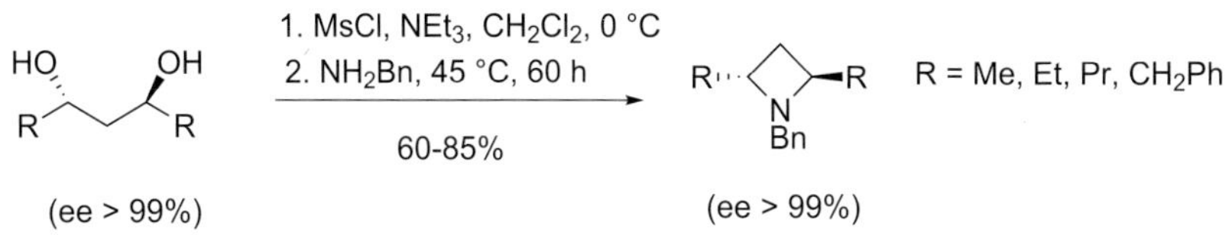

Scheme 3.14

3-Hydroxyazetidines have also been prepared by the reaction of epichlorohydrin with primary amines. The intermediate formation of γ-chloro-β-hydroxyamines or amino methyl oxiranes is a function of the reaction conditions and the nature of the amines. In some cases these two-step reactions were conducted as one-pot reactions (Scheme 3.15).

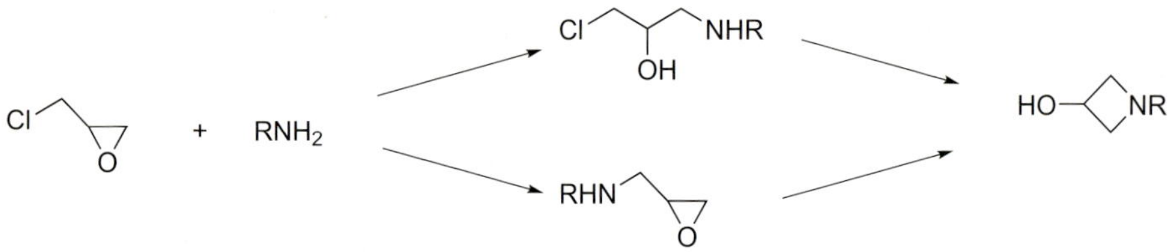

Cl, O, + RNH_2, Cl, NHR, OH, RHN, O, HO, NR

Scheme 3.15

For example, the reaction of epichlorohydrin with 1-silylalkylamine leads to the formation of 3-hydroxyazetidines in excellent yields. This method was used in a new preparation of 1-methyl-3-hydroxyazetidine (Scheme 3.16) [37]. N-Substituted azetidinols were also obtained by heating aminomethyloxiranes in the presence of triethylamine (Scheme 3.17) [38].

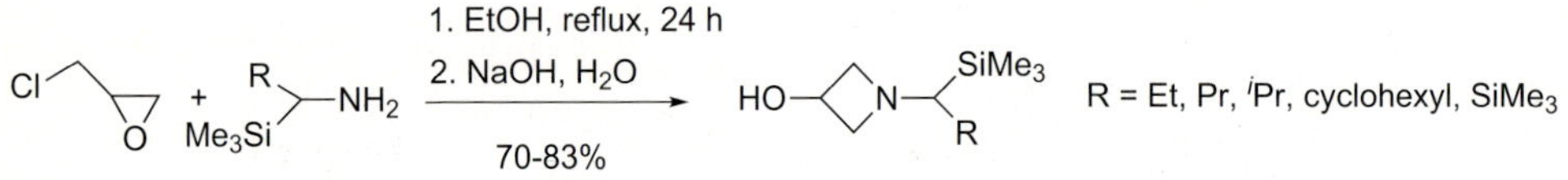

Cl, O, + R, Me_3Si, NH_2; 1. EtOH, reflux, 24 h; 2. NaOH, H_2O; 70-83%; HO, N, $SiMe_3$, R

R = Et, Pr, ^{i}Pr, cyclohexyl, $SiMe_3$

Scheme 3.16

Cl, O, + $R-NH_2$; ^{i}PrOH, rt, 6-24 h; 81-95%; RHN, O; Et_3N, MeCN; reflux, 12-24h; 26-85%; HO, N-R

R = ^{t}Bu, ^{i}Pr, n-Bu, $HOCH_2CH_2$-, allyl, $PhCH_2$, Ph_2CH

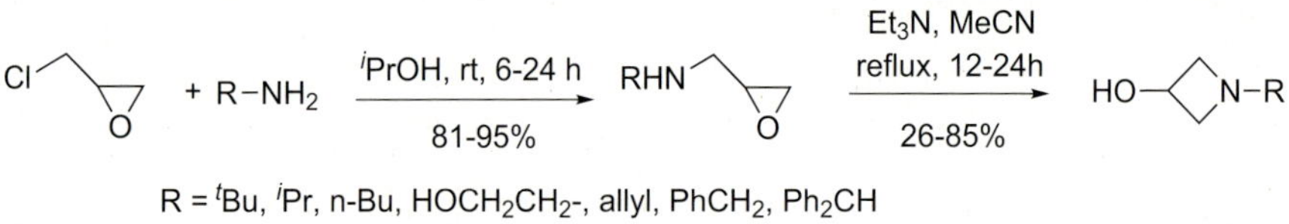

Scheme 3.17

Another approach to azetidines implies the addition of nucleophiles such as cyanide, hydride or organolithium reagents (Scheme 3.18) to β-chloroimines [39]. When the nucleophile is KO^{t}Bu, 2-methyleneazetidines are obtained if the work-ups are carried out in the absence of water (Scheme 3.18) [40].

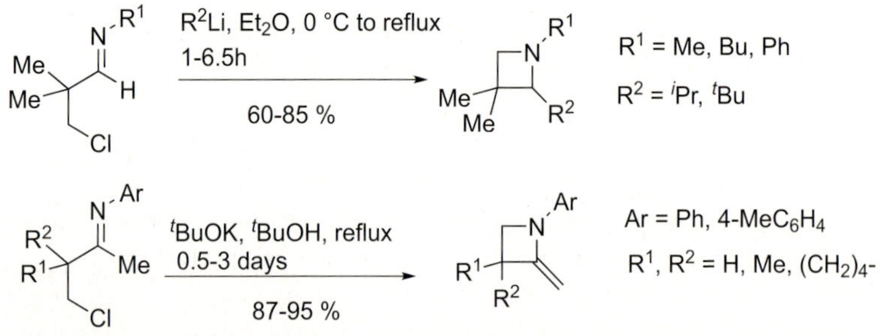

N, R^1, Me, Me, H, Cl; R^2Li, Et_2O, 0 °C to reflux; 1-6.5h; 60-85 %; N, R^1, Me, Me, R^2

R^1 = Me, Bu, Ph

R^2 = ^{i}Pr, ^{t}Bu

N, Ar, R^2, R^1, Me, Cl; ^{t}BuOK, ^{t}BuOH, reflux; 0.5-3 days; 87-95 %; N, Ar, R^1, R^2

Ar = Ph, 4-MeC_6H_4

R^1, R^2 = H, Me, $(CH_2)_4$-

Scheme 3.18

Electrophilic cyclization of N-cinnamyl tosylamides affords azetidines in good yields. These 4-*endo* mode cyclizations are diastereoselective (Scheme 3.19) [41].

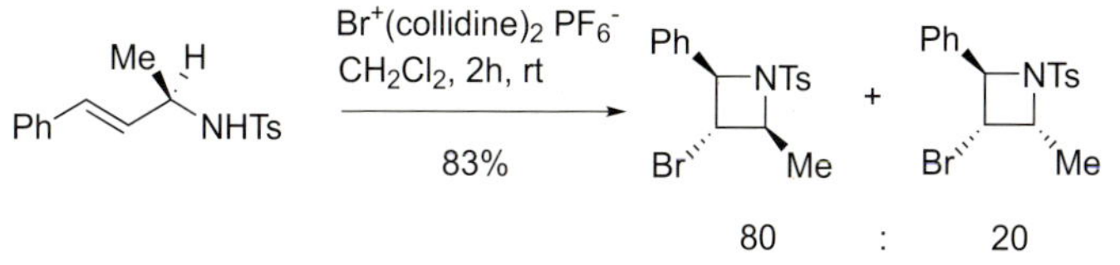

Scheme 3.19

4-*Exo* mode cyclizations of unsaturated amines have also been reported using iodine [42], NBS [43] and selenium reagents (Scheme 3.20) [44].

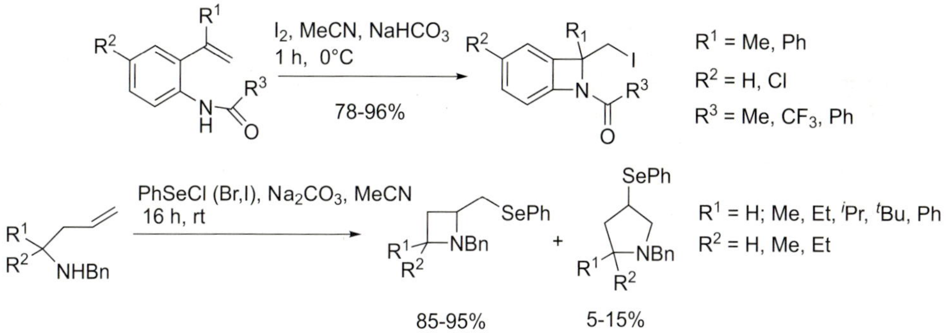

Scheme 3.20

An interesting method for the preparation of enantiopure azetidines is the intramolecular amination of β-aminoallenes catalyzed by $Pd(PPh_3)_4$ (Scheme 3.21). When the protecting group on the nitrogen atom was a tosyl, a mixture of alkenyl azetidine and tetrahydropyridine was obtained (R^2X = aryl iodide, e.g. PhI). However, with enol triflate as alkylating agent, the exclusive formation of azetidines was

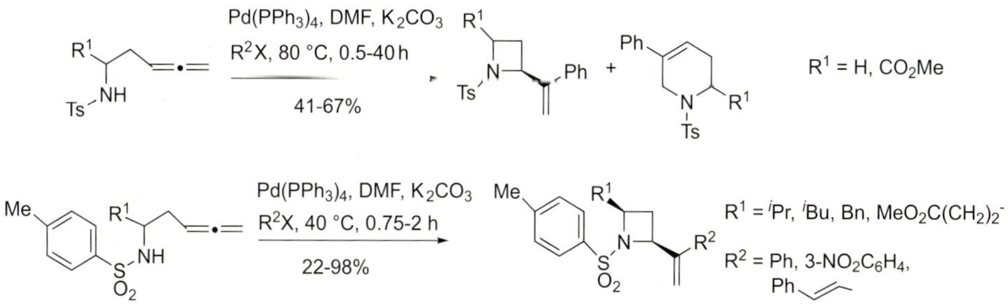

Scheme 3.21

observed (35–40%) [45]. This is also the case when the protecting group was a mesitylenesulfonyl (Scheme 3.21) [46].

3.1.3.1.2 **Formation of a C–C Bond** Intramolecular reaction of α-aminocarbanions with activated carbon–carbon double bonds is an excellent method by which to prepare functionalized azetidines. Mixtures of diastereomers have been obtained from α-aminonitriles. Better diastereoselectivities were observed from α-aminoesters (Scheme 3.22) [47, 48].

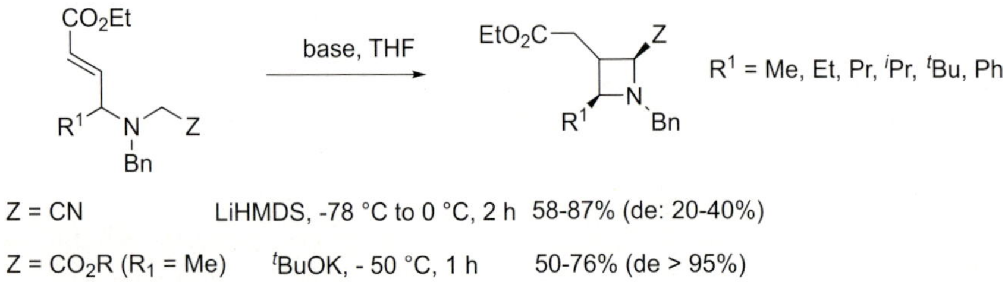

CO_2Et ; R^1 N Z ; Bn — base, THF → EtO_2C Z ; N ; R^1 Bn — R^1 = Me, Et, Pr, ^{i}Pr, ^{t}Bu, Ph

Z = CN   LiHMDS, -78 °C to 0 °C, 2 h   58-87% (de: 20-40%)

Z = CO_2R (R_1 = Me) ^{t}BuOK, - 50 °C, 1 h   50-76% (de > 95%)

Scheme 3.22

Intramolecular cyclization can also be observed if the carbon β of the amino group bears a leaving group (Scheme 3.23) [49]. This method has been applied to the preparation of *cis* and *trans* 3-phenyl-azetidine-2-carboxylic acids [49], azetidine-2-carbonitriles, [50] and diethyl azetidine-2-phosphonates (Scheme 3.24) [51].

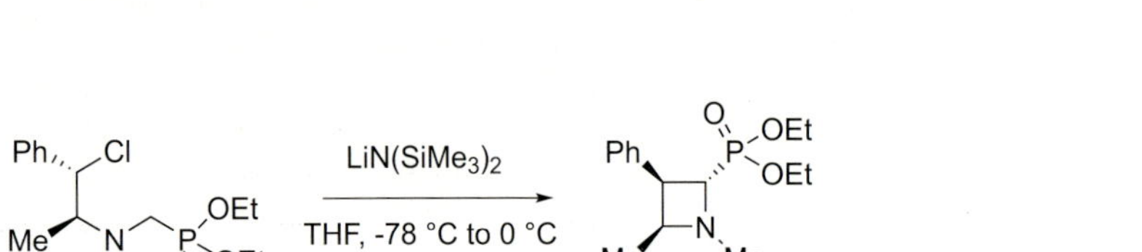

Ph ; Ph N Cl ; Ph $CO_2{}^tBu$ — $NaN(SiMe_3)_2$, THF, -78 °C, 69% → Ph $CO_2{}^tBu$; N ; Ph Ph — *trans*:*cis* = 82:18 → → Ph CO_2H ; N H

Scheme 3.23

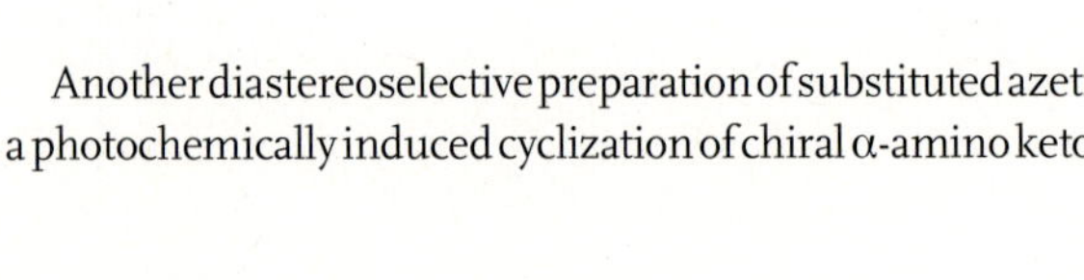

Ph,, Cl ; Me N P OEt OEt ; Me O — $LiN(SiMe_3)_2$, THF, -78 °C to 0 °C, 50% → O P OEt OEt ; Ph ; N ; Me Me

Scheme 3.24

Another diastereoselective preparation of substituted azetidines has been reported by a photochemically induced cyclization of chiral α-amino ketones (Scheme 3.25) [52, 53].

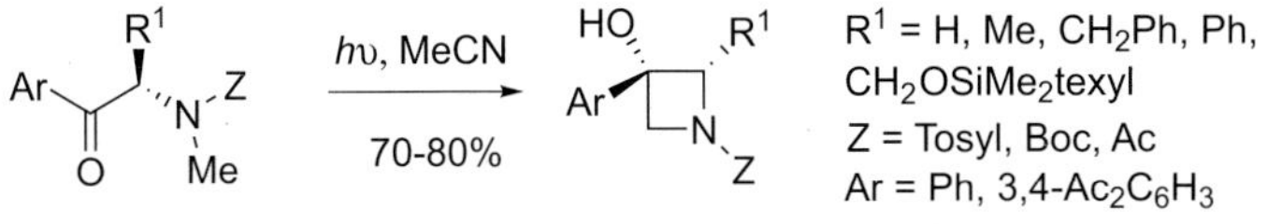

Scheme 3.25

Fused azetidine derivatives have been obtained by irradiation of dihydropyridines (Scheme 3.26) [54a]. This methodology has been applied to the formation of an analog of ABT-594 [54b]. However, this compound appeared less efficient than epibatidine as nicotine agonist.

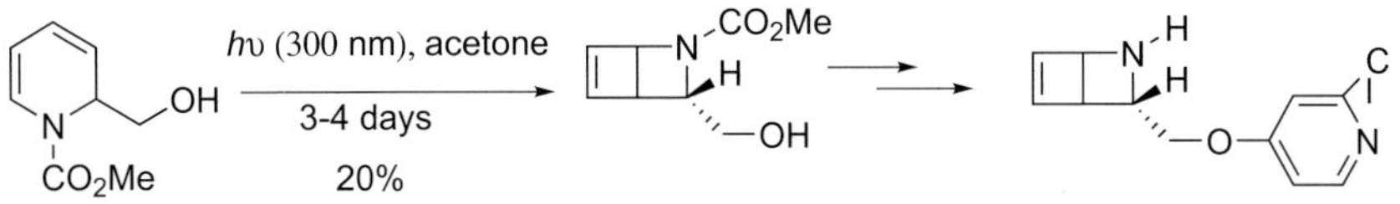

Scheme 3.26

Irradiation of 3-allyl or 3-benzyl-2-acyl perhydrobenzoxazine furnishes azetidin-3-ol derivatives in moderate chemical yields (50–60%). However, the diastereoselectivity of these cyclizations are good (up to 96%). The menthol part was then removed in low overall yields to give enantiopure azetidin-3-ol derivatives (Scheme 3.27) [55].

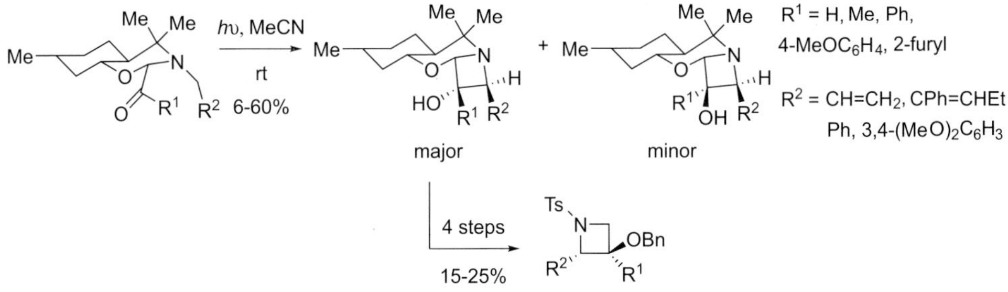

Scheme 3.27

Electroreduction by intramolecular coupling of chiral α–iminoesters in the presence of chlorotrimethylsilane affords *cis*-2,4-disubstituted azetidin-3-ones in good yields (Scheme 3.28) [56].

3.1.3.2 **Ring Transformations**

3.1.3.2.1 **Ring Expansions of Aziridines** Reactions of N-arenesulfonylaziridines with dimethyloxosulfonium methylide afford azetidines. These ring expansions are stereoselective. The *cis*-aziridines yield trans-azetidines, and reciprocally, via a

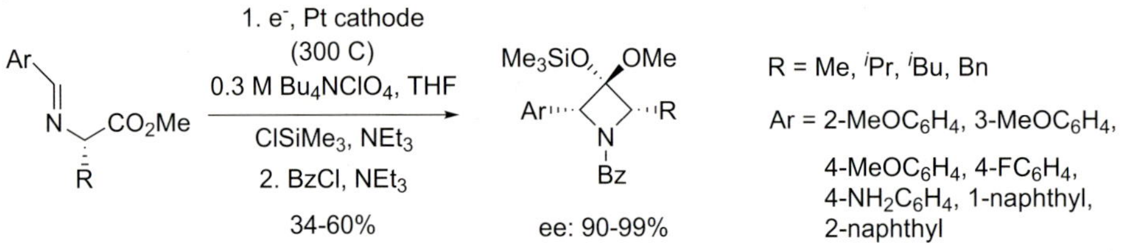

Scheme 3.28

S_N2-type reaction, followed by an intramolecular nucleophilic substitution [57]. This reaction has also been reported using dimethylsulfonium methoxycarbonyl methylide (Scheme 3.29) [58].

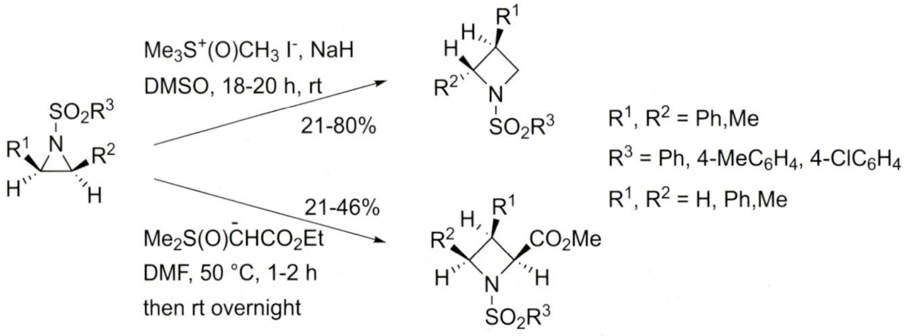

Scheme 3.29

1-Substituted 3-azabicyclo[1.1.0]butanes are easily prepared using the method first developed by Funke [59]. These compounds react with numerous electrophiles to afford azetidines (Scheme 3.30). More recent reports give improved conditions for the preparation of the bicycloaziridines, and their reactions with electrophilic reagents [60–62].

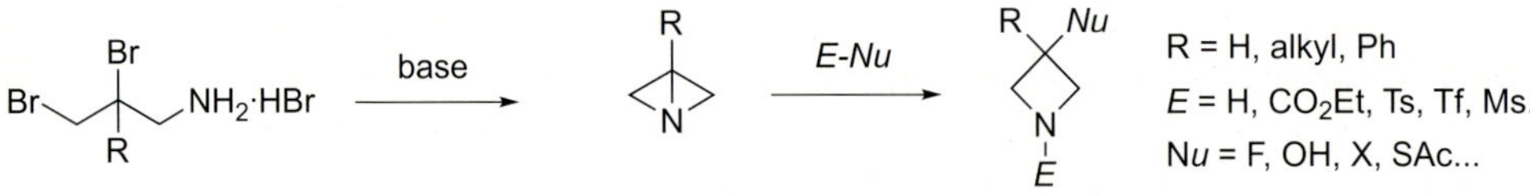

Scheme 3.30

3.1.3.2.2 **Ring Contractions** 5 → 4 Ring contractions appear rather rare. The reaction of a thiazolidine with Raney-nickel has been reported to lead to yield an azetidine (Scheme 3.31) [63].

Two interesting examples of 6 → 4 ring contractions have been also reported. They used the transformation of cyclic carbamates. Heating tetrahydro-3-benzyl-5,5-

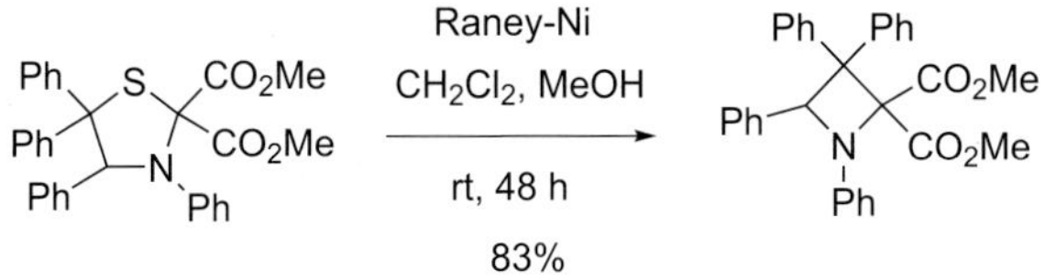

Scheme 3.31

dimethyl-1,3-oxazin-2-one at 250 °C in the presence of LiCl affords the corresponding azetidine [64]. 1,2,2,4-Tetrasubstituted azetidines have been prepared using this method. Reaction of a carbamate with a palladium(0) salt led to a comparable decarboxylation. This reaction has not been studied further (Scheme 3.32) [65].

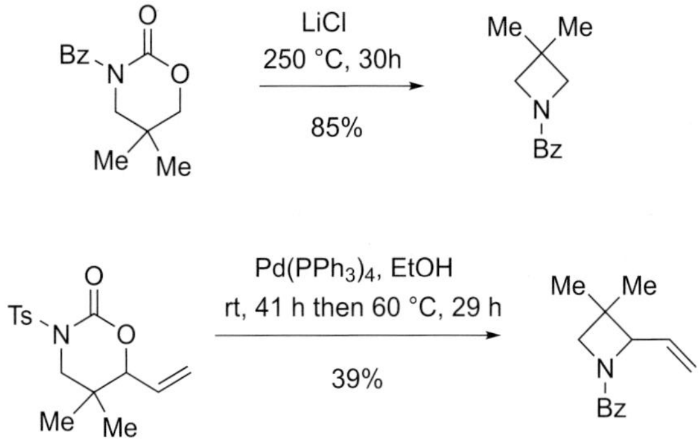

Scheme 3.32

3.1.3.2.3 From β-Lactams

Reductions Reductions of β-lactams into azetidines have been reported with common reducing agents such as $LiAlH_4$, $LiAlH_4$-$AlCl_3$, $ClAlH_2$, Cl_2AlH, AlH_3, BH_3 and DIBAL-H [66–69]. With optically active lactams, these reactions occur in general without loss of enantiomeric purity. However, these reagents were not compatible with the presence of ester groups. In this case, diphenylsilane in the presence of a rhodium salt (Scheme 3.33) could be efficient [70].

Scheme 3.33

Reduction of azetidine-2,4-dione by AlH_3 into azetidine has also been reported [71], while the reduction of 2-azetidinethiones is possible using Raney-nickel (Scheme 3.34) [72].

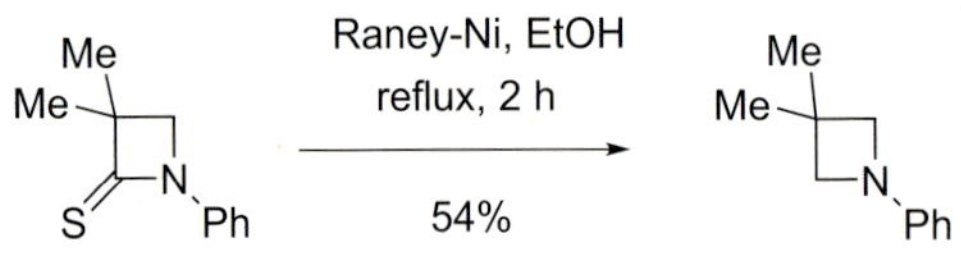

Me, Me, S, N, Ph — Raney-Ni, EtOH, reflux, 2 h, 54% — Me, Me, N, Ph

Scheme 3.34

Olefinations Stabilized ylides react with 2-azetidinones to afford 2-methyleneazetidines. This olefination does not take place with unstabilized ylides or when a substituent is present at the 3-position [73]. In such a case, methylenation is possible using dimethyltitanocene (Scheme 3.35) [74, 75].

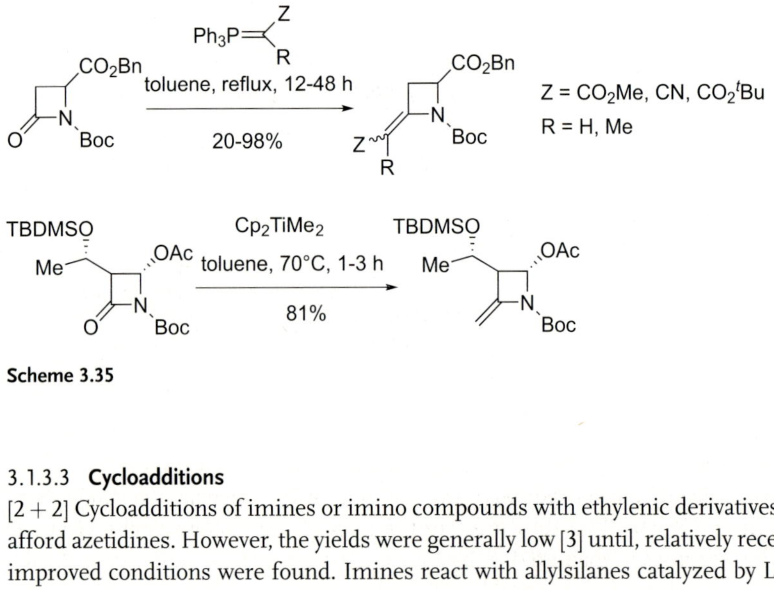

CO_2Bn, O, N, Boc — Ph_3P=C(Z)(R), toluene, reflux, 12-48 h, 20-98% — CO_2Bn, Z, R, N, Boc — Z = CO_2Me, CN, $CO_2{}^tBu$; R = H, Me

TBDMSO, Me, OAc, O, N, Boc — Cp_2TiMe_2, toluene, 70°C, 1-3 h, 81% — TBDMSO, Me, OAc, N, Boc

Scheme 3.35

3.1.3.3 Cycloadditions

[2 + 2] Cycloadditions of imines or imino compounds with ethylenic derivatives can afford azetidines. However, the yields were generally low [3] until, relatively recently, improved conditions were found. Imines react with allylsilanes catalyzed by Lewis acids to give azetidines in good yields (Scheme 3.36) [76]. Mixtures of diastereomers were obtained. N-Tosylimines have also been reported to react with α-allenic esters and α-allenic ketones in the presence of DABCO (Scheme 3.36) [77].

The reaction of α-imino esters with 1-methoxyallenylsilanes has been achieved by means of copper tetrafluoroborate. In the presence of (*R*)-Tol.BINAP, optically active azetidines with high enantiomeric excesses were obtained (Scheme 3.37) [78].

1-Aryl-4-phenyl-1-azadienes undergo [2 + 2] cycloadditions, after more than 40 days, with allenic esters at room temperature. Attempts to decrease the reaction time by heating gave rise to the exclusive formation of [4 + 2] adducts (Scheme 3.38) [79].

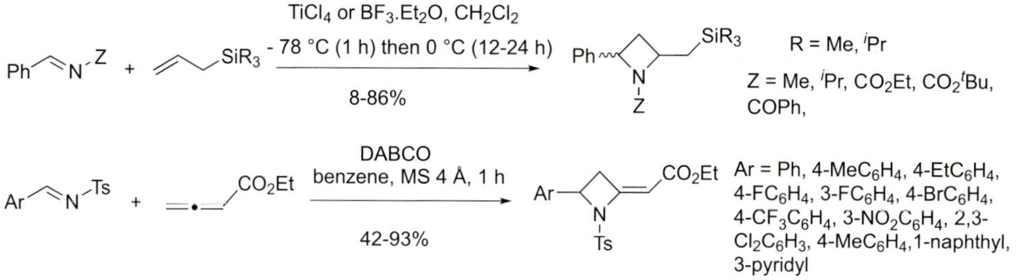

Scheme 3.36

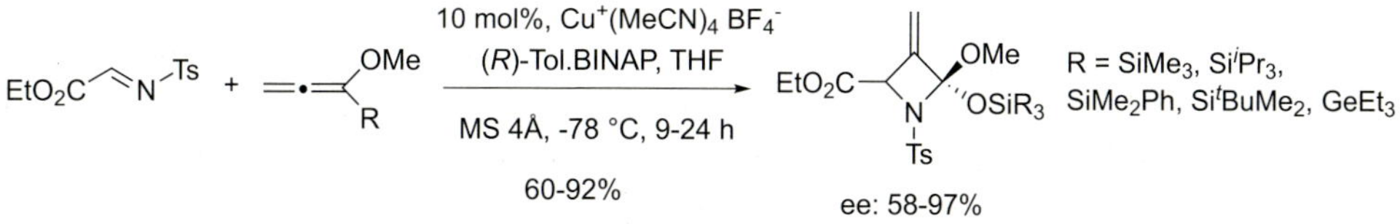

Scheme 3.37

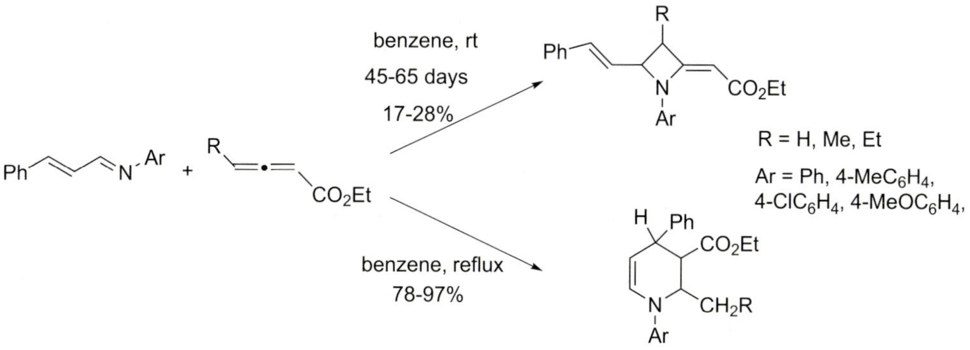

Scheme 3.38

The cycloaddition of benzyne with N-aryl imines occurs in good yields (Scheme 3.39) [80]. This is also the case for the [2 + 2] cycloadditions of alkoxy imines with phenyl ketene [81].

3.1.4 Reactivity and Useful Reactions

3.1.4.1 Reactions at the Nitrogen Atom

3.1.4.1.1 Reaction with Carboxylic Acid Derivatives Reactions of carboxylic acid chlorides with azetidine derivatives lead to N-acyl derivatives in high yields. These reactions were carried out in the presence of bases such as K_2CO_3 [82], NEt_3 or pyridine [3]. Recently, a solid-phase approach has been described for the preparation

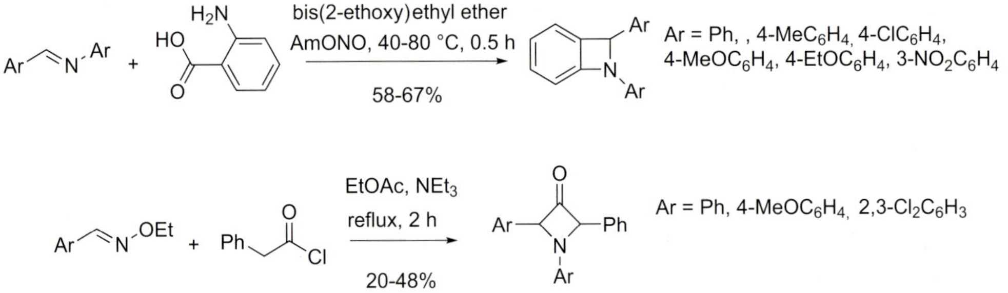

Scheme 3.39

of malondiamides. 2-Methoxyethylamine was fixed on a Bal-resin, and successive reactions with 2,2-dimethylmalonyl dichloride and azetidine lead to the malonamide. Cleavage with trifluoroacetic acid gives rise to the mixed malonamide in high yield (Scheme 3.40) [83].

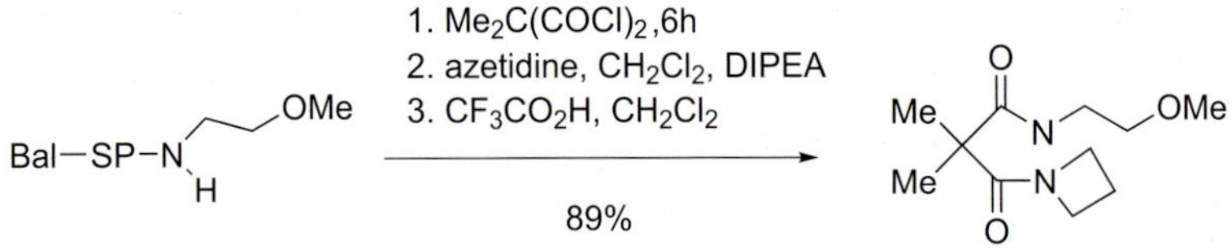

Scheme 3.40

N-Acetylazetidines can also be obtained by reaction with carboxylic anhydrides in the presence of DMAP (Scheme 3.41) [84], by reaction with ester in the presence of Horse Liver esterase [85], or by reaction with acid activated by reagents such as 1-[(3-dimethylamino)propyl]-3-ethylcarbodiimide (DMAP) [86] or 1*H*-1,2,3-benzotriazol-1-ol (HBTU) derivatives (Scheme 3.41) [87–89].

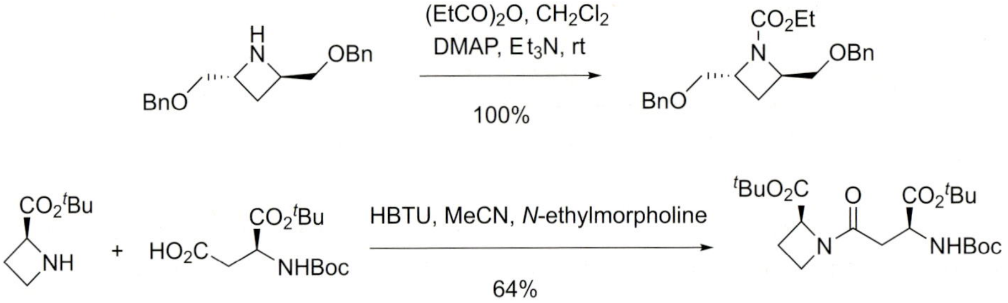

Scheme 3.41

3.1.4.1.2 **Reaction with Aldehydes and Ketones** Aldehydes react in the presence of $NaBH_3CN$ in acetic acid to furnish N-alkylazetidines (Scheme 3.42) [90]. Formic acid was sometimes preferred as the reducing agent [91].

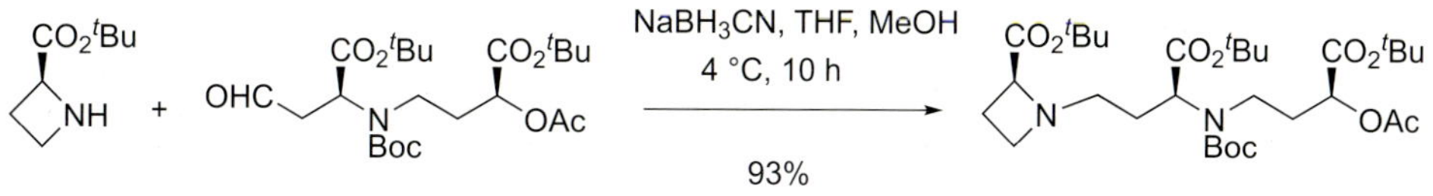

$NaBH_3CN$, THF, MeOH
4 °C, 10 h
93%

Scheme 3.42

These N-alkyl azetidines have been also obtained from the corresponding N-acyl compounds by carbonyl reduction after transformation into thioamides (Scheme 3.43) [88].

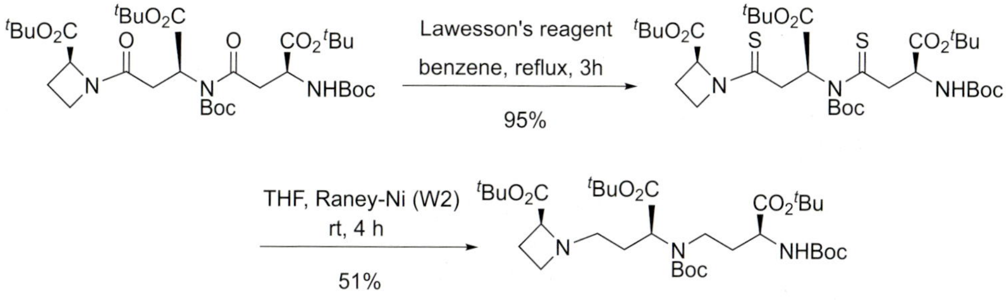

Lawesson's reagent
benzene, reflux, 3h
95%

THF, Raney-Ni (W2)
rt, 4 h
51%

Scheme 3.43

The reaction of (*S*)-azetidine 2-carboxylic acid with pivaldehyde gives a bicyclic compound. When this reaction is catalyzed by CF_3COOH, a racemic product is obtained [92]. With trimethylsilyl-trifluoromethanesulfonate as catalyst, though, racemization was not observed and enantiomerically pure product was obtained (Scheme 3.44) [93].

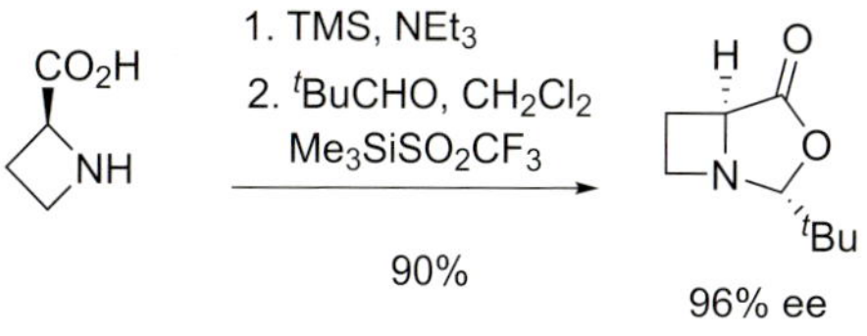

1. TMS, NEt_3
2. ^tBuCHO, CH_2Cl_2
$Me_3SiSO_2CF_3$
90%
96% ee

Scheme 3.44

The reaction of azetidines with ketones affords the corresponding enamines. Kinetic studies have shown that these amines are more reactive than pyrrolidine [94].

3.1.4.1.3 **N-Alkylations** N-Aryl azetidines can be obtained by the arylation of N-unsubstituted azetidines with aryl bromide, catalyzed by palladium salts (Scheme 3.45) [36a].

Another interesting method used the S_NAr substitution of the fluorine on fluoroaryl. These reactions were carried out under sonication of the reaction mixture (Scheme 3.46) [95].

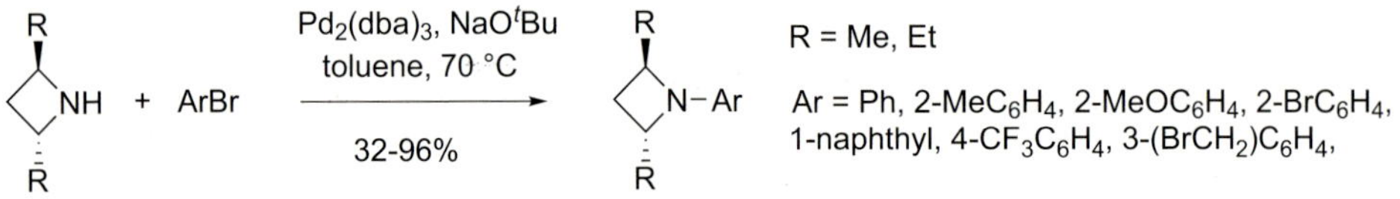

$Pd_2(dba)_3$, NaO^tBu, toluene, 70 °C, 32-96%

R = Me, Et

Ar = Ph, 2-MeC_6H_4, 2-$MeOC_6H_4$, 2-BrC_6H_4, 1-naphthyl, 4-$CF_3C_6H_4$, 3-$(BrCH_2)C_6H_4$,

Scheme 3.45

DMSO, K_2CO_3,))), 100 °C, 3-5 h, 24-81%

R = H, Me

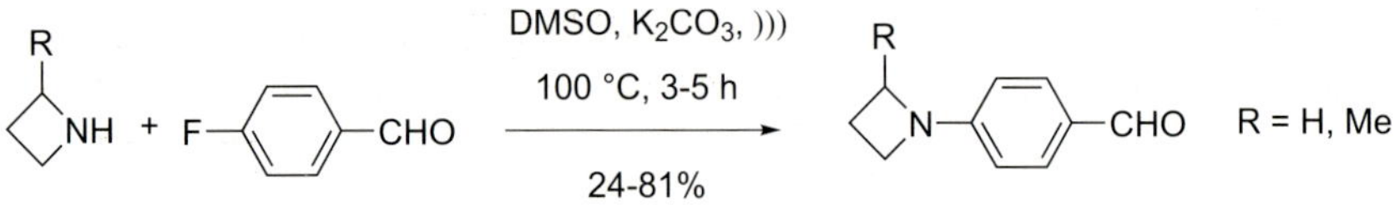

Scheme 3.46

An original preparation of (phosphorothioyl)oxycarbonyl azetidine has been reported by reaction of azetidines with diphenylphosphinothioic chloride [$Ph_2P(S)Cl$] in the presence of K_2CO_3. This reaction was greatly improved when carried out under carbon dioxide in the presence of crown ether (Scheme 3.47) [96].

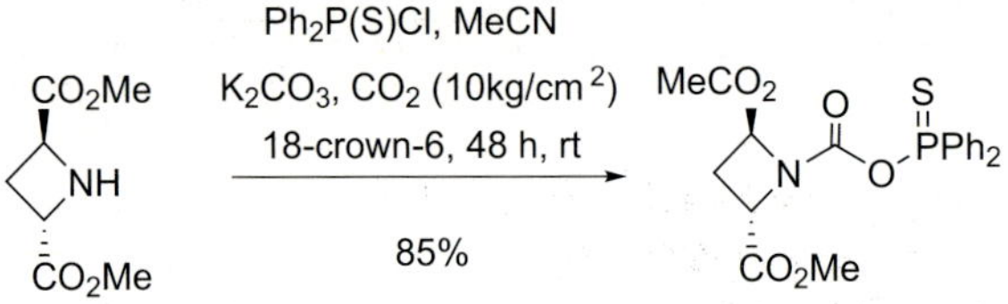

$Ph_2P(S)Cl$, MeCN, K_2CO_3, CO_2 (10kg/cm^2), 18-crown-6, 48 h, rt, 85%

Scheme 3.47

3.1.4.2 **Oxidizing Reactions**

Oxidations of azetidin-3-ols into azetidin-3-ones with common CrO_3, Swern or Dess-Martin reagents have been reported to occur in high yields. No cleavage or degradation of the four-membered ring was noticed [3, 97].

Microbial oxidations of azetidines lead to the formation of azetidin-3-ols (Scheme 3.48) [98, 99]. Oxidations α of the nitrogen-atom (formation of β-lactams) have been performed with a RuO_2-$NaIO_4$ mixture (Scheme 3.48) [100] and

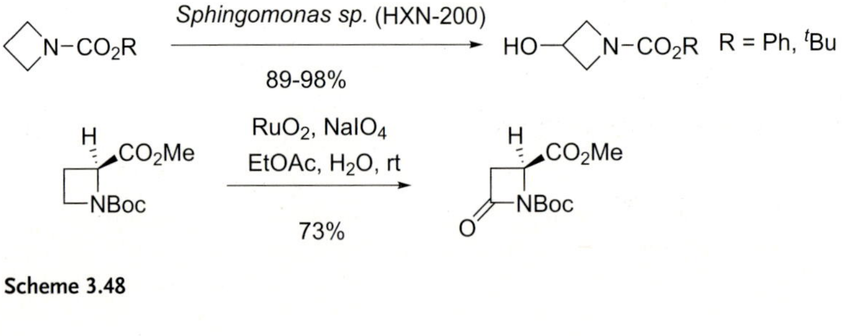

Sphingomonas sp. (HXN-200), 89-98%

R = Ph, ^tBu

RuO_2, $NaIO_4$, EtOAc, H_2O, rt, 73%

Scheme 3.48

$KMnO_4$ [101]. Transformations of azetidin-2-carboxylic acids and esters into of β-lactams using oxygen have been also reported [102].

3.1.4.3 **Reactions with Nucleophiles and Bases**

N-Chloroazetidines can react with DBU to furnish stable azetines (Scheme 3.49) [103]. Azetines were also obtained when the leaving group was in the 3-position. The reaction of 3-mesylazetidines with potassium ^{t}butylate leads to the formation of azetines that can be transformed by flash-thermolysis into 1-aza-1,3-dienes. By smooth heating, these later give rise to bicyclic compounds (Scheme 3.50) [104].

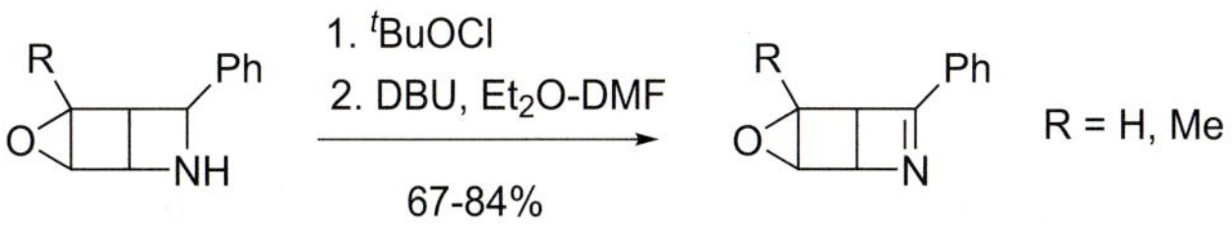

Scheme 3.49

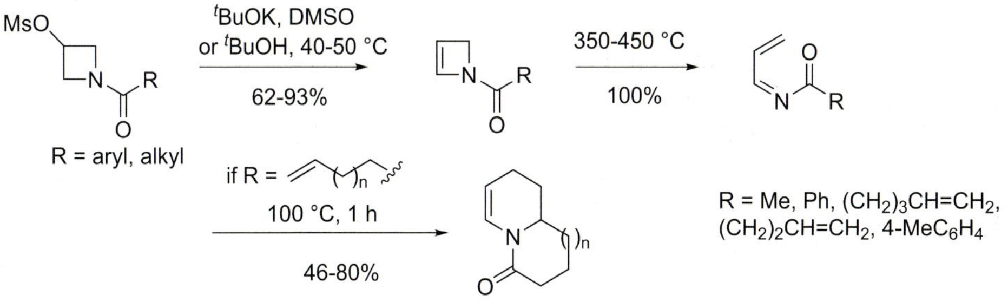

Scheme 3.50

The reaction of 3,3-dichloroazetidines with NaOMe or NaH also leads to azetines. The presence of an aryl substituent at the 2-position seems to increase their reactivities; subsequent reactions lead, finally, to aziridines (Scheme 3.51) [105].

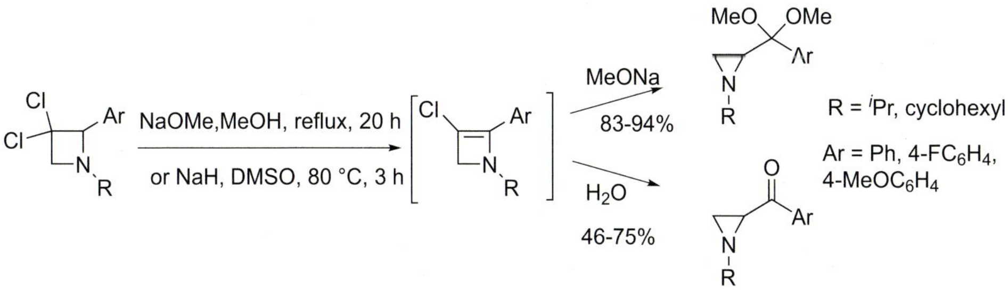

Scheme 3.51

Instead of elimination, substitution occurs when 3-arylsulfonate-azetidines are reacted with benzylamine (Scheme 3.52) [106]. These substitutions, which occur with *retention* of configurations (due to the participation of the ring nitrogen) were also reported with sodium cyanide [107].

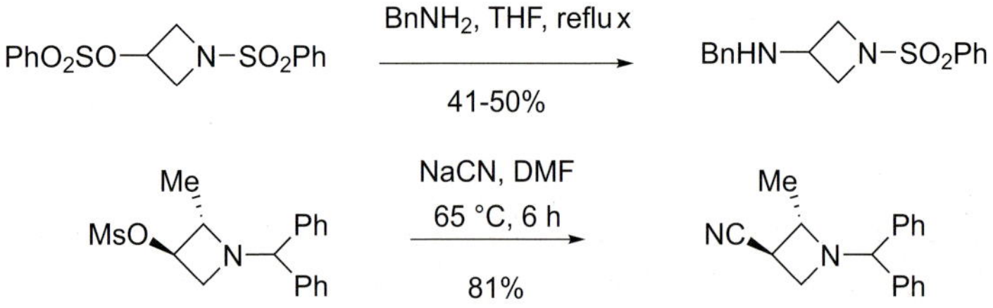

PhO_2SO–azetidine–N–SO_2Ph → ($BnNH_2$, THF, reflux; 41-50%) → BnHN–azetidine–N–SO_2Ph

MsO, Me, N, Ph, Ph → (NaCN, DMF, 65 °C, 6 h; 81%) → NC, Me, N, Ph, Ph

Scheme 3.52

3.1.4.4 Reactions of C-Metallated Azetidines

Lithiation of methyl azetidine-2-carboxylate in α of the ester function has been carried out with LiHMDS as base [108]. Metallation of the azetidine ring of a methyl azetidin-2-ylidenecarboxylate with KHMDS is also effective. Subsequent addition of electrophiles yields 3-substituted azetidines (Scheme 3.53) [109].

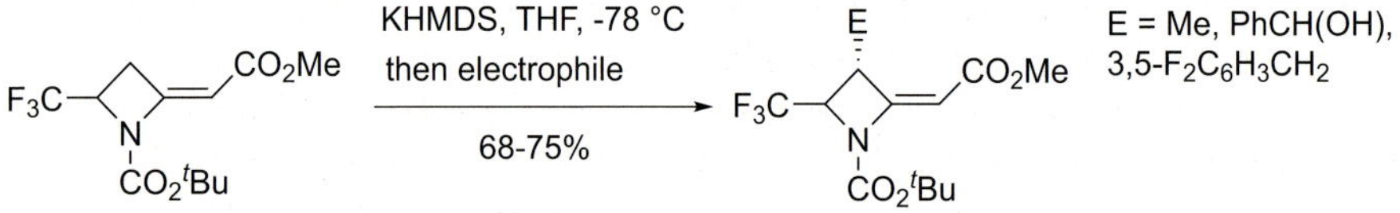

F_3C, CO_2Me, N, $CO_2{}^tBu$ → (KHMDS, THF, -78 °C then electrophile; 68-75%) → F_3C, E, CO_2Me, N, $CO_2{}^tBu$

E = Me, PhCH(OH), 3,5-$F_2C_6H_3CH_2$

Scheme 3.53

Organozinc compounds, formed by the reaction of 3-iodoazetidine with zinc, couple with aryl chlorides in the presence of palladium salts or with allyl bromide in the presence of CuCN to give 3-substituted azetidines in satisfactory yields (Scheme 3.54) [110].

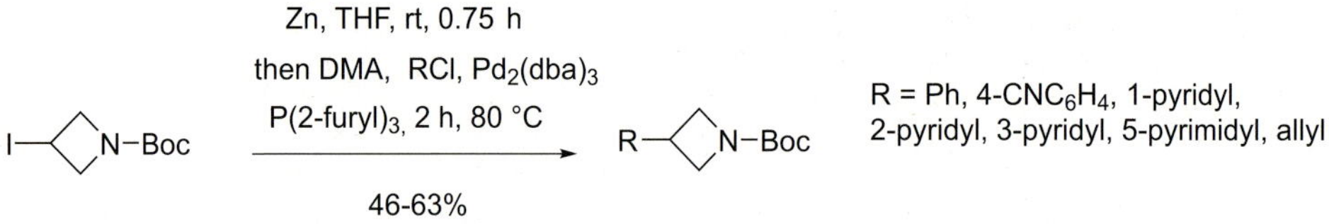

I–azetidine–N–Boc → (Zn, THF, rt, 0.75 h then DMA, RCl, $Pd_2(dba)_3$, $P(2\text{-furyl})_3$, 2 h, 80 °C; 46-63%) → R–azetidine–N–Boc

R = Ph, 4-CNC_6H_4, 1-pyridyl, 2-pyridyl, 3-pyridyl, 5-pyrimidyl, allyl

Scheme 3.54

3.1.4.5 Ring Expansions

Ring expansions of azetidines into five-, six- and eight-membered heterocycles have been reported. Pyrrolidin-2-ones have been isolated during the preparation of azetidin-2-carbonyl chlorides (Scheme 3.55) [111] or 2-(azetidin-3-yl)acetyl chloride [112].

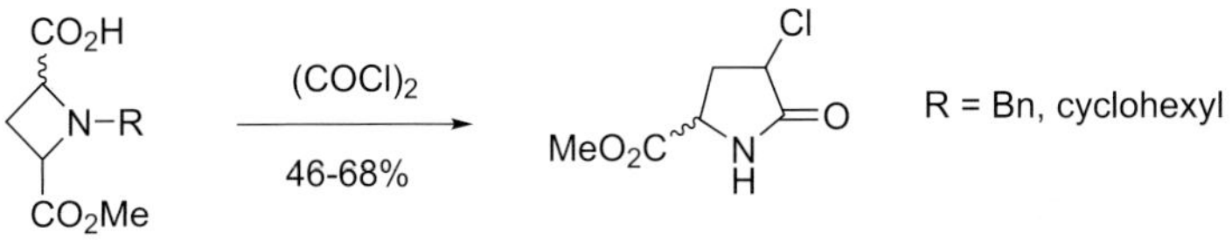

Scheme 3.55

Imidazolidines can also be obtained by Beckmann rearrangement of 3-(mesyloxyimino)- or 3-(tosyloxyimino)azetidines (Scheme 3.56) [113].

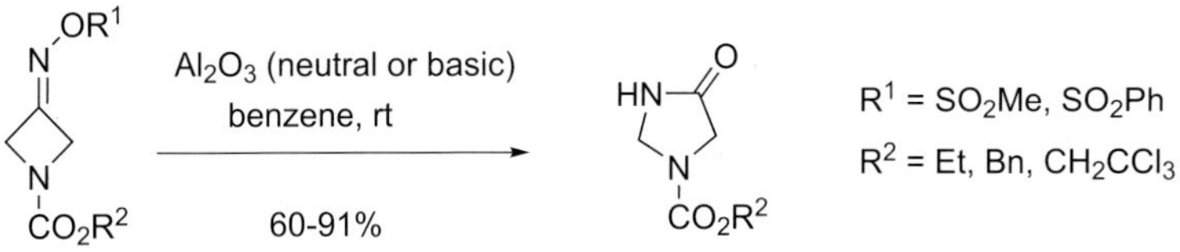

Scheme 3.56

Heating of 2-(phenylselenyl)methyl azetidines [114], 2-(methanesulfonyl)methylazetidines [115] or 2-(chloromethyl)azetidines (Scheme 3.57) [115] affords pyrrolidines in excellent yields. These rearrangements appeared highly diastereoselective. Ring expansions are also possible by treatment of azetidinium salts with bases [116], or by reaction of azetidines with cyclic olefins in the presence of Lewis acids [117].

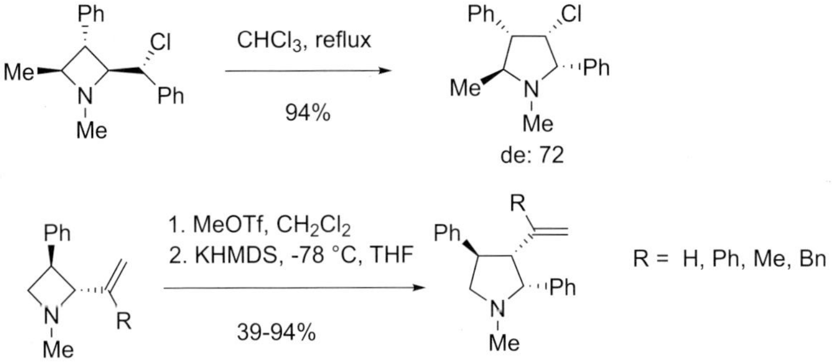

Scheme 3.57

Photochemical rearrangement of 3-benzoylazetidines leads to pyrroles (Scheme 3.58) [118]. Preparation of pyrroles is also possible by the reaction of 3-phenoxy (or 3-isopropylidene) 2-thioacetal-azetidines of cis-stereochemistry with $AlEt_2Cl$ (Scheme 3.58) [119]. When the substituents at the 3-position were phenyl or

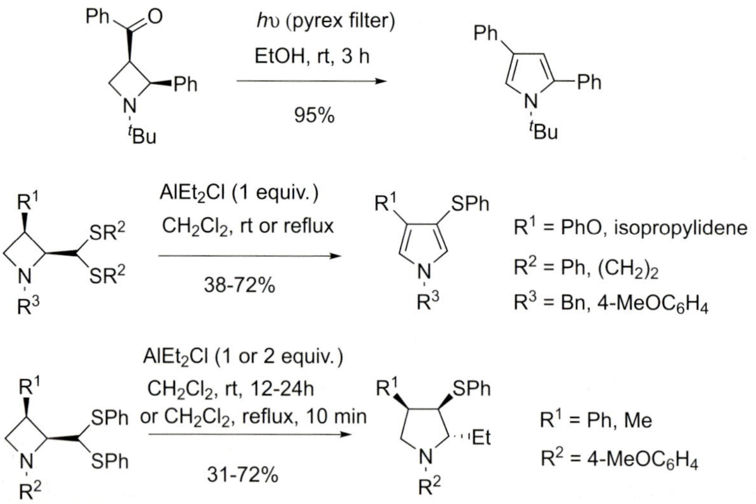

Scheme 3.58

methyl groups, the reactions yield pyrrolidines. With cyclic acetals or thioacetals, bicyclic pyrrolidines are obtained [119].

Another interesting preparation of pyrrolidinones is by the cobalt carbonyl catalyzed carbonylation of azetidines. The regioselectivity of the CO insertion depends on the substituent fixed to the nitrogen atom (Scheme 3.59) [120].

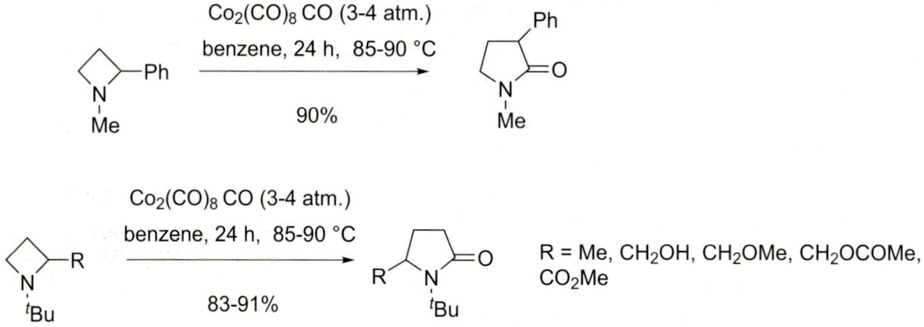

Scheme 3.59

2-Hydroxymethylazetidines react with *m*-chloroperbenzoic acid (*m*CPBA) to furnish the corresponding N-oxides with high diastereoselectivity. Upon heating, these N-oxides rearrange quantitatively into the 1,2-oxazinan-6-ol (Scheme 3.60) [121].

Piperidines, formed by [4 + 2] cycloadditions, have been isolated by reaction of azetidines with methylene-cycloalkanes (Scheme 3.61) [117]. [4 + 2] Cycloadditions have also been reported in the reaction of azetidines with nitriles (Scheme 3.61) [122].

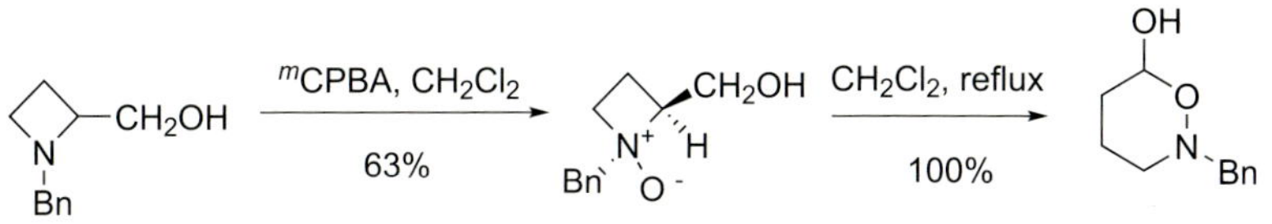

Scheme 3.60

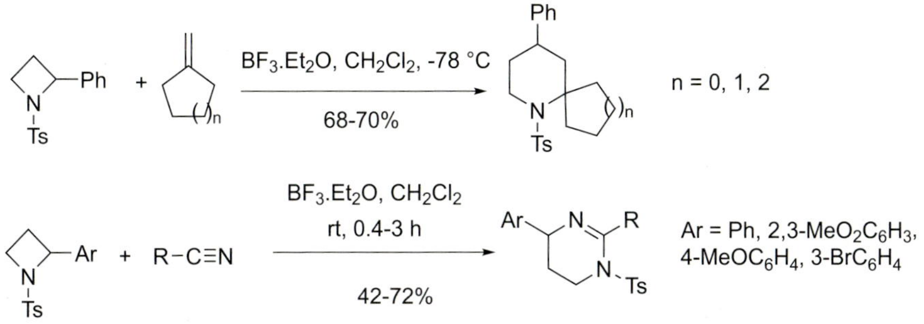

Scheme 3.61

Piperidines are also obtained by the reaction of 1,3-butadienyl-azetidines with a palladium(0) salt. The 1,3-butadienyl entity appears necessary for the rearrangement (Scheme 3.62) [123].

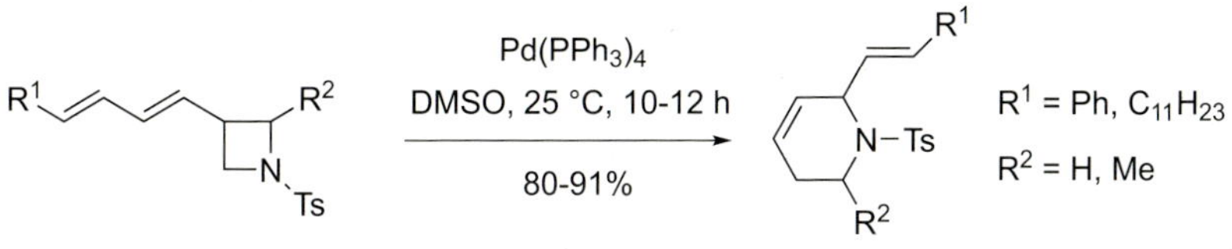

Scheme 3.62

The reaction of 2-alkenyl-azetidines with cobalt carbonyl leads to azepinones instead of piperidinones, as in the case of 2-unsaturated substituents (Scheme 3.63, cf. Scheme 3.59) [120].

Azocanes can be obtained by thermal rearrangement of 1,2-divinylazetidines (Scheme 3.64) [69].

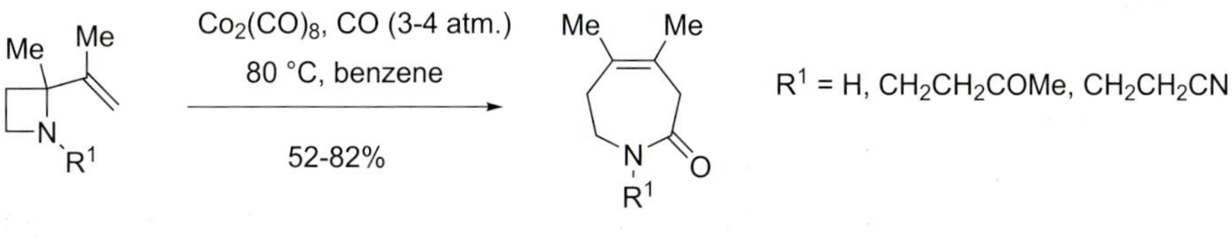

Scheme 3.63

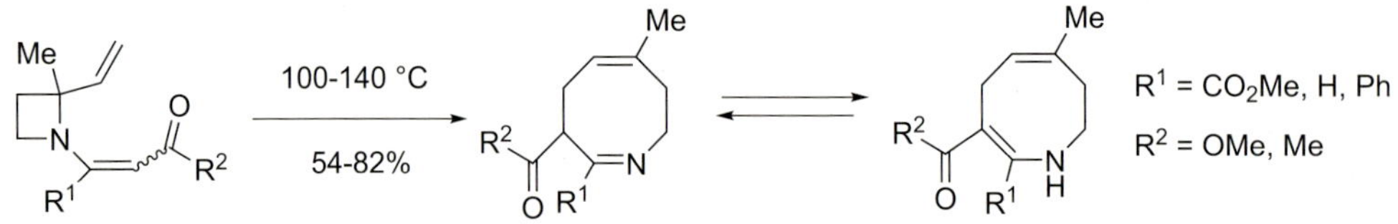

Scheme 3.64

3.1.4.6 Cleavage of the Azetidine Ring

3.1.4.6.1 Fragmentation Reactions Flash thermolysis at 900 °C of azetidine gives ethylene and *N*-methylenemethanamine [124]. However, heating under less drastic conditions leads to cleavage of only one of the N–C bonds (Scheme 3.65) [125] to afford unsaturated amines.

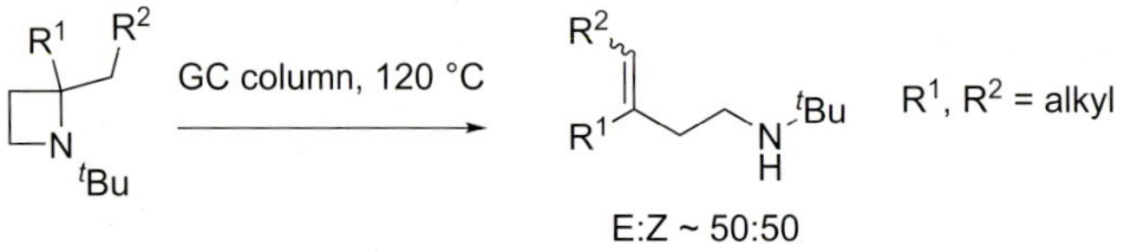

Scheme 3.65

3.1.4.6.2 Reaction with Nucleophilic Reagents Azetidin-2-carboxylic acid derivatives react with thiophenol (pH 8) to afford a mixture of 3-amino-2-phenyl- and 2-amino-3-phenylthiopropanoic acids [126]. Similarly, 3-hydroxyazetidines can be cleaved with phenols. These reactions are highly stereospecific if a substituent was present in the 2-position (Scheme 3.66) [127].

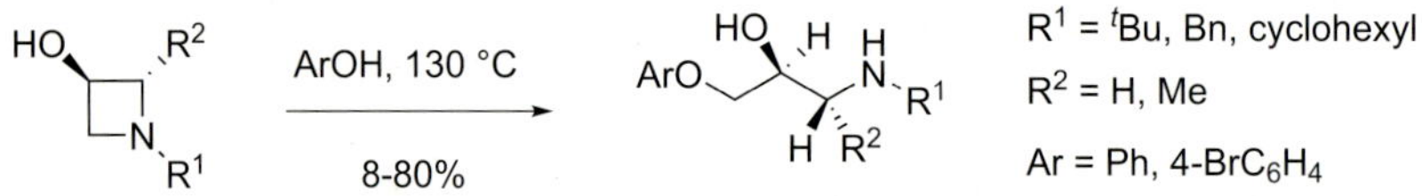

Scheme 3.66

The opening of azetidinium salts with nucleophiles such as NaN_3, butylamine or AcONa has also been examined. The regiochemistry of the cleavage appears to be a function of the substituents fixed on the azetidine ring (Scheme 3.67) [128].

3.1.4.6.3 Reaction with Electrophilic Reagents The reaction of N-alkylazetidines with N_2O_5 gives 1,3-nitroamine-nitrates (Scheme 3.68) [129].

3.1.4.7 Enzymatic Resolutions of Azetidines

Racemic methyl azetidin-2-carboxylate has been resolved using *Candida antartica* as lipase in ^{t}butanol saturated with ammonia. Excellent enantiomeric excesses were obtained for the remaining ester and the amide formed (Scheme 3.69) [130].

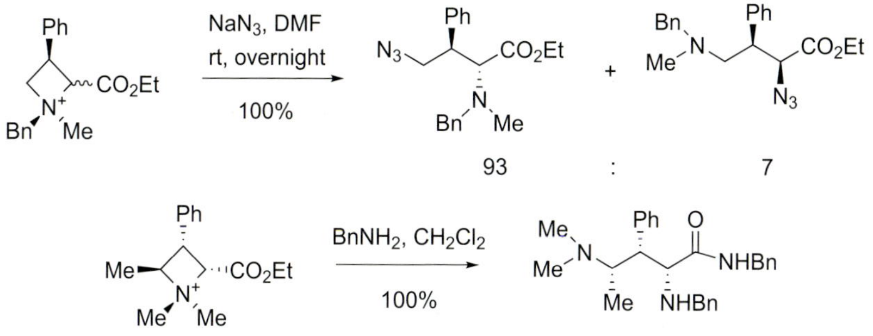

Ph; CO_2Et; N^+; Bn; Me — NaN_3, DMF, rt, overnight, 100% → N_3; Ph; CO_2Et; N; Bn; Me (93) + Bn; N; Me; Ph; CO_2Et; N_3 (7)

93 : 7

Ph; Me; CO_2Et; N^+; Me; Me — $BnNH_2$, CH_2Cl_2, 100% → Me; N; Me; Ph; O; NHBn; Me; NHBn

Scheme 3.67

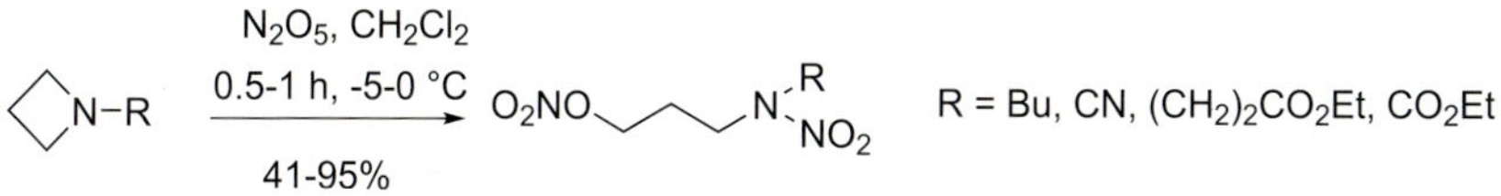

N–R — N_2O_5, CH_2Cl_2, 0.5-1 h, -5-0 °C, 41-95% → O_2NO; N; R; NO_2

R = Bu, CN, $(CH_2)_2CO_2Et$, CO_2Et

Scheme 3.68

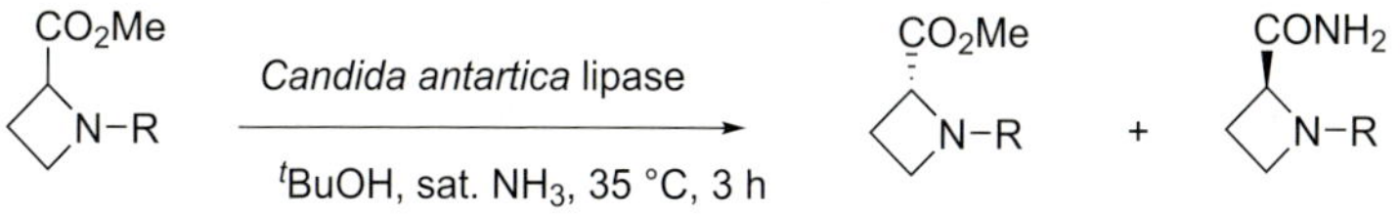

CO_2Me; N–R — *Candida antartica* lipase, ^tBuOH, sat. NH_3, 35 °C, 3 h → CO_2Me; N–R + $CONH_2$; N–R

R	conversion	ee (CO_2Me)	ee ($CONH_2$)
R = Bn	conversion: 56%	ee > 99%	ee = 80%
R = allyl	conversion: 50%	ee > 99%	ee = 97%
R = 4-$MeOC_6H_4CH_2$	conversion: 54%	ee > 99%	ee = 84%

Scheme 3.69

An azetidine with high enantiomeric excess has also been obtained by resolution of a meso-2,4-diol derivative. For a conversion of around 55%, a compound of high enantiomeric purity could be obtained (Scheme 3.70) [131] using porcine pancreas lipase (PPL) immobilized on Celite. Under these conditions, resolution of the racemic *trans*-2,4-diol gave a mixture of mono- and diacetate of lower enantiomeric excesses.

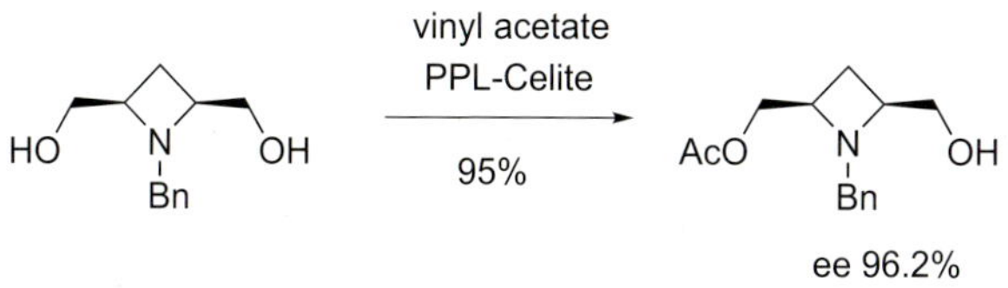

HO; N; Bn; OH — vinyl acetate, PPL-Celite, 95% → AcO; N; Bn; OH

ee 96.2%

Scheme 3.70

oxetane 2-oxetanone

Figure 3.2 Four-membered oxygenated heterocycles.

3.2 Oxetanes

3.2.1 Introduction

Oxetanes are four-membered oxygenated cycles. Their carbonyl-substituted analogues are named oxetanones and, in particular, the 2-oxetanone family is commonly referred as β-lactones (Figure 3.2).

The search for effective enantioselective methods of preparation is now the most common challenge. Nonetheless, new syntheses are still a topical question due to the frequent presence of these four-membered cyclic ethers in biologically active substances. These compounds have aroused much interest by their large range of reactivities. The ring strain common in both the oxetane and the oxetanone cycles causes significant susceptibility to thermal cleavage. The basicity of the ring oxygen makes the oxetanes sensitive to electrophile reagents and/or consequently to nucleophilic attacks at carbon. Oxetanes can be subject to polymerization. The 2-oxetanones mainly undergo nucleophilic addition at the carbonyl carbon, with or without electrophilic catalysis.

3.2.2 Physicochemical Data

Microwave, electron diffraction and X-ray diffraction methods have permitted a precise determination of bond lengths and angles in several oxetanes [132]. Table 3.1 reports the bond lengths and angles of oxetane and 2-oxetanone.

Notably, the C–O bond is longer than in other types of compounds: (C_{sp3}–O = 1.41 Å). The oxetane ring may be either coplanar or puckered, depending on the substituents present. A rocking motion allows the substituents to attain a less

Table 3.1 Bond lengths and angles of oxetane and 2-oxetanone.

Bond	Oxetane (Å)	2-Oxetanone (Å)
C4–O1	1.449	1.45
C3–C4	1.549	1.53
C2–O1–C4	91.8°	89°
C2–C3–C4	84.55°	83°

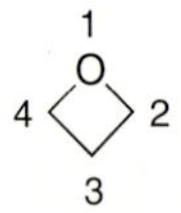

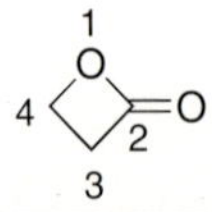

Table 3.2 Chemical shifts of various compounds in deuteriochloroform.

		H^A	H^B	H^C	H^D	H^E	H^F (ppm)
F O B / D / E A / C	Oxetane	4.73	4.73	2.72	2.72	4.73	4.73
	2-Me-oxetane	1.35(Me)	4.85	2.24	2.63	4.37	4.49
	4-Me-oxetanone	1.53(Me)	4.57	2.98	3.50	—	—

hindered conformation but is opposed by the resulting increase in bond angle deformation. The 2-oxetanone having a carbonyl group in place of a methylene group is strictly planar.

3.2.2.1 **NMR Data**

The proton NMR spectroscopy of oxetanes is largely understood [132]. Table 3.2 gives the chemical shifts of various compounds in deuterochloroform. A substituent can have special effects on the proton chemical shifts due to their proximity in the small ring. This could be also induced by changes in the puckering of the ring.

Table 3.3 gives some representative ^{13}C chemical shifts relative to TMS in deuterochloroform [133].

^{17}O NMR spectra of oxetanes and 2-oxetanones referenced to water show values of −13 ppm for the oxetane ether O and 347 ppm (C=O) and 241 ppm (ether O) for the two oxygens in 2-oxetanone [133].

3.2.2.2 **Infrared Spectroscopy**

A strong absorption band at about 980 cm^{-1} is characteristic of oxetane [132]. The far-IR spectrum of oxetane has been obtained at 205 K. The band of ring puckering transition was assigned at 52.92 cm^{-1}. 2-Oxetanones present an intense absorption at 1840–1820 cm^{-1} due to carbonyl stretching.

3.2.3
Natural or Bioactive Compounds

The oxetane ring appears in some biologically active compounds. We report in this chapter some representative examples. Taxol **12**, isolated from *Taxus brevifolia* and its derivative the taxotere, is an important drug in cancer chemotherapy [134].

Table 3.3 Representative ^{13}C chemical shifts relative to TMS in deuterochloroform.

		C^2	C^3	C^4	Me
1 O / 4 2 / 3	Oxetane	72.8	23.1	72.8	—
	2-Me-oxetane	78.3	29.0	66.6	24.1

Montanin D (**13**) has antifeedant activity. This substance has been extracted from the plant *Teucrium tomentosum* [135].

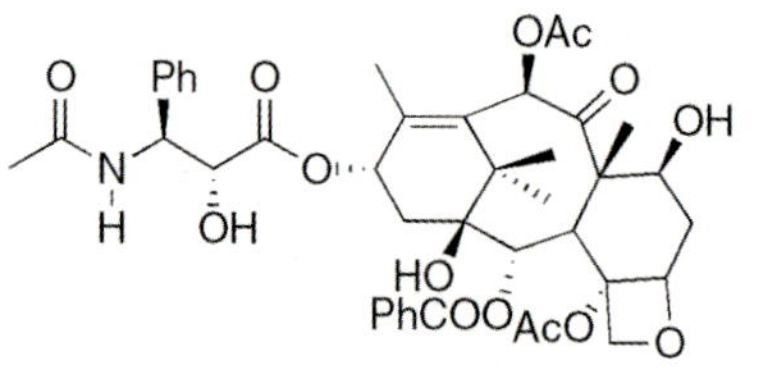

Taxol **12**

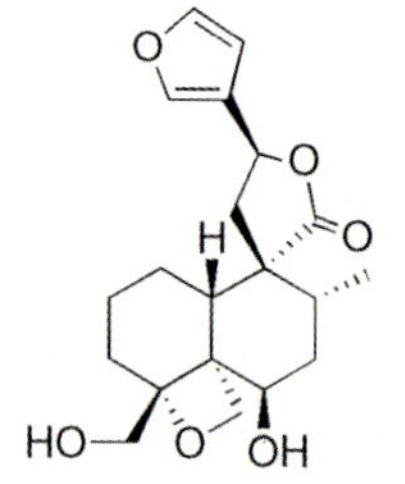

Montanin D **13**

(−)-Tetrahydrolipstatin (**14**) is a triglyceride mimic, an analogue of lisptatin isolated from *Streptomyces toxytricimi*. It is a potent and irreversible inhibitor of pancreatic lipase, used as an antiobesity agent under the name Xenical [136]. The antibiotic oxetanocin A (**15**) is a fermentation product of *Bacillus megaterium*. It is also known as an anti-HIV compound [137]. Curromycin A (**16**) is an antibiotic produced by a genetically modified strain of *Streptomyces hygroscopicus* [138].

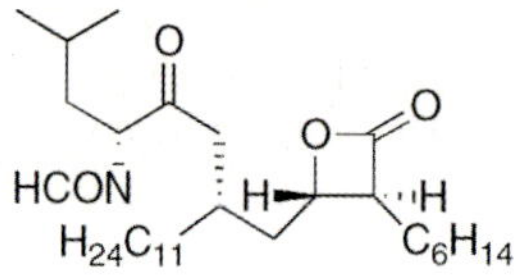

(-) tetrahydrolipstatin **14**

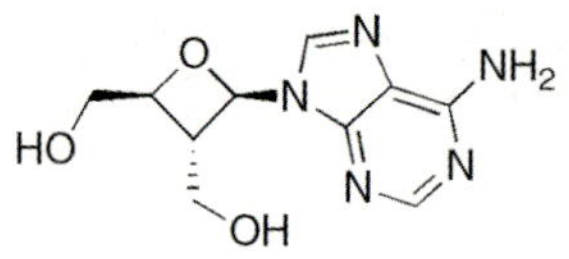

Oxetanocin A **15**

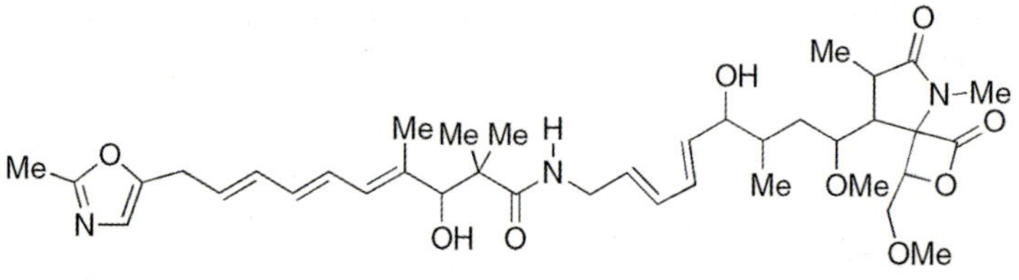

16

Anisatin **17** and neoanisatin **18** are convulsant principles isolated from the fruits of the toxic plant *Illicium anisatum L* [139]. Salinosporamide A (**19**) is a bioactive product of a marine microorganism [140].

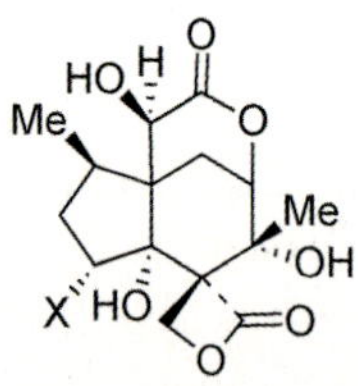

X = OH anisatin **17**

X = H neoanisatin **18**

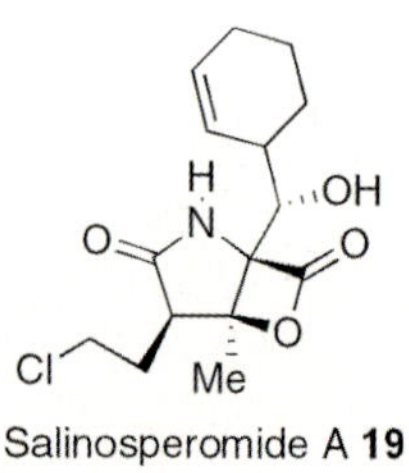

Salinosperomide A **19**

3.2.4
Synthesis of Oxetanes and Oxetan-2-ones

3.2.4.1 [2 + 2] Paterno–Büchi Cyclizations

The [2 + 2] photocycloaddition of carbonyl compounds with alkenes, the Paterno–Büchi reaction, is a useful method in organic synthesis. The challenge in this area is the regio and stereocontrol of the reaction. Bach *et al.* have reported significant facial diastereoselectivities in the reaction with silyl enol ethers carrying a chiral substituent. The carbonyl compound is directed to the less shielded face. With large and polar substituents at the stereogenic center, the facial selectivity leads to diastereomerically pure oxetanes (Scheme 3.71) [141].

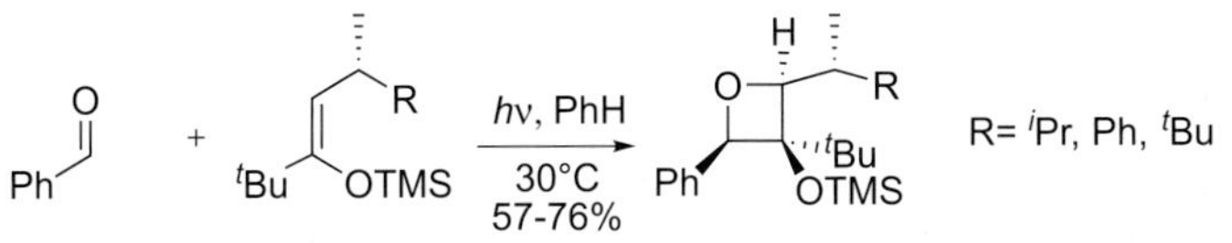

hν, PhH; 30°C; 57-76%; R= ^{i}Pr, Ph, ^{t}Bu

Scheme 3.71

Photocycloaddition of chiral N-acyl enamines with benzaldehyde produces the cis-diastereomers predominantly, with facial diastereomeric excess from 30 to 62% (Scheme 3.72) [142].

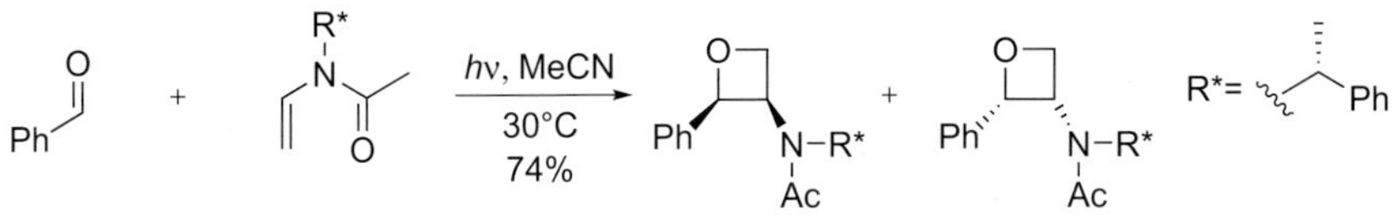

hν, MeCN; 30°C; 74%

Scheme 3.72

A N-benzylenecarbamate has been irradiated in the presence of various aldehydes. The cis-diastereoselectivity is always predominant [143]. This method has been applied to the synthesis of diastereomerically pure 1,2-amino alcohols. A diastereoselectivity of 62% has been observed in the reaction of benzaldehyde with an atropisomeric enamide (axial chirality) (Scheme 3.73) [144].

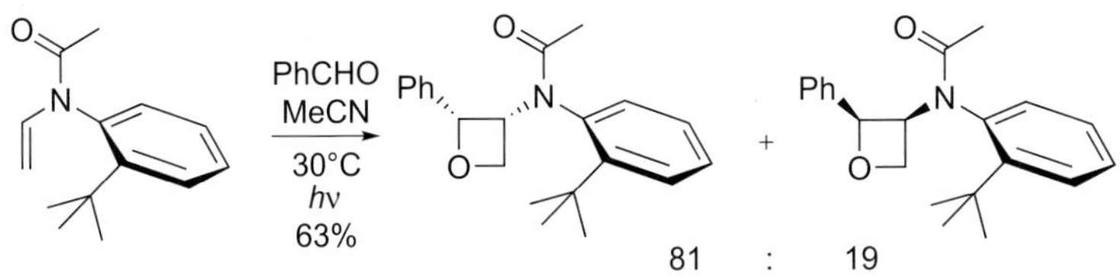

PhCHO, MeCN; 30°C, hν; 63%; 81 : 19

Scheme 3.73

A high facial diastereoselectivity in the photocyclization of a chiral aromatic aldehyde and an enamide has been induced by intermolecular hydrogen bonding. The simple diastereoselectivity led solely to cis-isomers and the facial diastereoselectivity was 95 : 5, when the reaction was carried out at low temperature in toluene (Scheme 3.74) [145]. The same reaction with enantiomerically pure aldehyde produced the oxetane with 95% of enantiomeric excess [146]. Bach *et al.* have studied the Paterno–Büchi reaction with 2,3-dihydropyrrole and α-alkylated enecarbamate as well [147, 148].

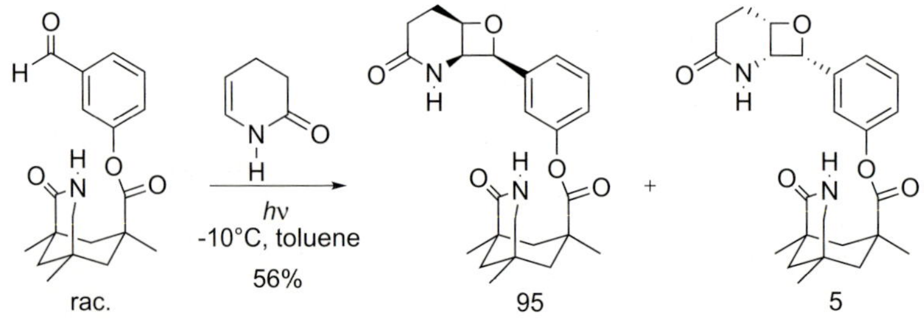

hν
-10°C, toluene
56%
rac.
95
5

Scheme 3.74

Griesbeck *et al.* have worked on the photocycloadditions of oxazole with carbonyl compounds [149]. The 4-unsubstituted 5-methyloxazole gave the cycloadducts with aromatic or aliphatic aldehydes, in high yield and excellent *exo*-diastereoselectivities (98 : 2) (Scheme 3.75). The cycloadducts are precursor of α-amino β-hydroxycarboxylic acid esters [150]. The [2 + 2] cycloaddition of methyl pyruvate with 4-alkylated 5-methoxyoxazoles led to the *exo*-adducts with high diastereoselectivity. When the reaction was run with phenylglyoxylate, oxetanes were formed with moderate diastereoselectivity (79 : 21) [151]. The same authors also studied the Paterno–Büchi reaction of enols with carbonyl compounds [152]. Hydrogen bonding effect has been explored in the reaction of allylic alcohols [153–155].

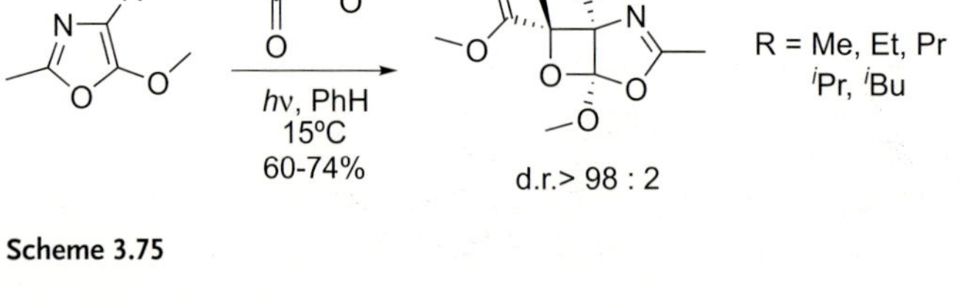

hν, PhH
15°C
86%
d.r.> 98 : 2

hν, PhH
15°C
60-74%
d.r.> 98 : 2
R = Me, Et, Pr, ^{i}Pr, ^{i}Bu

Scheme 3.75

Abe *et al.* have described the formation of 2-siloxy-2-alkoxyoxetanes, which can easily lead to aldol-type adducts, by photoreaction of cyclic ketene silyl acetals with 2-naphthaldehyde (Scheme 3.76) [156].

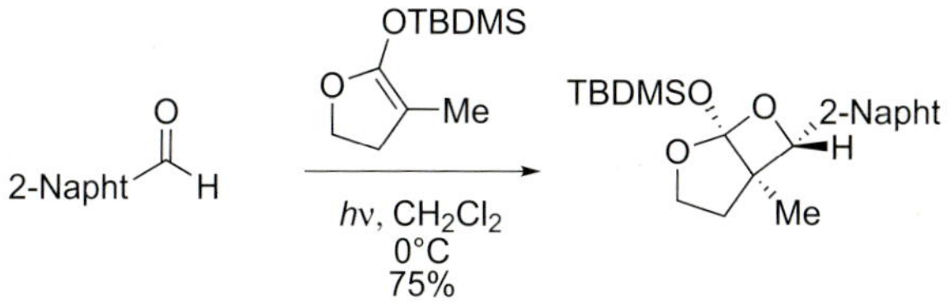

Scheme 3.76

Reaction of aromatic aldehydes with silyl O,S-ketene acetals produced 3-siloxyoxetanes with high regioselectivity due to a S-directed approach (Scheme 3.77) [157]. The same group has also studied the reactivity of furan derivatives in the Paterno–Büchi reaction [158, 159].

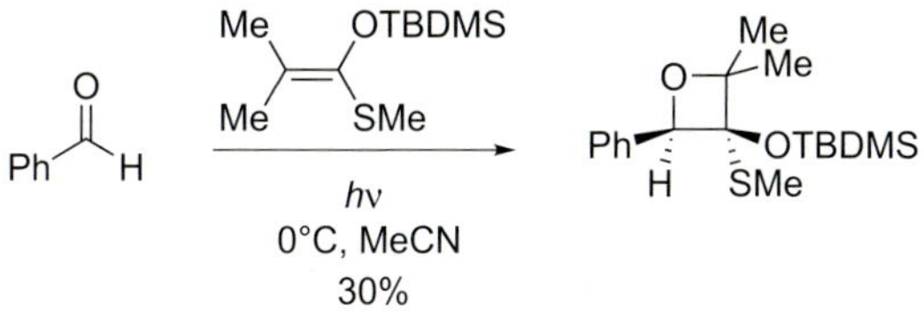

Scheme 3.77

3.2.4.2 **Catalyzed [2 + 2] Cyclizations**

[2 + 2] Cycloaddition can also be performed in a catalytic manner. Oxetanes have been prepared by zirconium(IV) chloride promoted cycloaddition of allylsilane to aldehydes (Scheme 3.78). Previous work was limited to activated carbonyl compounds such as α-ketoesters or α-diketones [160].

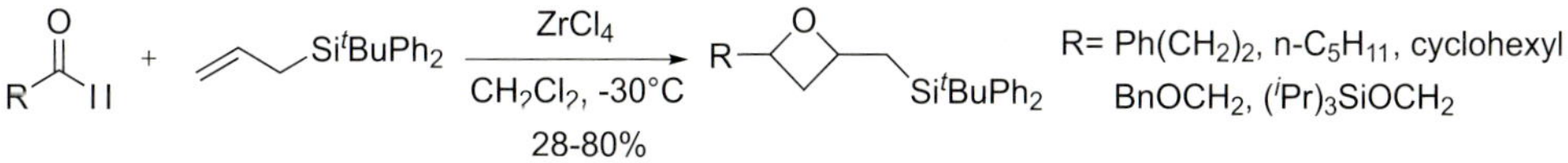

Scheme 3.78

Akiyama and Kirino have developed a stereoselective construction of oxetanes by titanium(IV) chloride promoted [2 + 2] cycloaddition of allylsilane to α-ketoesters. Best results were obtained with bulky substituents on the silicon and toluene as solvent. The diastereoselectivity was up to 96% (Scheme 3.79) [161].

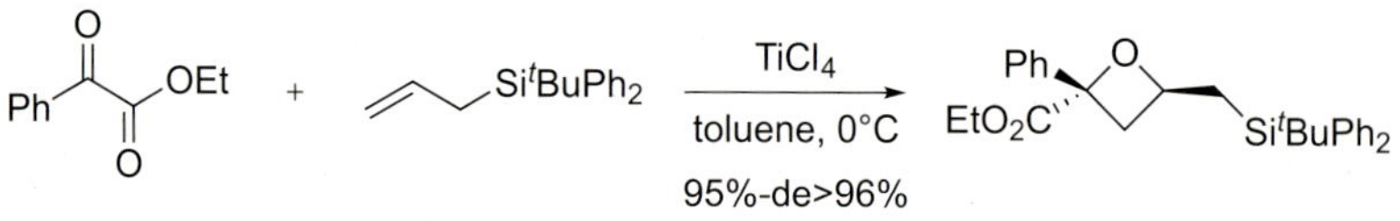

Ph–CO–CO–OEt + allyl–Si^{t}BuPh$_2$ → (TiCl$_4$; toluene, 0°C; 95%-de>96%) Ph, EtO$_2$C oxetane Si^{t}BuPh$_2$

Scheme 3.79

Polyfluorinated oxetanes have been prepared by the reaction between hexafluoroacetone and fluorinated ethylenic compounds. The reaction is catalyzed by aluminium chlorofluoride as Lewis acid. Hydrofluoroethylenes $HXC{=}CF_2$ (X = H, F, Cl, Br) led to only one regioisomer (Scheme 3.80) [162].

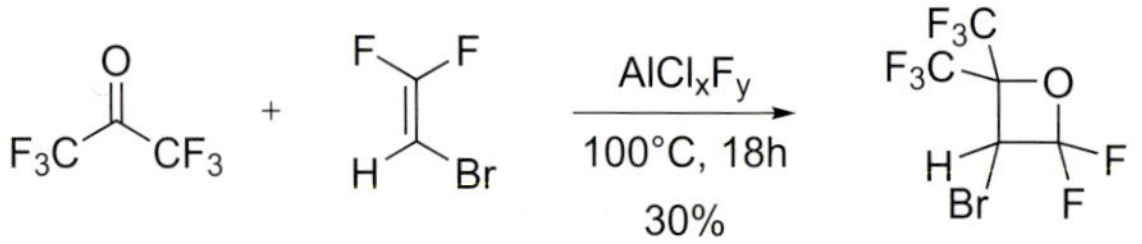

F_3C–CO–CF_3 + F, F, H, Br alkene → ($AlCl_xF_y$; 100°C, 18h; 30%) F_3C, F_3C, O, H, Br, F, F oxetane

Scheme 3.80

3.2.4.3 Ring Contraction of Butanolides

Ring contractions of γ-lactones into oxetanes have been studied and are used in the synthesis of oxetin or oxetanocin analogues (Scheme 3.81) [163, 164].

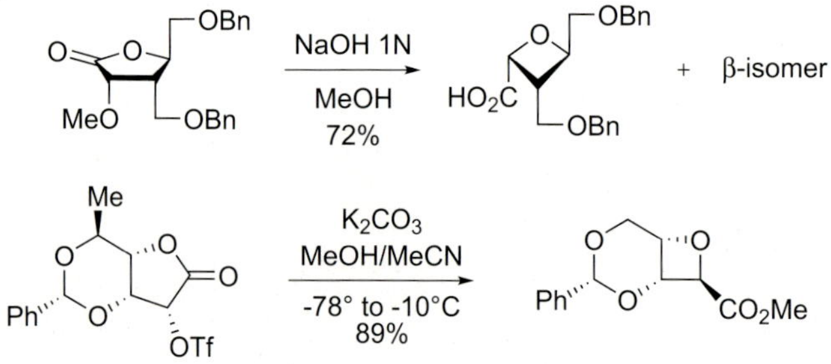

MeO, OBn, OBn lactone → (NaOH 1N; MeOH; 72%) HO_2C, OBn, OBn oxetane + β-isomer

Me, Ph, O, OTf lactone → (K_2CO_3; MeOH/MeCN; -78° to -10°C; 89%) Ph, O, CO_2Me oxetane

Scheme 3.81

Suzuki and Tomooka have found a new anionic ring contraction reaction. They obtained oxetanes by treating cyclic acetal systems with an excess of alkyl lithium (Scheme 3.82) [165]. The four-membered rings are obtained with high diastereoisomeric excesses.

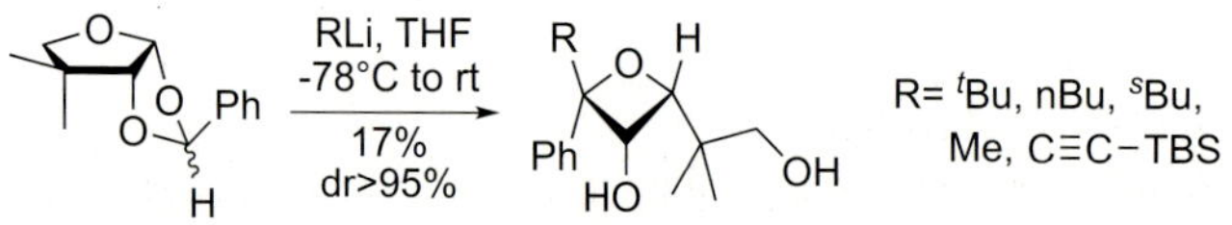

Ph, H acetal → (RLi, THF; -78°C to rt; 17%; dr>95%) R, H, Ph, HO, OH oxetane

R= ^{t}Bu, nBu, ^{s}Bu, Me, C≡C−TBS

Scheme 3.82

β-Lactones have been obtained by contraction of butanolides via an intramolecular nucleophilic displacement of an activated group by the carboxylate anion. De Angelis *et al.* have produced the total inversion of configuration of the (*S*)-β-hydroxy-γ-butyrolactone via a β-lactone intermediate. This last product was also presented as a versatile chiral synthon (Scheme 3.83) [166].

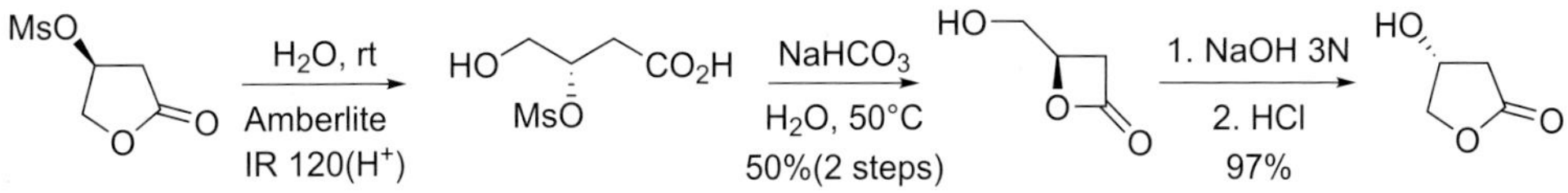

Scheme 3.83

3.2.4.4 **Oxirane Ring Opening by Carbanionic Attacks**

Mordini *et al.* have synthesized oxetanes by the isomerization of oxiranes using an equimolar mixture of butyllithium/diisopropylamine/potassium *tert*-butoxide (LIDAKOR) [167]. The reaction occurred when Y is a phenyl or a propargyl group, and the substituted oxetanes were produced in *anti*-(2,3)-configurations (Scheme 3.84) [168].

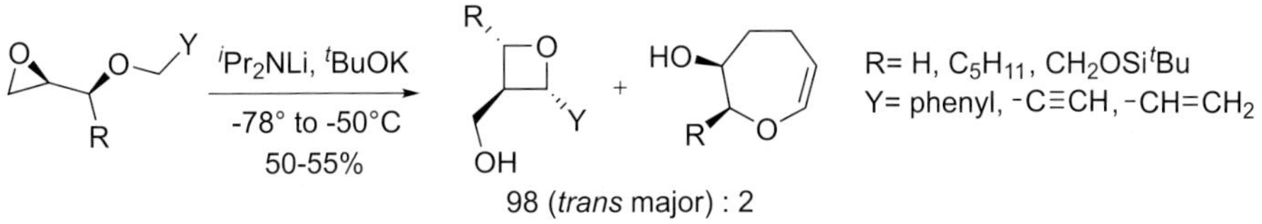

Scheme 3.84

3.2.4.5 **Williamson Reactions**

The intramolecular Williamson reaction has often been used for the preparation of cyclic ethers like oxetanes (Scheme 3.85). This S_N2 displacement of a leaving group (halogen, mesylate, tosylate, etc.) by the alkoxides moiety can be performed with considerable variation in the choice of base (DBU, KOH, ^{t}BuOK, NaHMDS, etc.) [132, 133]. No recent work has been published about this reaction in particular.

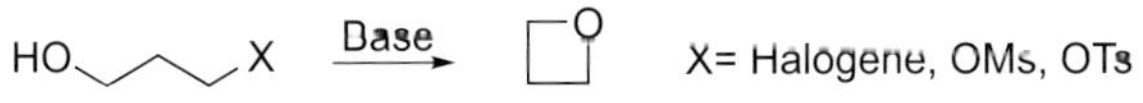

Scheme 3.85

3.2.4.6 **Isomerization of Oxiranyl Hydroxyls**

Since an epoxy oxygen atom can serve as a good leaving group, the isomerization of oxiranyl hydroxyls has been used for oxetane preparation in total syntheses (Scheme 3.86). This reaction can be performed under acidic ($BF_3{\cdot}Et_2O$) or basic

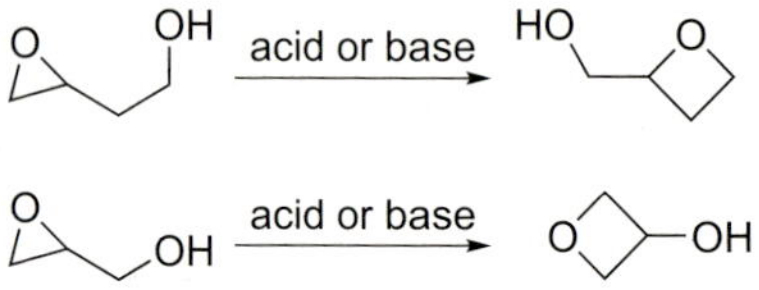

Scheme 3.86

(KOH) conditions [132, 133]. No new studies of this ring opening reaction have been published recently.

3.2.4.7 Oxirane Ring Expansions

Oxetanes can be obtained by treating oxiranes with dimethyloxosulfonium methylide (Scheme 3.87) [132, 133]. This well-known reaction has not receive renewed interest in recent years.

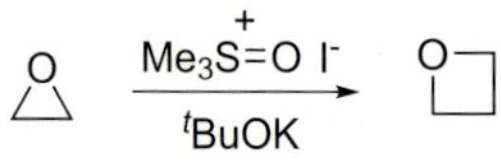

Scheme 3.87

3.2.4.8 Electrophilic Cyclizations

This reaction, which can be performed with mild experimental conditions, is often used in total synthesis. Few methodological studies have been done. Rousseau *et al.* have published the preparation of oxetanes via a 4-*endo* cyclization process on allylic alcohols [169, 170] and via a 4-*exo* cyclization process on homoallylic alcohols [171], using the electrophilic reagent bis(*sym*-collidine)bromine(I) hexafluorophosphate (HBB) or hexafluoroantimonate. This reaction was also used to prepare β-lactones starting from α,β-unsaturated acids (Scheme 3.88) [169].

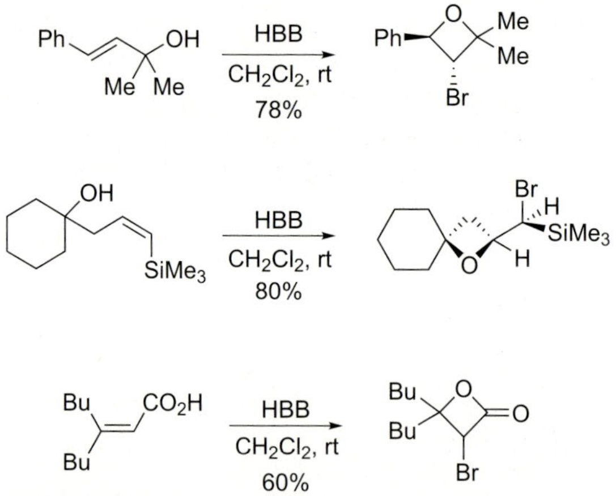

Scheme 3.88

3.2.4.9 [2 + 2] Cycloaddition of Ketene and Carbonyl Compounds

The [2 + 2] cycloaddition of ketenes with carbonyl compounds is an expedient way to substituted β-lactones. Silylketenes are reactant of choice due to their better stability. The most common reagent as Lewis acid is $BF_3{\cdot}Et_2O$, but the goal in this area is now the use of chiral Lewis acid catalysts. Asymmetric [2 + 2] cycloadditions of ketene with aliphatic aldehydes have been catalyzed by chiral aluminium Lewis acids to afford optically active 4-substituted oxetan-2-ones in moderate enantiomeric excesses (Scheme 3.89) [172].

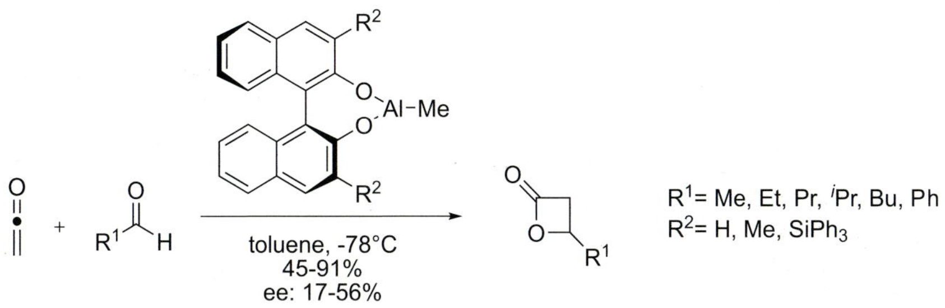

toluene, -78°C
45-91%
ee: 17-56%

R^1= Me, Et, Pr, ^{i}Pr, Bu, Ph
R^2= H, Me, $SiPh_3$

Scheme 3.89

Romo and Yang have developed Ti-TADDOL catalysts such as **20**, which provide good reactivity and moderate enantioselectivity (9–80% ee) in the asymmetric [2 + 2] cycloadditions of silylketenes and aldehydes [173].

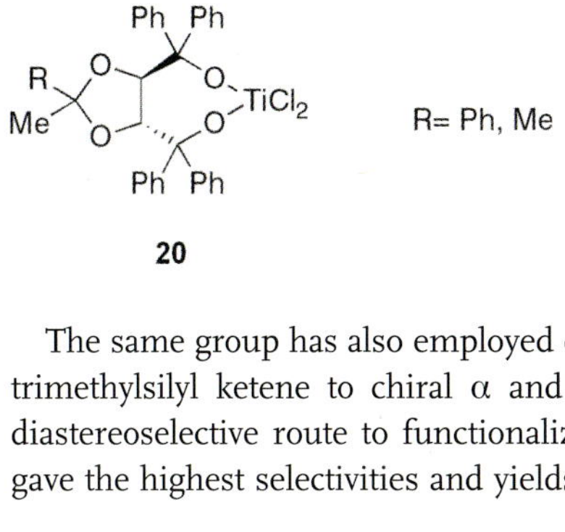

R= Ph, Me

20

The same group has also employed chelation controlled [2 + 2] cycloadditions of trimethylsilyl ketene to chiral α and β-benzyloxyaldehydes to provide a highly diastereoselective route to functionalized β-lactones. The Lewis acid $MgBr_2.Et_2O$ gave the highest selectivities and yields (Scheme 3.90) [174].

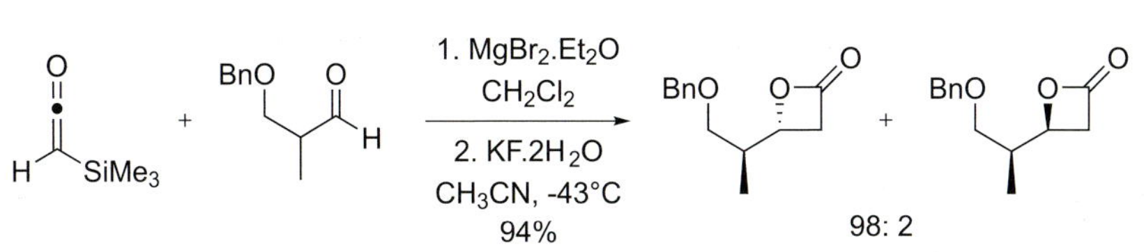

1. $MgBr_2.Et_2O$, CH_2Cl_2
2. $KF.2H_2O$, CH_3CN, -43°C
94%

98: 2

Scheme 3.90

Yamamoto *et al.* have disclosed a highly diastereoselective cyclization under the influence of bulky methyl aluminium bis(4-bromo-2,6-di-*tert*-butylphenoxide) (MABR) as Lewis acid (Scheme 3.91) [175]. No trace of the trans isomer could be detected.

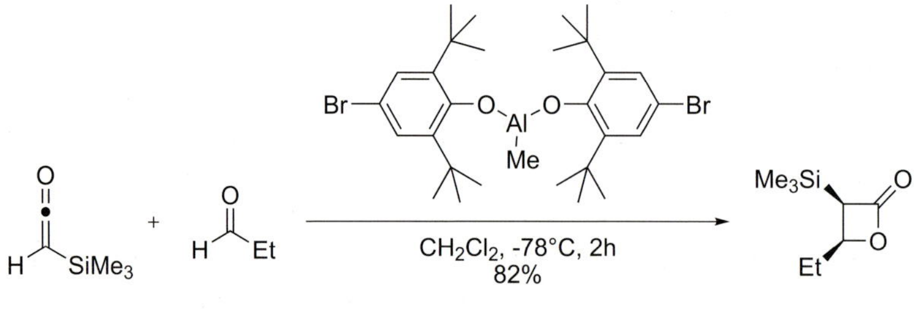

Scheme 3.91

Bis(oxazoline)-Cu(II) complexes have been used by Evans *et al.* to catalyze the enantioselective cycloaddition between silylketenes and chelating carbonyl substrates. β-Lactones have been produced in excellent yields and selectivities (Scheme 3.92) [176].

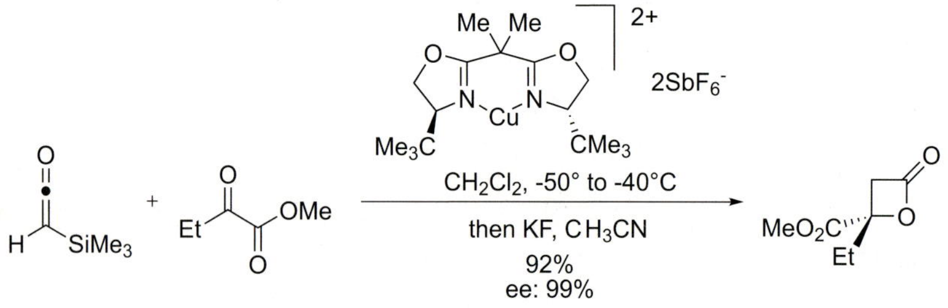

Scheme 3.92

3.2.4.10 **Acyl Halide–Aldehyde Cyclocondensations**

The acyl halide–aldehyde cyclocondensation can be classified between the [2 + 2] cycloaddition of ketenes to carbonyl compounds and the catalyzed aldol-lactonization reaction. Actually, the reaction can progress via *in situ* ketene formation or via an ammonium enolate (Scheme 3.93). This aldol-addition reaction equivalent has induced much interest, especially in its catalyzed asymmetric version. The use of tertiary amines as both base to effect dehydrochlorination and nucleophile to promote the reaction of ketenes and aldehydes has been demonstrated.

Romo and Tennyson have taken their inspiration from the Wynberg procedure [177] for the synthesis of optically active dichlorinated β-lactones via an *in situ* generated ketene. Di-chlorinated aldehydes were reacted with acetyl chloride in the

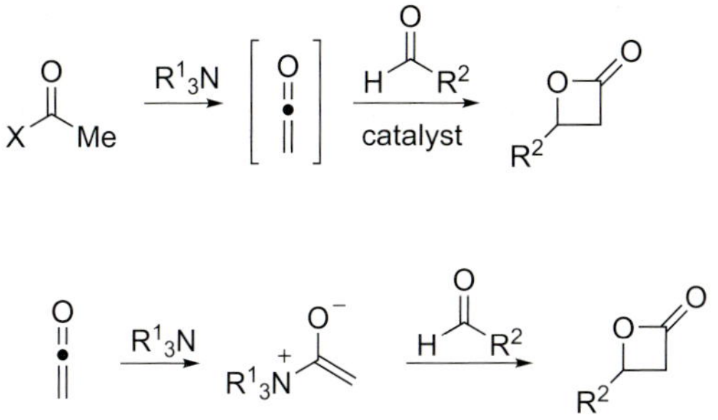

Scheme 3.93

presence of Hunig's base and 2 mol% of quinidine A (**21**). Dichlorinated β-lactones were obtained with 93–98% enantiomeric excesses (Scheme 3.94) [178].

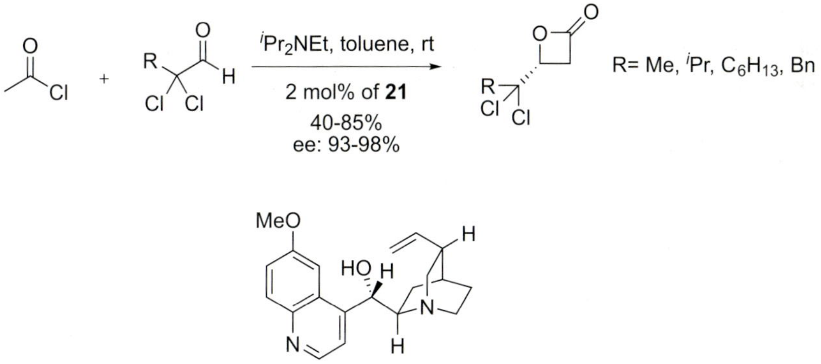

Scheme 3.94

Romo *et al.* have also investigated an intramolecular aldol-lactonization reaction. Chiral bicyclic β-lactones have been obtained with high enantiomeric excess from non-activated aldehydes in the presence of 10 mol% of catalyst **21** (Scheme 3.95) [179].

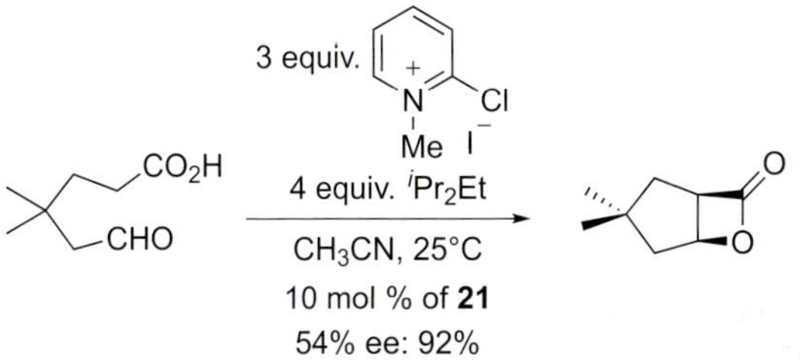

Scheme 3.95

In a first stage, Nelson *et al.* have presented an Al(III)-catalyzed acyl halide–aldehyde cyclocondensation reaction. A catalytic quantity of $Al(SbF_6)_3$ in concert with di (isopropyl)ethylamine constituted the most successful reaction promoter [180].They continued with asymmetric cyclocondensations catalyzed by chiral Al(III) triamine derivatives **22** or **23** [181, 182]. This methodology can be applied to a large range of aldehydes and to substituted ketenes. 3,4-Disubstituted β-lactones are accessible with excellent optical purity (Scheme 3.96).

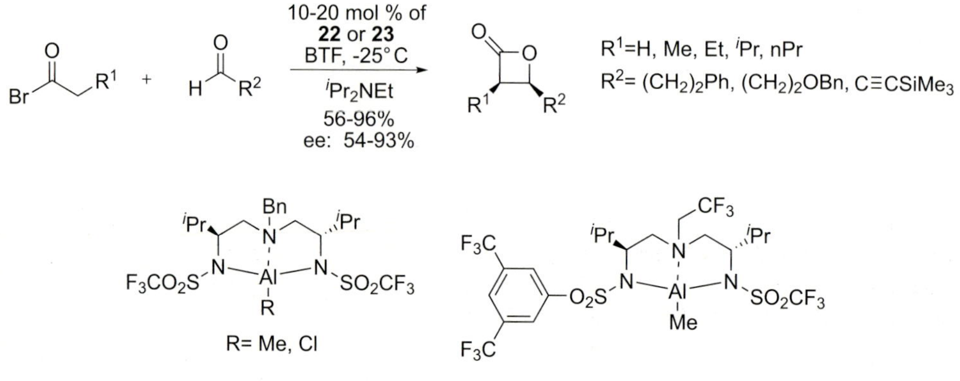

10-20 mol % of **22** or **23**, BTF, -25°C, $^iPr_2NEt$, 56-96%, ee: 54-93%

R^1=H, Me, Et, ^iPr, nPr

R^2= $(CH_2)_2Ph$, $(CH_2)_2OBn$, C≡CSiMe$_3$

R= Me, Cl

22 **23**

Scheme 3.96

In attempting to expand the scope of Wynberg's original ketene–aldehyde cycloadditions, Nelson *et al.* have developed a cinchona alkaloid–Lewis acid catalyzed acid chloride–aldehyde cyclocondensation. The TMSQ/$LiClO_4$ catalyst system allows the reaction with α-branched aldehydes or with methyl ketene. The 3,4-disubstituted β-lactones are obtained in excellent enantiomeric excess (Scheme 3.97) [183].

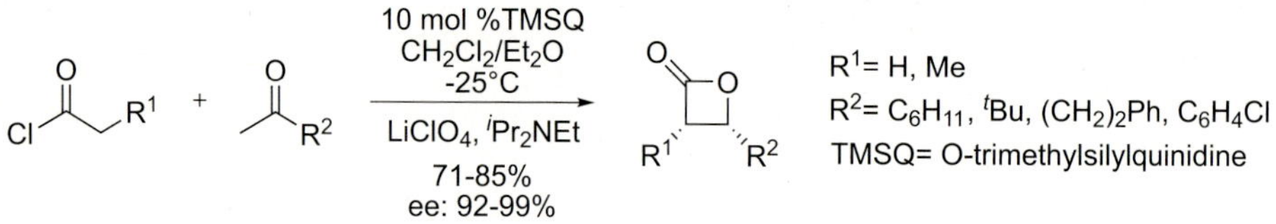

10 mol %TMSQ, CH_2Cl_2/Et_2O, -25°C, $LiClO_4$, $^iPr_2NEt$, 71-85%, ee: 92-99%

R^1= H, Me

R^2= C_6H_{11}, ^tBu, $(CH_2)_2Ph$, C_6H_4Cl

TMSQ= O-trimethylsilylquinidine

Scheme 3.97

3.2.4.11 C–H Insertions

Intramolecular C–H insertion in the catalytic decomposition of diazomalonic esters has often been used in the preparation of β- and γ-lactones. However, this reaction depends on the substitution pattern of insertion centers and the conformational bias of metallocarbenes, leading to a mixture of both β- and γ-lactones [184, 185]. Balaji and Chanda have shown that steric effects may play a major role in the C–H insertion of inactivated α-diazo-α-aroyl esters, catalyzed by rhodium(II) carboxylates. The reaction of α-diazo-α-benzoyl esters catalyzed by various rhodium carboxylates yielded β-lactones as the only products (Scheme 3.98) [186].

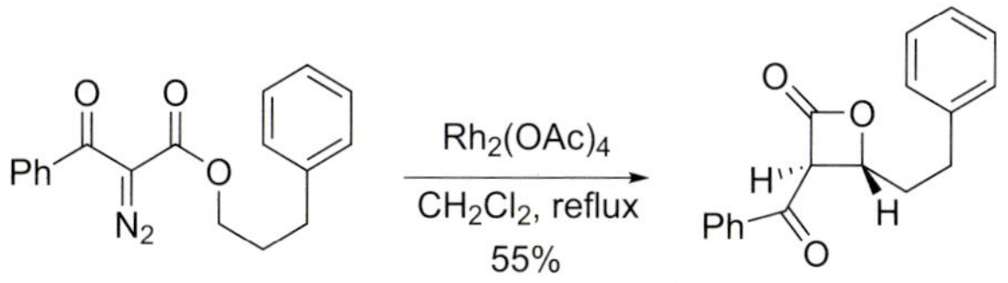
Ph
N2
O
Rh2(OAc)4
CH2Cl2, reflux
55%
H
Ph

Scheme 3.98

C–H insertion reactions catalyzed with immobilized dirhodium(II) salt having mixed chiral ligands have been reported by Doyle *et al.* They observed that the enantioselectivity was slightly increased with the most selective azetidinone-ligated homogeneous catalyst **25** (Scheme 3.99) [187].

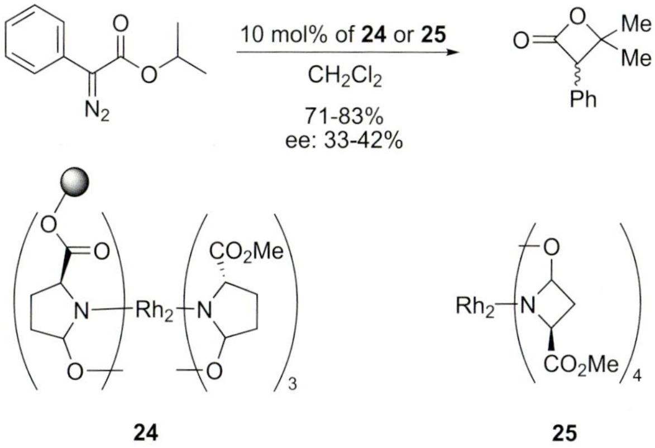
N2
10 mol% of 24 or 25
CH2Cl2
71-83%
ee: 33-42%
Me
Me
Ph
CO2Me
Rh2
N
3
4
24
25

Scheme 3.99

3.2.4.12 **Carbonylative Ring Expansion Reactions**

Alper *et al.* have described the carbonylation of epoxides with a new catalyst, $PPNCo(CO)_4$ [PPN=bis(triphenylphosphine)iminium], used in conjunction with $BF_3 \cdot Et_2O$. Carbonylations occurred selectively at the unsubstituted C–O bond of the epoxide rings. These reactions tolerate various functional groups such as alkenyls, halides, hydroxyls and alkyl ethers. The β-lactones are obtained in good yields without polymeric by-products (Scheme 3.100) [188].

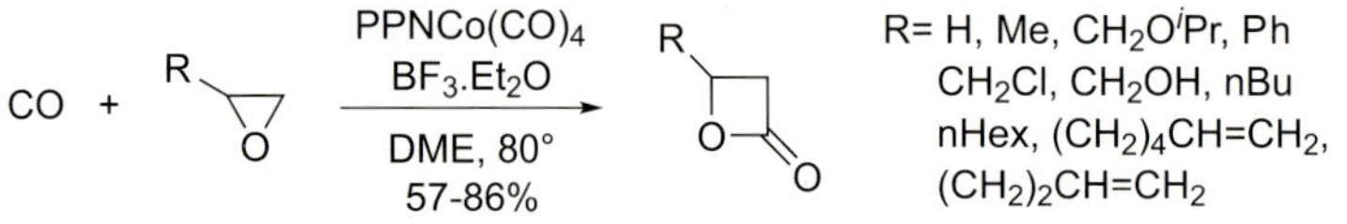

CO + R-epoxide → (PPNCo(CO)$_4$, BF$_3$.Et$_2$O, DME, 80°, 57-86%) β-lactone

R= H, Me, CH$_2$O^{i}Pr, Ph, CH$_2$Cl, CH$_2$OH, nBu, nHex, (CH$_2$)$_4$CH=CH$_2$, (CH$_2$)$_2$CH=CH$_2$

Scheme 3.100

Coates *et al.* have developed three catalyst for the carbonylation of substituted epoxides: [(salph)Al(THF)$_2$][Co(CO)$_4$] (**26**) [189], [Cp$_2$Ti(THF)$_2$][Co(CO)$_4$] (**27**) [190] and [(TPP)Cr(THF)$_2$][Co(CO)$_4$] [**28**, with THF in the axial positions (not shown)] [191]. Catalyst **28** is the most active and selective with a wide range of epoxides.

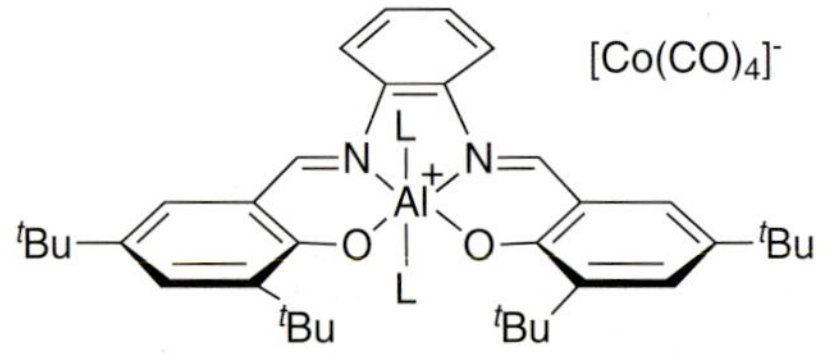

[Co(CO)$_4$]$^-$ **26** L= THF

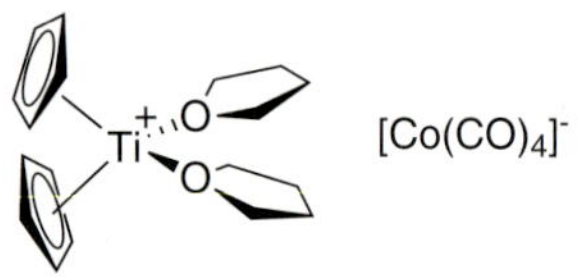

[Co(CO)$_4$]$^-$ **27**

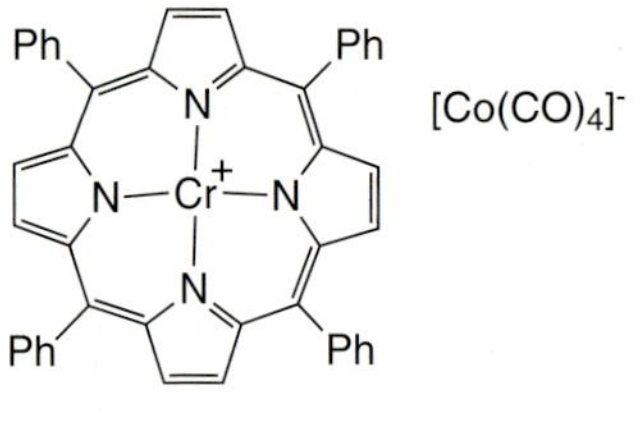

[Co(CO)$_4$]$^-$ **28**

3.2.4.13 β-Hydroxy Acid Cyclizations

The cyclization of β-hydroxy acids is largely used in the synthesis of molecules possessing a β-lactone moiety. This reaction can be executed following an addition–elimination process or a nucleophilic substitution process [132, 133]. In the addition–elimination way, Adam's method (pyridine/ArSO$_2$Cl; Ar = Ph, 4-MePh, 4-NO$_2$Ph) [192] is the most commonly used, but different reactants can also activate the carboxylic acid, such as Et$_3$N/BOPCl [bis-(2-oxo-3-oxazolidinyl)phosphonic chloride], DCC/HOBt, EDC/HOBt, etc. The intramolecular nucleophilic substitution can be pursued by the Mitsunobu reaction (PPh$_3$/DEAD or DMAD) [193].

3.2.5 Reactivity

3.2.5.1 β-Lactones

The reactivity of β-lactones has been already [194]. It can be broken down into four areas:

- nucleophilic attack with ring opening;
- Lewis acid promoted rearrangement;

- decarboxylation;
- enolate formation and reaction toward electrophiles.

3.2.5.1.1 **Nucleophilic Attacks** It has already been demonstrated that the ring opening of β-lactones can occur through two different pathways. Attack via an addition–elimination process at the carbonyl residue affords access to β-hydroxy adducts. A nucleophilic substitution reaction at the C4 atom can produce β-substituted carboxylic acid derivatives (Scheme 3.101).

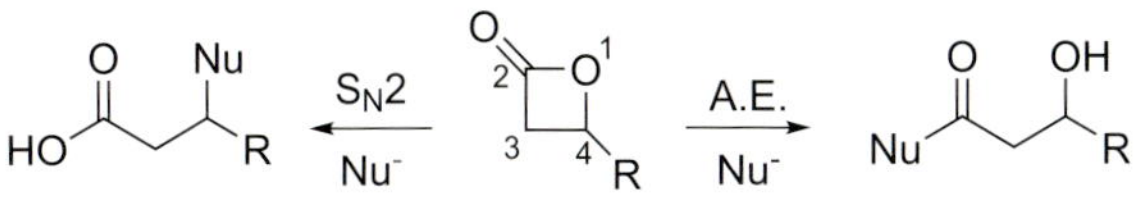

Scheme 3.101

Organomagnesium and organolithium reagents attack β-lactones at the carbonyl, leading to oxygen–acyl cleavage. Organocuprates induce the oxygen alkenyl cleavage. Ring opening of β-lactone by hetero-nucleophiles has been much studied. Goodman *et al.* [195], inspired by Vederas' results [196], have synthesized lanthiomine derivatives by ring opening of a protected serine β-lactone by the thiolate anion of methyl Boc-*(S)*-cysteinate derivatives (Scheme 3.102).

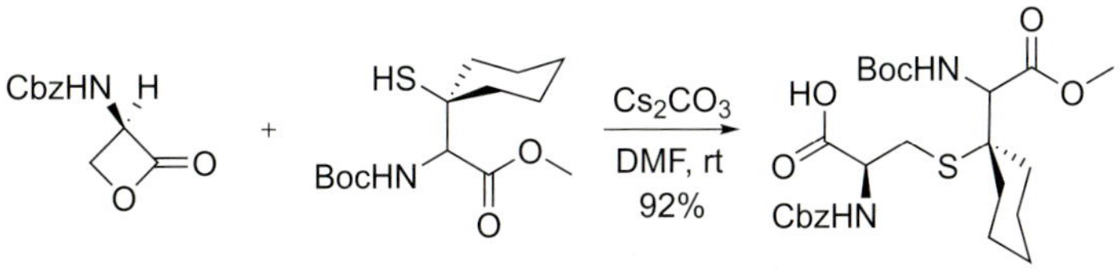

Scheme 3.102

Palomo *et al.* [197] have obtained dipeptides by coupling an α-dichloro β-lactone with *(S)*-leucine or *(S)*-phenylalanine and subsequent dechlorination (Scheme 3.103).

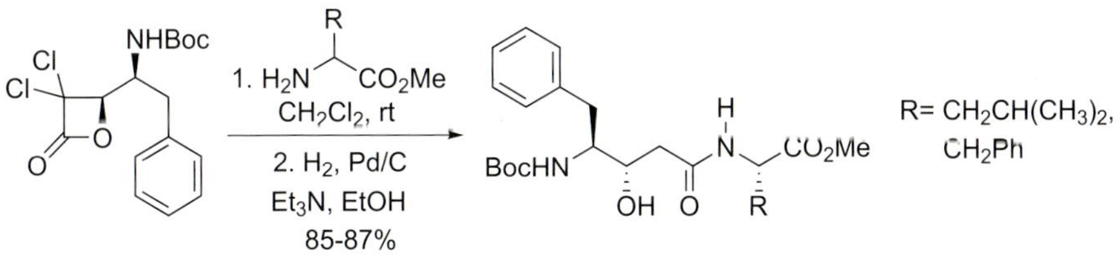

Scheme 3.103

Ring opening of β-lactones by hydrazone anions, followed by dehydroamination cyclizations of the β-ketohydrazone intermediates in the presence of amberlyst-15 acid resin, yields dihydropyrones (Scheme 3.104) [198].

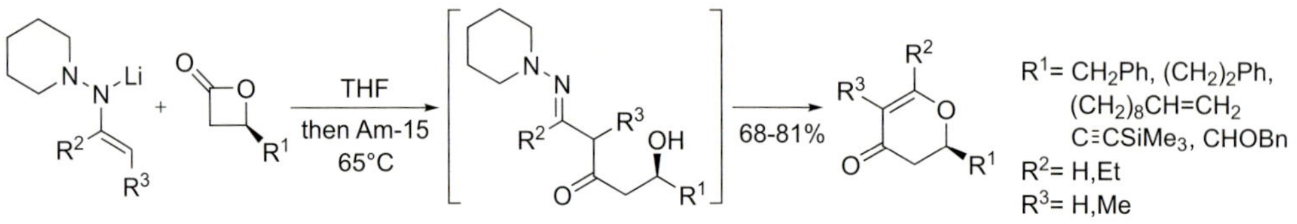
THF
then Am-15
65°C
68-81%
R1= CH2Ph, (CH2)2Ph,
(CH2)8CH=CH2
C≡CSiMe3, CHOBn
R2= H,Et
R3= H,Me

Scheme 3.104

Nelson *et al.* [199] have demonstrated that primary and secondary amines promote a ring opening by addition–elimination to deliver β-hydroxyamides (Scheme 3.105). Sodium azide reacts in an S_N2 manner to produce β-azido acids, and sulfonamide anions give rise to β-amino acids.

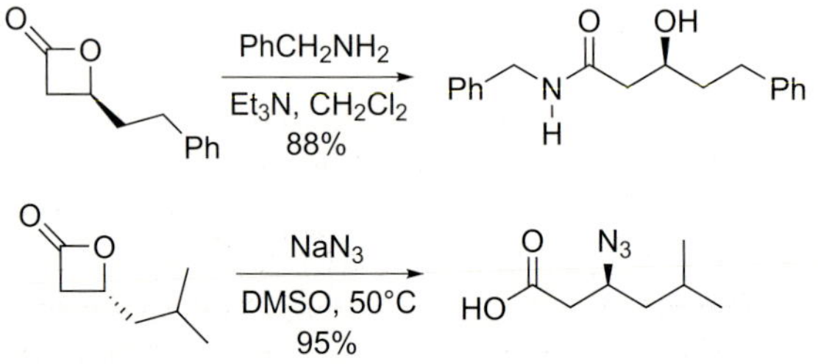
PhCH2NH2
Et3N, CH2Cl2
88%
NaN3
DMSO, 50°C
95%

Scheme 3.105

Acylation of aromatic compounds with chiral β-trichloromethyl β-propiolactone in the presence of Lewis acid has been investigated by Fujisawa *et al.* [200]. The Friedel–Crafts products retain completely the stereochemical integrity of the starting β-lactone. This method has been applied to the synthesis of a precursor of enalapril, an angiotensin converting enzyme (ACE) inhibitor (Scheme 3.106).

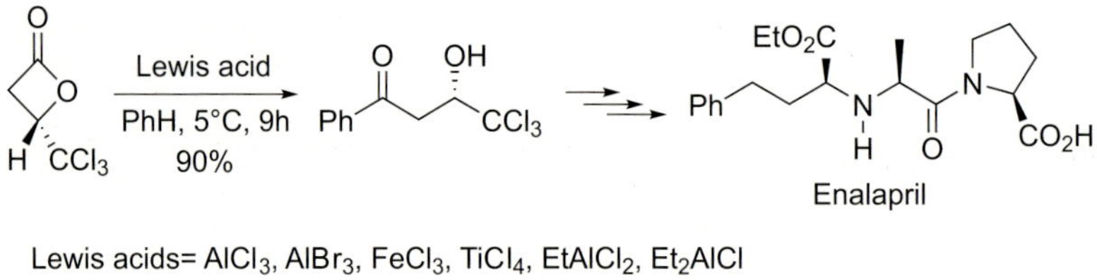
Lewis acid
PhH, 5°C, 9h
90%
Enalapril
Lewis acids= AlCl3, AlBr3, FeCl3, TiCl4, EtAlCl2, Et2AlCl

Scheme 3.106

The utility of β-lactones as intermediates for asymmetric synthesis has been limited by the difficulty for their preparation in enantiomerically pure form. Nelson and Spencer [201] have examined enzymatic resolution for the preparation of enantiomerically enriched β-lactones (Scheme 3.107). This method allows the preparation of 4-substituted β-propiolactones with high enantiomeric excesses simultaneously with enantiomerically enriched β-hydroxy esters.

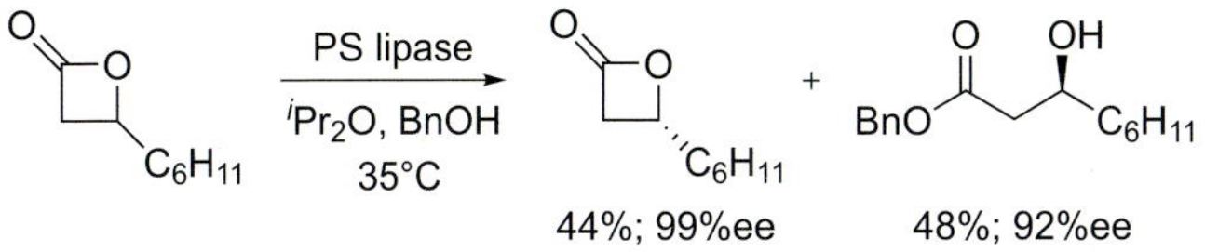

Scheme 3.107

3.2.5.1.2 **Lewis Acid Promoted Rearrangements** Ring expansion of β-lactones to afford γ-lactones can be performed with various Lewis acids, *p*.TsOH, $ZnCl_2$, BF_3.OEt or Ti(O-^{i}Pr)$_4$. The best catalyst appears to be $MgBr_2$ (Scheme 3.108) [202].

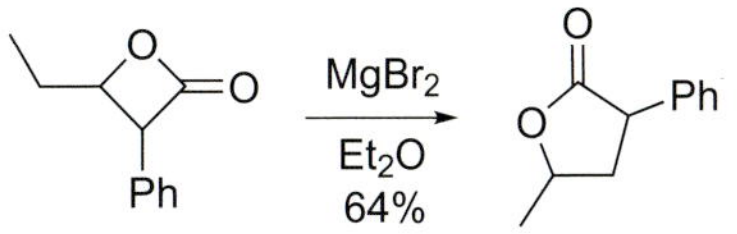

Scheme 3.108

This dyotropic rearrangement involves simultaneous positional interchange of two adjacent atoms having an anti-coplanar stereochemistry relationship. Inversion of stereochemistry at C4 atom is always observed [203]. Black *et al.* have prepared spiro [204] and cis-fused γ-lactones (Scheme 3.109) [205] via such β-lactones rearrangements. β-Elimination can occur in place of rearrangement when the C4 atom is tertiary [206].

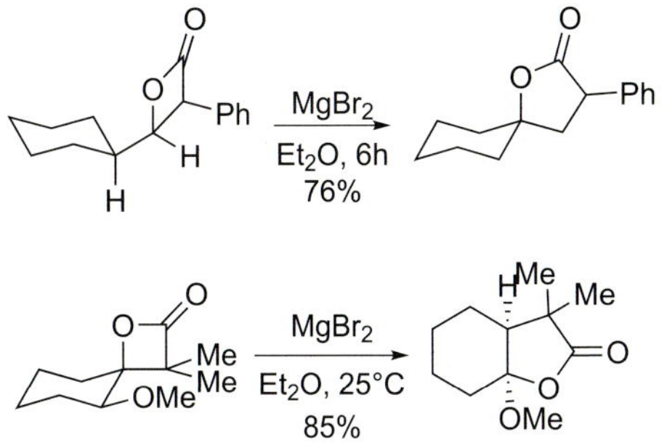

Scheme 3.109

3.2.5.1.3 **Decarboxylations** The thermal decomposition of β-lactones into alkenes takes place between 80 and 160 °C. The reaction is a stereospecific cis-elimination: cis-disubstituted β-lactones lead to a *(Z)*-olefins whereas trans-disubstituted β-lactones lead to *(E)*-olefins. These reactions can be regarded as alternative to Wittig reactions. Dolbier *et al.* have used this method to synthesize 1,1-difluoroalkenes via α,α-difluoro-β-lactones (Scheme 3.110) [207, 208].

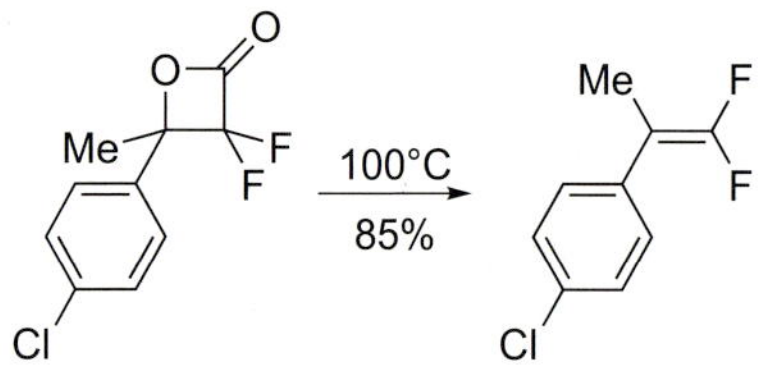

Scheme 3.110

Adam *et al.* [209] have obtained allyl amines and sulfides by conjugate addition of amine and thiol nucleophiles to α-methylene β-lactones and subsequent decarboxylation (Scheme 3.111).

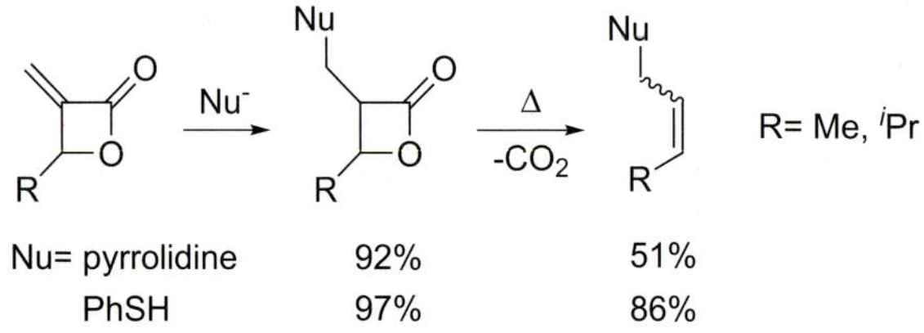

Scheme 3.111

With the same strategy the conjugate addition of ester and ketone enolates to α-methylene β-lactones followed by decarboxylation leads to γ,δ–unsaturated esters with complete stereoselectivity (Scheme 3.112) [210].

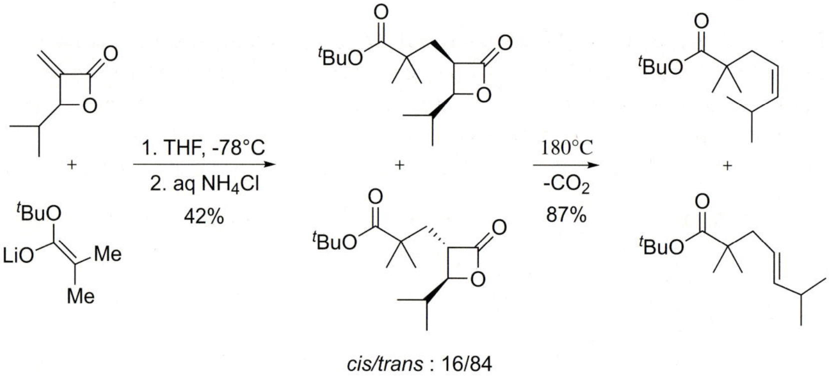

Scheme 3.112

3.2.5.1.4 **Enolate Formation and Reaction Towards Electrophiles** The enolate of β-lactones can be generated by treatment with lithium diisopropylamide and trapped with various electrophiles such as alkyl-, allyl- or propargyl halides, aldehydes, dimethyl maleate, etc. Diastereoselective reactions take place when oxetan-2-ones are β-substituted, leading to the trans-disubstituted products (Scheme 3.113) [194].

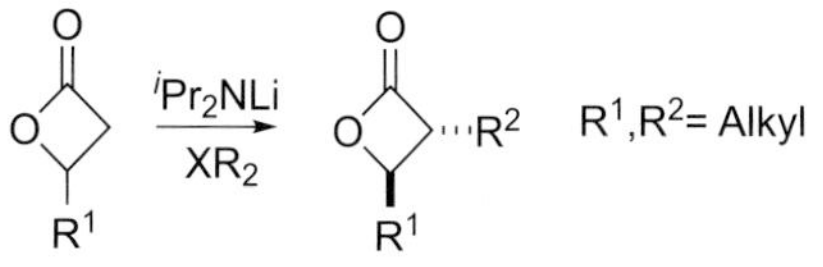

$^{i}Pr_2NLi$, XR_2 — R^1,R^2= Alkyl

Scheme 3.113

As the alkylation of β-lactones unsubstituted at the α-position leads to low yields, a desilylation–alkylation method has been introduced by Mead *et al.* [211]. Kocienski *et al.* have used this modified procedure, in the synthesis of the hypocholesterolemic agent 1233A, to introduce a hydroxymethyl group α to the carbonyl, via the tetrabutylammonium enolate of a β-lactone (Scheme 3.114) [212].

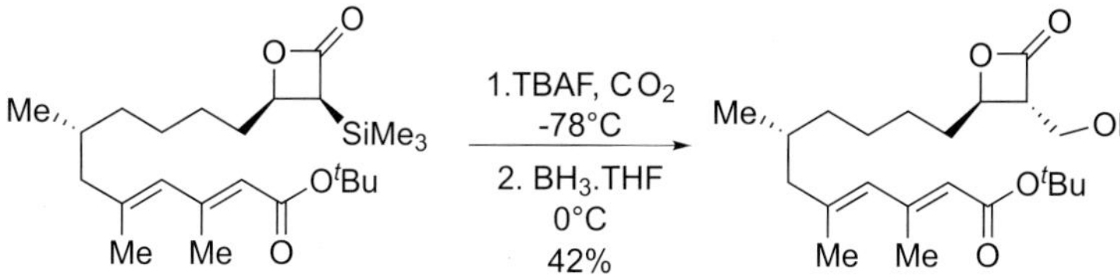

1.TBAF, CO_2, -78°C; 2. BH_3.THF, 0°C, 42%

Scheme 3.114

3.2.5.1.5 **Polymerization of β-Lactones** Polymerization of β-lactones has aroused interest as biodegradable polymer products can be applied in biomedical applications [213–215]. Pohl *et al.* [216] have synthesized chiral polyesters by protic acid-catalyzed polymerization of β-lactones. A stereogenic site repeated in each molecular unit can change the polymer properties, in comparison with the racemic polymers (Scheme 3.115).

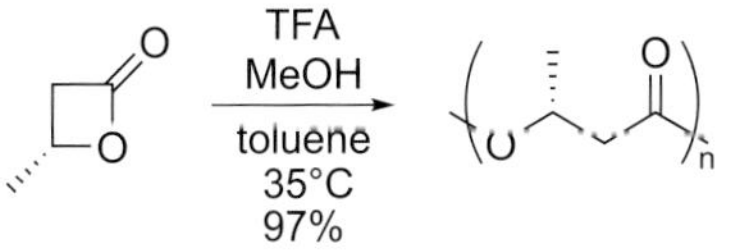

TFA, MeOH, toluene, 35°C, 97%

Scheme 3.115

3.2.5.1.6 **Miscellaneous** Methyl ketene dimer has been used to prepare β-ketoesters or β-ketoamides by nucleophilic attack on the carbonyl function by lithium amides or amines (Scheme 3.116) [217].

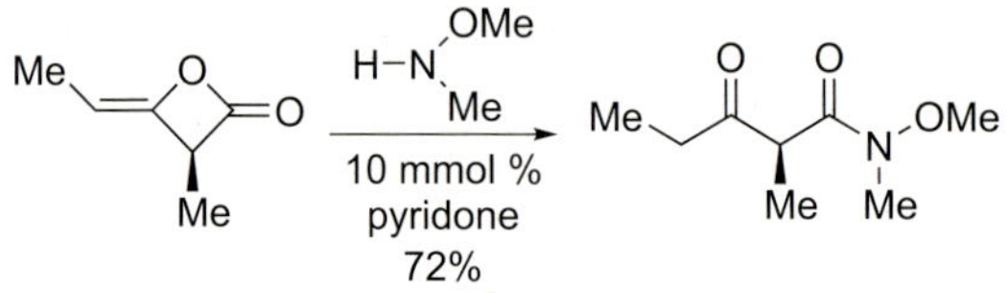

Scheme 3.116

3.2.5.2 **Oxetanes**

3.2.5.2.1 **Nucleophilic Attacks** Two positions (C2 and C4) in the oxetane are amenable to nucleophilic attack. Bach *et al.* [218] have studied intramolecular ring opening reactions of 2-phenyl-3-oxetanols. They have obtained tetrahydropyrans, thiotetrahydropyrans or piperidines by anionic attack at the C4 carbon atom of a heteroatom attached at the C3 position via an alkyl or aryl chain (Scheme 3.117).

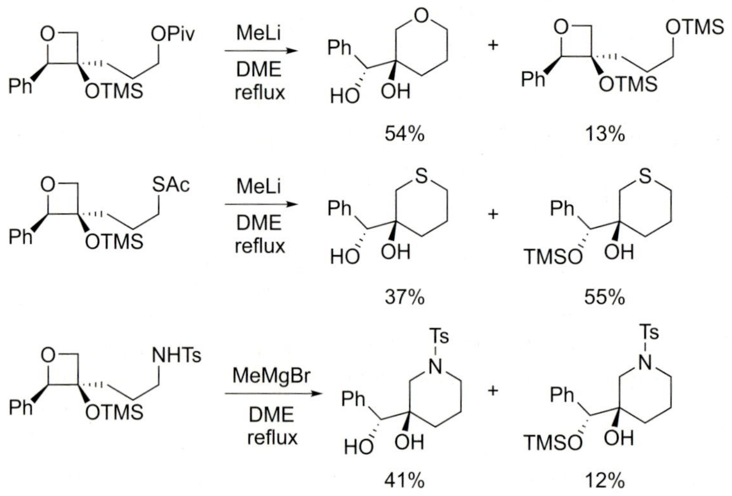

Scheme 3.117

A second ring-opening reaction proceeds by attack at the C2 position upon activation by acid reagents. Boc-protected 3-oxetanols have been transformed into cyclic carbonates as a mixture of diastereomers (Scheme 3.118) [218].

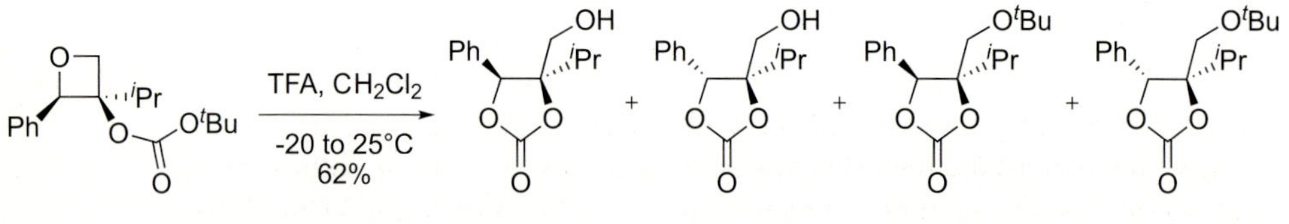

Scheme 3.118

Howell and Hashemzadeh [219] have carried out reductive cleavage of 3,3-dimethyl-2-methylene-4-phenyl oxetane with lithium and 4,4′-di-*tert*-butylbiphenyl (DTBB). The resulting dianion reacts with aldehydes and ketones to give aldol adducts (Scheme 3.119).

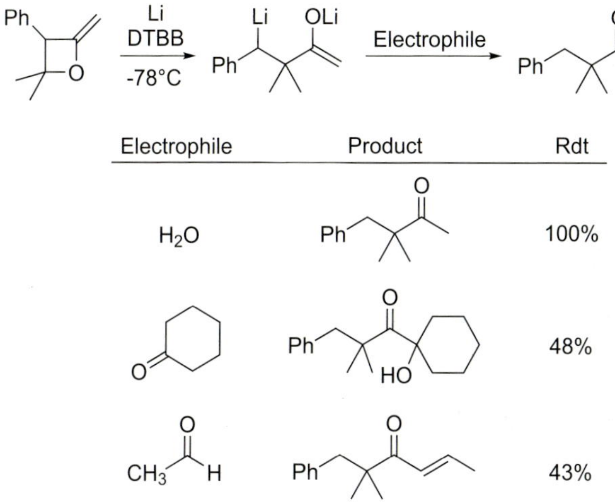

Li, DTBB, -78°C; Electrophile

Electrophile	Product	Rdt
H_2O		100%
	HO	48%
CH_3CHO		43%

Scheme 3.119

The ring opening of oxetanes by samarium diiodide and acyl chloride has been investigated, but the regioselectivity was not always satisfactory. In general, a mixture of iodo products was obtained. In the particular case of 2-phenyloxetane, only the deiodinated product was isolated (Scheme 3.120) [220].

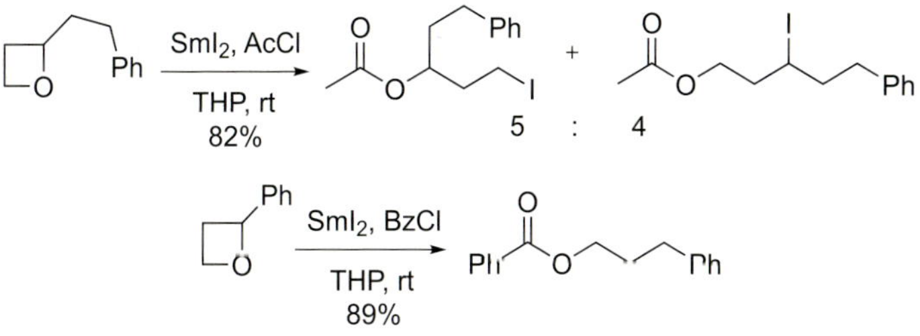

SmI_2, AcCl; THP, rt; 82%; 5 : 4

SmI_2, BzCl; THP, rt; 89%

Scheme 3.120

Kellogg *et al.* [221] have studied the acid catalyzed ring opening reactions of optically pure 2-aryl-3,3-dimethyloxetanes. The aqueous or alcoholic sulfuric acid catalyzed ring opening reaction occurs at the benzylic position with partial inversion of configuration (Scheme 3.121).

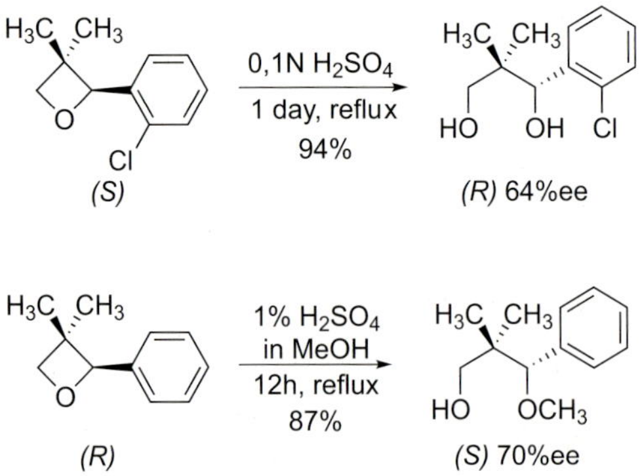

Scheme 3.121

Lewis acid catalyzed nucleophilic ring opening with alkyl lithiums or lithium thiolates occurs at the less hindered carbon with preservation of the stereochemical integrity. Benzyl alcohols are obtained in enantiomerically pure form, with good yields (Scheme 3.122) [221]. Grignard reagents or amines with or without acid catalyst were not successful in the ring opening.

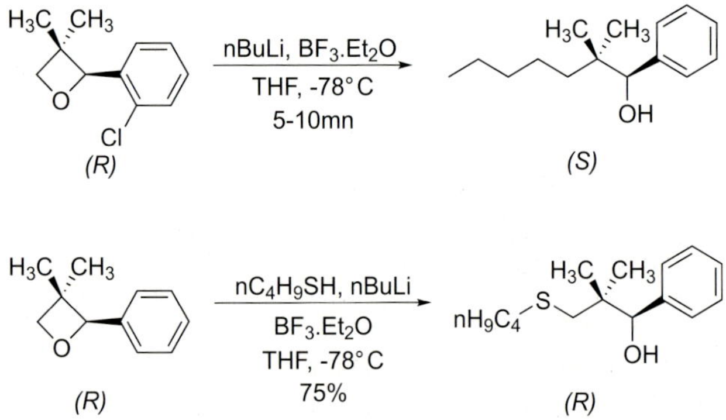

Scheme 3.122

Bach *et al.* [222] have shown that substituted oxetane rings could be opened at the more hindered carbon with $LiAlH_4$. To induce this regioselectivity they attached a hydroxyl group at the arene C2 substituent (Scheme 3.123).

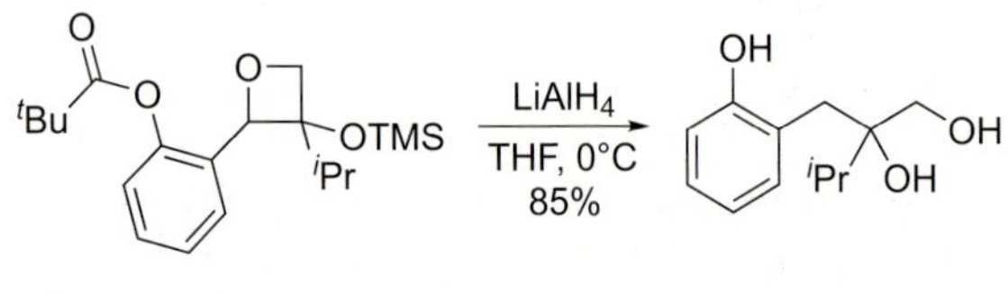

Scheme 3.123

3.2.5.2.2 **Ring Expansions** Nozaki and Noyori first reported in 1966 the asymmetric ring expansion of oxetanes to tetrahydrofurans using chiral copper catalyst (Scheme 3.124) [223].

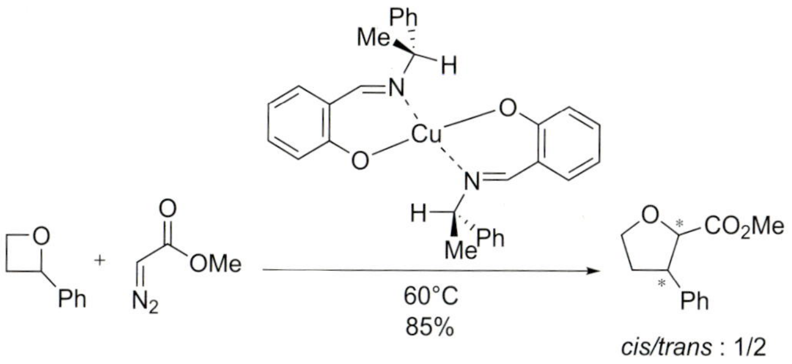

60°C
85%

cis/trans : 1/2

Scheme 3.124

Katzuki has established that this copper-catalyzed reaction proceeds with good stereoselectivity in the presence of bipyridine ligands [224]. Reaction of 2-aryl substituted oxetanes with *tert*-butyl diazoacetate in the presence of a chiral Cu complex furnishes 2,3-disubstituted furan derivatives with high enantiomeric excesses (Scheme 3.125).

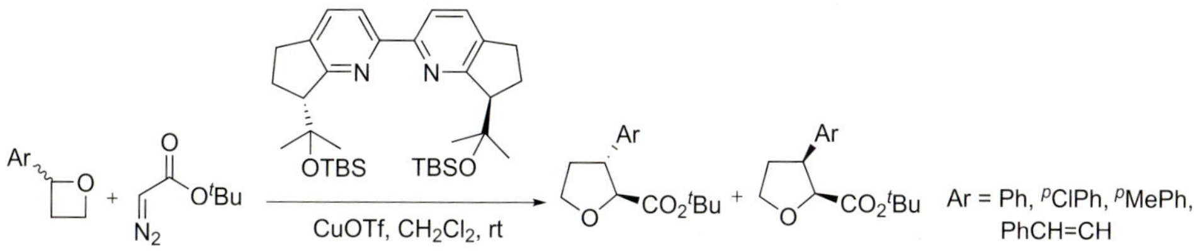

$CuOTf, CH_2Cl_2$, rt

Ar = Ph, ^{p}ClPh, ^{p}MePh, PhCH=CH

(*R*)-2-Phenyloxetane (89%ee)	Rdt= 35%	89 (92%ee)	11 (16%ee)
(*S*)-2-Phenyloxetane (85%ee)	Rdt= 30%	25 (11%ee)	75 (93%ee)

Scheme 3.125

This enantiospecific ring expansion method has been applied to formal syntheses of the natural products (−)-avenaciolide (**29**) and (−)-isoavenaciolide (**30**).

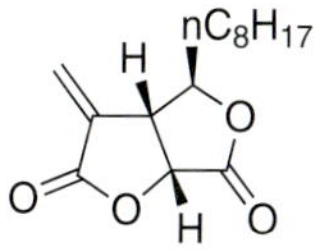

(-)-Avenaciolide

29

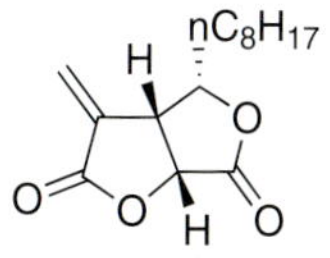

(-)-Isoavenaciolide

30

Fu *et al.* [225] have explored this reaction with a Cu(I) bisazaferrocene catalyst. Both trans- and cis-disubstituted tetrahydrofurans with good enantiomeric excesses are obtained when (*R,R*) or (*S,S*) catalysts **31** are used respectively (Scheme 3.126).

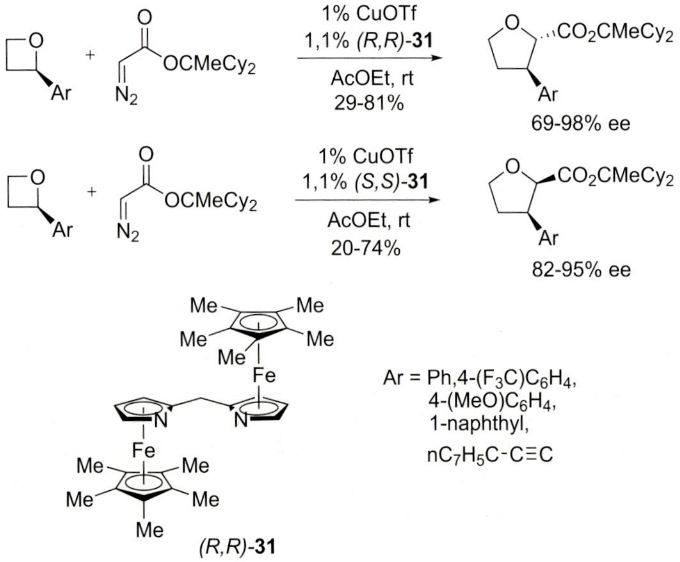

Scheme 3.126

An unusual ring expansion of 2-methyleneoxetane has been observed by Howell *et al.* [226] in the presence of lithium and 4,4′-di-*tert*-butylbiphenyl (Scheme 3.127). The resulting lactone was postulated to arise from a coupling between a radical enolate derived from 2-methyleneoxetane and the acetaldehyde enolate, a decomposition product of THF.

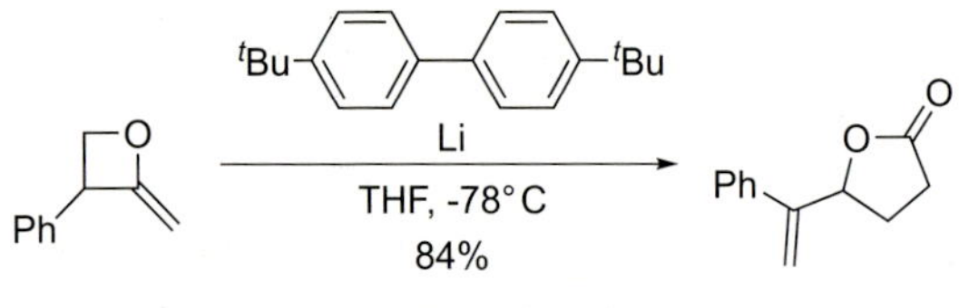

Scheme 3.127

Hara *et al.* [227] have focused on the stereoselective synthesis of fluorinated cyclic ethers, since their derivatives seemed to be involved in biochemical mechanisms. They have reported the ring expansion of oxetanes using 4-iodotoluene difluoride. Fluorinated furans were obtained in good yield and as a single stereoisomer from 2,4-substituted oxetanes (Scheme 3.128).

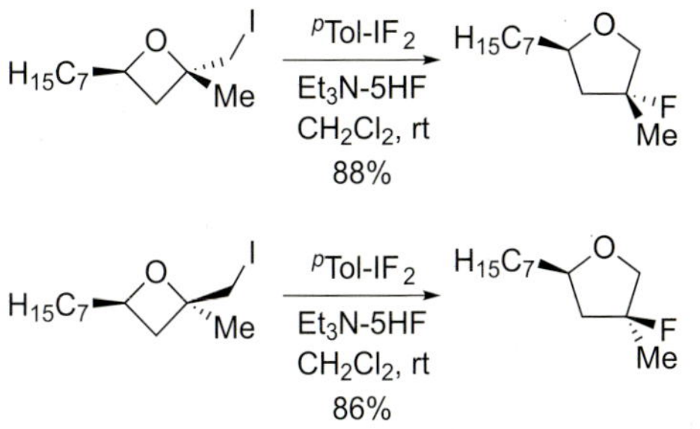

Scheme 3.128

3.2.5.2.3 **2-Methylene Oxetanes Ring Openings** Howell *et al.* [228] have worked on 2-methyleneoxetane derivatives. This ring system contains several potential reactive features: the ring strain, an exocyclic double bond, electron-rich enol ether and a latent enolate leaving group. They demonstrated that 2-methyleneoxetanes underwent regioselective ring opening at the C2 position when treated with trimethylaluminium and either phenyl- or butyllithium. The homopropargylic alcohol was then isolated as a single product in good yield. The reaction in the presence of LDA, the additive was not necessary (Scheme 3.129).

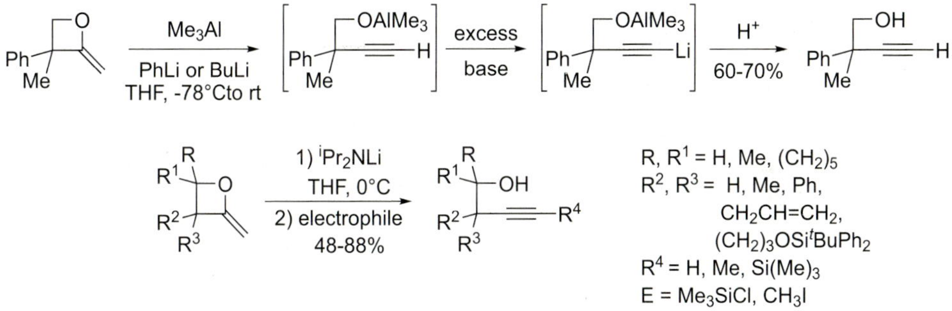

Scheme 3.129

The same group then achieved nucleophilic ring opening of 2-methyleneoxetanes at C4 [229]. Stabilized carbanionic nucleophiles and heteroatom nucleophiles provided C4 ring opening, leading to β-functionalized ketones (Scheme 3.130).

3.2.5.2.4 **Thermolysis and Photolysis** Thermolysis of oxetane derivatives can lead to multiple products from unsymmetrically substituted rings. The lack of regioselectivity has been attributed to similar bond dissociation energies for the carbon–carbon and carbon–oxygen bonds (Scheme 3.131) [132, 133].

Photolytic fragmentation of oxetane is the formal reverse of the Paterno–Büchi reaction. This reaction has attracted some interest as it appears to be involved in the

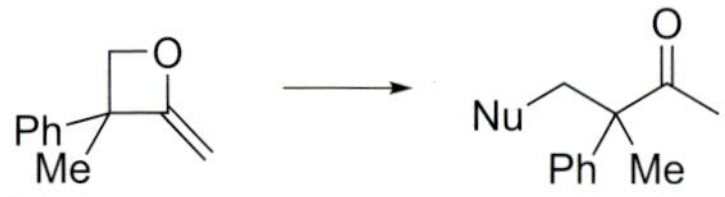

Conditions	Rdt	Nu
$PhCH_2Li$, BuLi, $PhCH_3$, TMEDA, 0°C	81%	$PhCH_2$
(allyl)MgCl, THF, reflux	86%	(allyl)CH_2
PhOH, NaH, DMF, 100°C	70%	PhO
NaI, DMF/H_2O, 100°C	40%	I

Scheme 3.130

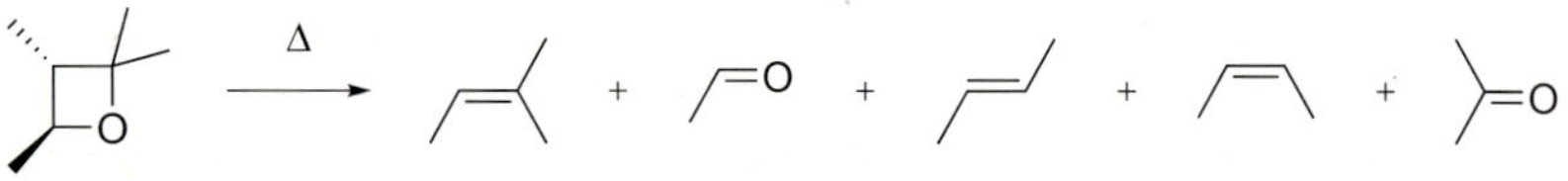

Scheme 3.131

photoenzymatic repair of the (6–4) photoproducts of DNA dipyrimidine sites by the enzyme photolyase [230].

3.2.5.2.5 **Polymerizations** Kakuchi *et al.* [231] have prepared a fused 15-crown-4 polymer, a novel ladder polymer, by a two-step polymerization of an oxetanyl oxirane. This polymer showed metal cation binding properties (Scheme 3.132).

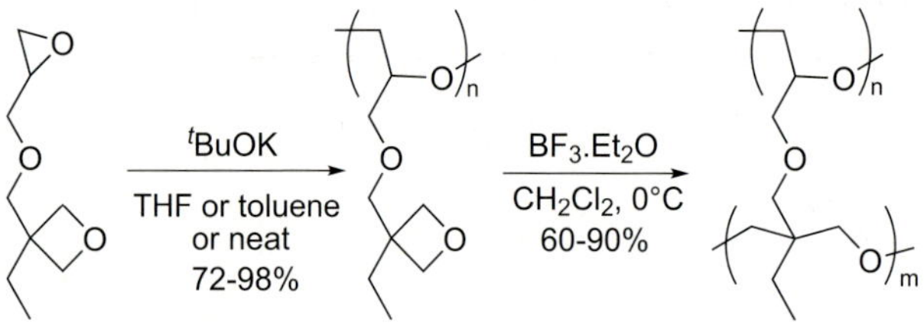

^{t}BuOK; THF or toluene or neat; 72-98%

$BF_3.Et_2O$; CH_2Cl_2, 0°C; 60-90%

Scheme 3.132

3.3 Thietanes

3.3.1 Introduction

Thietanes have received much less attention than azetidines and oxetanes. Some reviews have, though, been published on their chemistry [232–234]. The ring strain of thietane (80 kJ mol^{-1}) is comparable to that of thiirane (three-membered ring). This explains their difficult formation by ring closure and, conversely, their easy cleavage by electrophilic and nucleophilic reagents.

3.3.2 Physicochemical Data

Conformations of 3-substituted thietanes 1-oxide have been studied [235]. *Ab initio* SCF calculations (G-31G* level) of thietanes examined the change of C–C bond lengths and angular deformations [236] compared to sulfur heterocycles of higher ring sizes. Calculations on the reaction of thietane with NH_3 explained the greater reactivity of this compound compared to oxetane [237]. Calculations concerning the ring closure of $HS(CH_2)_3S$- to thietane were also reported [238].

Details concerning the NMR spectroscopy of thietane derivatives have been reported in previous reviews [232, 233]. The α-protons of thietane appear at δ 3.21 ppm (4.09 for thietane 1,1-dioxide). The ^{13}C NMR chemical shifts have been reported to be at 25 ppm for the α-carbons and 27 ppm for the β-carbon. The α-carbon shifts in thietanes appear in general at higher field than those of the β-carbon [233]. The ^{33}S NMR spectra of thietane, thietane 1-oxide and thietane 1,1-dioxide have been recorded [239]. In mass spectra, the main fragmentation includes retro [2 + 2] cycloaddition for thietanes, thiolactones and iminothietanes, and loss of, respectively, SO and SO_2 for thietane 1-oxides and thietane 1,1-dioxides [232].

3.3.3 Natural and Bioactive Compounds

Mono and dialkyl thietanes **32–35** have been detected in anal gland secretions of small mammals (ferrets, polecats, stoats, minks, weasels, kiores, voles, etc.).

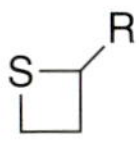

32 R = Me, Et, Pr, ^{i}Pr 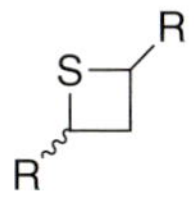 **33** R = Me, Et 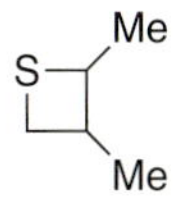 **34** 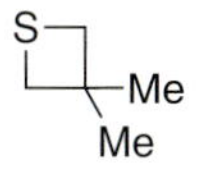 **35**

L-α-Aspartyl-*N*-(2,2,4,4-tetramethyl-3-thietanyl)-D-alaninamide (**36**), a sweetener that is 2000× more efficient than sugar, has been developed by Pfizer under the name Alitame [240] and commercialized in different countries.

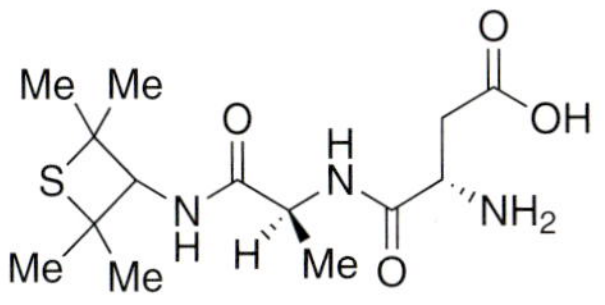

36

3.3.4
Synthesis of Thietanes

3.3.4.1 Synthesis by Formation of a S–C Bond

3.3.4.1.1 Formation from 3-Halo Thiol Derivatives The reaction of *S*-benzyl-*N*-phthaloylcysteinyl chloride with two equivalents of $AlCl_3$ affords the corresponding thiolactone (Scheme 3.133) [241]. Similar result was obtained for the free thiol function.

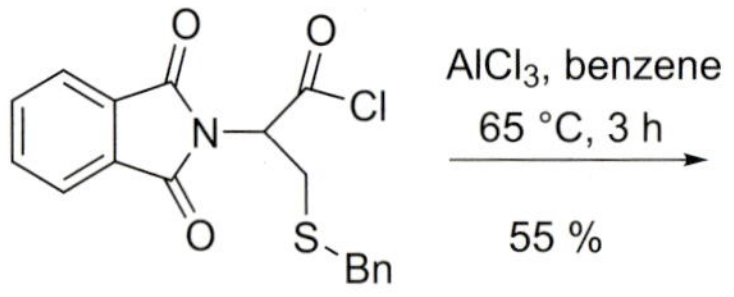

AlCl3, benzene
65 °C, 3 h
55 %

Scheme 3.133

Intramolecular reaction of thiolate, generated by base cleavage of thioacetate, on a secondary bromide leads to a bridged thietane [242] (Scheme 3.134). This approach to thietanes has been little studied.

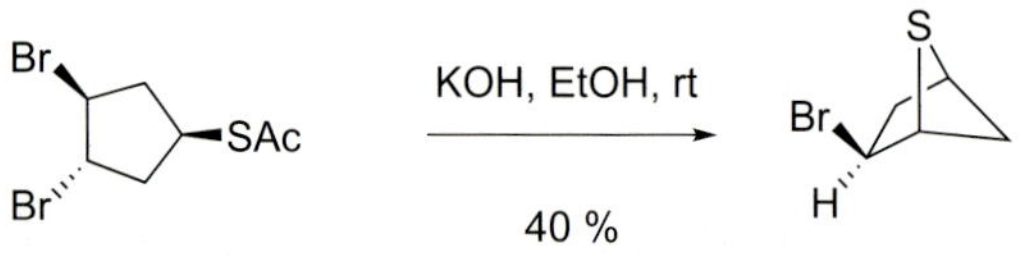

KOH, EtOH, rt
40 %

Scheme 3.134

3.3.4.1.2 Formation from 3-Hydroxy Thiol Derivatives Cyclization of 4-thio-4-methylbutanol into 2,2-dimethylthietane occurs in high yield when diethoxytriphenylphosphorane is used as reagent (Scheme 3.135) [243].

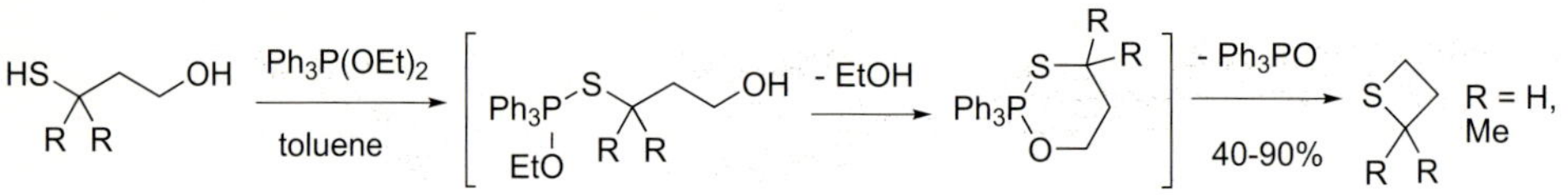

Ph3P(OEt)2, toluene; - EtOH; - Ph3PO, 40-90%; R = H, Me

Scheme 3.135

Sulfur analogues of thromboxane A_2 are obtained when the alcohol function is activated as mesylate [244] (Scheme 3.136). Nucleoside derivatives have also been obtained by a similar procedure [245].

Flash vacuum pyrolysis (500–750 °C) of (2-mercaptophenyl)methanol derivatives leads to the formation of thietanes, generally in good yields. For example, heating at

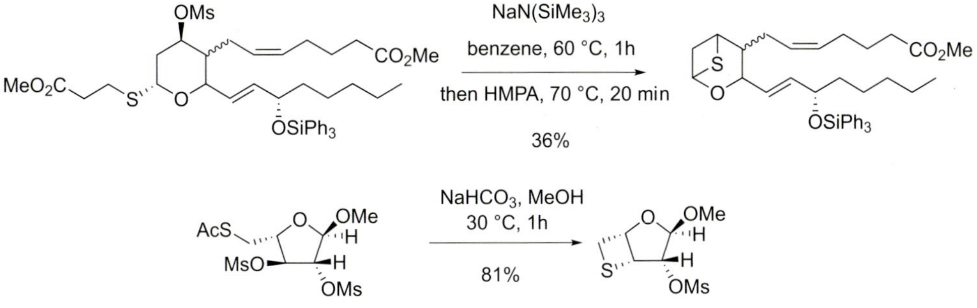

OMs
MeO2C S O CO2Me
OSiPh3
NaN(SiMe3)3 benzene, 60 °C, 1h then HMPA, 70 °C, 20 min
36%
S O CO2Me OSiPh3

AcS O OMe H MsO H OMs
NaHCO3, MeOH 30 °C, 1h
81%
O OMe H S H OMs

Scheme 3.136

750 °C of (3,6-dimercapto-1,2-phenylene)dimethanol furnishes 4,9-dithiatricyclo [6.2.0.0^{2,5}]deca-1,5,7-triene, which appears to be stable under 90 °C (Scheme 3.137) [246].

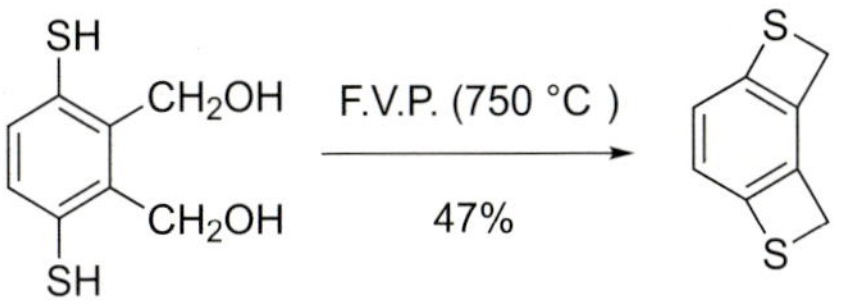

SH CH2OH CH2OH SH
F.V.P. (750 °C)
47%
S S

Scheme 3.137

Such thermal formation of thietanes was shown to be possible when benzotriazoles [247] or acids [248] were present as reaction groups instead of alcohol functions (Scheme 3.138).

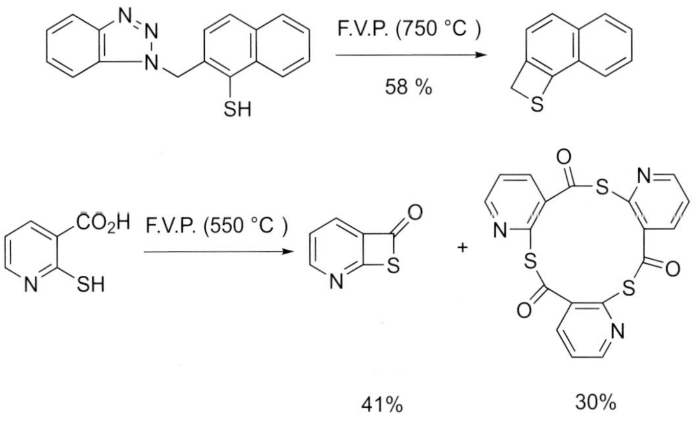

N N N SH
F.V.P. (750 °C)
58 %
S

CO2H N SH
F.V.P. (550 °C)
N S O
+
O S N N S S O O S N
41% 30%

Scheme 3.138

Thietan-2-ones are formed by the reaction of 3-mercaptocarboxylic acids with DCC [249], ^{i}butyl chloroformate [250], methyl chloroformate [251], Ac_2O [252], diethyl cyanophosphonate [253] or 1-ethyl-3-(3-dimethylaminopropyl)carbodiimide hydrochloride (EDAC) (Scheme 3.139) [254].

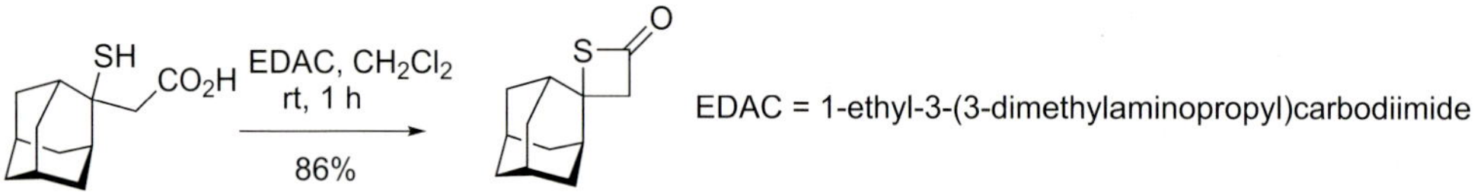

SH, CO_2H; EDAC, CH_2Cl_2, rt, 1 h, 86%; S, O

EDAC = 1-ethyl-3-(3-dimethylaminopropyl)carbodiimide

Scheme 3.139

An original preparation of thietan-2-one has also been reported: the reaction of 3-iodopropanoyl chloride with an ammonium tetrathiomolybdate salt (Scheme 3.140) [255].

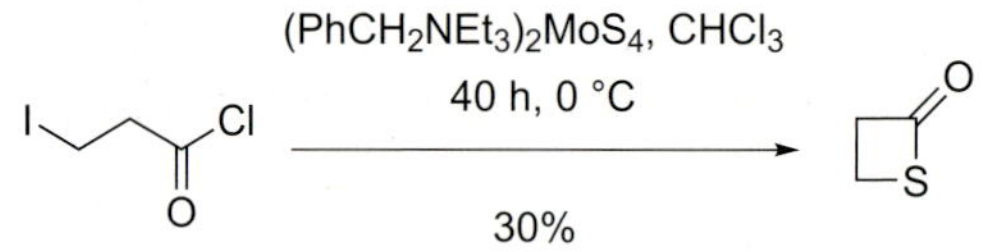

I, Cl, O; $(PhCH_2NEt_3)_2MoS_4$, $CHCl_3$, 40 h, 0 °C, 30%; O, S

Scheme 3.140

3.3.4.1.3 **Formation from Other 3-Functionalized Thiol Derivatives** α-Bromomethyl thioesters react with cobalt oximes to give the corresponding organocobalt compounds that, under photolysis, afford thietanes. This method appears synthetically useful only for the structure reported in Scheme 3.141 [256].

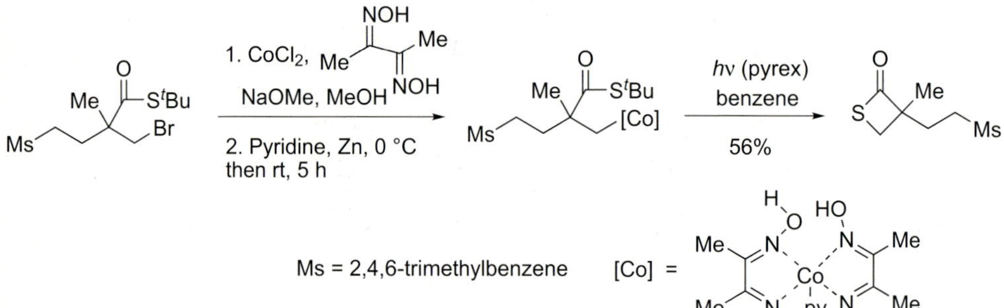

O, Me, S^tBu, Br, Ms; 1. $CoCl_2$, Me, NOH, Me, NOH, NaOMe, MeOH; 2. Pyridine, Zn, 0 °C then rt, 5 h; O, Me, S^tBu, [Co], Ms; $h\nu$ (pyrex) benzene, 56%; O, S, Me, Ms

Ms = 2,4,6-trimethylbenzene [Co] = H, O, HO, Me, N, N, Me, Co, Me, N, py, N, Me, O, HO, H

Scheme 3.141

Reaction of ethyl 2-mercaptoacetate with LDA followed by addition of a dichlorodiimine gives a 2,3-diiminothietane derivative (Scheme 3.142) [257]. When the R groups fixed at the nitrogen atoms were phenyl or 4-methoxyphenyl, the 2,3-diiminothietanes were not isolated, due to their ring opening in the reaction mixture.

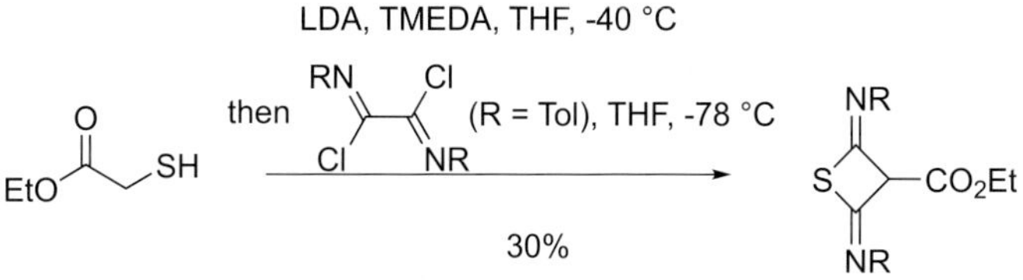

LDA, TMEDA, THF, -40 °C; then RN=C(Cl)–C(Cl)=NR (R = Tol), THF, -78 °C; 30%

Scheme 3.142

1,4-Addition of *O,O*-diethyl hydrogen dithiophosphate on the carbon–carbon double bond of chalcones affords a compound that then, under microwave irradiation in the presence of nucleophiles, gives thietanes (Scheme 3.143) [258]. This reaction was previously reported to occur by heating in the presence of a $NaH–NaBH_4$ mixture [259].

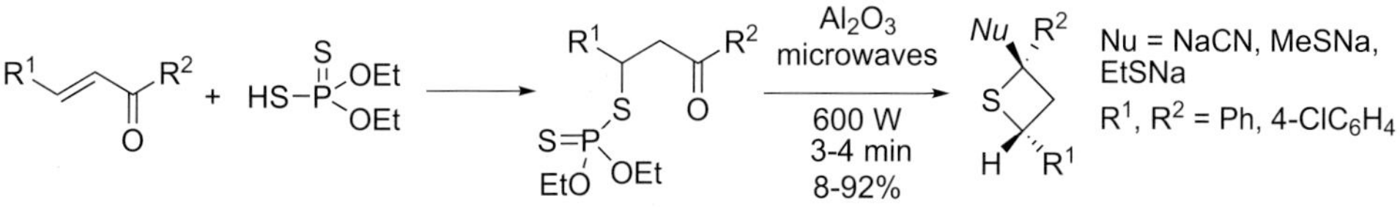

Al_2O_3, microwaves, 600 W, 3-4 min, 8-92%; Nu = NaCN, MeSNa, EtSNa; R^1, R^2 = Ph, 4-ClC_6H_4

Scheme 3.143

5-Substituted 2-thiooxoimidazolidin-4-ones react with sodium *tert*-butylate to yield bicyclothietenes (Scheme 3.144) [260].

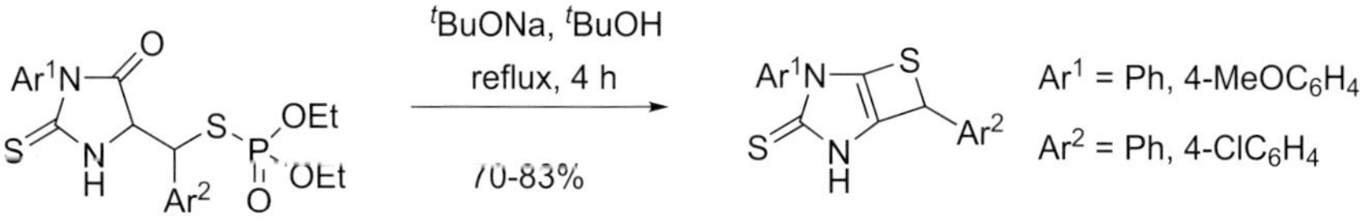

^tBuONa, ^tBuOH, reflux, 4 h, 70-83%; Ar^1 = Ph, 4-$MeOC_6H_4$; Ar^2 = Ph, 4-ClC_6H_4

Scheme 3.144

Thietanes are rarely formed by electrophile-induced cyclizations of unsaturated sulfides [261]. A thietane is obtained in mixture with other products by reaction of a γ-unsaturated thiol with bromine (Scheme 3.145). Utilization of the corresponding disulfide allowed exclusive formation of the expected thietane [262].

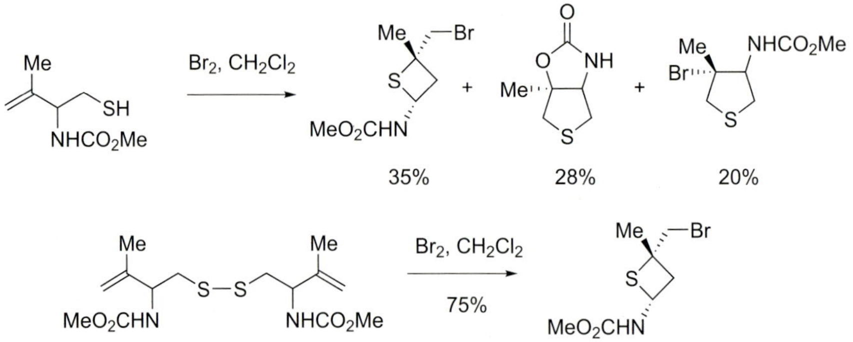

Scheme 3.145

The dianion of 1-(prop-2-ynyl)benzene reacts with phenyl thioisocyanate to give a thioimidate that, by reaction with potassium ^{t}butylate, leads to the formation of a thietane (Scheme 3.146) [263].

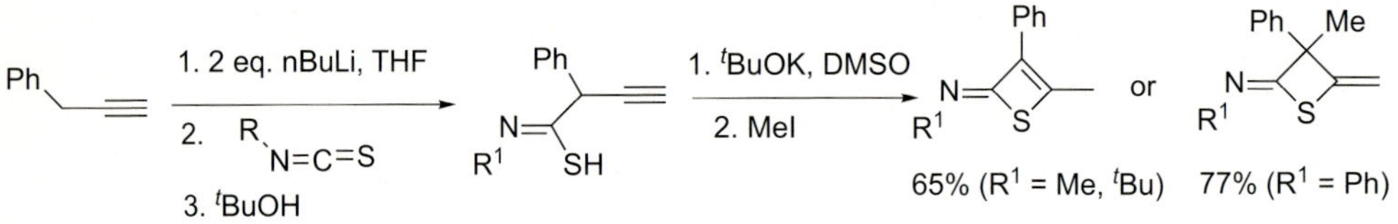

Scheme 3.146

An efficient preparation of thietanes in two steps occurs by reaction of 1,3-diols with dibenzoxazol-2-yl disulfide and tributylphosphine, followed by treatment of the resulting 2-(3-hydroxyalkylthio)benzoxazoles with KH (Scheme 3.147) [264]. The thio-Mitsunobu reaction applied to thioamides gives rise to the formation of α-iminothietanes (Scheme 3.147) [265].

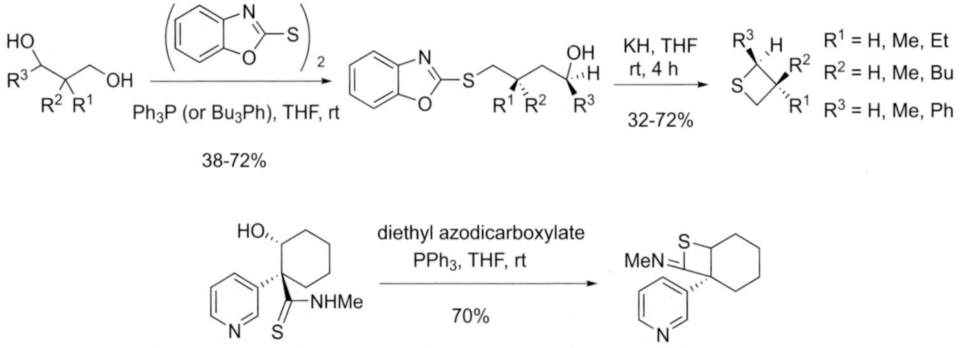

Scheme 3.147

The formation of thietanes is also possible by electroreduction in the presence of Ni(II) salem as catalyst (Scheme 3.148) [266].

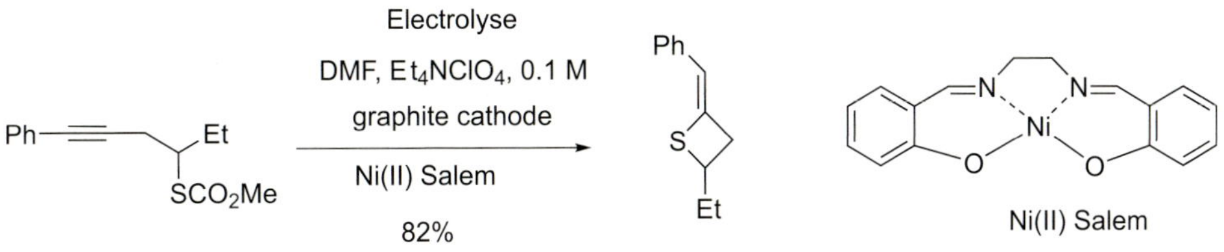

Electrolyse
DMF, Et_4NClO_4, 0.1 M
graphite cathode
Ni(II) Salem
82%

Ni(II) Salem

Scheme 3.148

3.3.4.2 **Synthesis by Formation of a C–C Bond**

Under phase transfer catalysis conditions, 1-bromomethylsulfonyl-2-phenylethane affords, in the presence of sodium hydroxide, 2-phenylthietane-1,1-dioxide, via intramolecular cyclization of the benzyl anion. The Ramberg–Bäcklund product is not observed (Scheme 3.149) [267].

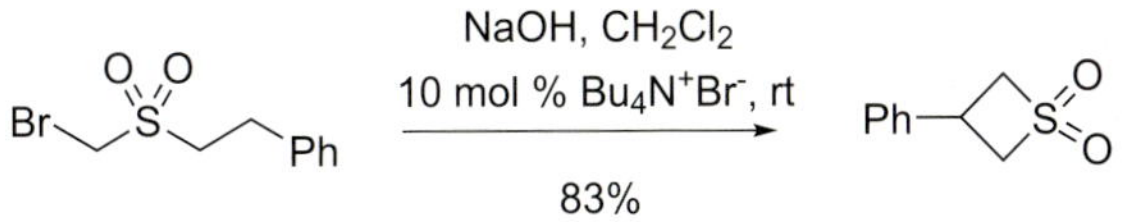

NaOH, CH_2Cl_2
10 mol % $Bu_4N^+Br^-$, rt
83%

Scheme 3.149

Thietanols can be formed by fluoride-mediated cyclization of (*Z*)-α-silyl vinylsulfides (Scheme 3.150) [268].

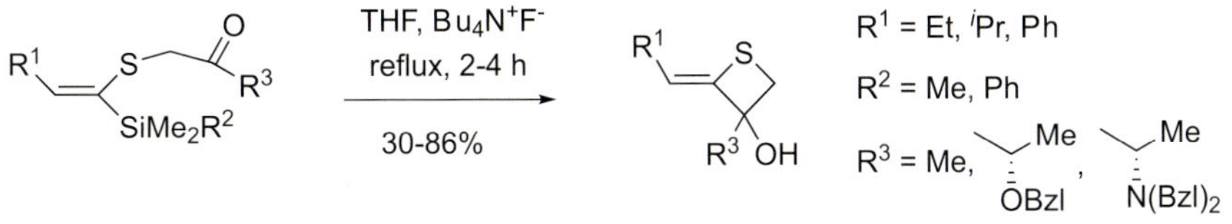

THF, $Bu_4N^+F^-$
reflux, 2-4 h
30-86%

R^1 = Et, ^{i}Pr, Ph
R^2 = Me, Ph
R^3 = Me, (Me, OBzl), (Me, $N(Bzl)_2$)

Scheme 3.150

3.3.4.3 **Synthesis by Formation of Two S–C Bonds**

The oldest method for the preparation of thietanes involves the reaction of 1,3-dihalides with sodium sulfide. Numerous conditions have been studied [232, 233]. Excellent results have been reported for phase transfer conditions (Scheme 3.151) [269]. A spiro thietane is formed upon reaction of a 1,3-dibromo derivative with Na_2S in DMF [270].

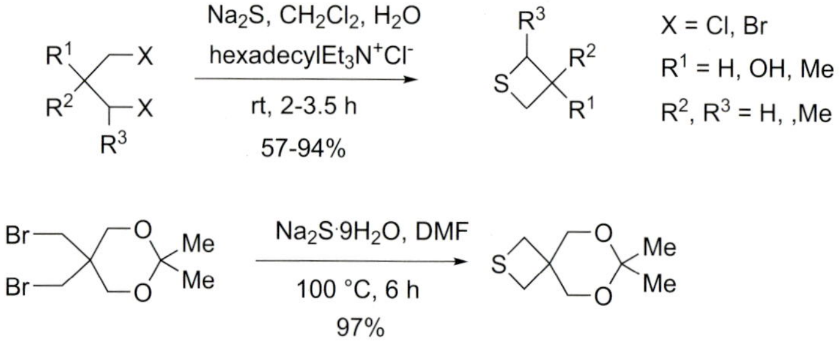

Scheme 3.151

Derivatives of 1,3-diol benzoates (Scheme 3.152) [271], mesylates [272] and tosylates (Scheme 3.152) [273] were also efficient in these preparations.

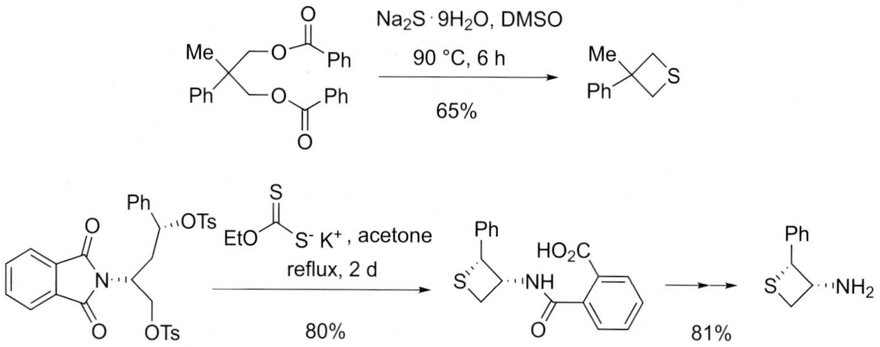

Scheme 3.152

1,3-Diols protected as carbonates can also lead to thietanes by reaction with KSCN at high temperature. This method has been reported for the preparation of (*S*)-2-propylthietane, which is a natural product (Section 3.3.3) (Scheme 3.153) [274]. This procedure was used more recently in the case of a 1,2,3-propane triol derivative [275].

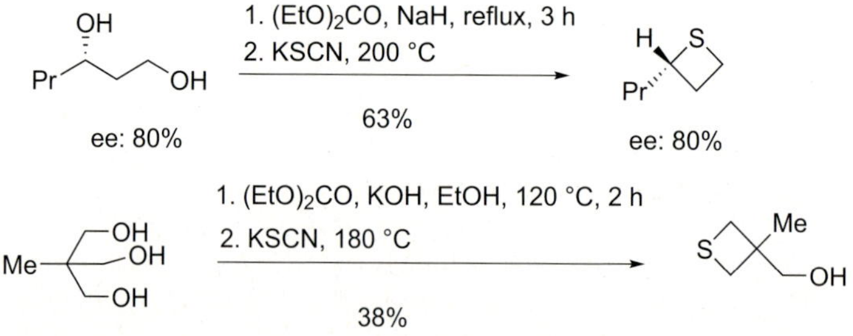

Scheme 3.153

Reactions of β-halo epoxides with different sulfur reagents give rise to 3-thietanols in good yields. The first sulfur reagent used was H_2S in the presence of sodium ethoxide (Scheme 3.154) [276], or $Ba(OH)_2$ [277].

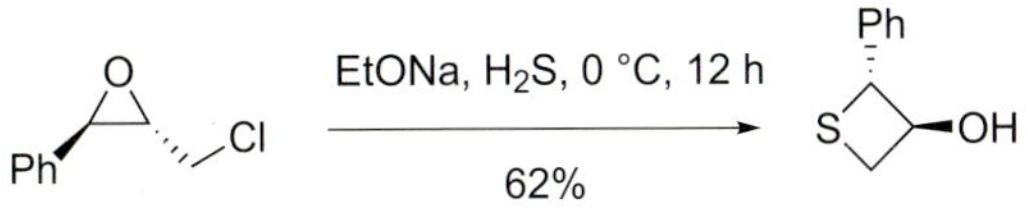

EtONa, H_2S, 0 °C, 12 h

62%

Scheme 3.154

Reactions of Li_2S [278] and benzyltriethylammonium tetrathiomolybdate [279], ammonium thiocarbamates (Scheme 3.155) [280] and thioacetate [281] with β-halooxiranes are very efficient.

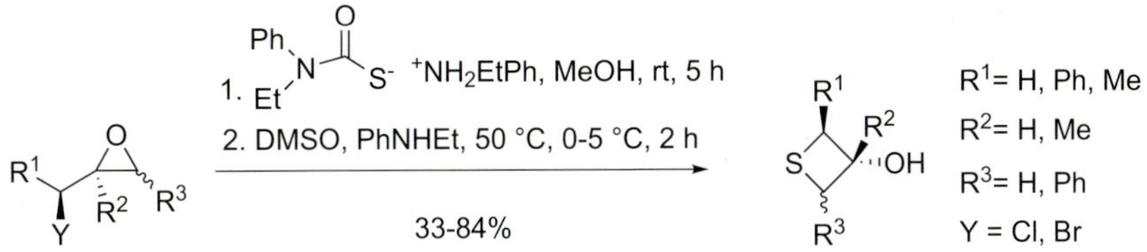

1. $^+NH_2EtPh$, MeOH, rt, 5 h

2. DMSO, PhNHEt, 50 °C, 0-5 °C, 2 h

33-84%

R^1 = H, Ph, Me

R^2 = H, Me

R^3 = H, Ph

Y = Cl, Br

Scheme 3.155

Thiols react with epichlorohydrin to give the corresponding addition products, which are then transformed into thietanes when reacted with sodium methoxide (Scheme 3.156) [282].

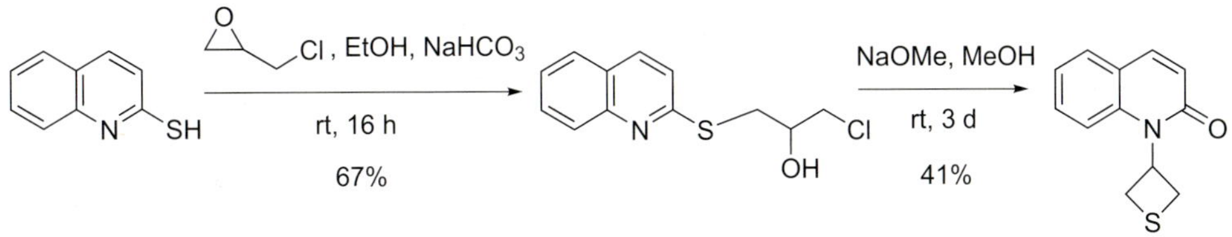

Cl, EtOH, $NaHCO_3$

rt, 16 h

67%

NaOMe, MeOH

rt, 3 d

41%

Scheme 3.156

The formation of thietanes by reaction of γ-hydroxyoxiranes with potassium thiocyanate has been also reported (Scheme 3.157) [283].

Thietanes can also be formed by radical addition of dimethyl disulfide on 1,3-dienes, by heating in the presence of iodine (Scheme 3.158) [284].

Electrophilic addition of SCl_2 [285], a mixture of $POCl_3$ (or $POBr_3$)/thiobismorpholine [286] or SO_2 (Scheme 3.159) [287] with norbornadiene produces thietane

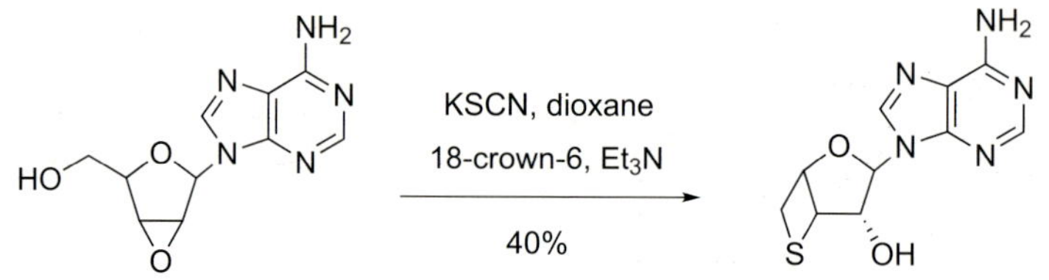

Scheme 3.157

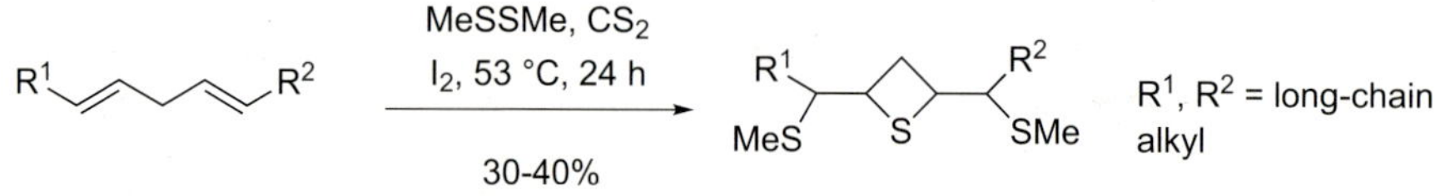

Scheme 3.158

derivatives. The reaction of SCl_2 with δ-diketones leads to thietanes in good yields (Scheme 3.159) [288].

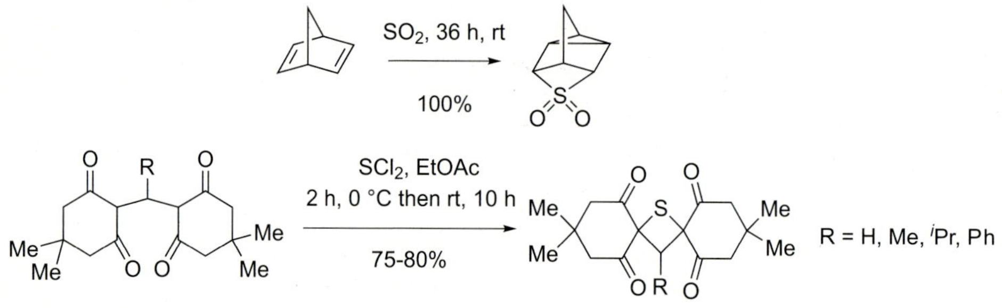

Scheme 3.159

3.3.4.4 Synthesis from Other Sulfur Heterocycles

3.3.4.4.1 Formation from Thiirane Derivatives The reaction of 2-(chloromethyl) thiirane with sodium acetate gives the corresponding thietane in 82% yield [289]. Similar ring expansions were reported using ammonium thiocyanate (Scheme 3.160)

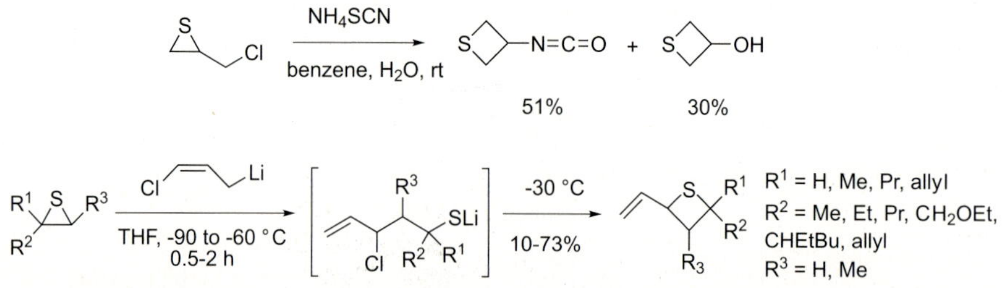

Scheme 3.160

[290], sulfonamides [291] and phenates [291], or from (thiiran-2-yl)methanol under the conditions of the Mitsunobu reaction [292]. Addition of allyl lithium compounds with thiiranes gives rise to thietanes in good yields (Scheme 3.160) [293].

An optically active thiirane has been transformed into a thietane, without epimerization of the chiral centers, in two steps by reaction with silver acetate followed by cyclization of the resultant 3-mercaptoacetal by camphor sulfonic acid (CSA) (Scheme 3.161) [294].

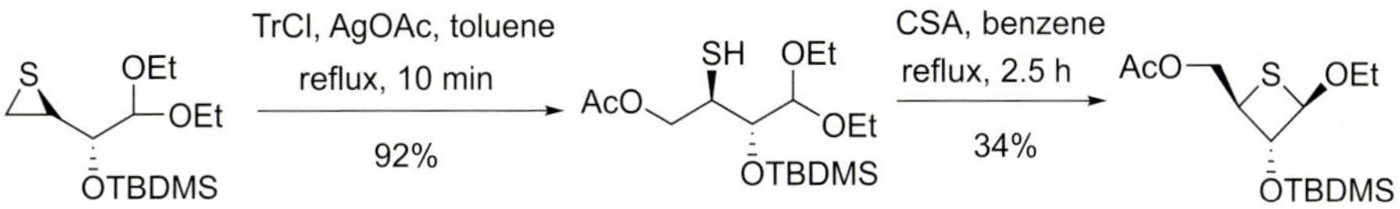

Scheme 3.161

Iminothiiranes react with isocyanates to afford 2,4-diiminothietanes (Scheme 3.162) [295]. This rearrangement of thiiranes into thietanes has also been reported during the reaction of diphenyl diazomethane with acyl isothiocyanate [296].

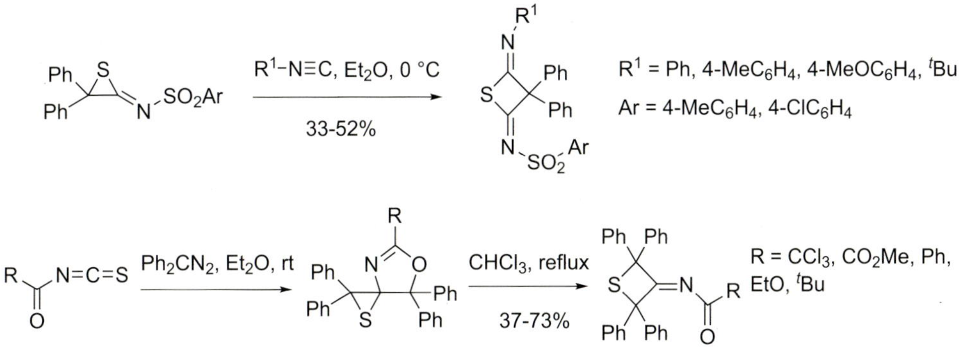

Scheme 3.162

3.3.4.4.2 **Formation from Thiolanes and Derivatives** 3-Chlorotetrahydro-2-methylthiophene reacts with water to give the 2-substituted thietane (Scheme 3.163) [297]. This reaction is reversible, since under acidic conditions (HCl) the tetrahydrothiophene was reformed.

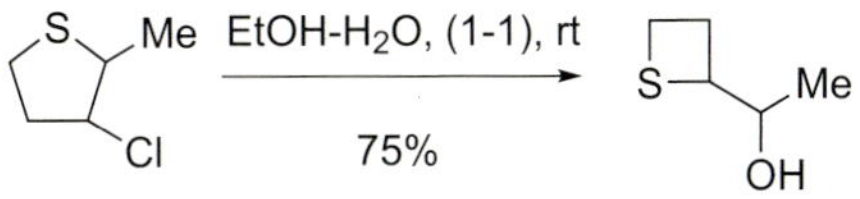

Scheme 3.163

4-Diazodihydrothiophen-3(2*H*)-one derivatives undergo ring contraction to thietan-2-ones upon heating in isooctane [298], irradiation in an alcohol [299] or in the presence of rhodium diacetate (Scheme 3.164) [300].

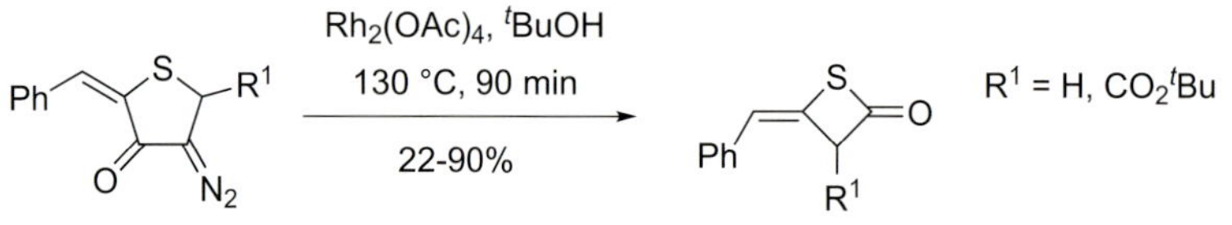

Scheme 3.164

Flash-thermolysis of benzothiophen-2(3*H*)-one (Scheme 3.165; X = 2H) [299] or benzothiophen-2,3-dione (Scheme 3.165: X = O) [301] gives rise to the corresponding thietanes in excellent yields.

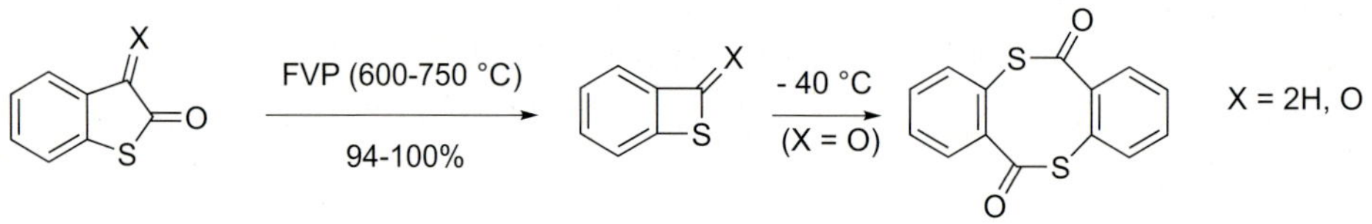

Scheme 3.165

Treatment of a 4-thiofuranose derivatives with diethylaminosulfur trifluoride (DAST) leads to the formation of the ring contraction product (Scheme 3.166) [302].

Rearrangement of thiazolotriazoliums, formed by electrophilic addition of bromine on 3-methallylthiotriazoles in basic conditions, affords thietanes (Scheme 3.167) [303]. The same rearrangement was observed with benzimidazole, benzothiazole and imidazole derivatives.

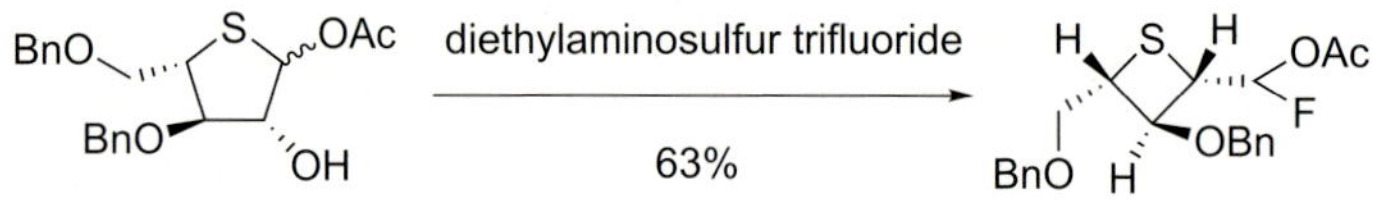

Scheme 3.166

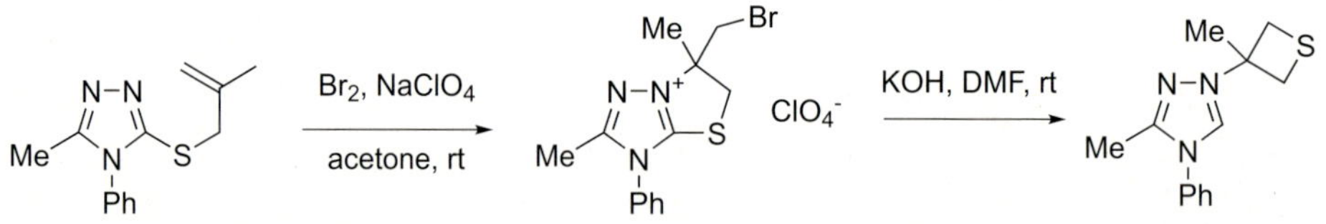

Scheme 3.167

Monodesulfurization of 1,2-dithiolanone with triphenylphosphine or tris(diethylamino)phosphine gives the thietanone in medium yields (Scheme 3.168) [304]. An attempt to use this method for the preparation of optically active thietanes was unsuccessful [305].

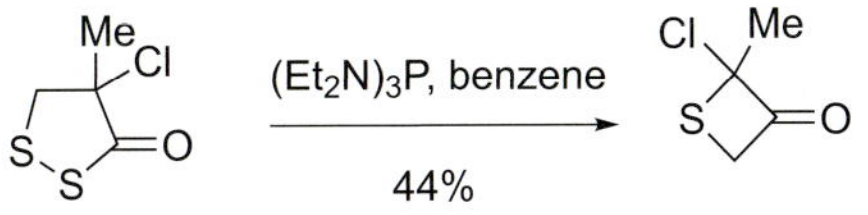

$(Et_2N)_3P$, benzene; 44%

Scheme 3.168

Reaction of benzyne with a thiophen-2,4-dithione gives a 2-thioacetal thietane in good yield (Scheme 3.169) [306].

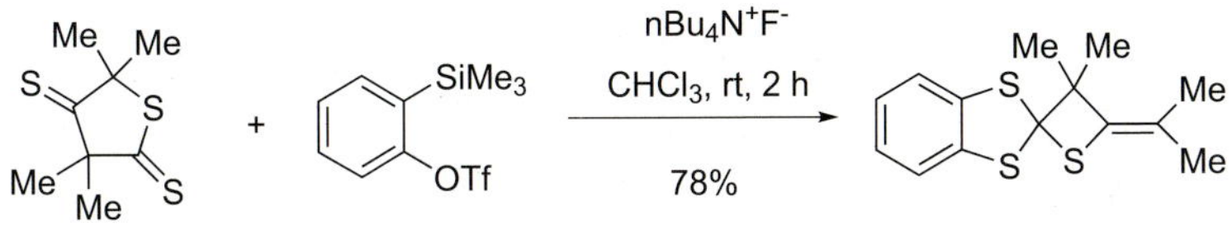

$nBu_4N^+F^-$, $CHCl_3$, rt, 2 h; 78%

Scheme 3.169

3.3.4.4.3 **Formation for Thiopyranes and Higher Ring Size Thio Heterocycles** Irradiation at low temperature of a 1,3-oxathian-6-one has been reported to lead to an unstable thianone, which leads to the formation of a dimer at room temperature (Scheme 3.170) [301].

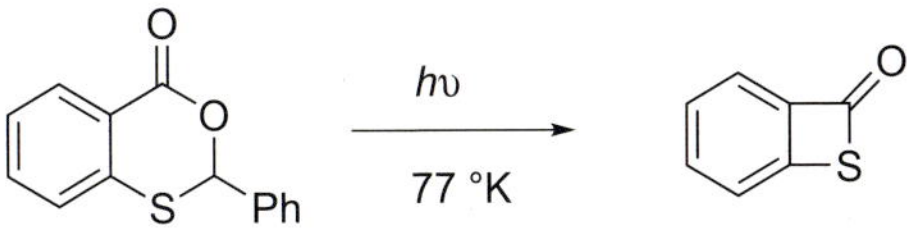

hυ, 77 °K

Scheme 3.170

CO_2 is extruded during the flash thermolysis of 1,3 oxathian 2 ones, affording thietanes that have been isolated in good yields (Scheme 3.171) [307].

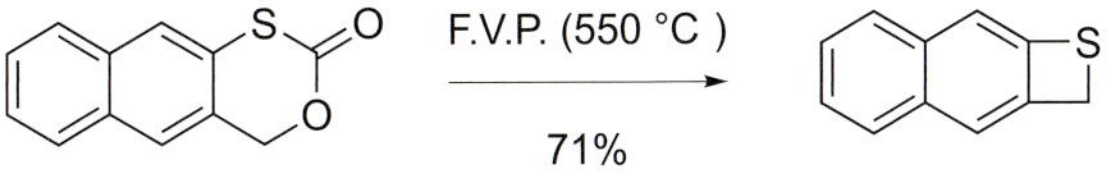

F.V.P. (550 °C); 71%

Scheme 3.171

Ultraviolet irradiation of 1-aza-8-thiacylo[4.2.1]non-2,4-diene-8,8-dioxide yields an unusual tricyclic compound (Scheme 3.172) [308].

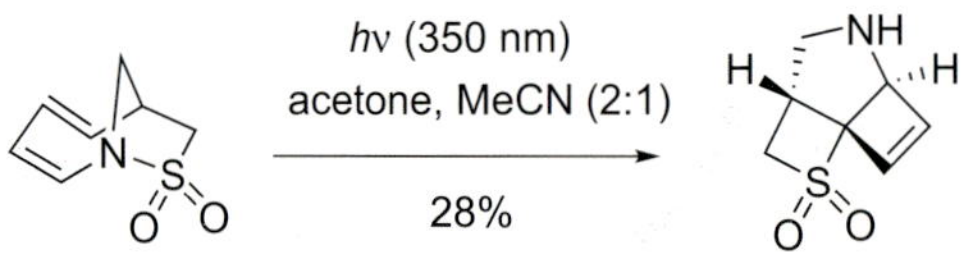
hν (350 nm)
acetone, MeCN (2:1)
28%
N
S
O
O
NH
H
H
S
O
O

Scheme 3.172

3.3.4.5 **Synthesis by [2 + 2] Cycloaddition**

Photochemical cycloadditions of thiones with unsaturated compounds have received considerable attention. Aromatic, aliphatic and α,β-unsaturated thioketones react as well with electron-poor rather than electron-rich olefins [232–234]. The more recent result concerns the reaction of silyl thioketones, which can be seen as an equivalent of thioaldehydes, with olefins. Irradiation of phenyl triphenylsilyl thioketone in acrylonitrile, methyl acrylate and *cis*- or *trans*-1,2-dichloroethene give the silylthietanes with high regio- and stereoselectivity (Scheme 3.173) [309]. Lower yields and selectivities are observed during the reaction with vinyl ether. Subsequent protiodesilylation reactions take place with predominant inversion of configuration at the carbon bearing the silicon group. The intramolecular cyclization of thiones was reported to be very efficient (Scheme 3.173) [310].

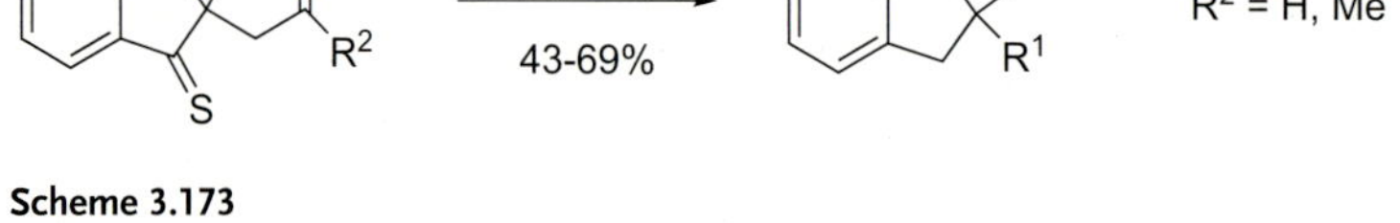
Ph
Ph3Si
S
EWG
in excess
hν, - 70 °C
58-60%
2-5 h
GWE
H
Ph
S
SiPh3
de: 84%
EWG = CN, CO2Me
R1
R2
S
hν, benzene
43-69%
S
R2
R1
R1 = H, allyl
R2 = H, Me

Scheme 3.173

Photochemical additions of olefins on the thiocarbonyl function of thioxo-3,4-dihydroisoquinolinones (Scheme 3.174) [311], benzooxazole-2-thione [312] and 5-thioxopyrrolidinones (Scheme 3.174) [313] have been reported. [2 + 2] Cycloadditions of thiopivalophenones were also reported by reaction with benzyne (Scheme 3.175) [314].

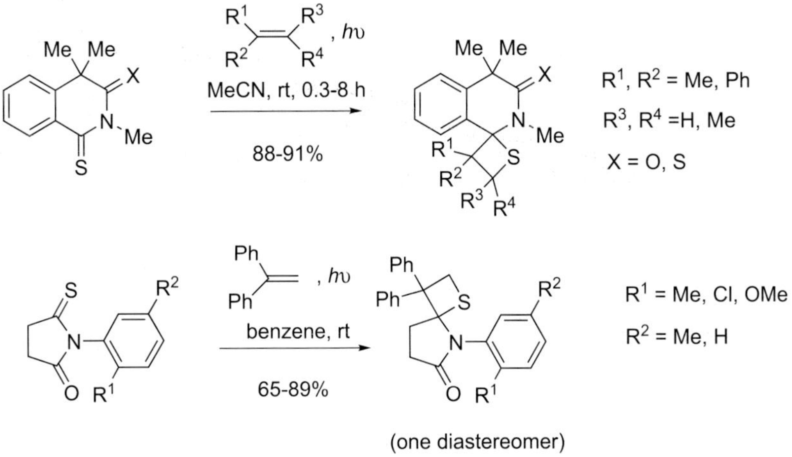

Scheme 3.174

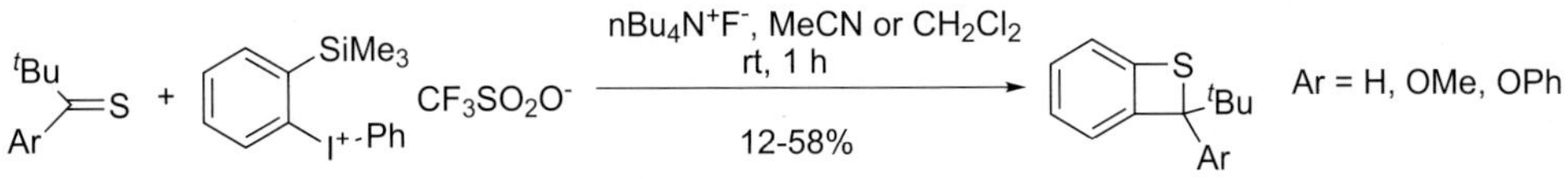

Scheme 3.175

Intramolecular cyclizations of unsaturated N-ethanethioylacetamide derivatives have also furnish tricyclic thietanes (Scheme 3.176) [315]. Cycloaddition of 2,4-dioxopenta-3-thione with allyltributyltin occurs at room temperature [316].

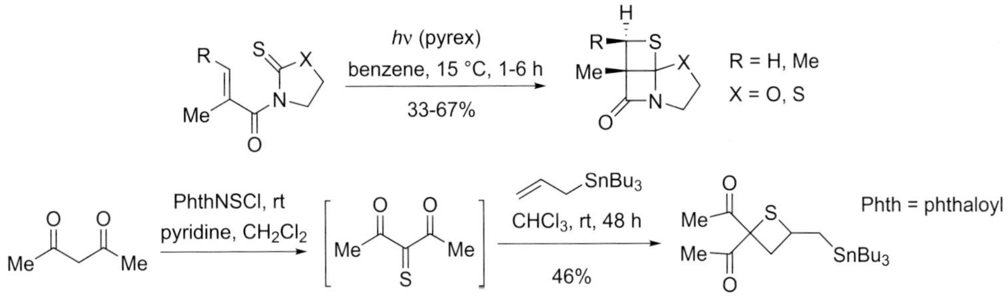

Scheme 3.176

Thietene-1,1-dioxides, which are precursors of thietanes, can be obtained by reaction of enamines with methanesulfonyl chloride in the presence of triethylamine (Scheme 3.177) [317]. Cycloadditions of acrylates with alkyl(chlorosulfonyl)acetate was reported to lead to 2-carboxylate thietanes [318].

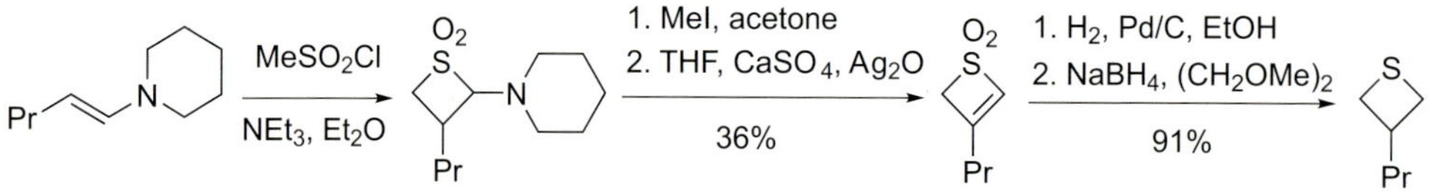

Scheme 3.177

3.3.4.6 Synthesis by Miscellaneous Methods

Isomerization of cyclobutane-1,3-dithiones into thietanones occurs in the presence of strong bases [319]. Utilization of tetrabutylammonium fluoride, allows a transformation under milder conditions (Scheme 3.178) [320]. The same product is obtained by reaction of the cyclobutane-1,3-dione with P_4S_{10} at reflux in pyridine [321].

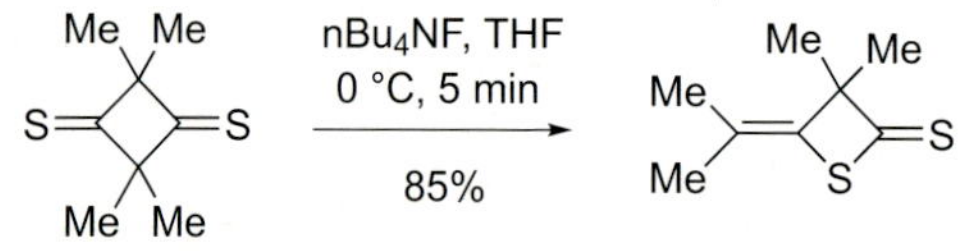

Scheme 3.178

Isoxazolidine-3-thiones react with ZnI_2 in chloroform to give 2-iminothietanes (Scheme 3.179) [322].

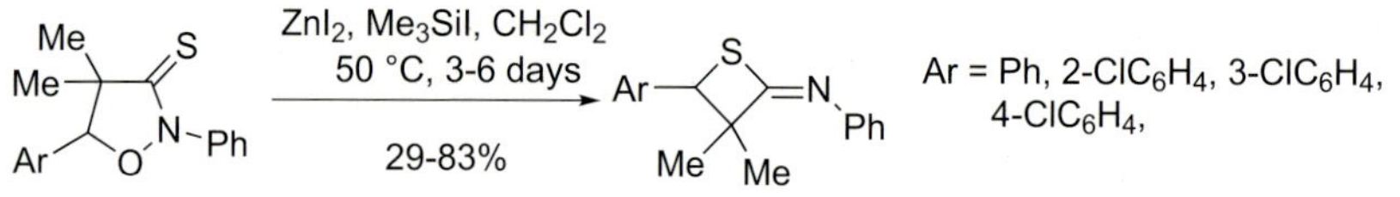

Scheme 3.179

An interesting preparation of thietan-1-oxides involves the reaction of ^{t}butyl sulfide with peroxydodecanoic acid (Scheme 3.180). Mixtures of diastereomers were obtained [323].

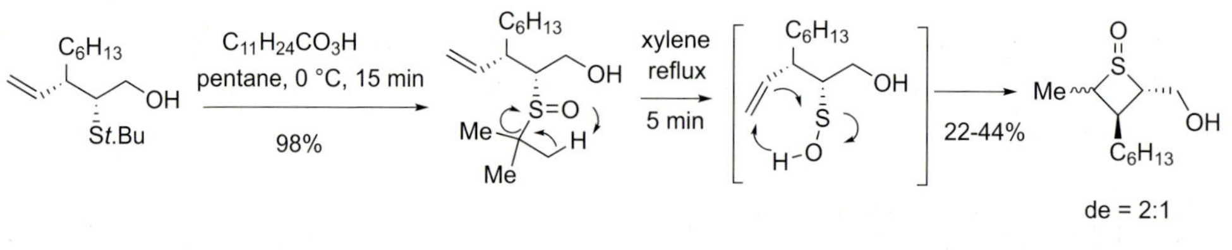

Scheme 3.180

3.3.5
Reactivity and Useful Reactions

3.3.5.1 Reactions with Electrophilic Reagents

Thietanes react with ethyl diazoacetate [324] or diethyl diazomalonate [325] in the presence of rhodium acetate to give ring expansion products. When the carbenes are generated from diaziridines, ring-opening compounds are obtained (Scheme 3.181) [326].

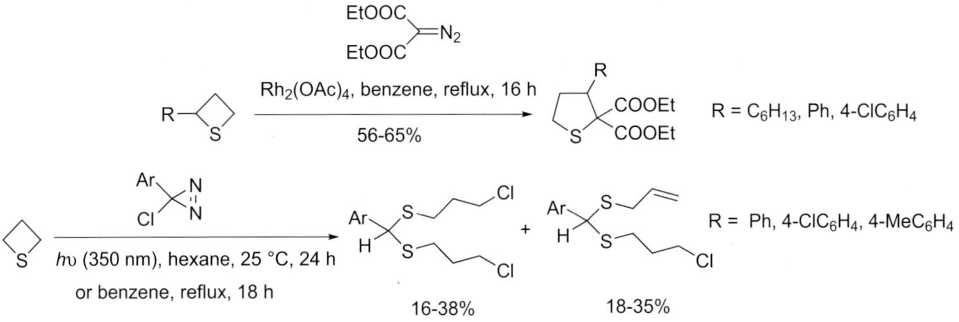

EtOOC, EtOOC =N2; $Rh_2(OAc)_4$, benzene, reflux, 16 h; 56-65%; COOEt, COOEt; $R = C_6H_{13}$, Ph, 4-ClC_6H_4

Ar, Cl, N, N; hυ (350 nm), hexane, 25 °C, 24 h or benzene, reflux, 18 h; 16-38%; 18-35%; R = Ph, 4-ClC_6H_4, 4-MeC_6H_4

Scheme 3.181

The reaction of 2- or 3-substituted thietanes with cobalt or ruthenium carbonyl [$Co_2(CO)_8$ or $Ru_3(CO)_{12}$] produces dihydrothiophenones in good yields [327]. More recently, it has been shown that platinum salts can catalyze this CO insertion (Scheme 3.182) [328]. However, these reactions were not observed with 2,4-disubstituted thietanes.

(dppe)MePt-Co$(CO)_4$; CO, THF, 100 °C, 2-5 h or 30 °C, 24 h; 89-99%; R = H, Me

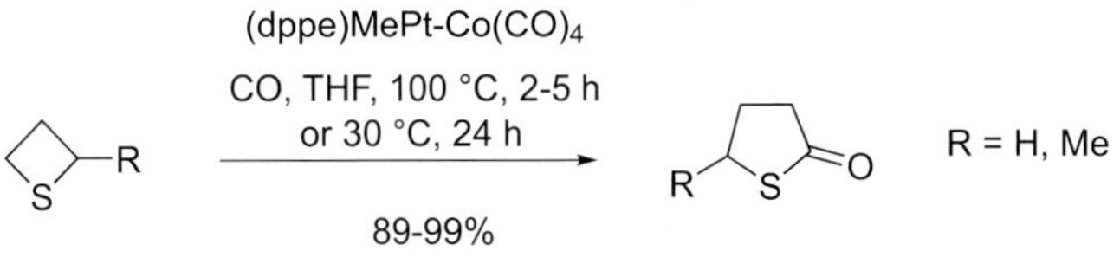

Scheme 3.182

Thietanes undergo ring opening by reaction with chloro reagents [232]. However, mono- or di-chlorinations and brominations of thietane-1,1-dioxides occur easily on C3 by irradiation [329]. 2-Iodination has also been reported by reaction with butyllithium, followed by addition of triethylaluminium and iodine (Scheme 3.183) [330].

3.3.5.2 Reactions with Oxidizing Agents

Oxidations of thietanes to thietanes-1-oxides and thietane-1,1-dioxides have been reported with the reagents generally used in the case of dialkyl sulfides: $KMnO_4$,

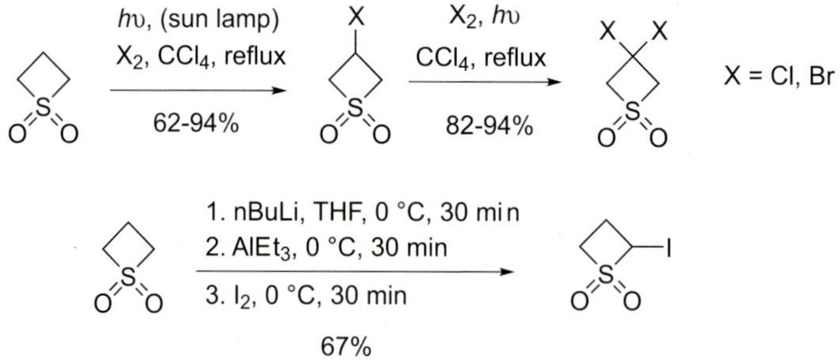

$h\nu$, (sun lamp) X_2, CCl_4, reflux 62-94%; X_2, $h\nu$ CCl_4, reflux 82-94%; X = Cl, Br

1. nBuLi, THF, 0 °C, 30 min 2. $AlEt_3$, 0 °C, 30 min 3. I_2, 0 °C, 30 min 67%

Scheme 3.183

*m*CPBA, $NaIO_4$, H_2O_2 in CH_3COOH or in presence of WO_3-H_2O, CrO_3 [232, 233]. Other reagents have been tested with success, such as cat. OsO_4-4-methylmorpholine *N*-oxide [331], 2-(phenylsulfonyl)-3-phenyl-oxaziridine [332], monoperphthalic acid [333], ^{t}pentyl hydroperoxide in the presence of $MoCl_5$ [334], oxone (Scheme 3.184) [309b] or 1-butyl-4-aza-1-azoniabicyclo[2.2.2]octane dichromate [335].

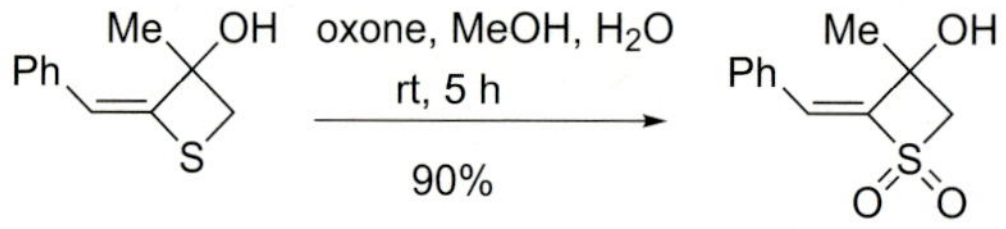

oxone, MeOH, H_2O rt, 5 h 90%

Scheme 3.184

The diastereoselectivity of the formation of thietane-1-oxides from thietanes has been examined using *m*CPBA [336]. Better selectivities have been reported using H_2O_2 in the presence of $TiCl_3$ (Scheme 3.185) [337].

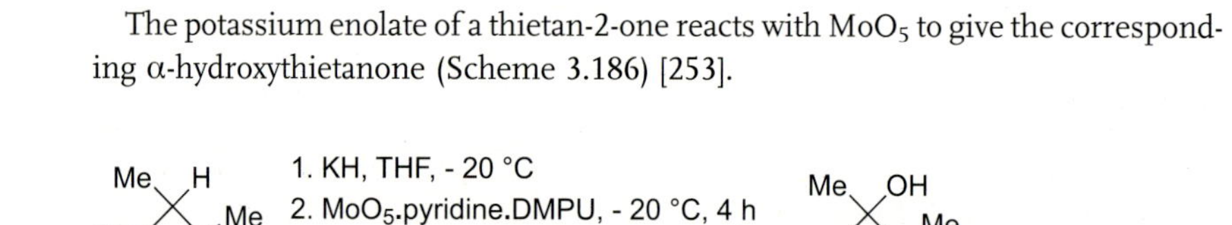

1. $TiCl_3$, H_2O-MeOH 2. H_2O_2, H_2O, MeOH 77-80%; minor (8-20%) : major (80-92%); Z = OH, NHAc

Scheme 3.185

The potassium enolate of a thietan-2-one reacts with MoO_5 to give the corresponding α-hydroxythietanone (Scheme 3.186) [253].

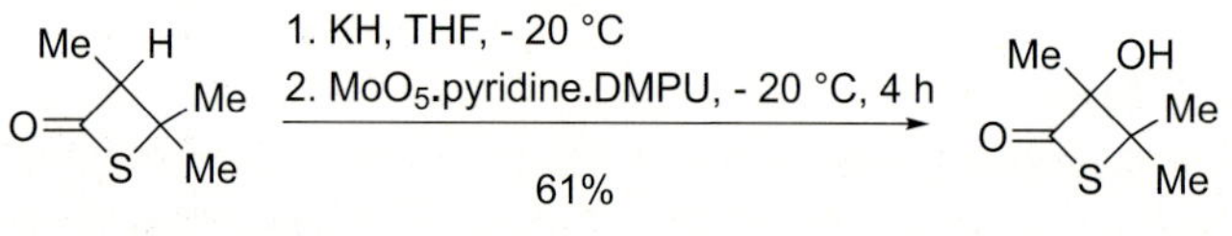

1. KH, THF, - 20 °C 2. MoO_5.pyridine.DMPU, - 20 °C, 4 h 61%

Scheme 3.186

3.3.5.3 Reactions with Nucleophilic Reagents

Thietane-1,1-dioxides have been reduced to thietanes by reaction with $LiAlH_4$ (Scheme 3.187) [317]. Reduction of thietane-1-oxides and sulfimides have been reported with a mixture $MeSiCl_2$-Zn [338].

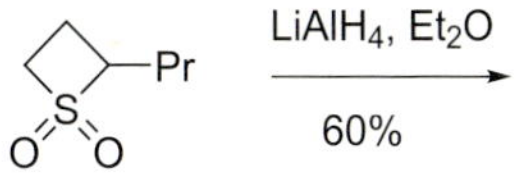

Scheme 3.187

2-Thietanones react with stabilized phosphoranes to give the corresponding *(E)*- and *(Z)*-2-alkylidenethietanes (Scheme 3.188) [339]. 3-Thietanones are easily transformed into the corresponding N-hydroxyimines; subsequent reaction with $NaBH_4$ led to 3-aminothietanes (Scheme 3.188) [340].

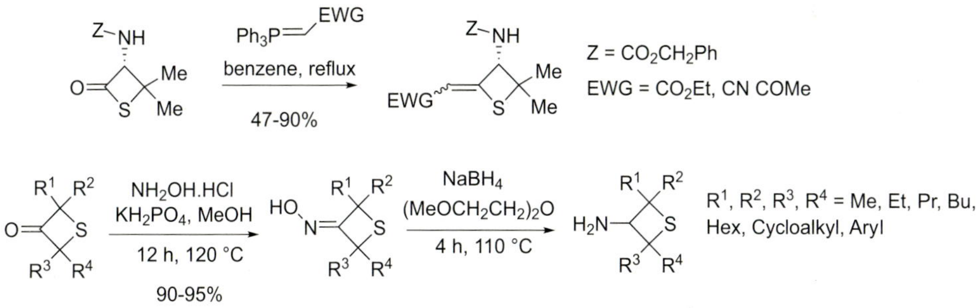

Scheme 3.188

Additions of nucleophiles or 3-chlorothieten-1,1-dioxide affords products without cleavage of the thietane ring. By reaction of sodium ethanoate the corresponding 3-acetal was obtained, while with sodium dimethyl malonate, the 3-alkylidenethietane was formed (Scheme 3.189) [329b].

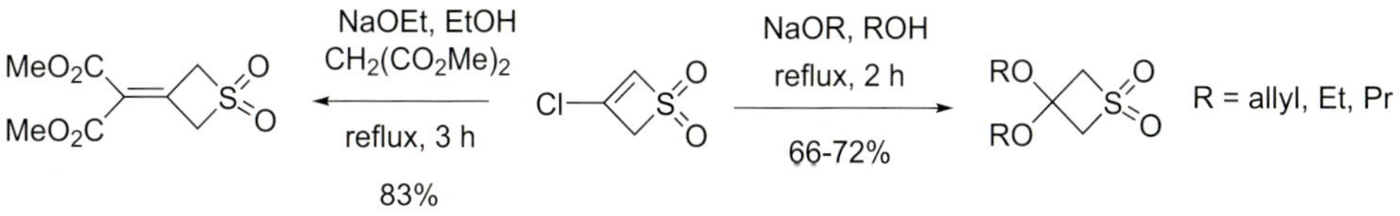

Scheme 3.189

3.3.5.4 Reactions with Bases

Thietanes are readily opened by reaction with sodium hydroxide to give linear sulfurs or polymerization products, depending on their structure [341]. With 3-hydroxythietane-1-oxides, ketones were obtained (Scheme 3.190).

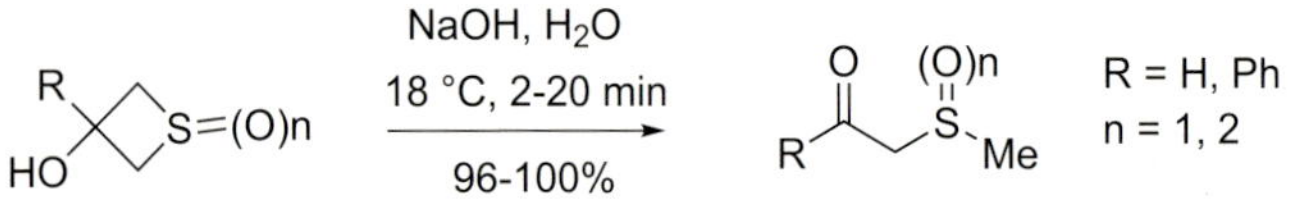

Scheme 3.190

Thiiranes have been formed by pyrolysis of lithium salts of hydrazones formed from thietan-3-ones (Scheme 3.191) [342].

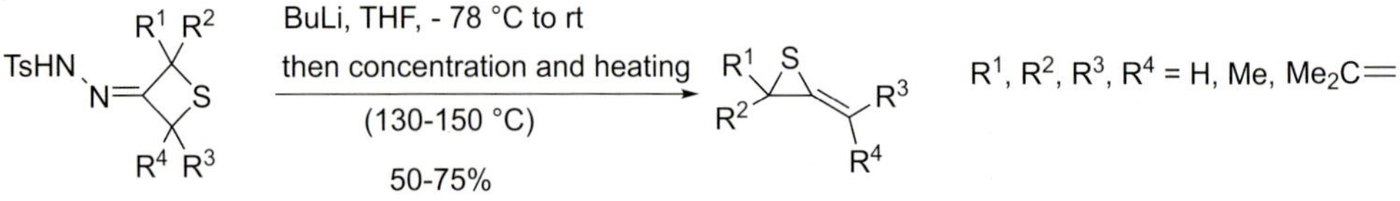

Scheme 3.191

Thietanes react with lithium diphenylphosphine to give the opened products in modest yields (Scheme 3.192) [275].

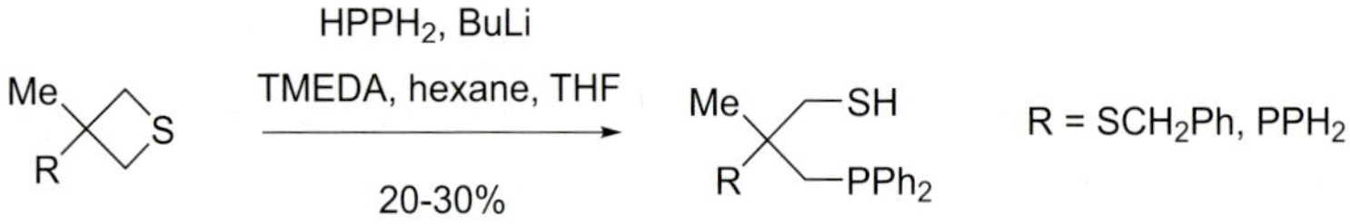

Scheme 3.192

Reaction of 2-phenylthietane with 4,4′-di-*t*-butylbiphenyllithium (DTBB-Li) has been reported to produce a dianion, which can react with electrophiles (Scheme 3.193) [343]. This reaction was not observed when starting with 2-methylthietane. With CO_2 as electrophile, dihydro-3-phenylthiophen-2(3*H*)-one was formed.

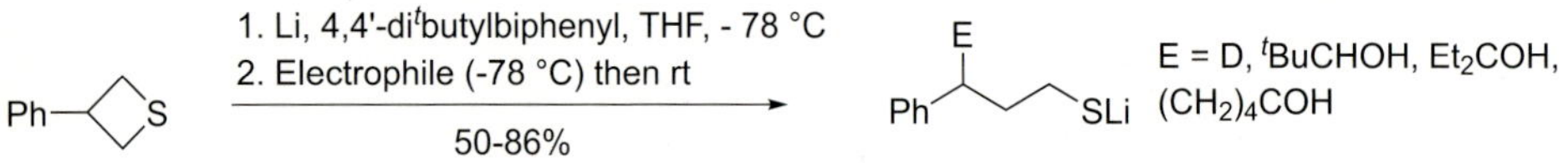

Scheme 3.193

Thietane-1-oxides undergo Pummerer rearrangements, when treated with Lewis acids in the presence of amino bases. This method has been used for the preparation of thietane nucleosides (Scheme 3.194) [264, 267, 344].

3.3.5.5 Reactions with Metal Complexes and Salts

Thietane is transformed into crown thio ethers (12S3, 16S4, 18S6, etc.) by the reaction of $Re_2(CO)_9$(thietane), $Os_4(CO)_{11}$(thietane) and $W(CO)_5$(thietane). Yields of 40%

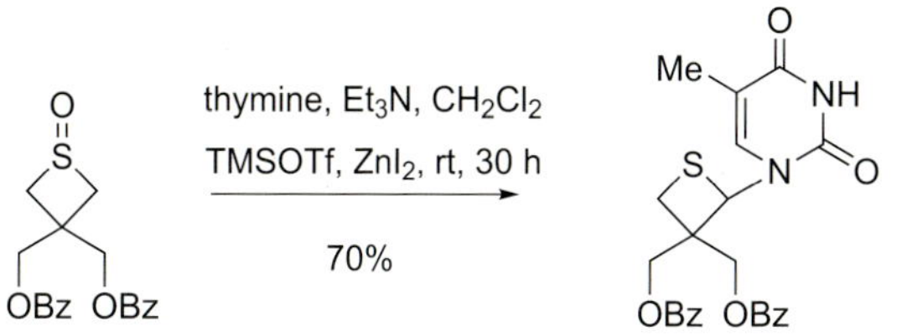

thymine, Et_3N, CH_2Cl_2
TMSOTf, ZnI_2, rt, 30 h
70%

OBz OBz

OBz OBz

Scheme 3.194

were reported for the formation of the 12S3 crown ether (Scheme 3.195) [345]. This transformation is possible with 2- and 3-methylthietanes and 3,3-dimethylthietanes. Starting from *(R)*-2-methylthietane, the corresponding optically active crown ether has been obtained [346].

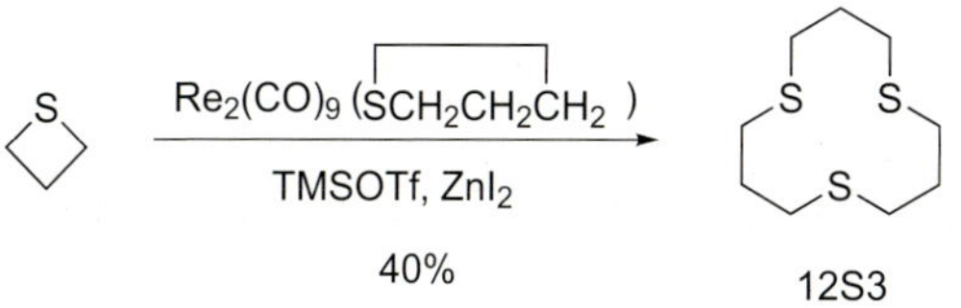

$Re_2(CO)_9$ ($SCH_2CH_2CH_2$)
TMSOTf, ZnI_2
40%

12S3

Scheme 3.195

Mannosyl iodide reacts with thietane in the presence of MgO to give mainly the thio β-anomer (Scheme 3.196) [347].

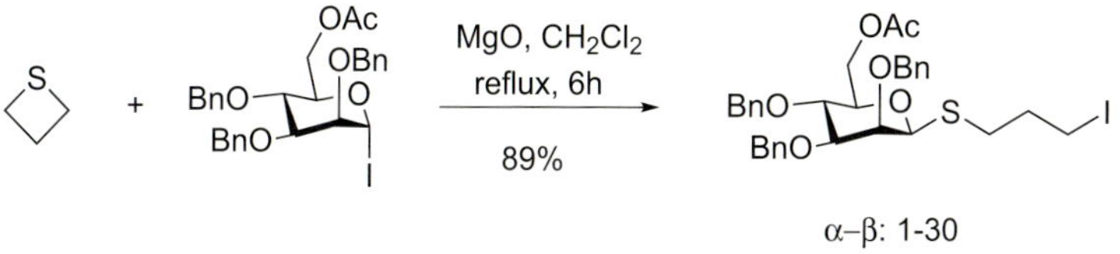

MgO, CH_2Cl_2
reflux, 6h
89%

α–β: 1-30

Scheme 3.196

3.3.5.6 **Electrocyclic Reactions**

Retro [2 + 2] cycloadditions of thietanes occur at high temperature to generate ethylenic compounds and thioaldehydes. This fragmentation is also possible by irradiation at low temperature. With thietan-3-one, the formation of ketene has been detected [2]. These retrocycloadditions were observed for substituted thietanes. For example, enol ethers are formed during irradiation of 3-alkoxythietanes (Scheme 3.197) [348]. These reactions also lead to the formation of thioketones or thioaldehydes.

Irradiations of thietan-2-ones in methanol lead, by Norrish I rearrangements, to five-membered heterocycles (Scheme 3.198) [349]. Dithiolactones give rise to the same kind of rearrangements [350].

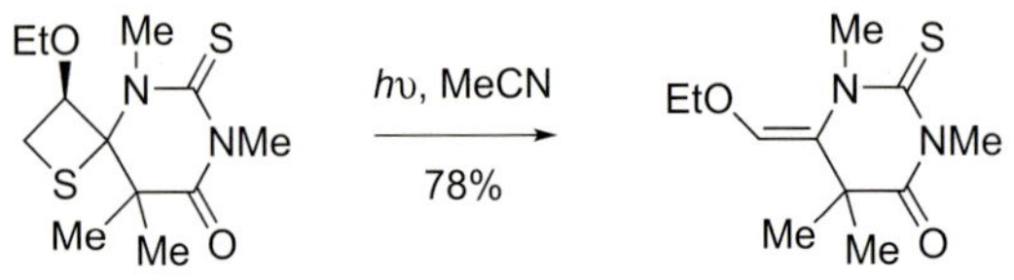

Scheme 3.197

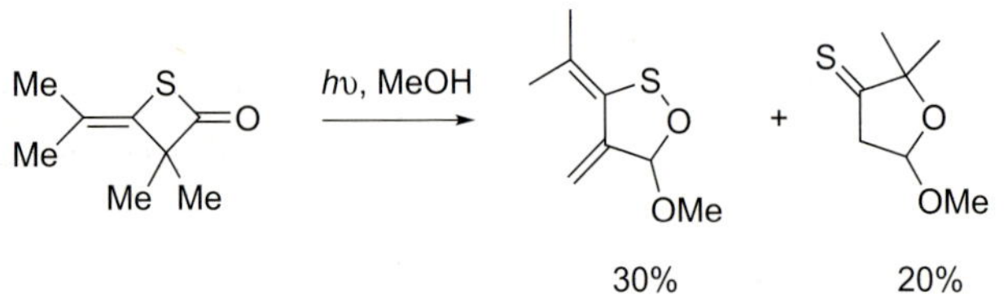

Scheme 3.198

2-Iminothietanes can be formed by cycloadditions of keteneimines with diphenyl thioketone in methylene chloride. Upon smooth heating, these compounds ($R^2 = H$) are transformed into unsaturated thioamides (Scheme 3.199) [351].

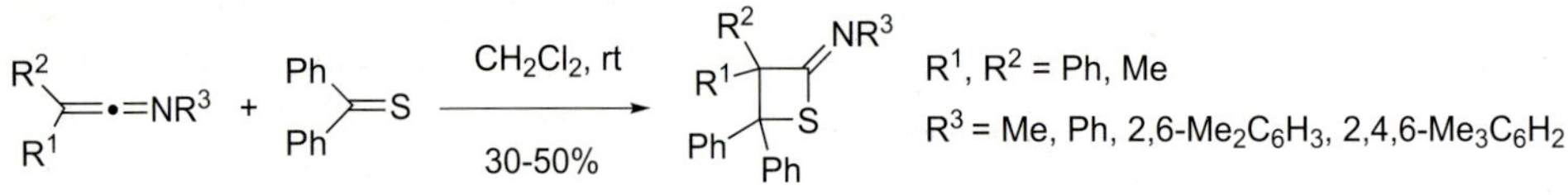

Scheme 3.199

Benzothiete undergoes, by heating or irradiation, an isomerization into methylenecyclohexadienethione. This compound can be trapped by various dienophiles (Scheme 3.200) [299]. In the absence of dienophile, a dimer of methylenecyclohexadienethione is formed. Reaction of this compound with C_{60} [352], imines and diazenes [353] has also been reported.

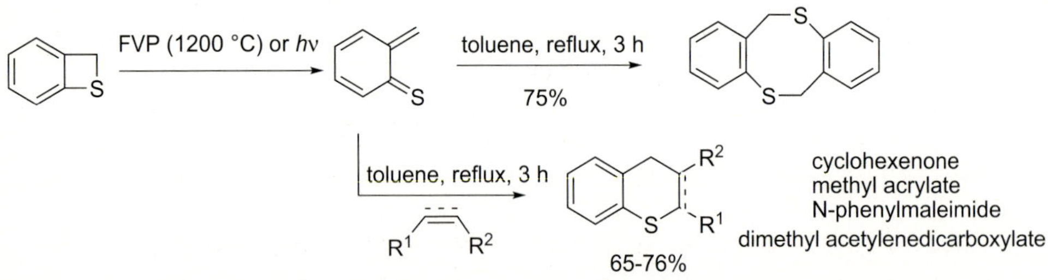

Scheme 3.200

3.3.5.7 Cleavage and Other Reactions

Thietane reacts with benzyl bromide in acetonitrile to give the ring cleavage product (Scheme 3.201) [354].

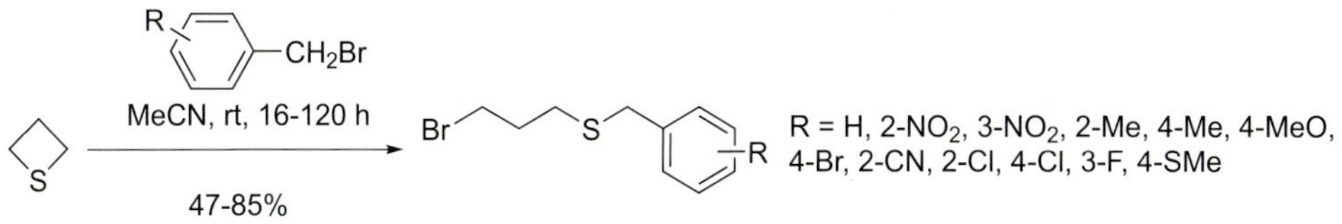

R = H, 2-NO_2, 3-NO_2, 2-Me, 4-Me, 4-MeO, 4-Br, 2-CN, 2-Cl, 4-Cl, 3-F, 4-SMe

MeCN, rt, 16-120 h; 47-85%

Scheme 3.201

Mercaptothiolic acids have been formed by the reaction of thietan-2-ones with H_2S [355]. These acids can be transformed into 1,2-dithiolan-3-ones by reaction with $FeCl_3$ [253, 356] or HIO_3 (Scheme 3.202) [357].

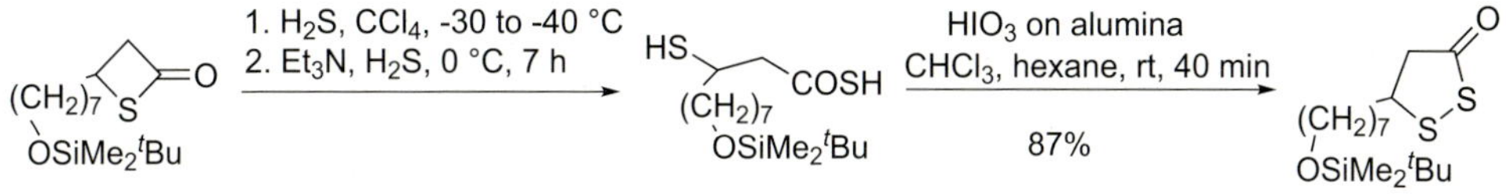

1. H_2S, CCl_4, -30 to -40 °C 2. Et_3N, H_2S, 0 °C, 7 h; HIO_3 on alumina, $CHCl_3$, hexane, rt, 40 min, 87%

Scheme 3.202

2,3-Diiminothietanes react with the dianion of ethyl mercaptoacetate to give bisthioles, while reaction with dimethyl acetylenedicarboxylate leads to thiopyran derivatives (Scheme 3.203) [358].

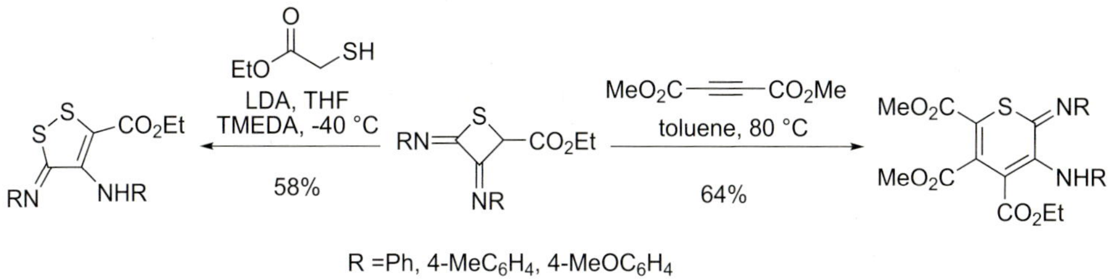

LDA, THF, TMEDA, -40 °C, 58%; MeO_2C—≡—CO_2Me, toluene, 80 °C, 64%

R =Ph, 4-MeC_6H_4, 4-$MeOC_6H_4$

Scheme 3.203

Intramolecular induced ring opening of 2-iminothietanes, in which the imino group bears a phenol function, occurs by simple heating (Scheme 3.204) [359].

Enzymatic resolution of 3-methylthietan-2-one has been studied in the presence of lipases. Best results were reported using *Pseudomonas cepacia* lipase ($E > 100$) (Scheme 3.205) [360].

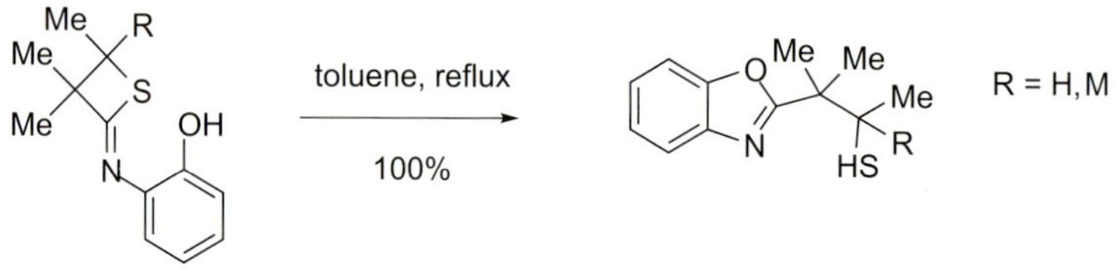

toluene, reflux
100%
R = H, M

Scheme 3.204

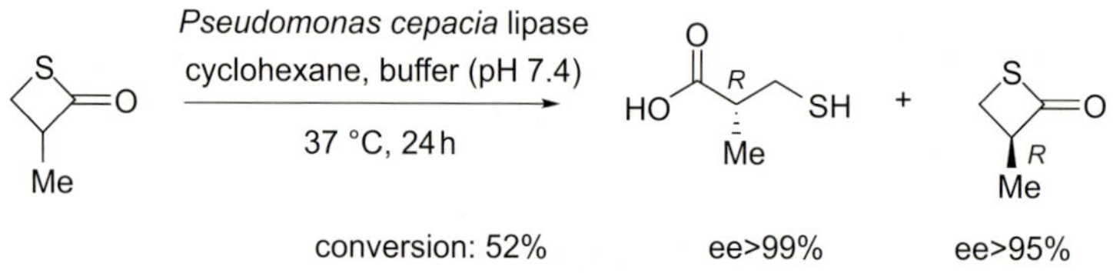

Pseudomonas cepacia lipase
cyclohexane, buffer (pH 7.4)
37 °C, 24 h
conversion: 52%
ee>99%
ee>95%

Scheme 3.205

3.4
Other Four-Membered Heterocycles

3.4.1
Selenetanes

3.4.1.1 Introduction

The chemistry of four-membered heterocyclic compounds containing one selenium atom has been until recently very little studied. These compounds appear to be stable enough to be synthesized and their reactivity examined, even if they have a relatively low thermal stability.

CNDO/2 calculations have been performed to determine the conformation of the parent compound [361]. Experiment values were measured by ^{1}H NMR in the case of 3,3-dimethylselenetane complexed by palladium(II) halides [362] IR and Raman spectra of this compound have been studied and the different vibrations assigned [363]. ^{1}H and ^{13}C NMR spectra of some selenetanes have been reported. Chemical shifts of protons and carbons appear comparable to those reported for thietanes. The ^{13}C-^{77}Se coupling has been used recently as a criterion of formation of selenetanes, even if the natural abundance of ^{77}Se is only 7% [364]. The X-ray structure of a selenetane has been reported [365]. In mass spectra, the parent ion corresponds to the retro [2 + 2] cycloaddition. All ions containing a Se atom are characteristic due to the particular isotopic abundance of selenium [366].

No natural product containing a selenetane ring has yet been reported. Derivative **37** has been tested as an antiviral compound [367].

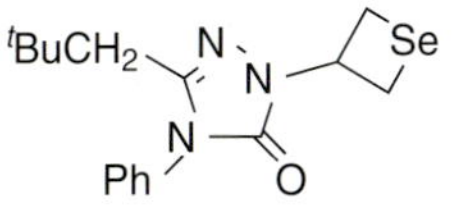

37

3.4.1.2 Synthesis of Selenetanes

Although much less studied than thietanes, preparations of these compounds have often been made by similar methods.

3.4.1.2.1 Synthesis by Formation of Two Se–C Bonds The first attempt to prepare selenetanes was by the reaction of 1,3-dibromopropane with Na_2Se [368]. The selenetane was not characterized due its easy polymerization. This method was reinvestigated and applied to the preparation of various selenetanes (Scheme 3.206) [369]. The parent compound was obtained in low yield (5%). This approach was recently used in the preparation of nucleosides (Scheme 3.206) [364].

Heating cyclic carbonates at high temperature (170–220 °C) in the presence of KSeCN gives selenetanes in satisfactory yields (Scheme 3.207) [370].

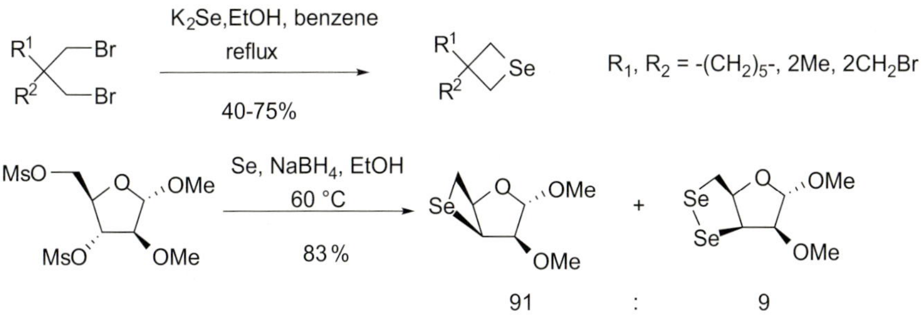

K_2Se, EtOH, benzene, reflux, 40-75%; R_1, R_2 = $-(CH_2)_5-$, 2Me, $2CH_2Br$

Se, $NaBH_4$, EtOH, 60 °C, 83 %; 91 : 9

Scheme 3.206

KSeCN, 170-220 °C, 35-50%; R = C_1-C_8, n = 0-8

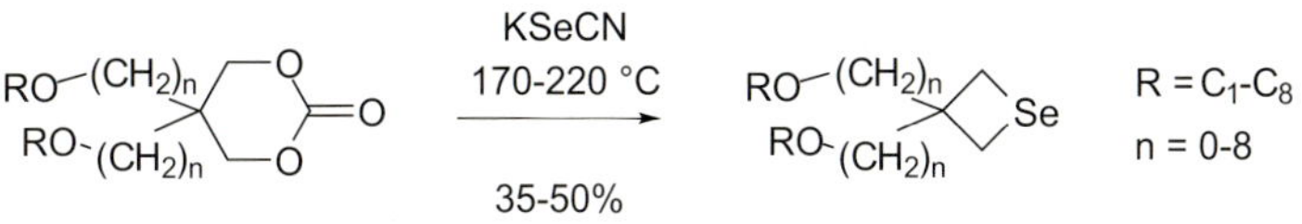

Scheme 3.207

Reaction of an excess of NaSeH with 2-hydroxy-1,3-dibromopropane gives the corresponding di-selenoate. This then leads to the formation of 3-hydroxyselenetane by reaction with $NaBH_4$ or a mixture $Hg(CN)_2$–Na_2S (Scheme 3.208) [371]. The reaction of 2-hydroxy-1,3-dibromopropane with NaSeH under phase transfer conditions has been reported to give 3-hydroxyselenetane in 56% yield [372].

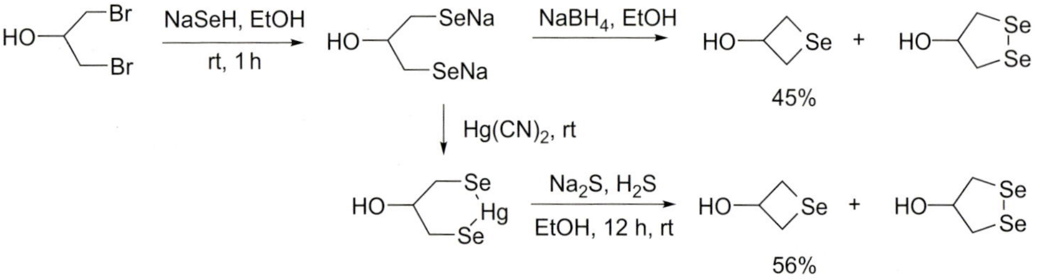

Scheme 3.208

3-Hydroxyseletanes have also been prepared by the reaction of 2-chlorooxiranes with Li_2Se, prepared *in situ* (Scheme 3.209) [372]. Utilization of H_2Se in the presence of $SnCl_4$ gives similar results [373]. The oxirane ring opening has been applied to the preparation of a taxol derivative [374] and nucleosides [375]. Opening of oxiranes with the formation of selenetanes is also possible if the leaving group is in the γ position (Scheme 3.209) [376]. 3-Hydroxyselenetane was also obtained by electrochemical opening of epichlorohydrin in the presence of selenium. The best results were obtained when the graphite electrode was doped with selenium [377].

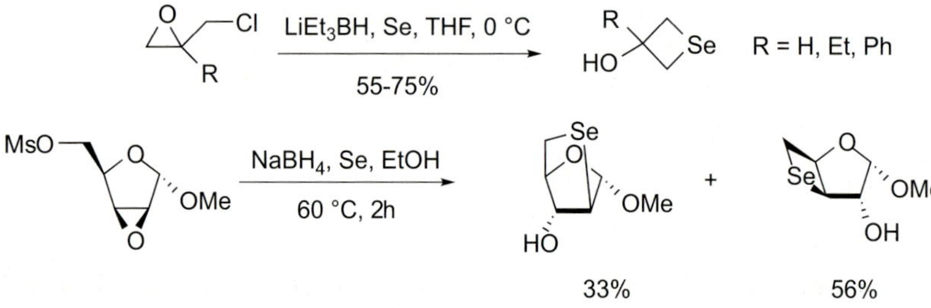

Scheme 3.209

Electrophilic addition of $SeBr_4$ to norbornadiene leads to the formation of a 1,1-dibromoselenetane (Scheme 3.210) [378].

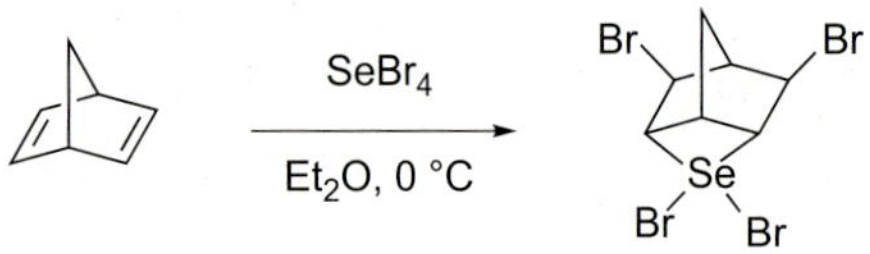

Scheme 3.210

3.4.1.2.2 **Synthesis by Formation of One Se–C Bond** 3,3-Dimethylselenetane-1,1-dioxide can be obtained by heating sodium 3-chloro-2,2-dimethylseleninoate (Scheme 3.211) [369].

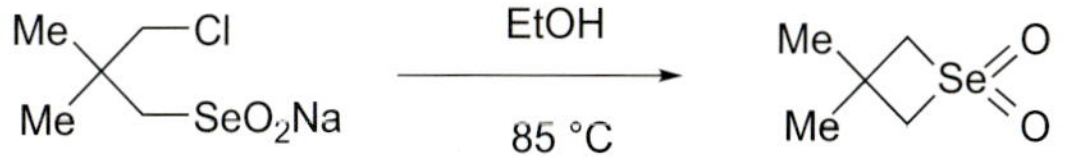

Scheme 3.211

2-Selenoalkenylbenzoimidazole derivatives lead to the formation of selenetanes by the sequence indicated in Scheme 3.212 [302].

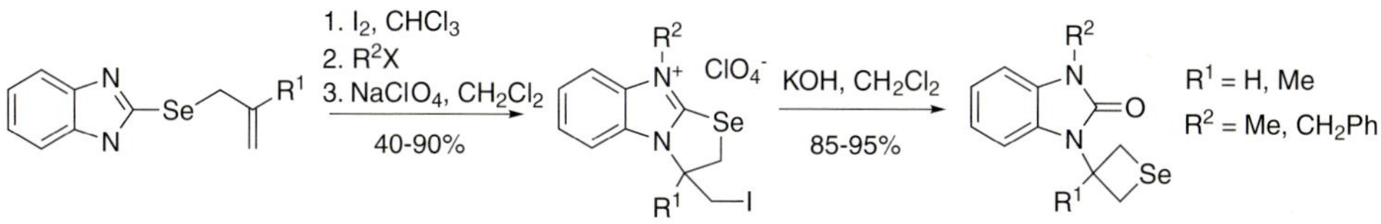

Scheme 3.212

Another interesting preparation of selenetanes uses the reaction of γ-seleno alcohols with KH; good yields were generally obtained (Scheme 3.213) [379].

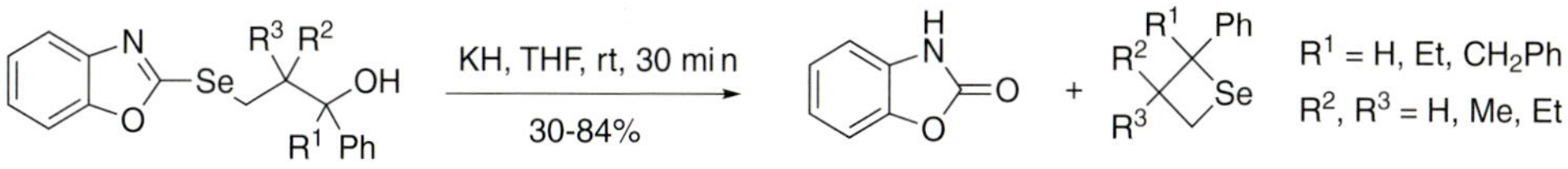

Scheme 3.213

3.4.1.2.3 **Formation by Ring Regression** A selenetane has been prepared in low yield by reaction of a diseleno compound with hexamethylphosphotriamine (Scheme 3.214) [380]. The 1,3-oxaselenetane derivative was obtained in 72% yield when the reaction was carried out in the presence of triphenylphosphine.

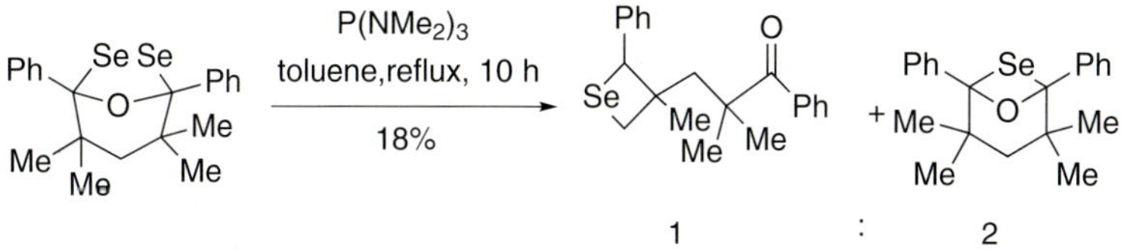

Scheme 3.214

An attempt to prepare benzoselenete by photolysis of 3-diazobenzoselenophenone has been reported. A dimer was isolated, the formation of which was postulated to occur by dimerization of the unstable benzoselenete. Calculations show that benzoselenete is 49.5 kcal mol^{-1} less stable than benzothiete (Scheme 3.215) [381].

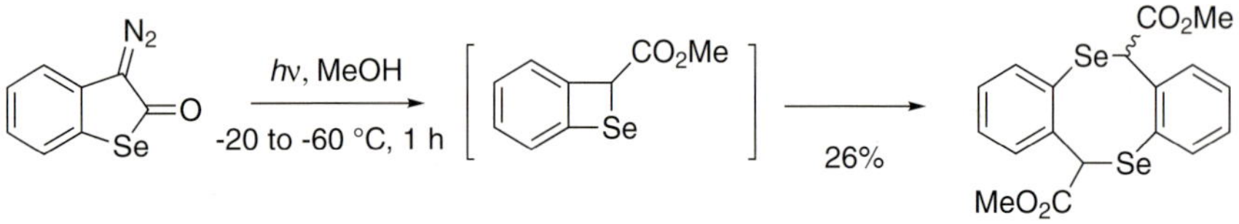

Scheme 3.215

3.4.1.2.4 **Formation by Cycloaddition** Stable benzoselenetes, characterized by X-ray crystallography, have been obtained by cycloaddition of sterically hindered seleno ketones with benzyne (Scheme 3.216) [365].

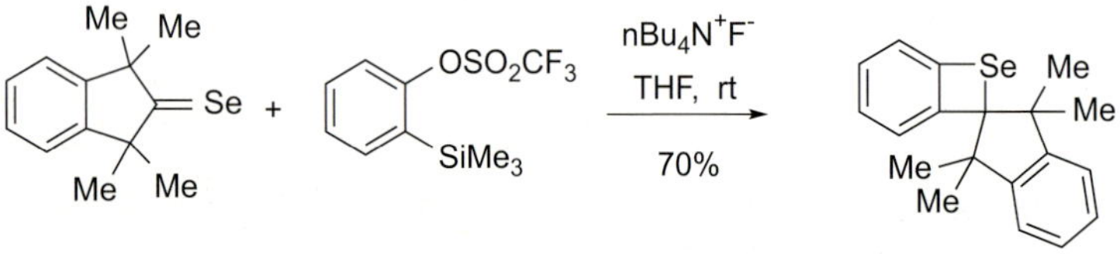

Scheme 3.216

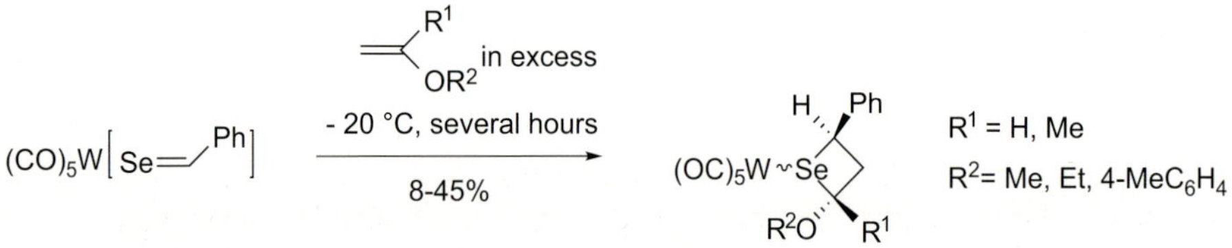

Scheme 3.217

Benzoselenoaldehyde stabilized as a tungsten complex reacts with enol ethers at low temperature to give stable selenetanes (Scheme 3.217) [382].

3.4.1.3 **Reactivity**

The reaction of 3,3-dimethylselenetane with electrophiles such as MeI, Br_2, I_2, SO_2Cl_2 gives ring cleavage products [383]. Additions of nucleophiles to the reaction mixture allow subsequent reactions. With ozone or hydrogen peroxide, insertions of an oxygen atom in the four-membered cycle are observed (Scheme 3.218).

Seleno crown ethers have been formed by the reaction of 3,3-dimethylselenetane with a rhenium carbonyl complex (Scheme 3.219) [384].

Oxidation of 3-hydroxyselenetane with the Dess-Martin reagent gives selenetan-3-one in quantitative yield [372]. However, this ketone was reported to be highly instable. In the presence of sodium hydroxymethanesulfonate (Rongalite), 3-phenyl-

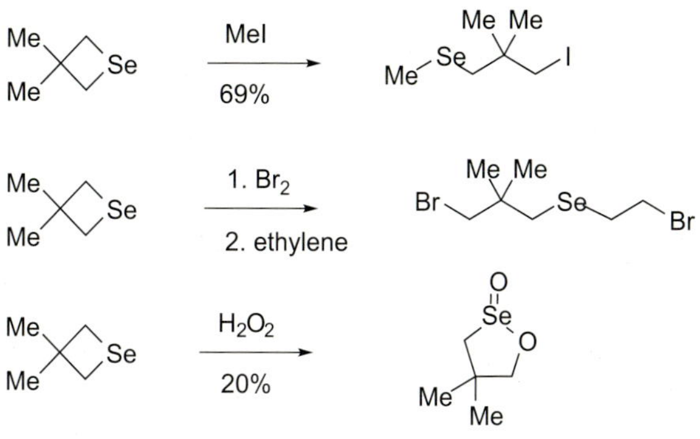

Scheme 3.218

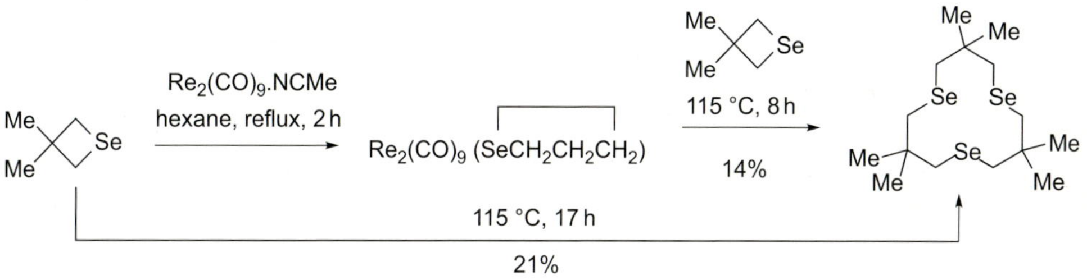

Scheme 3.219

3-hydroxyselenetane leads to the formation 2-phenylallylic alcohol [372]. The reactivity of tungsten-stabilized selenetanes with KSCN and KSeCN has been examined. In the first case a 1,3-selenazinone-2-thione is obtained (Scheme 3.220) [385], while in the second case a 1,2-diselenolane is formed [382b].

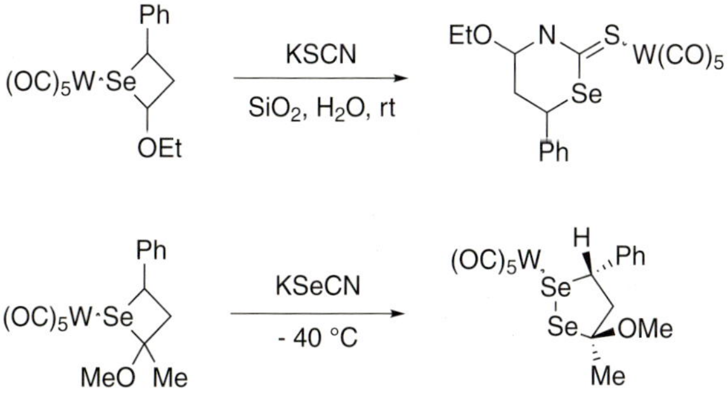

Scheme 3.220

3.4.2
Telluretanes

The chemistry of telluretanes is almost unknown, even though the first attempt to obtain the parent compound was reported in 1945 [386]. Reaction of 1,3-dibromo-propane with Na_2Te gave, apparently, the unstable tellurane and only a polymer was isolated during work-up of the reaction. A similar report was made on the reaction of a tellurate [387]. The reaction of epichlorohydrin derivatives with Na_2Te was also fruitless [372]. However, a telluretane derivative has been reported by reaction of norbornadiene with $TeBr_4$ (Scheme 3.221) [378].

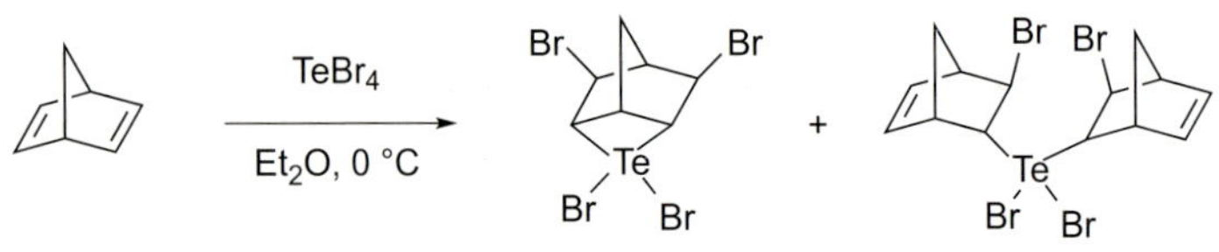

$TeBr_4$, Et_2O, 0 °C

Scheme 3.221

3.4.3
Phosphetanes

Four-membered phosphorus compounds have been known since 1957. Phosphorus (III) heterocycles have in general a low stability and are stabilized as P(V) derivatives (oxides, sulfurs, boranes, etc.). Chiral phosphetane derivatives have been studied as ligands in catalytic reactions (hydrosilylation, hydrogenation). Their synthesis and chemistry have been reviewed [388–390]. Chapter 23 of this book is devoted to phosphorous heterocycles.

3.4.4
Arsetanes

Arsetanes, also called arsacyclobutanes, have been known since 1977 [391]. Structural data on arsetene have been explored by PM3 semi-empirical studies [392]. These few compounds studied appear moderately stable at room temperature under inert atmosphere.

The first reported preparation of arsetanes involves reaction of (3-chloropropyl) iodo-arsines with sodium (Scheme 3.222) [391]. This method was subsequently modified, and it was reported that the ring closure of (3-chloropropyl)arsines could be carried out using potassium *tert*-butylate starting from iron(II) complexes of arsines [393]. 1-Phenylarsetane has been also prepared, by the reaction of dilithium phenyl arsenide with 1,3-dichloropropane [393].

An arsetene has been obtained by reaction of a methylenearsorane derivative complexed by iron with dimethyl fumarate. This arsetene has been characterized by

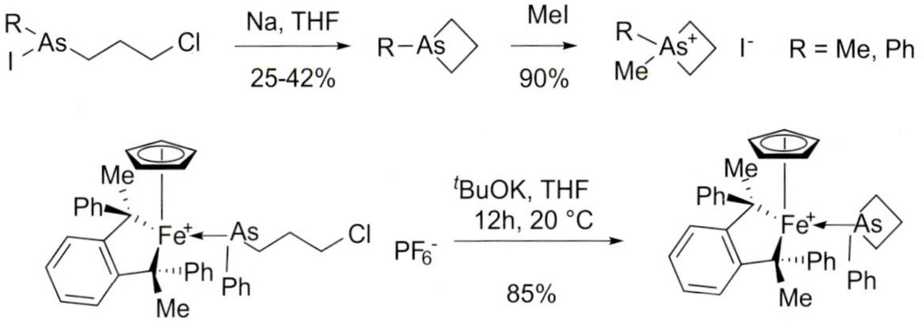

R, I–As–(CH2)3–Cl → (Na, THF; 25-42%) → R–As(ring) → (MeI; 90%) → R, Me–As⁺(ring) I⁻   R = Me, Ph

Me, Ph, Fe⁺←As, Ph, Ph, Me, Cl, PF_6^- → (^{t}BuOK, THF; 12h, 20 °C; 85%) → Me, Ph, Fe⁺←As(ring), Ph, Ph, Me

Scheme 3.222

spectroscopic analysis and by X-ray diffraction [394]. Similar cycloaddition, leading to an arsetane, has been reported from fumaronitrile (Scheme 3.223) [395].

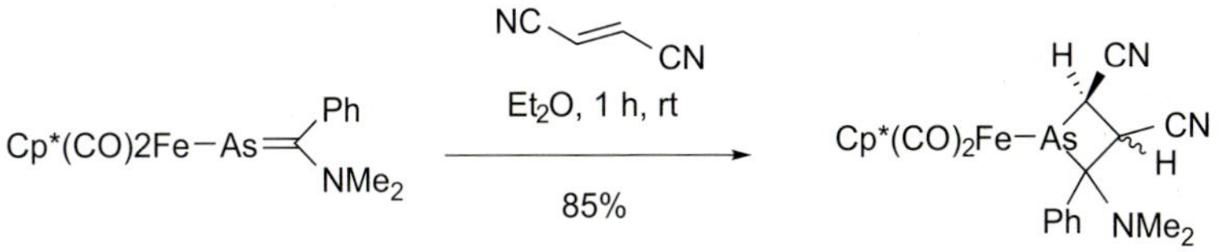

Cp*(CO)2Fe–As=C(Ph)(NMe2) → (NC–CH=CH–CN; Et_2O, 1 h, rt; 85%) → $Cp^*(CO)_2Fe$–As(ring) with H, CN, CN, H, Ph, NMe_2

Scheme 3.223

An original preparation of an arsetene involves metal exchange between titanium and arsenium, starting from a titanium cyclobutane derivative (Scheme 3.224) [396].

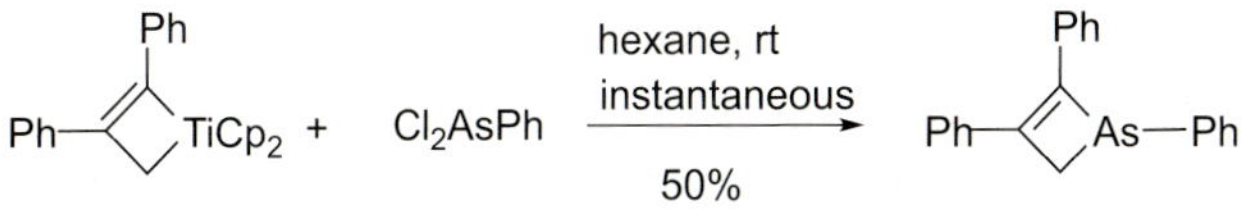

Ph, Ph–(ring)–$TiCp_2$ + Cl_2AsPh → (hexane, rt; instantaneous; 50%) → Ph, Ph–(ring)–As–Ph

Scheme 3.224

3.4.5 Siletanes

3.4.5.1 Introduction

Siletanes, also called silacyclobutanes, have been known since 1954 [397]. As this family of compounds has been reviewed [398], the present chapter focuses on new aspects of their preparation and reactivity. Siletanes have been studied by numerous theoretical methods. The geometry of 1,1-dimethylsiletane has been reported using gas electron diffraction [399]. *Ab initio* calculations shows that the barrier to inversion of 1-methylsiletane is comparable to that of methylcyclobutane [400]. NMR (^{1}H, ^{13}C and ^{32}Si), IR and UV spectra of substituted siletanes have been reported [398]. In mass spectra, the main fragmentation corresponds to [2 + 2] retrocycloadditions [398].

3.4.5.2 Preparation of Siletanes

3.4.5.2.1 Preparation from Chlorosilanes Cyclization of 1-chloro-(3-chloropropyl) silanes by the Wurtz reaction using magnesium is an efficient method to obtain siletanes (Scheme 3.225) [398]. Utilization of other metals (Li, Na, Na-K alloy) and electrochemical conditions extend this method [401].

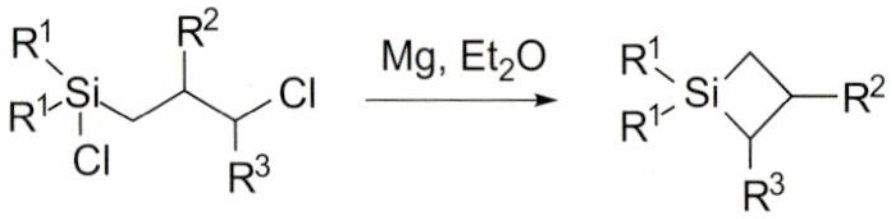

Mg, Et_2O

Scheme 3.225

Dichlorosilanes react similarly with 1,3-dimetallic species to form polycyclic silacyclobutanes (Scheme 3.226) [402].

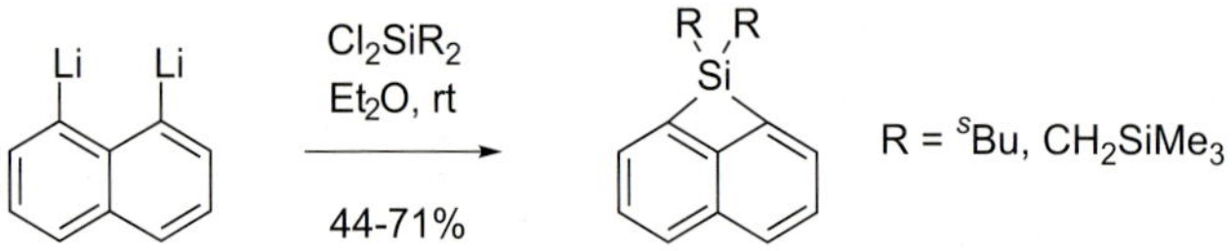

Cl_2SiR_2, Et_2O, rt, 44-71%

R = ^{s}Bu, CH_2SiMe_3

Scheme 3.226

3.4.5.2.2 Other Intramolecular Cyclizations 1,2-Bissilylalkanes lead, with alkenes, in the presence of palladium(II) catalyst to the formation of 1,2-bis(organosilyl)alkanes by the addition of a Si–Si bond across C–C double bonds. With 1,2-bissilylalkenes, intramolecular additions have been observed that give rise to silacycloalkanes. When the substituent fixed on one of the silicon atoms is a 3-butenyl chain, siletanes of low stability were formed by 4-*exo* cyclizations (Scheme 3.227) [403].

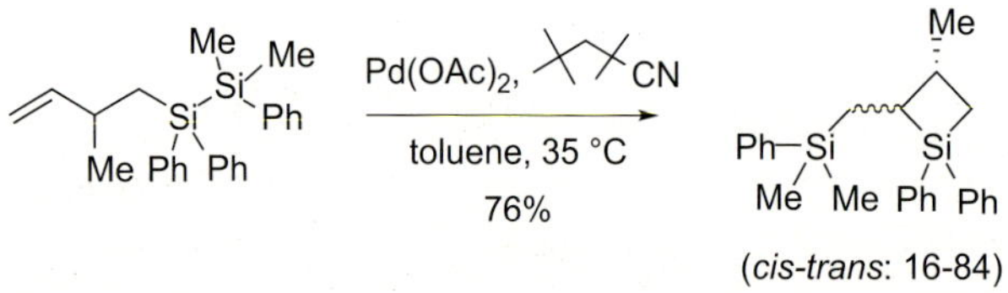

$Pd(OAc)_2$, toluene, 35 °C, 76%

(*cis-trans*: 16-84)

Scheme 3.227

Substitutions of silicon by sterically overcrowded substituents allow the preparation of stable silylenes. When one of the substituents is a silicon aryl substituted group, simple heating leads to rearrangement of these compounds into siletanes (Scheme 3.228) [404].

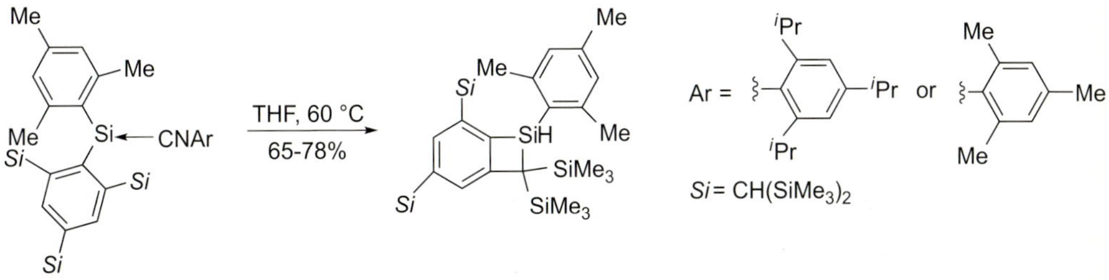

Scheme 3.228

Intramolecular insertion of carbenes into C–H bonds is also possible by irradiation of α-silyl-α-diazoacetates (Scheme 3.229) [405]. If one of the substituents of the silyl group is an allyl group, a bicyclic silacyclobutane is observed [406].

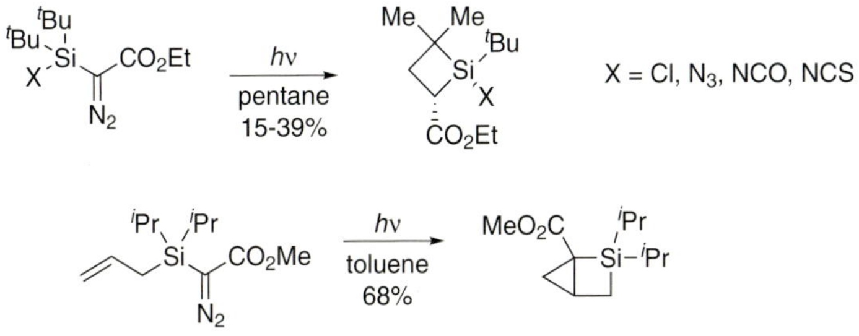

Scheme 3.229

3.4.5.2.3 **[2 + 2] Cycloadditions** Reaction of trichlorovinylsilane with *t*-butyl lithium produces, by 1,2 elimination of LiCl, reactive silenes (Si=C bond). In the presence of enol ethers, elimination does not take place, and the lithio intermediate undergoes an addition on the C=C double bond, leading by 1,4 cyclization to siletanes (Scheme 3.230) [407].

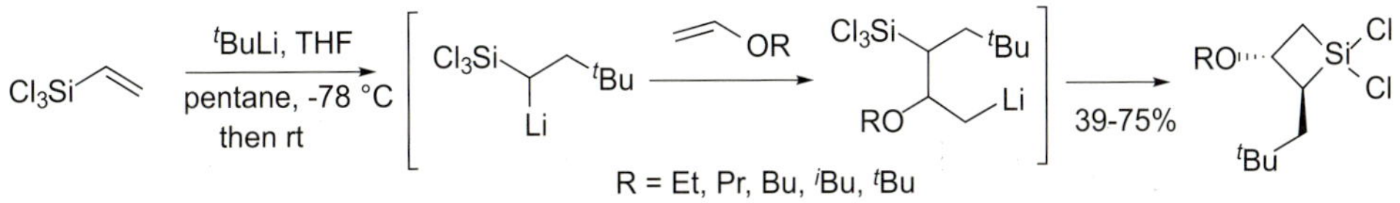

Scheme 3.230

3.4.5.2.4 **Preparation from Other Heterocyclic Compounds** Siliranes (silacyclopropanes) react with diazomethane to give siletanes in good yields [408]. These compounds give rise to similar ring expansions by reaction with isocyanate (Scheme 3.231) [409]. For monosubstituted silacyclopropanes a good regioselectivity in the formation of siletane is observed [409b].

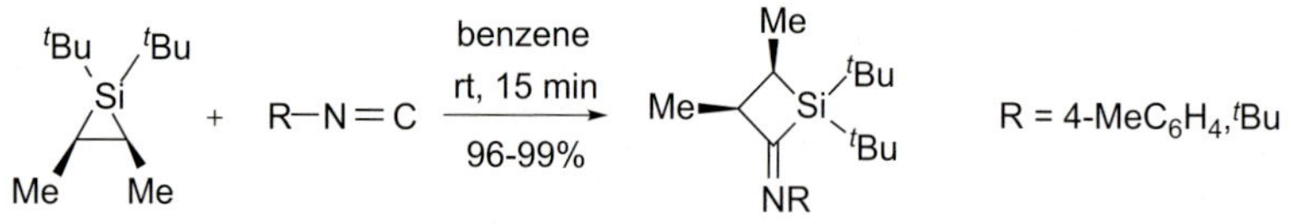

benzene, rt, 15 min, 96-99%; R = 4-MeC$_6$H$_4$,^{t}Bu

Scheme 3.231

Sterically hindered 1,2-disiletanes lead to the formation of silenes by heating at 60–100 °C. When the reaction is carried out in the presence of styrene, a siletane is formed in excellent yield (Scheme 3.232) [410].

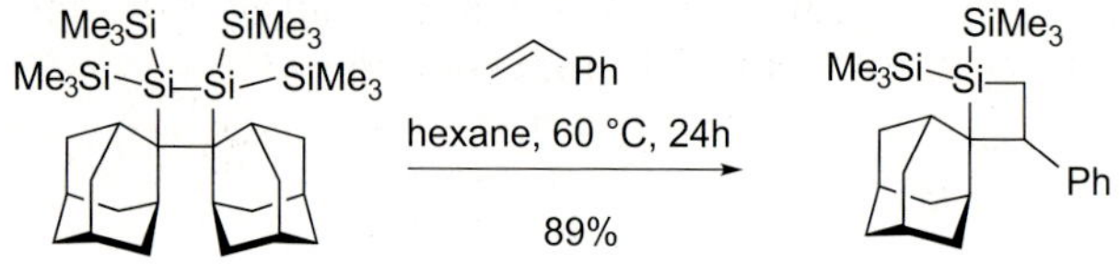

hexane, 60 °C, 24h, 89%

Scheme 3.232

Irradiation of 1,2-disilolane in the presence of isobutene furnishes a stable tricyclic siletane if one of the substituents fixed on the silicon atoms is a phenyl (Scheme 3.233) [411]. Irradiation of a sterically hindered 9-silaanthracene in benzene leads to the formation of a bicyclo[2.2.0] compound of low stability (Scheme 3.233) [412]. On standing at room temperature, the starting anthracene derivative is reformed.

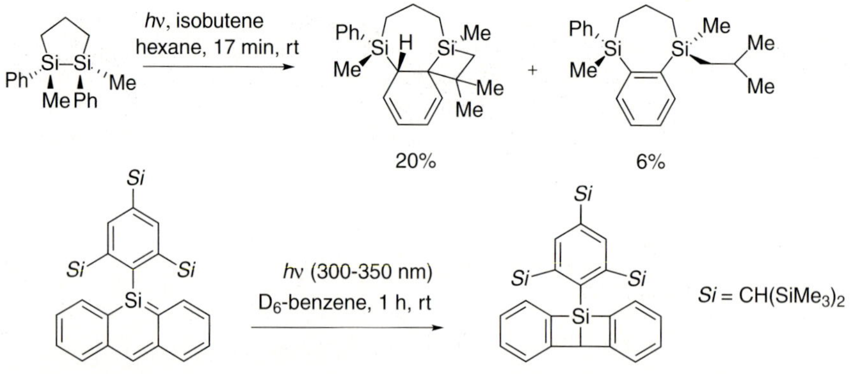

hν, isobutene, hexane, 17 min, rt; 20%; 6%

hν (300-350 nm), D$_6$-benzene, 1 h, rt; *Si* = CH(SiMe$_3$)$_2$

Scheme 3.233

3.4.5.3 Reactivity

Pyrolysis or photolysis of siletanes leads to the formation of silenes [413]. Polymerization of silacyclobutanes has also been studied intensively, and compounds with alternating silicon and trimethylene groups have been obtained [414]. Polymers in which the siletane unit was introduced have also been reported [415]. The presence of siletane cycles instead of trialkylsilanes in functionalized compounds increases their reactivity, allowing much easier reactions. For example, reaction of silyl ketene acetals with carbonyl compounds occurs without any catalyst if a siletane is present (Scheme 3.234) [416]. This increase in reactivity has also been reported for cross-coupling reactions of vinyl- and arylsilanes [417].

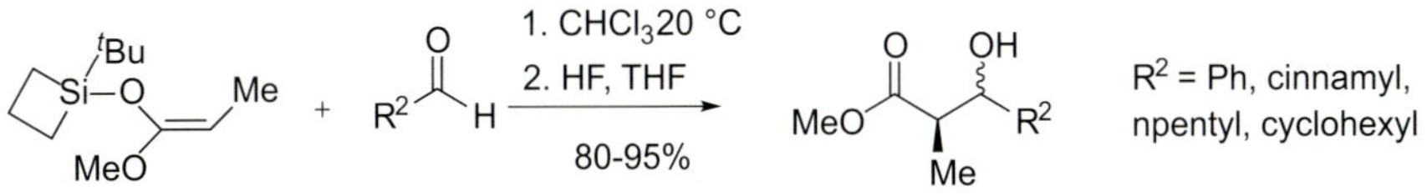

Scheme 3.234

Silacyclobutanes are opened by the action of nucleophilic and electrophilic reagents, sometimes with polymerization [398, 414]. This cleavage can be controlled [418, 419]. For example, in the presence of platinum salts and 1 equivalent of triethylsilane, opening of a benzosiletane occurs cleanly (Scheme 3.235). In the presence of a catalytic amount of triethylsilane, cyclic oligomers or polymers are obtained [418].

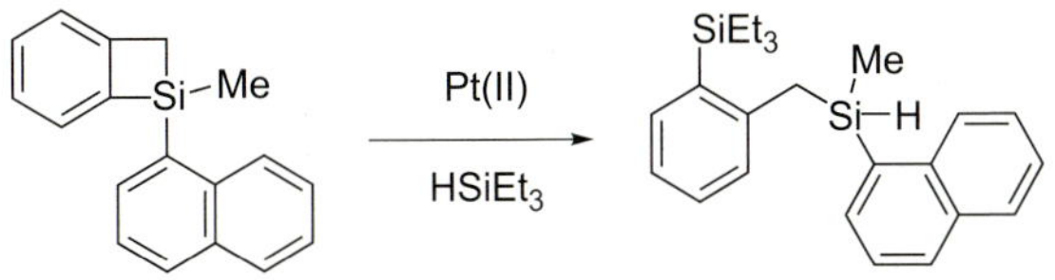

Scheme 3.235

Insertion of a functionalized chain in one of the C–H bonds in the 3-position of 1,1-dimethylsiletane is realized by reaction with diazoacetates in the presence of rhodium diacetate (Scheme 3.236) [420].

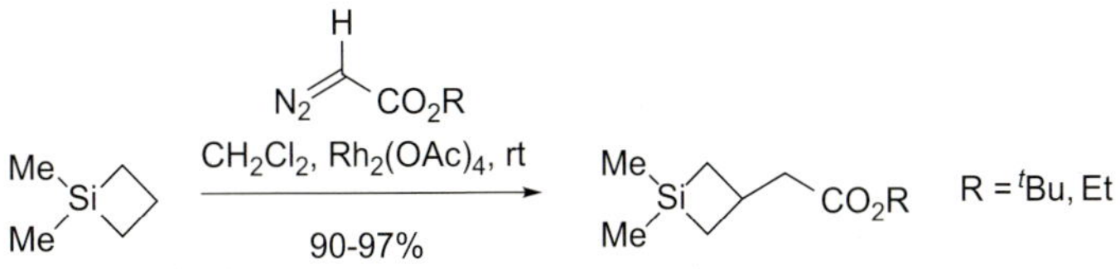

Scheme 3.236

Reactions of siletanes with carbenoids of type CHX_2Li lead to the formation of silacyclopentane derivatives (Scheme 3.237) [421]. Ring expansions are also observed by reaction of 1-(1-iodoalkyl)siletanes with MeLi, ^{t}BuOK or AgOAc and by reaction of 1-oxiranylsiletanes with MeLi (Scheme 3.237) [422]. The insertion of sulfur in 1,1-dimethysiletane is very efficient in the presence of KF and crown ethers, giving rise to 1,2-thiasilolanes [423].

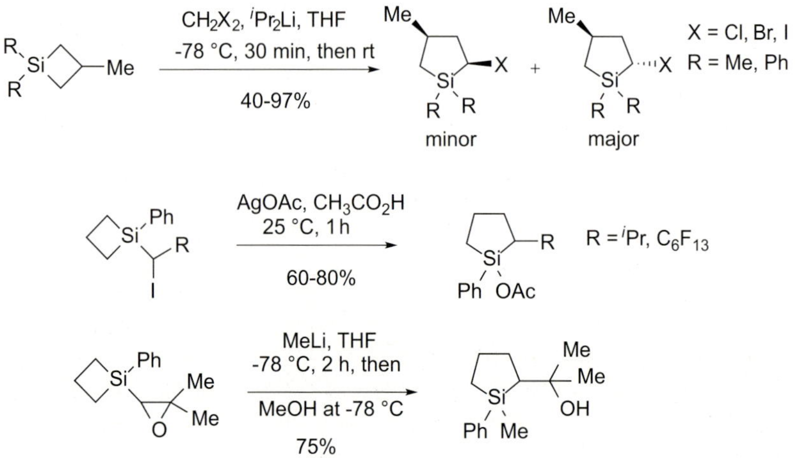

Scheme 3.237

Insertion of the carbonyl function of aldehydes in the cycle of siletanes is possible in the presence of ^{t}BuOK [424]. With 1,1-diphenyl-3-methylenesiletane, ^{t}BuOK is not necessary and the reaction occurs by simple heating. It was subsequently reported that this reaction is also possible with aryl ketones (Scheme 3.238) [425].

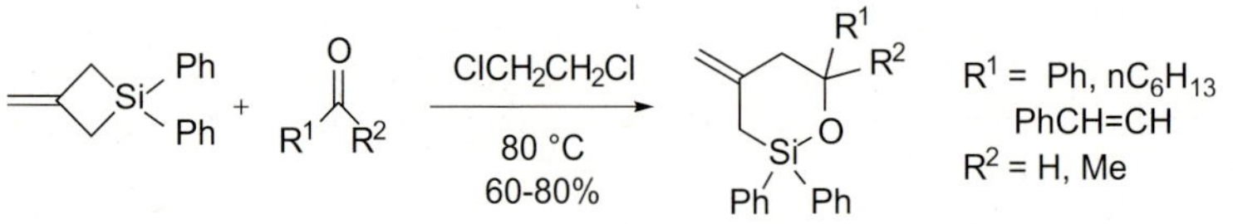

Scheme 3.238

Insertions of SO_2, SO_3 and phosphorus ylides similarly lead to six-membered heterocycles [398]. With acid chlorides and acetylenic compounds, these insertions occur only in the presence of palladium(II) catalysts (Scheme 3.239) [426, 427].

Insertion of the carbonyl function of CF_3COOH into the C–Si bond of 2-iminosiletanes, instead of the expected hydrolysis, has been reported (Scheme 3.240) [409b]. In the presence of an aqueous solution of $CuSO_4$, ring opening of these

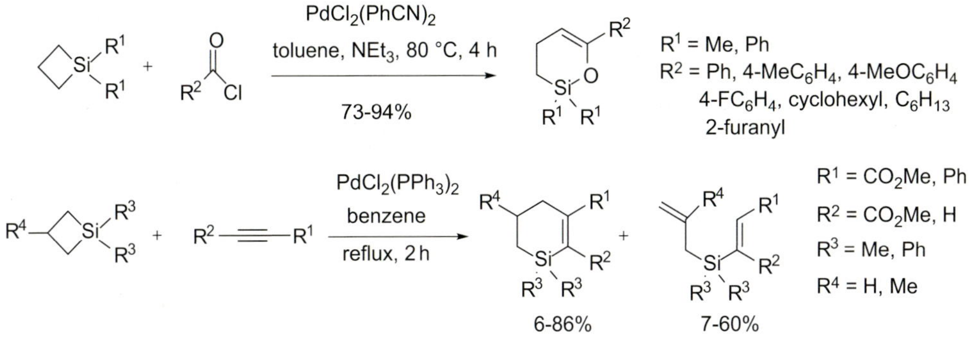

$PdCl_2(PhCN)_2$, toluene, NEt_3, 80 °C, 4 h, 73-94%

R^1 = Me, Ph
R^2 = Ph, 4-MeC_6H_4, 4-$MeOC_6H_4$, 4-FC_6H_4, cyclohexyl, C_6H_{13}, 2-furanyl

$PdCl_2(PPh_3)_2$, benzene, reflux, 2 h; 6-86% + 7-60%

R^1 = CO_2Me, Ph
R^2 = CO_2Me, H
R^3 = Me, Ph
R^4 = H, Me

Scheme 3.239

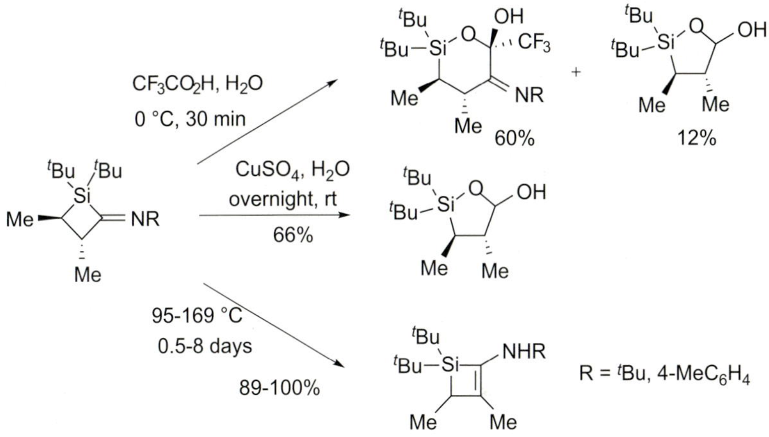

CF_3CO_2H, H_2O, 0 °C, 30 min: 60% + 12%

$CuSO_4$, H_2O, overnight, rt: 66%

95-169 °C, 0.5-8 days: 89-100%

R = ^{t}Bu, 4-MeC_6H_4

Scheme 3.240

substrates is observed. Heating of these 2-iminosiletanes leads to isomerization of the C–N double bond, to give the corresponding enamines [409b].

The silicon atom of silacyclobutanes has reactivity comparable to that of other silanes, and is not reported in this chapter. An interesting example concerning the reactivity of such compounds has been reported. Reactions of 1(γ-haloalkoxy)-1-methylsilacyclobutanes with magnesium leads to the formation of 1-(ω-hydroxyalkoxy)-1-methylsilacyclobutanes in good yields, by intramolecular Grignard reagent attack on the silicon atom, followed by alcoholate elimination (Scheme 3.241) [428].

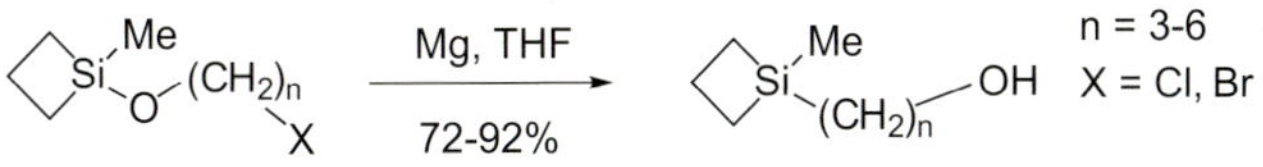

Mg, THF, 72-92%

n = 3-6
X = Cl, Br

Scheme 3.241

3.4.6
Germetanes

3.4.6.1 Introduction

The chemistry of germetanes has been less examined than that of siletanes, since the first report in 1966 [429]. Only limited spectra data are available concerning these compounds. In the IR spectrum, a vibration at 1120 cm^{-1} seems characteristic of this heterocycle [429]. These compounds appear to be stable at room temperature, and the mechanism of thermal decomposition of 1,1-dimethylgermetane has been examined. The formation of cyclopropane, propene and 1,2-digermacyclobutane was observed [430]. The structure of 1,1,3,3-tetramethylgermetane has been studied by gas electron diffraction and *ab initio* molecular orbital calculations (HF/6–31G* and MP2/6–31G*) [431]. An *ab initio* study has been made on the mechanism of formation of germetane by cycloaddition of alkylidenegermylene and ethylene [432].

3.4.6.2 Preparations

The first method reported for the preparation of germetanes consists in the cyclization of chloro-(3-chloropropyl)germanes with sodium or sodium/potassium alloy (Scheme 3.242) [429].

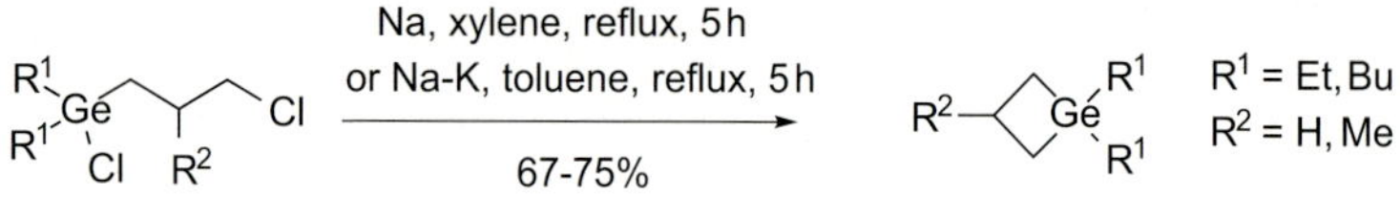

Na, xylene, reflux, 5 h
or Na-K, toluene, reflux, 5 h
67-75%
R^1 = Et, Bu
R^2 = H, Me

Scheme 3.242

In more recent studies, the reaction of 1,1-dichlorogermanes with 1,3-dimetallo compounds has been preferred [433]. This method has been applied to the preparation of benzo- [434] and tricyclic germetane derivatives (Scheme 3.243) [435].

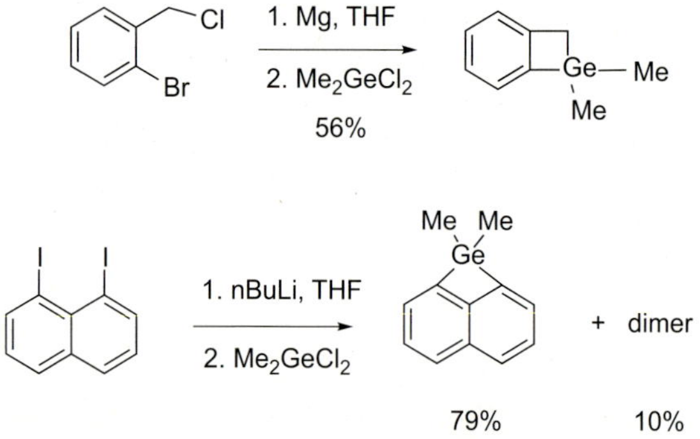

1. Mg, THF
2. Me_2GeCl_2
56%

1. nBuLi, THF
2. Me_2GeCl_2
+ dimer
79% 10%

Scheme 3.243

Intermolecular hydrogermanylation has been reported to afford 1,1-dimethylgermetane [436]. Heating of dihydrogermylprop-2-en-1-ols was reported to give 2-hydroxygermetanes in low yields, by intramolecular cyclization (Scheme 3.244) [437].

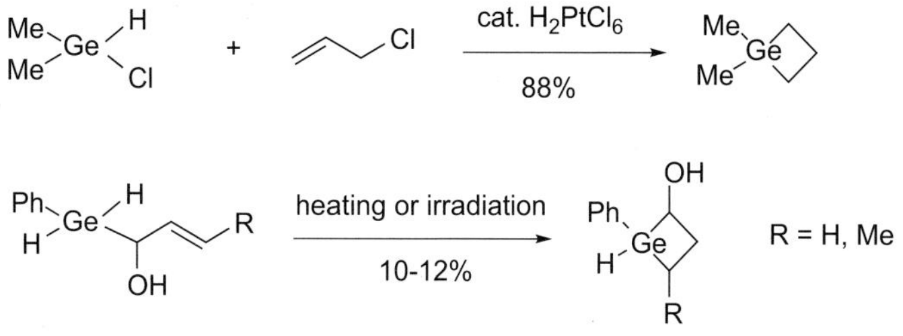

Scheme 3.244

Highly strained germetanes have been obtained by rearrangement of germanylenes [438] and ethylenidene germanes (Scheme 3.245) [439].

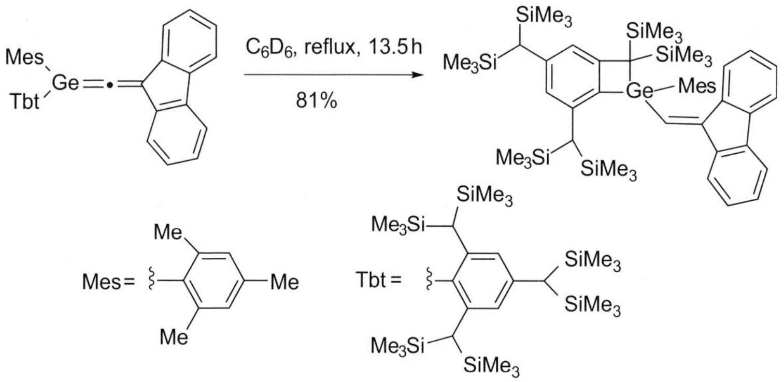

Scheme 3.245

3.4.6.3 **Reactivity**

Germatanes react with electrophilic or nucleophilic reagents to give ring cleavage products. Ring cleavages were reported with halogens, $LiAlH_4$, acids, bases [429], hydrosilanes (Scheme 3.246) [440], phenylphosphinidene [441] and alcohols [442].

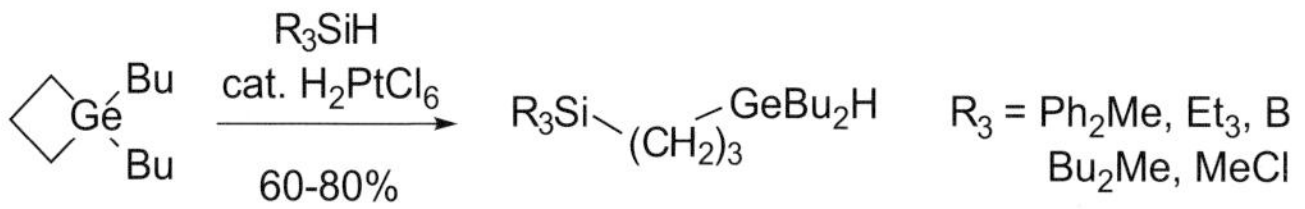

Scheme 3.246

Heating at 160 °C of 1,1-dimethylgermetane leads to formation of a polymer [436], while irradiation of 1,1-diphenylgermane in cyclohexane gives rise to the 1,3-digermanocyclobutane derivative (Scheme 3.247) [442].

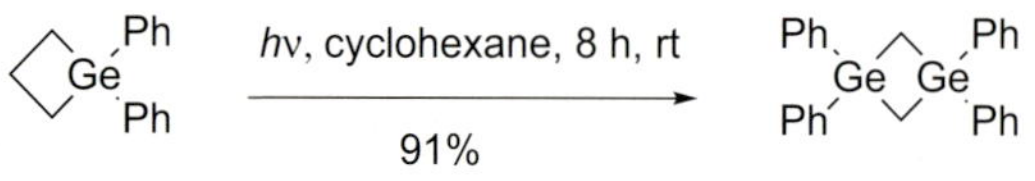

hν, cyclohexane, 8 h, rt
91%

Scheme 3.247

When heated (250 °C) in the presence of sulfur, an insertion in one Ge–C bond of a sulfur atom occurs (Scheme 3.248). The same insertion is observed on heating at 260 °C in the presence of selenium [443]. Comparable insertion was reported with dichlorocarbene, generated from the Seyferth reagent [444]. Reactions with SO_2 and SO_3 led, as with siletanes, to the formation of six-membered heterocycles [445].

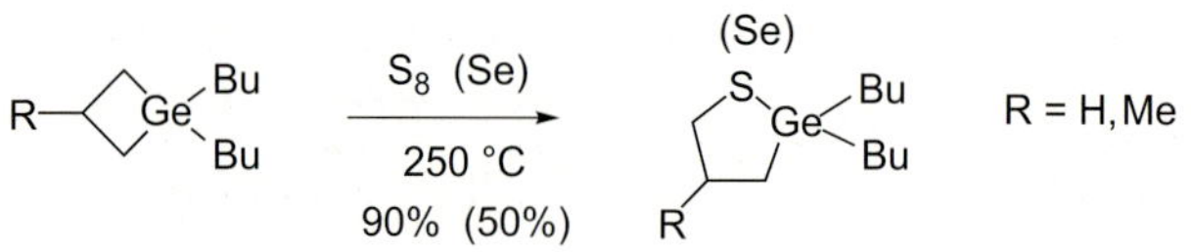

S_8 (Se)
250 °C
90% (50%)
R = H, Me

Scheme 3.248

3.4.7
Bismetanes and Stibetanes

These four-membered ring compounds are still unknown.

References

1 Cromwell, N.H. and Phillips, B. (1979) *Chemical Reviews*, **79**, 331–358.
2 Davies, D.E. and Storr, R.C. (1984) in *Comprehensive Heterocyclic Chemistry*, **7** (ed. W. Lwowski), Pergamon Press, Oxford, pp. 237–358.
3 De Kimpe, N. (1996) in *Comprehensive Heterocyclic Chemistry II*, **1B** (ed. A. Padwa), Pergamon Press, Oxford, pp. 507–589.
4 Hunt, S. (1985) in *Chemistry and Biochemistry of the amino acids* (ed. G.C. Barrett), Chapman and Hall, New York.
5 Rosenthal, D.A. (1982) *Plant Nonprotein Amino and Imino Acids*, Academic Press, New York.
6 Suzuki, K., Sffimada, K., Nozoe, S., Kazuhiko, T., and Ogita, T. (1996) *Journal of Antibiotics*, **49**, 1284–1285.
7 Sugiura, Y., Mino, Y., Iwashita, T., and Nomoto, K. (1985) *Journal of the American Chemical Society*, **107**, 4667–4669.
8 Koboyashi, J., Tsuda, M., Cheng, J.-f., Ishibashi, M., Takikawa, H., and Mori, K. (1996) *Tetrahedron Letters*, **37**, 6775–6776.

9 Isono, K., Funayama, S., and Suhadolnik, R.J. (1975) *Biochemistry*, **14**, 2992–2996.

10 Hanessian, S., Fu, J.-M., Tu, Y., and Isono, K. (1993) *Tetrahedron Letters*, **34**, 4153–4156.

11 Isono, K., Asahi, K., and Suzuki, S. (1969) *Journal of the American Chemical Society*, **91**, 7490–7505.

12 Kitajima, M., Kogure, N., Yamaguchi, K., Takayama, H., and Aimi, N. (2003) *Organic Letters*, **5**, 2075–2078.

13 Schummer, D., Forche, E., Wray, V., Domke, T., Reichenbach, H., and Hoefle, G. (1996) *Liebigs Annalen: Organic and Bioorganic Chemistry*, 971–9678.

14 Bannon, A.W., Decker, M.W., Holladay, M.W., Curzon, P., Donnelly-Roberts, D., Puttfarcken, P.S., Bitner, R.S., Diaz, A., Dickenson, A.H., Porsolt, R.D., Williams, M., and Arneric, S.P. (1999) *Science*, **279**, 77–81.

15 Holladay, M.W. and Decker, M.W. (2000) *Advances in Medicinal Chemistry*, **5**, 85–113.

16 Axenrod, T., Watnick, C., Yazdekhasti, H., and Dave, P.R. (1995) *The Journal of Organic Chemistry*, **60**, 1959–1964.

17 Di Martino, A., Galli, C., Gargano, P., and Mandolini, L. (1985) *Journal of the Chemical Society, Perkin Transactions 2*, 1345–1350.

18 Higgins, R.H., Faircloth, W.J., Baughman, R.G., and Eaton, Q.L. (1994) *The Journal of Organic Chemistry*, **59**, 2172–2178.

19 Crimaldi, K. and Lichter, R.L. (1980) *The Journal of Organic Chemistry*, **45**, 1277–1281.

20 Hiraki, T., Yamagiwa, Y., and Kamikawa, T. (1995) *Tetrahedron Letters*, **36**, 4841–4844.

21 Barluenga, J., Fernandez-Mari, F., Viado, A.L., Aguilar, E., and Olano, B. (1996) *The Journal of Organic Chemistry*, **61**, 5659–5662.

22 Liu, D.-G. and Lin, G.-Q. (1999) *Tetrahedron Letters*, **40**, 337–340.

23 Enders, D., Gries, J., and Kim, Z.-S. (2004) *European Journal of Organic Chemistry*, **69**, 4471–4482, see also the reference 123.

24 Hosono, F., Nishiyama, S., Yamamura, S., Izawa, T., Kato, K., and Terada, Y. (1994) *Tetrahedron*, **50**, 13335–13456.

25 Szmuszkovicz, J., Kane, M.P., Laurian, L.G., Chidester, C.G., and Scahill, T.A. (1981) *The Journal of Organic Chemistry*, **46**, 3562–3564.

26 Barluenga, J., Baragaña, B., and Concellón, J.M. (1997) *The Journal of Organic Chemistry*, **62**, 5974–5977.

27 Concellón, J.M., Riego, E., and Bernad, P.L. (2002) *Organic Letters*, **4**, 1299–1301.

28 Chowdhury, A.R., Kumar, V.V., Roy, R., and Bhaduri, A.P. (1997) *Journal of Chemical Research-S*, 254–255.

29 Podlech, J. and Seebach, D. (1995) *Helvetica Chimica Acta*, **78**, 1238–1246.

30 (a) Wang, J., Hou, Y., and Wu, P. (1999) *Journal of the Chemical Society, Perkin Transactions 1*, 2277–2280; (b) See also. Burtoloso, A.C.B. and Correia, C.R.D. (2005) *Journal of Organometallic Chemistry*, **690**, 5636–5646, and the references cited.

31 Karikomi, M., Arai, K., and Toda, T. (1997) *Tetrahedron Letters*, **38**, 6059–6062.

32 Breternitz, H.-J. and Schaumann, E. (1999) *Journal of the Chemical Society, Perkin Transactions 1*, 1927–1931.

33 Guanti, G. and Riva, R. (2001) *Tetrahedron Asymmetry*, **12**, 605–618.

34 Miller, R.A., Lang, F., Marcune, B., Zewge, D., Song, Z.J., and Karady, S. (2003) *Synthetic Communications*, **33**, 3347–3353.

35 Chong, J.M. and Sokoll, K.K. (1995) *Synthetic Communications*, **25**, 603–611.

36 (a) Marinetti, A., Hubert, P., and Genêt, J.-P. (2000) *European Journal of Organic Chemistry*, 1815–1820. (b) Sato, M., Gunji, Y., Ikeno, T., and Yamada, T. (2004) *Synthesis*, 1434–1438.

37 Constantieux, T., Grelier, S., and Picard, J.-P. (1998) *Synlett*, 510–512.

38 Oh, C.H., Rhim, C.Y., You, C.Y., and Cho, J.R. (2003) *Synthetic Communications*, **33**, 4297–4302.

39 (a) De Kimpe, N. and Stevens, C. (1993) *Synthesis*, 89–91. (b) For recent applications see: Salgado, A., Dejaegher, Y., Verniest, G., Boeykens, M., Gauthier, C., Lopin, C., Tehrani, K.A., and De Kimpe, N. (2003) *Tetrahedron*, **59**, 2231–2239, and the references cited therein.

40 Sulmon, P., De Kimpe, N., and Schamp, N. (1988) *The Journal of Organic Chemistry*, **53**, 4462–4465.

41 Robin, S. and Rousseau, G. (2000) *European Journal of Organic Chemistry*, 3007–3011.

42 Kobayashi, K., Miyamoto, K., Morikawa, O., and Konishi, H. (2005) *Bulletin of the Chemical Society of Japan*, **78**, 886–889.

43 Lemau de Talancé, V., Banide, E., Bertin, B., Comesse, S., and Kadouri-Puchot, C. (2005) *Tetrahedron Letters*, **46**, 8023–8025.

44 Pannecoucke, X., Outurquin, F., and Paulmier, C. (2002) *European Journal of Organic Chemistry*, 995–1006.

45 Rutjes, F.P.J.T., Tjen, K.C.M.F., Wolf, L.B., Karstens, W.F.J., Schoemaker, H.E., and Hiemstra, H. (1999) *Organic Letters*, **1**, 717–720.

46 Ohno, H., Anzai, M., Toda, A., Ohishi, S., Fujii, N., Tanaka, T., Takemoto, Y., and Ibuka, T. (2001) *The Journal of Organic Chemistry*, **66**, 4904–4914.

47 Carlin-Sinclair, A., Couty, F., and Rabasso, N. (2003) *Synlett*, 726–728.

48 Braüner-Osborne, H., Bunch, L., Chopin, N., Couty, F., Evano, G., Jensen, A.A., Kusk, M., Nielsen, B., and Rabasso, N. (2005) *Organic and Biomolecular Chemistry*, **3**, 3926–3936.

49 Blythin, D.J., Green, M.J., Lauzon, M.J.R., and Shue, H.-J. (1994) *The Journal of Organic Chemistry*, **59**, 6098–6100.

50 Agami, C., Couty, F., and Evano, G. (2002) *Tetrahedron Asymmetry*, **13**, 297–302.

51 Agami, C., Couty, F., and Rabasso, N. (2002) *Tetrahedron Letters*, **43**, 4633–4636.

52 Wessig, P. and Schwarz, J. (1998) *Helvetica Chimica Acta*, **81**, 1803–1814, and references cited.

53 Wessig, P., Lindemann, U., and Surygina, O. (2000) *Journal of Information Recording*, **25**, 245–249.

54 (a) For a review see: Krow, G.R. and Cannon, K.C. (2004) *Heterocycles*, **64**, 577–603; (b) Krow, G.R., Yuan, J., Fang, Y., Meyer, M.D., Anderson, M.J., and Campbell, J.E. (2001) *The Journal of Organic Chemistry*, **66**, 1811–1817.

55 Pedrosa, R., Andrés, C., Nieto, J., and del Pozo, S J. (2005) *Journal of Organic Chemistry*, **70**, 1408–1416.

56 Kise, N., Ozaki, H., Moriyama, N., Kitagishi, Y., and Ueda, N. (2003) *Journal of the American Chemical Society*, **125**, 11591–11596.

57 Nadir, U.K., Sharma, R.L., and Koul, V.K. (1989) *Tetrahedron*, **45**, 1851–1858.

58 Nadir, U.K. and Arora, A. (1995) *Journal of the Chemical Society, Perkin Transactions 1*, 2605–2609.

59 Funke, W. (1969) *Chemische Berichte*, **102**, 3148–3158.

60 Alvernhe, G., Laurent, A., Touhami, K., Bartnik, R., and Mloston, G. (1985) *Journal of Fluorine Chemistry*, **29**, 363–384.

61 Marchand, A.P., Rajagopal, O., Bott, S.G., and Archibald, T.G. (1994) *The Journal of Organic Chemistry*, **59**, 1608–1612.

62 Hayashi, H., Hiki, S., Kumagai, T., and Nagao, Y. (2002) *Heterocycles*, **56**, 433–442.

63 Mloston, G., Urbaniak, K., and Heimgartner, H. (2002) *Helvetica Chimica Acta*, **85**, 2056–2063.

64 (a) Renga, J.M. (1985) U.S Patent, 4,529,544; (b) (1986) *Chemical Abstracts*, **104**, 19498.

65 Tamaru, Y., Hojo, M., and Yoshida, Z.-i. (1988) *The Journal of Organic Chemistry*, **53**, 5731–5741.

66 Jackson, M.B., Mander, L.N., and Spotswood, T.M. (1983) *Australian Journal of Chemistry*, **36**, 779–788.

67 van Elburg, P.A. and Reinhoudt, D.N. (1987) *Heterocycles*, **26**, 437–445.

68 Ojima, I., Zhao, M., Yamato, T., Nakahashi, K., Yamashita, M., and Abe, R. (1991) *The Journal of Organic Chemistry*, **56**, 5263–5277.

69 Hassner, A. and Wiegand, N. (1986) *The Journal of Organic Chemistry*, **51**, 3652–3656.

70 Gerona-Navarro, G., Bonache, M.A., Alias, M., Perez de Vega, M.J.,

Garcia-López, M.T., Lopez, P., Cativiela, C., and González-Muñiz, R. (2004) *Tetrahedron Letters*, **45**, 2193–2196.

71 Testa, E., Fontanella, L., Gianfranco, C., and Mariani, L. (1959) *Helvetica Chimica Acta*, **42**, 2370–2379.

72 Verkoyen, C. and Rademacher, P. (1985) *Chemische Berichte*, **118**, 653–660.

73 Baldwin, J.E., Edwards, A.J., Farthing, C.N., and Russell, A.T. (1993) *Synlett*, 49–50.

74 Herdeis, C. and Heller, E. (1993) *Tetrahedron Asymmetry*, **4**, 2085–2094.

75 Martinez, I. and Howell, A.R. (2000) *Tetrahedron Letters*, **41**, 5607–5611.

76 Uyehara, T., Yuuki, M., Masaki, H., Matsumoto, M., Ueno, M., and Sato, T. (1995) *Chemistry Letters*, 789–790.

77 Zhao, G.-L., Huang, J.-W., and Shi, M. (2003) *Organic Letters*, **5**, 4737–4739.

78 Akiyama, T., Daidouji, K., and Fuchibe, K. (2003) *Organic Letters*, **5**, 3691–3693.

79 Ishar, M.P.S., Kumar, K., Kaur, S., Kumar, S., Girdhar, N.K., Sachar, S., Marwaha, A., and Kapoor, A. (2001) *Organic Letters*, **3**, 2133–2136.

80 Singal, K.K. and Kaur, J. (2001) *Synthetic Communications*, **31**, 2809–2815.

81 Adamu, H.M., Olagbemiro, T.O., Agho, M.O., and Kutama, I.U. (2002) *Journal of the Chemical Society-Nigeria*, **27**, 14–16.

82 Kiuchi, F., Nishizawa, S., Kawanishi, H., Kinoshita, S., Oshima, H., Uchitani, A., Shekino, N., Ishida, M., Kondo, K., and Tsuda, K. (1992) *Chemical & Pharmaceutical Bulletin*, **40**, 3234–3244.

83 Vögtle, M.M. and Marzinzik, A.L. (2005) *Synlett*, 496–500.

84 Hoshino, J., Hiraoka, J., Hata, Y., Sawada, S., and Yamamoto, Y. (1995) *Journal of the Chemical Society, Perkin Transactions 1*, 693–697.

85 Djeghaba, Z., Deleuze, H., Maillard, B., and De Jeso, B. (1995) *Bulletin des Sociétés Chimiques Belges*, **104**, 161–165.

86 Taylor, E.C. and Hu, B. (1997) *Heterocycles*, **45**, 241–253.

87 Klair, S.S., Mohan, H.R., and Kitahara, T. (1998) *Tetrahedron Letters*, **39**, 89–92.

88 Miyakoshi, K., Oshita, J., and Kitahara, T. (2001) *Tetrahedron*, **57**, 3355–3360.

89 Nichols, D.E., Frescas, S., Marona-Lewicka, D., and Kurrasch-Orbaugh, D.M. (2002) *Journal of Medicinal Chemistry*, **45**, 4344–4349.

90 Matsuura, F., Hamada, Y., and Shioiri, T. (1994) *Tetrahedron*, **50**, 9457–9470.

91 McFarland, J.W., Hecker, S.J., Jaynes, B.H., Jefson, M.R., Lundy, K.M., Vu, C.B., Glazer, E.A., Froshauser, S.A., Hayashi, S.F., Kamicker, B.J., Reese, C.P., and Olson, J.A. (1997) *Journal of Medicinal Chemistry*, **48**, 1041–1045.

92 Seebach, D., Boes, M., Naerf, R., and Schweizer, W.B. (1983) *Journal of the American Chemical Society*, **105**, 5390–5396.

93 Seebach, D., Vettiger, T., Mueller, H.M., Plattner, D.A., and Petter, W. (1990) *Liebigs Annalen der Chemie*, 687–695.

94 Thompson, H.W. and Swistok, J. (1981) *The Journal of Organic Chemistry*, **46**, 4907–4911.

95 Magdolen, P., Meciarová, M., and Toma, S. (2001) *Tetrahedron*, **57**, 4781–4785.

96 Shi, M., Jiang, J.-K., Shen, Y.-M., Feng, Y.-S., and Lei, G.-X. (2000) *The Journal of Organic Chemistry*, **65**, 3443–3448.

97 Ramtohul, Y.K., James, M.N.G., and Vederas, J.C. (2002) *The Journal of Organic Chemistry*, **67**, 3169–3178.

98 Olivo, H.F., Hemenway, M.S., and Gezginci, M.H. (1998) *Tetrahedron Letters*, **39**, 1309–1312.

99 Chang, D., Feiten, H.-J., Engesser, K.-H., van Beilen, J.B., Witholt, B., and Li, Z. (2002) *Organic Letters*, **4**, 1859–1862.

100 Tanaka, K.-i., Yoshifuji, S., and Nitta, Y. (1986) *Heterocycles*, **24**, 2539–2543.

101 Markgraf, J.H. and Stickney, C.A. (2000) *Journal of Heterocyclic Chemistry*, **37**, 109–110.

102 Wasserman, H.H., Lipshutz, B.H., Tremper, A.W., and Wu, J.S. (1981) *The Journal of Organic Chemistry*, **46**, 2991–2999.

103 Kurita, J., Iwata, K., and Tsuchiya, T. (1987) *Chemical & Pharmaceutical Bulletin*, **35**, 3166–3174.

104 Jung, M.E. and Choi, Y.M. (1991) *The Journal of Organic Chemistry*, **56**, 6729–6730.

105 Dejaegher, Y., Mangelinckx, S., and De Kimpe, N. (2002) *The Journal of Organic Chemistry*, **67**, 2075–2081.

106 Sum, F.-W., Wong, V., Han, S., Largis, E., Mulvey, R., and Tillett, J. (2003) *Bioorganic & Medicinal Chemistry Letters*, **13**, 2191–2194, and references cited.

107 Frigola, J., Torrens, A., Castrillo, J.A., Mas, J., Vañó, D., Berrocal, J.M., Calvet, C., Salgado, L., Redondo, J., Garcia-Granda, S., Valenti, E., and Quintana, J.R. (1994) *Journal of Medicinal Chemistry*, **37**, 4195–4210.

108 Ikeda, M., Kugo, Y., and Sato, T. (1996) *Journal of the Chemical Society, Perkin Transactions 1*, 1819–1824.

109 Jiang, J., Shah, H., and DeVita, R.J. (2003) *Organic Letters*, **5**, 4101–4103.

110 Billotte, S. (1998) *Synlett*, 379–380.

111 Wasserman, H.H., Han, W.T., Schaus, J.M., and Faller, J.W. (1984) *Tetrahedron Letters*, **25**, 3111–3114.

112 Bartholomew, D. and Stocks, M.J. (1991) *Tetrahedron Letters*, **32**, 4799–4800.

113 Nitta, Y., Yamaguchi, T., and Tanaka, T. (1986) *Heterocycles*, **24**, 25–28.

114 Outurquin, F., Pannecoucke, X., Berthe, B., and Paulmier, C. (2002) *European Journal of Organic Chemistry*, 1007–1014.

115 Couty, F., Durrat, F., and Prim, D. (2003) *Tetrahedron Letters*, **44**, 5209–5212.

116 Couty, F., Durrat, F., Evano, G., and Prim, D. (2004) *Tetrahedron Letters*, **45**, 7525–7528.

117 Ungureanu, I., Klotz, P., Schoenfelder, A., and Mann, A. (2001) *Chemical Communications*, 958–959.

118 Padwa, A. and Gruber, R. (1970) *Journal of the American Chemical Society*, **92**, 100–107.

119 Alcaide, B., Almendros, P., Aragoncillo, C., and Salgado, N.R. (1999) *The Journal of Organic Chemistry*, **64**, 9596–9604.

120 Roberto, D. and Alper, H. (1989) *Journal of the American Chemical Society*, **111**, 7539–7543.

121 O'Neil, I.A. and Potter, A.J. (1998) *Chemical Communications*, 1487–1488.

122 Prasad, B.A.B., Bisai, A., and Singh, V.K. (2004) *Organic Letters*, **6**, 4829–4831.

123 Fugami, K., Miura, K., Morizawa, Y., Oshima, K., Utimoto, K., and Nozaki, H. (1989) *Tetrahedron*, **45**, 3089–3098.

124 Rodler, M. and Bauder, A. (1983) *Journal of Molecular Structure*, **97**, 47–52.

125 De Kimpe, N., Tehrani, K.A., and Fonck, G. (1996) *The Journal of Organic Chemistry*, **61**, 6500–6503.

126 Hata, Y. and Watanabe, M. (1987) *Tetrahedron*, **43**, 3881–3888.

127 Higgins, R.H., Faircloth, W.J., Baughman, R.G., and Eaton, Q.L. (1994) *The Journal of Organic Chemistry*, **59**, 2172–2178.

128 Couty, F., Durrat, F., and Gwilherm, E. (2005) *Synlett*, 1666–1670.

129 Golding, P., Millar, R.W., Paul, N.C., and Richards, D.H. (1995) *Tetrahedron*, **51**, 5073–5082.

130 Starmans, W.A.J., Doppen, R.G., Thijs, L., and Zwanenburg, B. (1998) *Tetrahedron Asymmetry*, **9**, 429–435.

131 Guanti, G. and Riva, R. (2001) *Tetrahedron Asymmetry*, **12**, 605–618.

132 Searles, S. (1984) in *Comprehensive Heterocyclic Chemistry*, **7** (ed. W. Lwowski), Pergamon Press, Oxford, pp. 363–402.

133 Linderman, R.J. (1996) *Comprehensive Heterocyclic Chemistry II*, **1** (ed. A. Padwa), Pergamon Press, Oxford, pp. 721–753.

134 Chen, S.H. and Farina, V. (1995) Paclitaxel (Taxol®) chemistry and structure-activity relationships, in *The Chemistry and Pharmacology of Taxol® and Its Derivatives* (ed. V. Farina), Elsevier, pp. 165–253.

135 Malakov, P., Papanov, G., Mollov, N., and Spassov, S. (1978) *Zeitschrift fur Naturforschung Section B-A Journal of Chemical Sciences*, **33**, 1142–1144.

136 Chadha, N.K., Batcho, A.D., Tang, P.C., Courtney, L.F., Cook, C.M., Wovkulich, P.M., and Uskokovic, M.R. (1991) *The Journal of Organic Chemistry*, **56**, 4714–4718.

137 (a) Shimada, N., Hasegawa, S., Harada, T., Fujii, T., and Takita, T. (1986) *Journal of Antibiotics*, **39**, 1623–1625. (b) Gumina,

G. and Chu, C.C. (2002) *Organic Letters*, **4**, 1147–1149.
138 Ogura, M., Tanaka, T., Furihata, K., Shimazu, A., and Otaka, N. (1986) *Journal of Antibiotics*, **39**, 1443–1449.
139 Lane, J.F., Koch, W.T., Leeds, N.S., and Govin, G. (1952) *Journal of the American Chemical Society*, **74**, 3211–3215.
140 Feling, R.H., Buchanan, G.O., Mincer, T.J., Kauffman, C.A., Jensen, P.R., and Fenical, W. (2003) *Angewandte Chemie, International Edition*, **42**, 355–357.
141 Bach, T., Jödicke, K., Kather, K., and Fröhlich, R. (1997) *Journal of the American Chemical Society*, **119**, 2437–2445.
142 Bach, T., Shröder, J., Brandl, T., Hecht, J., and Harms, K. (1998) *Tetrahedron*, **54**, 4507–4520.
143 Bach, T. and Shröder, J. (1999) *The Journal of Organic Chemistry*, **64**, 1265–1273.
144 Bach, T., Shröder, J., and Harms, K. (1999) *Tetrahedron Letters*, **40**, 9003–9004.
145 Bach, T., Bergamnn, H., and Harms, K. (1999) *Journal of the American Chemical Society*, **121**, 10650–10651.
146 Bach, T., Bergamnn, H., Brummerhop, H., Lewis, W., and Harms, K. (2001) *Chemistry - A European Journal*, **7**, 4512–4521.
147 Bach, T., Brummerhop, H., and Harms, K. (2000) *Chemistry - A European Journal*, **6**, 3838–3848.
148 Bach, T. and Shröder, J. (2001) *Synthesis*, 1117–1124.
149 Griesbeck, A.G., Fiege, M., and Lex, J. (2000) *Chemical Communications*, 589–590.
150 (a) Griesbeck, A.G., Bondock, S., and Lex, J. (2003) *The Journal of Organic Chemistry*, **68**, 9899–9906. (b) Griesbeck, A.G. and Bondock, S. (2003) *Canadian Journal of Chemistry*, **81**, 555–559.
151 Griesbeck, A.G., Bondock, S., and Lex, J. (2004) *Organic and Biomolecular Chemistry*, **2**, 1113–1115.
152 (a) Buhr, S., Griesbeck, A.G., and Lex, J. (1996) *Tetrahedron Letters*, **37**, 1195–1196. (b) Griesbeck, A.G., Buhr, S., Fiege, M., Schmickler, H., and Lex, J. (1998) *The Journal of Organic Chemistry*, **63**, 3847–3854. (c) Griesbeck, A.G., Bondock, S., and Cygon, P. (2003) *Journal of the American Chemical Society*, **125**, 9016–9017.
153 Griesbeck, A.G. and Bondock, S. (2001) *Journal of the American Chemical Society*, **123**, 6191–6192.
154 Adam, W., Peter, K., Peters, E.M., and Stegmann, V.R. (2000) *Journal of the American Chemical Society*, **122**, 2958–2959.
155 D'Auria, M., Emanuele, L., Poggi, G., Racioppi, R., and Romaniello, G. (2002) *Tetrahedron*, **58**, 5045–5051.
156 Abe, M., Ikeda, M., Shirodai, Y., and Nojima, M. (1996) *Tetrahedron Letters*, **37**, 5901–5904.
157 Abe, M., Fujimoto, K., and Nojima, M. (2000) *Journal of the American Chemical Society*, **122**, 4005–4010.
158 Abe, M., Torii, E., and Nojima, M. (2000) *The Journal of Organic Chemistry*, **65**, 3426–3431.
159 Abe, M., Kawakami, T., Ohata, S., Nozaki, K., and Nojima, M. (2004) *Journal of the American Chemical Society*, **126**, 2838–2846.
160 Akiyama, T. and Yamanaka, M. (1996) *Synlett*, 1095–1096.
161 Akiyama, T. and Kirino, M. (1995) *Chemistry Letters*, 723–724.
162 Petrov, V.A., Davidson, F., and Smart, B.E. (1995) *The Journal of Organic Chemistry*, **60**, 3419–3422.
163 Gumina, G. and Cheu, C. (2002) *Organic Letters*, **4**, 1147–1149.
164 Johnson, S.W., Jenkinson, S.F., Angus, D., Jones, J., Fleet, G.W., and Taillefumier, C. (2004) *Tetrahedron Asymmetry*, **15**, 2681–2686.
165 Suzuki, M. and Tomooka, K. (2004) *Synlett*, 651–654.
166 De Angelis, F., De Fusco, E., Desiderio, P., Giannessi, F., Piccirilli, F., and Tinti, M.O. (1999) *European Journal of Organic Chemistry*, 2705–2707.
167 Mordini, A., Bindi, S., Pecchi, S., Delg'Innocenti, A., Reginato, G., and Serci, A. (1996) *The Journal of Organic Chemistry*, **61**, 4374–4378.
168 (a) Mordini, A., Bindi, S., Pecchi, S., Capperucci, A., Degl'Innocenti, A., and

Reginato, G. (1996) *The Journal of Organic Chemistry*, **61**, 4466–4468. (b) Mordini, A., Valacchi, M., Nardi, C., Bindi, S., Poli, G., and Reginato, G. (1997) *The Journal of Organic Chemistry*, **62**, 8557–8559. (c) Mordini, A., Bindi, S., Capperucci, A., Nistri, D., Reginato, G., and Valacchi, M. (2001) *The Journal of Organic Chemistry*, **66**, 3201–3205.

169 Homsi, F. and Rousseau, G. (1999) *The Journal of Organic Chemistry*, **64**, 81–85.

170 Albert, S., Robin, S., and Rousseau, G. (2001) *Tetrahedron Letters*, **42**, 2477–2479.

171 Rofoo, M., Roux, M.-C., and Rousseau, G. (2001) *Tetrahedron Letters*, **42**, 2481–2484.

172 Tamai, Y., Someya, M., Fukumoto, J., and Miyano, S. (1994) *Journal of the Chemical Society, Perkin Transactions 1*, 1549–1550.

173 Yang, H.W. and Romo, D. (1998) *Tetrahedron Letters*, **39**, 2877–2880.

174 Zemribo, R. and Romo, D. (1995) *Tetrahedron Letters*, **36**, 4159–4162.

175 Concepcion, A.B., Maruoka, K., and Yamamoto, H. (1995) *Tetrahedron*, **51**, 4011–4020.

176 Evans, D.A. and Janey, J.M. (2001) *Organic Letters*, **3**, 2125–2128.

177 (a) Wynberg, H. and Staring, E.G.J. (1982) *Journal of the American Chemical Society*, **104**, 166–168. (b) Wynberg, H. and Staring, E.G.J. (1985) *The Journal of Organic Chemistry*, **50**, 1977–1979.

178 Tennyson, R. and Romo, D. (2000) *The Journal of Organic Chemistry*, **65**, 7248–7252.

179 Cortez, G.S., Tennyson, R.L., and Romo, D. (2001) *Journal of the American Chemical Society*, **123**, 7945–7946.

180 Nelson, S.G., Wan, Z., Peelen, T.J., and Spencer, K.L. (1999) *Tetrahedron Letters*, **40**, 6535–6539.

181 Nelson, S.G., Peelen, T.J., and Wan, Z. (1999) *Journal of the American Chemical Society*, **121**, 9742–9743.

182 Nelson, S.G., Zhu, C., and Shen, X. (2004) *Journal of the American Chemical Society*, **126**, 14–15.

183 Zhu, C., Shen, X., and Nelson, S.G. (2004) *Journal of the American Chemical Society*, **126**, 5352–5353.

184 Eun, L., Kyung, W.J., and Yong, S.K. (1990) *Tetrahedron Letters*, **31**, 1023–1026.

185 Chelucci, G. and Saba, A. (1995) *Tetrahedron Letters*, **36**, 4673–4676.

186 Balaji, B.S. and Chanda, B.M. (1998) *Tetrahedron Letters*, **39**, 6381–6382.

187 Doyle, M.P., Yan, M., Gan, H.M., and Blossey, E.C. (2003) *Organic Letters*, **5**, 561–563.

188 Lee, J.T., Thomas, P.J., and Alper, H. (2001) *The Journal of Organic Chemistry*, **66**, 5424–5426.

189 Getzler, Y.D.Y.L., Mahadevan, V., Lobkovsky, E.B., and Coates, G.W. (2002) *Journal of the American Chemical Society*, **124**, 1174–1175.

190 Mahadevan, V., Getzler, Y.D.Y.L., and Coates, C.W. (2002) *Angewandte Chemie, International Edition*, **41**, 1784–2781.

191 Schmidt, J.A.R., Mahadevan, V., Getzler, Y.D.Y.L., and Coates, G.W. (2004) *Organic Letters*, **6**, 373–376.

192 Adam, W., Martinez, G., and Thompson, J. (1981) *The Journal of Organic Chemistry*, **46**, 3359–3361.

193 Kim, D.H., Park, J., Chung, S.J., Park, J.D., Park, N.-K., and Han, J.H. (2002) *Bioorganic and Medicinal Chemistry*, **10**, 2553–2560.

194 (a) Pommier, A. and Pons, J.M. (1993) *Synthesis*, 441–459. (b) Wang, Y., Tennyson, R.L., and Romo, D. (2004) *Heterocycles*, **64**, 605–658.

195 Shao, H., Wang, S.H., Lee, C.W., Osapay, G., and Goodman, M. (1995) *The Journal of Organic Chemistry*, **60**, 2956–2957.

196 Arnold, L.D., Kalantar, T.H., and Vederas, J.C. (1985) *Journal of the American Chemical Society*, **107**, 7105–7109.

197 Palomo, C., Miranda, J.I., and Linden, A.J. (1996) *Journal of Organic Chemistry*, **61**, 9196–9201.

198 Zipp, G.G., Hilfiker, M.A., and Nelson, S.G. (2002) *Organic Letters*, **4**, 1823–1826.

199 Nelson, S.G., Spencer, K.L., Cheung, W.S., and Mamie, S.J. (2002) *Tetrahedron*, **58**, 7081–7091.

200 Fujisawa, T., Ito, T., Fujimoto, K., Shimizu, M., Wynberg, H., and Staring, E.G.J. (1997) *Tetrahedron Letters*, **38**, 1593–1596.

201 Nelson, S.G. and Spencer, K.L. (2000) *The Journal of Organic Chemistry*, **65**, 1227–1230.

202 Black, T.H. and Fields, J.D. (1988) *Synthetic Communications*, **18**, 125–130.

203 Mulzer, J., Hoyer, K., and Müller-Fahrnow, A. (1997) *Angewandte Chemie, International Edition in English*, **36**, 1476–1478.

204 Black, T.H., Dubay, W., III, and Tully, P.S. (1988) *The Journal of Organic Chemistry*, **63**, 5922–5927.

205 Black, T.H., Smith, D.C., Eisenbeis, S.A., Peterson, K.A., and Harmon, M.S. (2001) *Chemical Communications*, 753–754.

206 Mulzer, J., Pointner, A., Strasser, R., and Hoyer, K. (1995) *Tetrahedron Letters*, **36**, 3679–3682.

207 Ocampo, R., Dolbier, W.R., Jr, and Paredes, R. (1998) *Journal of Fluorine Chemistry*, **88**, 41–50.

208 Dolbier, W.R., Jr, and Ocampo, R. (1995) *The Journal of Organic Chemistry*, **60**, 5378–5379.

209 Adam, W. and Nava-Salgado, V.O. (1995) *The Journal of Organic Chemistry*, **60**, 578–584.

210 Nava-Salgado, V.O., Peters, E.M., Peters, K., Von Schnering, H.G., and Adam, W. (1995) *The Journal of Organic Chemistry*, **60**, 3879–3886.

211 Mead, K.T. and Park, M. (1992) *The Journal of Organic Chemistry*, **57**, 2511–2514.

212 Dymock, B.W., Kocienski, P.J., and Pons, J.M. (1998) *Synthesis*, 1655–1661.

213 Bizzarri, R., Chiellini, F., Solaro, R., Chiellini, E., Cammas-Marion, S., and Guerin, P. (2002) *Macromolecules*, **35**, 1215–1223.

214 Kwon, Y., Faust, R., Chen, C.X., and Thomas, E.L. (2002) *Macromolecules*, **35**, 3348–3357.

215 Schreck, K.M. and Hillmyer, M.A. (2004) *Tetrahedron*, **60**, 7177–7185.

216 Jaipuri, F.A., Bower, B.D., and Pohl, N.L. (2003) *Tetrahedron Asymmetry*, **14**, 3249–3252.

217 Calter, M.A. and Bi, F.C. (2000) *Organic Letters*, **2**, 1529–1531.

218 Bach, T., Kather, K., and Krämer, O. (1998) *The Journal of Organic Chemistry*, **63**, 1910–1918.

219 Hashemzadeh, M. and Howell, A.R. (2000) *Tetrahedron Letters*, **41**, 1855–1858.

220 Kwon, D.W., Kim, Y.H., and Lee, K. (2002) *The Journal of Organic Chemistry*, **67**, 9488–9491.

221 Xianming, H. and Kellogg, R.M. (1995) *Tetrahedron Asymmetry*, **6**, 1399–1408.

222 Bach, T. and Lange, C. (1996) *Tetrahedron Letters*, **37**, 4363–4364.

223 Nozaki, H., Moriuti, S., Takaya, H., and Noyori, R. (1966) *Tetrahedron Letters*, **7**, 5239–5244.

224 (a) Ito, K., Yoshitake, M., and Katsuki, H. (1996) *Heterocycles*, **42**, 305–317. (b) Ito, K., Fukuda, T., and Katsuki, T. (1997) *Heterocycles*, **46**, 401–411.

225 Lo, M.M.-C. and Fu, G.C. (2001) *Tetrahedron*, **57**, 2621–2634.

226 Hashemzadeh, M. and Howell, A.R. (2000) *Tetrahedron Letters*, **41**, 1859–1862.

227 Inagaki, T., Nakamura, Y., Sawaguchi, M., Yoneda, N., Ayuba, S., and Hara, S. (2003) *Tetrahedron Letters*, **44**, 4117–4119.

228 Dollinger, L.M. and Howell, A.R. (1998) *The Journal of Organic Chemistry*, **63**, 6782–6783.

229 Wang, Y., Bekolo, H., and Howell, A.R. (2002) *Tetrahedron*, **58**, 7101–7107.

230 Miranda, M.A., Izguierdo, M.A., and Galindo, F. (2001) *Organic Letters*, **3**, 1965–1967.

231 Satoh, T., Ishihara, H., Sasaki, H., Kaga, H., and Kakuchi, T. (2003) *Macromolecules*, **36**, 1522–1525.

232 Block, E. and De Wang, M. (1996) *Comprehensive Heterocyclic Chemistry II*, **1B** (ed. A. Padwa), Pergamon Press, Oxford, pp. 773–821.

233 Block, E. (1984) in *Comprehensive Heterocyclic Chemistry*, **5** (ed. W. Lwowski), Pergamon Press, Oxford, pp. 403–447.

234 Sander, M. (1966) *Chemical Reviews*, **66**, 341–353.

235 Contreras, J.G., Hurtado, S.M., Gerli, L.A., and Madariaga, S.T. (2005) *Journal of Molecular Structure: THEOCHEM*, **713**, 207–213.

236 Mastryukov, V.S. and Boggs, J.E. (1995) *Journal of Molecular Structure: THEOCHEM*, **338**, 235–248.

237 Banks, H.D. (2003) *The Journal of Organic Chemistry*, **68**, 2639–2644.

238 Gronert, S. and Lee, J.M. (1995) *The Journal of Organic Chemistry*, **60**, 6731–6736.

239 Barbarella, G., Bongini, A., Chatgilialoglu, C., Rossini, S., and Tugnoli, V. (1987) *The Journal of Organic Chemistry*, **52**, 3857–3860.

240 Ellis, J.W. (1995) *Journal of Chemical Education*, **72**, 671–675.

241 Fles, D., Markovac-Prpic, A., and Tomasic, V. (1958) *Journal of the American Chemical Society*, **80**, 4654–4657.

242 Block, E. and Naganathan, S. (1993) *Heteroatom Chemistry*, **4**, 33–37.

243 Robinson, P.L., Kelly, J.W., and Evans, S.A., Jr, (1987) *Phosphorus Sulfur Silicon and the Related Elements*, **31**, 59–70.

244 Ohuchida, S., Hamanaka, N., Hashimoto, S., and Hayashi, M. (1982) *Tetrahedron Letters*, **23**, 2883–2886.

245 Schulze, O., Voss, J., Adiwidjaja, G., and Olbrich, F. (2004) *Carbohydrate Research*, **339**, 1787–1802.

246 Meier, H. and Rumpf, N. (1998) *Tetrahedron Letters*, **39**, 9639–9642, and references cited.

247 Rumpf, N., Groschl, D., Meier, H., Oniciu, D.C., and Katrizsky, A.R. (1998) *Journal of Heterocyclic Chemistry*, **35**, 1505–1508.

248 Chou, C.-H., Chiu, S.-J., and Liu, W.-M. (2002) *Tetrahedron Letters*, **43**, 5285–5286.

249 Al-Zaidi, S.M.R., Crilley, M.M.L., and Stoodley, R.J. (1983) *Journal of the Chemical Society, Perkin Transactions 1*, 2259–2266.

250 Lee, A.H.F., Chan, A.S.C., and Li, T. (2003) *Tetrahedron*, **59**, 833–839.

251 Lee, H.B., Park, H.-Y., Lee, B.-S., and Kim, Y.G. (2000) *Magnetic Resonance in Chemistry*, **38**, 468–471.

252 Ramirez, J., Yu, L., Li, J., Braunschweiger, P.G., and Wang, P.G. (1996) *Bioorganic & Medicinal Chemistry Letters*, **6**, 2575–2580.

253 Pattenden, G. and Shuker, A.J. (1992) *Journal of the Chemical Society, Perkin Transactions 1*, 1215–1221.

254 Lin, C.-E., Garvey, D.S., Janero, D.R., Letts, L.G., Marek, P., Richardson, S.K., Serebryanik, D., Shumway, M.J., Tam, S.W., Trocha, A.M., and Young, D.V. (2004) *Journal of Medicinal Chemistry*, **47**, 2276–2282.

255 Bhar, D. and Chandrasekaran, S. (1997) *Tetrahedron*, **53**, 11835–11842.

256 Tada, M., Nakamura, T., and Matsumoto, M. (1988) *Journal of the American Chemical Society*, **110**, 4647–4652.

257 Langer, P. and Döring, M. (1999) *Chemical Communications*, 2439–2440.

258 Yadav, L.D.S. and Kapoor, R. (2002) *Synthesis*, 1502–1504.

259 Ueno, Y., Yadav, L.D.S., and Okawara, M. (1981) *Synthesis*, 547–548.

260 Yadav, L.D.S. and Singh, S. (2003) *Synthesis*, 340–342.

261 Robin, S. and Rousseau, G. (2002) *European Journal of Organic Chemistry*, 3099–3414.

262 Abd Elall, E.H.M., Al Ashmawy, M.I., and Mellor, J.M. (1987) *Journal of the Chemical Society, Chemical Communications*, 1577–1578.

263 Brandsma, L., Spek, A.L., Trofimov, B.A., Tarasova, O.A., Nedolya, N.A., Afonin, A.V., and Zinshenko, S.V. (2001) *Tetrahedron Letters*, **42**, 4687–4689.

264 Ikemizu, D., Matsuyama, A., Takemura, K., and Mitsunobu, O. (1997) *Synlett*, 1247–1248.

265 Hart, T.W., Guillochon, D., Perrier, G., Sharp, B.W., and Vacher, B. (1992) *Tetrahedron Letters*, **33**, 5117–5120.

266 Ozaki, S., Matsui, E., Saiki, T., Yoshinaga, H., and Ohmori, H. (1998) *Tetrahedron Letters*, **39**, 8121–8124.

267 Scholz, D. (1983) *Liebigs Annalen der Chemie*, 98–106.

268 Bonini, B.F., Franchini, M.C., Fochi, M., Mangini, S., Mazzanti, G., and Ricci, A. (2000) *European Journal of Organic Chemistry*, 2391–2399.

269 Lancaster, M. and Smith, D.J.H. (1982) *Synthesis*, 582–583.

270 Nishizono, N., Koike, N., Yamagata, Y., Fuji, S., and Matsuda, A. (1996) *Tetrahedron Letters*, **37**, 7569–7572.

271 Buza, M., Andersen, K.K., and Pazdon, M.D. (1978) *The Journal of Organic Chemistry*, **43**, 3827–3834.

272 Ichikawa, E., Yamamura, S., and Kato, K. (1999) *Tetrahedron Letters*, **40**, 7385–7388.

273 Rozwadowska, M.D. (1994) *Tetrahedron Asymmetry*, **5**, 1327–1332.

274 Crump, D.R. (1982) *Australian Journal of Chemistry*, **35**, 1945–1948.

275 Reinhard, G., Rainer, H., Huttner, G., Barth, A., Walter, O., and Zsolnai, J. (1996) *Chemische Berichte*, **129**, 97–108.

276 Abbott, F.S. and Haya, K. (1978) *Canadian Journal of Chemistry*, **56**, 71–79.

277 Kozikowski, A.P. and Fauq, A.H. (1991) *Synlett*, 783–784.

278 Gunatilaka, A.A.L., Ramdayal, F.D., Sarragiotto, M.H., Kingston, D.G.I., Sackett, D.L., and Hamel, E. (1999) *The Journal of Organic Chemistry*, **64**, 2694–2703.

279 Devan, N., Sridhar, P.R., Prabhu, K.R., and Chandrasekaran, S. (2002) *The Journal of Organic Chemistry*, **67**, 9417–9420.

280 Karikomi, M., Narabu, S.-i., Yoshida, M., and Toda, T. (1992) *Chemistry Letters*, 1655–1658.

281 Mercklé, L., Dubois, J., Place, E., Thoret, S., Guéritte, F., Guénard, D., Poupat, C., Ahond, A., and Potier, P. (2001) *The Journal of Organic Chemistry*, **66**, 5058–5065.

282 Press, J.B., McNally, J.J., Hajos, Z.G., and Sawyers, R.A. (1992) *The Journal of Organic Chemistry*, **57**, 6335–6339.

283 Sivets, G.G., Kvasyuk, E.I., and Mikhailopulo, I.A. (1989) *Journal of Organic Chemistry USSR*, **25**, 172–177.

284 Carballeira, N.M., Shalabi, F., and Cruz, C. (1994) *Tetrahedron Letters*, **35**, 5575–5578.

285 Lautenschlaeger, F. (1966) *The Journal of Organic Chemistry*, **31**, 1679–1682.

286 Zyk, N.V., Beloglazkina, E.K., Vatsadze, S.Z., Titanyuk, I.D., and Dubinshaya, Y.A. (2000) *Russian Journal of Organic Chemistry*, **36**, 794–800.

287 De Lucchi, O. and de Lucchini, V. (1982) *Journal of the Chemical Society, Chemical Communications*, 1105–1106.

288 Ito, S. and Mori, J. (1978) *Bulletin of the Chemical Society of Japan*, **51**, 3403–3404.

289 Tabushi, I., Tamaru, Y., and Yoshida, Z. (1974) *Bulletin of the Chemical Society of Japan*, **47**, 1455–1459.

290 Allakhverdiev, M.A., Alekperov, R.K., Shirinova, N.A., and Akperov, N.A. (2000) *Russian Journal of Organic Chemistry*, **36**, 565–567.

291 Sokolov, V.V., Butkevitch, A.N., Yuskovets, V.N., Tomashevskii, A.A., and Potekhin, A.A. (2005) *Russian Journal of Organic Chemistry*, **41**, 1023–1035.

292 Gay, J. and Scherowsky, G. (1995) *Synthetic Communications*, **25**, 2665–2672.

293 Ongoka, P., Mauzé, B., and Miginiac, L. (1985) *Synthesis*, 1069–1070.

294 Uenishi, J., Motoyama, M., Kimura, Y., and Yonemitsu, O. (1998) *Heterocycles*, **47**, 439–451.

295 L'abbé, G., Dekerk, J.-P., Declercq, J.-P., Germain, G., and van Meerssche, M. (1979) *Tetrahedron Letters*, **20**, 3213–3216.

296 L'abbé, G., Francis, A., Dehaen, W., and Bosman, J. (1996) *Bulletin des Sociétés Chimiques Belges*, **105**, 253–258.

297 Cerny, J.V. and Polacik, J. (1966) *Collection of Czechoslovak Chemical Communication*, **31**, 1831–1838.

298 Bolster, J.M. and Kellog, R.M. (1982) *The Journal of Organic Chemistry*, **47**, 4429–4439.

299 Kanakarajan, K. and Meier, H. (1983) *The Journal of Organic Chemistry*, **48**, 881–883.

300 Stachel, H.-D., Poschenrieder, P., and Redlin, J. (1996) *Zeitschrift fur Naturforschung Section B-A Journal of Chemical Sciences*, **51**, 1325–1333.

301 Wentrup, C., Bender, H., and Gross, G. (1987) *The Journal of Organic Chemistry*, **52**, 3838–3847.

302 Jeong, L.S., Moon, H.R., Yoo, S.J., Lee, S.N., Chun, M.W., and Lim, Y.-H. (1998) *Tetrahedron Letters*, **39**, 5201–5204.

303 Korotkikh, N.I., Aslanov, A.F., Raenko, G.F., and Shvaika, O.P. (1999) *Russian*

Journal of Organic Chemistry, **35**, 730–740, and references cited.

304 Vasil'eva, T.P., Bystrova, V.M., Lin'Kova, M.G., Kil'disheva, O.V., and Knunyants, I.L. (1981) *Bulletin of the Academy of Sciences of the USSR, Division of Chemical Sciences*, **30**, 1324–1330.

305 Miyake, Y., Tanaka, H., Ohe, K., and Uemura, S. (2000) *Journal of the Chemical Society, Perkin Transactions 1*, 1595–1599.

306 Okuma, K., Tsubone, T., Shigetomi, T., Shioji, K., and Yokomori, Y. (2005) *Heterocycles*, **65**, 1553–1556.

307 Meier, H. and Mayer, A. (1996) *Synthesis*, 327–329.

308 Paquette, L.A., Barton, W.R.S., and Gallucci, J.C. (2004) *Organic Letters*, **6**, 1313–1315.

309 (a) Bonini, B.F., Franchini, M.C., Fochi, M., Mangini, S., Mazzanti, G., and Ricci, A. (2000) *European Journal of Organic Chemistry*, 2391–2399. (b) Bonini, B.F., Franchini, M.C., Fochi, M., Mangini, S., Mazzanti, G., and Ricci, A. (1995) *Journal of the Chemical Society, Perkin Transactions 1*, 2039–2044.

310 (a) Nishio, T., Okuda, N., and Kashima, C. (1996) *Liebigs Annalen: Organic and Bioorganic Chemistry*, 117–130. (b) See also Padwa, A., Jacquez, M.N., and Schmidt, A. (2004) *The Journal of Organic Chemistry*, **69**, 33–45.

311 Takechi, H., Machida, M., and Kanaoka, Y. (1992) *Synthesis*, 778–782.

312 Nishio, T., Shiwa, K., and Sakamoto, M. (2002) *Helvetica Chimica Acta*, **85**, 2383–2393.

313 Sakamoto, M., Shigekura, M., Saito, A., Ohtake, T., Mino, T., and Fujita, T. (2003) *Chemical Communications*, 2218–2219.

314 (a) Okuma, K., Shiki, K., Sonoda, S., Koga, Y., Shioji, K., Kitamura, T., Fujiwara, Y., and Yokomori, Y. (2000) *Bulletin of the Chemical Society of Japan*, **73**, 155–161. (b) See also Okuma, K., Tsubone, T., Shigetami, T., Shioji, T., and Yokomori, Y. (2005) *Heterocycles*, **65**, 1553–1556.

315 Sakamoto, M., Takahashi, M., Yoshiaki, M., Fujita, T., Watanabe, S., and Aoyama, H. (1994) *Journal of the Chemical Society-Perkin Transactions 1*, 2983–2986, and references cited.

316 Capozzi, G., Fragai, M., Menichetti, S., and Nativi, C. (1999) *European Journal of Organic Chemistry*, 3375–3379.

317 Woolhouse, A.D., Gainsford, G.J., and Crump, D.R. (1993) *Journal of Heterocyclic Chemistry*, **30**, 873–880.

318 Sayed, A.A.F. (1996) *Bulletin of the Polish Academy of Sciences. Chemistry*, **44**, 205–208.

319 Elam, E.U. and Davis, H.E. (1967) *The Journal of Organic Chemistry*, **32**, 1562–1566.

320 Mloston, G., Prakash, G.K.S., Olah, G.A., and Heimgartner, H. (2002) *Helvetica Chimica Acta*, **85**, 1644–1659.

321 Okuma, K., Shigetomi, T., Shibata, S., and Shioji, K. (2004) *Bulletin of the Chemical Society of Japan*, **77**, 187–189.

322 Yoon, K.S., Lee, S.J., and Kim, K. (1996) *Heterocycles*, **43**, 1211–1221.

323 Jones, D.N., Kogan, T.P., Murray-Rust, P., Murray-Rust, J., and Newton, R.F. (1982) *Journal of the Chemical Society, Perkin Transactions 1*, 1325–1332.

324 Meier, H. and Gröschl, D. (1995) *Tetrahedron Letters*, **36**, 6047–6050.

325 Nair, V., Nair, S.M., Mathai, S., Liebscher, J., Ziemer, B., and Narsimulu, K. (2004) *Tetrahedron Letters*, **45**, 5759–5762.

326 (a) Romashin, Y.N., Liu, M.T.H., and Bonneau, R. (2001) *Tetrahedron Letters*, **42**, 207–209. (b) See also Romashin, Y.N., Liu, M.T.H., Hill, B.T., and Platz, M.S. (2003) *Tetrahedron Letters*, **44**, 6519–6521.

327 Wang, M.-D., Calet, S., and Alper, H. (1989) *The Journal of Organic Chemistry*, **54**, 20–21.

328 Furuya, M., Tsutsuminai, S., Nagasawa, H., Komine, N., Hirano, M., and Komiya, S. (2003) *Chemical Communications*, 2046–2047.

329 (a) Dittmer, D.C. and Nelsen, T.R. (1976) *The Journal of Organic Chemistry*, **41**, 3044–3046. (b) Sedergran, T.C., Yokoyama, M., and Dittmer, D.C. (1984) *The Journal of Organic Chemistry*, **49**, 2408–2412.

330 Imamoto, T. and Koto, H. (1985) *Synthesis*, 982–983.

331 Ghosh, A.K., Lee, H.Y., Thompson, W.J., Culberson, C., Holloway, M.K., McKee, S.P., Munson, P.M., Duong, T.T., Smith, A.M., Darke, P.L.,

Zugay, J.A., Emini, E.A., Schleif, W.A., Huff, J.R., and Anderson, P.S. (1994) *Journal of Medicinal Chemistry*, **37**, 1177–1188.

332 Davis, F.A., Awad, S.B., Jenkins, R.H., Jr, Billmers, R.L., and Jenkins, L.A. (1983) *The Journal of Organic Chemistry*, **48**, 3071–3074.

333 Miljkovic, D., Popsavin, V., and Harangi, J. (1985) *Tetrahedron*, **41**, 2737–2743.

334 Tolstikov, G.A., Lerman, B.M., and Komissarova, N.G. (1985) *Zhurnal Organicheskoi Khimii*, **21**, 1915–1918.

335 (a) Hajipour, A.E., Bagheri, H.R., and Ruoho, A.E. (2003) *Phosphorus Sulfur Silicon and the Related Elements*, **178**, 2441–2446. (b) See also Hajipour, A.E. and Ruoho, A.E. (2003) *Sulfur Letters*, **26**, 83–87.

336 Glass, R.S., Singh, W.P., and Hay, B.A. (1994) *Tetrahedron Letters*, **35**, 5809–5812.

337 Glass, R.S., Singh, W.P., and Hay, B.A. (1994) *Sulfur Letters*, **17**, 281–286.

338 (a) Nagasawa, K., Umezawa, T., and Itoh, K. (1984) *Heterocycles*, **21**, 463–466. (b) Nagasawa, K., Yoneta, A., Umezawa, T., and Itoh, K. (1987) *Heterocycles*, **26**, 2607–2609.

339 Al-Zaidi, S. and Stoodley, R.J. (1982) *Journal of the Chemical Society, Chemical Communications*, 995–996.

340 Knaup, G., Retzow, S., Schwarm, M., and Drauz, K. (1996) Ger Offen DE 19505934 A1 Chemical Abstracts, 125, 221553.

341 Young, D.J. and Stirling, C.J.M. (1997) *Journal of the Chemical Society, Perkin Transactions 2*, 425–429.

342 (a) Furuhata, T. and Ando, W. (1986) *Tetrahedron*, **42**, 5301–5308. (b) Ando, W., Hanyu, Y., Kumamoto, Y., and Takata, T. (1986) *Tetrahedron*, **42**, 1989–1994.

343 Almena, J., Foubelo, F., and Yus, M. (1997) *Tetrahedron*, **53**, 5563–5572.

344 Nishizono, N., Koike, N., Yamagata, Y., Fujii, S., and Matsuda, A. (1996) *Tetrahedron Letters*, **37**, 7569–7572.

345 Adams, R.D., Cortopassi, J., and Falloon, S. (1993) *Journal of Organometallic Chemistry*, **463**, C5–C7.

346 For a review see: Adams, R.D. (2000) *Aldrichimica Acta*, **33**, 39–44.

347 Dabideen, D.R. and Gervay-Hague, J. (2004) *Organic Letters*, **6**, 973–975.

348 Takechi, H. and Machida, M. (1997) *Chemical & Pharmaceutical Bulletin*, **45**, 1–7, and references cited.

349 Muthuramu, K. and Ramamurthy, V. (1981) *Chemistry Letters*, 1261–1264.

350 Muthuramu, K., Sundari, B., and Ramamurthy, V. (1983) *Tetrahedron*, **39**, 2719–2724.

351 Dondini, A., Battaglia, A., and Giorgianni, P. (1980) *The Journal of Organic Chemistry*, **45**, 3766–3773.

352 Ohno, M., Kojima, S., Shirakawa, Y., and Eguchi, S. (1995) *Tetrahedron Letters*, **36**, 6899–6902.

353 Mayer, A., Rumpf, N., and Meier, H. (1995) *Liebigs Annalen: Organic and Bioorganic Chemistry*, 2221–2226.

354 Palmer, D.C. and Taylor, E.C. (1986) *The Journal of Organic Chemistry*, **51**, 846–850.

355 Carter, S.D., Kaura, A.C., and Stoodley, R.J. (1980) *Journal of the Chemical Society, Perkin Transactions 1*, 388–394.

356 Vasil'eva, T.P., Bystrova, V.M., Kil'disheva, O.V., and Knunyants, I.L. (1986) *Bulletin of the Academy of Sciences of the USSR, Division of Chemical Sciences*, **35**, 1180–1183.

357 Lee, A.H.F., Chen, J., Chan, A.S.C., and Li, T. (2003) *Phosphorus Sulfur Silicon and the Related Elements*, **178**, 1163–1174.

358 Langer, P. and Döring, M. (1999) *Chemical Communications*, 2439–2440.

359 Nishio, T., Iida, I., and Sugiyama, K. (2000) *Journal of the Chemical Society, Perkin Transactions 1*, 3039–3046.

360 Hwang, B.-Y., Lee, H.B., Kim, Y.G., and Kim, B.-G. (2000) *Biotechnology Progress*, **16**, 973–978.

361 Pousa, J.L., Sorarrain, O.M., and Maranon, J. (1981) *Journal of Molecular Structure*, **71**, 31–38.

362 De Silva, K.G., Monsef-Mirzai, Z., and McWhinnie, W.R. (1983) *Journal of The Chemical Society, Dalton Transactions*, 2143–2146.

363 Harvey, A.B., Durig, J.R., and Morrissey, A.C. (1969) *Journal of Chemical Physics*, **50**, 4949–4961.

364 Schultze, O., Voss, J., Adiwidjaja, G., and Olbrich, F. (2004) *Carbohydrate Research*, **339**, 1787–1802.

365 Okuma, K., Okada, A., Koga, Y., and Yokomori, Y. (2001) *Journal of the American Chemical Society*, **123**, 7166–7167.

366 Bird, C.W., Cheeseman, G.W.H., and Hornfeldt, A.-B. (1984) *Comprehensive Heterocyclic Chemistry*, 1st edn, **4**, 942.

367 (a) Korotkikh, N.I., Losev, G.A., and Shvaika, O.P. (2001) *Azotistye Geterotsikly I Alkaloidy*, **1**, 350–355. (b) (2001) *Chemical Abstracts*, **141**, 134472.

368 Morgan, G.T. and Burstrall, F.H. (1930) *Journal of the Chemical Society*, 1497–1502.

369 Backer, H.J. and Winter, H.J. (1937) *Recueil des Travaux Chimiques des Pays-Bas*, **56**, 492–509.

370 (a) Throckmorton, P.E. (1972) U.S. Patent 3,644,204. (b) (1972) *Chemical Abstracts*, **76**, 140766.

371 Arnold, A.P. and Candy, A.J. (1983) *Australian Journal of Chemistry*, **36**, 815–823.

372 Polson, G. and Dittmer, D.C. (1988) *The Journal of Organic Chemistry*, **53**, 791–794.

373 Kurbanov, S.B., Mamedov, E.S., Mishiev, R.D., Gusiev, N.K., and Agaeva, E.A. (1991) *The Journal of Organic Chemistry USSR*, **27**, 812–816.

374 Gunatilaka, A.A.L., Ramdayal, F.D., Sarragiotto, M.H., Kingston, D.G.I., Sackett, D.L., and Hamel, E. (1999) *The Journal of Organic Chemistry*, **64**, 2694–2703.

375 Schulze, O. and Voss, J. (1999) *Phosphorus Sulfur Silicon and the Related Elements*, **153**, 429–430.

376 Adiwidjaja, G., Schulze, O., Voss, J., and Wirsching, J. (2000) *Carbohydrate Research*, **325**, 107–119.

377 Akhmedov, A.M., Mamedov, E.S., Velieva, D.S., Guseinova, S.S., and Kulibekova, T.N. (2001) *Azerbaidzhanskii Khimicheskii Zhurnal*, 66–68, and references cited.

378 Migalina, Y.V., Lendel, V.G., Balog, I., and Staninets, V.I. (1981) *Ukrainskii Khimicheskii Zhurnal*, **47**, 1293–1295.

379 Takemura, K., Sakano, K., Takahashi, A., Sakamaki, T., and Mitsunobu, O. (1998) *Heterocycles*, **47**, 633–637.

380 Ding, M.-X., Ishii, A., Nakayama, J., and Hoshino, M. (1993) *Bulletin of the Chemical Society of Japan*, **66**, 1714–1721.

381 Yamazaki, S., Kohgami, K., Okazaki, M., Yamabe, S., and Arai, T. (1989) *The Journal of Organic Chemistry*, **54**, 240–243.

382 (a) Fischer, H., Kalbas, C., and Gerbing, U. (1992) *Journal of the Chemical Society. Chemical Communications*, 563–564. (b) Fischer, H., Kalbas, C., and Hofmann, J. (1992) *Journal of the Chemical Society, Chemical Communications*, 1050–1051.

383 Lindgren, B. (1980) *Chemica Scripta*, **16**, 24–27.

384 Adams, R.D., McBride, K.T., and Rogers, R.D. (1997) *Organometallics*, **16**, 3895–3901.

385 Fischer, H., Kalbas, C.Z., Troll, C., and Fluck, K.H. (1993) *Zeitschrift fur Naturforschung Section B-A Journal of Chemical Sciences*, **48**, 1613–1620.

386 Farrar, W.V. and Gulland, J.M. (1945) *Journal of the Chemical Society*, 11–14.

387 Abel, E.W., MacKenzie, T.E., Orrell, K.G., and Sik, V. (1986) *Journal of The Chemical Society, Dalton Transactions*, 205–211.

388 Kawashiwa, T. and Okazaki, R. (1996) in *Comprehensive Heterocyclic Chemistry II*, **1B** (eds A.R. Katritzky, C.W. Ress, and E.F.V. Scriven), Pergamon Press, pp. 833–866.

389 Kawashiwa, T. and Okazaki, R. (2001) in *Phosphorus-Carbon Heterocyclic Chemistry*, (ed. F. Mathey), Pergamon Press, pp. 105–165.

390 Marinetti, A. and Carmichael, D. (2002) *Chemical Reviews*, **102**, 201–230.

391 Mickiewicz, M. and Wild, S.B. (1977) *Journal of The Chemical Society, Dalton Transactions*, 704–708.

392 Berger, D.J., Gaspar, P.P., and Liebman, J.F. (1995) *Journal of Molecular Structure: THEOCHEM*, **338**, 51–71.

393 Bader, A., Kang, Y.B., Pabel, D.D., Pathak, D.D., Willis, A.C., and Wild, S.B. (1995) *Organometallics*, **14**, 1434–1441.

394 Weber, L., Kaminski, O., Stammler, H.-G., and Neumann, B. (1996) *Chemische Berichte*, **129**, 223–226.
395 Weber, L., Kleinbekel, S., Pumpenmeier, L., Stammler, H.-G., and Neumann, B. (2002) *Organometallics*, **21**, 1998–2005.
396 Tumas, W., Suriano, J.A., and Harlow, R.L. (1990) *Angewandte Chemie, International Edition in English*, **29**, 75–76.
397 Sommer, L.H., and Baum, G.A. (1954) *Journal of the American Chemical Society*, **76**, 5002–15002.
398 Lukevics, E. and Pudova, O. (1996) in *Comprehensive Heterocyclic Chemistry II*, **1B** (eds A.R. Katritzky, C.W. Ress, and E.F.V. Scriven), Pergamon Press, pp. 867–886.
399 Novikov, V.P., Tarasenko, S.A., Samdal, S., Shen, Q., and Vilkov, L.V. (1999) *Journal of Molecular Structure*, **477**, 71–89.
400 Durig, J.R., Zhen, P., Jin, Y., Gounev, T.K., and Guirgis, G.A. (1999) *Journal of Molecular Structure*, **477**, 31–47.
401 Jouikov, V. and Krasnov, V. (1995) *Journal of Organometallic Chemistry*, **498**, 213–219.
402 Ohshita, J., Matsushige, K., Kunai, A., Adachi, A., Sakamaki, K., and Okita, K. (2000) *Organometallics*, **19**, 5288–5582.
403 Murakami, M., Suginome, M., Fujimoto, K., Nakamura, H., Andersson, P.G., and Ito, Y. (1993) *Journal of the American Chemical Society*, **115**, 6487–6498.
404 Takeda, N., Kajiwara, T., Suzuki, H., Okazaki, R., and Tokitoh, N. (2003) *Chemistry - A European Journal*, **9**, 3530–3543, and references cited.
405 Maas, G. and Bender, S. (2000) *Chemical Communications*, 437–438.
406 Maas, G., Daucher, B., Maier, A., and Gettwert, G. (2004) *Chemical Communications*, 238–239.
407 (a) Auner, N., Heikenwälder, C.R., and Herrschaft, B. (2000) *Organometallics*, **19**, 2470–2476.
(b) Auner, N., Grasmann, M., Herrschaft, B., and Hummer, M. (2000) *Canadian Journal of Chemistry*, **78**, 1445–1458, and references cited.
408 Seyferth, D., Duncan, D.P., Schmidbaur, H., and Holl, P. (1978) *Journal of Organometallic Chemistry*, **159**, 137–145.
409 (a) Kroke, E., Willms, S., Weidenbruch, M., Saak, W., Pohl, S., and Marsmann, H. (1996) *Tetrahedron Letters*, **37**, 3675–3678.
(b) Nguyen, P.T., Palmer, W.S., and Woerpel, K.A. (1999) *The Journal of Organic Chemistry*, **64**, 1843–1848.
410 Apeloig, Y., Bravo-Zhivotovskii, D., Zharov, I., Panov, V., Leigh, W.J., and Sluggett, G.W. (1998) *Journal of the American Chemical Society*, **120**, 1398–1404.
411 Naka, A., Matsui, Y., Kobayashi, H., and Ishikawa, M. (2004) *Organometallics*, **23**, 1509–1518.
412 Shinohara, A., Takeda, N., and Tokitoh, N. (2003) *Journal of the American Chemical Society*, **125**, 10804–10805.
413 Gusel'nikov, L.E. (2003) *Coordination Chemistry Reviews*, **244**, 149–240.
414 (a) For leading references, see Matsumoto, K., Shimazu, H., Deguchi, M., and Yamaoka, H. (1997) *Journal of Polymer Science Part A-Polymer Chemistry*, **35**, 3207–3216.
(b) Sheikh, R.K., Tharanikkarasu, K., Imae, I., and Kawakami, Y. (2001) *Macromolecules*, **34**, 4384–4389.
415 Matsumoto, K., Hasegawa, H., and Matsuoka, H. (2004) *Tetrahedron*, **60**, 7197–7204.
416 Denmark, S.E., Griedel, B.D., Coe, D.M., and Schnute, M.E. (1994) *Journal of the American Chemical Society*, **116**, 7026–7043.
417 Denmark, S.E. and Sweis, R.F. (2002) *Accounts of Chemical Research*, **35**, 835–846.
418 Uenishi, K., Imae, I., Shirakawa, E., and Kawakami, Y. (2002) *Macromolecules*, **35**, 2455–2460.
419 Uenishi, K., Imae, I., Shirakawa, E., and Kawakami, Y. (2001) *Chemistry Letters*, 986–987.
420 Hatanaka, Y., Watanabe, M., Onozawa, S.-Y., Tanaka, M., and Sakurai, H. (1998) *The Journal of Organic Chemistry*, **63**, 422–423.

421 Matsumoto, K., Aoki, A., Oshima, K., Utimoto, K., and Rahman, N.A. (1993) *Tetrahedron*, **49**, 8487–8502.

422 Matsumoto, K., Takeyama, Y., Miura, K., Oshima, K., and Utimoto, K. (1995) *Bulletin of the Chemical Society of Japan*, **68**, 250–261, and references cited therein.

423 Boudjouk, P., Black, E., Kumarathasan, R., Samaraweera, U., Castellino, S., Oliver, J.P., and Krampf, J.W. (1994) *Organometallics*, **13**, 3715–3727.

424 Takeyama, Y., Oshima, K., and Utimoto, K. (1990) *Tetrahedron Letters*, **31**, 6059–6062.

425 Okada, K., Matsumoto, K., Oshima, K., and Utimoto, K. (1995) *Tetrahedron Letters*, **36**, 8067–8070.

426 Tanaka, Y., Yamashita, H., and Tanaka, M. (1996) *Organometallics*, **15**, 1524–1526.

427 Takeyama, Y., Nozaki, K., Matsumoto, K., Oshima, K., and Utimoto, K. (1991) *Bulletin of the Chemical Society of Japan*, **64**, 1461–1466.

428 Ushakov, N.V. and Fedorova, G.K. (1996) *Izvestiya Akademii Nauk, Seria Khimia*, 955–957.

429 Mazerolles, P., Dubac, J., and Lesbre, M. (1966) *Journal of Organometallic Chemistry*, **5**, 35–47, and references cited therein.

430 Namavari, M. and Conlin, R.T. (1992) *Organometallics*, **11**, 3307–3312.

431 Haaland, A., Samdal, S., Strand, T.G., Tafipolsky, M.A., Volden, H.V., Van de Heisteeg, B.J.J., Akkerman, O.S., and Bickelhaupt, F.B. (1997) *Journal of Organometallic Chemistry*, **536–537**, 217–221.

432 Lu, X., Xu, Y., Yu, H., and Wu, W. (2005) *Journal of Physical Chemistry A*, **109**, 6970–6973.

433 Seetz, J.W.F.L., Van de Heisteeg, B.J.J., Schat, G., Akkerman, O.S., and Bickelhaupt, F. (1984) *Journal of Organometallic Chemistry*, **277**, 319–322.

434 de Boer, H.J.R., Akkerman, O.S., and Bickelhaupt, F. (1987) *Journal of Organometallic Chemistry*, **321**, 291–306.

435 Tinga, M.A.G.M., Buisman, G.J.H., Schat, G., Akkerman, O.S., Bickelhaupt, F., Smeets, W.J.J., and Spek, A.L. (1994) *Journal of Organometallic Chemistry*, **484**, 137–145.

436 Nametkin, N.S., Kuz'min, O.V., Zav'yalov, V.I., Zueva, G.Y., Babich, E.D., Vdovin, V.M., and Chernysheva, T.I. (1969) *Izvestiya Akademii Nauk SSR, Seria Khimia*, 976–977.

437 Rivière, P. and Satgé, J. (1971) *Angewandte Chemie, International Edition in English*, **10**, 267–268.

438 Tokitoh, N., Matsumoto, T., and Okazaki, R. (1995) *Chemistry Letters*, 1087–1088.

439 Tokitoh, N., Kushikawa, K., and Okazaki, R. (1998) *Chemistry Letters*, 811–812.

440 Dubac, J. and Mazerolles, P. (1966) *Bulletin de la Societe Chimique de France*, 2153–12153.

441 Escudié, J., Couret, C., and Satgê, J. (1979) *Recueil des Travaux Chimiques des Pays-Bas*, **98**, 461–466.

442 Toltl, N.P. and Leigh, W.J. (1998) *Journal of the American Chemical Society*, **120**, 1172–1178.

443 Mazerolles, P., Dubac, J., and Lesbre, M. (1968) *Journal of Organometallic Chemistry*, **12**, 143–148.

444 Seyferth, D., Washburne, S.S., Jula, J.F., Mazerolles, P., and Dubac, J. (1969) *Journal of Organometallic Chemistry*, **16**, 503–506.

445 Dubac, J. and Mazerolles, P. (1969) *Bulletin de la Societe Chimique de France*, 3608–3609.

4
Five-Membered Heterocycles: Pyrrole and Related Systems

Jan Bergman and Tomasz Janosik

4.1
Introduction

The history of pyrrole **1** dates back to 1834, when Runge observed the presence of a compound that caused red coloration of a wood splinter moistened with mineral acid, in a fraction obtained through distillation of coal tar. He named the substance pyrrole [1] – a name maintained by Anderson, who later isolated a pure sample by distillation of bone oil [2]. Several years later, the correct structure was established by Baeyer and Emmerling [3]. The biological relevance and intriguing reactivity patterns of pyrrole derivatives have triggered intense interest in their chemistry, which has been exhaustively treated in several excellent monographs covering the advances of most essential aspects of the topic [4–6].

A general review with a practical perspective is included in *Science of Synthesis* [7]. Annual coverage detailing more recent achievements in synthetic pyrrole chemistry is provided in *Progress in Heterocyclic Chemistry* [8], whereas more comprehensive accounts on structure [9], ring synthesis [10], reactivity [11] and applications [12] are available in the second edition of *Comprehensive Heterocyclic Chemistry*. Since the scope of this chapter does not permit in-depth coverage of all aspects of pyrrole chemistry, and will be mainly restricted to 1*H*-pyrroles, which will hereinafter simply be referred to as pyrroles, readers may also want to consult the above mentioned reference works. Additional useful accounts highlighting special topics in pyrrole chemistry are cited in appropriate sections of this chapter.

4.1.1
Nomenclature

The IUPAC numbering convention for pyrrole (1*H*-pyrrole) is shown in structure **1**, including the commonly used designation of the 2(5)-positions as α, and the 3(4)-positions as β. The two remaining conceivable tautomeric forms **1a** and **1b** are known as 2*H*-pyrroles and 3*H*-pyrroles, respectively. Trivial names are relatively rare in the

Modern Heterocyclic Chemistry, First Edition.
Edited by Julio Alvarez-Builla, Juan Jose Vaquero, and José Barluenga.

pyrrole series, but are still used for some derivatives of natural origin. The general IUPAC rules are now generally applied to most pyrrole derivatives. Carbon containing substituents, such as carboxylic acid, nitrile, aldehyde, but also sulfonic acid, as well as derivatives thereof, should be incorporated into names as suffixes, although prefixes, for instance cyano-, may sometimes be encountered. Nomenclature that applies to special classes of pyrrole derivatives, such as partially saturated systems, will be evident from appropriate sections of the text.

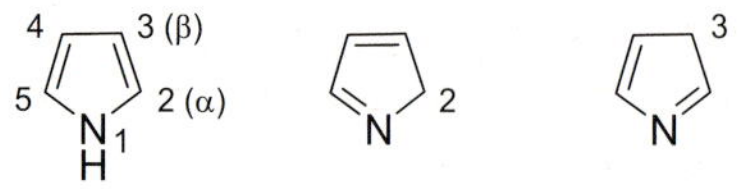

1 **1a** **1b**

4.2 General Reactivity

4.2.1 Relevant Physicochemical Data, Computational Chemistry, and NMR Data

Pyrrole is a colorless liquid [mp −23 °C, bp 130 °C (760 Torr)] that darkens in contact with air. It has limited water solubility, but is miscible with many common organic solvents. Although some simple pyrroles are oils, many derivatives with higher molecular weights are solids. Pyrrole displays weakly acidic properties ($pK_a = 17.25$ in aqueous medium [13], 17.51 in aqueous hydroxide solution [14] and 23.05 in DMSO [15]). The dipole moment (μ) of pyrrole is 1.74 ± 0.02 D, with the negative pole directed towards the ring carbon atoms [16].

The planar C_{2v} symmetric molecular structure of pyrrole has been determined based on microwave spectra [16]; Table 4.1 provides selected bond lengths and angles. Although it has for some time been possible to calculate quite accurate structural parameters for the structure of pyrrole [9], the increasing level of refinement of modern theoretical methods has enabled even better estimations. Some represen-

Table 4.1 Selected experimental and calculated bond lengths (Å) and angles (°) of pyrrole (**1**).

	N–C2 (Å)	C2–C3 (Å)	C3–C4 (Å)	C2–N–C5 (°)	N–C2–C3 (°)	C2–C3–C4 (°)	Reference
Experimental	1.370	1.382	1.417	109.8	107.7	107.4	[16]
MM3	1.380	1.382	1.417	110.1	107.1	107.8	[18]
B3P86	1.368	1.374	1.419	109.9	107.6	107.4	[19]
B3LYP/6-311G(2df,p)	1.370	1.373	1.421	109.8	107.7	107.4	[20]

tative values are included in Table 4.1. Computational methods involving complex pyrrole containing molecules are now commonplace, facilitating for instance studies of ligand–receptor interactions. The fundamental physicochemical properties of pyrroles, including the advances in spectroscopic and theoretical methods, have been compiled in several reviews [6, 9, 17].

As a "π-excessive" five-membered aromatic heterocycle with six π-electrons, pyrrole displays many features that are usually associated with such systems, such as high resonance energy, and a tendency to participate in substitution reactions. Both experimental and theoretical aspects of the aromaticity of pyrroles have been studied over the years, sometimes arousing controversy; this topic has been reviewed and discussed in considerable detail. The generally accepted relative aromaticity scale for the common five-membered heterocycles featuring one heteroatom versus benzene is: benzene > thiophene > selenophene > pyrrole > tellurophene > furan [21, 22].

Since detailed spectroscopic data, occasionally including complete assignments, are now included in virtually every research paper devoted to pyrroles, there is an immense wealth of information available on the subject [6, 9]. The chemical shifts of the protons attached to the carbon atoms of 1*H*-pyrroles reflect the aromatic character of this ring system, typically appearing in the range 5.5–7.8 ppm. Concentration and solvent dependence accounts for the considerable range of measured values for the proton attached to the nitrogen atom, which is often observed as a broad, sometimes even barely discernible peak.

Likewise, much information on ^{13}C NMR spectroscopy on pyrroles has been collected and evaluated in detail. The ^{13}C resonances of 1*H*-pyrroles lie in the aromatic region, and the shifts depend on the electronic effects transferred from the substituents. Alkyl- [9] or aryl [23] substituents at the nitrogen atom usually have only limited influence on the ring carbon chemical shifts regardless of their properties, whereas the presence of electron-withdrawing groups can cause relatively large downfield shifts of the resonances; the magnitude of this effect increases with the electron-withdrawing power of the N-substituent [24]. The tetrahedral carbon of 2*H*-pyrroles usually resonates at 78–98 ppm, in contrast to its counterpart in 3*H*-pyrroles, which appears in the range 49–70 ppm. The resonances originating from the C=N carbon in 2*H*- and 3*H*-pyrroles appear at 161–185 and 173–191 ppm, respectively [9].

The continuous development of two dimensional NMR experiments, such as gradient enhanced ^{15}N-HMBC, allows practical measurement of ^{15}N chemical shifts and ^{1}H$-^{15}$N couplings in pyrrole derivatives. The ^{15}N chemical shifts of pyrroles fall in the approximate δ range between −186 and −236 ppm, and may vary considerably due to influence from the solvent; this should be kept in mind when data recorded in different media are compared [25].

4.2.2
Fundamental Reactivity Patterns

The propensity of pyrrole to react by electrophilic substitution imparts a dominant effect on its general reactivity patterns. Pyrrole itself, as well as simple non-deactivated derivatives thereof, is susceptible to undergo reactions with electrophiles predominantly at C2 (α-position). Certain substituted pyrroles will, however, react with electrophiles selectively at C3, provided that an electron-withdrawing group is present at the nitrogen or at C2, or when both α-positions are blocked. Consequently, electrophilic substitution is a very useful tool for elaboration of pyrrole derivatives. An inspection of the Wheland intermediates resulting from attack on a suitable electrophile (E^+) at C2 (**2**) or C3 (**3**) gives an explanation to the preferred C2 substitution pathway observed for simple pyrroles, as the intermediate **2** is stabilized to a higher degree by more extensive delocalization of the positive charge (Scheme 4.1). Computational data on the differences in the total energy of pyrrole and the possible cationic σ-complexes formed upon its protonation, performed using for example the *ab initio* RHF/6-31G(d) and the DFT B3LYP/6-31G(d) methods [26], are in coherence with the experimental observations.

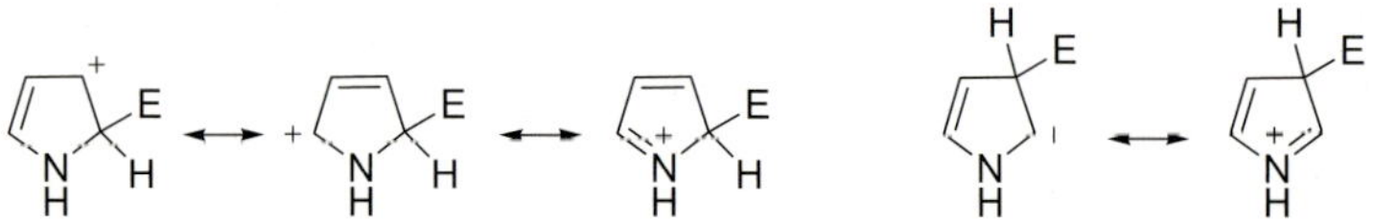

Scheme 4.1

The electron rich nature of most pyrroles is further manifested by their reluctance to participate in nucleophilic substitution reactions. The intrinsically low reactivity of pyrroles towards nucleophilic regents may, however, be enhanced upon protonation, or introduction of strongly electron-withdrawing substituents. Synthetically useful reactions of pyrroles may also be performed with radical reagents, leading to selective substitution at C2 under special conditions (Section 4.5.7).

Pyrrole reacts readily with strong bases giving the pyrrolyl anion **4** (Scheme 4.2). This ambident nucleophile is also of considerable synthetic importance, as it allows introduction of substituents in particular at the nitrogen atom, but also at the carbon atoms. In contrast, pyrroles possessing appropriate N-blocking substituents are usually metallated at C2, providing access to a wide variety of 2-substituted

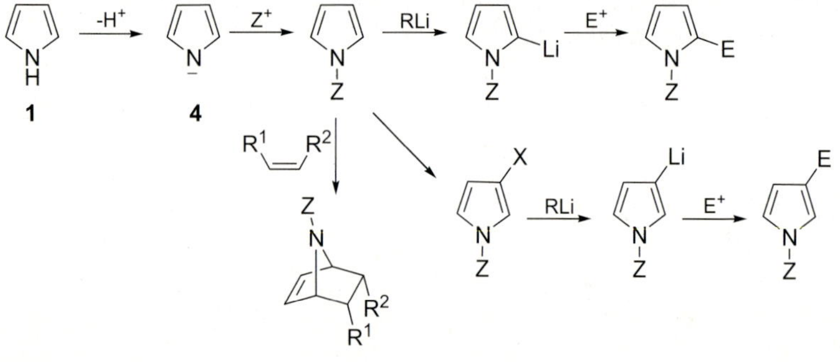

Scheme 4.2

derivatives upon quenching with suitable electrophiles. Metallation of pyrroles at C3 is conveniently accomplished by halogen–metal exchange using 3-halopyrroles incorporating a bulky N-protecting group, which effectively blocks access to C2. Owing to its aromatic character, pyrrole itself does not participate in Diels–Alder reactions, instead giving α-substitution products. However, this reactivity path may be precluded by introduction of an electron-withdrawing group on the nitrogen atom, thereby transforming the pyrrole nucleus into a useful diene component in Diels–Alder reactions. Examples of reactions where pyrroles act as dienophiles are quite rare, but have nevertheless found some applications.

Taken together, these fundamental reactions, combined with reductions, oxidations, classical functional group interconversions, pyrrole ring syntheses, as well as modern developments, such as transition metal catalyzed couplings, constitute a powerful arsenal of tools for the preparation and elaboration of a wide array of pyrrole derivatives. Additional aspects on the reactivity of the pyrrole nucleus are discussed in appropriate sections of this chapter.

4.3 Relevant Natural and/or Useful Compounds

The pyrrole nucleus is an essential component of several naturally occurring macrocyclic complexes of various metals of utmost importance for living systems by virtue of its ability to participate in coordination of metals. One molecule belonging to this class, chlorophyll-*a* (**5**), is a crucial prerequisite for sustaining life on our planet by its ability to participate in the conversion of carbon dioxide into carbohydrates with concomitant liberation of molecular oxygen by photosynthesis. The total synthesis of chlorophyll-*a* (**5**) conducted by Woodward constitutes one of the most prominent achievements in organic chemistry [27]. This, and several related pigments, is biosynthesized from the common building block porphobilinogen (**6**) [28–30]. Likewise, the amino acid L-proline is ubiquitous in biologically important peptides and proteins, as well as other natural products. In addition, numerous naturally occurring compounds incorporating derivatives of proline have been identified [31]. Detailed mechanisms for some of the intricate biosynthetic pathways responsible for pyrrole ring formation and incorporation of pyrrole units in natural products have been formulated [32].

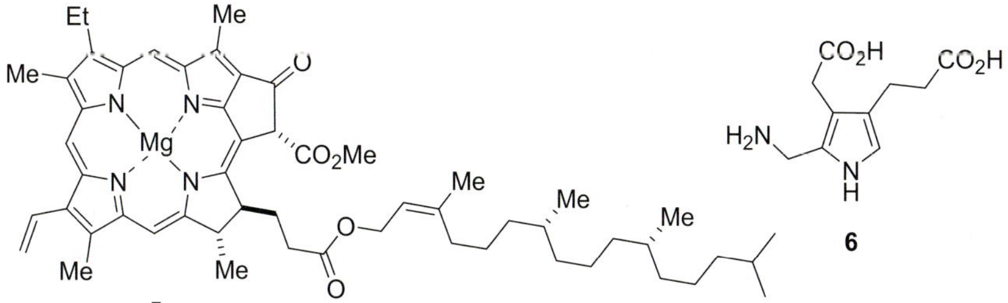

Et Me Me O N N Mg N N CO_2Me Me Me Me O O Me Me Me

5

CO_2H CO_2H H_2N N H

6

There are many other pyrrole based molecules of natural and of synthetic origin that exhibit various biological activities. The cytotoxic pyrrole alkaloid roseophilin (**7**) [33] is an excellent example of such a compound, and has also attracted considerable attention as a challenging target for total synthesis [34]. An increasing group of naturally occurring pyrrole derivatives feature the tetramic acid motif as the main structural element [35, 36], as illustrated by the plasmodial pigment fuligorubin A (**8**) isolated from the slime mold *Fuligo septica* [37]. The field of monopyrrolic natural products, including tetramic acid derivatives, has been comprehensively reviewed [38]; an account detailing recent synthetic strategies towards antitumor pyrroles bearing oxygenated aryl groups is also available [39]. Distamycin (**9**), an antiviral and antimitotic natural product isolated from a *Streptomyces* sp. [40], has served as a model compound for fruitful studies towards synthetic pyrrole containing polyamides for recognition [41–44] and sequence specific alkylation [45] of DNA. The development of ketorololac (**10**), a drug with potent anti-inflammatory and analgesic properties [46, 47], demonstrates the importance of synthetic pyrroles in medicinal chemistry.

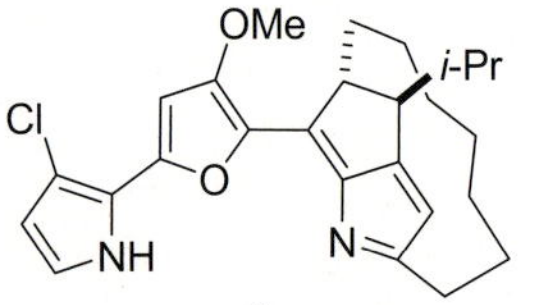

7

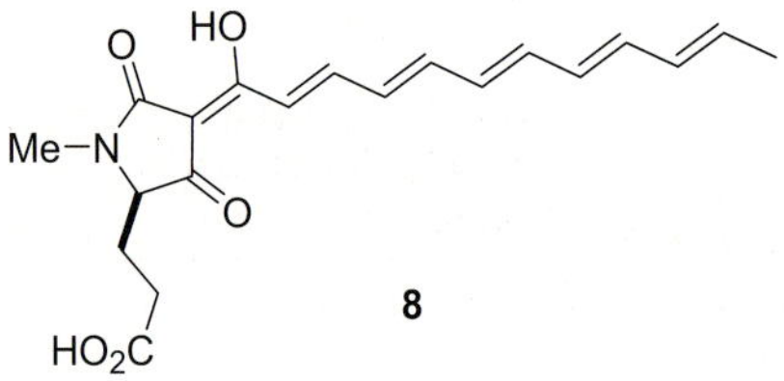

8

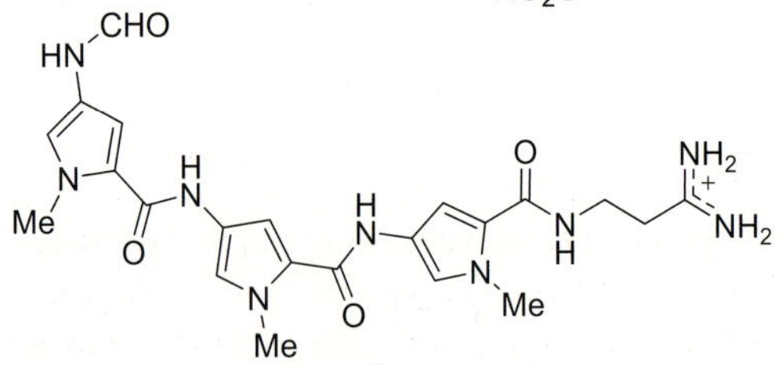

9

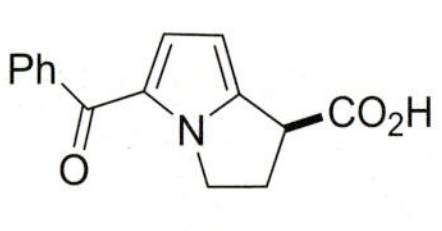

10

Pyrrole polymers constitute yet an additional group of derivatives that have captured considerable interest, for instance as new materials for electrocatalysis [48], or conducting polymer nanocomposites [49]. Accounts concerning syntheses (e.g., by electropolymerization [50]), properties and applications of polypyrroles are also available [51, 52].

4.4 Pyrrole Ring Synthesis

Numerous different approaches for the construction of pyrrole derivatives from acyclic materials have arisen from over one century of intense research activity in this particular field. Nevertheless, new developments, as well as further extensions of known methods still continue to attract the attention of synthetic chemists, providing additional effective routes to previously known pyrroles, as well as novel, and more exotic, derivatives. This section focuses particularly on general processes of practical importance. Selected syntheses of more specialized derivatives, such as oxypyrroles, aminopyrroles, pyrrolines and pyrrolidines, are incorporated in the sections devoted to these systems. A review detailing many recent advances in pyrrole ring synthesis since 1995 is available [52], whereas a more specialized account provides a survey of routes to pyrroles bearing two aryl or heteroaryl groups on adjacent positions [53].

4.4.1
Paal–Knorr Synthesis and Related Methods (4 + 1 Strategy)

The Paal–Knorr pyrrole synthesis [54, 55] deserves particular recognition as one the most valuable of all pyrrole ring forming reactions, as it relies on the condensation of 1,4-dicarbonyl compounds with primary amines or their equivalents, both of which are quite common and readily available materials. In an illustrative example, the 2,5-dioxohexanoate derivatives **11** are efficiently converted into the corresponding pyrroles **12** upon treatment with appropriate amines in the presence of acetic acid (Scheme 4.3) [56].

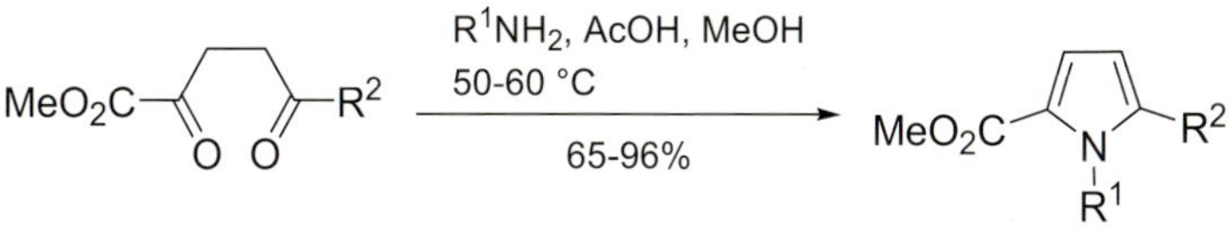

MeO_2C ... R^2 (**11**) → R^1NH_2, AcOH, MeOH, 50-60 °C; 65-96% → MeO_2C, N–R^1, R^2 (**12**)

Scheme 4.3

The versatility of the Paal–Knorr reaction is neatly demonstrated by the conversion of cyclododecane-1,4-dione into a pyrrole containing cyclophane [57], or transformation of the γ-ketoaldehyde **13** into the bicyclic system **14** in a key step of a synthesis of the bacterial tripyrrole pigment metacycloprodigiosin (Scheme 4.4) [58]. Modern applications emerge continuously, and allow for instance synthesis of 1-aminopyrrole derivatives by using monoprotected hydrazines [59] or *N*-aminophthalimide [60] as the amine components. A variant employing amine hydrobromides in refluxing pyridine is available [61], and an efficient synthesis of cyclopenta[*b*]pyrroles from suitable diketones with hexamethyldisilazane (HMDS) as the ammonia equivalent in

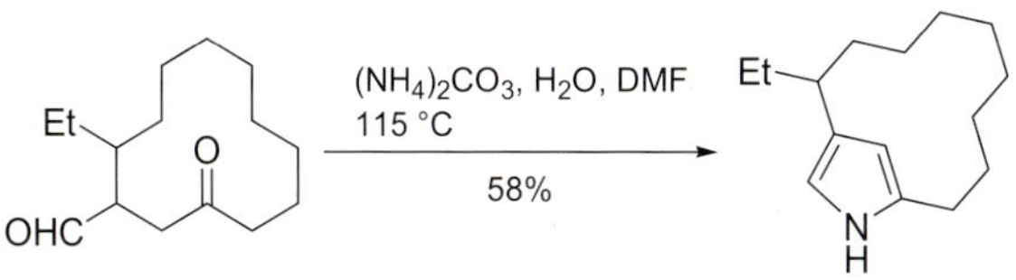

Et, OHC, O (**13**) → $(NH_4)_2CO_3$, H_2O, DMF, 115 °C; 58% → Et, N–H (**14**)

Scheme 4.4

the presence of Al_2O_3 has also been described [62]. Other useful extensions involve montmorillonite KSF clay [63], titanium isopropoxide [64] or iodine [65] as the catalysts. The Paal–Knorr synthesis has recently been performed under microwave irradiation [66–68], and has also been adapted to the solid phase employing immobilized 1,4-diketones [69]. A solution phase combinatorial approach has also been presented, involving construction of the 1,4-diketones from methyl esters by reaction with an excess of vinylmagnesium bromide in the presence of CuCN, followed by oxidation of the alkene unit in the resulting homoallylic ketones using $O_2/PdCl_2/CuCl$ in aqueous DMF [70].

The mechanism of the Paal–Knorr condensation has been scrutinized in detail, here exemplified by the conversion of the substituted 2,5-hexanediones **15** into the 2,5-dimethylpyrrole derivatives **16**, which appears to involve the intermediacy of the aminals **17**, which undergo cyclization to the diols **18**, followed by elimination of two equivalents of water (Scheme 4.5). These conclusions were supported by meticulous kinetic studies [71], as well as probing of the influence of the stereochemistry of the starting 1,4-dicarbonyl compounds [71, 72]. The reversibility of this series of events has recently been demonstrated by conversion of various pyrroles into the corresponding 1,4-dicarbonyl compounds by heating at pH 3, which allows exchange of the N-substituent [73].

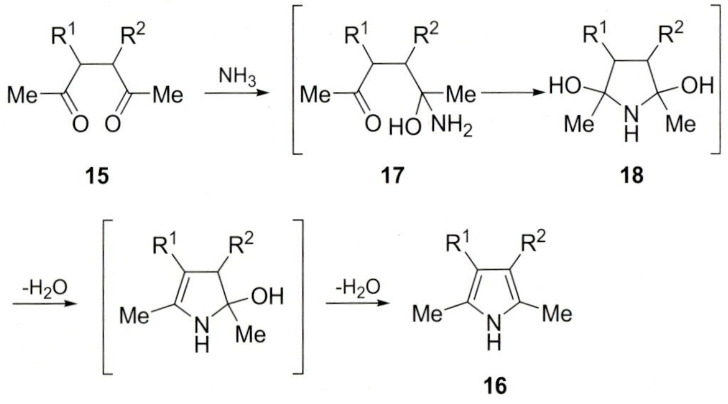

Scheme 4.5

This versatile method may also be utilized for the synthesis of 2*H*-pyrroles, as demonstrated by conversion of the 1,4-diketone **19** into the product **20** (Scheme 4.6). The series of events leading to this outcome involves the intermediacy of the

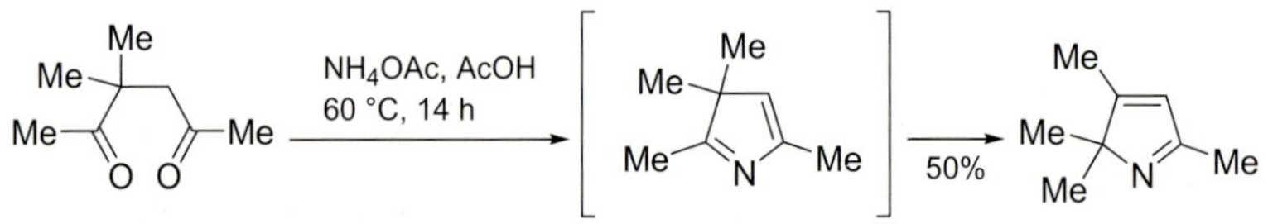

Scheme 4.6

3H-pyrrole **21**, which undergoes rearrangement to the 2H-pyrrole **20** due to the acidic reaction conditions in combination with heating. Nonetheless, this approach can in certain cases enable isolation of the 3H-tautomers [74]. A review detailing the early advances in the chemistry of 2H- and 3H-pyrroles is available [75].

A useful extension of the Paal–Knorr reaction is based on the cyclization of 2,5-dimethoxytetrahydrofuran (**22**) with primary amines, providing facile access to N-substituted pyrroles (e.g., **23**) (Scheme 4.7) [76, 77]. This process is further facilitated by using phosphorus pentoxide as the catalyst [78], or by heating in acetic acid under microwave conditions [79]. It has also been demonstrated that cyclizations involving **22** and amine components incorporating sensitive substituents proceeds in acceptable yields when carried out in a medium containing acetic acid and pyridine via a path featuring acid–base catalysis [80]. Application of arylsulfonamides as the amine synthons constitutes a useful route to 1-(arylsulfonyl)pyrroles [81]. Likewise, heating of 2,5-dimethoxytetrahydrofuran-3-carbaldehyde with ethyl carbamate [82] or *p*-toluenesulfonamide [83] under acidic conditions gives the corresponding N-substituted pyrrole-3-carboxaldehydes. Treatment of the related four-carbon precursor 2,5-dimethoxy-2,5-dihydrofuran with amines in 10% aqueous HCl gives the corresponding N-substituted 3-pyrroline-2-ones in good yield [84]. Initial hydrolysis of **22** in water to 2,5-dihydroxytetrahydrofuran, followed by reaction with primary amines in an acetate buffer, constitutes an additional modification that permits a broader range of N-substituents because of the less acidic conditions [85].

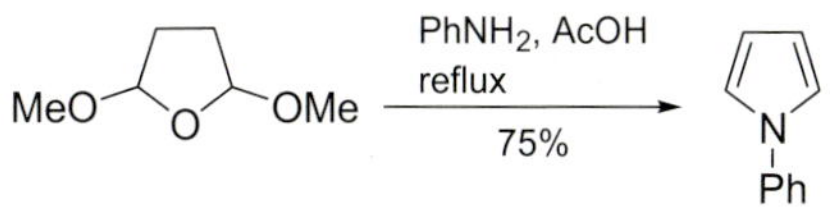

MeO, O, OMe — $PhNH_2$, AcOH, reflux, 75% → N, Ph

22 **23**

Scheme 4.7

Relatively mild conditions have also been employed in some related syntheses, wherein exposure of 1,4-dichloro-1,4-dimethoxybutane (**24**) to amino acids [86], or primary amides [87], led for example to the pyrrole **25**, or 1-acylpyrroles, respectively (Scheme 4.8).

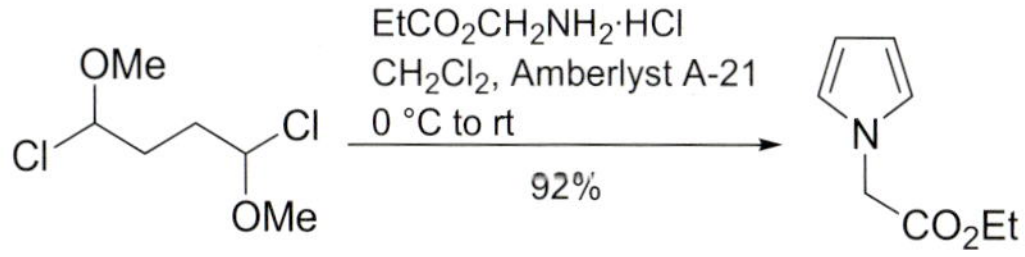

OMe, Cl, Cl, OMe — $EtCO_2CH_2NH_2·HCl$, CH_2Cl_2, Amberlyst A-21, 0 °C to rt, 92% → N, CO_2Et

24 **25**

Scheme 4.8

The Paal–Knorr condensation has also been incorporated as the key step in multicomponent approaches to pyrroles. A one-pot procedure, involving initial formation of the highly substituted 1,4-dicarbonyl compounds **26** from acylsilanes and a series of α,β-unsaturated ketones in the presence of the thiazolium salt **27** as the catalyst, is completed by ring closure using primary amines to the target pyrroles **28** (Scheme 4.9), featuring for example multiple aryl substituents [88]. A similar approach includes palladium-catalyzed coupling of aryl halides with propargylic alcohols, giving α,β-unsaturated ketones, which thereafter undergo thiazolium salt catalyzed Stetter reactions with aldehydes to provide the requisite 1,4-diketone precursors [89].

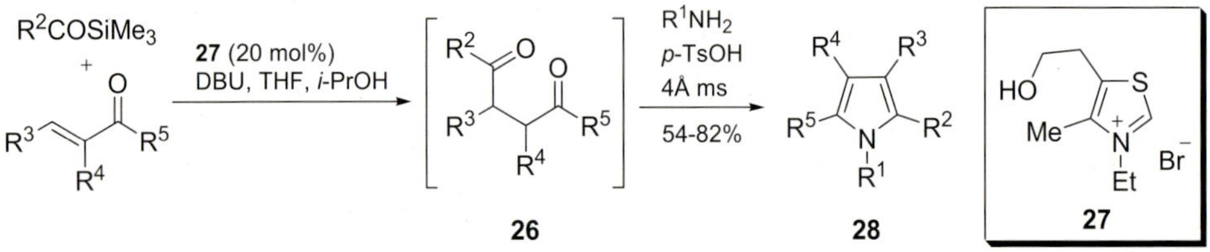

$R^2COSiMe_3$ + enone → [**27** (20 mol%), DBU, THF, *i*-PrOH] → **26** → [R^1NH_2, *p*-TsOH, 4Å ms, 54-82%] → **28**; catalyst **27**

Scheme 4.9

4.4.2
Other Cyclizations of Four-Carbon Precursors (5 + 0 and 4 + 1 Strategies)

Apart from the classical and modern variants of the Paal–Knorr reaction outlined above, several related approaches involving cyclization of four-carbon precursors are available. Generation of the γ-nitroketones **29** bearing an additional ester functionality geminal to the nitro group by Michael addition of ethyl nitroacetate to suitable enones, and subsequent cyclization thereof with formamidinesulfinic acid and triethylamine, gives the pyrrole-2-carboxylates **30** (Scheme 4.10), via the intermediacy of the oximes or imines **31** (X=NOH or NH) [90].

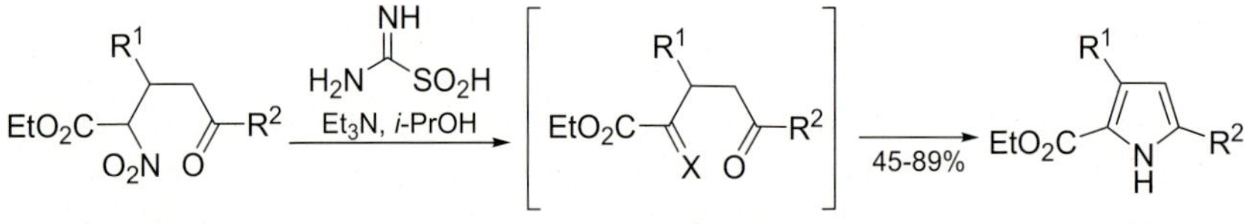

29 → [$H_2NC(=NH)SO_2H$, Et_3N, *i*-PrOH] → **31** → [45-89%] → **30**

Scheme 4.10

In contrast, reductive cyclization of the precursors **32**, available by conjugate addition of 2-nitropropane to α,β-unsaturated ketones, gives the pyrrolidines **33**, which may thereafter be converted into the corresponding 2*H*-pyrroles **34** upon dehydrogenation with 2,3-dichloro-5,6-dicyano-1,4-benzoquinone (DDQ) (Scheme 4.11) [91].

Suitable four-carbon precursors may also be prepared from the α-aminoaldehydes or -ketones **35**, which are readily available from *N*-Boc-α-amino acids by conversion

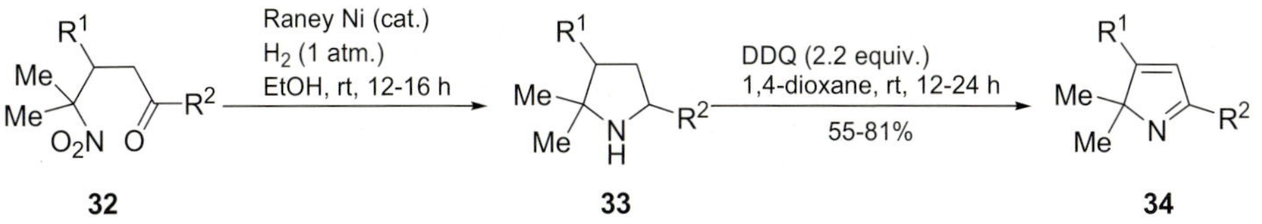

Scheme 4.11

into Weinreb amides, followed by reduction with $LiAlH_4$ or treatment with Grignard reagents, respectively. Thus exposure of **35** to lithium enolates of ketones, followed by cyclization of the resulting aldol adducts **36**, produced the set of pyrroles **37** (Scheme 4.12), including several fused derivatives, in low to moderate yields [92]. Reductive cyclization of similar aldol intermediates available from α-(*N*,*N*-dibenzyl) amino aldehydes or ketones has been utilized in a related, more high-yielding approach to *N*-benzylpyrroles [93]. Acid-induced cyclodehydration of Boc-protected γ-amino-α,β-enals or -enones derived from *N*-Boc-α-aminoaldehydes via Wittig reactions provides a route to various 1-(*tert*-butoxycarbonyl)pyrrole derivatives [94].

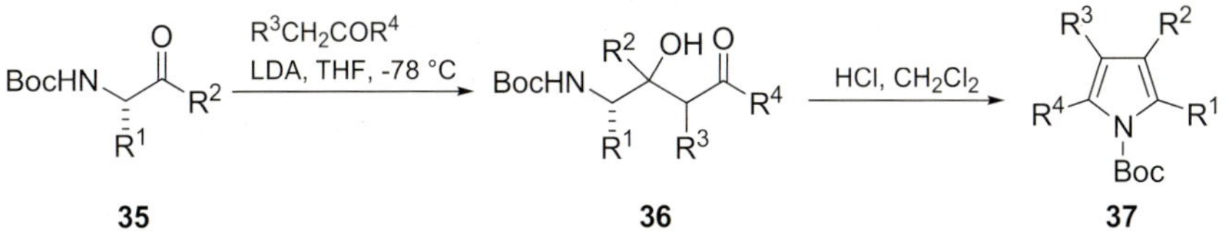

Scheme 4.12

An approach based on a microwave assisted domino process involving primary amines and the alkynoates **38**, which are derived from two equivalents of methyl propiolate and suitable aldehydes (R^1CHO), results in the pyrroles **39** (Scheme 4.13). The series of events leading to this outcome were suggested to involve rearrangement of 1,3-oxazolidine intermediates as the key feature [95]. A set of structurally related substrates has also been converted into pyrroles bearing multiple substituents by initial silver-catalyzed isomerization of the propargyl moiety to an allene, followed by condensation of the resulting intermediates with primary amines, and final gold-catalyzed 5-*exo-dig* cyclizations [96].

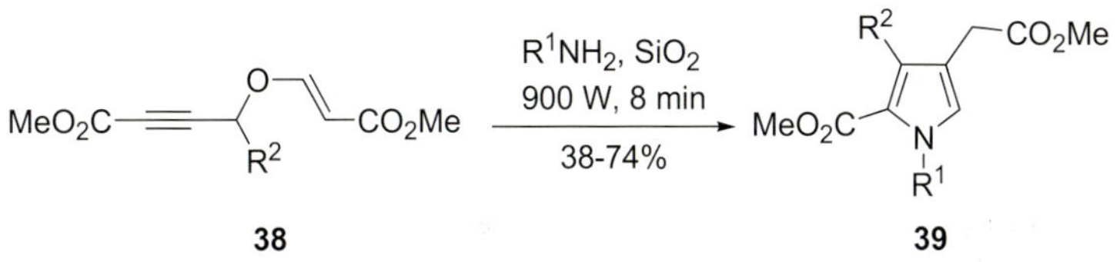

Scheme 4.13

It has been known for some time that pyrroles may be obtained from the reactions of azaallylic anions with suitable α-haloketones [97]. A new application of azaallylic anions in pyrrole synthesis has been realized by conversion of the α-haloimines **40** into the intermediates **41**, which in turn are cyclized to the 1-pyrrolines **42**, eventually giving the pyrroles **43** (Scheme 4.14), including the 3-chloro derivative (R^2=Ph, R^3=Cl) [98].

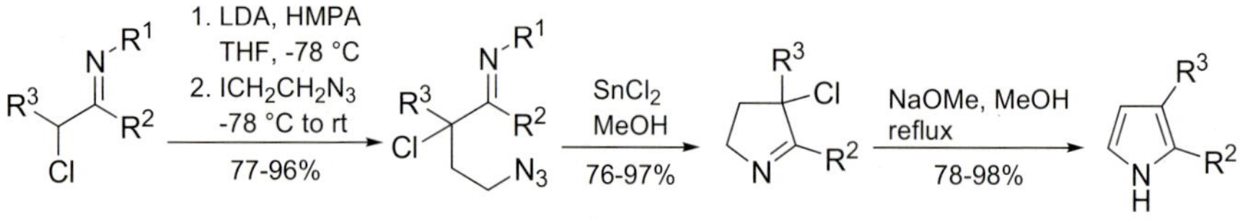

40 R^1 = *i*-Pr or *t*-Bu; 1. LDA, HMPA, THF, -78 °C; 2. $ICH_2CH_2N_3$, -78 °C to rt; 77-96% → **41**; $SnCl_2$, MeOH; 76-97% → **42**; NaOMe, MeOH, reflux; 78-98% → **43**

Scheme 4.14

Several approaches based on cyclization of propargylamines and homologues thereof have emerged in recent years. Base induced cyclization of the benzotriazol-1-yl (Bt) substituted precursors **44**, which are readily available from 1-propargylbenzotriazole, gives the corresponding pyrroles **45**, presumably via allene intermediates (Scheme 4.15) [99]. Rhodium-catalyzed hydroformylation of 1,3-disubstituted propargylamines affording 2,4-disubstituted pyrroles has also been accomplished [100].

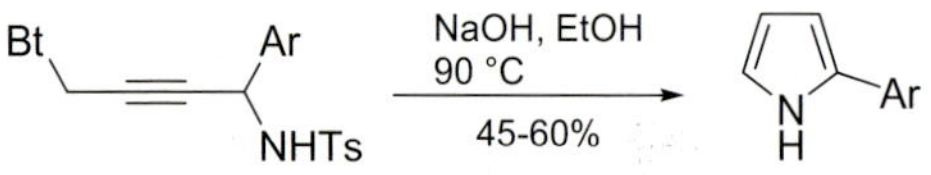

44 → NaOH, EtOH, 90 °C; 45-60% → **45**

Scheme 4.15

Homopropargylamines, which are available, for instance, by addition of propargylic Grignard reagents to Schiff bases, are also useful precursors, as exemplified by the silver(I) mediated conversion of **46** into the pyrroles **47** (Scheme 4.16) [101]. A synthetic route to pyrroles involving cyclization of homopropargylamines generated *in situ* by ring opening of ethynylepoxides with amines has also been described [102].

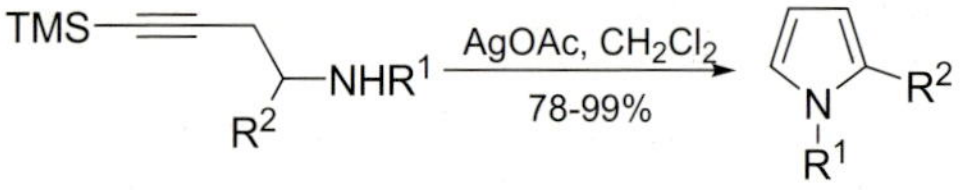

46 → AgOAc, CH_2Cl_2; 78-99% → **47**

Scheme 4.16

Annulation of the homopropargylic sulfonamides **48** (R=Ph, 2-furyl, 2-thienyl), which are prepared by alkylation of the benzophenone imine of methyl glycinate with propargyl bromide, followed by sequential hydrolysis, tosylation and Sonogashira coupling at the terminal acetylene unit, has been reported to give the substituted 4-iodo-2,3-dihydropyrrole derivatives **49** via a 5-*endo-dig* process, eventually leading to the β-iodopyrroles **50** (Scheme 4.17) [103, 104], whereas cyclization of related alkenyl derivatives provides access to β-iodopyrrolidine derivatives [105].

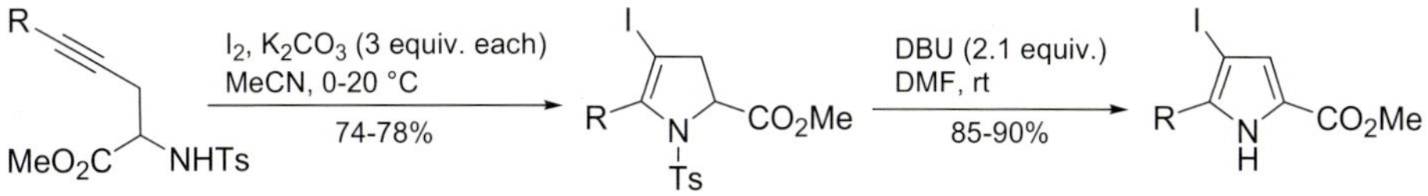

I_2, K_2CO_3 (3 equiv. each)
MeCN, 0-20 °C
74-78%

DBU (2.1 equiv.)
DMF, rt
85-90%

48 **49** **50**

Scheme 4.17

Azadienes **51** may also serve as starting materials for construction of pyrroles, as C-alkylation thereof provides the intermediates **52**, which undergo annulation to the pyrroles **53** upon heating in toluene or ethanol. A subsequent hydrolysis step completes this synthesis, resulting in the 3-acetylpyrrole derivatives **54** in excellent overall yields (Scheme 4.18) [106, 107]. An approach to, for instance, 1,2,3,5-tetra-substituted pyrroles, utilizing thermally induced cyclization of iminoalkyne intermediates derived from various substituted 4-pentynones and suitable amines, has also been reported [108, 109].

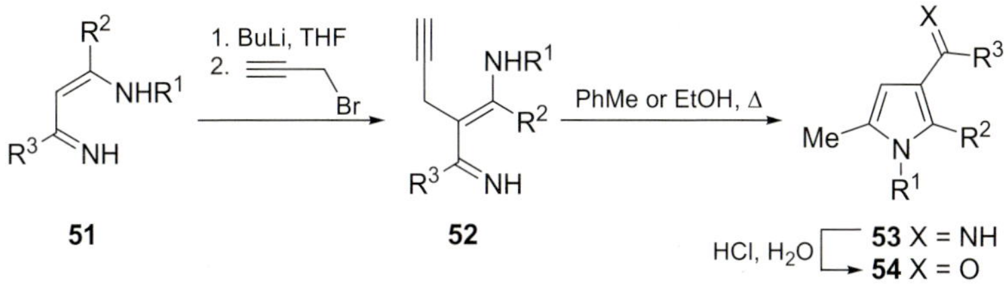

1. BuLi, THF
2. propargyl bromide

PhMe or EtOH, Δ

HCl, H_2O

51 **52** **53** X = NH **54** X = O

Scheme 4.18

4.4.3
Knorr Synthesis and Related Routes (3 + 2 Strategy)

The Knorr pyrrole synthesis [110], which relies on the condensation of an α-aminoketone with a carbonyl compound possessing acidic α-hydrogens (each contributing with a two-carbon fragment to the pyrrole ring) is also of considerable synthetic importance. Since α-aminoketones are rather reactive and difficult to handle, the practical procedures often involve generation thereof *in situ* from a suitable synthetic equivalent, as illustrated by the classical example below (Scheme 4.19), in which addition of one equivalent of sodium nitrite to two equivalents of ethyl acetoacetate (**55**) generates an oxime, which, upon reduction with zinc dust [110, 111] or

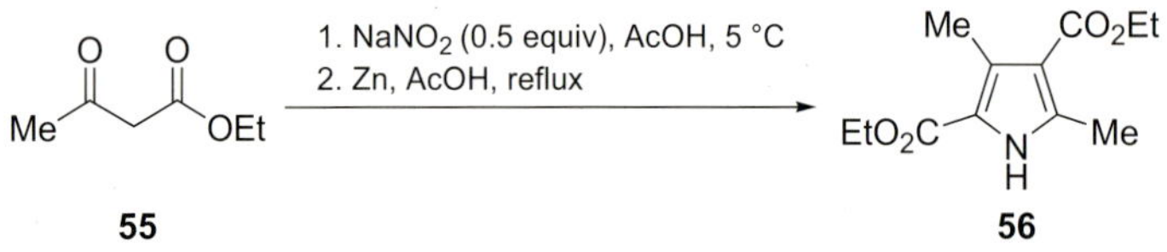

Scheme 4.19

dithionite [112], undergoes condensation with the remaining equivalent of **55** to furnish the pyrrole **56** in excellent yield (Scheme 4.19).

It has also been demonstrated that this reaction may follow a different path if a β-diketone is used as one of the reactants, as treatment of diethyl oximinomalonate (**57**) with 2,4-pentanedione (**58**) under reductive conditions will afford ethyl 3,5-dimethylpyrrole-2-carboxylate (**59**) [113]. Since this process involves the intermediacy of diethyl aminomalonate (**60**), the reductive conditions can be avoided by using this very reagent. The mechanism is considered to feature an initial formation of the aminal **61**, followed by elimination of water to form the enamine **62**, which will cyclize on the ketone carbonyl carbon to provide **63**, eventually leading to the final product **59** (Scheme 4.20) [114, 115]. The use of unsymmetrical 1,3-diketones instead of 2,4-pentanedione usually gives mixtures of regioisomeric pyrroles, unless relatively bulky groups (*i*-Pr, *t*-Bu, Ph) are present at the terminal carbon [116]. Many substituted pyrrole-2-carboxylate derivatives obtained using these procedures may be readily converted into further derivatives by decarboxylation (Section 4.6.2).

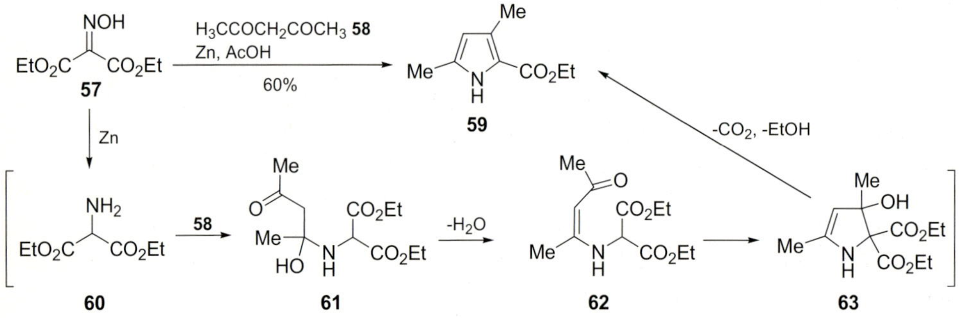

Scheme 4.20

A modern modification of the Knorr pyrrole synthesis involves elaboration of Weinreb amides, for instance **64**, derived from phenylalanine, giving the protected α-aminoketone **65**, which can subsequently be deprotected and condensed with 2,4-pentanedione to provide the pyrrole **66** (Scheme 4.21) [117]. Reaction of similar Weinreb amides lacking the Boc group with enamines gives *N*-methoxy-*N*-methyl-α-enaminocarboxamides, which take part in related conversions into α-enaminoketones, and ensuing annulations to pyrroles [118].

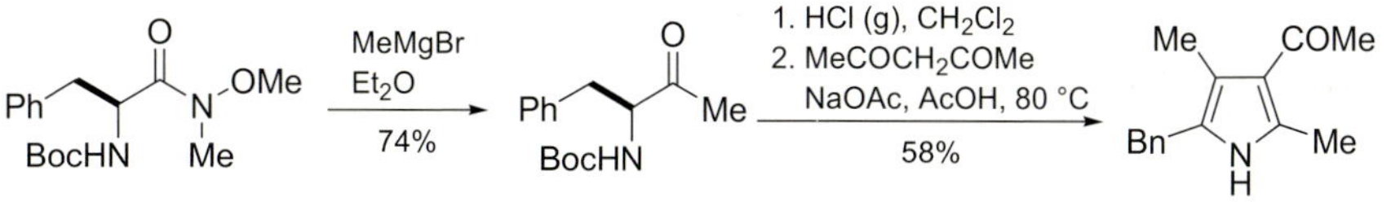

Scheme 4.21

In a related approach involving an initial C–N bond formation between two C_2-fragments, addition of the α-aminoketones **67** to dimethyl acetylenedicarboxylate (DMAD) yields the intermediates **68**, which finally undergo cyclization to give pyrroles **69** (Scheme 4.22) [119]. A related procedure involving initial reactions of α-amino acid esters with DMAD, and cyclization of the intermediate enamines with sodium methoxide in methanol rendering 3-hydroxypyrroles **69** (R^1=OH), has also been reported [120].

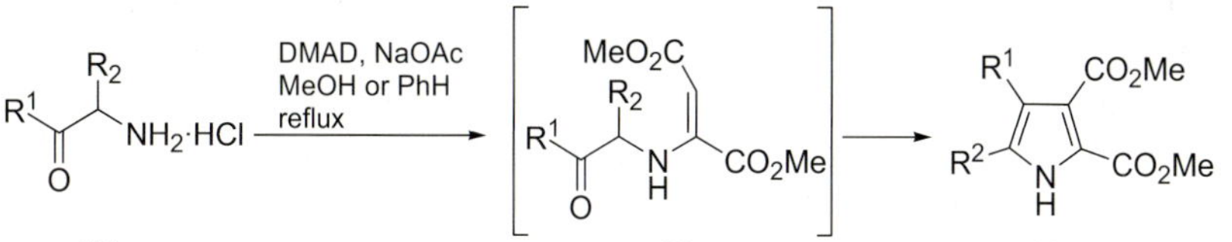

Scheme 4.22

A series of 2-substituted 3-nitropyrroles (**70**) has been prepared by displacement of a methylthio group of the nitroalkene **71**, followed by treatment of the resulting products **72** with aminoacetaldehyde dimethylacetal to furnish the intermediate enamines **73**, which underwent a final ring closure in acidic medium (Scheme 4.23) [121]. Intermediates similar to **73** featuring a trifluoroacetyl group instead of the nitro functionality have previously been prepared from β-(trifluoroacetylvinyl) ethers and aminoacetaldehyde dimethylacetal, and were cyclized to the corresponding 3-trifluoroacetylpyrroles in aqueous TFA [122].

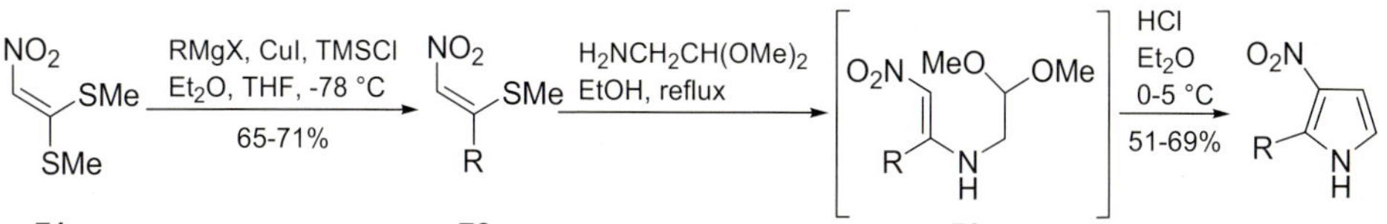

Scheme 4.23

4.4.4
Hantzsch Synthesis and Related Approaches (2 + 2 + 1 or 3 + 2 Strategy)

A conceptually related process involving two C_2-fragments and an amine component is the Hantzsch pyrrole synthesis (Scheme 4.24) [123]. In a typical procedure, a mixture consisting of an α-halo carbonyl compound (**74**) and ethyl acetoacetate (**75**) is treated with ammonia. This initially gives the enamine **76** derived from the β-ketoester, which will subsequently undergo cyclization with **74** to provide pyrrole **77** [124]. Application of β-aminoacrylonitriles as the enamine counterparts has been used to prepare 5-(trifluoromethyl)pyrrole-3-carbonitrile derivatives [125]. A solid phase variant utilizing an immobilized enamino component has also been developed [126].

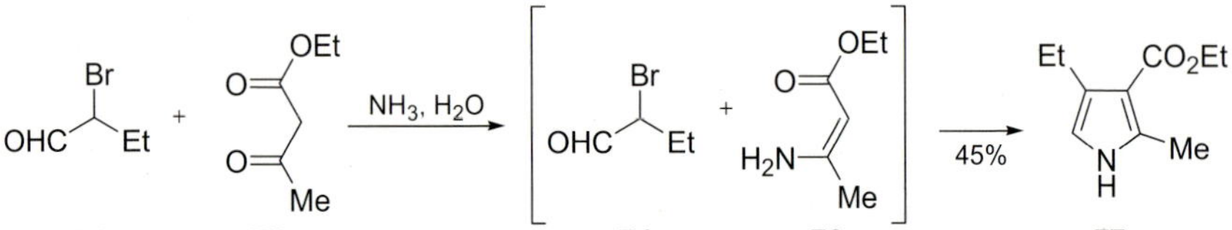

Scheme 4.24

It has also been demonstrated that titanium mediated annulation of substrate **78**, which is available in two steps from the appropriate 1,3-dicarbonyl compound, gives a good yield of 2,3,5-triphenylpyrrole (**79**) (Scheme 4.25). Similar precursors have also been cyclized in the presence of a preformed titanium–graphite reagent [127].

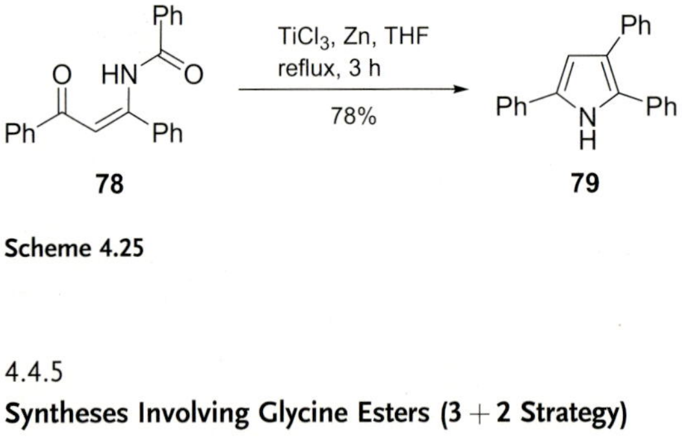

Scheme 4.25

4.4.5
Syntheses Involving Glycine Esters (3 + 2 Strategy)

Several useful routes to pyrroles are based on the reactions of glycine esters or related compounds with suitable C_3-synthons. For example, condensation of ethyl glycinate hydrochloride (**80**) with the 1,3-diketone **81** provides access to the pyrrole **82** [128] via the enaminoketone intermediate **83** (Scheme 4.26). Such intermediates may also be isolated in a stepwise approach involving milder conditions [129], whereas cyclization of related condensation products generated from 3-ethoxyacrolein derivatives and N-substituted glycine esters gives 2,4-disubstituted pyrroles [130].

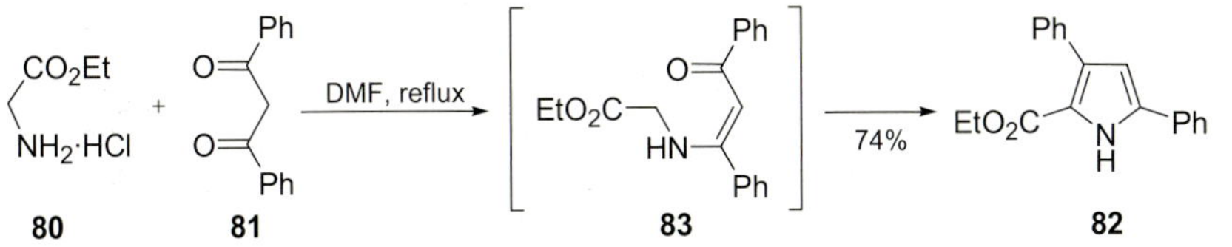

Scheme 4.26

Likewise, it has been demonstrated that *p*-toluenesulfonylglycine esters **84** undergo addition to α,β-unsaturated ketones **85** to render pyrrolidines **86**, which will eventually furnish pyrroles **87** by sequential elimination of water and *p*-toluenesulfinate, via the suggested 3-pyrroline intermediates **88** (Scheme 4.27) [131, 132].

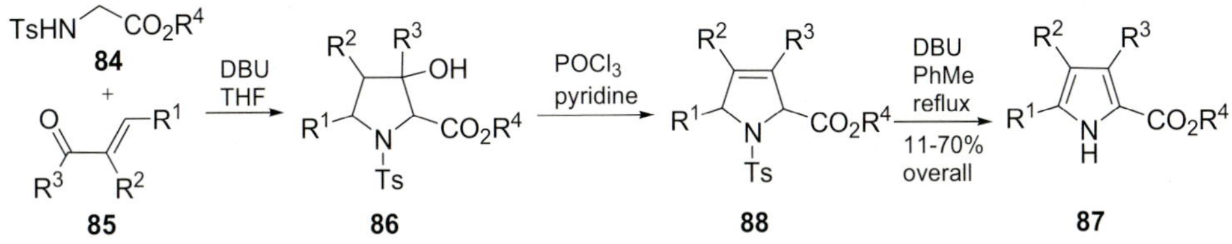

Scheme 4.27

Glycine derivatives may also give pyrroles upon treatment with various iminium salts. For example, the reaction of ethyl glycinate hydrochloride (**80**) with the vinamidinium perchlorates **89** provides 3-arylpyrrole-2-carboxylates (**90**) in good yields [133]. A related reaction involving the salt **91**, which is readily available from 3-acetylthiophene, leads to the thienylpyrrole **92** (Scheme 4.28) [134]. Similar chemistry has also been employed for the preparation of 5-arylpyrrole-2-carboxylates [135] and of various 3,4-disubstituted pyrrole-2-carboxylates [136].

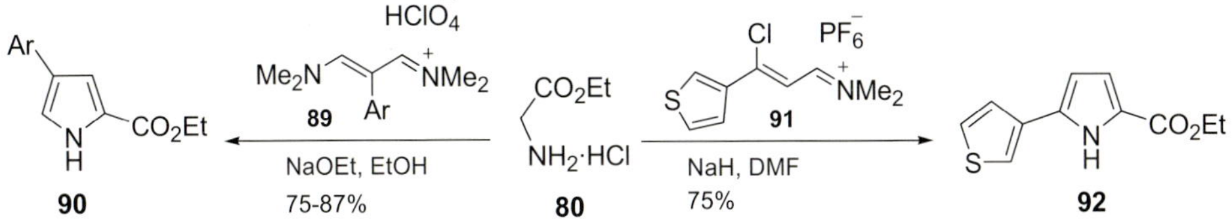

Scheme 4.28

4.4.6
Van Leusen Method (3 + 2 Strategy)

Since its introduction, the van Leusen pyrrole synthesis has enjoyed considerable popularity, as it provides convenient access to 3,4-disubstituted pyrroles from the readily available building blocks *p*-toluenesulfonylmethyl isocyanide (TosMIC) (**93**) and electron deficient alkenes. Treatment of the anion of TosMIC with the α,β-unsaturated ketones **94** initially gives the intermediate Michael adducts **95**. After cyclization to **96**, followed by tautomerization to **97**, *p*-toluenesulfinate is eliminated giving the 3*H*-pyrroles **98**, which eventually tautomerize to the final products **99** (Scheme 4.29) [137]. Application of methyl 3-arylacrylates in this approach gives methyl 4-arylpyrrole-3-carboxylates, which may be further converted into the corresponding 3-arylpyrroles by saponification and decarboxylation [138]. A variant involving aryl- or heteroarylalkenes, TosMIC and sodium *tert*-butoxide in DMSO allows direct access to 3-aryl- or heteroarylpyrroles, respectively, in moderate yields [139]. When acrylonitriles are used as the alkene reactants, pyrrole-3-carbonitriles are produced [137], whereas application of nitroalkenes [140, 141] or *tert*-butyl (*E*)-4,4,4-trifluorobutenoate [142] gives the corresponding β-nitropyrrole- or β-(trifluoromethyl)pyrrole derivatives, respectively. Extensions involving substituted TosMIC derivatives offer direct routes to 2,3,4-trisubstituted pyrrole derivatives [143], including 2-stannylpyrroles [144]. The closely related reagent benzotriazol-1-ylmethyl isocyanide (BetMIC) has also been evaluated in similar reactions, and may in some cases give better yields [145].

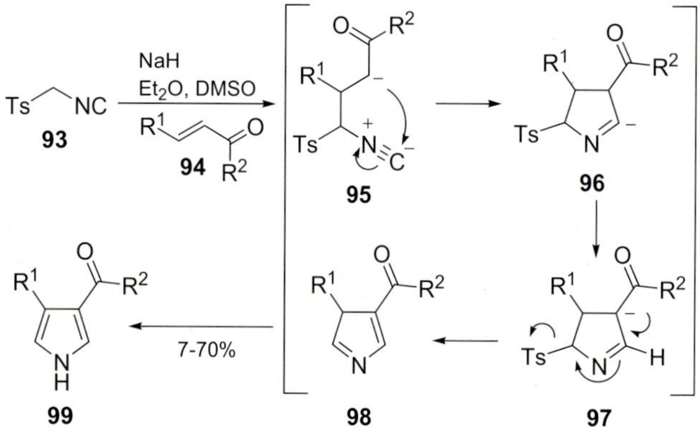

Scheme 4.29

In a further extension of this valuable method, reaction of the 1-isocyano-1-tosyl alkenes **100** with nitromethane in the presence of potassium *tert*-butoxide enables efficient preparation of the 3-nitropyrroles **101** (Scheme 4.30) [146]. Similar transformations involving suitable substituted ketones instead of nitromethane yield

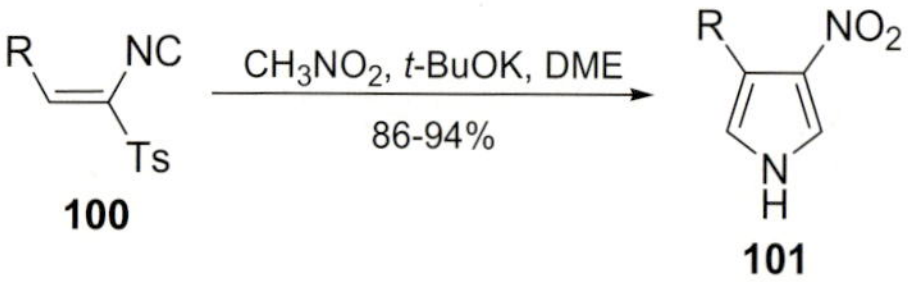

Scheme 4.30

various 3,4-disubstituted pyrroles, even such lacking electron-withdrawing substituents [147]. It is also noteworthy that alkenes generated from aldehydes and alkyl isocyanoacetates in the presence of DBU may react with an additional equivalent of the isocyanoacetate component, affording 2-substituted alkyl pyrrole-2,4-dicarboxylates in a convenient one-pot operation [148].

The reactions of TosMIC (**93**) or the methyl derivative **102** with dienes, for instance **103** [141] or **104** [149], furnish the corresponding substituted 3-vinylpyrroles, **105** and **106**, respectively (Scheme 4.31). Treatment of 1,4-disubstituted 2,3-dinitrobutadienes with TosMIC under similar conditions gives 3-alkynylpyrrole derivatives [150].

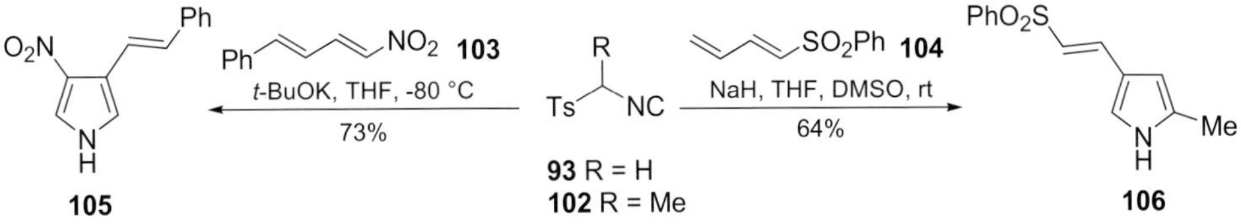

Scheme 4.31

4.4.7 Barton–Zard Synthesis (3 + 2 Strategy)

The equally versatile Barton–Zard synthesis features an initial conjugate addition of isocyanoacetate esters **107** to nitroolefins **108** in the presence of a base (e.g., DBU), generating the adducts **109**, which thereafter undergo cyclization to afford **110**. An ensuing isomerization of **110** to **111**, followed by elimination of nitrite, provides the 3*H*-pyrroles **112**, which finally tautomerize to the target pyrrole-2-carboxylate derivatives **113** (Scheme 4.32). Sensitive nitroolefins are preferably formed *in situ* from the corresponding β-nitroacetoxyalkanes [151, 152]. Application of benzyl

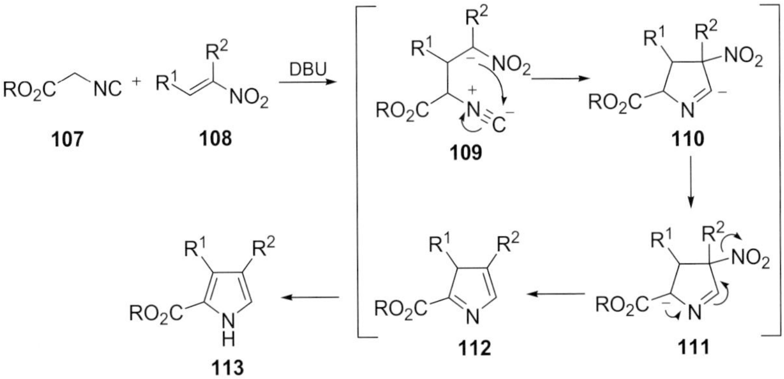

Scheme 4.32

isocyanoacetate in this approach allows efficient preparation of benzyl pyrrole-2-carboxylates [153, 154]. Alternatively, the nitroolefin components may be replaced by α,β-unsaturated sulfone derivatives [155–157] or by acetate precursors bearing a vicinal nitro group [158]. A variant of this reaction involving polymer supported reagents has also been developed [159].

Synthesis of the fused pyrrole derivatives **114** and **115** from 9-nitrophenanthrene (**116**) constitutes an interesting application of the Barton–Zard approach (Scheme 4.33) [160, 161]. Other condensed nitroaromatics, such as 3-nitrobenzothiophene, also give fused pyrrole derivatives under these conditions [162]. In cases where relatively unreactive nitroaromatics are involved, the use of a strong phosphazene base may give improved yields [163].

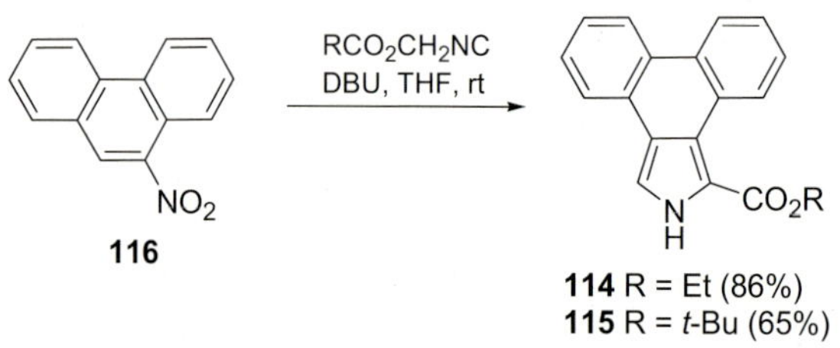

NO_2

116

RCO_2CH_2NC
DBU, THF, rt

N H CO_2R

114 R = Et (86%)
115 R = *t*-Bu (65%)

Scheme 4.33

4.4.8
Trofimov Synthesis (3 + 2 Strategy)

Various pyrroles have been prepared over the years using the Trofimov synthesis, which relies on cyclization of ketoximes with acetylenes in a strong basic medium [164, 165]. For example, exposure of oxime **117** to acetylene in the presence of KOH in DMSO at elevated pressure and temperature gives the fused pyrrole **118** in excellent yield (Scheme 4.34) [166]. However, the formation of mixtures of N-vinylated products and the corresponding parent pyrroles is a common outcome of this reaction. The N-vinylation may be suppressed by addition of water (about 5%) to the reaction mixture, whereas optimal conditions for synthesis of *N*-vinylpyrroles require the use of a large excess of KOH [165]. A recent application of this method provided access to 2,6-bis(pyrrol-2-yl)pyridines [167]. Several pyrroles incorporating sulfur containing moieties have been prepared using the Trofimov reaction [168].

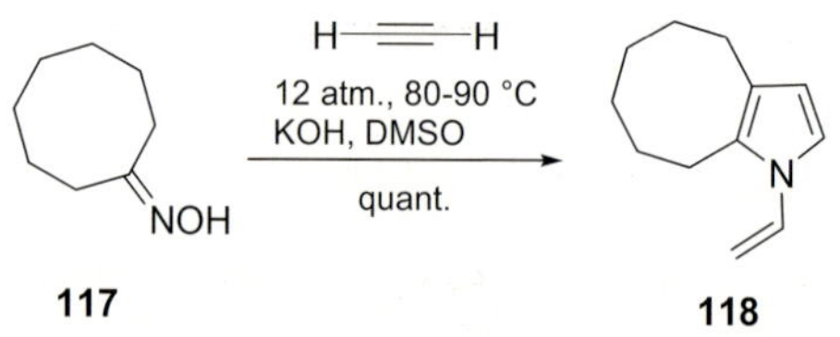

NOH

117

H—≡—H
12 atm., 80-90 °C
KOH, DMSO
quant.

N

118

Scheme 4.34

4.4.9 Cycloaddition Reactions and Related Approaches (3 + 2 Strategy)

1,3-Dipolar cycloadditions between mesoionic compounds and suitable dipolarophiles, for example alkynes, constitute another useful approach to pyrroles [169, 170]. Thus the alkyl- or aryl-substituted münchnones **119**, which are readily available from α-amino acids, participate in cycloadditions with acetylene derivatives (e.g., diesters) to provide pyrroles **120**, often in excellent yields, through expulsion of carbon dioxide from the intermediate adducts **121** (Scheme 4.35). Münchnones may also be generated *in situ* from *N*-acyl-α-amino acids and acetic anhydride. Preferably, one of the reactants should be symmetrically substituted, thus avoiding formation of mixtures containing isomeric pyrroles [171, 172]. Nevertheless, regiospecific reactions involving polyfluoro-2-alkynoic acid esters have been reported [173]. The use of *N*-acylmünchnones provides access to *N*-acylpyrroles as mixtures of isomers [174], whereas 2-arylthio- or alkylthio-substituted 5-amino-1,3-thiazolium salts give 2-arylthio- or 2-alkylthiopyrrole derivatives, respectively, upon reaction with dimethyl acetylenedicarboxylate (DMAD) via extrusion of isothiocyanates [175]. Modern, multi-component variants that presumably involve münchnones feature generation of the dipoles from imines, acid chlorides, and carbon monoxide via palladium catalysis [176], or by annulation of products derived from Ugi four-component reactions involving carboxylic acids, primary amines, aldehydes and 1-isocyanocyclohexene [177]. A solid phase version using polymer bound münchnones has also been described. [178] Various aspects concerning the regioselectivity of 1,3-dipolar cycloadditions involving münchnones have been discussed in detail; the outcome appears to be influenced by the electronic nature and location of the substituents on the dipole [179], as well as steric factors [180]. In connection with investigations of regioselective cycloadditions of a certain münchnone with a carbohydrate derived nitroolefin, it was concluded that predictions using frontier molecular orbital theory in combination with semi-empirical studies are not applicable for such processes, and that this particular example proceeded through a concerted, although somewhat asynchronous, transition state, as implied by results from *ab initio* MO calculations [181].

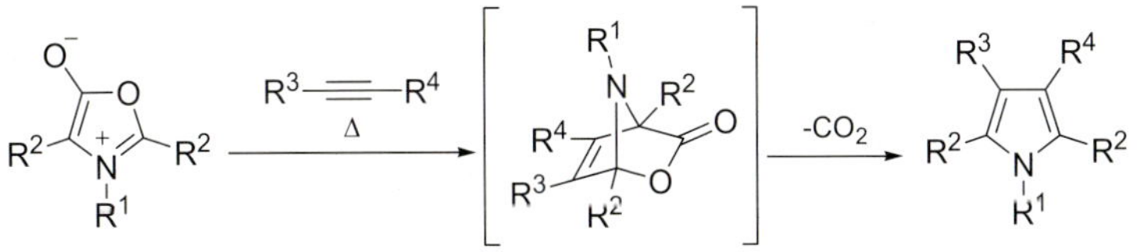

R^3—≡—R^4, Δ; $-CO_2$

119 **121** **120**

Scheme 4.35

Pyrroles may also be made by cycloaddition of dimethyl 1,2,4,5-tetrazine-3,6-dicarboxylate (**122**) with electron rich alkenes, followed by ring contraction of the resulting 1,2-diazines [182, 183]. In a representative procedure, **122** reacts with

1,1-dimethoxyethylene to give intermediate **123**, which is subsequently reduced to the pyrrole **124** (Scheme 4.36) [184]. The alkene components may also be replaced with acetylene derivatives [185]. Advances in ring contraction methodology for the construction of pyrroles have been discussed in detail in a specialized review [186].

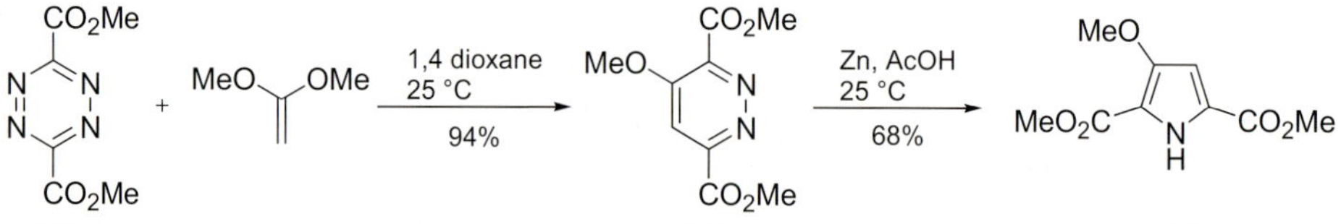

Scheme 4.36

Generation of dipoles from aziridines, and reaction thereof with suitable acetylenes, offers a route to pyrroles [187] that might be otherwise difficult to access. This method has been employed for the synthesis of the 3,4-disilylpyrrole **125** by dipolar cycloaddition of 2-cyano-1-trimethylsilylaziridine (**126**) with bis(trimethylsilyl)acetylene, followed by N-desilylation of the intermediate product **127** in methanol (Scheme 4.37) [188]. Azomethine ylides generated by desilylation of suitable immonium salts have been demonstrated to add to alkynes, giving pyrroles, or to alkenes to render 2-pyrrolines, which could in turn be further converted into the corresponding pyrroles by treatment with DDQ [189]. Based on a previously reported procedure [190], an approach to pyrroles has been devised that relies on reactions involving alkynes and imines in the presence of Ti(*i*-OPr)$_4$, *i*-PrMgCl and carbon monoxide at atmospheric pressure, via azatitanacyclopentene derivatives as intermediates [191].

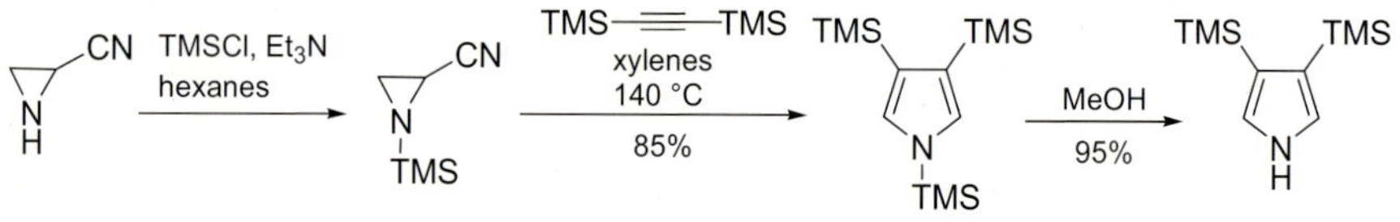

Scheme 4.37

An elegant route to pyrroles from isocyanides and electron deficient acetylenes has also become available. In an illustrative example, pyrrole **128** was obtained upon reaction of ethyl isocyanoacetate with alkyne **129** in the presence of dppp [192]. The regioselectivity may be reversed by changing the catalyst to Cu_2O, giving instead the product **130** (Scheme 4.38) [193]. A related synthesis of pyrroles involving addition of metallated isocyanides to acetylenes, featuring an intramolecular cycloaddition of an alkene unit with the isocyanide moiety in the initially formed intermediates as the key step, has also appeared [194].

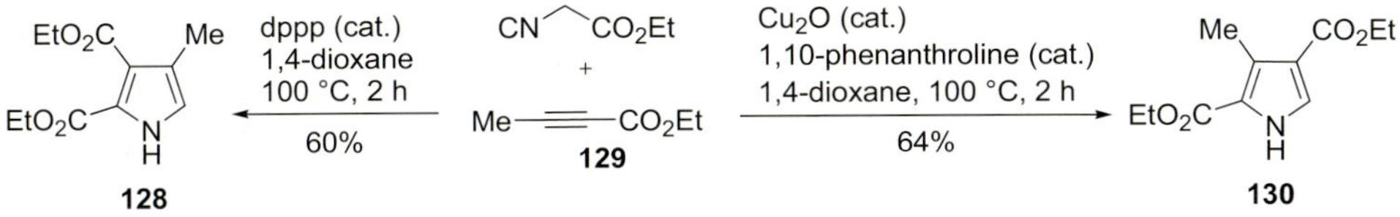

Scheme 4.38

4.4.10
Multi-Component Reactions (2 + 2 + 1 Strategy)

Multi-component processes, which are carried out in a one pot operation, have become increasingly popular tools for pyrrole synthesis in recent years. Some of these approaches employ well-known principles for pyrrole ring formation, for example the Paal–Knorr reaction [88], or dipolar cycloaddition of alkynes to münchnones [176, 177] (see above). Samarium-catalyzed three-component coupling of amines, aldehydes and nitroalkanes has been demonstrated to furnish modest to moderate yields of the pyrroles **131** (Scheme 4.39). Two aldehyde units are incorporated in the final products. The aldehydes may also be replaced by α,β-unsaturated aldehydes or ketones in a similar pyrrole ring forming reaction that does, however, not require the use of a catalyst [195]. Likewise, reactions between amines, α,β-unsaturated aldehydes or ketones and nitroethane in the presence of silica [196], or alternatively amines, aldehydes or ketones and nitroalkenes mediated by Al_2O_3 [197] under microwave irradiation, also produce useful yields of various pyrroles. The latter set of components may also be converted into pyrroles by heating in molten tetrabutylammonium bromide [198].

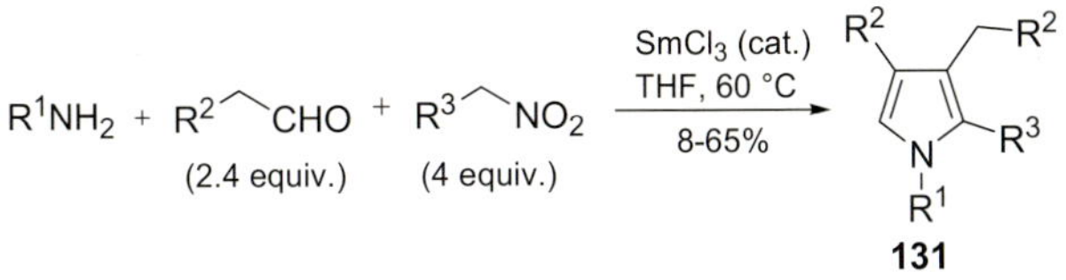

Scheme 4.39

4.4.11
Miscellaneous Transition Metal Catalyzed Methods (3 + 2 and 5 + 0 Strategies)

Transition metal catalyzed/mediated transformations of acyclic precursors to pyrroles have also attracted considerable attention. Although conceptually and mechanistically very interesting, some of these developments still appear to suffer from lack of practical synthetic applicability, involving rather complex starting materials and catalysts. Nevertheless, such procedures are now acknowledged as valuable tools for the preparation of exotic pyrroles having unusual substituents, substitution patterns or oxidation states, which are not easily available via the "classical" procedures. For example, the rhodium-catalyzed reaction of 3-fluoropentane-2,4-dione (**132**) with ethyl isocyanoacetate furnished the fluoropyrrole **133** (Scheme 4.40), whereas the use of unsymmetrical 1,3-diketones gave, in most cases, mixtures of regioisomeric pyrroles [199].

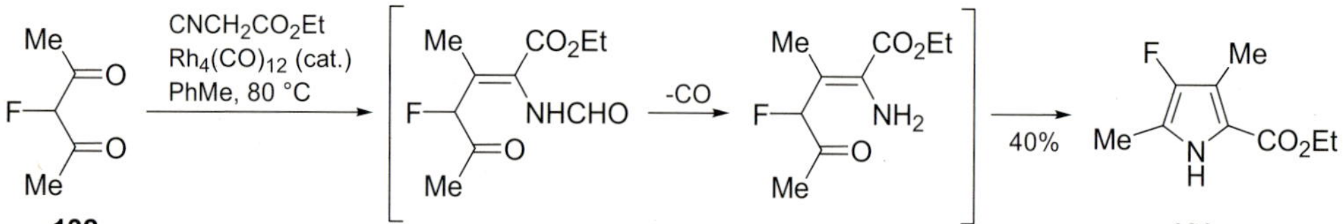

Scheme 4.40

Metallation of *N*-allylbenzotriazole **134**, followed by treatment with imines provided the intermediates **135**, which were thereafter converted into the 1,2-diarylpyrroles (**136**) by palladium-catalyzed annulation (Scheme 4.41) [200]. This approach resembles a previous route featuring lithiation of 1-(3-morpholinoprop-2-enyl)benzotriazole, wherein the final cyclization could be effected under acidic conditions [201].

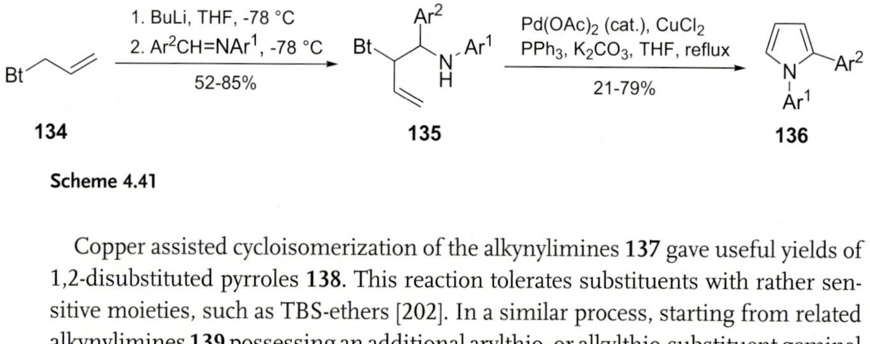

Scheme 4.41

Copper assisted cycloisomerization of the alkynylimines **137** gave useful yields of 1,2-disubstituted pyrroles **138**. This reaction tolerates substituents with rather sensitive moieties, such as TBS-ethers [202]. In a similar process, starting from related alkynylimines **139** possessing an additional arylthio- or alkylthio-substituent geminal to R^2, 2-alkyl-3-thio-substituted pyrroles **140** were produced in good yields via 1,2-migration of the thio group in intermediate thioallenylimines (Scheme 4.42) [203].

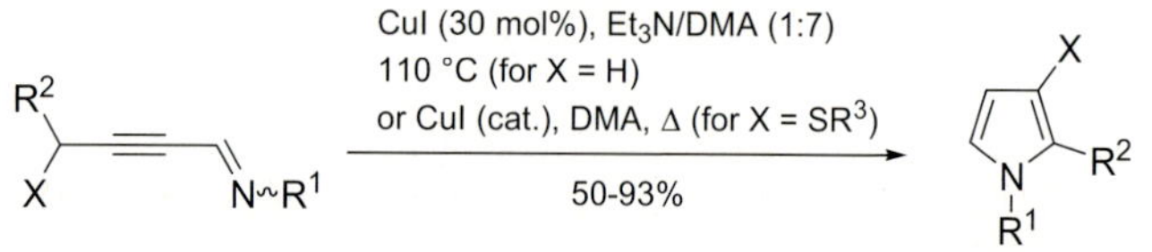

Scheme 4.42

The (*Z*)-(2-en-4-ynyl)amines **141** undergo Pd(II) [204, 205] or Cu(II)-catalyzed cycloisomerization to the pyrroles **142** (Scheme 4.43). The copper-catalyzed reactions require higher temperatures. Interestingly, the less stable of the substrates **141** (R^4=H, Ph, CH_2OTHP) underwent spontaneous cycloisomerization to the target heterocycles [205].

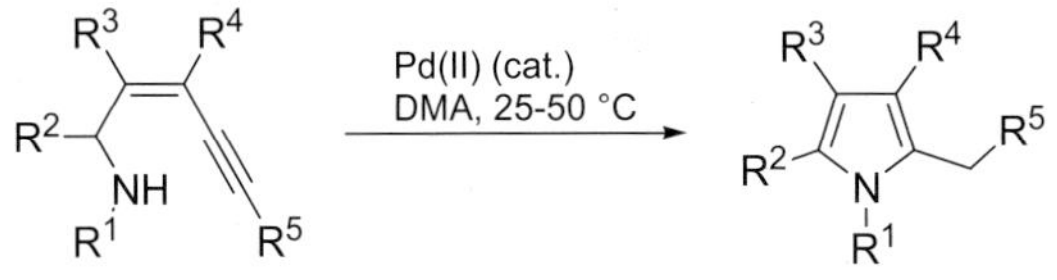

Pd(II) (cat.)
DMA, 25-50 °C

141 **142**

Scheme 4.43

4.5 Reactivity

4.5.1 Reactions with Electrophilic Reagents

4.5.1.1 General Aspects of Reactivity and Regioselectivity in Electrophilic Substitution

As indicated in Section 4.2.2, pyrrole is prone to undergo electrophilic substitution predominantly at the α-position (C2). Introduction of substituents alters both regioselectivity and reactivity by changing the electronic properties of the pyrrole nucleus by inductive effects, and sometimes also by steric interactions with the incoming electrophile. The transmission of electronic substituent effects in pyrroles appears to occur through the carbon atoms rather than via the ring nitrogen [206]. For example, it has been demonstrated that 2-methylpyrrole (R^2=Me; Figure 4.1) undergoes trifluoroacetylation at C5 some 23.8 times faster than pyrrole itself [207], whereas the reactivity of 1-methylpyrrole (R^1=Me; Figure 4.1) towards trifluoroacetic anhydride in 1,2-dichloroethane at 75 °C is only 1.9 times higher than that of the

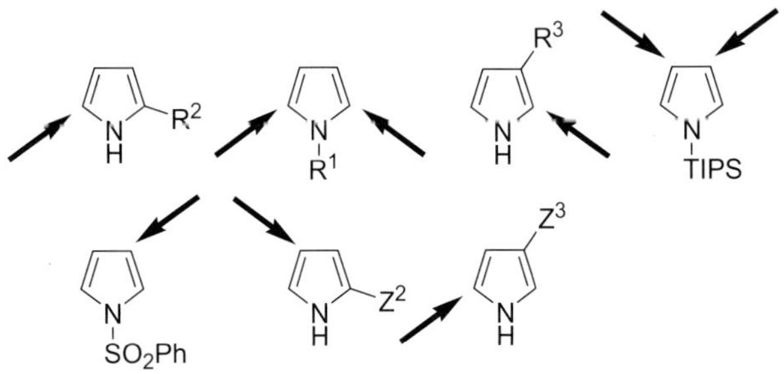

Figure 4.1 Transmission of electronic substituent effects in pyrroles.

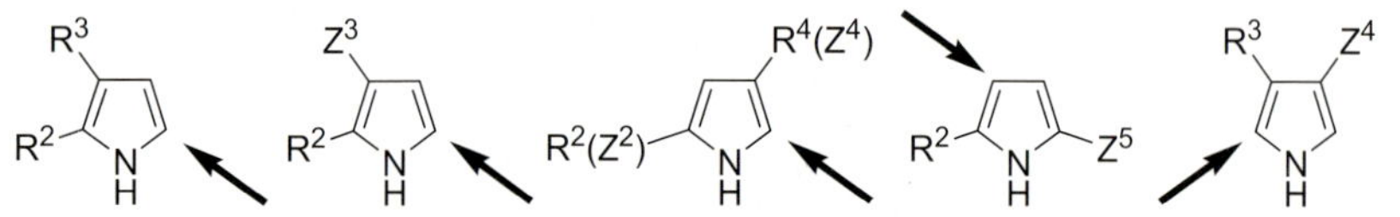

Figure 4.2 General trends of regioselectivity of electrophilic substitution of disubstituted pyrroles.

parent heterocycle [208]. Pyrroles having an electron releasing group at C3 (R^3) display enhanced reactivity both at C2 and C4, but the C2 position is still the most active because of the influence of the ring nitrogen. The introduction of a bulky N-substituent, for instance triisopropylsilyl (TIPS) [209] or trityl [210], gives access to C3 substituted or C3, C4 disubstituted products by blocking the intrinsic α-reactivity by steric interference. Substitution with high selectivity at C3 may also often be obtained with pyrroles having powerful electron-withdrawing substituents at the nitrogen, for example the phenylsulfonyl group [211]. Such electron deficient substrates are also considerably less reactive towards electrophiles than non-deactivated pyrroles. The presence of a strong "meta" directing, electron-withdrawing groups (Z^2) at C2 [212] or at C3 (Z^3) will direct substitution to C4 or C5, respectively. The general effects of various directing groups in monosubstituted pyrroles are summarized in Figure 4.1 (R=alkyl, Z=electron-withdrawing substituent). Additional examples, as well as special cases that deviate from the common pathways are discussed in appropriate sections below.

Prediction of the regioselectivity of electrophilic substitution of disubstituted pyrroles is more complicated, as the outcome is dependent on the combined influence of the substituents. There are also cases when the effects of sterically demanding substituents must be taken into account. Detailed studies of the reactivity of such systems have been conducted [213, 214]; Figure 4.2 depicts the general trends.

4.5.1.2 **Protonation**

In acid solution, pyrrole (**1**) undergoes reversible protonation, predominantly at C2, giving the thermodynamically favored 2*H*-pyrrolium cation **143**, which is stabilized by mesomeric delocalization of the charge (Scheme 4.44). The pK_a of −3.80 has been determined for protonation in dilute sulfuric acid solution [215]. It has also been demonstrated that protonation of pyrrole with the mild acids $C_4H_9^+$ and NH_4^+ in the gas phase occurs at C2, as well as at C3, giving the isomeric 3*H*-pyrrolium cation **144**, and that the affinity for protonation is higher at C2 [216]. The virtually non-existent N-basicity of pyrrole may be rationalized in terms of the absence of

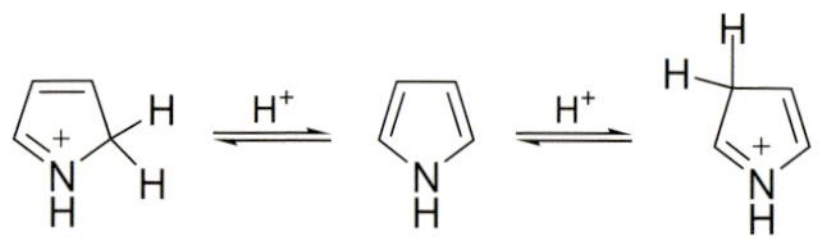

Scheme 4.44

mesomeric charge delocalization in the putative 1*H*-pyrrolyl cation, and the unavailability of the electron pair on the pyrrole ring nitrogen due to its contribution to the aromatic π-electron sextet. This is supported by the fact that the dipole moment of pyrrole is directed into the ring, as opposed to its tetrahydro derivative pyrrolidine, which displays a dipole moment direction towards the nitrogen atom [4].

Based on observations during protonation studies involving various substituted pyrrole derivatives [215, 217], it is clear that the basicity is markedly increased by introduction of alkyl groups by stabilizing the corresponding cations. The presence of *tert*-butyl groups even allows isolation of stable 2*H*-pyrrolium salts, for example **145**, which was obtained in quantitative yield as a crystalline solid from pyrrole **146** (Scheme 4.45) [218].

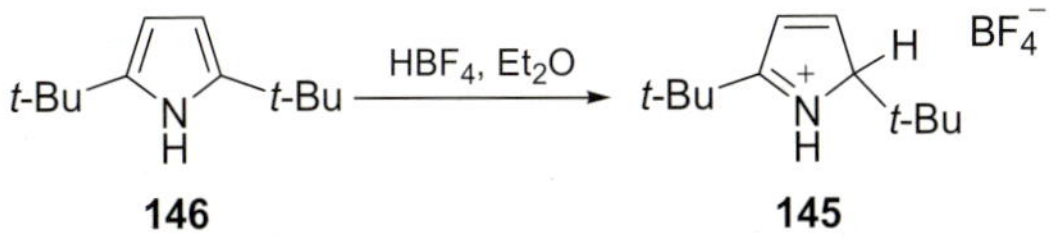

Scheme 4.45

Although the 3*H*-pyrrolium cation **144** is the less stabilized, and thus also the less abundant of the two C-protonated species, it is nevertheless very important, as it plays a major role in the acid-catalyzed oligomerization and polymerization of pyrrole because of its higher reactivity. The electrophilic cation **144** undergoes attack by pyrrole (**1**), thereby forming the unstable dimeric enamine **147** (Scheme 4.46). Protonation thereof generates a new electrophilic intermediate **148**, which reacts with an additional equivalent of pyrrole (**1**), rendering the isolable trimer **149** (ratio trans : cis of 2 : 1) [219], which may subsequently participate in further reactions, eventually giving polymeric products [220], for example fully aromatic polypyrrole [221], unless careful control of the conditions is maintained, and, therefore, the reaction allowing isolation of **149** is performed at 0 °C in 20% aqueous HCl [219]. The propensity of non-deactivated pyrroles to undergo polymerization makes such substrates unsuitable for reactions that involve strongly acidic conditions.

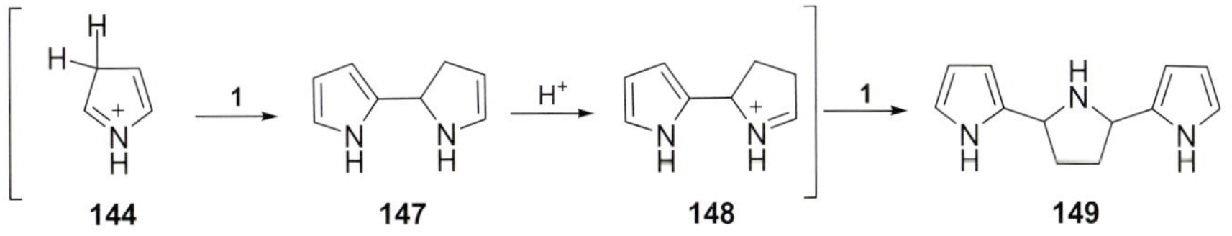

Scheme 4.46

4.5.1.3 Halogenation

Many electron rich halogenated pyrroles are rather unstable compounds that decompose readily upon exposure to air. Hence, in the early days, the availability of such pyrrole derivatives was severely limited, although syntheses of several relatively stable iodinated pyrroles, for example 2,3,4,5-tetraiodopyrrole [222], as well as some other bromo- or iodopyrroles featuring additional electron-withdrawing substituents were described [223, 224]. An early study on the bromination of methyl pyrrole-2-carboxylate and pyrrole-2-carboxaldehyde under various conditions clearly demonstrated that formation of mixtures containing several halogenation products is a common course of many of these reactions, thus illustrating yet another complicating factor [225]. The labile 1-chloropyrrole, generated in 65–72% yield by chlorination of pyrrole with NaOCl, was shown to rearrange readily to give mixtures containing several species, such as 2-choro- and 3-chloropyrrole, when subjected to acidic conditions or heated in methanol [226]. Previous findings indicated that both 2-chloro- and 2-bromopyrrole undergo rapid degradation, 2-chloropyrrole being somewhat more stable. Introduction of 1-alkyl substituents increases the stability of these compounds, whereas the presence of C-alkyl groups appears to lead to even faster decomposition. Since hydrogen chloride, which is formed during degradation of 2-chloropyrrole, also catalyzes the decomposition, the process is probably autocatalytic. Stabilization can be effected to some extent by storage in the presence of a suitable base [227]. Consequently, practical halogenations of pyrroles are generally performed under mild conditions that avoid generation of acidic by-products. Bromination of some 1-alkylpyrroles with one or two equivalents of *N*-bromosuccinimide (NBS) in THF provides the corresponding 2-bromo- or 2,5-dibromopyrrole derivatives, respectively. Chlorination with *N*-chlorosuccinimide (NCS) in THF gives similar results, albeit with lower selectivity [228]. Interestingly, treatment of 1-methylpyrrole with NBS at −78 °C to −10 °C in THF employing PBr_3 as the catalyst selectively gives 3-bromo-1-methylpyrrole, whereas the use of one equivalent of triethylamine as the additive allows regiospecific synthesis of 2-bromo-1-methylpyrrole [229]. Notably, however, treatment of 1-alkylpyrroles with NCS in chloroform with or without $NaHCO_3$ leads instead to introduction of the *N*-succinimide moiety at C2 [230]. Application of *N*-halosuccinimides in DMF provides convenient access to a series of 2,5-disubstituted or 2,4,5-trisubstituted 3-chloro-, 3-bromo- and 3-iodopyrrole derivatives [231]. The selective rearrangement of 1-benzyl-2,5-dibromopyrrole (**150**) with *p*-toluenesulfonic acid into the product **151** (Scheme 4.47) is also worth mentioning in this context [232]. In general, 3,4-dihalopyrroles, even such lacking additional electron-withdrawing groups, are stable compounds, which have been studied in detail [233].

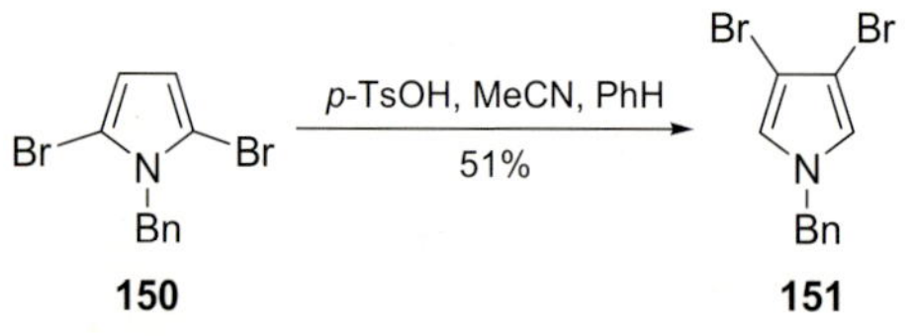

Br Br N Bn **150** —(*p*-TsOH, MeCN, PhH; 51%)→ Br Br N Bn **151**

Scheme 4.47

As implied above, the introduction of electron-withdrawing groups increases the stability of halogenated pyrroles. Halogenation of a series of substituted 3-acetylpyrroles with $CuBr_2$ in acetonitrile gave the corresponding 3-acetyl-4-bromopyrrole derivatives in moderate to high yields [234]. Similar bromination of rather densely substituted pyrrole-3-carboxylates **152** furnished the 4-bromo derivatives **153**, which could in turn be converted into the 3,4-dibromopyrroles **154** with concomitant decarboxylation (Scheme 4.48) [235]. 1-Methyl-2-(trichloroacetyl)pyrrole undergoes regioselective bromination upon treatment with NBS in chloroform at $-10\,°C$, rendering 4-bromo-1-methyl-2-(trichloroacetyl)pyrrole in 79% yield [236].

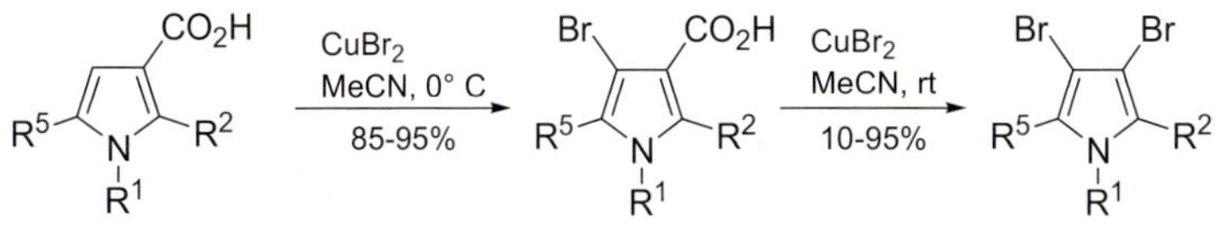

CO_2H R^5 N R^2 R^1 **152** — $CuBr_2$, MeCN, 0° C, 85-95% → Br CO_2H R^5 N R^2 R^1 **153** — $CuBr_2$, MeCN, rt, 10-95% → Br Br R^5 N R^2 R^1 **154**

Scheme 4.48

Stable and synthetically useful simple halogenated pyrroles have become readily available by introduction of the Boc protecting group (Scheme 4.49, cf. Section 4.5.1.6). Efficient preparation of 2-bromo-1-(*tert*-butoxycarbonyl)pyrrole **155**, as well as the related 2,5-dibromo derivative, has been accomplished by bromination of pyrrole with 1,3-dibromo-5,5-dimethylhydantoin **156**, followed by installation of the Boc-group on the intermediate 2-bromopyrrole **157** (Scheme 4.49) [237, 238]. On the other hand, 2,5-dibromo-1-(*tert*-butoxycarbonyl)pyrrole has also been obtained in 61% yield by exposure of **158** to NBS [239]. An alternative route to **155** encompasses conversion of 1-(*tert*-butoxycarbonyl)pyrrole (**158**) into the 2-stannyl derivative **159**, which thereafter undergoes a stannyl–bromo exchange reaction. 2-Bromo-1-(phenylsulfonyl)pyrrole may also be prepared in a similar manner [240]. The closely related 2-bromo-1-

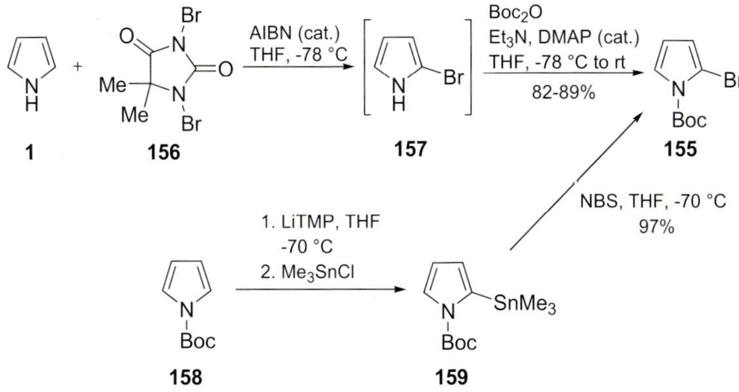

N H **1** + Br N O O N Me Me Br **156** — AIBN (cat.), THF, -78 °C → [N H Br] **157** — Boc_2O, Et_3N, DMAP (cat.), THF, -78 °C to rt, 82-89% → N Boc Br **155**

N Boc **158** — 1. LiTMP, THF, -70 °C; 2. Me_3SnCl → N Boc $SnMe_3$ **159** — NBS, THF, -70 °C, 97% → **155**

Scheme 4.49

(*p*-toluenesulfonyl)pyrrole is readily available in two steps from pyrrole via bromination with 1,3-dibromo-5,5-dimethylhydantoin, followed by N-tosylation [241].

The presence of a bulky, removable N-substituent on the pyrrole nucleus enables introduction of halogen atoms at C3 and C4 even in the absence of stabilizing/blocking groups at C2 and/or C5. Initial studies on the bromination of the sterically hindered 1-(triisopropylsilyl)pyrrole (**160**) with two equivalents of NBS gave 3,4-dibromo-1-(triisopropylsilyl)pyrrole (**161**), along with the 2,3-dibromo derivative **162** in a 1 : 1 ratio. The use of only one equivalent of NBS afforded predominantly the 3-brominated derivative **163**, together with minor amounts of the 2-bromo isomer (ratio 97 : 3) [209, 242]. Improved conditions, featuring a portion-wise addition of NBS at −78 °C, allow selective preparation of **163** in 78% yield [243]. It was also later emphasized that careful temperature control is essential to suppress the formation of side products [244]. Consequently, useful and high-yielding procedures for the synthesis of **163** are available [244, 245]. Investigation of the synthetic potential of **160** also revealed that attempted chlorination with NCS gives complex mixtures of products and unchanged starting material, whereas iodination with elemental iodine in the presence of mercuric acetate provides **164** in 61% yield [244]. Blocking of the normal α-substitution pathway by a bulky N-substituent has also been employed in the monobromination of 1-tritylpyrrole with pyridinium bromide perbromide, which proceeds cleanly to afford 3-bromo-1-tritylpyrrole (**165**) in 75% yield [210]. Treatment of the pyrrole **166** with NBS gives the 3-bromo derivative **167** via *ipso*-bromination. Similarly, *ipso*-iodination of **166** to provide **168** can be achieved employing iodine in the presence of silver trifluoroacetate [246, 247]. In addition, 3,4-dibromo-1-(*p*-toluenesulfonyl)pyrrole (**169**) has been prepared in 43% yield by treatment of 1-(*p*-toluenesulfonyl)pyrrole with bromine in refluxing acetic acid [248].

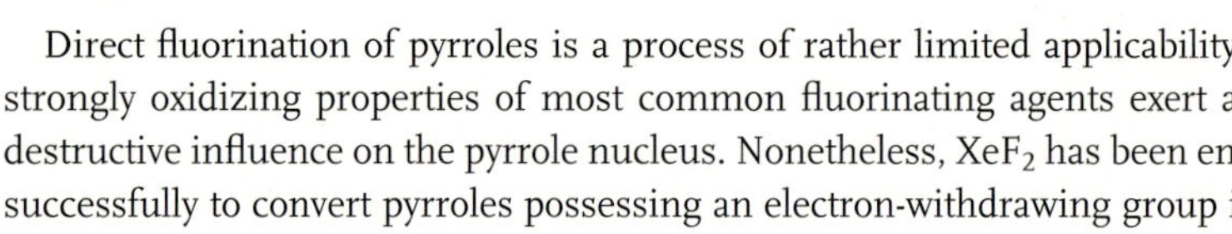

160 X = H, **163** X = Br, **164** X = I; **161** R = TIPS, **169** R = Ts; **162**; **165**; **166**; **167** X = Br, **168** X = I

Direct fluorination of pyrroles is a process of rather limited applicability, as the strongly oxidizing properties of most common fluorinating agents exert a highly destructive influence on the pyrrole nucleus. Nonetheless, XeF_2 has been employed successfully to convert pyrroles possessing an electron-withdrawing group into the corresponding α-fluoro derivatives in low to moderate yields [249, 250]. Other miscellaneous approaches include fluoro-decarboxylation of pyrrole-2-carboxylates with 1-chloromethyl-4-fluoro-1,4-diazoniabicyclo[2.2.2]octane bis(tetrafluoroborate) in modest yields [251], or photolysis of a pyrrole-β-diazonium tetrafluoroborate [252].

Halogenation on an α-methyl group of certain polysubstituted pyrroles with sulfuryl chloride in ether [253] or bromine in acetic acid [254] gives the corresponding α-(chloromethyl)- or α-(bromomethyl)pyrrole derivatives, respectively. Such halogenation processes have been suggested to occur via an initial electrophilic attack of the pyrrole ring, followed by a rearrangement rendering the final α-(halomethyl)pyrrole products [255, 256].

4.5.1.4 **Nitration**

Because of the sensitivity of simple pyrroles towards strongly acidic media, nitration must be conducted under relatively mild conditions, or involve deactivated substrates. Nitration of pyrrole itself with nitric acid in acetic anhydride normally gives mixtures of 2-nitropyrrole and 3-nitropyrrole in a ratio of approximately 4 : 1 over a wide temperature range. The reactivities at C2 and C3 are 1.3×10^5 and 3×10^4 times higher than for benzene, respectively [257]. A similar reactivity pattern was observed in the case of 1-alkyl- and 1-aryl-pyrroles, with C2/C3 nitration ratios ranging between 3.15 : 1 for 1-methylpyrrole and 0.25 : 1 for 1-(*tert*-butyl)pyrrole, depending on the electronic and steric properties of the N-substituent. The high reactivity of pyrrole is further manifested in the easy formation of di-, tri-, and even tetranitropyrrole derivatives [258]. Introduction of a deactivating acyl group either on the nitrogen atom or at C2 also proved to be insufficient for selective nitration [259]. Selective and efficient (80% yield) conversion of pyrrole into 2-nitropyrrole has more recently been accomplished using $(NH_4)_2Ce(NO_3)_5{\cdot}4H_2O$ in acetic anhydride [260].

Other practical procedures are based on pyrroles having strategically located, strongly deactivating or bulky N-substituents. For example, 2-(trichloroacetyl)pyrrole (Section 4.5.1.6) may be nitrated with 90% HNO_3 at −50 °C to provide the corresponding 4-nitro derivative as the major product in 77% yield [212]. In addition, β-acylated pyrroles are cleanly converted into the corresponding 4-acyl-2-nitropyrroles upon treatment with nitric acid in acetic anhydride at −15 °C [259]. Likewise, an extensive series of 1,5-dialkyl-4-nitropyrrole-2-carboxylates has been prepared involving nitration of suitable precursors at C4 [261]. Nitration of 1-(phenylsulfonyl)pyrrole (**170**) with nitric acid in acetic anhydride provides a selective route to 3-nitropyrrole **171** [211, 262]. The parent 3-nitropyrrole (**172**) can thereafter be obtained after removal of the phenylsulfonyl group with base [262]. In addition, the readily available 1-(triisopropylsilyl)pyrrole (**160**) can serve as an excellent precursor to 3-nitropyrrole (**172**), as demonstrated by nitration of **160** with cupric nitrate trihydrate in acetic anhydride to give **173** (77% yield), which may subsequently be efficiently desilylated [244].

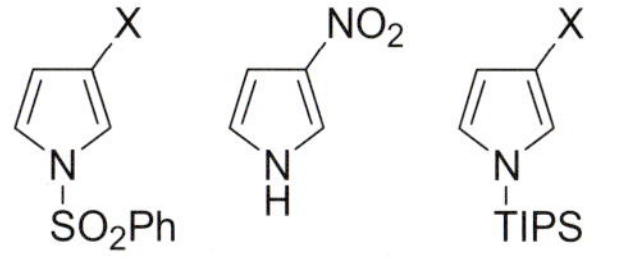

X NO2 X
N N N
SO2Ph H TIPS

170 X = H **172** **160** X = H
171 X = NO_2 **173** X = NO_2

4.5.1.5 **Reactions with Sulfur-Containing Electrophiles**

Sulfonation of pyrrole with the sulfur trioxide–pyridine complex has long been recognized to give pyridinium pyrrole-2-sulfonate (**174**) [263, 264]. A more recent reinvestigation of this reaction provided, however, strong evidence that substitution seems instead to occur at C3, leading to the isomeric pyridinium pyrrole-3-sulfonate (**175**), the formation of which was verified by detailed NMR studies of the corresponding sodium salt **176**, and preparation of several pyrrole-3-sulfonamides [265]. This intriguing preference for selective C3 substitution has yet to be rationalized in detail. When the pyrrole nucleus is already deactivated, sulfonation using acidic reagents is feasible, as illustrated by the transformation of the pyrrole **177** into the sulfonyl chloride **178** in 81% yield using chlorosulfonic acid [266]. Likewise, sulfonation of 1-(arylsulfonyl)pyrroles with chlorosulfonic acid in acetonitrile gives practical access to the corresponding pyrrole-3-sulfonyl chlorides in moderate yields [267]. Similar regioselectivity was observed upon treatment of 1-(phenylsulfonyl)pyrrole with dimethylsulfamoyl chloride in the presence of bismuth (III) trifluoromethanesulfonate, providing *N*,*N*-dimethyl-1-(phenylsulfonyl)pyrrole-3-sulfonamide in 49% yield. As noted in connection with studies of the latter reaction, one has to consider the possibility that the 3-substituted products may arise by a rearrangement of the conceivable 2-substituted products (see below), and further studies are required to gain deeper insight into the mechanistic pathways of such transformations [268].

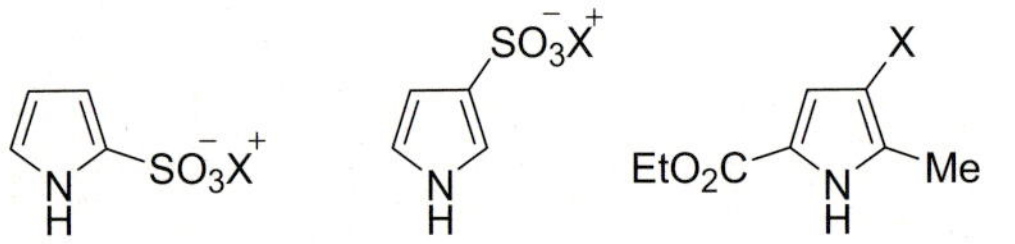

174 $X^+ = C_5H_6N^+$ **175** $X^+ = C_5H_6N^+$ **176** $X^+ = Na^+$ **177** X = H **178** $X = SO_2Cl$

The reaction of pyrrole and 1-methylpyrrole with alkyl- or aryl-sulfinyl chlorides at 0 °C offers a route to the 2-sulfinylpyrroles **179** (R^1=H or Me) in moderate to good yields, provided that the products are protected from the influence of the liberated hydrogen chloride. Without that precaution, rearrangement of the kinetic products **179** affords the corresponding C3 substituted isomers **180** as the major products. This outcome could be ascribed to an initial protonation of **179** to generate the intermediate **181**, which may thereafter undergo a sigmatropic rearrangement, followed by loss of a proton to give **180**. Crossover experiments also indicated the possibility of an intermolecular process involving dissociation of the complex **181**. Clean conversion of **179** into **180** can be effected by treatment with *p*-toluenesulfonic acid [269] or TFA [270]. Introduction of the phenylsulfinyl group at C2 may also be accomplished using *N*-(phenylsulfinyl)succinimide [269].

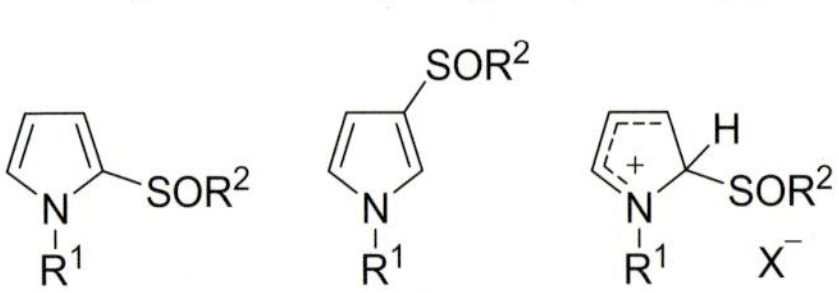

179 **180** **181**

Several different approaches are available for the synthesis of (alkylthio)- or (arylthio)pyrroles. Treatment of pyrrole with the *N*-chlorosuccinimide–dimethyl sulfide adduct **182** (formed *in situ*) afforded the salt **183**, which could then subsequently be converted, by thermal decomposition, into 2-(methylthio)pyrrole **184** in 58% overall yield [271]. Reaction of 1-alkylpyrroles with 1-(methylthio) morpholine in the presence of acid, or, even better, excess pyridine, gives access to 2,3,4,5-tetra(methylthio)pyrroles [272]. Exposure of 2-thiocyanatopyrrole (see below) to phenyl- [273] or alkylmagnesium bromides [274] provides useful routes to 2-(phenylthio)pyrrole or the corresponding 2-(alkylthio)pyrroles, respectively. Notably, the alkylthio unit serves as a protecting group for the α-position of pyrrole in a new approach to dipyrromethanes [274]. A synthesis of densely substituted 3,3′-dipyrrolyl sulfides possessing electron-withdrawing groups by treatment of suitable pyrroles having a vacant β-position with sulfur dichloride has also been reported [275].

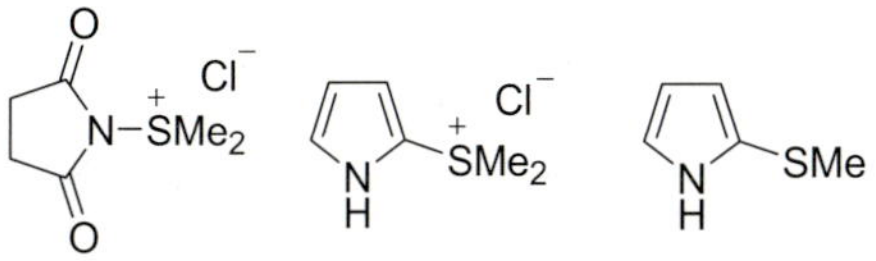

182 **183** **184**

In analogy to the behavior of the sulfoxides **179** (see above), the 2-(alkylthio) pyrroles **185** undergo rearrangement to furnish the isomeric C3 substituted derivatives **186** exclusively in good yields upon heating in a 1 : 1 mixture of TFA and 1,2-dichloroethane (Scheme 4.50) [276]. This behavior contrasts with the propensity of unprotected or N-methylated 2-(alkylthio)pyrroles and 3-(alkylthio)pyrroles to undergo acid induced equilibration under mild conditions [277]. The events leading to the conversion of **185** into **186** have been suggested to involve an initial protonation to provide the intermediate **187**, subsequent rearrangement via the episulfonium salt **188** to the C3-substituted intermediate **189**, and a final deprotonation. This mechanistic rationale was supported by crossover experiments, as no crossover products could be detected [276].

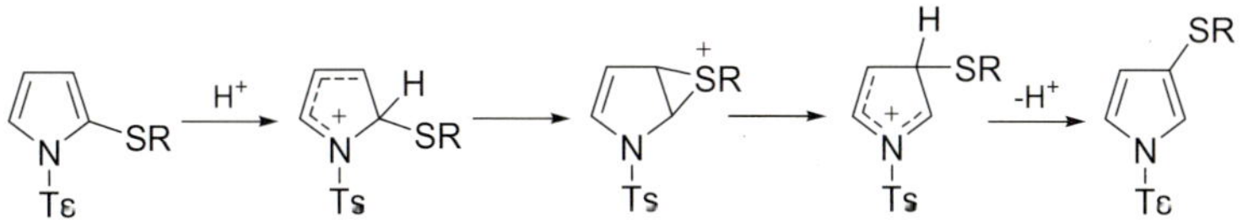

185 **187** **188** **189** **186**

Scheme 4.50

Thiocyanation of pyrroles with cupric thiocyanate [278, 279], thiocyanogen chloride [276] or, more conveniently, ammonium thiocyanate in the presence of iodine in methanol [280] or CAN [281], provides access to 2-thiocyanatopyrroles.

4.5.1.6 **Acylation**

Several useful procedures for N-acylation of pyrroles are available, for example by acyl transfer from 1-acetylimidazole, which offers an efficient route to 1-acetylpyrrole [282]. Alternative attractive methods for N-acylation rely on the use of acetic anhydride [283], or acyl chlorides in the presence of and triethylamine and DMAP as the catalyst [284]. Exposure of pyrroles to di(*tert*-butyl)dicarbonate in the presence of DMAP in acetonitrile solution gives access to many useful 1-(*tert*-butoxycarbonyl) pyrroles in high yields [285].

Formylation of pyrrole is most conveniently accomplished using the Vilsmeier–Haack reaction. Both pyrrole (**1**) [286, 287], 1-methylpyrrole (**190**) [287], as well as several C-methyl derivatives thereof [288], provide excellent yields of the corresponding pyrrole-2-carboxaldehydes **191** and **192** via the intermediates **193** and **194**, respectively, upon treatment with the reagent **195** [289], which is readily generated *in situ* from $POCl_3$ and DMF (Scheme 4.51). The presence of N-alkyl groups larger than methyl leads to mixtures of products formylated at C2 and C3, while sterically demanding N-substituents favor substitution at C3 over C2. Thus, for instance, for 1-*tert*-butylpyrrole the ratio of products is 14 : 1 in favor of 1-(*tert*-butyl)pyrrole-3-carboxaldehyde. The reactivity differences in the 1-arylpyrrole series are less pronounced (with prevalence for the C2 products), and are influenced by both steric and electronic effects. Formylation of 1-acetyl-, 1-benzoylpyrrole and ethyl pyrrole-2-carboxylate leads exclusively to substitution at C2 [290]. Interestingly, Friedel–Crafts acylation of intermediates such as **193** and **194**, followed by hydrolysis, provides a one-pot procedure to 4-acylpyrrole-2-carboxaldehydes due to the strong "meta"-directing properties of the iminium substituent [291]. Salts related to **193** have also been exploited in reactions with bromine or SO_2Cl_2, eventually yielding various pyrrole-2-carboxaldehyde derivatives halogenated at C4, or C4 and C5 [292]. Vilsmeier–Haack-type reagents generated from pyrophosphoryl chloride lead to increased preference for substitution at C3 due to more pronounced steric interaction with N-substituents [293]. Likewise, treatment of 1-(triisopropylsilyl)pyrrole with iminium salts gives substitution at C3 [294]. When extended to lactams, the Vilsmeier–Haack reaction allows preparation of imines such as **196** [295], whereas the use of *N,N*-dimethylamides, for instance DMA, provides an effective route to 2-acylpyrroles [296]. In addition, exposure of pyrroles to Vilsmeier–Haack reagents generated from aroylamides gives good yields of 2-aroylpyrroles [297]. An approach to 3,4-dialkylpyrrole-2,5-dicarboxaldehydes relying on treatment of 3,4-dialkylpyrrrole-2-carboxylic acids with triethyl orthoformate in TFA has also been described [298].

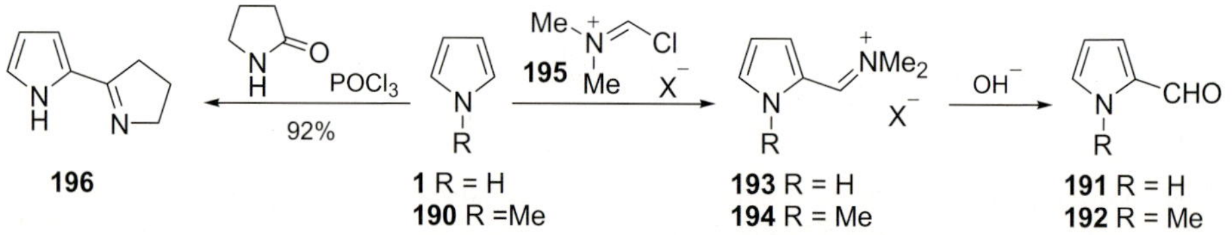

Scheme 4.51

Pyrroles are readily acylated at C2, as exemplified by the selective conversion of pyrrole itself into 2-(trichloroacetyl)pyrrole **197** (Scheme 4.52). This material provides easy access to pyrrole-2-carboxylic acid (**198**) upon alkaline hydrolysis [299]. Likewise, treatment of pyrrole with TFAA in the presence of *N*,*N*-dimethylaniline gave the corresponding trifluoroacetyl derivative, which could also be efficiently converted into **198** [300]. Alcoholysis of 2-(trichloroacetyl)pyrroles enables preparation of the corresponding esters, for example, methyl pyrrole-2-carboxylate (**199**). This reaction is, however, synthetically useful only with primary alcohols [301]. Acetylation with acetic anhydride is not practical, as it has been reported to give 2-acetylpyrrole in 39% yield, and minor amounts (8%) of the C3 acetylated isomer [302]. In contrast, acylation of pyrrole with cyanoacetic acid in acetic anhydride allows smooth introduction of a cyanoacetyl functionality, giving **200**. Alkylpyrroles are also cyanoacetylated at C2, unless both C2 and C5 are substituted, leading instead to 3-(cyanoacetyl)pyrrole derivatives in good yields [303]. Both pyrrole itself, as well as 1-methylpyrrole, are also effectively acylated at C2 using *N*-acylbenzotriazoles in the presence of $TiCl_4$ [304].

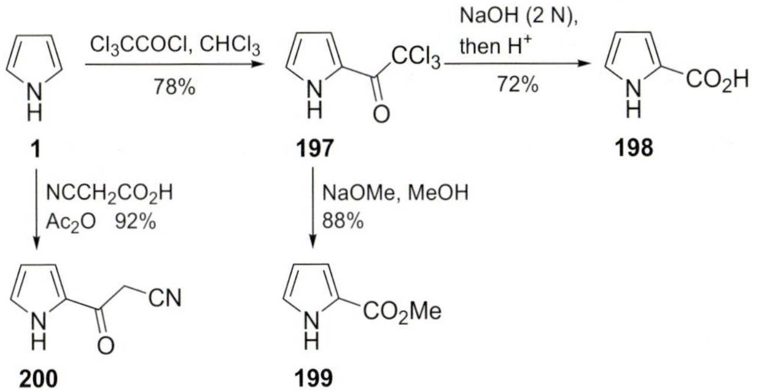

Scheme 4.52

Studies on the acid mediated rearrangements of acylpyrroles under anhydrous conditions revealed that 2-acylpyrroles possessing an N-methyl group (**201**) undergo conversion into the corresponding 3-acylpyrroles (**202**). In contrast, the parent 2-acylpyrroles (**203**) give equilibrium mixtures containing **203** and **204** under similar conditions (Scheme 4.53). Experiments aimed at explaining this discrepancy failed to

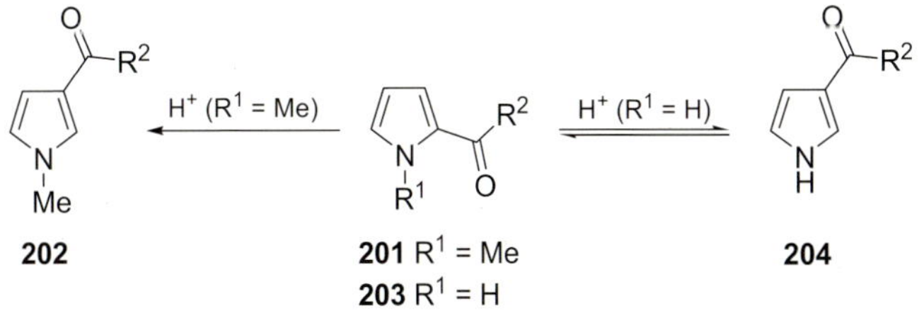

Scheme 4.53

give conclusive results, although it seems likely that a 1,2-acyl shift of a C2-protonated intermediate is involved [305].

Pyrroles acylated at C2 are valuable precursors for elaboration to more complex derivatives. As a result of the deactivating and "meta" directing properties of the trichloroacetyl group, 2-(trichloroacetyl)pyrrole **197** undergoes electrophilic substitution at C4 with various reagents, producing the substituted methyl pyrrole-2-carboxylates **205** (E=Cl, Br, I, COMe) in good overall yields after subsequent methanolysis [212]. The scope and limitations of Friedel–Crafts acylation reactions of ethyl pyrrole-2-carboxylate (**206**) using acyl chlorides have been evaluated in detail, leading to the conclusion that selective C4 acylation or aroylation is best achieved employing $AlCl_3$ as the catalyst, as the use of weaker Lewis acids, such as zinc chloride or boron trifluoride etherate, affords mixtures of C4 and C5 substituted products. This approach allows efficient syntheses of potentially useful pyrroles, for instance **207** [306]. Chlorination of **199** with two equivalents of SO_2Cl_2 in chloroform gives the corresponding 4,5-dichloro derivative, which may be subsequently iodinated at C3 using iodine in the presence of silver trifluoroacetate to provide methyl 4,5-dichloro-3-iodopyrrole-2-carboxylate, a useful partner for regioselective Suzuki reactions (Section 4.5.11) [307]. Pyrrole-2-carbonitrile, which is readily available by treatment of pyrrole with chlorosulfonyl isocyanate followed by warming in DMF [308], displays similar reactivity, affording 4-substituted pyrrole-2-carbonitriles [308, 309]. Likewise, ethyl pyrrole-2-thiolcarboxylate (**208**) (readily prepared by the action of ethyl chlorothiolformate on pyrrolyl magnesium halides) also takes part in electrophilic substitution reactions, providing good yields of 2,4-disubstituted pyrroles, for example **209**. These products can thereafter be converted into the corresponding 3-substituted pyrroles by treatment with Raney nickel [310] or to 3-alkylpyrroles by Wolff–Kishner reduction with concomitant decarboxylation [311]. A detailed study describing the applicability of β-acylpyrroles as useful substrates for Friedel–Crafts reactions leading to 2,4-diacylpyrroles is also available [312].

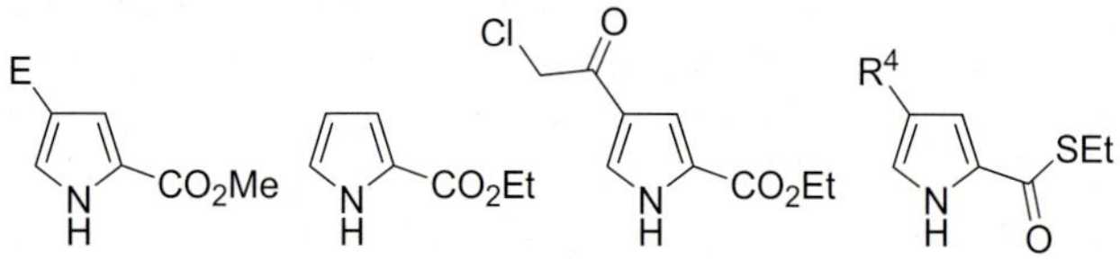

205 **206** **207** **208** R^4 = H
209 R^4 = COMe

In an interesting application of Friedel–Crafts chemistry, treatment of **206** with succinic anhydride in the presence of $AlCl_3$ gave the 2,4-disubstituted pyrrole **210**. Reduction of the ketone functionality of **210** afforded **211**, which was eventually subjected to an acid induced intramolecular acylation, leading to the fused system **212** in 60% overall yield (Scheme 4.54) [313].

The synthetic potential in acylation reactions of pyrrole derivatives possessing removable, strongly electron withdrawing, or bulky N-substituents, is nicely demonstrated by the selective and efficient conversion of 1-(phenylsulfonyl)pyrrole **170** (Section 4.5.1.4) into the corresponding C3 acylated products **213** (Scheme 4.55)

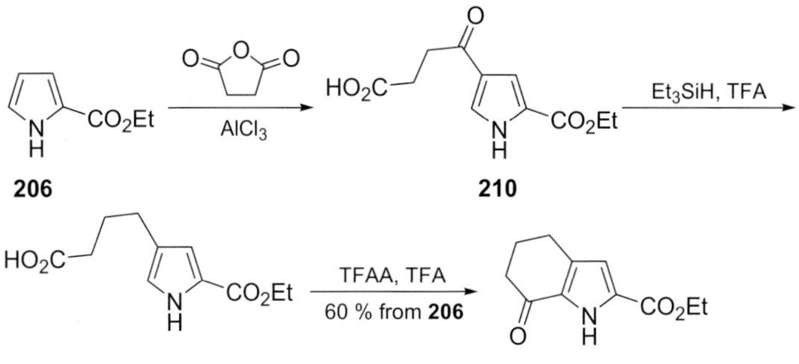

Scheme 4.54

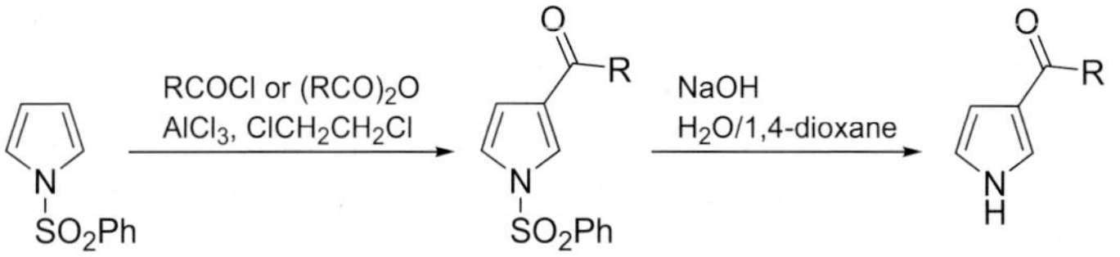

Scheme 4.55

upon treatment with acyl chlorides or carboxylic acid anhydrides in the presence of $AlCl_3$. An ensuing base-induced cleavage of the N-phenylsulfonyl group provides excellent overall yields of pyrroles **214** (R=alkyl or aryl) [211, 262, 314, 315]. However, there are cases when mixtures of C2 and C3 acylated products have been encountered in connection with related reactions involving relatively electron rich aroyl chlorides [316, 317]. Even though similar selectivity problems were noted during the reaction of **170** with 1-naphthoyl chloride in the presence of $AlCl_3$, selective C3 substitution was achieved by using nitromethane as the co-solvent [318]. Acylation at C3 may also be conveniently performed employing 1-(triisopropylsilyl) pyrrole as the substrate [244]. 1-(Triisopropylsilyl)pyrrole may be selectively acylated at C3 upon reaction with 1-acylbenzotriazoles assisted by $TiCl_4$ in refluxing CH_2Cl_2 [304].

In contrast, when $BF_3{\cdot}OEt_2$ instead of $AlCl_3$ is used as the catalyst during acylation of **170**, a dramatic change in regioselectivity is induced, as exclusive formation of the 2-substituted products **215** takes place (Scheme 4.56), providing an alternative route to the parent 2-acylpyrroles after removal of the protecting group under basic conditions. Although attempts to rationalize these differences in regioselectivity in terms of kinetic, steric or electronic factors have been made, no clear conclusions could be drawn [315]. A new contribution to this field allows selective α-acylation of 1-(*p*-toluenesulfonyl)pyrroles with carboxylic acids and TFAA, presumably involving mixed anhydrides as the acylating agents [319]. Friedel–Crafts acylation of 3-alkyl-1-(phenylsulfonyl)pyrroles with acetic anhydride also proceeds with pronounced

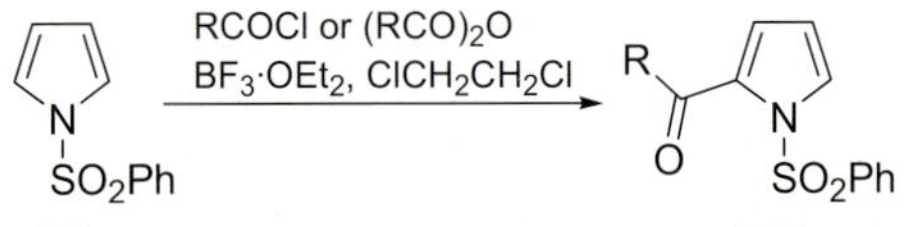

170
215

Scheme 4.56

selectivity, provided that the alkyl group at C3 is not too bulky, giving 2-substitution with $AlCl_3$, and 5-substitution with $BF_3{\cdot}OEt_2$ [320].

4.5.1.7 **Reactions with Aldehydes, Ketones, Nitriles and Iminium Ions**

The reactions of pyrroles with aldehydes and ketones have been studied extensively, as some of these transformations constitute powerful tools for the construction of the important porphyrin skeleton. Treatment of pyrrole with aldehydes in the presence of acid initially generates the intermediate carbinols **216** (Scheme 4.57), which readily lose water to provide the highly electrophilic 2-alkylidenepyrrolium (azafulvenium) ions **217** (Section 4.5.10).

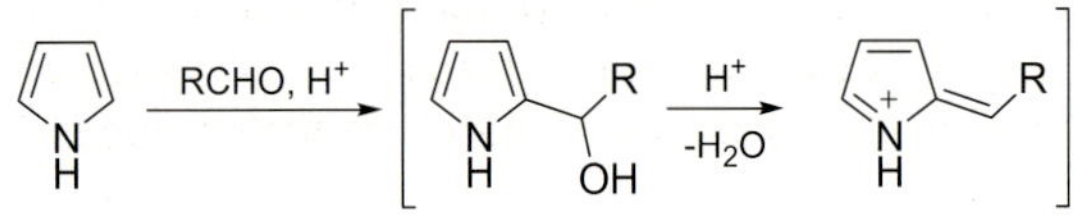

1
216
217

Scheme 4.57

Such condensation reactions eventually lead to the formation of porphyrins (Scheme 4.58) [321, 322], along with other oligomeric or polymeric products [323], for instance so-called "N-confused porphyrins," which are porphyrin isomers featuring a pyrrole unit linked through its α and β′ positions [324–327]. A widely used older procedure for the preparation of *meso*-tetraarylporphyrins **218** involves heating of pyrrole with an appropriate benzaldehyde in propionic acid (Scheme 4.58) [328]. This process involves the intermediacy of the porphyrinogens

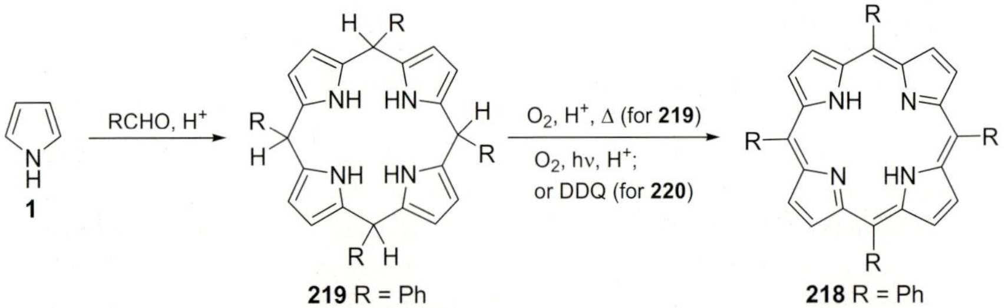

1
219 R = Ph
220 R = alkyl
218 R = Ph
221 R = alkyl

Scheme 4.58

219, which subsequently spontaneously undergo a rate limiting air oxidation step leading to **218** [329]. More recently it has been demonstrated that pyrrole and benzaldehydes react reversibly at ambient temperature in the presence of an acid catalyst to provide porphyrinogens, which may subsequently be irreversibly converted into the corresponding porphyrins by addition of an oxidant. Consequently, the method of choice relies on initial reaction of pyrrole with an appropriate benzaldehyde in anhydrous CH_2Cl_2 in the presence of $BF_3{\cdot}OEt_2$ or TFA at room temperature, followed by treatment of the resulting mixture with *p*-chloranil at reflux [330]. Reactions involving aliphatic aldehydes generate relatively stable *meso*-tetraalkylporphyrinogens (**220**), the conversion of which into the corresponding *meso*-tetraalkylporphyrins (**221**) requires an additional forced oxidation step [331, 332]. Likewise, condensation of pyrrole and acetone in the presence of acid gives a good yield of a cyclic tetramer [333, 334]. In contrast, the reactions between pyrrole and ortho esters under acidic conditions give rise to tris(pyrrol-2-yl)alkanes [335, 336]. Moreover, condensation reactions between suitable pyrrole fragments and pyrrole-2-carboxaldehyde derivatives constitute a common strategy to numerous dipyrrins and dipyrrinones [337].

Under carefully controlled conditions, the reaction of pyrrole with formaldehyde may give bis(pyrrol-2-yl)methane (dipyrromethane) (**222**), which is a useful precursor for the synthesis of 5,15-disubstituted porphyrins, for example **223** (Scheme 4.59) [335]. Exposure of aldehydes to excess pyrrole, in the presence of

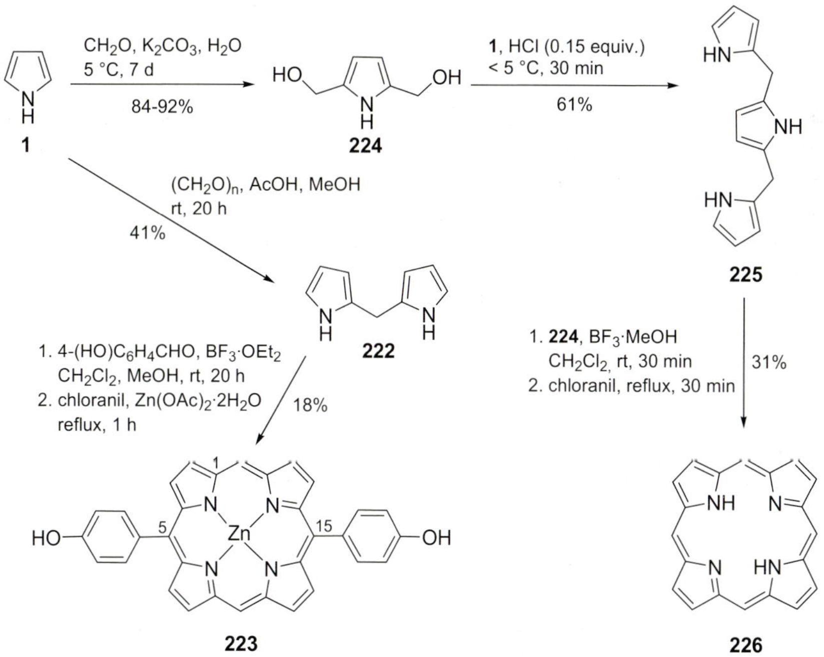

Scheme 4.59

TFA or $BF_3{\cdot}OEt_2$ as catalysts, constitutes an effective protocol (yields up to 86%) for the preparation of *meso*-substituted dipyrromethanes [338]. Alternatively, such systems are also effectively generated in 5 min from pyrrole and aldehydes (ratio 25 : 1) in the presence of TFA (0.1 equiv.) [339], or in water solution with catalytic amounts of HCl [340]. Treatment of pyrrole with formalin under basic conditions permits isolation of the dialcohol **224**, which can subsequently react with two equivalents of pyrrole to provide trimer **225**, which is an excellent precursor to the parent macrocycle **226** [341].

In a related process, the 2-acylpyrroles **227** (R=CF_3, CO_2Et) have been exposed to paraformaldehyde in the presence of HCl under anhydrous conditions, leading to efficient and selective chloromethylation to render the disubstituted pyrroles **228** (Scheme 4.60) [342].

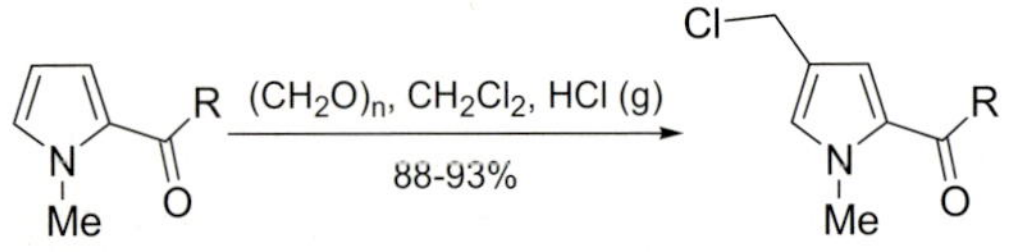

R $(CH_2O)_n$, CH_2Cl_2, HCl (g)
88-93%
N Me O **227**
Cl R N Me O **228**

Scheme 4.60

Application of a nitrile as the electrophilic reagent has been utilized in a concise one-pot synthesis of the monarch butterfly pheromone danaidone (**229**), wherein N-alkylation of 3-methylpyrrole (**230**) with acrylonitrile in the presence of DBU gave the intermediate **231**, which subsequently underwent intramolecular cyclization, followed by hydrolysis of the intermediate imine (Scheme 4.61) [343].

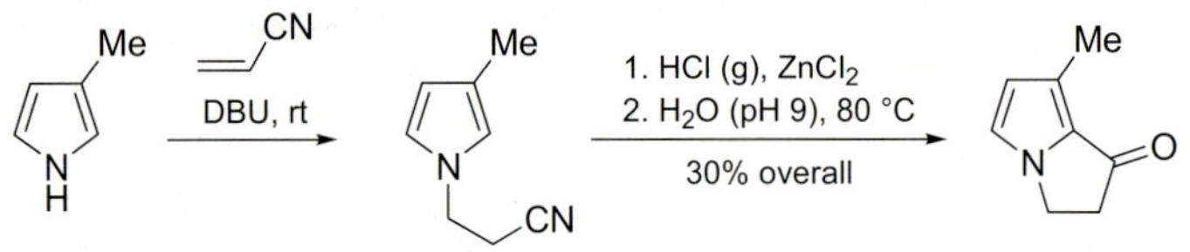

Me N H **230**
CN DBU, rt
Me N CN **231**
1. HCl (g), $ZnCl_2$
2. H_2O (pH 9), 80 °C
30% overall
Me N O **229**

Scheme 4.61

Pyrroles also react readily with iminium ions, generated *in situ* from formaldehyde and dialkylamines in acetic acid, to provide 2-(dialkylaminomethyl)pyrroles, which are useful synthetic intermediates (Section 4.5.10), or with the Vilsmeier–Haack reagent, affording pyrrole-2-carboxaldehydes (Section 4.5.1.6). Likewise, pyrrole (**1**) may be converted in high yield into 2-(dimethylaminomethyl)pyrrole **232** via the Mannich reaction (Scheme 4.62) [344]. It has also been found that pyrrole adds upon heating (neat) to 1-pyrroline **233**, leading to 2-(pyrrolidin-2-yl)pyrrole (**234**) [345].

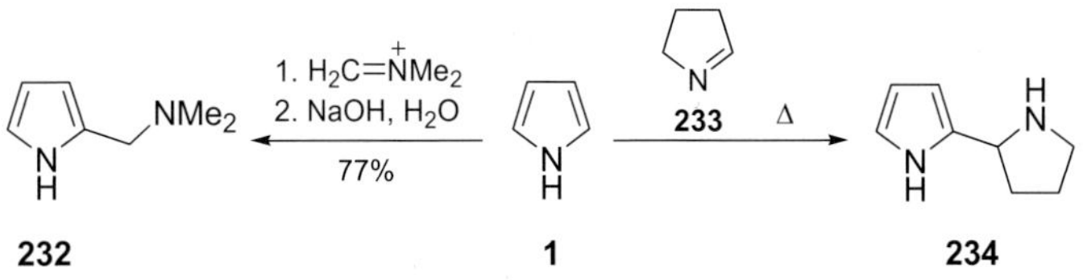

Scheme 4.62

4.5.1.8 **Conjugate Addition to α,β-Unsaturated Carbonyl Compounds**

Pyrroles are useful nucleophiles in conjugate addition reactions (Scheme 4.63). For instance, pyrrole has been alkenylated efficiently by treatment with (*E*)-4-(phenylsulfinyl)-3-buten-2-one to furnish the alkenyl derivative **235**. This reaction probably proceeds via a sequence featuring a Michael addition, followed by elimination of phenylsulfenic acid [346]. Addition of pyrrole to 1-acyl-2-bromoacetylenes in the presence of silica gives rise to modest yields of di(pyrrol-2-yl)ethenes, for example **236**, as the major products [347]. When the silica is replaced by alumina under solvent free conditions, pyrroles ethynylated at C2 are produced in useful yields [348]. Addition of pyrrole to methyl acrylate occurs in the presence of $BF_3{\cdot}OEt_2$, providing the diester **237** [349]. Other Lewis acids, such as $InCl_3$, can also catalyze such processes efficiently, as illustrated by the synthesis of **238** [350]. Silica supported zinc chloride has been demonstrated to promote conjugate addition of pyrrole to methyl α-acetamidoacrylate under microwave irradiation to give the amino acid derivative **239** in considerably better yield than under conventional thermal conditions [351]. Michael addition of pyrroles to a series of electron deficient alkenes under microwave irradiation has also been performed without solvent in the presence of silica gel only,

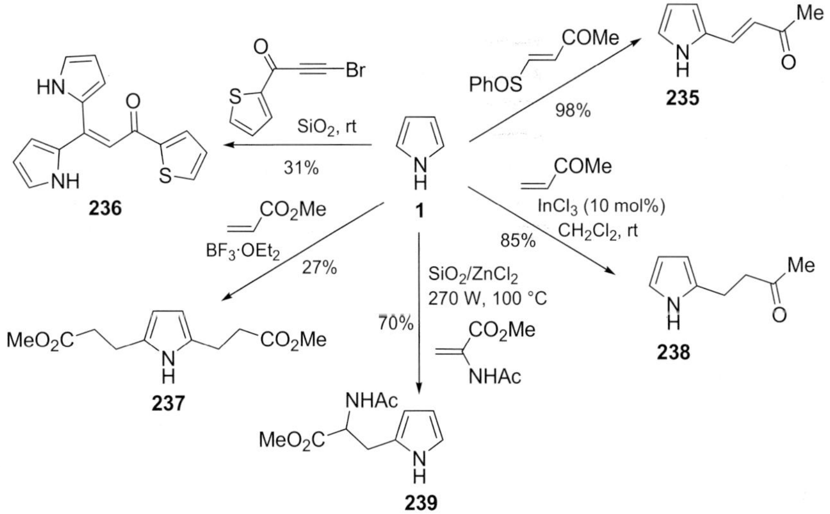

Scheme 4.63

whereas reactions involving some more sluggish substrates required the use of catalytic amounts of $BiCl_3$ [352]. A further development includes the efficient addition of pyrroles to electron deficient alkenes in water catalyzed by aluminium dodecyl sulfate trihydrate [353].

An interesting modern contribution to this field is the asymmetric Friedel–Crafts-type alkylation of pyrroles with α,β-unsaturated carbonyl compounds, which has, for example, been achieved employing organocatalysis by chiral imidazolinones, as illustrated by the conversion of 1-methylpyrrole (**190**) into the product **240** (Scheme 4.64) [354]. A study of additions of pyrroles to a set of α,β-unsaturated 2-acylimidazoles demonstrated that such reactions can also proceed with high yields and enantioselectivity using a bis(oxazolinyl)pyridine scandium(III) triflate complex as the catalyst [355].

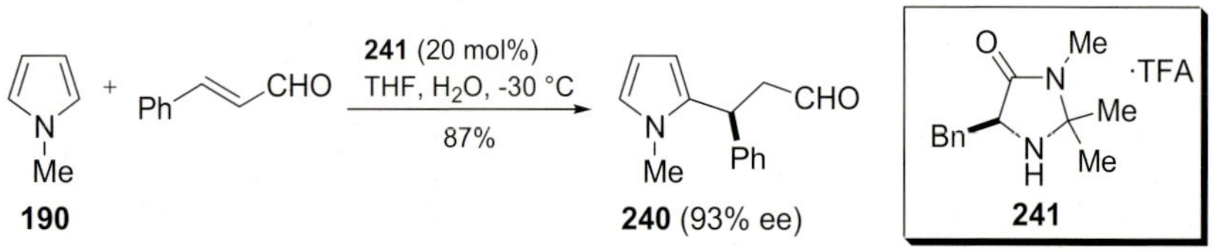

Scheme 4.64

4.5.2
Reactions with Oxidants

Powerful oxidants usually have a severely destructive effect on pyrroles, often leading to extensive decomposition or rather complex product mixtures [356]. Consequently, most useful transformations involving pyrroles and oxidizing agents require careful matching of substrates and reagents. A reaction belonging to this category is the oxidation of α-methylpyrroles to the corresponding carboxaldehydes, which can be effected by using $Pb(OAc)_4/PbO_2$ as the oxidant system [357]. Certain 2-methylpyrroles have also been successfully oxidized employing $Pb(OAc)_4$ [358]. The oxidation of pyrrole α-methyl groups with ceric ammonium nitrate (CAN) provides a reliable and selective route to the corresponding carboxaldehydes, as exemplified by the conversion **242** into the aldehyde **243** (Scheme 4.65) [359]. An extension of this approach allows, for example, preparation of the 2-(methoxymethyl)pyrrole derivative

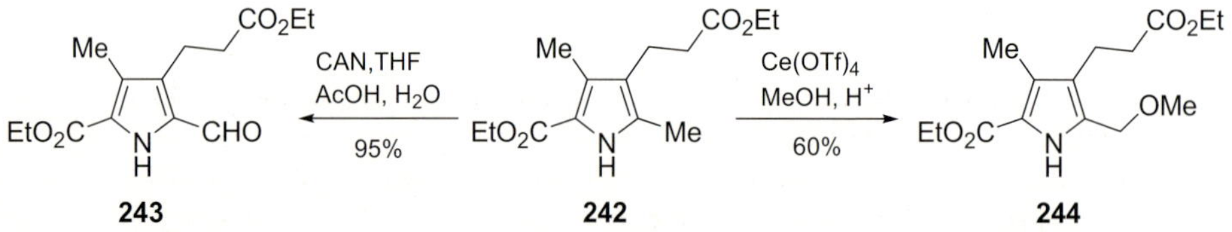

Scheme 4.65

244 using ceric triflate in methanol [360]. Substituted pyrrolecarboxaldehyde derivatives have also been accessed by oxidation of the corresponding α-methylpyrroles with IBX in DMSO [96]. An alternative route to pyrrole-2-carboxaldehydes involves oxidative cleavage of 2-(polyhydroxyalkyl)pyrroles with CAN [361]. In cases where the pyrrole nucleus is strongly deactivated, oxidation of an aldehyde functionality at C2 to the corresponding carboxylic acid may be achieved using the strong oxidant $KMnO_4$ [362].

A useful oxidative coupling reaction has been elaborated, providing access to quarter-, penta- and sexipyrrole derivatives. For instance, coupling of the 2,2′-bipyrrole **245** with K_3FeCN_6 afforded the system **246** (Scheme 4.66), which could thereafter be converted into its reduced form by treatment with NaBH$(OAc)_3$ [363].

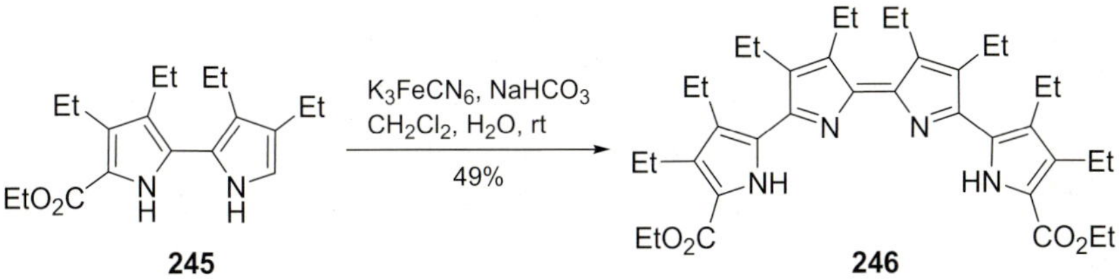

K_3FeCN_6, $NaHCO_3$
CH_2Cl_2, H_2O, rt
49%

245 **246**

Scheme 4.66

Treatment of the 1-(*p*-toluenesulfonyl)pyrroles **247** with phenyliodine bis(trifluoroacetate) (PIFA) and $BF_3{\cdot}OEt_2$ in the presence of TMSCN gives the pyrrole-2-carbonitrile derivatives **248** (Scheme 4.67). This cyanation reaction was suggested to involve initial formation of a pyrrole radical cations, which thereafter react with cyanide ions by a one-electron oxidation, giving the final products after deprotonation [364]. Similar radical cation intermediates are presumably involved in the PIFA/TMSBr (bromotrimethylsilane) mediated oxidative coupling of electron rich pyrroles to bipyrroles [365].

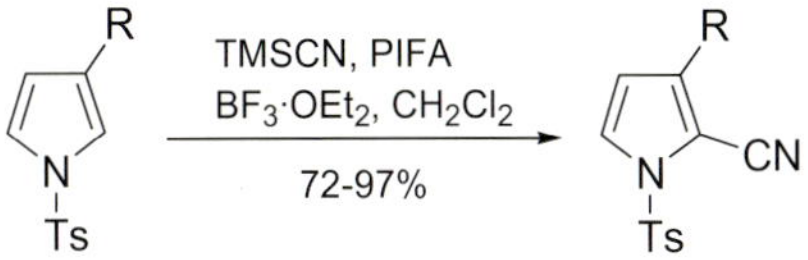

TMSCN, PIFA
$BF_3{\cdot}OEt_2$, CH_2Cl_2
72-97%

247 **248**

Scheme 4.67

The action of benzoyl peroxide on 1-alkyl- or 1-arylpyrroles leads to mixtures of the corresponding 2-hydroxy- and 2,5-dihydroxypyrrole-*O*-benzoates, whereas pyrrole itself gives only intractable mixtures under the same conditions [366]. Careful oxidation of pyrrole with hydrogen peroxide gives a modest yield of 3-pyrrolin-2-one (Section 4.6.3).

4.5.3
Reactions with Nucleophiles

Owing to their electron rich properties, most pyrroles are relatively unreactive towards nucleophilic reagents, with the exception of some derivatives or intermediates having strongly electron-withdrawing substituents, or carrying a positive charge resulting from, for instance, protonation (Section 4.5.1.2).

Treatment of the nitropyrroles **249** with ethylene oxide provides a route to the pyrrolo[2,1-*b*]oxazoles **250** via intramolecular displacement of the nitro group in the intermediates **251** (Scheme 4.68) [367]. Intramolecular displacement of methanesulfinate- or bromide ions at C2 in electron deficient pyrroles by sodium enolates constitutes an alternative approach to [1,2-*a*]-fused pyrrole derivatives [271].

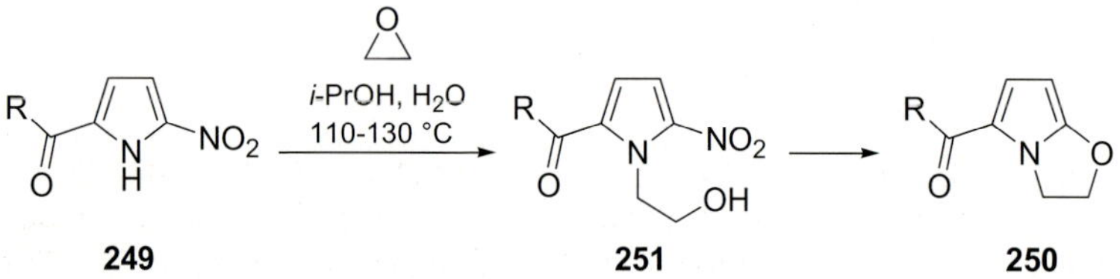

i-PrOH, H_2O, 110-130 °C

249 **251** **250**

Scheme 4.68

The rather electron deficient molecule 1-methyl-2,5-dinitropyrrole (**252**) reacts with piperidine in DMSO to produce the substitution product **253** (Scheme 4.69). A similar reaction occurs more readily with methoxide [368], and the rate is further enhanced by the presence of an additional electron-withdrawing substituent (NO_2, CN) at the adjacent C3 position [369]. Likewise, exposure of 1-methyl-2,3-dinitropyrrole to sodium methoxide in methanol furnishes 2-methoxy-1-methyl-3-nitropyrrole in 93% yield [370]. The treatment of 1-alkyl-2-nitropyrroles with Grignard reagents gives mixtures of C3- and C5 alkylated products [371], whereas treatment of the anion of chloromethyl phenyl sulfone affords the corresponding 5-substituted 1-alkyl-2-nitropyrrole derivatives via vicarious nucleophilic substitution [372]. It has also been established that the bromine atom in 2-acetyl-5-bromo-1-methyl-4-nitropyrrole may be displaced with various nucleophiles, such as azide, cyanide, alkoxides, thiophenols or amines [373].

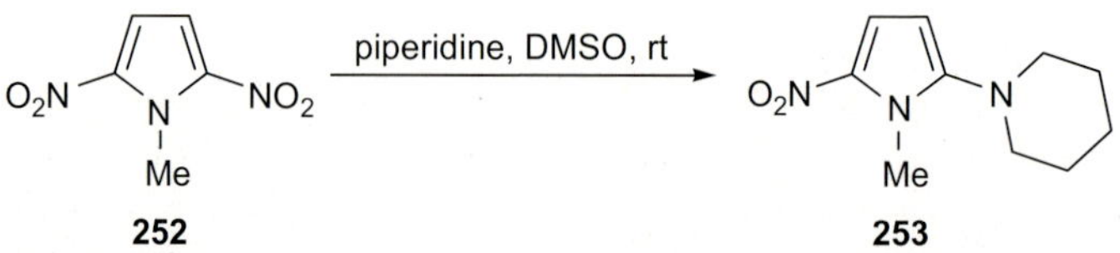

piperidine, DMSO, rt

252 **253**

Scheme 4.69

An unusual example of nucleophilic attack has been observed upon heating pyrrole with sodium hydrogen sulfite, affording the pyrrolidine derivative **254** (Scheme 4.70). This transformation presumably involves the intermediacy of a C3 protonated pyrrole [374].

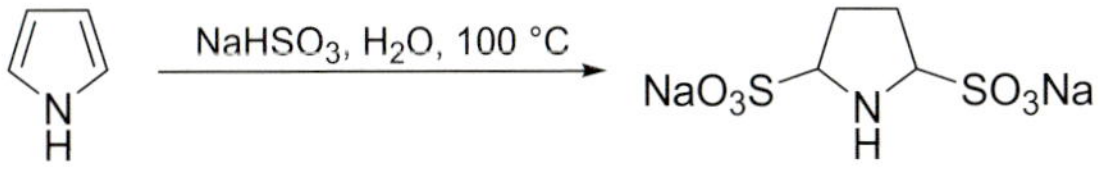

$NaHSO_3$, H_2O, 100 °C

1 → **254** (NaO_3S / SO_3Na)

Scheme 4.70

4.5.4
Reactions with Bases

4.5.4.1 N-Metallated Pyrroles

As a consequence of its weakly acidic properties [13–15], pyrrole will react readily with virtually any strong base to generate a reactive ambident pyrrolyl anion, which can either undergo reaction with electrophiles at the nitrogen atom or at C2/C3. The reaction site is dependent on the properties of the metal–nitrogen bond, and on the solvating power of the solvent used. In general, N-substitution is favored by increasing ionic character of the metal–nitrogen bond, and by stronger solvating power (polarity) of the solvent [375, 376]. Treatment of pyrrole (**1**) with potassium metal gives the salt **255** [377], which can thereafter be converted into a wide variety of N-substituted derivatives, for instance the useful 1-(arylsulfonyl)pyrroles **256** (Scheme 4.71) [378]. Such procedures have now in most cases been supplanted by modern methods, and the pyrrole anion is now usually generated by treatment of pyrrole with commercially available alkyl lithiums [377, 379]. N-Alkylation of **255** formed by deprotonation of pyrrole by KOH in DMSO offers a high-yielding route to 1-alkylpyrroles [380], although several more convenient and effective procedures rely on phase transfer conditions using 18-crown-6 as the catalyst, in combination with KOH in anhydrous benzene [381], or with potassium *tert*-butoxide in diethyl ether [382]. Phase transfer alkylation of pyrrole by alkyl halides with aqueous NaOH in CH_2Cl_2 in the presence of tetrabutylammonium bromide constitutes a useful alternative [383]. Interestingly, 2,5-dialkylpyrrolyl anions generated in the superbase system KOH–DMSO will instead react at C3 with carbon disulfide, providing a route

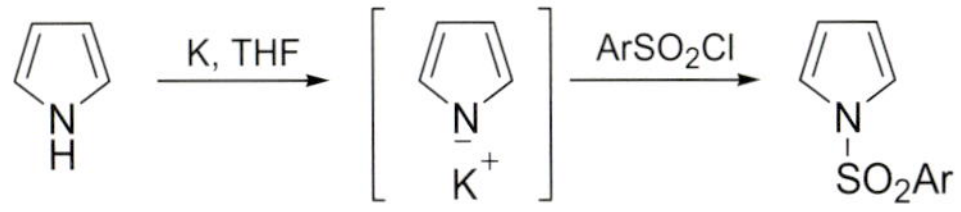

1 —K, THF→ [**255** (K^+)] —$ArSO_2Cl$→ **256** (SO_2Ar)

Scheme 4.71

to pyrrole-3-carbodithioates [384]. The pyrrolyl anion is also easily converted into the corresponding 1-silylated pyrroles by treatment with triisopropylsilyl chloride (TIPSCl) [244] or *tert*-butyldimethylsilyl chloride (TBSCl) [385], as well as with other trialkylsilyl chlorides [386]. The high reactivity of the pyrrolyl anion is neatly demonstrated by the reaction of pyrrolylsodium with hexafluorobenzene, which gives hexa(pyrrol-1-yl)benzene via a S_NAr mechanism [387]. Pyrrolylthallium(I), prepared from pyrrole and thallium(I) ethoxide, can also be utilized for N-alkylation of pyrroles [388], but should perhaps be avoided because of its toxicity.

Pyrrolylmagnesium halides, which are easily prepared by the action of alkylmagnesium halides on pyrrole, display more complex reactivity patterns towards most alkylating agents, rendering mixtures of C2 and C3 substituted products, as well as di- and tri-alkylpyrroles [383, 389]. In contrast, acylation of pyrrolylmagnesium halides is more selective, giving 2-acylpyrroles as the prevailing products [390]. Excellent selectivity for C2 acylation of pyrrolylmagnesium chloride **257** under mild conditions can be achieved by treatment with readily available 2-pyridylthiol esters, which gives high yields of the corresponding 2-acylpyrroles **258** (Scheme 4.72), probably as a result of coordination of the pyridine nitrogen atom to the magnesium [391]. The reaction of **257** with alkyl bromoacetates is an efficient way for selective synthesis of the alkyl (pyrrol-2-yl)acetates **259**, which is in this case mediated by coordination between the metal and the carbonyl oxygen of the alkylating agent [392]. A series of (pyrrole-2-yl)acetone derivatives have been prepared by employing a new heteroarylation reaction involving the pyrrolyl anion and suitable enolates in the presence of a Cu(II) oxidant [393]. Transmetallation of pyrrolylsodium with $ZnCl_2$ gives pyrrolylzinc chloride, which undergoes perfluoroalkylation at C2 upon exposure to perfluoroalkyl iodides in the presence of $PdCl_2(PPh_3)_2$ (10 mol.%) and PPh_3 [394].

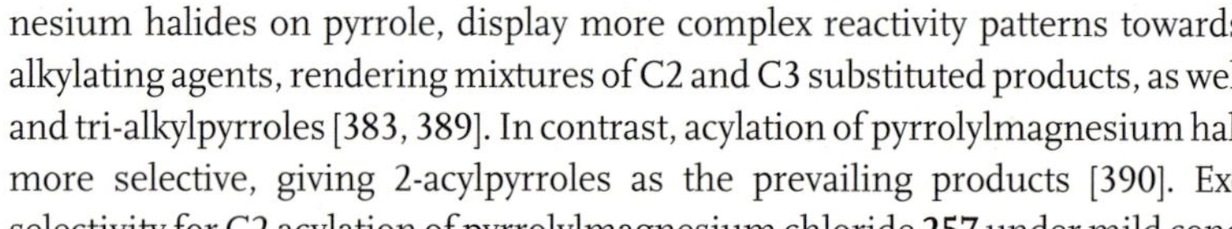

PhMe, -78 °C, 80-95%; $R^1CHBrCO_2R^2$, THF

258 **257** **259**

Scheme 4.72

4.5.4.2 C-Metallated Pyrroles

Pyrroles possessing suitable N-blocking substituents undergo C-metallation upon treatment with sufficiently strong bases. Initial studies early established that 1-methylpyrrole (**190**) is slowly metallated at C2 using BuLi in ether to give 2-lithio-1-methylpyrrole (**260**) (Scheme 4.73) [379], but the yield of the lithiopyrrole is improved significantly if *N,N,N′,N′*-tetramethylethylenediamine (TMEDA) is used as the additive [395]. It was later demonstrated that quantitative and selective generation of **260** is easily achieved using 2.5 equivalents of BuLi in hexane at ambient temperature in the presence of TMEDA [396], while a larger excess (4.5 equivalents) of BuLi and elevated temperatures promotes increasing formation of 2,5- and 2,4-dilithio- derivatives [396, 397]. Generation of **260** may also be carried out using BuLi at

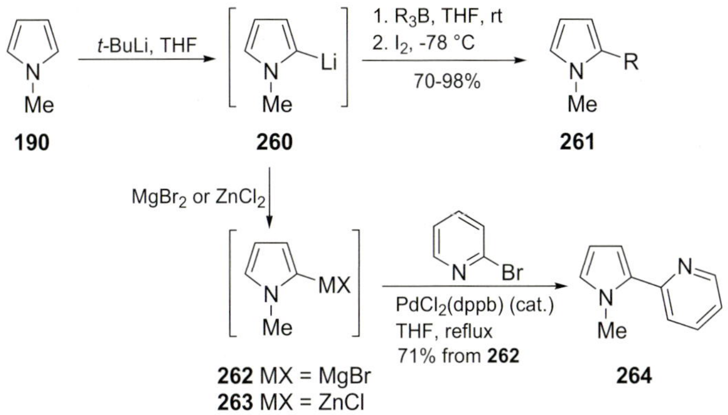

Scheme 4.73

ambient temperature in THF–hexane (2 : 1). This lithiopyrrole will subsequently react readily with a wide variety of electrophiles [398], and may for example also be converted into the 2-alkyl-1-methylpyrroles **261** by treatment with trialkylboranes, followed by addition of iodine or NCS [399, 400]. Quenching of 2-lithio-1-methyl-5-*n*-octylpyrrole with *N*-fluorodibenzenesulfonamide has been used to prepare the labile 2-fluoro-1-methyl-5-*n*-octylpyrrole [401]. Transmetallation of **260** formed from 1-methylpyrrole and *t*-BuLi with $MgBr_2$ or $ZnCl_2$ gives the corresponding metallated pyrroles **262** and **263**, which can subsequently participate in palladium-catalyzed cross-coupling reactions leading to 2-arylpyrroles, or to the 2-pyridylpyrrole **264** [402]. A procedure for selective C2 metallation and functionalization of 1-vinylpyrrole employing the base system BuLi/*t*-BuOK catalyzed by *i*-Pr_2NH, followed by addition of LiBr and suitable electrophiles, has also been described [403].

Selective C2 mercuration of pyrroles is effected by exposure of 1-acetyl- or 1-(phenylsulfonyl)pyrrole to mercuric chloride. The corresponding diorganomercury compounds, which are useful for the synthesis of pyrrole containing transition metal complexes, can thereafter be obtained after treatment with sodium iodide [404]. For N-H pyrroles, mercuration with mercury(II) acetate leads to N-mercuration, while various N-substituted pyrroles are mercurated at C2 or C3. Such C-mercurated pyrroles undergo Heck-type reactions with alkenes [405].

The development of directed metallation techniques has allowed facile access to a multitude of 2-substututed pyrroles that were previously difficult to prepare. The ideal N-protecting and directing group should be easily installed and cleaved, and induce high regioselectivity during metallation. The *tert*-butoxycarbonyl (Boc) group fulfils all these requirements, and has consequently proven to be extremely useful. For example, 1-(*tert*-butoxycarbonyl)pyrrole (**158**) [285] is efficiently lithiated at C2 by LiTMP, and subsequent quenching with aldehydes or acid chlorides gives the corresponding pyrrol-2-yl methanols **265** (Scheme 4.74) and 2-acylpyrroles, respectively. The Boc group may be removed by treatment with sodium methoxide in methanol [406], under thermal conditions [407] or using acid, which may, however, not always be compatible with electron rich substrates. Similar metallation of **158**

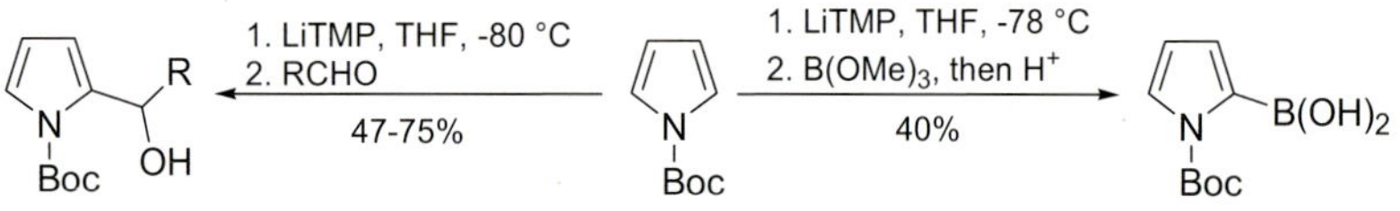
1. LiTMP, THF, -80 °C
2. RCHO
47-75%
1. LiTMP, THF, -78 °C
2. B(OMe)3, then H+
40%
R
OH
N
Boc
B(OH)2

265 158 266

Scheme 4.74

followed by quenching with trimethyl borate and subsequent hydrolysis provides a route to the useful boronic acid **266** [239]. An excellent recent review covers the advances in pyrrole protection strategies [408].

An alternative route permits introduction of two α-substituents by halogen–metal exchange on 2,5-dibromo-1-(*tert*-butoxycarbonyl)pyrrole (**267**) (Section 4.5.1.3), eventually producing the corresponding 2,5-disubstituted pyrroles, for instance **268** (Scheme 4.75) [237, 239]. Moreover, mono-lithiation of **267** offers access to 2-bromopyrroles having an additional substituent at C5 (e.g., **269**) [239].

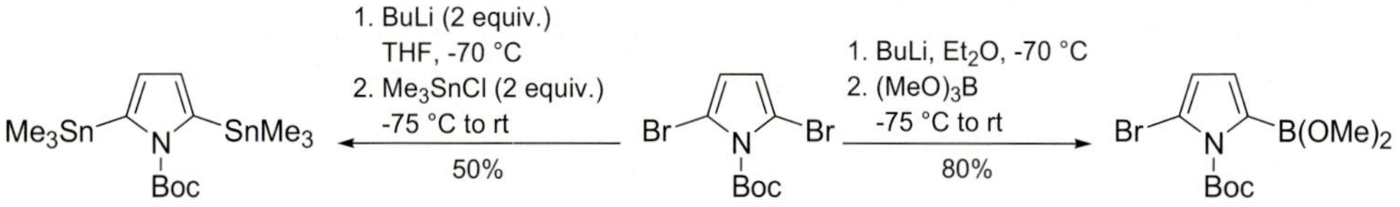
1. BuLi (2 equiv.)
THF, -70 °C
2. Me3SnCl (2 equiv.)
-75 °C to rt
50%
1. BuLi, Et2O, -70 °C
2. (MeO)3B
-75 °C to rt
80%
Me3Sn
SnMe3
Br
N
Boc
B(OMe)2

268 267 269

Scheme 4.75

Pyrroles that are N-protected with the *N-tert*-butylcarbamoyl group are also very useful substrates for C2 lithiation; the directing group can subsequently be removed by LiOH in MeOH–THF [409]. Generation of the dilithio species **270**, followed by quenching with suitable electrophiles, and final acidic work up which cleaves the carbamate, constitutes a one-pot protocol for preparation of C2 functionalized pyrroles **271** (Scheme 4.76) [410].

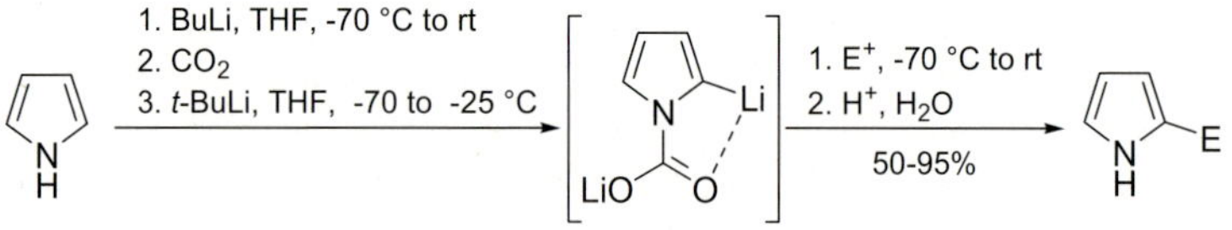
1. BuLi, THF, -70 °C to rt
2. CO2
3. t-BuLi, THF, -70 to -25 °C
1. E+, -70 °C to rt
2. H+, H2O
50-95%
N
H
Li
LiO
O
E

1 270 271

Scheme 4.76

Lithiation of 1-(*p*-toluenesulfonyl)pyrrole (**272**) also occurs at C2 with high selectivity, and subsequent quenching with $PhSSO_2Ph$ gives the sulfide **273** (Scheme 4.77) [276]. Similar techniques can also be utilized in synthesis of other chalcogen containing systems, as illustrated by the conversion of **272** into the

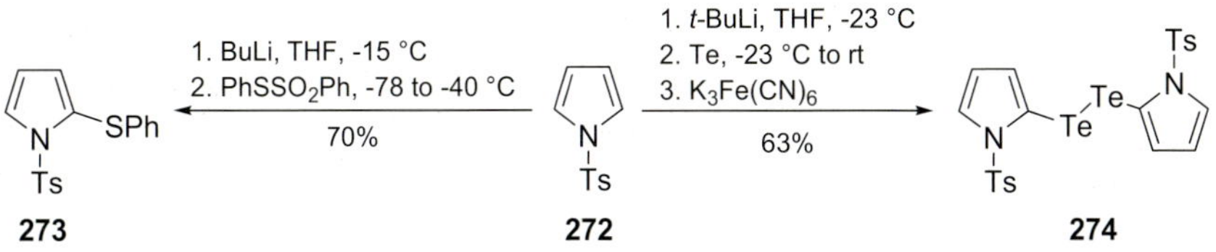

Scheme 4.77

ditelluride **274** [411]. Magnesium can be introduced at C2 in 1-(phenylsulfonyl) pyrrole by the action of 3 equivalents of *i*-PrMgBr in the presence of 5 mol.% *i*-PrNH; ensuing treatment with electrophiles gives the corresponding 2-substituted pyrroles in moderate yields [412].

The N-protected pyrroles discussed above are neatly complemented by 1-[2-(trimethylsilyl)ethoxymethyl]pyrrole (1-SEM-pyrrole) (**275**), which is available by deprotonation of pyrrole with NaH in DMF, followed by treatment with SEMCl [413]. Lithiation is also in this case directed to C2, rendering the lithiopyrrole **276**, treatment of which with appropriate electrophiles affords the 2-substituted N-protected pyrroles **277** in moderate to good yields (Scheme 4.78). The electron releasing SEM group may thereafter be conveniently cleaved using tetrabutylammonium fluoride (TBAF) under conditions that tolerate sensitive functionalities [414]. Even structurally rather complex and sterically congested electrophiles may be used in this approach, providing useful yields of unusual 2-substituted pyrroles [415]. Notably, lithiation of 1-(*N*,*N*-dimethylamino)pyrrole also occurs at C2, and this electron releasing directing group can subsequently be removed by treatment with $Cr_2(OAc)_4{\cdot}2H_2O$ [416].

Scheme 4.78

Metallation of pyrroles at C3 has enabled convenient preparation of derivatives that were prepared previously by cumbersome means [417]. Selective C3 metallation is achieved conveniently by halogen–metal exchange on 3-bromo-1-(triisopropylsilyl) pyrrole (**163**) by virtue of the steric bulk of the TIPS group (Scheme 4.79) [245]. Subsequent treatment of the so-obtained 3-lithiopyrrole **278** with suitable electrophiles, followed by desilylation with TBAF gives the corresponding 3-substitued pyrroles **279** via the 1-TIPS derivatives **280** [242]. Intermediate **278** may also be generated by lithiation of 3-iodo-1-(triisopropylsilyl)pyrrole (**164**), and converted into, for example, the boronic acid **281** or the stannyl derivative **282**, which are useful substrates in palladium-catalyzed coupling reactions (Section 4.5.11) [418]. Similar

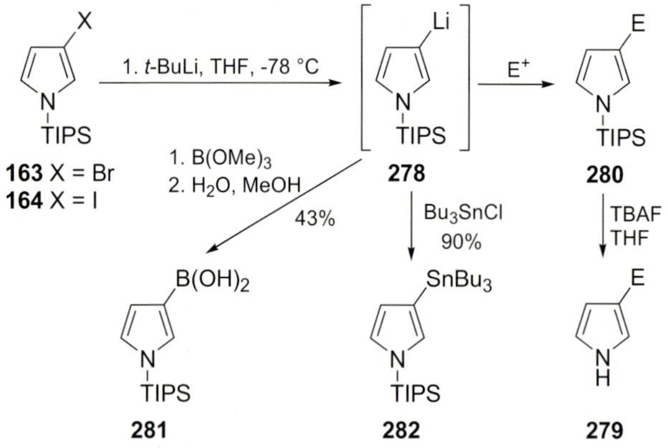

Scheme 4.79

techniques have been employed for the synthesis of 3-fluoro-1-TIPS-pyrrole in 50% yield by quenching of 3-lithio-1-TIPS-pyrrole by *N*-fluorobenzenesulfonimide [419]. Conversely, lithiations involving pyrroles having a trimethyl or triethylsilyl groups at the nitrogen are more difficult to control in terms of regioselectivity, and are under certain conditions further complicated by migration of the silyl groups [420]. Treatment of dimethoxyethyl protected 4-iodopyrrole-2-carbonitrile with *i*-PrMgCl and subsequent quenching with electrophiles provides a route to 4-substituted pyrrole-2-carbonitriles in good yields [421].

4.5.5
Reactions with Radical Reagents

Synthetically useful reactions of pyrroles with radical reagents were not been studied in much detail until it was demonstrated that effective and regioselective synthesis of 2-alkylpyrrole derivatives can be accomplished by radical substitution. The pyrrole-2-acetic acid derivatives **283** are readily available by treatment of the pyrroles **1** or **190** (Scheme 4.80) with radicals generated from α-carbonyl-, α,α′-dicarbonyl- and α-cyano-alkyl iodides [422]. Alternative procedures for the synthesis of pyrrole-2-acetic acid derivatives involve generation of radicals from various iodoacetates induced by stannanes [423] or under conditions avoiding stannanes, by irradiation

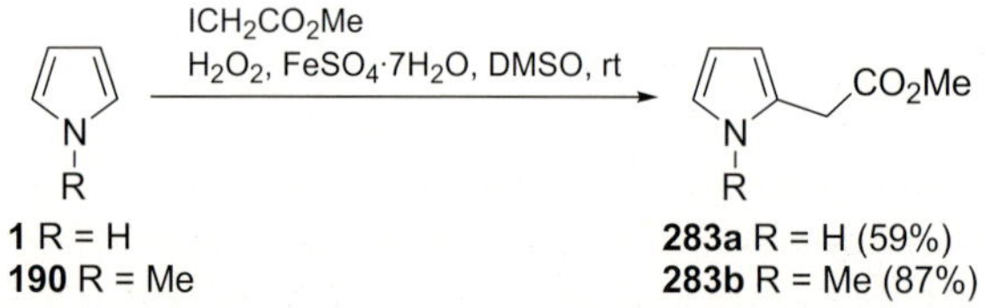

Scheme 4.80

in the presence of $Na_2S_2O_3$ as the I_2 reductant, Bu_4NI to aid in solubility, and propylene oxide [423, 424] or epoxydecane as the HI trap [425]. Pyrroles possessing electron-withdrawing groups at C2 undergo related alkylation reactions at C5 with α-acetyl or α-acetonyl radicals generated by exposure of suitable xanthate based precursors to dilauroyl peroxide [426].

Similar intramolecular processes provide routes to fused pyrroles, as illustrated by the manganese(III) acetate generated *in situ* induced conversion of the 2-aroylpyrrole **284** into the fused system **285** (Scheme 4.81), a useful precursor to the analgesic molecule ketorolac (**10**) [427]. In addition, [1,2-*a*]-fused pyrroles are also available via intramolecular radical cyclization of 1-(bromoalkyl)pyrroles induced by Bu_3SnH in the presence of AIBN [428, 429], annulation of 1-(iodoalkyl)pyrroles with the H_2O_2/Fe(II) system (*vide supra*) [430], or employing electroreduction with a Ni(II) complex as the electron-transfer catalyst [431]. Intramolecular cyclization of acyl radicals onto pyrroles leading to [1,2-*a*]-fused pyrrole derivatives in modest yields has also been reported [432].

Ph, O, N, $CH(CO_2Et)_2$ **284** — $Mn(OAc)_2{\cdot}4H_2O$, $AgNO_3$ (cat.), $Na_2S_2O_8$, NaOAc, Ac_2O, AcOH, 80 °C; 95% → Ph, O, N, CO_2Et, CO_2Et **285**

Scheme 4.81

It has been known for some time that 2-(perfluoroalkyl)pyrroles are formed in low yields upon heating of pyrroles in the presence of perfluoroalkyl iodides under forcing conditions [433]. Application of the H_2O_2/Fe(II) protocol offers a practical procedure for the preparation of, for example, the perfluoroalkylpyrroles **286** from pyrrole-2-carboxaldehyde (**191**) and perfluoroalkyl iodides via a homolytic substitution process (Scheme 4.82) [434]. In contrast, perfluoroalkylation of pyrroles at C2 using bis(perfluoroalkanoyl) peroxides follows a different mechanistic path, which appears to proceed via coupling of perfluoroalkyl radicals with pyrrole cation radicals [435].

OHC, N, H **191** — $I(CF_2)_nCF_3$, H_2O_2, $FeSO_4{\cdot}7H_2O$, DMSO, rt → OHC, N, H, $(CF_2)_nCF_3$ **286a** n = 2 (64%), **286b** n = 3 (55%)

Scheme 4.82

Addition of the 3-pyridyl radical generated from *N*-nitroso-*N*-(3-pyridyl)-isobutyramide to ethyl pyrrole-1-carboxylate gives **287** in 23% yield [436]. More efficient radical C2 arylations of pyrroles having an electron-withdrawing substituent at the nitrogen atom have been afforded using anilines in the presence of amyl nitrite in warm acetic acid [437]. Intramolecular radical arylations involving pyrroles are useful tools for the construction of more complex systems, such as the tricyclic derivative **288**, which was derived from the precursor **289** by exposure thereof to tributyltin hydride in the presence of AIBN in refluxing toluene [438, 439]. Related cyclization reactions may also produce spiropyrrolidinyloxindoles, depending on the properties of the pyrrole protecting group [440].

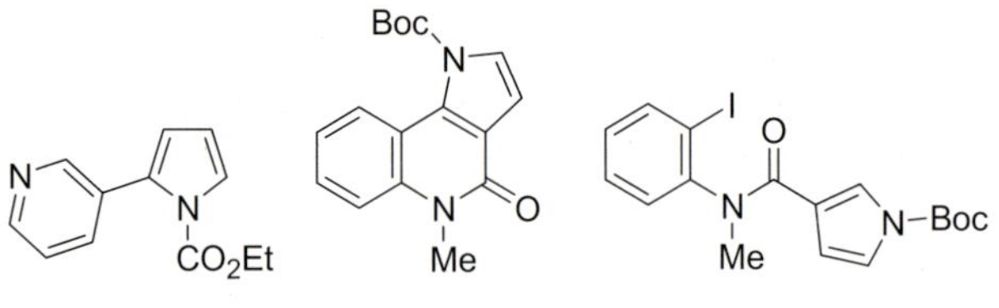

Boc N N N O N Me O I N Me N–Boc CO_2Et

287 **288** **289**

A radical displacement reaction at C2 of 1-phenylsulfonyl-2-(*p*-toluenesulfonyl) pyrrole with stannyl radicals generated by the reagent combination Bu_3SnH/AIBN in refluxing benzene, resulting in the corresponding 2-stannylpyrrole derivative, has also been described [441].

4.5.6
Reactions with Reducing Agents

It has long been established that reduction of pyrroles **290** with zinc in hydrochloric acid gives 2,5-dihydropyrroles (3-pyrrolines) **291** (Scheme 4.83) [442, 443], although it was later found that the material obtained from the reduction of pyrrole itself is frequently contaminated with pyrrolidine, which is difficult to separate from the desired product [444]. The synthetic utility is further compromised by the fact that the formation of both *cis* and *trans* products occurs upon reduction of 2,5-dimethylpyrrole [445]. Nevertheless, this procedure seems to be useful in certain applications, and a modified version thereof has been applied successfully for the reduction of some 1,2-disubstituted pyrroles to give the corresponding 3-pyrroline derivatives *en route* to the antibiotic (±)-anisomycin [446]. Treatment of pyrrole-2-carboxamide with phosphonium iodide in fuming hydroiodic acid affords 3,4-dehydroprolinamide as the major product, provided that careful control of the reaction conditions is maintained [447]. A similar reduction using the combination H_3PO_2/HI in acetic acid has been used for the conversion of pyrrole-2-carboxylic acid into 3,4-dehydroproline in 74% yield on a large scale [448]. Reduction of C2 or C3 substituted

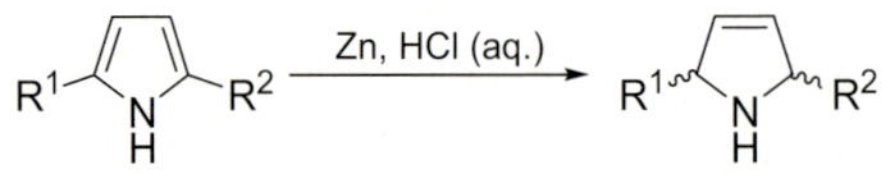

R^1 N H R^2 → Zn, HCl (aq.) → R^1 N H R^2

290 **291**

Scheme 4.83

1-phenylsulfonylpyrroles with $NaCNBH_3$ in TFA leads to the corresponding 3-pyrrolines in good yields, offering a convenient complement to the procedures discussed above [449].

Catalytic hydrogenation of pyrroles leads to pyrrolidines, and can be performed under various conditions, often proceeding exclusively with cis-stereoselectivity. Useful catalysts for this application are 5% Rh on Al_2O_3 [450–452], PtO_2 or 10% Pd/C in 6 N aqueous HCl [453] or alternatively Pd/C in the presence of catalytic amounts of H_2SO_4 [454]. Moreover, catalytic hydrogenation of pyrroles at atmospheric pressure offers a convenient procedure for the synthesis of *cis*-2,5-disubstituted pyrrolidines, as illustrated by the conversion of pyrrole **292** into pyrrolidine **293** (Scheme 4.84) [455].

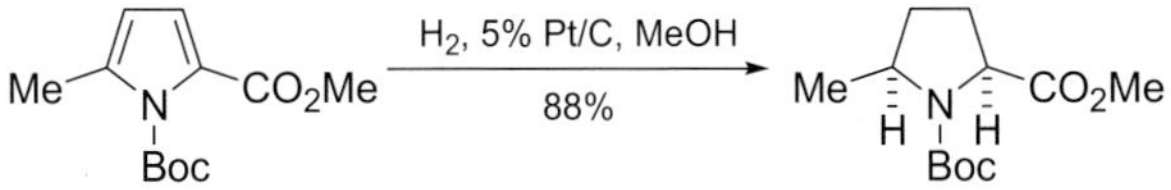

Scheme 4.84

Birch reduction of pyrroles has not been explored until recently, but has already attracted considerable attention as a tool for preparation of interesting 3-pyrrolines and 2,3-dihydropyrroles (2-pyrrolines). Initial studies demonstrated that pyrroles possessing electron-withdrawing groups, for instance **294**, undergo efficient conversion into 3-pyrrolines **295** upon Birch reduction and subsequent alkylation (Scheme 4.85) [456, 457]. Diastereoselectivity at C2 may be induced in alkylation [458] or protonation [459], by using for example pyrrole-2-carboxylates or pyrrole-2-carboxamides, respectively, containing chiral moieties as substrates. A variant featuring a reductive aldol reaction has also been developed [460]. It was later established that such reductions may also be conveniently performed by employing lithium in THF in the presence of bis(methoxyethylamine) and naphthalene [461, 462], or 4,4′-di-*tert*-butylbiphenyl (**296**, see below) [463], thus supplanting the classical Birch conditions. Further extensions of this methodology allow reductive acylation [462], and aldol reactions with excellent *anti*-selectivity [464, 465].

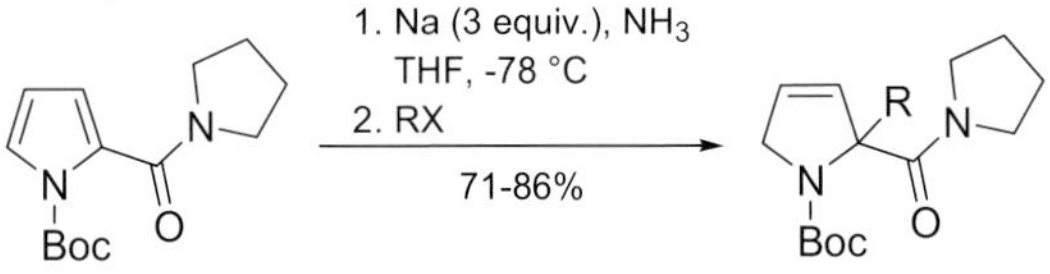

Scheme 4.85

A recent contribution to this field concerns the enantioselective partial reduction of pyrroles involving chiral protonation using (−)-ephedrine. Thus, diester **297** underwent conversion into a 1 : 1 mixture of the cis- and trans-diastereomers **298a** and **298b**, the latter of which was found to be formed with an enantiomeric excess of 74%. A simple recrystallization of **298b** provided material with >94% ee (Scheme 4.86) [466].

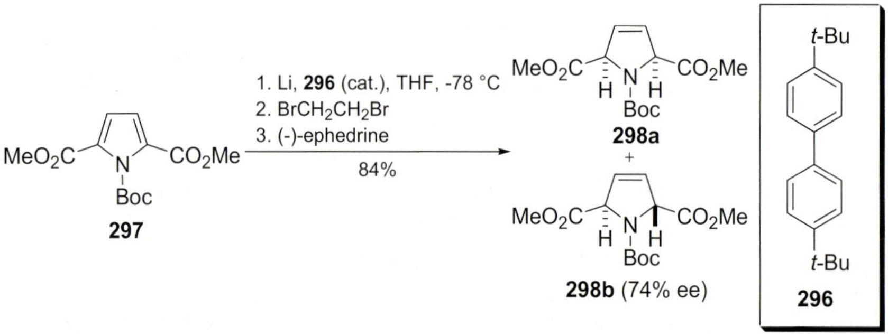

MeO$_2$C–(N-Boc pyrrole)–CO$_2$Me **297** → 1. Li, **296** (cat.), THF, -78 °C; 2. BrCH$_2$CH$_2$Br; 3. (-)-ephedrine; 84% → **298a** + **298b** (74% ee); **296**: 4,4'-di-*t*-Bu-biphenyl

Scheme 4.86

Pyrroles bearing amide or ester functionalities at the β-position also undergo Birch reduction, and ensuing alkylation provides the corresponding 4,4-disubstituted 2-pyrrolines [467]. The 3,4-disubstituted pyrrole **299** (Adoc=adamantyloxycarbonyl) is converted into *cis*-pyrrolidines **300** under similar conditions (Scheme 4.87). Sequential alkylation with two different electrophiles gives access to unsymmetrically substituted derivatives [468].

299 → 1. Li (6 equiv.), NH$_3$, THF, (MeOCH$_2$CH$_2$)$_2$NH, -78 °C; 2. RI; 70-82% → **300**

Scheme 4.87

4.5.7
Cycloaddition Reactions

Cycloaddition reactions involving pyrroles constitute a powerful tool for crafting rather complex heterocyclic structures from simple precursors. It was early recognized that some pyrroles, for example 1-methylpyrrole, give mainly C2 substituted products resulting from Michael-type addition to maleic anhydride [469]. Interestingly, the reaction of 1-methylpyrrole with dimethyl acetylenedicarboxylate (DMAD) gives a product for which an indolic structure was proposed [470] – the true identity was later elucidated by Acheson [471]. Further studies of these reactions suggested that Michael additions are favored in the presence of a proton source, whereas under neutral conditions and at elevated temperatures Diels–Alder additions will take place. The adducts, however, being rather unstable, undergo retro Diels–Alder reactions to the starting materials, yield further pyrrole derivatives via extrusion of acetylene derivatives, or react further with DMAD to give more complex indolic products [472]. Based on the results of *ab initio* calculations on the reaction of 1-methylpyrrole with DMAD, which support the preferential formation of Diels–Alder products instead of Michael adducts, a step-wise mechanism involving a zwitterionic intermediate was suggested, rather than a concerted pericyclic process [473].

Even though elimination of acetylene from intermediate Diels–Alder adducts of the deactivated methyl pyrrole-1-carboxylate with DMAD has also been observed [474, 475], stable addition products could nevertheless be isolated in very low yields from the reactions of 1-benzylpyrrole [476] or methyl pyrrole-1-carboxylate [477] with acetylenedicarboxylic acid. A different decomposition path is in operation during reactions of 1-aminopyrroles with DMAD, which give substituted benzene derivatives as the final products [478]. Conditions for practical Diels–Alder reactions involving deactivated pyrroles soon became available, permitting the synthesis of adduct **301** in a moderate yield upon heating of the pyrrole **302** in neat DMAD (Scheme 4.88) [479], whereas the same reaction performed in CH_2Cl_2 in the presence of five equivalents of $AlCl_3$ at 40 °C is much faster and gives the adduct in 93% yield [480]. The use of high pressure (15 kbar) also leads to high yields of **301**, but, even under these conditions, pyrrole itself undergoes mainly C2-substitution [481].

Scheme 4.88

Deactivated pyrroles also give good yields of Diels–Alder adducts with alkenes at elevated pressure, as demonstrated by the synthesis of the *endo*-adduct **303** from 1-acetylpyrrole (**304**) and *N*-methylmaleimide (Scheme 4.89). In some cases,

Scheme 4.89

mixtures of *exo*- and *endo*-adducts are produced, and it has been proposed that the *endo*-isomers are in some cases formed under kinetic control and isomerize to the *exo*-products [482]. Mixtures of *exo*- and *endo*-adducts have also been encountered during high pressure reactions of methyl 3-methylthio- or 3-phenylthiopyrrole-1-carboxylate with *N*-phenylmaleimide or methyl acrylate [483]. However, it appears that 1-acylpyrroles do not participate in Diels–Alder reactions with alkenes at atmospheric pressure [484].

An intramolecular cycloaddition reaction involving the precursor **305** (Ns=2-nitrophenylsulfonyl), incorporating a deactivated pyrrole as the diene component has been used in an elegant synthesis of the tricyclic system **306** (Scheme 4.90). This study was extended to the efficient construction of several conformationally strained analogues, featuring stereoselective generation of multiple stereogenic centers [485].

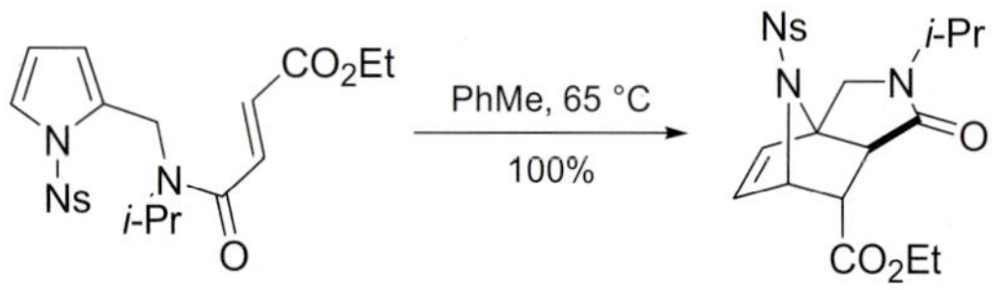

Scheme 4.90

Diels–Alder reactions involving pyrroles and acetylenes are the key feature of several total syntheses of the powerful analgesic natural product epibatidine (**307**) (Scheme 4.91). The common 4π-component 1-(*tert*-butoxycarbonyl)pyrrole (**158**) undergoes efficient conversion into the bicyclic system **308** upon heating with the electron deficient dienophile (*p*-toluenesulfonyl)acetylene [486]. A similar reaction involving **158** and methyl pyrrole-1-carboxylate has previously been demonstrated to give excellent results [487, 488]. The intermediate **308** proved to be a useful precursor for the synthesis of epibatidine (**307**) [489, 490]. Other related approaches to **307** employ Diels–Alder reactions of methyl pyrrole-1-carboxylate with (*p*-toluenesulfonyl)acetylene [491], or a (6-chloro-3-pyridyl)acetylene derivative [492]. Likewise, heating of the pyrrole **158** with the dienophile methyl 3-bromopropiolate gives the adduct **309**, which is also a useful vehicle for further manipulations that eventually lead to **307** [493, 494].

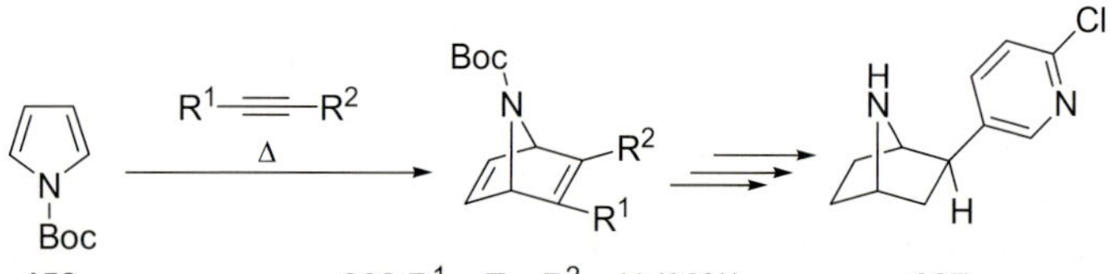

Scheme 4.91

Allenes may also participate in Diels–Alder reactions with electron deficient pyrroles, and high selectivity can be attained, depending on the choice of starting materials and conditions. For example, 1-(*tert*-butoxycarbonyl)pyrrole (**158**) gives exclusively the *endo*-isomer **310** with diethyl allene-1,3-dicarboxylate **311** (Scheme 4.92). Under similar conditions, 1-methyl- or 1-benzylpyrrole give only C2 substitution products [495]. Variants involving *endo*-selective Diels–Alder addition of **158** to an optically active allene-1,3-dicarboxylate [496], or 1-(benzenesulfonyl)-1,2-propadiene to provide **312** [497], have also been disclosed.

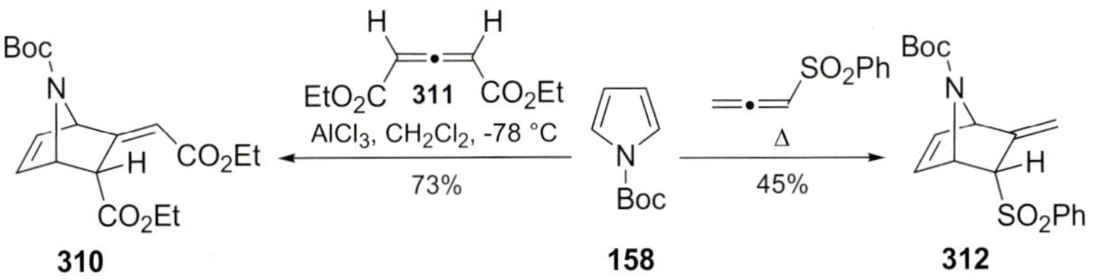

Scheme 4.92

The reactions of 1-methylpyrrole or 1-benzylpyrrole with benzyne give rather unstable adducts that could only be isolated in low yields as the methiodide or the picrate, respectively, as they readily react further with benzyne, finally rendering carbazole derivatives [498, 499], or rearrange to 2-naphthylamines [500]. In contrast, 1-(*tert*-butoxycarbonyl)pyrrole (**158**) reacts with benzyne, to afford a stable adduct (**313**) in moderate yield (Scheme 4.93) [501], as do 1-alkyl-2,3,4,5-tetramethylpyrroles [502, 503]. Tetrafluorobenzyne also adds readily to for, example, 1-methylpyrrole to form a stable adduct [504], whereas the reaction of the benzyne generated from 1-bromo-2,5-difluorobenzene with 1-(trimethylsilyl)pyrrole (**314**) gives **315** with concomitant desilylation [505]. Related addition products have also been obtained upon addition of tetrachlorobenzyne to 1-alkylpyrroles [506].

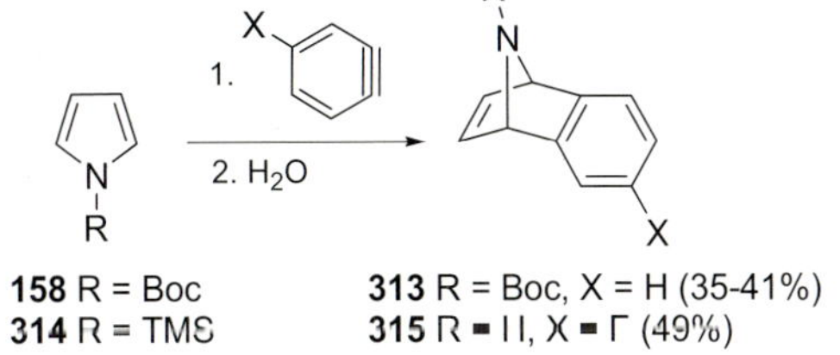

Scheme 4.93

Although the photooxygenation of pyrroles was discussed already in 1912 [507], evidence for the mechanism was not available until almost 70 years later, when it was demonstrated that addition of singlet oxygen to 1-methylpyrrole (**190**) provides the unstable endoperoxide **316**, which undergoes a subsequent rearrangement or reacts with water to furnish, among other products, the oxypyrrole **317** (Scheme 4.94). The intermediacy of **316** was suggested on the basis of low-temperature NMR measurements [508].

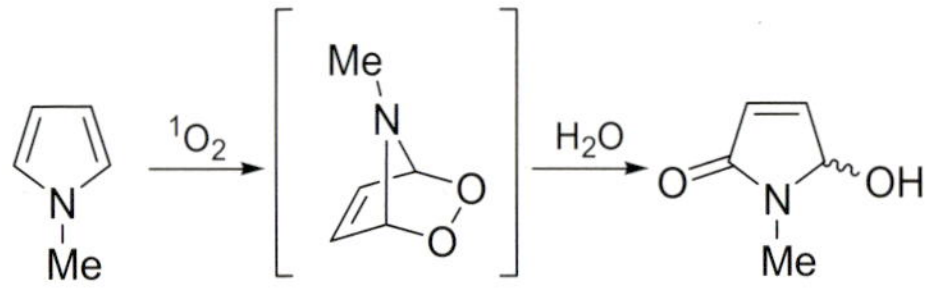

Scheme 4.94

Owing to their aromatic character, the applicability of pyrroles as the 2π reactants in normal electron demand Diels–Alder reactions has been quite limited, requiring harsh conditions and giving only moderate yields of products as mixtures of regioisomers [509]. Recently, however, it was demonstrated that pyrroles possessing the strong electron-withdrawing trifluoromethylsulfonyl (Tf) group plus an acetyl group can indeed act as dienophiles at elevated pressure, as illustrated by the conversion of **318** into **319** (Scheme 4.95) [510, 511].

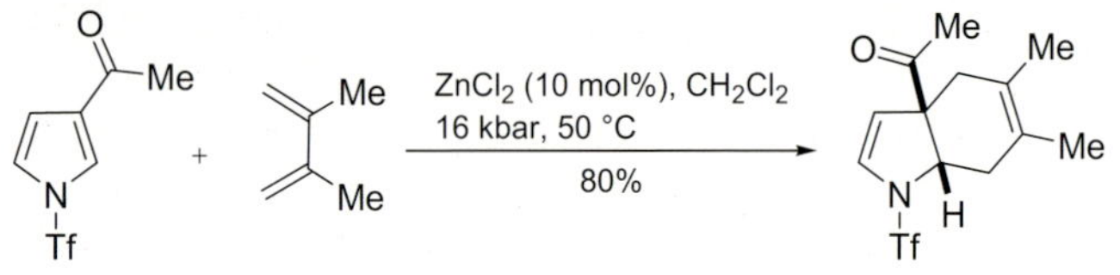

Scheme 4.95

Pyrroles may also play the role of dipolarophiles in cycloaddition reactions. Thus for instance, the reactions of certain 1-alkylpyrroles with nitrileimines have been demonstrated to give pyrrolo-fused pyrazoles [512, 513], whereas intramolecular cyclization of nitrile oxide functionalities onto pyrroles involved unstable pyrrolo-fused isoxazole adducts that underwent ring opening of the isoxazole ring [514]. It has also been established that a series of osmium pyrrole complexes, for example **320** [Os(II)=[Os(NH_3)$_5$](OTf)$_2$], participate as dipoles in cycloadditions with alkenes, for instance providing the *endo*-adduct **321** in good yield along with minor amounts of the *exo*-isomer (Scheme 4.96) [515, 516].

Inverse electron demand Diels–Alder reactions have been used successfully for the synthesis of the pyrrolo[2,3-*d*]pyrimidines **322** from the 2-amino-4-cyanopyrroles **323** and the 1,3,5-triazine **324** (Scheme 4.97). This process is remarkable, as it proceeds readily even at ambient temperature [517]. Based on a theoretical study of the reaction between 2-aminopyrrole and 1,3,5-triazine, it appears that these transformations involve an initial nucleophilic attack of the aminopyrrole on the triazine to form a zwitterionic intermediate, which is in equilibrium with a neutral species, eventually

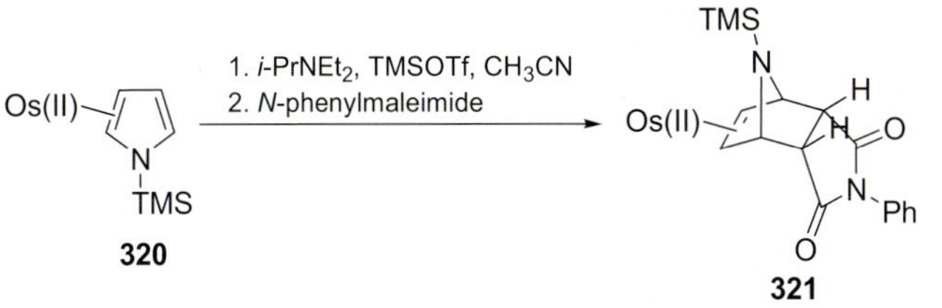

Scheme 4.96

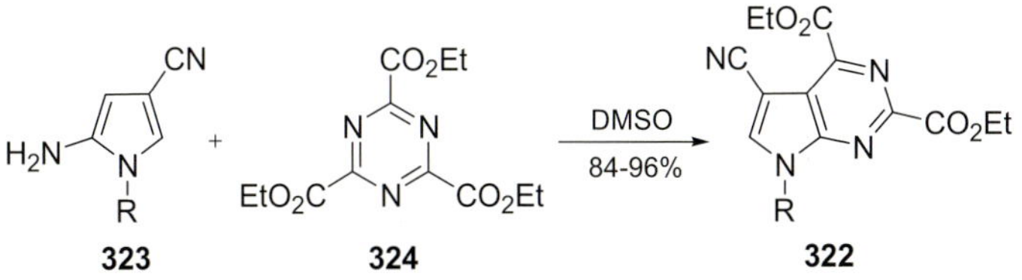

Scheme 4.97

undergoing a rate determining cyclization step [518]. The related reaction between 1-methylpyrrole and 4,5-dicyanopyridazine gave only a low yield of 5,6-dicyano-1-methylindole [519].

An intramolecular inverse electron demand Diels–Alder reaction involving pyrrole as the dienophile component has been employed for conversion of the pyrrole-tethered 1,2,4-triazine **325** into the tricyclic system **326** in good yield (Scheme 4.98) [520]. Likewise, 1-acyl- or 1-benzoylpyrroles have been used as 2π-components in cycloadditions to masked *o*-benzoquinones [521].

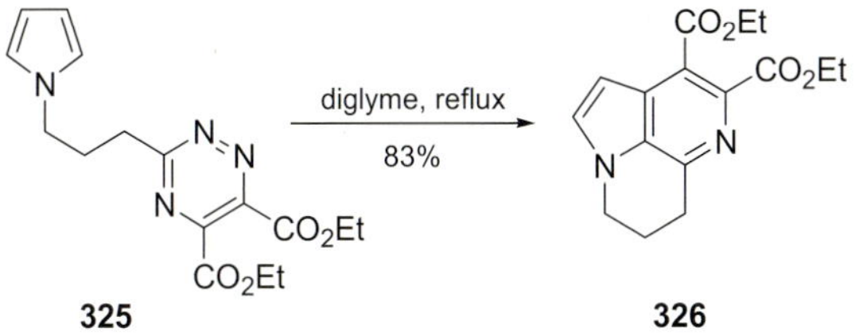

Scheme 4.98

Since the initial reports that 1-alkyl- and 1-aryl-2(3)-vinylpyrroles may serve as diene components in Diels–Alder reactions leading to indoles with interesting substitution patterns [522, 523], several studies exploiting this strategy have emerged. Heating of 1-triisopropylsilyl-(*E*)-2-(2-phenylsulfinylvinyl)pyrrole with suitable acetylenecarboxylic acid derivatives furnished TIPS protected 4-acylindoles in good yields [524]. A more practical approach involves the readily available 3-vinylpyrrole **327**, which gives the adduct **328** upon reaction with *N*-phenylmaleimide and subsequent isomerization (Scheme 4.99). Similar reactions yielding isomeric adducts were performed using 1-(phenylsulfonyl)-2-vinylpyrrole [525]. Other related indole syntheses encompass the use of silyl enol ethers of 2-acylpyrroles [526, 527], or a diene generated by S-alkylation of 1-methyl-3-thioacetylpyrrole [528] as the 4π-components. It is also noteworthy that some sterically hindered N-substituted 2-vinylpyrroles give mainly Michael-type addition at C5, as the required *cisoid* conformation is disrupted by steric interaction between the group at the pyrrole nitrogen and the substituted vinyl moiety [529].

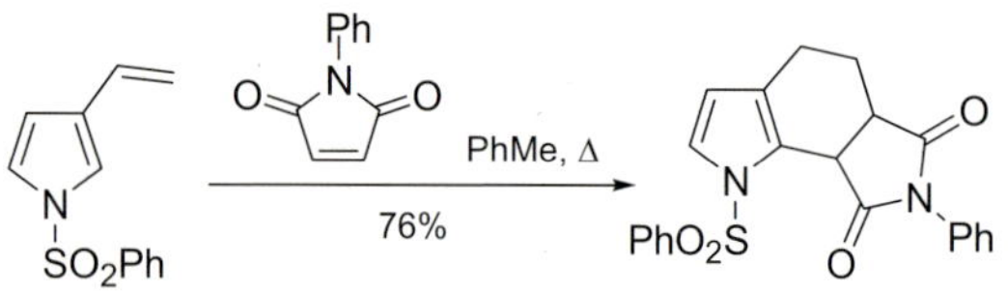

PhMe, Δ

76%

SO_2Ph

327

PhO_2S

328

Scheme 4.99

4.5.8
Reactions with Carbenes and Carbenoids

Although it was established early on that pyrroles readily undergo substitution reactions with carbenes [530] the synthetic utility of this reaction appears to be severely limited, as simple pyrroles, for example 1-methylpyrrole, give mixtures of C2 and C3 substituted products [531–533]. Interestingly, in this context, the reactivity rate of pyrrole versus naphthalene towards thermally generated (ethoxycarbonyl) carbene at 150 °C is 23 times higher [534]. A useful ring expansion reaction occurs when 2,5-dimethylpyrrole (**329**) is exposed to dichlorocarbene under neutral aprotic conditions, rendering the pyridine **330**. Under basic protic conditions, very little of **330** is formed, along with even lower amounts the 2*H*-pyrrole **331** [535]. A similar mixture of products, although in considerably higher total yield, is obtained under basic conditions in the presence of a phase transfer catalyst [536]. The pyridine product may result from a rearrangement of the dichlorocyclopropane intermediate **332**, in analogy with observations during studies on similar reactions of 2-*tert*-butyl-5-methylpyrrole [537].

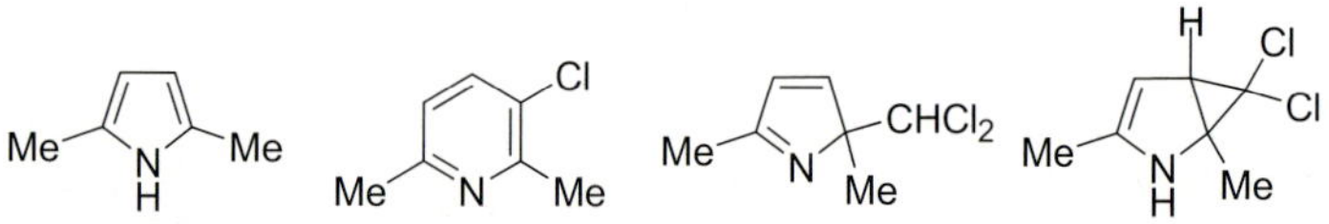

329 **330** **331** **332**

Addition of carbenes to pyrroles possessing electron-withdrawing groups at the nitrogen atom is recognized to lead to cyclopropanation, as illustrated by the transformation of pyrrole **302** into the bicyclic system **333**, with co-formation of the product **334** (Scheme 4.100) [538, 539]. An analogous outcome giving cyclopropanated products can be observed upon CuCl-catalyzed decomposition of diazomethane in the presence of **302** [539, 540]. Similar conditions involving CuOTf $^{1}/_{2}C_6H_6$ as the catalyst allow cyclopropanation of 1-acylpyrroles with methyl diazoacetate in up to 44% yield [541].

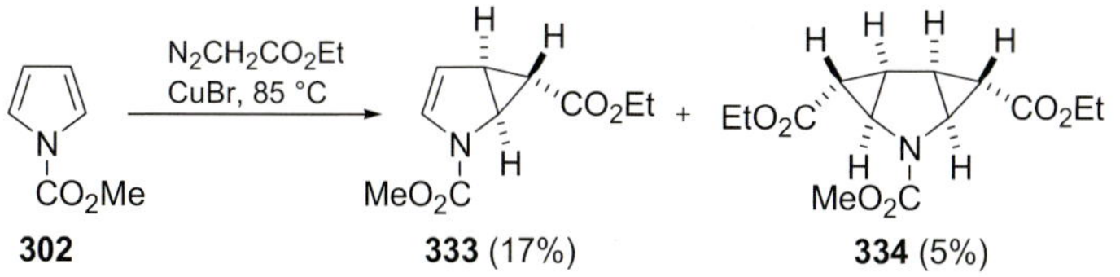

Scheme 4.100

Further studies also indicated that **333** undergoes rearrangement to the dihydropyridine derivative **335** by heating at 285 °C. This is also true for the diazomethane adduct **336**, which is completely converted into **337** within 30 min under the same conditions (Scheme 4.101) [539]. Based on detailed mechanistic studies, these transformations were suggested to proceed through the intermediate acyclic azatrienes **338**/**339**, which give the dihydropyridines **335**/**337** after final 6π-electrocyclization [539, 542].

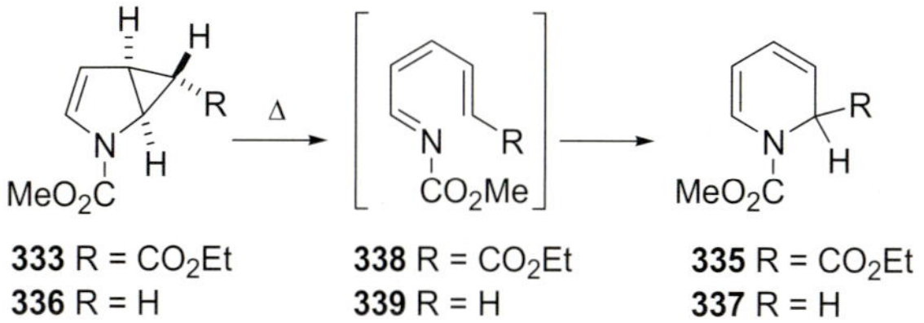

Scheme 4.101

Stabilized vinyl carbenes also add readily to deactivated pyrroles, for example **340**, providing the expected cyclopropanated intermediates **341**, which, however, can not be isolated, as an ensuing Cope rearrangement leads directly to the tropane skeleton **342** (Scheme 4.102) [543, 544]. In analogy with previous findings (see above), pyrrole

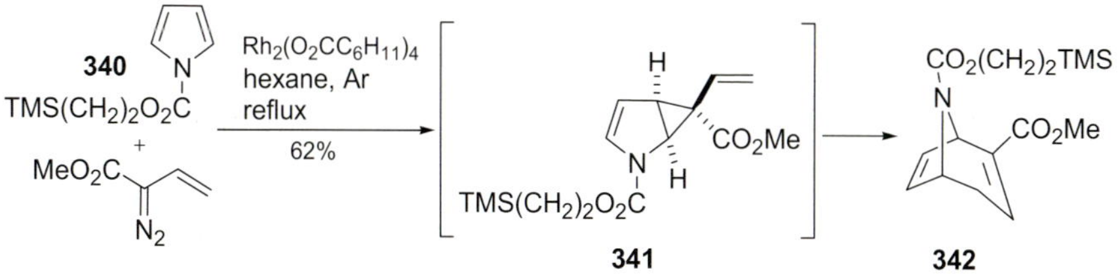

Scheme 4.102

itself gives instead a mixture of C2 and C3 substitution products under similar conditions [543].

4.5.9
Photochemical Reactions

The photochemistry of pyrroles has been studied relatively scantily, and only few synthetically useful procedures have appeared over the years. Interestingly, during the irradiation of pyrrole-2-carbonitrile (**343**) in methanol solution the isomeric pyrrole-3-carbonitrile **344** was isolated in 55% yield as the major product (Scheme 4.103) [545]. This intriguing isomerization has been suggested to encompass initial generation of the unstable Dewar pyrrole **345**, and a subsequent rearrangement involving the aziridine nitrogen to form intermediate **346** [546]. The *N*-ethoxycarbonyl derivative of the parent ring system of **345** has, interestingly, been generated and trapped as an adduct with 1,3-diphenylisobenzofuran [547], although the formation and reactions of tetrakis(trifluoromethyl) derivatives thereof had been reported previously [548, 549].

hν, MeOH 55%

343 345 346 344

Scheme 4.103

The photo-induced reaction of pyrroles with aldehydes or ketones provides a route to C3 substituted pyrroles, as illustrated by the conversion of **190** into **347** (Scheme 4.104). Based on NMR studies of the reaction mixture, the oxetane **348** was proposed as a conceivable intermediate [550]. Similar reactions of 1-benzoylpyrrole with 3- or 4-benzoylpyridine led to the isolation of low yields of related bis-adducts [551].

Me_2CO, hν 40%

190 347 348

Scheme 4.104

Light-induced cyclizations of suitable pyrrole containing precursors offer useful routes to more complex fused pyrroles. Thus the heterocyclic stilbene analog **349** undergoes conversion into the system **350** upon irradiation in the presence of Pd/C

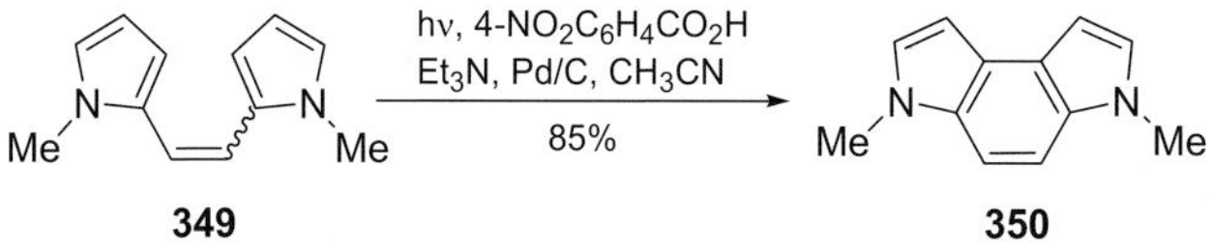

hν, 4-$NO_2C_6H_4CO_2H$
Et_3N, Pd/C, CH_3CN
85%

349 → 350

Scheme 4.105

as the dehydrogenating catalyst (Scheme 4.105) [552]. Photochemical annulation reactions of related precursors may also be performed in the presence of iodine with access to air, or alternatively by employing iodine and excess propylene oxide under an argon atmosphere [553].

4.5.10
Pyrryl-C-X Compounds: Synthesis and Reactions

Pyrrolyl-C-X compounds constitute an important class of derivatives, as pyrroles possessing aminomethyl- or hydroxymethyl substituents are excellent substrates in reactions with nucleophilic reagents. Conversion of the readily available 2-(dimethylamino)pyrrole **232** into the methoiodide **351**, followed by treatment with NaCN provides convenient access to (pyrrol-2-yl)acetonitrile (**352**, Scheme 4.106) [554], whereas a similar displacement with the anion of diethyl phosphite gives the useful diethyl (pyrrol-2-yl)methylphosphonate in 91% yield [555]. This type of elimination–addition processes with N-substituted pyrroles has been implied to involve the intermediacy of azafulvenium ions [287] (Section 4.5.1.7), or in the case of pyrrole itself the azafulvene **353** [556, 557], both of which are prone to attack by nucleophiles at the *exo*-cyclic carbon. Likewise, 2,5-bis(dimethylaminomethyl)pyrrole may, for example, be converted into the corresponding 2,5-bis(phenylthiomethylene)pyrrole via an intermediate quaternization with iodomethane in good overall yield [558]. However, products that presumably resulted from nucleophilic attack at C5 of 1-methylazafulvenium ions have also been observed [555, 559].

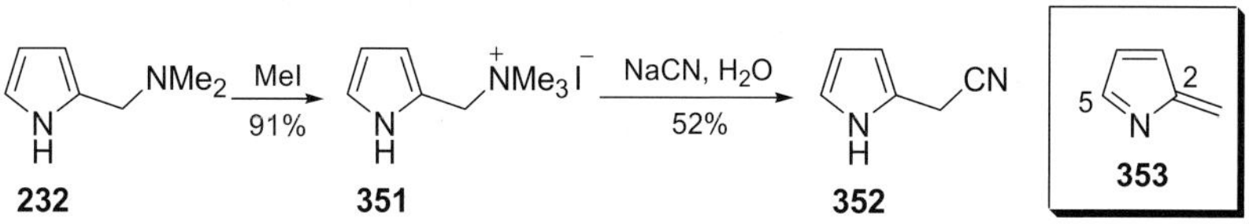

232 → (MeI, 91%) → 351 → (NaCN, H_2O, 52%) → 352; 353

Scheme 4.106

Pyrroles containing acetoxymethyl groups at C2 are particularly useful for the construction of di(pyrrol-2-yl)methanes [560–562]. The reaction of the 2-(acetoxymethyl)pyrrole derivative **354** with **355** leading to the molecule **356** (Scheme 4.107) serves as an excellent illustration of the applicability of such approaches [561]. Related transformations may also be carried out using 2-(hydroxymethyl)pyrroles [341, 563], as well as 3-(hydroxymethyl)pyrroles, which display similar behavior, and may for instance be converted into the corresponding 3-(cyanomethyl)pyrroles upon

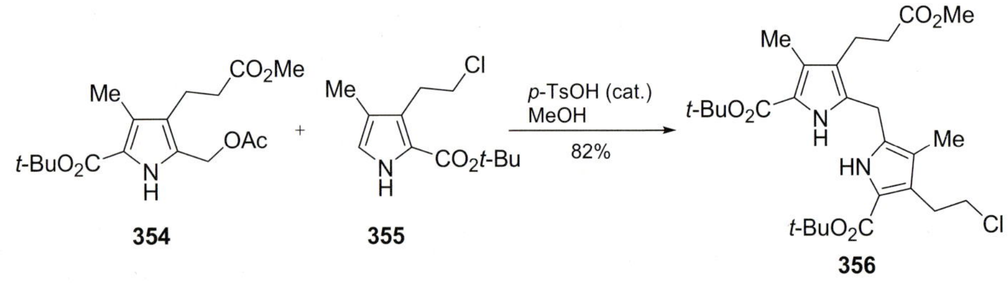

Scheme 4.107

treatment with NaCN [564]. Introduction of the strong electron-withdrawing (trifluoromethyl)sulfonyl group at the nitrogen of 2-(hydroxymethyl)pyrroles blocks the pathway involving azafulvenium ions, thus enabling Mitsunobu-type reactions at the hydroxymethyl moiety [565].

Di(pyrrol-2-yl)methanes have also been prepared under non-acidic conditions, as demonstrated by the reaction of a magnesium derivative of **357** with the 2-(chloromethyl)pyrrole **358** to afford the product **359** (Scheme 4.108). The compound **358** is readily available from pyrrole-2-carboxaldehyde by N-protection, followed by reduction, yielding a 2-(hydroxymethyl)pyrrole, and treatment thereof with methanesulfonyl chloride in the presence of Hünig's base [566]. 2-(Haloalkyl)pyrroles may also be used in displacement reactions with, for example, azide ions [567], pyridine or alkoxide ions [568], giving additional useful synthetic intermediates.

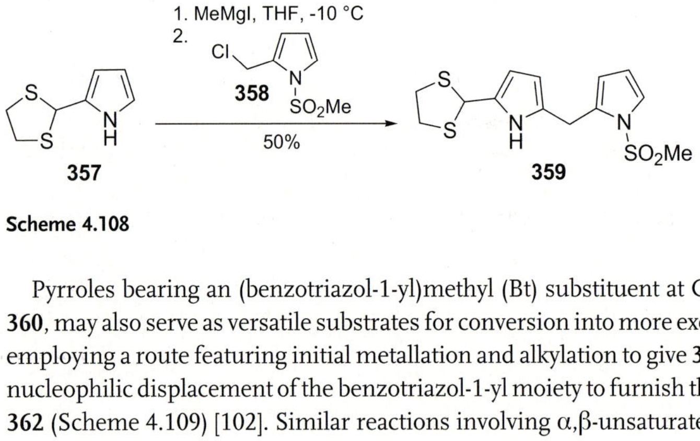

Scheme 4.108

Pyrroles bearing an (benzotriazol-1-yl)methyl (Bt) substituent at C2, for example **360**, may also serve as versatile substrates for conversion into more exotic derivatives, employing a route featuring initial metallation and alkylation to give **361**, followed by nucleophilic displacement of the benzotriazol-1-yl moiety to furnish the final product **362** (Scheme 4.109) [102]. Similar reactions involving α,β-unsaturated aldehydes or

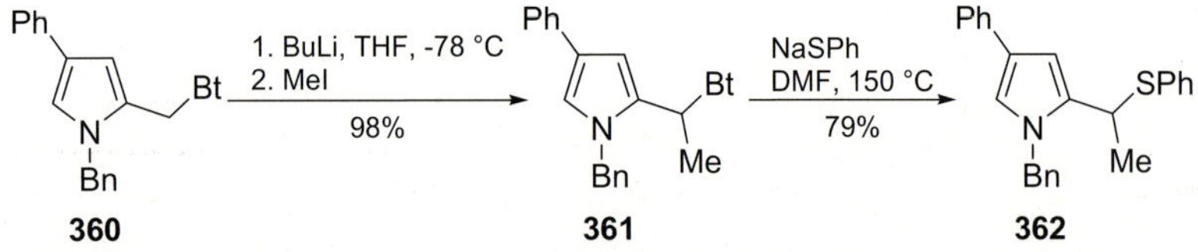

Scheme 4.109

ketones instead of alkyl halides as the electrophiles, eventually leading to indole derivatives, have also been reported [569].

Reduction of 3-acyl-1-(*p*-toluenesulfonyl)pyrroles with 0.5 equivalents of $NaBH_4$ in refluxing dioxane containing 1 equivalent of *i*-PrOH gives the corresponding intermediate 1-(pyrrol-3-yl)methanol derivatives, which undergo dehydration in hot DMSO, rendering 3-vinylpyrroles [570]. C-Vinylpyrroles are useful in various applications, for example cycloaddition reactions (Section 4.5.7), and comprehensive reviews highlighting the preparation [571] and synthetic uses of these compounds have appeared quite recently [572].

4.5.11 Transition Metal Catalyzed Coupling Reactions

The availability of stable halopyrroles, stannylpyrroles and pyrroleboronic acids has opened new possibilities for functionalization of the pyrrole nucleus by transition metal catalyzed reactions, enabling the synthesis of derivatives otherwise difficult to access [573]. A few reactions involving transition metal catalyzed C–H activation in pyrroles have also emerged.

There are only a few examples of N-arylation of pyrrole itself using aryl bromides, performed in the presence of *t*-BuONa and catalytic amounts of $Pd(OAc)_2$ and diphenylphosphinoferrocene (DPPF) and a suitable base [574], or employing the combination $Pd(dba)_2/P(t\text{-}Bu)_3/Cs_2CO_3$ [575]. Pyrrole has also been N-arylated with an aryl iodide using CuI/*trans*-1,2-cyclohexanediamine in the presence of K_3PO_4 as the catalytic system [576]. 2-Acetylpyrrole, as well as pyrrole-2-carboxaldehydes possessing an additional electron-withdrawing substituent at C4, undergoes efficient N-arylation with arylboronic acids at ambient temperature using stoichiometric amounts of $Cu(OAc)_2$ [577]. Treatment of pyrroles with vinyl triflates in the presence of the system $Pd_2(dba)_3$/Xphos/K_3PO_4 constitutes a new route to 1-vinylpyrroles [578]. It has also been reported that N-alkynylation of electron deficient pyrroles can be achieved using alkynyl bromides in the presence of catalytic amounts of $CuSO_4{\cdot}5H_2O$ and 1,10-phenanthroline [579].

The readily available N-protected 2-bromopyrrole **155** is an excellent partner for Suzuki couplings, offering a convenient route to the 2-arylpyrroles **363** (Scheme 4.110) [580, 581]. An alternative approach is based on coupling of the pyrrole-2-boronic acid **266** with aryl bromides or iodides [582]. Suzuki reactions have also been performed using 1-(phenylsulfonyl)pyrrole-2-boronic acid [583], 2-bromo-1-(*p*-toluenesulfonyl)pyrrole [241] and 2-iodo-1-(phenylsulfonyl)pyrrole [412]. In

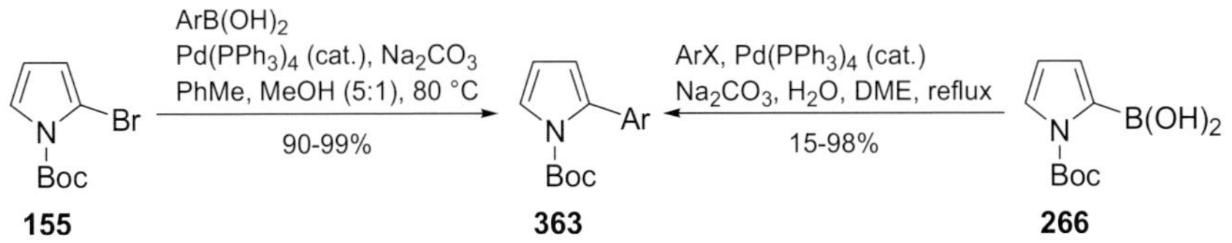

ArB(OH)$_2$, Pd(PPh$_3$)$_4$ (cat.), Na$_2$CO$_3$, PhMe, MeOH (5:1), 80 °C, 90-99%: **155** (N-Boc-2-bromopyrrole) → **363** (N-Boc-2-arylpyrrole)

ArX, Pd(PPh$_3$)$_4$ (cat.), Na$_2$CO$_3$, H$_2$O, DME, reflux, 15-98%: **266** (N-Boc-pyrrole-2-B(OH)$_2$) → **363**

Scheme 4.110

addition, it has been demonstrated that electron deficient di- or tribrominated pyrroles undergo selective Suzuki reactions at the α-position with phenylboronic acid derivatives [584]. Heck reactions at the α-position of iodinated pyrroles involving vinylbenzene derivatives [585, 586], as well as Pd(0)-catalyzed couplings with alkynes [587], have also been reported.

Suzuki coupling of the triflate **364** with the boronic acid **266** has been employed as the final step in an elegant synthesis of undecylprodigiosin (**365**) (Scheme 4.111) [588], as well as during preparation of a series of analogues thereof [589].

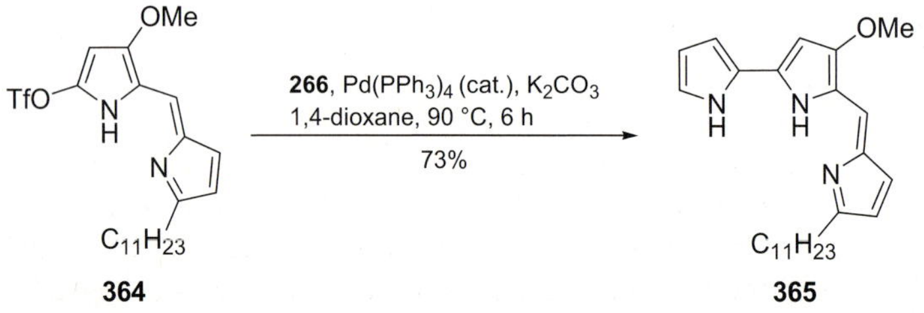

OMe
TfO
N
H
N
$C_{11}H_{23}$
364

266, $Pd(PPh_3)_4$ (cat.), K_2CO_3
1,4-dioxane, 90 °C, 6 h
73%

OMe
N H
N H
N
$C_{11}H_{23}$
365

Scheme 4.111

A new arylation reaction of pyrrolylsodium (**366**) (Section 4.5.4.1) has been developed. In a representative set of conditions, exposure of **366** to various aryl chlorides or bromides and $ZnCl_2$ in the presence of $Pd(OAc)_2$ and 2-(di-*tert*-butylphosphino)biphenyl (**367**), afforded 2-arylpyrroles **368** in high yields (Scheme 4.112). The products **368** may also be subjected to further arylation under similar conditions at the remaining free α-position, providing access to various 2,5-diarylpyrroles [590]. It has also been known for some time that palladium mediated arylation of 1-acylpyrroles with arenes occurs at the α-position. However, this requires stoichiometric amounts of the palladium source [591]. A recent interesting contribution involving C–H activation features regioselective palladium-catalyzed oxidative alkenylation of N-protected pyrroles at C2 or C3 using alkenes under aerobic conditions. The regioselectivity is highly dependent on the steric and electronic properties of the N-substituent of the substrate. For instance, the use of 1-(*tert*-butoxycarbonyl)pyrrole gives C2 alkenylated products, whereas implementation of 1-TIPS-pyrrole leads to functionalization at C3 [592].

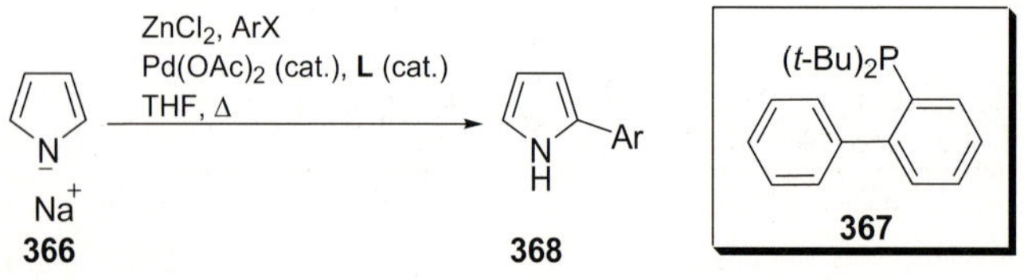

N
Na^+
366

$ZnCl_2$, ArX
$Pd(OAc)_2$ (cat.), **L** (cat.)
THF, Δ

N
H
Ar
368

$(t\text{-Bu})_2P$
367

Scheme 4.112

Cross-coupling techniques may also be applied to the preparation of various 3-substituted pyrroles. The pyrrole-3-boronic acid **281**, which is derived from 1-triisopropylsilyl-3-iodopyrrole, takes part in Suzuki couplings with both electron rich and electron deficient aryl halides (X=Br or I) to provide useful yields of the 3-arylpyrroles **369** (Scheme 4.113). Stille reactions of 1-triisopropylsilyl-3-(tributyl-stannyl)pyrrole with suitable aryl halides constitute an alternative route to pyrroles of type **369**, whereas palladium-catalyzed coupling of 1-triisopropylsilyl-3-iodopyrrole with terminal acetylenes gives high yields of the corresponding 3-ethynylpyrroles. The TIPS group in all products may be removed efficiently by treatment with Bu_4NF [418]. In connection with studies on Suzuki couplings involving ethyl 4-bromopyrrole-2-carboxylate, a competing dehalogenation of the halopyrrole was observed. This side reaction was, however, suppressed by using the corresponding N-Boc-protected derivative, leading to good yields of ethyl 4-arylpyrrole-2-carboxy-lates with a concomitant removal of the Boc-group during the process [593]. Other useful developments in this area encompass palladium-catalyzed coupling of 1-(*p*-toluenesulfonyl)-4-(tributylstannyl)pyrrole-2-carboxaldehyde with aryl- and hetero-aryl halides giving the corresponding 4-arylpyrrole-2-carboxaldehydes [594], and synthesis of 3-vinylpyrroles by Stille reactions between various 3-iodopyrroles and vinyltributyltin [595]. In addition, Suzuki couplings involving the triflate derived from 1-benzylpyrrolidine-3-one are accompanied by concomitant dehydrogenation, giving access to 3-aryl-1-benzyl-pyrroles [596]. Finally, both 2-iodo-1-(phenylsulfonyl) pyrrole and its 3-iodo isomer are efficiently cyanated using CuCN in the presence of catalytic amounts of $Pd_2(dba)_3$ and dppf [597].

$B(OH)_2$ / N-TIPS **281** — ArX, $Pd(PPh_3)_4$ (cat.), Na_2CO_3; PhH, H_2O, MeOH, reflux; 59-96% → Ar / N-TIPS **369**

Scheme 4.113

In an interesting approach towards unsymmetrically 3,4-disubstituted pyrroles, the 3,4-disilylated pyrrole derivative **125** was N-protected, followed by an *ipso*-iodination to provide the key intermediate **370**, which was in turn subjected to various cross-couplings, providing, for example, the products **371** via the Sonoga-shira reaction (Scheme 4.114) [246, 247], or the corresponding aryl derivatives

TMS, TMS / N-H **125** — 1. Boc_2O, DMAP, CH_2Cl_2; 2. I_2, CF_3CO_2Ag, THF, -78 °C; 88% → TMS, I / N-Boc **370** — ≡–R, $PdCl_2(PPh_3)_2$, CuI, Et_2NH, rt; 83-92% → TMS, C≡C–R / N-Boc **371**

Scheme 4.114

employing Suzuki conditions. A second *ipso*-iodination and subsequent palladium-catalyzed coupling reactions give access to further derivatives [247]. Moreover, phosphine-free Suzuki reactions involving methyl 4,5-dichloro-3-iodopyrrole-2-carboxylate occur selectively at C3, giving access to 3-arylpyrrole-2-carboxylates after final removal of the chlorine atoms by catalytic hydrogenation [307].

4.6 Pyrrole Derivatives

4.6.1 Alkyl Derivatives

N-Alkylation of pyrroles is a well documented process that is conveniently accomplished by treatment of pyrrolyl anions with suitable alkylating agents under various conditions (Section 4.5.4.1).

Direct C-alkylation is often not particularly practical, as it has been demonstrated that treatment of pyrrolyl anions with allyl-, crotyl- and benzyl-halides gives mixtures of N- and C- monoalkylated products, along with disubstituted derivatives [375]. The methods of choice for the synthesis of alkylpyrroles rely on pyrrole ring formation from acyclic precursors (Section 4.4), or reduction of readily available 2- or 3-acylpyrroles (Section 4.5.1.6). Reduction of 2-benzoylpyrrole (**372**) with $NaBH_4$ in boiling 2-propanol gives a high yield of 2-benzylpyrrole (**373**) (Scheme 4.115) [598]. An alternative procedure employs a *tert*-butylamine–borane complex in the presence of $AlCl_3$ as the reducing system, enabling for example effective transformation of the N-protected 2-acylpyrrole **374** into the corresponding 2-alkylpyrrole **375**. This approach is also useful for the reduction of 1-phenylsulfonylpyrrole-2-carboxaldehyde to the corresponding 2-methylpyrrole derivative [599].

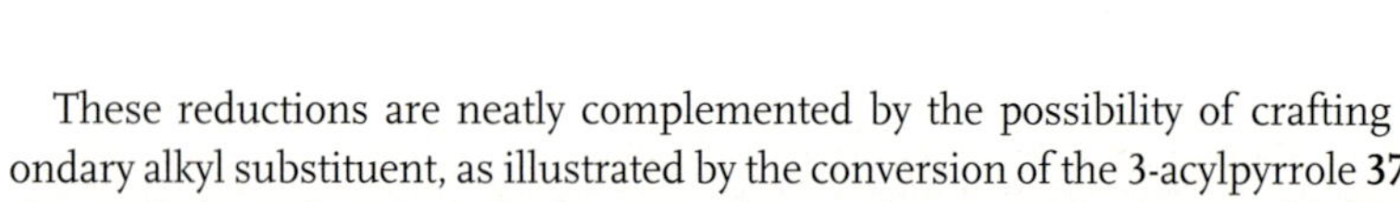

$NaBH_4$, *i*-PrOH, reflux, 99%; *t*-$BuNH_2 \cdot BH_3$, $AlCl_3$, CH_2Cl_2, rt, 87%

Bn; R_2; R^1; O; *n*-Pr; SO_2Ph

373

372 R^1 = H, R^2 = Ph
374 R^1 = SO_2Ph, R^2 = Et

375

Scheme 4.115

These reductions are neatly complemented by the possibility of crafting a secondary alkyl substituent, as illustrated by the conversion of the 3-acylpyrrole **376** into the 3-alkylpyrrole **377**, which proceeds via the unstable tertiary alcohol **378** (Scheme 4.116) [598]. A somewhat related approach involving addition of organometallic reagents to 2-acylpyrroles, followed by reduction with lithium in liquid ammonia, has also been described [600].

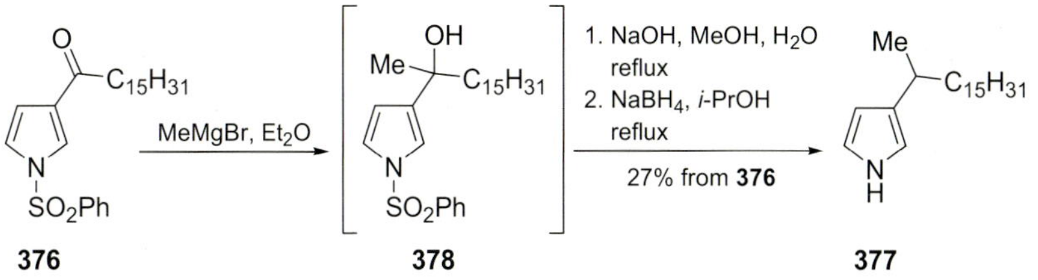

Scheme 4.116

4.6.2
Pyrrole Carboxylic Acids and Carboxylates

Many pyrrole carboxylic acids are readily available compounds, which are quite prone to decarboxylation. This particular characteristic is very attractive from a synthetic point of view, extending the scope of those pyrrole ring syntheses that give pyrrole carboxylates (Section 4.4). Pyrrole-2-carboxylates are useful substrates for further functionalization by means of electrophilic substitution, as the substituent is strongly "meta" directing, thereby allowing selective synthesis of 2,4-disubstituted pyrroles (Section 4.5.1.6). It is also worth mentioning that amides derived, for instance, from pyrrole-2,5-dicarboxylic acids have recently attracted interest as anion receptors and membrane transport agents for HCl [601].

Decarboxylation of pyrrole-3-carboxylic acids may, for example, be effected by heating [602], whereas ethyl pyrrole-3-carboxylates can be hydrolyzed and decarboxylated in one pot by heating with aqueous NaOH at 175 °C in a sealed vessel [124]. Based on kinetic studies, the mechanism of the decarboxylation of pyrrole-2-carboxylic acid **198** in acidic media has been suggested to proceed through the intermediate **379**, which eventually releases carbon dioxide (Scheme 4.117) [603]. A striking feature during saponification of pyrrolecarboxylates is the relatively high rate constant for pyrrole-2-carboxylates compared to that of the C3 substituted isomers. This behavior has been ascribed to the possibility of intramolecular hydrogen bonding in the intermediate resulting from the attack of a hydroxide ion on the carbonyl carbon in the C2 isomers [604, 605].

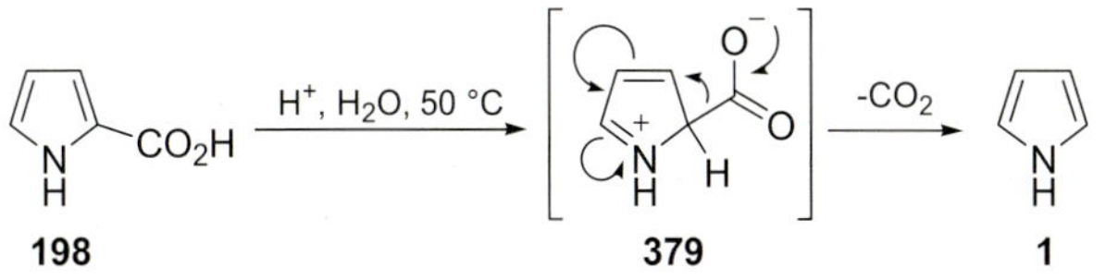

Scheme 4.117

4.6.3
Oxy Derivatives

1-Hydroxypyrrole derivatives are available by several routes, for example employing hydroxylamine in the Paal–Knorr pyrrole synthesis [606], by thermolysis of 1-(*tert*-butyl)-3-pyrrolin-1-oxide to 1-hydroxy-3-pyrroline [607] or via treatment of suitable monooximes derived from 1,2-dicarbonyl compounds with sodium hydride, followed by vinyltriphenylphosphonium bromide, which gives 2,3-disubstituted 1-hydroxypyrroles [608]. Deuterium exchange studies with D_2O in $CDCl_3$ performed on 1-hydroxy-2,3-diphenylpyrrole (**380**) indicated incorporation of deuterium at C4 and C5 along with the expected deuterium exchange at the oxygen, suggesting contributions from the tautomeric forms **381** and **382** (Scheme 4.118) [608].

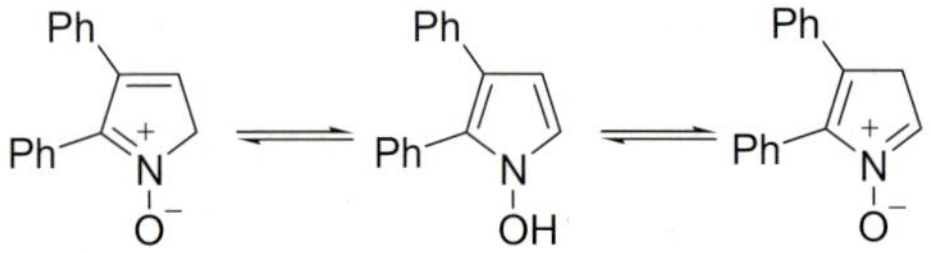

Ph Ph N+ O− **381** ⇌ Ph Ph N OH **380** ⇌ Ph Ph N+ O− **382**

Scheme 4.118

Oxidation of pyrrole employing hydrogen peroxide gives a modest yield of the tautomeric 2-oxypyrroles **383** and **384**, the former being the prevalent component as judged from NMR data (ratio **383** : **384** = 9 : 1 in acetone-d_6) (Scheme 4.119) [609]. Purified samples of **383** remain rather stable for several weeks upon storage at −10 °C, whereas the isomer **384** undergoes much faster isomerization [610]. After initial N-protection of **383**, access to a useful synthetic intermediate is gained by trapping of the corresponding 2-hydroxypyrrole tautomer as the silyl ether **385**, which reacts with aldehydes, rendering for example the 5-substituted pyrroline-2-one **386** [611, 612]. The oxypyrrole **385** has also been efficiently prepared on multi-kilogram scale via cyclization of racemic 4-amino-3-hydroxybutyric acid with HMDS

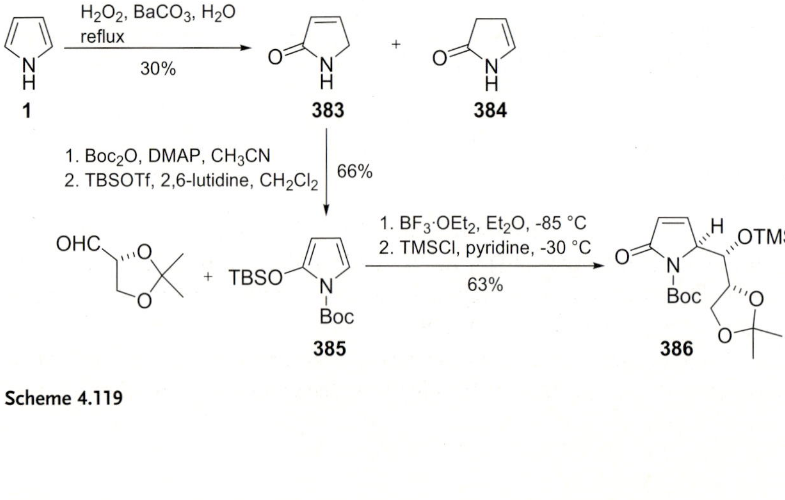

1 → (H_2O_2, $BaCO_3$, H_2O, reflux; 30%) **383** + **384**

383 → (1. Boc_2O, DMAP, CH_3CN; 2. TBSOTf, 2,6-lutidine, CH_2Cl_2; 66%) **385**

OHC-dioxolane + TBSO-pyrrole N-Boc **385** → (1. $BF_3{\cdot}OEt_2$, Et_2O, -85 °C; 2. TMSCl, pyridine, -30 °C; 63%) **386** (H, OTMS, N-Boc)

Scheme 4.119

in the presence of pyridine to 4-(trimethylsiloxy)pyrrolidine-2-one, which underwent subsequent Boc-protection, desilylation and elimination to yield 1-(*tert*-butoxycarbonyl)-3-pyrroline-2-one. This material could then be efficiently converted into **385** using TBSOTf in the presence of triethylamine [613]. It is also noteworthy that electrolytic fluorination of 2-cyano-1-methylpyrrole with $Et_3N{\cdot}3HF$ in acetonitrile, followed by treatment with water, gives 5,5-difluoro-1-methyl-3-pyrrolin-2-one, a useful starting material for the construction of some *gem*-difluorinated heterocycles [614].

The reaction of ethyl *N*-ethoxycarbonyl glycinate with ethyl fumarate in the presence of sodium in benzene solution, followed by decarboxylation, provides convenient access to the 3-oxypyrrole derivative **387** [615]. This material may for instance be further converted into the substituted 3-methoxypyrrole **388** by ketalization and dehydrogenation over Pd/C with concomitant cleavage of the carbamate functionality (Scheme 4.120) [616]. 3-Alkoxypyrroles have also been prepared by cyclization of alkyl 4-bromo-3-alkoxy-2-butenoates with suitable amines to 4-alkoxy-3-pyrrolin-2-ones, and subsequent treatment thereof with diisobutylaluminium hydride [617].

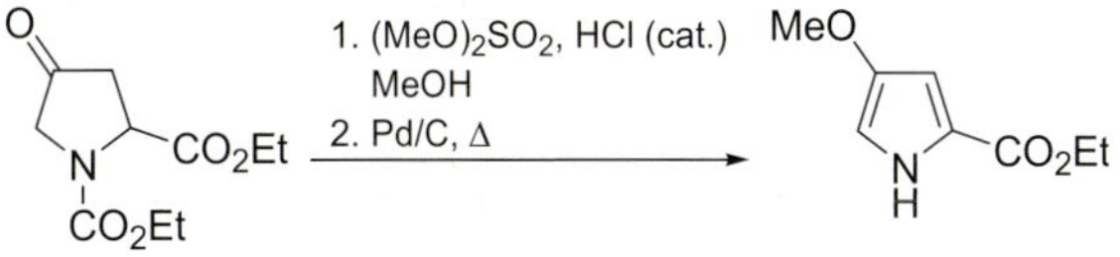

O
N
CO_2Et
CO_2Et
387
1. $(MeO)_2SO_2$, HCl (cat.)
MeOH
2. Pd/C, Δ
MeO
N
H
CO_2Et
388

Scheme 4.120

Cyclization of the precursors **389**, which are readily available from glycine esters and diethyl ethoxymethylenemalonate, provides a route to the stable 3-hydroxypyrrole derivatives **390** (Scheme 4.121). The C4 substituent may be selectively removed by alkaline hydrolysis, followed by decarboxylation [618].

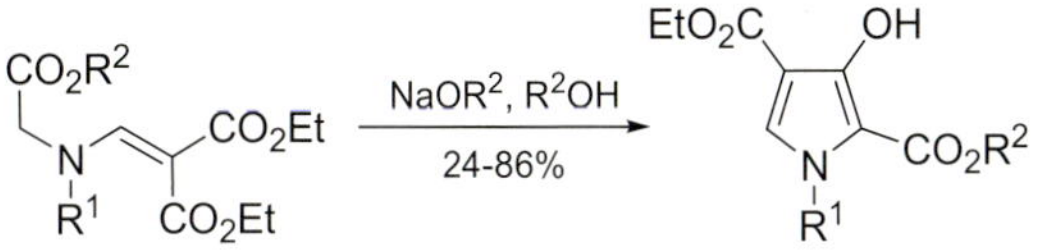

CO_2R^2
N
R^1
CO_2Et
CO_2Et
389
$NaOR^2$, R^2OH
24-86%
EtO_2C
OH
N
R^1
CO_2R^2
390

Scheme 4.121

In an approach based on intramolecular Wittig olefination, alkylation of, for example, *N*-acetylacetamide (**391**) with the reagents **392** (X = Br or Cl) afforded the precursor **393**, which was in turn cyclized to the 3-oxypyrrole derivative **394** under thermal conditions (Scheme 4.122) [619].

Apart from the examples mentioned above, 3-oxypyrroles are also available by for instance flash vacuum pyrolysis of N,N-disubstituted aminomethylene derivatives of

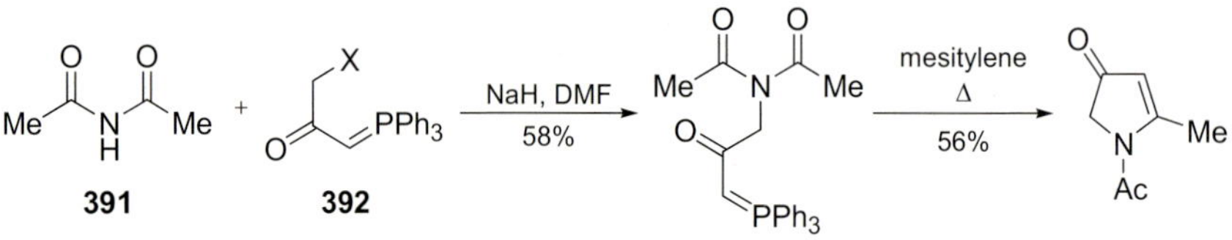

Scheme 4.122

Meldrum's acid [620]. A representative member of this class, 1-phenyl-1*H*-pyrrol-3(2*H*)-one, exists as mixtures of the tautomeric forms **395** or **396** depending on the medium. In general, polar solvents favor the enol form **396**, whereas in the solid state the keto tautomer **395** alone was detected (Scheme 4.123) [621]. The enol form was also detected as the prevalent species in DMSO solution in the case of the parent 3-hydroxypyrrole [622]. In addition, preference for the enol tautomers has been observed in connection with studies of derivatives containing an ester functionality at the adjacent C4 position [623]. Regiospecific O-alkylation of 3-hydroxypyrroles can be accomplished in polar aprotic solvents such as dimethylimidazolidinone (DMI), using the hard alkylating agent methyl *p*-toluenesulfonate to give, for example, pyrrole **397**. In contrast, the use of soft alkylating agents, such as iodomethane in relatively nonpolar solvents, leads to increased amounts of C-alkylated products [624].

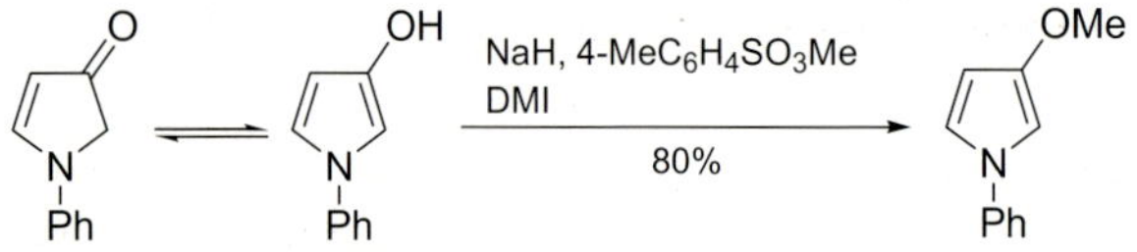

Scheme 4.123

Derivatives of tetramic acid (**398**) constitute a relatively large class of oxygenated pyrroles, which has been studied in considerable detail [35, 625]. The parent compound **398** is a relatively weak acid, (pK_a=6.4 in water), which exists in the keto from in the solid state, whereas in aqueous solution a minor contribution from the enol **399** may be discerned [626]. The enolization behavior of 3-acetyltetramic acids is more complex; studies involving for instance the 3-acetyl-5-isopropyl derivative revealed a considerable contribution from the *exo*-enol tautomer **400** [627].

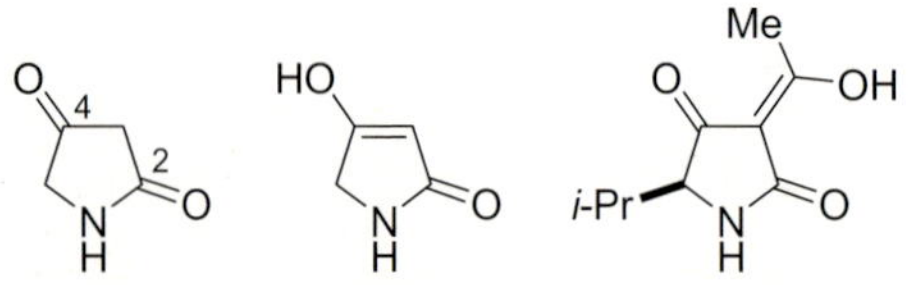

The most practical and widely used approach to tetramic acid derivatives has been developed by Lacey, and involves Dieckmann cyclization of *N*-acyl-α-amino esters. For example, the precursor **401**, which is available by treatment of ethyl glycinate with diketene, gives 3-acetyltetramic acid **402** upon treatment with sodium methoxide (Scheme 4.124) [628]. Application of these conditions to substrates derived from optically active amino acids may cause racemization, which can, however, be avoided by conducting the cyclization in the presence of TBAF in THF, or potassium *tert*-butoxide in *tert*-butanol during short periods of time. These modified routes also involve generation of the acyclic precursors from β-ketothioesters and amino acids [629]. Suitable precursors to tetramic acids may also be prepared by treatment of hippuric acid [630] or aceturic acid [631] derivatives with anions of active methylene compounds. The Dieckmann cyclization strategy has also been utilized in a solid phase approach starting from amino acid derivatives attached to the resin by an ester linkage [632]. A recent contribution to this field encompasses preparation of 5-substituted teramic acid derivatives by cyclocondensation of amidines with DMAD, followed by alkaline hydrolysis of the intermediate 5-amino-4-pyrrolin-3-ones [633].

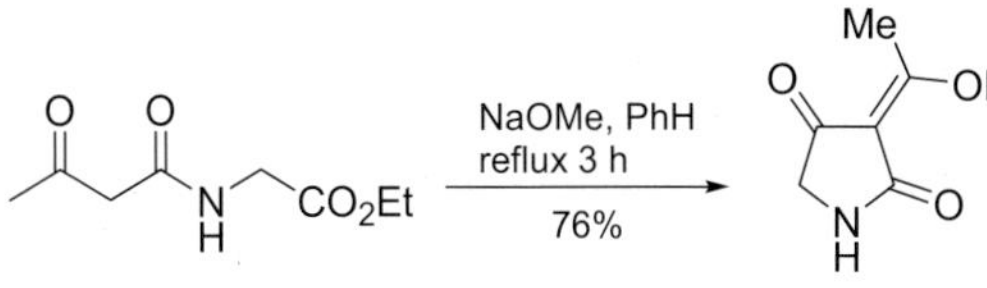

NaOMe, PhH, reflux 3 h, 76%

401 → **402**

Scheme 4.124

The methyl tetramate **403**, as well as several similar compounds, is available by treatment of methyl 4-bromo-3-methoxy-2-butenoate **404** with methylamine [634], and may be converted into the corresponding 3-alkoxypyrroles, for instance **405**, by treatment with diisobutylaluminium hydride (Scheme 4.125) [617]. Furthermore, the representative tetramate **403** undergoes metallation at C5 upon exposure to butyllithium, and the resulting lithio derivative gives access to various C5 substituted products after subsequent treatment with suitable electrophiles [635]. It has also been established that methyl tetramates of type **403** having an isopropyl moiety [636] or two substituents [637] at C5 may be acylated at C3 via lithiation, followed by introduction of aldehydes, and oxidation of the intermediate alcohols. The active methylene unit of tetramates may also participate in condensation reactions with aldehydes in the presence of sodium hydroxide in aqueous DMSO [588].

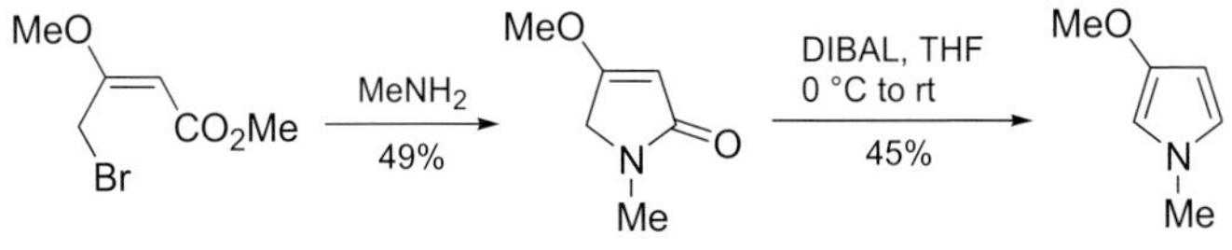

404 → ($MeNH_2$, 49%) → **403** → (DIBAL, THF, 0 °C to rt, 45%) → **405**

Scheme 4.125

4.6.4 Aminopyrroles

Simple aminopyrroles are highly electron rich and thus often labile species, but the stability may be improved considerably by the presence of electron-withdrawing substituents. A series of 1-aminopyrroles, including **406**, has been prepared by N-amination of the corresponding NH-pyrroles with anhydrous ethereal NH_2Cl in the presence of NaH [638]. Recently, a more convenient N-amination protocol for pyrroles has been realized under phase transfer conditions using *in situ* generated chloramine as the electrophilic aminating agent [639]. Interestingly, 1-(*N*,*N*-dimethylamino)pyrrole (**407**) can be easily prepared by heating 2,5-dimethoxytetrahydrofuran and *N*,*N*-dimethylhydrazine in acetic acid, and is cleanly lithiated at C2 [416].

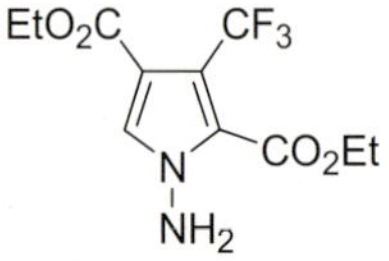

EtO_2C CF_3 CO_2Et N NH_2 **406**

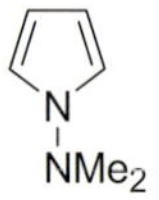

N NMe_2 **407**

The parent 2-aminopyrrole (**408**) was generated from the (pyrrole-2-yl)phthalimide **409** [640], which is in turn available by treatment of 1-trimethylsilylpyrrole with chlorophthalimide followed by aqueous workup (Scheme 4.126) [230, 641]. The aminopyrrole **408**, as well as 1-alkyl- and 1-aryl derivatives thereof, have to be kept in acid solution. A fast proton exchange at C5 in this series was observed in glacial acetic acid [640], and further NMR studies indicated that the conjugate acids of the 2-aminopyrroles exist as protonated imines under these conditions [641]. It has also been demonstrated that 2-aminopyrroles incorporating an electron-withdrawing substituent at the adjacent β-carbon can undergo protonation at either the exocyclic nitrogen atom or C5, depending on the conditions, but react at the exocyclic nitrogen only with acylating agents, thus behaving like typical aromatic amines rather than enamines [642].

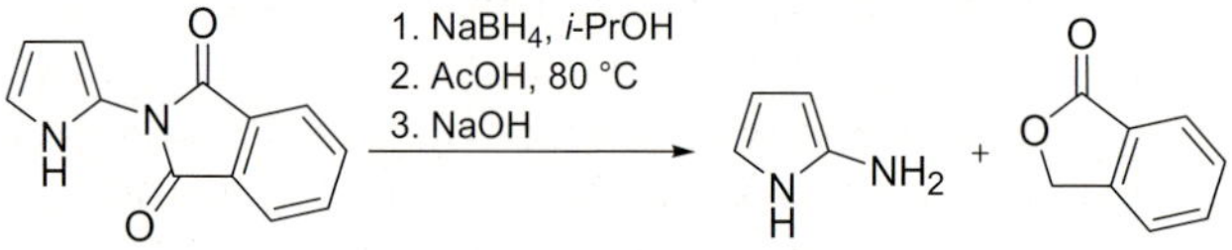

409 → 1. $NaBH_4$, *i*-PrOH; 2. AcOH, 80 °C; 3. NaOH → **408** (NH_2) + phthalide

Scheme 4.126

A useful approach featuring a pyrrole ring synthesis involves treatment of the readily available aminoacetaldehyde dimethyl acetals **410** with malononitrile under acidic conditions to provide facile access to the 2-aminopyrrole-3-carbonitriles **411** in moderate yields (Scheme 4.127). Similar products may also be prepared from ethyl 3-cyanopyrrole-2-carboxylates using a route based on the Curtius rearrangement [643]. 2-Aminopyrroles have also been obtained by base induced condensation reactions

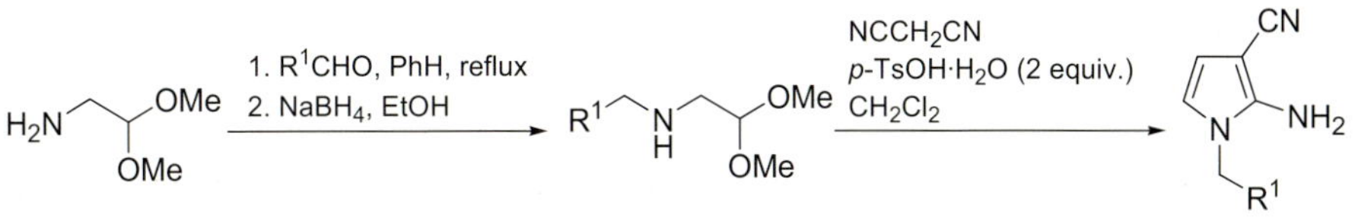

Scheme 4.127

between acetylaminoacetone and substituted acetonitriles [644], or from *N*-acetyl-α-aminoketones and malononitrile [645].

Yet another contribution to aminopyrrole chemistry is represented by a procedure for palladium-catalyzed amination of 2-acetyl-5-bromo-1-methylpyrrole (**412**), which can be performed using various primary and secondary amines, giving for example the 2-aminopyrrole derivative **413** (Scheme 4.128) [646]. An effective amidation of methyl 4-bromo-1-methylpyrrole-2-carboxylate with *tert*-butyl carbamate in the presence of the combination $CuI/K_3PO_4/N,N'$-dimethylethylenediamine has also been reported [236].

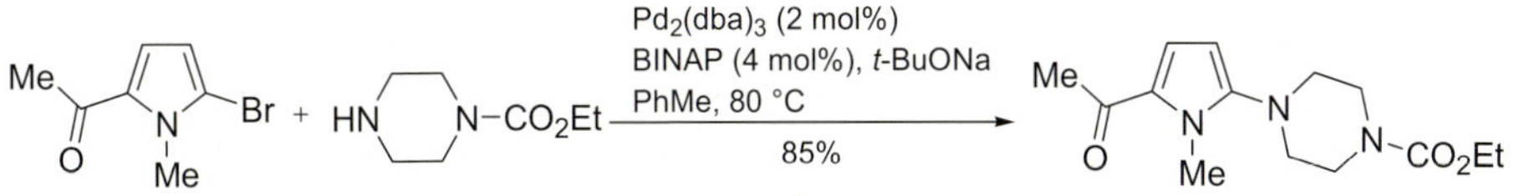

Scheme 4.128

3-Aminopyrroles featuring electron-withdrawing moieties have attracted some interest as building blocks for pyrrolo[2,3-*d*]pyrimidines [647, 648]. Consequently, several synthetic approaches to such derivatives have been described. The α-cyanoaldehydes **414**, which are derived from suitable aldehydes by base induced condensation with 3,3-dimethoxypropionitrile, followed by hydrolysis of the acetal and catalytic hydrogenation, serve as excellent substrates for the generation of enamine intermediates **415** by treatment with diethyl aminomalonate (Scheme 4.129). A final cyclization step afforded the 3-aminopyrrole derivatives **416** in moderate to good overall yields [648]. A route based on cyclization of similar enamine intermediates has been used for the synthesis of methyl 3-amino-4-arylpyrrole-2-carboxylates [649].

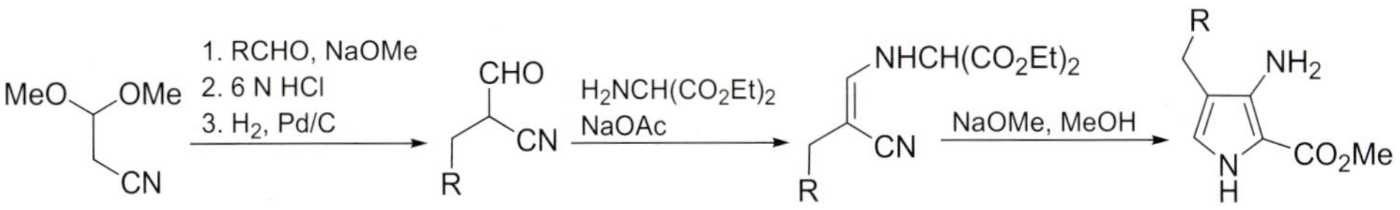

Scheme 4.129

A different substitution pattern is displayed in the pyrroles **417**, which were prepared by tosylation of the cyanoacetyl compounds **418** rendering the precursors **419** (Scheme 4.130). These intermediates were in turn converted into the target heterocycles by reaction with diethyl aminomalonate hydrochloride in the presence of ethoxide [650]. Treatment of benzyl 4-oxoproline-2-carboxylate derivatives protected at the nitrogen by the 9-(9-phenylfluorenyl) group with primary or secondary amines in the presence of catalytic amounts of *p*-TsOH provides an efficient route to 4-aminopyrrole-2-carboxylates [651].

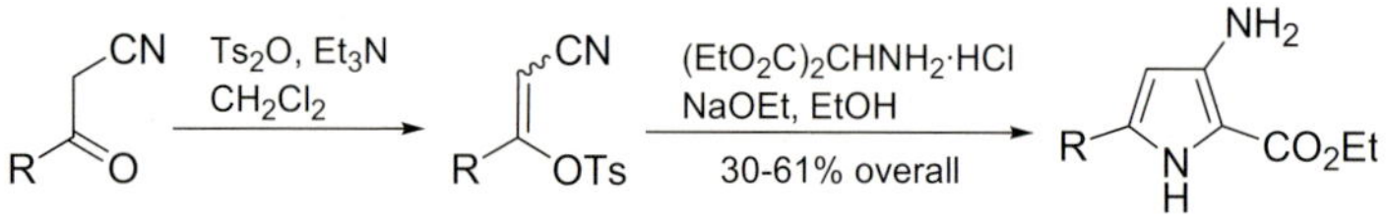

Ts_2O, Et_3N, CH_2Cl_2; $(EtO_2C)_2CHNH_2{\cdot}HCl$, NaOEt, EtOH, 30-61% overall

418 **419** **417**

Scheme 4.130

3-Amino-1-tritylpyrrole (**420**) has been prepared from the corresponding pyrrole-3-carboxylic acid in several steps using a route involving the Curtius rearrangement, and was demonstrated to exist exclusively as the 3-imino tautomer **421** in $CDCl_3$ solution [210]. A series of 1-arylpyrroles, as well as 1-methylpyrrole, has recently been shown to undergo amination, providing the interesting derivatives **422** in good yields using *N*-(*p*-toluenesulfonyl)imino-phenyliodinane (PhI=NTs) in the presence of a ruthenium(II) porphyrin catalyst [652].

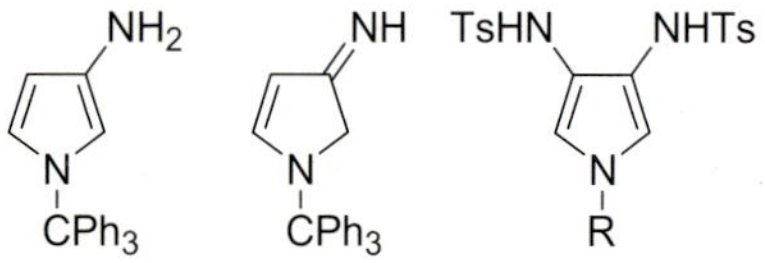

420 **421** **422**

4.6.5 Dihydro- and Tetrahydro-Derivatives

The scope of this chapter only allows inclusion of selected examples of procedures involving dihydro- and tetrahydro derivatives, but it is important to emphasize that many significant compounds belong to these thoroughly studied systems, for instance the amino acid L-proline. Both 2- and 3-pyrrolines (2,3-dihydro- and 2,5-dihydropyrroles), as well as pyrrolidines (tetrahydropyrroles), are available by reduction of pyrrole derivatives (Section 4.5.6). The reverse transformation, that is, conversion of pyrrolidines into pyrroles, may be accomplished for instance by dehydrogenation with MnO_2 in refluxing THF [653].

In more recent years, several useful ring syntheses of partially, or completely saturated pyrrole derivatives have emerged, for example 2,5-disubstituted pyrrolidines [654]. Several practical approaches rely on the cyclization of suitable diol derivatives, for instance **423**, which gives the *trans*-pyrrolidine **424** upon treatment with benzylamine (Scheme 4.131) [655]. Base induced annulation of mesylates derived from suitable chiral γ-aminoalcohols constitutes an alternative procedure for the stereoselective preparation of *cis*- or *trans*-2,5-disubstituted pyrrolidines [656].

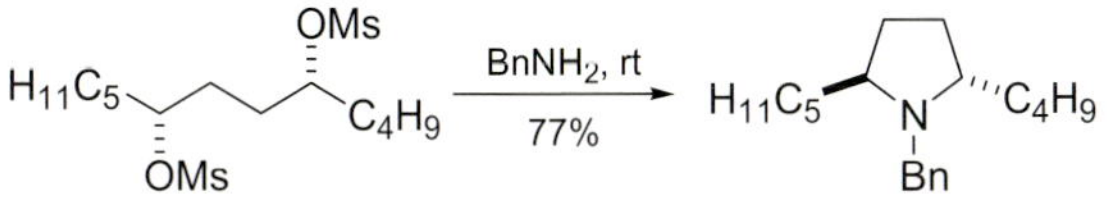

Scheme 4.131

Reductive amination of 1,4-diketones with ammonium acetate [657] or amines in the presence of $NaCNBH_3$ provides routes to pyrrolidines as mixtures of cis- and trans-isomers, the former being favored by increasing size of the amine reactant [658]. A related one-pot approach giving the 1-(2-naphthyl)methylene- or 1-(3,4-dimethoxybenzyl)pyrrolidines **425** involves reductive cyclization of the four-carbon precursors **426**, which are in turn available by conjugate addition of the α-(alkylideneamino)nitriles **427** to the α,β-unsaturated ketones **428** (Scheme 4.132) [659].

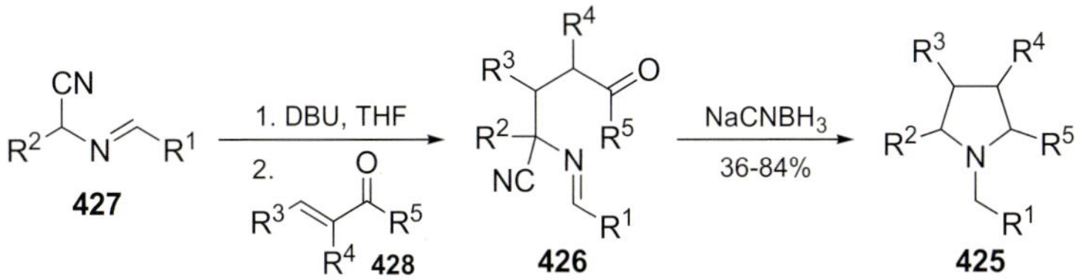

Scheme 4.132

Pyrrolidines containing sensitive substituents have been prepared under mild conditions using a tandem cationic aza-Cope rearrangement–Mannich process. Thus, treatment of the substituted 3-butenamines **429** with appropriate aldehydes in benzene or toluene affords moderate to excellent yields of the substituted 3-acetylpyrrolidines **430** (Scheme 4.133) [660].

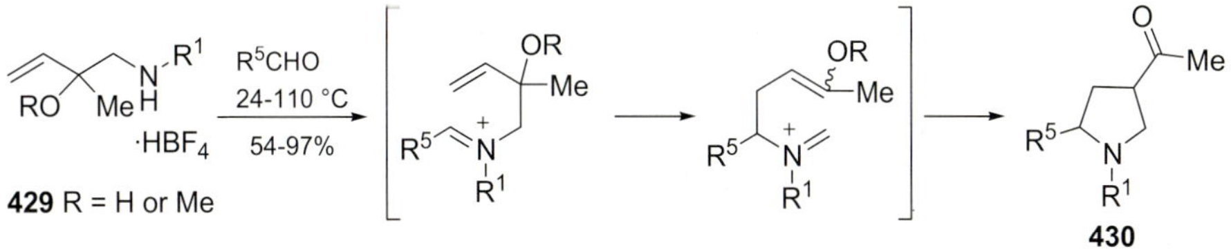

Scheme 4.133

A general approach to pyrrolidines involves 1,3-dipolar cycloadditions of alkenes and azomethine ylides, which are for instance available by desilylation of α-silyl iminium salts [661]. This strategy has been exploited in construction of the system **431**, which resulted from a reaction of the ylide **432** with *N*-methylmaleimide (Scheme 4.134) [662]. The various routes to enantiopure pyrrolidine derivatives based on cycloaddition reactions involving azomethine ylides have been recently summarized in a review [663].

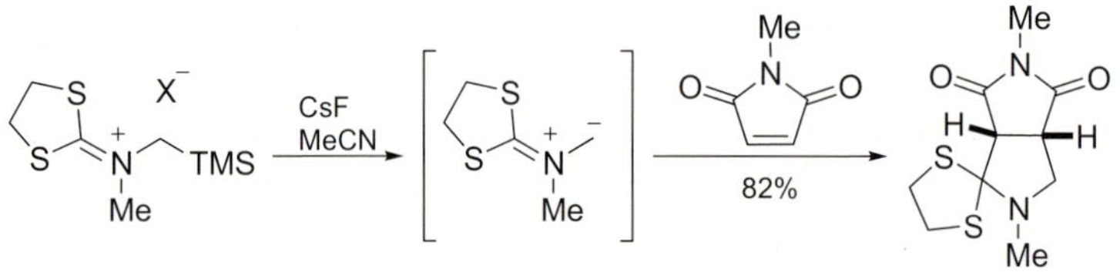
S
X⁻
S
N
TMS
Me
CsF
MeCN
S
S
N
Me
Me
N
O
O
82%
Me
N
O
O
H
H
S
S
N
Me

432 **431**

Scheme 4.134

The pyrrolidines **433** incorporating an exocyclic double bond have been synthesized by cycloaddition of the *N*-tosylimines **434** with a palladium complex derived from the precursor **435** (Scheme 4.135). Similar syntheses of various pyrrolidine derivatives may also be performed starting from aliphatic *N*-tosylimines (readily generated from the corresponding aldehydes by treatment with chloramine-T and elemental selenium), nitrimines or other electron deficient imines [664].

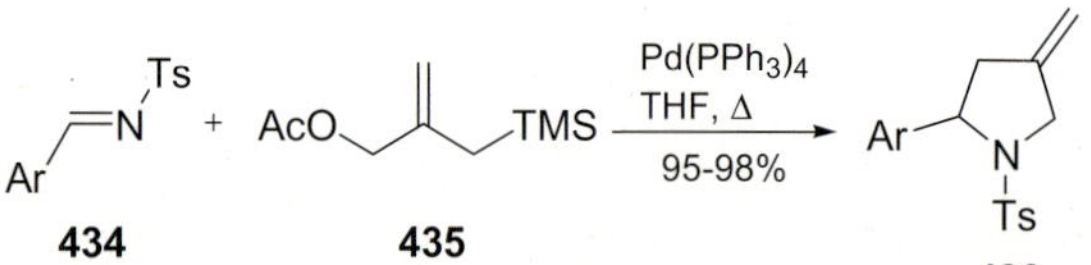
Ts
N
Ar
+
AcO
TMS
Pd(PPh3)4
THF, Δ
95-98%
Ar
N
Ts

434 **435** **433**

Scheme 4.135

Based on a previous approach for palladium-catalyzed cyclization of 3-alkynylamines yielding 1-pyrrolines [665], the precursors **436**, derived from a propargylglycine derivative via Sonogashira coupling, were deprotected under acidic conditions, followed by silver-catalyzed cyclization, providing a route to the 1-pyrrolines **437** (Scheme 4.136) [666]. Palladium-catalyzed ring closure of related precursors may instead give 2-pyrrolines, whereas the application of 4-alkynylamines in similar

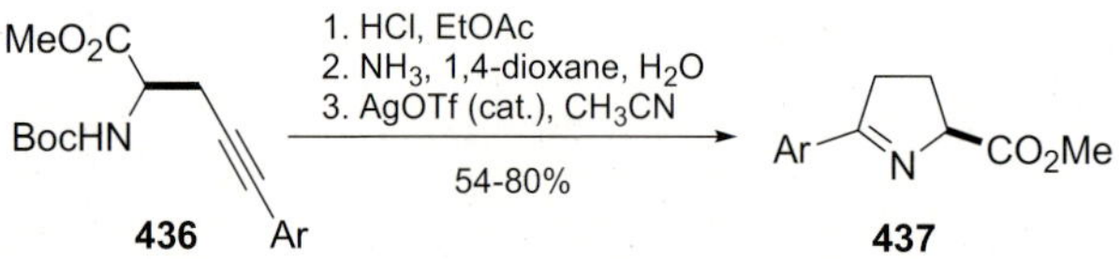
MeO2C
BocHN
Ar
1. HCl, EtOAc
2. NH3, 1,4-dioxane, H2O
3. AgOTf (cat.), CH3CN
54-80%
Ar
N
CO2Me

436 **437**

Scheme 4.136

processes gives pyrrolidines incorporating an exocyclic double bond [667]. Densely substituted 1-pyrrolines may also be accessed by 1,3-dipolar cycloaddition reactions between acrylamide derivatives and nitrile ylides generated from imidoyl chlorides [668]. Preparative routes and transformations involving 1-pyrrolines have been reviewed [669].

A traditional approach to 1-pyrrolines relies on addition of Grignard reagents to γ-halonitriles, which gives rise to the 2-substituted systems [670, 671]. It was later demonstrated that application of hydrocarbon/ether solvent mixtures for such reactions constitutes a practical modification [672]. Addition of but-3-enylmagnesium bromide to benzonitriles **438**, followed by treatment with NBS, leads to formation of 1-pyrrolines **439** (Scheme 4.137) [673], whereas chlorination using NCS gives the related 5-(chloromethyl)-1-pyrroline derivatives [674].

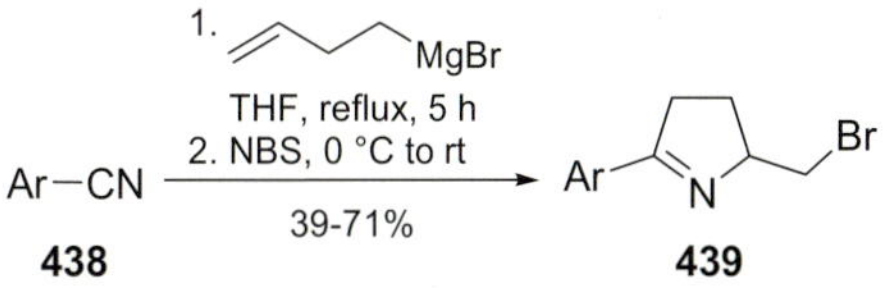

Scheme 4.137

Ring closing metathesis (RCM) of the enamides **440** (R^1=Ts, Bz, CO_2Et) has been employed for the preparation of the 2-pyrrolines **441** using the catalyst **442** (Scheme 4.138). In some cases, better yields were obtained using a related, Ru–imidazoline RCM catalyst [675].

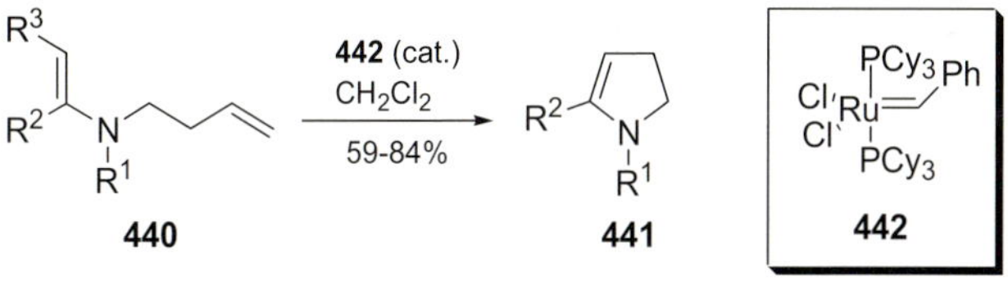

Scheme 4.138

Likewise, RCM methodology is also applicable for the construction of 3-pyrrolines, as illustrated by the synthesis of **443** from the diene **444** (Scheme 4.139) [676]. In a

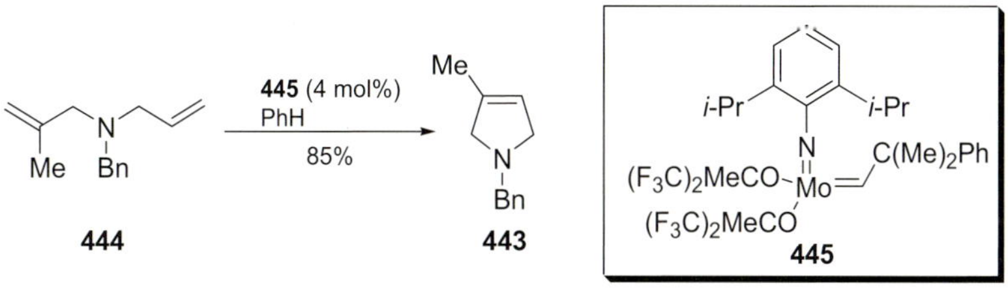

Scheme 4.139

related RCM approach to 3-pyrrolines, a suitable set of N-SES protected dienes were constructed from 2-(trimethylsilylethane)sulfonamide, aldehydes and methyl acrylate in an aza-Baylis–Hillman reaction [677]. On the other hand, the cyclization of diallylamines using the second generation Grubbs' catalyst (10 mol.%) in combination with $RuCl_3 \cdot H_2O$ (2 mol.%) gives rise to pyrroles in moderate yields [678]. An alternative route to 3-pyrrolines relies on triphenylphosphine-catalyzed [3 + 2] cycloaddition reactions of methyl 2,3-butadienoate with *N*-tosyl aldimines [679].

As a typical secondary amine, pyrrolidine displays pronounced basic character, and reacts readily as an N-nucleophile, affording, for example, amides. A well known application of pyrrolidines is its condensation with carbonyl groups to give enamines, which may subsequently be alkylated or acylated, providing an excellent and versatile route to α-substituted carbonyl compounds after a final hydrolysis step [680].

Reactions at the carbon atoms of simple pyrrolidines have been less studied. Nevertheless, several synthetically useful transformations have been described. It has for example been established that deprotonation and subsequent silylation of *N*-Boc-pyrrolidine (**446**) gives the intermediate **447** (Scheme 4.140) [681]. This material was subjected to a second lithiation/silylation cycle to provide the pyrrolidine derivative **448**. After removal of the Boc-group, followed by N-benzylation, the resulting pyrrolidine **449** was converted into the azamethine ylide **450**, which was trapped with ethyl propiolate to furnish adduct **451** [682]. An enantioselective cycloaddition between **450** and a chiral alkene has also been described [683]. Deprotonation of methyl 3-pyrroline-1-carboxylate with LDA at C2, and subsequent quenching with suitable electrophiles, provides an example of the functionalization of 3-pyrrolines [684].

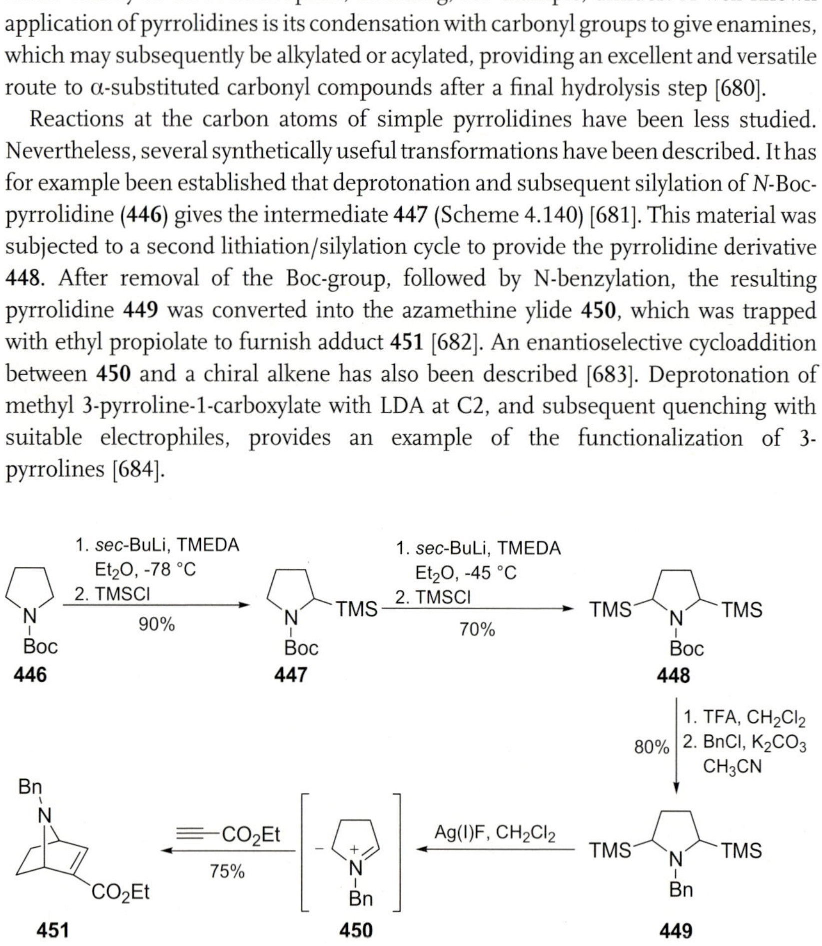

Scheme 4.140

In an interesting example of enantioselective functionalization at C2, *N*-Boc-pyrrolidine **446** was treated with *sec*-BuLi in the presence of (−)-sparteine (**452**) to produce the chiral lithio derivative **453**, which upon quenching with suitable electrophiles gave the 2-substituted derivatives **454** (Scheme 4.141) [685, 686]. Two

sec-BuLi, 452, Et$_2$O, -78 °C; E$^+$, 12-99%; 446; 453; 454 (ee up to 96%); 452

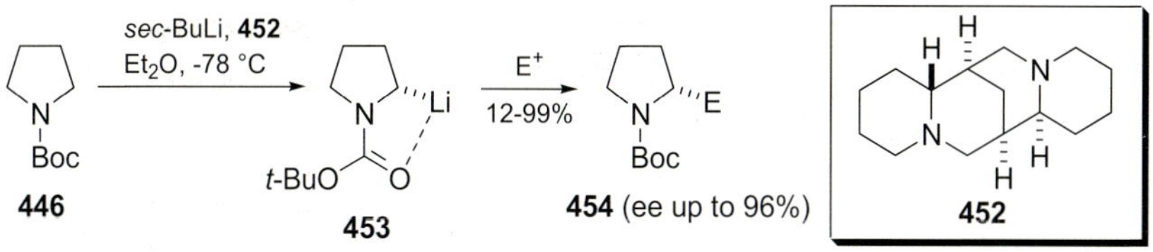

Scheme 4.141

sequential lithiations of **446**, each followed by quenching with dimethyl sulfate, furnished the corresponding *trans*-2,5-dimethylpyrrolidine derivative [686]. The species **453** may also be used for enantioselective ring opening of epoxides, provided that one equivalent of $BF_3 \cdot OEt_2$ is added directly after the electrophile [687]. Transmetallation of **453** with CuCN·2LiCl generates a corresponding cuprate with retention of configuration, and subsequent reactions thereof with, for example, vinyl iodides or triflates give vinylated products with excellent enantioselectivity [688].

Pyrrolidine derivatives may also be functionalized via conversion into *N*-acyliminium ions, as illustrated by the conversion of the pyrrolidine **455** into the product **456** (cis : trans ratio 7 : 3) employing $SnCl_4$ and TMSCN, via the intermediate **457** (Scheme 4.142) [689].

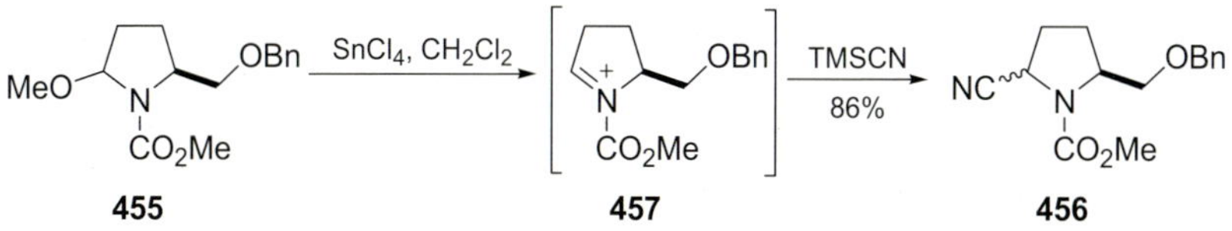

$SnCl_4$, CH_2Cl_2; TMSCN, 86%; 455; 457; 456

Scheme 4.142

4.7 Addendum

Even a rather superficial glance at the most recent literature of organic chemistry clearly reflects the tremendous amount of effort currently invested in research activities focusing on pyrrole based molecules. This addendum highlights some selected new developments reported during the production process of this book, as well as some related relevant studies. As usual, an annual summary of the most important recent advances in *Progress in Heterocyclic Chemistry* provides an excellent source of information [690]. In addition, the new edition of *Comprehensive Heterocyclic Chemistry* has appeared, covering various aspects of pyrrole chemistry explored during the last decade [691–694]. Several specialized reviews have also emerged, discussing for example some synthetic aspects of pyrroles bearing multiple substituents [695], or asymmetric synthesis of pyrrolidines by [3 + 2] cycloadditions of azomethine ylides [696]. Because much novel pyrrole chemistry is directed towards synthesis and applications of macrocycles, topics such as carbaporphyrins and related porphyrinoids [697], acyclic oligopyrroles [698], expanded porphyrins [699], and their transition metal complexes [700], and nonlinear optical properties of porphyrins [701], as well as synthetic work towards porphyrins involving the Barton–Zard reaction [702], have received treatment. In addition, the pyrrole alkaloids lamellarins and their relatives have been discussed in detail [703].

Mechanistic aspects of the Paal–Knorr synthesis have been addressed in a density functional theory study [704], supporting the previously suggested pathway [71] that involves a cyclization of a hemiaminal intermediate in the rate-limiting step (Section 4.4.1). Although the extensive arsenal of well-established routes for pyrrole ring synthesis gives access to a wide variety of products, there is a continuous stream of new approaches that attempt to target pyrroles with rare substitution patterns, or serve as complementary methods for construction of known groups of useful derivatives.

Numerous routes rely on reactions of substrates containing all the necessary carbon atoms. It has been shown that $PtCl_4$ catalyzes cyclization of homopropargyl azides in the presence of a bulky pyridine as the base, affording for instance 2,5-substituted pyrroles, or tetrahydroindole derivatives [705]. The recent surge in gold- or silver-catalyzed organic transformations has also exerted some impact on heterocyclic chemistry, as illustrated by the conversion precursor **458** into the pyrrole **459** by exposure to benzyl amine in the presence of silver trifluoromethanesulfonate (Scheme 4.143). Such transformations can also be performed using the catalytic system AuCl/AgOTf/PPh_3 [706]. Moreover, palladium-catalyzed reactions between N-protected γ-aminoalkenes and functionalized aryl bromides in the presence of a phosphine ligand and a base have furnished a set of multiply substituted pyrrolidines [707]. A series of potentially useful pyrroles has also been prepared by CuI-catalyzed cyclization of 1,4-dihalo-1,3-dienes with *tert*-butyl carbamate [708].

BocHN … O … CO_2Me **458** → ($BnNH_2$, AgOTf (5 mol%), $ClCH_2CH_2Cl$, 50 °C, 18 h; 78%) → BocHN … N-Bn … CO_2Me **459**

Scheme 4.143

Azidodienes are readily available by condensation of azidoacetic acid esters with α,β-unsaturated aldehydes, and contain all the atoms necessary for a construction of a pyrrole ring. This fact was exploited in the conversion of precursor **460** into the pyrrole-2-carboxylate **461** in good yield upon treatment with catalytic amounts of zinc iodide, providing a representative illustration of this route (Scheme 4.144). The

Cl … CO_2Me, N_3 **460** → (ZnI_2 (5 mol%), CH_2Cl_2, 25 °C, 15 h; 80%) → Cl … N-H … CO_2Me **461**

Scheme 4.144

procedure could be applied for the preparation of various pyrrole-2-carboxylates bearing aryl, heteroaryl, and alkyl groups [709]. Likewise, pyrrole derivatives have also been constructed by initial reactions of 1,3-dicarbonyl anions with α-azidoketones, and ensuing annulation of the resulting intermediates employing the Staudinger–aza-Wittig reaction [710]. Precursors for similar reductive cyclizations may also be assembled from 1,3-bis-silyl enol ethers and 1-azido-2,2-dimethoxyethane in the presence of TMSOTf [711].

The elaboration of new [3 + 2] strategies, as well as the development of new aspects of the classical methods, continues, as illustrated by a new variation of the Barton–Zard pyrrole synthesis (Section 4.4.7), which has been applied for construction of pyrrolic Weinreb amides *en route* to pyrrole-2-carboxaldehydes and pyrroline-3-ones. For example, the isocyanide **462**, which is available in four steps from Boc-glycine, was converted upon reaction with the β-nitroacetate **463** into the Weinreb amide **464** (Scheme 4.145) [712]. The use of ketene *S,S*- or *S,N*-acetals in Barton–Zard reactions provides a route to substituted pyrroles bearing methylthio- or amino groups at C3. Such systems could also be prepared employing the related van Leusen method (Section 4.4.6) for the pyrrole ring formation [713]. A practical one-pot route to 4-substituted pyrrole-3-carboxylates on a multi-kilogram scale has been presented, featuring a Horner–Wadsworth–Emmons reaction of aliphatic or aromatic aldehydes with trimethyl phosphonoacetate, followed by a van Leusen pyrrole synthesis involving TosMIC [714].

Scheme 4.145

It has been demonstrated that the vinyl azides **465** may serve as an excellent starting point for pyrrole synthesis, as heating of such substrates with acetylacetone in toluene produced pyrroles **466** (Scheme 4.146). Interestingly, a related copper-catalyzed reaction involving instead ethyl acetoacetate proceeded with different regioselectivity, to provide the ethyl pyrrole-3-carboxylates **467**, thereby widening the scope of this strategy [715].

Scheme 4.146

Palladium-catalyzed pyrrole ring synthesis from the halogenated aminoester **468** with various acetylenes has been examined. For instance, reaction between **468** and diphenylacetylene in the presence of $Pd(OAc)_2$ affords pyrrole **469** with concomitant loss of the acyl group (Scheme 4.147). However, some similar cyclizations gave products where the acyl group was untouched. The use of certain unsymmetrical alkynes (e.g., 1-phenyl-2-trimethylsilylacetylene) can give rise to regioselective formation of pyrroles [716].

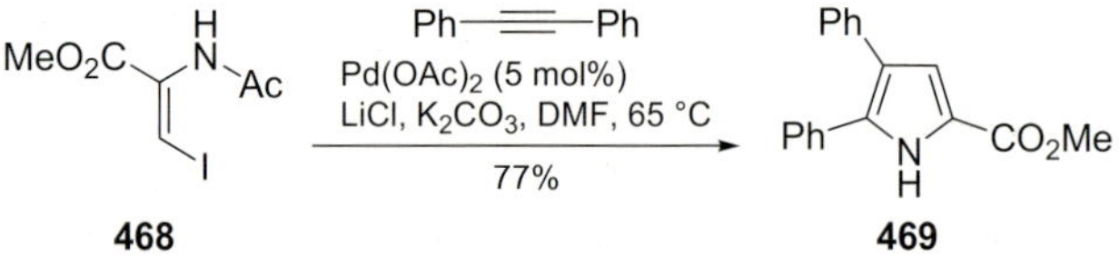

Ph—≡—Ph, $Pd(OAc)_2$ (5 mol%), LiCl, K_2CO_3, DMF, 65 °C, 77%; **468** → **469**

Scheme 4.147

Pyrrolidines (Section 4.6.5) may be produced efficiently by cycloaddition reactions involving azomethine ylides [696]. An organocatalytic application of this approach has now became available, utilizing the catalyst **470**, which could for instance mediate the efficient and enantioselective [3 + 2] cycloaddition of components **471** and **472**, affording the pyrrolidine **473** (Scheme 4.148) [717]. It should also be mentioned that application of α-(alkylideneamino)nitriles in reactions with nitroalkenes provides a useful pyrrole synthesis, which relies on elimination of HCN and HNO_2 as the driving force for aromatization. This pathway involves a stepwise annulation mechanism rather than a cycloaddition [718]. In addition, a series of 4-hydroxypyrrole-2,3-dicarboxylates have been prepared by reactions of α-amino acids with acetylenedicarboxylates in the presence of cyclohexyl isocyanide or *N,N′*-dicyclohexylcarbodiimide as the coupling reagents [719].

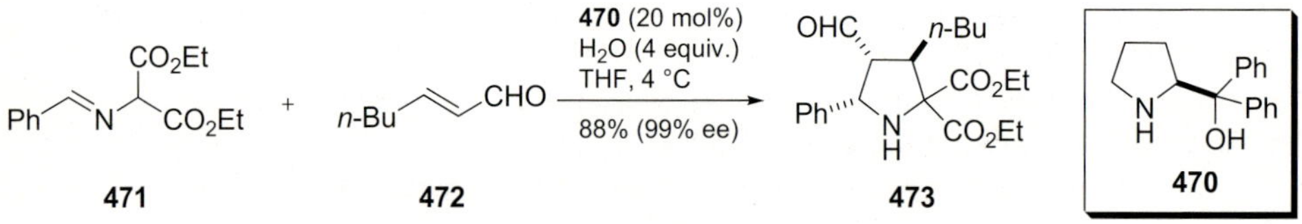

471 + **472**: **470** (20 mol%), H_2O (4 equiv.), THF, 4 °C, 88% (99% ee) → **473**

Scheme 4.148

The largely neglected Piloty–Robinson [720–722] pyrrole synthesis has been adapted to microwave conditions [723], providing a route to 3,4-dialkylpyrroles, in particular 3,4-diethylpyrrole, a building block for construction of octaethylporpyrins [724, 725]. Thus, the intermediate azine **474** was generated by exposure of butyraldehyde to hydrazine hydrate in diethyl ether. Subsequent heating of **474** in the presence of benzoyl chloride in pyridine in a microwave apparatus gave the N-substituted product **475** (Scheme 4.149), which could subsequently be hydrolyzed to the desired 3,4-diethylpyrrole [723].

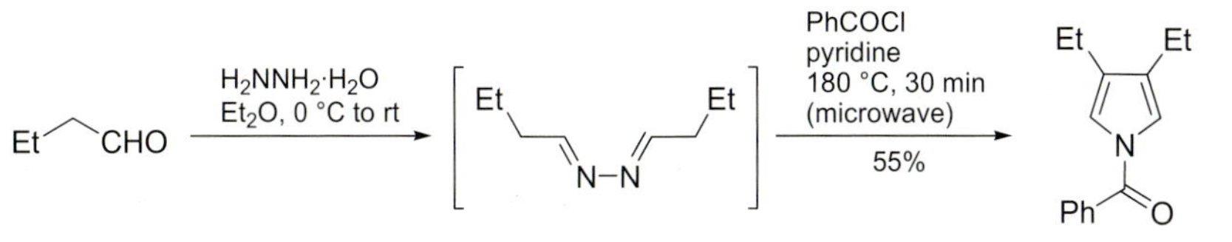

Scheme 4.149

Several new direct pyrrole syntheses based on multicomponent reactions (Section 4.4.10) have emerged recently, featuring for instance imines, diazoacetonitrile, and alkynes as the reactants in the presence of a rhodium(II) catalyst. The sequence of events leading to pyrroles presumably involves initial decomposition of the diazo compound, formation of an intermediate azomethine ylide upon reaction with the imine, and finally cycloaddition with the alkyne [726]. Similar sets of starting compounds, namely, imines, acid chlorides, and alkynes, may also be converted into pyrroles in the presence of either isocyanides [727] or phosphines [728]. The latter approaches have been suggested to proceed via cycloaddition reactions between mesoionic intermediates with the alkyne components [727, 728]. Intermediate münchnones can also be generated by a palladium-catalyzed reaction between certain α-amidoethers with carbon monoxide, eventually affording pyrroles via cycloadditions with suitable alkynes [729]. Additional efforts resulted in conversion of 1,3-dicarbonyl compounds, arylglyoxals, and ammonium acetate into 2-alkyl-5-aryl-4-hydroxypyrroles in water as the reaction medium [730], and preparation of 4,5-dimethylpyrrole-2,3-dicarboxylates by cyclizations of butane-2,3-dione with ylides derived from acetylenedicarboxylates and ammonium acetate in the presence of triphenylphosphine [731].

Some useful developments for modification of existing pyrrole rings have also appeared, such as an efficient procedure for CuI-catalyzed arylation of pyrroles with aromatic or heteroaromatic halides in the presence of simple *N*-hydroxyimides [732]. Moreover, a protocol for assembly of various 2,2′-bipyrrole-5,5′-dicarboxaldehydes by homocoupling of 5-iodopyrrole-2-carboxaldehyde precursors employed palladium on carbon and activated zinc dust as the catalyst [733]. Classical cross-coupling techniques have also found new applications, further demonstrating their extraordinary synthetic potential in pyrrole chemistry. For example, tetramic acid triflates have been demonstrated to participate in Suzuki couplings at C4, giving access to 3,4-diarylpyrrolin-2-ones [734], whereas Suzuki reactions have been employed in regioselective conversion of 1-methyltetrabromopyrrole into its 5-aryl-2,3,4-tribromo- or 2,5-diaryl-3,4-dibromo- derivatives [735], or transformations involving 1-phenylsulfonyl-3,4-dibromopyrrole [736]. A reaction sequence featuring an initial iridium-catalyzed borylation of 1-*tert*-butoxycarbonyl-2-trimethylsilylpyrrole at C4, followed by Suzuki coupling, as well as an intramolecular palladium-catalyzed C–H bond functionalization at C5, has been implemented in an elegant synthesis of the alkaloid rhazinicine [737]. Efficient conditions for generation of pyrrole-3-boronate esters by palladium-catalyzed reactions of pinacol borane with 3-bromopyrroles, and their subsequent Suzuki reactions in the presence of a monophosphine based catalyst, have also been established [738].

A series of 2,2′-bipyrroles has been obtained by regioselective oxidative coupling of pyrroles in the presence of phenyliodine(III) bis(trifluoroacetate) (PIFA) and bromotrimethylsilane (TMSBr), whereas for instance *N*-benzylpyrrole gave a 2,3′-coupled product in good yield under different conditions where $BF_3{\cdot}OEt_2$ was used instead of TMSBr [739]. Intermolecular radical alkylation of some 3-substituted pyrroles with xanthates mediated by dilauroyl peroxide occurs at C2 in moderate to good yields with high regioselectivity, as shown by preparation of the product **476** from 3-phenylpyrrole **477** (Scheme 4.150) [740].

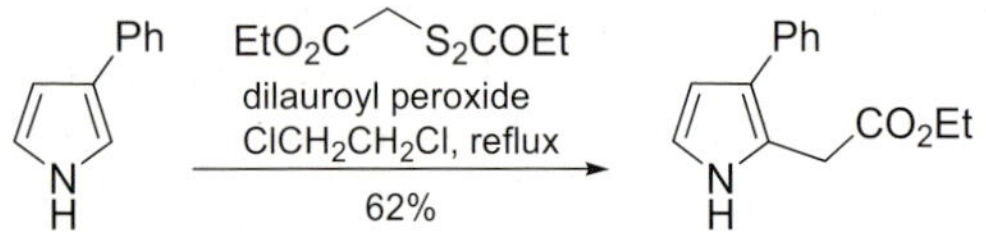

Ph
EtO_2C S_2COEt
dilauroyl peroxide
$ClCH_2CH_2Cl$, reflux
62%
477
476
CO_2Et

Scheme 4.150

Catalytic hydrogenation of 2- or 3-nitropyrroles in the presence of carboxylic acid anhydrides has resulted in a new useful route to the corresponding series of pyrrolylamides or pyrrolylimides [741]. Alternatively, similar chemistry may also be accomplished using tin or indium as the reducing agents [742], while application of 1,4-diketones instead of anhydrides has provided efficient access to 1,2′- and 1,3′-bipyrroles [743].

Finally, it should also be mentioned that determination of the second-order rate constants of the reactions of a series of pyrroles with benzhydrylium ions in acetonitrile provided the basis for a nucleophilicity scale, where 1-triisopropylsilylpyrrole was the least nucleophilic member and 3-ethyl-2,4-dimethylpyrrole was the strongest nucleophile, comparable to enamines in its reactivity [744]. A systematic reinvestigation of an acylation, where a solution of 1-(*p*-toluenesulfonyl)pyrrole and $AlCl_3$ as the Lewis acid in 1,2-dichloroethane as the solvent was quenched with an acyl halide, led to the conclusion that this process may involve the initial formation of pyrrolic organoaluminium species, which thereafter react with the highly reactive electrophile at C3 via a reactant-like transition state [745]. This is in contrast with earlier findings, which gave no evidence of complex formation of the closely related 1-(phenylsulfonyl)pyrrole with $AlCl_3$ in CH_2Cl_2 [315]. On the other hand, reactions featuring weaker Lewis acids such as $EtAlCl_2$ or Et_2AlCl, or less than equimolar amounts of $AlCl_3$, gave considerable amounts of products substituted at C2, proceeding via a normal Friedel–Crafts mechanism, where the pyrrole undergoes acylation by a complex generated from the Lewis acid and an acyl chloride [745]. Clearly, these very useful, but mechanistically complex, reactions are not yet completely understood, and further studies are necessary to provide a complete picture accounting for the observed outcomes.

References

1 Runge, F.F. (1834) *Annalen Der Physik*, **31**, 65–78.

2 Anderson, T. (1857) *Transactions of the Royal Society of Edinburgh*, **21**, 571–595.

3 Baeyer, A. and Emmerling, A. (1870) *Chemische Berichte*, **3**, 514–517.

4 Gossauer, A. (1974) *Die Chemie der Pyrrole*, Springer-Verlag, Berlin.

5 Jones, R.A. and Bean, G.P. (1977) *The Chemistry of Pyrroles*, Academic Press, London.

6 Jones, R.A. (1990) *Pyrroles*, John Wiley & Sons, New York.

7 Black, D.S.C. (2002) in *Science of Synthesis*, Vol. 9 (ed. G. Maas), Thieme, Stuttgart, pp. 441–552.

8 Pelkey, E.T. (2005) in *Progress in Heterocyclic Chemistry*, Vol. 17 (eds G.W. Gribble and J.A. Joule), Elsevier, Amsterdam, pp. 109–141.

9 Jones, G.B. and Chapman, B.J. (1996) in *Comprehensive Heterocyclic Chemistry II*, Vol. 2 (eds A.R. Katritzky, C.W. Rees and E.F.V. Scriven), Elsevier, Amsterdam, pp. 1–38.

10 Sundberg, R.J. (1996) in *Comprehensive Heterocyclic Chemistry II*, Vol. 2 (eds A.R. Katritzky, C.W. Rees, and E.F.V. Scriven), Elsevier, Amsterdam, pp. 119–206.

11 Black, D.S.C. (1996) in *Comprehensive Heterocyclic Chemistry II*, Vol. 2 (eds A.R., Katritzky, C.W. Rees and E.F.V. Scriven), Elsevier, Amsterdam, pp. 39–117.

12 Gribble, G.W. (1996) in *Comprehensive Heterocyclic Chemistry II*, Vol. 2 (eds A.R. Katritzky, C.W. Rees, and E.F.V. Scriven), Elsevier, Amsterdam, pp. 205–257.

13 Balón, M., Carmona, M.A., Muñoz, M.A., and Hidalgo, J. (1989) *Tetrahedron*, **45**, 7501–7504.

14 Yagil, G. (1967) *Tetrahedron*, **23**, 2855–2861.

15 Bordwell, F.G., Zhang, X., and Cheng, J.-P. (1991) *The Journal of Organic Chemistry*, **56**, 3216–3219.

16 Nygaard, L., Nielsen, J.T., Kirchheiner, J., Maltesen, G., Rastrup-Andersen, J., and Sørensen, G.O. (1969) *Journal of Molecular Structure*, **3**, 491–506.

17 Jones, R.A. (1970) *Advances in Heterocyclic Chemistry*, **11**, 383–472.

18 Tai, J.C., Yang, L., and Allinger, N.L. (1993) *Journal of the American Chemical Society*, **115**, 11906–11917.

19 Geidel, E. and Billes, F. (2000) *Journal of Molecular Structure (THEOCHEM)*, **507**, 75–87.

20 Lee, S.Y. and Boo, B.H. (1996) *The Journal of Physical Chemistry*, **100**, 15073–15078.

21 Katritzky, A.R., Jug, K., and Oniciu, D.C. (2001) *Chemical Reviews*, **101**, 1421–1449.

22 Balaban, A.T., Oniciu, D.C., and Katritzky, A.R. (2004) *Chemical Reviews*, **104**, 2777–2812.

23 Lee, C.K., Jun, J.H., and Yu, J.S. (2000) *Journal of Heterocyclic Chemistry*, **37**, 15–24.

24 Thompson, A., Gao, S., Modzelewska, G., Hughes, D.S., Patrick, B., and Dolphin, D. (2000) *Organic Letters*, **2**, 3587–3590.

25 Claramunt, R.M., Sanz, D., López, C., Jiménez, J.A., Jimeno, M.L., Elguero, J., and Fruchier, A. (1997) *Magnetic Resonance in Chemistry*, **35**, 35–75.

26 Belen'kii, L.I., Kim, T.G., Suslov, I.A., and Chuvylkin, N.D. (2005) *Russian Chemical Bulletin*, **54**, 853–863.

27 Woodward, R.B., Ayer, W.A., Beaton, J.M., Bickelhaupt, F., Bonnett, R., Buchschacher, P., Closs, G.L., Dutler, H., Hannah, J., Hauck, F.P., Ito, S., Langemann, A., Le Goff, E., Leimgruber, W., Lwowski, W., Sauer, J., Valenta, Z., and Volz, H. (1990) *Tetrahedron*, **46**, 7599–7659.

28 Battersby, A.R. (1987) *Natural Product Reports*, **4**, 77–87.

29 Battersby, A.R. and McDonald, E. (1979) *Accounts of Chemical Research*, **12**, 14–22.

30 Battersby, A.R. (1993) *Accounts of Chemical Research*, **26**, 15–21.

31 Mauger, A.B. (1996) *Journal of Natural Products*, **59**, 1205–1211.

32 Walsh, C.T., Garneau-Tsodikova, S., and Howard-Jones, A.R. (2006) *Natural Product Reports*, **23**, 517–531.

33 Hayakawa, Y., Kawakami, K., Seto, H., and Furihata, K. (1992) *Tetrahedron Letters*, **33**, 2701–2704.

34 Fürstner, A. (2003) *Angewandte Chemie-International Edition*, **42**, 3582–3603.
35 Royles, B.J.L. (1995) *Chemical Reviews*, **95**, 1981–2001.
36 Ghisalberti, E.L. (2003) in *Studies in Natural Products Chemistry*, Vol. 28 (ed. Atta-Ur-Rahman), Elsevier, Amsterdam, pp. 109–163.
37 Casser, I., Steffan, B., and Steglich, W. (1987) *Angewandte Chemie-International Edition*, **26**, 586–587.
38 Gossauer, A. (2003) in *Progress in the Chemistry of Organic Natural Products*, Vol. 86 (eds W. Herz, H. Falk, and G.W.Kirby), Springer, Wien, pp. 1–188.
39 Gupton, J.T. (2006) *Topics in Heterocyclic Chemistry*, **2**, 53–92.
40 Arcamone, A., Penco, S., Orezzi, P., Nicolella, V., and Pirelli, A. (1964) *Nature*, **203**, 1064–1065.
41 Dervan, P.B. (1986) *Science*, **232**, 464–471.
42 Marques, M.A., Doss, R.M., Urbach, A.R., and Dervan, P.B. (2002) *Helvetica Chimica Acta*, **85**, 4485–4517.
43 Dervan, P.B., Doss, R.M., and Marques, M.A. (2005) *Current Medicinal Chemistry – Anti-Cancer Agents*, **5**, 373–387.
44 Dervan, P.B., Poulin-Kerstien, A.T., Fechter, E.J., and Edelson, B.S. (2005) *Topics in Current Chemistry*, **253**, 1–31.
45 Bando, T. and Sugiyama, H. (2006) *Accounts of Chemical Research*, **39**, 935–944.
46 Muchowski, J.M., Unger, S.H., Ackrell, J., Cheung, P., Cooper, G.F., Cook, J., Gallegra, P., Halpern, O., Koehler, R., Kluge, A.F., Van Horn, A.R., Antonio, Y., Carpio, P., Franco, F., Galeazzi, E., Garcia, I., Greenhouse, R., Guzmàn, A., Iriarte, J., Leon, A., Peña, A., Peréz, V., Valdéz, D., Ackerman, N., Ballaron, S.A., Murthy, D.V.K., Rovito, J.R., Tomolonis, A.J., Young, J.M., and Rooks, W.H., II (1985) *Journal of Medicinal Chemistry*, **28**, 1037–1049.
47 Guzmàn, A., Yuste, F., Toscano, R.A., Young, J.M., Van Horn, A.R., and Muchowski, J.M. (1986) *Journal of Medicinal Chemistry*, **29**, 589–591.
48 Deronzier, A. and Moutet, J.-C. (1989) *Accounts of Chemical Research*, **22**, 249–255.
49 Gangopadhyay, R. and De, A. (2000) *Chemistry of Materials*, **12**, 608–622.
50 Sadki, S., Schottland, P., Brodie, N., and Sabouraud, G. (2000) *Chemical Society Reviews*, **29**, 283–293.
51 Wang, L.-X., Li, X.-G., and Yang, Y.-L. (2001) *Reactive & Functional Polymers*, **47**, 125–139.
52 Ferreira, V.F., de Souza, M.C.B.V., Cunha, A.C., Pereira, L.O.R., and Ferreira, M.L.G. (2001) *Organic Preparations and Procedures International*, **33**, 411–454.
53 Bellina, F. and Rossi, R. (2006) *Tetrahedron*, **62**, 7213–7256.
54 Paal, C. (1884) *Chemische Berichte*, **17**, 2756–2767.
55 Knorr, L. (1884) *Chemische Berichte*, **17**, 2863–2870.
56 Thompson, W.J. and Buhr, C.A. (1983) *The Journal of Organic Chemistry*, **48**, 2769–2772.
57 Nozaki, H., Koyama, T., and Mori, T. (1969) *Tetrahedron*, **25**, 5357–5364.
58 Wasserman, H.H., Keith, D.D., and Nadelson, J. (1976) *Tetrahedron*, **32**, 1867–1871.
59 McLeod, M., Boudreault, N., and Leblanc, Y. (1996) *The Journal of Organic Chemistry*, **61**, 1180–1183.
60 Jacobi, P.A., Buddhu, S.C., Fry, D., and Rajeswari, S. (1997) *The Journal of Organic Chemistry*, **62**, 2894–2906.
61 Lynn, D.G., Jaffe, K., Cornwall, M., and Tramontano, W. (1987) *Journal of the American Chemical Society*, **109**, 5858–5859.
62 Rousseau, B., Nydegger, F., Gossauer, A., Bennua-Skalmowski, B., and Vorbrüggen, H. (1996) *Synthesis*, 1336–1340.
63 Samajdar, S., Becker, F.F., and Banik, B.K. (2001) *Heterocycles*, **55**, 1019–1022.
64 Yu, S.-X. and Le Quesne, P.W. (1995) *Tetrahedron Letters*, **36**, 6205–6208.
65 Banik, B.K., Samajdar, S., and Banik, I. (2004) *The Journal of Organic Chemistry*, **69**, 213–216.
66 Danks, T.N. (1999) *Tetrahedron Letters*, **40**, 3957–3960.
67 Minetto, G., Raveglia, L.F., and Taddei, M. (2004) *Organic Letters*, **6**, 389–392.

68 Minetto, G., Raveglia, L.F., Sega, A., and Taddei, M. (2005) *European Journal of Organic Chemistry*, 5277–5288.

69 Raghavan, S. and Anuradha, K. (2003) *Synlett*, 711–713.

70 Hansford, K.A., Zanzarova, V., Dörr, A., and Lubell, W.D. (2004) *Journal of Combinatorial Chemistry*, **6**, 893–898.

71 Amarnath, V., Anthony, D.C., Amarnath, K., Valentine, W.M., Wetterau, L.A., and Graham, D.G. (1991) *The Journal of Organic Chemistry*, **56**, 6924–6931.

72 Szakál-Quin, G., Graham, D.G., Millington, D.S., Maltby, D.A., and McPhail, A.T. (1986) *The Journal of Organic Chemistry*, **51**, 621–624.

73 Zamora, R. and Hidalgo, F.J. (2006) *Synlett*, 1428.

74 Lui, K.-H. and Sammes, M.P. (1990) *Journal of the Chemical Society, Perkin Transactions 1*, 457–468.

75 Sammes, M.P. and Katritzky, A.R. (1982) *Advances in Heterocyclic Chemistry*, **32**, 233–284.

76 Elming, N. and Clauson-Kaas, N. (1952) *Acta Chemica Scandinavica*, **6**, 867–874.

77 Josey, A.D. and Jenner, E.L. (1962) *The Journal of Organic Chemistry*, **27**, 2466–2470.

78 Fang, Y., Leysen, D., and Ottenheijm, H.C.J. (1995) *Synthetic Communications*, **25**, 1857–1861.

79 Dumoulin, H., Rault, S., and Robba, M. (1995) *Journal of Heterocyclic Chemistry*, **32**, 1703–1707.

80 D'Silva, C. and Walker, D.A. (1998) *The Journal of Organic Chemistry*, **63**, 6715–6718.

81 Wasley, J.W.F. and Chan, K. (1973) *Synthetic Communications*, **3**, 303–304.

82 Plieninger, H., El-Berins, R., and Hirsch, R. (1973) *Synthesis*, 422–423.

83 Karousis, N., Liebscher, J., and Varvounis, G. (2006) *Synthesis*, 1494–1498.

84 Baussanne, I., Chiaroni, A., Husson, H.-P., Riche, C., and Royer, J. (1994) *Tetrahedron Letters*, **35**, 3931–3934.

85 Gourlay, B.S., Molesworth, P.P., Ryan, J.H., and Smith, J.A. (2006) *Tetrahedron Letters*, **47**, 799–801.

86 Chan, T.H. and Lee, S.D. (1983) *The Journal of Organic Chemistry*, **48**, 3059–3061.

87 Lee, S.D., Brook, M.A., and Chan, T.H. (1983) *Tetrahedron Letters*, **24**, 1569–1572.

88 Bharadwaj, A.R. and Scheidt, K.A. (2004) *Organic Letters*, **6**, 2465–2468.

89 Braun, R.U., Zeitler, K., and Müller, T.J.J. (2001) *Organic Letters*, **3**, 3297–3300.

90 Quiclet-Sire, B., Thévenot, I., and Zard, S.Z. (1995) *Tetrahedron Letters*, **36**, 9469–9470.

91 Cheruku, S.R., Padmanilayam, M.P., and Vennerstrom, J.L. (2003) *Tetrahedron Letters*, **44**, 3701–3703.

92 Nagafuji, P. and Cushman, M. (1996) *The Journal of Organic Chemistry*, **61**, 4999–5003.

93 Lagu, B., Pan, M., and Wachter, M.P. (2001) *Tetrahedron Letters*, **42**, 6027–6030.

94 Paulus, O., Alcaraz, G., and Vaultier, M. (2002) *European Journal of Organic Chemistry*, 2565–2572.

95 Tejedor, D., Gonzàlez-Cruz, D., García-Tellado, F., Marrero-Tellado, J.J., and López Rodríguez, M. (2004) *Journal of the American Chemical Society*, **126**, 8390–8391.

96 Binder, J.T. and Kirsch, S.F. (2006) *Organic Letters*, **8**, 2151–2153.

97 Wittig, G., Röderer, R., and Fischer, S. (1973) *Tetrahedron Letters*, **14**, 3517–3520.

98 Aelterman, W., De Kimpe, N., Tyvorskii, V., and Kulinkovich, O. (2001) *The Journal of Organic Chemistry*, **66**, 53–58.

99 Katritzky, A.R., Li, J., and Gordeev, M.F. (1994) *Synthesis*, 93–96.

100 Campi, E.M., Jackson, W.R., and Nilsson, Y. (1991) *Tetrahedron Letters*, **32**, 1093–1094.

101 Agarwal, S. and Knölker, H.-J. (2004) *Organic and Biomolecular Chemistry*, **2**, 3060–3062.

102 Katritzky, A.R. and Li, J. (1996) *The Journal of Organic Chemistry*, **61**, 1624–1628.

103 Knight, D.W., Redfern, A.L., and Gilmore, J. (1998) *Chemical Communications*, 2207–2208.

104 Knight, D.W., Redfern, A.L., and Gilmore, J. (2002) *Journal of the Chemical Society, Perkin Transactions 1*, 622–628.

105 Knight, D.W. Redfern, A.L., and Gilmore, J. (1998) *Synlett*, 731–732.

106 Barluenga, J., Tomás, M., and Suárez-Sobrino, A. (1990) *Synlett*, 351–352.

107 Barluenga, J., Tomás, M., Kouznetsov, V., Suárez-Sobrino, A., and Rubio, E. (1996) *The Journal of Organic Chemistry*, **61**, 2185–2190.

108 Arcadi, A. and Rossi, E. (1997) *Synlett*, 667–668.

109 Arcadi, A. and Rossi, E. (1998) *Tetrahedron*, **54**, 15253–15272.

110 Knorr, L. (1886) *Liebigs Annalen der Chemie*, **236**, 290–332.

111 Fischer, H. (1943) *Organic Syntheses, Collective Volumes II*, 202–204.

112 Treibs, A., Schmidt, R., and Zinsmeister, R. (1957) *Chemische Berichte*, **90**, 79–84.

113 Kleinspehn, G.G. (1955) *Journal of the American Chemical Society*, **77**, 1546–1548.

114 Paine, J.B. III and Dolphin, D. (1985) *The Journal of Organic Chemistry*, **50**, 5598–5604.

115 Paine, J.B. III, Brough, J.R., Buller, K.K., and Erikson, E.E. (1987) *The Journal of Organic Chemistry*, **52**, 3986–3993.

116 Fujii, H., Yoshimura, T., and Kamada, H. (1997) *Tetrahedron Letters*, **38**, 1427–1430.

117 Hamby, J.M. and Hodges, J.C. (1993) *Heterocycles*, **35**, 843–850.

118 Alberola, A., González Ortega, A., Sádaba, M.L., and Sañudo, C. (1999) *Tetrahedron*, **55**, 6555–6566.

119 Hendrickson, J.B., Rees, R., and Templeton, J.F. (1964) *Journal of the American Chemical Society*, **86**, 107–111.

120 Kolar, P. and Tišler, M. (1994) *Synthetic Communications*, **24**, 1887–1893.

121 Terang, N., Mehta, B.K., Ila, H., and Junjappa, H. (1998) *Tetrahedron*, **54**, 12973–12984.

122 Okada, E., Masuda, R., Hojo, M., and Yoshida, R. (1992) *Heterocycles*, **34**, 1435–1441.

123 Hantzsch, A. (1890) *Chemische Berichte*, **23**, 1474–1476.

124 Roomi, M.W. and MacDonald, S.F. (1970) *Canadian Journal of Chemistry*, **48**, 1689–1697.

125 Kameswaran, V. and Jiang, B. (1997) *Synthesis*, 530–532.

126 Trautwein, A.W., Süßmuth, R.D., and Jung, G. (1998) *Bioorganic & Medicinal Chemistry Letters*, **8**, 2381–2384.

127 Fürstner, A., Weintritt, H., and Hupperts, A. (1995) *The Journal of Organic Chemistry*, **60**, 6637–6641.

128 Mataka, S., Takahashi, K., Tsuda, Y., and Tashiro, M. (1982) *Synthesis*, 157–159.

129 Hombrecher, H.K. and Horter, G. (1990) *Synthesis*, 389–391.

130 Walizei, G.H. and Breitmaier, E. (1989) *Synthesis*, 337–340.

131 Terry, W.G., Jackson, A.H., Kenner, G.W., and Kornis, G. (1965) *Journal of the Chemical Society*, 4389–4393.

132 Lash, T.D. and Hoehner, M.C. (1991) *Journal of Heterocyclic Chemistry*, **28**, 1671–1676.

133 Gupton, J.T., Krolikowski, D.A., Yu, R.H., Riesinger, S.W., and Sikorski, J.A. (1990) *The Journal of Organic Chemistry*, **55**, 4735–4740.

134 Gupton, J.T., Petrich, S.A., Hicks, F.A., Wilkinson, D.R., Vargas, M., Hosein, K.N., and Sikorski, J.A. (1998) *Heterocycles*, **47**, 689–702.

135 Gupton, J.T., Krolikowski, D.A., Yu, R.H., Vu, P., Sikorski, J.A., Dahl, M.L., and Jones, C.R. (1992) *The Journal of Organic Chemistry*, **57**, 5480–5483.

136 Gupton, J.T., Krumpe, K.E., Burnham, B.S., Dwornik, K.A., Petrich, S.A., Du, K.X., Bruce, M.A., Vu, P., Vargas, M., Keertikar, K.M., Hosein, K.N., Jones, C.R., and Sikorski, J.A. (1998) *Tetrahedron*, **54**, 5075–5088.

137 van Leusen, A.M., Siderius, H., Hoogenboom, B.E., and van Leusen, D. (1972) *Tetrahedron Letters*, **13**, 5337–5340.

138 Pavri, N.P. and Trudell, M.L. (1997) *The Journal of Organic Chemistry*, **62**, 2649–2651.

139 Smith, N.D., Huang, D., and Cosford, N.D.P. (2002) *Organic Letters*, **4**, 3537–3539.

140 Ono, N., Muratani, E., and Ogawa, T. (1991) *Journal of Heterocyclic Chemistry*, **28**, 2053–2055.

141 ten Have, R., Leusink, F.R., and van Leusen, A.M. (1996) *Synthesis*, 871–876.

142 Leroy, J. (1991) *Journal of Fluorine Chemistry*, **53**, 61–70.

143 Possel, O. and van Leusen, A.M. (1977) *Heterocycles*, **7**, 77–80.

144 Dijkstra, H.P., ten Have, R., and van Leusen, A.M. (1998) *The Journal of Organic Chemistry*, **63**, 5332–5338.

145 Katritzky, A.R., Cheng, D., and Musgrave, R.P. (1997) *Heterocycles*, **44**, 67–70.

146 Van Leusen, D., Flentge, E., and van Leusen, A.M. (1991) *Tetrahedron*, **47**, 4639–4644.

147 van Leusen, D., van Echten, E., and van Leusen, A.M. (1992) *The Journal of Organic Chemistry*, **57**, 2245–2249.

148 Suzuki, M., Miyoshi, M., and Matsumoto, K. (1974) *The Journal of Organic Chemistry*, **39**, 1980.

149 Carter, P., Fitzjohn, S., Halazy, S., and Magnus, P. (1987) *Journal of the American Chemical Society*, **109**, 2711–2717.

150 Dell'Erba, C., Giglio, A., Mugnoli, A., Novi, M., Petrillo, G., and Stagnaro, P. (1995) *Tetrahedron*, **51**, 5181–5192.

151 Barton, D.H.R. and Zard, S.Z. (1985) *Journal of the Chemical Society, Chemical Communications*, 1098–1100.

152 Barton, D.H.R., Kervagoret, J., and Zard, S.Z. (1990) *Tetrahedron*, **46**, 7587–7598.

153 Lash, T.D., Bellettini, J.R., Bastian, J.A., and Couch, K.B. (1994) *Synthesis*, 170–172.

154 Ono, N., Katayama, H., Nisyiyama, S., and Ogawa, T. (1994) *Journal of Heterocyclic Chemistry*, **31**, 707–710.

155 Uno, H., Inoue, K., Inoue, T., Fumoto, Y., and Ono, N. (2001) *Synthesis*, 2255–2258.

156 Haake, G. Struve, D., and Montforts, F.-P. (1994) *Tetrahedron Letters*, **35**, 9703–9704.

157 Abel, Y., Haake, E., Haake, G., Schmidt, W., Struve, D., Walter, A., and Montforts, F.-P. (1998) *Helvetica Chimica Acta*, **81**, 1978–1996.

158 Bobál, P. and Lightner, D.A. (2001) *Journal of Heterocyclic Chemistry*, **38**, 527 530.

159 Caldarelli, M., Habermann, J., and Ley, S.V. (1999) *Journal of the Chemical Society, Perkin Transactions 1*, 107–110.

160 Lash, T.D., Novak, B.H., and Lin, Y. (1994) *Tetrahedron Letters*, **35**, 2493–2494.

161 Novak, B.H. and Lash, T.D. (1998) *The Journal of Organic Chemistry*, **63**, 3998–4010.

162 Ono, N., Hironaga, H., Ono, K., Kaneko, S., Murashima, T., Ueda, T., Tsukamura, C., and Ogawa, T. (1996) *Journal of the Chemical Society, Perkin Transactions 1*, 417–423.

163 Lash, T.D., Thompson, M.L., Werner, T.M., and Spence, J.D. (2000) *Synlett*, 213–216.

164 Trofimov, B.A. (1990) *Advances in Heterocyclic Chemistry*, **51**, 177–301.

165 Trofimov, B.A. and Mikhaleva, A.I. (1994) *Heterocycles*, **37**, 1193–1232.

166 Vasil'tsov, A.M., Polubentsev, E.A., Mikhaleva, A.I., and Trofimov, B.A. (1990) *Izvestiya Akademii Nauk SSSR, Seriya Khimia*, 864.

167 Trofimov, B.A., Vasil'tsov, A.M., Schmidt, E.Y., Zorina, N.V., Afonin, A.V., Mikhaleva, A.I., Petrushenko, K.B., Ushakov, I.A., Krivdin, L.B., Belsky, V.K., and Bruyukvina, L.I. (2005) *European Journal of Organic Chemistry*, 4338–4345.

168 Trofimov, B.A. (1994) *Phosphorus, Sulfur Silicon Related Elements*, **95–96**, 145–163.

169 Huisgen, R., Gotthardt, H., and Bayer, H.O. (1964) *Angewandte Chemie-International Edition in English*, **3**, 135–136.

170 Huisgen, R., Gotthardt, H., Bayer, H.O., and Schaefer, F.C. (1964) *Angewandte Chemie-International Edition in English*, **3**, 136–137.

171 Bayer, H.O., Gotthardt, H., and Huisgen, R. (1970) *Chemische Berichte*, **103**, 2356–2367.

172 Huisgen, R., Gotthardt, H., Bayer, H.O., and Schaefer, F.C. (1970) *Chemische Berichte*, **103**, 2611–2624.

173 Funabiki, K., Ishihara, T., and Yamanaka, H. (1995) *Journal of Fluorine Chemistry*, **71**, 5–7.

174 Wilde, R.G. (1988) *Tetrahedron Letters*, **29**, 2027–2030.

175 Berrée, F. and Morel, G. (1995) *Tetrahedron*, **51**, 7019–7034.

176 Dhawan, R. and Arndtsen, B.A. (2004) *Journal of the American Chemical Society*, **126**, 468–469.

177 Keating, T.A. and Armstrong, R.W. (1996) *Journal of the American Chemical Society*, **118**, 2574–2583.

178 Mjalli, A.M.M., Sarshar, S., and Baiga, T.J. (1996) *Tetrahedron Letters*, **37**, 2943–2946.

179 Padwa, A., Burgess, E.M., Gingrich, H.L., and Roush, D.M. (1982) *The Journal of Organic Chemistry*, **47**, 786–791.

180 Texier, F., Mazari, M., Yebdri, O., Tonnard, F., and Carrié, R. (1990) *Tetrahedron*, **46**, 3515–3526.

181 Avalos, M., Babiano, R., Cabanillas, A., Cintas, P., Jiménez, J.L., Palcios, J.C., Aguilar, M.A., Corchado, J.C., and Espinosa-García, J. (1996) *The Journal of Organic Chemistry*, **61**, 7291–7297.

182 Boger, D.L., Coleman, R.S., Panek, J.S., and Yohannes, D. (1984) *The Journal of Organic Chemistry*, **49**, 4405–4409.

183 Boger, D.L., Panek, J.S., and Patel, M. (1992) *Organic Synthesis*, **70**, 79–92.

184 Boger, D.L. and Patel, M. (1988) *The Journal of Organic Chemistry*, **53**, 1405–1415.

185 Boger, D.L., Boyce, C.W., Labroli, M.A., Sehon, C.A., and Jin, Q. (1999) *Journal of the American Chemical Society*, **121**, 54–62.

186 Joshi, U., Pipelier, M., Naud, S., and Dubreuil, D. (2005) *Current Organic Chemistry*, **9**, 261–288.

187 Uchida, T. (1978) *Journal of Heterocyclic Chemistry*, **15**, 241–248.

188 Liu, J.-H., Chan, H.-W., Xue, F., Wang, Q.-G., Mak, T.C.W., and Wong, H.N.C. (1999) *The Journal of Organic Chemistry*, **64**, 1630–1634.

189 Padwa, A., Haffmanns, G., and Tomas, M. (1984) *The Journal of Organic Chemistry*, **49**, 3314–3322.

190 Buchwald, S.L., Wannamaker, M.W., and Watson, B.T. (1989) *Journal of the American Chemical Society*, **111**, 776–777.

191 Gao, Y., Shirai, M., and Sato, F. (1996) *Tetrahedron Letters*, **37**, 7787–7790.

192 Kamijo, S., Kanazawa, C., and Yamamoto, Y. (2005) *Tetrahedron Letters*, **46**, 2563–2566.

193 Kamijo, S., Kanazawa, C., and Yamamoto, Y. (2005) *Journal of the American Chemical Society*, **127**, 9260–9266.

194 Larionov, O.V. and de Meijere, A. (2005) *Angewandte Chemie-International Edition*, **44**, 5664–5667.

195 Shiraishi, H., Nishitani, T., Sakaguchi, S., and Ishii, Y. (1998) *The Journal of Organic Chemistry*, **63**, 6234–6238.

196 Ranu, B.C., Hajra, A., and Jana, U. (2000) *Synlett*, 75–76.

197 Ranu, B.C. and Hajra, A. (2001) *Tetrahedron*, **57**, 4767–4773.

198 Ranu, B.C. and Dey, S.S. (2003) *Tetrahedron Letters*, **44**, 2865–2868.

199 Takaya, H., Kojima, S., and Murahashi, S.-I. (2001) *Organic Letters*, **3**, 421–424.

200 Katritzky, A.R., Zhang, L., Yao, J., and Denisko, O.V. (2000) *The Journal of Organic Chemistry*, **65**, 8074–8076.

201 Katritzky, A.R., Chang, H.-X., and Verin, S.V. (1995) *Tetrahedron Letters*, **36**, 343–346.

202 Kel'in, A.V., Sromek, A.W., and Gevorgyan, V. (2001) *Journal of the American Chemical Society*, **123**, 2074–2075.

203 Kim, J.T., Kel'in, A.V., and Gevorgyan, V. (2003) *Angewandte Chemie-International Edition*, **42**, 98–101.

204 Gabriele, B., Salerno, G., Fazio, A., and Bossio, M.R. (2001) *Tetrahedron Letters*, **42**, 1339–1341.

205 Gabriele, B., Salerno, G., and Fazio, A. (2003) *The Journal of Organic Chemistry*, **68**, 7853–7861.

206 Fringuelli, F., Marino, G., and Savelli, G. (1969) *Tetrahedron*, **25**, 5815–5818.

207 Clementi, S. and Marino, G. (1972) *Journal of the Chemical Society, Perkin Transactions 2*, 71–73.

208 Clementi, S. and Marino, G. (1969) *Tetrahedron*, **25**, 4599–4603.

209 Muchowski, J.M. and Solas, D.R. (1983) *Tetrahedron Letters*, **24**, 3455–3456.

210 Chadwick, D.J. and Hodgson, S.T. (1983) *Journal of the Chemical Society, Perkin Transactions 1*, 93–102.

211 Xu, R.X., Anderson, H.J., Gogan, N.J., Loader, C.E., and McDonald, R. (1981) *Tetrahedron Letters*, **22**, 4899–4900.

212 Bélanger, P. (1979) *Tetrahedron Letters*, **20**, 2505–2508.

213 Treibs, A. and Fritz, G. (1958) *Liebigs Annalen der Chemie*, **611**, 162–193.

214 Merino, G. (1971) *Advances in Heterocyclic Chemistry*, **13**, 235–314.

215 Chiang, Y. and Whipple, E.B. (1963) *Journal of the American Chemical Society*, **85**, 2763–2767.

216 Nguyen, V.Q. and Tureek, F. (1996) *Journal of Mass Spectrometry*, **31**, 1173–1184.

217 Chiang, Y., Hinman, R.L., Theodoropulos, S., and Whipple, E.B. (1967) *Tetrahedron*, **23**, 745–759.

218 Gassner, R., Krumbholz, E., and Steuber, F.W. (1981) *Liebigs Annalen der Chemie*, 789–791.

219 Zhao, Y., Beddoes, R.L., and Joule, J.A. (1997) *Journal of Chemical Research (S)*, 42–43.

220 Smith, G.F. (1963) *Advances in Heterocyclic Chemistry*, **2**, 287–309.

221 Garsuch, A., Sattler, R.R., and Pickup, P.G. (2004) *Chemical Communications*, 344–345.

222 Ciamician, G.L. and Dennstedt, M. (1882) *Chemische Berichte*, **15**, 2579–2585.

223 Kleinspehn, G.G. and Corwin, A.H. (1953) *Journal of the American Chemical Society*, **75**, 5295–5298.

224 Treibs, A. and Kolm, H.G. (1958) *Liebigs Annalen der Chemie*, **614**, 176–198.

225 Anderson, H.J. and Lee, S.-F. (1965) *Canadian Journal of Chemistry*, **43**, 409–414.

226 De Rosa, M. (1982) *The Journal of Organic Chemistry*, **47**, 1008–1010.

227 Cordell, G.A. (1975) *The Journal of Organic Chemistry*, **40**, 3161–3169.

228 Gilow, H.M. and Burton, D.E. (1981) *The Journal of Organic Chemistry*, **46**, 2221–2225.

229 Dvornikova, E. and Kamienska-Trela, K. (2002) *Synlett*, 1152–1154.

230 De Rosa, M., Cabrera Nieto, G., and Ferrer Gago, F. (1989) *The Journal of Organic Chemistry*, **54**, 5347–5350.

231 Aiello, E., Dattolo, G., Cirrincione, G., Almerico, A.M., and D'Asdia, I. (1982) *Journal of Heterocyclic Chemistry*, **19**, 977–979.

232 Choi, D.-S., Huang, S., Huang, M., Barnard, T.S., Adams, R.D., Seminario, J.M., and Tour, J.M. (1998) *The Journal of Organic Chemistry*, **63**, 2646–2655.

233 Motekaitis, R.J., Heinert, D.H., and Martell, A.E. (1970) *The Journal of Organic Chemistry*, **35**, 2504–2511.

234 Petruso, S., Caronna, S., and Sprio, V. (1990) *Journal of Heterocyclic Chemistry*, **27**, 1209–1211.

235 Petruso, S., Caronna, S., Sferlazzo, M., and Sprio, V. (1990) *Journal of Heterocyclic Chemistry*, **27**, 1277–1280.

236 Jaramillo, D., Liu, Q., Aldrich-Wright, J., and Tor, Y. (2004) *The Journal of Organic Chemistry*, **69**, 8151–8153.

237 Chen, W. and Cava, M.P. (1987) *Tetrahedron Letters*, **28**, 6025–6026.

238 Chen, W., Stephenson, E.K., Cava, M.P., and Jackson, Y.A. (1992) *Organic Synthesis*, **70**, 151–156.

239 Martina, S., Enkelmann, V., Wegner, G., and Schlüter, A.-D. (1991) *Synthesis*, 613–615.

240 Groenendaal, L., Van Loo, M.E., Vekemans, J.A.J.M., and Meijer, E.W. (1995) *Synthetic Communications*, **25**, 1589–1600.

241 Knight, L.W., Huffman, J.W., and Isherwood, M.L. (2003) *Synlett*, 1993–1996.

242 Muchowski, J.M. and Naef, R. (1984) *Helvetica Chimica Acta*, **67**, 1168–1172.

243 Shum, P.W. and Kozikowski, A.P. (1990) *Tetrahedron Letters*, **31**, 6785–6788.

244 Bray, B.L., Mathies, P.H., Naef, R., Solas, D.R., Tidwell, T.T., Artis, D.R., and Muchowski, J.M. (1990) *The Journal of Organic Chemistry*, **55**, 6317–6328.

245 Kozikowski, A.P. and Cheng, X.-M. (1984) *The Journal of Organic Chemistry*, **49**, 3239–3240.

246 Chan, H.-W., Chan, P.-C., Liu, J.-H., and Wong, H.N.C. (1997) *Chemical Communications*, 1515–1516.

247 Liu, J.-H., Chan, H.W., and Wong, H.N.C. (2000) *The Journal of Organic Chemistry*, **65**, 3274–3283.

248 Zonta, C., Fabris, F., and De Lucchi, O. (2005) *Organic Letters*, **7**, 1003–1006.

249 Wang, J. and Scott, A.I. (1994) *Tetrahedron*, **50**, 6181–6192.

250 Wang, J. and Scott, A.I. (1994) *Tetrahedron Letters*, **35**, 3679–3682.

251 Wang, J. and Scott, A.I. (1995) *Journal of the Chemical Society, Chemical Communications*, 2399–2400.

252 Onda, H., Toi, H., Aoyama, Y., and Ogoshi, H. (1985) *Tetrahedron Letters*, **26**, 4221–4224.

253 Fischer, H., Sturm, E., and Friedrich, H. (1928) *Liebigs Annalen der Chemie*, **461**, 244–277.

254 Fischer, H. and Scheyer, H. (1923) *Liebigs Annalen der Chemie*, **434**, 237–251.

255 Angelini, G., Illuminati, G., Monaci, A., Sleiter, G., and Speranza, M. (1980) *Journal of the American Chemical Society*, **102**, 1377–1382.

256 Angelini, G., Giancaspro, C., Illuminati, G., and Sleiter, G. (1980) *The Journal of Organic Chemistry*, **45**, 1786–1790.

257 Cooksey, A.R., Morgan, K.J., and Morrey, D.P. (1970) *Tetrahedron*, **26**, 5101–5111.

258 Doddi, G., Mencarelli, P., Razzini, A., and Stegel, F. (1979) *The Journal of Organic Chemistry*, **44**, 2321–2323.

259 Morgan, K.J. and Morrey, D.P. (1971) *Tetrahedron*, **27**, 245–253.

260 Tanemura, K., Suzuki, T., Nishida, Y., Satsumabayashi, K., and Horaguchi, T. (2003) *J Chem Res (S)*, 497–499.

261 Molins-Pujol, A.M., Moranta, C., Arroyo, C., Rodríguez, M.T., Meca, M.C., Pujol, M.D., and Bonal, J. (1996) *Journal of the Chemical Society, Perkin Transactions 1*, 2277–2289.

262 Anderson, H.J., Loader, C.E., Xu, R.X., Lê, N., Gogan, N.J., McDonald, R., and Edwards, L.G. (1985) *Canadian Journal of Chemistry*, **63**, 896–902.

263 Terent'ev, A.P. and Yanovski, L.A. (1949) *Journal of General Chemistry of the USSR (English Translation)*, **19**, 487–491.

264 Terentyev, A.P., Yanovskaya, L.A., and Yashunsky, V.G. (1950) *Journal of General Chemistry of the USSR (English Translation)*, **20**, 539–542.

265 Mizuno, A., Kan, Y., Fukami, H., Kamei, T., Miyazaki, K., Matsuki, S., and Oyama, Y. (2000) *Tetrahedron Letters*, **41**, 6605–6609.

266 Moranta, C., Molins-Pujol, A.M., Pujol, M.D., and Bonal, J. (1998) *Journal of the Chemical Society, Perkin Transactions 1*, 3285–3291.

267 Janosik, T., Shirani, H., Wahlström, N., Malky, I., Stensland, B., and Bergman, J. (2006) *Tetrahedron*, **62**, 1699–1707.

268 Grossie, D.A., Malwitz, D.J., and Ketcha, D.M. (2006) *Acta Crystallographica Section E Structure Reports Online*, **E62**, o980–o982.

269 Carmona, O., Greenhouse, R., Landeros, R., and Muchowski, J.M. (1980) *The Journal of Organic Chemistry*, **45**, 5336–5339.

270 Ortiz, C. and Greenhouse, R. (1985) *Tetrahedron Letters*, **26**, 2831–2832.

271 Franco, F., Greenhouse, R., and Muchowski, J.M. (1982) *The Journal of Organic Chemistry*, **47**, 1682–1688.

272 Gilow, H.M., Brown, C.S., Copeland, J.N., and Kelly, K.E. (1991) *Journal of Heterocyclic Chemistry*, **28**, 1025–1034.

273 Campiani, G., Nacci, V., Bechelli, S., Ciani, S.M., Garofalo, A., Fiorini, I., Wikström, H., de Boer, P., Liao, Y., Tepper, P.G., Cagnotto, A., and Mennini, T. (1998) *Journal of Medicinal Chemistry*, **41**, 3763–3772.

274 Thamyongkit, P., Bhise, A.D., Taniguchi, M., and Lindsey, J.S. (2006) *The Journal of Organic Chemistry*, **71**, 903–910.

275 Chen, Q. and Dolphin, D. (2001) *Synthesis*, 40–42.

276 Kakushima, M. and Frenette, R. (1984) *The Journal of Organic Chemistry*, **49**, 2025–2027.

277 De Sales, J., Greenhouse, R., and Muchowski, J.M. (1982) *The Journal of Organic Chemistry*, **47**, 3668–3672.

278 Gronowitz, S., Hörnfeldt, A.-B., Gestblom, B., and Hoffman, R.A. (1961) *The Journal of Organic Chemistry*, **26**, 2615–2616.

279 Gronowitz, S., Hörnfeldt, A.-B., Gestblom, B., and Hoffman, R.A. (1961) *Arkiv foer Kemi*, **18**, 151–163.

280 Yadav, J.S., Reddy, B.V.S., Shubashree, S., and Sadashiv, K. (2004) *Tetrahedron Letters*, **45**, 2951–2954.

281 Nair, V., George, T.G., Nair, L.G., and Panicker, S.B. (1999) *Tetrahedron Letters*, **40**, 1195–1196.

282 Reddy, G.S. (1965) *Chemistry & Industry*, 1426–1427.

283 Nickisch, K., Klose, W., and Bohlmann, F. (1980) *Chemische Berichte*, **113**, 2036–2037.

284 D'Silva, C. and Iqbal, R. (1996) *Synthesis*, 457–458.

285 Grehn, L. and Ragnarsson, U. (1984) *Angewandte Chemie-International Edition*, **23**, 296–297.

286 Smith, G.F. (1954) *Journal of the Chemical Society*, 3842–3846.

287 Silverstein, R.M., Ryskiewicz, E.E., Willard, C., and Koehler, R.C. (1955) *The Journal of Organic Chemistry*, **20**, 668–672.

288 de Groot, J.A., Gorter-La Roy, G.M., van Koeveringe, J.A., and Lugtenburg, J. (1981) *Organic Preparations and Procedures International*, **13**, 97–101.

289 Jugie, G., Smith, J.A.S., and Martin, G.J. (1975) *Journal of the Chemical Society, Perkin Transactions 2*, 925–927.

290 Candy, C.F., Jones, R.A., and Wright, P.H. (1970) *Journal of the Chemical Society (C)*, 2563–2567.

291 Anderson, H.J., Loader, C.E., and Foster, A. (1980) *Canadian Journal of Chemistry*, **58**, 2527–2530.

292 Sonnet, P.E. (1972) *The Journal of Organic Chemistry*, **37**, 925–929.

293 Downie, I.M., Earle, M.J., Heaney, H., and Shuhaibar, K.F. (1993) *Tetrahedron*, **49**, 4015–4034.

294 Bray, B.L. and Muchowski, J.M. (1988) *The Journal of Organic Chemistry*, **53**, 6115–6118.

295 Rapoport, H. and Castagnoli, N. Jr (1962) *Journal of the American Chemical Society*, **84**, 2178–2181.

296 Alonso Garrido, D.O., Buldain, G., and Frydman, B. (1984) *The Journal of Organic Chemistry*, **49**, 2619–2622.

297 White, J. and McGillivray, G. (1977) *The Journal of Organic Chemistry*, **42**, 4248–4251.

298 Tardieux, C., Bolze, F., Gros, C.P., and Guilard, R. (1998) *Synthesis*, 267–268.

299 Treibs, A. and Kreuzer, F.-H. (1969) *Liebigs Annalen der Chemie*, **721**, 105–115.

300 Sonnet, P.E. (1972) *Journal of Medicinal Chemistry*, **15**, 97–98.

301 Harbuck, J.W. and Rapoport, H. (1972) *The Journal of Organic Chemistry*, **37**, 3618–3622.

302 Anderson, A.G. Jr and Exner, M.M. (1977) *The Journal of Organic Chemistry*, **42**, 3952–3955.

303 Slätt, J., Romero, I., and Bergman, J. (2004) *Synthesis*, 2760–2765.

304 Katritzky, A.R., Suzuki, K., Singh, S.K., and He, H.-Y. (2003) *The Journal of Organic Chemistry*, **68**, 5720–5723.

305 Carson, J.R. and Davis, N.M. (1981) *The Journal of Organic Chemistry*, **46**, 839–843.

306 Tani, M., Ariyasu, T., Nishiyama, C., Hagiwara, H., Watanabe, T., Yokoyama, Y., and Murakami, Y. (1996) *Chemical & Pharmaceutical Bulletin*, **44**, 48–54.

307 Smith, J.A., Ng, S., and White, J. (2006) *Organic and Biomolecular Chemistry*, **4**, 2477–2482.

308 Loader, C.E. and Anderson, H.J. (1981) *Canadian Journal of Chemistry*, **59**, 2673–2676.

309 Anderson, H.J., Riche, C.R., Costello, T.G., Loader, C.E., and Barnett, G.H. (1978) *Canadian Journal of Chemistry*, **56**, 654–657.

310 Loader, C.E. and Anderson, H.J. (1969) *Tetrahedron*, **25**, 3879–3885.

311 Groves, J.K., Anderson, H.J., and Nagy, H. (1971) *Canadian Journal of Chemistry*, **49**, 2427–2432.

312 Cadamuro, S., Degani, I., Dughera, S., Fochi, R., Gatti, A., and Piscopo, L. (1993) *Journal of the Chemical Society, Perkin Transactions 1*, 273–283.

313 Tani, M., Ariyasu, T., Ohtsuka, M., Koga, T., Ogawa, Y., Yokoyama, Y., and Murakami, Y. (1996) *Chemical & Pharmaceutical Bulletin*, **44**, 55–61.

314 Rokach, J., Hamel, P., Kakushima, M., and Smith, G.M. (1981) *Tetrahedron Letters*, **22**, 4901–4904.

315 Kakushima, M., Hamel, P., Frenette, R., and Rokach, J. (1983) *The Journal of Organic Chemistry*, **48**, 3214–3219.

316 Ezaki, N.S.S. and Sakai, S. (1984) *Yakugaku Zasshi*, **104**, 238–245.

317 Nicolaou, I. and Demopoulos, V.J. (1998) *Journal of Heterocyclic Chemistry*, **35**, 1345–1348.

318 Lainton, J.A.H., Huffman, J.W., Martin, B.R., and Compton, D.R. (1995) *Tetrahedron Letters*, **36**, 1401–1404.

319 Song, D., Knight, D.W., and Whatton, M.A. (2004) *Tetrahedron Letters*, **45**, 9573–9576.

320 Xiao, D., Schreier, J.A., Cook, J.H., Seybold, P.G., and Ketcha, D.M. (1996) *Tetrahedron Letters*, **37**, 1523–1526.

321 Rothemund, P. (1936) *Journal of the American Chemical Society*, **58**, 625–627.

322 Rothemund, P. (1939) *Journal of the American Chemical Society*, **61**, 2912–2915.

323 Treibs, A. and Häberle, N. (1968) *Liebigs Annalen der Chemie*, **718**, 183–207.

324 Chmielewski, P.J., Latos-Grażyńyski, L., Rachlewicz, G., and Głowiak, T. (1994) *Angewandte Chemie-International Edition in English*, **33**, 779–781.

325 Furuta, H., Asano, T., and Ogawa, T. (1994) *Journal of the American Chemical Society*, **116**, 767–768.

326 Ghosh, A. (2004) *Angewandte Chemie-International Edition*, **43**, 1918–1931.

327 Maeda, H. and Furuta, H. (2006) *Pure and Applied Chemistry*, **78**, 29–44.

328 Adler, A.D., Longo, F.R., Finarelli, J.D., Goldmacher, J., Assour, J., and Korsakoff, L. (1967) *The Journal of Organic Chemistry*, **32**, 476.

329 Kim, J.B., Leonard, J.J., and Longo, F.R. (1972) *Journal of the American Chemical Society*, **94**, 3986–3992.

330 Lindsey, J.S., Schreiman, I.C., Hsu, H.C., Kearney, P.C., and Marguerettaz, A.M. (1987) *The Journal of Organic Chemistry*, **52**, 827–836.

331 Rocha Gonsalves, A.M.d'A., Varejão, J.M.T.B., and Pereira, M.M. (1991) *Journal of Heterocyclic Chemistry*, **28**, 635–640.

332 Rocha Gonsalves, A.M.d'A. and Pereira, M.M. (1985) *Journal of Heterocyclic Chemistry*, **22**, 931–933.

333 Rothemund, P. and Gage, C.L. (1955) *Journal of the American Chemical Society*, **77**, 3340–3342.

334 Corwin, A.H., Chivvis, A.B., and Storm, C.B. (1964) *The Journal of Organic Chemistry*, **29**, 3702–3703.

335 Wang, Q.M. and Bruce, D.W. (1995) *Synlett*, 1267–1268.

336 Reese, C.B. and Yan, H. (2001) *Tetrahedron Letters*, **42**, 5545–5547.

337 Boiadjiev, S.E. and Lightner, D.A. (2006) *Organic Preparations and Procedures International*, **38**, 347–399.

338 Lee, C.-H. and Lindsey, J.S. (1994) *Tetrahedron*, **50**, 11427–11440.

339 Littler, B.J., Miller, M.A., Hung, C.-H., Wagner, R.W., O'Shea, D.F., Boyle, P.D., and Lindsey, J.S. (1999) *The Journal of Organic Chemistry*, **64**, 1391–1396.

340 Sobral, A.J.F.N., Rebanda, N.G.C.L., da Silva, M., Lampreia, S.H., Silva, M.R., Beja, A.M., Paixão, J.A., and Rocha Gonsalves, A.M.d'A. (2003) *Tetrahedron Letters*, **44**, 3971–3973.

341 Taniguchi, S., Hasegawa, H., Nishimura, M., and Takahashi, M. (1999) *Synlett*, 73–74.

342 Barker, P.L. and Bahia, C. (1990) *Tetrahedron*, **46**, 2691–2694.

343 Murtagh, J.E., McCooey, S.H., and Connon, S.J. (2005) *Chemical Communications*, 227–229.

344 Herz, W., Dittmer, K., and Cristol, S.J. (1947) *Journal of the American Chemical Society*, **69**, 1698–1700.

345 Fuhlhage, D.W. and VanderWerf, C.A. (1958) *Journal of the American Chemical Society*, **80**, 6249–6254.

346 Hayakawa, K., Yodo, M., Ohsuki, S., and Kanematsu, K. (1984) *Journal of the American Chemical Society*, **106**, 6735–6740.

347 Stepanova, Z.V., Sobenina, L.N., Mikhaleva, A.I., Ushakov, I.A., Chipanina, N.N., Elokhina, V.N., Voronov, V.K., and Trofimov, B.A. (2004) *Synthesis*, 2736–2742.

348 Trofimov, B.A., Stepanova, Z.V., Sobenina, L.N., Mikhaleva, A.I., and Ushakov, I.A. (2004) *Tetrahedron Letters*, **45**, 6513–6516.

349 Treibs, A. and Michl, K.-H. (1954) *Liebigs Annalen der Chemie*, **589**, 163–173.

350 Yadav, J.S., Abraham, S., Reddy, B.V.S., and Sabitha, G. (2001) *Tetrahedron Letters*, **42**, 8063–8065.

351 de la Hoz, A., Díaz-Ortiz, A., Gómez, M.V., Mayoral, J.A., Moreno, A., Sánchez-Migallón, A.M., and Vásquez, E. (2001) *Tetrahedron*, **57**, 5421–5428.

352 Zhan, Z.-P., Yang, W.-Z., and Yang, R.-F. (2005) *Synlett*, 2425–2428.

353 Firouzabadi, H., Iranpoor, N., and Nowrouzi, F. (2005) *Chemical Communications*, 789–791.

354 Paras, N.A. and MacMillan, D.W.C. (2001) *Journal of the American Chemical Society*, **123**, 4370–4371.

355 Evans, D.A. and Fandrick, K.R. (2006) *Organic Letters*, **8**, 2249–2252.

356 Gardini, G.P. (1973) *Advances in Heterocyclic Chemistry*, **15**, 67–98.

357 Battersby, A.R., Dutton, C.J., and Fookes, C.J.R. (1988) *Journal of the Chemical Society, Perkin Transactions 1*, 1569–1576.

358 Siedel, W. and Winkler, F. (1943) *Liebigs Annalen der Chemie*, **554**, 162–201.

359 Thyrann, T. and Lightner, D.A. (1995) *Tetrahedron Letters*, **36**, 4345–4348.

360 Thyrann, T. and Lightner, D.A. (1996) *Tetrahedron Letters*, **37**, 315–318.

361 Moreno-Vargas, A.J., Robina, I., Fernández-Bolaños, J.G., and Fuentes, J. (1998) *Tetrahedron Letters*, **39**, 9271–9274.

362 Moranta, C., Pujol, M.D., Molins-Pujol, A.M., and Bonal, J. (1999) *Synthesis*, 447–452.

363 Sessler, J.L., Aguilar, A., Sanchez-Garcia, D., Seidel, D., Köhler, T., Arp, F., and Lynch, V.M. (2005) *Organic Letters*, **7**, 1887–1890.

364 Dohi, T., Morimoto, K., Kiyono, Y., Tohma, H., and Kita, Y. (2005) *Organic Letters*, **7**, 537–540.

365 Dohi, T., Morimoto, K., Maruyama, A., and Kita, Y. (2006) *Organic Letters*, **8**, 2007–2010.

366 Aiura, M. and Kanaoka, Y. (1975) *Chemical & Pharmaceutical Bulletin*, **23**, 2835–2841.

367 Vecchietti, V., Dradi, E., and Lauria, F. (1971) *Journal of the Chemical Society (C)*, 2554–2557.

368 Doddi, G., Mencarelli, P., and Stegel, F. (1975) *Journal of the Chemical Society, Chemical Communications*, 273–274.

369 Bazzano, F., Mencarelli, P., and Stegel, F. (1984) *The Journal of Organic Chemistry*, **49**, 2375–2377.

370 Di Lorenzo, A., Mencarelli, P., and Stegel, F. (1986) *The Journal of Organic Chemistry*, **51**, 2125–2126.

371 Ballini, R., Bartoli, G., Bosco, M., Dalpozzo, R., and Marcantoni, E. (1988) *Tetrahedron*, **44**, 6435–6440.

372 Mąkosza, M. and Kwast, E. (1995) *Tetrahedron*, **51**, 8339–8354.

373 Cirrincione, G., Almerico, A.M., Passannanti, A., Diana, P., and Mingoia, F. (1997) *Synthesis*, 1169–1173.

374 Treibs, A. and Zimmer-Galler, R. (1963) *Liebigs Annalen der Chemie*, **664**, 140–145.

375 Hobbs, C.F., McMillin, C.K., Papadopoulos, E.P., and VanderWerf, C.A. (1962) *Journal of the American Chemical Society*, **84**, 43–51.

376 Nunomoto, S., Kawakami, Y., Yamashita, Y., Takeuchi, H., and Eguchi, S. (1990) *Journal of the Chemical Society, Perkin Transactions 1*, 111–114.

377 Treibs, A. and Dietl, A. (1958) *Liebigs Annalen der Chemie*, **619**, 80–95.

378 Papadopoulos, E.P. and Haidar, N.F. (1968) *Tetrahedron Letters*, **9**, 1721–1723.

379 Shirley, D.A., Gross, B.H., and Roussel, P.A. (1955) *The Journal of Organic Chemistry*, **20**, 225–231.

380 Heaney, H. and Ley, S.V. (1973) *Journal of the Chemical Society, Perkin Transactions 1*, 499–500.

381 Santaniello, E., Farachi, C., and Ponti, F. (1979) *Synthesis*, 617–618.

382 Guida, W.C. and Mathre, D.J. (1980) *The Journal of Organic Chemistry*, **45**, 3172–3176.

383 Wang, N.-C., Teo, K.-E., and Anderson, H.J. (1977) *Canadian Journal of Chemistry*, **55**, 4112–4116.

384 Trofimov, B.A., Sobenina, L.N., Mikhaleva, A.I., Demenev, A.P., Tarasova, O.A., Ushakov, I.A., and Zinchenko, S.V. (2000) *Tetrahedron*, **56**, 7325–7329.

385 Simchen, G. and Majchrzak, M.W. (1985) *Tetrahedron Letterss*, **26**, 5035–5036.

386 Birkofer, L., Richter, P., and Ritter, A. (1960) *Chemische Berichte*, **93**, 2804–2809.

387 Biemans, H.A.M., Zhang, C., Smith, P., Kooijman, H., Smeets, W.J.J., Spek, A.L., and Meijer, E.W. (1996) *The Journal of Organic Chemistry*, **61**, 9012–9015.

388 Candy, C.F. and Jones, R.A. (1971) *The Journal of Organic Chemistry*, **36**, 3993–3994.

389 Skell, P.S. and Bean, G.P. (1962) *Journal of the American Chemical Society*, **84**, 4655–4660.

390 Bean, G.P. (1965) *Journal of Heterocyclic Chemistry*, **2**, 473–474.

391 Nicolaou, K.C., Claremon, D.A., and Papahatjis, D.P. (1981) *Tetrahedron Letters*, **22**, 4647–4650.

392 Schloemer, G.C., Greenhouse, R., and Muchowski, J.M. (1994) *The Journal of Organic Chemistry*, **59**, 5230–5234.

393 Baran, P.S., Richter, J.M., and Lin, D.W. (2004) *Angewandte Chemie-International Edition*, **43**, 2–4.

394 Filippini, L. Gusmeroli, M., and Riva, R. (1992) *Tetrahedron Letters*, **33**, 1755–1758.

395 Gjøs, N. and Gronowitz, S. (1971) *Acta Chemica Scandinavica*, **25**, 2596–2608.

396 Chadwick, D.J. and Willbe, C. (1977) *Journal of the Chemical Society, Perkin Transactions 1*, 887–893.

397 Chadwick, D.J. (1974) *Journal of the Chemical Society, Chemical Communications*, 790–791.

398 Brittain, J.M., Jones, R.A., Sepulveda Arques, J., and Aznar Saliente, T. (1982) *Synthetic Communications*, **12**, 231–248.

399 Sotoyama, T., Hara, S., and Suzuki, A. (1979) *Bulletin of the Chemical Society of Japan*, **52**, 1865–1866.

400 Marinelli, E.R., and Levy, A.B. (1979) *Tetrahedron Letters*, **20**, 2313–2316.

401 Dvornikova, E., Bechcicka, M., Kamienska-Trela, K., and Krówczynski, A. (2003) *Journal of Fluorine Chemistry*, **124**, 159–168.

402 Minato, A., Tamao, K., Hayashi, T., Suzuki, K., and Kumada, M. (1981) *Tetrahedron Letters*, **22**, 5319–5322.

403 Mal'kina, A.G., Tarasova, O.A., Verkruijsse, H.D., van der Kerk, A.C.H.T.M., Brandsma, L., and Trofimov, B.A. (1995) *Recueil des Travaux Chimiques des Pays-Bas*, **114**, 18–21.

404 Clark, G.R., Ng, M.M.P., Roper, W.R., and Wright, L.J. (1995) *Journal of Organometallic Chemistry*, **491**, 219–229.

405 Ganske, J.A., Pandey, R.K., Postich, M.J., Snow, K.M., and Smith, K.M. (1989) *The Journal of Organic Chemistry*, **54**, 4801–4807.

406 Hasan, I., Marinelli, E.R., Lin, L.-C.C., Fowler, F.W., and Levy, A.B. (1981) *The Journal of Organic Chemistry*, **46**, 157–164.

407 Rawal, V.H. and Cava, M.P. (1985) *Tetrahedron Letters*, **26**, 6141–6142.

408 Jolicoeur, B., Chapman, E.E., Thompson, A., and Lubell, W.D. (2006) *Tetrahedron*, **62**, 11531–11563.

409 Gharpure, M., Stoller, A., Bellamy, F., Firnau, G., and Snieckus, V. (1991) *Synthesis*, 1079–1082.

410 Katritzky, A.R. and Akutagawa, K. (1988) *Organic Preparations and Procedures International*, **20**, 585–590.

411 Engman, L. and Cava, M.P. (1982) *Organometallics*, **1**, 470–473.

412 Dinsmore, A., Billing, D.G., Mandy, K., Michael, J.P., Mogano, D., and Patil, S. (2004) *Organic Letters*, **6**, 293–296.

413 Muchowski, J.M. and Solas, D.R. (1984) *The Journal of Organic Chemistry*, **49**, 203–205.

414 Edwards, M.P., Doherty, A.M., Ley, S.V., and Organ, H.M. (1986) *Tetrahedron*, **42**, 3723–3729.

415 Edwards, M.P., Ley, S.V., Lister, S.G., Palmer, B.D., and Williams, D.J. (1984) *The Journal of Organic Chemistry*, **49**, 3503–3516.

416 Martinez, G.R., Grieco, P.A., and Srinivasan, C.V. (1981) *The Journal of Organic Chemistry*, **46**, 3760–3761.

417 Anderson, H.J. and Loader, C.E. (1985) *Synthesis*, 353–364.

418 Alvarez, A., Guzmàn, A., Ruiz, A., Velarde, E., and Muchowski, J.M. (1992) *The Journal of Organic Chemistry*, **57**, 1653–1656.

419 Barnes, K.D., Hu, Y., and Hunt, D.A. (1994) *Synthetic Communications*, **24**, 1749–1755.

420 Chadwick, D.J. and Hodgson, S.T. (1982) *Journal of the Chemical Society, Perkin Transactions 1*, 1833–1836.

421 Bergauer, M. and Gmeiner, P. (2001) *Synthesis*, 2281–2288.

422 Baciocchi, E., Muraglia, E., and Sleiter, G. (1992) *The Journal of Organic Chemistry*, **57**, 6817–6820.

423 Byers, J.H., Campbell, J.E., Knapp, F.H., and Thissell, J.G. (1999) *Tetrahedron Letters*, **40**, 2677–2680.

424 Byers, J.H., Duff, M.P., and Woo, G.W. (2003) *Tetrahedron Letters*, **44**, 6853–6855.

425 Byers, J.H., DeWitt, A., Nasveschuk, C.G., and Swigor, J.E. (2004) *Tetrahedron Letters*, **45**, 6587–6590.

426 Osornio, Y.M., Cruz-Almanza, R., Jiménez-Montaño, V., and Miranda, L.D. (2003) *Chemical Communications*, 2316–2317.

427 Artis, D.R., Cho, I.-S., and Muchowski, J.M. (1992) *Canadian Journal of Chemistry*, **70**, 1838–1842.

428 Aldabbagh, F., Bowman, W.R., and Mann, E. (1997) *Tetrahedron Letters*, **38**, 7937–7940.

429 Aldabbagh, F., Bowman, W.R., Mann, E., and Slawin, A.M.Z. (1999) *Tetrahedron*, **55**, 8111–8128.

430 Artis, D.R., Cho, I.-S., Jaime-Figueroa, S., and Muchowski, J.M. (1994) *The Journal of Organic Chemistry*, **59**, 2456–2466.

431 Ozaki, S., Mitoh, S., and Ohmori, H. (1996) *Chemical & Pharmaceutical Bulletin*, **44**, 2020–2024.

432 Allin, S.M., Barton, W.R.S., Bowman, W.R., and McInally, T. (2001) *Tetrahedron Letters*, **42**, 7887–7890.

433 Cantacuzène, D., Wakselman, C., and Dorme, R. (1977) *Journal of the Chemical Society, Perkin Transactions 1*, 1365–1371.

434 Baciocchi, E. and Muraglia, E. (1993) *Tetrahedron Letters*, **34**, 3799–3800.

435 Yoshida, M., Yoshida, T., Kobayashi, M., and Kamigata, N. (1989) *Journal of the Chemical Society, Perkin Transactions 1*, 909–914.

436 Rapoport, H. and Look, M. (1953) *Journal of the American Chemical Society*, **75**, 4605–4607.

437 Saeki, S., Hayashi, T., and Hamana, M. (1984) *Chemical & Pharmaceutical Bulletin*, **32**, 2154–2159.

438 Jones, K., Ho, T.C.T., and Wilkinson, J. (1995) *Tetrahedron Letters*, **36**, 6743–6744.

439 Ho, T.C.T. and Jones, K. (1997) *Tetrahedron*, **53**, 8287–8294.

440 Escolano, C. and Jones, K. (2002) *Tetrahedron*, **58**, 1453–1464.

441 Aboutayab, K., Caddick, S., Jenkins, K., and Khan, S.J.S. (1996) *Tetrahedron*, **52**, 11329–11340.

442 Knorr, L. and Rabe, P. (1901) *Chemische Berichte*, **34**, 3491–3502.

443 Andrews, L.H. and McElvain, S.M. (1929) *Journal of the American Chemical Society*, **51**, 887–892.

444 Hudson, C.B. and Robertson, A.V. (1967) *Tetrahedron Letters*, **8**, 4015–4017.

445 Lemal, D.M. and McGregor, S.D. (1966) *Journal of the American Chemical Society*, **88**, 1335–1336.

446 Schumacher, D.P. and Hall, S.S. (1982) *Journal of the American Chemical Society*, **104**, 6076–6080.

447 Robertson, A.V. and Witkop, B. (1962) *Journal of the American Chemical Society*, **84**, 1697–1701.

448 Scott, J.W., Focella, A., Hengartner, U.O., Parrish, D.R., and Valentine, D. Jr (1980) *Synthetic Communications*, **10**, 529–540.

449 Ketcha, D.M., Carpenter, K.P., and Zhou, Q. (1991) *The Journal of Organic Chemistry*, **56**, 1318–1320.

450 Overberger, C.G., Palmer, L.C., Marks, B.S., and Byrd, N.R. (1955) *Journal of the American Chemical Society*, **77**, 4100–4104.

451 Bates, H.A. and Rapoport, H. (1979) *Journal of the American Chemical Society*, **101**, 1259–1265.

452 Turner, W.W. (1986) *Journal of Heterocyclic Chemistry*, **23**, 327–328.

453 Jefford, C.W., Tang, Q., and Zaslona, A. (1991) *Journal of the American Chemical Society*, **113**, 3513–3518.

454 Bond, T.J., Jenkins, R., Ridley, A.C., and Taylor, P.C. (1993) *Journal of the Chemical Society, Perkin Transactions 1*, 2241–2242.

455 Kaiser, H.-P. and Muchowski, J.M. (1984) *The Journal of Organic Chemistry*, **49**, 4203–4209.

456 Donohoe, T.J. and Guyo, P.M. (1996) *The Journal of Organic Chemistry*, **61**, 7664–7665.

457 Donohoe, T.J., Guyo, P.M., Beddoes, R.L., and Helliwell, M. (1998) *Journal of the Chemical Society, Perkin Transactions 1*, 667–676.

458 Donohoe, T.J., Guyo, P.M., and Helliwell, M. (1999) *Tetrahedron Letters*, **40**, 435–438.

459 Schäfer, A. and Schäfer, B. (1999) *Tetrahedron*, **55**, 12309–12312.

460 Donohoe, T.J., Ace, K.W., Guyo, P.M., Helliwell, M., and McKenna, J. (2000) *Tetrahedron Letters*, **41**, 989–993.

461 Donohoe, T.J., Harji, R.R., and Cousins, R.P.C. (2000) *Tetrahedron Letters*, **41**, 1327–1330.

462 Donohoe, T.J., Harji, R.R., and Cousins, R.P.C. (2000) *Tetrahedron Letters*, **41**, 1331–1334.

463 Donohoe, T.J. and House, D. (2002) *The Journal of Organic Chemistry*, **67**, 5015–5018.

464 Donohoe, T.J. and House, D. (2003) *Tetrahedron Letters*, **44**, 1095–1098.

465 Donohoe, T.J., House, D., and Ace, K.W. (2003) *Organic and Biomolecular Chemistry*, **1**, 3749–3757.

466 Donohoe, T.J., Freestone, G.C., Headley, C.E., Rigby, C.L., Cousins, R.P.C., and

Bhalay, G. (2004) *Organic Letters*, **6**, 3055–3058.

467 Donohoe, T.J., Guyo, P.M., Harji, R.R., Helliwell, M., and Cousins, R.P.C. (1998) *Tetrahedron Letters*, **39**, 3075–3078.

468 Donohoe, T.J., Harji, R.R., and Cousins, R.P.C. (1999) *Chemical Communications*, 141–142.

469 Diels, O. and Alder, K. (1931) *Liebigs Annalen der Chemie*, **486**, 211–225.

470 Diels, O. and Alder, K. (1931) *Liebigs Annalen der Chemie*, **490**, 267–276.

471 Acheson, R.M. and Vernon, J.M. (1962) *Journal of the Chemical Society*, 1148–1157.

472 Noland, W.E. and Lee, C.K. (1980) *The Journal of Organic Chemistry*, **45**, 4573–4582.

473 Domingo, L.R., Picher, M.T., and Zaragozá, R.J. (1998) *The Journal of Organic Chemistry*, **63**, 9183–9189.

474 Acheson, R.M. and Vernon, J.M. (1961) *Journal of the Chemical Society*, 457–459.

475 Gabel, N.W. (1962) *The Journal of Organic Chemistry*, **27**, 301–303.

476 Mandell, L. and Blanchard, W.A. (1957) *Journal of the American Chemical Society*, **79**, 6198–6201.

477 Acheson, R.M. and Vernon, J.M. (1963) *Journal of the Chemical Society*, 1008–1011.

478 Schultz, A.G. and Shen, M. (1979) *Tetrahedron Letters*, **20**, 2969–2972.

479 Kitzing, R., Fuchs, R., Joyeux, M., and Prinzbach, H. (1968) *Helvetica Chimica Acta*, **51**, 888–895.

480 Bansal, R.C., McCulloch, A.W., and McInnes, A.G. (1969) *Canadian Journal of Chemistry*, **47**, 2391–2394.

481 Kotsuki, H., Mori, Y., Nishizawa, H., Ochi, M., and Matsuoka, K. (1982) *Heterocycles*, **19**, 1915–1920.

482 Drew, M.G.B., George, A.V., Isaacs, N.S., and Rzepa, H.S. (1985) *Journal of the Chemical Society, Perkin Transactions 1*, 1277–1284.

483 Keijsers, J., Hams, B., Kruse, C., and Scheeren, H. (1989) *Heterocycles*, **29**, 79–86.

484 Corey, E.J. and Loh, T.-P. (1993) *Tetrahedron Letters*, **34**, 3979–3982.

485 Paulvannan, K. (2004) *The Journal of Organic Chemistry*, **69**, 1207–1214.

486 Chen, Z. and Trudell, M.L. (1994) *Tetrahedron Letters*, **35**, 9649–9652.

487 Altenbach, H.-J., Blech, B., Marco, J.A., and Vogel, E. (1982) *Angewandte Chemie-International Edition in English*, **21**, 778.

488 Altenbach, H.-J., Constant, D., Martin, H.-D., Mayer, B., Müller, M., and Vogel, E. (1991) *Chemische Berichte*, **124**, 791–801.

489 Giblin, G.M.P., Jones, C.D., and Simpkins, N.S. (1997) *Synlett*, 589–590.

490 Giblin, G.M.P., Jones, C.D., and Simpkins, N.S. (1998) *Journal of the Chemical Society, Perkin Transactions 1*, 3689–3697.

491 Clayton, S.C. and Regan, A.C. (1993) *Tetrahedron Letters*, **34**, 7493–7496.

492 Huang, D.F. and Shen, T.Y. (1993) *Tetrahedron Letters*, **34**, 4477–4480.

493 Zhang, C. and Trudell, M.L. (1996) *The Journal of Organic Chemistry*, **61**, 7189–7191.

494 Zhang, C. and Trudell, M.L. (1998) *Tetrahedron*, **54**, 8349–8354.

495 Nishide, K., Ichihashi, S., Kimura, H., Katoh, T., and Node, M. (2001) *Tetrahedron Letters*, **42**, 9237–9240.

496 Node, M., Nishide, K., Fujiwara, T., and Ichihashi, S. (1998) *Chemical Communications*, 2363–2364.

497 Pavri, N.P. and Trudell, M.L. (1997) *Tetrahedron Letters*, **38**, 7993–7996.

498 Wittig, G. and Behnisch, W. (1958) *Chemische Berichte*, **91**, 2358–2365.

499 Wittig, G. and Reichel, B. (1963) *Chemische Berichte*, **96**, 2851–2858.

500 Wolthuis, E., Vander Jagt, D., Mels, S., and De Boer, A. (1965) *The Journal of Organic Chemistry*, **30**, 190–193.

501 Carpino, L.A. and Barr, D.E. (1966) *The Journal of Organic Chemistry*, **31**, 764–767.

502 Wolthuis, E. and De Boer, A. (1965) *The Journal of Organic Chemistry*, **30**, 3225–3227.

503 Wolthuis, E., Cady, W., Roon, R., and Weidenaar, B. (1966) *The Journal of Organic Chemistry*, **31**, 2009–2011.

504 Callander, D.D., Coe, P.L., Tatlow, J.C., and Uff, A.J. (1969) *Tetrahedron*, **25**, 25–36.

505 Anderson, P.S., Christy, M.E., Engelhardt, E.L., Lundell, G.F., and Ponticello, G.S. (1977) *Journal of Heterocyclic Chemistry*, **14**, 213–218.

506 Ahmed, M. and Vernon, J.M. (1976) *Journal of the Chemical Society, Chemical Communications*, 462–463.

507 Ciamician, G. and Silber, P. (1912) *Chemische Berichte*, **45**, 1842–1845.

508 Lightner, D.A., Bisacchi, G.S., and Norris, R.D. (1976) *Journal of the American Chemical Society*, **98**, 802–807.

509 Wenkert, E., Moeller, P.D.R., and Piettre, S.R. (1988) *Journal of the American Chemical Society*, **110**, 7188–7194.

510 Chrétien, A., Chataigner, I., and Piettre, S.R. (2005) *Chemical Communications*, 1351–1353.

511 Chrétien, A., Chataigner, I., and Piettre, S.R. (2005) *Tetrahedron*, **61**, 7907–7915.

512 Ruccia, M., Vivona, N., and Cusmano, G. (1972) *Tetrahedron Letters*, **13**, 4703–4706.

513 Ruccia, M., Vivona, N., and Cusmano, G. (1978) *Journal of Heterocyclic Chemistry*, **15**, 293–296.

514 Dehaen, W. and Hassner, A. (1991) *The Journal of Organic Chemistry*, **56**, 896–900.

515 Hodges, L.M., Gonzalez, J., Koontz, J.I., Myers, W.H., and Harman, W.D. (1993) *The Journal of Organic Chemistry*, **58**, 4788–4790.

516 Gonzalez, J., Koontz, J.I., Hodges, L.M., Nilsson, K.R., Neely, L.K., Myers, W.H., Sabat, M., and Harman, W.D. (1995) *Journal of the American Chemical Society*, **117**, 3405–3421.

517 Dang, Q. and Gomez-Galeno, J.E. (2002) *The Journal of Organic Chemistry*, **67**, 8703–8705.

518 Yu, Z.-X., Dang, Q., and Wu, Y.-D. (2001) *The Journal of Organic Chemistry*, **66**, 6029–6036.

519 Giomi, D. and Cecchi, M. (2002) *Tetrahedron*, **58**, 8067–8071.

520 Li, J.-H. and Snyder, J.K. (1993) *The Journal of Organic Chemistry*, **58**, 516–519.

521 Hsieh, M.-F., Peddinti, R.K., and Liao, C.-C. (2001) *Tetrahedron Letters*, **42**, 5481–5484.

522 Jones, R.A., Marriott, M.T.P., Rosenthal, W.P., and Sepulveda Arques, J. (1980) *The Journal of Organic Chemistry*, **45**, 4515–4519.

523 Jones, R.A. and Sepulveda Arques, J. (1981) *Tetrahedron*, **37**, 1597–1599.

524 Muchowski, J.M. and Scheller, M.E. (1987) *Tetrahedron Letters*, **28**, 3453–3456.

525 Xiao, D. and Ketcha, D.M. (1995) *Journal of Heterocyclic Chemistry*, **32**, 499–503.

526 Ohno, M., Shimizu, S., and Eguchi, S. (1990) *Tetrahedron Letters*, **31**, 4613–4616.

527 Ohno, M., Shimizu, S., and Eguchi, S. (1991) *Heterocycles*, **32**, 1199–1202.

528 Murase, M., Yoshida, S., Hosaka, T., and Tobinaga, S. (1991) *Chemical & Pharmaceutical Bulletin*, **39**, 489–492.

529 Jones, R.A., Aznar Saliente, T., and Sepúlveda Arques, J. (1984) *Journal of the Chemical Society, Perkin Transactions 1*, 2541–2543.

530 Nenitzescu, C.D. and Solomonica, E. (1931) *Chemische Berichte*, **64**, 1924–1931.

531 Maryanoff, B.E. (1977) *Journal of Heterocyclic Chemistry*, **14**, 177–178.

532 Maryanoff, B.E. (1979) *The Journal of Organic Chemistry*, **44**, 4410–4419.

533 Maryanoff, B.E. (1982) *The Journal of Organic Chemistry*, **47**, 3000–3002.

534 Pomerantz, M. and Rooney, P. (1988) *The Journal of Organic Chemistry*, **53**, 4374–4378.

535 Jones, R.L. and Rees, C.W. (1969) *Journal of the Chemical Society (C)*, 2249–2251.

536 De Angelis, F., Gambacorta, A., and Nicoletti, R. (1976) *Synthesis*, 798–800.

537 Gambacorta, A., Nicoletti, R., Cerrini, S., Fedeli, W., and Gavuzzo, E. (1978) *Tetrahedron Letters*, **19**, 2439–2442.

538 Fowler, F.W. (1969) *Journal of the Chemical Society, Chemical Communications*, 1359–1360.

539 Tanny, S.R., Grossman, J., and Fowler, F.W. (1972) *Journal of the American Chemical Society*, **94**, 6495–6501.

540 Fowler, F.W. (1971) *Angewandte Chemie-International Edition*, **10**, 135.

541 Voigt, J., Noltemeyer, M., and Reiser, O. (1997) *Synlett*, 202–204.

542 Biellmann, J.F. and Goeldner, M.P. (1971) *Tetrahedron*, **27**, 2957–2965.

543 Davies, H.M.L., Young, W.B., and Smith, H.D. (1989) *Tetrahedron Letters*, **30**, 4653–4656.

544 Davies, H.M.L., Saikali, E., and Young, W.B. (1991) *The Journal of Organic Chemistry*, **56**, 5696–5700.

545 Hiraoka, H. (1970) *Journal of the Chemical Society. Chemical Communications*, 1306.

546 Barltrop, J.A., Day, A.C., and Ward, R.W. (1978) *Journal of the Chemical Society, Chemical Communications*, 131–133.

547 Warrener, R.N., Amarasekara, A.S., and Russell, R.A. (1996) *Chemical Communications*, 1519–1520.

548 Kobayashi, Y., Kumadaki, I., Ohsawa, A., and Ando, A. (1977) *Journal of the American Chemical Society*, **99**, 7350–7351.

549 Kobayashi, Y., Ando, A., Kawada, K., and Kumadaki, I. (1980) *The Journal of Organic Chemistry*, **45**, 2966–2968.

550 Jones, G. II, Gilow, H.M., and Low, J. (1979) *The Journal of Organic Chemistry*, **44**, 2949–2951.

551 Rivas, C. and Bolivar, R.A. (1976) *Journal of Heterocyclic Chemistry*, **13**, 1037–1040.

552 Rawal, V.H., Jones, R.J., and Cava, M.P. (1985) *Tetrahedron Letters*, **26**, 2423–2426.

553 Antelo, B., Castedo, L., Delamano, J., Gómez, A., López, C., and Tojo, G. (1996) *The Journal of Organic Chemistry*, **61**, 1188–1189.

554 Herz, W. (1953) *Journal of the American Chemical Society*, **75**, 483.

555 Berlin, A., Bradamante, S., Ferraccioli, R., Pagani, G.A., and Sannicolò, F. (1987) *Journal of the Chemical Society, Perkin Transactions 1*, 2631–2635.

556 Battersby, A.R., Fookes, C.J.R., Gustafson-Potter, K.E., McDonald, E., and Matcham, G.W.J. (1982) *Journal of the Chemical Society, Perkin Transactions 1*, 2427–2444.

557 Barock, R.A., Moorcroft, N.A., Storr, R.C., Young, J.H., and Fuller, L.S. (1993) *Tetrahedron Letters*, **34**, 1187–1190.

558 Kim, I.T. and Elsenbaumer, R.L. (1998) *Tetrahedron Letters*, **39**, 1087–1090.

559 Carson, J.R., Hortenstine, J.T., Maryanoff, B.E., and Molinari, A.J. (1977) *The Journal of Organic Chemistry*, **42**, 1096–1098.

560 Rocha Gonsalves, A.M.d'A., Kenner, G.W., and Smith, K.M. (1972) *Tetrahedron Letters*, **13**, 2203–2206.

561 Cavaleiro, J.A.S., Kenner, G.W., and Smith, K.M. (1974) *Journal of the Chemical Society, Perkin Transactions 1*, 1188–1194.

562 Jackson, A.H., Pandey, R.K., Rao, K.R.N., and Roberts, E. (1985) *Tetrahedron Letters*, **26**, 793–796.

563 Tietze, L.F., Kettschau, G., and Heitmann, K. (1996) *Synthesis*, 851–857.

564 De Leon, C.Y. and Ganem, B. (1996) *The Journal of Organic Chemistry*, **61**, 8730–8731.

565 Abell, A.D., Nabbs, B.K., and Battersby, A.R. (1998) *Journal of the American Chemical Society*, **120**, 1741–1746.

566 Abell, A.D., Nabbs, B.K., and Battersby, A.R. (1998) *The Journal of Organic Chemistry*, **63**, 8163–8169.

567 Treibs, A. and Jacob, K. (1970) *Liebigs Annalen der Chemie*, **737**, 176–178.

568 Hayes, A., Kenner, G.W., and Williams, N.R. (1958) *Journal of the Chemical Society*, 3779–3788.

569 Katritzky, A.R., Levell, J.R., and Li, J. (1996) *Tetrahedron Letters*, **37**, 5641–5644.

570 Settambolo, R., Lazzaroni, R., Messeri, T., Mazzetti, M., and Salvadori, P. (1993) *The Journal of Organic Chemistry*, **58**, 7899–7902.

571 Sobenina, L.N., Demenev, A.P., Mikhaleva, A.I., and Trofimov, B.A. (2002) *Russian Chemical Reviews*, **71**, 563–591.

572 Trofimov, B.A., Sobenina, L.N., Demenev, A.P., and Mikhaleva, A.I. (2004) *Chemical Reviews*, **104**, 2481–2506.

573 Banwell, M.G., Goodwin, T.E., Ng, S., Smith, J.A., and Wong, D.J. (2006) *European Journal of Organic Chemistry*, 3043–3060.

574 Mann, G., Hartwig, J.F., Driver, M.S., and Fernández-Rivas, C. (1998) *Journal of the American Chemical Society*, **120**, 827–828.

575 Hartwig, J.F., Kawatsura, M., Hauck, S.I., Shaughnessy, K.H., and Alcazar-Roman, L.M. (1999) *The Journal of Organic Chemistry*, **64**, 5575–5580.

576 Klapars, A., Antilla, J.C., Huang, X., and Buchwald, S.L. (2001) *Journal of the American Chemical Society*, **123**, 7727–7729.

577 Yu, S., Saenz, J., and Srirangam, J.K. (2002) *The Journal of Organic Chemistry*, **67**, 1699–1702.

578 Movassaghi, M. and Ondrus, A.E. (2005) *The Journal of Organic Chemistry*, **70**, 8638–8641.

579 Zhang, Y., Hsung, R.P., Tracey, M.R., Kurtz, K.C.M., and Vera, E.L. (2004) *Organic Letters*, **6**, 1151–1154.

580 Thoresen, L.H., Kim, H., Welch, M.B., Burghart, A., and Burgess, K. (1998) *Synlett*, 1276–1278.

581 Burghart, A., Kim, H., Welch, M.B., Thoresen, L.H., Reibenspies, J., Burgess, K., Bergström, F., and Johansson, L.B.-Å. (1999) *The Journal of Organic Chemistry*, **64**, 7813–7819.

582 Johnson, C.N., Stemp, G., Anand, N., Stephen, S.C., and Gallagher, T. (1998) *Synlett*, 1025–1027.

583 Grieb, J.G. and Ketcha, D.M. (1995) *Synthetic Communications*, **25**, 2145–2153.

584 Schröter, S. and Bach, T. (2005) *Synlett*, 1957–1959.

585 Tietze, L.F. and Nordmann, G. (2001) *Synlett*, 337–340.

586 Tietze, L.F., Kettschau, G., Heuschert, U., and Nordmann, G. (2001) *Chemistry – A European Journal*, **7**, 368–373.

587 O'Neal, W.G., Roberts, W.P., Ghosh, I., and Jacobi, P.A. (2005) *The Journal of Organic Chemistry*, **70**, 7243–7251.

588 D'Alessio, R. and Rossi, A. (1996) *Synlett*, 513–514.

589 D'Alessio, R., Bargiotti, A., Carlini, O., Colotta, F., Ferrari, M., Gnocchi, P., Isetta, A., Mongelli, N., Motta, P., Rossi, A., Rossi, M., Tibolla, M., and Vanotti, E. (2000) *Journal of Medicinal Chemistry*, **43**, 2557–2565.

590 Rieth, R.D., Mankad, N.P., Calimano, E., and Sadighi, J.P. (2004) *Organic Letters*, **6**, 3981–3983.

591 Itahara, T. (1985) *The Journal of Organic Chemistry*, **50**, 5272–5275.

592 Beck, E.M., Grimster, N.P., Hatley, R., and Gaunt, M.J. (2006) *Journal of the American Chemical Society*, **128**, 2528–2529.

593 Handy, S.T., Bregman, H., Lewis, J., Zhang, X., and Zhang, Y. (2003) *Tetrahedron Letters*, **44**, 427–430.

594 Wang, J. and Scott, A.I. (1996) *Tetrahedron Letters*, **37**, 3247–3250.

595 Wang, J. and Scott, A.I. (1995) *Tetrahedron Letters*, **36**, 7043–7046.

596 Lee, C.-W. and Chung, Y.J. (2000) *Tetrahedron Letters*, **41**, 3423–3425.

597 Sakamoto, T. and Ohsawa, K. (1999) *Journal of the Chemical Society, Perkin Transactions 1*, 2323–2326.

598 Greenhouse, R., Ramirez, C., and Muchowski, J.M. (1985) *The Journal of Organic Chemistry*, **50**, 2961–2965.

599 Ketcha, D.M., Carpenter, K.P., Atkinson, S.T., and Rajagopalan, H.R. (1990) *Synthetic Communications*, **20**, 1647–1655.

600 Schumacher, D.P. and Hall, S.S. (1981) *The Journal of Organic Chemistry*, **46**, 5060–5064.

601 Gale, P.A. (2005) *Chemical Communications*, 3761–3772.

602 Lancaster, R.E. Jr and VanderWerf, C.A. (1958) *The Journal of Organic Chemistry*, **23**, 1208–1209.

603 Dunn, G.E. and Lee, G.K.J. (1971) *Canadian Journal of Chemistry*, **49**, 1032–1035.

604 Khan, M.K.A. and Morgan, K.J. (1965) *Tetrahedron*, **21**, 2197–2204.

605 Williams, A. and Salvadori, G. (1972) *Journal of the Chemical Society, Perkin Transactions 2*, 883–889.

606 Ramasseul, R. and Rassat, A. (1970) *Bulletin de la Societe Chimique de France*, 4330–4341.

607 Kreher, R. and Pawelczyk, H. (1976) *Zeitschrift fur Naturforschung*, **31b**, 599–604.

608 Schweizer, E.E. and Kopay, C.M. (1972) *The Journal of Organic Chemistry*, **37**, 1561–1564.

609 Bocchi, V., Chierici, L., Gardini, G.P., and Mondelli, R. (1970) *Tetrahedron*, **26**, 4073–4082.

610 Baker, J.T. and Sifniades, S. (1979) *The Journal of Organic Chemistry*, **44**, 2798–2800.

611 Casiraghi, G., Rassu, G., Spanu, P., and Pinna, L. (1992) *The Journal of Organic Chemistry*, **57**, 3760–3763.

612 Casiraghi, G., Rassu, G., Spanu, P., and Pinna, L. (1994) *Tetrahedron Letters*, **35**, 2423–2426.

613 Tian, Z., Rasmussen, M. and Wittenberger, S.J. (2002) *Organic Process Research & Development*, **6**, 416–418.

614 Tajima, T., Nakajima, A., and Fuchigami, T. (2006) *The Journal of Organic Chemistry*, **71**, 1436–1441.

615 Kuhn, R. and Osswald, G. (1956) *Chemische Berichte*, **89**, 1423–1442.

616 Rapoport, H. and Willson, C.D. (1962) *Journal of the American Chemical Society*, **84**, 630–635.

617 Kochhar, K.S. and Pinnick, H.W. (1984) *The Journal of Organic Chemistry*, **49**, 3222–3224.

618 Momose, T., Tanaka, T., Yokota, T., Nagamoto, N., and Yamada, K. (1978) *Chemical & Pharmaceutical Bulletin*, **26**, 2224–2232.

619 Flitsch, W. and Hohenhorst, M. (1990) *Liebigs Annalen der Chemie*, 397–399.

620 McNab, H. and Monahan, L.C. (1988) *Journal of the Chemical Society, Perkin Transactions 1*, 863–868.

621 Blake, A.J., McNab, H., and Monahan, L.C. (1988) *Journal of the Chemical Society, Perkin Transactions 2*, 1455–1458.

622 Capon, B. and Kwok, F.-C. (1989) *Journal of the American Chemical Society*, **111**, 5346–5356.

623 Momose, T., Tanaka, T., Yokota, T., Nagamoto, N., and Yamada, K. (1979) *Chemical & Pharmaceutical Bulletin*, **27**, 1448–1453.

624 Hunter, G.A., McNab, H., Monahan, L.C., and Blake, A.J. (1991) *Journal of the Chemical Society, Perkin Transactions 1*, 3245–3251.

625 Henning, H.-G. and Gelbin, A. (1993) *Advances in Heterocyclic Chemistry*, **57**, 139–185.

626 Mulholland, T.P.C., Foster, R., and Haydock, D.B. (1972) *Journal of the Chemical Society, Perkin Transactions 1*, 2121–2128.

627 Nolte, M.J., Steyn, P.S., and Wessels, P.L. (1980) *Journal of the Chemical Society, Perkin Transactions 1*, 1057–1065.

628 Lacey, R.N. (1954) *Journal of the Chemical Society*, 850–854.

629 Ley, S.V., Smith, S.C., and Woodward, P.R. (1992) *Tetrahedron*, **48**, 1145–1174.

630 Igglessi-Markopoulou, O. and Sandris, C. (1982) *Journal of Heterocyclic Chemistry*, **19**, 883–890.

631 Igglessi-Markopoulou, O. and Sandris, C. (1985) *Journal of Heterocyclic Chemistry*, **22**, 1599–1606.

632 Matthews, J. and Rivero, R.A. (1998) *The Journal of Organic Chemistry*, **63**, 4808–4810.

633 Erden, I., Ozer, G., Hoarau, C., and Cao, W. (2006) *Journal of Heterocyclic Chemistry*, **43**, 395–399.

634 Kochhar, K.S., Carson, H.J., Clouser, K.A., Elling, J.W., Gramens, L.A., Parry, J.L., Sherman, H.L., Braat, K., and Pinnick, H.W. (1984) *Tetrahedron Letters*, **25**, 1871–1874.

635 Jones, R.C.F. and Bates, A.D. (1986) *Tetrahedron Letters*, **27**, 5285–5288.

636 Jones, R.C.F. and Peterson, G.E. (1983) *Tetrahedron Letters*, **24**, 4751–4754.

637 Jones, R.C.F. and Patience, J.M. (1990) *Journal of the Chemical Society, Perkin Transactions 1*, 2350–2351.

638 Hynes, J. Jr, Doubleday, W.W., Dyckman, A.J., Godfrey, J.D. Jr, Grosso, J.A., Kiau, S., and Leftheris, K. (2004) *The Journal of Organic Chemistry*, **69**, 1368–1371.

639 Bhattacharya, A., Patel, N.C., Plata, R.E., Peddicord, M., Ye, Q., Parlanti, L., Palaniswamy, V.A., and Grosso, J.A. (2006) *Tetrahedron Letters*, **47**, 5341–5343.

640 De Rosa, M., Issac, R.P., and Houghton, G. (1995) *Tetrahedron Letters*, **36**, 9261–9264.

641 De Rosa, M., Issac, R.P., Marquez, M., Orozco, M., Luque, F.J., and Timken, M.D. (1999) *Journal of the Chemical Society, Perkin Transactions 2*, 1433–1437.

642 Almerico, A.M., Cirrincione, G., Diana, P., Grimaudo, S., Dattolo, G., Aiello, E., and Mingoia, F. (1995) *Journal of Heterocyclic Chemistry*, **32**, 985–989.

643 Chien, T.-C., Meade, E.A., Hinkley, J.M., and Townsend, L.B. (2004) *Organic Letters*, **6**, 2857–2859.

644 Wamhoff, H. and Wehling, B. (1976) *Synthesis*, 51.

645 Johnson, R.W., Mattson, R.J., and Sowell, J.W. Sr (1977) *Journal of Heterocyclic Chemistry*, **14**, 383–385.

646 Castellote, I., Vaquero, J.J., and Alvarez-Builla, J. (2004) *Tetrahedron Letters*, **45**, 769–772.

647 Lim, M.-I., Ren, W.-Y., Otter, B.A., and Klein, R.S. (1983) *The Journal of Organic Chemistry*, **48**, 780–788.

648 Elliott, A.J., Morris, P.E. Jr, Petty, S.L., and Williams, C.H. (1997) *The Journal of Organic Chemistry*, **62**, 8071–8075.

649 Rochais, C., Lisowski, V., Dallemagne, P., and Rault, S. (2004) *Tetrahedron*, **60**, 2267–2270.

650 Chen, N., Lu, Y., Gadamasetti, K., Hurt, C.R., Norman, M.H., and Fotsch, C. (2000) *The Journal of Organic Chemistry*, **65**, 2603–2605.

651 Marcotte, F.-A. and Lubell, W.D. (2002) *Organic Letters*, **4**, 2601–2603.

652 He, L., Chan, P.W.H., Tsui, W.-M., Yu, W.-Y., and Che, C.-M. (2004) *Organic Letters*, **6**, 2405–2408.

653 Bonnaud, B. and Bigg, D.C.H. (1994) *Synthesis*, 465–467.

654 Pichon, M. and Figadère, B. (1996) *Tetrahedron: Asymmetry*, **7**, 927–964.

655 Bloch, R., Brillet-Fernandez, C., and Mandville, G. (1994) *Tetrahedron: Asymmetry*, **5**, 745–750.

656 Bäckvall, J.-E., Schink, H.E., and Renko, Z.D. (1990) *The Journal of Organic Chemistry*, **55**, 826–831.

657 Jones, T.H., Franko, J.B., Blum, M.S., and Fales, H.M. (1980) *Tetrahedron Letters*, **21**, 789–792.

658 Boga, C., Manescalchi, F., and Savoia, D. (1994) *Tetrahedron*, **50**, 4709–4722.

659 Meyer, N. Werner, F., and Opatz, T. (2005) *Synthesis*, 945–956.

660 Overman, L.E., Kakimoto, M., Okazaki, M.E., and Meier, G.P. (1983) *Journal of the American Chemical Society*, **105**, 6622–6629.

661 Vedejs, E. and West, F.G. (1986) *Chemical Reviews*, **86**, 941–955.

662 Fishwick, C.W.G., Foster, R.J., and Carr, R.E. (1995) *Tetrahedron Letters*, **36**, 9409–9412.

663 Pandey, G., Banerjee, P., and Gadre, S.R. (2006) *Chemical Reviews*, **106**, 4484–4517.

664 Trost, B.M. and Marrs, C.M. (1993) *Journal of the American Chemical Society*, **115**, 6636–6645.

665 Fukuda, Y., Matsubara, S., and Utimoto, K. (1991) *The Journal of Organic Chemistry*, **56**, 5812–5816.

666 van Esseveldt, B.C.J., Vervoort, P.W.H., van Delft, F.L., and Rutjes, F.P.J.T. (2005) *The Journal of Organic Chemistry*, **70**, 1791–1795.

667 Wolf, L.B., Tjen, K.C.M.F., ten Brink, H.T., Blaauw, R.H., Hiemstra, H., Schoemaker, H.E., and Rutjes, F.P.J.T. (2002) *Advanced Synthesis & Catalysis*, **344**, 70–83.

668 Yoo, C.L., Olmstead, M.M., Tantillo, D.J., and Kurth, M.J. (2006) *Tetrahedron Letters*, **47**, 477–481.

669 Shvekhgeimer, M.-G.A. (2003) *Chemistry of Heterocyclic Compounds (English Translation)*, **39**, 405–448.

670 Cloke, J.B. (1929) *Journal of the American Chemical Society*, **51**, 1174–1187.

671 Maginnity, P.M. and Cloke, J.B. (1951) *Journal of the American Chemical Society*, **73**, 49–51.

672 Fry, D.F., Fowler, C.B., and Dieter, R.K. (1994) *Synlett*, 836–838.

673 Dechoux, L., Jung, L., and Stambach, J.-F. (1995) *Synthesis*, 242–244.

674 Verniest, G., Claessens, S., and De Kimpe, N. (2005) *Tetrahedron*, **61**, 4631–4637.

675 Kinderman, S.S., van Maarseveen, J.H., Schoemaker, H.E., Hiemstra, H., and Rutjes, F.P.J.T. (2001) *Organic Letters*, **3**, 2045–2048.

676 Fu, G.C. and Grubbs, R.H. (1992) *Journal of the American Chemical Society*, **114**, 7324–7325.

677 Declerck, V., Ribière, P., Martinez, J., and Lamaty, F. (2004) *The Journal of Organic Chemistry*, **69**, 8372–8381.

678 Dieltiens, N., Stevens, C.V., De Vos, D., Allaert, B., Drozdzak, R., and Verpoort, F. (2004) *Tetrahedron Letters*, **45**, 8995–8998.

679 Xu, Z. and Lu, X. (1997) *Tetrahedron Letters*, **38**, 3461–3464.

680 Stork, G., Brizzolara, A., Landesman, H., Szmuszkovicz, J., and Terrell, R. (1963) *Journal of the American Chemical Society*, **85**, 207–222.

681 Beak, P. and Lee, W.K. (1993) *The Journal of Organic Chemistry*, **58**, 1109–1117.

682 Pandey, G., Bagul, T.D., and Sahoo, A.K. (1998) *The Journal of Organic Chemistry*, **63**, 760–768.

683 Pandey, G., Laha, J.K., and Lakshmaiah, G. (2002) *Tetrahedron*, **58**, 3525–3534.

684 Macdonald, T.L. (1980) *The Journal of Organic Chemistry*, **45**, 193–194.

685 Kerrick, S.T. and Beak, P. (1991) *Journal of the American Chemical Society*, **113**, 9708–9710.

686 Beak, P., Kerrick, S.T., Wu, S., and Chu, J. (1994) *Journal of the American Chemical Society*, **116**, 3231–3239.

687 Deng, X. and Mani, N.S. (2005) *Tetrahedron: Asymmetry*, **16**, 661–664.

688 Dieter, R.K., Oba, G., Chandupatla, K.R., Topping, C.M., Lu, K., and Watson, R.T. (2004) *The Journal of Organic Chemistry*, **69**, 3076–3086.

689 Langlois, N. and Rojas, A. (1993) *Tetrahedron*, **49**, 77–82.

690 Pelkey, E.T. and Russel, J.S. (2008) *Progress in Heterocyclic Chemistry*, Vol. 19 (eds G.W. Gribble and J.A. Joule), Elsevier, Oxford, pp. 135–175.

691 D'Ischia M. and Napolitano Pezzella, A. (2008) *Comprehensive Heterocyclic Chemistry III*, Vol. 3 (eds A.R. Katritzky, C.A. Ramsden, E.F.V. Scriven, and R.J.K. Taylor), Elsevier, Oxford, pp. 1–43.

692 Trofimov B.A. and Nedolya, N.A. (2008) *Comprehensive Heterocyclic Chemistry III*, Vol. 3 (eds A.R. Katritzky, C.A. Ramsden, E.F.V. Scriven, and R.J.K. Taylor), Elsevier, Oxford, pp. 45–268.

693 Bergman J. and Janosik, T. (2008) in *Comprehensive Heterocyclic Chemistry III*, Vol. 3 (eds A.R. Katritzky, C.A. Ramsden, E.F.V. Scriven, and R.J.K. Taylor), Elsevier, Oxford, pp. 269–351.

694 D'Ischia M. and Napolitano Pezzella, A. (2008) in *Comprehensive Heterocyclic Chemistry III*, Vol. 3 (eds A.R. Katritzky, C.A. Ramsden, E.F.V. Scriven, and R.J.K. Taylor), Elsevier, Oxford, pp. 353–388.

695 Schmuck C. and Rupprecht, D. (2007) *Synthesis*, 3095–3110.

696 Pandey, G., Banerjee, P., and Gadre, S.R. (2006) *Chem. Rev.*, **106**, 4484–4517.

697 Lash, T.D. (2007) *Eur. J. Org. Chem.*, 5461–5481.

698 Maeda, H. (2007) *Eur. J. Org. Chem.*, 5313–5325.

699 Misra, R. and Chandrashekar, T.K. (2008) *Acc. Chem. Res.*, **41**, 265–279.

700 Sessler, J.L. and Tomat, E. (2007) *Acc. Chem. Res.*, **40**, 371–379.

701 Senge, M.O., Fazekas, M., Notaras, E.G.A., Blau, W.J., Zawadzka, M., Locos, O.B., and Ni Mhuircheartaigh, E.M. (2007) *Adv. Mater.*, **19**, 2737–2774.

702 Ono, N. (2008) *Heterocycles*, **75**, 243–284.

703 Fan, H., Peng, J., Hamann, M.T., and Hu, J.-F. (2008) *Chem. Rev.*, **108**, 264–287.

704 Mothana, B. and Boyd, R.J. (2007) *J. Mol. Struct. (Theochem)*, **811**, 97–107.

705 Hiroya, K., Matsumoto, S., Ashikawa, M., Ogiwara, K., and Sakamoto, T. (2006) *Org. Lett.*, **8**, 5349–5352.

706 Harrison, T.J., Kozak, J.A., Corbella-Pané, M., and Dake, G.R. (2006) *J. Org. Chem.*, **71**, 4525–4529.

707 Bertrand, M.B., Leathen, M.L., and Wolfe, J.P. (2007) *Org. Lett.*, **9**, 457–460.

708 Martín, R., Larsen, C.H., Cuenca, A., and Buchwald, S.L. (2007) *Org. Lett.*, **9**, 3379–3382.

709 Dong, H., Shen, M., Redford, J.E., Stokes, B.J., Pumphrey, A.L., and Driver, T.G. (2007) *Org. Lett.*, **9**, 5191–5194.

710 Freifeld, I., Shojaei, H., and Langer, P. (2006) *J. Org. Chem.*, **71**, 4965–4968.

711 Bellur, E., Görls, H., and Langer, P. (2005) *J. Org. Chem.*, **70**, 4751–4761.

712 Coffin, A.C., Roussell, M.A., Tserlin, E., and Pelkey, E.T. (2006) *J. Org. Chem.*, **71**, 6678–6681.

713 Misra, N.C., Panda, K., Ila, H., and Junjappa, H. (2007) *J. Org. Chem.*, **72**, 1246–1251.

714 Chang, J.H. and Shin, H. (2008) *Org. Proc. Res. Dev.*, **12**, 291–293.

715 Chiba, S., Wang, Y.-F., Lapointe, G., and Narasaka, K. (2008) *Org. Lett.*, **10**, 313–316.

716 Crawley, M.L., Goljer, I., Jenkins, D.J., Mehlmann, J.F., Nogle, L., Dooley, R., and Mahaney, P.E. (2006) *Org. Lett.*, **8**, 5837.

717 Vicario, J.L., Reboredo, S., Badía, D., and Carillo, L. (2007) *Angew. Chem. Int. Ed.*, **46**, 5168–5170.

718 Bergner, I. and Opatz, T. (2007) *J. Org. Chem.*, **72**, 7083–7090.

719 Alizadeh, A., Hosseinpour, R., and Rostamina, S. (2008) *Synthesis*, 2462–2466.

720 Piloty, O. (1910) *Chem. Ber.*, **43**, 489–498.

721 Robinson, G.M. and Robinson, R. (1918) *J. Chem. Soc.*, 639–645.

722 Baldwin, J.E. and Bottaro, J.C. (1982) *J. Chem. Soc., Chem. Commun.*, 624–625.

723 Milgram, B.C., Eskildsen, K., Richter, S.M., Scheidt, W.R., and Scheidt, K.A. (2007) *J. Org. Chem.*, **72**, 3941–3944.

724 Sessler, J.L., Mozaffari, A., and Johnson, M.R. (1992) *Org. Synth.*, **70**, 68–78.

725 Barkigia, K.M., Berber, M.D., Fajer, J., Medforth, C.J., Renner, M.W., and Smith, K.M. (1990) *J. Am. Chem. Soc.*, **112**, 8851–8857.

726 Galliford, C.V. and Scheidt, K.A. (2007) *J. Org. Chem.*, **72**, 1811–1813.

727 St. Cyr., D.J., Martin, N., and Arndtsen, B.A. (2007) *Org. Lett.*, **9**, 449–452.

728 St. Cyr, D.J. and Arndtsen, B.A. (2007) *J. Am. Chem. Soc.*, **129**, 12366–12367.

729 Lu, Y. and Arndtsen, B.A. (2008) *Angew. Chem.*, **120**, 5510–5513.

730 Khalili, B., Jajarmi, P., Eftekhari-Sis, B., and Hashemi, M.M. (2008) *J. Org. Chem.*, **73**, 2090–2095.

731 Kassaee, M.Z., Masrouri, H., Movahedi, F., and Partovi, T. (2008) *Helv. Chim. Acta*, **91**, 227–231.

732 Ma, H.-C. and Jiang, X.-Z. (2007) *J. Org. Chem.*, **72**, 8943–8946.

733 Jiao, L., Hao, E., Vicente, M.G.H., and Smith, K.M. (2007) *J. Org. Chem.*, **72**, 8119–8122.

734 Dorward, K.M., Guthrie, N.J., and Pelkey, E.T. (2007) *Synthesis*, 2317–2322.

735 Dang, T.T., Ahmad, R., Dang, T.T., Reinke, H., and Langer, P. (2008) *Tetrahedron Lett.*, **49**, 1698–1700.

736 Fukuda, T., Sudo, E., Shimokawa, K., and Iwao, M. (2008) *Tetrahedron*, **64**, 328–338.

737 Beck, E.M., Hatley, R., and Gaunt, M.J. (2008) *Angew. Chem. Int. Ed.*, **47**, 3004–3007.

738 Billingsley, K. and Buchwald, S.L. (2007) *J. Am. Chem. Soc.*, **129**, 3358–3366.

739 Dohi, T., Morimoto, K., Ito, M., and Kita, Y. (2007) *Synthesis*, 13–19.

740 Guadarrama-Morales, O., Méndez, F., and Miranda, L.D. (2007) *Tetrahedron Lett.*, **48**, 4515–4518.

741 Fu, L. and Gribble, G.W. (2007) *Tetrahedron Lett.*, **48**, 9155–9158.

742 Fu, L. and Gribble, G.W. (2008) *Synthesis*, 788–794.

743 Fu, L. and Gribble, G.W. (2008) *Tetrahedron Lett.*, **49**, 3545–3548.

744 Nigst, T.A., Westermaier, M., Ofial, A.R. and Mayr, H. (2008) *Eur. J. Org. Chem.*, 2369–2374.

745 Huffman, J.W., Smith, V.J., and Padgett, L.W. (2008) *Tetrahedron*, **64**, 2104–2112.

5
Five-Membered Heterocycles: Indole and Related Systems

José Barluenga and Carlos Valdés

5.1
Introduction

5.1.1
General Introduction

Indole (1*H*-indole) (**1**) is the benzopyrrole with the ring fusion through the 2 and 3 positions of pyrrole. It is one of the most abundant heterocycles found in natural products and biologically active molecules. In fact, it can be regarded as the most important of all the privileged structures in medicinal chemistry [1]. For this reason, research in the different areas of indole chemistry has been, and continuous to be, extraordinarily intense. Many excellent reviews, covering advances in specific topics in the chemistry of indoles, are available and will be referred to throughout this chapter. The present chapter covers the developments in indole chemistry that appeared in the literature until mid-2006 (some subsequent developments are given in the Addendum). For further information the monographs cited in References [2, 3] are recommended.

Indole
1*H*-Indole

1

The discovery and structure elucidation of indole dates from 1866, when Adolf von Baeyer synthesized indole by zinc-dust pyrolysis of oxindole (**2**), which had been obtained by reduction of isatin (**3**), a product of the oxidation of the natural blue pigment Indigo (**4**) [4]. Consequently, the name Indole derives from that of *Ind*igo.

Modern Heterocyclic Chemistry, First Edition.
Edited by Julio Alvarez-Builla, Juan Jose Vaquero, and José Barluenga.

oxindole
2

isatin
3

Indigo
4

5.1.2
System Isomers and Nomenclature

The tautomers of 1*H*-indole are 2*H*-indole and 3*H*-indole (Figure 5.1). Both systems are highly unstable, although 3*H*-indole has been characterized spectroscopically, and its derivatives have been isolated [5]. High level quantum chemical DFT calculations predict an energy difference of 5.20 and 24.1 kcal mol^{-1} between 1*H*-indole and 3*H*-indole, and 1*H*-indole and 2*H*-indole respectively [6].

The other isomeric benzopyrroles are isoindole and indolizine. Indoline is the name for 2,3-dihydro-1*H*-indole.

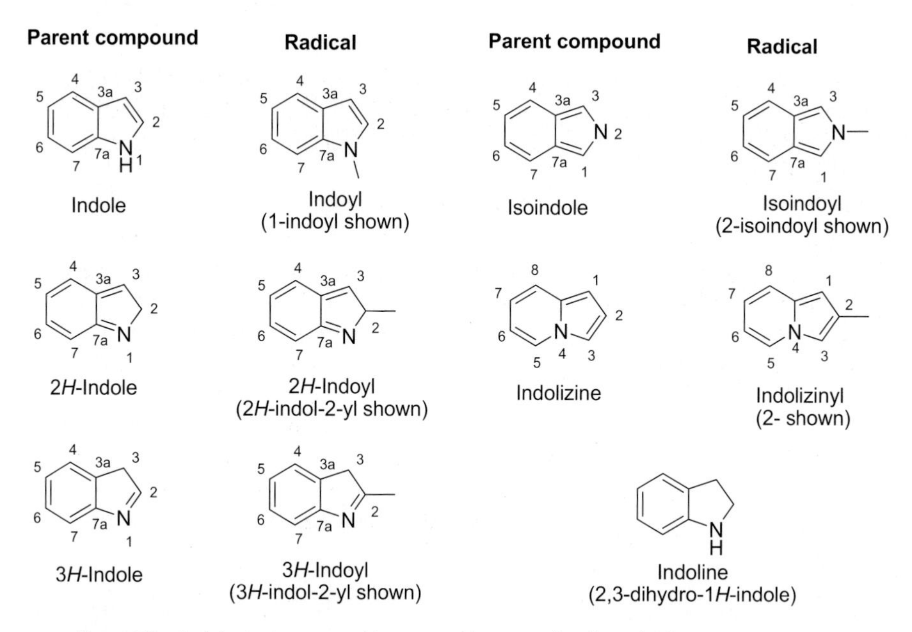

Figure 5.1 Indole, tautomers and isomers with conventional numbering.

5.2 General Properties

5.2.1 Physicochemical Data

Indole is a crystalline solid (mp = 54–54 °C, bp = 253–254 °C) with a fecal smell. The main commercial source of indole comes from the 220–260 °C fraction of coal-tar distillation. It is soluble in organic solvents such as diethyl ether, ethanol and benzene, and also in hot water.

The crystal structure of indole [7] and of several simple derivatives is available [8]. In addition, state of the art quantum chemical DFT calculations provide very accurate results regarding structural [9], electronic [10] and chemical properties of indole and indole derivatives (Figure 5.2) [11]. DFT calculations of the magnetic properties, to estimate the aromaticity of the indole ring, reveal a stabilization due to the π-molecular orbital delocalization of 10 electrons between the two aromatic rings [12].

The ^{1}H NMR spectra of indole feature all the resonances for the hydrogens in the aromatic region, and corroborate the aromaticity of the ring (Figure 5.3). The upfield shifts observed for H3 and C3 in the ^{1}H and ^{13}C NMR spectra indicate the higher electron density around C3. Substitution on the indole ring may cause important variations in the chemical shifts of H2 and H3, which can be rationalized in terms of resonance and inductive effects (Table 5.1).

5.2.2 General Reactivity

Indole is a π-excessive aromatic heterocycle with ten π-electrons. The lone pair of the nitrogen atom (which features sp^2 hybridization) completes the ten π-electrons delocalized across the ring. As in pyrrole, the π-excessive nature of the aromatic ring governs its reactivity and chemical properties.

Indole is a weak base (pK_a = −2,4 for the conjugated acid), as protonation of the nitrogen atom would disrupt the aromaticity of the five-membered ring. In contrast, as a π-excessive aromatic heterocycle, electrophilic aromatic substitution is one of the most characteristic reactions. Unlike pyrrole, addition of electrophiles takes place preferentially at C3. A simple explanation for this can be deduced by analysis of the Wheland intermediates resulting from the attack of a nucleophile at C2 and C3 (Scheme 5.1).

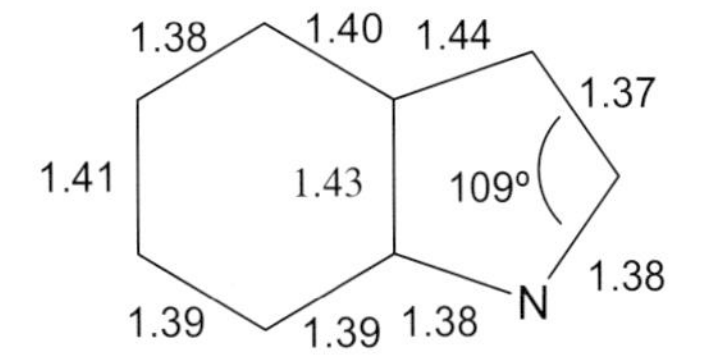

Figure 5.2 Indole structural parameters at the B3LYP/6-311 + G(2p,d) level of theory.

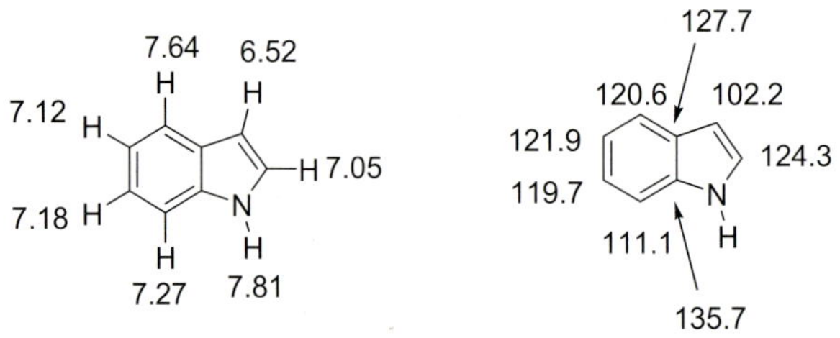

Figure 5.3 ^{1}H and ^{13}C NMR chemical shifts in $CDCl_3$ (δ, ppm) for indole.

The intermediate of the attack at C3 is stabilized by delocalization of the positive charge. However, no delocalization is possible in the intermediate derived from attack at C2 without disrupting the aromaticity of the six-membered ring.

Frontier Molecular Orbital theory considerations provide an alternative theoretical explanation for this reactivity trend. Indole features a relative high-energy HOMO, with the highest value at C3 (Figure 5.4). Moreover, the condensed Fukui function for electrophilic attack f_q^- [17], takes values of 0.08, 0.05 and 0.15 for N, C2 and C3, respectively, pointing to the higher reactivity of C3 towards soft electrophiles [12].

Table 5.1 Chemical shifts for H-1 and H-2 in substituted indoles ($CDCl_3$).

Substituent	δ H2 (ppm)	δ H3 (ppm)
None	7.05	6.52
1-Me	6.90	6.43
2-Me	—	6.20
3-Me	6.85	—
1-CO_2Me [13]	7.55	6.57
2-CO_2Me [14]	—	7.15
3-CO_2Me [15]	7.93	—
2-Cl [16]	—	6.41
3-Cl	7.44	—

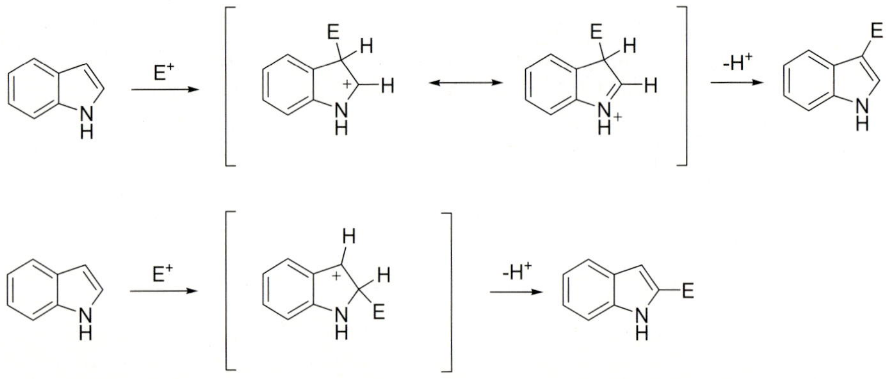

Scheme 5.1 Possible regioisomers in the electrophilic attack on the indole ring.

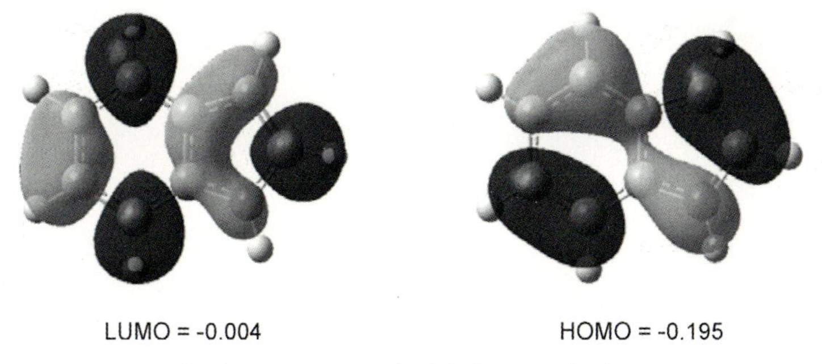

Figure 5.4 Graphical representation of indole frontier orbitals.

Typical electrophilic aromatic substitution reactions that allow for the introduction of functionalized side-chains at C3 are Friedel–Crafts acylations, Vilsmeier–Haack reaction, Mannich type alkylations and halogenations (Scheme 5.2).

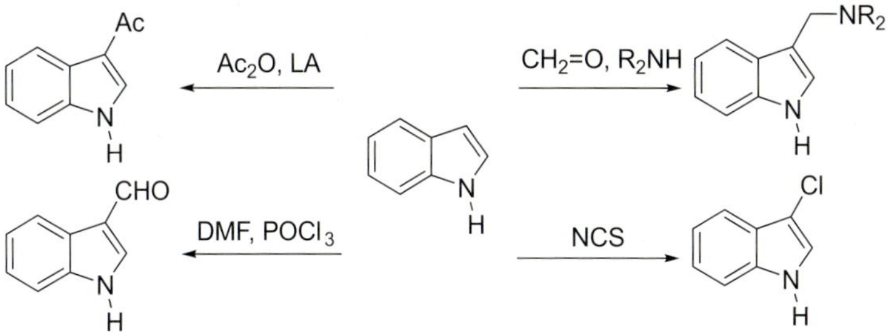

Ac; Ac_2O, LA; CH_2=O, R_2NH; NR_2; CHO; DMF, $POCl_3$; NCS; Cl

Scheme 5.2 Some typical electrophilic substitution reactions of indole.

Electrophilic substitution at C2 can be achieved in 3-substituted indoles, although the reaction usually starts with electrophilic attack at C3, followed by rearrangement or reversal of the reaction to produce the substitution at C2 (Scheme 5.3).

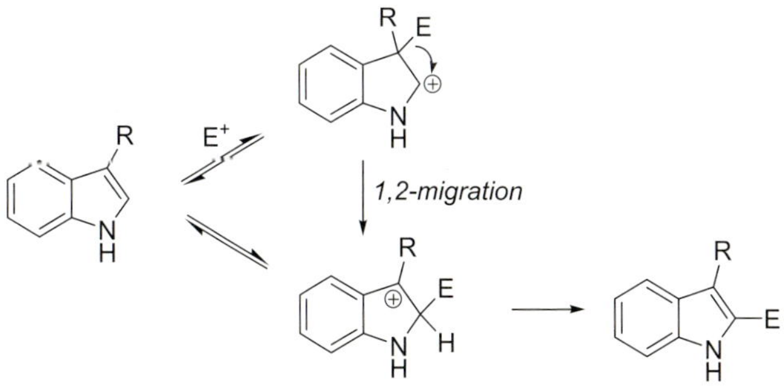

E^+; *1,2-migration*

Scheme 5.3

The indole N–H is weakly acidic and, thus, can be deprotonated by strong bases (pK_a 16.7 in water) to provide the indolyl anion. Therefore, substitution at the nitrogen can be achieved through base-promoted processes, such as alkylations, acylations and, more recently, transition metal catalyzed arylations (Scheme 5.4).

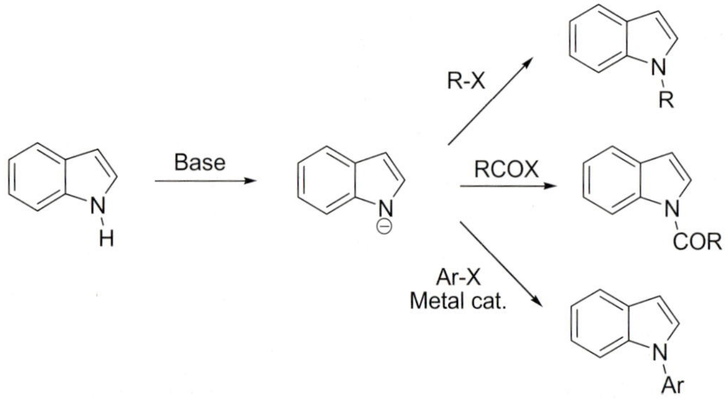

Scheme 5.4

The most reliable method to carry out the substitution at C2 is the heteroatom assisted metallation at C2 of *N*-acyl or *N*-sulfonylindoles, followed by reaction with an electrophile (Scheme 5.5).

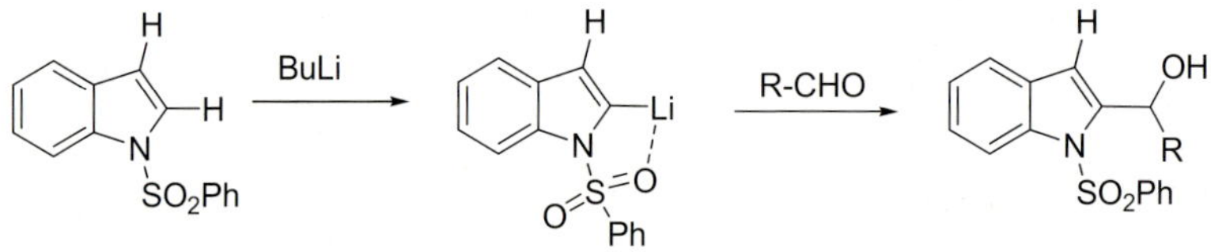

Scheme 5.5

Metal catalyzed cross-coupling reactions represent nowadays one of the main ways to modify the substituents in both rings of indole. All kinds of Pd catalyzed processes (Stille, Suzuki and Buchwald–Hartwig) can be achieved successfully from properly substituted indole species (Scheme 5.6).

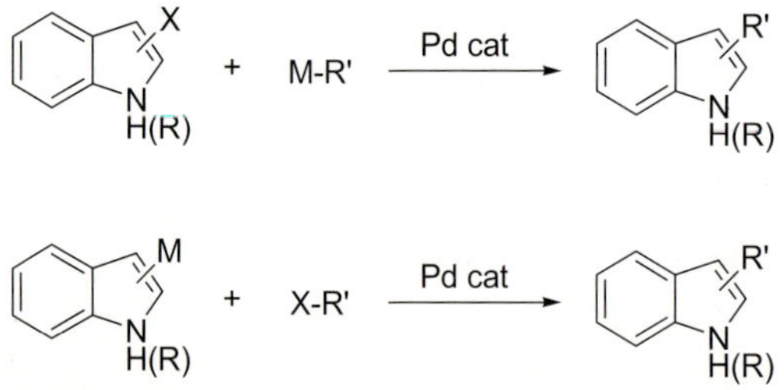

Scheme 5.6

5.3
Relevant Natural and/or Useful Compounds

The indole nucleus is present in the essential amino acid tryptophan (**5**), in many metabolites derived from tryptophan and also in natural molecules with high structural complexity.

Among the naturally occurring molecules that feature the indole ring in their structure, two worth noting: (i) the family of *tryptamines* [18], such as serotonin **6**, a very important neurotransmitter with numerous functions in the human body, and melatonin **7**, a hormone that participates in the regulation of the circadian rhythms (sleep–wake); (ii) the *auxins*, a group of plant growth substances, such as the natural auxin indole 3-acetic acid (**8**) and the synthetic auxin indole-3-butyric acid (**9**) (Figure 5.5) .

The indole structure is also present in structurally complex indole alkaloids with biological activity. To name a few: the hallucinogen D-lysergic acid diethyl amide (LSD) (**10**); the strichnous family of alkaloids (e.g., strychnine, **11**); the family of marine indole alkaloids isolated from blue-algae such as Fischer-indole I (**12**) [19], the bisindole alkaloids vinblastine (**13**) and vincristine (**14**), which are extremely potent cytotoxic agents, used in the therapy of leukemia and lymphoma tumor types [20, 21]; and reserpine **15**, a pentacyclic alkaloid that is a central nervous system depressant employed in the treatment of hypertension and psychiatric disorders (Figure 5.6) [22].

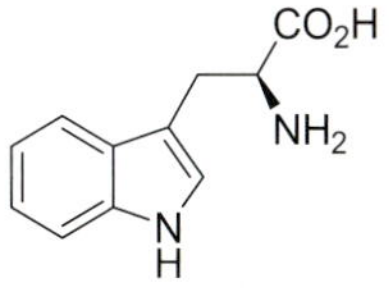

CO_2H NH_2 N H
5
L-Tryptophan

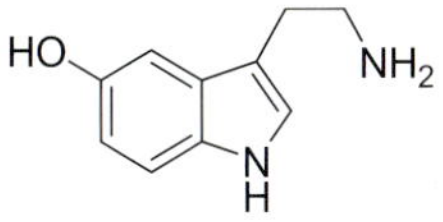

HO NH_2 N H
serotonin
(5-hydroxytryptamine)
6

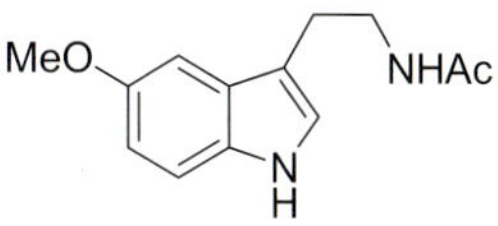

MeO NHAc N H
melatonin
7

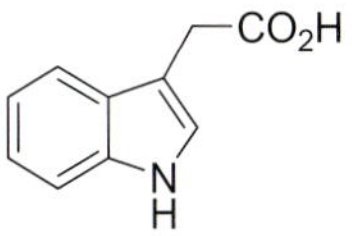

CO_2H N H
Indole-3-acetic acid
8

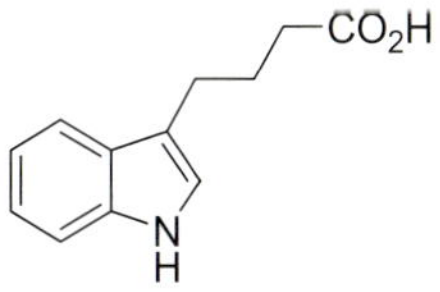

CO_2H N H
Indole-3-butyric acid
9

Figure 5.5 Some important simple indole natural products.

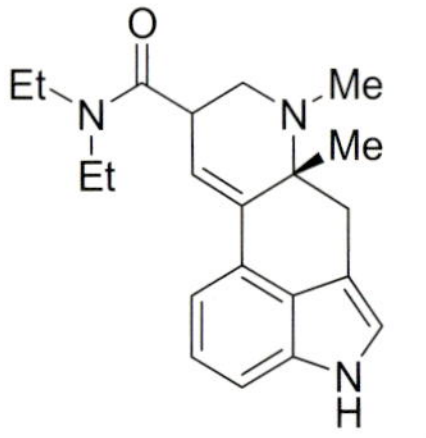

D-Lysergic acid diethyl amide (LSD)
10

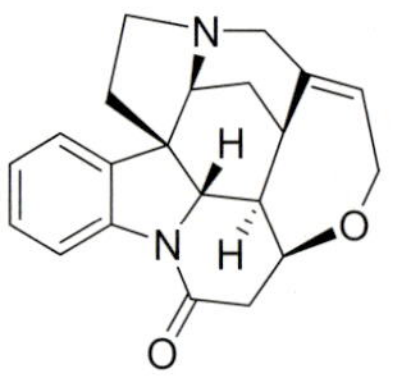

Strychnine
11

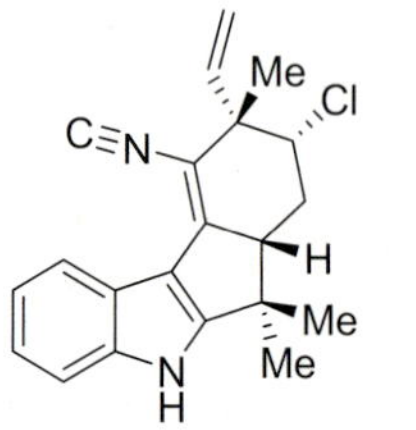

12 Fischer-indole I

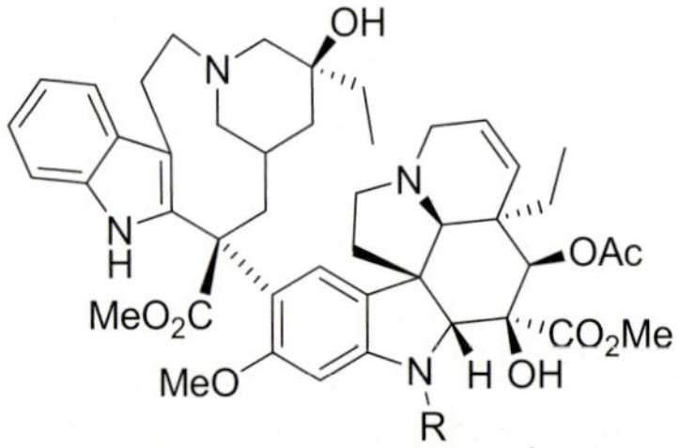

13 R = CH_3, vinblastine
14 R = CHO, vincristine

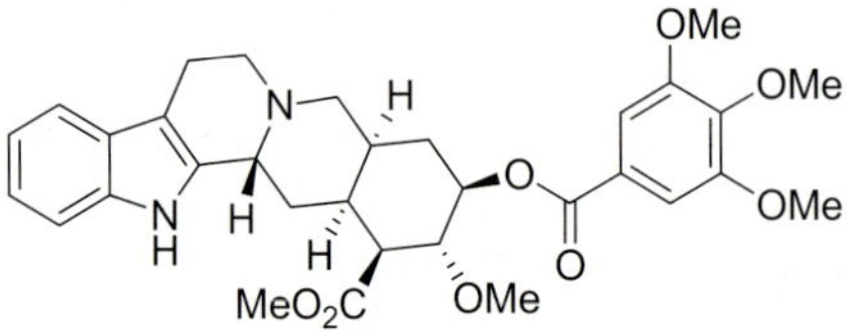

15 (-)-Reserpine

Figure 5.6 Some important indole alkaloids.

In contrast, several synthetic drugs currently in use contain the indole nucleus, for instance Sumatriptan, a synthetic tryptamine used in the treatment of migraine [23], and the non-steroidal anti-inflammatory drugs Indomethacin and Etodolac (Figure 5.7).

5.4
Indole Synthesis

5.4.1
Introduction

Indole is one of the most important heterocycles, as a result of its abundance in natural products and pharmaceuticals. For this reason, the synthesis of the indole

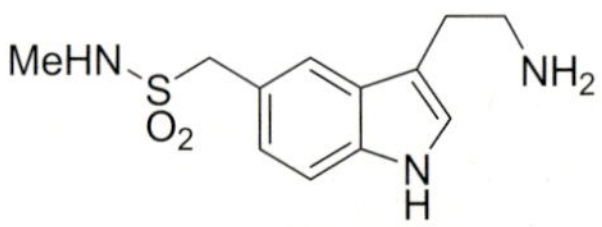

Sumatryptan

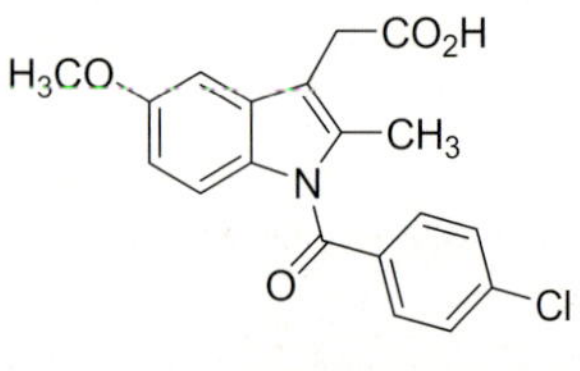

Indomethacin

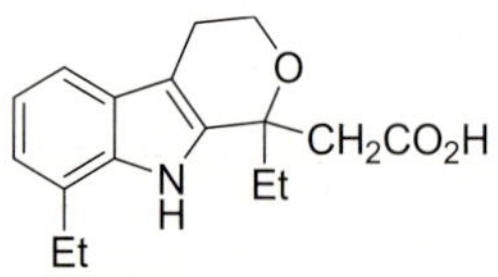

Etodolac

Figure 5.7 Indole-containing drugs currently in use.

ring has attracted great attention in synthetic organic chemistry, and, in turn, an extraordinary large number of different approaches to the synthesis of the indole ring have been devised over the years [24]. Despite the ample repertoire of methods available to build the indole ring, it continues to be an area of active research due to the enormous interest in the indole structure. Moreover, while most classic approaches to the indole ring usually rely on a final cyclization through a condensation reaction, the development of new methodologies for transition catalyzed C–C and C–N bond-forming reactions has led to a new family of methods in which the cyclization step is a metal-catalyzed process. In this area, the Pd-catalyzed reactions have a prominent position. This subject has been reviewed and monographs dealing with this particular subject are available [25, 26]. In addition, the rise of combinatorial approaches to drug discovery has motivated the development of new methodologies, and the adaptation of solution phase chemistries into solid phase synthesis. Specific reviews on this topic are also available [27]. The variety of different approaches to the indole ring is enormous, and an exhaustive coverage would largely exceed the aim of this book. For this reason, this chapter is restricted to the classical methods that still find application in the preparation of indoles and the more recent advances in the area, with particular attention to transition metal catalyzed processes.

From a retrosynthetical point of view, the indole ring can be constructed by two main strategies: formation of the pyrrole ring onto a properly substituted benzene precursor and formation of the benzene ring by annelation of a substituted pyrrole. By far, the strategies that start from a substituted benzene have been more extensively used (Scheme 5.7).

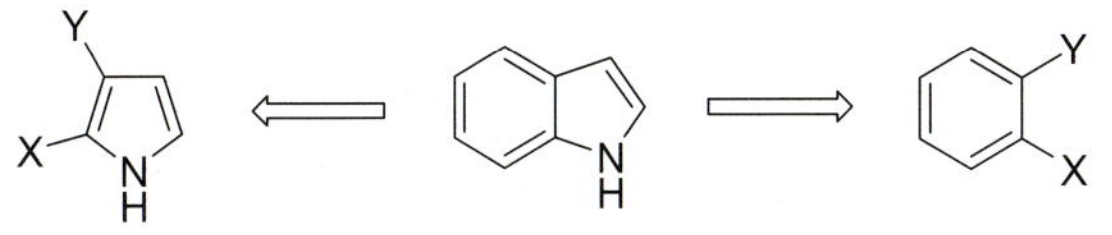

Scheme 5.7

5.4.2 Synthesis of the Indole Ring from a Benzene Ring

In an attempt to organize the many methods, they have been classified according to the formation of the last bond in the cyclization process (Scheme 5.8). Of particular importance in the construction of the indole ring are methods involving a sigmatropic rearrangement: the Fischer indole synthesis and related process, which will be covered in a specific section.

5.4.2.1 Indole Synthesis Involving a Sigmatropic Rearrangement

Indole ring syntheses that include a sigmatropic rearrangement are particularly appealing strategies since no *o*-substitution is required in the starting aniline (Scheme 5.9). The C–C bond is formed during the rearrangement. The most prominent member of this family of methods is the Fischer indole synthesis.

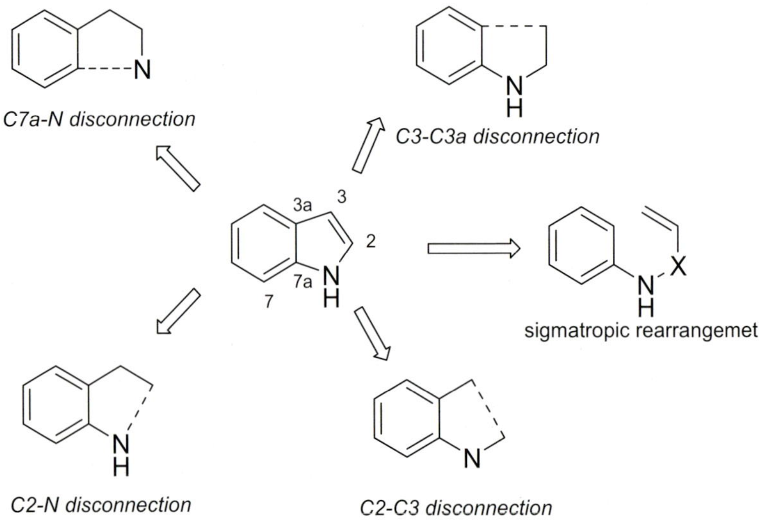

C7a-N disconnection

C3-C3a disconnection

3a 3 2 7a 7

sigmatropic rearrangemet

C2-N disconnection

C2-C3 disconnection

Scheme 5.8 General strategies for the construction of the five-membered ring of indole.

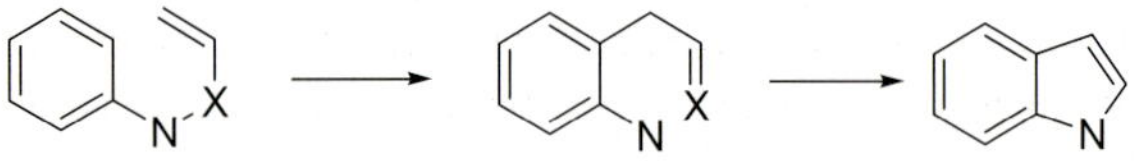

Scheme 5.9

5.4.2.1.1 **Fischer Synthesis** The Fischer indole synthesis [28], which was first discovered in 1883, is still considered as the most popular, and one of the most general and efficient approaches to the indole ring. It consists of the acid-catalyzed cyclization of aryl hydrazones **18** with loss of ammonia. The aryl hydrazones are easily obtained by condensation of a ketone (**17**) with an aryl hydrazine (**16**) (Scheme 5.10).

H^+ or LA

+ NH_3

16 **17** **18** **19**

Scheme 5.10 Fischer indole synthesis.

The mechanism accepted for the overall reaction was already formulated by Robinson and Robinson back in 1924. It involves a [3,3] sigmatropic rearrangement of the ene-hydrazine tautomer **20** of the arylhydrazone **18**, with cleavage of the N–N bond and formation of a C–C bond. Aromatization of **21** to give the intermediate **22**, followed by cyclization and NH_3 elimination provides the indole **19** (Scheme 5.11).

The Claisen-like sigmatropic rearrangement is strongly accelerated by an acid catalyst. Both protic and Lewis acids are effective catalyst for the Fischer indolizidation.

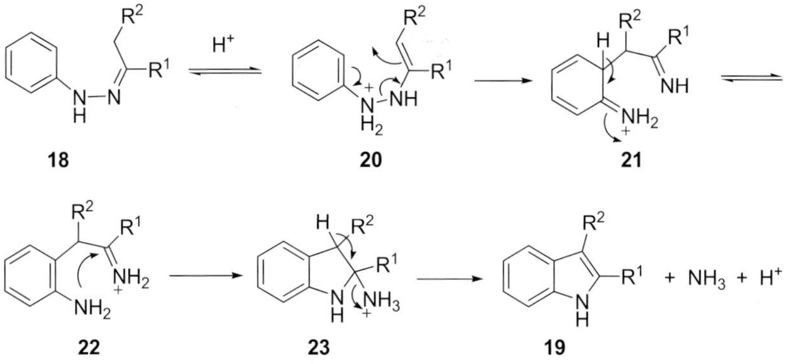

Scheme 5.11 Accepted mechanism for the Fischer indole synthesis.

Sulfuric and hydrochloric aqueous, alcoholic or acetic acid solutions have been used to promote the Fischer cyclization, as well as *p*-toluenesulfonic acid and phosphorous trichloride. Among the Lewis acids, $ZnCl_2$ is the most frequently used catalyst. Very recently, solid supports such as montmorillonite AK10/$ZnCl_2$ have been used to promote the indolizidation, in combination with microwave heating. This technique allows for the preparation of indoles unavailable through the standard conditions [29].

The nature of the substituents on the aromatic ring exerts a remarkable influence on the rate of the overall reaction: electron-donating substituents accelerate the reaction, while electron-releasing substituents slow the cyclization. For further information, detailed reviews covering all these topics are available [3].

The regioselectivity in the indole synthesis is an issue of major importance when unsymmetrical ketones or *m*-substituted aromatic rings are involved in the cyclization. For instance, the cyclization of hydrazones of type **24**, derived from methyl alkyl ketones, can deliver regioisomeric indoles **25** and **26** (Scheme 5.12).

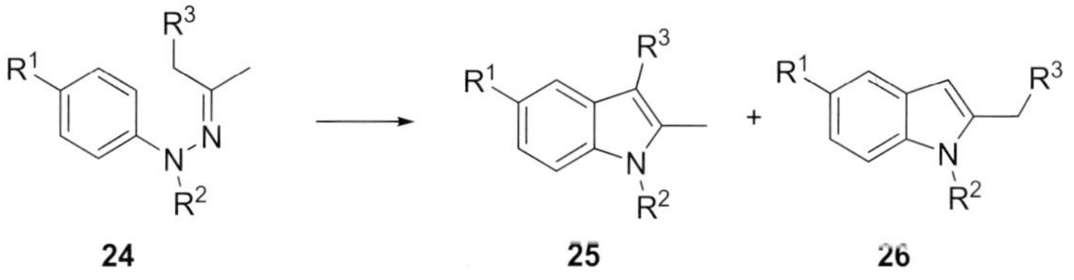

Scheme 5.12

Usually, under standard conditions, the major (or unique) product obtained is indole **25**, the one derived from the more highly substituted ene-hydrazine **27**, which is considered the thermodynamically controlled product. Therefore, for indole **25** to be the major isomer, the [3,3] rearrangement must be the rate-determining step of the whole sequence (Scheme 5.13).

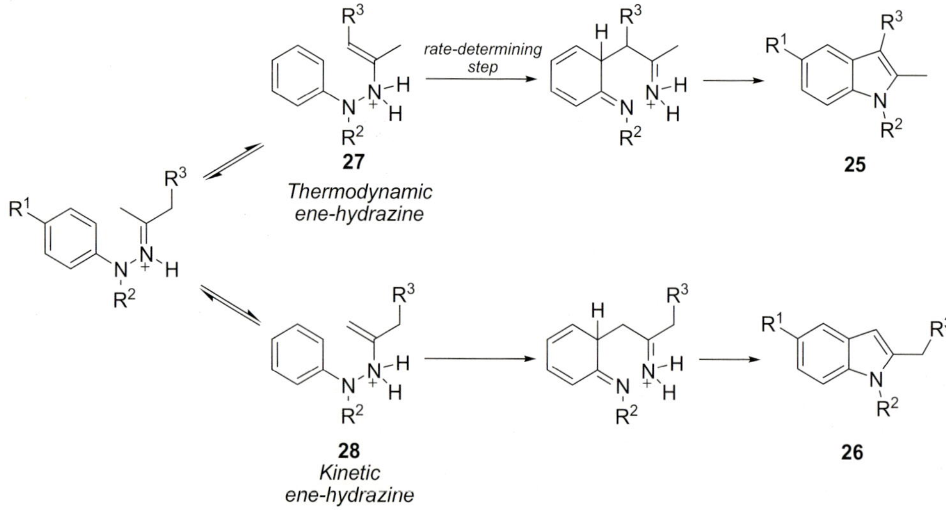

Scheme 5.13 The problem of the regioselectivity in the Fischer indole synthesis.

Nevertheless, it has been possible to obtain selectively 3-unsubstituted indoles such as **31** by reacting hydrazine **29** with the unsymmetrical methylketone **30** under kinetic conditions by employing very strong acids such as Eaton's acid ($P_2O_5/MeSO_3$) (Scheme 5.14) [30].

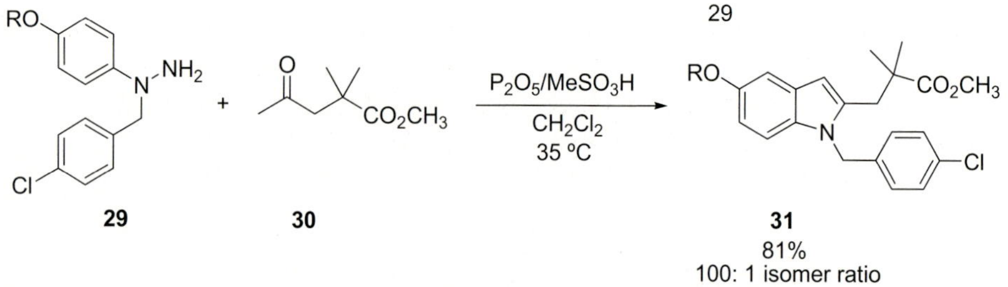

Scheme 5.14 Fischer indole synthesis under kinetically controlled conditions.

Mechanistic studies have suggested that under kinetic conditions the reaction proceeds through the doubly protonated intermediate **32**. The activation barrier for the [3,3] rearrangement in this case must be lower (since no aromaticity is lost during the rearrangement) and therefore, the formation of the ene-hydrazine intermediate (and not the rearrangement) becomes rate determining (Scheme 5.15) [31].

In some instances the cyclization can be carried out thermally, in the absence of acid catalyst, although it requires much harsher conditions and a protic solvent as a proton source. In contrast, preformed *N*-trifluoroacetyl enehydrazines **33** undergo cyclization without the need acid catalyst or very high temperatures to provide 2,3-disubstituted indole **34** [32] (Scheme 5.16).

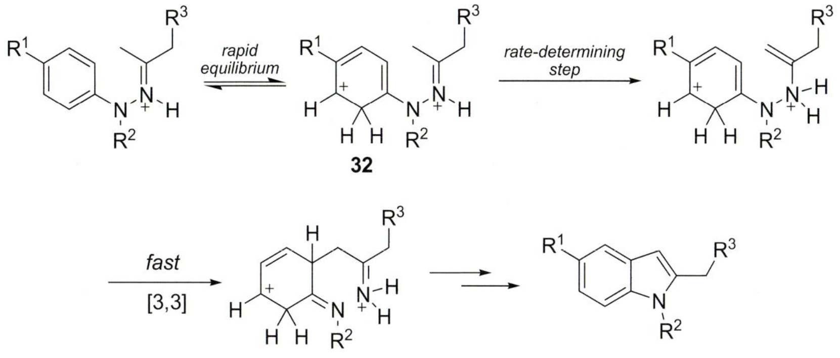

Scheme 5.15

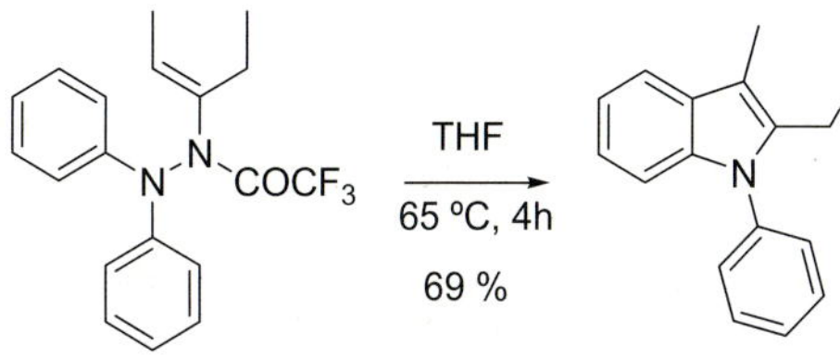

Scheme 5.16

The Fischer indolizidation is a very general process that proceeds with yields ranging from moderate to quantitative, depending on the substrate. Moreover, a large array of functional groups are tolerated by the reaction conditions, including the relatively sensitive, amide, ester or hydroxy. Schemes 5.17–5.20 depict some representative recent examples of the application of the Fischer indole synthesis.

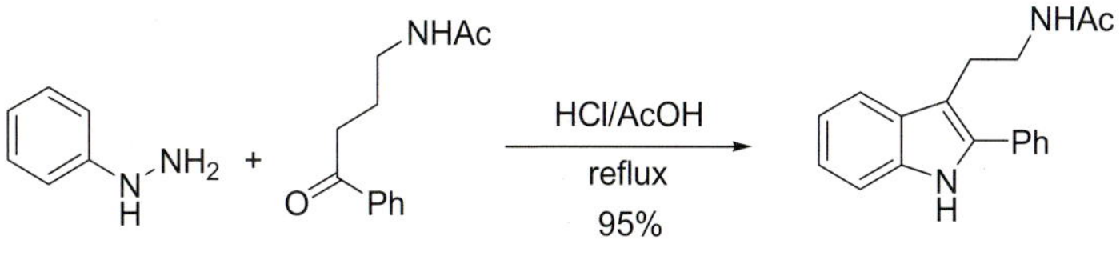

Scheme 5.17 [33].

An interesting variation of the Fischer indolizidation gives rise directly to tryptamines **36**, a particularly important type of indole, due to their biological activity. This so-called Grandberg indole synthesis employs 4-halobutenals **35** as carbonyl components. The nitrogen atom, which is usually liberated as ammonia in the Fischer indolizidation, is incorporated to the molecule, likely through the pathway represented in Scheme 5.20 [36].

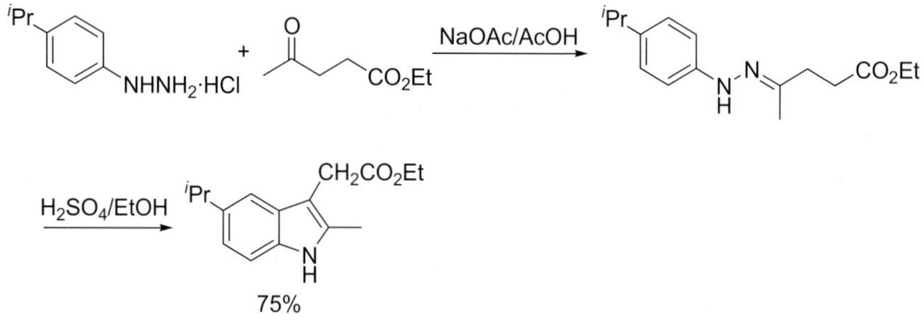

Scheme 5.18 [34].

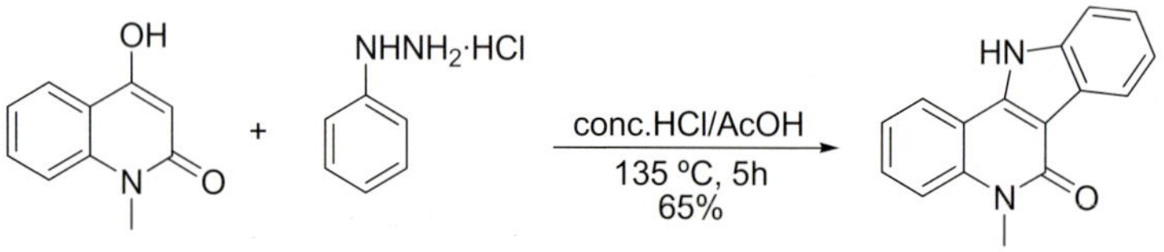

Scheme 5.19 [35].

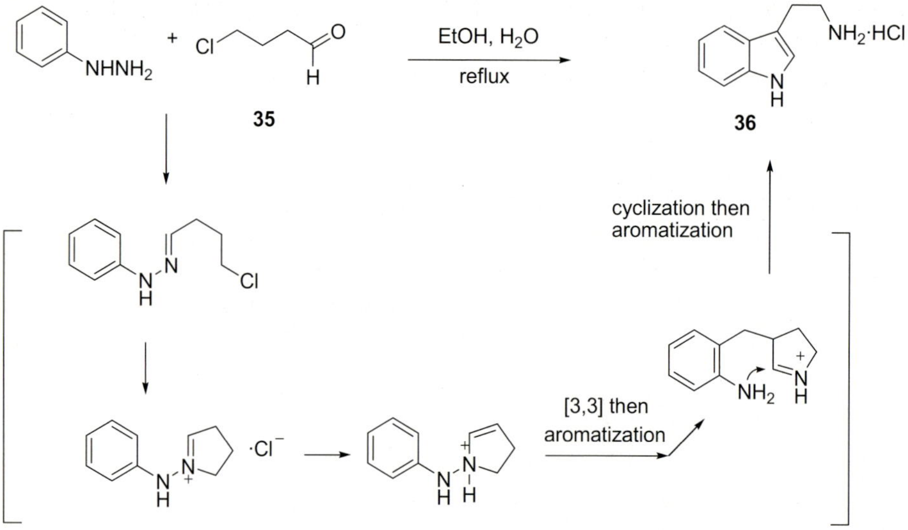

Scheme 5.20 Grandberg indole synthesis.

Several examples of the Fischer indole synthesis in the solid phase have been described [37]. A very elegant traceless synthesis has been reported that employs a solid supported hydrazine (**37**) [38], which participates in the Fischer indolizidation upon treatment with a ketone in the presence of TFA. Interestingly, during the cyclization step, indole **38** is released from the resin in a traceless manner (Scheme 5.21).

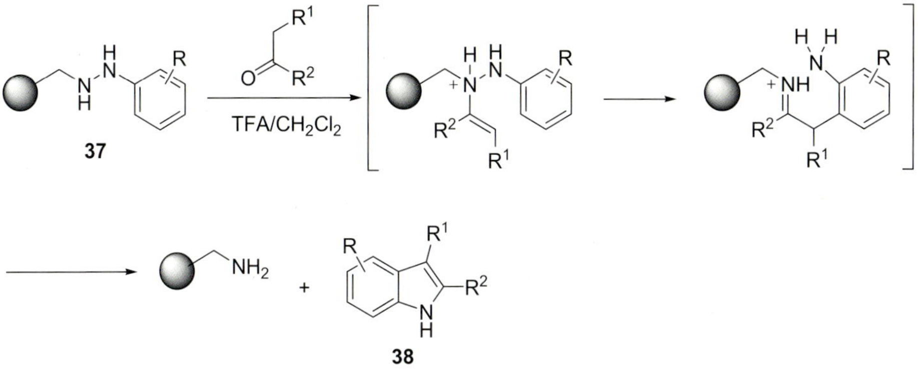

Scheme 5.21 A traceless solid-phase Fischer indole synthesis.

A library of structurally diverse indomethacine analogs (**45**) have been prepared by a "resin-capture-release" strategy, a technique that combines solid-supported and solution chemistry. Aryl hydrazine **39** is "captured" by an aldehyde functionalized resin (**40**), and the unprotected N−H is acylated, to build an array of solid supported protected acylhydrazines (**42**). Treatment with trifluoroacetic acid (**46**) releases the hydrazine **43**, which then reacts with a ketone, to provide hydrazone **44**, which suffers the indolizidation to deliver the corresponding indole **45** (Scheme 5.22) [39].

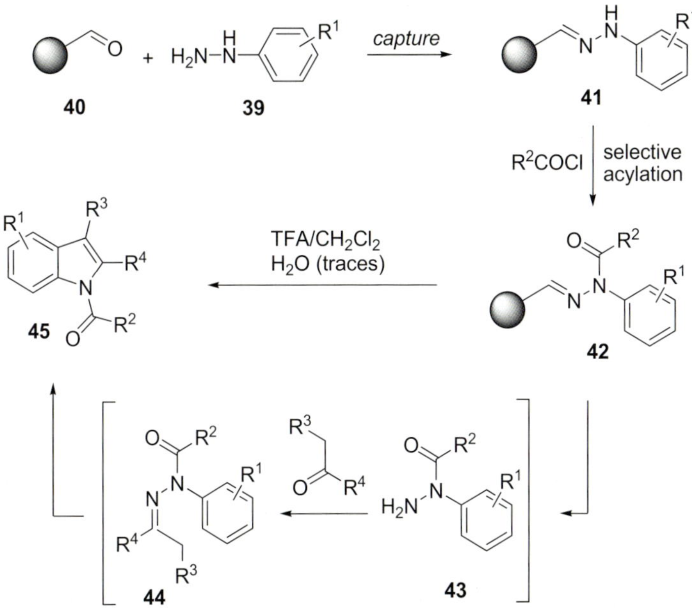

Scheme 5.22 Solid-supported synthesis of indomethacine analogs.

Cyclic enol-ethers and enol-lactones **47** have been used as synthetic equivalents of aldehydes and ketones, giving rise directly to functionalized indoles **48** (Scheme 5.23) [40].

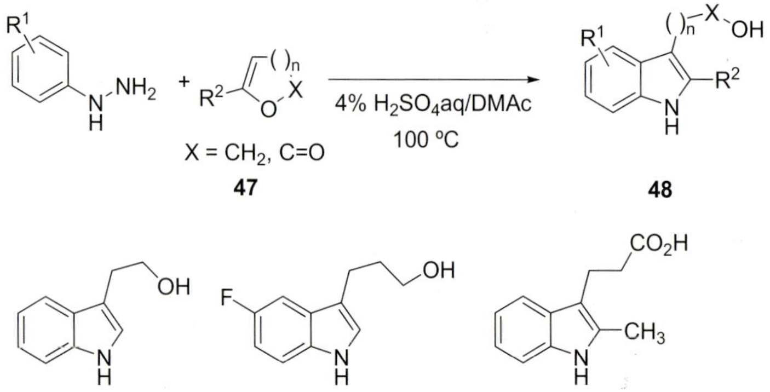

Scheme 5.23 Fischer indole synthesis with enol ethers and enol lactones.

5.4.2.1.2 **Japp–Klingemann Reaction** The condensation of carbonyls (or synthetic equivalents) with aryl hydrazines is not the only route to the aryl hydrazones required for the Fischer indolizidation. The coupling of aryldiazonium salts (**49**) with the enolates of ketones **50** – the Japp–Klingemann reaction [41] – represents an alternative that has been employed extensively. Hydrazone **52** is formed after a deacylation step of the intermediate diazo compound **51** (Scheme 5.24).

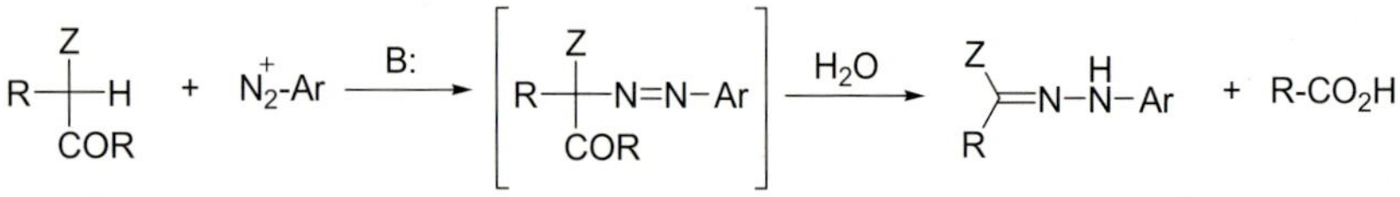

Scheme 5.24 General scheme of the Japp–Klingemann reaction.

β-Ketoesters are the usual substrates for the Japp–Klingemann reaction. For instance, ketoester **53** reacts with benzenediazonium chloride in an alkaline solution to form hydrazone **54**. Indolizidation of hydrazone **54** under standard Fischer conditions affords 2,3-disubstituted indole **55**, which can be decarboxylated to give 3-substituted indole **56** (Scheme 5.25) [42].

When cyclic β-ketoesters (e.g., **57**) are used, deacylation by ring opening occurs under the basic conditions of the reaction, to provide the carboxylic acid **58**. Fischer indolization of **58** gives rise to 2,3-disubstituted indoles **59** [43] (Scheme 5.26).

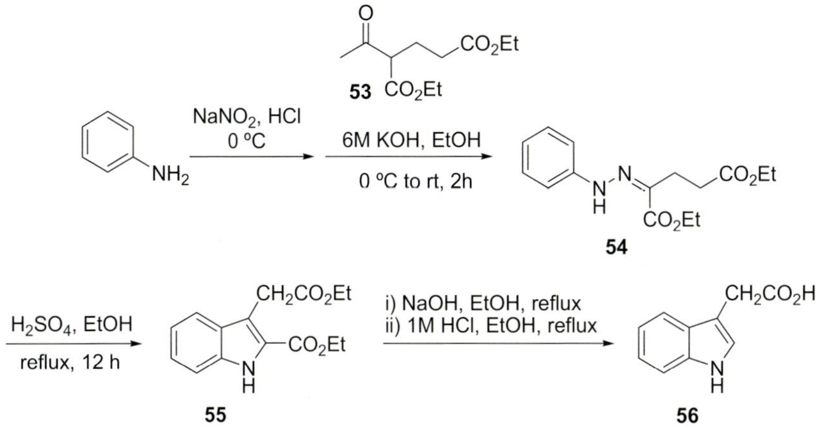

Scheme 5.25

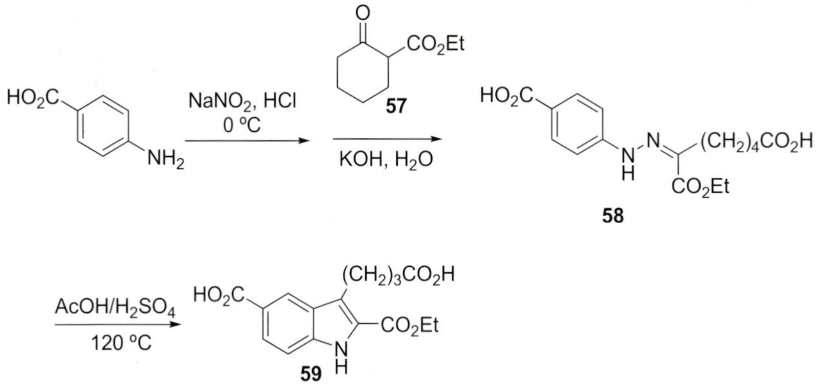

Scheme 5.26

5.4.2.1.3 **Hydroamination-Based Fischer Indole Synthesis** Recently, an interesting variation to the traditional Fischer approach has been introduced that uses alkyne hydroamination as an alternative way to access the intermediate arylhydrazone. Scheme 5.27 presents a *one pot* approach to the indole framework based on intermolecular titanium amide-catalyzed hydroamination reactions of alkynes **61** with 1,1-disubstituted hydrazines **60**. Subsequent addition of 3–5 equiv $ZnCl_2$ is necessary to convert the generated hydrazone **62** into the corresponding indole **63** [44].

The Grandberg strategy has been combined with the hydroamination to prepare tryptamine analogs **66** (Scheme 5.28) [45]. The Ti-catalyzed hydroamination of terminal chloroalkynes **64** gives rise directly the hydrochloric salt of aminoindoles **65**.

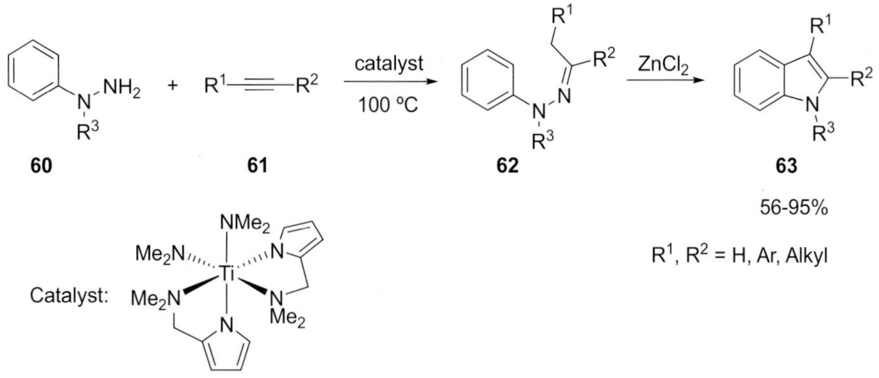

Scheme 5.27 Hydroamination-based Fischer indole synthesis.

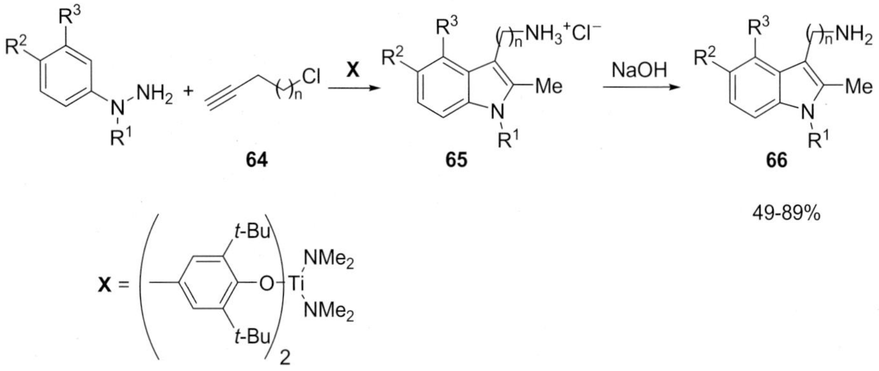

Scheme 5.28 Hydroamination-based Grandberg indole synthesis.

This one pot process involves a titanium-catalyzed Markovnikov hydroamination to furnish the corresponding hydrazone, which can follow the same reaction pathway as that proposed in the original Grandberg route.

Another interesting modification of the hydroamination makes use of inexpensive $TiCl_4$ as precatalyst for the hydroamination reaction, and it also serves as Lewis acid for the cyclization process. This methodology allows for the preparation of 1,2,3-trisubstituted indoles **70** from unsymmetrical alkynes **68** (Scheme 5.29) [46].

The aryl hydrazones required for the Fischer cyclization have also been prepared through a novel rhodium-catalyzed hydroaminomethylation of olefins [47]. Thus, reaction of aliphatic olefins **71** with synthesis gas ($CO : H_2$, 1 : 1) and aryl hydrazines **72** in the presence of rhodium phosphine catalysts leads to the corresponding hydrazones **73**, which can be subsequently transformed into indole **74** by treatment with $ZnCl_2$ in a *one pot* process (Scheme 5.30) [48].

As an alternative method, the Buchwald–Hartwig amination has been applied for the synthesis of the arylhydrazines required for the Fischer synthesis [49]. Thus,

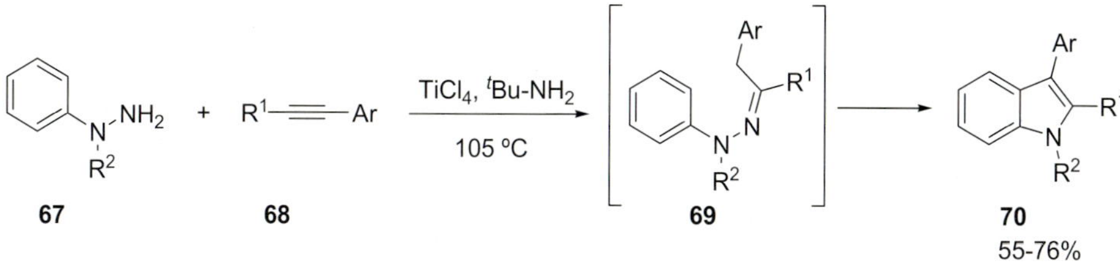

Scheme 5.29

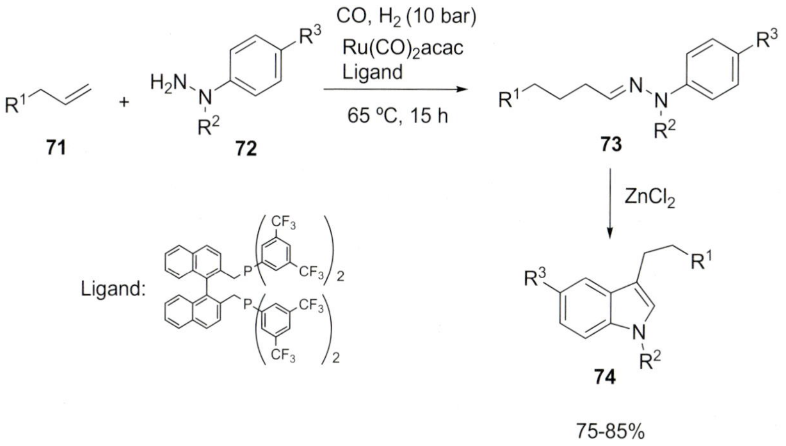

Scheme 5.30

Pd-catalyzed cross-coupling of aryl bromides **76** with benzophenone hydrazone (**75**) furnishes *N*-arylbenzophenone hydrazone **77**, which is hydrolyzed to the arylhydrazine **78** by treatment with acid. If the hydrolysis is carried out in the presence of a carbonyl compound, indole **79** is formed in a one pot process, without isolation of any of the intermediates (Scheme 5.31).

5.4.2.1.4 **Gassman Synthesis** A [2,3] sigmatropic shift of anilinosulfonium ylide **80** is the key step in the Gassman synthesis [50]. The sulfur substituted indoles **81** so-obtained can be easily reduced with Raney Ni (Scheme 5.32).

In the original Gassman procedure the indole is formed directly from the corresponding aniline **82** in a *one pot–three step* procedure that involves the formation of chloramine **83**, which then reacts with a α-thioketone to form the anilinosulfonium salt **84**. Low temperature rearrangement of **84** yields the 3-thioindole **85** (Scheme 5.33).

The anilinosulfonium salt **86** can be accessed also by a modified procedure that avoids the use of *t*BuOCl and provides slightly better yields. This strategy has been employed in the synthesis of oxyindoles **87** (Scheme 5.34) [51].

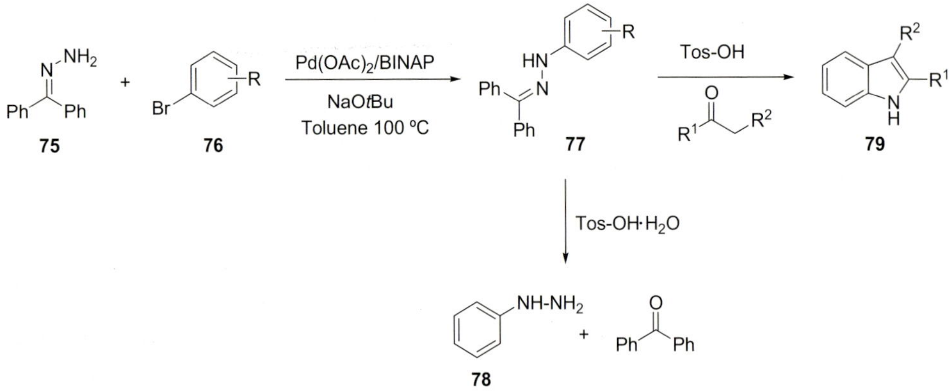

Scheme 5.31

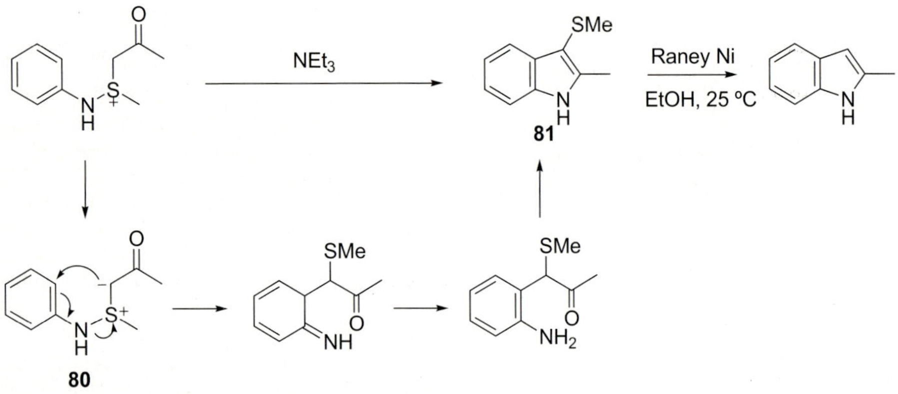

Scheme 5.32

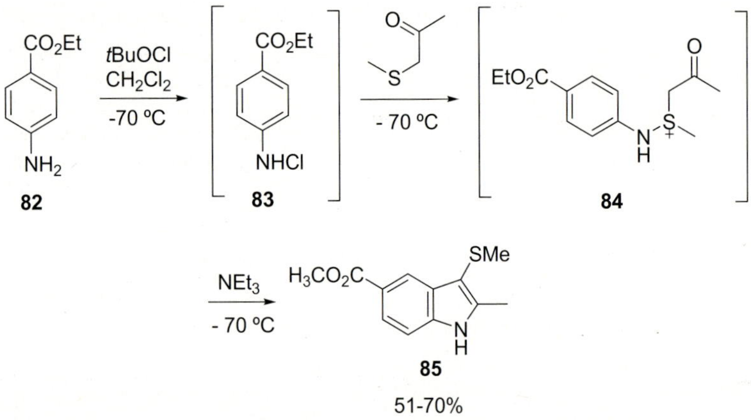

Scheme 5.33

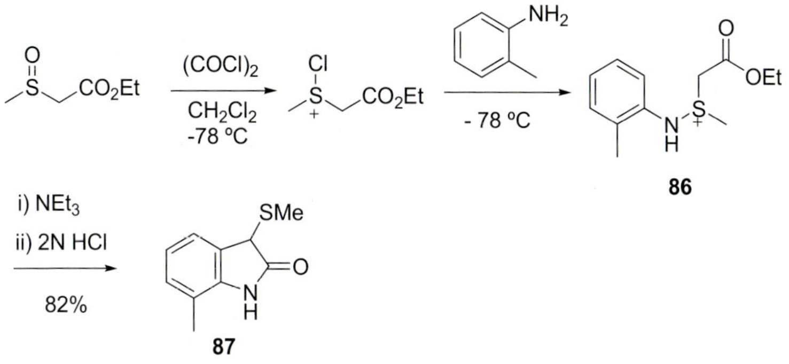

Scheme 5.34

5.4.2.1.5 **Bartoli Synthesis** In the Bartoli synthesis, 7-substituted indoles **89** are obtained from 2-substituted nitrobenzene **88** upon treatment with 3 equivalents of vinylmagnesium chloride (Scheme 5.35) [52]. The availability of the starting materials and its simplicity make this reaction one of the most efficient methods for the preparation of 7-substituted indoles. The necessity of an ortho-substituent on the aromatic ring is the main limitation. Nevertheless, Br is a very suitable substituent that can enforce the sigmatropic rearrangement as requested by the mechanism (see below), and can be easily removed or transformed thereafter.

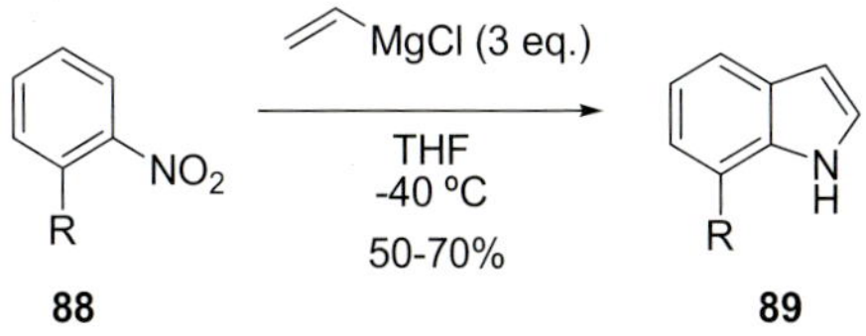

Scheme 5.35

Scheme 5.36 depicts the mechanism proposed for the Bartoli reaction. The reaction starts with the addition of the first equivalent of vinyl Grignard to an oxygen

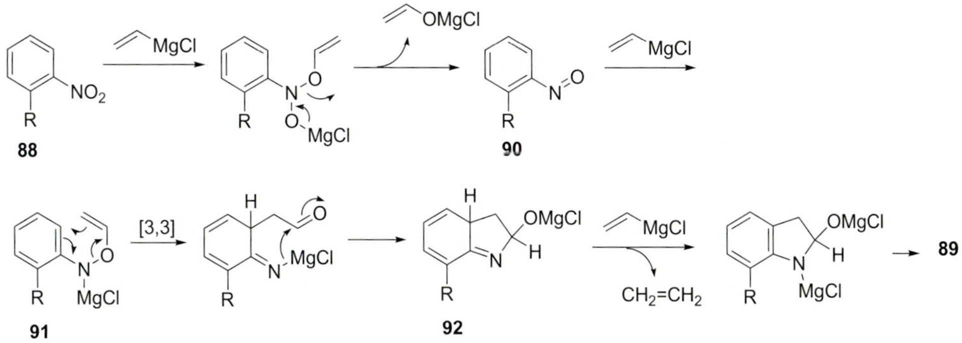

Scheme 5.36

atom of the nitro group of **88**, with subsequent elimination of acetaldehyde enolate, to generate nitrosobenzene derivative **90**. Then, the second equivalent attacks the oxygen atom of the nitroso functionality. The intermediate **91** generated suffers a Claisen-like [3,3]-sigmatropic rearrangement, with cleavage of the N–O bond, followed by heterocyclization to form the five-membered ring. A third equivalent of the Grignard is required to abstract a proton on **92** before the final elimination takes place.

Slightly modified Bartoli protocols [53], including a solid phase version [54], have been implemented more recently.

Other examples of [3,3] sigmatropic rearrangement-based indoles synthesis have been described [55, 56].

5.4.2.2 Cyclization by Formation of the N–C2 Bond

The construction of the indole ring with formation of the N–C2 bond includes the most popular approaches other than the Fischer synthesis.

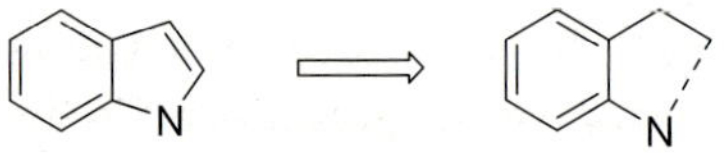

N ⟹ N

An important class of methods for indole ring synthesis involves a cyclization of an aminoketone (**93**) with formation of the C2–N bond (Scheme 5.37). Usually, the precursor that suffers the intramolecular condensation has to be preformed in a preliminary step, and must contain a carbonyl or masked carbonyl (imine, enamine, enolether) functionality.

R O NH_2 **93** → N –R

Scheme 5.37

Many different approaches have been investigated to generate the cyclization precursor. Among them, those that rely on the reductive cyclization of *o*-substituted nitroaryls are noteworthy.

5.4.2.2.1 Reissert Indole Synthesis

The Reissert indole synthesis is a very reliable method for the preparation of benzene-substituted indole-2-carboxylates [57]. In the first step, condensation of *o*-nitrotoluene (**94**) with ethyl oxalate under basic media affords the potassium salt of *o*-nitrophenylpyruvate **95** (Scheme 5.38). The key step in the Reissert synthesis is the reductive cyclization of this intermediate, which generates the aminoketone **96** that undergoes cyclization to provide the indole-2-carboxylate (**97**). The reductive cyclization has been effected under different catalytic hydrogenation conditions (Pt/AcOH, Pd-C/EtOH [58]) and also with various low oxidation state metal salts ($SnCl_2$-$TiCl_3$) [59].

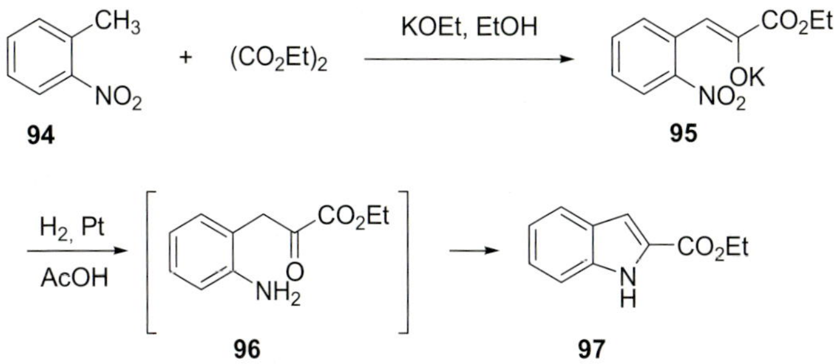

Scheme 5.38 Classic Reissert indole synthesis.

Many variations on the classic Reissert synthesis are known, which differ in the way the intermediate *o*-aminobenzyl ketones or aldehydes, ready for the cyclization, are obtained [60].

Nucleophilic substitution on nitroarenes has been applied extensively in several approaches, a theme that has been reviewed [61]. Silylenol ethers **99** activated with fluoride anion [tris(dimethylamino)sulfonium difluorotrimethylsiliconate (TASF)] behave as strong C-nucleophiles and add to nitroarenes **98** in the ortho position to the nitro group. Subsequent aromatization of the intermediate generated with DDQ leads to *o*-(2-nitroaryl)alkyl ketones **100**. Finally, reductive cyclization under standard conditions provides the indole **101** (Scheme 5.39) [62].

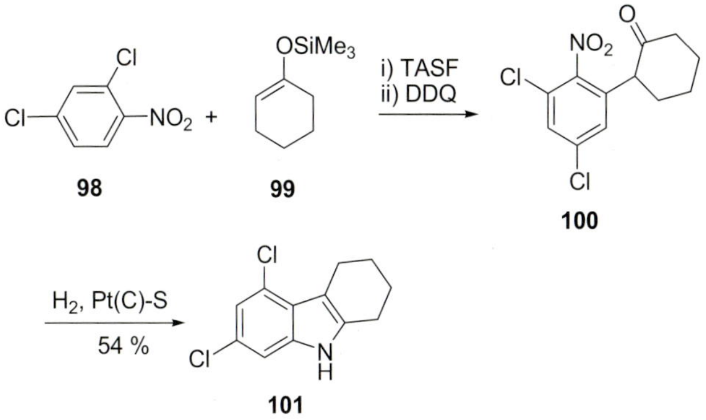

Scheme 5.39

The vicarious nucleophilic substitution (VNS) of hydrogen [63] in nitroarenes **102** with α-chloroalkyl ketones **103** gives rise to nitroaryl ketones **104**, which are converted into indoles **105** under classical Reissert conditions (Scheme 5.40).

A similar valuable approach to indoles consists of the reductive cyclization of (*o*-nitroaryl)acetonitriles **108** (Scheme 5.41). The cyanomethyl group is efficiently introduced in nitroarenes **106** by VNS of hydrogen with chloroacetonitrile **107** or aryloxyacetonitrile. Catalytic hydrogenation transforms the cyano group into an

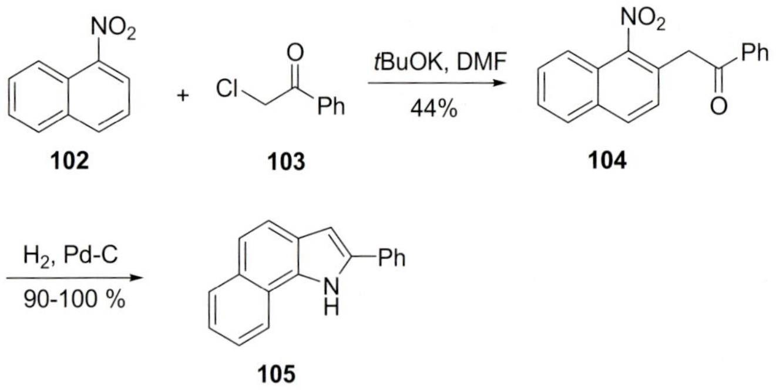

Scheme 5.40

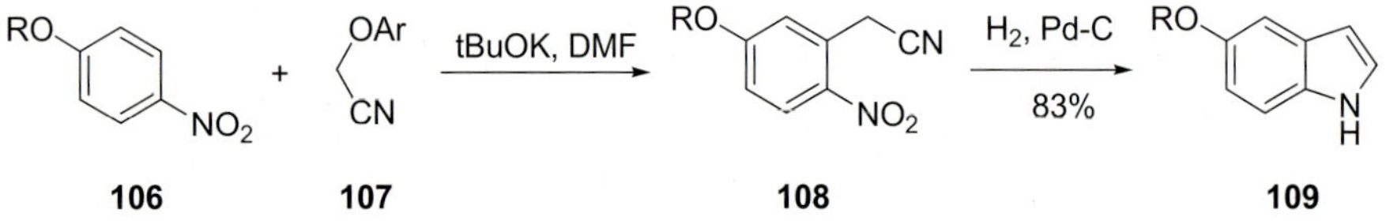

Scheme 5.41

imine and the nitro group to an amine, giving rise to **108**, which cyclizes spontaneously to the indole **109** [64].

Carbocyclic-fused indoles have been prepared by several other alternative routes that involve the cyclization of a α-(*o*-nitrophenyl) ketone (Scheme 5.42). For instance, the intermediate nitroketones **112** can be obtained by arylation of cyclic silylenolether **111** with (*o*-nitrophenyl)phenyliodonium fluoride **110** [65]. This methodology was employed in the total synthesis of (−)-tabersonine [66]. In a different approach, an Ullman-type cross-coupling of *o*-halonitroarenes **113** with α-haloenones **114** has been also employed to obtain the *o*-nitroarylketones **115** [67]. In both cases, reductive cyclization affords the expected indole ring.

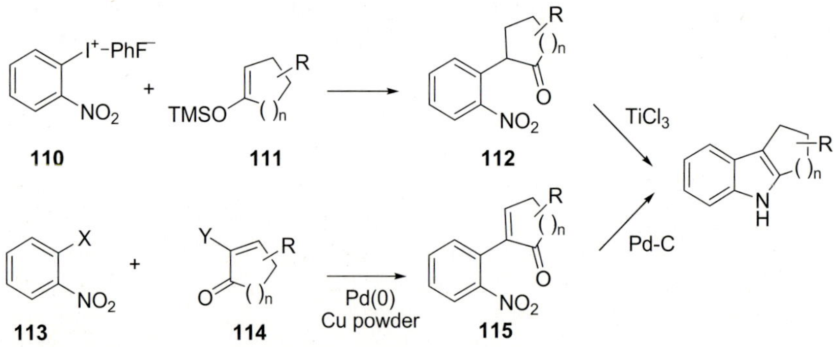

Scheme 5.42

In Shibasaki's total synthesis of strychnine [68] an advanced intermediate is also an α-(*o*-nitrophenyl)ketone (**118**), which is prepared by Stille cross-coupling of *o*-nitro-stannylbenzene (**116**) with a vinyl iodide (**117**) (Scheme 5.43).

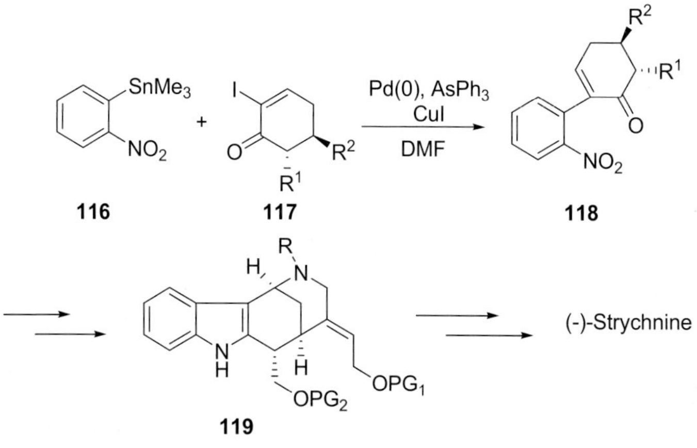

Scheme 5.43

A remarkable innovation has been uncovered by Buchwald – applying the Pd-catalyzed arylation of ketone enolates [69] to transform *o*-bromonitrobenzene derivatives **120** into *o*-nitroketones **122** (Scheme 5.44). Subsequent reductive cyclization provides the corresponding indoles **123** [70]. The reaction is very general and has been utilized successfully in the synthesis of indoles bearing both electron-withdrawing and electron-donating substituents on the benzene ring and a wide variety of ketones **121**. Moreover, additional substitution can be introduced if the intermediate

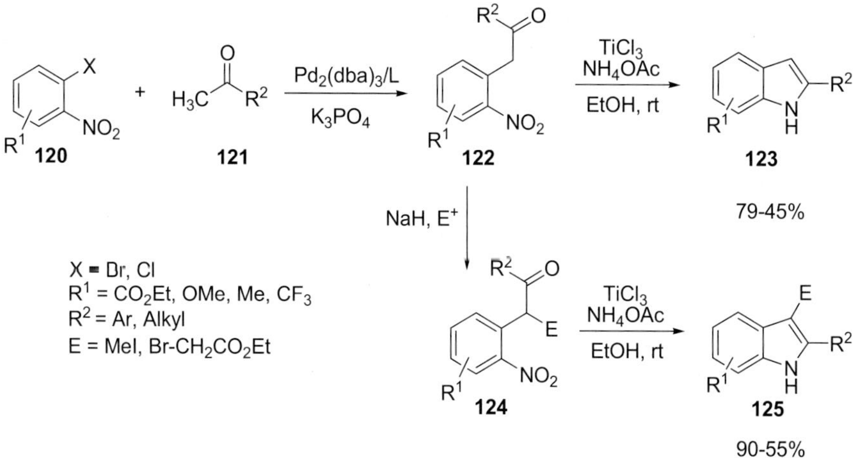

Scheme 5.44

nitroketone **122** is deprotonated and alkylated before the reductive cyclization step to furnish substituted nitroketones **124**. Therefore, this important development allows for the preparation of very highly substituted indoles **125**.

5.4.2.2.2 **Leimgruber–Batcho Synthesis** The Leimgruber–Batcho synthesis is a very convenient method to prepare indoles with substitution only in the benzene ring [71]. The two-step procedure starts with the three-components reaction of an *o*-methylnitroaryl (**126**) with dimethylformamide dimethylacetal in the presence of pyrrolidine, to provide *o*-nitro-β-pyrrolidinostyrene **127** (Scheme 5.45). Reductive cyclization on **127** furnishes indoles **128**, usually in very high yields.

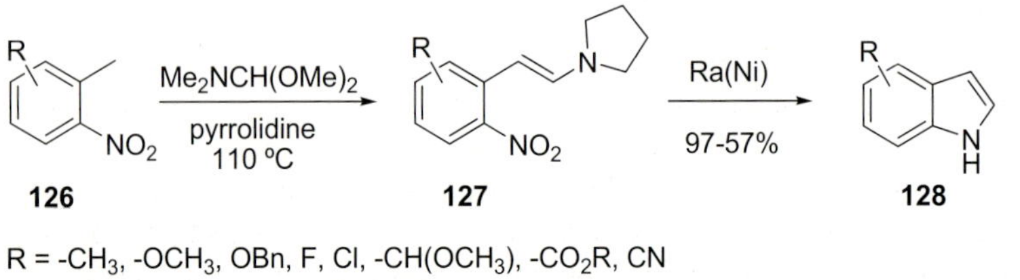

R = $-CH_3$, $-OCH_3$, OBn, F, Cl, $-CH(OCH_3)$, $-CO_2R$, CN

Scheme 5.45

Strong points of this approach are the compatibility with many functional groups, and the ready availability of the starting materials, which make this method a very interesting entry to polysubstituted indoles [72–74]. Enhanced conditions for a Lewis acid catalyzed version of the reaction using microwave acceleration have been described recently [75]. This modification has been applied to the synthesis of a wide variety of substituted nitroenamines, including several examples of heteroaromatics such as **129**, which expand the scope of the Leimbruger–Batcho synthesis (Scheme 5.46).

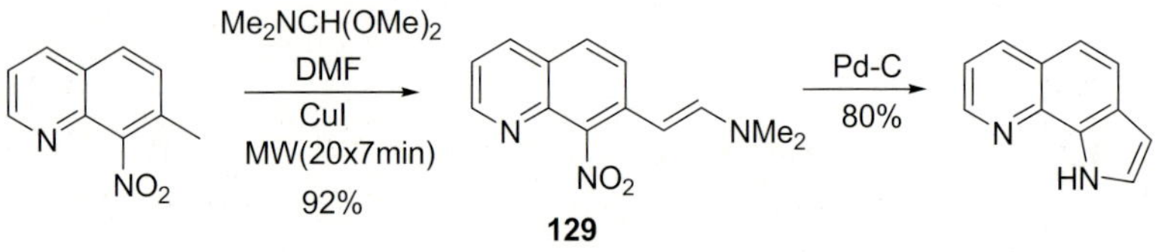

Scheme 5.46

5.4.2.2.3 **Reductive cyclizations *o*-Nitrostyrenes** The reductive cyclization *o*,β-dinitrostyrenes is another two step synthesis of indoles closely related with the Leimbruger–Batcho approach. The formation of the *o*,β-dinitrostyrenes **131** is usually achieved by Henry condensation [76, 77] of nitromethane with *o*-nitrobenzaldehydes **130**, or by nitration of benzaldehydes [78]. The reduction step has been carried out with several classes of reducing agents, including different metal/acid combinations [79] and catalytic hydrogenation conditions (Scheme 5.47) [80]. This

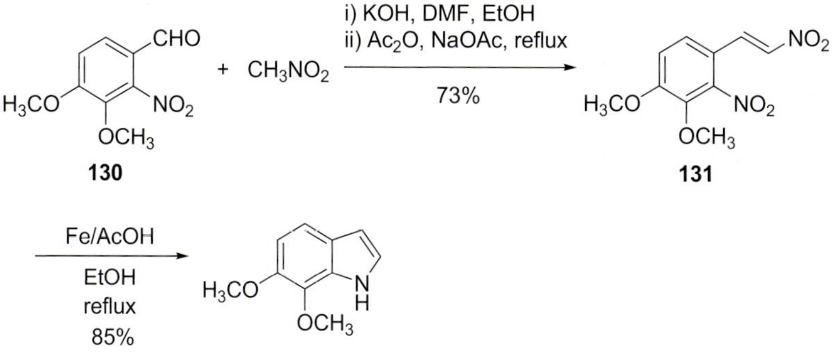

Scheme 5.47

methodology has been applied as starting point in the synthesis of several natural products containing the indole skeleton [81–84].

A versatile methodology, related with the methods described above, is the palladium-catalyzed reductive N-heteroannulation of *o*-nitrostyrenes **132** [85, 86]. The reaction is carried out under CO pressure and in the presence of a Pd/ligand catalytic system (Scheme 5.48) [87].

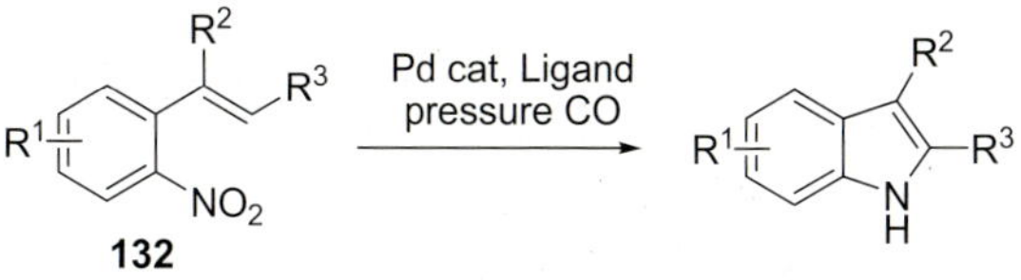

Scheme 5.48

The required precursors, the *o*-nitrostyrenes, are readily available from *o*-bromonitrobenzenes, through metal-catalyzed reactions, such as Stille couplings [88] or Heck reactions [89]. Examples of both approaches are represented in Scheme 5.49.

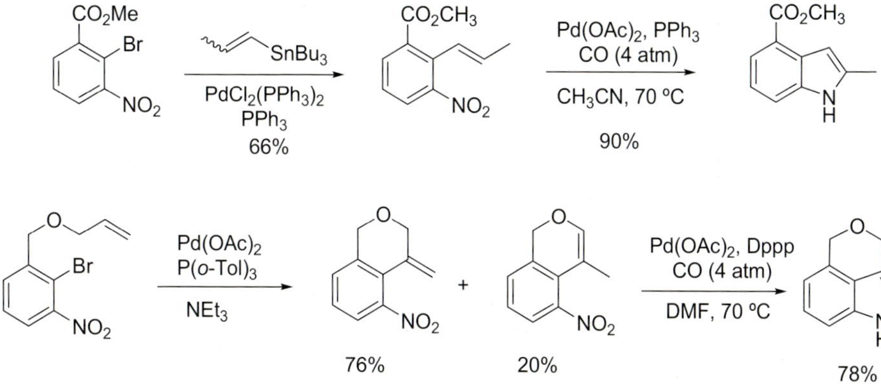

Scheme 5.49

5.4.2.2.4 **Sugasawa Synthesis** Another synthesis of the indole ring from anilines is based on the *o*-chloroacetylation of aniline, the so-called Sugasawa reaction (Scheme 5.50) [90, 91]. Treatment of an aniline (**133**) with chloroacetonitrile in the presence of BCl_3 and another Lewis acid furnishes *o*-chloroacetylaniline **134**, through the intermediate **135**. Chloroacetylaniline **134** can be cyclized to the indole **136** by reduction with $NaBH_4$. The Sugasawa synthesis has been used very efficiently for the preparation of N–H unprotected indoles with substitution in the benzene ring [92, 93].

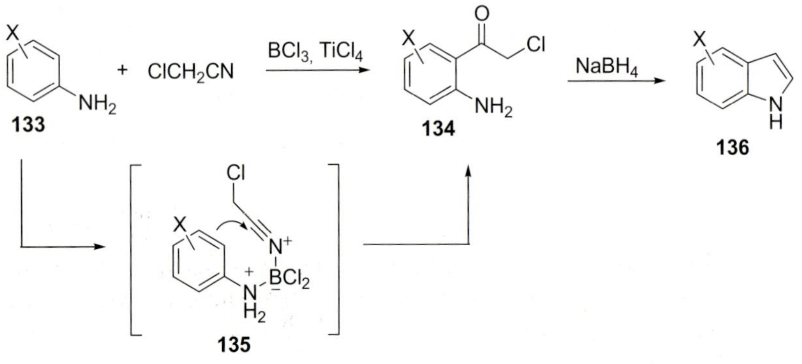

Scheme 5.50

5.4.2.2.5 **Indoles from 5-Aminodihydronaphthalenes: Plieninger Indole Synthesis** The oxidative cleavage of 5-aminodihydronapthalenes **137** provides the amino carbonyl **138** required for the N–C2 condensation reaction to form 7-substituted indoles **139** (Scheme 5.51) [94].

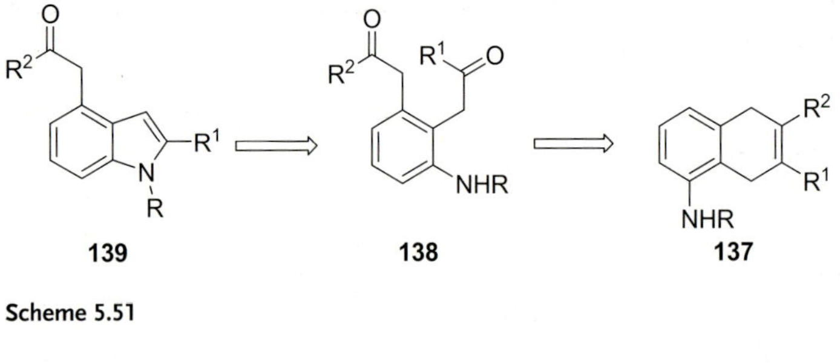

Scheme 5.51

The key intermediate in this synthesis, the 2-aminodihydronaphthalene **141**, is best obtained by [4 + 2] cycloaddition reaction of a *p*-benzoquinone mono- or di-imine, such as **140** with electron-rich dienes [95]. The great potential of this approach is exemplified by the synthesis of the polysubstituted indole **142**, *en route* to the total synthesis of the antitumor agent (+)-yatakemycin (Scheme 5.52) [96].

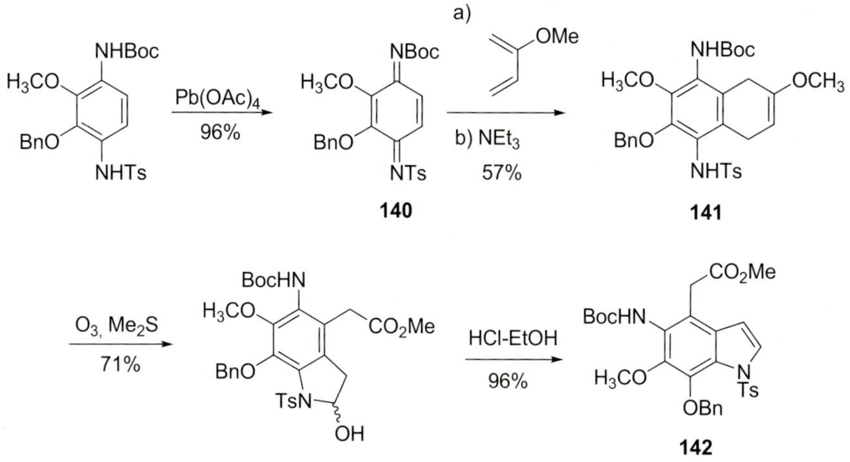

Scheme 5.52

5.4.2.2.6 **Cyclizations of Nitrenes** Another type of cyclization with formation of the N–C2 bond involves the insertion of nitrenes in a vinylic C–H bond to form indoles (Scheme 5.53). Obviously, the key in this type of methodology is the generation of the unstable nitrene species.

Scheme 5.53

The intermediate nitrenes can be formed by thermolysis of β-substituted-*o*-azidostyrenes **143**, which undergo insertion to form the indole [97, 98]. Scheme 5.54 shows, as an example, the synthesis of 2-nitroindole **144**, which is not easily available through other procedures [99].

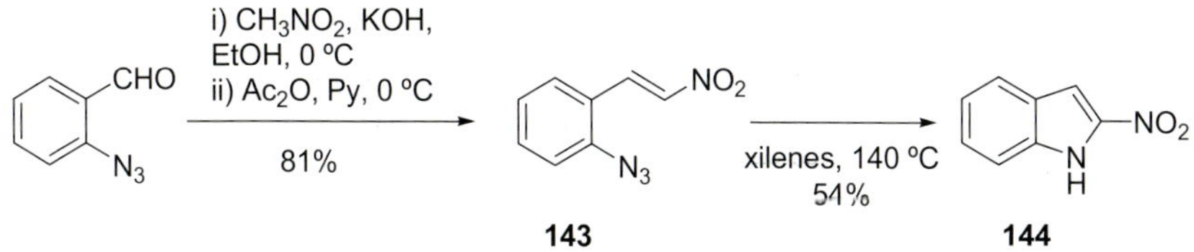

Scheme 5.54

Nitrenes have also been generated from *o*-nitrostyrenes or *o*-nitrostilbenes by deoxygenation of the nitro group with triethyl phosphite [100, 101]. In a more recent modification of this approach, the nitrene is generated by treatment of the nitroarene **145** with phenylmagnesium chloride under very mild conditions (Scheme 5.55) [102].

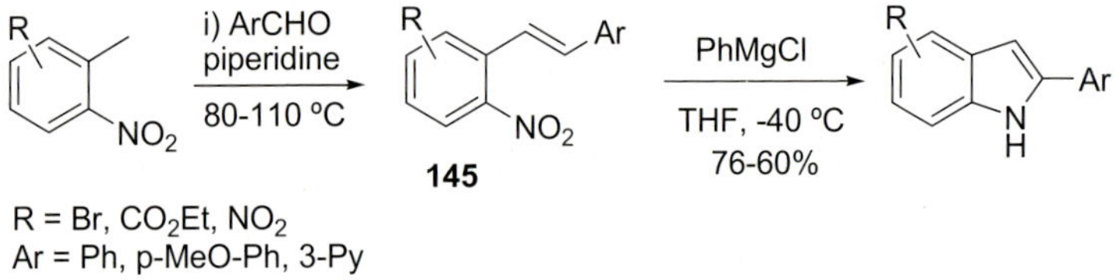

R = Br, CO_2Et, NO_2
Ar = Ph, p-MeO-Ph, 3-Py

Scheme 5.55

Scheme 5.56 shows the mechanism proposed for this novel nitrene generation reaction.

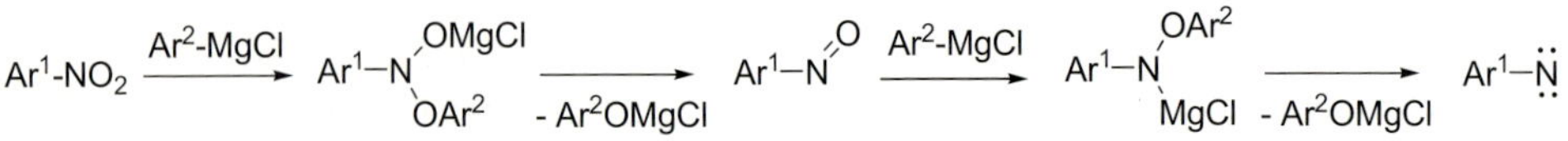

Scheme 5.56

5.4.2.2.7 **Synthesis from *o*-Allylanilines or *o*-Vinylanilines** Palladium-catalyzed intramolecular C–N bond formation is one of the most powerful methods for the synthesis of the indole ring. The first contributions in this area were due to Hegedus, who described the cyclization of *o*-allylanilines **146** using stoichiometric amounts of Pd(II) in a Wacker-type process (Scheme 5.57) [103]. The reaction tolerates functional groups as well as substitution in the allyl moiety.

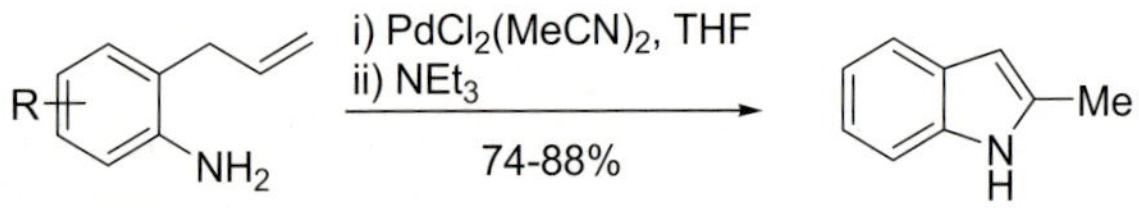

Scheme 5.57

During the reaction the Pd(II) species is reduced to Pd(0). To avoid the use of stoichiometric Pd, benzoquinone is used in the catalytic process as reoxidant. This methodology has been applied in the synthesis of 3-hydroxyindoles **147** (Scheme 5.58) [104].

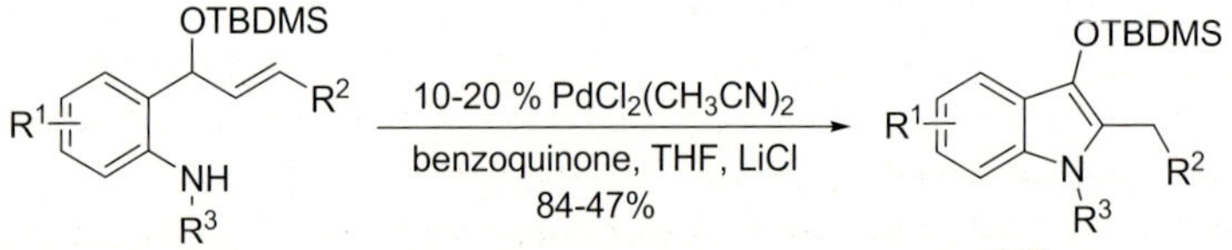

Scheme 5.58

5.4.2.2.8 **Synthesis from *o*-Alkynylanilines or *o*-Alkynylanilides** Intramolecular hydroamination of *o*-alkynylanilines **148**, through a 5-*endo*-dig cyclization, is one of the most popular modern methods for the construction of the indole ring (Scheme 5.59). The required alkynylanilines are usually synthesized by a Pd-catalyzed Sonogashira cross-coupling of an *o*-iodoaniline with a terminal acetylene [105]. The cyclization reaction has been studied extensively and applied in numerous synthetic efforts. Several metal-based catalytic systems are able to promote the cyclization. Among them, Pd and Cu have been the most studied. Detailed accounts of this methodology have been published [106]. Typically, the cyclization is carried out in the presence of a Pd(II) catalyst, such as $PdCl_2$ (Scheme 5.60) [107].

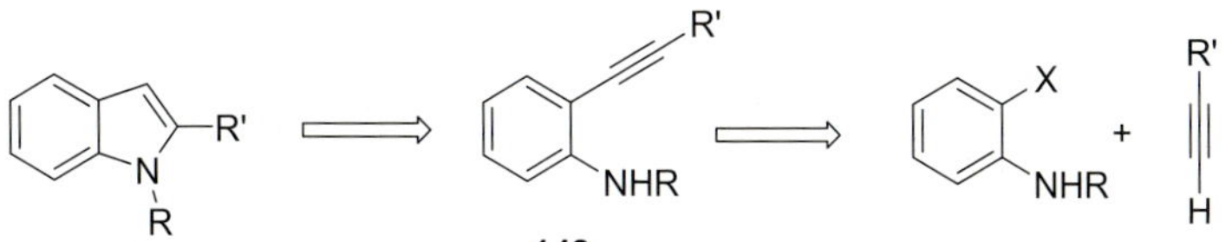

R' N R NHR 148 X NHR + R' H

Scheme 5.59

R^2 NHR^1 cat. $PdCl_2$, MeCN reflux 52-83% R^2 N R^1

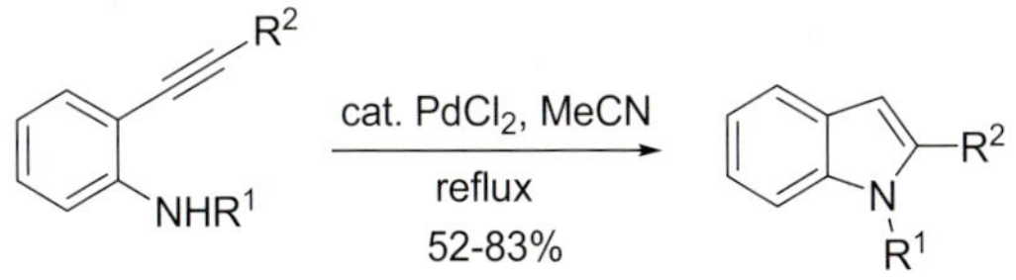

R^1 = H, -COR; R^2 = Alkyl, Ar

Scheme 5.60

The mechanism of the reaction is postulated to proceed via an intramolecular aminopalladation of the alkynylaniline **148**, which produces the a σ-indolylpalladium species **149**, followed by protonolysis of the C–Pd bond, which releases the indole **150** and recovers the Pd(II) catalyst (Scheme 5.61).

The *one pot* synthesis of 2-substituted indoles from *o*-haloaniline derivatives **151** and terminal acetylenes **152** can be achieved by employing a combination of Pd(II) and Cu(I) catalyst (Scheme 5.62). In this process, the Sonogashira coupling to form alkynylaniline **153** and the intramolecular hydroamination occur consecutively and are promoted by the same catalytic system [108, 109].

Indoles are also prepared in *one pot* fashion from *o*-chloroalkynylbenzene **154** derivatives and primary amines **155** (Scheme 5.63) [110]. In a first step, a Pd-catalyzed aryl amination takes place to furnish the *o*-alkynylaniline **156**, which then cyclizes to give the indole **157**, in a process promoted by the same Pd catalyst.

An interesting variation of this methodology is the "aminopalladation/reductive elimination" domino reaction that provides 2,3-disubstituted indoles **160** from *o*-alkynyl trifluoroacetanilides **158** and organic halides **159** (Scheme 5.64) [111].

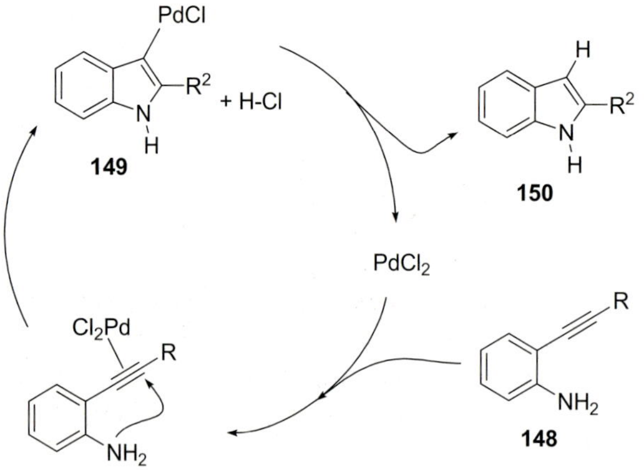

Scheme 5.61

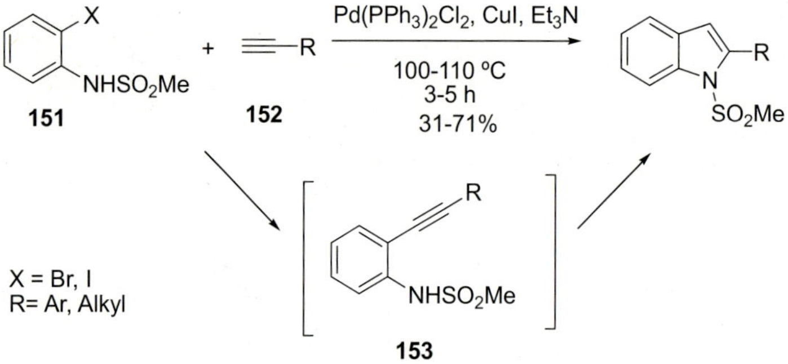

Scheme 5.62

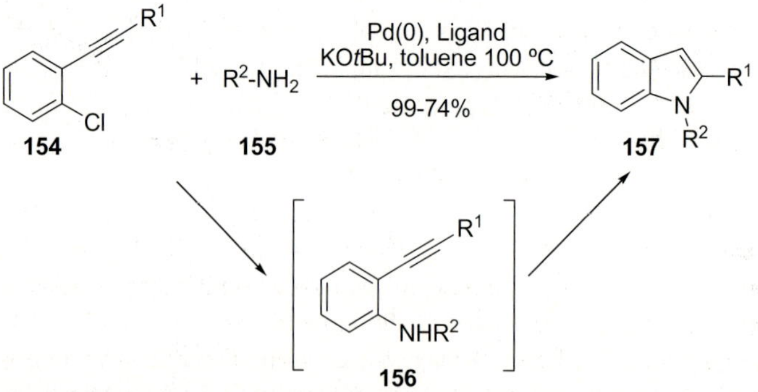

Scheme 5.63

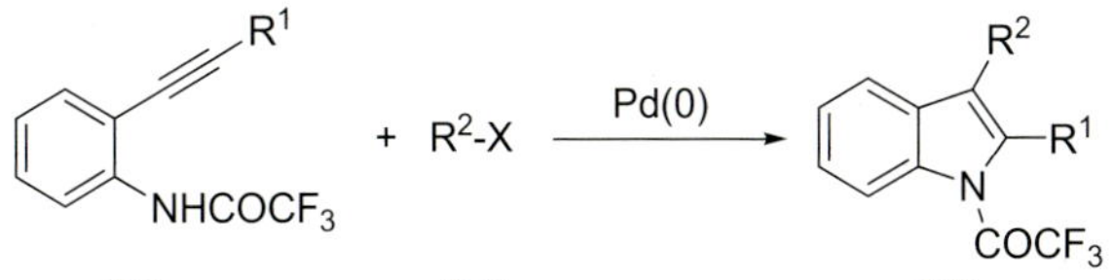

Scheme 5.64

A plausible mechanism for this Pd(0)-catalyzed domino reaction involves: (i) oxidative addition of the halide to the Pd(0) species to form Pd complex **161**; (ii) coordination of the Pd(II) species **161** to the triple bond of **158** to form the (η^2-alkyne) organopalladium complex **162**; (iii) intramolecular nucleophilic attack of the nitrogen across the triple bond to form the σ-indolylpalladium intermediate **163**; and (iv) reductive elimination that regenerates the Pd(0) catalyst and releases the 2,3-disubstituted indole **160** (Scheme 5.65).

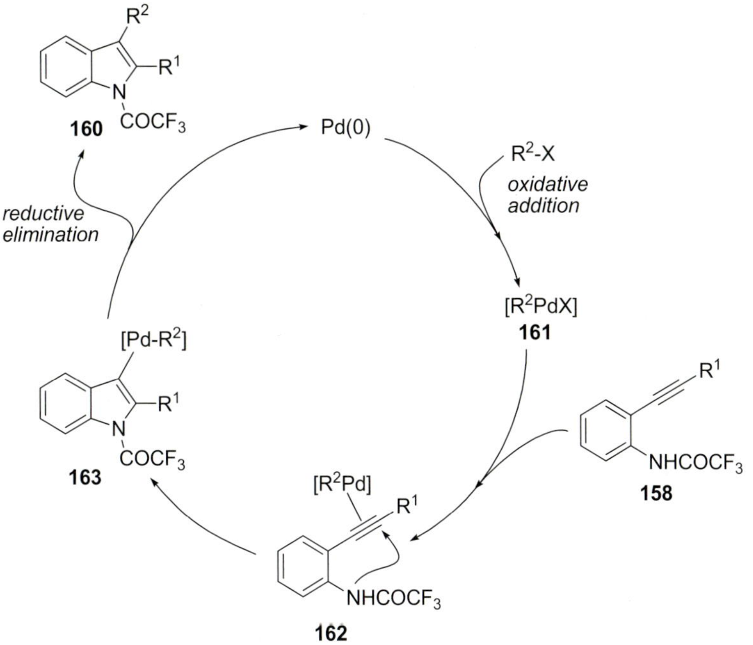

Scheme 5.65

The reaction has been carried out employing aryl halides [112], alkenyl triflates, alkynyl bromides [113], and allylic carbonates [114] to provide the corresponding 2,3-disubstituted indoles. Moreover, if the process is carried out under CO atmosphere, a three-component reaction takes place with incorporation of the carbon monoxide molecule, to yield 3-acyl-substituted indoles (Scheme 5.66) [115].

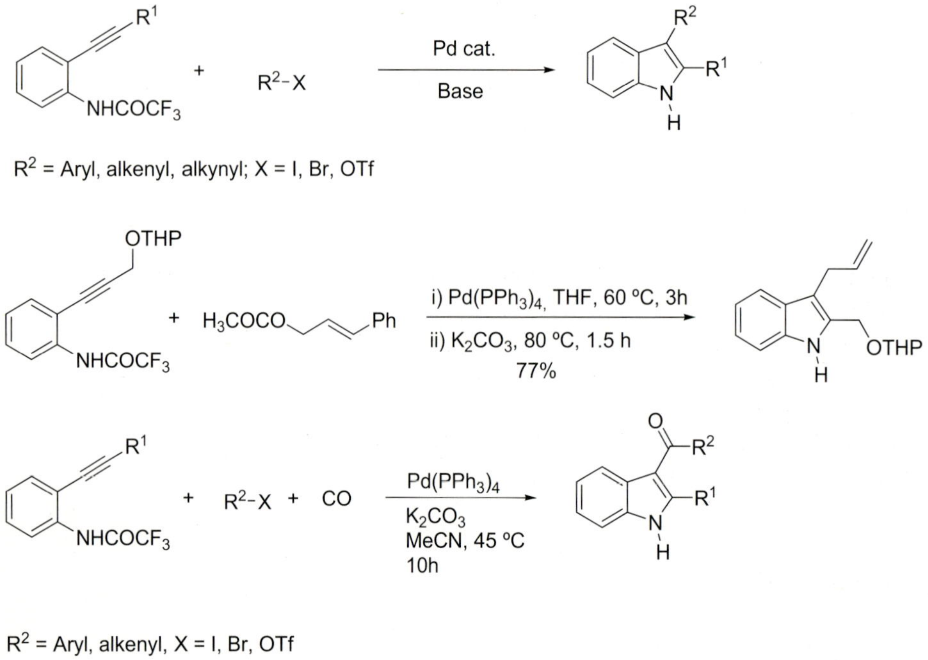

R^1 NHCOCF$_3$ + R^2–X → (Pd cat., Base) → indole (R^1, R^2, N–H)

R^2 = Aryl, alkenyl, alkynyl; X = I, Br, OTf

OTHP, NHCOCF$_3$ + H_3COCO…Ph → i) $Pd(PPh_3)_4$, THF, 60 °C, 3h; ii) K_2CO_3, 80 °C, 1.5 h; 77% → N–H, OTHP

R^1 NHCOCF$_3$ + R^2–X + CO → $Pd(PPh_3)_4$, K_2CO_3, MeCN, 45 °C, 10h → indole (O, R^2, R^1, N–H)

R^2 = Aryl, alkenyl, X = I, Br, OTf

Scheme 5.66

This domino chemistry has been also carried out with a solid support and applied to the preparation of a combinatorial library of indoles **164** with three points of diversity (Scheme 5.67) [116].

A very impressive extension of this approach is the recently described Pt-catalyzed carboamination of *o*-alkynylanilides **165** [117]. A representative example is presented in Scheme 5.68. During the reaction, the acyl group migrates from the N-atom to C3 to form acylated indole **166**. Therefore, this methodology constitutes a very attractive method for the preparation of 2,3-disubstituted indoles.

The indolization of *o*-alkynylanilines has been performed with alternative promoters other than the classic transition metal catalysts. For instance, K and Cs bases promote successfully the cyclization (Scheme 5.69) [118]. This simple procedure appears to be very general regarding the substituents in the alkyne and the aromatic system, and has been adapted to solid-supported synthesis. Moreover, a similar strategy has been applied to the preparation of oxindoles.

Iodinating reagents, such as bis(pyridine)iodonium(I) tetrafluoroborate (IPy_2BF_4) [119], can also promote the cyclization of *o*-alkynylanilines **167** (Scheme 5.70). Interestingly, the reaction affords 3-iodoindoles **168**, offering the opportunity of further transformation by substitution of the iodine substituent by well known cross-coupling protocols. Moreover, the reaction has also been carried out on a solid support.

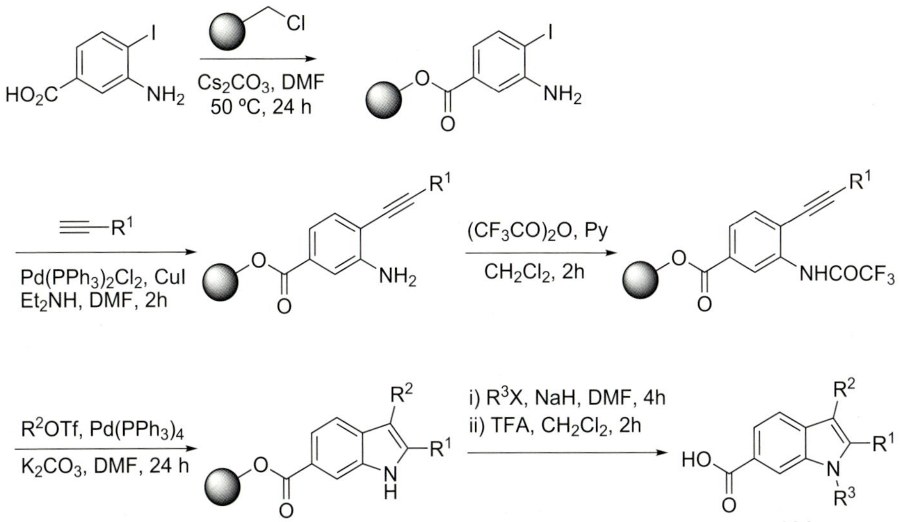

Scheme 5.67

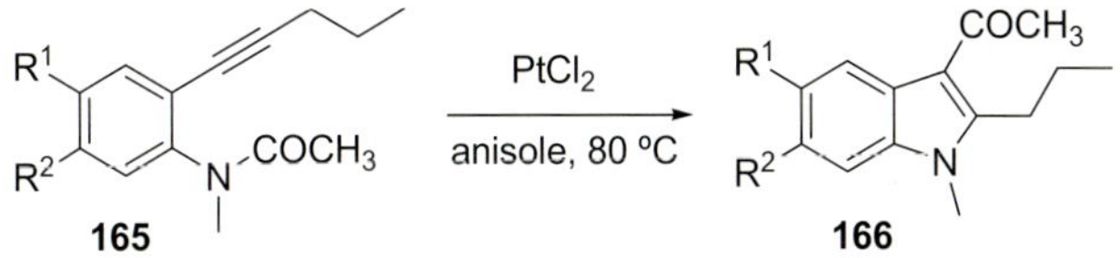

R^1 = OMe, R^2 = H: 89%
R^1 = H, R^2 = OMe: 99%

Scheme 5.68

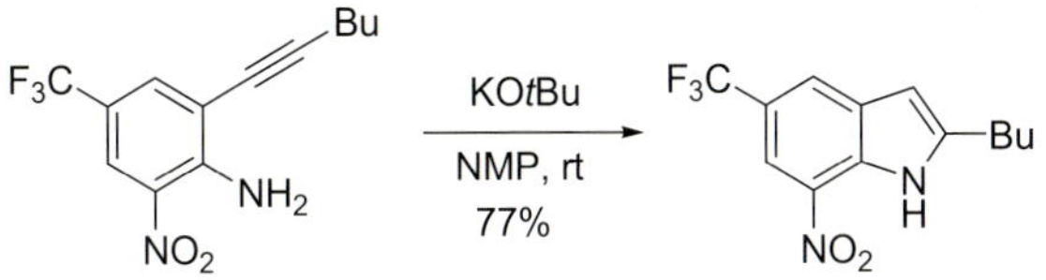

Scheme 5.69

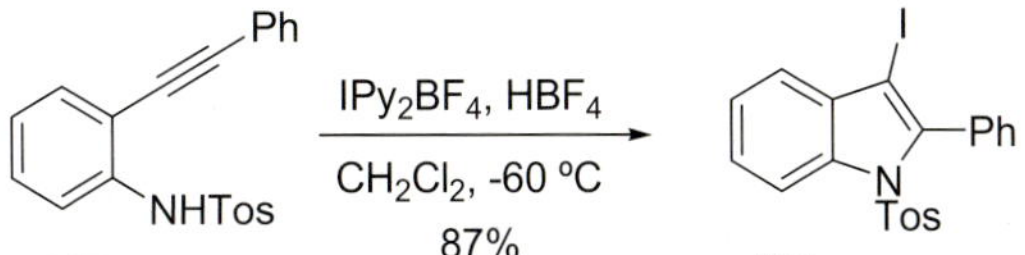

Scheme 5.70

5.4.2.2.9 **Synthesis from *o*-Alkynyl isocyanides or *o*-Alkynyl Isocyanates** *N*-Cyanoindoles **171** are prepared by a very elegant Pd-catalyzed three-component coupling reaction of *o*-alkynylphenylisocyanide **169**, allyl carbonate **170** and trimethylsilyl azide (Scheme 5.71) [120].

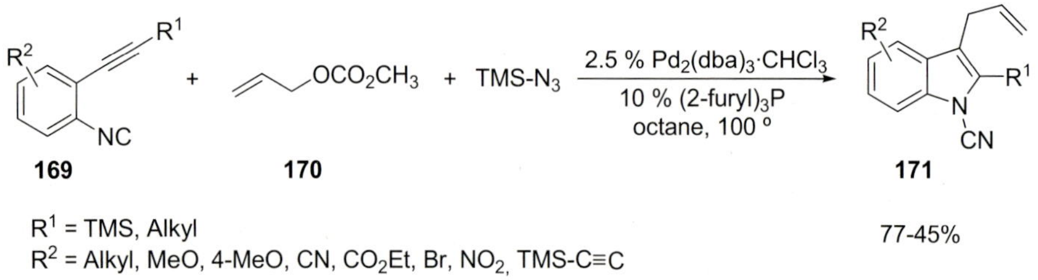

R^1 = TMS, Alkyl
R^2 = Alkyl, MeO, 4-MeO, CN, CO_2Et, Br, NO_2, TMS-C≡C

2.5 % $Pd_2(dba)_3 \cdot CHCl_3$, 10 % $(2\text{-furyl})_3P$, octane, 100 °; 77-45%

Scheme 5.71

The catalytic cycle proposed for this remarkable cascade process (Scheme 5.72) starts with the reaction of the Pd(0) species with allyl carbonate **170** and $TMSN_3$ to give the π-allylpalladium azide complex **172**. Insertion of the isocyanide in the Pd–N_3 bond would give a new π-allylpalladium complex (**173**). Elimination of N_2, with 1,2-migration of the π-allylpalladium moiety from C to N, would provide π-allylpalladium carbodiimide complex **174**. This rearrangement can be envisioned as a π-allylpalladium

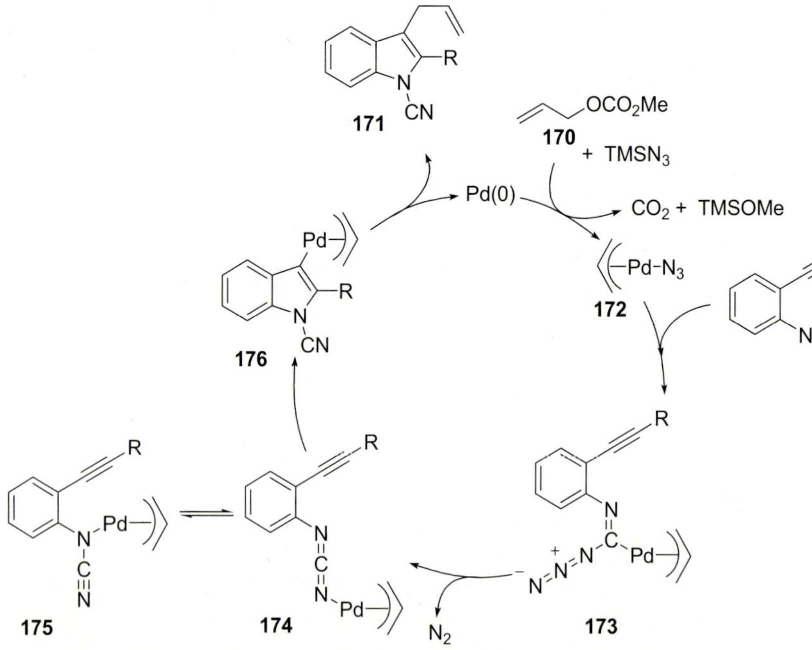

Scheme 5.72

mimic of the Curtius rearrangement. Complex **174** can be in equilibrium with the π-allylpalladium cyanamide complex **175**. Insertion of the alkyne in the Pd–N bond of π-allylpalladium **175** would provide the 2-indolyl-π-allylpalladium complex **176**, which upon reductive elimination yields the cyanoindole **171** and the Pd(0) catalyst.

A related reaction can be applied to prepare *N*-methoxycarbonylindoles **178** from *o*-alkynylisocyanates **177** and allyl carbonate **170**. In this case a Pd^0-Cu^I bimetallic catalyst is required (Scheme 5.73) [121].

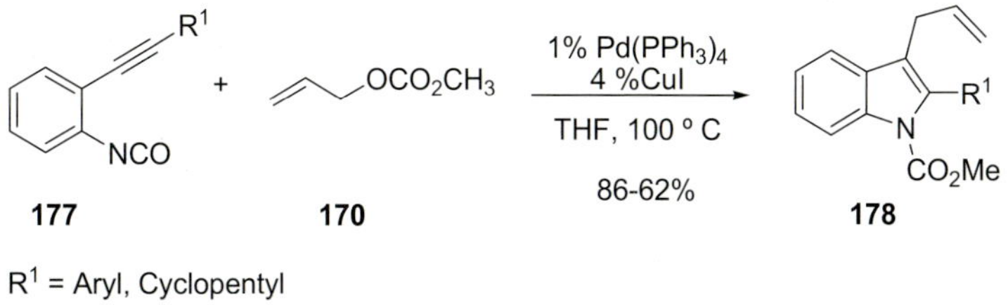

R1; NCO; +; OCO_2CH_3; 1% $Pd(PPh_3)_4$; 4 %CuI; THF, 100 °C; 86-62%; R1; N; CO_2Me

177 **170** **178**

R^1 = Aryl, Cyclopentyl

Scheme 5.73

5.4.2.2.10 **Synthesis from *o*-Haloanilines and Acetylenes: The Larock Indole Synthesis**

The Pd-catalyzed synthesis of indoles from *o*-iodoanilines **179** and internal alkynes **180** is known as the Larock indole synthesis, and stands as one of the more powerful methods for the preparation of 2,3-disubstituted indoles **181** (Scheme 5.74) [122].

When unsymmetrical alkynes are used, the annulation reaction is regioselective, with the bulkiest substituent being placed at C2 in the indole. In particular, silylated alkynes afford exclusively the C2 silylated indole.

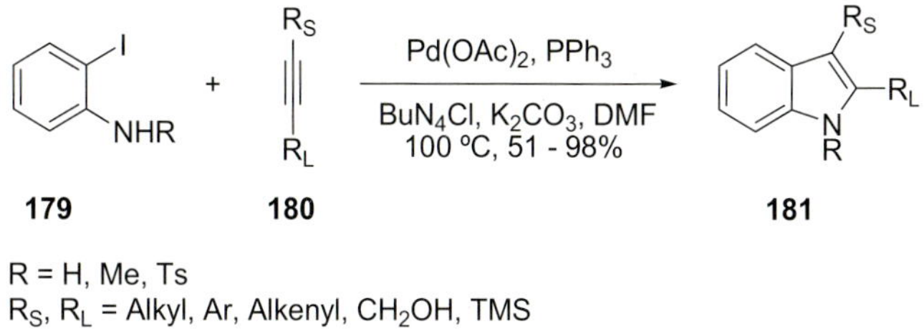

I; NHR; +; R_S; R_L; $Pd(OAc)_2$, PPh_3; BuN_4Cl, K_2CO_3, DMF; 100 °C, 51 - 98%; R_S; R_L; N; R

179 **180** **181**

R = H, Me, Ts

R_S, R_L = Alkyl, Ar, Alkenyl, CH_2OH, TMS

Scheme 5.74

It is postulated that the mechanism of the annulation involves (Scheme 5.75): (i) oxidative addition of the aryl iodide **179** to the Pd(0) species to form arylpalladium complex **182**; (ii) coordination of the alkyne **180** to the arylpalladium complex to form complex **183**; (iii) insertion of the alkyne in the Pd–C_{Ar} bond, and coordination of the nitrogen with ligand displacement to generate palladacycle intermediate **184**; and (iv) reductive elimination to form the indole and regenerate the Pd(0). The regioselectivity of the reaction is determined by the insertion of the alkyne in the Pd–C bond.

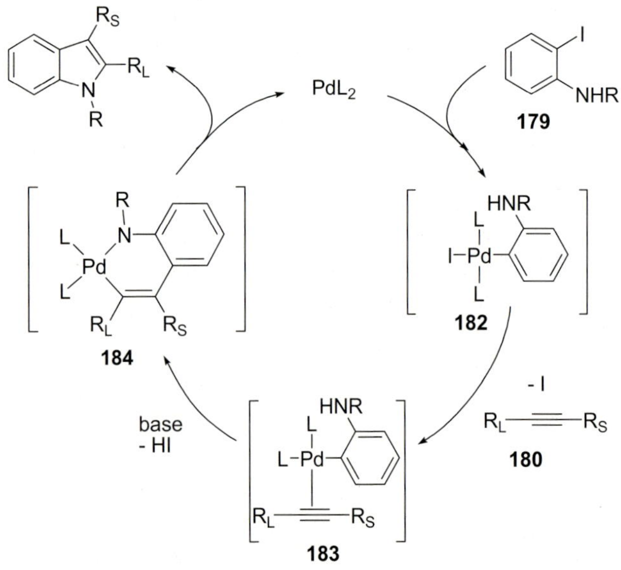

Scheme 5.75

The bulkier substituent is situated next to the Pd, to avoid steric interactions with the aromatic ring.

Larock's heteroannulation has been applied as the key step in the preparation of several indole alkaloids and heterocycles. For instance, 7-methoxytryptophan **185**, an early intermediate in the preparation of some indole alkaloids, has been efficiently synthesized with this methodology (Scheme 5.76) [123]. Moreover, recently, the scope

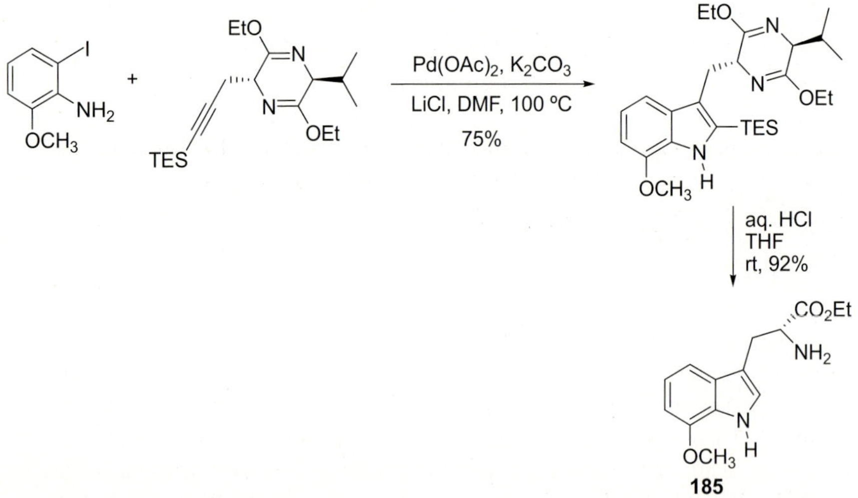

Scheme 5.76

of the reaction has been extended to *o*-bromo- and *o*-chloroanilines, by using an appropriate supporting ligand for the Pd, which enhances its reactivity towards oxidative addition [124].

A different Pd-catalyzed tandem process, which shares some mechanistic features with Larock's indolization, has led to indoles from *o*-(2,2-dibromovinyl)anilines **186** [125]. In this case, the tandem process takes advantage of the different reactivity of the two C–Br bonds towards oxidative addition. Thus, in a first step, Suzuki coupling with substitution of the more reactive trans-bromine atom gives the bromovinyl aniline **187**, which undergoes Pd-catalyzed intramolecular C–N bond formation to provide the indole **188** with high yields (Scheme 5.77).

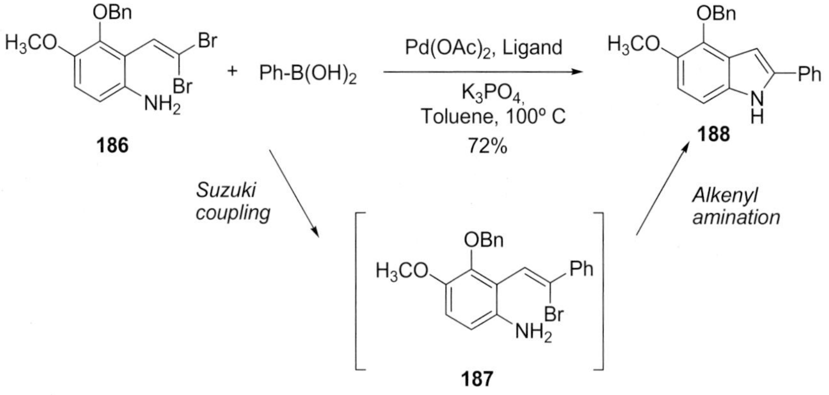
OBn
H3CO
Br
Br
NH2
186
+ Ph-B(OH)2
Pd(OAc)2, Ligand
K3PO4,
Toluene, 100° C
72%
OBn
H3CO
Ph
N
H
188
Suzuki
coupling
Alkenyl
amination
OBn
H3CO
Ph
Br
NH2
187

Scheme 5.77

5.4.2.3 Ring Synthesis by Formation of the C3–C3a Bond

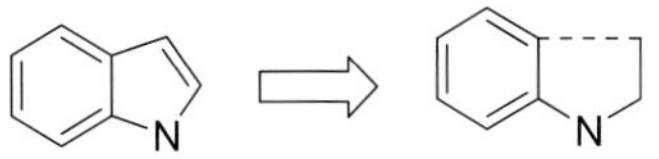
N
N

5.4.2.3.1 Electrophilic Cyclizations of the Aromatic Ring: The Bischler Synthesis This section includes those indole syntheses in which the key step is a cyclization by electrophilic attack of the aromatic ring to a nucleophilic center in a 5-exo-trig type of cyclization (Scheme 5.78).

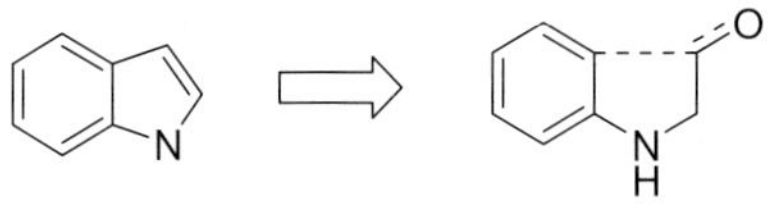
N
O
N
H

Scheme 5.78

Cyclization of α-(*N*-arylaminoketones), known as the Bischler synthesis, and discovered over 100 years ago [126], is the most characteristic representative of this approach. In the original protocol, arylaminoketone **190** is obtained from *N*-methylaniline and α-haloketones **189**. Acid-catalyzed 5-exo-trig cyclization then leads to indole **191** (Scheme 5.79).

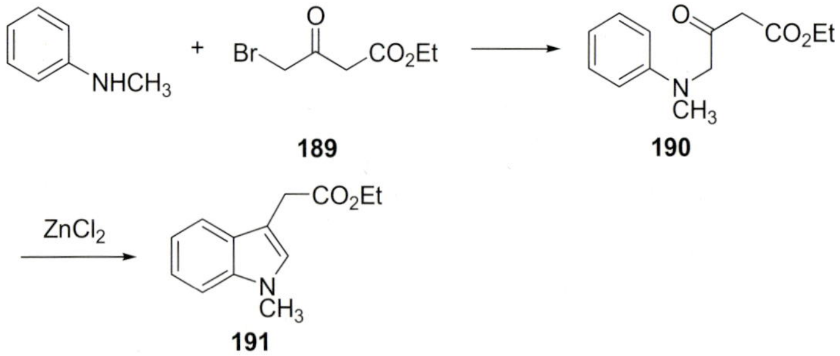

Scheme 5.79

Notable among the most relevant modifications of this procedure are the use of an acetal as a masked aldehyde functionality and the acylation or sulfonation [127] of the nitrogen, which allows a much more controlled cyclization. The Nordlander synthesis [128] exemplifies such variations (Scheme 5.80).

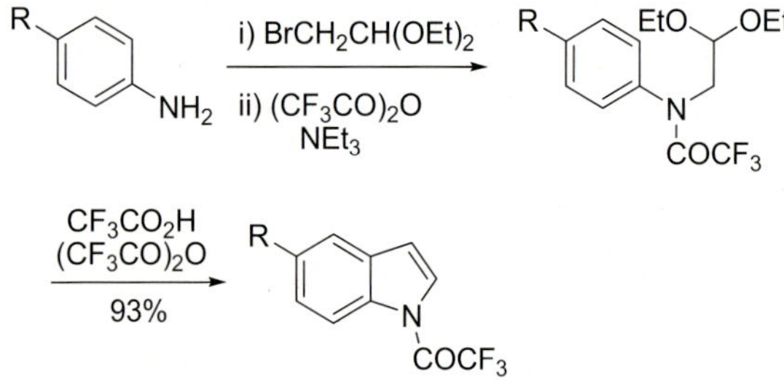

Scheme 5.80

In another modification of the Bischler synthesis, the intermediate α-(*N*-arylaminoketones) **194** are prepared by the N−H insertion reaction of anilines **192** with rhodium carbenoids [129], generated from diazocarbonyl compounds **193** [130]. The final cyclization is carried out using the acidic ion-exchange resin Amberlyst 15, which provides better results than the usual Lewis acid catalysts (Scheme 5.81). A further refinement of the reaction by the same authors allows for the preparation of N−H indoles [131].

Hydroamination of propargylic alcohols with anilines represents another transition metal catalyzed alternative to the classic Bischler synthesis [132]. In a *one pot* process, the hydroxyimine **196**, which comes from the ruthenium-catalyzed hydroamination of the propargylic alcohol **195**, undergoes hydrogen migration to the Bischler intermediate **197**, followed by cyclization to provide a mixture of regioisomeric 2,3-substituted indoles (Scheme 5.82).

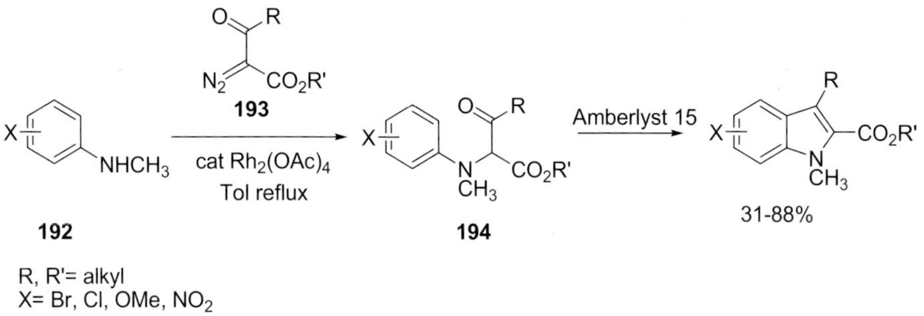

Scheme 5.81

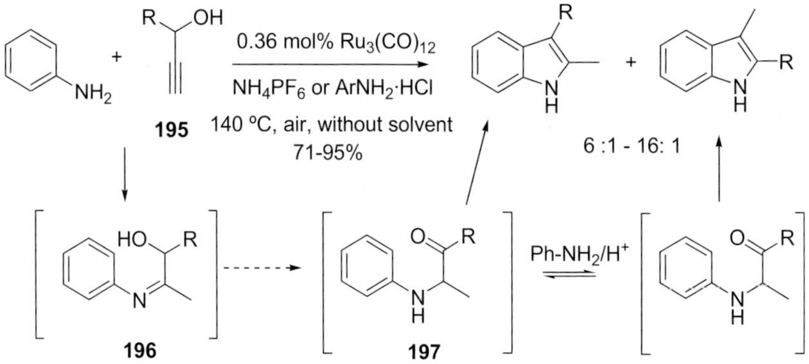

Scheme 5.82

5.4.2.3.2 **Carbometallation of *N*-(2-Lithioallyl)anilines** The intramolecular carbometallation of lithiated double bonds has been used to prepare several types of functionalized indoles [133]. For instance, *N*-bromoallyl-*o*-fluoroanilines **198** are converted into indoles by treatment with 3 equivalents of *t*BuLi. Cyclization of the benzyne intermediate **199** generated gives rise to C(4)-lithiated 3-methyleneindoline derivative **200**. Quenching of the lithiated species **200** with selected electrophiles allows functionalization at this position to provide methyleneindoline **201**, which isomerizes on workup or on silica gel chromatography to the corresponding indole derivatives **202** (Scheme 5.83) [134].

Interestingly, 3-methyleneindolines such as **201** are known to participate in Alder-ene reactions with activated enophiles to furnish 3,4-disubstituted indoles **203** (Scheme 5.84) [135].

Taking advantage of this reaction, addition of an enophile to the reaction mixture provides 3,4-disubstituted indoles in a *one-pot* sequence. Schemes 5.85 and 5.86 show, as representative examples, the one-pot synthesis of a tryptamine analog **204** and a substituted carbazole **205**, respectively.

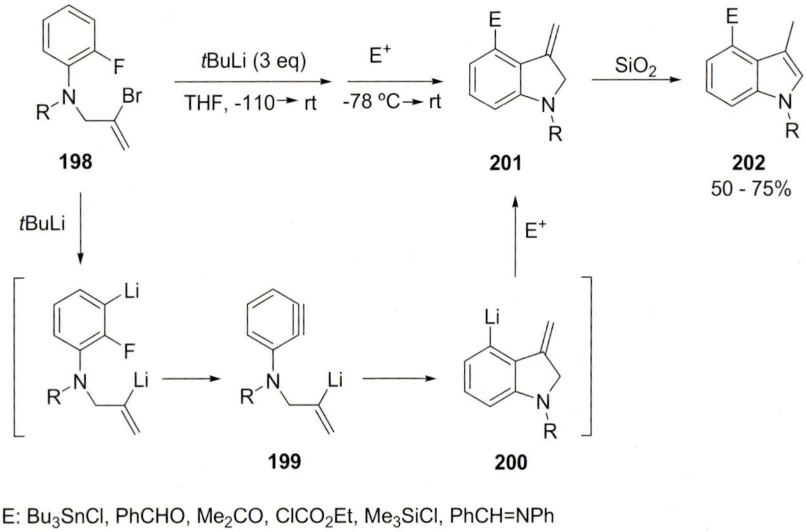

F R N Br **198** → tBuLi (3 eq), THF, -110 → rt; E^+, -78 °C → rt → E N R **201** → SiO_2 → E N R **202** 50 - 75%

tBuLi ↓

[Li F R N Li → R N Li **199** → Li N R **200**] → E^+ → **201**

E: Bu_3SnCl, PhCHO, Me_2CO, $ClCO_2Et$, Me_3SiCl, PhCH=NPh

Scheme 5.83

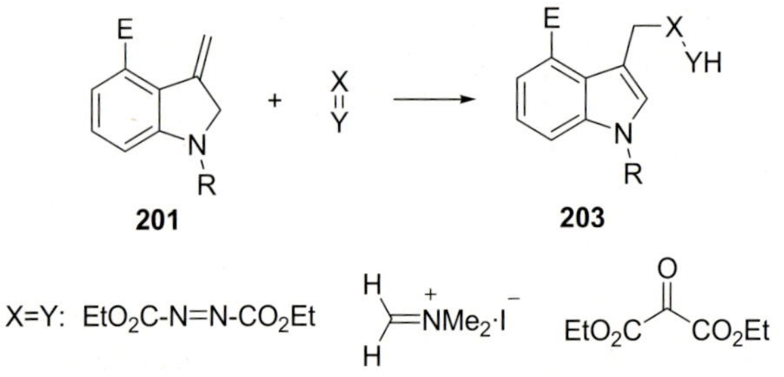

E N R **201** + X=Y → E X YH N R **203**

X=Y: EtO_2C-N=N-CO_2Et H_2C=$\overset{+}{N}Me_2$·I^- EtO_2C-CO-CO_2Et

Scheme 5.84

5.4.2.3.3 **Palladium-Catalyzed Heck Reactions** The intramolecular Heck reactions of *o*-halo-*N*-allylanilines **206** and *o*-halo-*N*-vinylanilines **207** are very effective methods for the construction of the indole ring (Scheme 5.87).

The Pd-catalyzed cyclization of *N*-allyl-*o*-haloanilines, such as **208** [136–138], is a very reliable method for the preparation of indoles. The intramolecular Heck reaction

H_3CO Cl N Br → i) tBuLi, THF, -110 °C to rt; ii) H_2O, -78 to 20 °C; iii) CH_2=NMe_2·I, 67 °C; 55% → H_3CO NMe_2 N **204**

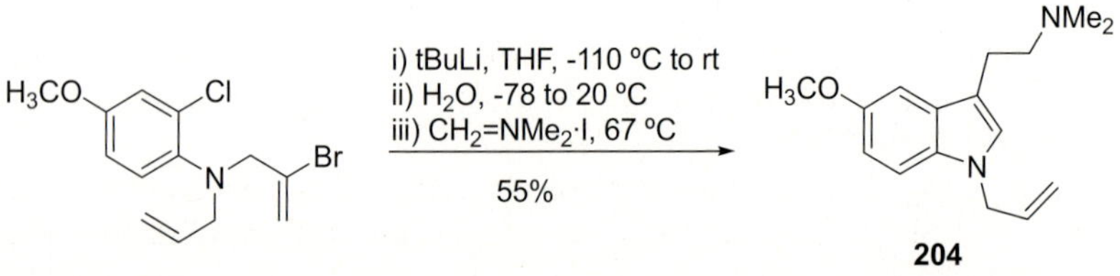

Scheme 5.85

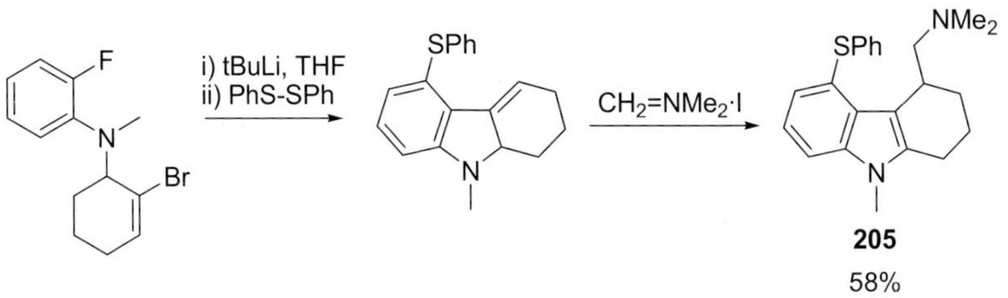

Scheme 5.86

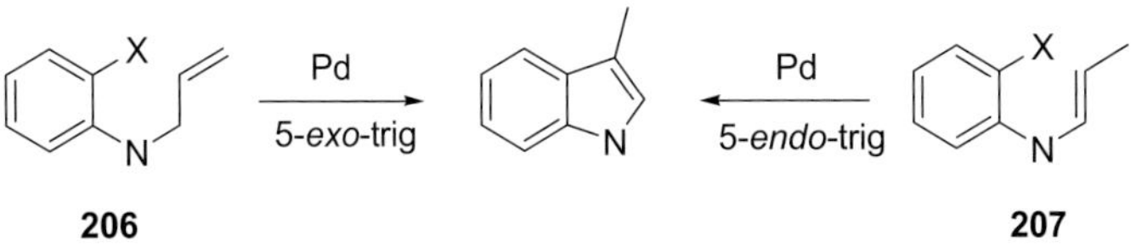

Scheme 5.87

is usually achieved under Pd ligandless conditions and in the presence of a base (Scheme 5.88) [139, 140].

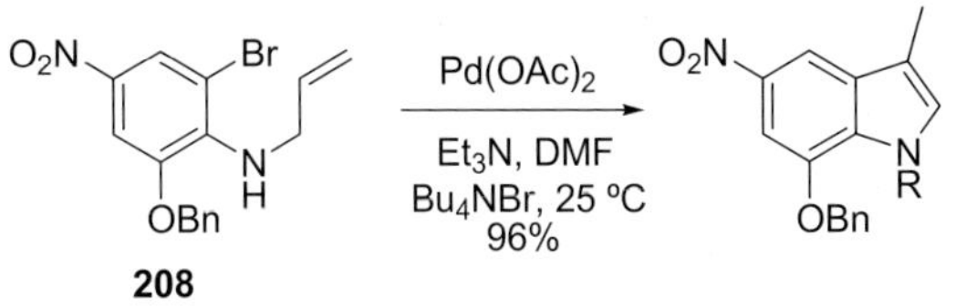

Scheme 5.88

The formation of the indole can be explained by the accepted mechanism of the Heck reaction. Oxidative addition of the aryl halide to form arylpalladium complex **209**, followed by cyclization by olefin insertion, gives rise to the Pd(II) substituted indoline **210**. β-Elimination regenerates the Pd(0) species and produces a 3-methylenindoline **211**, which isomerizes spontaneously to indole **212** (Scheme 5.89).

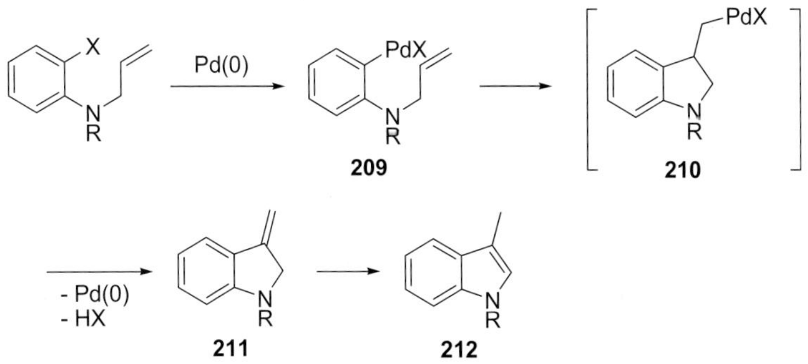

Scheme 5.89

Nevertheless, it is possible to isolate the intermediate indoline under certain reaction conditions, which include the use of silver carbonate as base [141].

Many examples have appeared in the literature of the application of this methodology in the synthesis of complex molecules [142], including adaptation to solid-phase synthesis [143]. Scheme 5.90 shows a solid-phase synthesis of trisubstituted indoles **213** that allows for the introduction of several points of diversity in the indole scaffold [144].

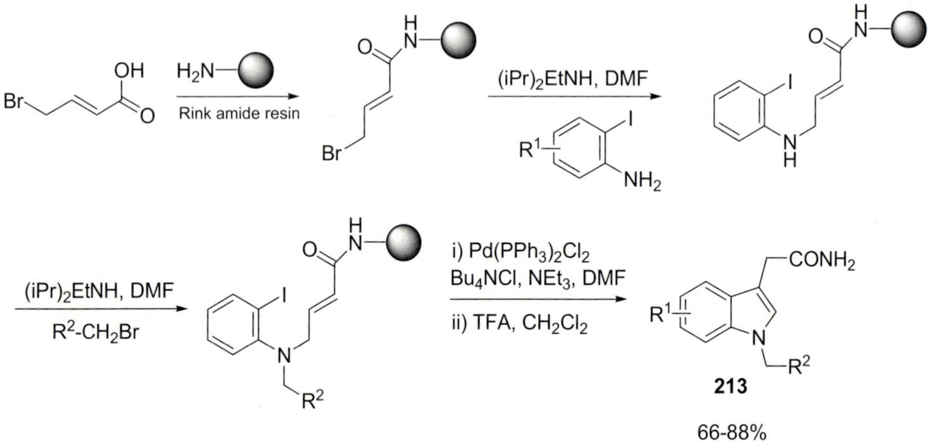

Scheme 5.90

The Heck reaction of *o*-haloenamines has attracted comparatively less attention than the previously discussed reaction with *N*-allylanilines. The instability of the enamines when compared with the allylamines, as well as the more difficult 5-endo-trig cyclization when compared with the 5-exo-trig, may account for this difference. The first examples of this approach were carried out with acylenamines **214**, easily prepared through different procedures, which undergo Pd-catalyzed cyclization to provide 2,3-disubstituted indoles **215** (Scheme 5.91) [145].

Direct annulation of *o*-iodoanilines with ketones is a very convenient variation of this route [146]. The reaction has been extended recently to the more easily available chloroanilines **216** [147]. In a first step, condensation of the amine with the ketone forms the enamine **217**, which then undergoes Pd-catalyzed cyclization (Scheme 5.92). The use of a very active catalytic system is crucial.

Enamine **217** required for the cyclization has also been prepared by acetylene hydroamination. Thus, the indole ring has been formed from acetylenes **219** and *o*-chloroaniline (**218**), in a *one pot* process that requires two different metal catalysts: a Ti catalyst for the hydroamination, and a Pd catalyst for the Heck reaction (Scheme 5.93) [148].

In a related process, indoles have been synthesized from *o*-haloanilines **220** and alkenyl bromides **221**. Notably, in this process the same Pd catalyst promotes two

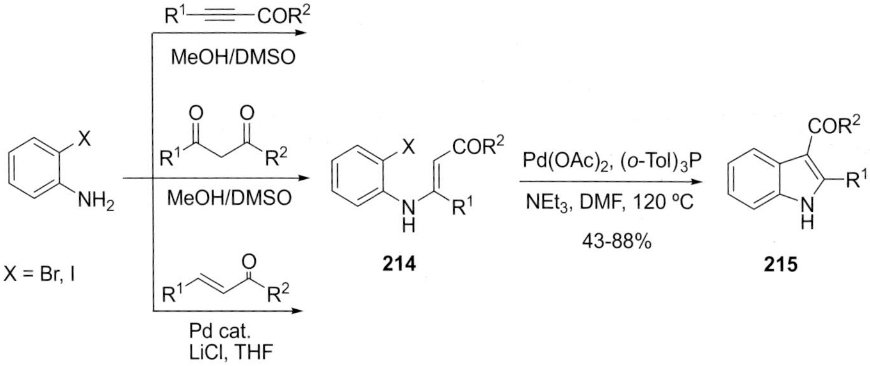

Scheme 5.91

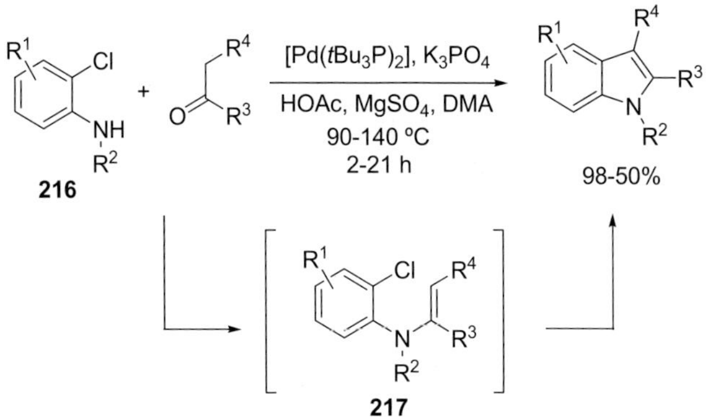

Scheme 5.92

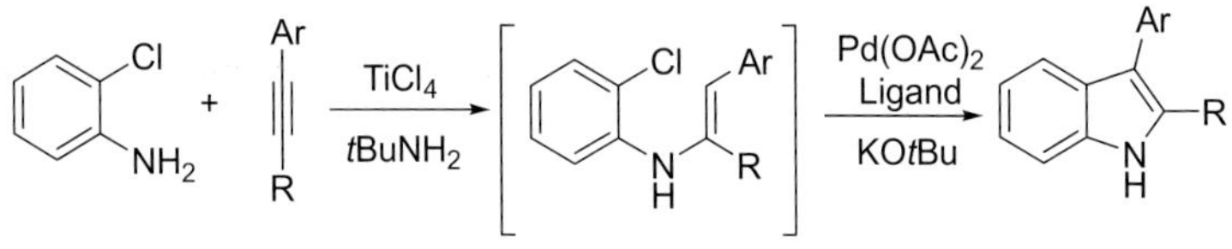

Scheme 5.93

different reactions: the alkenyl amination, which forms the intermediate enamine 222, and the subsequent Heck reaction to form the indole (Scheme 5.94) [149].

5.4.2.4 Ring Synthesis by Formation of the C2–C3 Bond

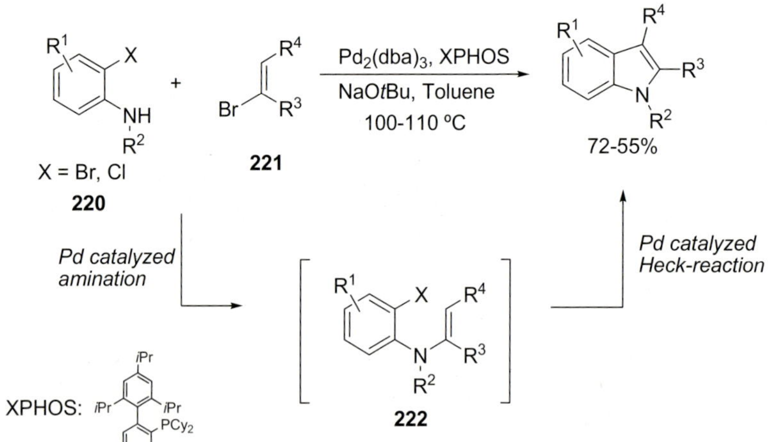

Scheme 5.94

5.4.2.4.1 **Madelung Synthesis** The construction of the indole ring by cyclocondensation of 2-methylanilides (**223**) is known as the Madelung indole synthesis (Scheme 5.95). In its original formulation the reaction used bases such as $NaNH_2$ and KO*t*Bu, and required extremely harsh reaction conditions with temperatures over 250 °C (check out the procedure published in *Organic Synthesis Collective*) [150]. The replacement of these bases by alkyllithiums or LDA allows the reaction to take place under much milder reaction conditions (Scheme 5.96) [151, 152].

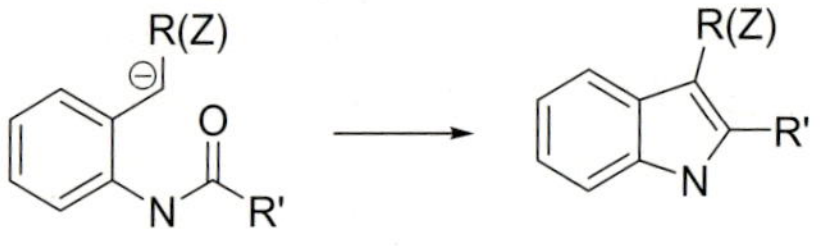

Scheme 5.95

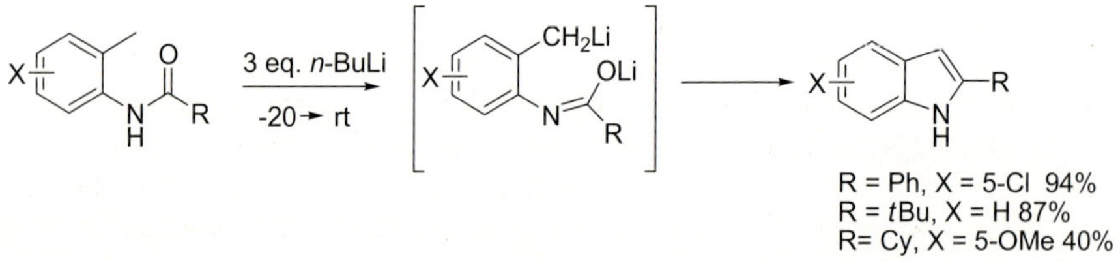

Scheme 5.96

Introduction of electron-withdrawing substituents at the ortho-methyl group makes the benzylic proton of derivatives **224** more acidic and thus facilitates the overall process [153]. This approach is illustrated by the solid-phase synthesis of a library of 2,3-disubstutited-indoles **225** (Scheme 5.97) [154].

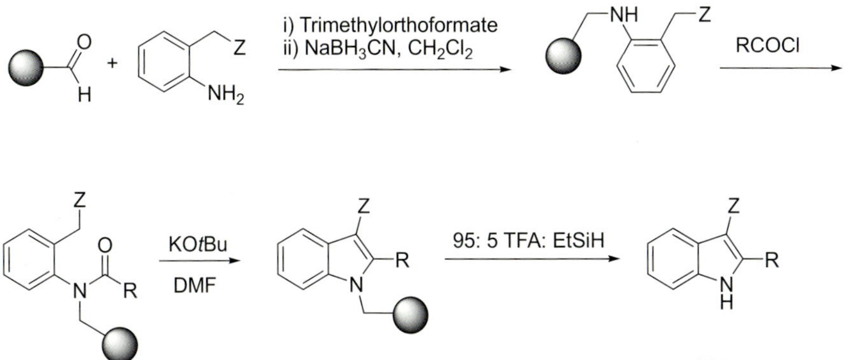

Z = CN, CO_2tBu, $CONH_2$

Scheme 5.97

In another variation of the Madelung synthesis, phosphonium salts **227** led to the corresponding indoles through a Wittig-like intramolecular reaction in the presence of base [155] or under thermal conditions [156]. The required phosphonium salts can be easily obtained from *o*-nitrobenzyl bromide (**226**) in a sequence that includes substitution with triphenylphosphine, reduction of the nitro group and acylation [157]. This procedure has been applied to the preparation of 2-alkyl, 2-alkenyl and 2-arylindoles **228** (Scheme 5.98). A solid-phase version of this procedure using a polymer-bound triphenylphosphine has also been developed [158].

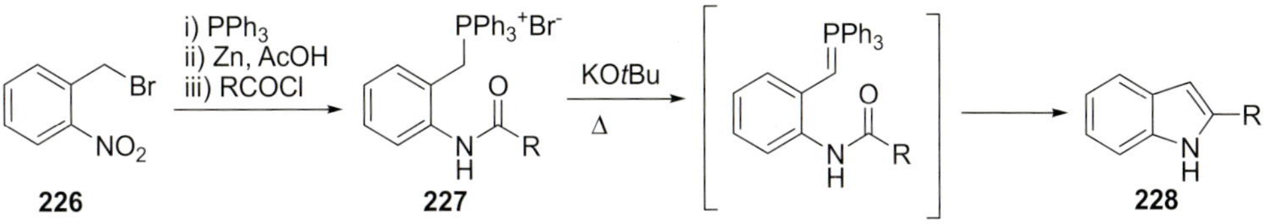

Scheme 5.98

2,3-Disubstituted indoles **231** have been prepared by a very original strategy from *o*-aminoketones **229** through an intramolecular Horner–Wadsworth–Emmons reaction. The key step in this synthesis is the generation of the intermediate phosphonate **230**, required for the olefination reaction, which was prepared by insertion in a N−H bond of an *in situ* generated Rh carbenoid (Scheme 5.99). The whole sequence can be conducted in a one-pot process [159].

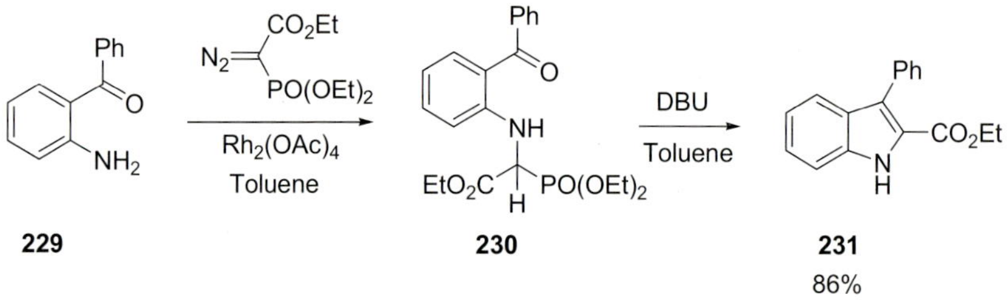

Scheme 5.99

5.4.2.4.2 **Fürstner Synthesis** The intramolecular reductive cyclization of oxo-amides **232** promoted by low-valent titanium (an intramolecular McMurry coupling) [160] is known as the Fürstner indole synthesis [161] (Scheme 5.100).

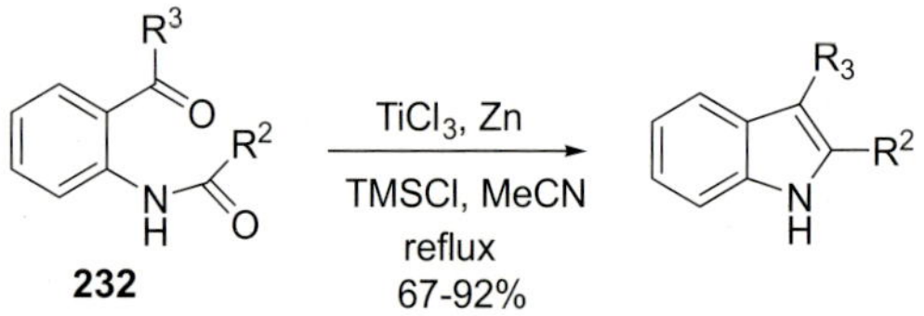

Scheme 5.100

This coupling reaction is a very powerful method for the preparation of 2,3-disubstituted indoles – in particular of 2-arylindoles – and has been applied in the total synthesis of several indole alkaloids and biologically active molecules [162]. Scheme 5.101 shows a particular example of this methodology, applied to the preparation of an endothelin receptor antagonist [163]. The oxo-amide **235** required for the cyclization reaction is easily prepared from *o*-nitrobenzaldehyde (**233**). Addition of an aryl Grignard followed by oxidation of the resulting alcohol installs the ketone functionality to obtain **234**. Then, catalytic hydrogenation of the nitro group followed by acylation provides the amido functionality of **235**, which is transformed into indole **236** under the standard low-valent-titanium reduction.

5.4.2.4.3 **Radical Cyclizations** A wide range of 2,3-disubstituted indoles can be prepared by the tin-promoted radical cyclizations of 2-alkenylphenylisonitriles [164] and 2-alkenylthioanilides **237** devised by Fukuyama [165] (Scheme 5.102).

The cyclizations are carried out in the presence of tributyltin hydride and a radical initiator. In a first step, the tin radical adds to the thioamide **237** to form a sp^3 radical **238** or an imidoyl radical **239**, which cyclizes to yield the indole after a tautomeric equilibrium (Scheme 5.103).

The tin-promoted cyclizations are compatible with a wide range of functional groups. Moreover, the required 2-alkenylthioamides are easily available through various procedures. All this factors make this approach a very useful method for the preparation of 2,3-disubstituted indoles. For instance, this reaction has been applied by Fukuyama to build both indole rings present in the alkaloid (+)-vinblastine, a

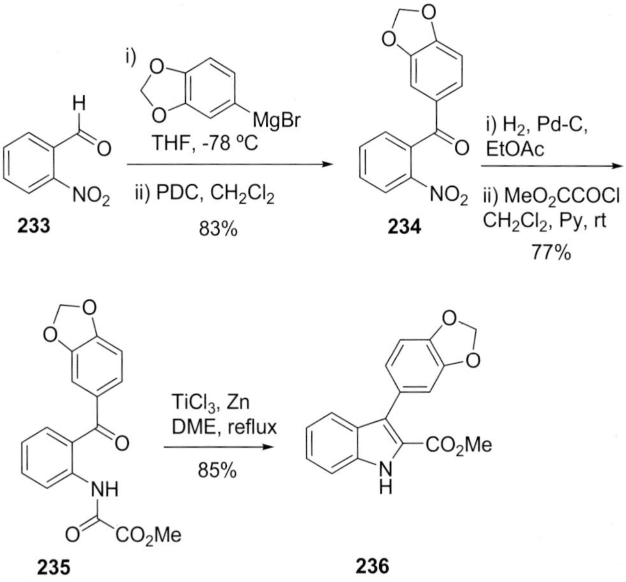

i) (3,4-methylenedioxyphenyl)MgBr, THF, -78 °C; ii) PDC, CH_2Cl_2, 83%

233 → 234

i) H_2, Pd-C, EtOAc; ii) MeO_2CCOCl, CH_2Cl_2, Py, rt, 77%

$TiCl_3$, Zn, DME, reflux, 85%

235 → 236

Scheme 5.101

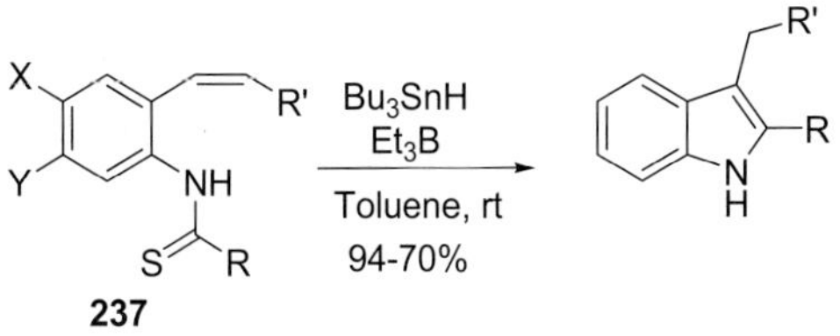

Bu_3SnH, Et_3B, Toluene, rt, 94-70%

237

Scheme 5.102

237 → $\cdot SnBu_3$ → 238 → - $HSSnBu_3$ → 239

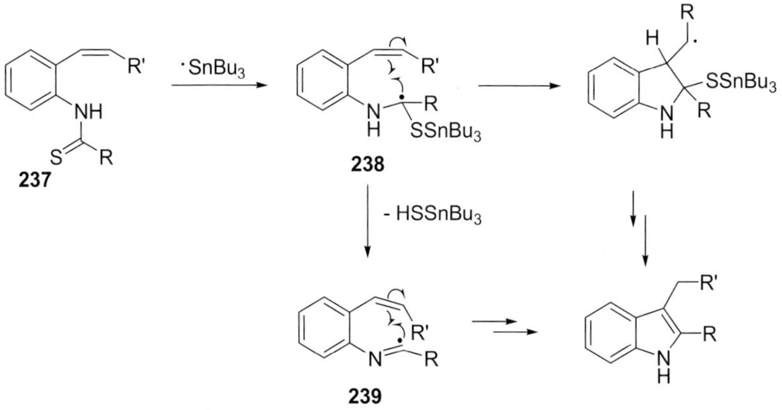

Scheme 5.103

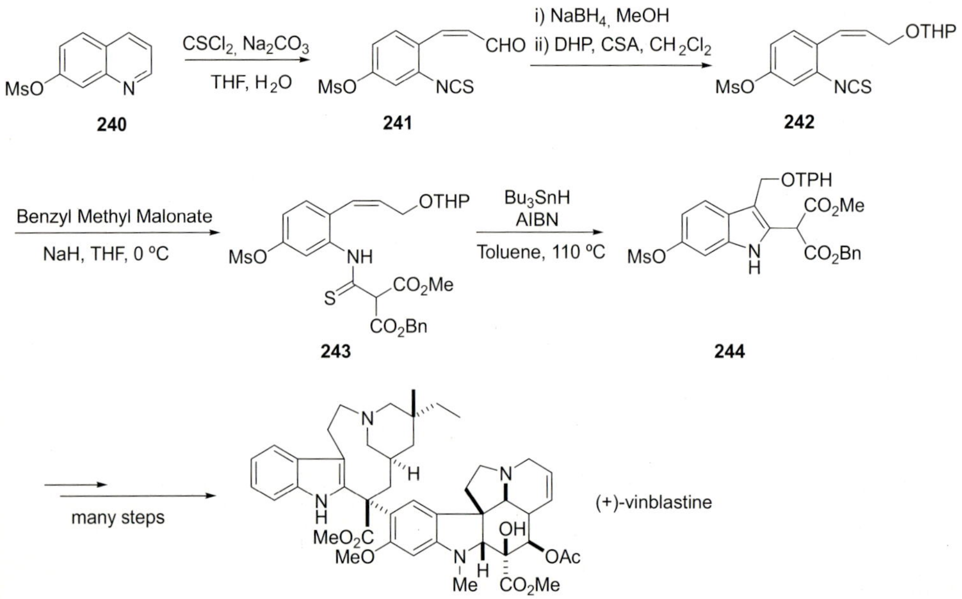

Scheme 5.104

potent agent for cancer therapy (Scheme 5.104) [166]. In this example, the 2-alkenylthioamide is prepared in several steps from quinoline **240**. Ring opening of quinoline **240** promoted by thiophosgene produces **241**. Reduction the of the aldehyde functionality, followed by protection of the hydroxyl group with dihydropyran, leads to the isothiocyanide **242**, which is transformed into the thioamide **243** by addition of a nucleophile. Finally, radical cyclization under standard conditions gives rise to the highly functionalized indole **244**.

5.4.2.4.4 **Palladium-Catalyzed Cyclizations** The Pd-catalyzed cyclization of 2-(1-alkynyl)-*N*-alkylideneanilines **245** gives rise to 3-alkenylindoles **246** [167] (Scheme 5.105).

According to the authors' proposal, the reaction most probably proceeds through the regioselective insertion of a Pd hydride in the triple bond of **245**, to form vinylpalladium intermediate **247**. From this intermediate, both an "oxidative

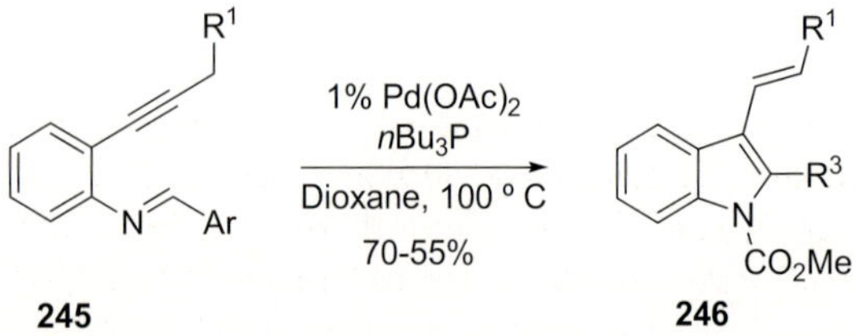

Scheme 5.105

addition/reductive elimination" sequence (path a) and carbopalladation of the imine followed by β-elimination (path b) can explain the formation of the indole (Scheme 5.106).

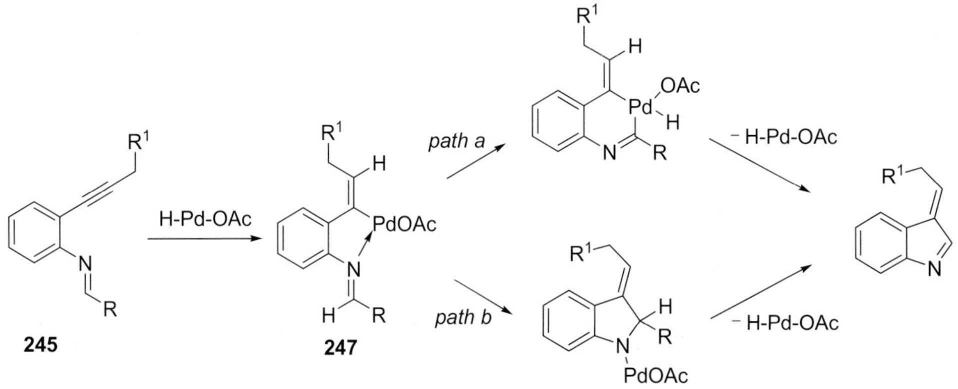

Scheme 5.106

5.4.2.5 **Cyclizations with Formation of the N–C7a Bond**

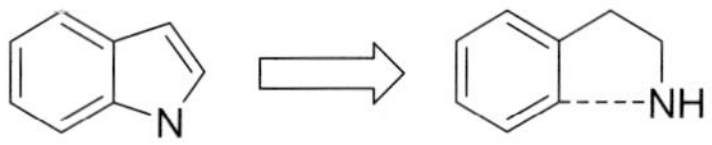

5.4.2.5.1 **Nenitzescu Indole Synthesis** The preparation of 5-hydroxyindole **250** derivatives from 1,4-benzoquinone (**248**) and a β-enaminoester **249** is known as the Nenitzescu reaction and was first reported in 1929 (Scheme 5.107) [168].

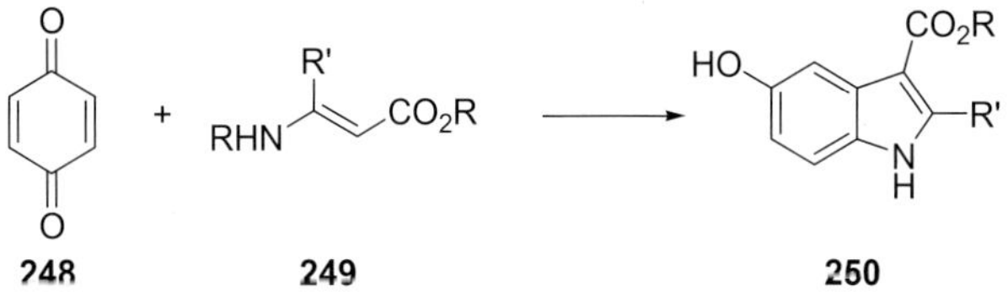

Scheme 5.107

Although the mechanism of the reaction remains somewhat obscure, it is presumed that it involves an internal oxidation–reduction process within the following steps: Michael-type addition of the enamine **249** to the quinone **248**, oxidation of the hydroquinone **251** into a substituted benzoquinone **252**, condensation to form the quinoimmonium cation **253** and reduction to form 5-hydroxyindole **250** (Scheme 5.108).

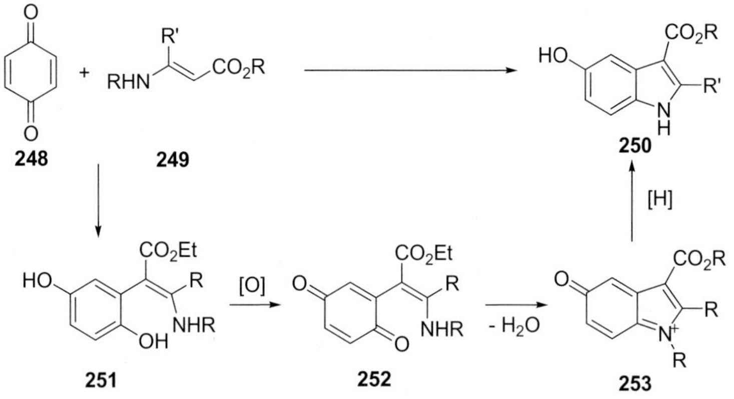

Scheme 5.108

The Nenitzescu synthesis has been widely used for the preparation of 5-hydroxyindole derivatives with additional substituents in both rings [169]. An example of its implementation to solid-phase synthesis is given in Scheme 5.109 [170].

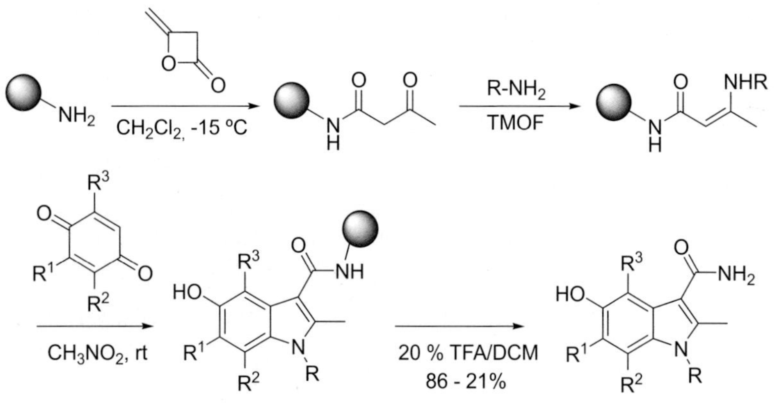

Scheme 5.109

The scope of the Nenitzescu reaction has been recently extended to substrates different than the classical β-enaminocarbonyls. Pyrimidine-2,4,6-triamines react with *p*-benzoquinones to provide hydroxypyrimido[4,5-*b*]indoles with moderate yields [171].

Benzylimines of simple ketones (**254**), which are in tautomeric equilibrium with the corresponding enamines **255**, are also good substrates for the indolization. This new reaction has been applied to simple cyclic imines, and also to more complex systems such as **256** to prepare new hydroxyindolomorphinans **257** with potent biological activity (Scheme 5.110) [172].

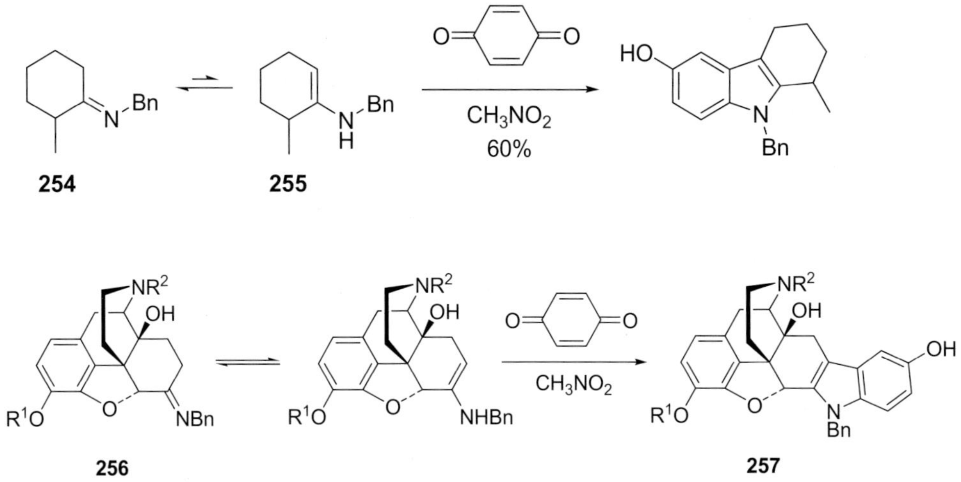

254 ⇌ 255 → (quinone, CH_3NO_2, 60%)

256 ⇌ → 257 (quinone, CH_3NO_2)

Scheme 5.110

5.4.2.5.2 **Synthesis by Nitrene Insertion: The Hemetsberger Synthesis** The indole ring can be built by insertion of a nitrene placed on the side-chain (**260**). In this approach, commonly known as the Hemetsberger indole synthesis, cyclization is achieved by thermolysis of the corresponding azides **259** [173]. Although the reaction may be understood as a nitrene insertion into a C–H bond, an azirine intermediate **261** has been isolated at lower temperatures [174], which provides the indole **262** after a subsequent rearrangement (Scheme 5.111).

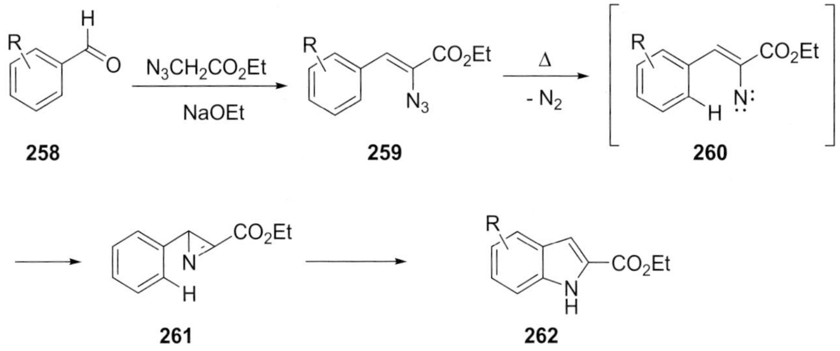

258 → ($N_3CH_2CO_2Et$, NaOEt) 259 → (Δ, $-N_2$) [260] → 261 → 262

Scheme 5.111

The vinylazides are best obtained by condensation of azidoacetate with aryl aldehydes **258**. Thus, this methodology is particularly useful for the preparation of 2-carboxyindoles **262** from aromatic aldehydes [175]. This protocol has been employed successfully in the preparation of polycyclic indole derivatives. For instance, application of the Hemetsberger sequence to the formylbenzoxazine **263** gives oxazinindole **264** in moderate overall yield (Scheme 5.112) [176].

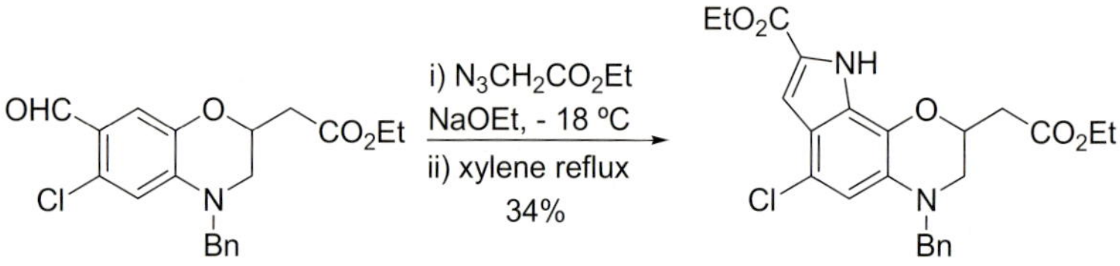

Scheme 5.112

5.4.2.5.3 **Intramolecular Buchwald–Hartwig Amination** The Pd-catalyzed cross-coupling of aryl halides with amines is nowadays one of the most powerful methods for the formation of $C_{aryl}{-}N$ bonds [177]. The intramolecular version of the reaction leads to indolines, which can be converted into indoles by treatment with Pd/C. Scheme 5.113 shows the cyclization of the *in situ* generated 1-(*o*-bromophenyl)-2-ethylamine **265** to furnish the corresponding indole **267** after the oxidation step of the intermediate indoline **266** [178]. Further optimization of the reaction showed that modification of the ligand and the base provides better yields under milder conditions [179]. Moreover, a similar copper-catalyzed cyclization has also been reported [180, 181].

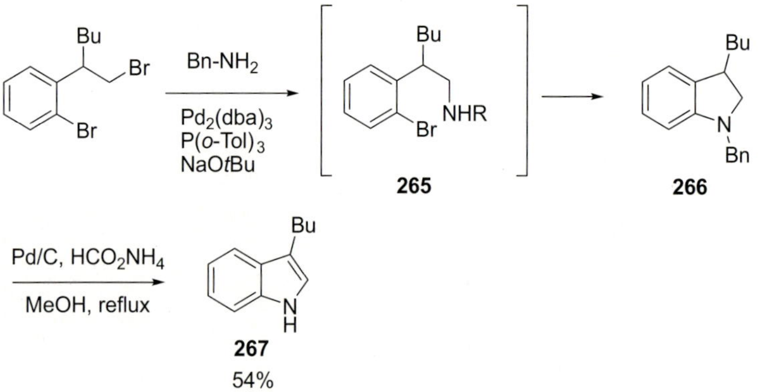

Scheme 5.113

Intramolecular N-arylation of enamines has also been applied in the synthesis of indoles. Enamine **270** (in tautomeric equilibrium with the imine **269**), which is ready for cyclization, is obtained by Ti-catalyzed hydroamination of the *o*-chloro-substituted alkynylbenzene **268**. Then, the Pd-catalyzed C–N bond-forming reaction furnishes *N*,2-disubstituted indoles **271** with good yields in a one-pot process (Scheme 5.114) [182].

Indoles have been prepared by two consecutive Pd-catalyzed amination reactions on the enoltriflate **272** in a closely related strategy that features two reactive positions towards Pd-catalyzed amination (Scheme 5.115) [183]. First, amination of the more reactive enoltriflate occurs, to produce the enamine **273**, which then undergoes intramolecular aryl amination to give indole **274**.

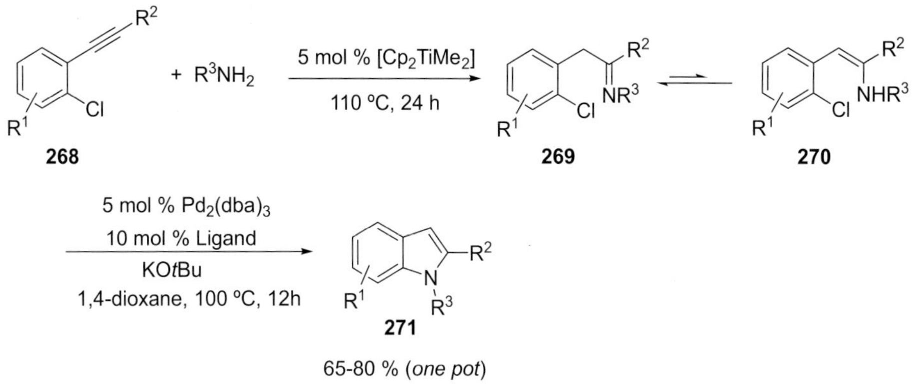

Scheme 5.114

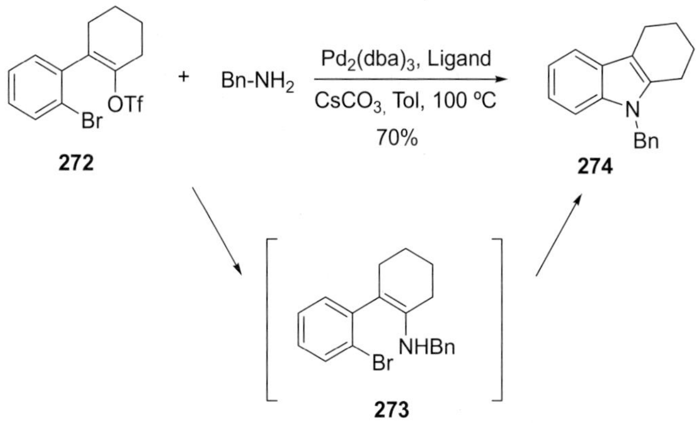

Scheme 5.115

5.4.3
Synthesis of the Indole Ring by Annelation of Pyrroles

Construction of the indole ring by annelation of pyrroles has been much less exploited. Nevertheless, several efficient methods have been disclosed during the last two decades. Most of the existing methods lie in one of the retrosynthetic schemes represented in Scheme 5.116.

5.4.3.1 Synthesis by Electrophilic Cyclization

The most popular type of methods for synthesis of the indole ring from pyrroles involves an electrophilic cyclization, with formation of either the C7–C7a or C3a–C4 bond (Scheme 5.117). In both cases, annelation takes place by an intramolecular electrophilic attack of the pyrrole to a carbonyl function. To facilitate aromatization, a leaving group is usually present in the carbon chain. The different methods described in the literature differ in the way to synthesize the intermediate carbonyl pyrrole ready for the cyclization.

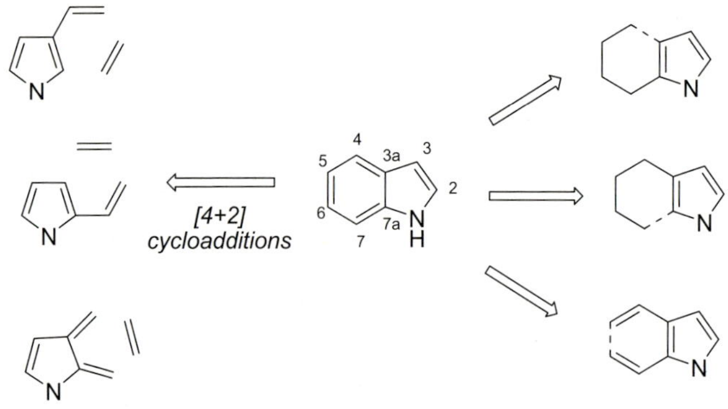

N
N
N
4
3a
3
5
2
6
7a
N
H
7
[4+2]
cycloadditions
N
N
N

Scheme 5.116

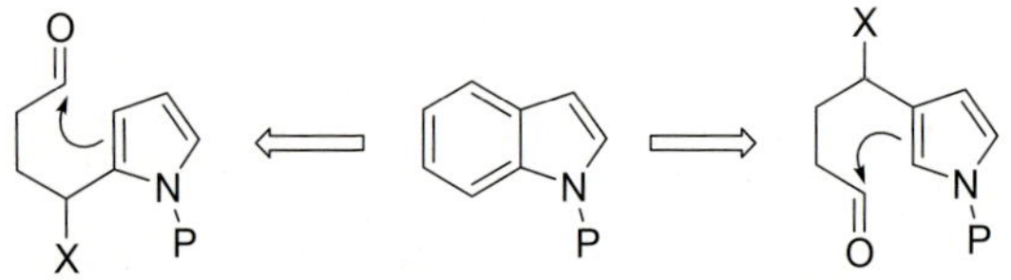

O
N
X P
N
P
X
N
O P

Scheme 5.117

5.4.3.1.1 **Natsume Synthesis** In the Natsume approach, the intermediate for cyclization is prepared by addition of a organometallic bearing a masked carbonyl functionality (**276**) to a pyrrolyl ketone (**275**). Acid-catalyzed cyclization on the resulting functionalized pyrrole **277**, with concomitant aromatization, provides indole **278**. This method is a very powerful strategy for the preparation of alkyl-substituted indoles in the benzene portion. Scheme 5.118 shows the preparation of

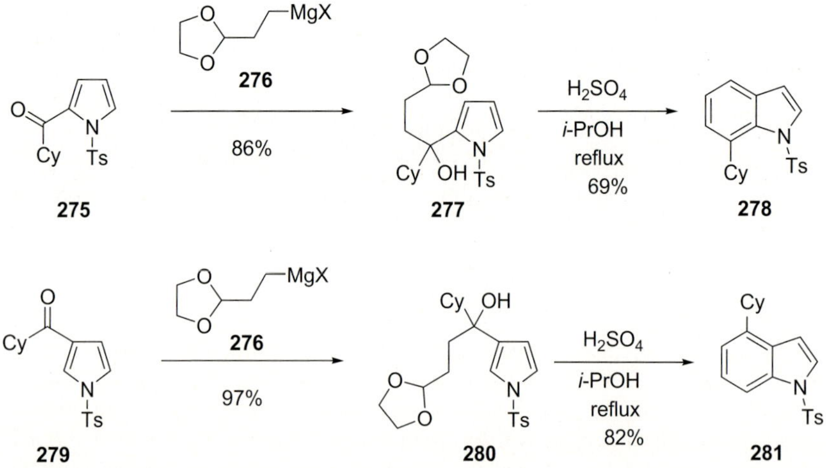

O
O
MgX
276
O
N
Cy Ts
86%
O
O
N
Cy OH Ts
H_2SO_4
i-PrOH
reflux
69%
N
Cy Ts
275 **277** **278**

O
O
MgX
O
Cy
276
N
Ts
97%
Cy OH
O
O
N
Ts
H_2SO_4
i-PrOH
reflux
82%
Cy
N
Ts
279 **280** **281**

Scheme 5.118

both a 7-substituted (**278**) and a 4-substituted indole (**281**) by application of the same methodology but starting from 2-acyl (**275**) and 3-acyl (**279**) substituted pyrrole, respectively [184].

Modifications of this strategy have been applied to the preparation of several natural indole alkaloids, characterized by complex alkyl-substitution in the benzene ring, such as herbindoles and trikentrins (Scheme 5.119) [185].

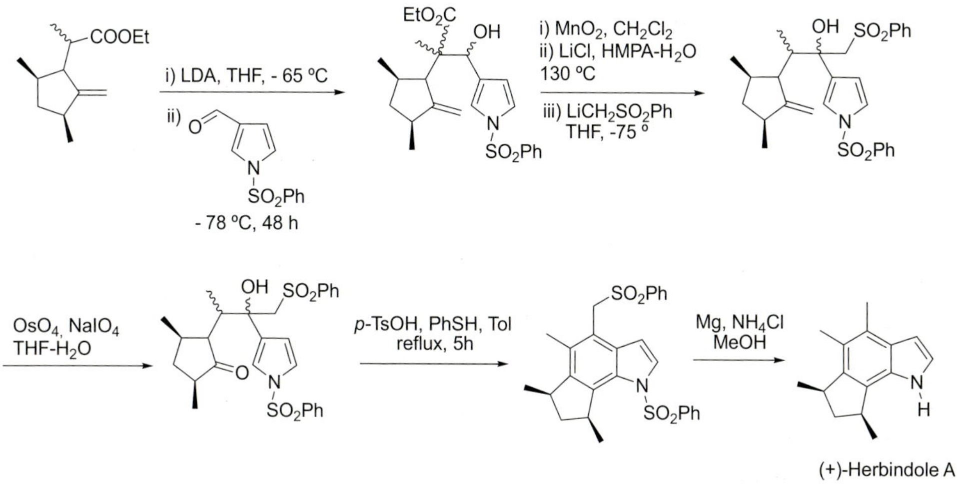

COOEt
i) LDA, THF, - 65 °C
ii) (N-SO$_2$Ph pyrrole-3-carbaldehyde), - 78 °C, 48 h
EtO_2C OH
N SO$_2$Ph
i) MnO_2, CH_2Cl_2
ii) LiCl, HMPA-H_2O 130 °C
iii) $LiCH_2SO_2Ph$ THF, -75 °
OH SO$_2$Ph
OsO_4, $NaIO_4$ THF-H_2O
p-TsOH, PhSH, Tol reflux, 5h
SO$_2$Ph
Mg, NH_4Cl MeOH
N H
(+)-Herbindole A

Scheme 5.119

5.4.3.1.2 **Katritzky Synthesis** In the method by Katritzky, the carbonyl substituted pyrrole **283** ready for the cyclization, is generated by a Michael-type addition of the carbanion generated from a benzotriazole (Bt)-substituted pyrrole (**282**) with a α,β-unsaturated ketone (Scheme 5.120) [186]. The benzotriazolyl functionality acts as both an anion-stabilizing group and leaving group in the overall transformation. Both approaches, cyclization by formation of the C3a–C4 bond (Scheme 5.120) or the C7–C7a bond (Scheme 5.121), have been conducted [187].

5.4.3.2 Palladium-Catalyzed Cyclizations

The six-membered ring of the indole skeleton has also been constructed by Pd-catalyzed intramolecular Heck reactions. In the example shown in Scheme 5.122, a 6-exo-trig cyclization of a bromo-substituted pyrrole (**284**), bearing a double bond in an appropriate position, provides an advanced intermediate for the synthesis of the antitumor antibiotic duocarmycin SA [188].

The intramolecular Heck reaction has also been employed to build the benzene ring of the indole system with formation of the C5–C6 bond [189].

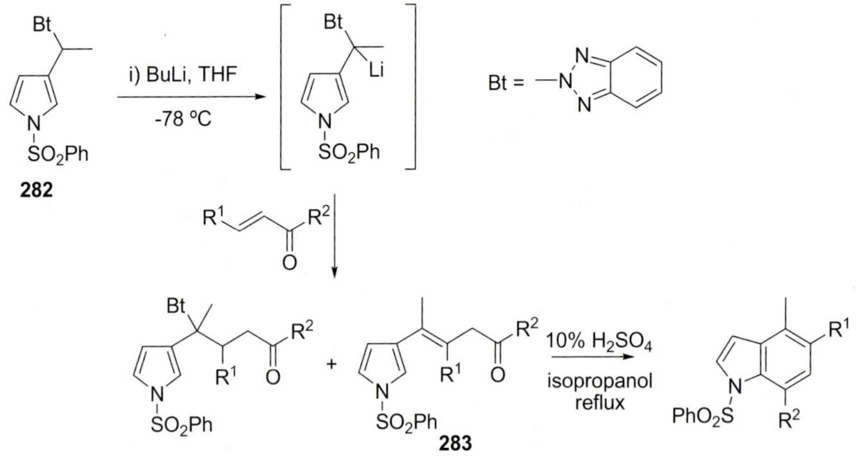

Scheme 5.120

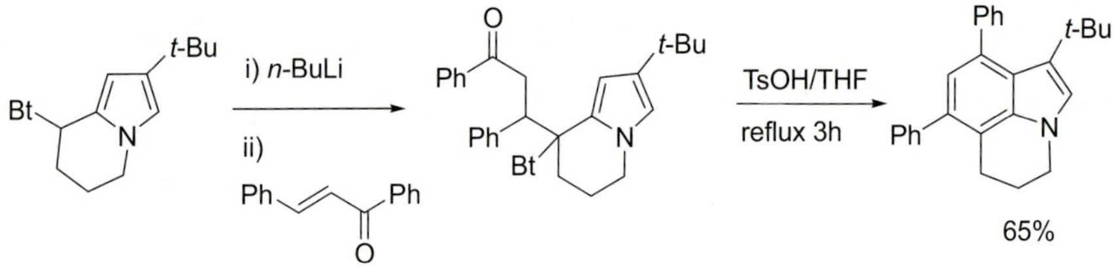

Scheme 5.121

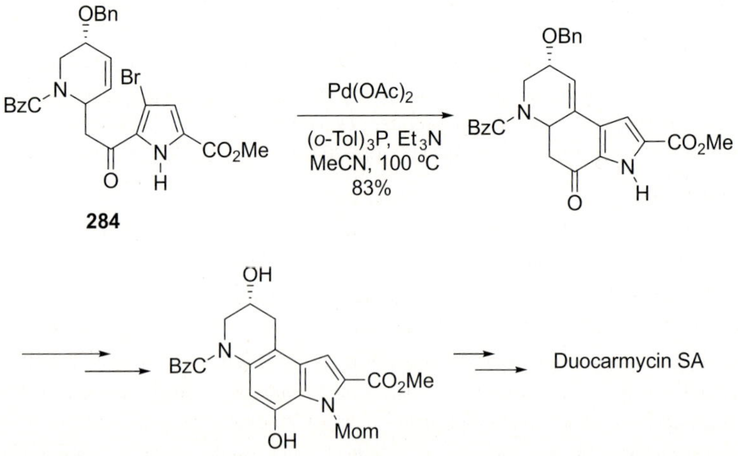

Scheme 5.122

5.4.3.3 **Electrocyclizations**

Electrocyclization of 2,3-dialkenyl-4-nitropyrroles **285** gives rise to 3-nitroindoles **286** – compounds that are difficult to prepare selectively through other routes. After the thermal 6π-electrocyclization, aromatization of the intermediate dihydroindoles occurs to provide the indole directly (Scheme 5.123) [190].

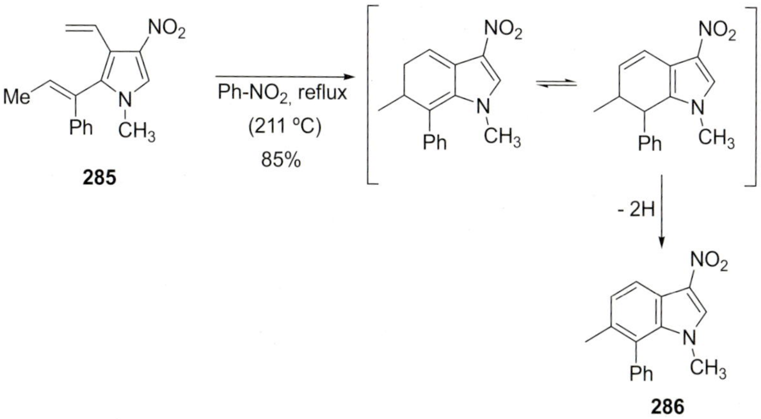

Scheme 5.123

5.4.3.4 **[4 + 2] Cycloadditions**

The cycloaddition of 2- and 3-vinyl pyrroles should allow for the preparation highly functionalized indoles in the benzene ring [191]. In a particularly interesting approach, 4,5-η^2-Os(II)pentaammine-3-vinylpyrrole complexes **287** readily undergo Diels–Alder reactions with activated dienophiles such as *N*-phenylmaleimide to generate the 5,6,7,7a-tetrahydroindole nucleus. The tetrahydroindole complexes **288** can be decomplexed and oxidized with DDQ to generate highly functionalized indoles **289** (Scheme 5.124) [192].

In contrast, the cyclic pyrrolo-2,3-quinodimethanes **290** undergo Diels–Alder cycloaddition with alkynes. After CO_2 loss, indoles with high substitution in the benzene ring (**291**) are obtained (Scheme 5.125) [193].

5.4.3.5 **Indoles from 3-Alkynylpyrrole-2-Carboxaldehydes**

Benzannulation of readily available 3-alkynylpyrrole-2-carboxaldehydes **292** with alkenes, promoted by iodonium ions, yields indoles **293** with a high level of substitution and functionalization in the benzene ring (Scheme 5.126) [194].

The mechanism proposed for this unusual transformation is represented in Scheme 5.127. Interaction of the iodonium ion with the triple bond of **292** would promote the formation of intermediate **294**. Nucleophilic attack of the alkene to the electrophilic carbon of **294**; subsequent intramolecular cyclization would then provide **296**. The simple loss of a proton to give a conjugated double bond then yields **297**. Finally, aromatization by elimination of HI gives the indoles **293**. This

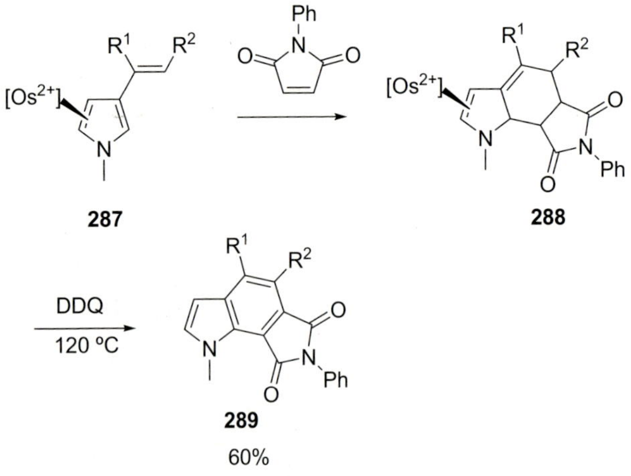

Scheme 5.124

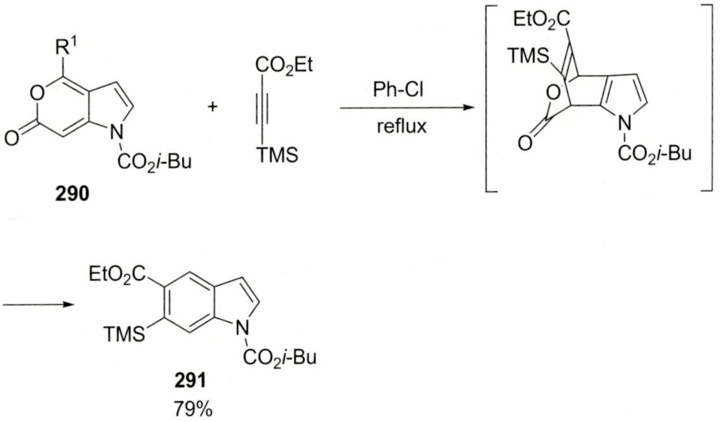

Scheme 5.125

proposal is supported by detailed mechanistic and spectroscopic studies, which allowed the isolation of analogues of the cationic intermediates **294** and **296** [195].

5.5
Reactivity of Indole

5.5.1
Reactions with Electrophiles

The chemistry of indole is dominated by the strong electrophilic character of the π-electron excessive heterocycle. Electrophilic aromatic substitution takes place at C3, and is one of the most efficient methods for the introduction of substituents on

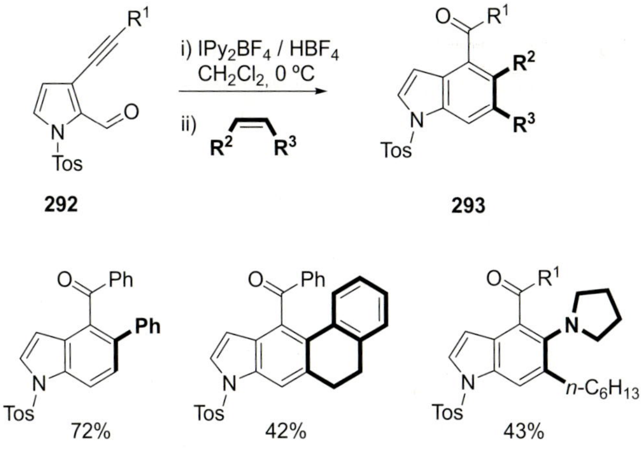

R1
i) IPy_2BF_4 / HBF_4
CH_2Cl_2, 0 °C
ii) R^2 R^3
O
N
Tos
292

O R1
R2
R3
N
Tos
293

O Ph
Ph
N
Tos
72%

O Ph
N
Tos
42%

O R1
N
n-C_6H_{13}
N
Tos
43%

Scheme 5.126

R1
O
N
Tos
292

$I^+BF_4^-$
R3 R2
+

O R1
R2
R3
N
Tos
293

$I^+BF_4^-$

BF_4^- I^+ R1
O
N
Tos

HI

R1
O I
H
R2
R3
N
Tos
297

HBF4

BF_4^-
I
R1
N
O+
Tos
294

R2
+
R3

I
R1
N
O:
Tos
R3
+
R2
BF_4^-
295

R2
R3
H
I
R1
BF_4^-
N
O+
Tos
296

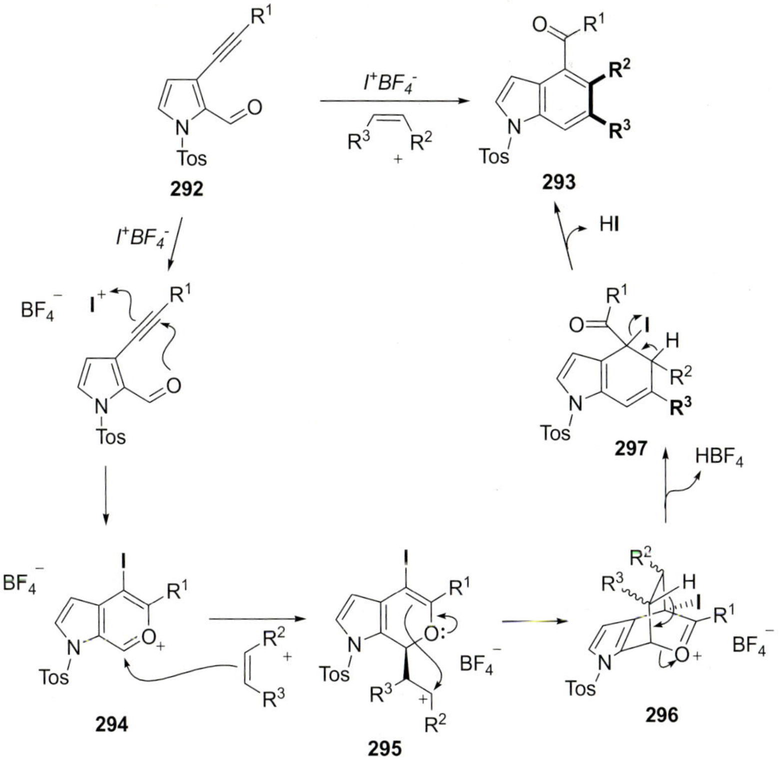

Scheme 5.127

the indole ring at that position. The regioselectivity of the electrophilic aromatic substitution is easily explained by the different stability of the intermediate indolium cations generated. The positive charge at C2, generated by electrophilic attack at C3, can be delocalized between C2 and the N atom without compromising the aromaticity of the benzene ring. In contrast, electrophilic attack at C2 generates an indolium cation with the positive charge at C3, which cannot be delocalized without breaking the aromaticity of the benzene ring. Electrophilic aromatic substitution on the benzene ring occurs only on indoles strongly deactivated on the five-membered ring.

The method of choice to introduce electrophiles at the C2 position of indoles consists in a stepwise procedure, involving N-protection, lithiation at C2, reaction with the electrophile and N deprotection.

5.5.1.1 **Protonation**

Indole and substituted indoles are weak bases, with pK_a values ranging from -2.4 for protonated indole to −0.3 for the protonated electron-richer 2-methylindole [196]. As expected, C3 is the main site for protonation of indole, to produce the 3-*H*-indolium cation. Although the N-protonated indole (1-*H*-indolium cation) is not detected, even spectroscopically, it is believed that N-protonation occurs rapidly and equilibrates into the more stable C3 protonated species.

Under weak acidic media indole undergoes dimerization and oligomerization by attack of a molecule of non-protonated indole to the strong electrophilic C2 position of the 3-*H*-indolium cation (Scheme 5.128).

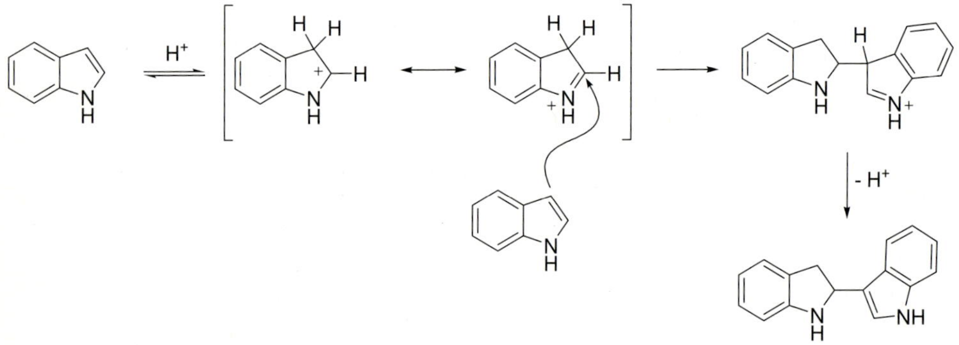

Scheme 5.128

5.5.1.2 **Friedel–Crafts Alkylations of Indole**

Indolyl compounds can be regioselectively alkylated at C3 through different Friedel–Crafts (FC) type of reactions (Scheme 5.129). Alkyl halides, epoxides and aziridines, carbonyl compounds and imines, α,β-unsaturated compounds, alkenes and alkynes, and allylic acetates or carbonates have all been employed as electrophiles. The amount of literature regarding this topic is enormous. These types of reaction usually proceed smoothly in the presence of a Lewis or protic acid as catalyst. Thus, the most recent advances have focused mainly on developing catalytic systems

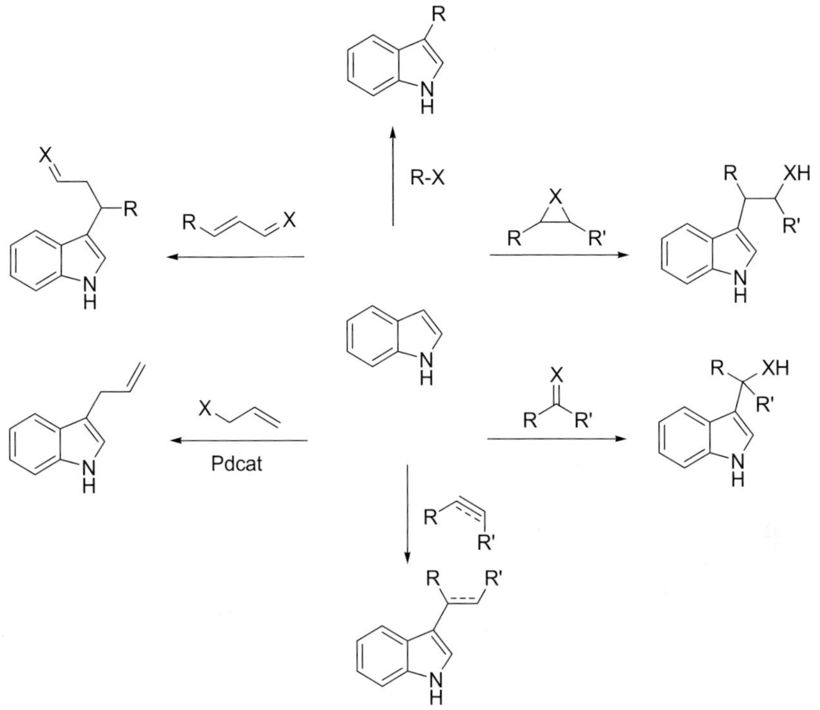

Scheme 5.129

to perform the electrophilic aromatic substitution reaction in a milder, regio- and stereoselective manner [197].

5.5.1.2.1 **Michael Additions** The reactions of indoles with α,β-unsaturated compounds, such as α,β-unsaturated ketones, nitriles and nitroolefins, usually require activation of the electrophile by an acid, and proceed with either protic [198] or Lewis acids [199] and clays [200] to provide the corresponding 3-substituted indoles (Scheme 5.130).

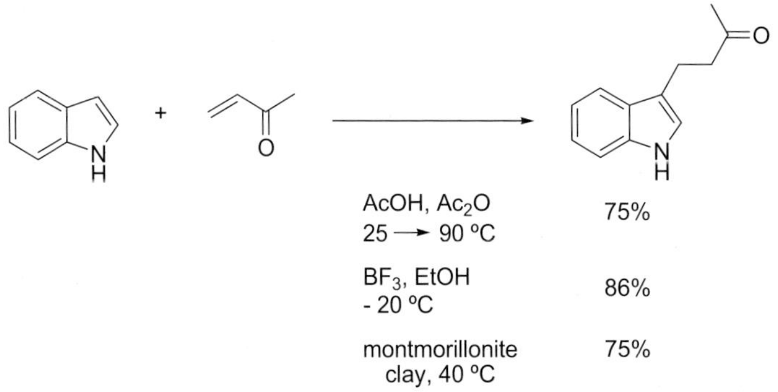

Conditions	Yield
AcOH, Ac_2O, 25 → 90 °C	75%
BF_3, EtOH, - 20 °C	86%
montmorillonite clay, 40 °C	75%

Scheme 5.130

These transformations have attracted much attention in recent years, and several different new Lewis acids have been developed, including bismuth(III) salts [201], cerium salts [202], Pd salts [203] and iodine [204]. For instance, the use of lanthanide-based Lewis acids allows for the alkylation of indoles at the 3-position with various Michael acceptors and under mild conditions (Scheme 5.131) [205].

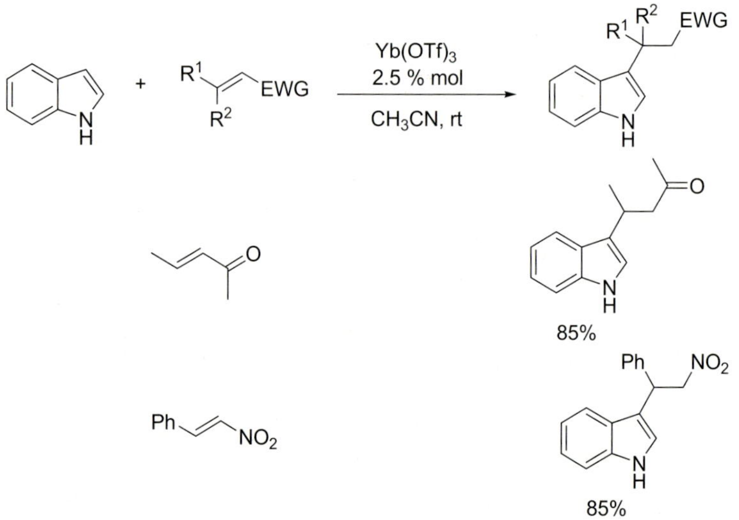

Yb(OTf)$_3$ 2.5 % mol, CH_3CN, rt

85%

85%

Scheme 5.131

Michael additions have been carried out in water at room temperature using a scandium salt of an anionic surfactant – $Sc(DS)_3$ (scandium dodecyl sulfate) (Scheme 5.132) [206].

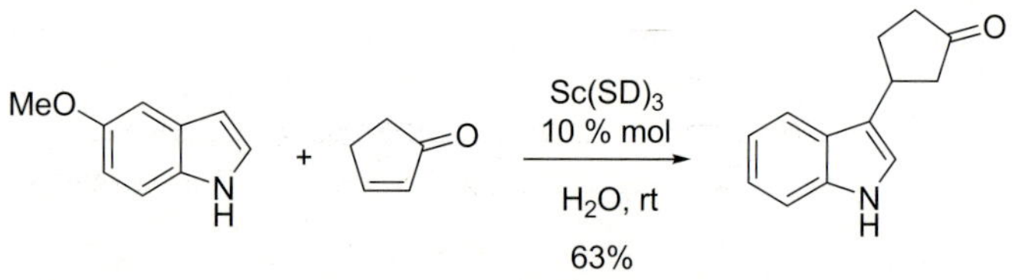

Sc(SD)$_3$ 10 % mol, H_2O, rt, 63%

Scheme 5.132

Indium salts are amongst the most general and effective Lewis acids to promote the Michael addition, which can be performed with relatively hindered enones and 2-substituted indoles [207, 208]. The same catalyst promotes the reaction with nitroalkenes, under aqueous conditions, with excellent yields in most instances (Scheme 5.133) [209].

Gold(III) salts are also excellent catalysts for this reaction [210]. Interestingly, when the Michael addition is carried out with 3-substituted indoles, substitution occurs at C2 although with relatively lower yields (Scheme 5.134).

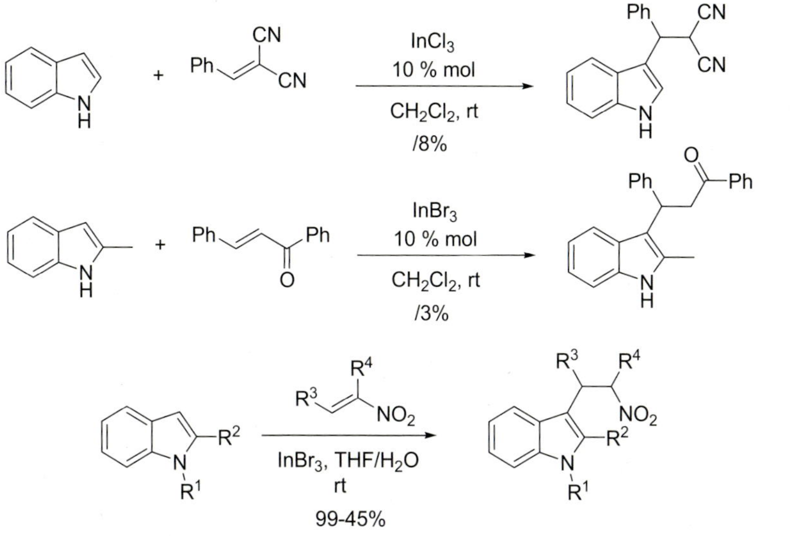

Scheme 5.133

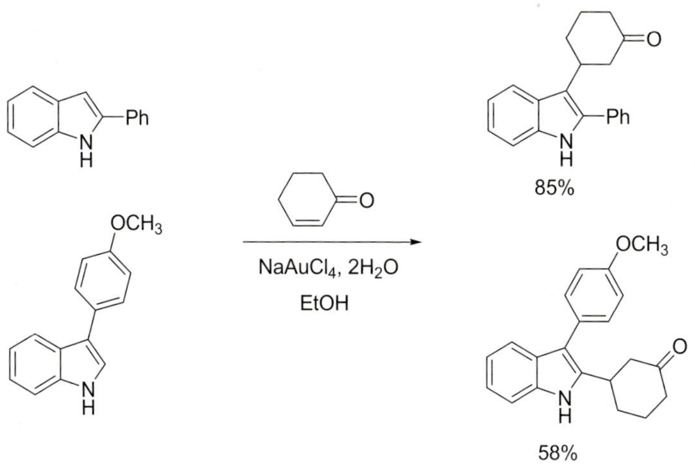

Scheme 5.134

In the reaction of indoles with β-substituted α,β-unsaturated compounds one or two new stereogenic centers can be created. Control of the enantioselectivity of such processes has aroused much interest in recent years (Scheme 5.135).

Few examples have appeared of the catalytic asymmetric Michael reaction of indoles. To achieve good enantioselectivities the use of bidentate chelating Michael acceptors is required, to keep fixed the chiral environment provided by the ligand (Scheme 5.136). Good to excellent enantioselectivities have been achieved in reactions of alkylidene malonates **298** [211], β,γ-unsaturated α-ketoesters **299** [212], acyl

Scheme 5.135

298 X = CO_2R', Y = OR'
299 X = H, Y = CO_2R'
300 X = H, Y = $P(O)OMe_2$
301 X = H, Y = $C(Me)_2OH$
302 X = H, Y = (N-methylimidazol-2-yl)

monodentate Michael acceptor: poor enantioselectivity

bidentate Michael acceptor: good enantioselectivity

Scheme 5.136

phosphonates **300** [213], α′-hydroxyenones **301** [214] and α,β-unsaturated-2-acylimidazoles **302** [215, 216], using appropriate chelating C_2 symmetric chiral ligands combined with Cu or lanthanide Lewis acids.

Scheme 5.137 shows the asymmetric Friedel–Crafts alkylation of indoles with α′-hydroxyenone **301**, yielding 3-substituted indole **303** with very high ee.

301, cat., CH_2Cl_2, rt, 2h

303: 96% yield, 97% ee

steps

X = H, R, OMe

Scheme 5.137

In a totally different and extremely elegant approach, the asymmetric Michael addition of indoles to α,β-unsaturated aldehydes has been achieved using a chiral imidazolidinone (**304**) as an asymmetric organocatalyst [217]. The imidazolidinone plays a double catalytic role: to activate the aldehyde **305** by the formation of the highly reactive iminium species **306**, and to create a chiral environment that differentiates both enantiotopic faces of the Michael acceptor (Scheme 5.138).

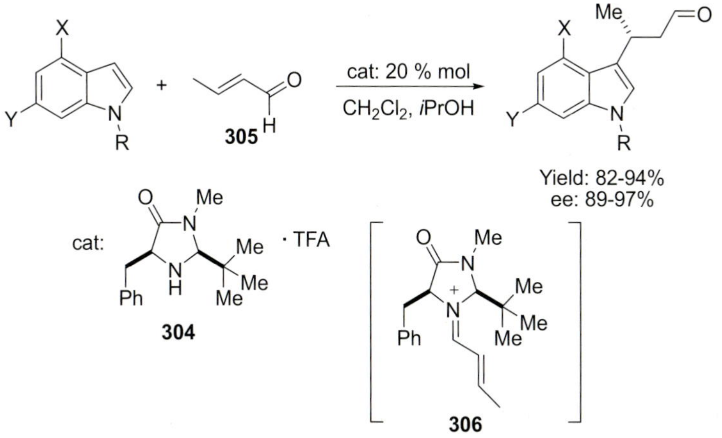

cat: 20 % mol
CH_2Cl_2, *i*PrOH

Yield: 82-94%
ee: 89-97%

cat: **304** · TFA

306

Scheme 5.138

Thiourea-based organocatalyst **307** promotes the asymmetric Friedel–Crafts alkylation of indoles with nitroalkenes with high yields and enantiomeric excesses. The stereochemical course of the reaction is controlled by the asymmetric platform provided by the chiral thiourea organocatalyst, which forms hydrogen bonds simultaneously with both reactants (Scheme 5.139) [218].

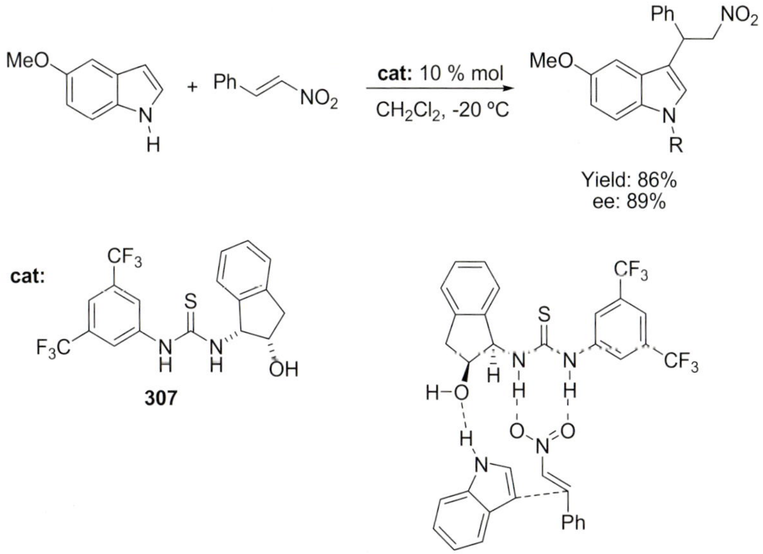

cat: 10 % mol
CH_2Cl_2, -20 °C

Yield: 86%
ee: 89%

cat: **307**

Scheme 5.139

5.5.1.2.2 **Reactions with Unactivated Olefins** The intramolecular cyclization of alkenylindoles **308** can be promoted by $PtCl_2$ to give tetrahydrocarbazole derivatives **309** [219]. The cyclization proceeds through nucleophilic attack of the indole on the Pt(II) complexed olefin in intermediate **310**, followed by protonolysis of the C–Pt bond in **311** (Scheme 5.140). The same authors have recently reported an extension of this methodology to the use of Pd(II) and Cu(II) catalysts [220].

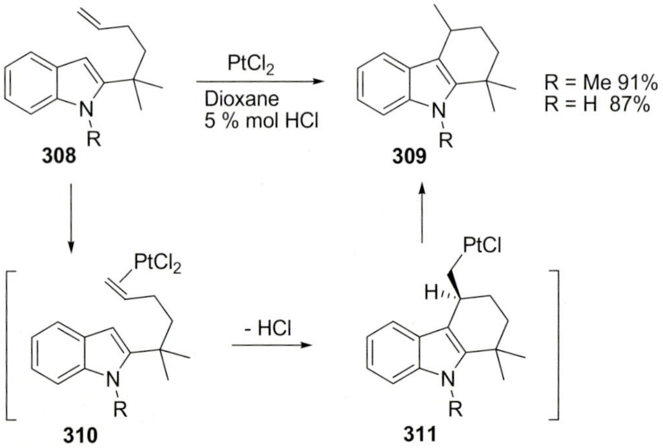

Scheme 5.140

5.5.1.2.3 **Reactions with Carbonyl Compounds** Friedel–Crafts alkylation of indoles with aldehydes and ketones takes place under Brønsted [221] or Lewis acid catalysis. The initially formed indole-3-yl-carbinols **312** are usually not isolated and evolve to produce the azafulvenium salts **313**. Finally, addition of a second molecule of indole to the azafulvenium salt gives rise to bis(indoylmethanes) **314** (Scheme 5.141).

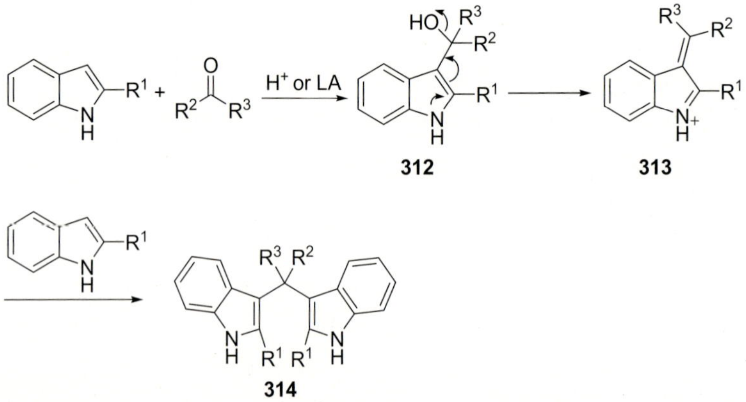

Scheme 5.141

Several different types of protic and Lewis acid catalysts have been employed for this transformation. For instance, reaction of indole with aldehydes or ketones such as **315** and **317**, in the presence of $LiClO_4$, affords the bis(indoylmethanes) **316** and **318**, respectively, in excellent yields (Scheme 5.142) [222].

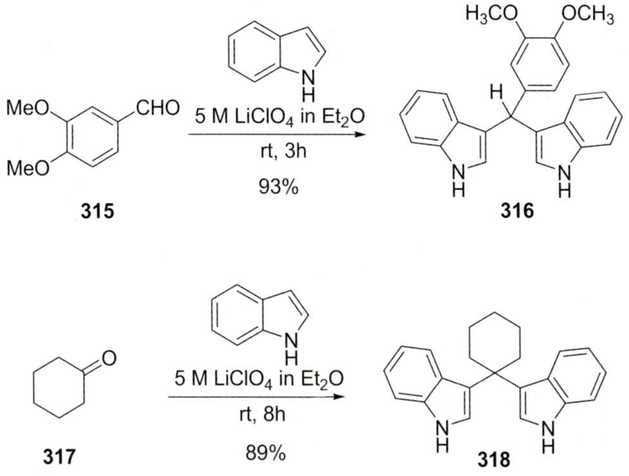

Scheme 5.142

When the reaction is carried out with 2-oxoesters such as glyoxalate **319**, the indoylcarbinol **320** does not undergo elimination and can be isolated (Scheme 5.143) [223].

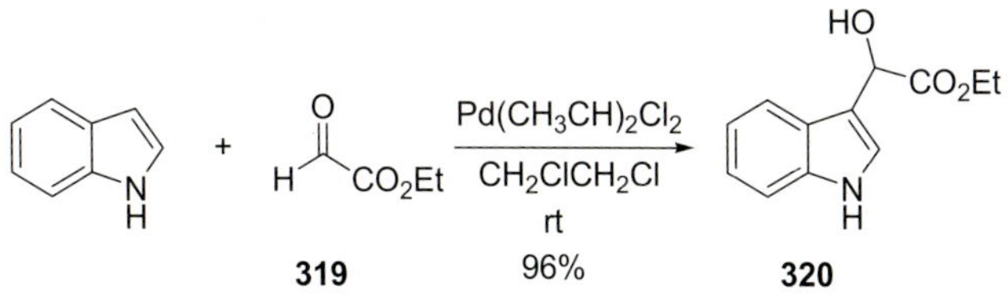

Scheme 5.143

The enantioselective version of this reaction has been successfully carried out both with copper-based chiral Lewis acids **321** [224] and with *Cinchona* alkaloid organic catalysts **322** (Scheme 5.144) [225]. The organocatalyzed reaction, although not well understood yet, is remarkable, as the appropriate choice of alkaloid (chinchonidine, CD or Chinchonine, CN) provides either enantiomer in quantitative yield and with very high ee.

5.5.1.2.4 **Reactions with Imines and Imminium Ions: Mannich Reaction** Under typical Mannich conditions indole undergoes alkylation at C3. This reaction leads to the synthesis of gramines **323**, which are important intermediates for the

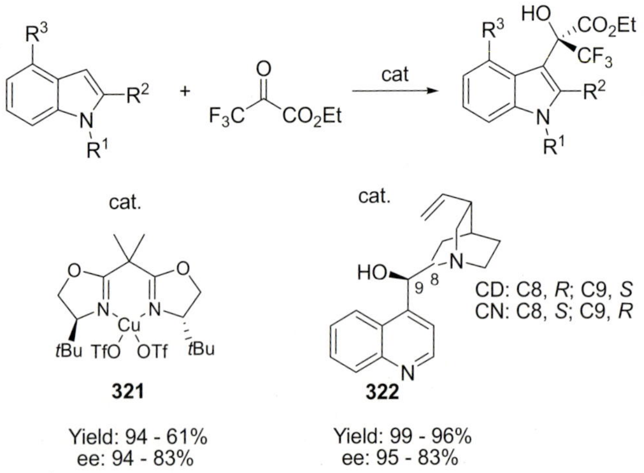

Yield: 94 - 61%
ee: 94 - 83%

Yield: 99 - 96%
ee: 95 - 83%

Scheme 5.144

preparation of substituted indoles (Section 5.6.1). When the reactions are conducted in water at low temperatures, the kinetically controlled N-alkylation product **324** is obtained [226]. The resulting N-aminal indoles are relatively stable but convert into the thermodynamically more stable C3-substituted aminomethyl indoles upon heating at neutral pH or acid treatment at room temperature (Scheme 5.145).

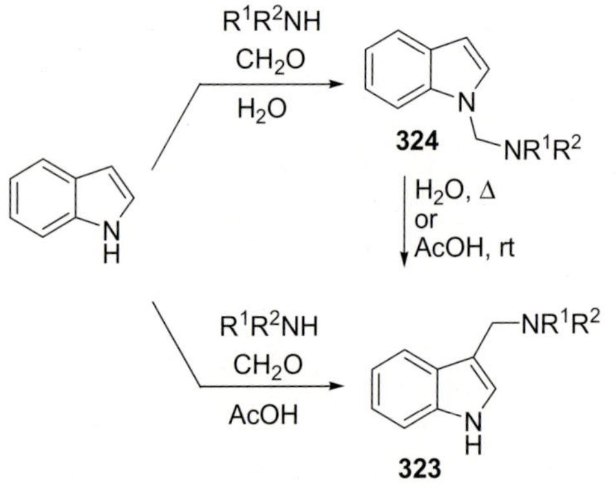

Scheme 5.145

Gramines can undergo a second Mannich reaction, yielding 1,3-disubstituted indoles **325**. This reaction has been applied to the parallel synthesis of an indole-based library (Scheme 5.146) [227].

Indole reacts with imines under acidic conditions to give substituted gramines. For instance the reaction of glyoxylimine **326** with indoles employing $Yb(OTf)_3$ as Lewis acid leads to gramine derivatives **327** (Scheme 5.147) [228, 229].

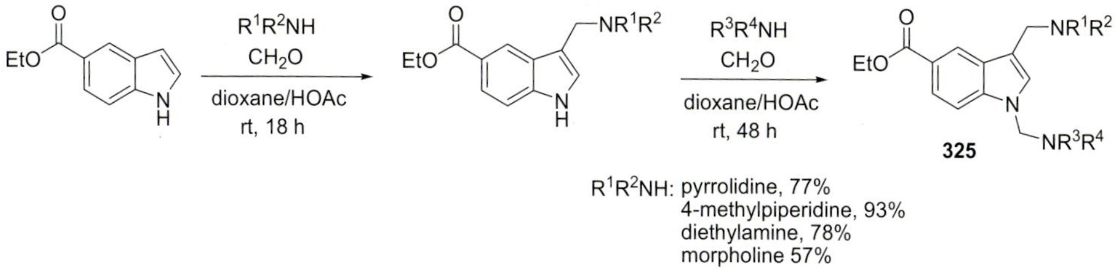

Scheme 5.146

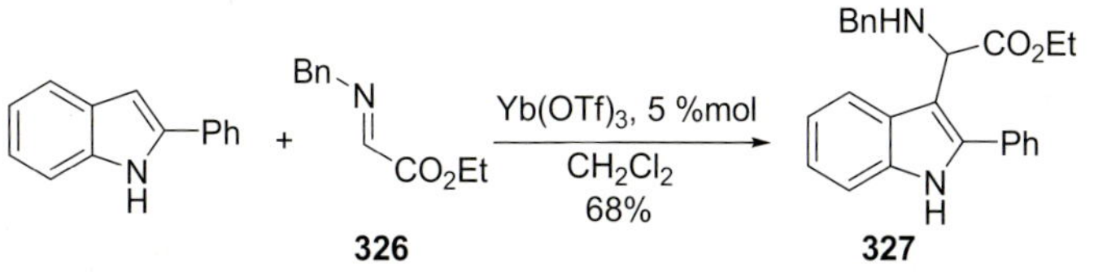

Scheme 5.147

Asymmetric versions of this reaction have been carried out employing Cu-based chiral catalysts [230] and also organic catalysts. Scheme 5.148 presents the Friedel–Crafts alkylation of indoles with *N*-tosylimines, such as **328**, promoted by the quinine-based organic catalyst **329** [231].

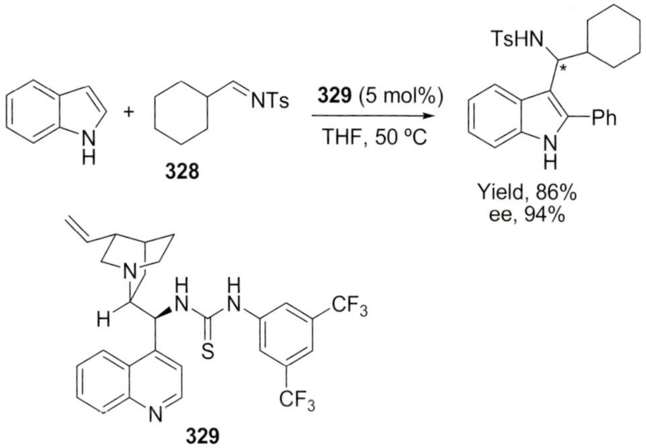

Scheme 5.148

5.5.1.2.5 **Epoxide and Aziridine Ring Opening** Epoxides and aziridines are versatile alkylating agents for indole. The ring opening of these strained heterocycles by indole proceeds with Lewis acid, bases and solid acids, giving rise to triptophols and tryptamine derivatives respectively (Scheme 5.149). Moreover, the ready availability of enantioenriched cis and trans epoxides makes this approach a very valuable entry

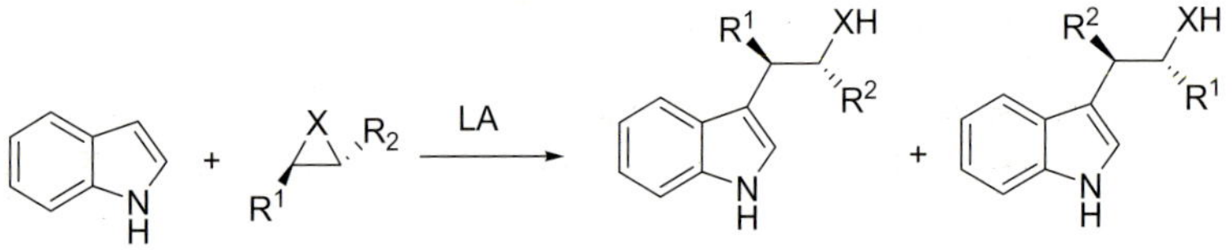

Scheme 5.149

into enantiomerically pure indoles. The regiochemistry and stereochemistry of the ring opening are the main issues that have to be addressed during the reaction.

The alkylation of indole with enantiomerically pure styrene oxide is catalyzed by most of the common Lewis acids, with $InBr_3$ being the most efficient in terms of regioselectivity, enantioselectivity and yield (Scheme 5.150) [232, 233]. Other catalysts such as $Sc(OTf)_3$ and $SnCl_4$ [234] also promote the ring opening without racemization.

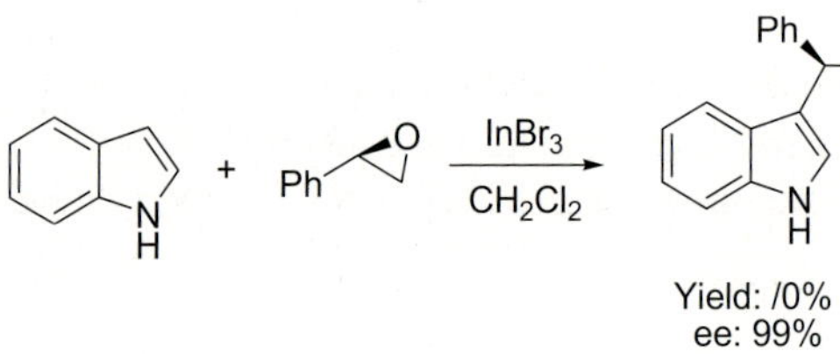

Scheme 5.150

Enantioenriched chiral triptophols **330** can also be prepared by kinetic resolution of racemic epoxides, or by desymmetrization of *meso* epoxides, employing Cr(Salen) Cl complex **331** as chiral catalyst (Scheme 5.151) [235].

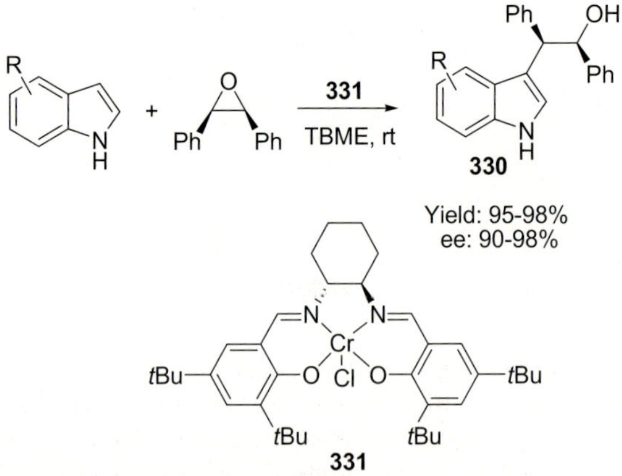

Scheme 5.151

The ring opening of aziridines with indoles provides tryptamine derivatives. Lewis acids based on Zn [236], Sc, Yb, and indium [237], and SiO_2 promote efficiently the FC reaction. Scheme 5.152 shows the alkylation of methyl indole-2-carboxylate with chiral aziridine **332** to produce the enantiomerically pure substituted indole **333**. The alkylation occurs cleanly simply by adsorbing both reagents in SiO_2 and heating the mixture at 70 °C [238].

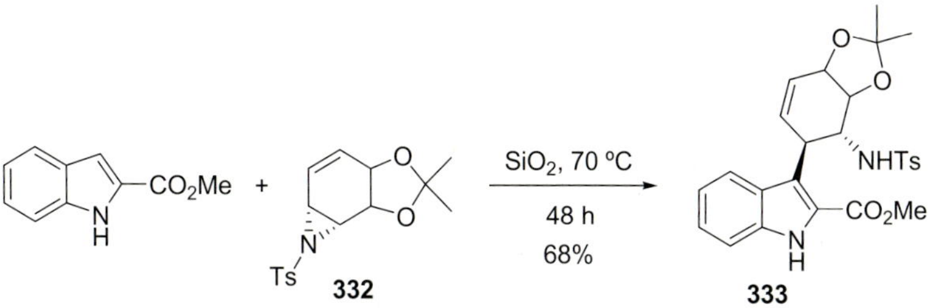

Scheme 5.152

5.5.1.2.6 **Indole as Nucleophile in Palladium-Catalyzed Allylic Alkylations** Indoles are also competent nucleophiles for Pd-catalyzed allylic substitutions (Tsuji–Trost reaction) [239]. The reaction between allyl carbonate **334** and indole leads regioselectively to the C3-allylated indole **335** or the *N*-allylated indole **336**, depending on the reaction conditions applied (Scheme 5.153) [240].

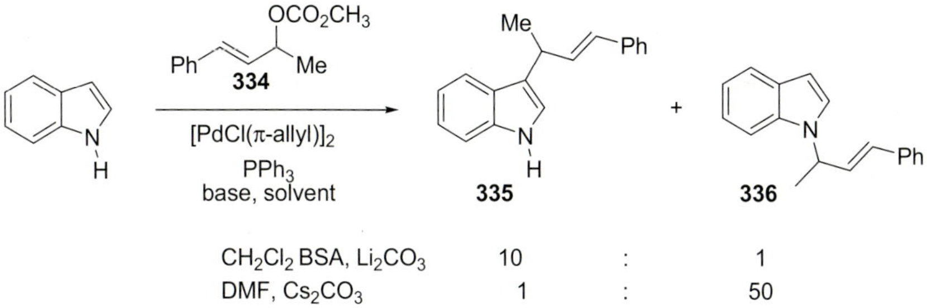

Scheme 5.153

The intramolecular variant of this reaction provides a new entry into polycyclic fused indoles. Remarkably, use of the appropriate chiral ligand allows for the asymmetric allylic alkylation (AAA) reaction [241] to take place with very high ee, giving rise to tetrahydro-β-carbolines **338** (Scheme 5.154) [242]. Moreover, when the pedant group is attached at C3, such as in **339**, the allylic alkylation takes place at C2, also with very high regio- and enantioselectivity, to produce tetrahydro-γ-carbolines **340**.

5.5.1.3 **Nitration**

Indoles are highly sensitive to acids, and for this reason the nitration of the indole ring requires carefully designed experimental conditions. Although C3 is the

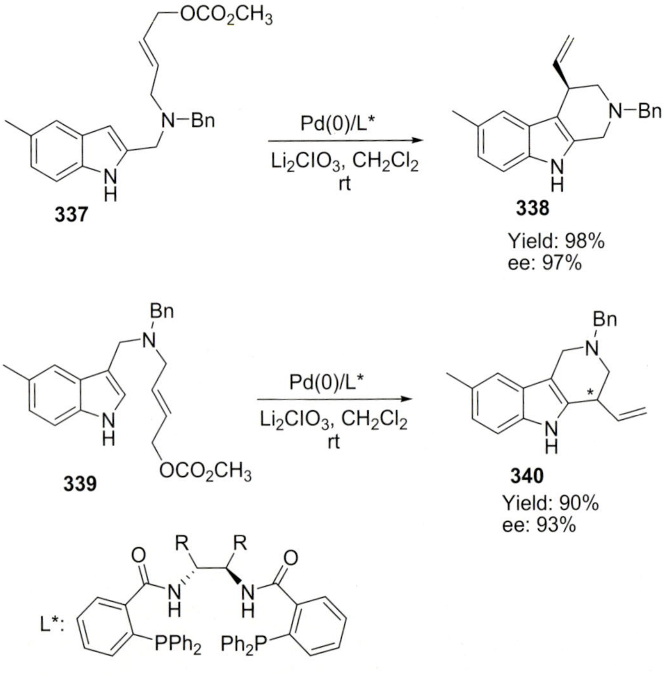

Scheme 5.154

preferred position for electrophilic attack on indole, nitration under strong acid conditions proceeds through the 3-*H*-indolium cation, and nitration occurs mainly at C5, owing to the directing influence of the iminium substituent. Indole itself can be nitrated at C3 with moderate yield with benzoyl nitrate [243], and 2-arylindoles can be successfully nitrated at C3 by treatment with 2-cyano-2-propyl nitrate under phase transfer catalysis conditions [244]. More interestingly, N-protected indoles can be successfully nitrated at C3 by treatment with *in situ* generated acetyl nitrate at very low temperature (Scheme 5.155) [245].

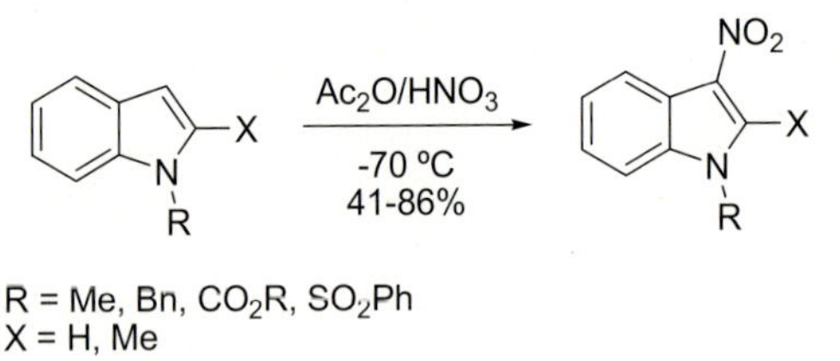

Scheme 5.155

In contrast, 2-nitroindoles **342** can be synthesized by a lithiation-nitration protocol on *N*-protected indoles **341** at very low temperature (Scheme 5.156) [246], and by treatment of N-protected-2-bromoindoles **343** with silver nitrite (Scheme 5.157) [247].

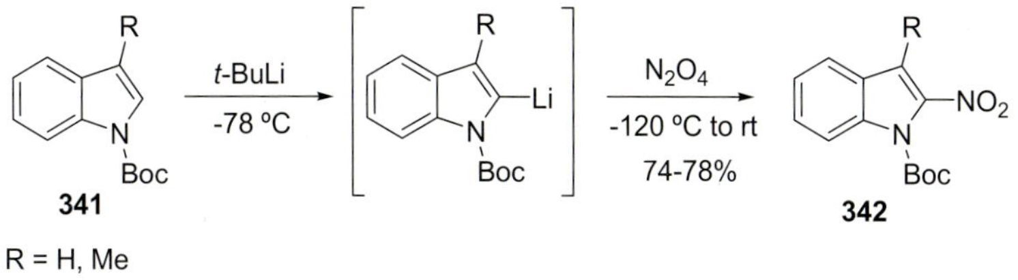

R = H, Me

Scheme 5.156

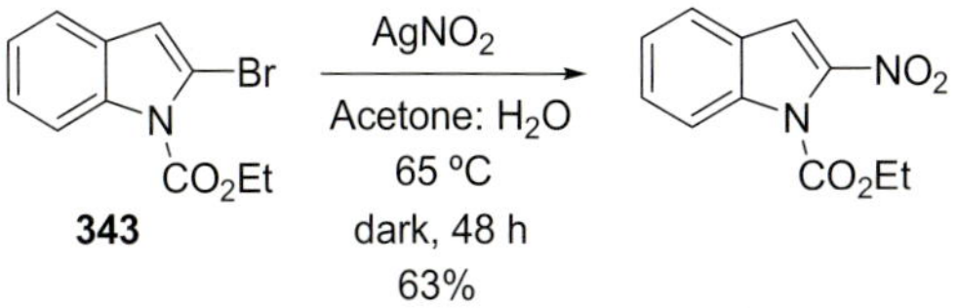

Scheme 5.157

5.5.1.4 Acylation

Indoles can be acylated at C3 by the Vilsmeier–Haack reaction [248], by Friedel–Crafts acylations [249] and also by reactions of indole Grignard reagents [250] or indole zinc chloride salts with acid chlorides (Scheme 5.158) [251].

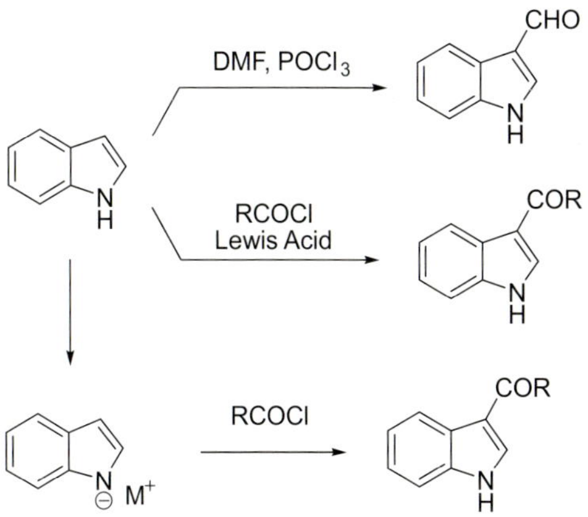

Scheme 5.158

The Vilsmeier–Haack reaction is the classical method for the preparation of 3-formylindole and other 3-acylindoles, starting from tertiary amides. This reaction provides good yields for the acylation of indoles but is limited to formamide and alkylcarboxamides (Scheme 5.159).

Friedel–Crafts acylation is a very convenient process for electron-withdrawing-substituted indoles and for N-protected indoles. However, the reaction of N–H indoles requires a fine tuning of the catalyst and the reaction conditions to avoid

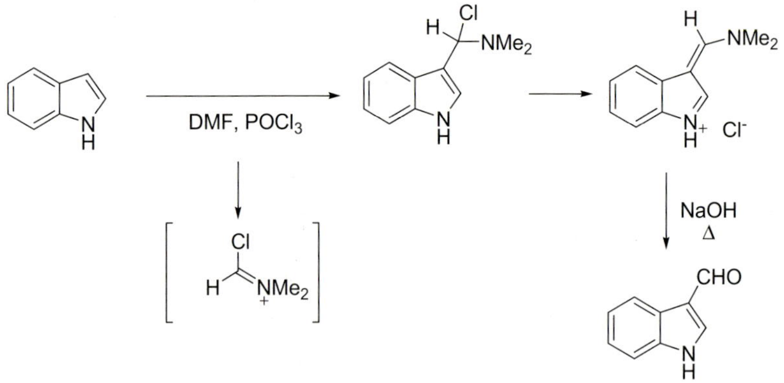

DMF, $POCl_3$

NaOH
Δ

Scheme 5.159

N-acylation and polymerization reactions [252]. The use of an excess of dialkylaluminium chlorides as Lewis acids represents a very general method for the Friedel–Crafts acylation of indoles [253]. The excess dialkylaluminum reagent is required to quench the HCl liberated in the acylation process, thereby avoiding side reactions, such as dimerization and polymerization, that might be originated by the presence of the acid (Scheme 5.160).

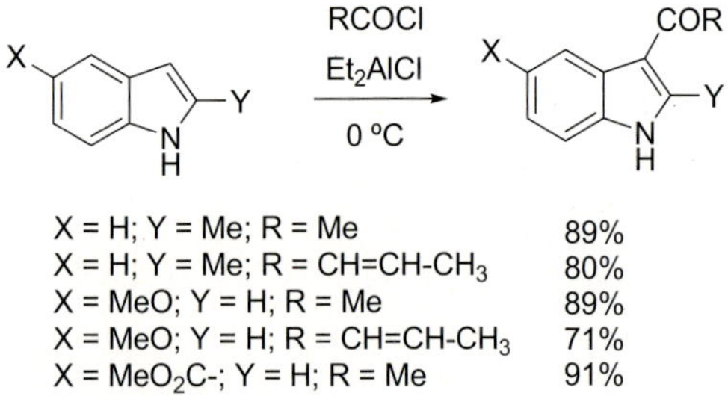

RCOCl
Et_2AlCl
0 °C

X = H; Y = Me; R = Me 89%
X = H; Y = Me; R = $CH{=}CH{-}CH_3$ 80%
X = MeO; Y = H; R = Me 89%
X = MeO; Y = H; R = $CH{=}CH{-}CH_3$ 71%
X = MeO_2C-; Y = H; R = Me 91%

Scheme 5.160

The use of *N*-acylbenzotriazoles **344** represents an alternative to acid chlorides, eliminating the complications associated with the release of HCl [254]. This procedure permits the acylation of both *N*-H and *N*-alkylindoles (Scheme 5.161).

5.5.1.5 Halogenation

The indole ring can be easily halogenated at C3 by employing bromine and iodine in DMF (the presence of potassium hydroxide is required for the iodination reaction), providing nearly quantitative yields of the corresponding 3-haloindoles [255], while 3-chloroindoles are best prepared with *N*-chlorosuccinimide (NCS) (Scheme 5.162) [256, 257]. Many other reagents have been described to promote these transformations [258, 259].

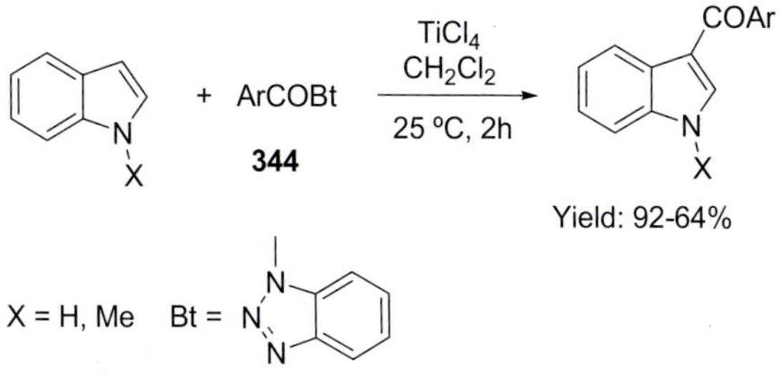
+ ArCOBt
344
TiCl4
CH2Cl2
25 °C, 2h
COAr
N
X
Yield: 92-64%
X = H, Me
Bt =
N
N
N

Scheme 5.161

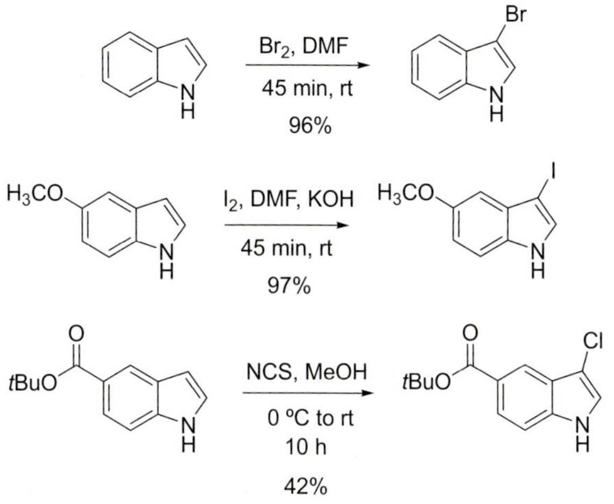
Br2, DMF
45 min, rt
96%
Br
H3CO
I2, DMF, KOH
45 min, rt
97%
I
tBuO
NCS, MeOH
0 °C to rt
10 h
42%
Cl

Scheme 5.162

Halogens are best introduced at C2 by the metallation–halogenation sequence discussed in Section 5.5.2.2.

5.5.2
Reactions with Bases

5.5.2.1 *N*-Metallation of Indoles

Indole is a weak acid (pK_a = 16.7 and 20.9 in water [260] and DMSO [261], respectively) and therefore undergoes deprotonation by strong bases to provide a reactive anion. Most of the reactions involving substitution at nitrogen – alkylation, acylation, and sulfonation – are carried out through the indolyl anion.

The usual methods for the N-alkylation of indole involve the use of alkali metal hydroxides [262], alkali metal hydrides [263] or $NaNH_2$ in polar aprotic solvents (Scheme 5.163). Alternatives include the use of potassium hydroxide in acetone under phase transfer catalysis conditions [264], caesium carbonate in either DMPU [265] or in the presence of a crown ether [266].

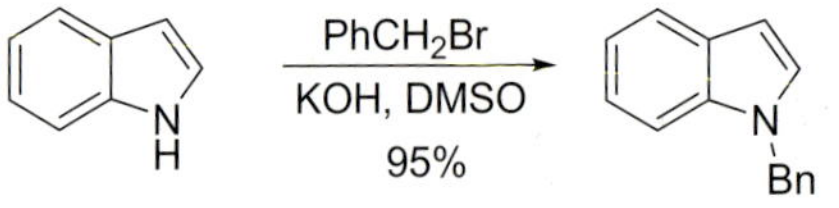

Scheme 5.163

Both N-acylation and N-sulfonation of indoles are synthetically relevant transformations, not only for the interest in *N*-acyl and *N*-sulfonyl indoles themselves, but also because the protection of the indole N−H in a synthetic sequence almost always involves an acylation or sulfonation step [267]. The use of acid chlorides and anhydrides in the presence bases make up the most common reaction conditions. In particular, the Boc group has been introduced using di-*tert*-butyl dicarbonate, phenyl-*tert*-butyl carbonate and $BocN_3$ [268]. Scheme 5.164 depicts some representative reactions for the N−H acylation of indole with different acylating agents [269–271].

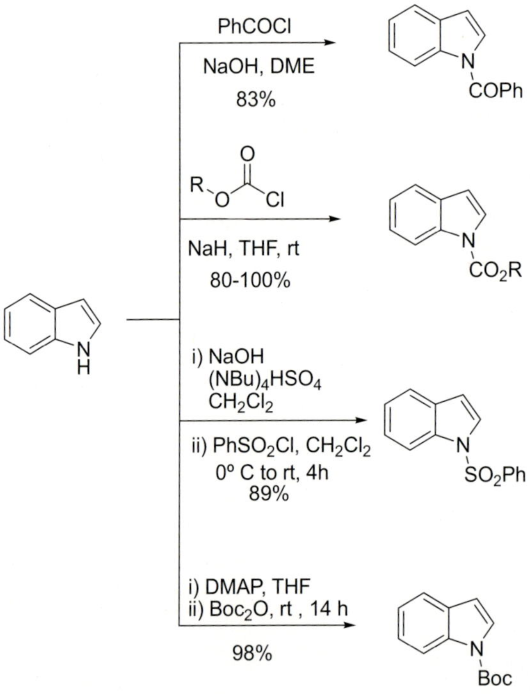

Scheme 5.164

Direct acylation of indoles with aromatic carboxylic acids can be achieved using DCC as coupling agent (Scheme 5.165) [272].

5.5.2.2 C-Metallation of Indoles

N-Substituted indoles undergo direct lithiation at C2, and in certain cases at C3 upon treatment with organolithium reagents [273]. In fact, lithiation followed by reaction with an electrophile is the most common strategy to introduce substituents at C2.

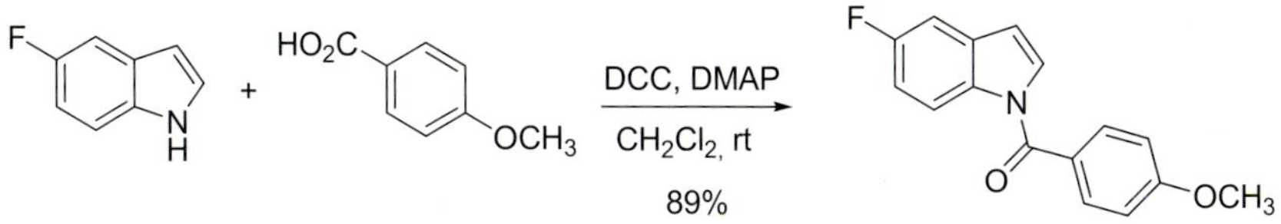

Scheme 5.165

Both *N*-acyl and *N*-sulfonyl indoles can be selectively lithiated at C2 by treatment with organolithium reagents. The presence of the coordinating group at N1 assists the lithiation at C2 and governs the regioselectivity of the process. Recent applications of this strategy are found in Gribble's syntheses of 2-nitroindole **345** [274] and 2-iodoindole **346** [275] from, respectively, *N*-Boc-indole and *N*-phenylsulfonylindole (Scheme 5.166) and in the preparation of the important synthetic intermediates 2-indolylborates **347** (Scheme 5.167) [276].

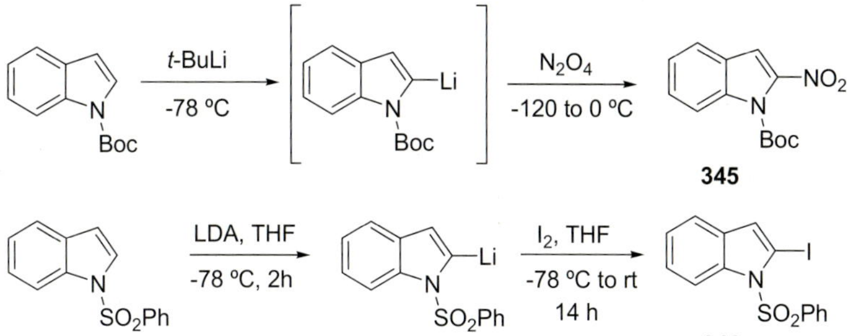

Scheme 5.166

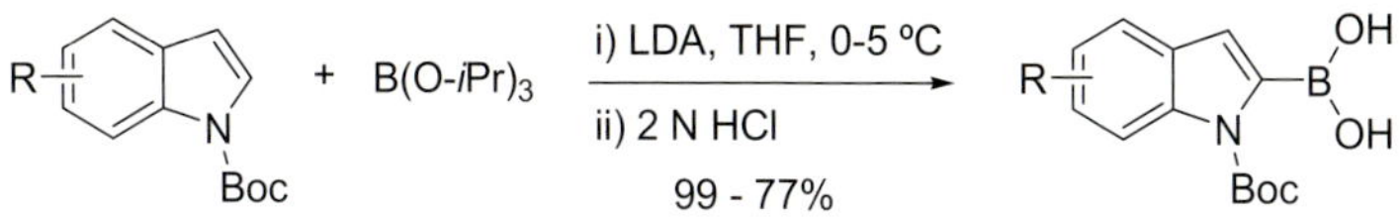

Scheme 5.167

Substitution at C2 on NH indoles can be achieved by employing Katritzky's elegant indole C2 lithiation protocol [277]. In this sequential process the dilithiated intermediate **348** is formed, which provides the C2 substituted indole **349** after reaction with an electrophile and aqueous workup (Scheme 5.168). A typical example of the application of this methodology is Bergman's synthesis of 2-bromoindole **350** (Scheme 5.169) [278].

The nature of the N-substituent can direct the position of the lithiation in *N*-alkyl indoles. While indoles substituted with non-bulky alkyl groups are lithiated at C2, the

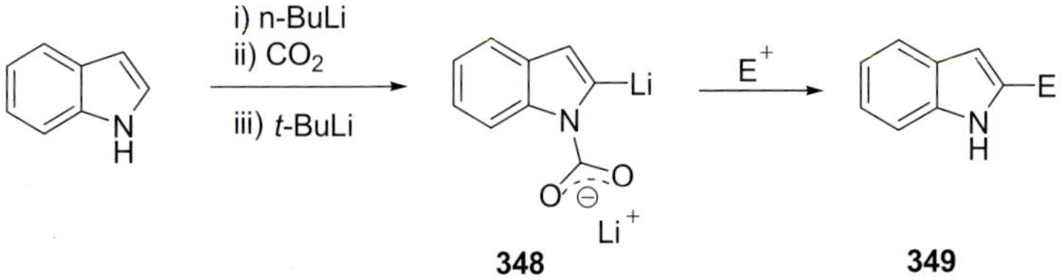

Scheme 5.168

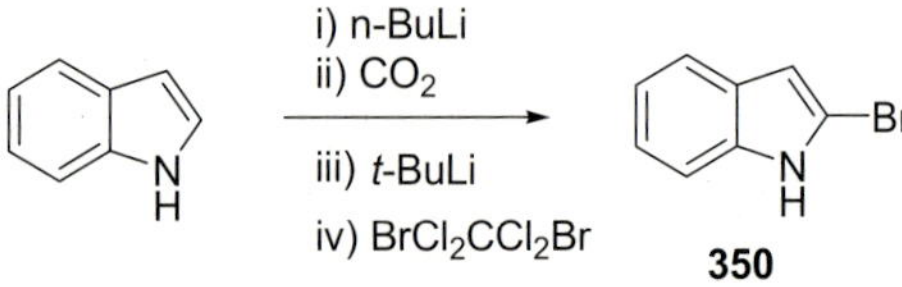

Scheme 5.169

presence of a bulky non-coordinating group drives the lithiation at C3. For instance, *N*-methyl-2-stannylindole **352** is selectively synthesized from *N*-methylindole (**351**) by deprotonation at C2 with BuLi, followed by reaction with tributylstannyl chloride (Scheme 5.170).

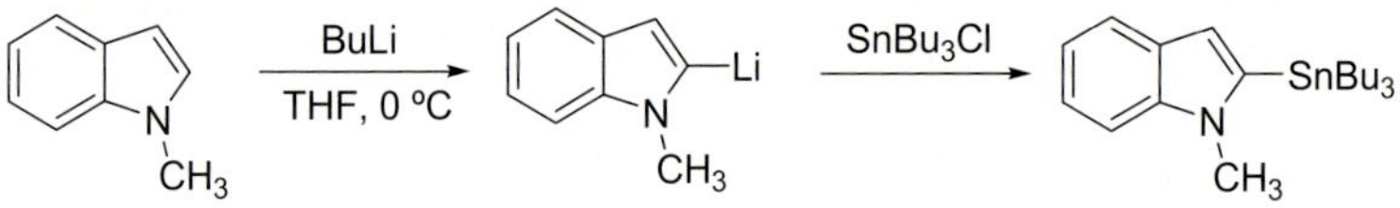

Scheme 5.170

In contrast, indole **353**, bearing the very bulky triisopropylsilyl N-substituent, is lithiated selectively at C3 upon treatment with *t*-BuLi (Scheme 5.171) [279].

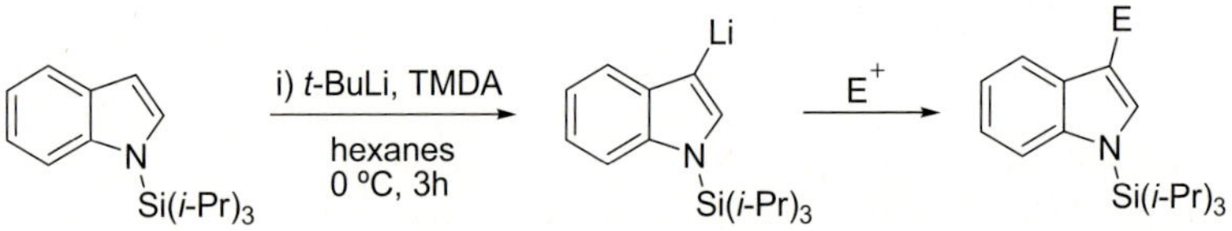

Scheme 5.171

The directing effect of the N-protecting group has been also applied to effect regioselective ortho-metallation at the C7 position [280]. For this purpose, it is necessary to protect the C2 position with a removable group, such as TMS. Then, treatment of *N*-$CONEt_2$ protected indole **354** with *t*-BuLi/THF at −78 °C followed by quench with TMSCl gives rise to the indole silylated at C2 **355**. Treatment of **355** with *s*-BuLi/TMEDA/THF at −78 °C produces the regioselective metallation at C7, which provides the corresponding C7 functionalized indole **356** upon reaction with an electrophile (Scheme 5.172). Interestingly, the whole process can be conducted in a *one pot* fashion.

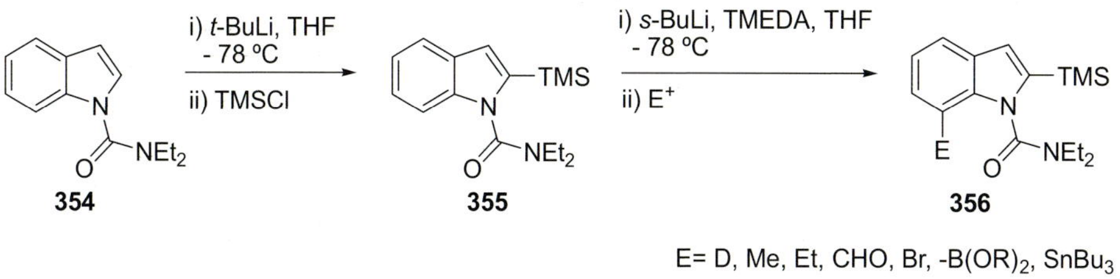

i) *t*-BuLi, THF - 78 °C; ii) TMSCl

i) *s*-BuLi, TMEDA, THF - 78 °C; ii) E^+

354 **355** **356**

E= D, Me, Et, CHO, Br, $-B(OR)_2$, $SnBu_3$

Scheme 5.172

5.5.3 Transition Metal Catalyzed Reactions

The most reliable synthetic methods to create C−C and C−X bonds from C-sp^2 are currently transition metal catalyzed reactions. The well-developed Pd-catalyzed Heck and cross-coupling reactions are doubtless the most prominent methods, and have been applied successfully to the incorporation of new substituents in the indole ring. Moreover, alternative methodologies that make use of different transition metals are also noteworthy, and are discussed in this section.

5.5.3.1 General Considerations on Palladium-Catalyzed Cross-Coupling Reactions

Indolyl halides or triflates behave as regular aryl halides in Pd-catalyzed cross-coupling reactions. Thus, 2- and 3-halo or triflate substituted indoles have been employed to synthesize numerous indole derivatives. The process starts with the oxidative addition of indolyl halide **357** to the Pd(0) catalyst to form indolyl-Pd(II) complex **358**. The nature of the coupling partner, alkene, alkyne, stannane, boronate, organozinc, amine, determines the subsequent steps of the cross-coupling process. In a prototypical cross-coupling reaction, transmetallation of the organometallic partner followed by reductive elimination produces the cross-coupling product **360** and releases the Pd(0) catalyst (Scheme 5.173).

5.5.3.2 Reactions with Alkenes and Alkynes: Heck Reactions

The reaction of indolyl halides or pseudohalides with alkenes under standard Heck conditions gives rise to vinylindoles [281]. For instance, *N*-protected-4-bromoindole **361** can be transformed into 4-alkenylsubstituted indoles **362** under standard Heck conditions (Scheme 5.174) [282].

The higher reactivity of iodide than bromide towards oxidative addition to Pd, allows for the sequential substitution of 3-iodo-4-bromoindole **363** through two consecutive Heck reactions. In the first step substitution of the more reactive iodine occurs to yield **364**. A second Heck reaction with substitution of the bromine leads to **365** (Scheme 5.175). This strategy was employed by Hegedus in the synthesis of ergot alkaloids [283].

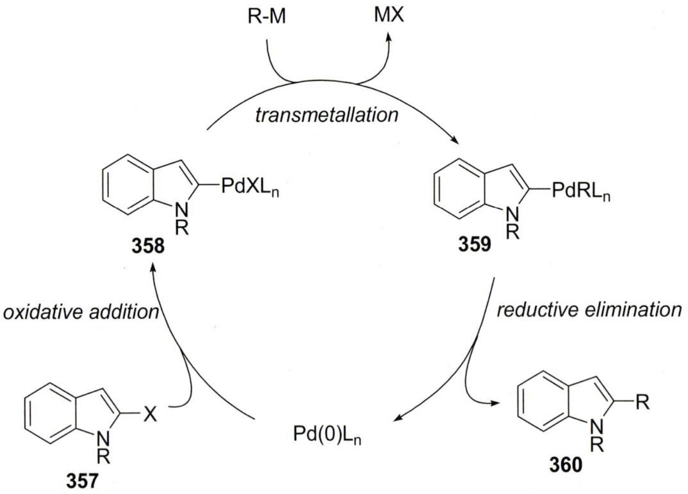

Scheme 5.173

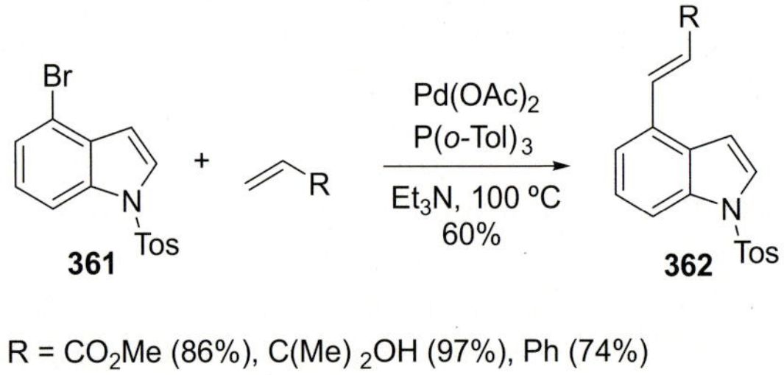

R = CO_2Me (86%), $C(Me)_2OH$ (97%), Ph (74%)

Scheme 5.174

In a similar manner to other Pd-catalyzed cross-couplings, the reaction is general regarding the position of the leaving halogen. For instance, 3-substituted 2-iodoindole **366** reacts with methyl acrylate to furnish 2-alkenylindole **367** (Scheme 5.176) [284].

The intramolecular Heck reaction is particularly appealing, as it leads to polycyclic indole derivatives that might be difficult to synthesize through other strategies [285]. For instance, *N*-allylindoles **368** undergo cyclization to give 3*H*-pyrroloquinoline **369** under typical Heck reaction conditions (Scheme 5.177) [286], and carbolines **371** are easily prepared from 3-iodoindoles **370** (Scheme 5.178) [287].

5.5.3.3 **Sonogashira Reaction**

Indolyl halides or triflates react under typical Sonogashira [288] conditions with terminal alkynes to give rise to the corresponding alkynes (Scheme 5.179) [289]. The alkynylation can be employed to prepare 2-alkynylindoles **372**, 3-alkynylindoles **373**, and also to introduce the alkyne in the benzene ring [290] from the corresponding indolyl halides or triflates.

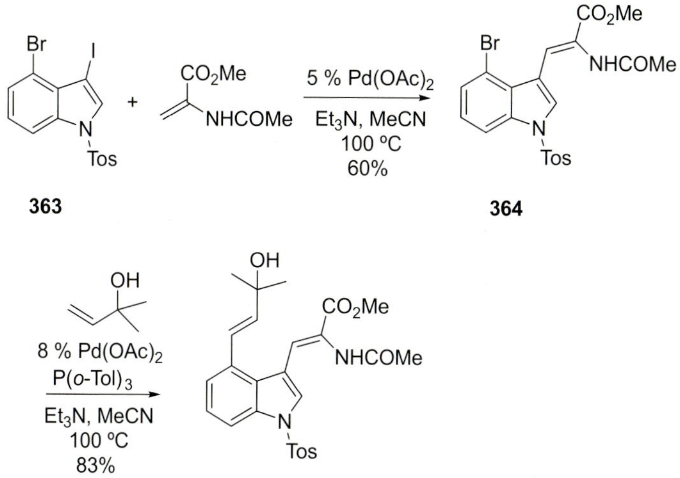

363 + methyl 2-acetamidoacrylate (CO_2Me, NHCOMe) → 364

5 % $Pd(OAc)_2$
Et_3N, MeCN
100 °C
60%

364 → 365

2-methylbut-3-en-2-ol (OH)
8 % $Pd(OAc)_2$
$P(o\text{-}Tol)_3$
Et_3N, MeCN
100 °C
83%

Scheme 5.175

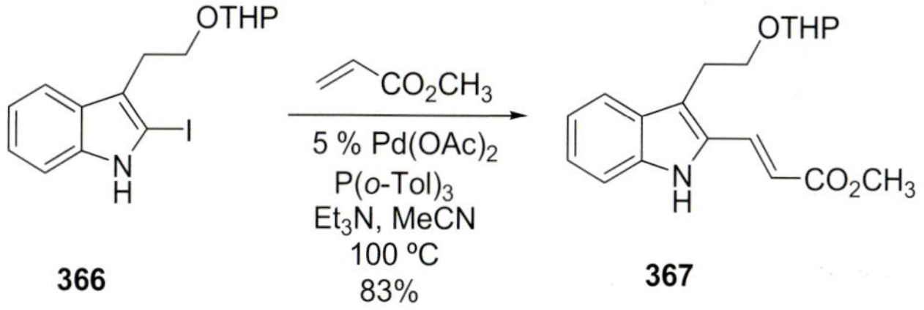

366 (OTHP) → 367 (CO_2CH_3)

methyl acrylate (CO_2CH_3)
5 % $Pd(OAc)_2$
$P(o\text{-}Tol)_3$
Et_3N, MeCN
100 °C
83%

Scheme 5.176

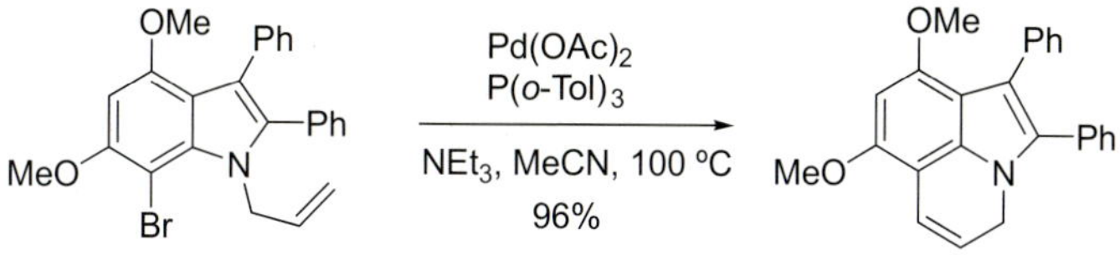

368 (OMe, Ph, Ph, MeO, Br) → 369

$Pd(OAc)_2$
$P(o\text{-}Tol)_3$
NEt_3, MeCN, 100 °C
96%

Scheme 5.177

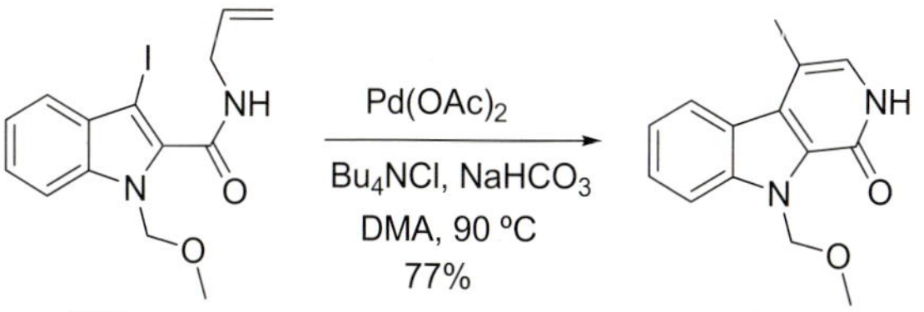

370 → 371

$Pd(OAc)_2$
Bu_4NCl, $NaHCO_3$
DMA, 90 °C
77%

Scheme 5.178

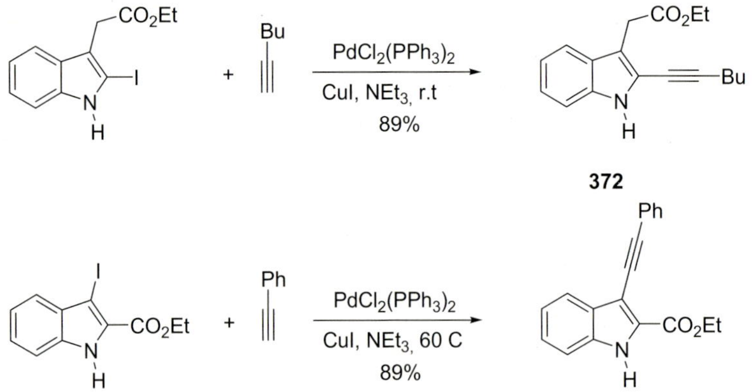

Scheme 5.179

5.5.3.4 Cross-Coupling Reactions with Organometallic Reagents

Typical Pd-catalyzed cross-couplings consist of the reaction of an organic halide with an organometallic reagent. The most popular methods involve the use of organotin (Stille reaction), organoboron (Suzuki–Miyaura reaction) [291] and organozinc (Negishi reaction) [292] compounds. Two different strategies are possible to conduct a cross-coupling reaction involving an indole derivative: (i) couple an indolyl halide or triflate with an organometallic reagent and (ii) react an indolylmetal derivative with an organic halide. Regardless of the strategy chosen, in most of the cases, cross-coupling reactions involving indolyl species can be performed efficiently under the typical conditions developed for cross-couplings with benzenoid systems.

5.5.3.4.1 Suzuki–Miyaura Cross-Coupling

The Pd-catalyzed reaction of a boronic acid and an aryl or alkenyl halide or sulfonate is one of the most popular C–C bond-forming reactions involving aromatic species, and has been extensively applied to the functionalization of indoles on every position of the ring. Indolyl halides and triflates have been employed in the coupling reaction. For instance, the reactions of 2-indolyl triflates with aryl boronic acids afford the corresponding aryl substituted indoles in high yields (Scheme 5.180) [293].

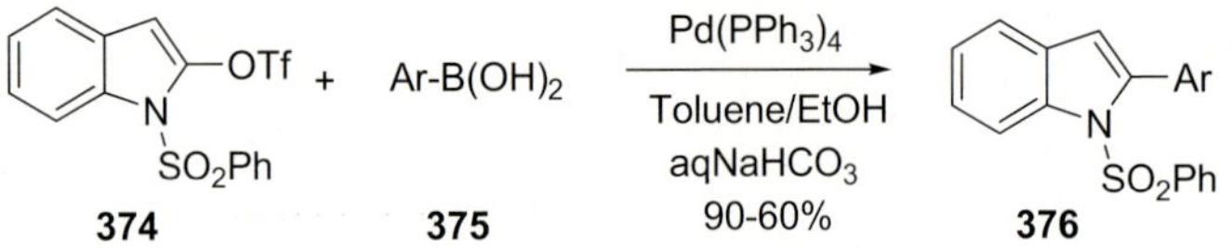

Scheme 5.180

The Suzuki coupling has been also applied to the preparation of vinylindoles. One example is the reaction of vinylboronate **378** with 3-iodoindole **377**, which leads to 3-vinylindole **379** (Scheme 5.181) [294].

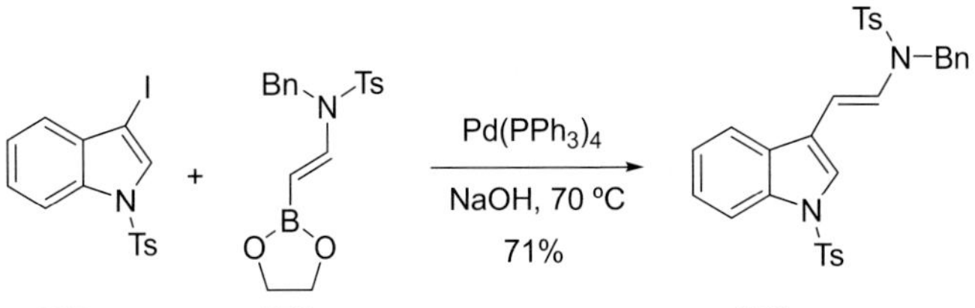

Pd(PPh3)4
NaOH, 70 °C
71%

377 378 379

Scheme 5.181

Thiophen-3-trifluoroborates **381** have been employed in a coupling reaction with 7-bromoindole **380**, giving rise to the corresponding thiophene-substituted indole **382** (Scheme 5.182) [295].

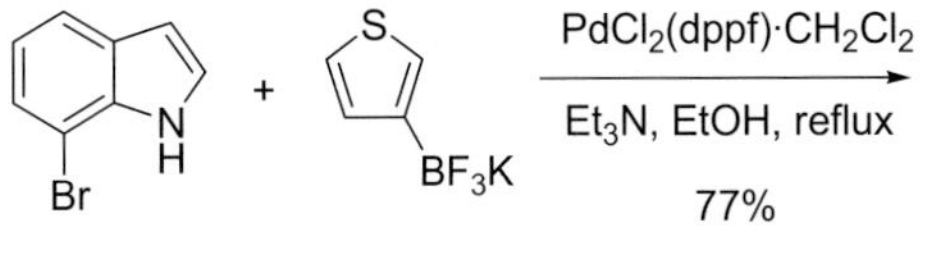

PdCl2(dppf)·CH2Cl2
Et3N, EtOH, reflux
77%

380 381 382

Scheme 5.182

Indoylboronic acids are also appropriate coupling partners in cross-coupling reactions [296]. The sometimes difficult purification of indolylboronic acids recommends its utilization as crude products. An example of this methodology is represented in the synthesis of **386** (Scheme 5.183), a precursor of a tyrosine kinase

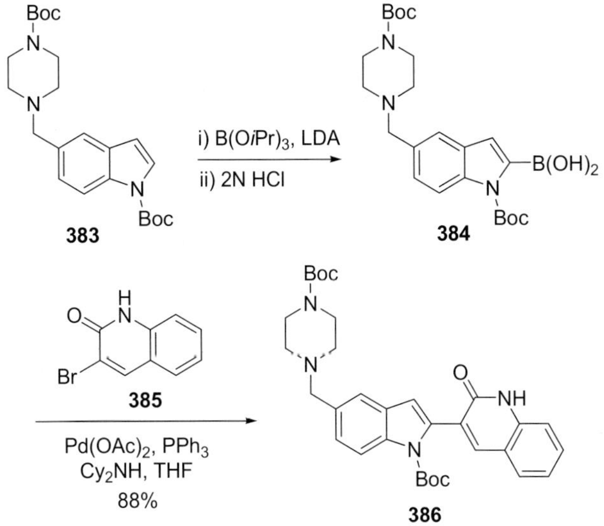

i) B(OiPr)3, LDA
ii) 2N HCl

383 384

385
Pd(OAc)2, PPh3
Cy2NH, THF
88%

386

Scheme 5.183

inhibitor. The indolylboronic acid **384** required is prepared by metallation of indole **383**, followed by reaction with triisopropylborate. The coupling reaction with the corresponding bromide **385** then affords **386** [297]. Remarkably, the sequence can be conducted on a multi-kilogram scale.

7-Arylsubstituted indoles can be synthesized from the corresponding boronates **387**, which can be prepared by the synthetic sequence discussed in Scheme 5.167. Subsequent Suzuki reaction leads to the 7-aryl substituted indoles **388** (Scheme 5.184) [298].

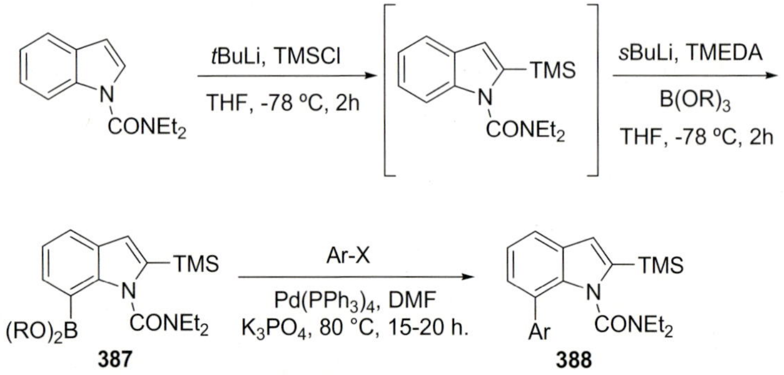

tBuLi, TMSCl; THF, -78 °C, 2h; TMS; $CONEt_2$; sBuLi, TMEDA; $B(OR)_3$; THF, -78 °C, 2h

$(RO)_2B$; TMS; $CONEt_2$; **387**; Ar-X; $Pd(PPh_3)_4$, DMF; K_3PO_4, 80 °C, 15-20 h.; Ar; TMS; $CONEt_2$; **388**

Scheme 5.184

5.5.3.4.2 **Stille Cross-Coupling** The Pd-catalyzed coupling of organostannanes with halides or pseudohalides is generally known as the Stille reaction. This coupling reaction has been widely employed in the modification of indoles. Both indolyl triflates (Scheme 5.185) [299, 300] and halides have been employed in the coupling reaction.

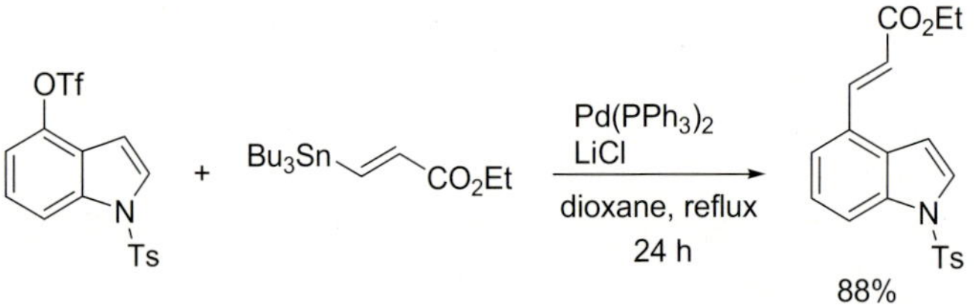

OTf; Ts; + Bu_3Sn CO_2Et; $Pd(PPh_3)_2$; LiCl; dioxane, reflux; 24 h; CO_2Et; Ts; 88%

Scheme 5.185

The Stille coupling has been employed in many syntheses of biologically active indole alkaloids [301]. In the example represented in Scheme 5.186, 2-vinylindole **391** (an intermediate in the synthesis of the natural alkaloid tabersonine) was prepared from 2-iodoindole **389** and vinylstannane **390** [302].

Interestingly, when the Stille cross coupling conditions are applied under a CO atmosphere, the corresponding α,β-unsaturated ketones are isolated (Scheme 5.187) [303].

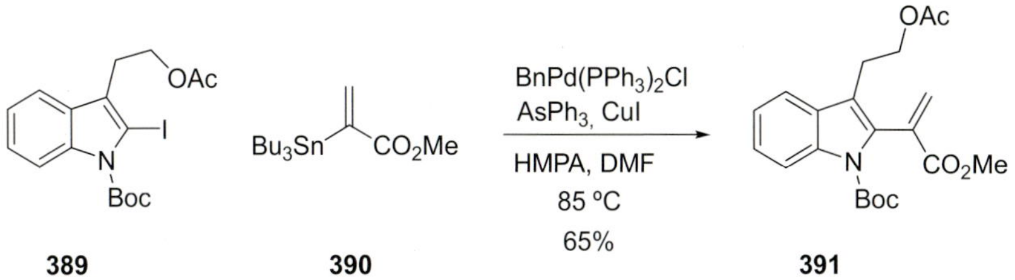

389 390 391

Scheme 5.186

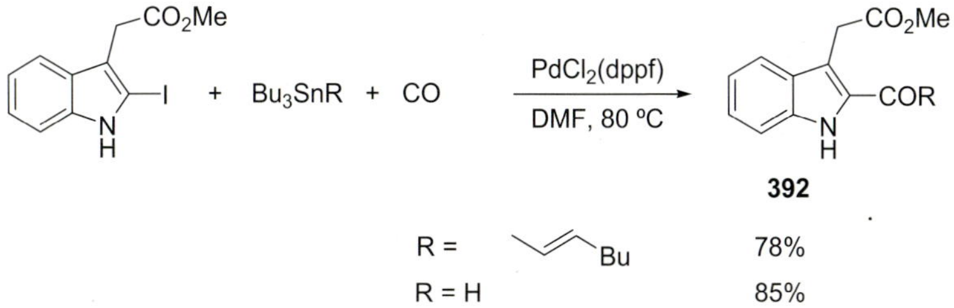

392

Scheme 5.187

Indolylstannanes [304] have also been widely utilized for the functionalization of indoles, mostly at C2 [305] and C3 [306]. Scheme 5.188 presents the synthesis of 2-arylindoles by coupling of *N*-protected-2-indolylstannane **393** with aryl halides under typical Stille conditions. A variation of this reaction has also been adapted to a solid-phase synthesis [307].

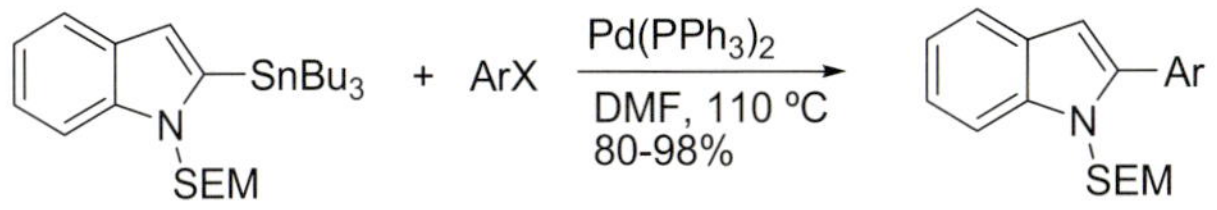

393

SEM: $-CH_2OCH_2CH_2TMS$

Scheme 5.188

5.5.3.5 **C–N Bond-Forming Reactions**

Application of the Buchwald–Hartwig arylation to N–H indoles allows for the preparation of *N*-arylindoles from N–H indoles and aryl halides [308–310]. The reaction is very general regarding the structure of both coupling partners, the indole and the aryl halide. Aryl bromides, chlorides, and triflates can be employed successfully upon selection of the proper combination of ligand and base. Scheme 5.189 gives an example of a coupling reaction between indole **394** and *m*-bromoacetophenone (**395**). The reaction proceeds in the presence of potentially sensitive functional groups to give arylated indole **396** in quantitative yield. In a similar procedure,

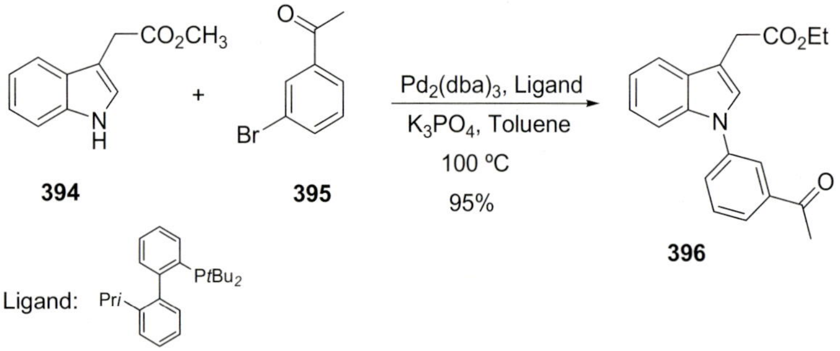

CO_2CH_3 + Br ... 394 395 → $Pd_2(dba)_3$, Ligand / K_3PO_4, Toluene / 100 °C / 95% → CO_2Et 396

Ligand: Pri ... PtBu2

Scheme 5.189

treatment with vinyl halides allows for the preparation of the corresponding *N*-vinylindoles [311].

On the other hand, Buchwald–Hartwig amination can be applied to incorporate an amino group into the indole structure. Thus, coupling of haloindoles with amines provides the corresponding aminoindoles. The coupling reaction can even be conducted with NH-containing indoles such as **397** to give amino substituted indole **398**, provided that an excess of base is employed (Scheme 5.190) [312].

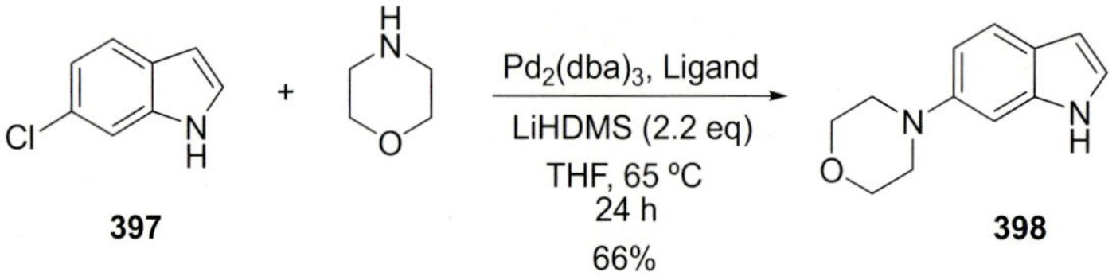

Cl ... 397 + morpholine → $Pd_2(dba)_3$, Ligand / LiHDMS (2.2 eq) / THF, 65 °C / 24 h / 66% → 398

Scheme 5.190

5.5.3.6 Transition Metal Catalyzed C–H Activation

Modification of the indole ring via substitution of a C–H bond by a C–C bond is a very desirable reaction, as it does not require the previous presence of a reactive functionalization such as C–halogen, C–triflate, or C–metal bond in the indole ring. Nevertheless, this challenging transformation has been comparatively much less studied than the cross-coupling reactions, and at the present suffers from some limitations, such as lack of selectivity and generality.

Many examples exist of intramolecular Pd(II)-catalyzed oxidative cyclizations of indoles with a proper pedant aryl or alkenyl group [313]. For instance, the oxidative cyclization of bisindolylmaleimides **399** gives rise to indolo[2,3-*a*]pyrrolo[3,4-*c*]carbazoles **400** [314], a substructure that is present in a large number of natural products (Scheme 5.191).

One approach for the direct substitution of the C–H bond is the reaction with metal salts that usually starts with the electrophilic metallation of the indole (Scheme 5.192).

An interesting recent example is the Pd-catalyzed oxidative annulation of alkenyl indoles **401** (Scheme 5.193), which gives rise to tricyclic derivatives **402** [315].

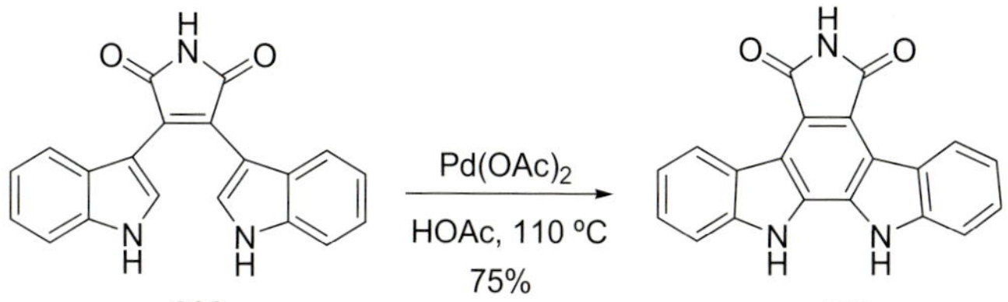

Scheme 5.191

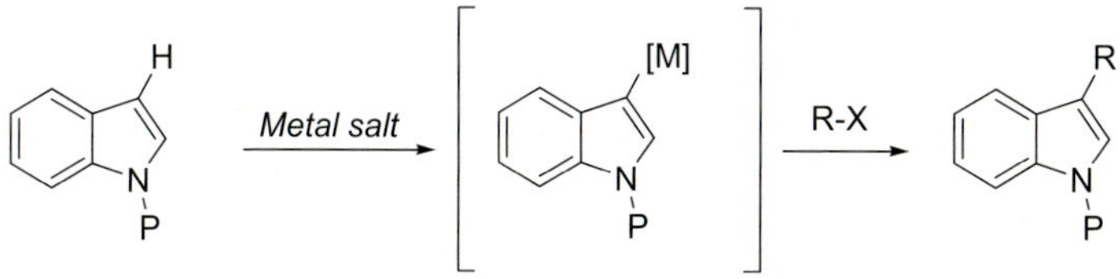

Scheme 5.192

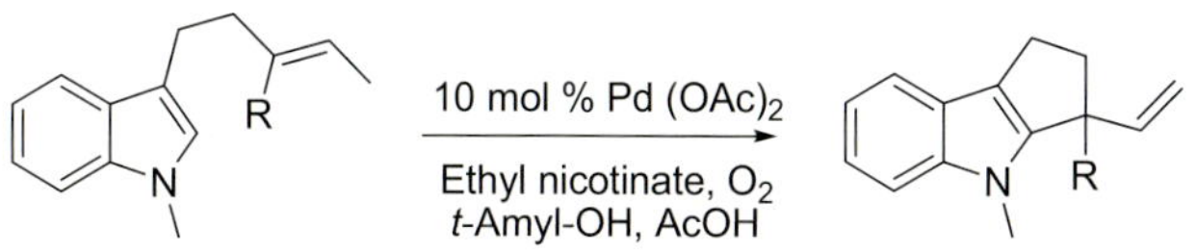

Scheme 5.193

The mechanism proposed for this annulation includes (i) electrophilic palladation of the indole to form indolyl palladium complex **403**; (ii) intramolecular olefin insertion to give palladated complex **404**; and (iii) β-hydrogen elimination, which gives rise to the annulated indole **402** and releases a Pd(0) species. To obtain a catalytic reaction, reoxidation of the Pd(0) to Pd(II) is necessary. In this example, a pyridine derivative (ethyl nicotinate) in an oxygen atmosphere is responsible for reoxidation of the Pd catalyst (Scheme 5.194).

Some examples of the intermolecular oxidative coupling of indoles with olefins and alkynes have been also described [316, 317]. The intermolecular oxidative Heck reaction of indoles with alkenes allows for the selective alkenylation at C2 or C3, depending on the solvent and reaction conditions chosen [318]. Thus, when the reaction is carried out using a mixture of DMF and DMSO as solvent, and $Cu(OAc)_2$ as oxidant, the expected C3 oxidative alkenylation is produced, giving rise to the C3-alkenylindole **405** (Scheme 5.195). However, if the reaction is performed under acidic conditions (dioxane : AcOH), employing *tert*-butyl benzoyl peroxide as oxidant, the

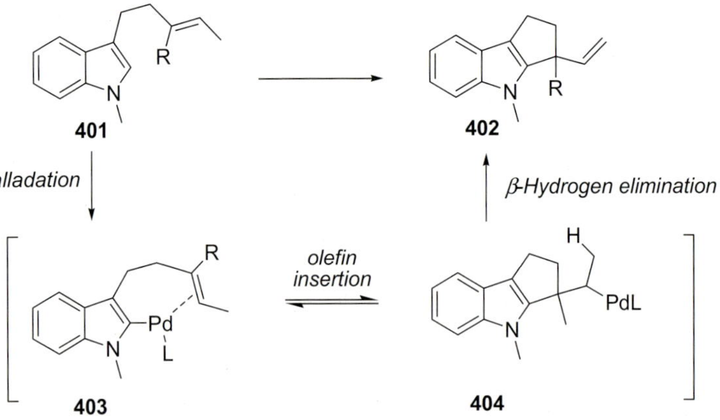

401 → 402

palladation

β-Hydrogen elimination

403 olefin insertion 404

Scheme 5.194

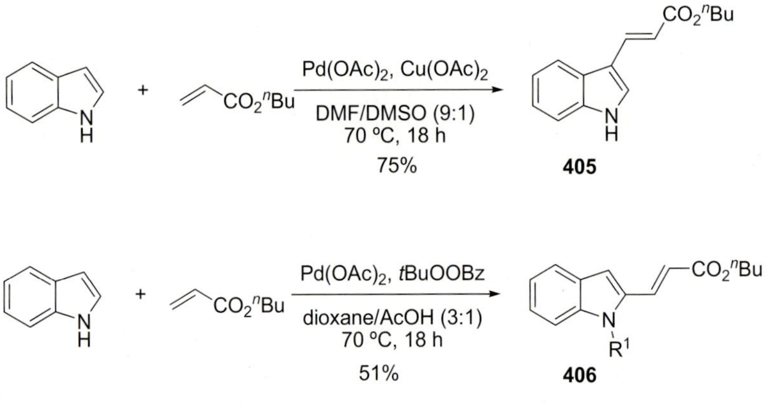

Indole + $CH_2=CHCO_2{}^nBu$ → (Pd(OAc)$_2$, Cu(OAc)$_2$; DMF/DMSO (9:1), 70 °C, 18 h; 75%) → 405

Indole + $CH_2=CHCO_2{}^nBu$ → (Pd(OAc)$_2$, tBuOOBz; dioxane/AcOH (3:1), 70 °C, 18 h; 51%) → 406

Scheme 5.195

regioselectivity of the alkenylation is switched to the 2 position to give alkenylindole **406** as the major product.

An explanation of this switch in the regioselectivity can be found in the mechanisms proposed for each process as represented in Scheme 5.196. The reaction starts with the electrophilic palladation at the electron-richer C3 position, to provide cationic indole complex **407**. At this point, aromatization by loss of a proton leads to indolyl palladium complex **408** (left-hand cycle), which then follows the path of a typical Heck reaction. The catalytic cycle is completed by oxidation of the Pd(0) species liberated. When the reaction is carried out under acidic media, the deprotonation of complex **407** is disfavored (right-hand cycle). Instead, 1,2-Pd migration takes place to give the new cationic complex **409**, which then follows the reaction path of a Heck reaction to produce alkenylation at C2.

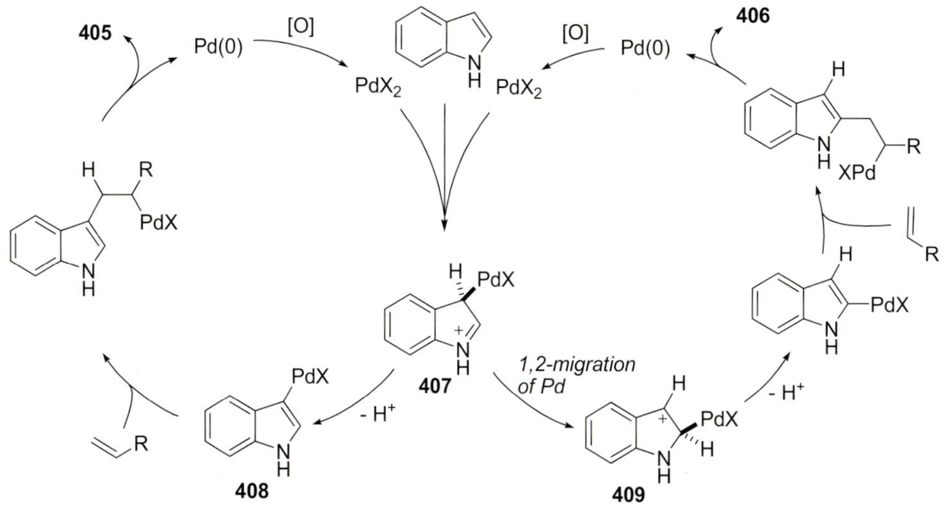

Scheme 5.196

In a similar way, indoles add regioselectively to alkynoates **410** in the presence of $Pd(OAc)_2$ and acetic acid [319]. Indole itself gives rise to 3-alkenylindoles **411**, while 3-methylindole provides 2-alkenylindoles **412**, also with good yields (Scheme 5.197).

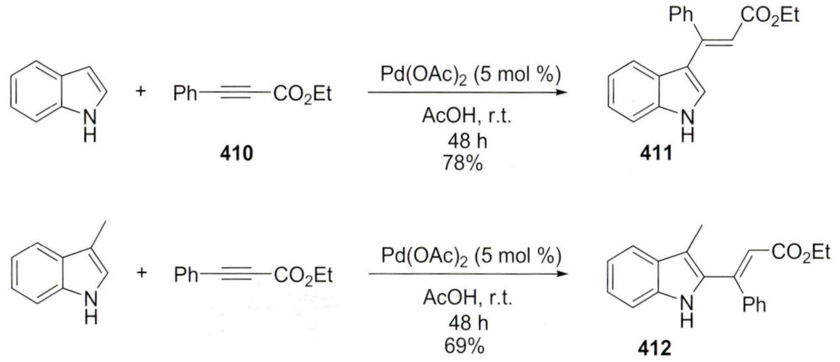

Scheme 5.197

The mechanism proposed for the reaction involves (i) electrophilic palladation of the indole to form indolyl palladium complex; (ii) complexation of the alkyne followed by regioselective carbopalladation of the triple bond to provide vinyl palladium complex; and (iii) protonolysis of the C–Pd bond by action of the AcOH, affording alkenylated indole **411** and liberating the Pd(II) species (Scheme 5.198) .

Direct Pd-catalyzed C–H substitution has also been accomplished by reaction of indolyl anions **413** with organopalladium complexes **414**, generated *in situ* from Pd(0) complexes and aryl halides (Scheme 5.199). Arylation of NH-indoles can be carried out selectively either at C2 [320, 321] or C3 [322], depending on the reaction

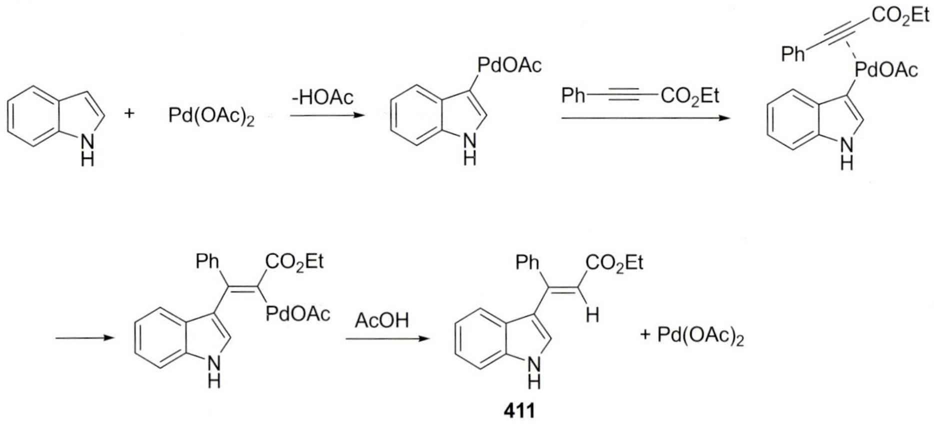

Scheme 5.198

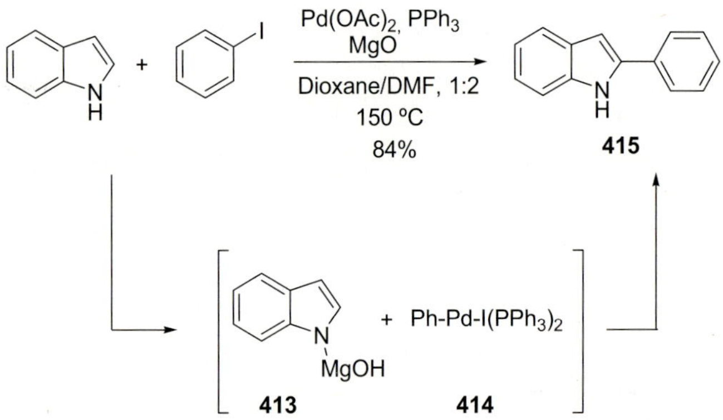

Scheme 5.199

conditions (Scheme 5.200). Thus, reaction with MgO as base gives rise exclusively the C2 arylated product **415**, while the use of larger bases such as $Mg(HMDS)_2$ provides the C3 arylated indole **417** with very high regioselectivity.

A rationale for this reactivity trend has been found after detailed mechanistic investigations (Scheme 5.201). Electrophilic palladation of the indolylmagnesium salt with the arylpalladium complex **414** gives rise to the cationic indolinylarylpalladium complex **418**. Aromatization by deprotonation gives indolylaryl palladium complex **419**, and reductive elimination produces the C3 arylated indole **417**. In contrast, palladium migration on complex **418** leads to the C2 palladated indole **420**, which will evolve through 2-indolylaryl palladium complex to furnish the C2 arylated indole **415**. The driving force for the migration step has been related to stabilization of the carbon–palladium bond by the adjacent nitrogen atom on **420**. However, bulky ligands on the magnesium destabilize the C2 palladated indole **420** and, therefore, the migration does not occur and so the arylation takes place at C3 (Scheme 5.201).

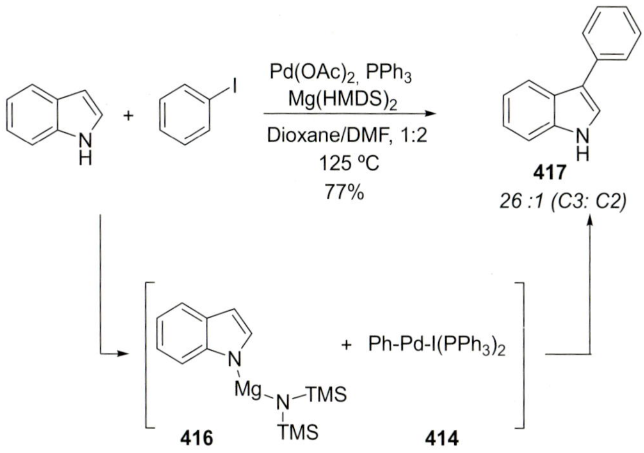

Scheme 5.200

In contrast, when the steric interactions are minimized with small bases, such as MgO, the migration step is favored and the arylation occurs at C2. In agreement with this hypothesis, bulkier ligands on the Pd also drive the reaction to the C3 arylation product.

Regioselective indole C2 arylation has also been conducted employing Rh(III) complexes as catalysts. The coupling process is achieved in the presence of a catalyst that is assembled *in situ* from a rhodium species, a phosphine ligand and CsOPiv as base (Scheme 5.202) [323].

The proposed catalytic cycle involves (i) oxidative addition of the aryl halide, to form Rh(III) complex **422**, (ii) complexation of the indole to form complex **423**, followed by

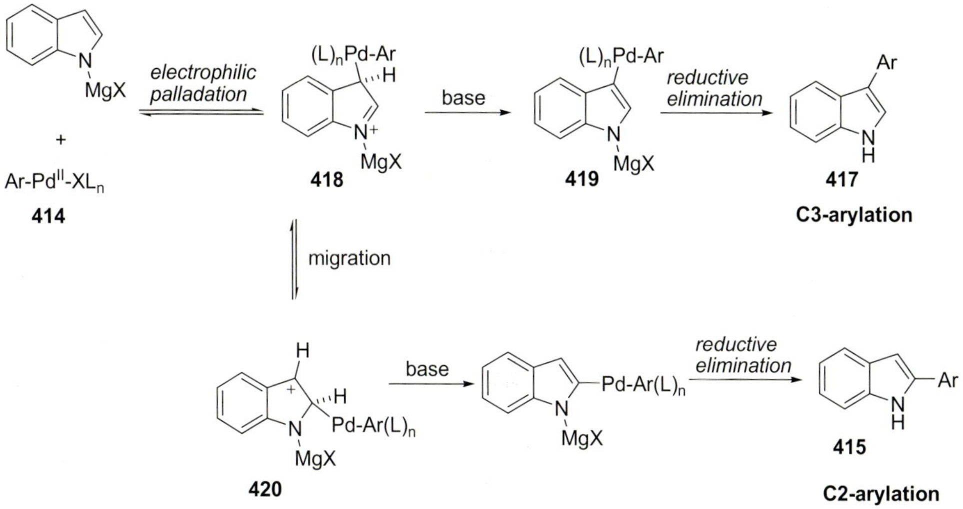

Scheme 5.201

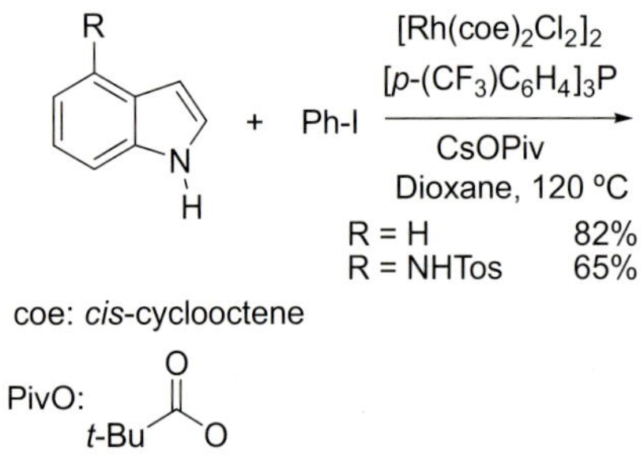

$[Rh(coe)_2Cl_2]_2$, $[p\text{-}(CF_3)C_6H_4]_3P$, CsOPiv, Dioxane, 120 °C

R = H 82%
R = NHTos 65%

coe: *cis*-cyclooctene

PivO: t-Bu

Scheme 5.202

pivalate promoted C–H bond metallation to give indolyl rhodium complex **424** and (iii) reductive elimination to release the C2 arylated indole **421** (Scheme 5.203).

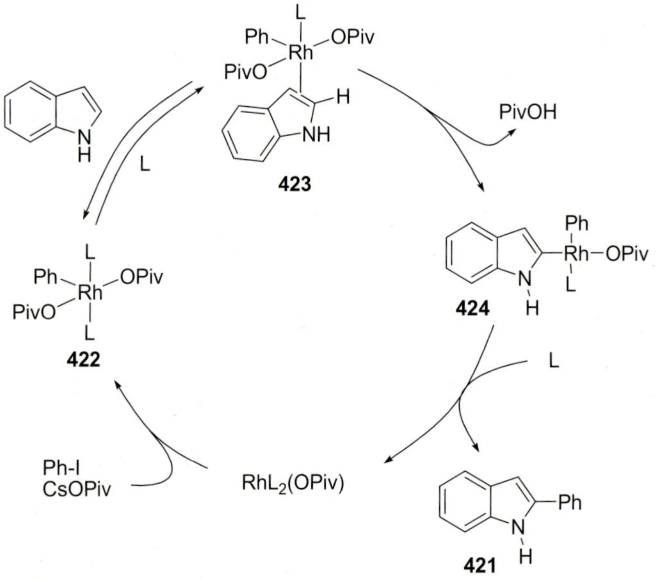

PivOH
Ph-I
CsOPiv
$RhL_2(OPiv)$

Scheme 5.203

5.5.4
Radical Reactions

Intermolecular radical aromatic substitution on indole can be effected at C2 with electrophilic carbon-centered radicals. For instance, indole reacts with radicals generated from iodoacetates or bromomalonates **425** to give the alkylated indole **426** (Scheme 5.204) [324, 325]. The reactions proceed with high regioselectivity, although a large excess of indole is required to avoid polysubstitution reactions.

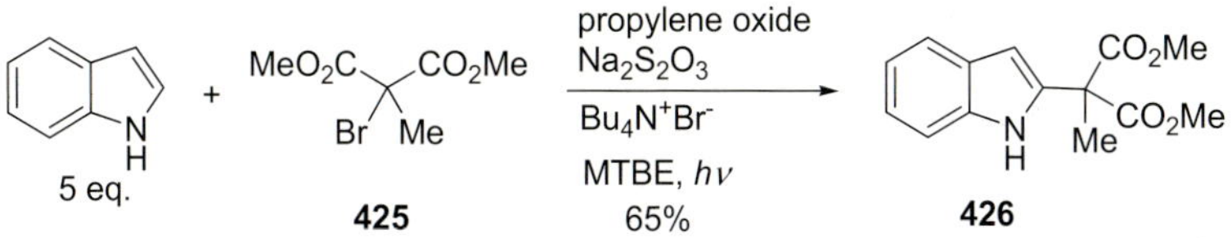

Scheme 5.204

A more efficient procedure for the oxidative radical alkylation of indoles employs α-acetyl or α-acetonyl radicals generated from the corresponding xantenes **427**. In this way, radical alkylation can be accomplished to prepare 2-indolyl acetate **428** in preparatively useful yields without the need of excess indole (Scheme 5.205) [326, 327].

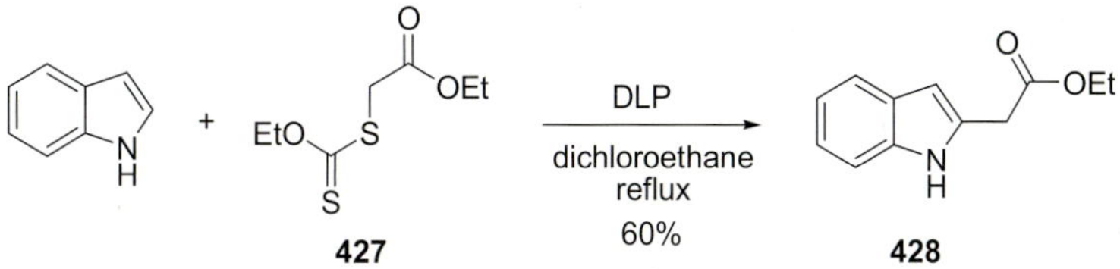

DLP: dilauryol peroxide

Scheme 5.205

While the intermolecular reactions of indoles with radicals have found limited application, there are many examples of intramolecular radical additions to the indole ring that have been employed as an entry into more complex heterocyclic structures [328–331]. For instance, spirocyclic dearomatized indole derivatives **430**, **432**, and **434** are formed via 5-exo-trig cyclizations from aryl, vinyl and alkyl radical precursors **429**, **431**, and **433**, respectively (Scheme 5.206) [332].

In these examples the reaction proceeds by attack of the intermediate radical **435** to the C2 position of the indole ring (Scheme 5.207). The additional stability of the benzylic radical **436** might account for the preference of the C2 attack instead of C3 attack.

Nevertheless, it has been observed that the nature of the pedant group may influence the course of the cyclization, and, in some cases, the addition of the radical at C3 is the main or unique reaction pathway [333, 334]. This is the case of the tandem radical sequence that has been applied to the preparation of functionalized indolenine **441** from amidoindole **437** (Scheme 5.208) [335]. Reaction of indole amide **437** with the tributylstannyl radical produces the aryl radical **438** by bromine atom abstraction. The aryl radical undergoes [1,5]-hydrogen atom abstraction, which generates the amido radical **439**. Intramolecular addition of the radical to the C3 position of the indole produces indolyl radical **440**, which leads to the indolenine **441** after hydrogen atom abstraction of the tributylstannyl hydride. The same authors have further extended the radical sequence to the preparation of tetracyclic structures [336].

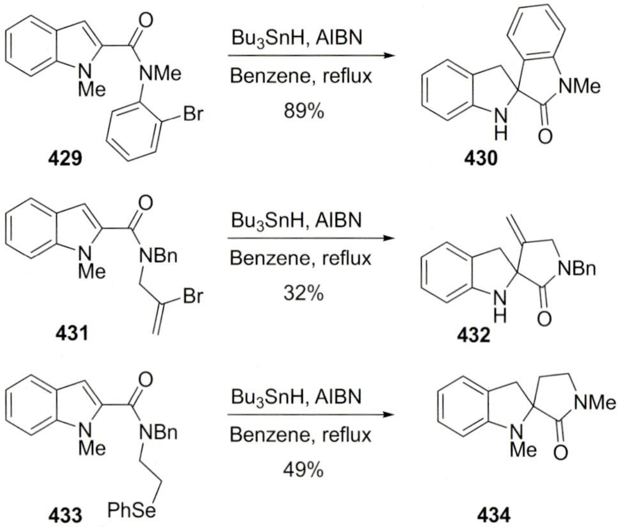

Scheme 5.206

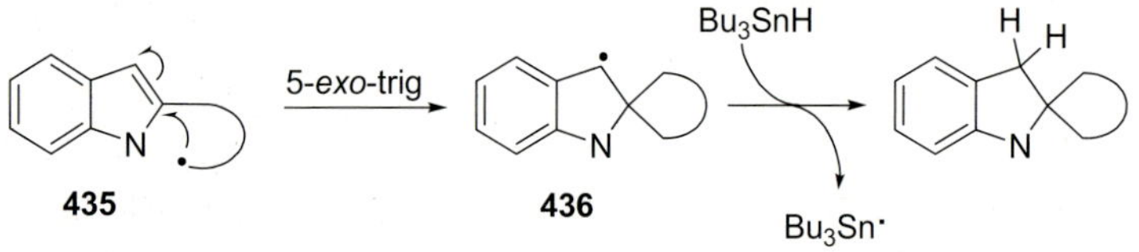

Scheme 5.207

On the other hand, indolyl radicals can be generated from the corresponding haloindoles by treatment with tributyltin hydride in the presence of a radical initiator. The indolyl radical can be trapped in an intermolecular [337] or in an intramolecular sense by reaction with a suitable radical acceptor. Scheme 5.209 shows the generation of 2-indolyl radical **443** from a 2-bromoindole (**442**) and the subsequent intramolecular cyclization to provide the tricyclic indole **444** [338].

Indolyl radicals such as **446**, generated from bromoindole **445**, which cannot undergo a direct intramolecular annulation, evolve through a [1,5]-hydrogen atom abstraction reaction to generate transient radical **447**. Then, intramolecular radical addition to the indole ring leads to the tetracyclic radical **448**, and finally to indoline **449** as a single diastereoisomer (Scheme 5.210) [339]. The reaction proceeds with moderate yield, as a 42% yield of dehalogenated indole **450** is also recovered.

A 3-indolyl radical can be also generated by oxidation of the corresponding anion. This strategy has been applied to devise an extremely potent coupling between unprotected indoles and ketone enolates [340]. Thus, simultaneous deprotonation of

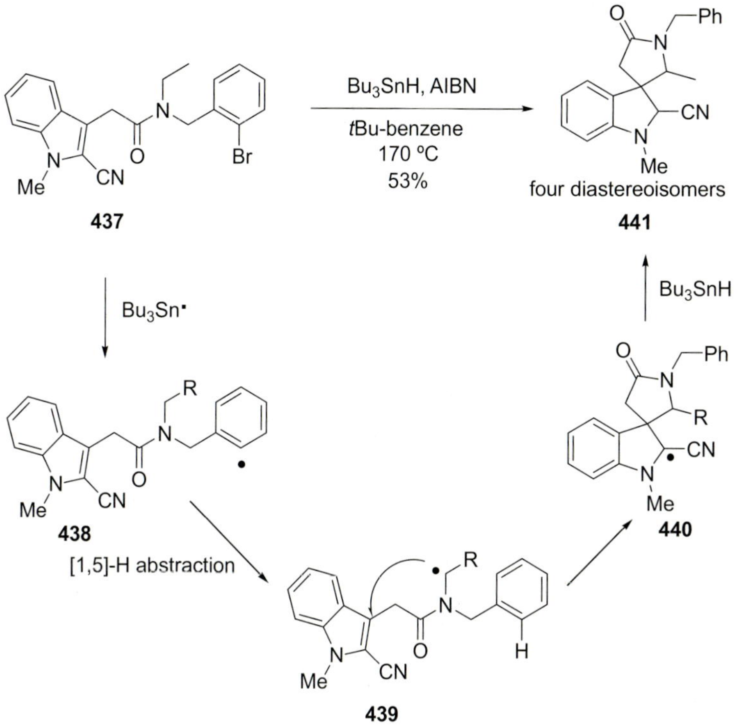

Bu$_3$SnH, AIBN
*t*Bu-benzene
170 °C
53%

four diastereoisomers

437 441 438 439 440

Bu$_3$Sn$^{\bullet}$
Bu$_3$SnH
[1,5]-H abstraction

Scheme 5.208

Bu$_3$SnH
Toluene
reflux

442 443 444
79%

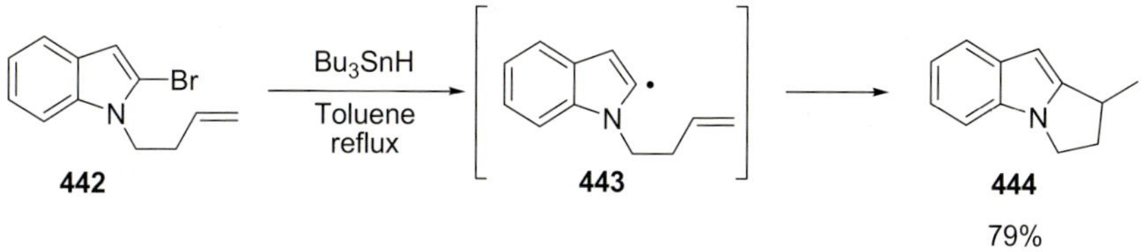

Scheme 5.209

indole and the ketone **451** gives indole anion **452** and ketone enolate **453**. Oxidation with a Cu(II) salt provides 3-indolyl radical **454** and ketone radical **455**, which upon radical coupling lead to substituted indole **456** (Scheme 5.211).

Importantly, the application of this methodology to ketones bearing stereogenic centers leads to the corresponding coupling products as single diastereoisomers. This strategy has been applied in the crucial step of a highly convergent total synthesis of various indole alkaloids such as Fischer-indoles I and G and welwitindolinone A [341]. Scheme 5.212 shows the coupling of ketone **457** with indole to give enantiomerically pure substituted indole **458**.

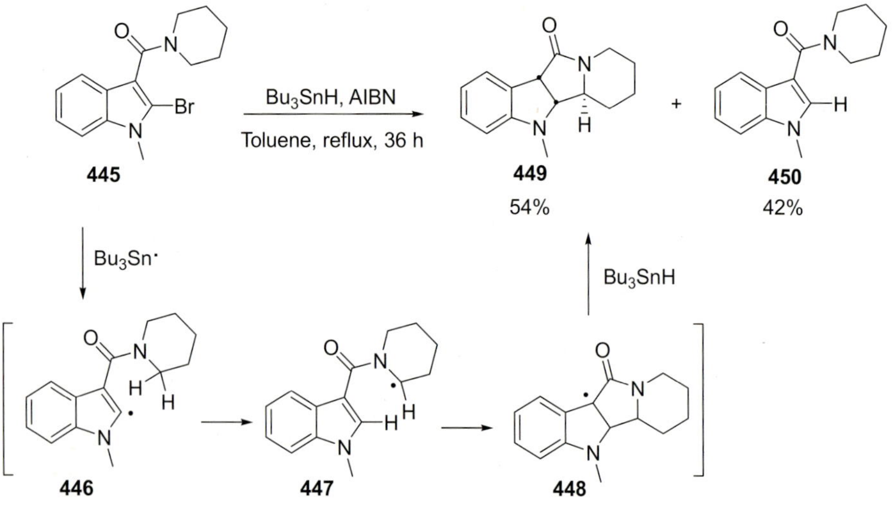

445 → (Bu_3SnH, AIBN; Toluene, reflux, 36 h) → 449 (54%) + 450 (42%)

445 → ($Bu_3Sn^{\cdot}$) → [446 → 447 → 448] → (Bu_3SnH) → 449

Scheme 5.210

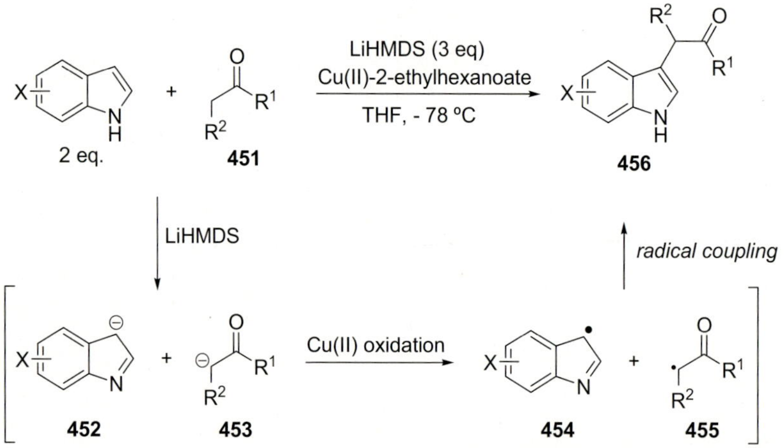

X-indole (2 eq.) + 451 → (LiHMDS (3 eq), Cu(II)-2-ethylhexanoate; THF, - 78 °C) → 456

LiHMDS → [452 + 453 → (Cu(II) oxidation) → 454 + 455] → (*radical coupling*) → 456

Scheme 5.211

457 → (indole, LiHMDS, Cu(II)2-ethylhexanoate; THF, -78 °C, 30 min; 55%) → 458 → → (-)-Fischer-indole G

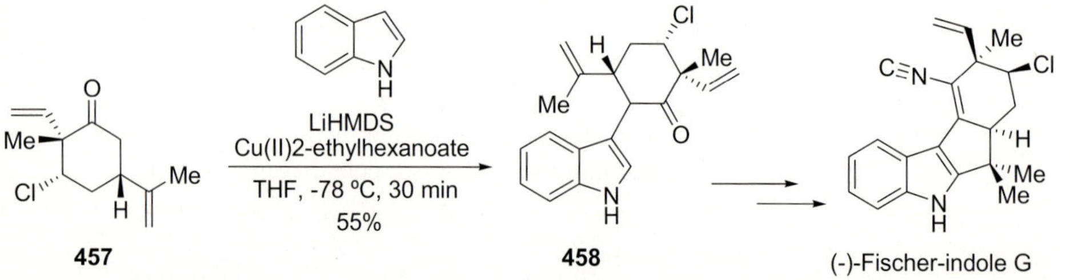

Scheme 5.212

5.5.5 Oxidation Reactions

Indoles are highly sensitive to oxidation and, thus, undergo aerial autooxidation to produce 3-hydroperoxo-3*H*-indoles, which are rarely isolated but decompose to produce complex mixtures of degradation and polymerization derivatives.

The transformation of *N*-tosylindole into indoxyl **459** has been described with oxodiperoxomolybdenum(IV) [342, 343], and by catalytic oxidation with dichlororuthenium(IV)*meso*-tetrakis(2,6-dichlorophenyl)porphyrin complex $[Ru^{IV}(2,6\text{-}Cl_2tpp)Cl_2]$. The latter process requires the presence of 2,6-dichloropyridine *N*-oxide as stoichiometric oxidant (Scheme 5.213) [344].

$[Ru^{IV}(2,6\text{-}Cl_2tpp)Cl_2]$, 2,6-dichloropyridine N-oxide, CH_2Cl_2, 40 °C; N-Ts indole → **459** (95%)

Scheme 5.213

Indoles can be converted into oxindoles through several different oxidative strategies. A general reaction is the treatment of indoles with conc. HCl in dimethyl sulfoxide, which provides cleanly the corresponding oxindoles **460** [345]. The reaction is likely to proceed through C3 protonation of indole to give indolenium intermediate **461**, followed by electrophilic addition of the DMSO to give intermediate **462** followed by elimination of dimethylsulfane (Scheme 5.214).

conc. HCl, DMSO, 20 °C, 0.5-4h → **460**; HCl → **461**; **461** + DMSO → **462** ($O\text{-}\overset{+}{S}Me_2$) → **460**

R
$-CO_2H$ (82%)
$-CH_2CH_2NH_2$ (61%)
$-CH_2(NH_2)CO_2H$ (54%)

Scheme 5.214

Oxidation of indoles to oxindoles **460** can also be produced by treatment of the indole with a halogenating agent, followed by hydrolysis. Thus, the intermediate indolenium cation **463** formed gives 2-hydroxy-3-halodihydroindole **464**, which suffers dehydrohalogenation to give oxindole **460** (Scheme 5.215) [346]. A closely related procedure allows for the substitution of indoles at C2. This oxidative-coupling procedure involves reaction of the indole with a suitable electrophile [X^+]. Addition of a nucleophile [Nu^-] to the transient indolenium cation **463** formed gives **465**, and aromatization by elimination of HX furnishes the functionalized indole **466**.

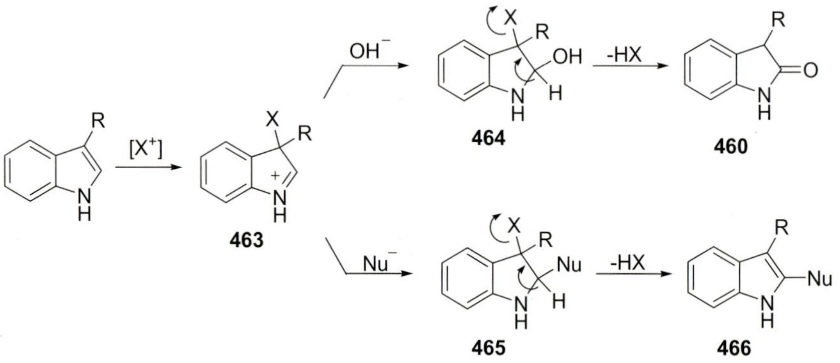

Scheme 5.215

Halogenating agents, such as *t*BuOCl, NCS or NBS are the electrophiles of choice. The oxidative-coupling strategy has been performed with various carbon nucleophiles, such as allyl boranes and stannanes, enol ethers, enamines, acetylide and even indole. Scheme 5.216 presents the synthesis of C2 substituted tryptophan **469** by nucleophilic addition of a silylenol ether (**467**) to the indolenium cation generated from protected tryptophan **468** [347].

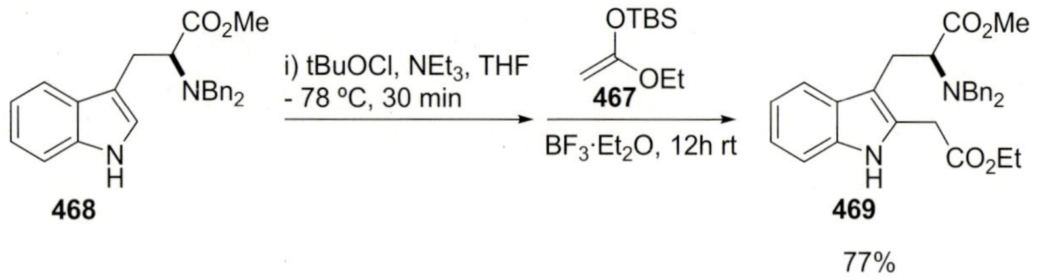

Scheme 5.216

Heteronucleophiles such as alcohols [348], anilines, phenols and thiophenols [349] are also appropriate reagents for the oxidative coupling, providing the corresponding C2 functionalized indoles. For instance, 2-aminoindole **471** can be prepared from indole **470** by employing this sequence (Scheme 5.217).

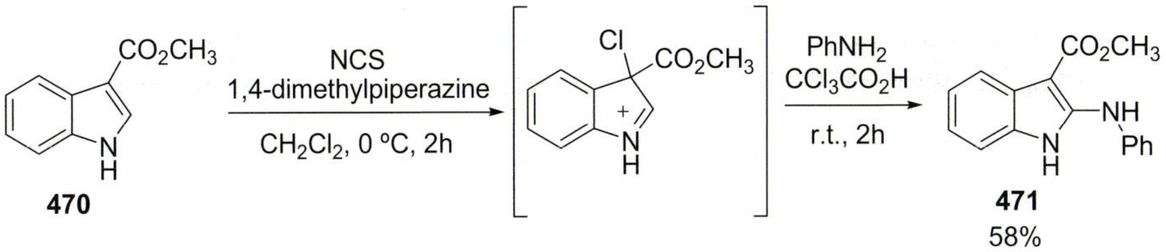

Scheme 5.217

Moreover, the versatility and functional group compatibility of this oxidative-coupling procedure has been shown in the coupling of a tryptophan derived peptide **472** with the imidazole ring of the histidine-containing peptide **473**, which represents the key step in the synthesis of the right-hand ring of the natural octapeptide celogentin (**474**) (Scheme 5.218) [350].

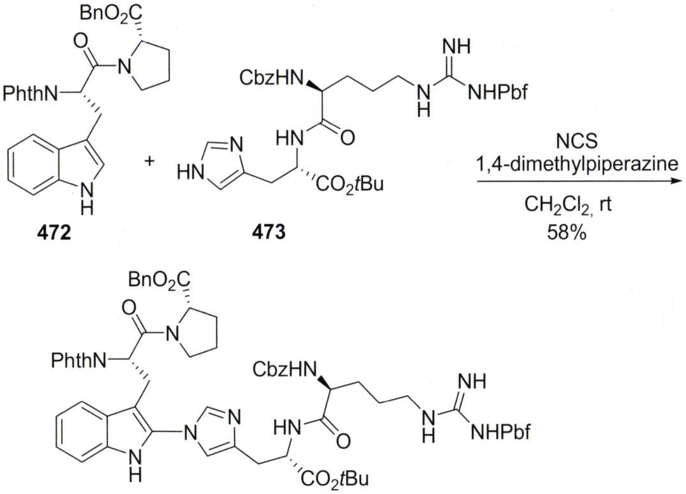

Scheme 5.218

A very valuable synthetic transformation is the oxidation of indoles with dimethyldioxirane, which produces the corresponding epoxides **475** [351]. Although the epoxides are unstable in most cases, their generation in the presence of a nucleophile gives rise to 3-hydroxyindoline derivatives **476** (Scheme 5.219).

The application of this strategy to protected tryptophan **477** led to pyrrolo[2,3-*b*]indole **478**, an early key intermediate in Danishefsky's total synthesis of himastatin (Scheme 5.220) [352].

In contrast, aqueous workup of the preformed epoxide leads to 3-hydroxyoxindoles **479** (Scheme 5.221) [353].

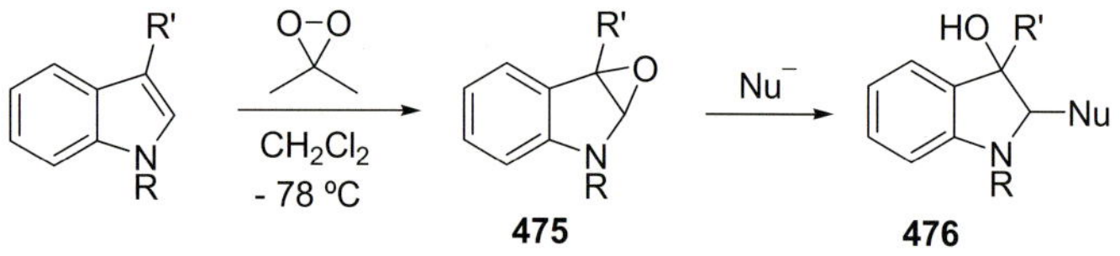

Scheme 5.219

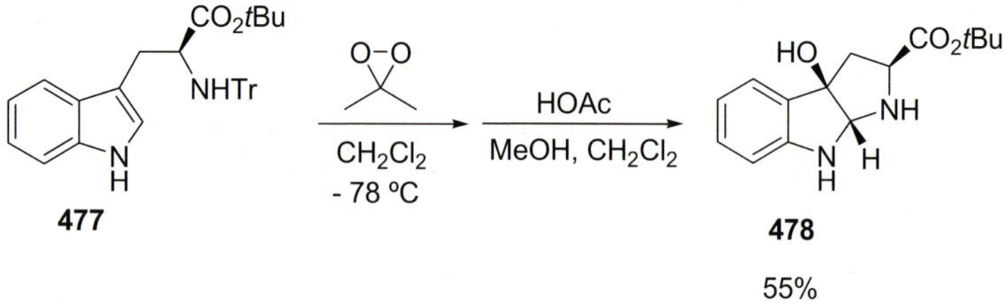

Scheme 5.220

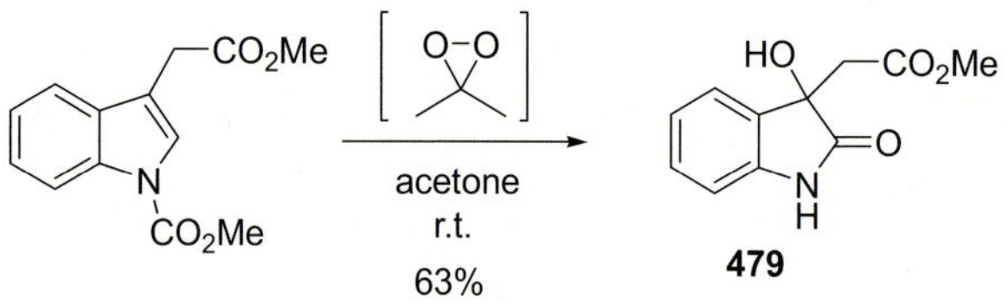

Scheme 5.221

5.5.6
Reduction of the Heterocyclic Ring

Indoles are reduced to indolines by several methods, including catalytic hydrogenation, dissolving metals and metal hydrides. Some detailed reviews on this subject are available [354].

5.5.6.1 Catalytic Hydrogenation

Catalytic hydrogenation of the heterocyclic ring has been achieved employing various heterogeneous catalysts, but usually requires harsh conditions and sometimes is not very selective. In the presence of strong acids, relatively milder conditions can be used, as the reaction proceeds through the C3 protonated indole [355]. For instance, *N*-Boc-2,3-disubstituted indoles **480** are hydrogenated to the corresponding *cis*-2,3-disubstituted indolines **481** over a rhodium-alumina catalyst in a EtOH/AcOH mixture (Scheme 5.222) [356].

The catalytic asymmetric hydrogenation of *N*-acylindoles **482** has been performed employing as catalyst a rhodium complex bearing the chiral diphosphine ligand **484**. In this way, optically active substituted indolines **483** can be obtained in high yield and enantiomeric excesses (Scheme 5.223) [357].

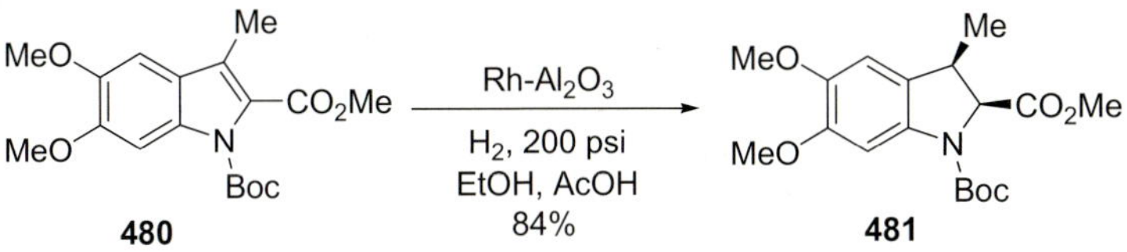
MeO
MeO
Me
CO_2Me
N
Boc
480
$Rh\text{-}Al_2O_3$
H_2, 200 psi
EtOH, AcOH
84%
MeO
MeO
Me
CO_2Me
N
Boc
481

Scheme 5.222

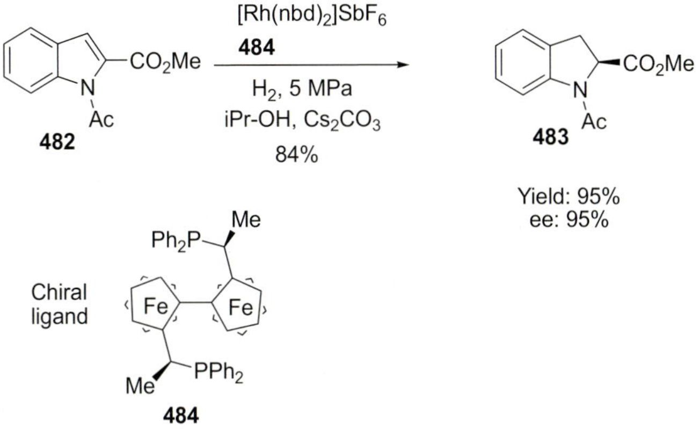
CO_2Me
N
Ac
482
$[Rh(nbd)_2]SbF_6$
484
H_2, 5 MPa
iPr-OH, Cs_2CO_3
84%
CO_2Me
N
Ac
483
Yield: 95%
ee: 95%
Chiral
ligand
Ph_2P
Me
Fe
Fe
PPh_2
Me
484

Scheme 5.223

5.5.6.2 **Metal-Promoted Reductions**

The reduction of indoles with dissolving metals may give partial reduction of both the benzene and the heterocyclic ring, depending on the particular substrate and reaction conditions applied. In certain examples, high chemo- and stereoselectivity can be achieved. For instance, *N*-phenylindoles **485** are converted into the corresponding indolines **486** upon treatment with Na/NH_3 in THF (Scheme 5.224) [358].

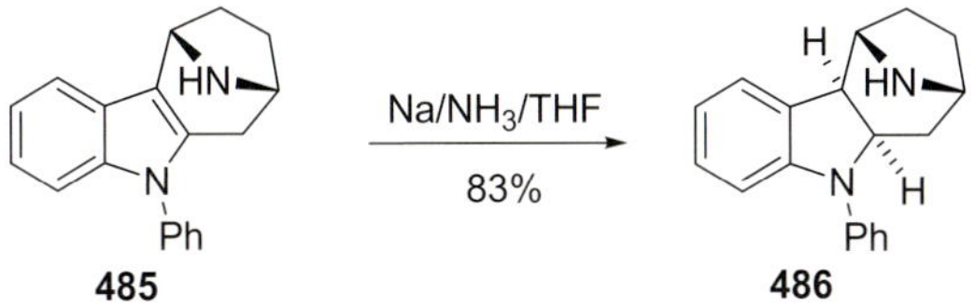
HN
N
Ph
485
Na/NH_3/THF
83%
H
HN
N
H
Ph
486

Scheme 5.224

On the other hand, 2-acylindoles can be reduced to the indolines through several protocols involving metal–acid combinations, such as Mg/MeOH [359] and Sn/HCl [360].

5.5.6.3 **Metal Hydride Complexes**

A very general method for the reduction of indoles to indolines **488** consists of treatment with a hydride reagent under acidic conditions (Scheme 5.225). The

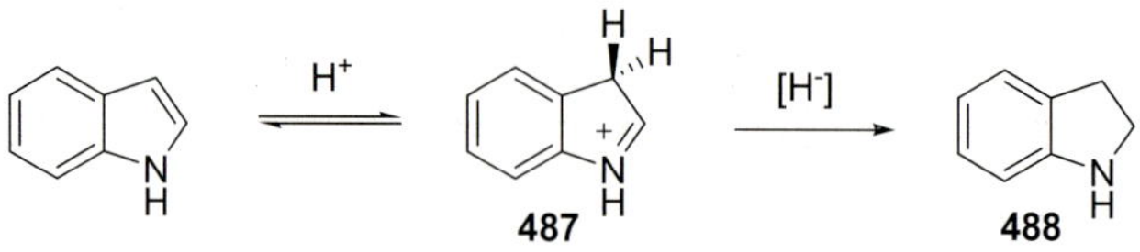

Scheme 5.225

reaction proceeds via initial protonation at C3 to generate indolenium ion **487** which is subsequently reduced by the hydride donor [361].

Hence, the best results have been obtained with hydride sources that are stable to acid, such as $NaBH_3CN$ in AcOH [362], BH_3/THF in TFA [363] and triethylsilane in TFA [364]. Representative examples are given in Scheme 5.226.

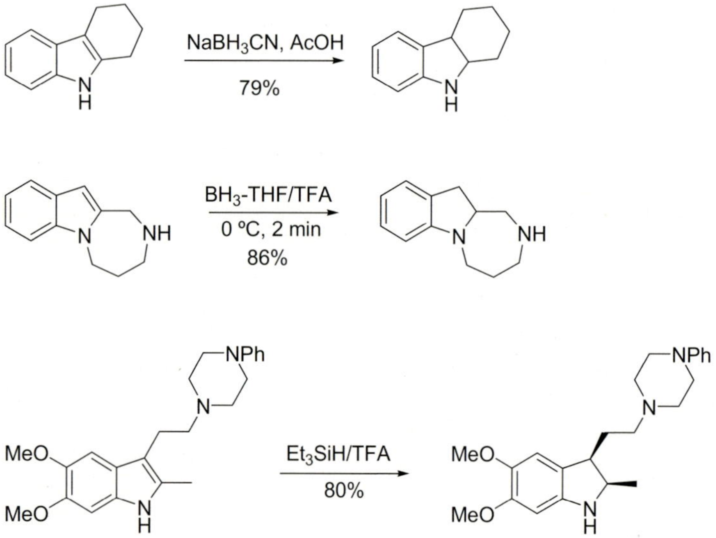

Scheme 5.226

5.5.7 Pericyclic Reactions Involving the Heterocyclic Ring

5.5.7.1 Cycloaddition Reactions

The C2–C3 double bond of indole can participate in different types of cycloaddition processes as a 2π or 4π component. As a 2π component it can behave as a dienophile in [4 + 2] cycloadditions, and as dipolarophile in dipolar [3 + 2] cycloadditions. On the other hand, 3-vinylindoles **489** and 2-vinylindoles **490** take part as dienes in [4 + 2] cycloadditions. Moreover, orthoquinodimethane indole derivatives **491** are highly reactive dienes in Diels–Alder reactions, and methylene indolines **492** and oxindoles can react as 2π components in [4 + 2] and dipolar cycloadditions (Scheme 5.227).

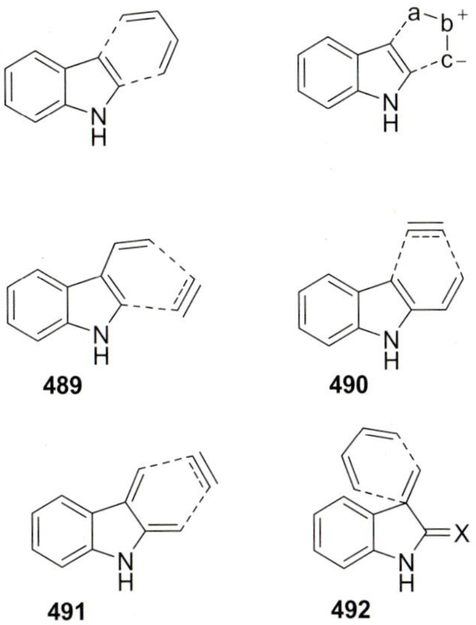

489 490 491 492

Scheme 5.227

Owing to the high electron density of the C2–C3 double bond, indole derivatives have been mostly employed as electron-rich dienophiles in inverse-electron-demand Diels–Alder reactions [365]. In particular, the use of aromatic heterodienes such as 1,2,4,5-tetrazines, 1,2,4-triazines and pyridazines represents a versatile route into more complex polycyclic structures [366]. Scheme 5.228 shows the synthesis of a 3,9b-dihydro-5*H*-pyridazino[4,5-*b*]indole **494** by cycloaddition of 3-methylindole with the 1,2,4,5-tetrazine **493**. The final compound **494** is obtained after loss of N_2 and hydrogen transposition of the initially formed cycloadduct **495** [367].

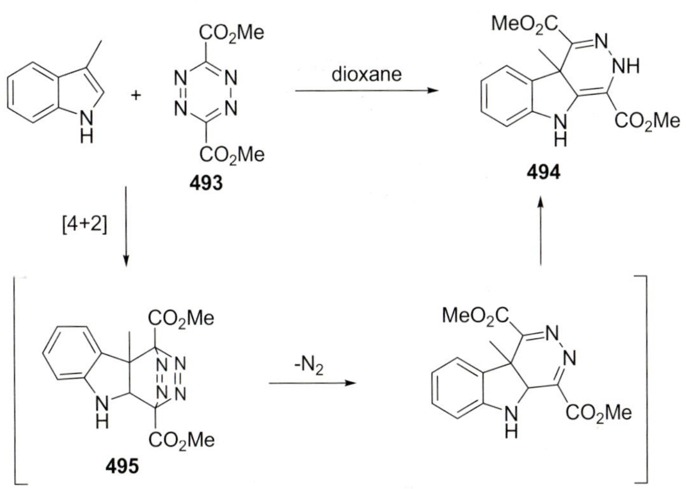

493 494 495

Scheme 5.228

The application of [4 + 2] cycloaddition methodology in an intramolecular version leads to polycyclic skeletons in very short synthetic sequences [368], and has been employed in several approaches to complex indole alkaloids. For instance, transannular cycloaddition of the pyridazinoindolophane **496** gives polycyclic adduct **497**, which provides the final dihydrocarbazole **498** after loss of N_2 via a retro-Diels–Alder reaction [369]. This particular sequence is the key step in a total synthesis of strychnine (Scheme 5.229) [370].

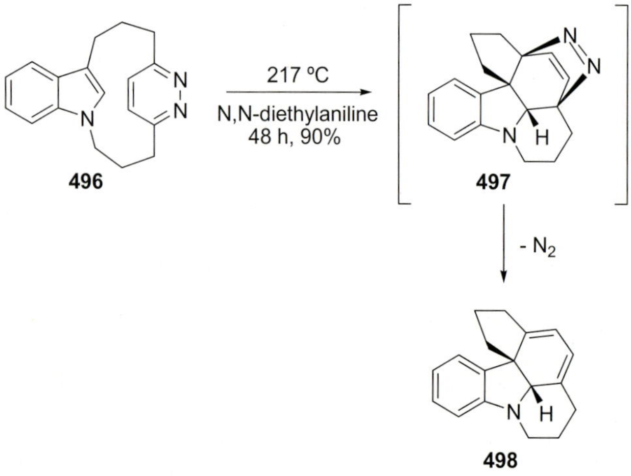

Scheme 5.229

Another example of an intramolecular inverse-electron-demand Diels–Alder reaction is the cycloaddition of **499**, a molecule that features a substituted indole and an aza-*o*-xylylene, a very reactive heterodiene moiety [371]. Cycloaddition gives rise to **500**, which features the heptacyclic ring system present in the natural alkaloid communesin B (Scheme 5.230).

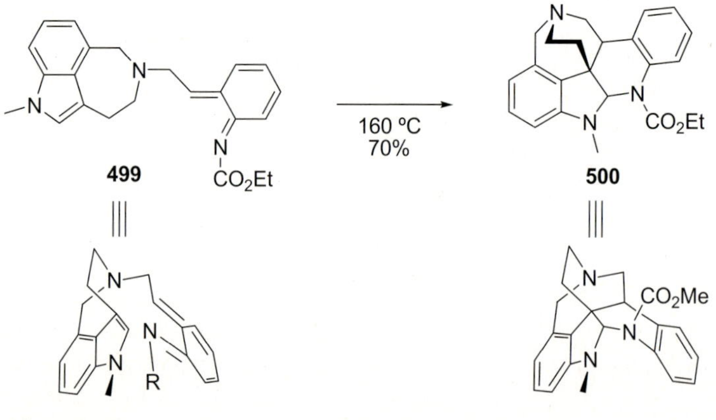

Scheme 5.230

Indoles substituted with electron-withdrawing groups in positions 1 and 3, such as **501**, can react with electron-rich dienes like Danishefsky's diene **502**, in normal electron-demand cycloadditions [372]. Very high temperatures are required unless activation by Lewis acids or high pressure are employed (Scheme 5.231) [373, 374].

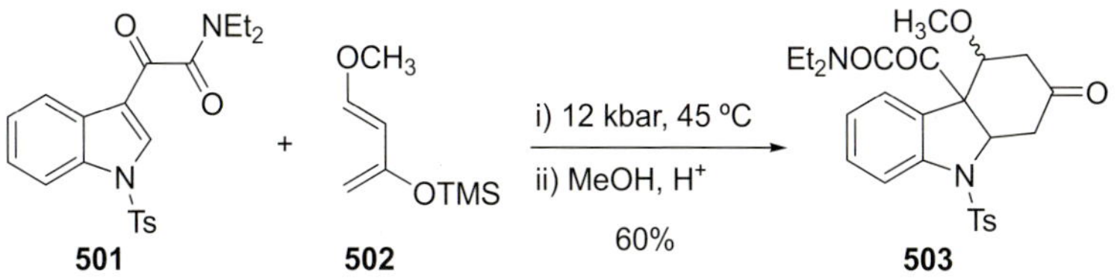

Scheme 5.231

The intramolecular [4 + 2] cycloaddition of **504**, which features an amidofuran moiety tethered onto an indole, can be regarded also as an example of a normal electron-demand cycloaddition [375]. In this process, the cycloadduct **505** undergoes spontaneous rearrangement onto tetracycle **506** [376]. The reactions proceed at very high temperatures and only substrates substituted with an electron-withdrawing group at the indole nitrogen undergo the cycloaddition (Scheme 5.232).

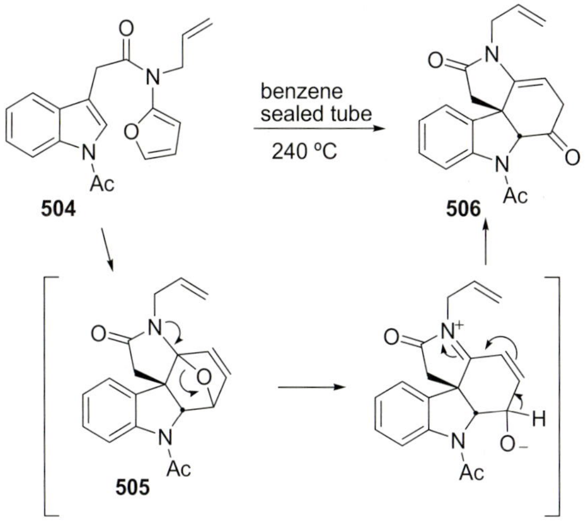

Scheme 5.232

The electron-rich C2–C3 bond of indole can also participate as dipolarophile in 1,3-dipolar cycloaddition reactions. Nitrones [377], azomethine ylides [378], azides [379] and carbonyl ylides are among the dipoles successfully employed, in most cases in an intramolecular fashion. In the example presented below, the carbonyl ylide, generated by cyclization of a rhodium carbenoid generated from diazoimide **507**, reacts with the tethered indole giving rise to **508**, which presents the

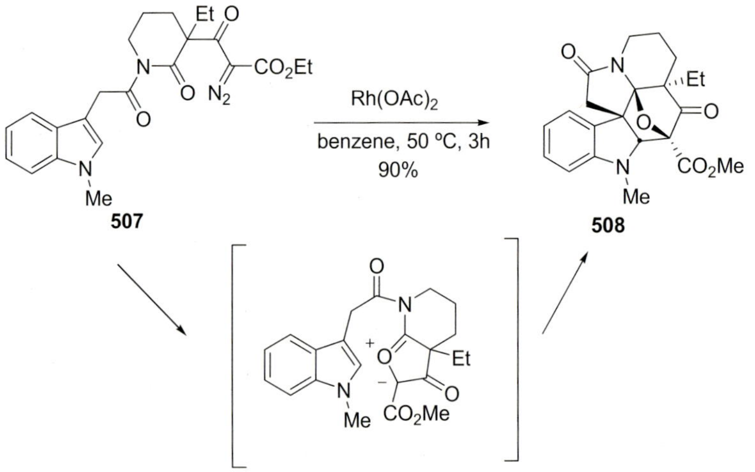

Scheme 5.233

pentacyclic skeleton of Aspidosperma alkaloids (Scheme 5.233) [380]. The carbonyl ylide–indole cycloaddition can be also conducted in an intermolecular fashion [381].

The power of intramolecular cycloadditions across the C2−C3 double bond has been shown in a remarkable application of cycloaddition cascades employed by Boger *et al.* in the total synthesis of (–)-vindoline (Scheme 5.234) [382]. The reaction is initiated by the [4 + 2] cycloaddition of the tethered enol ether with the 1,3,4-oxadiazole on **509**, to give a cycloadduct. Loss of N_2 by a retro-dipolar cycloaddition then generates the intermediate carbonyl ylide **510**. Intramolecular [1,3] dipolar

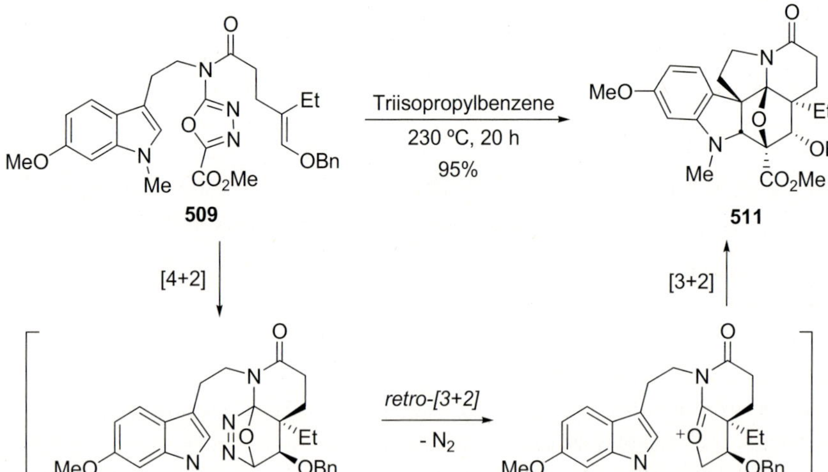

Scheme 5.234

cycloaddition across the indole C2–C3 double bond provides the polycycle **511** with total control of the stereochemistry on the six new stereogenic centers generated along the cascade process.

Among the very few intermolecular 1,3-dipolar cycloadditions of indoles known, noteworthy is the reaction of 2- and 3-nitroindoles with the mesoionic münchnones **512** [383]. The reaction gives rise to pyrrolo[3,4-*b*]indoles **514** in a highly regioselective manner, and is thought to proceed through a 1,3-dipolar cycloaddition to form the unstable cycloadduct **513**, which gives the final compounds after loss of CO_2 and NO_2 (Scheme 5.235).

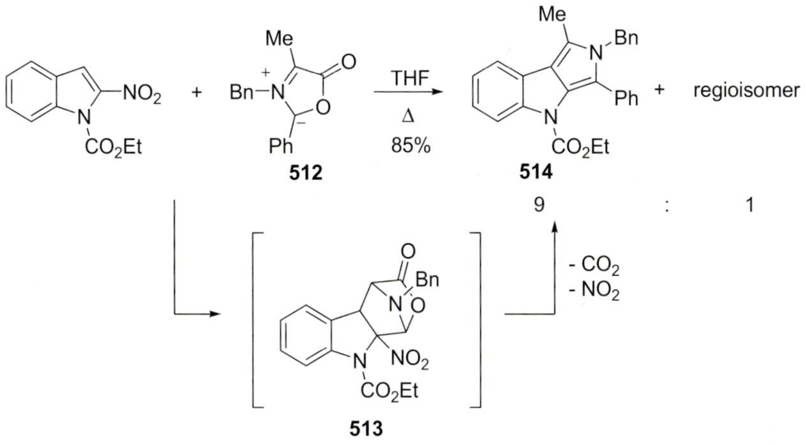

Scheme 5.235

Many examples have been described of [4 + 2] cycloadditions of 2- and 3-vinylindoles as 4π-electron components. Vinylindoles can be seen as electron-rich dienes, and therefore will react preferentially with electron-poor dienophiles in normal electron-demand cycloadditions. Thus, intermolecular reactions involve typical dienophiles such as methyl acetylenedicarboxylate [384], *N*-phenylmaleimide [385], and benzoquinones [386]. When asymmetric dienophiles are employed, the regiochemistry can be predicted by employing frontier molecular orbital theory principles. For instance, the regiochemistry of the reactions of 3-vinylindole **515** with ethyl acrylate [387], and the cycloaddition of 2-vinylindole **516** with methyl propiolate can be both explained assuming the directing interaction of the HOMO of the vinylindole with the LUMO of the dienophile (Scheme 5.236) [388, 389].

The intramolecular version of the [4 + 2] cycloaddition leads to complex structures in a convergent manner. In a classical example, the 2-vinylindole **518** generated *in situ* from indoline **517** undergoes cycloaddition with the enaminic double bond to give **519**, a very advanced intermediate in the total synthesis of pseudotabersonine (Scheme 5.237) [390, 391].

Intramolecular Diels–Alder reactions of 3-vinylindoles usually involve a dienophile tethered through the nitrogen atom. For instance, the intramolecular

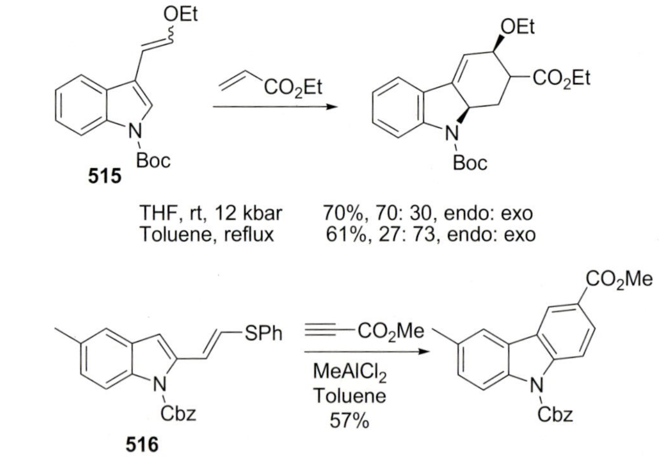

Scheme 5.236

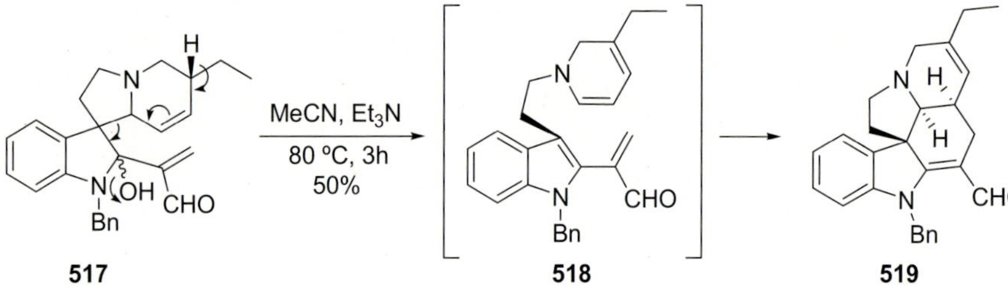

Scheme 5.237

cycloaddition of the nitrovinylindole with a terminal alkyne **520** provides the tetracyclic structure **521** after loss of HNO_2 (Scheme 5.238) [392].

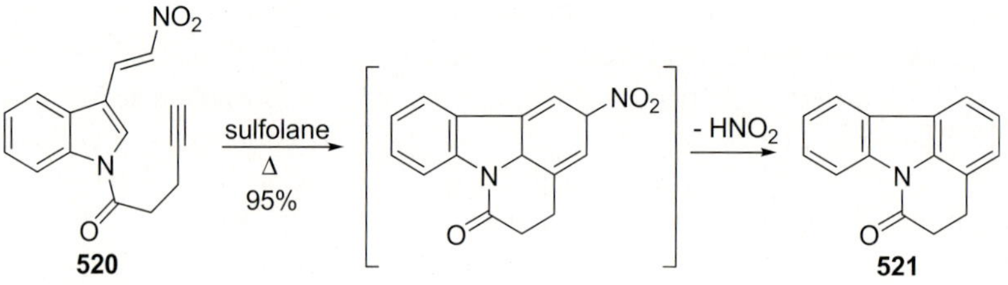

Scheme 5.238

The oxime-substituted indole **522** behaves as an 1-aza-1,3-butadiene in an intramolecular hetero-Diels–Alder reaction with a tethered alkyne, to give tetracycle **523**, which features the skeleton of Canthine alkaloids (Scheme 5.239) [393].

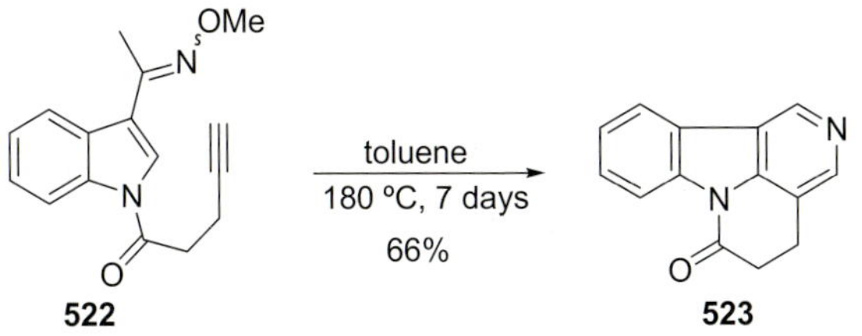

Scheme 5.239

Indole-2,3-quinodimethanes are highly reactive dienes in [4 + 2] cycloadditions and represent a very valuable entry into the carbazole ring structure [394]. They can be generated *in situ* by fluoride or iodide ion induced 1,4-elimination of silylated indolyl ammonium salts [395], by double elimination on *N*-protected-2,3-bis(dibromomethyl)indoles [396], by thermal fragmentation of 2-substituted 3-aminomethylindoles [397] and by [1,5]-H shift on 3-cyanomethyl-2-vinylindoles [398]. Moreover, the anionic indole-2,3-dienolate can be formed by deprotonation of 1,2-dimethylindole-3-carboxaldehyde [399].

For instance, disubstituted indole-*o*-quinodimethane **525** is generated by thermal decomposition of gramine **524** in the presence of the dienophile; it then reacts to give the tetracyclic adduct **526** in very high yield (Scheme 5.240).

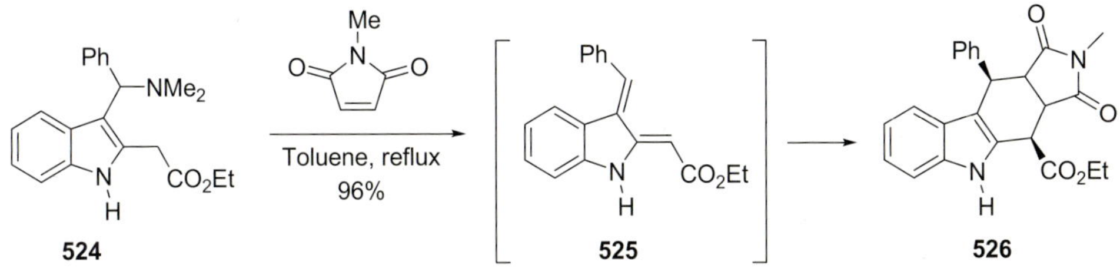

Scheme 5.240

Alternatively, reactive indole-2,3-quinodimethane intermediates **528** can be generated from *o*-allenylanilines **527**, and trapped with dienophiles to obtain directly the corresponding carbazole derivatives **529** in moderate yields (Scheme 5.241) [400].

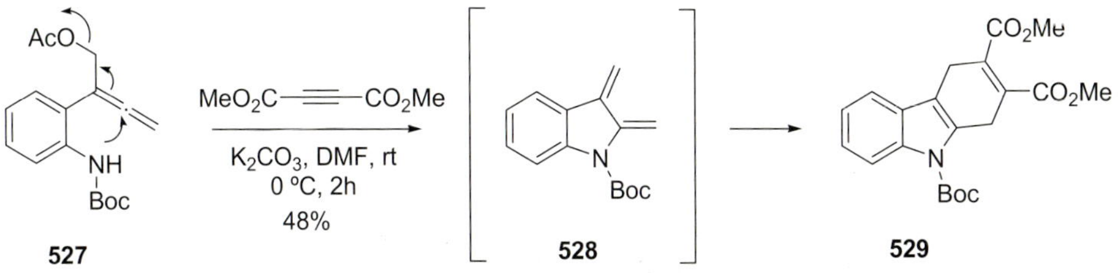

Scheme 5.241

5.5.7.2 **Electrocyclizations**

The C2–C3 double bond of indole can participate in electrocyclic reactions. Although the 6π-electron electrocyclization of 2,3-divinylindoles is known, the participation of allene intermediates is more efficient. Thus, a versatile route to carbazoles is based on an allene-mediated electrocyclic reaction of a 6π-electron system involving the indole 2,3-bond [401]. This strategy has been applied in the preparation of numerous alkaloids containing the carbazole moiety. Scheme 5.242 depicts the preparation of oxygenated carbazole **532** from propargyl ether containing indole **530**. Treatment with KO*t*Bu produces the acetylene–allene isomerization to give intermediate **531**, which suffers electrocyclization leading to the carbazole.

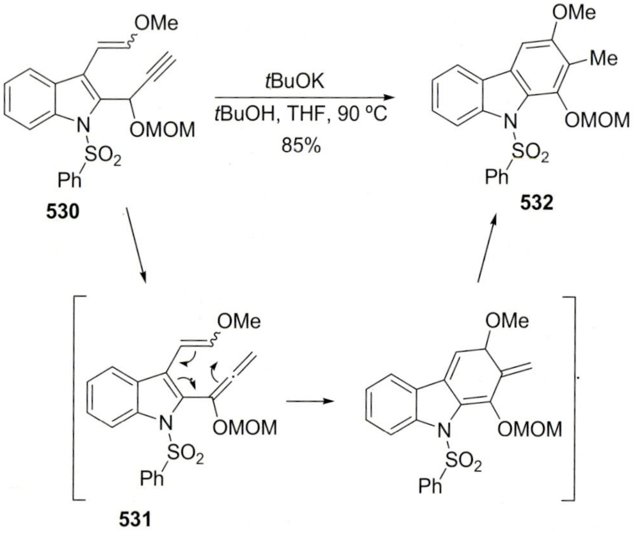

OMe
OMOM
N
SO_2
Ph
530
*t*BuOK
*t*BuOH, THF, 90 °C
85%
OMe
Me
OMOM
N
SO_2
Ph
532
OMe
OMOM
N
SO_2
Ph
531
OMe
OMOM
N
SO_2
Ph

Scheme 5.242

5.5.7.3 **Sigmatropic Rearrangements**

The indole C2–C3 bond can participate in [3,3] sigmatropic rearrangements, and this strategy has been employed in many synthetic efforts to introduce additional substitution at C2 or C3 [402–404].

The Claisen rearrangement of 2-allyloxiindoles leads to oxindoles with creation of a quaternary center. For example, allyloxyindole **534**, generated *in situ* by olefination of 2-allyloxiindoxyl **533**, suffers the [3,3]-rearrangement to provide oxindole **535** (Scheme 5.243) [405]. The required 2-allyloxiindoles can be also generated by oxidative coupling of indoles with allyl alcohols (Section 5.5.5) [406].

Another type of [3,3]-rearrangement with functionalization at C3 is found in the conversion of 1-vinylaminoindole **536** into tricyclic compound **538** under thermolysis conditions [407]. The formation of **538** can be explained by [3,3]-rearrangement to form imine **537**, which generates the pyrrolo[2,3-*b*]indole by intramolecular

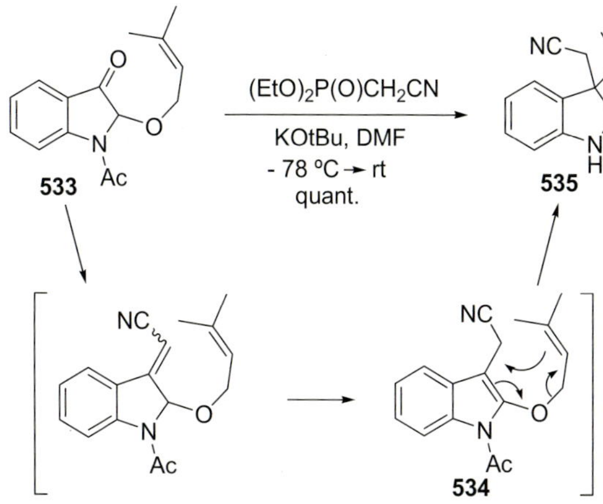

Scheme 5.243

cyclization (Scheme 5.244). On the other hand, the incorporation of a substituent at C2 by a Claisen rearrangement has been also described [408].

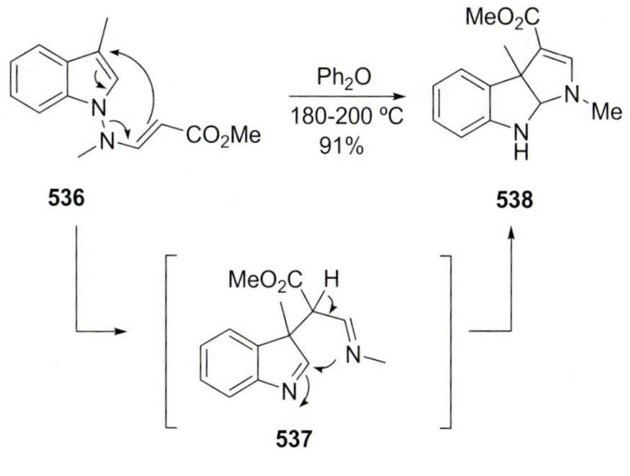

Scheme 5.244

5.5.8
Photochemical Reactions

Synthetic applications of indole photochemical reactions are very limited, and therefore this technique has been scarcely employed. Ultraviolet light irradiation promotes the [2 + 2] cycloaddition of N-protected indoles with alkenes and alkynes [409, 410]. For instance, *N*-acylindoles **539** undergo [2 + 2] photocycloaddition with monosubstituted alkenes to form the cyclobutane ring regioselectively regardless of the nature of the substituent on the alkene [411]. The reaction is postulated to proceed through the biradical intermediate **540**, which is formed by bonding the 2-position of the indole with the less substituted position of the alkene (Scheme 5.245).

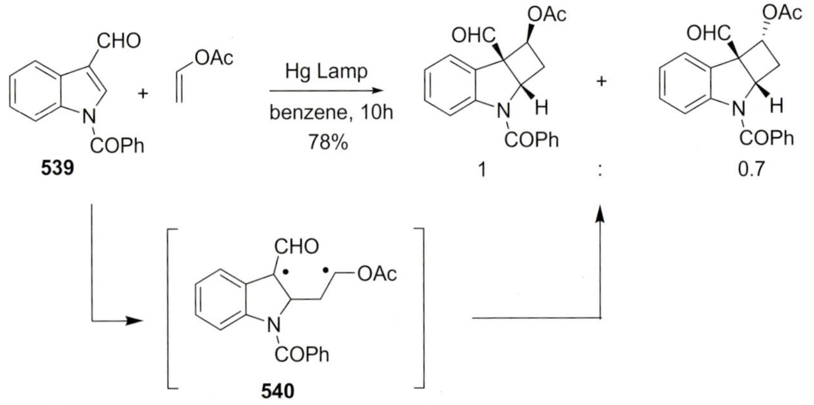

Scheme 5.245

Several examples have been reported of the intramolecular version of the [2 + 2] photocycloaddition [412]. Interestingly, the intramolecular reaction with the double bond tethered through the N atom leads to the opposite regiochemistry in the [2 + 2] cycloadduct [413].

The [4 + 2] cycloaddition reaction of indoles with electron-rich dienes, which does not proceed thermally, can be promoted photochemically by electron-transfer catalysis. Nevertheless, at present, these procedures are far from being synthetically practical [414].

Another synthetically useful photochemical transformation is the photocyclization of indole-containing stilbenes **541** and related systems to give polycycles **543** featuring the carbazole moiety. The reaction occurs through a 6π-electron conrotatory electrocyclization to give **542**, followed by oxidation to furnish directly the aromatic system. Thus, the photocyclization is usually carried out in the presence of an oxidant, such as iodine or Pd/C (Scheme 5.246) [415].

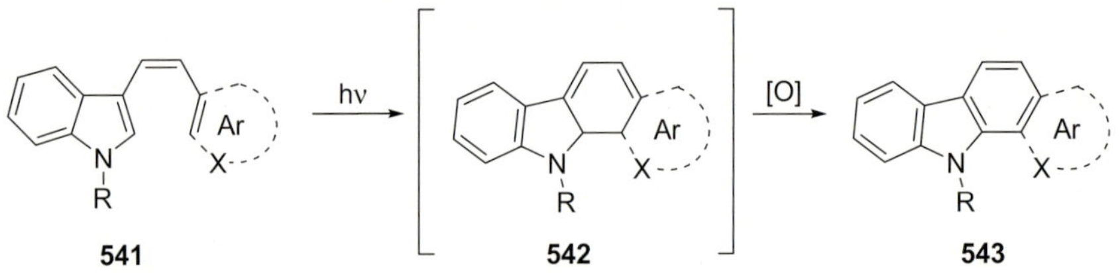

Scheme 5.246

The main application of this methodology is in the preparation of [2,3-*a*]pyrrolo [3,4-*c*]carbazoles, a substructure that is present in several naturally occurring alkaloids with potent biological activities [416], such as rebeccamycin [417] and staurosporin [418]. In a recent example, naphtho[2,3-*a*]carbazole **545** is obtained by photooxidation of 2-naphthylindolyl maleimide **544** in nearly quantitative yield (Scheme 5.247) [419, 420].

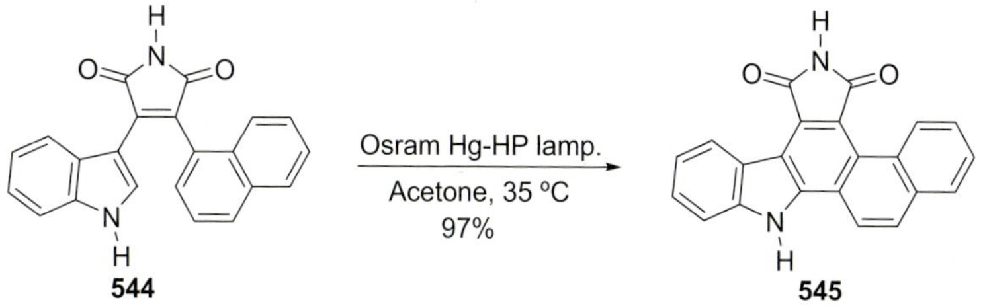

Osram Hg-HP lamp.
Acetone, 35 °C
97%

544 **545**

Scheme 5.247

5.5.9 Reactions with Carbenes and Carbenoids

The reaction of indole with carbenes or carbenoids does not lead to cyclopropanation but, instead, insertion in the C3–H double bond occurs, to give the corresponding substitution product [421]. In the example shown in Scheme 5.248, the rhodium carbenoid generated from the cyclic diazo compound **546** reacts with indole to give the C3-substituted indole **547** [422]. This chemistry has been applied in the preparation of natural indolocarbazole alkaloids.

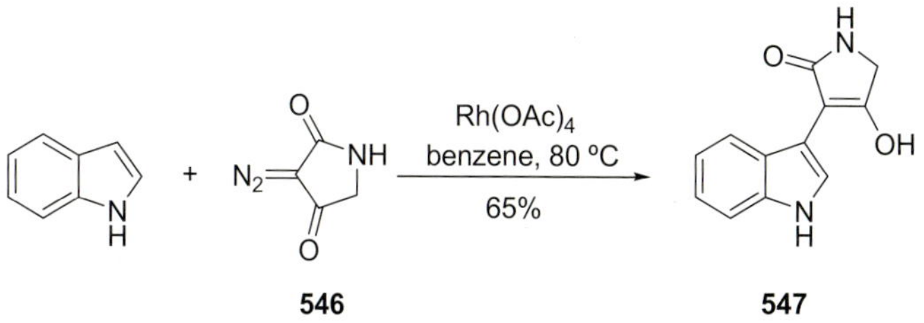

$Rh(OAc)_4$
benzene, 80 °C
65%

546 **547**

Scheme 5.248

5.6 Chemistry of Indole Derivatives

5.6.1 Alkylindoles

Deprotonation of the alkyl chain of indoles is usually difficult. Nevertheless, NH-2-alkylindoles **548** can be lithiated at the α-position through a one pot sequence that involves formation of the lithium carbonate **549**, by NH deprotonation with BuLi and quench with CO_2, followed by treatment with *t*-BuLi to obtain dianion **550**. Subsequent reaction with an electrophile yields the 2-(substituted a1kyl)indole **551** after thermally induced loss of CO_2 (Scheme 5.249) [423].

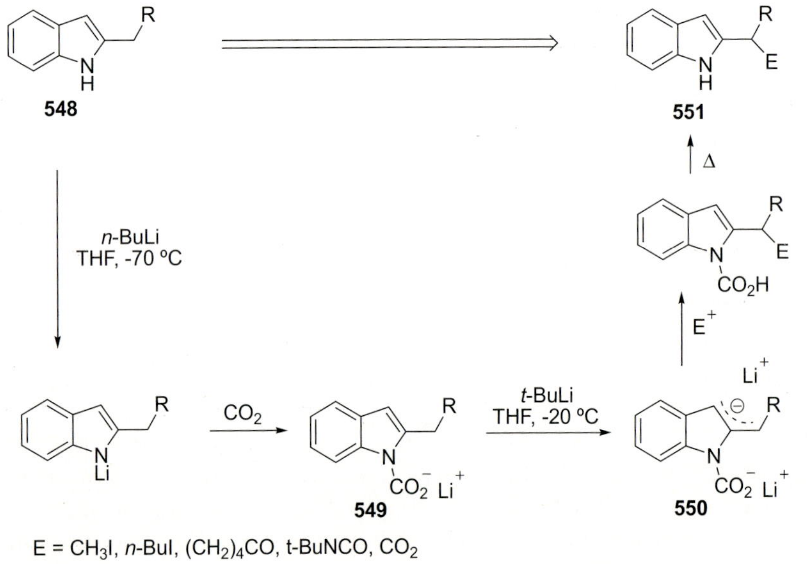

548 *n*-BuLi, THF, -70 °C; CO_2 **549** *t*-BuLi, THF, -20 °C **550** E^+; Δ **551**

E = CH_3I, *n*-BuI, $(CH_2)_4CO$, t-BuNCO, CO_2

Scheme 5.249

Alternatively, NH-2,3-dialkylindoles **552** can be directly deprotonated by treatment with an excess of base that consists of a combination of BuLi and KO*t*-Bu. Treatment with an electrophile provides the 2-substituted indole **553**, indicating that the lithiation is produced exclusively at the C2 side-chain (Scheme 5.250).

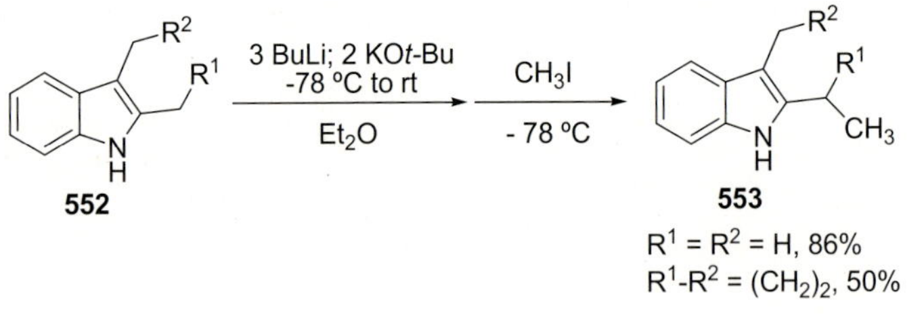

552 3 BuLi; 2 KO*t*-Bu, -78 °C to rt, Et_2O; CH_3I, - 78 °C **553**

$R^1 = R^2 = H$, 86%
R^1-R^2 = $(CH_2)_2$, 50%

Scheme 5.250

Bromination of the side-chain of N-protected indoles such as **554** can be carried out by treatment with NBS in the presence of a radical initiator. Interestingly, while the reaction under radical conditions provides the indole brominated on the side-chain **555**, the same reaction in the absence of the radical initiator gives rise to the C2-brominated indole **556** (Scheme 5.251) [424].

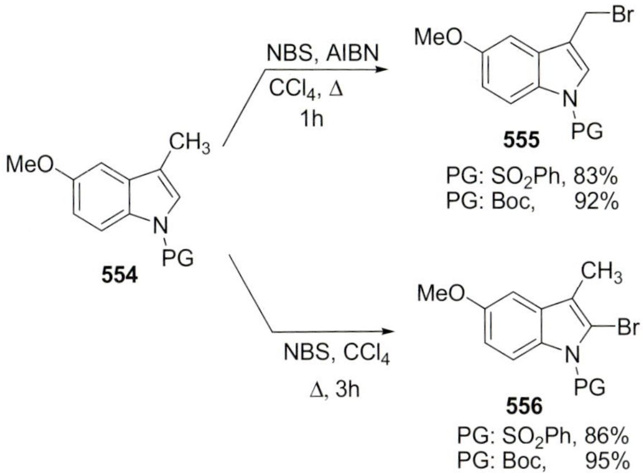

Scheme 5.251

Among functionalized alkylindoles, the chemistry of gramines **557** is noteworthy. The elimination of the dimethylamino group can be promoted either by quaternization or by treatment with bases, generating the highly reactive electrophile **558**, which undergoes the addition of a nucleophile. The overall reaction is the substitution of the dimethylamino group by the nucleophile (Scheme 5.252). This strategy has been employed for the introduction of different types of carbon [425], nitrogen [426] and sulfur nucleophiles [427].

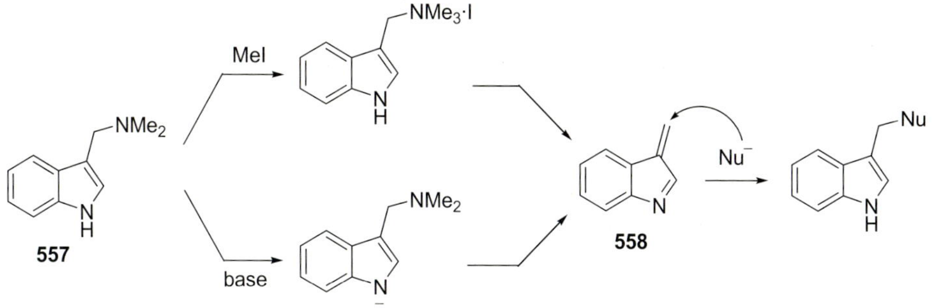

Scheme 5.252

Scheme 5.253 presents the use of gramine **557** as electrophile in a typical acetylacetate alkylation, leading to substituted indole **559** [428].

This chemistry has been applied in an original method for the preparation of vinylindoles, in a substitution/Wittig olefination tandem sequence. Thus, treatment of gramine with an aromatic aldehyde in the presence of an excess of PBu_3, affords directly the vinylindole **561**, through the *in situ* generated ylide **560** (Scheme 5.254) [429].

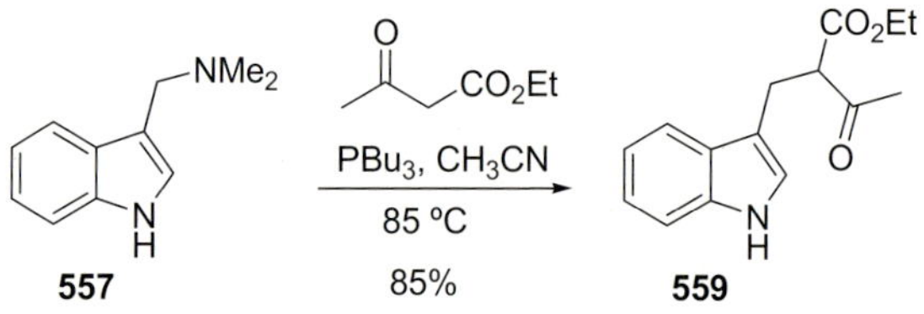

Scheme 5.253

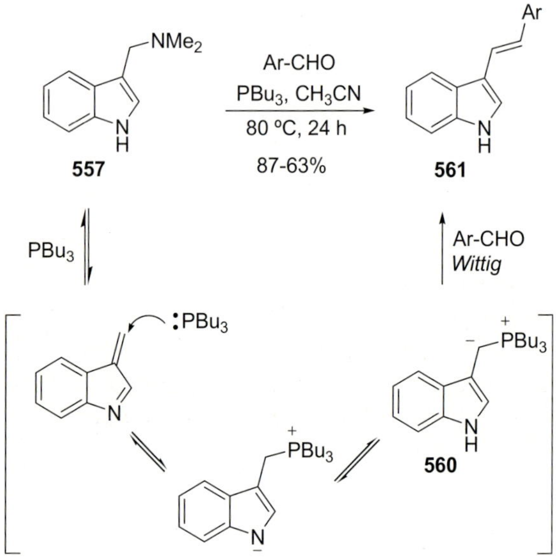

Scheme 5.254

5.6.2
Oxiderivatives

Oxindole and indoxyl are the stable tautomeric forms of 2-hydroxy- and 3-hydroxyindole respectively (Figure 5.8). The aromatic 2-hydroxyindole is unstable and undetectable. In contrast, 3-hydroxyindole contributes in the tautomeric equilibrium, and in some 2-substituted-indoxyls is the thermodynamically controlled product.

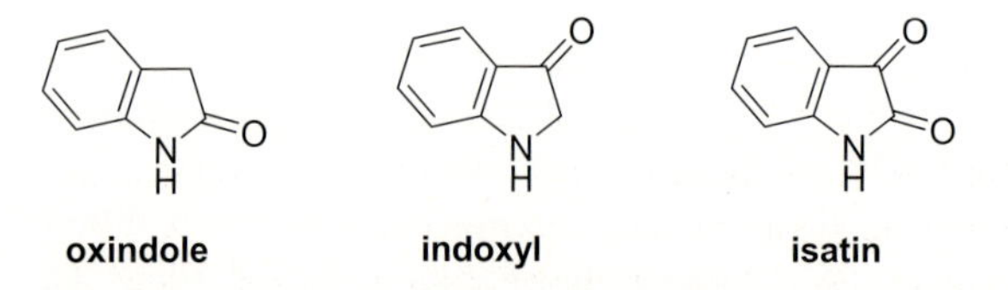

Figure 5.8 Structures of oxi-derivatives of indole.

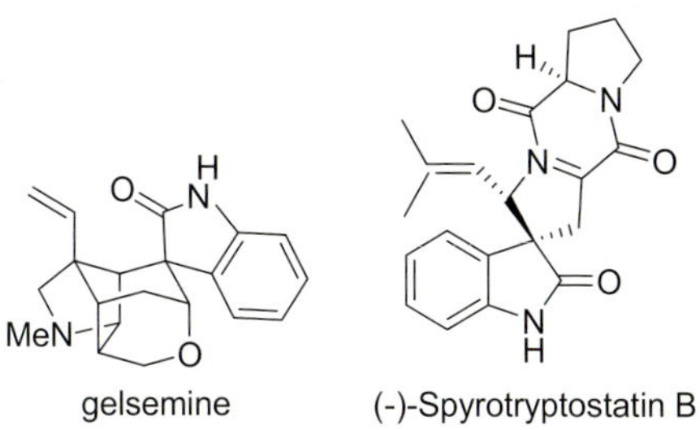

gelsemine (-)-Spyrotryptostatin B

Figure 5.9 Some important oxindole alkaloids.

5.6.2.1 **Oxindole**

The substructure of oxindole is present in numerous naturally occurring alkaloids and pharmaceutically active compounds. Substituted oxindoles and, in particular, spirocyclic oxindoles featuring a quaternary center at C3 are attractive targets in organic synthesis due to their usefulness as drug candidates and as intermediates in alkaloid synthesis. Figure 5.9 shows some natural oxindole-containing alkaloids such as the well-known hexacyclic cage-like alkaloid gelsemine [430] and the highly potent cytotoxic agent spyrotryprostatin B [431].

Numerous methodologies for the preparation of oxindoles are available, and detailed coverage would largely exceed the aim of this chapter. Oxindoles can be prepared from indole or indole derivatives, by derivatization of isatin (Figure 5.8), and by cyclization processes.

5.6.2.1.1 **Synthesis of Oxindoles from Indoles** Indoles can be transformed into oxindoles under the oxidation protocols discussed in Section 5.5.5. Additionally, *N*-Boc-indoles can be cleanly converted into *N*-Boc-oxindoles **562** by a two-step sequence that involves formation of a 2-indolylborate [432] and oxidation with oxone (Scheme 5.255) [433]. Moreover, as discussed in Section 5.5.7.3, oxindoles can be prepared by Claisen rearrangement of 2-allyloxiindoles.

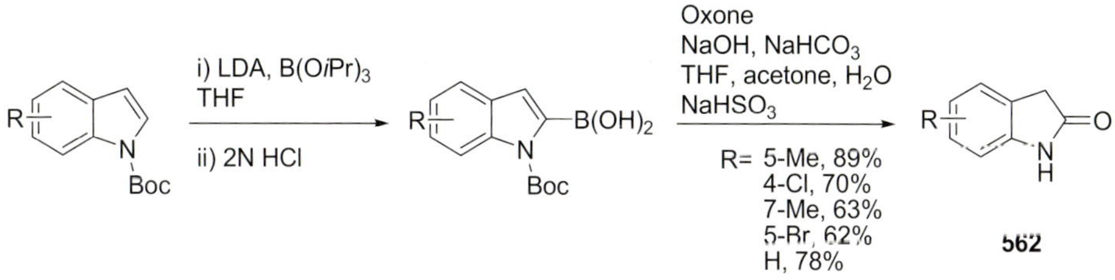

i) LDA, $B(O\textit{i}Pr)_3$, THF; ii) 2N HCl

Oxone, NaOH, $NaHCO_3$, THF, acetone, H_2O; $NaHSO_3$

R= 5-Me, 89%; 4-Cl, 70%; 7-Me, 63%; 5-Br, 62%; H, 78%

562

Scheme 5.255

5.6.2.1.2 **Synthesis of Oxindoles from Isatins** Isatin (Figure 5.8) features two carbonyl groups, a ketone and an amide carbonyl. The higher reactivity of the C3 ketone carbonyl can be exploited to introduce functionalization and prepare 3-substituted oxindoles. Thus, reductions [434], aldol reactions [435], additions of nucleophiles [436] and Wittig olefinations [437] lead to the corresponding substituted oxindoles. Notably, the rhodium-catalyzed asymmetric addition of arylboronic acids to isatin **563** gives rise to enantiomerically enriched 3-substituted-3-hydroxyoxindoles **564** (Scheme 5.256) [438].

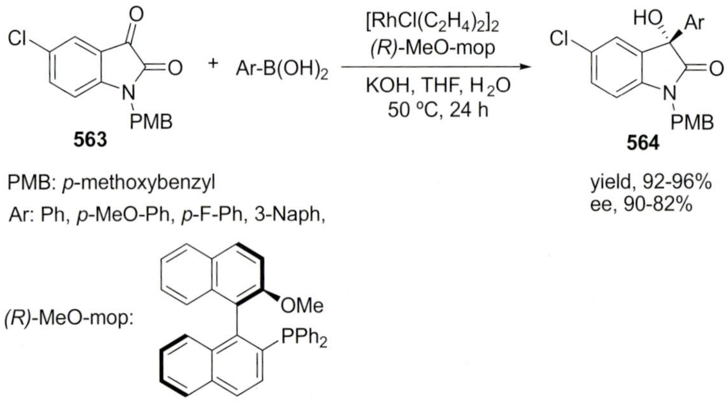

Cl, O, =O, N, PMB, **563** + Ar-B(OH)$_2$ → [RhCl(C$_2$H$_4$)$_2$]$_2$ / (*R*)-MeO-mop / KOH, THF, H$_2$O / 50 °C, 24 h → Cl, HO, Ar, =O, N, PMB, **564**

PMB: *p*-methoxybenzyl

Ar: Ph, *p*-MeO-Ph, *p*-F-Ph, 3-Naph,

yield, 92-96%

ee, 90-82%

(*R*)-MeO-mop: OMe, PPh$_2$

Scheme 5.256

5.6.2.1.3 **Synthesis of Oxindoles by Cyclization Reactions** Cyclization strategies are usually reminiscent of the methods employed for the synthesis of indoles, such as the Fischer synthesis [439], cyclization of *o*-aminophenylacetic acid derivatives [440], intramolecular Friedel–Crafts alkylations [441], radical cyclizations [442], intramolecular Heck reactions [443], intramolecular Buchwald–Hartwig amidations [444] and intramolecular α-arylation of amide enolates [445]. Extensive coverage of the methods of synthesis of oxindoles is beyond the scope of this chapter, and the reader is referred to the original papers. Nevertheless, some relevant modern alternatives to the general approaches depicted in Figure 5.10 is briefly presented.

Cyclization of *o*-aminophenylacetic acid derivatives is one common approach to the synthesis of oxindoles. Several different strategies can be applied to prepare the intermediate amino acid [446]. The [3,3]-sigmatropic rearrangement of the enolate **568**, generated from *N,O*-diacylated phenyl hydroxylamine **567**, is the key step in a three-step synthesis of oxindoles from *N*-acylhydroxylamines **565** and carboxylic acids **566** [447]. The *N*-protected amino acid **569** generated in the rearrangement is condensed to give the spirocyclic oxindole **570** (Scheme 5.257).

Samarium iodide reductive coupling of isocyanates with a tethered α,β-unsaturated ketone (**571**) gives rise to oxindoles **572** (Scheme 5.258) [448]. In this original approach, the oxindole is built by formation of the C2–C3 bond, and has been employed in the preparation of advanced intermediates towards the synthesis of welwitindolinone alkaloids.

Intramolecular Friedel–Crafts alkylation of α-chloroacetanilides is one of the most classical methods for the synthesis of oxindoles. A variant of this reaction has been recently disclosed by Buchwald, avoiding the harsh reaction conditions required in

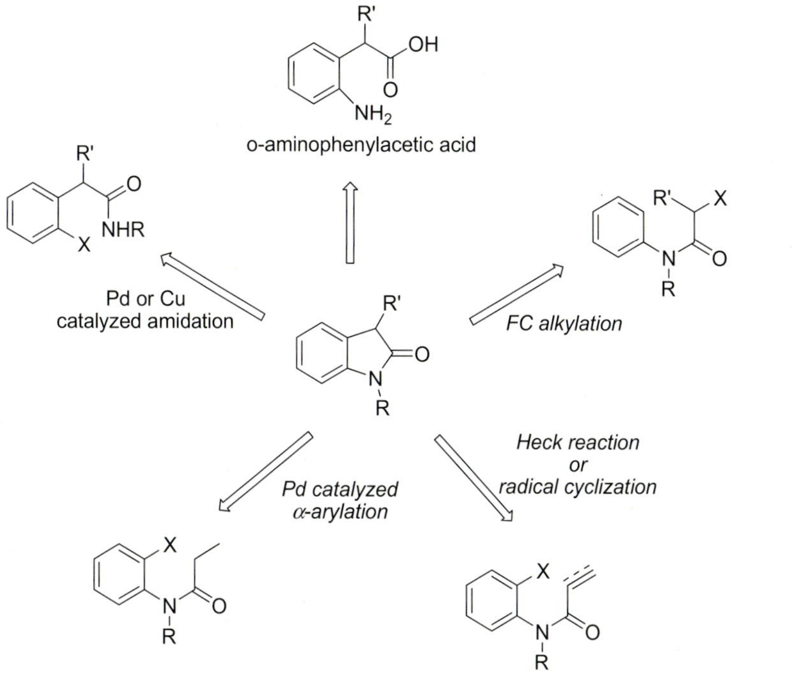

Figure 5.10 General approaches to the synthesis of oxindoles through cyclization reactions.

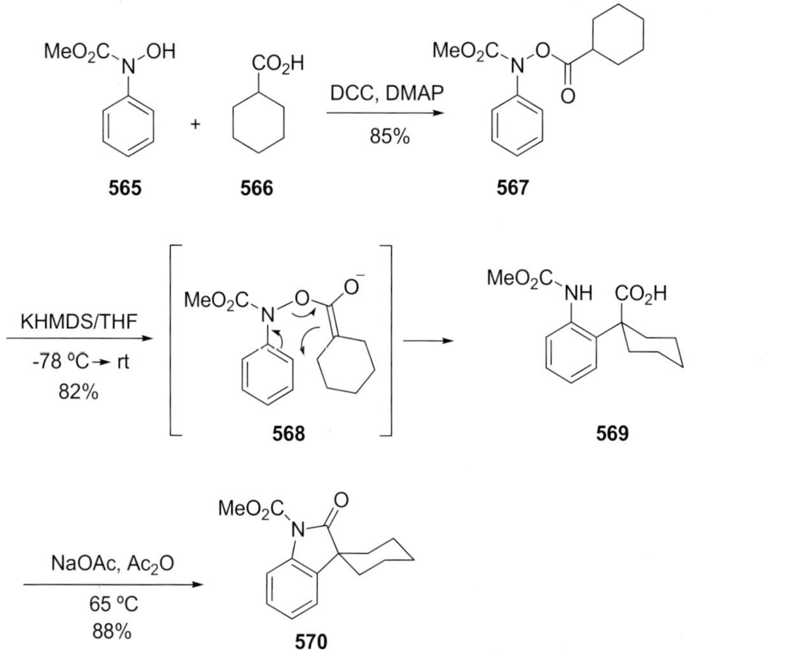

Scheme 5.257

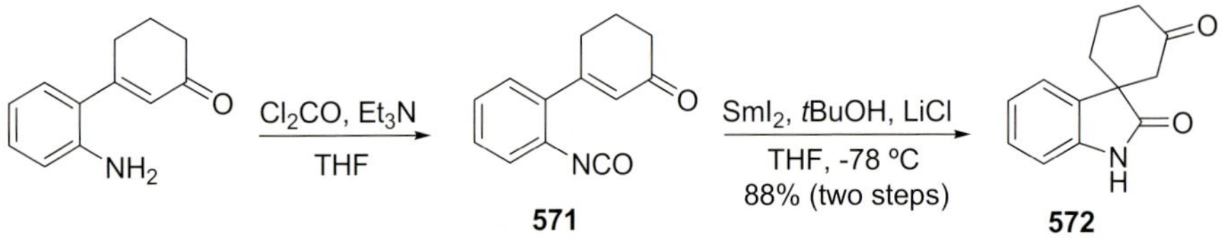

NH_2 — Cl_2CO, Et_3N / THF → NCO **571** — SmI_2, *t*BuOH, LiCl / THF, -78 °C / 88% (two steps) → **572**

Scheme 5.258

the FC alkylation. Thus, treatment of α-chloroacetanilides **573** with a Pd(0)-based catalytic system leads to the oxindoles **574** in excellent yields, under milder conditions and with high functional group tolerance (Scheme 5.259) [449]. Interestingly, the overall process results in a Pd-catalyzed aromatic C−H activation. Nevertheless, at present there is no unambiguous mechanistic proposal.

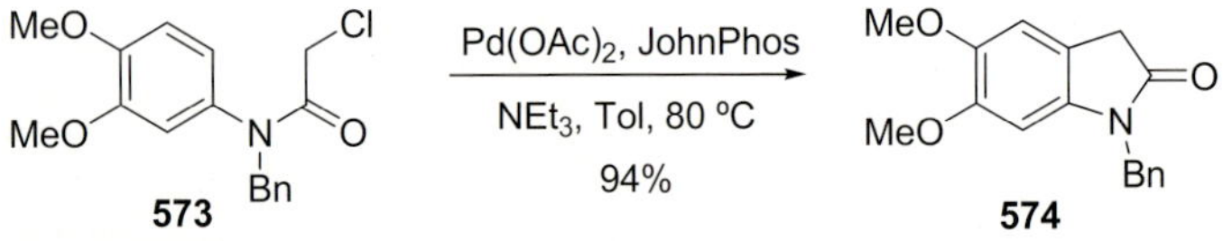

MeO, MeO, Cl, N, Bn, O **573** — $Pd(OAc)_2$, JohnPhos / NEt_3, Tol, 80 °C / 94% → MeO, MeO, N, Bn, O **574**

Scheme 5.259

5.6.2.1.4 **Oxindole Reactivity** The chemistry of oxindole resembles a typical five-membered ring lactam. Deprotonation at the β-carbon occurs readily, as the resulting anion is stabilized by the aromatic character of the oxindole enolate. Thus, the oxindole anion can react with electrophiles in alkylation, acylation and condensation reactions.

On the other hand, oxindoles can be transformed into the corresponding indolyl-triflates **575** (Scheme 5.260) [450], which can be further employed in metal-catalyzed cross-coupling reactions (Section 5.5.3.6).

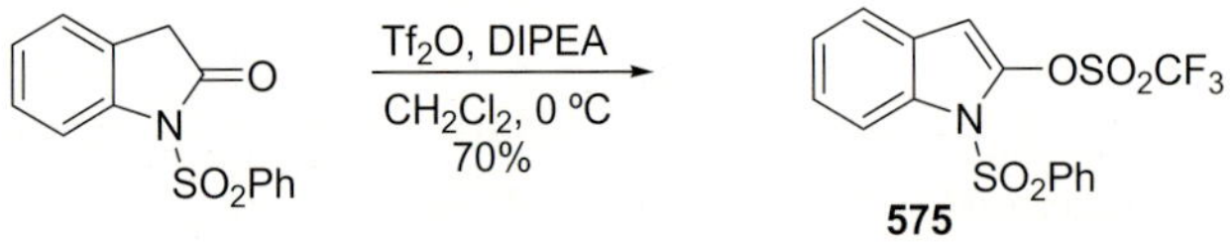

N, SO_2Ph, O — Tf_2O, DIPEA / CH_2Cl_2, 0 °C / 70% → OSO_2CF_3, N, SO_2Ph **575**

Scheme 5.260

5.6.2.2 ***N*-Hydroxyindoles**

The structure of *N*-hydroxyindole is also present in a considerable number of biologically molecules. Moreover, several *N*-hydroxylated analogues of biologically inactive indoles have shown biological activity [451].

The most common approach to the preparation of *N*-hydroxyindoles is the reductive cyclization of *o*-nitrobenzylketones **576** or aldehydes in the presence of a metal reducing agent [452]. A fairly general and chemoselective protocol is the reaction with the Pb/TEAF system (TEAF: triethylammonium formate), which provides *N*-hydroxyindoles **577** in very high yields (Scheme 5.261) [453].

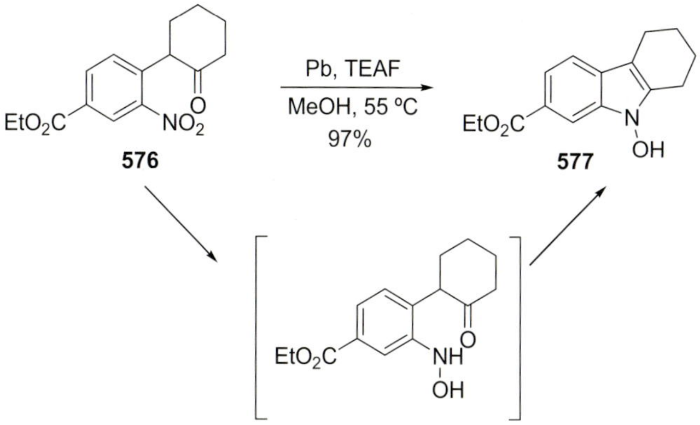

Scheme 5.261

In a closely related reaction, structurally diverse substituted *N*-hydroxyindoles **580** can be prepared in a tandem process from nitroaromatic α,β-unsaturated keto-esters **578** [454]. The initially formed nitrone **579** is trapped by a hetero- or carbonucleophile to provide substituted *N*-hydroxyindoles **580** (Scheme 5.262).

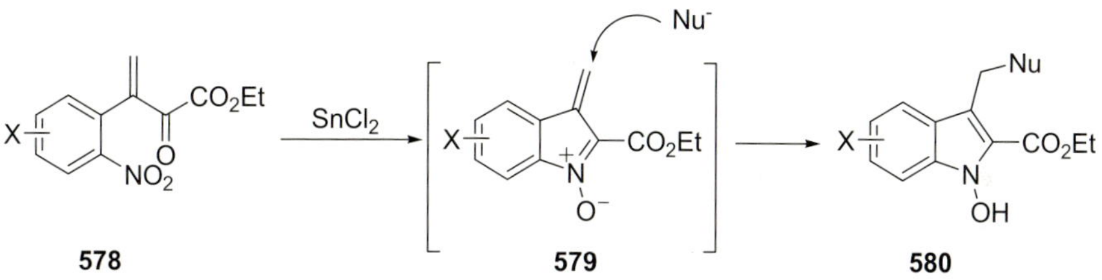

Scheme 5.262

For instance, the reaction in the presence of silylenol ethers gives rise to the corresponding C-alkylated derivatives **581** (Scheme 5.263).

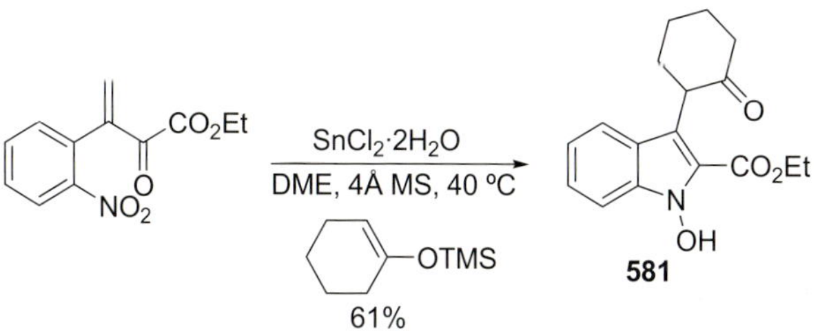

Scheme 5.263

5.6.3 Aminoindoles

Although most amino-substituted heterocycles exist mainly in the amino tautomer, 2-aminoindole exists predominantly as the 3*H*-tautomer, which is stabilized by the amidine resonance form (Scheme 5.264).

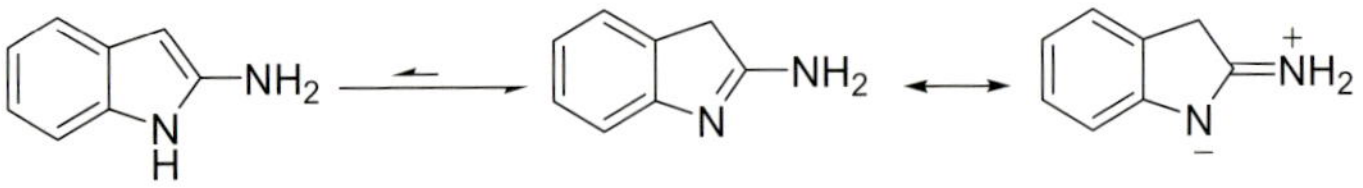

Scheme 5.264

Regarding the synthesis, the most reliable modern method to introduce an amino group in the heterocyclic indole ring is Pd- or Cu-catalyzed amination (Section 5.5.3.5).

N-Amination of indole can be accomplished with different "NH_2^+" type of reagents such as monochloramine (NH_2Cl) [455] or hydroxylamine *o*-sulfonic acid [456] under basic media (Scheme 5.265).

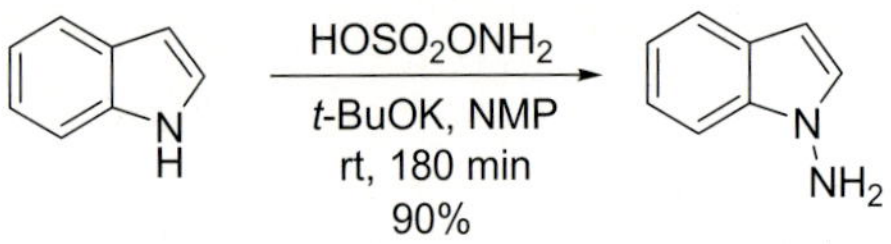

Scheme 5.265

On the other hand, *N*-aminoindoles **583** can be prepared by the Pd-catalyzed intramolecular cyclization of *o*-chloroarylacetaldehyde hydrazones **582** (Scheme 5.266) [457].

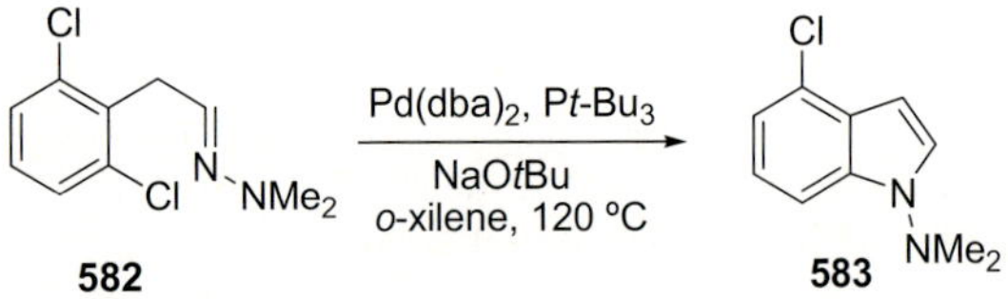

Scheme 5.266

5.6.4 Indole Carboxylic Acids

Decarboxylation of indole-3-carboxylic acid, and also indoyl-2-acetic acid, takes place under reflux of water (Scheme 5.267). The reaction proceeds through the protonated 3*H*-indolium cation **584** [458].

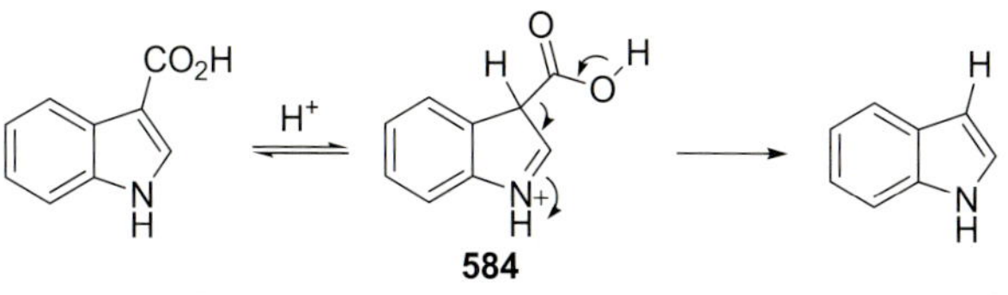

Scheme 5.267

Decarboxylation of indole-2-carboxylic acid requires much harsher conditions, such as heating in the presence of either mineral acids or copper powder [459]. The latter is a valuable synthetic transformation that has been employed to prepare 2-unsubstituted indoles **586** from the more readily available 2-indole carboxylates **585**. Scheme 5.268 gives an example of the application of this sequence [460, 461].

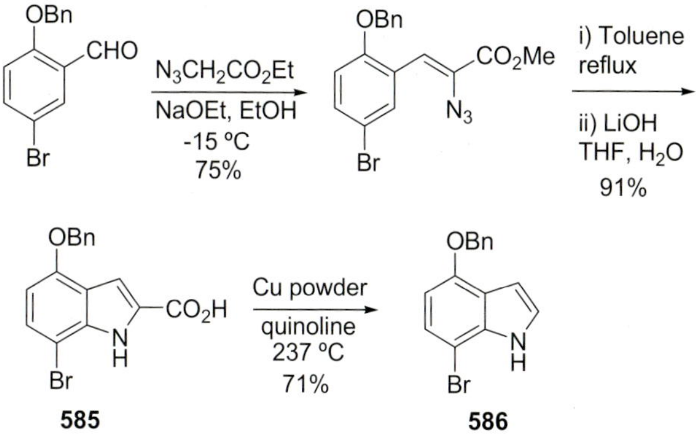

Scheme 5.268

5.7 Addendum

During the production of this book, and since the initial elaboration of this chapter, remarkable advances have occurred in the field of indole chemistry, which indicate the high interest in this particular type of heterocyclic structure. This addendum is not intended to be a comprehensive revision of the most recent literature, and collects only some important advances that have not been mentioned above. First of all, several reviews covering synthesis and different aspects of indole reactivity have appeared recently [462].

5.7.1 Ring Synthesis

5.7.1.1 Fischer Indole Synthesis

The direct synthesis of N-Cbz indoles from the corresponding N-Aryl-N-Cbz hydrazide has been disclosed [463].

An improved methodology for the sequence hydrohydrazination of alkynes/Fischer indolization has been described employing inexpensive and environmentally friendly Zn salts as catalysts [464].

5.7.1.1.1 **Cyclizations by Formation of the N–C2 Bond** The synthesis of indoles by cyclization of nitrenes, generated from aryl azides, has been carried out employing rhodium catalysts [465].

The aryl amination/hydroamination sequence on *o*-chloroalkynyl benzenes has been implemented for the preparation of indoles with sterically demanding N-substituents [466]. Similar routes have been employed for the synthesis of 2-aminoindoles [467] and 4-alkoxyindoles [468]. A multicomponent domino process for the preparation of 2-aminomethyl indoles, involving the cyclization of an *o*-alkynylaniline, has been reported [469].

A new approach for the synthesis of indoles is the cycloisomerization of 2-propargylanilines (**591**). This reaction has been effected employing Pt-based and Brønsted acid catalysts, and leads to functionalized indole skeletons **592** (Scheme 5.269). The mechanism proposed for the acid-catalyzed reaction involves a 5-*exo*-dig cyclization followed by an aza-Cope rearrangement [470].

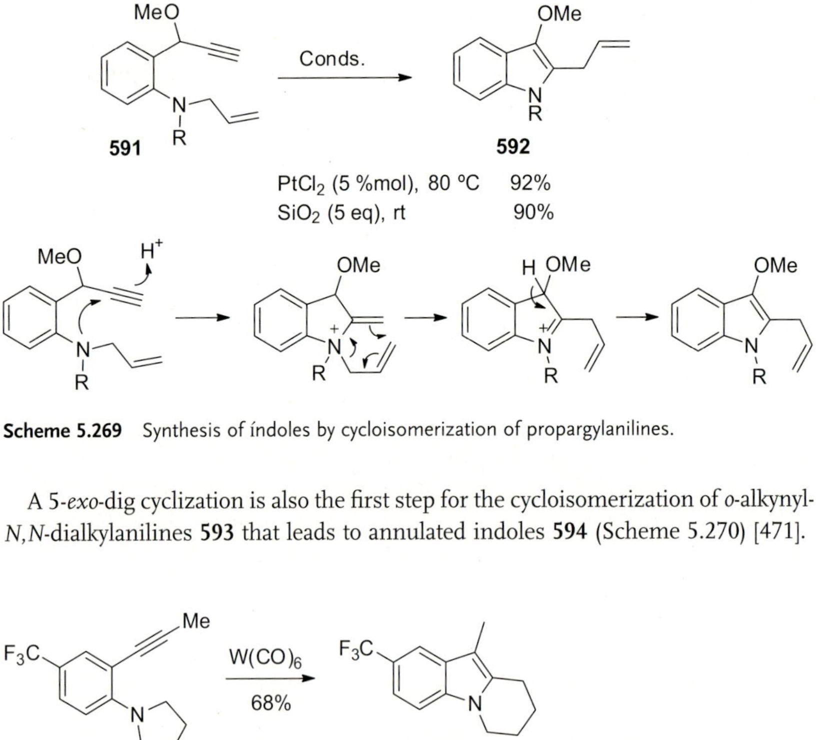

Scheme 5.269 Synthesis of índoles by cycloisomerization of propargylanilines.

A 5-*exo*-dig cyclization is also the first step for the cycloisomerization of *o*-alkynyl-*N*,*N*-dialkylanilines **593** that leads to annulated indoles **594** (Scheme 5.270) [471].

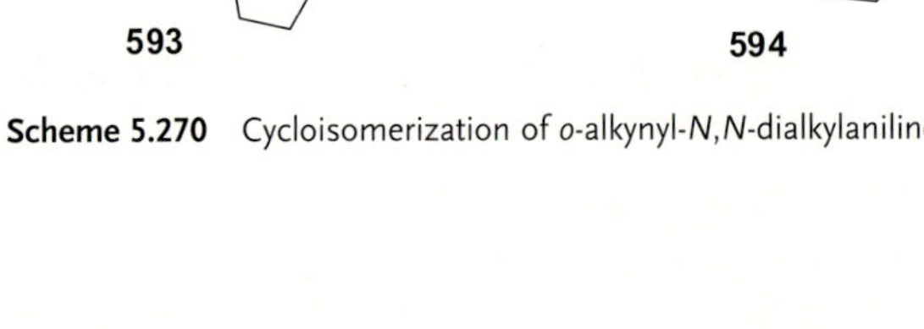

Scheme 5.270 Cycloisomerization of *o*-alkynyl-*N*,*N*-dialkylanilines.

The reaction involves activation of the alkyne by the action of the metal catalyst, followed by 5-*exo*-dig cyclization to produce a metal-containing ammonium ylide (**595**). Then, a [1,2]-Stevens rearrangement, followed by a [1,2]-alkyl migration, renders the fused indole structure (Scheme 5.271).

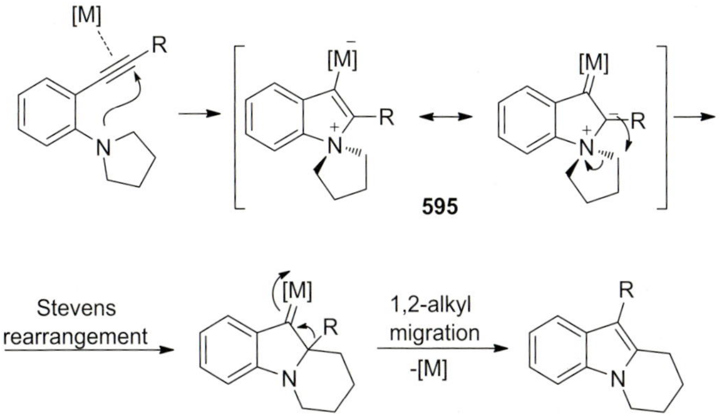

Scheme 5.271 Mechanism proposed for the cycloisomerization of *o*-alkynyl-*N*,*N*-dialkylanilines.

Indoles have been synthesized from *o*-iodobenzoic acids and alkynes in a sequential process that involves a Curtius rearrangement followed by a Pd-catalyzed Larock-type indolization [472]. Quite similarly, *o*-alkynylamides have been employed to synthesize indoles through a Hoffmann-rearrangement/alkyne hydroamination sequence [473].

The Pd-catalyzed synthesis of indoles from *o*-amino-*gem*-dihalostyrenes has been studied in detail, and represents a very powerful methodology for the synthesis of substituted indoles [474].

A new approach to tryptamines from readily available *N*-Boc anilines **596** and *N*-Boc-3-pyrrolidinone **597** has been developed (Scheme 5.272). The process, which

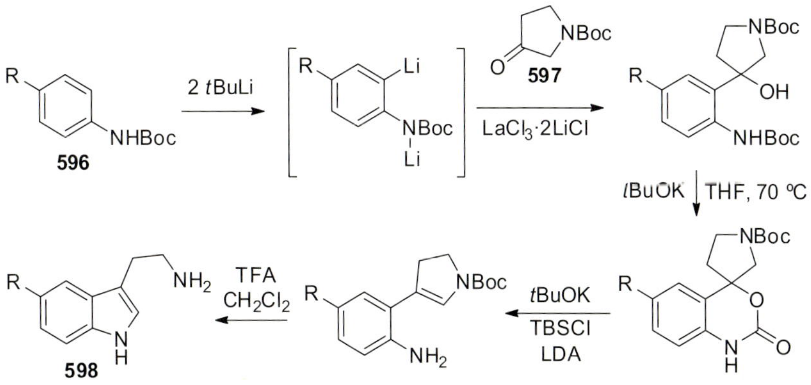

Scheme 5.272 Stepwise synthesis of triptamines.

consists of various steps, as represented in the scheme, allows for the synthesis of a large variety of tryptamines **598** substituted in the benzene ring [475].

5.7.1.1.2 **Cyclizations by Formation of the C2–C3 Bond** The synthesis of cyclic-ketone fused indoles has been achieved by a Pt-catalyzed cycloisomerization of *N*-(2-alkynylphenyl)lactams [476].

A structural variety of 2-substituted indoles **601** can be prepared from *o*-amino-2-chloroacetophenones **599** and Grignard reagents [477]. The reaction involves a [1,2]-migration of the R group from the intermediate magnesium alcoholate **600** (Scheme 5.273).

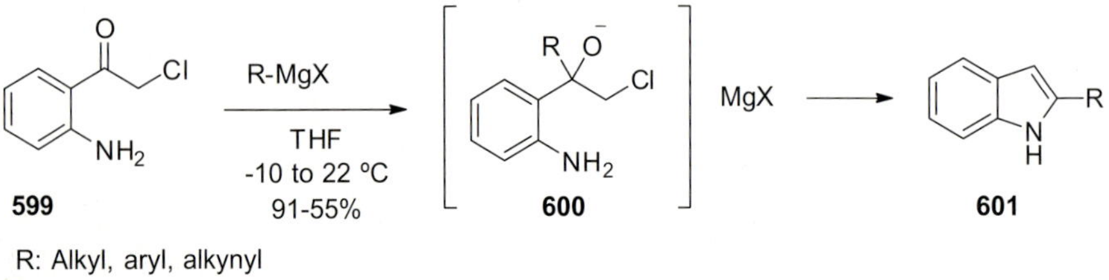

599 → (R-MgX; THF, -10 to 22 °C, 91-55%) → [600] MgX → 601

R: Alkyl, aryl, alkynyl

Scheme 5.273

N-fused indoles have been synthesized through a catalytic carbenoid C–H insertion approach. In this novel reaction, a niobium carbenoid (**603**) is generated from a CF_3 group of the *o*-trifluoromethylaniline derivative **602** (Scheme 5.274). Then, an intramolecular C–H insertion leads to a mixture of indole **605** and indoline **604**. Subsequent oxidation leads to the N-fused indoles **605** with good yields.

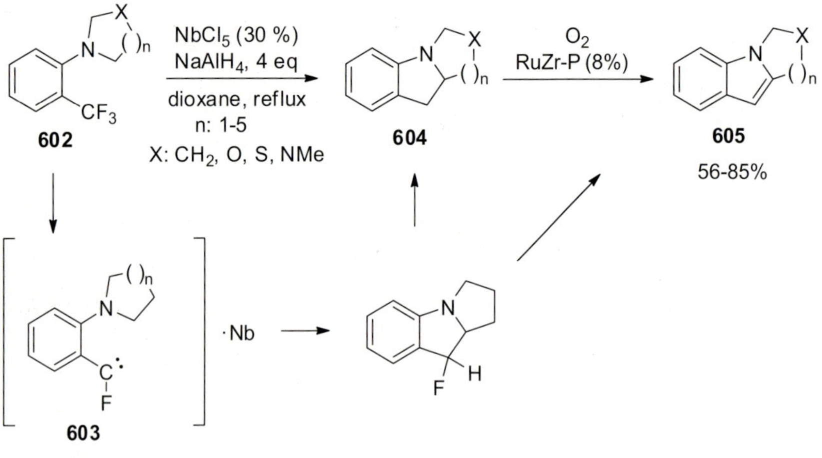

602 → ($NbCl_5$ (30 %), $NaAlH_4$, 4 eq; dioxane, reflux; n: 1-5; X: CH_2, O, S, NMe) → 604 → (O_2, RuZr-P (8%)) → 605 (56-85%)

602 → [603] ·Nb → (fluoroindoline) → 604 / 605

Scheme 5.274

5.7.1.1.3 **Cyclizations by Formation of the C3–C4 Bond** A very versatile new methodology for indole ring synthesis is the Pd-catalyzed oxidative cyclization of

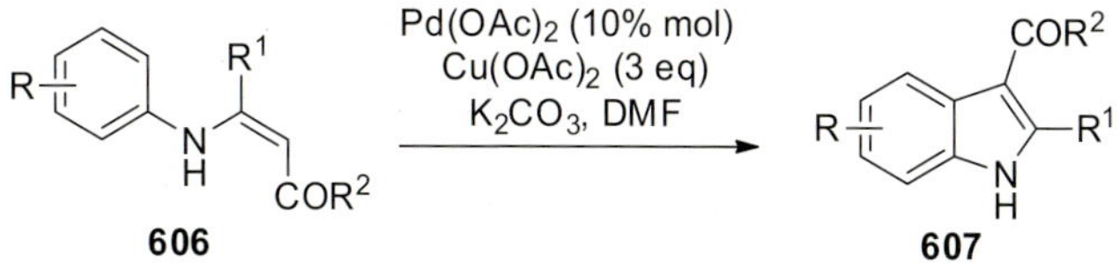

R R[1] N H COR[2] **606**

Pd(OAc)$_2$ (10% mol)
Cu(OAc)$_2$ (3 eq)
K$_2$CO$_3$, DMF

COR[2] R R[1] N H **607**

Scheme 5.275

N-arylenaminones **606** (Scheme 5.275) [478]. This methodology, which leads to 3-acyl substituted indoles **607**, is clearly advantageous when compared with strategies based on intramolecular Heck reactions, which require *o*-haloanilines as starting materials.

The mechanism proposed for this reaction (Scheme 5.276) starts with the electrophilic palladation of the enaminone **606**, to give acylpalladium intermediate **608**, followed by formation of palladacycle **609** by a σ-bond metathesis or a base-assisted deprotonation. Reductive elimination furnishes the indole **607** and releases a Pd(0) species that is reoxidized to the active Pd(II) by the stoichiometric Cu(II) salt.

More recently, a similar reaction, but employing CuI as a catalyst [479], and a metal-free version of this transformation, mediated by phenyl-iodide diacetate, have been reported [480].

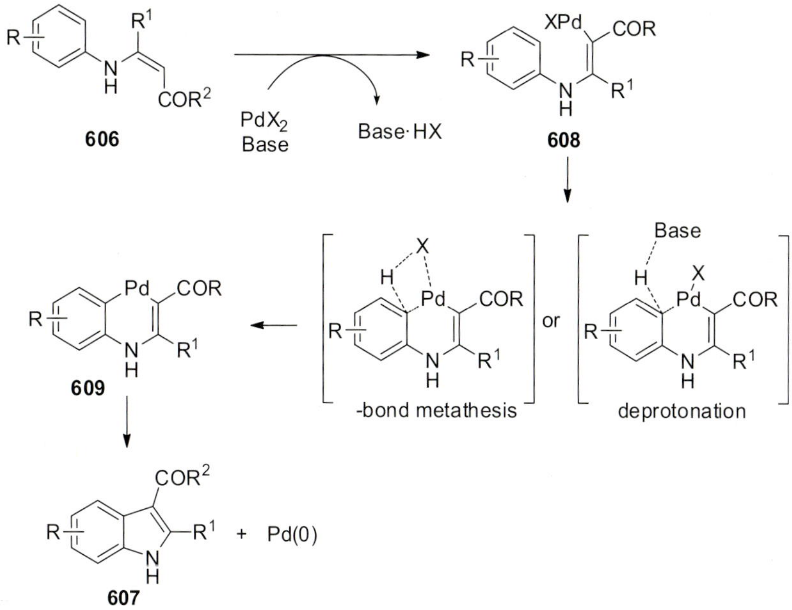

R R[1] N H COR[2] **606**

PdX$_2$
Base

Base·HX

R XPd COR N H R[1] **608**

R Pd COR N H R[1] **609**

R H X Pd COR N H R[1]
-bond metathesis

or

R Base H X Pd COR N H R[1]
deprotonation

R COR[2] R[1] N H **607** + Pd(0)

Scheme 5.276

Moreover, this type of cyclization has been adapted to a domino process in which indoles are prepared directly from electrophilic alkynes and anilines in a Pd(II)-catalyzed process [481].

The same type of products obtained from the cyclization of enaminones has been obtained by a metal-free cascade reaction between N-arylamides and ethyl diazoacetate [482].

Some other Heck based cyclizations have been reported recently: a Pd-catalyzed Suzuki–Heck sequence [483], a Pd-catalyzed aryl amination-Heck sequence [484], and a Cu-catalyzed cyclization of 2-iodoenaminones [485].

Acetanilides have been employed in a rhodium-catalyzed oxidative indole synthesis (Scheme 5.277). In this impressive reaction, 2,3-disubstituted indoles **612** are built from acetanilides **610** and internal alkynes **611** [486]. As in the oxidative cyclization of enaminones described above, the reaction does not need an ortho substituent to enable the cyclization.

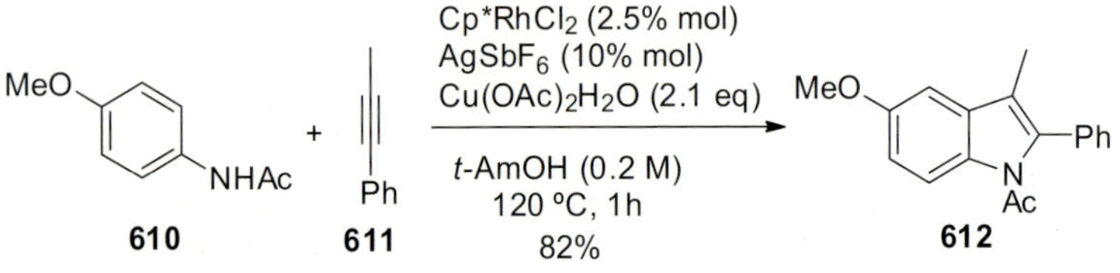

MeO NHAc **610** + Ph **611** → Cp*$RhCl_2$ (2.5% mol), $AgSbF_6$ (10% mol), $Cu(OAc)_2H_2O$ (2.1 eq), *t*-AmOH (0.2 M), 120 °C, 1h, 82% → MeO, N, Ac, Ph **612**

Scheme 5.277

5.7.1.1.4 **Cyclizations with Formation of the N−C7a Bond** A Pd-catalyzed N−C7a bond-forming reaction is the last step in the Pd-catalyzed cascade synthesis of indoles from 1,2-dihalobenzene **613** derivatives and imines **614** (Scheme 5.278). This synthesis consists of two independent reactions catalyzed by the same Pd catalyst: the imine α-arylation and the intramolecular C−N bond-forming reaction. It is worth noting the high modularity of this synthesis of indoles, which are prepared from three fragments: the aromatic system and the carbonyl compound and the amine employed in the preparation of the imine [487].

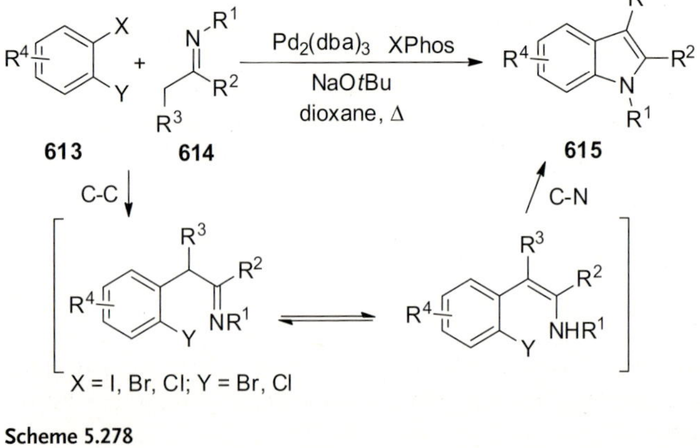

R^4, X, Y **613** + N, R^1, R^2, R^3 **614** → $Pd_2(dba)_3$ XPhos, NaO*t*Bu, dioxane, Δ → R^4, R^3, R^2, N, R^1 **615**

C-C ↓ ; C-N ↑

[R^4, R^3, R^2, NR^1, Y ⇌ R^4, R^3, R^2, NHR^1, Y]

X = I, Br, Cl; Y = Br, Cl

Scheme 5.278

This methodology has been extended to employ *o*-chlorononafluorosulfonates (*o*-chlorononaflates) **616** (Scheme 5.279). The reaction is totally regioselective, with the C–C bond being formed between imine α-carbon and the C-atom that supports the nonaflate. This is an important improvement, because *o*-chlorononaflates are readily prepared from phenols, and moreover is particularly adequate for the preparation of 6-substituted and 4,6-disubstituted indoles, which are not easily prepared by other conventional methods [488].

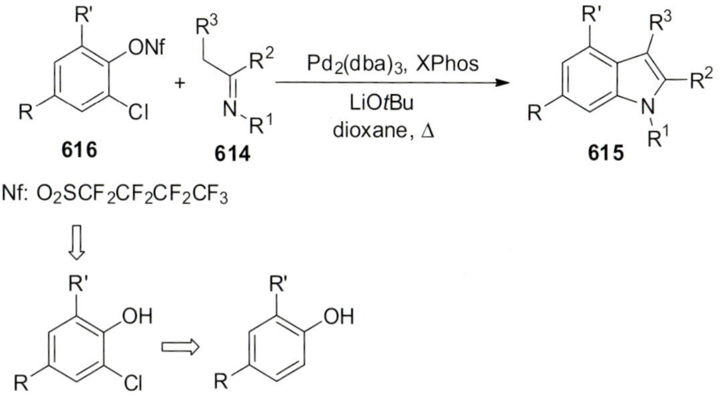
R'
ONf
R
Cl
616
R^3
R^2
N
R^1
614
$Pd_2(dba)_3$, XPhos
LiO*t*Bu
dioxane, Δ
615
Nf: $O_2SCF_2CF_2CF_2CF_3$
OH

Scheme 5.279

5.7.1.1.5 **Synthesis of Indoles by a [4 + 2] Cycloaddition** A very elegant construction of the functionalized indole skeleton has been carried out by the sequence presented in Scheme 5.280 [489]. The key step is the MW-promoted intramolecular [4 + 2] cycloaddition of the aminofuran intermediate **617**, which forms both rings at the same time. The method is particularly useful for the synthesis of 4-substituted indoles (**618**).

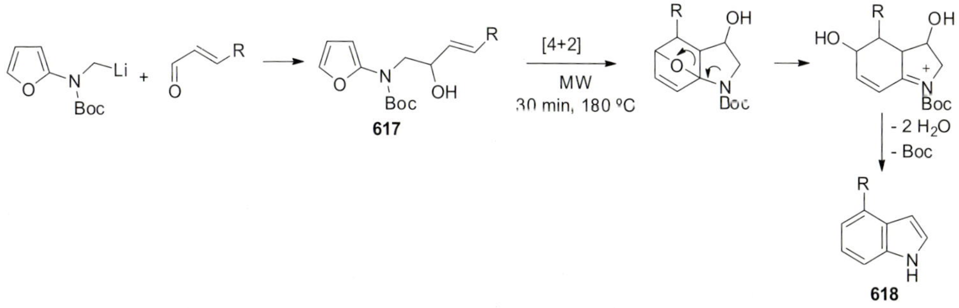
Li
Boc
617
[4+2]
MW
30 min, 180 °C
HO
OH
- 2 H_2O
- Boc
618

Scheme 5.280

5.7.2 Reactivity

5.7.2.1 Reactions with Electrophiles

5.7.2.1.1 Michael Additions and Reactions with Nitroolefins The recent explosion of the field of asymmetric organocatalysis has had an important impact in the chemistry of indoles, in particular in the reactions of indoles with electrophiles such as α,β-unsaturated compounds, nitroolefins, carbonyl compounds, and imines. A review on this field is available [490].

The Friedel–Crafts alkylation of indoles with α,β-unsaturated carbonyl compounds has been carried out employing chiral primary amines as organocatalyst [491].

A chiral binol *N*-triflylphosphoramide derivative has been employed as a Brønsted acid catalyst in the Friedel–Crafts alkylation of indoles with β,γ-unsaturated-α-ketoesters [492]. An intramolecular version has also been reported [493].

Regarding Lewis acid based asymmetric catalysis, highly enantioselective Friedel–Crafts alkylations with α,β-unsaturated phosphonates have been uncovered employing bis(oxazolinyl)pyridine-scandium(III) triflate complexes [494].

Several organocatalytic approaches for the asymmetric addition of indoles to nitroolefins have been reported, employing thiourea based organocatalysts [495], and promoted by a chiral phosphoric acid derivative [496]. Also noteworthy is a recent catalytic asymmetric version employing a Cu(I) chiral catalyst [497].

5.7.2.1.2 Asymmetric Organocatalyzed Pictet–Spengler Reactions Several asymmetric variants of the Pictet–Spengler (Scheme 5.281) reaction of have been reported [498].

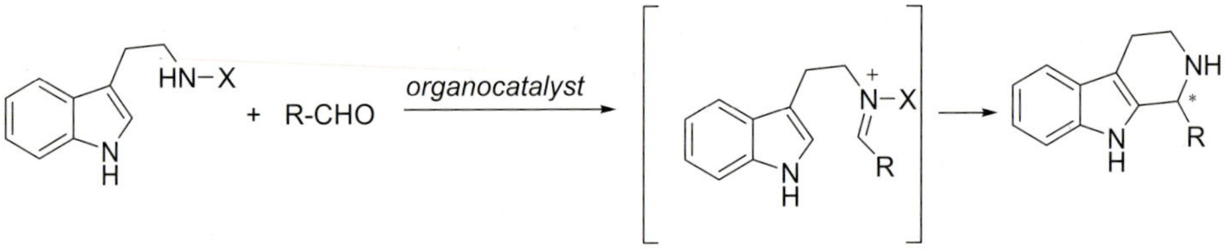

HN–X + R-CHO —organocatalyst→ [N⁺–X iminium intermediate] → NH, *, R

Scheme 5.281

For instance, carboline **621** was obtained in the chiral Brønsted acid catalyzed reaction between a properly functionalized tryptamine (**619**) and aliphatic aldehyde (**620**) (Scheme 5.282) [499].

5.7.2.1.3 Indole as Nucleophile in Pd-Catalyzed Allylic Alkylations The Pd-catalyzed allylic alkylation reaction has been applied in the allylation of 3-substituted indoles, leading to indolenines featuring a C3-quaternary center [500].

The electrophilic arylation of 3-substituted indoles employing diaryl λ^3-iodanes as electrophilic aryl transfer reagents leads to indolenines **622**, which also feature a C3-

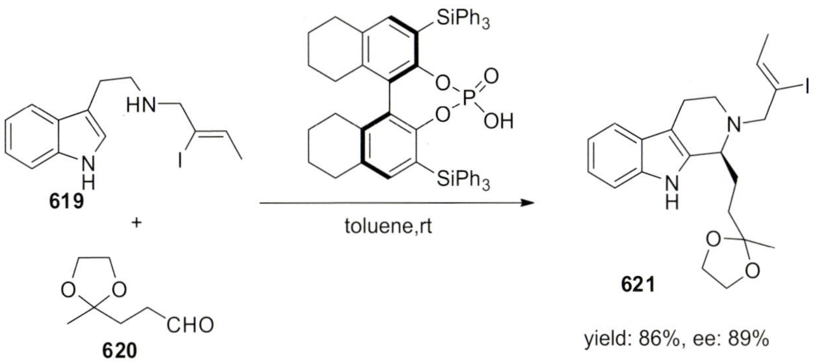

yield: 86%, ee: 89%

Scheme 5.282

quaternary center (Scheme 5.283) [501]. The relative unstable indolenines **622** are isolated upon reduction to the corresponding indolines **623**.

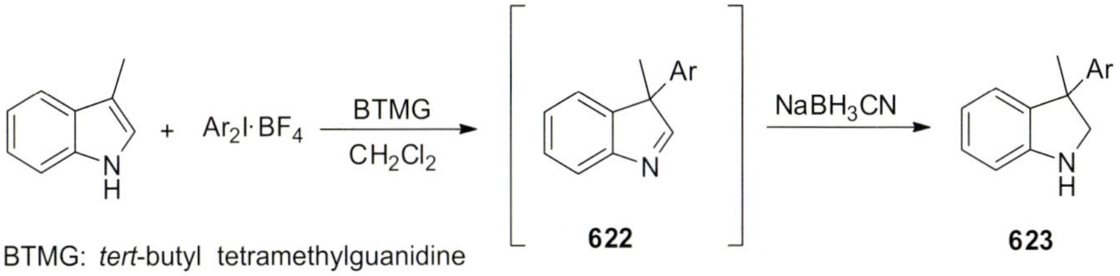

BTMG: *tert*-butyl tetramethylguanidine

Scheme 5.283

5.7.2.2 Transition Metal Catalyzed Reactions

5.7.2.2.1 Direct Alkenylation, Alkynylation, and Arylation reactions The indole ring has served as a platform to develop new C–C bond-forming cross-coupling from C–H bonds. Indeed, much effort has been made in very recent years in the study of the direct arylation of indoles, which has led to very exciting achievements in transition metal catalysis. A review in this field is available [502].

Palladium(II)-catalyzed C2-alkenylation has been reported that employs *N*-(2-pyridyl)sulfonyl indoles [503].

Very recently, an intermolecular Au(I)-catalyzed alkynylation of indoles has been disclosed [504].

Palladium-catalyzed selective arylation of indoles at C2 under mild conditions has been accomplished employing a $Pd^{II/IV}$ catalytic cycle [505]. A different strategy for the C–H arylation, employing boronic acids, has also been reported [506].

Site-selective arylation at either C2 or C3 has been described that employs a Cu(II) catalyst and diaryl-iodine(III) reagents as electrophiles (Scheme 5.284) [507]. Aryla-

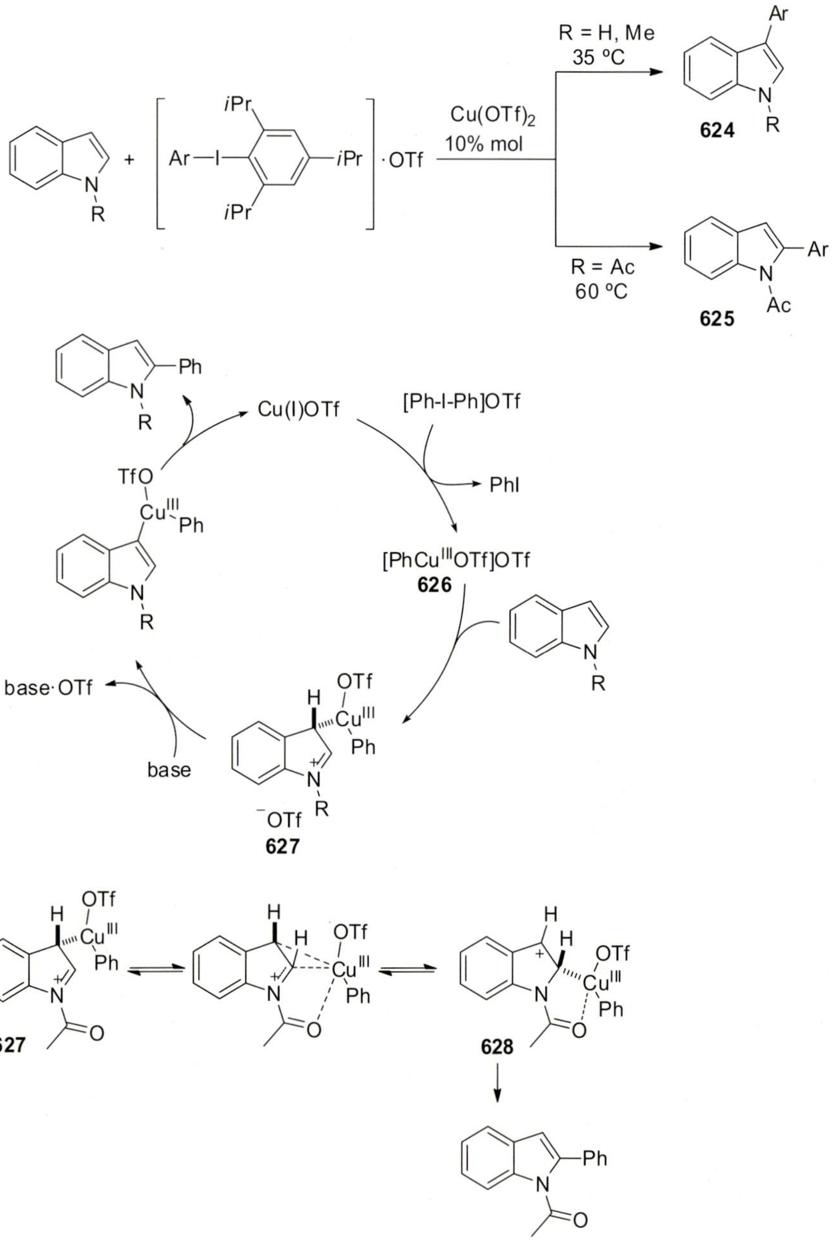

Scheme 5.284

tion at C3 to give **624** occurs on N-H and N-Me indoles, while N-acetylated indoles undergo arylation at C2 leading to **625**. The catalytic cycle proposed for this reaction involves a Cu^{I}/Cu^{III} system (Scheme 5.284), and involves oxidative addition to give a highly electrophilic Cu(III) species (**626**) that undergoes attack at the 3 position of the indole to form intermediate **627**. Rearomatization and reductive elimination releases the arylated product and regenerates the Cu(I) catalyst.

For the N-acetylated indole, the intermediate **627** suffers a C3–C2 migration directed by complexation with the oxygen atom of the carbonyl, giving rise to the C2-metallated system **628**. Reductive elimination and rearomatization provide the C2-arylates indole.

A truly remarkable achievement, carried out by the group of the late Keith Fagnou, is the arylation of indoles with unactivated arenes (Scheme 5.285) [508]. The reactions take place in the presence of a Pd(II) catalyst and a stoichiometric oxidant. C3 or C2 selectivity can be controlled by modification of several parameters: the stoichiometric metal oxidant, the substitution at the N atom of the indole, and the additives present in the reaction. Typically, N-acetyl indoles **629** give C3 arylation, while N-pivaloyl indoles **630** suffer arylation at C2.

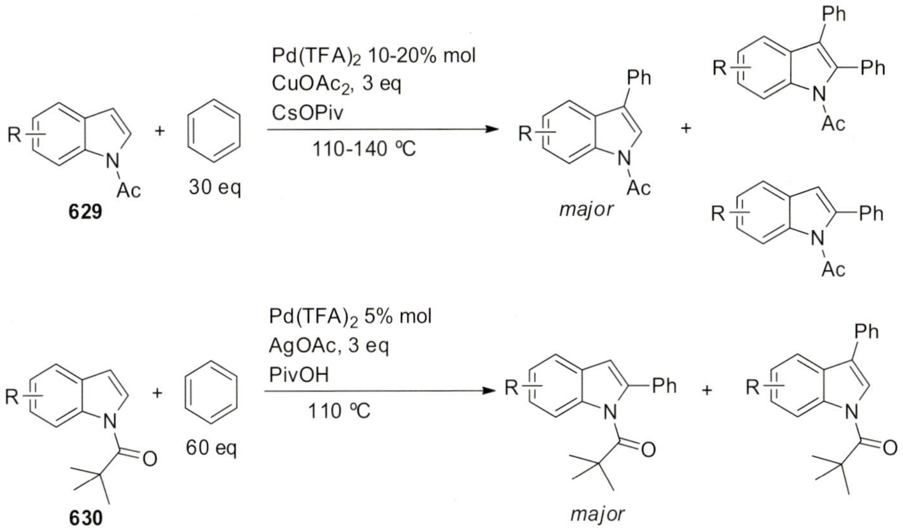

Scheme 5.285

5.7.2.2.2 **C—N Bond-Forming Reactions** The introduction of secondary and tertiary substituents at the N position of the indole is a challenging transformation. Important advances in this sense have been achieved recently.

A one-step *tert*-prenylation of indoles has been devised employing a Pd(II) catalyst (Scheme 5.286) [509]. This is an important transformation as the prenyl group is present in a large number of indole natural products and intermediates.

Various contributions have appeared regarding the asymmetric N-allylation of indoles [510]. For instance, a method has been described employing a chiral metallacyclic iridium phosphoramidite complex (**631**, Scheme 5.287) [511]. This important reaction provides the corresponding N-allylated indoles **632** with total regioselectivity (no C3-allylation occurs) for a wide range of 2,3-disubstituted, 3-

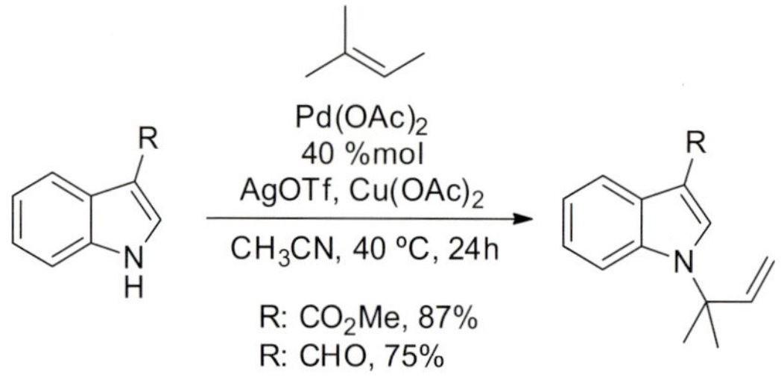
Pd(OAc)2
40 %mol
AgOTf, Cu(OAc)2
CH3CN, 40 °C, 24h
R: CO2Me, 87%
R: CHO, 75%

Scheme 5.286

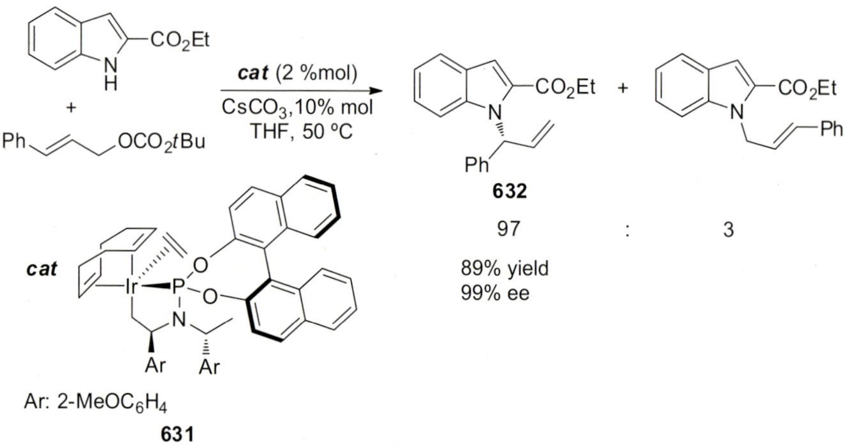
CO2Et
Ph
OCO2tBu
cat (2 %mol)
CsCO3,10% mol
THF, 50 °C
632
97 : 3
89% yield
99% ee
cat
Ar: 2-MeOC6H4
631

Scheme 5.287

substituted, and also 2-substituted indoles. In the last case, the presence of an electron-withdrawing group at C2 is necessary to reduce the reactivity of the C3 position and direct the reaction to the N-position. The branched regioisomer **632** is always obtained as major or exclusively with very high enantiomeric excesses.

5.7.2.2.3 **Pericyclic Reactions** A Claisen rearrangement involving the C2–C3 bond of the indole is the key step in the synthesis of kojic acid derivatives **634** (Scheme 5.288) [512]. The intermediate pyranone substituted indole **633** is prepared employing the Larock indole synthesis.

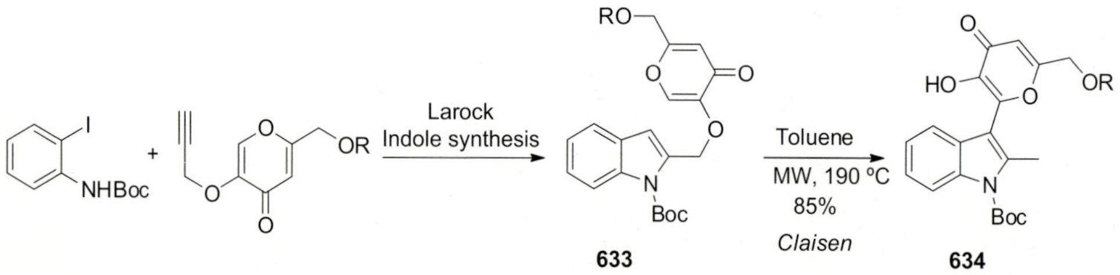
NHBoc
OR
Larock
Indole synthesis
633
Toluene
MW, 190 °C
85%
Claisen
634

Scheme 5.288

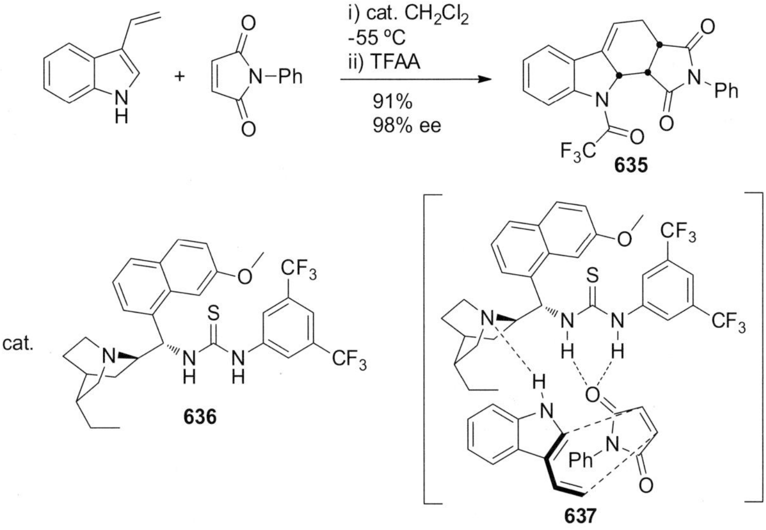

Scheme 5.289

A highly enantioselective organocatalytic [4 + 2] cycloaddition of 3-vinylindoles with maleimides has been reported recently (Scheme 5.289) [513]. The reaction gives rise to indolenines **634** with very high yields and enantioselectivities. The chiral thiourea **635** is the catalyst for the reaction. The nearly complete enantioselectivity obtained has been justified by the simultaneous interaction of the organocatalyst with diene and dienophile, as presented in the transition state model **636**.

References

1. Evans, B.E., Rittle, K.E., Bock, M.G., DiPardo, R.M., Freidinger, R.M., Whitter, W.L., Lundell, G.F., Veber, D.F., Anderson, P.S., Chang, R.S.L., Lotti, V.J., Cerino, D.J., Chen, T.B., Kling, P.J., Kunkel, K.A., Springer, J.P., and Hirshfield, J. (1988) *Journal of Medicinal Chemistry*, **31**, 2235; Nicolaou, K.C., Pfefferkorn, J.A., Roecker, A.J., Cao, G.-Q., Barluenga, S., and Mitchell, H.J. (2000) *Journal of the American Chemical Society*, **122**, 9939; Kleeman, A., Engel, J., Kutscher, B., and Reichert, D. (2001) *Pharmaceutical Substances*, 4th edn, Thieme, New York.
2. Bird, C.W. (ed.) (1995) *Comprehensive Heterocyclic Chemistry II*, Vol. **2**, Elsevier.
3. Sundberg, R.J. (1996) *Indoles*, Academic Press, San Diego CA.
4. Baeyer, A. and Emmerling, A. (1869) *Chemische Berichte*, **2**, 679.
5. Gut, I.G. and Wirz, J. (1994) *Angewandte Chemie, International Edition in English*, **33**, 1153.
6. Smith, B.J. and Liu, R. (1999) *Journal of Molecular Structure: THEOCHEM*, **491**, 211; Dubnikova, F. and Lifshitz, A. (2001) *Journal of Physical Chemistry A*, **105**, 3605.
7. Roychowdhury, P. and Basak, B.S. (1975) *Acta Crystallographica*, **B31**, 1559.
8. Williams, I.D. and Kurtz, S.K. (1992) *Acta Crystallographica, Section C*, **48**, 724; Smith, G., Wermutha, U.D., and Healy, P.C. (2003) *Acta Crystallographica, Section E*, **59**, o1766; Morzyk-Ociepaa, B.,

Michalskab, D., and Pietraszko, A. (2004) *Journal of Molecular Structure*, **688**, 79.

9 Catalán, J. and de Paz, J.L.G. (1997) *Journal of Molecular Structure: THEOCHEM*, **401**, 189; Koeppe, R.E., II, Sun, H., van der Wel, P.C.A., Scherer, E.M., Pulay, P., and Greathouse, D.V. (2003) *Journal of the American Chemical Society*, **125**, 12268

10 Kettle, L.J., Bater, S.P., and Mount, A.R. *Physical Chemistry Chemical Physics*, **2** 2000, 195.

11 Alagona, G., Ghio, C., and Monti, S. (1998) *Journal of Molecular Structure: THEOCHEM*, **433**, 203; Walden, S.E. and Wheeler, R.A. (1996) *Journal of the Chemical Society-Perkin Transactions 2*, 2653.

12 Martínez, A., Vázquez, M.-V., Carreón-Macedo, J.L., Sansores, L.E., and Salcedo, R. (2003) *Tetrahedron*, **59**, 6415.

13 Shieh, W.-C., Dell, S., Bach, A., Repic, O., and Blacklock, T.J. (2003) *The Journal of Organic Chemistry*, **68**, 1954.

14 Sechi, M., Derudas, M., Dallocchio, R., Dessi, A., Bacchi, A., Sannia, L., Carta, F., Palomba, M., Ragab, O., Chan, C., Shoemaker, R., Sei, S., Dayam, R., and Neamati, N. (2004) *Journal of Medicinal Chemistry*, **47**, 5298.

15 Yamazaki, K., Nakamura, Y., and Kondo, Y. (2003) *The Journal of Organic Chemistry*, **68**, 6011.

16 Bergman, J. and Venemalm, L. (1992) *The Journal of Organic Chemistry*, **57**, 2495.

17 Méndez, F. and Gázquez, J.L. (1994) *Journal of the American Chemical Society*, **116**, 9298.

18 Hibino, S. and Choshi, T. (2002) *Natural Product Reports*, **19**, 148, and earlier reviews in this series; Szantay, C. (1990) *Pure and Applied Chemistry*, **62**, 1299.

19 Stratmann, K., Moore, R.E., Bonjouklian, R., Deeter, J.B., Patterson, G.M.L., Shaffer, S., Smith, C.D., and Smitka, T.A. (1994) *Journal of the American Chemical Society*, **116**, 9935.

20 Brossi, A. (ed.) (1990) *The Alkaloids*, Vol. **37**, Academic Press, New York.

21 Yokoshima, S., Ueda, T., Kobayashi, S., Sato, A., Kuboyama, T., Tokuyama, H., and Fukuyama, T. (2002) *Journal of the American Chemical Society*, **124**, 2137; Schneider, C. (2002) *Angewandte Chemie, International Edition*, **41**, 4217.

22 Schlitter, E. (1965) in *The Alkaloids: Chemistry and Physiology*, Vol. **VIII** (ed. R.H.F. Manske), Academic Press, New York, pp. 287–334.

23 Meng, C.Q. (1997) *Current Medicinal Chemistry*, **4**, 385; Dahloef, C. (2005) *Therapy*, **2**, 349.

24 For more detailed reviews on the synthesis of the indole ring the reader is referred to: Pindur, U. and Adam, R. (1988) *Journal of Heterocyclic Chemistry*, **25**, 1; Moody, C.J. (1994) *Synlett*, 681; Sundberg, R.J. (1996) *Indoles*, Academic Press, San Diego;Gribble, G.W. (2000) *Journal of the Chemical Society, Perkin Transactions 1*, 1045.

25 Li, J.J. and Gribble, G.W. (2000) *Palladium in Heterocyclic Chemistry*, Pergamon, Oxford.

26 Cacchi, S. and Fabrizzi, C. (2005) *Chemical Reviews*, **105**, 2873.

27 Tois, J., Franzen, R., and Koskinen, A. (2003) *Tetrahedron*, **59**, 5395.

28 Robinson, B. (1963) *Chemical Reviews*, **63**, 373; Robinson, B. (1969) *Chemical Reviews*, **69**, 227; Robinson, B. (1982) *The Fischer Indole Synthesis*, Wiley-Interscience, New York;Hughes, D.L. (1993) *Organic Preparations and Procedures International*, **25**, 609.

29 Lipinska, T. (2004) *Tetrahedron Letters*, **45**, 8831.

30 Zhao, D., Hughes, D.L., Bender, D.R., DeMarco, A.M., and Reider, P.J. (1991) *The Journal of Organic Chemistry*, **56**, 3001.

31 Hughes, D.L. and Zhao, D. (1993) *The Journal of Organic Chemistry*, **58**, 228.

32 Miyata, O., Kimura, Y., Muroya, K., Hiramatsu, H., and Naito, T. (1999) *Tetrahedron Letters*, **40**, 3601.

33 Nenajdenko, V.G., Zakurdaev, E., Presov, E.V., and Balenkova, E.S. (2004) *Tetrahedron*, **60**, 11719.

34 Menciu, C., Duflos, M., Fouchard, F., Le Baut, G., Emig, P., Achterrath, U., Szelenyi, I., Nickel, B., Schmidt, J.,

Kutscher, B., and Günther, E. (1999) *Journal of Medicinal Chemistry*, **42**, 638.

35 Dhanabal, T., Sangeetha, R., and Mohan, P.S. (2005) *Tetrahedron Letters*, **46**, 4509.

36 Hugel, H.M. and Kennaway, D.J. (1995) *Organic Preparations and Procedures International*, **27**, 1.

37 Hutchins, S.M. and Chapman, K.T. (1996) *Tetrahedron Letters*, **37**, 4869; Cheng, Y. and Chapman, K.T. (1997) *Tetrahedron Letters*, **38**, 1497; Ohno, H., Tanaka, H., and Takahashi, T. (2004) *Synlett*, 508; Mun, H.-S., Ham, W.-H., and Jeong, J.-H. (2005) *Journal of Combinatorial Chemistry*, **7**, 130.

38 Rosenbaum, C., Katzka, C., Marzinzik, A., and Waldmann, H. (2003) *Chemical Communications*, 1822.

39 Rosenbaum, C., Müller, O., Baumhof, P., Mazitschek, R., Giannis, A., and Waldmann, H. (2004) *Angewandte Chemie, International Edition*, **43**, 224; Rosenbaum, C., Röhrs, S., Müller, O., and Waldmann, H. (2005) *Journal of Medicinal Chemistry*, **48**, 1179.

40 Campos, K.R., Woo, J.C.R., Lee, S., and Tillyer, R.D. (2004) *Organic Letters*, **6**, 79.

41 Phillips, R.R. (1959) *Organic Reactions*, **10**, 143.

42 Menciu, C., Duflos, M., Fouchard, F., Le Baut, G., Emig, P., Achterrath, U., Szelenyi, I., Nickel, B., Schmidt, J., Kutscher, B., and Günther, E. (1999) *Journal of Medicinal Chemistry*, **42**, 638.

43 Böttcher, H., Barnickel, G., Hausberg, H.H., Haase, A.F., Seyfried, C.A., and Eiermann, V. (2002) *Journal of Medicinal Chemistry*, **42**, 4020; Heinrich, T. and Böttcher, H. (2004) *Bioorganic & Medicinal Chemistry Letters*, **14**, 2681.

44 Cao, C., Shi, Y., and Odom, A.L. (2002) *Organic Letters*, **4**, 2853.

45 Khedkar, V., Tillack, A., Michalik, M., and Beller, M. (2004) *Tetrahedron Letters*, **45**, 3123.

46 Ackermann, L. and Born, R. (2004) *Tetrahedron Letters*, **45**, 9541.

47 Seayad, A., Ahmed, M., Klein, H., Jackstell, R., Gross, T., and Beller, M. (2002) *Science*, **297**, 1676; Seayad, A., Ahmed, M., Jackstell, R., and Beller, M. (2003) *Journal of the American Chemical Society*, **125**, 10311.

48 Ahmed, M., Jackstell, R., Seayad, A.M., Klein, H., and Beller, M. (2004) *Tetrahedron Letters*, **45**, 869; Köhling, P., Schmidt, A.M., and Eilbracht, P. (2003) *Organic Letters*, **18**, 3213.

49 Wagaw, S., Yang, B.H., and Buchwald, S.L. (1998) *Journal of the American Chemical Society*, **120**, 6621; Chae, J. and Buchwald, S.L. (2004) *The Journal of Organic Chemistry*, **69**, 3336.

50 Gassman, P.G., van Bergen, T.J., Gilbert, D.P., and Cue, B.W. Jr, (1974) *Journal of the American Chemical Society*, **96**, 5495; Gassman, P.G. and van Bergen, T.J. (1988) *Organic Synthesis*, **6**, 601.

51 Wright, S.W., McClure, L.D., and Hageman, D.L. (1996) *Tetrahedron Letters*, **37**, 4631.

52 Bartoli, G., Palmieri, G., Bosco, M., and Dalpozzo, R. (1989) *Tetrahedron Letters*, **30**, 2129; Bartoli, G., Bosco, M., Dalpozzo, R., Palmieri, G., and Marcantoni, E. (1991) *Journal of the Chemical Society, Perkin Transactions 1*, 2757; Bosco, M., Dalpozzo, R., Bartoli, G., Palmieri, G., and Petini, M. (1991) *Journal of the Chemical Society, Perkin Transactions 2*, 657; Dobson, D.R., Gilmore, J., and Long, D.A. (1992) *Synlett*, 79.

53 Dobbs, A.P., Voyle, M., and Whittall, N. (1999) *Synlett*, 1594; Dobbs, A.P. (2001) *The Journal of Organic Chemistry*, **66**, 638; Pirrung, M.C., Wedel, M., and Zao, Y. (2002) *Synlett*, 143.

54 Knepper, K. and Brässe, S. (2003) *Organic Letters*, **5**, 2829.

55 Thyagarajan, B.S., Hillard, J.B., Reddy, K.V., and Majumdar, K.C. (1974) *Tetrahedron Letters*, 1999, Majumdar, K.C., Jana, G.H., and Das, U. (1996) *Chemical Communications*, 517.

56 Baudin, J.-B., Commenil, M.-G., Julia, S.A., Lorne, R., and Mauclaire, L. (1996) *Bulletin de la Société chimique de France*, **133**, 329.

57 Noland, W.E. and Baude, F.J. (1973) *Organic Synthesis*, **5**, 567–571.

58 Suzuki, H., Gyoutoku, H., Yokoo, H., Shinba, M., Sato, Y., Yamada, H., and Murakami, Y. (2000) *Synlett*, 1196.

59 Katayama, S., Ae, N., and Nagata, R. (2001) *The Journal of Organic Chemistry*, **66**, 3474–3483.

60 Tsuji, Y., Kotachi, S., Huh, K.-T., and Watanabe, Y. (1990) *The Journal of Organic Chemistry*, **55**, 580.

61 Makosza, M. and Wojciechowski, K. (2004) *Chemical Reviews*, **104**, 2631.

62 RajanBabu, T.V., Chenard, B.L., and Petti, M.A. (1986) *The Journal of Organic Chemistry*, **51**, 1704.

63 Makosza, M. and Winiarski, J. (1987) *Accounts of Chemical Research*, **20**, 282.

64 Marino, J.P. and Hurt, C.R. (1994) *Synthetic Communications*, **24**, 839.

65 Kozmin, S.A. and Rawal, V.H. (1998) *Journal of the American Chemical Society*, **120**, 13523; Iwama, T., Birman, V.B., Kozmin, S.A., and Rawal, V.H. (1999) *Organic Letters*, **1**, 673.

66 Kozmin, S.A., Iwama, T., Huang, Y., and Rawal, V.H. (2002) *Journal of the American Chemical Society*, **124**, 4628.

67 Banwell, M.G., Kelly, B.D., Kokas, O.J., and Lupton, D.W. (2003) *Organic Letters*, **5**, 2497.

68 Ohshima, T., Xu, Y., Takita, R., Shimizu, S., Zhong, D., and Shibasaki, M. (2002) *Journal of the American Chemical Society*, **124**, 14546.

69 Kawatsura, M. and Hartwig, J.F. (1999) *Journal of the American Chemical Society*, **121**, 1473; Fox, J.M., Huang, X., Chieffi, A., and Buchwald, S.L. (2000) *Journal of the American Chemical Society*, **122**, 1360.

70 Rutherford, J.L., Rainka, M.P., and Buchwald, S.L. (2002) *Journal of the American Chemical Society*, **124**, 15168.

71 Batcho, A.D. and Leimgruber, W. (1990) *Organic Syntheses Collection Volume 7*, 34.

72 Ochi, M., Kataoka, K., Ariki, S., Iwatsuki, C., Kodama, M., and Fukuyama, Y. (1998) *Journal of Natural Products*, **61**, 1043.

73 Showalter, H.D.D., Sun, L., Sercel, A.D., Winters, R.T., Denny, W.A., and Palmer, B.D. (1996) *The Journal of Organic Chemistry*, **61**, 1155.

74 Fetter, J., Bertha, F., Poszavacz, L., and Simig, G. (2005) *Journal of Heterocyclic Chemistry*, **42**, 137.

75 Siu, J., Baxendale, I.R., and Ley, S.V. (2004) *Organic & Biomolecular Chemistry*, **2**, 160.

76 Benington, F., Morin, R.D., and Clark, L.C., Jr, (1959) *The Journal of Organic Chemistry*, **24**, 917–919.

77 Rogers, C.B., Blue, C.A., and Murphy, B.P. (1987) *Journal of Heterocyclic Chemistry*, **24**, 941.

78 Sinhababu, A.K. and Borchardt, R.T. (1983) *The Journal of Organic Chemistry*, **48**, 3347.

79 Novellino, l., d'Ischia, M., and Prota, G. (1999) *Synthesis*, **5**, 793.

80 Murphy, B.P. (1985) *The Journal of Organic Chemistry*, **50**, 5873.

81 Fukuyama, T. and Chen, X. (1994) *Journal of the American Chemical Society*, **116**, 3125.

82 Yang, L.-M., Chent, C.-F., and Lee, K.-H. (1995) *Bioorganic & Medicinal Chemistry Letters*, **5**, 465.

83 He, F., Bo, Y., Altom, J.D., and Corey, E.J. (1999) *Journal of the American Chemical Society*, **121**, 6771.

84 Magnus, P. and Westlund, M. (2000) *Tetrahedron Letters*, **41**, 9369.

85 Akazome, M., Kondo, T., and Watanabe, Y. (1994) *The Journal of Organic Chemistry*, **59**, 3375–3380.

86 Söderberg, B.C. and Shriver, J.A. (1997) *The Journal of Organic Chemistry*, **62**, 5838–5845.

87 Davies, I.W., Smitrovich, J.H., Sidler, R., Qu, C., Gresham, V., and Bazaral, C. (2005) *Tetrahedron*, **61**, 6425.

88 Söderberg, B.C., Chisnell, A.C., O'Neil, S.N., and Shriver, J.A. (1999) *The Journal of Organic Chemistry*, **64**, 9731–9734.

89 Söderberg, B.C.G., Hubbard, J.W., Rector, S.R., and O'Neil, S.N. (2005) *Tetrahedron*, **61**, 3637.

90 Sugasawa, T., Toyoda, T., Adachi, M., and Sasakura, K. (1978) *Journal of the American Chemical Society*, **100**, 4842.

91 Douglas, A.W., Abramson, N.L., Houpis, I.N., Karady, S., Molina, A., Xavier, L.C., and Yasuda, N. (1994) *Tetrahedron Letters*, **35**, 6807.

92 Sugasawa, T., Adachi, M., Sasakura, K., and Kitagawa, A. (1979) *The Journal of Organic Chemistry*, **44**, 578; Adachi, M., Sasakura, K., and Sugasawa, T. (1985) *Chemical & Pharmaceutical Bulletin*, **33**, 1826; Sasakura, K., Adachi, M., and Sugasawa, T. (1988) *Synthetic Communications*, **18**, 265.

93 Jiang, B., Smallheer, J.M., Amaral-Ly, C., and Wounola, M.A. (1994) *The Journal of Organic Chemistry*, **59**, 6823.

94 Plieninger, H. and Voekl, A. (1976) *Chemische Berichte*, **109**, 2121; Plieninger, H., Suhr, K., Werst, G., and Kiefer, B. (1956) *Chemische Berichte*, **89**, 270.

95 England, D.B. and Kerr, M.A. (2005) *The Journal of Organic Chemistry*, **70**, 6519.

96 Tichenor, M.S., Kastrinsky, D.B., and Boger, B.L. (2004) *Journal of the American Chemical Society*, **126**, 8396.

97 Sundberg, R.J., Russell, H.F., Ligon, W.V., and Lin, L.-S. (1980) *The Journal of Organic Chemistry*, **45**, 4767.

98 Molina, P., Alcántara, J., and López-Leonardo, C. (1995) *Tetrahedron Letters*, **36**, 953.

99 Pelkey, E.T. and Gribble, G.W. (1997) *Tetrahedron Letters*, **38**, 5603.

100 Cadogan, J.I.G. and Todd, M.J. (1967) *Journal of the Chemical Society. Chemical Communications*, 178; Cadogan, J.I.G. and Kulik, S. (1970) *Journal of the Chemical Society. Chemical Communications*, 233

101 Holzapfel, C.W. and Dwyer, C. (1998) *Heterocycles*, **48**, 1513.

102 Dohle, W., Staubitz, A., and Knochel, P. (2003) *Chemistry - A European Journal*, **9**, 5323.

103 Hegedus, L.S., Allen, G.F., and Waterman, E.L. (1976) *Journal of the American Chemical Society*, **98**, 2674; Hegedus, L.S., Allen, G.F., Bozell, J.J., and Waterman, E.L. (1978) *Journal of the American Chemical Society*, **100**, 5800; Hegedus, L.S., Allen, G.F., and Olsen, D.J. (1980) *Journal of the American Chemical Society*, **102**, 3583.

104 Gowan, M., Caillé, A.S., and Lau, C.K. (1997) *Synlett*, 1312.

105 Takahashi, S., Kuroyama, Y., Sonogashira, K., and Hagihara, N. (1980) *Synthesis*, 627; Sakamoto, T., Shiraiwa, M., Kondo, Y., and Yamanaka, H. (1983) *Synthesis*, 312; Ezquerra, J., Pedregal, C., Lamas, C., Barluenga, J., Perez, M., Garcia-Martín, M.A., and González, J.M. (1996) *The Journal of Organic Chemistry*, **61**, 5804; Litke, A.F. and Fu, G.C. (1999) *Angewandte Chemie, International Edition*, **38**, 2411.

106 Zeni, G. and Larock, R.C. (2004) *Chemical Reviews*, **104**, 2285; Alonso, F., Beletskaya, I.P., and Yus, M. (2004) *Chemical Reviews*, **104**, 3079.

107 Iritani, K., Matsubara, S., and Utimoto, K. (1988) *Tetrahedron Letters*, **29**, 1799.

108 Sakamoto, T., Kondo, Y., Iwashita, S., Nagano, T., and Yamanaka, H. (1988) *Chemical & Pharmaceutical Bulletin*, **36**, 1305.

109 Rudisill, D.E. and Stille, J.K. (1989) *The Journal of Organic Chemistry*, **54**, 5856.

110 Ackermann, L. (2005) *Organic Letters*, **7**, 439.

111 Battistuzzi, G., Cacchi, S., and Fabrizi, G. (2002) *European Journal of Organic Chemistry*, 2671; Cacchi, S., Fabrizi, G., and Parisi, L.M. (2004) *Synthesis*, 1889

112 Arcadi, A., Cacchi, S., and Marinelli, F. (1992) *Tetrahedron Letters*, **33**, 3915; Cacchi, S., Fabrizi, G., Lamba, D., Marinelli, F., and Parisi, L.M. (2003) *Synthesis*, 728.

113 Arcadi, A., Cacchi, S., Fabrizi, G., Marinelli, F., and Parisi, L.M. (2005) *The Journal of Organic Chemistry*, **70**, 6213.

114 Cacchi, S., Fabrizi, G., and Pace, P. (1998) *The Journal of Organic Chemistry*, **63**, 1001. 116.

115 Arcadi, A., Cacchi, S., Carnicelli, V., and Marinelli, F. (1994) *Tetrahedron*, **50**, 437; Cacchi, S., Fabrizi, G., Pace, P., and Marinelli, F. (1999) *Synlett*, 620.

116 Collini, M.D. and Ellingboe, J.W. (1997) *Tetrahedron Letters*, **38**, 7963.

117 Shimada, T., Nakamura, I., and Yamamoto, Y. (2004) *Journal of the American Chemical Society*, **126**, 10546.

118 Rodríguez, A.L., Koradin, C., Dohle, W., and Knochel, P. (2000) *Angewandte Chemie, International Edition*, **39**, 2488; Koradin, C., Dohle, W., Rodríguez, A.L., Schmid, B., and Knochel, P. (2003) *Tetrahedron*, **59**, 1571.

119 Barluenga, J., Trincado, M., Rubio, E., and González, J.M. (2003) *Angewandte Chemie, International Edition*, **42**, 2406.

120 Kamijo, S. and Yamamoto, Y. (2002) *Journal of the American Chemical Society*, **124**, 11940.

121 Kamijo, S. and Yamamoto, Y. (2002) *Angewandte Chemie, International Edition*, **41**, 3230; Kamijo, S. and Yamamoto, Y. (2003) *The Journal of Organic Chemistry*, **68**, 474.

122 Larock, R.C. and Yum, E.K. (1991) *Journal of the American Chemical Society*, **113**, 6689; Larock, R.C., Yum, E.K., and Refvik, M.D. (1998) *The Journal of Organic Chemistry*, **63**, 7652.

123 Zhou, H., Liao, X., and Cook, J.M. (2004) *Organic Letters*, **6**, 249.

124 Shen, M., Li, G., Lu, B.Z., Hossain, A., Roschangar, F., Farina, V., and Senanayake, C.H. (2004) *Organic Letters*, **6**, 4129.

125 Thielges, S., Meddah, E., Bisseret, P., and Eustache, J. (2004) *Tetrahedron Letters*, **45**, 907; Fang, Y.-Q. and Lautens, M. (2005) *Organic Letters*, **16**, 3549.

126 Bischler, A. and Brion, H. (1892) *Chemische Berichte*, **25**, 2860; Bischler, A. and Firemann, P. (1893) *Chemische Berichte*, **26**, 1336.

127 Sundberg, R.J. and Laurino, J.P. (1984) *The Journal of Organic Chemistry*, **49**, 249.

128 Nordlander, J.E., Catalane, D.B., Kotian, K.D., Stevens, R.M., and Haky, J.E. (1981) *The Journal of Organic Chemistry*, **46**, 778.

129 Aller, E., Buck, R.T., Drysdale, M.J., Ferris, L., Haigh, D., Moody, C.J., Pearson, N.D., and Sanghera, J.B. (1996) *Journal of the Chemical Society, Perkin Transactions 1*, 2879.

130 Moody, C.J., and Swann, E. (1998) *Synlett*, 135.

131 Bashford, K.E., Cooper, A.L., Kane, P.D., Moody, C.J., Muthusamy, S., and Swann, E. (2002) *Journal of the Chemical Society, Perkin Transactions 1*, 1672.

132 Tokunaga, M., Ota, M., Haga, M., and Wakatsuki, Y. (2001) *Tetrahedron Letters*, **42**, 3865.

133 Fañanás, F.J., Granados, A., Sanz, R., Ignacio, J.M., and Barluenga, J. (2001) *Chemistry - A European Journal*, **7**, 2896.

134 Barluenga, J., Fañanás, F.J., Sanz, R., and Fernández, Y. (1999) *Tetrahedron Letters*, **40**, 4865; Barluenga, J., Fañanás, F.J., Sanz, R., and Fernández, Y. (2037) *Chemistry - A European Journal*, **2002, 8**.

135 Tidwell, J.H., Senn, D.R., and Buchwald, S.L. (1991) *Journal of the American Chemical Society*, **113**, 4685; Tidwell, J.H. and Buchwald, S.L. (1992) *The Journal of Organic Chemistry*, **57**, 6380; Tidwell, J.H. and Buchwald, S.L. (1994) *Journal of the American Chemical Society*, **116**, 11797.

136 Mori, M., Chiba, K., and Ban, Y. (1977) *Tetrahedron Letters*, **18**, 1037.

137 Terpko, M.O. and Heck, R.F. (1979) *Journal of the American Chemical Society*, **101**, 5281.

138 Odle, R., Blevins, B., Ratcliff, M., and Hegedus, L.S. (1980) *The Journal of Organic Chemistry*, **45**, 2709.

139 Larock, R.C. and Babu, S. (1987) *Tetrahedron Letters*, **28**, 5291.

140 Sundberg, R.J. and Pitts, W.J. (1991) *The Journal of Organic Chemistry*, **56**, 3048.

141 Sakamoto, T., Kondo, Y., Uchiyama, M., and Yamanaka, H. (1993) *Journal of the Chemical Society, Perkin Transactions 1*, 1941.

142 Tietze, L.F., Hannemann, R., Buhr, W., Lögers, M., Menningen, P., Lieb, M., Starck, D., Grote, T., Döring, A., and Schubert, I. (1996) *Angewandte Chemie International Edition in English*, **35**, 2674; Macor, J.E., Ogilvie, R.J., and Wythes, M.J. (1996) *Tetrahedron Letters*, **37**, 4289; Wensbo, D. and Gronowitz, S. (1996) *Tetrahedron*, **52**, 14975; Li, J.J. (1999) *The Journal of Organic Chemistry*, **64**, 8425; Caddick, S. and Kofie, W. (2002) *Tetrahedron Letters*, **43**, 9347.

143 Yun, W. and Mohan, R. (1996) *Tetrahedron Letters*, **37**, 7189.

144 Zhang, H.-C. and Maryanoff, B.E. (1997) *The Journal of Organic Chemistry*, **62**, 1804.

145 Sakamoto, T., Nagano, T., Kondo, Y., and Yamanaka, H. (1990) *Synthesis*, 215.

146 Chen, C., Lieberman, D.R., Larsen, R.D., Verhoeven, T.R., and Reider, P.J. (1997) *The Journal of Organic Chemistry*, **62**, 2676Chen, C. and Larsen, R.D. *Organic Synthesis*, **10**, 683.

147 Nazaré, M., Schneider, C., Lindenschmidt, A., and Will, D.W. *Angewandte Chemie, International Edition*, **43** (2004), 4526.
148 Ackermann, L., Kaspar, L.T., and Gschrei, C.J. (2004) *Chemical Communications*, 2824.
149 Barluenga, J., Fernández, M.A., Aznar, F., and Valdés, C. (2005) *Chemistry - A European Journal*, **11**, 2276.
150 Allen, C.F.H. and Vanallan, J. (1955) *Organic Synthesis*, **3**, 597.
151 Houlihan, W.J., Parrino, V.A., and Uike, Y. (1981) *The Journal of Organic Chemistry*, **46**, 4511.
152 Spadoni, G., Stankov, B., Duranti, A., Biella, G., Lucini, V., Salvatori, A., and Fraschini, F. (1993) *Journal of Medicinal Chemistry*, **36**, 4069.
153 Orlemans, E.O.M., Schreuder, A.H., Conti, P.G.M., Verboom, W., and Reinhoudt, D.N. (1987) *Tetrahedron*, **43**, 3817.
154 Wacker, D.A. and Kasireddy, P. (2002) *Tetrahedron Letters*, **43**, 5189.
155 Li, J.P., Newlander, K.A., and Yellin, T.O. (1988) *Synthesis*, 73; Le Corre, M., Hercouet, A., and Le Baron, H. (1981) *Journal of the Chemical Society. Chemical Communications*, 14; Etiel, C.M. and Pindur, U. (1989) *Synthesis*, 364.
156 Miyashita, K., Tsuchiya, K., Kondoh, K., Miyabe, H., and Imanishi, T. (1996) *Heterocycles*, **42**, 513–516; Miyashita, K., Tsuchiya, K., Kondoh, K., Miyabe, H., and Imanishi, T. (1996) *Journal of the Chemical Society, Perkin Transactions 1*, 1261.
157 Lyle, R.E. and Skarlos, L. (1966) *Journal of the Chemical Society, Chemical Communications*, 644.
158 Hughes, I. (1996) *Tetrahedron Letters*, **37**, 7595.
159 Nakamura, Y. and Ukita, T. (2002) *Organic Letters*, **4**, 2317.
160 McMurry, J.E. (1989) *Chemical Reviews*, **89**, 1513; Fürstner, A. (1998) *Chemistry - A European Journal*, **4**, 567.
161 Fürstner, A. and Hupperts, A. (1995) *Journal of the American Chemical Society*, **117**, 4468; Fürstner, A. and Bogdanovich, B. (1996) *Angewandte Chemie International Edition in English*, **35**, 2442.
162 Fürstner, A., Hupperts, A., Ptock, A., and Janssen, E. (1994) *The Journal of Organic Chemistry*, **59**, 5215; Fürstner, A. and Ernst, A. (1995) *Tetrahedron*, **51**, 773.
163 Fürstner, A., Ernst, A., Krause, H., and Ptock, A. (1996) *Tetrahedron*, **52**, 7329.
164 Fukuyama, T., Chen, X., and Peng, G. (1994) *Journal of the American Chemical Society*, **116**, 3127.
165 Tokuyama, H., Yamashita, T., Reding, M.T., Kaburagi, Y., and Fukuyama, T. (1999) *Journal of the American Chemical Society*, **121**, 3791.
166 Yokoshima, S., Ueda, T., Kobayashi, S., Sato, A., Kuboyama, T., Tokuyama, H., and Fukuyama, T. (2002) *Journal of the American Chemical Society*, **124**, 2137.
167 Takeda, A., Kamijo, S., Kamijo, S., and Yamamoto, Y. (2000) *Journal of the American Chemical Society*, **122**, 5662.
168 Allen, G.R. Jr, (1973) *Organic Reactions*, **20**, 337.
169 Kinuwaga, M., Arai, H., Nishikawa, H., Sakaguchi, T., Ogasa, T., Tomioka, S., and Kasai, M. (1995) *Journal of the Chemical Society, Perkin Transactions 1*, 2677; Pawlak, J.M., Khau, V.V., Hutchinson, D.R., and Martinelli, M.J. (1996) *The Journal of Organic Chemistry*, **61**, 9055.
170 Ketcha, D.M., Wilson, L.J., and Portlock, D.E. (2000) *Tetrahedron Letters*, **41**, 6253.
171 Dotzauer, B. and Troschütz, R. (2004) *Synlett*, 1039.
172 Shefali, S., Srivastava, S.K., Husbands, S.M., and Lewis, J.W. (2005) *Journal of Medicinal Chemistry*, **48**, 635.
173 Hemetsberger, H., Knittel, D., and Weidmann, H. (1970) *Monatshefte fur Chemie*, **101**, 161; Hemetsberger, H. and Knittel, D. (1972) *Monatshefte fur Chemie*, **103**, 194.
174 Nittel, D. (1985) *Synthesis*, 186.
175 Kiselov, A.S., Van Aken, K., Gulevich, Y. and Strekovsky, L. (1994) *Journal of Heterocyclic Chemistry*, **31**, 1299; Kiselov, A.S., Van Aken, K., Gulevich, Y., and Strekovsky, L. (1997) *Chemical & Pharmaceutical Bulletin*, **45** 1739.
176 Mayer, S., Mérour, J.-Y., Joseph, B., and Guillaumet, G. (2002) *European Journal of Organic Chemistry*, 1643.

177 Wolfe, J.P., Wagaw, S., Marcoux, J.F., and Buchwald, S.L. (1998) *Accounts of Chemical Research*, **31**, 805; Hartwig, J.F. (1998) *Accounts of Chemical Research*, **31**, 852; Hartwig, J.F. (2000) in *Modern Amination Methods* (ed. A. Ricci), Wiley-VCH Verlag GmbH, Weinheim. Muci, A.R. and Buchwald, S.L. (2002) *Topics in Current Chemistry*, **219**, 133Jiang, L. and Buchwald, S.L. (2004) in *Metal-Catalyzed Cross-Coupling Reactions* 2nd edn (eds A. deMeijere and F. Diedrich), Vol. **2**, Wiley-VCH Verlag GmbH, Weinheim, pp. 699–760.

178 Wolfe, J.P., Rennels, R.A., and Buchwald, S.L. (1996) *Tetrahedron*, **52**, 7525; Wolfe, J.P., Rennels, R.A., and Buchwald, S.L. (1997) *Journal of the American Chemical Society*, **119**, 8451; Aoki, K., Peat, A.J., and Buchwald, S.L. (1998) *Journal of the American Chemical Society*, **120**, 3068.

179 Yang, B.H. and Buchwald, S.L. (1999) *Organic Letters*, **1**, 35.

180 Yamada, K., Kubo, T., Tokuyama, H., and Fukuyama, T. (2002) *Synlett*, 231.

181 Kwong, F.Y. and Buchwald, S.L. (2003) *Organic Letters*, **6**, 793.

182 Siebeneicher, H., Bytschkov, I., and Doye, S. (2003) *Angewandte Chemie, International Edition*, **42**, 3042.

183 Willis, M.C., Brace, G.N., and Holmes, I.P. (2005) *Angewandte Chemie, International Edition*, **44**, 403.

184 Muratake, K. and Natsume, M. (1990) *Heterocycles*, **31**, 683.

185 Muratake, H., Mikawa, A., and Natsume, M. (1992) *Tetrahedron Letters*, **33**, 4595; Muratake, H., Mikawa, A., and Natsume, M. (1993) *Tetrahedron Letters*, **34**, 4815.

186 Katritzky, A.R., Ledoux, S., and Nair, S.K. (2003) *The Journal of Organic Chemistry*, **68**, 5728.

187 Katritzky, A.R., Levell, J.R., and Li, J. (1996) *Tetrahedron Letters*, **27**, 5641; Katritzky, A.R., Fali, C.N., and Li, J. (1997) *The Journal of Organic Chemistry*, **62**, 4148; Katritzky, A.R., Li, J., and Xie, L. (1999) *Tetrahedron*, **55**, 8263.

188 Muratake, H., Abe, I., and Natsume, M. (1994) *Tetrahedron Letters*, **35**, 2573.

189 Muratake, K., Tonegawa, M., and Natsume, M. (1996) *Chemical & Pharmaceutical Bulletin*, **44**, 1631; Muratake, K., Tonegawa, M., and Natsume, M. (1998) *Chemical & Pharmaceutical Bulletin*, **46**, 400.

190 ten Have, R. and van Leusen, A.M. (1998) *Tetrahedron*, **54**, 1913.

191 Trofimov, A.B., Sobenina, L.N., Demenev, A.P., and Mikhaleva, A.I. (2004) *Chemical Reviews*, **104**, 2481.

192 Hodges, L.M., Moody, M.W., and Harman, W.D. (1994) *Journal of the American Chemical Society*, **116**, 7931; Hodges, L.M., Spera, M.L., Moody, M.W., and Harman, W.D. (1996) *Journal of the American Chemical Society*, **118**, 7117.

193 Andrews, J.F.P., Jackson, P.M., and Moody, C.J. (1993) *Tetrahedron*, **49**, 7353; Harrison, C.-A., Jackson, P.M., Moody, C.J., and Williams, J.M.J. (1995) *Journal of the Chemical Society, Perkin Transactions 1*, 9131.

194 Barluenga, J., Fernández-Villa, H., Ballesteros, A., and González, J.M. (2005) *Advanced Synthesis and Catalysis*, **347**, 526.

195 Barluenga, J., Vázquez-Villa, H., Merino, I., Ballesteros, A., and González, J.M. (2006) *Chemistry - A European Journal*, **12**, 5790.

196 Hinman, R.L. and Lang, J. (1964) *Journal of the American Chemical Society*, **86**, 3796. 198.

197 Bandini, M., Melloni, A., and Umani-Ronchi, A. (2004) *Angewandte Chemie, International Edition*, **43**, 550; Bandini, M., Melloni, A., Tomáis, S., and Umani-Ronchi, A. (2005) *Synlett*, 1199.

198 Szmuszkovicz, J. (1957) *Journal of the American Chemical Society*, **79**, 2819.

199 Dujardin, G. and Poirier, J.-M. (1994) *Bulletin de la Société chimique de France*, **131**, 900.

200 Iqbal, Z., Jackson, A.H., and Rao, K.R.N. (1988) *Tetrahedron Letters*, **29**, 2577.

201 Srivastava, N. and Banik, B.K. (2003) *The Journal of Organic Chemistry*, **68**, 2109; Alam, M.M., Varala, R., and Adapa, S.R. (2003) *Tetrahedron Letters*, **44**, 5115; Reddy, A.V., Ravinder, K., Goud, T.V., Krishnaiah, P., Raju, T.V., and Venkateswarlu, Y. (2003) *Tetrahedron Letters*, **44**, 6257; Yadav, J.S., Reddy, B.V.S., and Swamy, T. (2003) *Tetrahedron Letters*, **44**, 9121.

202 Bartoli, G., Bartolacci, M., Bosco, M., Foglia, G., Giuliani, A., Marcantoni, E., Sambri, L., and Torregiani, E. (2003) *The Journal of Organic Chemistry*, **68**, 4594; Ji, S.-J., Wang, S.-Y. (2003) *Synlett*, 2074.

203 Li, W.-J., Lin, X.-F., Wang, J., Li, G.-L., and Wang, Y.-G. (2005) *Synlett*, 2003.

204 Lin, C., Hsu, J., Sastry, M.N.V., Fang, H., Tu, Z., Liu, J.-T., and Ching-Fa, Y. (2005) *Tetrahedron*, **611**, 1751.

205 Harrington, P.E. and Kerr, M.A. (1996) *Synlett*, 1047.

206 Manabe, K., Aoyama, N., and Kobayashi, S. (2001) *Advanced Synthesis and Catalysis*, **343**, 174; Firouzabadi, H., Iranpoor, N., and Nowrouzi, F. (2005) *Chemical Communications*, 789.

207 Yadav, J.S., Abraham, S., Reddy, B.V.S., and Sabitha, G. (2001) *Synthesis*, 2165.

208 Bandini, M., Cozzi, P.G., Giacomini, M., Melchiorre, P., Selva, S., and Umani-Ronchi, A. (2002) *The Journal of Organic Chemistry*, **67**, 3700.

209 Bandini, M., Melchiorre, P., Melloni, A., and Umani-Ronchi, A. (2002) *Synthesis*, 1110.

210 Arcadi, A., Bianchi, G., Chiarini, M., D'Anniballe, G., and Marinelli, F. (2004) *Synlett*, 944. 212.

211 Zhuang, W., Hausen, T., and Jørgensen, K.A. (2001) *Chemical Communications*, 347; Zhou, J. and Tang, Y. (2002) *Journal of the American Chemical Society*, **124**, 9030–9031; Zhou, J. and Tang, Y. (2004) *Chemical Communications*, 432–433; Zhou, J., Ye, M.-C., Huang, Z.-Z., and Tang, Y. (2004) *The Journal of Organic Chemistry*, **69**, 1309–1320.

212 Jensen, K.B., Thorhauge, J., Mazell, R.-G., and Jørgensen, K.A. (2001) *Angewandte Chemie, International Edition*, **40**, 160.

213 Evans, D.A., Scheidt, K.A., Frandrick, K.R., Lam, H.W., and Wu, J. (2003) *Journal of the American Chemical Society*, **125**, 10780.

214 Palomo, C., Oiarbide, M., Kardak, B.G., García, J.M., and Linden, A. (2005) *Journal of the American Chemical Society*, **127**, 4154.

215 Evans, D.A., Fandrick, K.R., and Song, H.-J. (2005) *Journal of the American Chemical Society*, **127**, 8942.

216 Bandini, M., Fagioli, P., Garavelli, M., Melloni, A., Trigari, V., and Umani-Ronchi, A. (2004) *The Journal of Organic Chemistry*, **69**, 7511.

217 Austin, J.F. and MacMillan, D.W.C. (2002) *Journal of the American Chemical Society*, **124**, 1172; Austin, J.F., Kim, S.-G., Sinz, C.J., Xiao, W.-J., and MacMillan, D.W.C. (2004) *Proceedings of the National Academy of Sciences of the United States of America*, **101**, 5482.

218 Herrera, R.P., Sgarzani, V., Bernardi, L., and Ricci, A. (2005) *Angewandte Chemie, International Edition*, **44**, 6576.

219 Liu, C., Han, X., Wang, X., and Widenhoefer, R.A. (2004) *Journal of the American Chemical Society*, **126**, 3700.

220 Liu, C. and Widenhoefer, R.A. (2006) *Chemistry - A European Journal*, **12**, 2371.

221 Roomi, M. and MacDonald, S. (1970) *Canadian Journal of Chemistry*, **48**, 139; Gregorovich, B., Liang, K., Clugston, D., and MacDonald, S. (1968) *Canadian Journal of Chemistry*, **46**, 3291.

222 Yadav, J.S., Subba Reddy, B.V., Murthy, Ch.V.S.R., Mahesh Kumar, G., and Madan, Ch. (2001) *Synthesis*, 783. 224.

223 Pindur, U. and Kim, M.-H. (1989) *Tetrahedron*, **45**, 6427; Hao, J., Taktak, S., Aikawa, K., Yusa, Y., Hatano, M., and Mikami, K. (2001) *Synlett*, 1443.

224 Zhuang, W., Gathergood, N., Hazell, R.G., and Jørgensen, K.A. (2001) *The Journal of Organic Chemistry*, **66**, 1009; Lyle, M.P.A., Draper, N.D., and Wilson, P.D. (2005) *Organic Letters*, **7**, 901.

225 Török, B., Abid, M., London, G., Esquibel, J., Török, M., Mhadgut, S.C., Yan, P., and Prakash, G.K.S. (2005) *Angewandte Chemie, International Edition*, **44**, 3086.

226 Katritzky, A.R., Lue, P., and Chen, Y.-X. (1990) *The Journal of Organic Chemistry*, **55**, 3688.

227 Lindquist, C., Ersoy, U., and Somfai, P. (2006) *Tetrahedron*, **63**, 3439.

228 Janczuk, A., Zhang, W., Xie, W., Lou, S., Cheng, J., and Wang, P.G. (2002) *Tetrahedron*, **43**, 4271.
229 Esquivias, J., Arrayas, R.M., and Carretero, J.C. (2006) *Angewandte Chemie, International Edition*, **45**, 629.
230 Jia, Y., Xie, J., Duan, H., Wang, L., and Zhou, Q. (2006) *Organic Letters*, **8**, 1621.
231 Wang, Y.-Q., Song, J., Hong, R., Li, H., and Deng, L. (2006) *Journal of the American Chemical Society*, **128**, 8156.
232 Bandini, M., Cozzi, P.G., Melchiorre, P., and Umani-Ronchi, A. (2002) *The Journal of Organic Chemistry*, **67**, 5386.
233 Yadav, J.S., Reddy, B.V.S., Abraham, S., and Sabitha, G. (2002) *Synlett*, 1550.
234 Deechongkit, S., You, S.-L., and Kelly, J.W. (2004) *Organic Letters*, **6**, 497.
235 Bandini, M., Cozzi, P.G., Melchiorre, P., and Umani-Ronchi, A. (2004) *Angewandte Chemie, International Edition*, **43**, 84–87.
236 Sato, K. and Kozikowski, A.P. (1989) *Tetrahedron Letters*, **30**, 4073.
237 Bennani, Y.L., Zhu, G.D., and Freeman, J.C. (1998) *Synlett*, 754; Nishikawa, T., Kajii, S., Wada, K., Ishikawa, M., and Isobe, M. (2002) *Synthesis*, 1658
238 Rinner, U., Hudlicky, T., Gordon, H., and Pettit, G.R. (2004) *Angewandte Chemie, International Edition*, **43**, 5342.
239 Trost, B.M. and Van Vraken, D.L. (1996) *Chemical Reviews*, **96**, 395.
240 Bandini, M., Melloni, A., and Umani-Ronchi, A. (2004) *Organic Letters*, **6**, 3199.
241 Trost, B.M. and Crawley, M.L. (2003) *Chemical Reviews*, **103**, 2921.
242 Bandini, M., Melloni, A., Piccinelli, F., Sinisi, R., Tommasi, S., and Umani-Ronchi, A. (2006) *Journal of the American Chemical Society*, **128**, 1424.
243 Berti, G., Da Settimo, A., and Nannipieri, E. (1968) *Journal of the Chemical Society (C)*, 2145.
244 González, A. and Gálvez, C. (1983) *Synthesis*, 212.
245 Pelkey, E.T. and Gribble, G.W. (1999) *Synthesis*, 1117.
246 Jiang, J. and Gribble, G.W. (2002) *Tetrahedron Letters*, **43**, 4115.
247 Roy, S. and Gribble, G.W. (2005) *Tetrahedron Letters*, **46**, 1325.
248 Anthony, W.C. (1960) *The Journal of Organic Chemistry*, **25**, 2049; James, P.N. and Snyder, H.R. (1963) *Organic Synthesis*, **4**, 539.
249 Ketcha, D.M. and Gribble, G.W. (1985) *The Journal of Organic Chemistry*, **50**, 5451.
250 Heathcock, R.A. and Kasparek, S. (1969) *Advances in Heterocyclic Chemistry*, **10**, 61.
251 Bergman, J. and Venemalm, L. (1990) *Tetrahedron*, **46**, 6061; Faul, M.M. and Winneroski, L.L. (1997) *Tetrahedron Letters*, **37**, 4749.
252 Ottoni, O., Neder, A.V.F., Dias, A.K.B., Cruz, R.P.A., and Aquino, L.B. (2001) *Organic Letters*, **3**, 1005. 254.
253 Okauchi, T., Itonaga, M., Minami, T., Owa, T., Kitoh, K., and Yoshino, H. (2000) *Organic Letters*, **2**, 1485; Wynne, J.H., Lloyd, C.T., Jensen, S.D., Boson, S., and Stalick, W.M. (2004) *Synthesis*, 2277.
254 Katritzky, A.R., Suzuki, K., Singh, S.K., and He, H.-Y. (2003) *The Journal of Organic Chemistry*, **68**, 5720.
255 Bocchi, V. and Palla, G. (1982) *Synthesis*, 1096.
256 Brennan, M.R., Ericksson, K.L., Szmalc, F.S., Tansey, M.J., and Thornton, J.M. (1986) *Heterocycles*, **24**, 2879.
257 Ludwig, J., Bovens, S., Brauch, C., Elfringhoff, A.S., and Lehr, M. (2006) *Journal of Medicinal Chemistry*, **49**, 2611.
258 Piers, K., Meimaroglou, C., Jardine, R.V., and Brown, R.K. (1963) *Canadian Journal of Chemistry*, **41**, 2399; Calo, V., Ciminale, T., López, L., Naso, P., and Tudesco, P. (1972) *Journal of the Chemical Society, Perkin Transactions 1*, 2567.
259 Barluenga, J., Gonzalez, J.M., Garcia-Martin, M.A., Campos, P.J., and Asensio, G. (1993) *The Journal of Organic Chemistry*, **58**, 2058.
260 Balón, M., Carmona, M.C., Muñoz, M.A., and Hidalgo, J. (1989) *Tetrahedron*, **45**, 7501.
261 Bordwell, F.G., Zhang, X., and Cheng, J.-P. (1991) *The Journal of Organic Chemistry*, **56**, 3216.
262 Rubottom, G.M. and Chabala, J.C. (1972) *Synthesis*, 566; Heaney, H. and Ley, S.V. (1973) *Journal of the Chemical Society, Perkin Transactions 1*, 499; Kikugawa, Y. and Miyake, Y. (1981) *Synthesis*, 461
263 Muchowsky, J.M. and Solas, D.R. (1984) *The Journal of Organic Chemistry*, **49**, 203; Benson, S.C., Li, J.-H., and Snyder, J.K.

(1992) *The Journal of Organic Chemistry*, **57**, 5285.
264 Barco, A., Benetti, S., and Pollini, G.P. (1976) *Synthesis*, 124; Guida, W.C. and Mathre, D.J. (1980) *The Journal of Organic Chemistry*, **45**, 3172; Galons, H., Miocque, M., Combet- Farnoux, C., Bensaid, Y., Decodts, G., and Bram, G. (1985) *Chemical & Pharmaceutical Bulletin*, **33**, 5108.
265 Fink, D.M. (2004) *Synlett*, 2394.
266 Schmolka, S.J. and Zimmer, H. (1984) *Synthesis*, 29.
267 Greene, T.W. and Wuts, P.G.M. (1999) *Protective Groups in Organic Synthesis*, 3rd edn, John Wiley & Sons, New York, p. 615.
268 Grehn, L. and Ragnarsson, U. (1984) *Angewandte Chemie, International Edition in English*, **23**, 296.
269 Kikugawa, Y. (1981) *Synthesis*, 460.
270 Jacquemard, U., Bénéteau, V., Lefoix, M., Routier, S., Mérour, J.-Y., and Coudert, G. (2004) *Tetrahedron*, **60**, 10039.
271 Roy, S. and Gribble, G.W. (2005) *Tetrahedron Letters*, **46**, 1325.
272 Bremner, J.B., Samosorn, S., and Ambrus, J.I. (2004) *Synthesis*, 2653.
273 Chinchilla, R., Najera, C., and Yus, M. (2004) *Chemical Reviews*, **104**, 2667.
274 Jiang, J. and Gribble, G.W. (2002) *Tetrahedron Letters*, **43**, 4115.
275 Roy, S. and Gribble, G.W. (2005) *Tetrahedron Letters*, **46**, 1325.
276 Vazquez, E., Davies, I.W., and Payack, J.F. (2002) *The Journal of Organic Chemistry*, **67**, 7551.
277 Katritzky, A.R. and Akutagava, K. (1985) *Tetrahedron Letters*, **26**, 5935.
278 Bergman, J. and Venemalm, L. (1992) *The Journal of Organic Chemistry*, **57**, 2495.
279 Matsuzono, M., Fukuda, T., and Iwao, M. (2001) *Tetrahedron Letters*, **42**, 7621.
280 Hartung, C.G., Fecher, A., Chapell, B., and Snieckus, V. (2003) *Organic Letters*, **5**, 1899.
281 Frank, W.C., Kim, Y.C., and Heck, R.F. (1978) *The Journal of Organic Chemistry*, **43**, 2947.
282 Harrington, P.J. and Hegedus, L.S. (1984) *The Journal of Organic Chemistry*, **49**, 2657.
283 Harrington, P.J., Hegedus, L.S., and McDaniel, K.F. (1987) *Journal of the American Chemical Society*, **109**, 4335.
284 Tokuyama, H., Kaburagi, Y., Chen, X., and Fukuyama, T. (2000) *Synthesis*, 429.
285 Sundberg, R.J. and Cherney, R.J. (1990) *The Journal of Organic Chemistry*, **55**, 6028; Yokohama, Y., Kondo, K., Mitsuhashi, M., and Murakami, Y. (1996) *Tetrahedron Letters*, **37**, 9309.
286 Black, D.St.C., Keller, P.A., and Kumar, N. (1992) *Tetrahedron*, **48**, 7601.
287 Beccalli, E.M., Broggini, G., Marchesini, A., and Rossi, E. (2002) *Tetrahedron*, **58**, 6673. 289.
288 Sonogashira, K., Tohda, Y., and Hagihara, N. (1975) *Tetrahedron Letters*, **16**, 4467; Sonogashira, K. (2002) in *Handbook of Organopalladium Chemistry for Organic Synthesis*, Vol. 1, (ed. E. Negishi), John Wiley & Sons, New York, p. 493.
289 Sakamoto, T., Nagano, T., Kondo, Y., and Yamanaka, H. (1988) *Chemical & Pharmaceutical Bulletin*, **36**, 2248.
290 Tidwell, J.H., Peat, A.J., and Buchwald, S.L. (1994) *The Journal of Organic Chemistry*, **59**, 7164.
291 Miyaura, T.N. and Suzuki, A. (1995) *Chemical Reviews*, **95**, 2457.
292 Negishi, E., King, A.O., and Okukado, N. (1977) *The Journal of Organic Chemistry*, **42**, 1821–1823.
293 Malapel-Andrieu, B. and Merour, J.-Y. (1998) *Tetrahedron*, **54**, 11079; Joseph, B., Malapel-Andrieu, B., and Merour, J.-Y. (1996) *Synthetic Communications*, **26**, 3289.
294 Witulski, B., Buschmann, N., and Bergsträsser, U. (2000) *Tetrahedron*, **56**, 8473.
295 Molander, G.A. and Biolatto, B. (2003) *The Journal of Organic Chemistry*, **68**, 4302. 297.
296 Johnson, C.N., Stemp, G., Anand, N., Stephen, S.C., and Gallagher, T. (1998) *Synlett*, 1025; Merlic, C.A., McInness, D.M., and You, Y. (1997) *Tetrahedron Letters*, **38**, 6787; Witulski, B., Azcon, J.R., Alayrac, C., Arnautu, A., Collot, V., and Rault, S. (2005) *Synthesis*, 771.
297 Payack, J.F., Vazquez, E., Matty, L., Kress, M.H., and McNamara, J. (2005) *The Journal of Organic Chemistry*, **70**, 175.

298 Hartung, C.G., Fecher, A., Chapell, B., and Snieckus, V. (2003) *Organic Letters*, **5**, 1899.

299 Krolski, M.E., Renaldo, A.F., Rudisill, D.E., and Stille, J.K. (1988) *The Journal of Organic Chemistry*, **53**, 1170.

300 Joseph, B., Malapel, B., and Merour, J.-Y. (1996) *Synthetic Communications*, **26**, 3289; Malapel, B. and Merour, J.-Y. (1998) *Tetrahedron*, **54**, 11079.

301 Somei, M., Sayama, S., Naka, K., and Yamada, F. (1988) *Heterocycles*, **27**, 1585; Choshi, T., Yamada, S., Sugino, E., Kuwada, T., and Hibino, S. (1995) *The Journal of Organic Chemistry*, **60**, 6218; Choshi, T., Sada, T., Fujimoto, H., Nagayama, C., Sugino, E., and Hibino, S. (1996) *Tetrahedron Letters*, **37**, 2593.

302 Kobayashi, S., Peng, G., and Fukuyama, T. (1999) *Tetrahedron Letters*, **40**, 1519.

303 Tokuyama, H., Kaburagi, Y., Chen, X., and Fukuyama, T. (2000) *Synthesis*, 429. 305.

304 Hodson, H.F., Madge, D.J., and Widdowson, D.A. (1992) *Synlett*, 131; Hodson, H.F., Madge, D.J., Slawin, A.N.Z., Widdowson, D.A., and Williams, D.J. (1994) *Tetrahedron*, **50**, 1899; Amat, M., Hadida, S., Sathyanarayana, R., and Bosch, J. (1994) *The Journal of Organic Chemistry*, **59**, 10.

305 Palmisano, G. and Santagostino, M. (1993) *Helvetica Chimica Acta*, **76**, 2356; Palmisano, G. and Santagostino, M. (1993) *Synlett*, 771.

306 Ciattini, P.G., Morera, E., and Ortar, G. (1994) *Tetrahedron Letters*, **35**, 2405.

307 Kraxner, J., Arlt, M., and Gmeiner, P. (2000) *Synlett*, 125.

308 Mann, G., Hartwig, J.F., Driver, M.S., and Fernández-Rivas, C. (1998) *Journal of the American Chemical Society*, **120**, 827.

309 Old, D.W., Harris, M.C., and Buchwald, S.L. (2000) *Organic Letters*, **2**, 1404.

310 Grasa, G.A., Viciu, M.S., Huang, J., and Nolan, S.P. (2001) *The Journal of Organic Chemistry*, **66**, 7729.

311 Lebedev, A.Y., Izmer, V.V., Kazyul'kin, D.N., Beletskaya, I.P., and Voskoboynikov, A.Z. (2002) *Organic Letters*, **4**, 623.

312 Charles, M.D., Schultz, P., and Buchwald, S.L. (2005) *Organic Letters*, **7**, 3965.

313 Itahara, T. (1985) *The Journal of Organic Chemistry*, **50**, 5272; Itahara, T. (1985) *The Journal of Organic Chemistry*, **50**, 5546; Black, D., St, C., Keller, P.A., and Kumar, N. (1993) *Tetrahedron Letters*, **49**, 151.

314 Harris, W., Hill, C.H., Keech, E., and Malsher, P. (1993) *Tetrahedron Letters*, **34**, 8361; Ohkubo, M., Nishimura, T., Jona, H., Honma, T., and Morishima, H. *Tetrahedron*, **52**, 8099; Faul, M.M., Winnerosky, L.L., and Krumrich, C.A. (1998) *The Journal of Organic Chemistry*, **63**, 6053.

315 Ferreira, E.M. and Stoltz, B.M. (2003) *Journal of the American Chemical Society*, **125**, 9578. 317.

316 Itahara, T., Kawasaki, K., and Ouseto, F. (1984) *Synthesis*, 236; Yokoyama, Y., Matsumoto, T., and Murakami, Y. (1995) *The Journal of Organic Chemistry*, **60**, 1486.

317 Capito, E., Brown, J.M., and Ricci, A. (2005) *Chemical Communications*, 1854.

318 Grimster, N.P., Gauntlett, C., Godfrey, C.R.A., and Gaunt, M.J. (2005) *Angewandte Chemie, International Edition*, **44**, 3125.

319 Lu, W., Jia, C., Kitamura, T., and Fujiwara, Y. (2000) *Organic Letters*, **2**, 2927.

320 Sezen, B. and Sames, D. (2003) *Journal of the American Chemical Society*, **125**, 5274.

321 Touré, B.B., Lane, B.S., and Sames, D. (2006) *Organic Letters*, **10**, 1979.

322 Lane, B.S., Brown, M.A., and Sames, D. (2005) *Journal of the American Chemical Society*, **127**, 8051.

323 Wang, X., Lane, B.S., and Sames, D. (2005) *Journal of the American Chemical Society*, **127**, 4996.

324 Baciocchi, E., Muraglia, E., and Sleiter, G. (1992) *The Journal of Organic Chemistry*, **57**, 6817; Baciocchi, E. and Muraglia, E. (1993) *The Journal of Organic Chemistry*, **58**, 7610.

325 Bryers, J.H., Campbell, J.E., Knapp, F.H., and Thisell, J.G. (1999) *Tetrahedron Letters*, **40**, 2677.

326 Osornio, Y.M., Cruz-Almanza, R., Jiménez-Montaño, V., and Miranda, L.D. (2003) *Chemical Communications*, 2316.

327 Guerrero, M.A. and Miranda, L.D. (2006) *Tetrahedron Letters*, **47**, 2517.

328 Ziegler, F.E. and Berlin, M.Y. (1998) *Tetrahedron Letters*, **39**, 2455.

329 Tanino, H., Fukuishi, K., Ushiyama, M., and Okada, K. (2004) *Tetrahedron*, **60**, 3273.

330 Stevens, C.V., Van Meenen, E., Eeckhout, Y., Vanderhoydonck, B., and Hooghe, W. (2005) *Chemical Communications*, 4827.

331 Bremner, J.B. and Sengpracha, W. (2005) *Tetrahedron*, **61**, 941.

332 Kyei, A.S., Tchabanenko, K., Baldwin, J.E., and Adlington, R.M. (2004) *Tetrahedron Letters*, **45**, 8931.

333 Flanagan, S.R., Harrowven, D.C., and Bradley, M. (2003) *Tetrahedron Letters*, **44**, 1795.

334 Bennasar, M.-L., Roca, T., Griera, R., and Bosch, J. (2001) *The Journal of Organic Chemistry*, **66**, 7547.

335 Hilton, S.T., Ho, T.C.T., Pljevaljcic, G., and Jones, K. (2000) *Organic Letters*, **2**, 2639.

336 Hilton, S.T., Ho, T.C.T., Pljevaljcic, G., Schulte, M., and Jones, K. (2001) *Chemical Communications*, 209.

337 Fiumana, A. and Jones, K. (1999) *Chemical Communications*, 1761.

338 Dobbs, A.P., Jones, K., and Veal, K.T. (1995) *Tetrahedron Letters*, **36**, 4857; Dobbs, A.P., Jones, K., and Veal, K.T. (1998) *Tetrahedron*, **54**, 2149.

339 Gribble, G.W., Fraser, H.L., and Badenock, J.C. (2001) *Chemical Communications*, 805.

340 Baran, P.S. and Richter, J.M. (2004) *Journal of the American Chemical Society*, **126**, 7450.

341 Baran, P.S. and Richter, J.M. (2005) *Journal of the American Chemical Society*, **127**, 15394.

342 Chien, C.S., Suzuki, T., Kawasaki, T., and Sakamoto, M. (1984) *Chemical & Pharmaceutical Bulletin*, **32**, 3945; Chien, C.S., Kawasaki, T., and Sakamoto, M. (1985) *Chemical & Pharmaceutical Bulletin*, **33**, 5071.

343 Altinis-Kiraz, C.I., Emge, T.J., and Jimenez, L.S. (2004) *The Journal of Organic Chemistry*, **69**, 2200.

344 Zhang, J.-L. and Che, C.-M. (2005) *Chemistry - A European Journal*, **11**, 3899.

345 Szabó-Pusztay, K. and Szabó, L. (1979) *Synthesis*, 276.

346 Hinman, R.L. and Bauman, C.P. (1964) *The Journal of Organic Chemistry*, **29**, 1206.

347 Schkeryantz, J.M., Woo, J.C.G., Siliphaivanh, P., Depew, K.M., and Danishefsky, S.J. (1999) *Journal of the American Chemical Society*, **121**, 11964.

348 Booker-Milburn, K.I., Fedouloff, M., Paknoham, S.J., Strachan, J.B., Melville, J.L., and Voyle, M. (2000) *Tetrahedron Letters*, **41**, 4657.

349 Engqvist, R. and Bergman, J. (2003) *Tetrahedron*, **59**, 9649.

350 He, L., Yang, L., and Castle, S.L. (2006) *Organic Letters*, **8**, 1165.

351 Zhang, X. and Foote, C.S. (1993) *Journal of the American Chemical Society*, **115**, 8867; Adam, W., Ahrweiler, M., Peters, K., and Schmiedeskamp, B. (1994) *The Journal of Organic Chemistry*, **59**, 2733.

352 Schkeryantz, J.M., Woo, J.C.G., and Danishefsky, S.J. (1995) *Journal of the American Chemical Society*, **117**, 7025–7026;Schkeryantz, J.M., Woo, J.C.G., Siliphaivanh, P., Depew, K.M., and Danishefsky, S.J. (1999) *Journal of the American Chemical Society*, **121**, 11964–11975;Kamenecka, T.M. and Danishefsky, S.J. (2001) *Chemistry - A European Journal*, **7**, 41–63.

353 Suárez-Castillo, O.R., Sánchez-Zavala, M., Meléndez-Rodríguez, M., Castelán-Duarte, L.E., Morales-Ríos, M.S., and Joseph-Nathan, P. (2006) *Tetrahedron*, **62**, 3040.

354 Gribble, G.W. (1991) in *Comprehensive Organic Synthesis*, Vol. **8** (eds B.M. Trost and I. Fleming) Pergamon, Oxford, p. 603;Donohue, T.J., Garg, R., and Stevenson, C.A. (1996) *Tetrahedron: Asymmetry*, **7**, 317.

355 Ames, D.E., Ansari, H.R., France, A.D.G., Lovesey, A.C., Novitt, B., and Simpson, R. (1971) *Journal of the Chemical Society C*, 3088; Watanabe, Y., Ohta, T., Tsuji, Y., and Hiyoshi, T. (1984) *Bulletin of the Chemical Society of Japan*, **57**, 2440.

356 Coulton, S., Gilchrist, T.L., and Graham, K. (1997) *Tetrahedron*, **53**, 791.

357 Kuwano, R., Sato, K., Kurokawa, T., Karube, D., and Ito, Y. (2000) *Journal of the American Chemical Society*, **122**, 7614; Kuwano, R. and Kashiwabara, M. (2006) *Organic Letters*, **8**, 2653.

358 Berger, J.G., Teller, S.R., Adams, C.D., and Guggenberger, L.J. (1975) *Tetrahedron Letters*, **16**, 1807; Repic, O. and Long, D.J. (1983) *Tetrahedron Letters*, **24**, 1115.

359 Fagan, G.P., Chapleo, C.B., Lane, A.C., Myers, M., Roach, A.G., Smith, C.F.C., Stillings, M.R., and Wellbourn, A.P. (1988) *Journal of Medicinal Chemistry*, **31**, 944.

360 Corey, E.J., McCaully, R.J., and Sachdev, H.S. (1970) *Journal of the American Chemical Society*, **92**, 2476.

361 Gribble, G.W., Lord, P.D., Skotnicki, J., Dietz, S.E., Eaton, J.T., and Johnson, J.L. (1974) *Journal of the American Chemical Society*, **96**, 7812.

362 Gribble, G.W. and Hoffmann, J.H. (1977) *Synthesis*, 859.

363 Maryanoff, B.E. and McComsey, D.F. (1978) *The Journal of Organic Chemistry*, **43**, 2733.

364 Lanzilotti, A.E., Litte, R., Fanshawe, U.W.J., McKenzie, T.C., and Lovell, F.M. (1979) *The Journal of Organic Chemistry*, **44**, 4809.

365 Fleming, I. (1976) *Frontier Orbitals and Organic Chemical Reactions*, John Wiley & Sons, Ltd., London.

366 Lee, L. and Snyder, J.K. (1999), in *Advances in Cycloaddition Chemistry*, Vol. 6 (ed. M. Harmata), JAI Press, Stamford, CT, p. 119–171.

367 Benson, S.C., Lee, L., and Snyder, J.K. (1996) *Tetrahedron Letters*, **37**, 5061.

368 Wan, Z.-K. and Snyder, J.K. (1998) *Tetrahedron Letters*, **39**, 2487; Benson, S.C., Lee, L., Yang, L., and Snyder, J.K. (2000) *Tetrahedron*, **56**, 1165.

369 Bodwell, G.J. and Li, J. (2002) *Organic Letters*, **4**, 127.

370 Bodwell, G.J. and Li, J. (2002) *Angewandte Chemie, International Edition*, **41**, 3261.

371 Crawley, S.L. and Funk, R.L. (2003) *Organic Letters*, **5**, 3169.

372 Wenkert, E., Moeller, P.D.R., and Piettre, S.R. (1988) *Journal of the American Chemical Society*, **110**, 7188; Kraus, G.A., Raggon, J., Thomas, P.J., and Bougie, D. (1988) *Tetrahedron Letters*, **29**, 5605; Kraus, G.A. and Bougie, D. (1989) *The Journal of Organic Chemistry*, **55**, 2425; Biolatto, B., Kneeteman, M., and Mancini, P. (1999) *Tetrahedron Letters*, **40**, 3343.

373 Chataigner, I., Hess, E., Toupet, L., and Piettre, S.R. (2001) *Organic Letters*, **3**, 515; Chrétien, A., Chataigner, I., L'Hélias, N., and Piettre, S.R. (2003) *The Journal of Organic Chemistry*, **68**, 7990.

374 Biolatto, B., Kneeteman, M., Paredes, E., and Mancini, P.M.E. (2001) *The Journal of Organic Chemistry*, **66**, 3906.

375 Padwa, A., Brodney, M.A., Lynch, S.M., Rashatasakhon, P., Wang, Q., and Zhang, H. (2004) *The Journal of Organic Chemistry*, **69**, 3735.

376 Padwa, A., Brodney, M.A., Satake, K., and Straub, C.S. (1999) *The Journal of Organic Chemistry*, **64**, 4617.

377 Dehaen, W. and Hassner, A. (1991) *The Journal of Organic Chemistry*, **56**, 896.

378 Subramaniyan, G., Jayashankaran, J., Manian, D.R.S., and Raghunathan, R. (2005) *Synlett*, 1167.

379 de la Mora, M.A., Cuevas, E., Muchowskib, J.M., and Cruz-Almanza, R. (2001) *Tetrahedron Letters*, **42**, 5351.

380 Padwa, A. and Price, A.T. (1998) *The Journal of Organic Chemistry*, **63**, 556.

381 Muthusamy, S., Gunanathan, C., and Suresh, E. (2004) *Tetrahedron*, **60**, 7885.

382 Wilkie, G.D., Elliott, G.I., Blagg, B.S.J., Wolkenberg, S.E., Soenen, D.R., Miller, M.M., Pollack, S., and Boger, D.L. (2002) *Journal of the American Chemical Society*, **124**, 11292; Choi, Y., Ishikawa, H., Velcicky, J., Elliott, G.I., Miller, M.M., and Boger, D.L. (2005) *Organic Letters*, **7**, 4539.

383 Gribble, G.W., Pelkey, E.T., Simon, W.M., and Trujillo, H.A. (2000) *Tetrahedron*, **56**, 10133.

384 Pindur, U. and Eitel, M. (1988) *Helvetica Chimica Acta*, **71**, 1060; Jones, R.A., Fresneda, P.M., Saliente, T.A., and Arques, J.S. (1984) *Tetrahedron*, **40**, 4837; Murase, M., Hosaka, T., Koike, T., and Tobinaga, S. (1989) *Chemical & Pharmaceutical Bulletin*, **37**, 1999.

385 Pfeuffer, L. and Pindur, U. (1988) *Helvetica Chimica Acta*, **71**, 467; Pindur, U. and Otto, C. (1992) *Tetrahedron*, **48**, 3515.

386 Etiel, M. and Pindur, U. (1988) *Heterocycles*, **27**, 2353; Saroja, B. and Srinivasan, P.C. (1986) *Synthesis*, 748.

387 Le Strat, F. and Maddaluno, J. (2002) *Organic Letters*, **4**, 2791.

388 Back, T.G., Bethell, R.J., Parvez, M., and Taylor, J.A. (2001) *The Journal of Organic Chemistry*, **66**, 8599; Back, T.G., Pandyra, A., and Wulff, J.E. (2003) *The Journal of Organic Chemistry*, **68**, 3299.

389 Etiel, M. and Pindur, U. (1990) *The Journal of Organic Chemistry*, **55**, 5368.

390 Carroll, W.A. and Grieco, P.A. (1993) *Journal of the American Chemical Society*, **115**, 1164.

391 Pilarčík, T., Havlíček, J., and Hájíček, J. (2005) *Tetrahedron Letters*, **46**, 7909.

392 Markgraf, J.H., Finkelstein, M., and Cort, J.R. (1996) *Tetrahedron Letters*, **52**, 461.

393 Markgraf, J.H., Snyder, S.A., and Vosburg, D.A. (1998) *Tetrahedron Letters*, **39**, 1111; Snyder, S.A., Vosburg, D.A., Jarvis, M.G., and Markgraf, J.H. (2000) *Tetrahedron*, **56**, 5329.

394 Pindur, U. and Erfanian-Abdoust, H. (1989) *Chemical Reviews*, **89**, 1681.

395 Marinelli, E.R. (1982) *Tetrahedron Letters*, **23**, 2745.

396 Saroja, B. and Srinivasan, P.C. (1984) *Tetrahedron Letters*, **25**, 5429; Vice, S.F., de Carvalho, H.N., Taylor, N.G., and Dmitrienko, G.I. (1989) *Tetrahedron Letters*, **30**, 7289; Terzidis, M., Tsoleridis, C.A., and Stephanidou-Stephanatou, J. (2005) *Tetrahedron Letters*, **46**, 7239.

397 Diker, K., Döé de Maindreville, M., and Lévy, J. (1999) *Tetrahedron Letters*, **40**, 7459; Diker, K., Döé de Maindreville, M., Roger, D., LeProvost, F., and Lévy, J. (1999) *Tetrahedron Letters*, **40**, 7463.

398 Laronze, M. and Sapi, J. (2002) *Tetrahedron Letters*, **43**, 7925.

399 Basaveswara Rao, M.V., Satyanarayana, J., Ha, H., and Junjappa, H. (1995) *Tetrahedron Letters*, **36**, 3385.

400 Kuroda, N., Takahashi, Y., Yoshinaga, K., and Mukai, C. (2006) *Organic Letters*, **8**, 1843.

401 Choshi, T., Sada, T., Fujimoto, T.H., Nagayama, C., Sugino, E., and Hibino, E.S. (1996) *Tetrahedron Letters*, **15**, 2593; Choshi, T., Sada, T., Fujimoto, H., Nagayama, C., Sugino, E., and Hibino, S. (1997) *The Journal of Organic Chemistry*, **62**, 2535; Hagiwara, H., Choshi, T., Fujimoto, H., Sugino, E., and Hibino, S. (2000) *Tetrahedron*, **56**, 5807; Tohyama, S., Choshi, T., Matsumoto, K., Yamabuki, A., Ikegata, K., Nobuhiro, J., and Hibino, S. (2005) *Tetrahedron Letters*, **46**, 5263.

402 Martín Castro, A.M. (2004) *Chemical Reviews*, **104**, 2939.

403 Novikov, A.V., Kennedy, A.R., and Rainier, J.D. (2003) *The Journal of Organic Chemistry*, **68**, 993.

404 Kawasaki, T., Nonaka, Y., Watanabe, K., Ogawa, A., Higuchi, K., Terashima, R., Masuda, K., and Sakamoto, M. (2001) *The Journal of Organic Chemistry*, **66**, 1200.

405 Kawasaki, T., Terashima, R., Sakaguchi, K., Sekiguchi, H., and Sakamoto, M. (1996) *Tetrahedron Letters*, **37**, 7525; Kawasaki, T., Ogawa, A., Terashima, R., Sekiguchi, H., and Sakamoto, M. (2003) *Tetrahedron Letters*, **44**, 1591–1593;Kawasaki, T., Ogawa, A., Terashima, R., Saheki, T., Ban, N., Sekiguchi, H., Sakaguchi, K., and Sakamoto, M. (2005) *The Journal of Organic Chemistry*, **70** 2957.

406 Kawasaki, T., Ogawa, A., Terashima, R., Saheki, T., Ban, N., Sekiguchi, H., Sakaguchi, K., and Sakamoto, M. (2000) *Tetrahedron Letters*, **41**, 4657.

407 Santos, P.F., Lobo, A.M., and Prabhakar, S. (1995) *Tetrahedron Letters*, **36**, 8099; Santos, P.F., Srinivasan, N., Almeida, P.S., Lobo, A.M., and Prabhakar, S. (2005) *Tetrahedron*, **61**, 9147.

408 Roucher, S. and Klein, P. (1986) *The Journal of Organic Chemistry*, **51**, 123.

409 Weedon, A.C. and Zhang, B. (1992) *Synthesis*, 95.

410 Ito, Y. and Fujita, H. (2000) *Chemistry Letters*, 288.

411 Ikeda, M., Ohno, K., Mohri, S.-i., Takahashi, M., and Tamura, Y. (1984) *Journal of the Chemical Society, Perkin Transactions 1*, 405.

412 Winkler, J.D., Scott, R.D., and Williard, P.G. (1990) *Journal of the American Chemical Society*, **112**, 8971.

413 Oldroyd, D.L. and Weedon, A.C. (1994) *The Journal of Organic Chemistry*, **59**, 1333.

414 Haberl, U., Steckhan, E., Blechert, S., and Wiest, O. (1999) *Chemistry - A European*

Journal, **5**, 2859; Pérez-Prieto, J., Stiriba, S.-E., González-Béjar, M., Domingo, L.R., and Miranda, M.A. (2004) *Organic Letters*, **6**, 3905.
415 Rawal, V.H., Jones, R.J., and Cava, M.P. (1985) *Tetrahedron Letters*, **26**, 2423.
416 Omura, S., Sasaki, Y., Iwai, Y., and Takeshima, H. (1995) *The Journal of Antibiotics*, **48**, 535; Gribble, G.W. and Berthel, S.J. (1993) *Studies in Natural Products Chemistry*, **12**, 365.
417 Gallant, M., Link, T.J., and Danishefsky, S.J. (1993) *The Journal of Organic Chemistry*, **58**, 343; Link, T.J., Gallant, M., Danishefsky, S.J., and Huber, S. (1993) *Journal of the American Chemical Society*, **115**, 3782.
418 Xie, G. and Lown, J.W. (1994) *Tetrahedron Letters*, **35**, 5555.
419 Sanchez-Martínez, C., Faul, M.M., Shih, C., Sullivan, K.A., Grushch, J.L., Cooper, J.T., and Kolis, S.P. (2003) *The Journal of Organic Chemistry*, **68**, 8008.
420 Fedorova, O.A., Fedorov, Y.V., Andryukhina, E.N., Gromov, S.P., Alfimov, M.V., and Lapouyade, R. (2003) *Organic Letters*, **5**, 4533.
421 Gillespie, R.J. and Porter, A.E.A. (1979) *Journal of the Chemical Society, Chemical Communications*, 50.
422 Wood, J.L., Stoltz, B.M., and Dietrich, H.-J. (1995) *Journal of the American Chemical Society*, **117**, 10413; Wood, J.L., Stoltz, B.M., Dietrich, H.-J., Pflum, D.A., and Petsch, D.T. (1997) *Journal of the American Chemical Society*, **119**, 9641.
423 Katritzky, A.R. and Akutawaga, K. (1986) *Journal of the American Chemical Society*, **108**, 6809.
424 Liu, R., Zhang, P., Gan, T., and Cook, J.M. (1997) *The Journal of Organic Chemistry*, **62**, 7447.
425 Thesing, J. and Schülde, F. (1952) *Chemische Berichte*, **85**, 324.
426 Howe, E.E., Zambito, A.J., Snyder, H.R., and Tishler, M. (1945) *Journal of the American Chemical Society*, **67**, 38.
427 Gill, N.S., James, K.B., Lions, F., and Potts, K.T. (1952) *Journal of the American Chemical Society*, **74**, 4923.
428 Novikov, A.V., Sabahi, A., Nyong, A.M., and Rainier, J.D. (2003) *Tetrahedron: Asymmetry*, **14**, 911.
429 Low, K.H. and Magomedov, N.A. (2005) *Organic Letters*, **7**, 2003.
430 Madin, A., O'Donnell, C.J., Oh, T., Old, D.W., Overman, L.E., and Sharp, M.J. (1999) *Angewandte Chemie, International Edition*, **38**, 2934; Yokoshima, S., Tokuyama, H., and Fukuyama, T. (2000) *Angewandte Chemie, International Edition*, **39**, 4073.
431 Cui, C.B., Kakeya, H., and Osada, H. (1996) *The Journal of Antibiotics*, **49**, 832–835;Cui, C.B., Kakeya, H., and Osada, H. (1996) *Tetrahedron*, **52**, 12651–12666;Edmondson, S.D. and Danishefsky, S.J. (1998) *Angewandte Chemie, International Edition*, **37**, 1138–1140;Marti, C. and Carreira, E.C. (2005) *Journal of the American Chemical Society*, **127**, 11505.
432 Vazquez, E., Davies, I.W., and Payack, J.F. (2002) *The Journal of Organic Chemistry*, **67**, 7551.
433 Vazquez, E. and Payack, J.F. (2004) *Tetrahedron Letters*, **45**, 6549.
434 Crestini, C. and Saladino, R. (1994) *Synthetic Communications*, **24**, 2835; Sonderegger, O.J., Bürgi1, T., Limbach, L.K., and Baiker, A. (2004) *Journal of Molecular Catalysis A-Chemical*, **217**, 93.
435 Garden, S.J., da Silva, R.D., and Pinto, A.C. (2002) *Tetrahedron*, **58**, 8399.
436 Boissard, C.G., Post-Munson, D.J., Gao, Q., Huang, S., Gribkoff, V.K., and Meanwell, N.A. (2002) *Journal of Medicinal Chemistry*, **45**, 1487; Amiri-Attou, O., Terme, T., and Vanelle, P. (2005) *Synlett*, 3047.
437 Lathourakis, G.E. and Litinas, K.E. (1996) *Journal of the Chemical Society, Perkin Transactions 1*, 491; Jiang, T., Kuhen, K.L., Wolff, K., Yin, H., and Bieza, K. (2006) *Bioorganic & Medicinal Chemistry Letters*, **16**, 2105.
438 Shintani, R., Inoue, M., and Hayashi, T. (2006) *Angewandte Chemie, International Edition*, **45**, 3353; Toullec, P.Y., Jagt, R.B.C., de Vries, J.G., Feringa, B.L., and Minnaard, A.J. (2006) *Organic Letters*, **8**, 2715.

439 Moore, R.F. and Plant, S.G.P. (1951) *Journal of the Chemical Society*, 3457.

440 Cravotto, G., Giovenzana, G.B., Pilati, T., Sisti, M., and Palmesano, G. (2001) *The Journal of Organic Chemistry*, **66**, 8447.

441 Acemoglu, M., Allmendinger, T., Calienni, J., Cercus, J., Loiseleur, O., Sedelmeiera, G.H., and Xu, D. (2004) *Tetrahedron*, **60**, 11571.

442 Cossy, J., Cases, M., and Gomez Pardo, D. (1998) *Tetrahedron Letters*, **39**, 2331; Reiko Yanada, R., Obika, S., Oyama, M., and Takemoto, Y. (2004) *Organic Letters*, **6**, 2825.

443 Sakamoto, T., Nagano, Y., Kondo, Y., and Yamanaka, H. (1990) *Synthesis*, 215; Arumugam, V., Routledge, A., Abell, C., and Balasubramanian, S. (1997) *Tetrahedron Letters*, **38**, 6473; Dounay, A.B., Hatanaka, K., Kodanko, J.J., Oestreich, M., Overman, L.E., Pfeifer, L.A., and Weis, M.M. (2003) *Journal of the American Chemical Society*, **125**, 6261; Yanada, R., Obika, S., Inokuma, T., Yanada, K., Yamashita, M., Ohta, S., and Takemoto, Y. (2005) *The Journal of Organic Chemistry*, **70**, 6972; Madin, A., O'Donnell, C.J., Oh, T., Old, D.W., Overman, L.E., and Sharp, M.J. (2005) *Journal of the American Chemical Society*, **127**, 18054.

444 Yang, B.H. and Buchwald, S.L. (1999) *Organic Letters*, **1**, 35; Poondra, R.R. and Turner, N.J. (2005) *Organic Letters*, **7**, 863.

445 Shaughnessy, K.H., Hamann, B.C., and Hartwig, J.F. (1998) *The Journal of Organic Chemistry*, **63**, 6546; Lee, S. and Hartwig, J.F. (2001) *The Journal of Organic Chemistry*, **66**, 3402.

446 Bella, M., Kobbelgaard, S., and Jørgensen, K.A. (2005) *Journal of the American Chemical Society*, **127**, 3670.

447 Mao, Z. and Baldwin, S.W. (2004) *Organic Letters*, **6**, 2425.

448 Ready, J.M., Reisman, S.E., Hirata, M., Weiss, M.M., Tamaki, K., Ovaska, V., and Wood, J.L. (2004) *Angewandte Chemie, International Edition*, **73**, 1270.

449 Hennessy, E.J. and Buchwald, S.L. (2003) *Journal of the American Chemical Society*, **125**, 12084.

450 Conway, S.C. and Gribble, G.W. (1992) *Synthetic Communications*, **22**, 2987; Bourlot, A.S., Desarbre, E., and Mérour, J.Y. (1994) *Synthesis*, 411.

451 Somei, M. (2002) *Advances in Heterocyclic Chemistry*, **82**, 101; Somei, M. (1999) *Heterocycles*, **50**, 1157.

452 Somei, M., Inoue, S., Tokutake, S., Yamada, F., and Kaneko, C. (1981) *Chemical & Pharmaceutical Bulletin*, **29**, 726; Somei, M. (1986) *Chemical & Pharmaceutical Bulletin*, **34**, 4109; Reboredo, F.J., Treus, M., Estévez, J.C., Castedo, L., and Estévez, R.J. (2002) *SynLett*, 999; Myers, A.G. and Herzon, S.B. (2003) *Journal of the American Chemical Society*, **125**, 12080.

453 Wong, A., Kuethe, J.T., and Davies, I.W. (2003) *The Journal of Organic Chemistry*, **68**, 9865.

454 Nicolaou, K.C., Lee, S.H., Estrada, A.A., and Zak, M. (2005) *Angewandte Chemie, International Edition*, **44**, 3736; Nicolaou, K.C., Estrada, A.A., Lee, S.H., and Freestone, G.C. (2006) *Angewandte Chemie, International Edition*, **45**, 5364.

455 Hynes, J., Jr, Doubleday, W.W., Dyckman, A.J., Godfrey, J.D., Jr, Grosso, J.A., Kiau, S., and Leftheris, K. (2004) *The Journal of Organic Chemistry*, **69**, 1368.

456 Weiberth, F., Lee, G.E., Hanna, R.G., Dubberke, S., Utz, R., and Mueller-Lehar, J. (2005) PTC Int. Appl. 2005035496.

457 Watanabe, M., Yamamoto, T., and Nshiyama, M. (2000) *Angewandte Chemie, International Edition*, **39**, 2501.

458 Challis, B.C. and Rzepa, H.S. (1977) *Journal of the Chemical Society-Perkin Transactions 2*, 281.

459 Abramovitch, R.A., Kress, A.O., Pillay, K.S., and Thompson, W.M. (1985) *The Journal of Organic Chemistry*, **50**, 2066.

460 Fresneda, P.M., Molina, P., and Bleda, J.A. (2001) *Tetrahedron*, **57**, 2355; Djalil Coowar, D., Bouissac, J., Hanbali, M., Paschaki, M., Mohier, E., and Luu, B. (2004) *Journal of Medicinal Chemistry*, **47**, 6270.

461 Itoh, S., Takada, N., Ando, T., Haranou, S., Xin Huang, X., Uenoyama, Y., Ohshiro, Y., Komatsu, M., and Fukuzumi, S. (1997) *The Journal of Organic Chemistry*, **62**, 5898.

462 General: (a) Russel, J.S. and Pelkey, E.T. (2009) *Progr. Heterocycl. Chem.*, **20**, 122. Synthesis: (b) Humphrey, G.R. and Kuethe, J.T. (2006) *Chem. Rev.*, **106**, 2875. (c) Barluenga, J., Rodríguez, F., and Fañanás, F.J. (2009) *Chem. Asian J.*, **4**, 1036. (d) Krueger, K., Tillack, A., and Beller, M. (2008) *Adv. Synth. Cat.*, **350**, *2153*. Reactivity: (e) Bandani, M. and Eichholzer, A. (2009) *Angew. Chem. Int. Ed.*, **48**, 9608.

463 Park, I.-K., Suh, S.-E., Lim, B.-Y., and Cho, C.-G. (2009) *Org. Lett.*, **11**, 5454.

464 Alex, K., Tillack, A., Schwarz, N., and Beller, M. (2008) *Angew. Chem. Int. Ed.*, **47**, 2304.

465 Shen, M., Leslie, B.E., and Driver, T.G. (2008) *Angew. Chem. Int. Ed.*, **38**, 5056.

466 Ackermann, L., Sandmann, R., and Kondrashov, M.V. (2009) *Synlett*, 1219.

467 Yao, P.-Y., Zhang, Y., Hsung, R.P., and Zhao, K. (2008) *Org. Lett.*, **10**, 4275.

468 Roberto Sanz, R., Castroviejo, M.P., Guilarte, V., Pérez, A., and Fañanás, F.J. (2007) *J. Org. Chem.*, **72**, 5113.

469 Ohno, H., Ohta, Y., Oishi, S., and Fujii, N. (2007) *Angew. Chem. Int. Ed.*, **37**, 3173.

470 Cariou, K., Ronan, B., Mignani, S., Fensterbank, L., and Malacria, M. (2007) *Angew. Chem. Int. Ed.*, **46**, 1881.

471 Takaya, J., Udagawa, S., Kusama, H., and Iwasawa, N. (2008) *Angew. Chem. Int. Ed.*, **38**, 4906.

472 Leogane, O. and Lebel, H. (2008) *Angew. Chem. Int. Ed.*, **38**, 350.

473 Okamoto, N., Miwa, Y., Minami, H., Takeda, K., and Yanada, R. (2009) *Angew. Chem. Int. Ed.*, **48**, 9693.

474 Fang, Y.-Q. and Lautens, M. (2008) *J. Org. Chem.*, **73**, 538.

475 Nicolaou, K.C., Krasovskiy, A., Trépanier, V.E., David, Y.-K., and Chen, D.Y.-K. (2008) *Angew. Chem. Int. Ed.*, **47**, 4217.

476 Li, G., Huang, X., and Zhang, L. (2008) *Angew. Chem. Int. Ed.*, **47**, 346.

477 Pei, T., Chen, C.-Y., Dormer, P.G., and Davies, I.W. (2008) *Angew. Chem. Int. Ed.*, **47**, 4231.

478 Würtz, S., Rakshit, S., Neumann, J.J., Dröge, T., and Glorius, F. (2008) *Angew. Chem. Int. Ed.*, **38**, 7230.

479 Bernini, R., Fabrizi, G., Sferrazza, A., and Cacchi, S. (2009) *Angew. Chem. Int. Ed.*, **48**, 8078.

480 Yu, W., Du, Y., and Zhao, K. (2009) *Org. Lett.*, **11**, 2417.

481 Shi, Z., Zhang, C., Li, S., Pan, D., Ding, S., Cui, Y., and Jiao, N. (2009) *Angew. Chem. Int. Ed.*, **48**, 4572.

482 Cui, S.-L., Wang, J., and Wang, Y.-G. (2008) *J. Am. Chem. Soc.*, **130**, 13526.

483 Fuwa, H. and Sasaki, M. (2009) *J. Org. Chem.*, **74**, 212.

484 Jensen, T., Pedersen, H., Bang-Andersen, B., Madsen, R., and Jørgensen, M. (2008) *Angew. Chem. Int. Ed.*, **47**, 888.

485 Bernini, R., Cacchi, S., Fabrizi, G., Filisti, E., and Sferrazza, A. (2009) *Synlett*, 1480.

486 Stuart, D.R., Bertrand-Laperle, M., Burgess, K.M.N., and Fagnou, K. (2008) *J. Am. Chem. Soc.*, **130**, 16474.

487 Barluenga, J., Jiménez-Aquino, A., Valdés, C., and Aznar, F. (2007) *Angew. Chem. Int. Ed.*, **46**, 1529.

488 Barluenga, J., Jiménez-Aquino, A., Aznar, F., and Valdés, C. (2009) *J. Am. Chem. Soc.*, **131**, 4101.

489 Petronijevic, F., Timmons, C., Cuzzupey, A., and Wipf, P. (2009) *Chem. Commun.*, 104.

490 Marques-Lopez, E., Diez-Martinez, A., Merino, P., and Herrera, R.P. (2009) *Curr. Org. Chem.*, **13**, 1585.

491 Bartoli, G., Bosco, M., Carlone, C., Pesciaioli, F., Sambri, L., and Melchiorre, P. (2007) *Org. Lett.*, **9**, 1403.

492 Rueping, M., Nachtsheim, B.J., Moreth, S.A., and Bolte, M. (2008) *Angew. Chem. Int. Ed.*, **47**, 593.

493 Cai, Q., Zhao, Z.-A., and You, S.-L. (2009) *Angew. Chem. Int. Ed.*, **48**, 7428.

494 Evans, D.A., Fandrick, K.R., Song, H.-J., Scheidt, K.A., and Xu, R. (2007) *J. Am. Chem. Soc.*, **129**, 10029.

495 Ganesh, M., and Seidel, D. (2008) *J. Am. Chem. Soc.*, **130**, 16464.

496 Itoh, J., Fuchibe, K., and Akiyama, T. (2008) *Angew. Chem. Int. Ed.*, **47**, 4016.

497 Arai, T. and Yokoyama, N. (2008) *Angew. Chem. Int. Ed.*, **47**, 4989.

498 (a) Taylor, M.S. and Jacobsen, E.N. (2004) *J. Am. Chem. Soc.*, **126**, 10558. (b) Seayad, J., Seayad, A.M., and List, B. (2006) *J. Am. Chem. Soc.*, **128**, 1086. (c) Wanner, M.J., van der Haas, R.N.S., de Cuba, K.R., van Maarseveen, J.H., and Hiemstra, H. (2007) *Angew. Chem. Int. Ed.*, **46**, 7485. (d) Bou-Hamdan, F.R. and Leighton, J.L. (2009) *Angew. Chem. Int. Ed.*, **48**, 2403. (e) Muratore, M.E., Holloway, C.A., Pilling, A.W., Storer, R.-I., Graham Trevitt, G., and Dixon, D.J. (2009) *J. Am. Chem. Soc.*, **131**, 10796.

499 Wanner, M.J., Boots, R.N.A., Eradus, B., de Gelder, R., van Maarseveen, J.H., and Hiemstra, H. (2009) *Org. Lett.*, **11**, 2579.

500 (a) Trost, B.M. and Quancard, J.J. (2006) *Am. Chem. Soc.*, **128**, 6314. (b) Kagawa, N., Malerich, J.P., and Rawal, V.H. (2008) *Org. Lett.*, **10**, 2381.

501 Eastman, K. and Baran, P.S. (2009) *Tetrahedron*, **65**, 3149.

502 Joucla, L. and Djakovitch, L. (2009) *Adv. Synth. Cat.*, **351**, 673.

503 García-Rubia, A., Gómez Arrayás, R., and Carretero, J.C. (2009) *Angew. Chem. Int. Ed.*, **48**, 6511.

504 Brand, J.P., Charpentier, J., and Waser, J. (2009) *Angew. Chem. Int. Ed.*, **48**, 9346.

505 (a) Deprez, N.R., Kalyani, D., Krause, A., and Sanford, M.S. (2006) *J. Am. Chem. Soc.*, **128**, 4972. (b) Lebrasseur, N. and Larrosa, I. (2008) *J. Am. Chem. Soc.*, **130**, 2926.

506 Yang, S., Sun, C., Fang, Z., Li, B., Li, Y., and Shi, Z. (2008) *Angew. Chem., Int. Ed.*, **47**, 1473.

507 Philipps, R.J., Grimster, N.P., and Gaunt, M.J. (2008) *J. Am. Chem. Soc.*, **130**, 8174.

508 (a) Stuart, D.R. and Fagnou, K. (2007) *Science*, **316**, 1172. (b) Stuart, D.R., Villemure, E., and Fagnou, K. (2007) *J. Am. Chem. Soc.*, **129**, 12072.

509 Luzung, M.R., Lewis, C.A., and Baran, P.S. (2009) *Angew. Chem. Int. Ed.*, **48**, 7025.

510 (a) Bandini, M., Eichholzer, A., Tragni, M., and Umani-Ronchi, A. (2008) *Angew. Chem. Int. Ed.*, **47**, 3238. (b) Cui, H.-L., Feng, X., Peng, J., Lei, J., Jiang, K., and Chen, Y.-C. (2009) *Angew. Chem. Int. Ed.*, **48**, 5737.

511 Stanley, L.M. and Hartwig, J.F. (2009) *Angew. Chem. Int. Ed.*, **48**, 7841.

512 Xion, X. and Pirrung, M.C. (2008) *Org. Lett.*, **10**, 1151.

513 Claudio Gioia, C., Hauville, A., Bernardi, L., Fini, F., and Ricci, A. (2008) *Angew. Chem. Int. Ed.*, **47**, 9236.

6
Five-Membered Heterocycles: Furan

Henry N.C. Wong, Xue-Long Hou, Kap-Sun Yeung, and Hui Huang

6.1
Introduction

Furan (**1**) is prepared by a gas-phase decarbonylation procedure starting from furfural, which, in turn, is produced in large amounts via an acid treatment of vegetable residues after industrial production of porridge oats and cornflakes [1]. In addition, acid-promoted dehydration of saccharides such as D-fructose also leads to the formation of hydroxymethylfurfural (**2**) [2, 3]. In this connection, furan (**1**) and hydroxymethylfurfural (**2**) are generally regarded as compounds that are readily accessible from renewable resources.

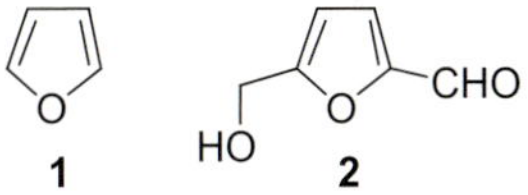

The structure of **1** is closely related to those of pyrrole and thiophene. There are two electron lone pairs on the oxygen atom, one being conjugated with the two double bonds to form a sextet, and the other located in the molecular plane in an sp^2 hybrid orbital.

Furan molecular frameworks are found in many naturally occurring molecules, and play a very significant role in the field of heterocyclic chemistry. Furans have been applied to various commercially important products such as pharmaceuticals, flavors and fragrant products, and functional polymers. They are also versatile precursors and synthetic intermediates in the preparation of cyclic and acyclic molecules. For example, furans are latent 1,4-dicarbonyl units and are also widely employed as 1,3-dienes in Diels–Alder reactions. Efficient synthesis of polysubstituted furans therefore continues to be of great interest to synthetic chemists due to their widespread applications and frequent occurrence in Nature [4].

Modern Heterocyclic Chemistry, First Edition.
Edited by Julio Alvarez-Builla, Juan Jose Vaquero, and José Barluenga.

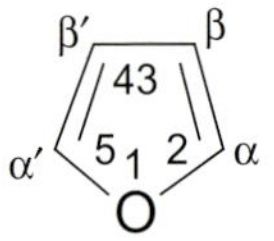

β′ β
43
α′ 5 1 2 α
O

Figure 6.1 Numbering of the furan ring system.

6.1.1
Nomenclature

The positions next to the oxygen atom are indicated as α and α′, while those away from the oxygen are called β and β′. Figure 6.1 shows the numbering of the furan ring system [5]. Reduced furans are called 2,3-dihydrofuran (**3**), 2,5-dihydrofuran (**4**) and 2,3,4,5-tetrahydrofuran (**5**).

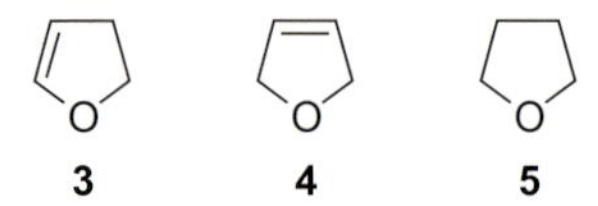

6.1.2
General Reactivity

Sections 6.3 and 6.4 give a much more thorough discussion on the reactions of furan **1** and its derivatives. Here we concentrate on the overall reactivity of **1** [1, 6]. In general, furans are rather stable towards weak aqueous acids. However, in concentrated sulfuric acid or Lewis acids the furan framework will be decomposed. Of the three types of five-membered heterocycles that contain NH, S or O, furans are the least aromatic [7] and would therefore react like dienes. Electrophilic substitution reactions with **1** are regiospecific and lead to mostly α-substituted furans **6** via a general addition–elimination pathway (Scheme 6.1) [1, 6, 8]. β-Substitution reactions only occur when both the α and α′ positions are occupied by substituents. Scheme 6.2 depicts an example of β-acylation, in which **7** is converted into **8** [9].

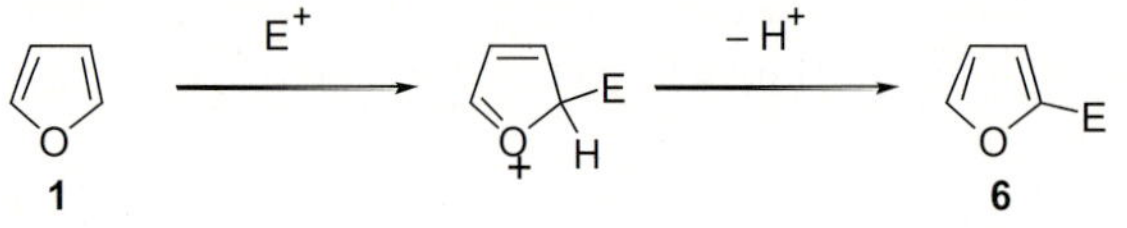

E^+ $-H^+$
1 E H 6

Scheme 6.1

Me MeCOCl $AlCl_3$ O Me Me
Me N CS_2 Me N
7 O reflux, 6 hr 8 O
65%

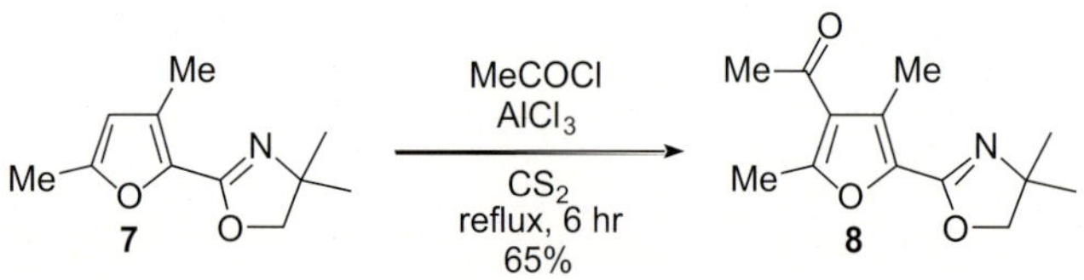

Scheme 6.2

In a nitration reaction, furan **1** reacts with acetyl nitrate to give, initially, an addition product, which undergoes subsequent elimination to offer 2-nitrofuran (**9**) (Scheme 6.3).

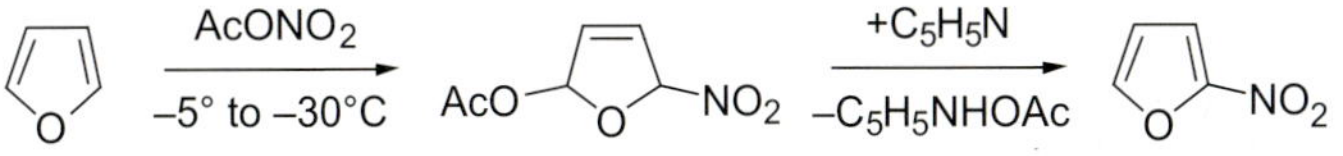

Scheme 6.3

In sulfonation reactions, furan (**1**) is again sulfonated at C2, with a pyridine–sulfur trioxide complex, followed by acidification with hydrochloric acid to afford **10** (Scheme 6.4).

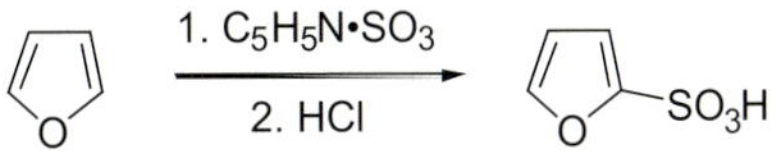

Scheme 6.4

Halogenation reactions are trickier with furans. Thus, polyhalogenated products are obtained from **1** through reactions with chlorine and bromine at room temperature. Pure 2-chlorofuran and 2-bromomfuran can only be obtained when **1** reacts with chlorine in dichloromethane at −40 °C, and bromine in dioxane at 0 °C, respectively. 3-Methylfuran (**11**), in contrast, reacts with *N*-bromosuccinimide (NBS) to furnish 2-bromo-3-methylfuran (**12**) (Scheme 6.5) [10].

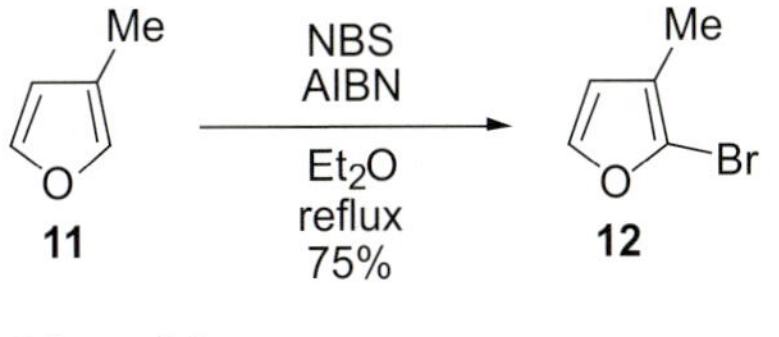

Scheme 6.5

Acylation under Vilsmeier–Haack conditions with a catalyst such as boron trifluoride etherate or phosphoric acid converts **1** into 2-acylfurans. As shown in Scheme 6.6, treatment of 3,4-bis(trimethylsilyl)furan (**13**) with acetyl chloride and titanium(IV) chloride affords the α-acetylfuran **14** [11].

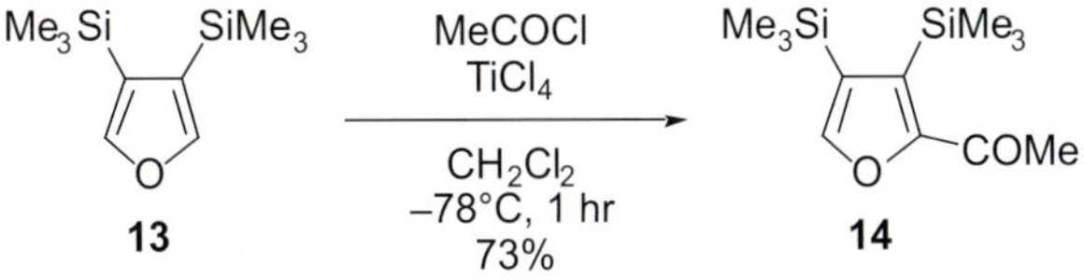

Scheme 6.6

Scheme 6.7 gives an appropriate example of a Lewis acid catalyzed alkylation of a furan moiety (**15**) [12]. This reaction is stereospecific, leading only to the trans-fused diastereomer **16**.

BF_3•Et_2O
84%
15 **16**

Scheme 6.7

Condensation of **1** with acetone in the presence of an acid, intriguingly, gives 16-membered macrocycle **17** (Scheme 6.8) [13].

$Me_2C=O$
$MeSO_3H$
EtOH
reflux
1 **17**

Scheme 6.8

Metallation such as mercuration with mercury (II) chloride and sodium acetate leads to α-mercurated furans. As demonstrated in Scheme 6.9, 3-methylfuran (**11**) undergoes mercuration to give the corresponding furanmercuric chloride **18** [14]. Lithiation with alkyllithium in refluxing diethyl ether proceeded smoothly at C2, providing a 2-lithiofuran (**19**). In the presence of tetramethyl ethylenediamine (TMEDA) in hexane, 2,5-bis(lithio)furan (**20**) is formed instead (Scheme 6.10) [15].

$HgCl_2$
H_2O-MeOH
r.t.
95%
11 **18**

Scheme 6.9

n-BuLi
Et_2O
n-BuLi (excess)
TMEDA
C_6H_{14}
19 **1** **20**

Scheme 6.10

Lithium chemistry opens up an avenue for the realization of some useful α-- substituted furans. An example is shown in Scheme 6.11 [16]. As can be seen, an excess of butyllithium presumably generates a 2,5-bis(lithio)furan, which is then silylated at the less hindered side to give furan **22** with a 2,4-disubstituted pattern.

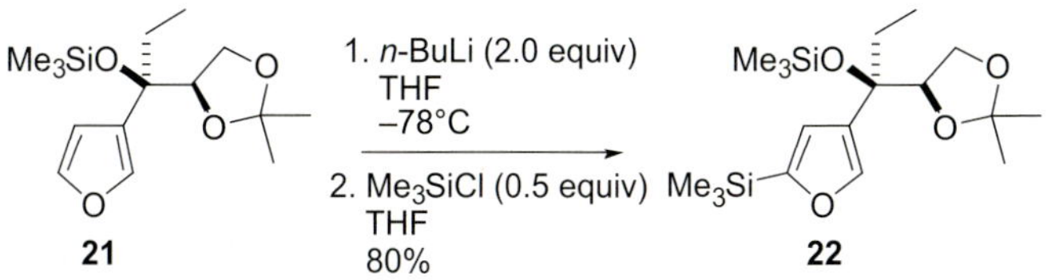

Scheme 6.11

Furan (**1**) undergoes catalytic hydrogenation to afford the very commercially important solvent tetrahydrofuran. As mentioned above, furans behave like a 1,3-diene, so the reaction of 2,5-disubstituted furan **23** with methanolic bromine produces 2,5-dimethoxy-2,5-dihydrofuran **24** (Scheme 6.12) [17].

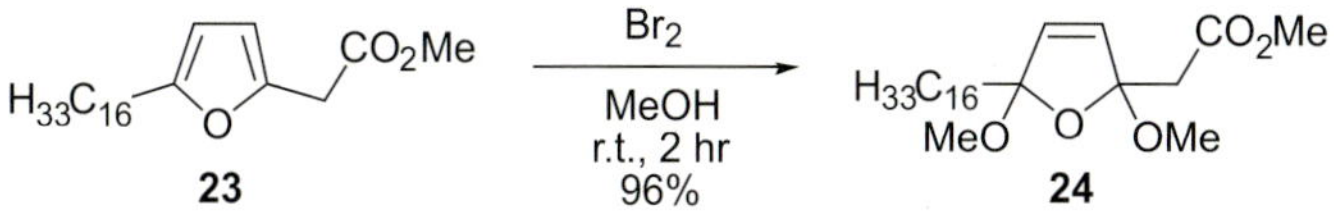

Scheme 6.12

The diene character of furan can be further demonstrated by its Diels–Alder cycloaddition reaction with maleic anhydride, forming the *endo*-adduct **25** at a lower temperature as the kinetic product, and an *exo*-product **26** as the thermodynamic product after much longer reaction times or at a higher temperature (Scheme 6.13). Notably, this cycloaddition reaction was one of the reactions that led Otto Diels and Kurt Alder to report their now famous Diels–Alder reaction [18]. A recent example utilizing an intramolecular Diels–Alder reaction between the furan unit **27** and maleic anhydride, leading to the Diels–Alder adduct **29** via **28**, is shown in Scheme 6.14 [19].

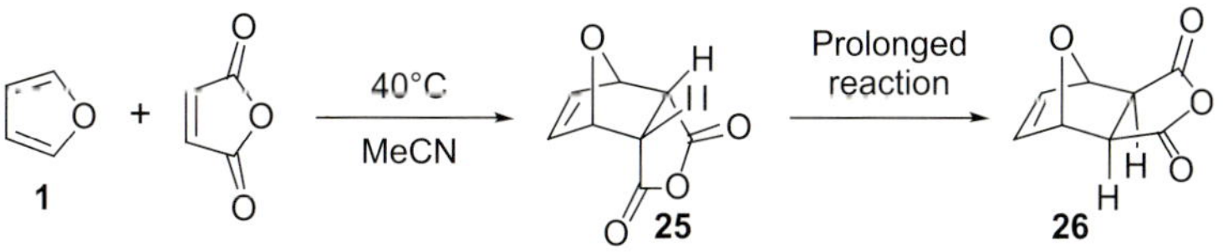

Scheme 6.13

Under the conditions of the Paterno–Büchi reaction, hydroxymethylfuran (**30**) reacts readily with benzophenone to form the [2 + 2] adducts oxetanes **31** (65%) and

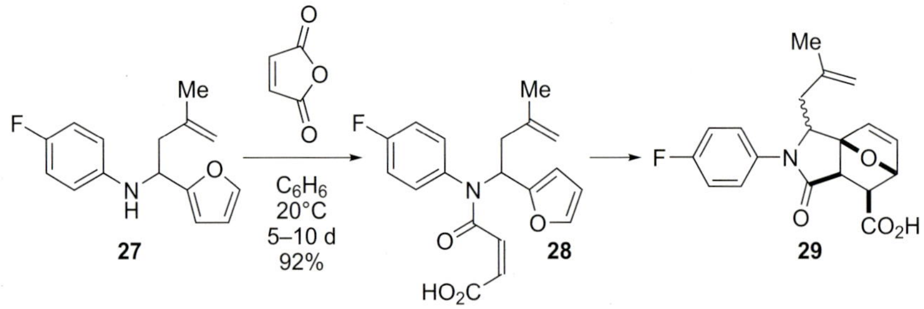

Scheme 6.14

32 (20%) (Scheme 6.15) [20]. Reductive methylation of furan **33** with sodium in liquid ammonia and subsequent addition of methyl iodide, on the other hand, afforded 2,5-dihydrofuran **34** (Scheme 6.16) [21].

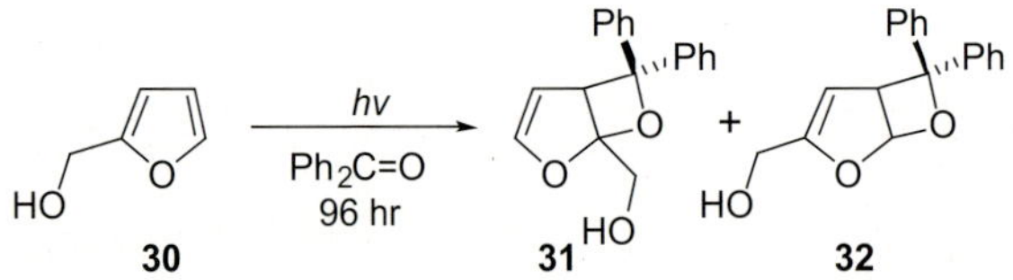

Scheme 6.15

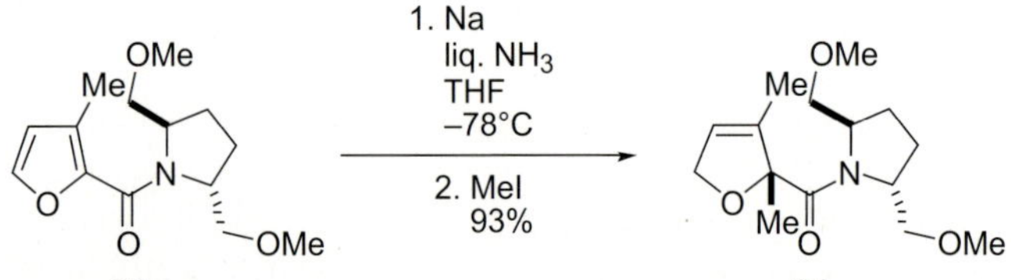

Scheme 6.16

6.1.3
Relevant Physicochemical Data [8]

Furan (**1**) shows only limited solubility in water (1 part in 35 at 20 °C) [8]. The hydrogen bonding effect of its oxygen atom with the water protons is the reason why it is more soluble in water than thiophene (1 part in 700 at 20 °C).

Microwave spectroscopy of **1** and its deuterated analogs in the gas phase has been reinvestigated, and highly accurate structure parameters have been obtained [22]. The dimensions are shown in Figure 6.2. In contrast, the solid-state structural dimensions of furan-2-carboxylic acid [23] and furan-3,4-dicarboxylic acid [24] have been determined by an X-ray diffraction study; the values obtained are in good agreement with those found in the gas phase study [22].

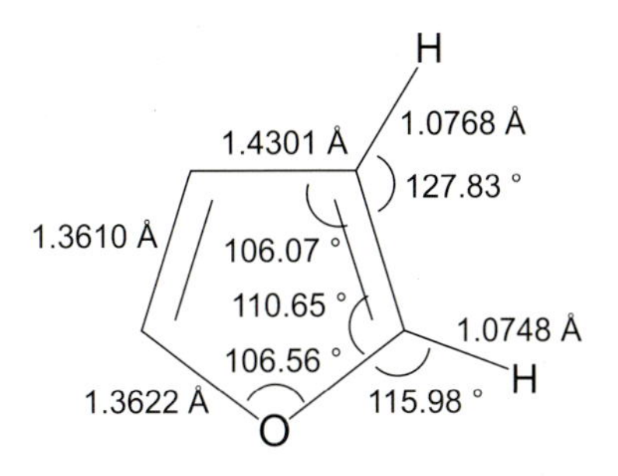

Figure 6.2 Geometry of furan.

The UV/Visible spectroscopic absorption maximum of furan (**1**) in ethanol is λ_{max} 208 nm (log ε 3.9). This value is the smallest amongst pyrrole (210 nm) and thiophene (235 nm), and is therefore a good indication that **1** behaves more like a diene than a conjugated system [25].

Chemical shifts of protons are related to the electron-density of the carbons they are attached to, with lower field shifts corresponding to electron-deficient carbon centers. Thus, the α-proton signals of furan (**1**) at δ 7.29 are at a lower field than those of the β-protons (δ 6.24), which is primarily due to the inductive electron-withdrawing effect of the oxygen atom [26]. Coupling constants between protons on **1** are $J_{2,3} = 1.75$ Hz, $J_{2,4} = 0.85$ Hz, $J_{2,5} = 1.40$ Hz and $J_{3,4} = 3.30$ Hz [27]. ^{13}C NMR chemical shifts also show the electron-withdrawing effect of the oxygen atom, giving signals at δ 142.7 ppm of the α-carbons and δ 109.6 ppm for the β-carbons [28].

In its mass spectrum, the fragmentation pathway of furan **1** is like that of other five-membered heterocycles (Scheme 6.17). For **1**, the strongest peak (70%) is the molecular ion [29].

The He(1α) photoelectron spectrum of furan (**1**) [30] reveals the energy of the third molecular orbital (π_3) as 8.89 eV, which is also the energy of the highest occupied

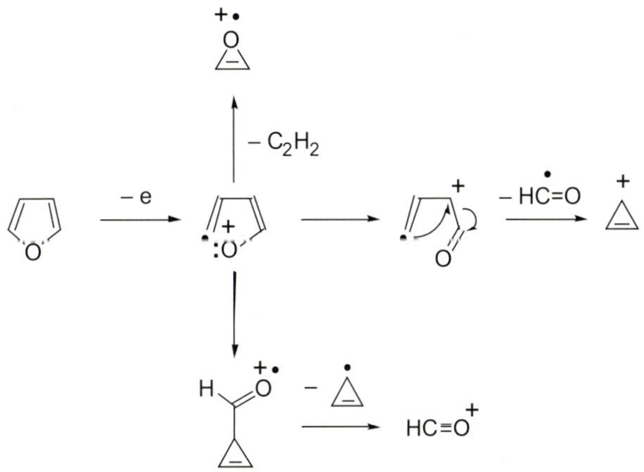

Scheme 6.17

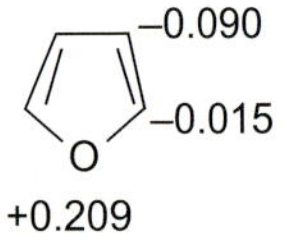

−0.090
−0.015
O
+0.209

Figure 6.3 π-Electron excess at positions 2 and 3 of furan.

molecular orbital as well as close to the first ionization potential (IP_1) of **1** (8.99 eV) [31]. Other molecular orbital energies π_1 and π_2 are measured at 14.4 and 10.2 eV, respectively. Since a high π-donation character of a compound always leads to substantial π-electron excess, the best measure of π-donation is the value of the first ionization potential (IP_1). In this connection, furan and thiophene ($\pi_3 = 8.92$ eV) [32] show almost equal π-donor properties, while that of pyrrole is significantly larger ($\pi_3 = 8.2$ eV) [32]. HMO calculations lead to the same conclusion, namely that the total π-electron excess of **1** is smaller than that of pyrrole [8]. As can be seen in Figure 6.3, the β-carbon atoms carry significantly larger negative charges than the α-carbons. Such a distribution of π-electron excess is also in good agreement with the larger shielding of H_β- and C_β-nuclei in the proton and ^{13}C NMR spectra, respectively [8].

The Pariser–Parr–Pople (PPP) approximation method reproduces well the electronic spectral features of furan (**1**) [33], while the simple Hückel method has also been employed to compute the resonance energy of **1** in a quantitative manner, providing a value of 18 kcal mol^{-1}, which is less stable than thiophene (resonance energy 29 kcal mol^{-1}) [34]. Satisfactory molecular geometry and heat of formation for **1** have been obtained from MINDO/3 semi-empirical MO calculations [35]. In addition, MINDO/3 calculated vibration frequencies for **1** [36] agree well with values obtained experimentally [37]. In addition, satisfactory dipole moment [38] and ionization energies [39] for **1** have been obtained from an *ab initio* calculation. The electronic spectrum [40] and electronic structure [41] of **1** have also been computed by other *ab initio* methods.

6.1.4
Relevant Natural and Useful Compounds

As mentioned in Section 6.1, the furan molecular skeleton is found widely in many naturally occurring molecules. Shown below are several furan-containing natural products (**35–40**) identified in 2004 and 2005. Thus, shinsonefuran (**35**), a sesterterpene exhibiting cytotoxicity against HeLa cells with an IC_{50} of 16 μg mg^{-1}, was obtained from the deep-sea sponge *Stoeba extensa* [42]. A limonoid containing also a 3-substituted furan, namely xyloccensin L (**36**), was isolated from the stem bark of *Xylocarpus granatum* [43]. The new furanocembranolide **37** was isolated from the octocorals *Leptogorgia alba* and *Leptogorgia ridida* collected on the Pacific coast of Panama [44]. The linderazulene **38**, a novel azulenoid showing moderate activity against the PANC-1 pancreatic cell line with an IC_{50} of 18.7 μg ml^{-1}, was obtained from a deep-sea gorgonian *Paramuricea* sp. [45]. In a chemical and genetic study of *Ligularia tongolensis* in the Hengduan Mountains of China, a strongly Ehrlich-positive furanoeremophilane compound (**39**) was identified [46]. During automated screening for small-molecule agonists to peroxisome proliferator-activated receptor-γ (PPAR-γ), a new biologically active linear triterpene **40** was isolated and characterized from the bark of *Cupaniopsis trigonocarpa* [47].

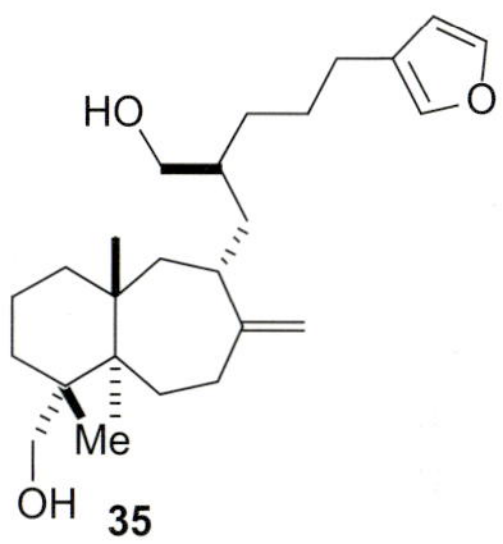

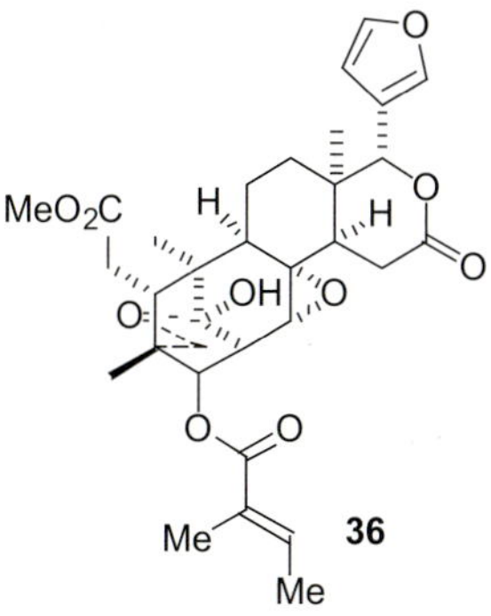

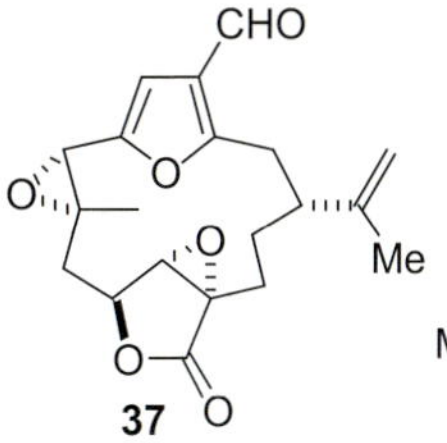

Furan-containing molecules are also useful as catalysts, pharmaceuticals and electronic devices. For example, in the synthesis of a highly potent and selective serotonin reuptake inhibitor (BMS-594726), MacMillan's furan-containing imidazolidinone catalyst **41** [48] has been employed in the key enantioselective indole alkylation step, leading to a desired product with an enantiomeric excess of 84% [49]. Ranitidine (**42**) (Zantac©), a histamine H_2-receptor antagonist that inhibits gastric acid secretion, possesses a furan framework as the pivotal structural unit [50]. Anther furan-containing oligoarylcyclophanene (**43**) – showing a strong broad luminescence at 499 nm, due presumably to an intramolecular interaction between the chromophores and the presence of a double bond – has potential use in opto-electronic devices [51].

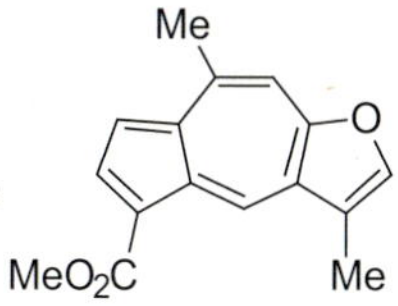

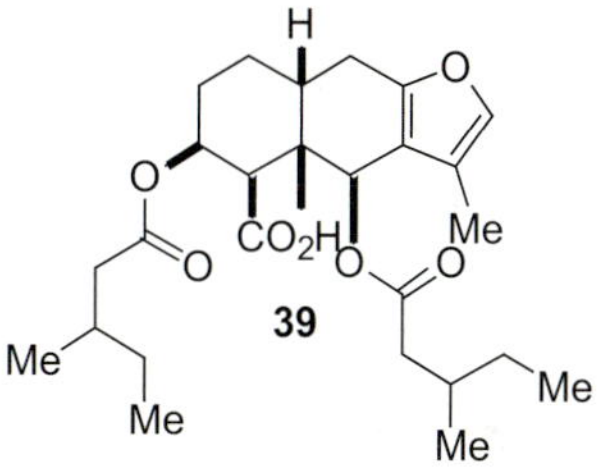

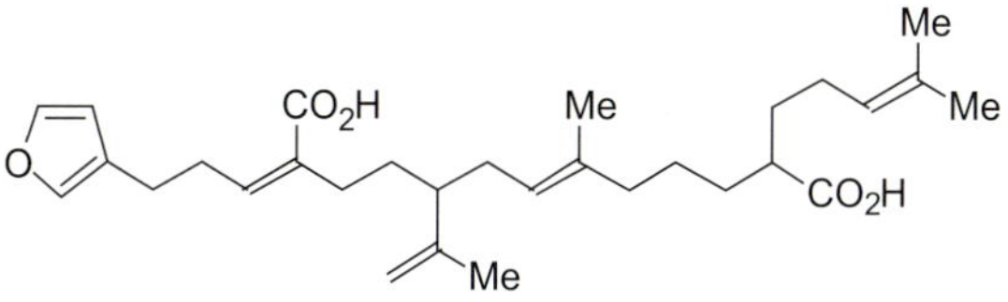

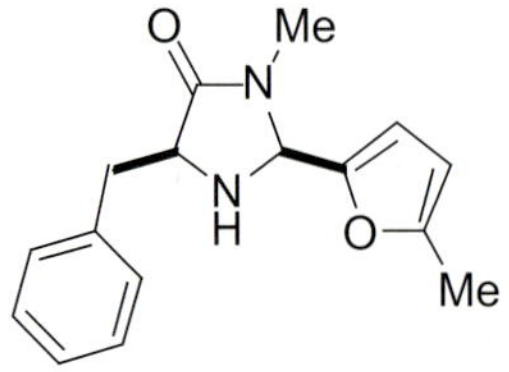

41 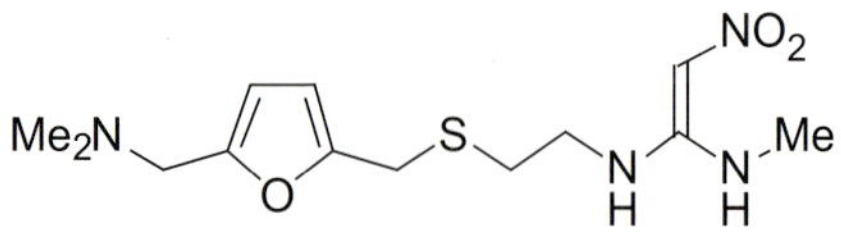 **42**

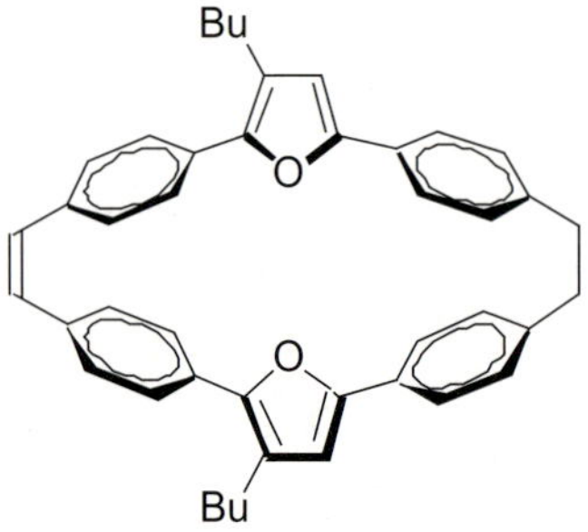

43

6.2
Synthesis of Furans

6.2.1
Introduction

The synthesis of furans has continued to attract the attention of synthetic chemists during the past several decades [4]. In this connection, many efficient procedures have been documented. Some of these methods have become standard procedures and have been extremely useful in the synthesis of furans with different substituted patterns. One of these classical methods is the acid-catalyzed dehydration reaction of 1,4-diketones to form furans, known as the Paal–Knorr synthesis [52]. Not only 1,4-ketones but also masked diketones, such as chloroallyl ketones and oxiranes with appropriate substituents, as well as 1,4-dicarboxylic acid derivatives, including esters and nitriles, are suitable substrates. Many variants have also been recorded. Furthermore, cyclization of γ-hydroxy ketones or aldehydes in the presence of an acid catalyst can also provide the corresponding substituted furans. Various acids, such as sulfuric acid, *p*-toluenesulfonic acid, oxalic acid, Amberlyst 15 and zinc bromide, have been used as catalyst. Some examples (**44** → **45** and **46** → **47**) are shown in Scheme 6.18 [53].

Reaction of oxazole derivative **48** and a dienophile via a Diels–Alder cycloaddition–retro-Diels–Alder reaction strategy provides a useful protocol to produce furan **49** with different substitution patterns, provided these starting materials are carefully chosen (Scheme 6.19). The procedure has been employed in the total synthesis of (−)-teubrevin G (**50**) [54]. Furans themselves may also serve as dienes in the reaction with electron-deficient alkynes, leading to 7-oxabicyclic compounds, which undergo a retro-Diels–Alder reaction to provide the corresponding furans. The advantages of this strategy lie in the functional group compatibility as well as the starting material

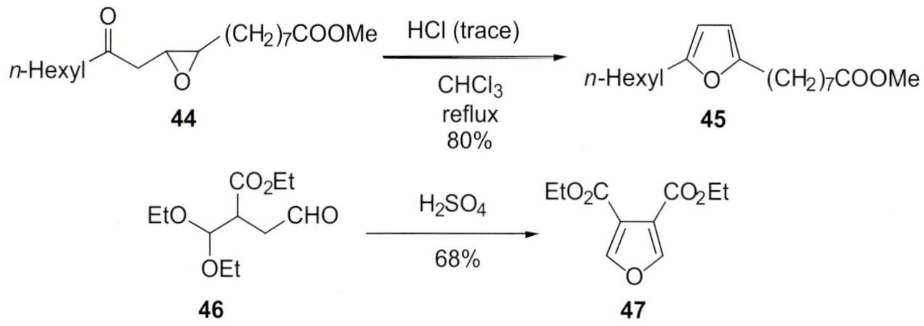

Scheme 6.18

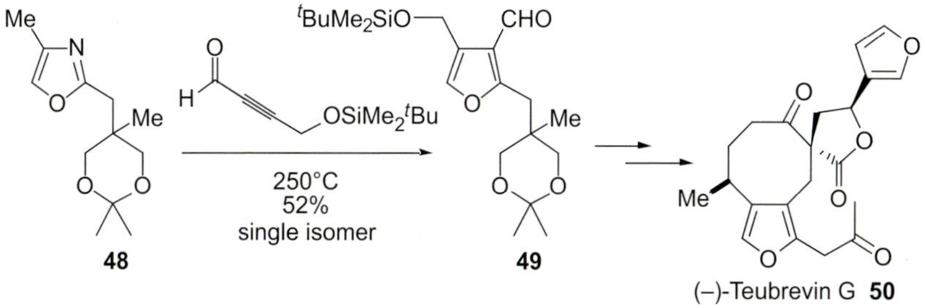

Scheme 6.19

accessibility. If the availability of starting materials is not a problem, this procedure may be regarded as the simplest and the most efficient.

Organometallic compounds have found significant application in furan synthesis in recent years. A lithiation–electrophile trapping strategy has been widely applied to the synthesis of 2,3-disubstituted and 2,3,5-trisubstituted furans. These furyllithium species have been obtained by various means, such as the direct metallation of furans at the α-position using *n*-BuLi or LDA (Scheme 6.10) [15] or using lithium magnesates [55] and metal–halogen exchange methodology. All these furyllithium compounds can be converted into the corresponding substituted furans upon quenching with various electrophiles such as halides or carbonyls. As shown in Scheme 6.20, 3-bromofuran **51** can be lithiated and alkylated at the α-carbon to afford **52**, which undergoes further reactions to give rosefuran (**53**) [56]. Furyllithium species can also be converted into other organometallic species, such as titanium and manganese derivatives through a transmetallation procedure. In addition to titanium and manganese derivatives, organocopper, zinc and tin have all been utilized in the preparation of polysubstituted furans [57].

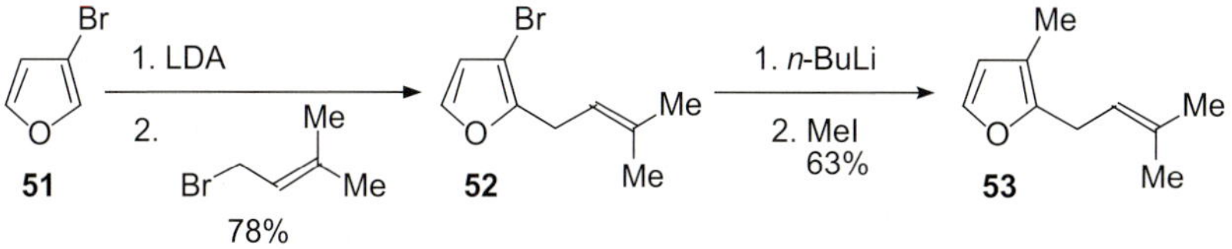

Scheme 6.20

Despite the fact that a lot of strategies and procedures for the synthesis of furans have been recorded, the quest for efficient and reliable synthetic methodologies, especially with acceptable regioselectivity, is still an active research field. Some progress in this area is reviewed in the following section.

6.2.2 Monosubstituted Furans

Transition metal catalyzed coupling reactions of arenes and halides and amines have been very popular in the synthesis of arene derivatives. These protocols can also be used in the synthesis of furans with substituents at different positions. C–N cross-coupling reaction of a bromo-substituted furan with various amides, carbamates and lactams catalyzed by CuI furnishes 2- and 3-substituted amidofurans in 45–95% yield [58]. Arylboronic acid has been employed as an aryl source in the synthesis of 2-arylfuran under Mn(II) acetate-promoted radical reaction conditions. Although the yields are not high, they are better than those of the phase-transfer Gomberg–Bachmann synthesis using arenediazonium ions [59]. Arylation of 3-furoate with 3-bromonitrobenzene using $Pd(PPh_3)_4$ as catalyst in toluene affords 2,3-disubstituted furan while 5-aryl products are generated predominantly when Pd/C is used as the catalyst and NMP as solvent [60].

2-Substituted furan **56** has been prepared using the phosphorylated allenic glycol **54** through a cyclization pathway under basic conditions, followed by dehydration of intermediate **55** in the presence of a catalytic amount of *p*-TsOH (Scheme 6.21) [61].

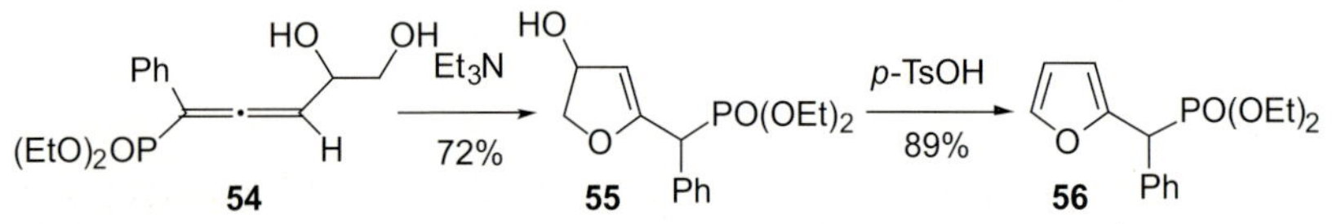

Scheme 6.21

2-Substituted furans can be prepared under mild conditions. Thus, reaction of 2,5-dimethoxy-2,5-dihydrofuran (**57**) and an appropriate vinyl ether in the presence of a catalytic amount of $MgBr_2{\cdot}Et_2O$ affords a functionalized 2-alkylfuran **58** in good yields (Scheme 6.22). The reaction might proceed through a concerted mechanism, in which the $MgBr_2$-activated dihydrofuran reacts with the vinyl ether via a cyclic intermediate [62].

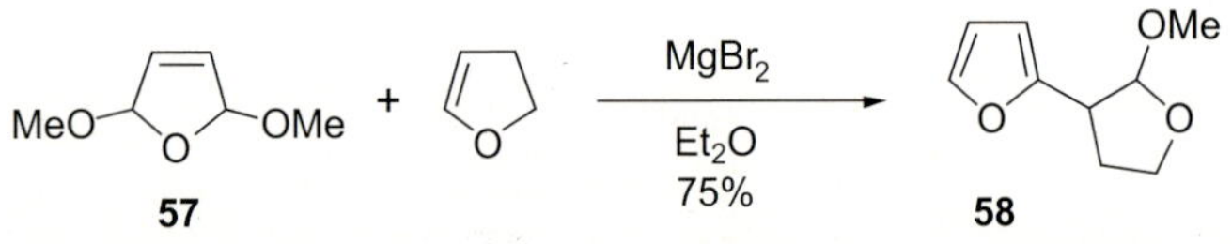

Scheme 6.22

Furyldifluoromethyl aryl ketones **59** are formed when furan (**1**) is allowed to react with difluoroenol silyl ethers in the presence of $Cu(OTf)_2$ as depicted by an example in Scheme 6.23 [63]. If 2-furfurylcarboxylate is used, the corresponding substituted furan is also obtained [63].

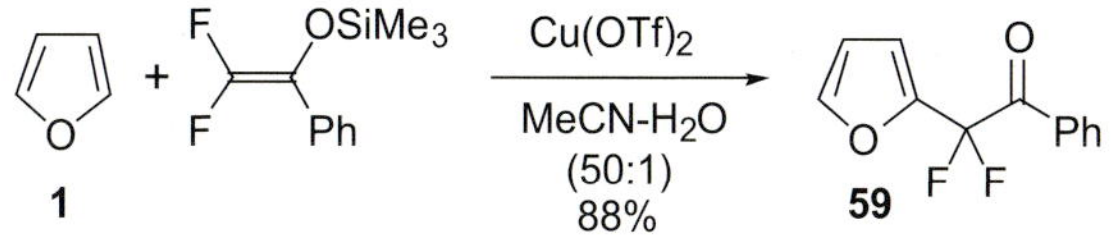

Scheme 6.23

The Yb(fod)$_3$-catalyzed carbonyl-ene reaction of 3-methylene-2,3-dihydrofuran (**60**) and aldehydes is a mild, efficient way to provide 2-substituted furans, which can be transformed into furano[2,3-*c*]pyrans **61** in good yield via the oxa-Pictet–Spengler procedure (Scheme 6.24) [64]. Similarly, **60**, produced from the Wolff–Kishner reduction of 2-furylhydrazone, reacts with aldehydes in the presence of Yb (fod)$_3$ or Ti(OPr-*i*)$_4$, to afford the same 2-substituted furans. Notably, an optically active product is realized when Ti(OPr-*i*)$_4$/(*S*)-BINOL is the catalyst [65].

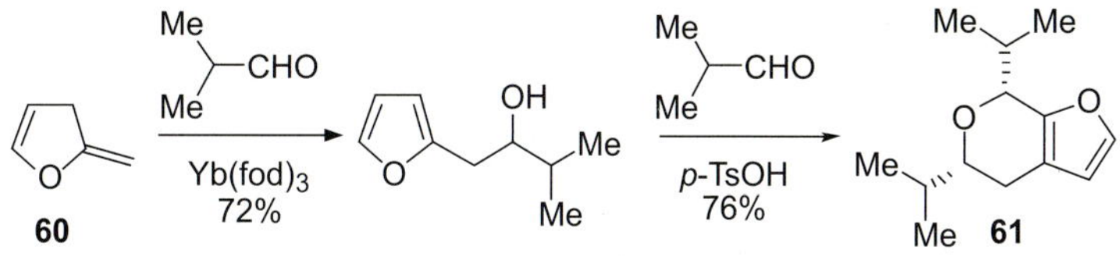

Scheme 6.24

The 2-position of furan is usually derivatized by a simple electrophilic aromatic substitution or metallation. However, the introduction of a substituent to the 3-position of furans requires special strategies. Several such procedures have been developed, one of which is the synthesis of furan-3-carboxylic acid (**63**) via aromatization of 3-trichloroacetyl-4,5-dihydrofuran (**62**) followed by a nucleophilic displacement with hydroxide, alcohols and amines (Scheme 6.25) [66].

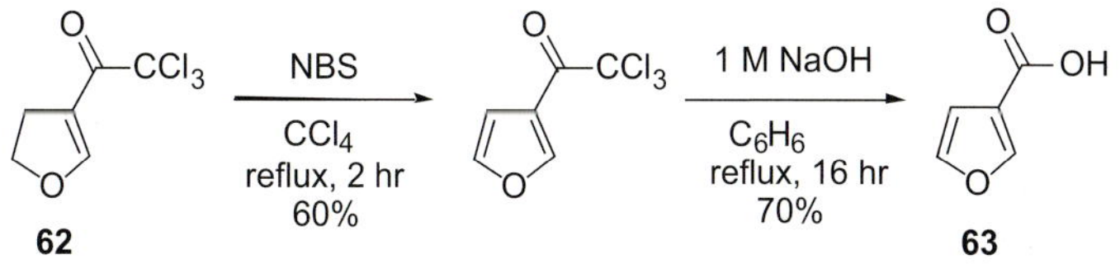

Scheme 6.25

Reductive annulation of 1,1,1-trichloroethyl propargyl ether **64** in the presence of a catalytic amount of Cr(II) regenerated by Mn/Me_3SiCl provides another entry to a 3-substituted furan (**65**) in high yields (Scheme 6.26). The reaction conditions proved compatible with most other common functional groups. Some natural products, such as perillene and dendrolasin, have been prepared by utilizing this procedure [67].

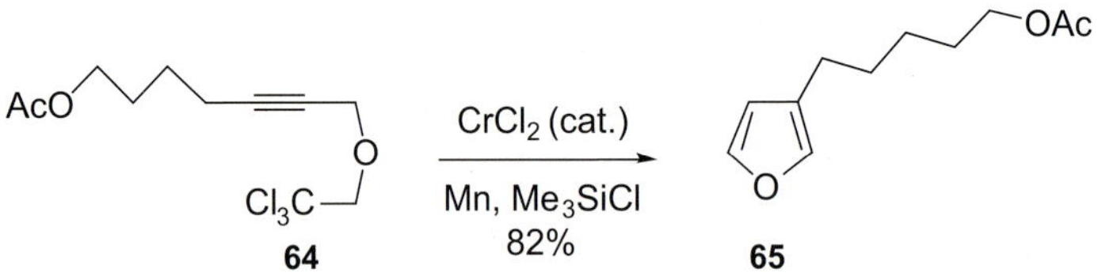

Scheme 6.26

The reaction of furan with *N*-tosylimine produced *in situ* from TsN=S=O and aldehydes in the presence of $ZnCl_2$ gives no Diels–Alder reaction products. Instead, furyl sulfonamides have been separated in high yields (Scheme 6.27) [68]. This procedure provides an efficient synthesis of 2-substituted furans such as **67** from **66**, and is general with respect to aldehydes. Moreover, it is possible to synthesize 2,3-disubstituted furans by an intramolecular aromatic substitution of *N*-tosylimines at the 3-position of furans.

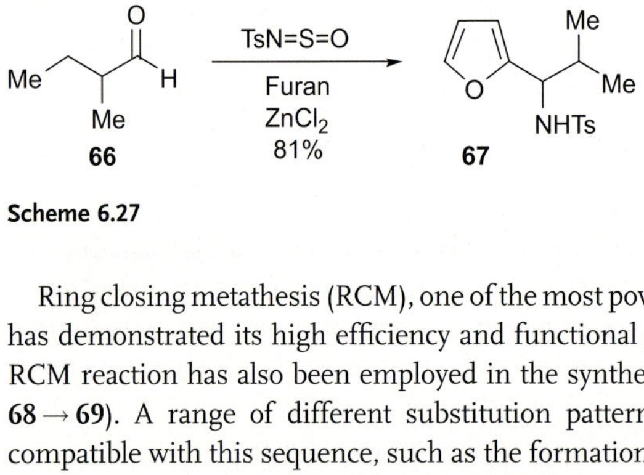

Scheme 6.27

Ring closing metathesis (RCM), one of the most powerful tools for ring-formation, has demonstrated its high efficiency and functional group tolerance. Recently, the RCM reaction has also been employed in the synthesis of substituted furans (e.g., **68** → **69**). A range of different substitution patterns and functional groups are compatible with this sequence, such as the formation of both **70** and **72**, employing also the Grubbs catalyst **71**(Scheme 6.28) [69].

6.2.3
Disubstituted Furans

Direct metallation of 2-substituted furan with *t*-BuLi in the presence of TMEDA followed by treatment of the lithiated furan species with electrophiles provides a simple and efficient way to the corresponding 2,3-disubstituted furans [70]. 2-Furancarboxamide reacts with vinylsilanes catalyzed by $Ru_3(CO)_{12}$ or RuHCl $(CO)(PPh_3)_3$ to deliver 3-trialkylsilyl-2-furancarboxamide also in high yield [71]. A general protocol has been developed towards a mild, regioselective arylation of

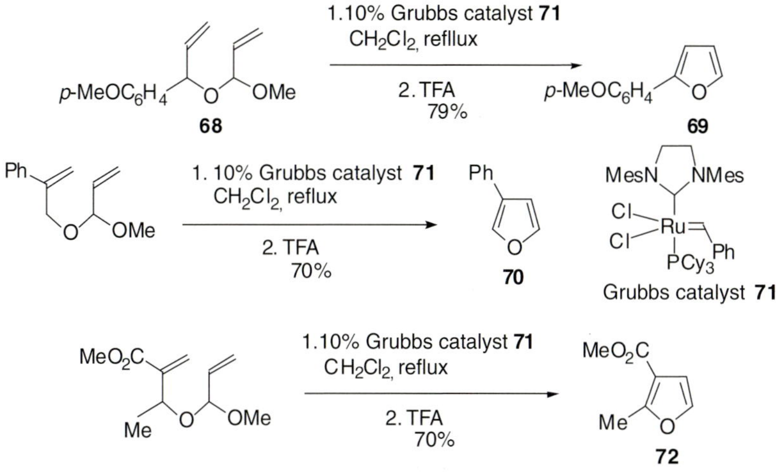

Scheme 6.28

2-furaldehyde using functionalized aryl halides in the presence of a catalytic amount of $PdCl_2$-PCy_3. Slow addition of aryl halides to the reaction mixture avoids the homocoupling efficiently [72]. A one-pot Suzuki coupling of aryl halides with *in situ* generated 5-(diethoxymethyl)-2-furylboronic acid catalyzed by palladium(0) also gave 5-aryl-2-furaldehydes in high yields [73]. A simple procedure to prepare 5-aryl- and 5-pyridyl-2-furaldehydes from the inexpensive and commercially available 2-furaldehyde diethyl acetal through direct lithiation–transmetallation–electrophile trapping has been reported. The reaction proceeded in a four-step, one-pot manner and the yields of the coupling step were usually 58–91% [74].

Allene derivatives are useful starting materials in the synthesis of furans containing different substitution patterns. In 1990, Marshall found that Rh(I) and Ag(I) can catalyze the isomerization of allene **73** to the corresponding substituted furan **74** in high yield (Scheme 6.29) [75]. Later, this strategy was successfully applied to the synthesis of furanocembranes and other related natural products [76].

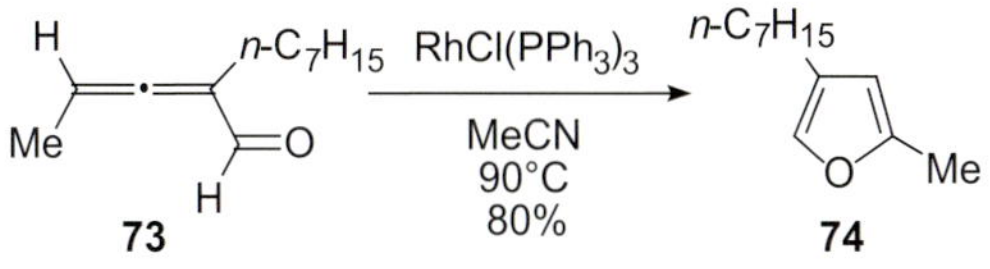

Scheme 6.29

Several reports have discussed the isomerization between allenes and propargyl groups. One example is the intermolecular reaction of allenyl ketone **75** and an α,β-unsaturated ketone catalyzed by $AuCl_3$, which gives rise to 2,5-disubstituted furan **76** in good yield (Scheme 6.30) [77]. Both ethyl propargyl ketone and ethyl allenyl ketone lead to the same 2,5-disubstituted furan.

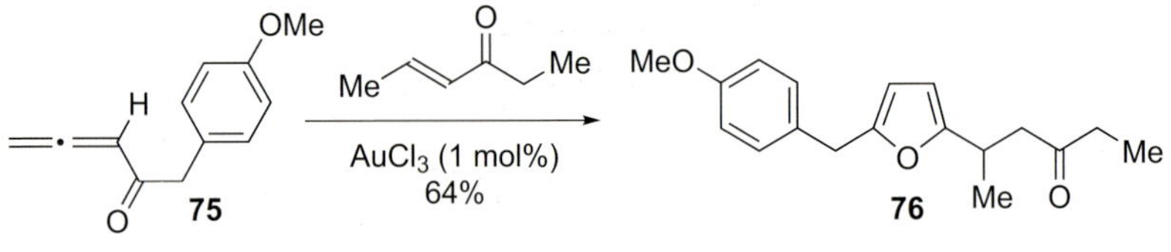

Scheme 6.30

A general, efficient procedure for the preparation of 2,5-disubstituted furans containing acid- and base-labile groups via CuI-catalyzed cycloisomerization of alkynyl ketones such as the conversion of **77** into **78** has appeared (Scheme 6.31), for which a plausible mechanism has been proposed [78]. The allenyl ketone produced from the triethylamine-Cu(I)-catalyzed isomerization of the starting material should be the intermediate. Coordination of copper to the terminal double bond of allene followed by an intramolecular nucleophilic attack of the oxygen lone pair and subsequent isomerization eventually leads to furans as final products. Indeed, 2-phenylfuran was afforded in 33% yield when allenyl phenyl ketone was treated with CuI in DMA.

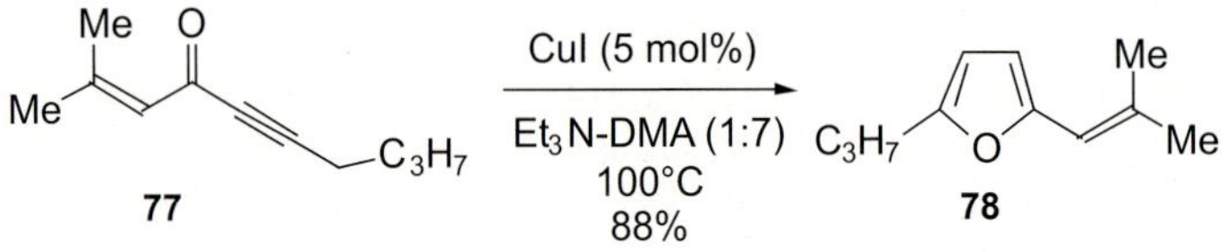

Scheme 6.31

2,4-Disubstituted furan **81** has been obtained in high yield through a novel oxidative cyclization–dimerization reaction between the two different allenes **79** and **80** (Scheme 6.32) [79].

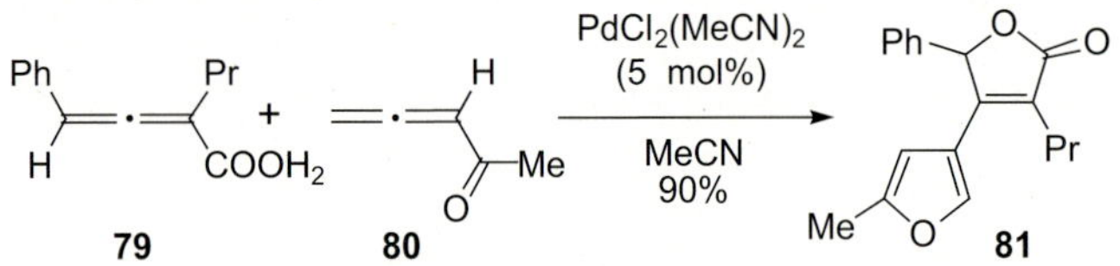

Scheme 6.32

As shown in Scheme 6.33, 2,5-disubstituted furan **84**, with a cyclopropane subunit at the 5-position, has been synthesized in acceptable yield by a metal-catalyzed

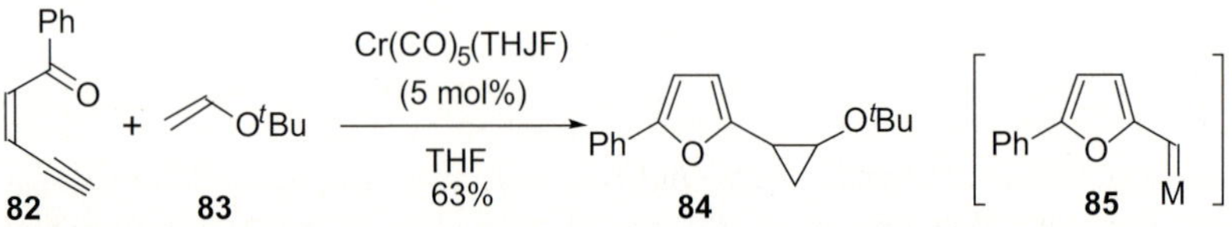

Scheme 6.33

cyclization reaction of 1-benzoyl-*cis*-1-buten-3-yne (**82**) and the enol ether **83** via (2-furyl)carbene complex **85**. Many types of metal complexes, such as Mo, W, Ru, Rh, Pd, and Pt, complexes are suitable catalysts [80].

Reaction of 2-alkynal acetal with a divalent titanium reagent and aldehydes provides, after an acid work up, 2,3-disubstituted furans in good to excellent yields [81]. Michael addition–aldol condensation reactions of α,β-unsaturated enones with an organocopper reagent and (tetrahydropyranyloxy)acetaldehyde have been followed by treatment of the products with *p*-TsOH to afford 2,3-disubstituted furans in moderate to good yields [82]. α,β-Unsaturated carbonyl compounds with an appropriate leaving group undergo 1,5-electrocyclization reactions to yield 2,5-disubstituted furans upon heating in the presence of an acid, presumably through an intermediate formed from 1-oxapentadienyl cations, whose conformational and energy properties were also studied by DFT calculations (B3LYP/6-31 + G*) [83]. A 2,3-disubstituted furan, such as **88**, with different functionalities has been synthesized through an acid-catalyzed elimination reaction of 2-alkylidenetetrahydrofuran **87**, which can be prepared from the cyclization of **86** in high regioselectivity (Scheme 6.34) [84]. 2-Substituted furans and bicyclic furans can be synthesized by this methodology.

MeO O O Cl OEt **86** — DBU (2 equiv), THF, 20°C, 80% → MeO O OEt **87** — TFA, CH_2Cl_2, 20°C, 100% → O O OEt **88**

Scheme 6.34

Aldol reaction of aziridine and α,β-epoxyaldehydes (e.g., **89**) followed by an intramolecular enol cyclization in the presence of Bu_2BOTf/DIPEA furnishes 5-substituted-2-furyl amines and carbinols **90** in good yields. The reaction proceeds in a one-pot manner (Scheme 6.35) [85].

Me X CHO **89** + Me O Me Me — Bu_2BOTf, DIPEA, CH_2Cl_2, 0-25°C → Me X OH O Me Me → Me Me O **90** X Me

X = NHBoc, 72% yield
X = O, 75% yiel

Scheme 6.35

Mercury triflate-catalyzed cyclization of 1-alkyn-5-one **91** produces 2-methyl-5-phenylfuran (**92**) (Scheme 6.36). The reaction may involve a protodemercuration of a vinylmercury intermediate generated *in situ*. When the substituent is at the

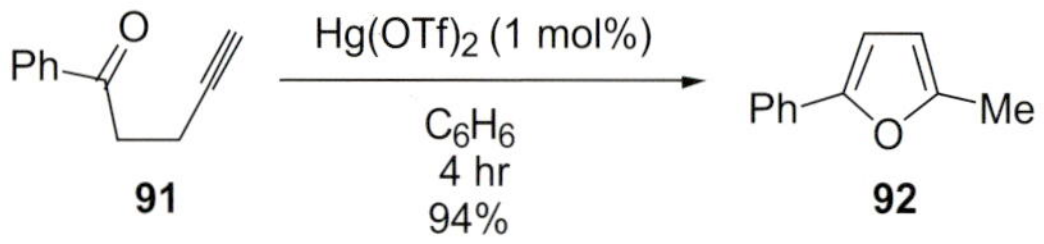

Ph · O · 91 — $Hg(OTf)_2$ (1 mol%), C_6H_6, 4 hr, 94% → Ph · O · Me · 92

Scheme 6.36

α-position of the carbonyl group, 2-methyl-4,5-disubstituted furans are formed. A plausible reaction mechanism has also been provided [86].

Trifluoromethylsulfonamidofuran **94** has been prepared in high yield through the reaction of a cyclic carbinol amide **93** with triflic anhydride (Scheme 6.37). Many lactams have been tested and the reaction proceeds under mild conditions. Keto-amides are also suitable substrates in these reactions [87].

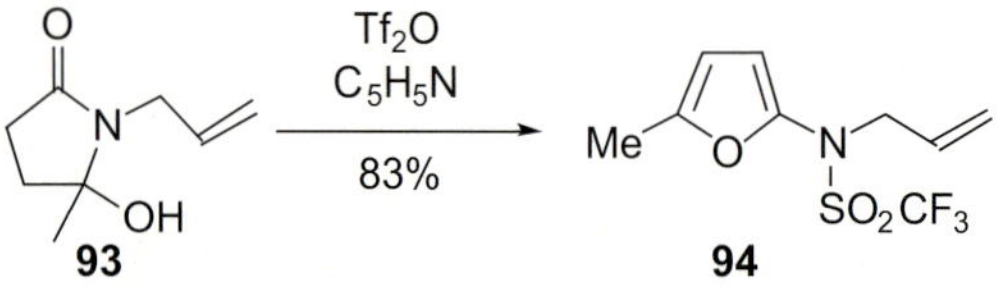

O · N · OH · 93 — Tf_2O, C_5H_5N, 83% → Me · O · N · SO_2CF_3 · 94

Scheme 6.37

A two-step electrochemical annulation directed to polycyclic systems containing annulated furans has been developed. This pathway involves an initial conjugate addition of a furylethyl cuprate to cyclopentenone (**95**) and trapping of the enolate as the corresponding silyl enol ether **96**. The second step involves anodic coupling of the furan and the silyl enol ether to form the cis-fused six-membered ring **97** with high stereoselectivity (Scheme 6.38) [88]. The reaction can be extended to the formation of seven-membered ring fused furan **98** [89]. However, the substituent on the 3-position of cyclopentanone is important for the formation of a seven-membered ring fused

O · 95 — (O, MgBr), CuI, THF, TMEDA, Me_3SiCl → Me_3SiO · O · 96 — 1. carbon anode, ^{i}PrOH, $LiClO_4$, 2,6-lutidine, MeCN; 2. 1N HCl, 64% (two steps) → O · H · O · 97

Me · Me_3SiO · O — 1. carbon anode, ^{i}PrOH, $LiClO_4$, 2,6-lutidine, MeCN; 2. 1N HCl, 61% → O · H · O · 98

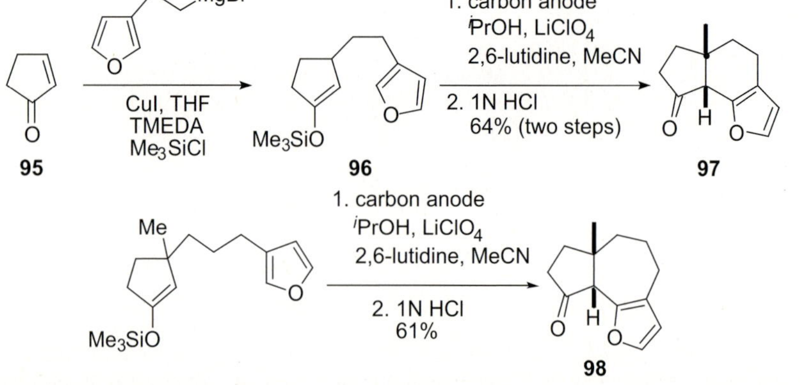

Scheme 6.38

furans. If the substituent is H, no desired cyclization product is provided. The results show that the gem-dialkyl effect plays a crucial role in this electron transfer reaction.

6.2.4
Trisubstituted Furans

Many procedures mentioned above can also be utilized effectively to the synthesis of 2,3,4- and 2,3,5-trisubstituted furans. In addition, several novel procedures have also been reported.

3-Trifluoromethylfuran **100** has been obtained from (*Z*)-2-alkynyl-3-trifluoromethyl allylic alcohol **99** through a palladium-catalyzed cyclization-isomerization procedure (Scheme 6.39) [90].

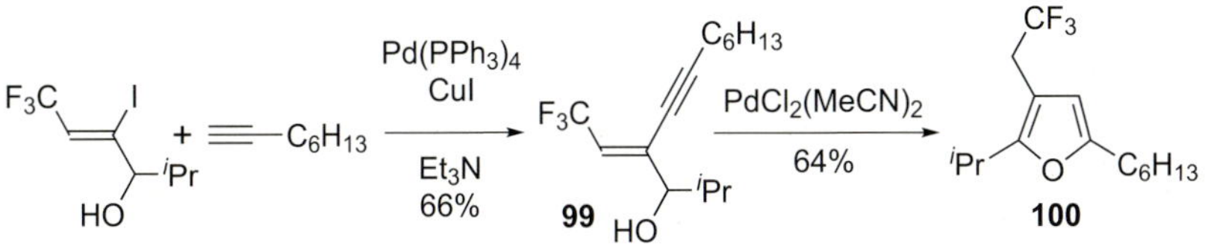

Scheme 6.39

A mild, simple reaction of dimethyl acetylenedicarboxylate with an ammonium ylide supplies trisubstituted furan **101** in good yield (Scheme 6.40) [91].

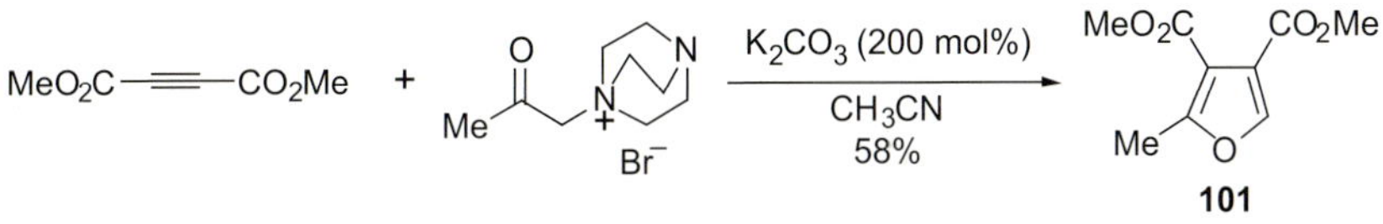

Scheme 6.40

The reaction of 1,4-diarylbut-3-yne-1-one **102** with NBS, NIS or ICl affords, via a 5-endo-dig electrophilic cyclization, 3-halofuran **103** in high regioselectivity and high yield (Scheme 6.41) [92].

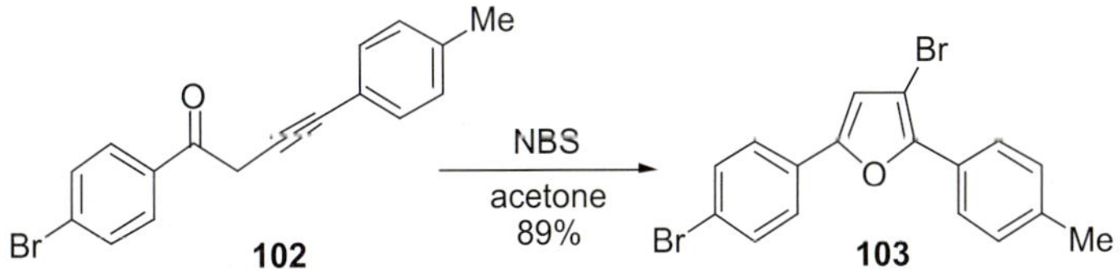

Scheme 6.41

Müller has reported the synthesis of 3-halofurans [93]. Thus, cross-coupling of acid chlorides with THP-protected propargyl alcohols gives rise to the corresponding alkynones, which undergo an acid-assisted electrophilic addition of hydrogen halide with concomitant deprotection and cyclization to afford 3-chlorofuran **104**. If RB $(OH)_2$ is added to the reaction system before workup, this reaction can provide Suzuki-coupling products in moderate yields (Scheme 6.42).

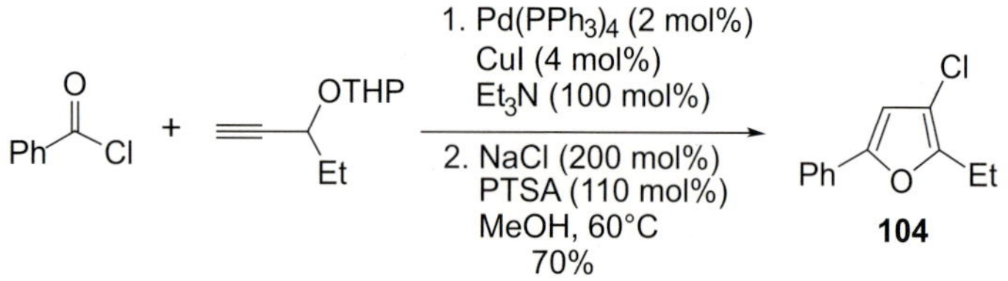

1. $Pd(PPh_3)_4$ (2 mol%), CuI (4 mol%), Et_3N (100 mol%); 2. NaCl (200 mol%), PTSA (110 mol%), MeOH, 60°C, 70%; **104**

Scheme 6.42

Organic molecules can also be used as catalysts in furan ring formation reactions. Krische has reported an example of organophosphine serving as catalyst in the construction of substituted furans, in which various γ-acyloxybutynoates such as **105** were converted into 2,3-disubstituted, 2,4-disubstituted, and 2,3,5-trisubstituted furans (e.g., **106**) (Scheme 6.43) [94].

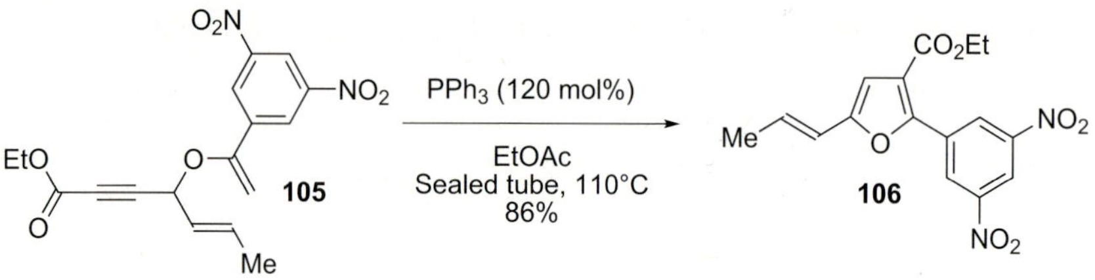

105; PPh_3 (120 mol%), EtOAc, Sealed tube, 110°C, 86%; **106**

Scheme 6.43

Reaction of 2-penten-4-yn-1-one **107** with benzaldehyde initiated by tributylphosphine delivers 2-vinyl substituted furan **108** in a good yield. The reaction might proceed through a 1,6-addition of phosphine to the enynes, and is followed by ring closure and Wittig reaction between the ylide resulting from cyclization and an aldehyde (Scheme 6.44) [95].

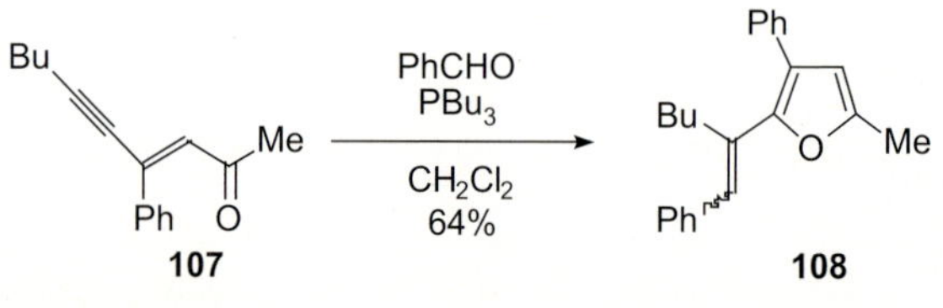

107; PhCHO, PBu_3, CH_2Cl_2, 64%; **108**

Scheme 6.44

3-Aminofuran-2-carboxylate **109** has been prepared in good yield through the reaction of a cyanoketone with glycolate under Mitsunobu conditions followed by treatment with NaH (Scheme 6.45). 4-Pyridylcarbinol and 4-nitrobenzyl alcohol can also react with cyanoketone to furnish 3-aminofurans [96].

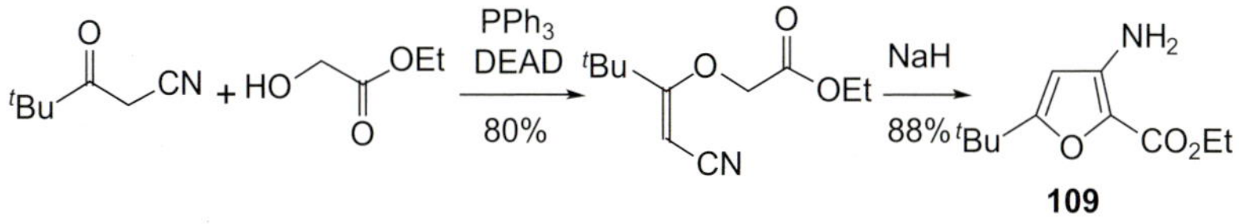

Scheme 6.45

2,3,5-Trisubstituted furan **112** has been synthesized from a two-step one-pot reaction from epoxyalkyne **110**. A facile SmI_2-mediated reduction generates 2,3,4-trien-1-ol **111**, and the reduction is followed by a Pd(II)-catalyzed cycloisomerization (Scheme 6.46) [97]. An attractive variant of this reaction has been extended to the preparation of tetrasubstituted furans. Thus, when electrophilic Pd(II) complexes are generated *in situ* by an oxidative addition of aryl halides or triflates to Pd(0), the oxypalladation process is followed by a reductive elimination and, as a result, tetrasubstituted furans are formed [98].

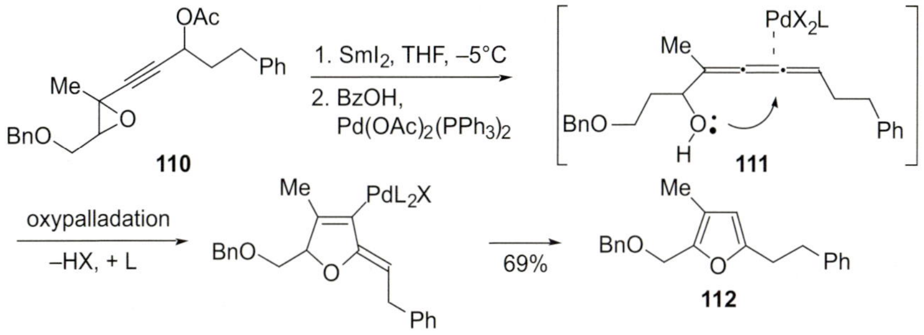

Scheme 6.46

Nucleophilic substitution of sulfinylfuran **113** with acetylacetone and allyl tin reagent via a Pummerer-type reaction has been used in the formation of 2,3,5-trisubstituted furan **114**, as well as with other 2,3-disubstituted furans, in high regioselectivity (Scheme 6.47) [99].

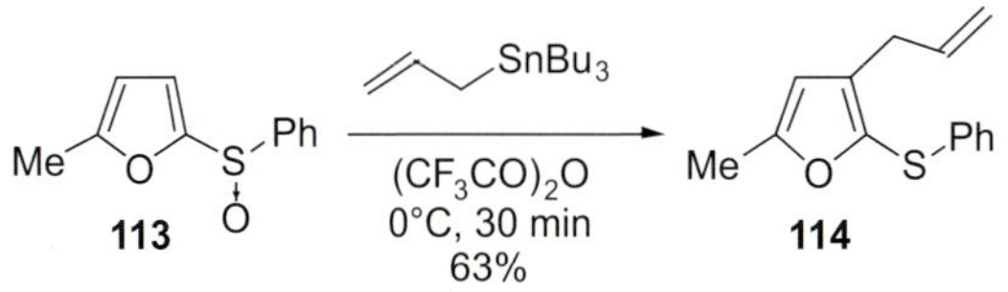

Scheme 6.47

Regioselective gold-catalyzed cyclization of 2-(1-alkynyl)-2-alken-1-one **115** in the presence of a nucleophile affords 2,3,5-trisubstituted furan **116** in an efficient, atom-economical manner. Various alcohols, 1,3-diketones as well as some indoles and amines can serve as nucleophiles. This reaction provides another good example of using gold as catalyst in furan synthesis (Scheme 6.48) [100].

O
115
1 mol% $AuCl_3$
1.5 eq MeOH
CH_2Cl_2
1 hr
80%
OMe
O
116

Scheme 6.48

Yamamoto has demonstrated that, in the presence of CuBr as a catalyst, trisubstituted furan **118** is formed from **117** and isopropanol (Scheme 6.49) [101], while Liu has shown that tetrasubstituted 3-iodofurans are afforded on employing similar starting materials and I_2/K_3PO_4 as reagents (see below) [102].

O
Ph
117
+
Me OH
Me
CuBr (10 mol%)
DMF
80°C
62%
Ph
O
Me
O
Me
118

Scheme 6.49

Conjugated ene-yne-carbonyl **119** has been employed as a precursor of 2-furylcarbene **121**. In this way, **121** was allowed to react with an allyl sulfide to form an *S*-ylide followed by [2,3]sigmatropic rearrangement to give the corresponding trisubstituted furan **120** in an excellent yield (Scheme 6.50) [103].

O
H
Bz
119
+
SPh
$[Rh(OAc)_2]_2$
(2.5 mol%)
CH_2Cl_2
reflux
98%
O
SPh
Bz
120
O
=[Rh]
Bz
121

Scheme 6.50

Palladium-catalyzed cyclization of α-propargyl α-keto ester **122** provided 2-alkenyl-4,5-disubstituted furan **123a**. High *E*/*Z*-selectivity is realized when the Me_3Si- group is introduced to the α′-position of the triple bond. The geometry of the double bond in **123a** is almost completely inverted by reaction with a catalytic amount of diphenyl diselenide, providing **123b**. (Scheme 6.51) [104].

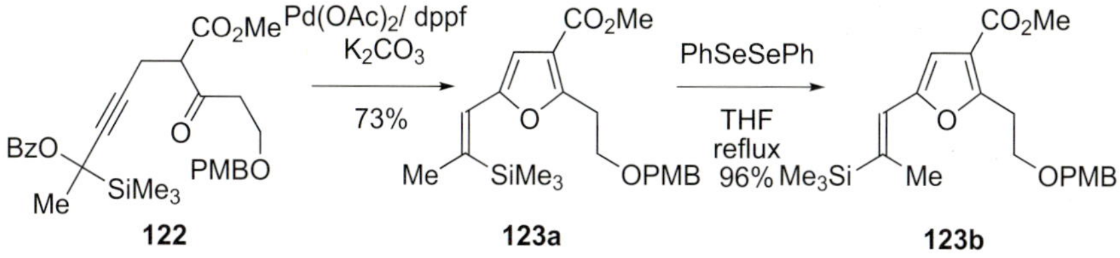

Scheme 6.51

Ma has reported an elegant regioselective synthesis of 2,3,4-trisubstituted furan **126** using a Cu-catalyzed ring-opening cycloisomerization reaction of cyclopropenyl ketone **124** (Scheme 6.52). Notably, the regioselectivity of this reaction can be tuned using different catalysts. 2,3,5-Trisubstituted furan **125** is produced from the same starting materials with excellent regiocontrol using a Pd catalyst [105].

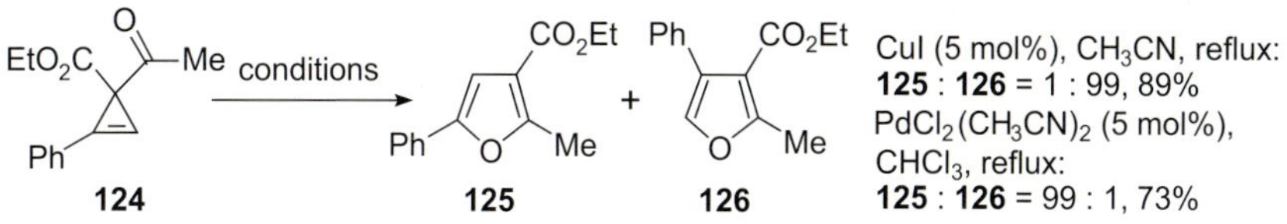

Scheme 6.52

Alkylidenecyclopropyl ketones, analogs of allene ketones, have been used as a starting material in the synthesis of polysubstituted furans. Ma has given another example in which highly regiocontrolled transformation of an alkylidenecyclopropyl ketone (**127**), easily prepared by the regioselective cyclopropanation of an allene or the reaction of alkylidenecyclopropanyllithium with *N,N*-dimethyl carboxylic acid amide, into 2,3,4-trisubstituted furan **128** is realized in the presence of NaI. Interestingly, the same starting material **127** gives rise to 2,3,4,5-tetrasubstituted furan **129** if Pd $(PPh_3)_4$ or $PdCl_2(MeCN)_2$ is the catalyst (Scheme 6.53) [106].

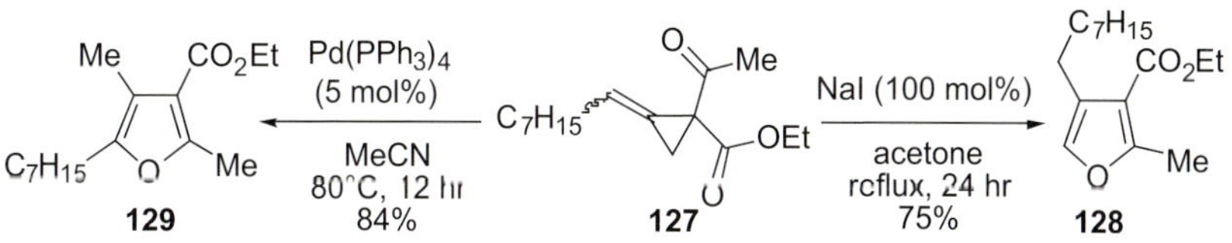

Scheme 6.53

Allene derivatives are important precursors in the regioselective synthesis of substituted furans. One example is the reaction of thioallenyl ketone via a 1,2-migration of the thio group from an sp^2 carbon atom in allenyl sulfide **130** catalyzed by CuI to afford trisubstituted furan **131** in high yield (Scheme 6.54) [107]. Propargyl sulfides led to similar results. If thiopropargyl aldehydes are used as starting materials, 2,3-disubstituted furans are obtained. This route therefore provides a simple entry for the preparation of disubstituted and trisubstituted furans. Similarly, reaction of propargylic dithioacetals with an organocopper reagent followed by treatment with an aldehyde and then with acid also furnishes 2,3,5-trisubstituted furans, in moderate to good yields [108].

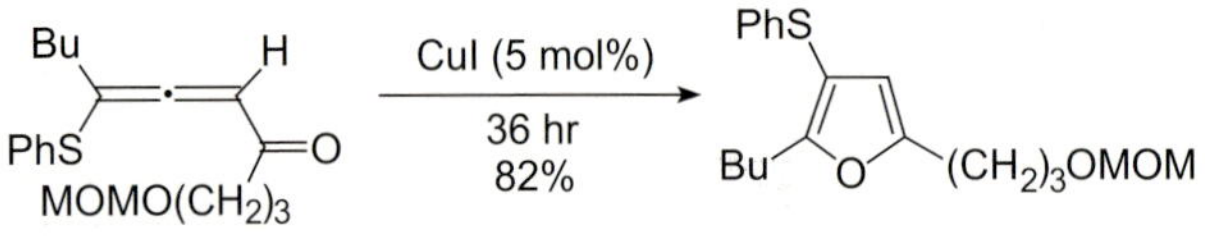

Bu, H, PhS, O, $MOMO(CH_2)_3$ **130** → CuI (5 mol%), 36 hr, 82% → PhS, Bu, O, $(CH_2)_3OMOM$ **131**

Scheme 6.54

Isomerization of α-allenylcyclopentenone **132**, obtained from propargyl ether and morpholino unsaturated amide, in the presence of Hg-catalyst gives furylcyclopentenone **133**, therefore providing another example of the conversion of allenyl ketones into furans (Scheme 6.55) [109].

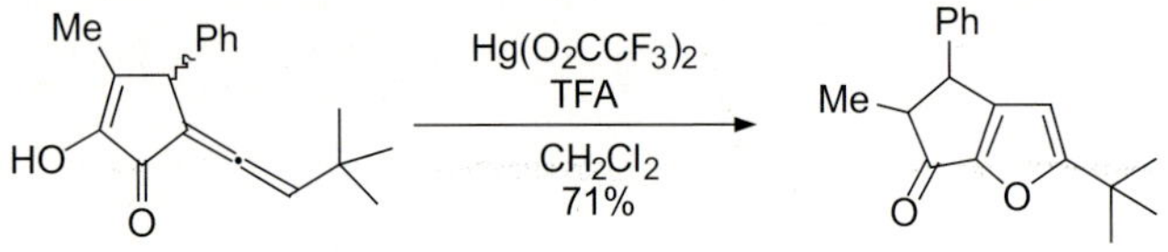

Me, Ph, HO, O **132** → $Hg(O_2CCF_3)_2$, TFA, CH_2Cl_2, 71% → Ph, Me, O, O **133**

Scheme 6.55

Gevorgyan has reported a regio and divergent synthesis of halofuran **135** via 1,2-halogen migration of haloallenyl aldehyde **134** catalyzed by $AuCl_3$ (Scheme 6.56) [110]. The procedure furnished 3-halofurans, some of which are otherwise difficult to access.

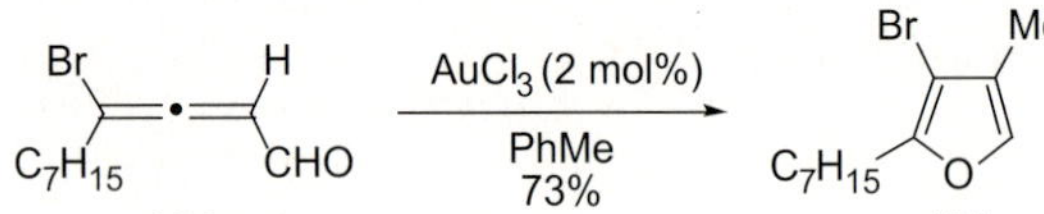

Br, H, C_7H_{15}, CHO **134** → $AuCl_3$ (2 mol%), PhMe, 73% → Br, Me, C_7H_{15}, O **135**

Scheme 6.56

Reaction of aryl or alkyl-1-enyl halides with allenyl ketone **136** in the presence of Pd-catalyst followed by cyclization affords substituted furans with aryl or alk-1-enyl as substituent at the 3-position. This reaction is a new route to 3-substituted furan **137** (Scheme 6.57) [111].

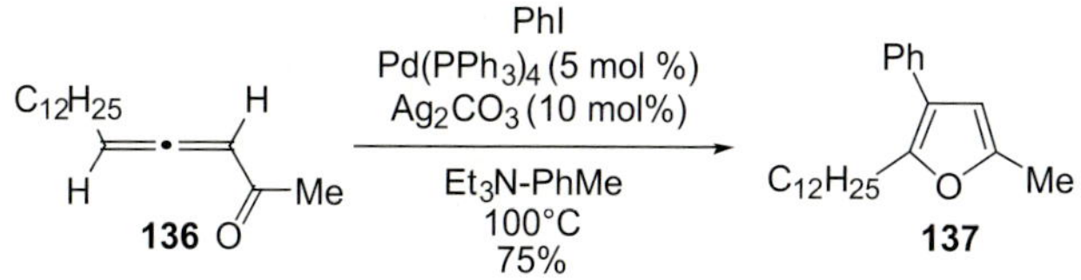

Scheme 6.57

6.2.5 Tetrasubstituted Furans

Many procedures discussed above are also suitable for the synthesis of tetrasubstituted furans if there are substituents at appropriate positions in the starting materials. However, some special methodologies have also been developed solely for the synthesis of tetrasubstituted furans.

Liu has shown that cyclization of 2-(1-alkynyl)-2-alken-1-one **117** in the presence of a nucleophile affords fully substituted furan **138** in high regioselectivity and a good yield if I_2 and K_3PO_4 are also used. This reaction provides a mild, efficient route to 3-iodofurans. The reaction might be initiated by the formation of iodonium through coordination of the triple bond to an iodine cation, followed by cyclization and, finally, a nucleophilic attack (Scheme 6.58) [102].

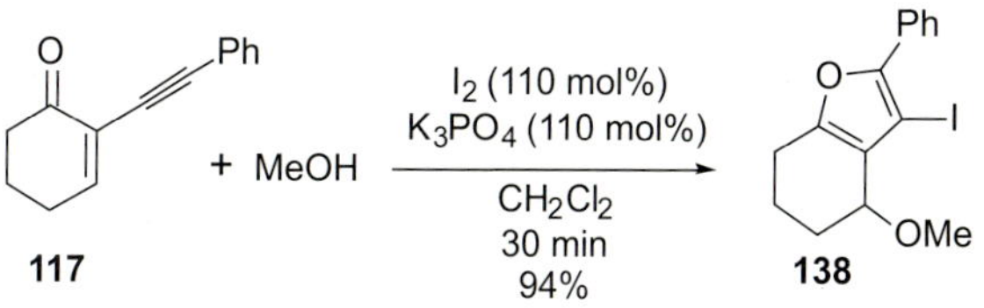

Scheme 6.58

Ruthenium- and platinum-catalyzed sequential reaction of propargylic alcohols (e.g., **139**) and cyclohexanone affords tri- and tetrasubstituted furans such as **140** (Scheme 6.59) [112]. In this reaction, two different kinds of catalysts sequentially promote each catalytic cycle in the same medium and give the products in high regioselectivity and good yields.

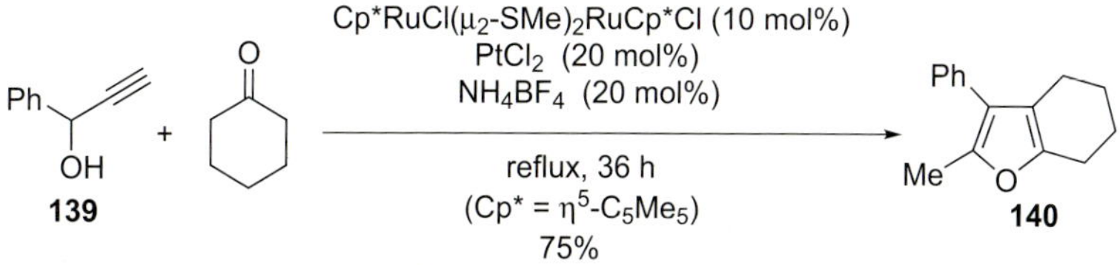

Scheme 6.59

Another example of the regioselective synthesis of substituted furan **142** from the reaction of allenyl ketone **141** with organic halides initiated by Pd-catalyzed nucleophilic attack of the aryl group to allene followed by cyclization reaction has been reported (Scheme 6.60) [113]. This methodology shows a high substituent-loading capacity and functional group tolerance, as well as generality and versatility. If one of the substituents of the allenyl ketones is H, 2,3,4- and 2,3,5-trisubstituted furans can also be generated.

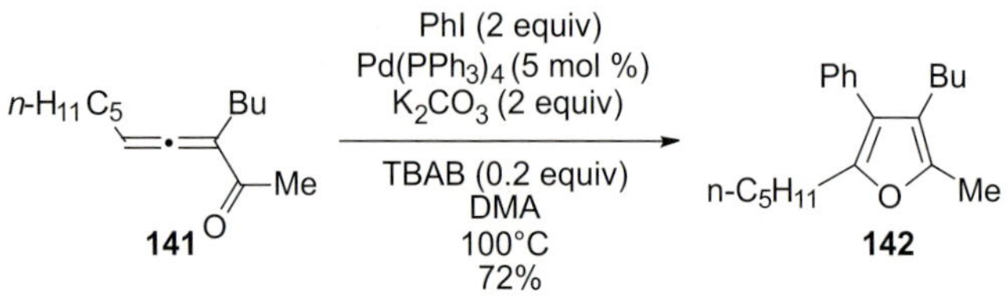

Scheme 6.60

Gold-catalyzed reactions of propargyl vinyl ether **143** furnish tetrasubstituted furan **145** in high yield. The reaction proceeds through cyclization of the 2-allenyl-1,3-dicarbonyl intermediate **144** produced from a propargyl-Claisen rearrangement (Scheme 6.61) [114].

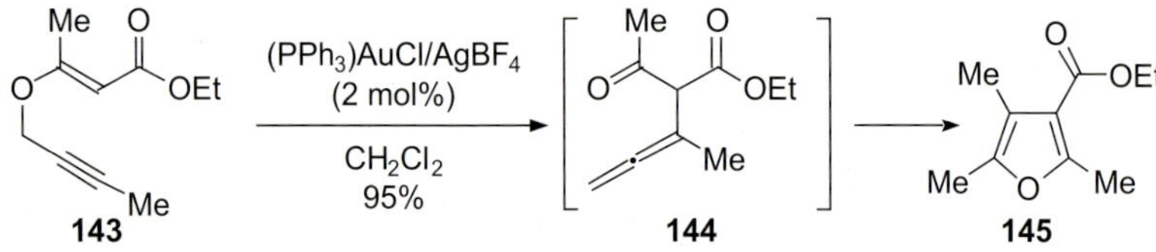

Scheme 6.61

Palladium-mediated sequential cross-coupling Sonogashira reaction–Wacker-type heteroannulation and deprotection reactions of pyridones, alkynes and organic halides furnish substituted furo[2,3-*b*]pyridones (e.g., **146**) in a one-pot operation (Scheme 6.62) [115]. The coupling products of pyridones and alkynes can be separated and a single palladium catalyst intervenes in three different transformations.

A palladium-catalyzed three-component cyclization–coupling reaction of acetoacetate, propargyl bromide or carbonate and aryl halides gives tetrasubstituted furan **147** in high regioselectivity and a good yield (Scheme 6.63) [116].

Several examples concerning the synthesis of tetrasubstituted furans utilizing acetylenecarboxylate, aldehydes and nitrile or isonitrile have been reported. One such example is a one-pot reaction, affording tetrahydrofuro[2,3-*c*]pyridine **148** in high yield (Scheme 6.64) [117].

Scheme 6.65 shows a further example, in which the reaction is carried out in an ionic liquid under mild conditions, leading to tetrasubstituted furan **149** in high

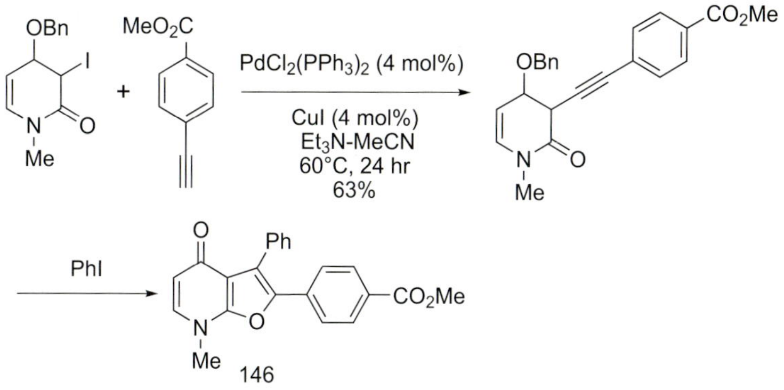

Scheme 6.62

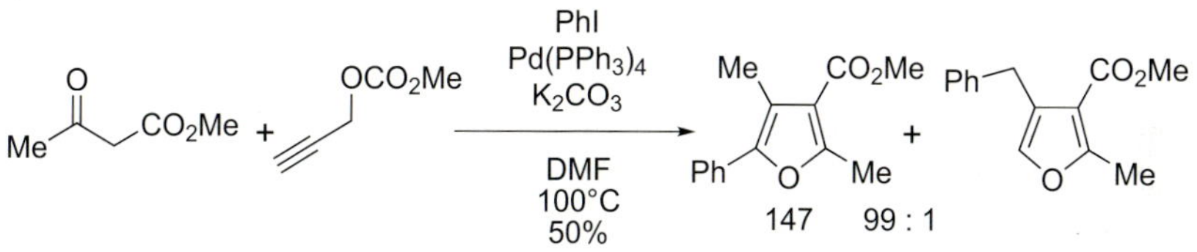

Scheme 6.63

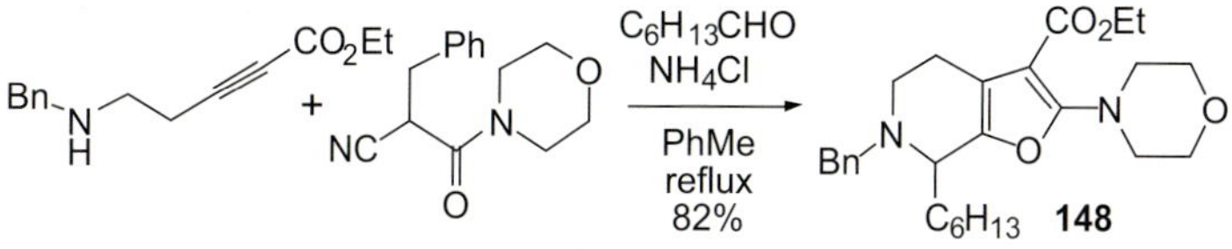

Scheme 6.64

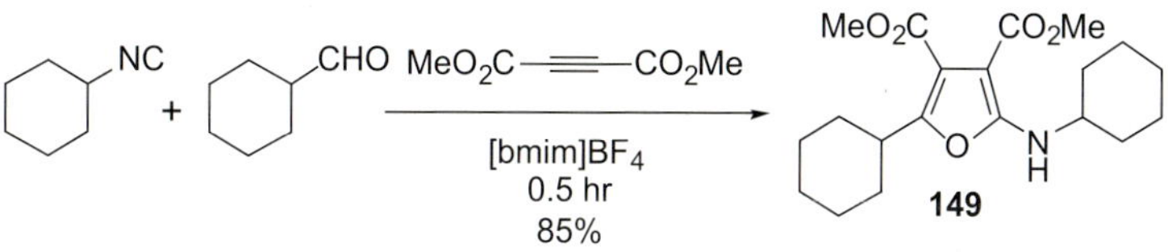

Scheme 6.65

yield [118]. A similar reaction employing 2-furyl-2-oxoacetamides instead of aldehydes in the preparation of substituted furylfurans was also recorded [119].

A detailed description of the synthesis of furylcyclopropane **150** from the reaction of an alkene and previously reported 2-furylcarbenoid by a metal-catalyzed cyclization of enyne ketone has been disclosed (Scheme 6.66) [120]. This protocol has been expanded by the same authors to the synthesis of furylcyclopropane-containing

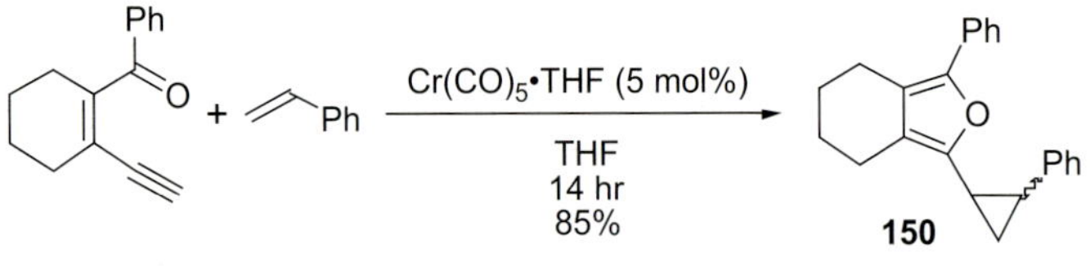

Scheme 6.66

polymers as well as furfurylidene-containing polymers when phenyl enyne ketones with a vinyl and formyl group at the *ortho* position of the benzene ring are employed [120].

Cycloisomerization of acyloxy-, phosphatyloxy- and sulfonyloxy-substituted alkynylketones gives tri- and tetrasubstituted furans (e.g., **151**) with good regioselectivities; the allene intermediates are produced *in situ* via migration of the substituents, catalyzed by CuCl or $AgBF_4$ (Scheme 6.67) [121].

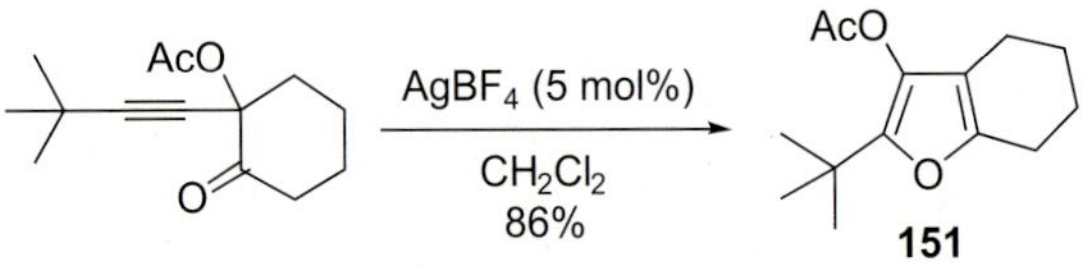

Scheme 6.67

N-Heterocyclic carbenes (NHC) have found use as reagents in the synthesis of polysubstituted aminofurans. Multicomponent reaction of an imidazolinium salt with an aldehyde and an acetylenecarboxylate in the presence of NaH leads to tetrasubstituted furan **152** in a good yield (Scheme 6.68). A plausible reaction mechanism has been proposed [122]. A similar procedure for the synthesis of tetrasubstituted 3-aminofurans, using a thiazolium salt, aldehydes and acetylenedicarboxylate, has also been reported [123].

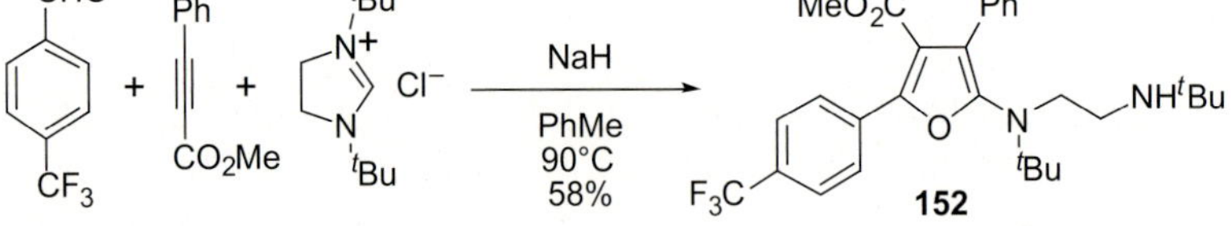

Scheme 6.68

6.3 Reactivity

This section primarily deals with the reactivity of furans that results in an overall transformation of the ring. Reactions of furans, particularly reactions of *C*-metallated furans, that lead to substituted furan derivatives have been discussed in Section 6.2.1. Most examples in this section are extracted from recent literature so as to document the progress made in each reaction category [4].

6.3.1
Reactions with Electrophilic Reagents

Furans are reactive π-nucleophiles. Nucleophilic additions of furans to electrophiles, from either the 2-position or 3-position, often involve regeneration of the furan ring from the oxonium ion intermediate by loss of a proton, as shown, for example, in Scheme 6.7, Section 6.1.2. An interesting example of direct hydrolytic cleavage that occurs subsequent to furan addition to an *N*-acyliminium ion is shown in Scheme 6.69. The initially formed oxonium ion likely undergoes a 1,5-proton shift to generate an intermediate that is then hydrolyzed to the dicarbonyl compound **153** [124].

HCO_2H, C_6H_{12}, r.t., 30 min; n = 1, 75%; n = 2, 64%

Scheme 6.69

2-Silyloxyfurans **154** are chemically equivalent to "cyclic" vinyl silyl ketene acetals, and have been employed for vinylogous additions to electrophiles under Lewis acid catalyzed conditions to provide γ-butenolides **155**, which are common structures present in natural products. These include vinylogous Mannich [125, 126], aldol [126] and Michael reactions. Consistent with vinylogous reactions of acyclic silyloxy dienolates, additions generally occur at the 5-position of 2-silyloxyfurans **154** (Scheme 6.70). Further synthetic manipulations of these adducts can lead to carbocycles, piperidines, sugars and azasugars. A subsequent intramolecular Michael/hetero Michael addition to the α,β-unsaturated system of butenolide can also be exploited to form polycyclic ring structures [127]. Although the factors and the transition state structures (Diels–Alder like versus open chain) that govern the stereoselectivity have not been well-defined, *anti* (*erythro*) adducts are in general isolated as the major isomers for additions to iminium ions, except cyclic acylimi-

154 X = NR^3; X = O; Lewis acid; **155a** *syn*-isomer; **155b** *anti*-isomer

Scheme 6.70

nium ions, from which the *syn*-isomers are obtained preferentially. For reactions with aldehydes, *syn* (*threo*) products generally predominate.

The synthetic utility of vinylogous additions using 2-siyoxyfurans (e.g. **156**) is exemplified by the total synthesis of the plant alkaloid croomine (**157**), in which two vinylogous Mannich reactions of 2-silyloxyfuran to a pyrrolinium ion constitute the key steps for assembling the carbon framework (Scheme 6.71) [128].

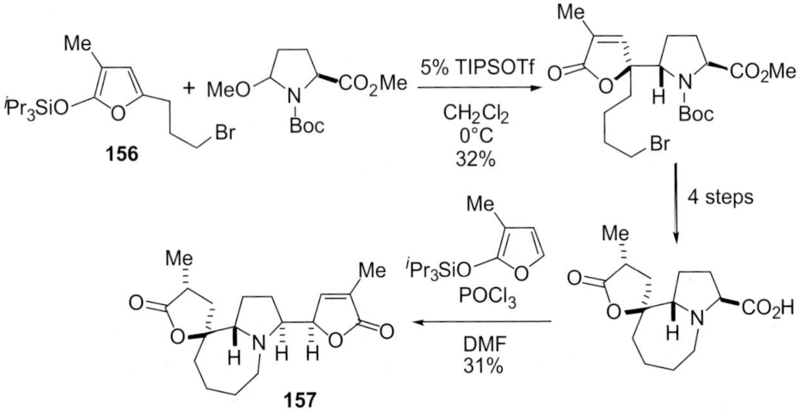

Scheme 6.71

Vinylogous Michael additions of 2-silyloxyfurans to 3-alkenyl-2-oxazolidinones are *anti*-selective, and are highly enantioselective in the presence of a BINAP-Lewis acid catalyst [129]. As illustrated in Scheme 6.72, a complementary, *syn*-selective, organocatalytic, enantioselective vinylogous Michael addition of 2-silyloxyfuran **158** with an α,β-unsaturated aldehyde to produce γ-butenolide **159** has been achieved by using the chiral amine catalyst **160** [130].

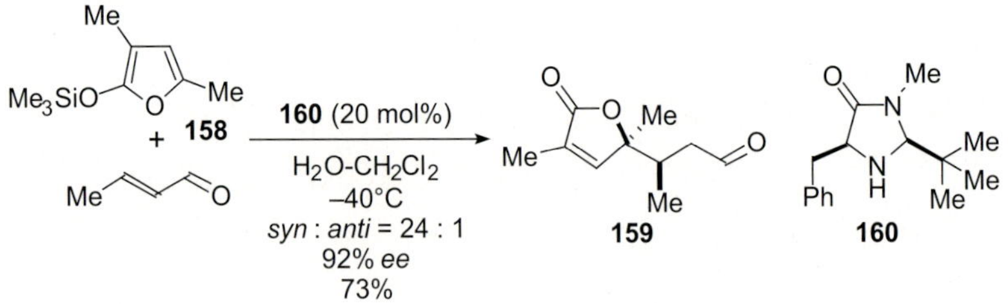

Scheme 6.72

As shown in the example in Scheme 6.73, 2-trimethylsilyloxyfuran (**161**) also reacts with Morita–Baylis–Hillman acetate **162** to provide the interesting γ-butenolide **163**. These triphenylphosphine-catalyzed substitutions proceed regio- and diastereoselectively. However, the reaction mechanism (vinylogous Michael versus Diels–Alder) has not been elucidated [131].

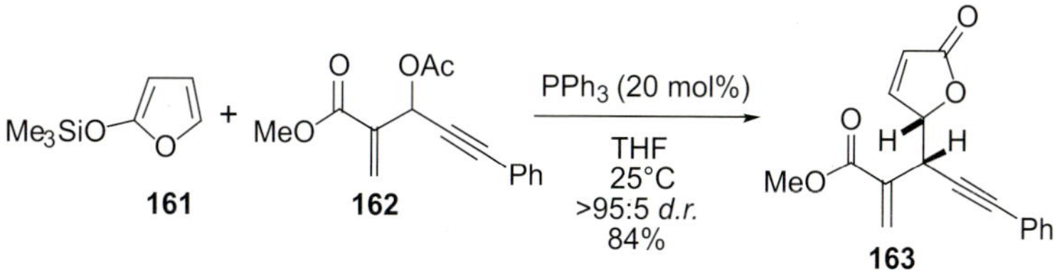

Scheme 6.73

Similar to 2-silyloxyfuran, the relatively unexplored 3-silyloxyfuran also participates in an aldol addition manner with aldehydes under Lewis acidic conditions. High *syn*-diastereoselectivity is obtained with bulky α-branched aldehydes [132]. In addition, 2-methoxy, 2-aryloxy and 2-phenylthiofurans also serve as nucleophiles [133].

The nucleophilicity of furans can be modulated by complexation to transition metals. When furan is dihapto-coordinated to a rhenium π-base, the nucleophilicity of the uncoordinated C3-position is enhanced. In this manner, furan acts as an enol ether [134].

6.3.2 Reactions with Nucleophilic Reagents

Electron-deficient furans, for example 2-nitrofuran (**9**), can undergo nucleophilic substitutions. Various Grignard reagents react with **9** in a Michael addition fashion, providing predominantly *trans*-2,3-disubstituted 2,3-dihydrofurans such as **164** (Scheme 6.74) [135].

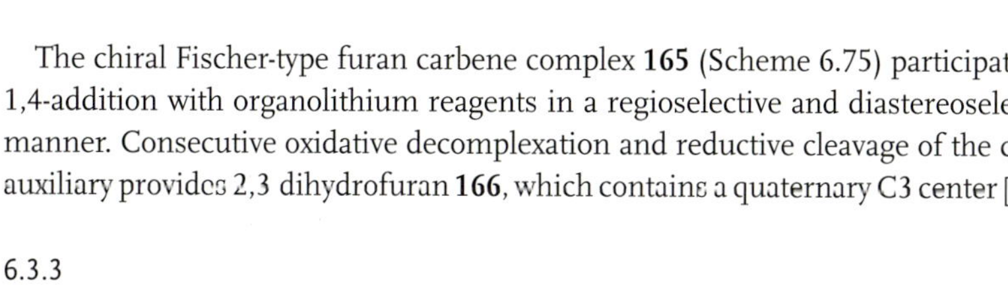

Scheme 6.74

The chiral Fischer-type furan carbene complex **165** (Scheme 6.75) participates in 1,4-addition with organolithium reagents in a regioselective and diastereoselective manner. Consecutive oxidative decomplexation and reductive cleavage of the chiral auxiliary provides 2,3 dihydrofuran **166**, which contains a quaternary C3 center [136].

6.3.3 Reactions with Oxidizing Reagents

Furans are valuable precursors to 1,4-dicarbonyl compounds, which are not directly accessible from the reactions between electrophiles and nucleophiles, such as those employed for the synthesis of 1,3- and 1,5-dicarbonyl compounds. Furans are

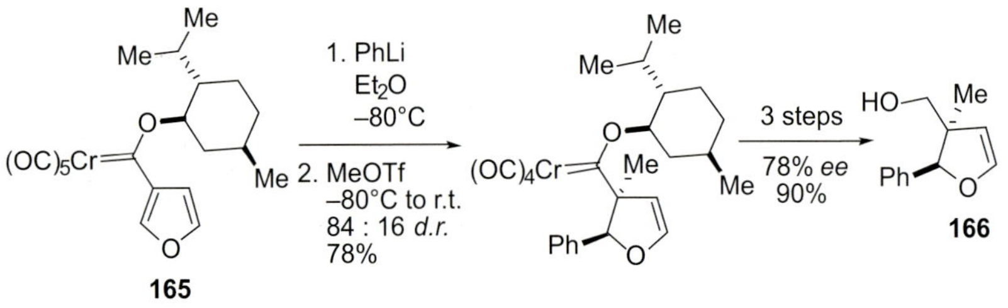

Scheme 6.75

hydrolytically cleaved to saturated 1,4-dicarbonyl compounds under reflux in strongly acidic conditions (Section 6.3.5). The 1,4-dicarbonyl moiety can also be revealed by oxidation of furans using various oxidizing reagents [4, 137].

2,5-Dialkoxy-2,5-dihydrofurans formed from the oxidation of furans in a methanolic solution of bromine (e.g., Scheme 6.12, Section 6.1.2) can be hydrolyzed to *cis*-α,β-unsaturated-1,4-dicarbonyls. Oxidation of 2-substituted or 2,5-disubstituted furans to *cis*-α,β-unsaturated-1,4-dicarbonyls **167** can be performed using *m*-CPBA, PCC or NBS [4, 137], magnesium monoperoxyphthalate (MMPP) [138], dimethyldioxirane [139], methyltrioxorhenium/urea hydrogen peroxide [140] and buffered sodium chlorite [141] (Scheme 6.76). The corresponding *trans* isomers **168** are obtained by *in situ* isomerization, especially in the presence of an amine base, or directly by $Mo(CO)_6$-catalyzed oxidation using cumyl hydroperoxide [142]. Furans also serve as surrogates to carboxylic acids, which are obtained by oxidative disassembly of the aromatic ring using ruthenium tetroxide [143].

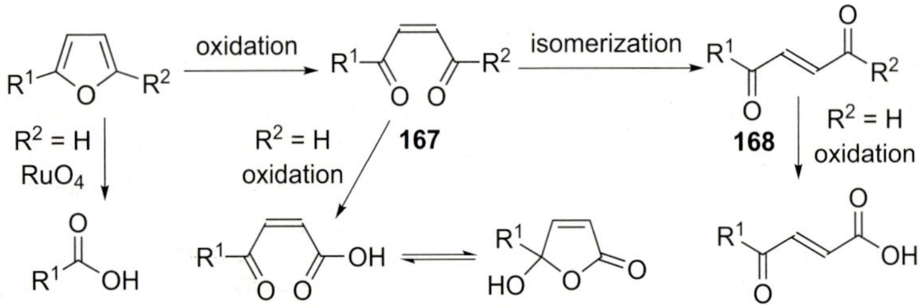

Scheme 6.76

The 1,4-dicarbonyl compounds are useful precursors towards the syntheses of cyclopentenones and cyclopentanones [137]. Interception of the transient oxonium ion during oxidation of furans (e.g., by NBS) by pendant nucleophiles leads to interesting spiro ring systems [144]. Scheme 6.77 depicts an application of furan oxidation in the total synthesis of the indolizidine alkaloid monomorine (**169**) [145].

Regioselective oxidation of unsymmetrical 3-substituted furan **170** to butenolide **172** has been accomplished by using alcoholic bromine or NBS and controlling the acid-catalyzed hydrolysis of the 2,5-dialkoxy intermediate **171** in acetone–water mixture [146] (Scheme 6.78).

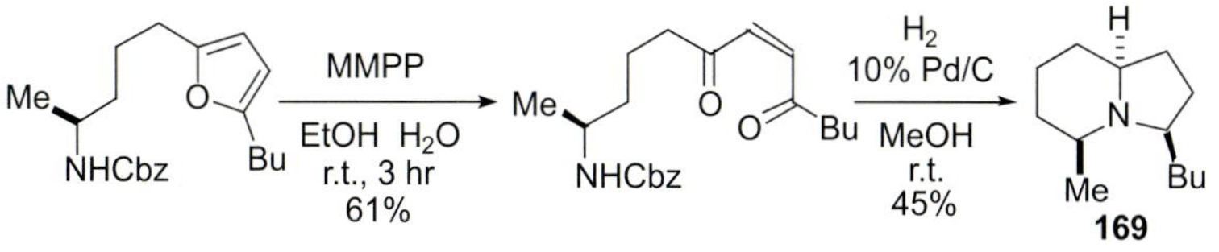

Scheme 6.77

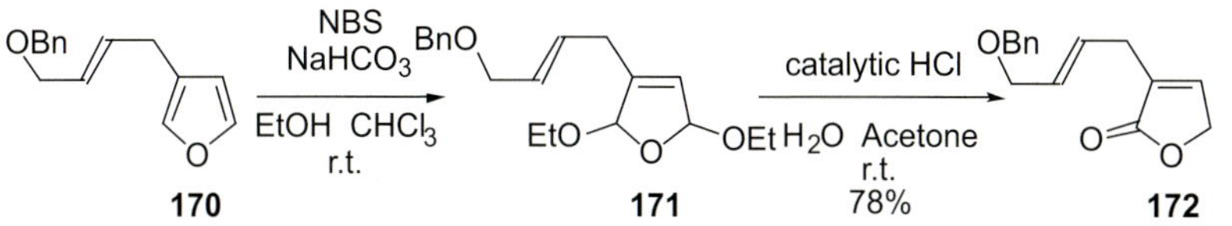

Scheme 6.78

Alternatively, regioselectivity can be controlled by the incorporation of a silyl group in the furan ring (Scheme 6.79), producing **173** in a regiospecific manner [147].

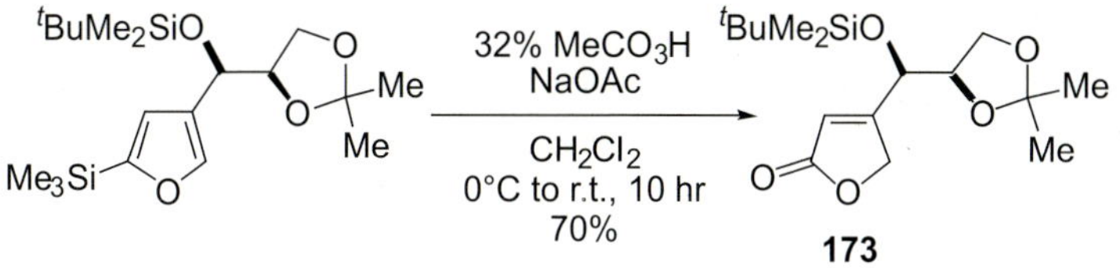

Scheme 6.79

Photooxidations of furans by singlet oxygen, sensitized by, for example, methylene blue, Rose Bengal or tetraphenylporphyrin, generate endoperoxides [148] (e.g., **174**, Scheme 6.80) via a Diels–Alder cycloaddition. Intermediate **174** is prone to Baeyer–Villiger type rearrangement to provide carboxylate **175** (R including silyl).

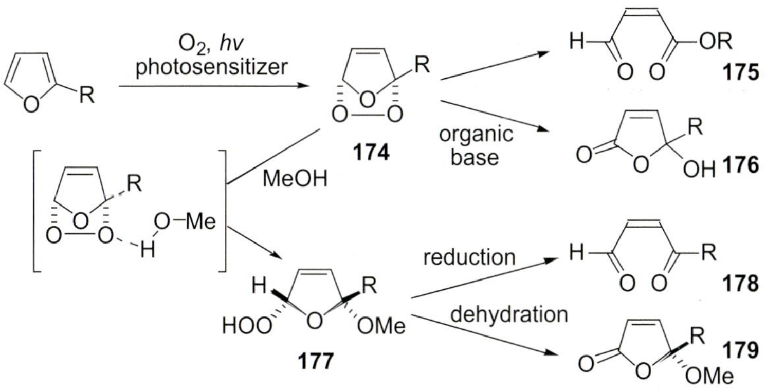

Scheme 6.80

Deprotonation at the bridgehead carbon of **174** by an organic base provides hydroxybutenolide **176**. This provides a useful method for the regioselective oxidation of unsymmetrical 3-substituted furans by using a hindered base (e.g., *i*-Pr_2NEt) [149]. Reaction of **174** with an alcohol (MeOH is often used as a solvent) at higher temperature forms hydroperoxide **177** regio and stereoselectively due to a hydrogen bonding assisted front side attack of the alcohol on the most stabilized carbocation. Reduction of hydroperoxide **177** and subsequent eliminative ring opening provides 1,4-dicarbonyl compound **178**, while dehydration of **177** gives butenolide **179**. Photooxidations of furans have the benefit of being performed selectively in the presence of alkenes.

Scheme 6.81 shows a novel example of taking advantage of the singlet-oxygen photooxidation of furan in the presence of two trisubstituted alkenes in the side chain (R group) during the total synthesis of litseaverticillols **180** [150]. The unusual regio and diastereochemistry obtained from the methylene blue (MB) sensitized oxidation in MeOH is, presumably, the result of a backside attack of MeOH on the intermediate endoperoxide.

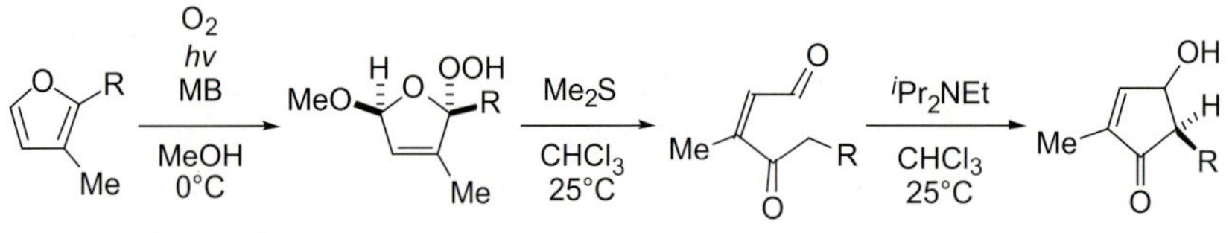

R = geranyl or neryl

Scheme 6.81

Oxidation of furfuryl alcohols under similar conditions leads to ring expansion to form dihydropyranones. These Achmatowicz oxidations are often accomplished by using buffered NBS, $VO(acac)_2$/*t*-BuOH and singlet oxygen. Dimethyldioxirane [139] and methyltrioxorhenium/urea hydrogen peroxide [140] are also effective oxidants. The corresponding aza-Achmatowicz oxidation of furfurylamines [151], frequently by NBS or *m*-CPBA, provides a novel method for the synthesis of azasugars and piperidine-containing compounds [152]. An interesting example of simultaneously applying both reaction variants in the synthesis of aza-C-linked disaccharide, such as **182** from **181**, is shown in Scheme 6.82 [153].

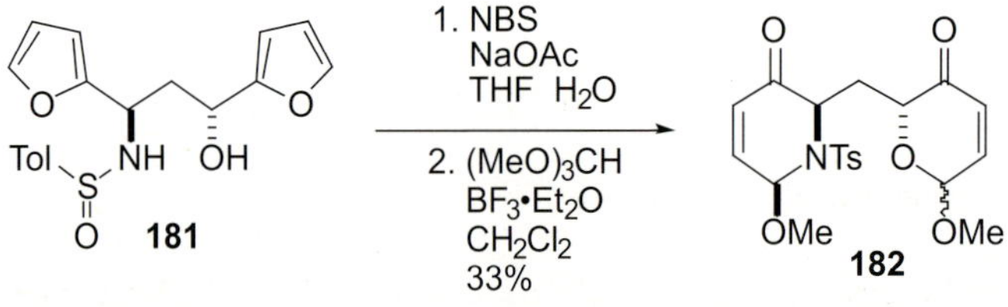

Scheme 6.82

A novel route for the construction of the daphnane BC-ring in the quest for resiniferatoxin (Scheme 6.83) employed furan **183** in an Achmatowicz rearrangement via, presumably, oxidopyrylium ion **184**, which participates in an intramolecular [5 + 2] cycloaddition to provide dihydropyranone **185** [154, 155].

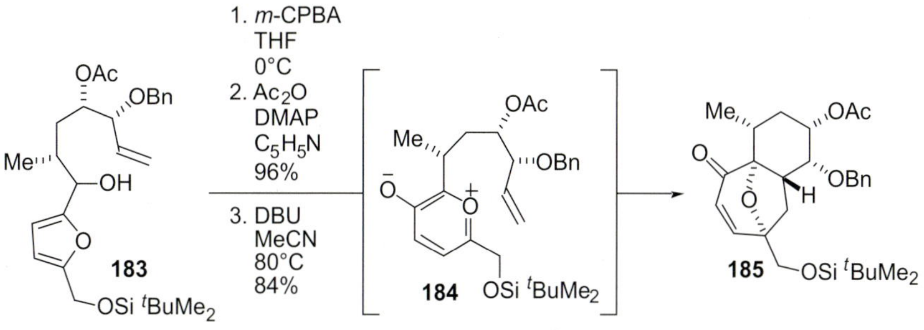

1. *m*-CPBA THF 0°C 2. Ac_2O DMAP C_5H_5N 96% 3. DBU MeCN 80°C 84%

Scheme 6.83

The furan nucleus can be oxidized electrochemically, and the radical cations generated at the anode can be trapped by the reaction medium, for example, methanol, to give 2,5-dioxygenated-2,5-dihydrofurans. Intramolecular electrochemical annulation of furans provides an interesting avenue for the synthesis of polycyclic ring systems. Anodic oxidation of furans that contain pendant vinylsulfide, methyl or silyl enol ethers generates radical cation intermediates that cyclize to form six- and seven-membered rings [156]. Scheme 6.84 shows a recent example of the construction of the complex tetracyclic core **186** of guanacastepenes, obtained as a single diastereoisomer, by such an electron transfer reaction [157]. A *gem*-dialkyl effect has been identified as essential for the efficient formation of a seven-membered ring in this type of reaction [89, 158].

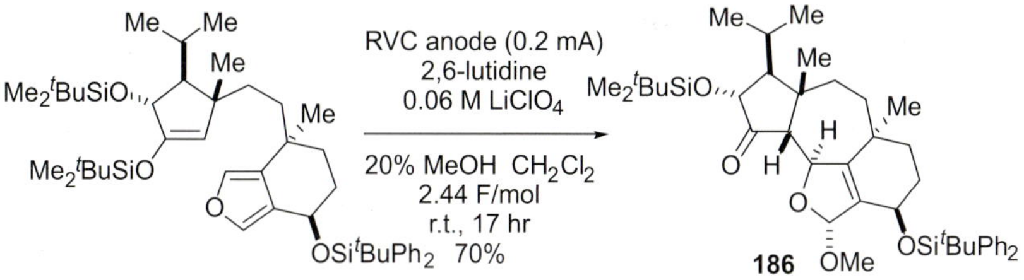

RVC anode (0.2 mA) 2,6-lutidine 0.06 M $LiClO_4$ 20% MeOH CH_2Cl_2 2.44 F/mol r.t., 17 hr 70%

Scheme 6.84

6.3.4
Reactions with Reducing Reagents

Furans are reduced by catalytic hydrogenation (Pd/C, Raney Ni and Rh), dissolving metals (Birch reduction) and alkylsilanes to dihydrofurans and tetrahydrofurans [159]. 2,5-*cis*-Reduction products **187a** and **187b** are obtained from 2,5-disubstituted furans, an example that is essential in the total synthesis of tetranactin (Scheme 6.85) [160].

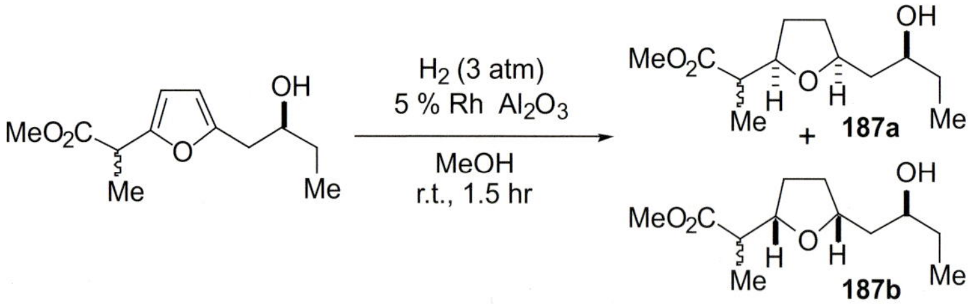

H_2 (3 atm)
5 % Rh Al_2O_3
MeOH
r.t., 1.5 hr

187a + **187b**

Scheme 6.85

In the stereoselective reduction of electron-deficient chiral furoic amides (Scheme 6.16, Section 6.1.2), under Birch-type reductive alkylation to form dihydrofuran derivatives, methyl and trimethylsilyl substituents at the 3-position of the furan moiety are essential for achieving high diastereoselectivity in the alkylation step, presumably by controlling the enolate geometry [161].

6.3.5
Reactions with Acids or Bases

Rearrangement of furfuryl alcohols in aqueous acidic medium at *p*H 4.0–6.0 is a useful reaction for the preparation of hydroxycyclopentenones on a preparative scale (Scheme 6.86). 5-Unsubstituted and 5-methylfurfuryl alcohols having a range of R groups are generally good substrates for this type of reaction. 5-Nitrofurfuryl alcohol, however, does not undergo the rearrangement [137]. As illustrated in Scheme 6.86, this kind of rearrangement provides *trans*-4-hydroxy-5-substituted 2-cyclopentenones, which can isomerize to 2-substituted 2-cyclopentenones **188** under mild basic conditions, for example, using alumina and phosphate buffer *p*H 7.9 [137], as well as amine bases [162]. When R is phenyl or 2-thienyl, a reaction in refluxing water without the use of any acid leads directly to isomer **188** [163].

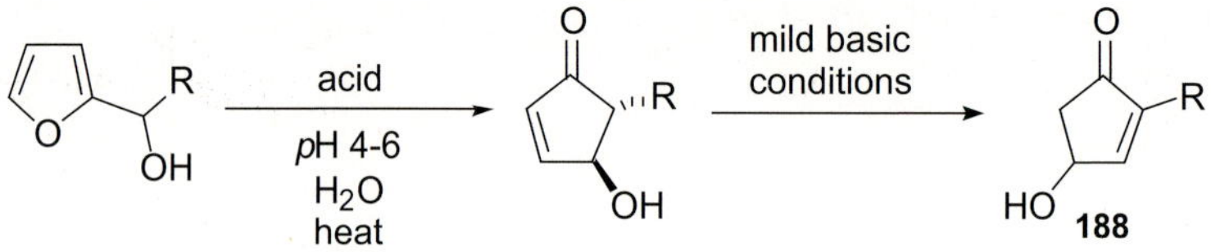

acid
*p*H 4-6
H_2O
heat

mild basic conditions

188

Scheme 6.86

A 2-acylfuran derivative reacts with aqueous ammonia to give 3-hydroxypyridine **189** (Scheme 6.87). An initial attack of ammonia at the C5-position of the furan followed by a subsequent ring opening/ring closing sequence has been proposed as the reaction mechanism [164].

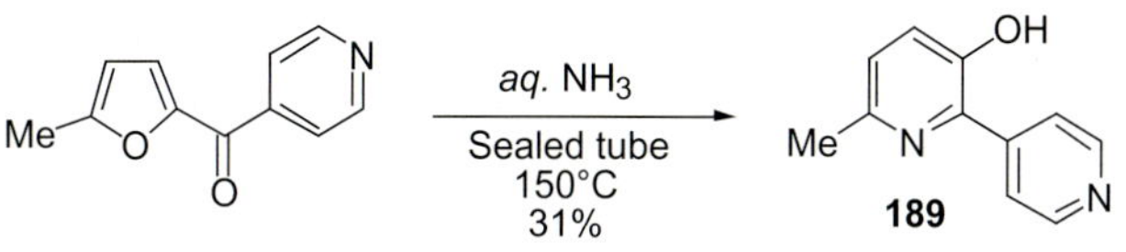

Scheme 6.87

6.3.6
Reactions of C-Metallated Furans

C-Metallated furans are primarily used for the synthesis of substituted furans (Section 6.2.1). 2-Furylcuprate **190**, as shown in Scheme 6.88, was recently discovered to undergo a 1,2-metalate rearrangement to form **191** [165]. A similar dyotropic rearrangement of 2-furylzirconocene complexes has also been reported [166].

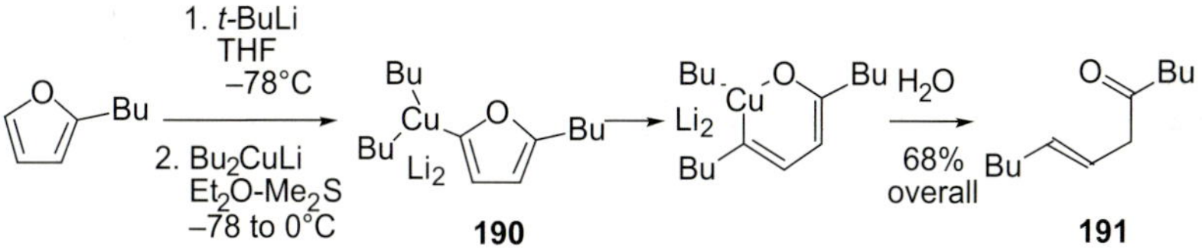

Scheme 6.88

6.3.7
Reactions with Radical Reagents

Furans react with radical reagents at the α-positions. For example, an acetyl radical, generated from a xanthate, reacts with 2-acetylfuran to provide the α-substituted product [167]. As illustrated in Scheme 6.89, addition of the alkenyl radical generated from the cyclohexenyl bromide **192** to the pendant furan moiety gives the spiro-dihydrofuran radical intermediate **193**. Radical fragmentation in **193** then provides a cyclohexyl radical, leading to unsaturated ketone **194** [168]. Similar methodology has been examined in the synthesis of polycyclic ring system [169].

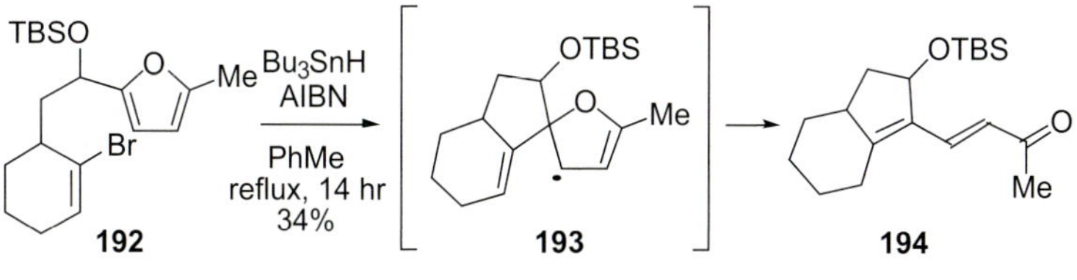

Scheme 6.89

6.3.8 Electrocyclic Reactions

There is a large volume of literature on both inter- and intramolecular Diels–Alder reactions of furans with dienophiles, for example, alkenes, alkynes, allenes and benzynes, under thermal, Lewis acid promoted or high-pressure conditions. These [4 + 2] cycloadditions, which form six-membered rings, have been applied to the total synthesis of natural products, and are well documented and reviewed [4, 170]. Diels–Alder reaction of furans often requires elevated reaction temperature, as furans are generally poor dienes, unless the dienophiles are reactive or Lewis acids are used. However, structural elements can be incorporated into the starting materials such that these cycloadditions proceed at or below ambient temperature. For example, the cycloaddition of **195** which has an unactivated double bond, occurs at room temperature to furnish **196** (Scheme 6.90). This reaction is a consequence of **195** being populated in a reactive conformation imparted by the amide carbonyl of the tether [171].

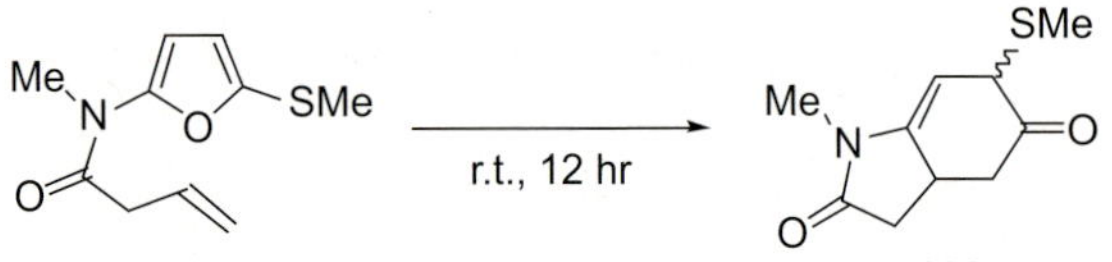

Me N O SMe O → r.t., 12 hr → Me N SMe O O

195 **196**

Scheme 6.90

Examples of transannular Diels–Alder reactions of furanophanes are very rare. In such an approach, macrocyclic conformational control can offer high diastereoselectivity, as demonstrated in the synthesis of the chatancin core **198** from **197** (Scheme 6.91) [172].

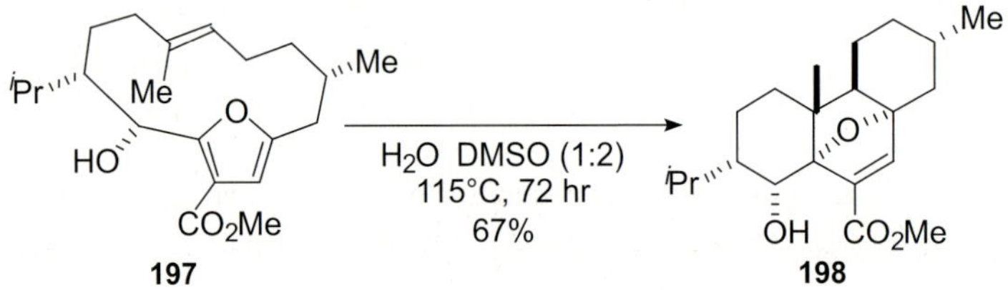

iPr Me Me O HO CO_2Me → H_2O DMSO (1:2) 115°C, 72 hr 67% → Me O iPr OH CO_2Me

197 **198**

Scheme 6.91

The [4 + 3] cycloadditions of furans have been used to form seven-membered rings, which are widely present in natural products. Reactions with oxyallyl cations are the most explored [4 + 3] cycloadditions of furans [173], although reactions with oxyallyl equivalents, for example, silyloxyacroleins [174] and cyclopropanone hemiacetals [175], and aminoallyl cations [176], have also been reported. These cycloadditions of furans with oxyallyl cations and their equivalents are generally considered as concerted processes. However, theoretical calculations support an alternative stepwise mechanism for certain examples [174, 177]. An example of employing such a reaction as the key step in the total synthesis of colchicine is shown in Scheme 6.92 [178]. Thus, coupling of the highly substituted furan **199** with the oxyallyl cation generated from silyl enol ether **200** produces the desired *endo*-adduct **201** as a single isomer.

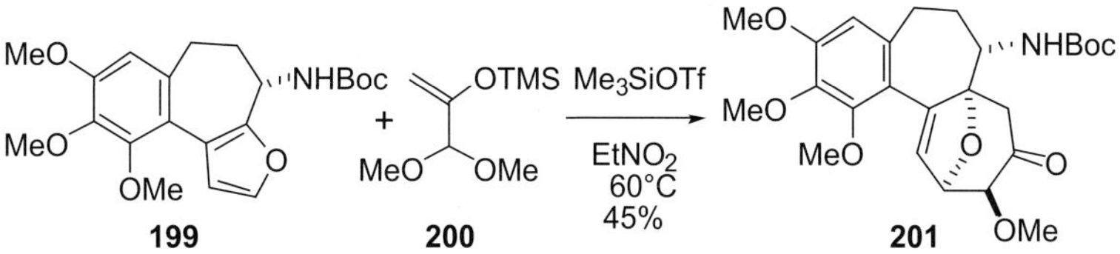

Scheme 6.92

An enantioselective organocatalytic [4 + 3] cycloaddition of furan using the chiral amine catalyst **160** has been realized [179]. As shown in Scheme 6.93, the *endo* selective [4 + 3] cycloaddition between 3-methylfuran (**11**) and the nitrogen stabilized oxyallyl cation **202** derived from an allenamide can also be rendered highly enantioselective by using a 1,2-cyclohexanediamine derived C_2 symmetric salen(Cu) complex as the catalyst, leading to the formation of **203** [180].

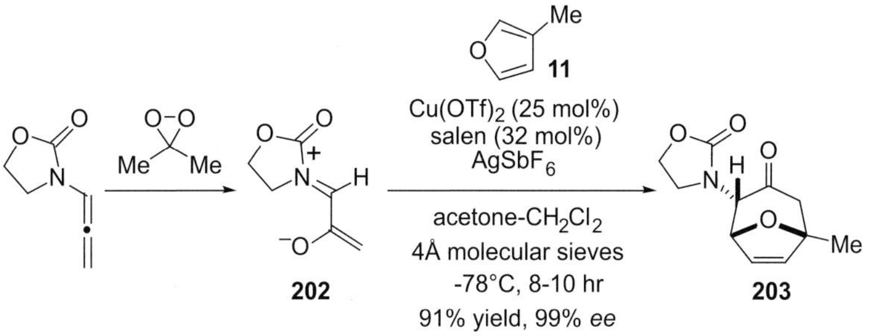

Scheme 6.93

The [6 + 4] cycloaddition between furan and tropone, a previously unsuccessful transformation, has recently been realized in an intramolecular conversion of **204** into **205** during the assembly of the ABC-ring of ingenol (Scheme 6.94) [181].

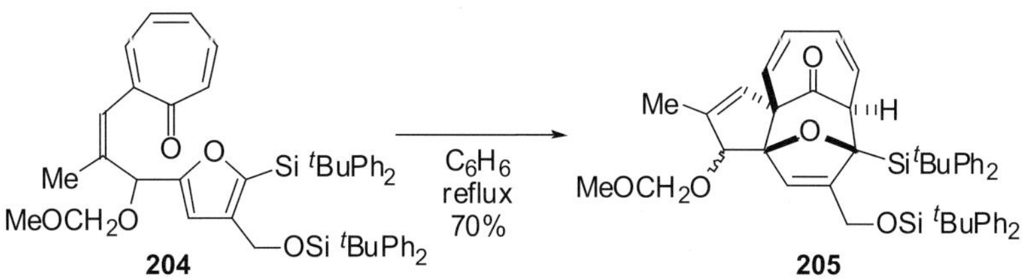

Scheme 6.94

Gold-catalyzed intramolecular cycloisomerization of furans with a pendant terminal alkyne provide phenol product **207** (Scheme 6.95) [182]. Although the reaction may be a result of a [4 + 2] or [2 + 2] process, a viable mechanism has not been identified. The formation of arene oxide intermediate **206**, however, has been characterized experimentally [183].

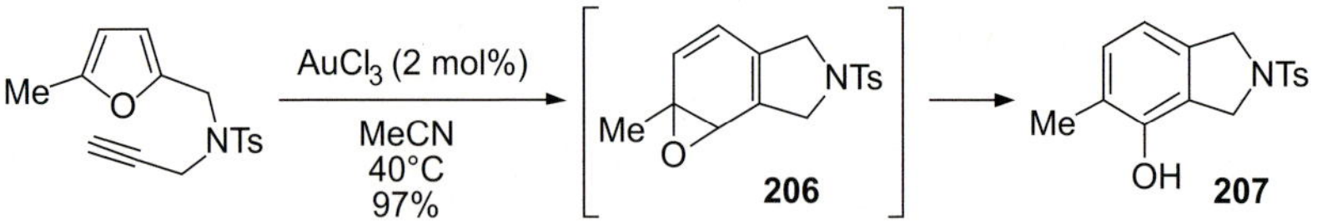

Scheme 6.95

2-Vinylfurans participate in extra annular [4 + 2] cycloadditions in which the vinyl group and the furan 2,3-π bond function as the 4π component to form tetrahydrobenzofurans [184]. 2-Butadienylfuran **208** has recently been shown to behave as an 8π-component in cycloaddition with dimethyl acetylenedicarboxylate, providing an oxygen-bridged ten-membered ring (**209**, Scheme 6.96) [185].

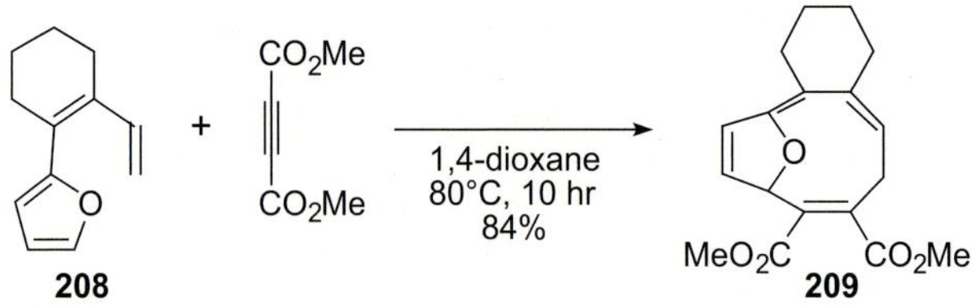

Scheme 6.96

Furans also behave as dienophiles and dipolarophiles. The 2,3-π bonds of furans react with *o*-quinodimethide [186] and *o*-benzoquinones. Scheme 6.97 shows an example of a regio- and diastereoselective cyclization involving (*R*)-furfuryl alcohol **211** and a masked *o*-benzoquinone **210**, producing the *ortho, endo* adduct **212** [187]. The α-hydroxyl group controlled the facial selectivity of the Diels–Alder type reaction. Recent studies, however, suggest a stepwise, double Michael addition as the mechanism for this type of reaction [188].

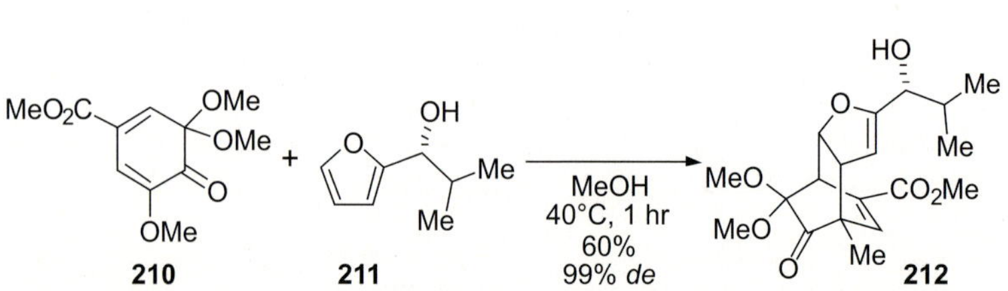

Scheme 6.97

Furans are less reactive dipolarophiles. However, 1,3-dipolar cycloaddition between nitrile oxides and furans to give furoisoxazolines has been developed as a novel entry to amino sugars and amino acids [189]. Recently, intramolecular cycloaddition of a furan with a carbonyl ylide dipole was also shown to proceed under microwave promoted conditions, providing the cycloadduct in modest yield [190].

The furan 2,3 π-bond also undergoes cyclopropanation with metal carbenoids. Asymmetric cyclopropanation of furan-2-carboxylate with ethyl diazoacetate to provide the *exo*-diastereoisomer has been achieved under copper(I)-bisoxazoline catalyzed conditions [191]. A retro-Claisen type rearrangement often accompanies cyclopropanation with diazocarbonyl compounds, leading to a 2,4-diene-1,6-dicarbonyl structural motif. Intramolecular versions are attractive methods for synthesizing [6,7]-, [6,6]-, [6,5]- and even [6,4]-fused ring systems [192]. Scheme 6.98 gives an example of a rhodium-catalyzed reaction as applied in the synthesis of guanacastepene core structure **213** [193].

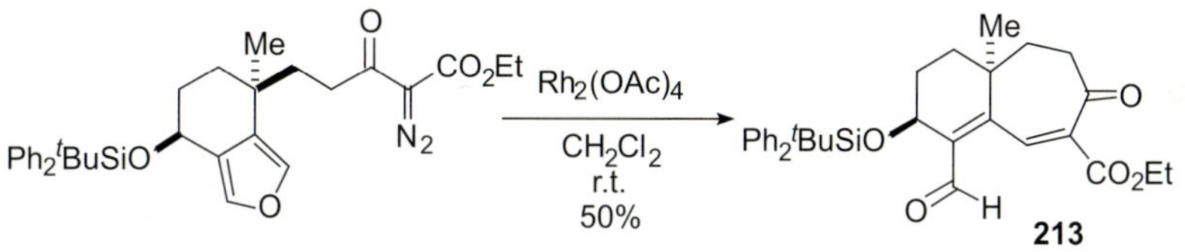

Scheme 6.98

6.3.9
Photochemical Reactions

In addition to furan behaving as a 4π-component in photochemical reactions with aromatic compounds, the furan 2,3-π bond also reacts photochemically. The most synthetically interesting photochemical reaction involving furans is the Paterno–Büchi [2 + 2] cycloaddition with carbonyl compounds. The oxetane products obtained are useful intermediates for the synthesis of natural products. Regio- and stereoselectivities of the reaction are determined by the conformational stability of the triplet diradical intermediate [194]. As illustrated in a study with 2-silyloxyfurans (Scheme 6.99) [195], reaction with ketones provided higher substituted products

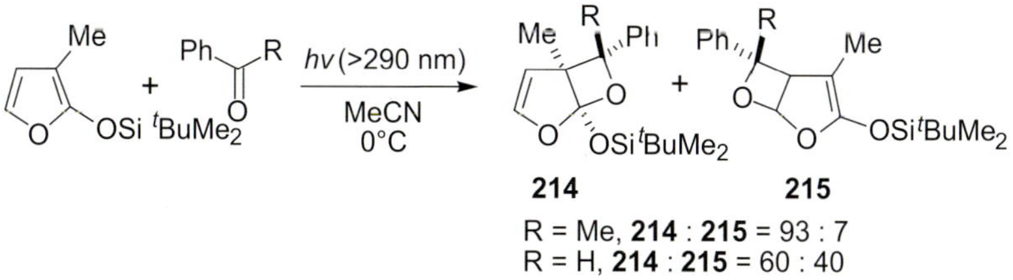

Scheme 6.99

(e.g. **214**) regioselectively, while those with aldehydes are regio-random. As usual, *exo*-oxetanes were produced predominantly in both examples.

The [2 + 2] photocycloaddition of furans with alkenes is also synthetically useful. A remarkable example is the pivotal intramolecular cyclization to form **216** (Scheme 6.100), as employed in the total synthesis of ginkgolide B [196].

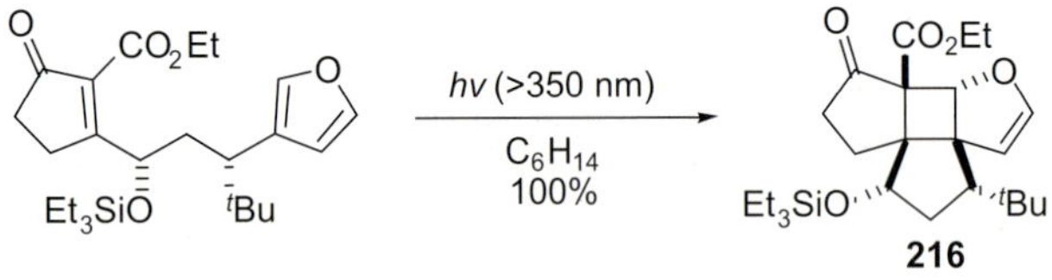

Scheme 6.100

6.4 Oxyfurans and Aminofurans

6.4.1 Oxyfurans

Hydroxyfurans exhibit fairly low stability that is quite different from their benzenoid counterparts. This phenomenon can be explained by a careful structure investigation on 2-furanol (**217**), which can be treated as the enol form of 2- or 3-butenolide. Estimations of resonance energy reveal that 2-furanol is of much higher energy than the corresponding furanone structure. Therefore, 2(3*H*)-furanone (**218**) and 2(5*H*)-furanone (**219**) dominate at equilibrium in the absence of additional stabilizing effect for enolization (Scheme 6.101) [197, 198].

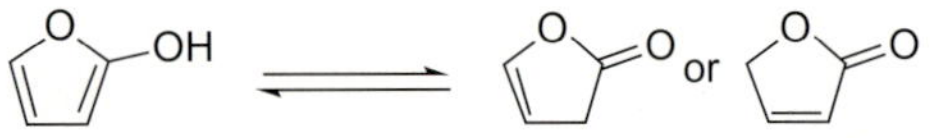

Scheme 6.101

Both the 2(5*H*)-, and 2(3*H*)-furanone structures widely occur in biologically active natural products. They are, in general, also called "butenolides," as derivatives of 4-hydroxybutenoic acids, and not as furan derivatives. Thus, 3-butenolide corresponds to 2(3*H*)-furanone (**218**) and 2-butenolide corresponds to 2(5*H*)-furanone (**219**) [199].

2(5*H*)-Furanones are important synthetic intermediates in the construction of substituted γ-butyrolactones. In addition, they also serve as useful building blocks in the syntheses of various organic compounds [200]. Several excellent reviews dealing with the synthesis of these unsaturated lactones have been published [199, 201–203]. Instead of providing a thorough literature survey, only recent advances are discussed here.

Palladium-catalyzed cyclocarbonylation of propargyl alcohol is an excellent method for the construction of 2(5*H*)-furanones. Thus, a combination of $Pd(dba)_2$ and dppb has been used to catalyze the transformation of **220** into 5,5-disubstituted 2(5*H*)-furanone **221**. The reaction mechanism involves the insertion of a Pd(0) species into the C–O bond of the substrate, followed by rearrangement to the allenylpalladium intermediate. Insertion of CO and subsequent reductive elimination then lead to the 2,3-dienoic acid, which then affords **221** (Scheme 6.102) [204].

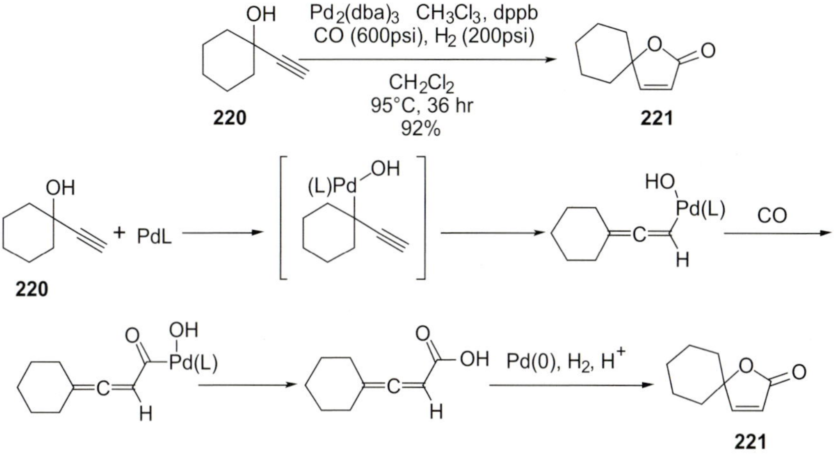

OH Pd$_2$(dba)$_3$ CH$_3$Cl$_3$, dppb CO (600psi), H$_2$ (200psi) CH$_2$Cl$_2$ 95°C, 36 hr 92% **220** **221**

OH + PdL (L)Pd OH HO Pd(L) C H CO **220**

O OH Pd(L) C H O OH Pd(0), H$_2$, H$^+$ C H **221**

Scheme 6.102

A highly functionalized 2(5*H*)-furanone (**222**) has been prepared from a triphenylphosphine-catalyzed reaction of activated carbonyl compound with dimethyl acetylenedicarboxylate (Scheme 6.103) [205].

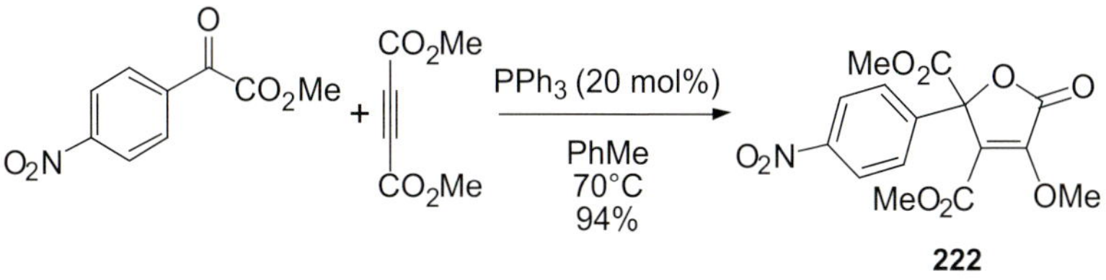

O CO$_2$Me O$_2$N + CO$_2$Me CO$_2$Me PPh$_3$ (20 mol%) PhMe 70°C 94% MeO$_2$C O O O$_2$N MeO$_2$C OMe **222**

Scheme 6.103

Besides acyclic substrates, 2(5*H*)-furanones can also be synthesized from cyclobutenone precursors. Thus, when hydroxycyclobutenone **223** was subjected to reaction with a hypervalent iodine reagent, an oxidative ring enlargement took place to give 5-acetoxy-2(5*H*)-furanone **224** (Scheme 6.104) [206].

The presence of an α,β-unsaturated carbonyl moiety renders the 2(5*H*)-furanones highly reactive towards different reagents (Section 6.3).

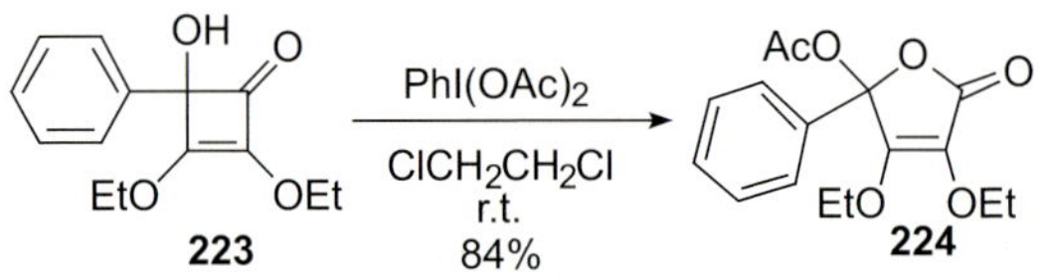

Scheme 6.104

Attempts to synthesize simple 2(3*H*)-furanones usually result in a mixture containing also 2(5*H*)-furanones [207]. Generally, 2(3*H*)-furanone (**218**) is thermodynamically less stable than 2(5*H*)-furanone (**219**). Computational (SCF-MO) results reveal that the energy of 2(3*H*)-furanone (**218**) is of 53 kJ mol^{-1} higher than that of 2(5*H*)-furanone (**219**) [198]. Experimentally, isomerization can be achieved by amine bases or at elevated temperature (Scheme 6.105) [208].

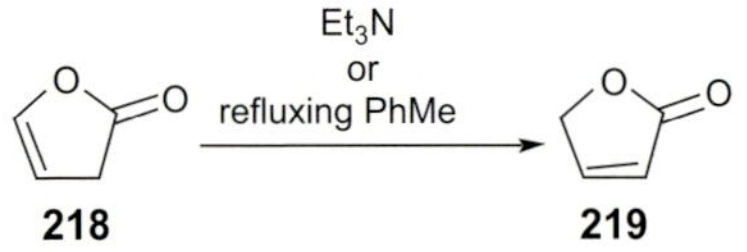

Scheme 6.105

The reactively low stability of 2(3*H*)-furanones makes them susceptible to nucleophilic attack to give ring-opening products, which may be employed to construct other important heterocyclic systems [209].

As illustrated in Scheme 6.106, a hydrazide **226** is formed on treatment of furanone **225** with hydrazine hydrate. Further reaction with a mixture of $NaNO_2$ and HCl promotes ring closure to afford **227** with a pyridazinone structure [210].

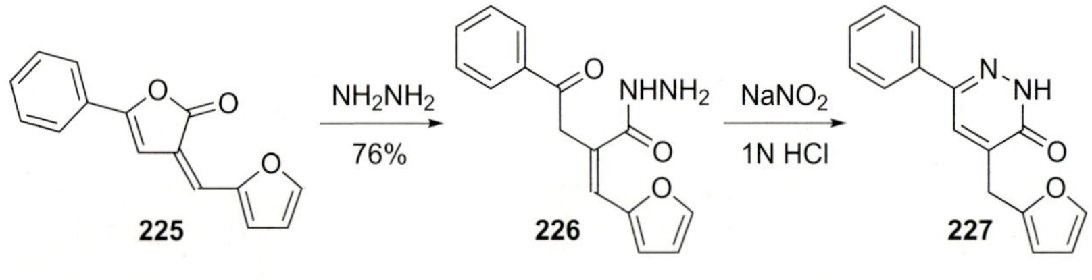

Scheme 6.106

Similar to their 2-hydroxyl isomers, 3-hydroxyfurans tend to exist in the keto form as 3(2*H*)-furanones [198]. As depicted in Scheme 6.107, free 3(2*H*)-furanone **228** can be synthesized from 3-bromofuran (**51**) [211].

Although the 3(2*H*)-furanone motif is less common than 2(5*H*)-furanones and 2(3*H*)-furanones in bioactive naturally occurring molecules, a series of 4,5-diaryl substituted 3(2*H*)-furanones (**229–233**) have been studied as cyclooxygenase-2 inhibitors and show excellent anti-inflammatory activities [212].

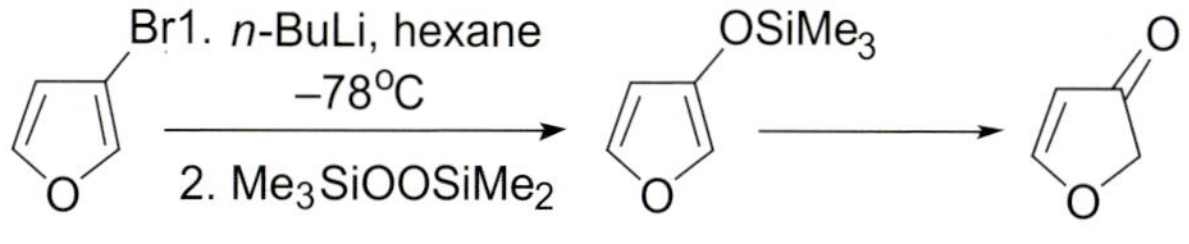

51 228

Scheme 6.107

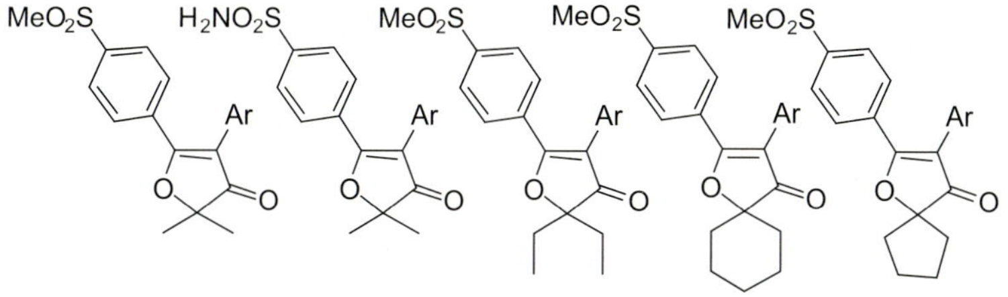

229 230 231 232 233

6.4.2
Aminofurans

3-Aminofuran is stable only in an inert atmosphere or *in vacuo*. It polymerizes rapidly upon contact with air. 2-Aminofuran is also a reactive species. However, as illustrated in Scheme 6.108, **234** is one of the exceptions, whose stability may be attributed to the presence of a conjugated electron-withdrawing group on the furan ring. When **234** is treated with a dienophile, the rearranged cycloadduct **235** is obtained in high yield [213].

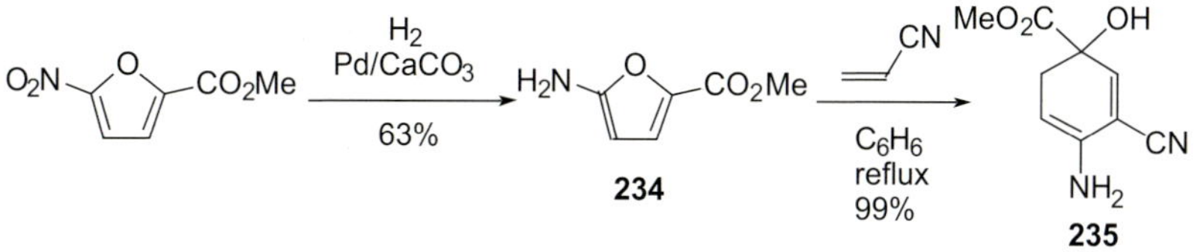

234 235

Scheme 6.108

6.5
Addendum

6.5.1
Additional Syntheses of Furans

In the past few years, several procedures for the synthesis of substituted furans in high regioselectivity and efficiency have appeared. Amongst them, reports on the using of Au-catalysts have increased substantially. In all cases the starting materials contain an alkynyl or allenyl group. Because of the exceptionally alkynophilic, but not as oxophilic, property of Au-catalysts, the Au-catalyzed reactions are oxygen-, water-, and alcohols-tolerated and thus do not need air- and moisture-free conditions. Also noteworthy is that in Au-catalyzed reactions non-classical carbocation or carbenoid intermediates are very often involved so that the selectivity of these reactions can be controlled. A few examples are shown below. Cyclization of allenones in the presence of Au(III)-porphyrin gave rise to the corresponding substituted furan in good to high yields. The catalyst can be recycled several times and its catalytic activity remained intact [Scheme 6.109, reaction (1)] [214a]. Another example of Au catalysis has been

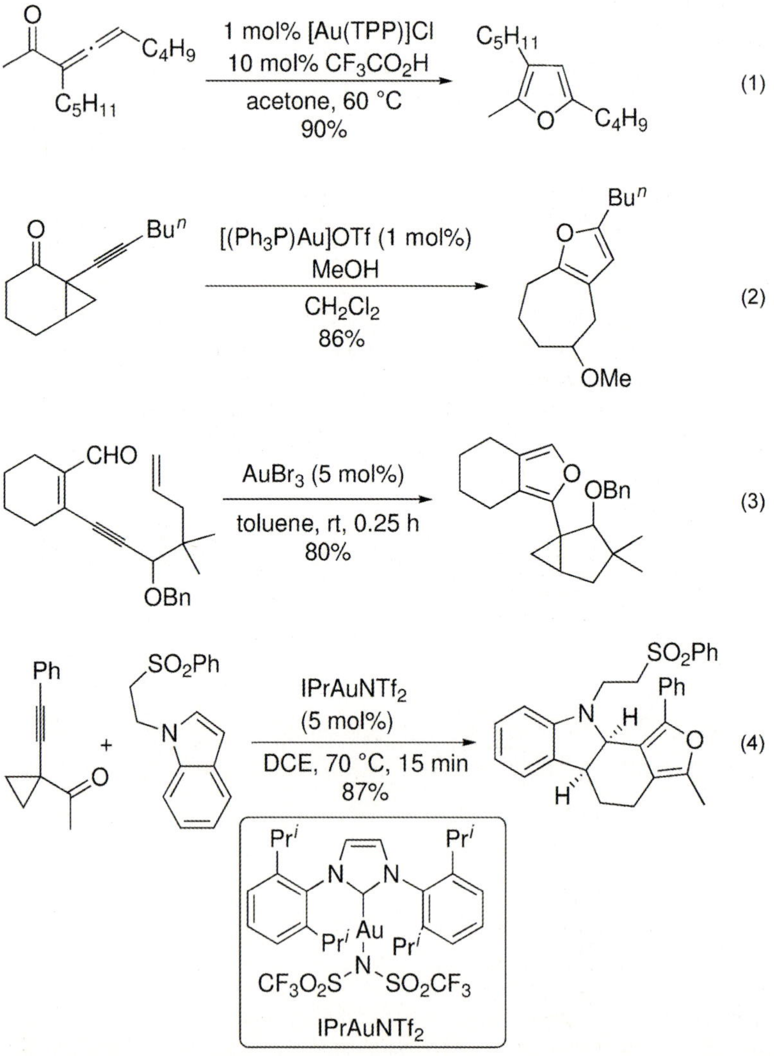

(1) 1 mol% [Au(TPP)]Cl, 10 mol% CF_3CO_2H, acetone, 60 °C, 90%

(2) [(Ph_3P)Au]OTf (1 mol%), MeOH, CH_2Cl_2, 86%

(3) $AuBr_3$ (5 mol%), toluene, rt, 0.25 h, 80%

(4) $IPrAuNTf_2$ (5 mol%), DCE, 70 °C, 15 min, 87%

$IPrAuNTf_2$

Scheme 6.109

reported using alkynyl cyclopropyl ketones as a starting material. Trisubstituted furans were afforded in high yields under mild conditions via a domino reaction process [Scheme 6.109, reaction (2)] [214b]. A carbonyl-ene-yne compound is also a suitable starting material in Au-catalyzed furan formation. In the presence of Au-catalyst, 2-alkynyl-1-cycloalkenecarbaldehydes were converted into trisubstituted furans via an Au-carbene intermediate [Scheme 6.109, reaction (3)] [214c]. Alkynyl cyclopropyl ketones have also been employed as starting material, as 1,4-dipoles in an Au-catalyzed [4 + 2] annulation reaction, providing fully substituted furans in high yields [Scheme 6.109, reaction (4)] [214d].

In addition to Au-catalysis, transition metals were still used widely as efficient catalysts in the synthesis of furans with full control of regioselectivity. Employing ruthenium complexes as catalysts, trisubstituted furans have been provided through an unprecedented 1,4-shift of the sulfanyl group of allenyl sulfides in high yields. Furan products have also been afforded in a one-pot reaction from α-diazocarbonyls and propargyl sulfide using both Rh- and Ru-complexes or only Ru-complexes as catalysts [Scheme 6.110, reaction (1)] [215a]. The second reaction in Scheme 6.110 provides another example of furan synthesis using metal catalysts. In the presence of $In(OTf)_3$, terminal disubstituted allenyl ketones were smoothly converted into

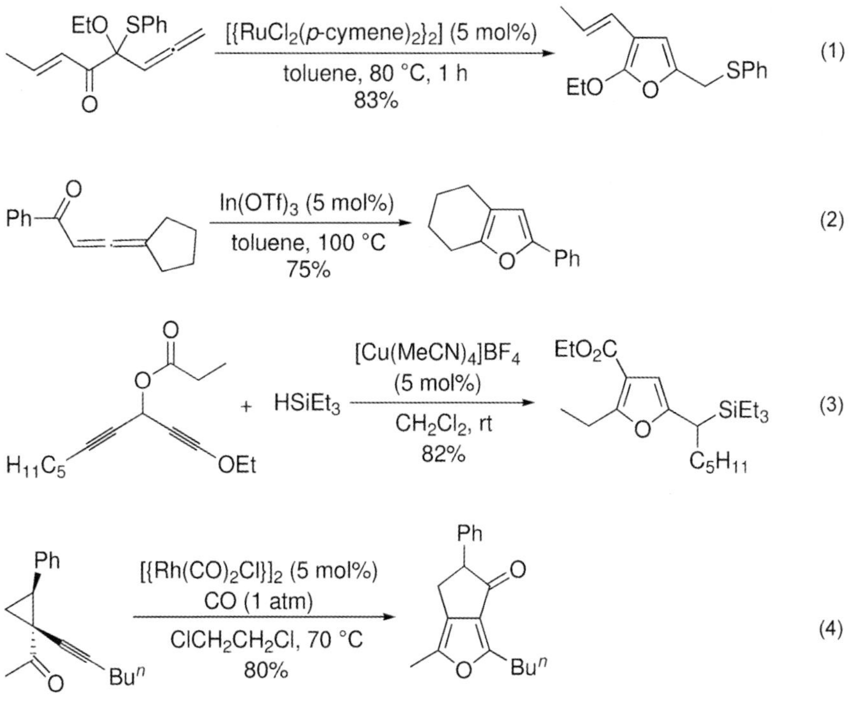

EtO SPh — [{$RuCl_2$(p-cymene)$_2$}$_2$] (5 mol%), toluene, 80 °C, 1 h, 83% → EtO, O, SPh (1)

O, Ph — $In(OTf)_3$ (5 mol%), toluene, 100 °C, 75% → O, Ph (2)

O, O, $H_{11}C_5$, OEt + $HSiEt_3$ — [$Cu(MeCN)_4$]BF_4 (5 mol%), CH_2Cl_2, rt, 82% → EtO_2C, O, $SiEt_3$, C_5H_{11} (3)

Ph, O, Bu^n — [{Rh(CO)$_2$Cl}]$_2$ (5 mol%), CO (1 atm), $ClCH_2CH_2Cl$, 70 °C, 80% → Ph, O, O, Bu^n (4)

Scheme 6.110

tri- and tetra-substituted furans in high yields. The 1,2-shift of the terminal alkyl group was a key step in this reaction [215b]. In addition to In salt, some other Lewis acids, such as $Sn(OTf)_2$, AgOTf, and [Au(PPh3)]OTf, could also be used in similar reactions. Control of regioselectivity in the synthesis of multi-substituted furans has been realized by using Cu-catalyst in the reaction of *bis*-propargylic ester via Cu-carbene intermediate [Scheme 6.110, reaction (3)]. The reaction is suitable for the preparation of tetra-substituted furans, and silane is not necessary when CuBr or CuCl is the catalyst [215c]. 1-(1-Alkynyl)-cyclopropyl ketones have proved useful as a building block in the synthesis of multi-substituted furans by using an Au-catalyst [215b,c]. They are also useful in transition-metal catalyzed furan-formation reactions. In the presence of Rh-catalyst, carbonylation takes place and thus the procedure was developed for the synthesis of tetra-substituted furans in high regioselectivity via a Rh-catalyzed carbonylation/cyclization [Scheme 6.110, reaction (4)] [215d].

A domino-reaction is a powerful strategy that has been adopted by many groups to synthesize furans. Some examples are shown below (Scheme 6.111). Thermally induced cascade cyclizations under metal-free conditions using epoxyhexene as acid scavenger provided polycyclic furans in high yields [Scheme 6.111, reaction (1)] [216a]. Another Pd-catalyzed cascade reaction using conjugated enynals as starting materials has been reported. Here, good to high yields of 2,3,4-trisubstituted furans were realized [Scheme 6.111, reaction (2)] [216b]. Tetrasubstituted furans have been obtained in the three-component Michael addition–cyclization–cross coupling reaction by using a Pd-complex as catalyst. [Scheme 6.111, reaction (3)] [216c].

Scheme 6.111

6.5.2
Additional Reactions of Furans

Noteworthy and interesting transformations of furan nucleus that are related to the types of furan reactions as described in Section 6.3 have been reported between 2006 and 2009.

The reaction of a furan tethered at the 2-position to an iminium ion gave a spiro-2,5-dihydrofuran derivative as the sole diastereoisomer. This spirocyclization has been used in forming the ABC tricyclic core of manzamine A [217]. Several new catalytic asymmetric addition reactions of silyloxyfurans to electrophiles using various chiral catalysts have been developed and reviewed in detail [218]. The addition of 2-trimethylsilyloxyfuran to Morita–Baylis–Hillman acetates to form γ-butenolides has been rendered enantioselective by using a chiral phosphine catalyst [219].

As exemplified in Scheme 6.112, regioselective addition of 2-methoxyfuran or 2-trimethylsilyloxyfuran to chromium(0) alkynylcarbene complexes furnished interesting dienyne and dienediyne carboxylates by proceeding through a formal vinylogous Michael intermediate [220]. 2-Methoxyfuran reacted with chiral tungsten (0) alkenylcarbene complexes in a similar fashion [221].

MeO $=Cr(CO)_5$ Me_3Si + OMe THF 0 °C to r.t. 89% MeO Me_3Si CO_2Me

Scheme 6.112

Scheme 6.113 shows a reaction of 2-furaldehyde with a secondary amine nucleophile in a lanthanide-catalyzed condensation/ring-opening/electrocyclization process to provide *trans*-4,5-diaminocyclopenten-2-ones, presumably via a ring-opened, deprotonated Stenhouse salt [222].

H O + HN O $Dy(OTf)_3$ (10 mol%) 4Å m.s. MeCN r.t., 18 h quantitative

Scheme 6.113

The opposite regioselectivity was obtained in the photooxidation of 3-bromofuran to bromo-γ-hydroxybutenolides by using DBU and phosphaxene [223]. The regioselectivity provided by the commonly used Hünig's base was reversed by using n-Bu_4NF in the photooxidation of unprotected furan Baylis–Hillman adducts to α-substituted γ-butenolides [224].

2-Furanyl carbamates undergo iodine-promoted oxidative rearrangement to form 5-methoxypyrrol-2(5*H*)-ones, which have been used as intermediates for the synthesis of 2,4-disubstituted pyrroles [225]. Enantiomeric enriched dihydropyranones can be obtained from Achmatowicz oxidation of furfuryl alcohols under Sharpless kinetic resolution conditions. This approach was adopted in a recent total synthesis of the acetogenin pyranicin [226]. As illustrated in Scheme 6.114, a "homologous" Achmatowicz oxidation of 2,5-disubstituted, 2-(β-hydroxyalkyl)furans by singlet oxygen produced 3-keto-tetrahydrofurans, presumably *via* a Michael addition to an intermediate 1,4-enedione [227].

1. O_2, *hv*, Rose bengal, MeOH, 2 min; 2. DMS, CH_2Cl_2, 18 h, then *p*-TsOH, 85%

Scheme 6.114

As depicted in Scheme 6.115, electrochemical oxidation of furans tethered to silyl enol ethers at the 2-position leads to a spiroannulation product as a result of the higher nucleophilicity of the furan 2-position [228].

carbon anode, $LiClO_4$, 2,6-lutidine, MeOH–^{i}PrOH, 85%, *d.r.* = 3 : 1

Scheme 6.115

An enantioselective hydrogenation has been developed in which 2-substituted furans using chiral iridium/pyridine-phosphinite complexes as catalysts provide tetrahydrofurans with ee up to 93% [229]. Another interesting example is the Rh/butiphane-catalyzed enantioselective *cis*-hydrogenation of a furylnucleoside to form the reduced product with 72% ee [230].

The presence of a halogen or methoxy substituent at the 2-position of furan enhances the rate of intramolecular Diels–Alder reactions and the yield of cycloadduct. This phenomenon is attributed to the decreased activation energy and a greater stabilization of the cycloadduct imparted by the substitution as determined by CBS-QB3 calculations. The same substitution at the 3-position can also provide a similar effect although to a lesser extent than that of the 2-position [231]. In the example shown in Scheme 6.116, the Diels–Alder reaction occurred mainly at the 3-chloro-

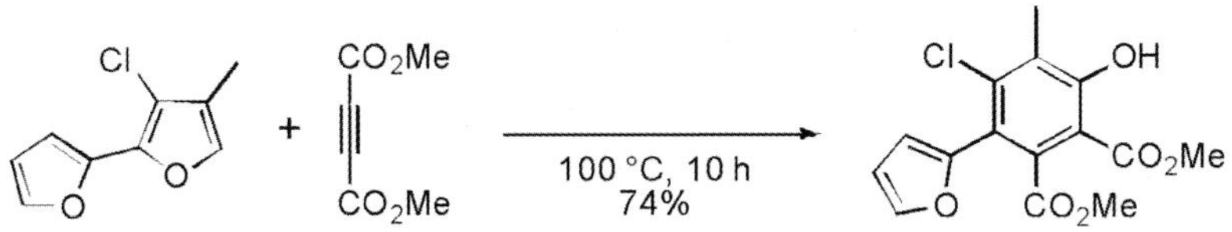

Scheme 6.116

furan ring, which suggests a dominant effect of a 3-halo substituent on intermolecular cycloaddition [232].

The [4 + 3] cycloaddition of furans with epoxy enol silane derived oxyallyl cations (Scheme 6.117) has been rendered a viable process by optimization of the reaction conditions and use of a bulky triethylsilyl group [233]. Dioxines were used as oxyallyl cation equivalents in a Au/Ag-catalyzed [4 + 3] cycloaddition with furan [234].

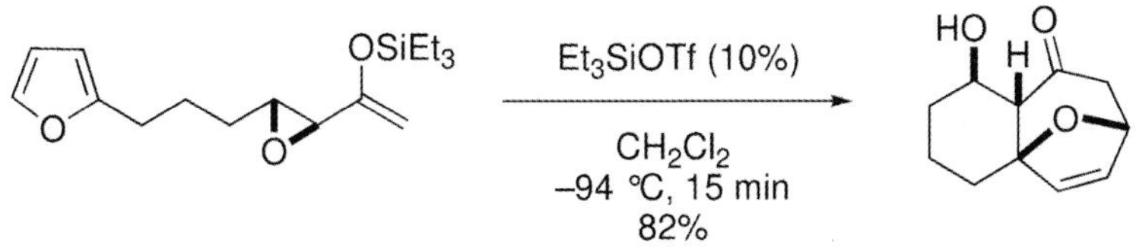

Scheme 6.117

A novel Nazarov cyclization of silyloxyfuran as catalyzed by a strong Lewis acidic iridium complex (Scheme 6.118) was the pivotal step in a total synthesis of the sesquiterpene merrilactone A [235].

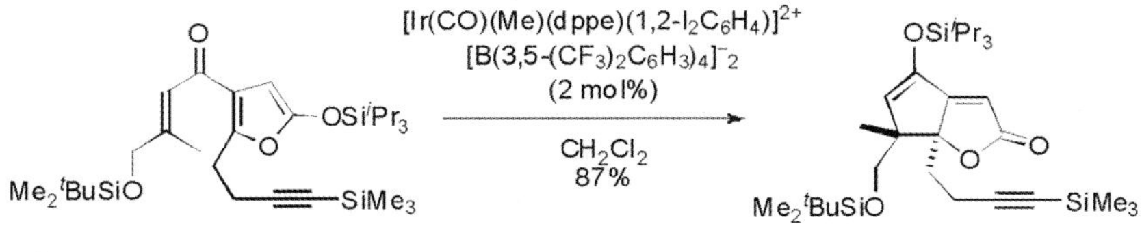

Scheme 6.118

Besides reacting with rhodium carbenoids, furans also react regioselectively with ruthenium and platinum carbenoids derived from tertiary propargyl carboxylates [236] and *sec*-O-propargyl thiocarbamates [237], leading to interesting triene systems.

References

1 Joule, J.A. and Mills, K. (2000) *Heterocyclic Chemistry*, 4th edn, Blackwell Science, Oxford, pp. 296–318.

2 Lichtenthaler, F.W., Cuny, E., Martin, D., Rönninger, S., and Weber, T. (1991) in *Carbohydrates as Organic Raw Materials*

(ed. F.W. Lichtenthaler), VCH, Weinheim, pp. 207–246.

3 Daub, J., Rapp, K.M., Salbeck, J., and Schöberl, U. (1991) in *Carbohydrates as Organic Raw Materials* (ed. F.W. Lichtenthaler), VCH, Weinheim, pp. 323–350.

4 Dunlop, A.P. and Peters, F.N. (1953) *The Furan*, Reinhold, New York. Bosshard, P. and Eugster, C.H. (1966) *Advances in Heterocyclic Chemistry*, **7**, 377–490. Gschwend, H.W. and Rodriguez, H.R. (1979) *Organic Reactions*, **26**, 1–360; Dean, F.M. (1982) *Advances in Heterocyclic Chemistry*, **30**, 167–238. Dean, F.M. (1982) *Advances in Heterocyclic Chemistry*, **31**, 237–344; Sargent, M.V. and Dean, F.M. (1984) in *Comprehensive Heterocyclic Chemistry, Vol 3* (eds C.W. Bird and G.W.H. Cheeseman), Oxford, Pergamon, pp. 599–656. Donnelly, D.M.X. and Meegan, M.J. (1984) in *Comprehensive Heterocyclic Chemistry, Vol. 4* (eds Bird, C.W. and Cheeseman, G.W.H.), Pergamon, Oxford, pp. 657–712.Lipshutz, B.H. (1986) *Chemical Reviews*, **86**, 795–819; Hou, X.L., Cheung, H.Y., Hon, T.Y., Kwan, P.L., Lo, T.H., Tong, S.Y., and Wong, H.N.C. (1998) *Tetrahedron*, **54**, 1955–2020; Keay, B.A. (1999) *Chemical Society Reviews*, **28**, 209–215; Wright, D.L. (2005) in *Progress in Heterocyclic Chemistry, Vol. 17* (eds G.W. Gribble and J.A. Joule), Elsevier, Amsterdam, Chapter 1, pp. 1–32.

5 McNaught, A.D. (1976) *Advances in Heterocyclic Chemistry*, **20**, 175–319.

6 Eicher, T. and Hauptmann, S. (1995) *The Chemistry of Heterocycles*, Thieme, Stuttgart, pp. 52–62, translated by H. Suschitzky and J. Suschitzky.

7 Simkin, B.Y., Minkin, V.I., and Glukhovtsev, M.N. (1993) *Advances in Heterocyclic Chemistry*, **56**, 303–428.

8 Pfleiderer, W. (1963) in *Physical Methods in Heterocyclic Chemistry, Vol. 1* (ed. A.R. Katritzky), Academic Press, New York, pp. 177–188. Katritzky, A.R. and Pozharskii, A.F. (2000) *Handbook of Heterocyclic Chemistry*, 2nd edn, Pergamon Press, Amsterdam.

9 Domínguez, C., Csáky, A.G., and Plumet, J. (1992) *Tetrahedron*, **48**, 149–158.

10 Bock, I., Bornowski, H., Rauft, A., and Theis, H. (1990) *Tetrahedron*, 1199–1210.

11 Song, Z.Z., Ho, M.S., and Wong, H.N.C. (1994) *The Journal of Organic Chemistry*, **59**, 3917–3926.

12 Nasipuri, D. and Das, G. (1979) *Journal of the Chemical Society, Perkin Transactions 1*, 2776–2778.

13 Chastrette, M. and Chastrette, F. (1973) *Journal of the Chemical Society, Chemical Communications*, 534–535.

14 Kutney, J.P., Hanssen, H.W., and Nair, G.V. (1971) *Tetrahedron*, **27**, 3323–3330.

15 Gilman, H. and Breuer, F. (1934) *Journal of the American Chemical Society*, **56**, 1123–1127.

16 Lee, H.K. and Wong, H.N.C. (2002) *Chemical Communications*, 2114–2115.

17 Al-Busafi, S. and Whitehead, R.C. (2000) *Tetrahedron Letters*, **41**, 3467–3470.

18 Diels, O. and Alder, K. (1928) *Annalen Der Chemie-Justus Liebig*, **460**, 98–122.Diels, O., Alder, K., and Naujoks, E. (1929) *Chemische Berichte*, **62B**, 554–562.

19 Boltulchina, E.V., Zubkov, F.I., Nikitina, E.V., and Varlamov, A.V. (2005) *Synthesis*, 1859–1875.

20 D'Auria, M., Racioppi, R., and Romaniello, G. (2000) *European Journal of Organic Chemistry*, 3265–3272.

21 Donohoe, T.J., Guillermin, J.-B., Frampton, C., and Walter, D.S. (2000) *Chemical Communications*, 465–466.

22 Mata, F., Martin, M.C., and Sørensen, G.O. (1978) *Journal of Molecular Structure*, **48**, 157–163.

23 Hudson, P. (1962) *Acta Crystallographica*, **15**, 919–920.

24 Williams, D.E. and Rundle, R.E. (1964) *Journal of the American Chemical Society*, **86**, 1660–1666.

25 Armarego, W.L.F. (1971) in *Physical Methods in Heterocyclic Chemistry, Vol. 3* (ed. A.R. Katritzky), Academic Press, New York, pp. 67–222. Bowden, K., Braude, E.A., and Jones, E.R.H. (1946) *Journal of the Chemical Society*, 948–952.Ott, D.G., Hayes, F.N., Hansbury, E., and Kerr, V.N. (1957) *Journal of the American Chemical Society*, **79**, 5448–5454. Grigg, R., Knight, J.A.,

and Sargent, M.V. (1966) *Journal of the Chemical Society (C)*, 976–981. Horváth, G. and Kiss, Á.I. (1967) *Spectrochimica Acta Part A-Molecular and Biomolecular Spectroscopy*, **23**, 921–924.

26 Batterham, T.J. (1973) *NMR Spectra of Simple Heterocycles*, John Wiley & Sons, Inc., New York, pp. 370–382.Jackman, L.M. and Sternhell, S. (1969) *Applications of Nuclear Magnetic Resonance Spectroscopy in Organic Chemistry*, 2nd edn, Pergamon Press, Oxford, pp. 201–214.White, R.F.M. (1963) in *Physical Methods in Heterocyclic Chemistry, Vol. 2* (ed. A.R. Katritzky), Academic Press, New York, pp. 103–159.Gronowitz, S., Sorlin, G., Gestblom, B., and Hoffman, R.A. (1962) *Arkiv för Kemi*, **19**, 483–497. Pascal, Y., Morizur, J.P., and Wiemann, J. (1965) *Bulletin de la Société chimique de France*, 2211–2219.

27 Read, J.M., Jr, Mathis, C.T., and Goldstein, J.H. (1965) *Spectrochim Acta*, **21**, 85–93.

28 Silverstein, R.M., Webster, F.X., and Kiemle, D.J. (2005) *Spectrometric Idenification of Organic Compounds*, 7th edn, Ch 4, John Wiley & Sons Inc., New York, pp. 204–244. White, R.F.M. and Williams, H. (1971) in *Physical Methods in Heterocyclic Chemistry, Vol. 4* (ed. A.R. Katritzky), Academic Press, New York, pp. 121–235; Reddy, G.S. and Goldstein, J.H. (1962) *Journal of the American Chemical Society*, **84**, 583–585. Page, T.F., Jr, Alger, T., and Grant, D.M. (1965) *Journal of the American Chemical Society*, **87**, 5333–5339; Weigert, F.J. and Roberts, J.D. (1968) *Journal of the American Chemical Society*, **90**, 3543–3549.

29 Spiteller, G. (1971) in *Physical Methods in Heterocyclic Chemistry, Vol. 3* (ed. A.R. Katritzky), Academic Press, New York, pp. 223–296; Collin, J. (1960) *Bulletin des Sociétés Chimiques Belges*, **69**, 449–465.

30 Turner, D.W., Baker, A., Baker, A.D., and Brundle, C.R. (1970) *Molecular Photoelectron Spectroscopy*, John Wiley & Sons Inc., New York, pp. 329.Eland, J.H.D. (1969) *International Journal of Mass Spectrometry and Ion Physics*, **2**, 471–484.

31 Distefano, G., Pignataro, S., Innorta, G., Fringuelii, F., Marino, G., and Taticchi, A. (1973) *Chemical Physics Letters*, **22**, 132–136.

32 Sell, J.A. and Kuppermann, A. (1979) *Chemical Physics Letters*, **61**, 355–362.

33 Fabian, J., Mehlhorn, A., and Zahradník, R. (1968) *Theoretica Chimica Acta*, **12**, 247–255.

34 Julg, A. and Sabbah, R. (1977) *Comptes Rendus de l'Academie des Sciences Paris C*, **285**, 421–424.

35 Bingham, R.C., Dewar, M.J.S., and Lo, D.H. (1975) *Journal of the American Chemical Society*, **97**, 1302–1306.

36 Dewar, M.J.S. and Ford, G.P. (1977) *Journal of the American Chemical Society*, **99**, 1685–1691.

37 Rico, M., Barrachina, M., and Orza, J.M. (1967) *Journal of Molecular Spectroscopy*, **24**, 133–148.

38 Palmer, M.H., Findlay, R.H., and Gaskell, A.J. (1974) *Journal of the Chemical Society, Perkin Transactions 2*, 420–428.

39 Nakatsuji, H., Kitao, O., and Yonezawa, T. (1985) *Journal of Chemical Physics*, **83**, 723–734.

40 Serrano-Andrés, L., Merchán, M., Nebot-Gil, I., Roos, B.O., and Fülscher, M. (1993) *Journal of the American Chemical Society*, **115**, 6184–6197.

41 Cooper, D.L. and Wright, S.C. (1989) *Journal of the Chemical Society, Perkin Transactions 2*, 263–267.

42 Phuwapraisirisan, P., Matsunaga, S., van Soest, R.W.M., and Fusetani, N. (2004) *Tetrahedron Letters*, **45**, 2125–2128.

43 Wu, J., Zhang, S., Xiao, Q., Li, Q.-X., Huang, J.-S., Long, L.-J., and Huang, L.-M. (2004) *Tetrahedron Letters*, **45**, 591–593.

44 Gutiérrez, M., Capson, T.L., Guzmán, H.M., González, J., Ortega-Barría, E., Quiño,á E., and Riguera, R. (2005) *Journal of Natural Products*, **68**, 614–616.

45 Reddy, N.S., Reed, J.K., Longley, R.E., and Wright, A.E. (2005) *Journal of Natural Products*, **68**, 248–250.

46 Hanai, R., Gong, X., Tori, M., Kondo, S., Otose, K., Okamoto, Y., Nishihama, T., Murota, A., Shen, Y.-M., Wu, S.-G., and Kuroda, C. (2005) *Bulletin of the Chemical Society of Japan*, **78**, 1302–1308.

47 Bousserouel, H., Litaudon, M., Morleo, B., Martin, M.-T., Thoison, O., Nosjean, O., Boutin, J.A., Renard, P., and Sévenet, T. (2005) *Tetrahedron*, **61**, 845–851.

48 Austin, J.F. and MacMillan, D.W.C. (2002) *Journal of the American Chemical Society*, **124**, 1172–1173.

49 King, H.D., Meng, Z.-X., Denhart, D., Mattson, R., Kimura, R., Wu, D.-D., Gao, Q., and Macor, J.E. (2005) *Organic Letters*, **7**, 3437–3440.

50 Garcia, A.A. and Martinez, J.L.O. (1982) Span. ES 506,422. Clitherow, J.W. (1985) U.S. Patent 4,497,961. Price, B.J., Clitherow, J.W., and Bradshaw, J. (1984) Patentschrift CH 640,846.

51 Tseng, J.-C., Huang, S.-L., Lin, C.-L., Lin, H.-C., Jin, B.-Y., Chen, C.-Y., Yu, J.-K., Chou, P.-T., and Luh, T.-Y. (2003) *Organic Letters*, **5**, 4381–4384.

52 Krasnoslobodskaya, L.D. and Gol'dfarb, Ya.L. (1969) *Uspekhi Khimii*, **38**, 854–891, (Russ.).

53 Ranganathan, S., Ranganathan, D., and Mehrotra, M.M. (1977) *Synthesis*, 838.Kornfeld, E.C. and Jones, R.G. (1954) *The Journal of Organic Chemistry*, **19**, 1671–1680.

54 Efremov, I. and Paquette, L.A. (2000) *Journal of the American Chemical Society*, **122**, 9324–9325.

55 Mongin, F., Bucher, A., Bazureau, J.P., Bayh, O., Awad, H., and Trécourt, F. (2005) *Tetrahedron Letters*, **46**, 7989–7992.

56 Bock, I., Bornowski, H., Rault, A., and Theis, H. (1990) *Tetrahedron*, **46**, 1199–1210.

57 Cahiez, G., Chavant, P.-Y., and Metais, E. (1992) *Tetrahedron Letters*, **33**, 5245–5248; Kojima, Y., Wakita, S., and Kato, N. (1979) *Tetrahedron Letters*, **20**, 4577–4580.Ennis, D.S. and Gilchrist, T.L. (1990) *Tetrahedron*, **46**, 2623–2632;Yang, Y. and Wong, H.N.C. (1994) *Tetrahedron*, **50**, 9583–9608.

58 Padwa, A., Crawford, K.R., Rashatasakhon, P., and Rose, M. (2003) *The Journal of Organic Chemistry*, **68**, 2609–17.

59 Demir, A.S., Reis, Ö., and Emrullahoglu, M. (2003) *The Journal of Organic Chemistry*, **68**, 578–580.

60 Glover, B., Harvey, K.A., Liu, B., Sharp, M.J., and Tymoschenko, M.F. (2003) *Organic Letters*, **5**, 301–304.

61 Brel, V.K. (2001) *Synthesis*, 1539–1545.

62 Malanga, C. and Mannucci, S. (2001) *Tetrahedron Letters*, **42**, 2023–2025.

63 Uneyama, K., Tanaka, H., Kobayashi, S., Shioyama, M., and Amii, H. (2004) *Organic Letters*, **6**, 2733–2736.

64 Miles, W.H., Heinsohn, S.K., Brennan, M.K., Swarr, D.T., Eidam, P.M., and Gelato, K.A. (2002) *Synthesis*, 1541–1545.

65 Miles, W.H., Dethoff, E.A., Tuson, H.H., and Ulas, G. (2005) *The Journal of Organic Chemistry*, **67**, 2862–2865.

66 Zanatta, N., Faoro, D., Silva, S.C., Bonacorso, H.G., and Martins, M.A.P. (2004) *Tetrahedron Letters*, **45**, 5689–5691.

67 Barma, D.K., Kundu, A., Baati, R., Mioskowski, C., and Falck, J.R. (2002) *Organic Letters*, **4**, 1387–1389.

68 Padwa, A., Zanka, A., Cassidy, M.P., and Harris, J.M. (2003) *Tetrahedron*, **59**, 4939–4944.

69 Donohoe, T.J., Orr, A.J., Gosby, K., and Bingham, M. (2005) *European Journal of Organic Chemistry*, 1969–1971.

70 Grimaldi, T., Romero, M., and Pujol, M.D. (2000) *Synlett*, 1788–1792.

71 Barma, D.K., Kundu, A., Baati, R., Mioskowski, C., and Falck, J.R. (2000) *Chemistry Letters*, 750–751.

72 McClure, M.S., Glover, B., McSorley, E., Millar, A., Osterhout, M.H., and Roschangar, F. (2001) *Organic Letters*, **3**, 1677–1680.

73 McClure, M.S., Roschangar, F., Hodson, S.J., Millar, A., and Osterhout, M.H. (2001) *Synthesis*, 1681–1685.

74 Gauthier, D.R., Jr, Szumigala, R.H., Jr, Dormer, P.G., Armstrong, J.D., III, Volante, R.P., and Reider, P.J. (2002) *Organic Letters*, **4**, 375–378.

75 Marshall, J.A. and Robinson, E.D. (1990) *The Journal of Organic Chemistry*, **55**, 3450–3451.

76 Marshall, J.A. and Sehon, C.A. (1997) *The Journal of Organic Chemistry*, **59**, 4313–4320.

77 Hashimi, A.S.K., Schwarz, L., Choi, J.-H., and Frost, T.M. (2000) *Angewandte Chemie, International Edition*, **39**, 2285–2588.

78 Kel'in, A.V. and Gevorgyan, V. (2002) *The Journal of Organic Chemistry*, **67**, 95–98.

79 Ma, S. and Yu, Z. (2002) *Angewandte Chemie, International Edition*, **41**, 1775; Ma, S., Gu, Z., and Yu, Z. (2005) *The Journal of Organic Chemistry*, **67**, 6291–6294.

80 Miki, K., Nishino, F., Ohe, K., and Uemura, S. (2002) *Journal of the American Chemical Society*, **124**, 5260–5261.

81 Teng, X., Wada, T., Okamoto, S., and Sata, F. (2001) *Tetrahedron Letters*, **42**, 5501–5503.

82 Méndez-Andino, J. and Paquette, L.A. (2000) *Organic Letters*, **2**, 4095–4097.

83 Alickmann, D., Fröhlich, R., Maulitz, A.H., and Würthwein, E.-U. (2002) *European Journal of Organic Chemistry*, 1523–1537.

84 Bellur, E., Görls, H., and Langer, P. (2005) *European Journal of Organic Chemistry*, 2074–2090.

85 Righi, G., Antonioletti, R., Ciambrone, S., and Fiorini, F. (2005) *Tetrahedron Letters*, **46**, 5467–5469.

86 Imagawa, H., Kurisaki, T., and Nishizawa, M. (2004) *Organic Letters*, **6**, 3679–3681.

87 Padwa, A., Rashatasakhon, P., and Rose, M. (2003) *The Journal of Organic Chemistry*, **68**, 5139–5146. Rashatasakhon, P. and Padwa, A. (2003) *Organic Letters*, **5**, 189–191.

88 Whitehead, C.R., Sessions, E.H., Ghiviriga, I., and Wright, D.L. (2002) *Organic Letters*, **4**, 3763–3765.

89 Sperry, J.B. and Wright, D.L. (2005) *Journal of the American Chemical Society*, **127**, 8034–8035.

90 Qing, F.-L., Gao, W.-Z., and Ying, J.-W. (2000) *The Journal of Organic Chemistry*, **65**, 2003–2006.

91 Fan, M., Guo, L., Liu, X., and Liang, Y. (2005) *Synthesis*, 391–396.

92 Sniady, A., Wheeler, K.A., and Dembinski, R. (2005) *Organic Letters*, **7**, 1769–1772.

93 Karpov, A.S., Merkul, E., Oeser, T., and Müller, T.J.J. (2005) *Chemical Communications*, 2581–2583.

94 Jung, C.-K., Wang, J.-C., and Krische, M.J. (2004) *Journal of the American Chemical Society*, **126**, 4118–4119.

95 Kuroda, H., Hanaki, E., and Kawakami, M. (1999) *Tetrahedron Letters*, **40**, 3753–3756; Kuroda, H., Hanaki, E., Izawa, H., Kano, M., and Itahashi, H. (2004) *Tetrahedron*, **60**, 1913–1920.

96 Redman, A.M., Dumas, J., and Scott, W.J. (2000) *Organic Letters*, **2**, 2061–2063.

97 Aurrecoechea, J.M., Pérez, E., and Solay, M. (2001) *The Journal of Organic Chemistry*, **66**, 564–569.

98 Aurrecoechea, J.M. and Pérez, E. (2001) *Tetrahedron Letters*, **42**, 3839–3841.

99 Akai, S., Kawashita, N., Satoh, H., Wada, Y., Kakiguchi, K., Kuriwaki, I., and Kita, Y. (2004) *Organic Letters*, **6**, 3793–3796.

100 Yao, T., Zhang, X., and Larock, R.C. (2004) *Journal of the American Chemical Society*, **126**, 11164–11165.

101 Patil, N.T., Wu, H., and Yamamoto, Y. (2005) *The Journal of Organic Chemistry*, **70**, 4531–4534.

102 Liu, Y. and Zhou, S. (2005) *Organic Letters*, **7**, 4609–4611.

103 Kato, Y., Miki, K., Nishino, F., Ohe, K., and Uemura, S. (2003) *Organic Letters*, **5**, 2619–2621.

104 Wipf, P. and Soth, M.J. (2002) *Organic Letters*, **4**, 1787–1790.

105 Ma, S. and Zhang, J. (2003) *Journal of the American Chemical Society*, **125**, 12386–12387.

106 Ma, S., Lu, L., and Zhang, J. (2004) *Journal of the American Chemical Society*, **126**, 9645–9660.

107 Kim, J.T., Kel'in, A.V., and Gevorgyan, V. (2003) *Angewandte Chemie, International Edition*, **42**, 98–101.

108 Lee, C.-F., Yang, L.-M., Hwu, T.-Y., Feng, A.-S., Tseng, J.-C., and Luh, T.-Y. (2000) *Journal of the American Chemical Society*, **122**, 4992–4993.

109 Leclerc, E. and Tius, M.A. (2003) *Organic Letters*, **5**, 1171–1174.

110 Sromek, A.W., Rubina, M., and Gevorgyan, V. (2005) *Journal of the American Chemical Society*, **127**, 10500–10501.

111 Ma, S. and Zhang, J. (2000) *Chemical Communications*, 117–118.

112 Nishibayashi, Y., Yoshikawa, M., Inada, Y., Milton, M.D., Hidai, M., and Uemura, S. (2003) *Angewandte Chemie, International Edition*, **42**, 2681–2684.

113 Ma, S., Zhang, J., and Lu, L. (2003) *Chemistry – A European Journal*, **9**, 2447–2456.

114 Suhre, M.H., Reif, M., and Kirsch, S.F. (2005) *Organic Letters*, **7**, 3925–3927.

115 Bossharth, E., Desbordes, P., Monteiro, N., and Balme, G. (2003) *Organic Letters*, **5**, 2441–2444.

116 Duan, X.-H., Liu, X.-Y., Guo, L.-N., Liao, M.-C., Liu, W.-M., and Liang, Y.-M. (2005) *The Journal of Organic Chemistry*, **67**, 6980–6983.

117 Fayol, A. and Zhu, J. (2004) *Organic Letters*, **6**, 115–118.

118 Yadav, J.S., Reddy, B.V.S., Shubashree, S., Sadashiv, K., and Naidu, J.J. (2004) *Synthesis*, 2376–2380.

119 Yavari, I., Nasiri, F., Moradi, L., and Djahaniani, H. (2004) *Tetrahedron Letters*, **45**, 7099–7101.

120 Miki, K., Yokoi, T., Nishino, F., Kato, Y., Washitake, Y., Ohe, K., and Uemura, S. (2004) *The Journal of Organic Chemistry*, **66**, 1557–1564; Miki, K., Washitake, Y., Ohe, K., and Uemura, S. (2004) *Angewandte Chemie, International Edition*, **43**, 1857–1860.

121 Sromek, A.W., Kel'in, A.V., and Gevorgyan, V. (2004) *Angewandte Chemie, International Edition*, **43**, 2280–2282.

122 Nair, V., Streekumar, V., Bindu, S., and Suresh, E. (2005) *Organic Letters*, **7**, 2297–2300.

123 Ma, C. and Yang, Y. (2005) *Organic Letters*, **7**, 1343–1345.

124 Tanis, S.P., Deaton, M.V., Dixon, L.A., McMills, M.C., Raggon, J.W., and Collins, M.A. (1998) *The Journal of Organic Chemistry*, **63**, 6914–6928.

125 Bur, S.K. and Martin, S.F. (2001) *Tetrahedron*, **57**, 3221–3242.

126 Rassu, G., Zanardi, F., Battistini, L., and Casiraghi, G. (1999) *Synlett*, 1333–1350; Rassu, G., Zanardi, F., Battistini, L., and Casiraghi, G. (2000) *Chemical Society Reviews*, **29**, 109–118; Casiraghi, G., Zanardi, F., Appendino, G., and Rassu, G. (2000) *Chemical Reviews*, **100**, 1929–1972.

127 For examples Carreño, C., Luzón, C.G., and Ribagorda, M. (2002) *Chemistry – A European Journal*, **8**, 208–216.Brimble, M.A., Davey, R.M., and McLeod, M.D. (2002) *Synlett*, 1318–1322.

128 Martin, S.F., Barr, K.J., Smith, D.W., and Bur, S.K. (1999) *Journal of the American Chemical Society*, **121**, 6990–6997.

129 Suga, H., Kitamura, T., Kakechi, A., and Baba, T. (2004) *Chemical Communications*, 1414–1415; Kitajima, H., Ito, K., and Katsuki, T. (1997) *Tetrahedron*, **53**, 17015–17028.

130 Brown, S.P., Goodwin, N.C., and MacMillan, D.W.C. (2003) *Journal of the American Chemical Society*, **125**, 1192–1194.

131 Cho, C.-W. and Krische, M.J. (2004) *Angewandte Chemie, International Edition*, **43**, 6689–6691.

132 Winkler, J.D., Oh, K., and Asselin, S.M. (2005) *Organic Letters*, **7**, 387–389.

133 Naito, S., Escobar, M., Kym, P.R., Liras, S., and Martin, S.F. (2002) *The Journal of Organic Chemistry*, **67**, 4200–4208.

134 Friedman, L.A., You, F., Sabat, M., and Harman, W.D. (2003) *Journal of the American Chemical Society*, **125**, 14980–14981; Chen, H., Liu, R., Myers, W.H., and Harman, W.D. (1998) *Journal of the American Chemical Society*, **120**, 509–520.

135 Hwu, J.R., Sambaiah, T., and Chakraborty, S.K. (2003) *Tetrahedron Letters*, **44**, 3167–3169.

136 Barluenga, J., Nandy, S.K., Laxmi, Y.R.S., Suárez, J.R., Merino, I., Flórez, J., García-Granda, S., and Montedo-Bernardo, J. (2003) *Chemistry – A European Journal*, **9**, 5725–5736.

137 Pianacatelli, G., D'Auria, M., and D'Onofrio, F. (1994) *Synthesis*, 867–889.

138 Csáky, A.G. and Plumet, J. (1990) *Tetrahedron Letters*, **31**, 7669–7670.

139 Adger, B.M., Barrett, C., Brennan, J., MaKervey, M.A., and Murray, R.W. (1991) *Journal of the Chemical Society, Chemical Communications*, 1553–1554.

140 Finlay, J., MaKervey, M.A., and Gunaratne, H.Q.N. (1998) *Tetrahedron Letters*, **39**, 5651–5654.

141 Annangudi, S.P., Sun, M., and Salomon, R.G. (2005) *Synlett*, 1468–1470; Clive,

D.L.J. and Minaruzzaman, Ou, L.-G. (2005), *The Journal of Organic Chemistry*, **70**, 3318–3320.

142 Massa, A., Acocella, M.R., De Rosa, M., Soriente, A., Villano, R., and Scettri, A. (2003) *Tetrahedron Letters*, **44**, 835–837.

143 Giovannini, R. and Petrini, M. (1997) *Tetrahedron Letters*, **38**, 3781–3784.

144 McDermott, P.J. and Stockman, R.A. (2005) *Organic Letters*, **7**, 27–29.

145 Kim, G., Jung, S.-D., Lee, E.-J., and Kim, N. (2003) *The Journal of Organic Chemistry*, **68**, 5395–5398.

146 Cenñal, J.P., Carreras, C.R., Tonn, C.E., Padrón, J.I., Ramírez, M.A., Díaz, D.D., García-Tellado, F., and Martín, V.S. (2005) *Synlett*, 1575–1578; Ley, S.V. and Mahon, M. (1983) *Journal of the Chemical Society, Perkin Transactions 1*, 1379–1380.

147 Yu, P., Yang, Y., Zhang, Z.Y., Mak, T.C.W., and Wong, H.N.C. (1997) *The Journal of Organic Chemistry*, **62**, 6539–6366.

148 Gollnick, K. and Griesbeck, A. (1985) *Tetrahedron*, **41**, 2057–2068.

149 Kerman, M.R. and Faulkner, D.J. (1988) *The Journal of Organic Chemistry*, **53**, 2773–2776.

150 Vassilikogiannakis, G. and Stratakis, M. (2003) *Angewandte Chemie, International Edition*, **42**, 5465–5468. Vassilikogiannakis, G., Margaros, I., and Montagnon, T. (2004) *Organic Letters*, **6**, 2039–2042.

151 Ciufolini, M.A., Hermann, C.Y.W., Dong, Q., Shimizu, T., Swaminathan, S., and Xi, N. (1997) *Synlett*, 105–114.

152 Haukaas, M.H. and O'Doherty, G.A. (2001) *Organic Letters*, **3**, 401–404; Cassidy, M.P. and Padwa, A. (2004) *Organic Letters*, **6**, 4029–4031.

153 Kennedy, A., Nelson, A., and Perry, A. (2005) *Chemical Communications*, 1646–1648.

154 Katritzky, A.R. and Dennis, N. (1989) *Chemical Reviews*, **89**, 827–861.

155 Wender, P.A., Jesudason, C.D., Nakahira, H., Tamura, N., Tebbe, A.L., and Ueno, Y. (1997) *Journal of the American Chemical Society*, **119**, 12976–12977.

156 Moeller, K.D. (2000) *Tetrahedron*, **56**, 9527–9554.

157 Hughes, C.C., Miller, A.K., and Trauner, D. (2005) *Organic Letters*, **7**, 3425–3428.

158 Yim, H.K., Liao, Y., and Wong, H.N.C. (2003) *Tetrahedron*, **59**, 1877–1884.

159 Gribble, G.W. (1991) In *Comprehensive Organic Synthesis, Vol. 8* (eds B.M. TrostC and I. Fleming), Oxford, Pergamon, pp. 606–608; Donohoe, T.J., Garg, R., and Stevenson, C.A. (1996) *Tetrahedron Asymmetry*, **7**, 317–344.

160 Schmidt, U. and Werner, J. (1986) *Journal of the Chemical Society, Chemical Communications*, 996–998.

161 Donohoe, T.J., Calabrese, A.A., Stevenson, C.A., and Ladduwahetty, T. (2000) *Journal of the Chemical Society, Perkin Transactions 1*, 3724–3731; Donohoe, T.J., Calabrese, A.A., Guillermin, J.-B., Frampton, C.S., and Walter, D. (2002) *Journal of the Chemical Society, Perkin Transactions 1*, 1748–1756.

162 Rodríguez, A., Nomen, M., Spur, B.W., and Godfroid, J.-J. (1999) *European Journal of Organic Chemistry*, 2655–2662.

163 D'Auria, M. (2000) *Heterocycles*, **52**, 185–194.

164 Chubb, R.W.J., Bryce, M.R., and Tarbit, B. (2001) *Chemical Communications*, 1853–1854.

165 Pommier, A., Stepanenko, V., Jarowicki, K., and Kocienski, P.J. (2003) *The Journal of Organic Chemistry*, **68**, 4008–4013.

166 Erker, G., Petrenz, R., Krüger, C., Lutz, F., Weiss, A., and Werner, S. (1992) *Organometallics*, **11**, 1646–1655.

167 Osornio, Y.M., Cruz-Almanza, R., Jiménez-Montaño, V., and Miranda, L.D. (2003) *Chemical Communications*, 2316–2317.

168 Demircan, A. and Parsons, P.J. (2003) *European Journal of Organic Chemistry*, 1729–1732; Demircan, A. and Parsons, P.J. (1998) *Synlett*, 1215–1216;Parsons, P.J., Penverne, M., and Pinto, I.L. (1994) *Synlett*, 721–722.

169 Jones, P., Li, W.-S., Pattenden, G., and Thomson, N.M. (1997) *Tetrahedron Letters*, **38**, 9069–9072.

170 Kappe, O., Murphree, S., and Padwa, A. (1997) *Tetrahedron*, **53**, 14179–14233.

171 Padwa, A., Ginn, J.D., Bur, S.K., Eidell, C.K., and Lynch, S.M. (2002) *The Journal of Organic Chemistry*, **67**, 3412–3424.

172 Toró, A. and Deslongchamps, P. (2003) *The Journal of Organic Chemistry*, **68**, 6847–6852.
173 Harmata, M. (2001) *Accounts of Chemical Research*, **34**, 595–605; Harmata, M. (1997) *Tetrahedron*, **53**, 6235–6280;Rigby, J.H. and Pigge, F.C. (1997) in *Organic Reactions, Vol 51* (ed. Paquette, L.A.), John Wiley & Sons Inc., New York, pp. 351–478.
174 Sáez, J.A., Arnó, M., and Domingo, L.R. (2003) *Organic Letters*, **5**, 4117–4120, and references cited therein.
175 Cho, S.Y., Lee, H.I., and Cha, J.K. (2001) *Organic Letters*, **3**, 2891–2893.
176 Prié G., Prévost, N., Twin, H., Fernandes, S.A., Hayes, J.F., and Shipman, M. (2004) *Angewandte Chemie, International Edition*, **43**, 6517–6519.
177 Harmata, M. and Schreiner, P.R. (2001) *Organic Letters*, **3**, 3663–3665.
178 Lee, J.C. and Cha, J.K. (2000) *Tetrahedron*, **56**, 10175–10184.
179 Hamata, M., Ghosh, S.K., Hong, X., Wacharasindhu, S., and Kirchhoefer, P. (2004) *Journal of the American Chemical Society*, **125**, 2058–2059.
180 Huang, J. and Hsung, R.P. (2005) *Journal of the American Chemical Society*, **127**, 50–51.
181 Rigby, J.H. and Chouraqui, G. (2005) *Synlett*, 2501–2503.
182 Hashmi, A.S.K., Frost, T.M., and Bats, J.W. (2000) *Journal of the American Chemical Society*, **122**, 11553–11554.
183 Hashmi, A.S.K., Rudolph, M., Weyrauch, J.P., Wölfle, M., Frey, W., and Bats, J.W. (2005) *Angewandte Chemie, International Edition*, **44**, 2798–2801.
184 Drew, M.G.B., Jahans, A., Harwood, L.M., and Apoux, S.A.B.H. (2002) *European Journal of Organic Chemistry*, 3589–3594, and references cited therein.
185 Zhang, L., Wang, Y., Buckingham, C., and Herndon, J.W. (2005) *Organic Letters*, **7**, 1665–1667.
186 For two recent examples see: Anderson, E.A., Alexanian, E.J., and Sorensen, E.J. (2004) *Angewandte Chemie, International Edition*, **43**, 1998–2001. Toyooka, N., Nagaoka, M., Kakuda, H., and Nemoto, H. (2001) *Synlett*, 1123–1124.
187 Chou, Y.-Y., Peddinti, R.K., and Liao, C.-C. (2003) *Organic Letters*, **5**, 1637–1640.
188 Avalos, M., Babiano, R., Cabello, N., Cintas, P., Hursthouse, M.B., Jimenez, J., Light, M.E., and Palacios, J.C. (2003) *The Journal of Organic Chemistry*, **68**, 7193–7203.
189 Jäger, V. and Müller, I. (1985) *Tetrahedron*, **41**, 3519–3528;Zimmermann, P.J., Lee, J.Y., Hlobilova, I., Endermann, R., Häbich, D., and Jäger, V. (2005) *European Journal of Organic Chemistry*, 3450–3460.
190 Mejia-Oneto, J.M. and Padwa, A. (2004) *Organic Letters*, **6**, 3241–3244.
191 Chhor, R.B., Nosse, B., Sörgel, S., Böhm, C., Seitz, M., and Reiser, O. (2003) *Chemistry – A European Journal*, **9**, 260–270.
192 Curini, M., Epifano, F., Marcotullio, M.C., Rosati, O., Guo, M., Guan, Y., and Wenkert, E. (2005) *Helvetica Chimica Acta*, **88**, 330–338.
193 Hughes, C.C., Kennedy-Smith, J.J., and Trauner, D. (2003) *Organic Letters*, **5**, 4113–4115.
194 Abe, M., Kawakami, T., Ohata, S., Nozaki, K., and Nojima, M. (2004) *Journal of the American Chemical Society*, **126**, 2838–2846.
195 Abe, M., Torri, E., and Nojima, M. (2000) *The Journal of Organic Chemistry*, **65**, 3426–3431.
196 Crimmins, M.T., Pace, J.M., Nantermet, P.G., Kim-Meade, A.S., Thomas, J.B., Watterson, S.H., and Wagman, A.S. (2000) *Journal of the American Chemical Society*, **122**, 8453–8463.
197 Capon, B. and Kwok, F.-C. (1986) *Tetrahedron Letters*, **27**, 3275–3278.
198 Bodor, N., Dewar, M.J.S., and Harget, A.J. (1970) *Journal of the American Chemical Society*, **92**, 2929–2936.
199 Rao, Y.S. (1976) *Chemical Reviews*, **76**, 625–694.
200 Fariña, F., Maestro, M.C., Martin, M.R., Martin, M.V., Sánchez, F., and Soria, M.L. (1986) *Tetrahedron*, **42**, 3715–3722.
201 Rao, Y.S. (1964) *Chemical Reviews*, **64**, 353–388.
202 Laduwahetty, T. (1995) *Contemporary Organic Synthesis*, **2**, 133–179.

203 Hashem, A. and Kleinpeter, E. (2001) *Advances in Heterocyclic Chemistry*, **81**, 107–165.

204 Yu, W.-Y. and Alper, H. (1997) *The Journal of Organic Chemistry*, **62**, 5684–5687.

205 Nozaki, K., Sato, N., Ikeda, K., and Takaya, H. (1996) *The Journal of Organic Chemistry*, **61**, 4516–4519.

206 Ohno, M., Oguri, I., and Eguchi, S. (1999) *The Journal of Organic Chemistry*, **64**, 8995–9000.

207 Jan-Anders, H., Nasman, K., and Pensar, G. (1985) *Synthesis*, **8**, 786–788.

208 Eberhard, G., Walter, K., Wolfgang, W., and Volker, J. (1986) *Synthesis*, **9**, 921–926.

209 Hashem, A. and Senning, A. (1999) *Advances in Heterocyclic Chemistry*, **73**, 275–293.

210 Hashm, A.I. and Shaban, M.E. (1981) *Journal für Praktische Chemie*, **323**, 164–168.

211 Camici, L., Ricci, A., and Taddei, M. (1986) *Tetrahedron Letters*, **27**, 5155–5158.

212 Shin, S.S., Byun, Y., Lim, K.M., Choi, J.K., Lee, K.-W., Moh, J.H., Kim, J.K., Jeong, Y.S., Kim, J.Y., Choi, Y.H., Koh, H.-J., Park, Y.-H., and Oh, Y.I. (2004) *Journal of Medicinal Chemistry*, **47**, 792–804.

213 Cochran, J.E., Wu, T., and Padwa, A. (1996) *Tetrahedron Letters*, **37**, 2903–2906.

214 (a) Zhou, C.Y., Hong, P.W. and Che, C.M. (2006) *Organic Letters*, **8**, 325–328. (b) Zhang, J. and Schmalz, H.-G. (2006) *Angewandte Chemie, International Edition*, **45**, 6704–6707. (c) Oh, C.H., Lee, S.J., Lee, J.H. and Na, Y.J. (2008) *Chemical Communications* 5794–5796. (d) Zhang, G., Huang, X., Li, G. and Zhang, L. (2008) *Journal of the American Chemical Society*, **130**, 1814–1815.

215 (a) Peng, L., Zhang, X., Ma, M. and Wang, J. (2007) *Angewandte Chemie, International* Edition, **46**, 1905–1908. (b) Dudnik, A.S. and Gevorgyan, V. (2007) *Angewandte Chemie, International Edition*, **46**, 5195–5197. (c) Barluenga, J., Riesgo, L., Vicente, R., López, L.A. and Tomás, A. (2008) *Journal of the American Chemical Society*, **130**, 13528–13529. (d) Zhang, Y., Chen, Z., Xiao, Y. and Zhang, J. (2009) *Chemistry – A European Journal*, **15**, 5208–5211.

216 (a) Parsons, P., Waters, A.J., Walter, D.S. and Board, J. (2007) *The Journal of Organic Chemistry*, **72**, 1395–1398. (b) Ho, C.H., Park, H.M. and Park, D.I. (2007) *Organic Letters*, **9**, 1191–1193. (c) Xiao, Y. and Zhang, J. (2008) *Angewandte Chemie, International Edition*, **47**, 1903–1906.

217 Tokumaru, K., Arai, S. and Nishida, A. (2006) *Organic Letters*, **8**, 27–30.

218 Casiraghi, G. Zanardi, F., Battistini, L. and Rassu, G. (2009) *Synlett*, 1525–1542; other new examples: Wieland, L.C., Vieira, E.M. Snapper, M.L. and Hoveyda, A.H. (2009) *Journal of the American Chemistry Society*, **131**, 570–576;Frings, M., Atodiresei, I., Runsink, J., Raabe, G. and Bolm, C. (2009) *Chemistry–A European Journal*, **15**, 1566–1569;Yuan, Z.-J., Jiang, J.-J. and Shi, M. (2009) *Tetrahedron*, **65**, 6001–6007.

219 Jiang, Y.-Q., Shi, Y.-L. and Shi, M. (2008) *Journal of the American Chemistry Society*, **130**, 7202–7203.

220 Barluenga, J., García-García, P., de Sáa, D. and Fernández-Rodríguez, M.A. (2007) *Angewandte Chemie, International Edition*, **46**, 2610–2612.

221 Barluenga, J., de Prado, A., Santamaría, J. and Tomás, M. (2007) *Chemistry–A European Journal*, **13**, 1326–1331; Barluenga, J. de Prado, A., Santamaría, J. and Tomás, M. (2005) *Angewandte Chemie, International Edition*, **44**, 6583–6585.

222 Li, S.-W. and Batey, R.A. (2007) *Chemical Communications*, 3759–3761.

223 Aquino, M., Bruno, I., Riccio, R. and Gomez-Paloma, L. (2006) *Organic Letters*, **8**, 4831–4834.

224 Patril, S.N. and Liu, F. (2007) *The Journal of Organic Chemistry*, **72**, 6305–6308.

225 Kiren, S., Hong, X., Leverett, C.A. and Padwa, A. (2009) *Organic Letters*, **11**, 1233–1235.

226 Griggs, N.D. and Phillips, A.J. (2008) *Organic Letters*, **10**, 4955–4957.

227 Tofi, M., Koltsida, K. and Vassilikogiannakis, G. (2009) *Organic Letters*, **11**, 313–316.

228 Sperry, J.B., Ghiviriga, I. and Wright, D.L. (2006) *Chemical Communications*, 194–196.

229 Kaiser, S., Smidt, S.P. and Pfaltz, A. (2006) *Angewandte Chemie, International Edition*, **45**, 5194–5197.

230 Feiertag, P., Albert, M., Nettekoven, U. and Spindler, F. (2006) *Organic Letters*, **8**, 4133–4135.

231 Pieniazek, S.N. and Houk, K.N. (2006) *Angewandte Chemie, International Edition*, **45**, 1442–1445; Padwa, A., Crawford, K.R., Straub, C.S., Pieniazek, S.N. and Houk, K.N. (2006) *The Journal of Organic Chemistry*, **71**, 5432–5439.

232 Ram, R.N. and Kumar, N. (2008) *Tetrahedron Letters*, **49**, 799–802;232.Ram, R.N. and Kumar, N. (2008) *Tetrahedron*, **64**, 10267–10271.

233 Chung, W.K., Lam, S.K., Lo, B., Liu, L.L., Wong, W.-T. and Chiu, P. (2009) *Journal of the American Chemistry Society*, **130**, 4556–4557.

234 Harmata, M. and Huang C. (2009) *Tetrahedron Letters*, **50**, 5701–5703.

235 He, W., Huang, J., Sun, X. and Frontier, A.J. (2007) *Journal of the American Chemistry Society*, **129**, 498–499; He, W., Huang, J., Sun, X. and Frontier, A.J. (2008) *Journal of the American Chemistry Society*, **130**, 300–308.

236 Miki, K., Fujita, M., Uemura, S. and Ohe, K. (2006) *Organic Letters*, **8**, 1741–1743; Miki, K., Senda, Y., Kowada, T. and Ohe, K. (2009) *Synlett*, 1937–1940.

237 Ikeda, Y., Murai, M., Abo, T., Miki, K. and Ohe, K. (2007) *Tetrahedron Letters*, **48**, 6651–6654.

7
Five-Membered Heterocycles: Benzofuran and Related Systems

Jie Wu

7.1
Introduction

Benzofuran is a very important heterocycle and broadly found in natural [1] and biologically important [2] molecules and is frequently used as a building block in materials science [3a] and in organic synthesis [3b]. Several recent mini-reviews cover the investigation of benzo[*b*]furans in natural products, bioactivity and synthesis [4]. Two general approaches are commonly used for the preparation of substituted benzofurans: (i) functionalization of existing benzofuran-containing precursors by introduction of new substituents [5] and (ii) the formation of a new benzofuran ring by cyclization of acyclic substrates [6–10]. The methods based on the first approach are not general. Derivatization of benzofuran via an electrophilic substitution is not easy due to poor regioselectivity and the low stability of benzofurans under strongly acidic conditions, whereas protocols involving metallation of benzofuran derivatives followed by trapping of the benzofuryl anion with electrophiles are limited to base-stable benzofuran substrates and, in the case of alkylation, to primary electrophiles only. Among cyclization approaches, the classical oxidative cyclocondensation of phenol or related precursor remains the most powerful method for the construction of some naturally occurring benzofurans. Significant attention has been paid to the development of metal-catalyzed approaches aimed at cascade cyclization with substituted phenols or iodophenols under rather mild or neutral conditions [10]. In this chapter, in consideration of the limited space, we first discuss recent progress in the synthetic methods for metal-catalyzed benzofuran synthesis based on the reaction classification, and then discuss their important uses in terms of drug discovery and material science. For other synthetic methods regarding the synthesis of benzofurans, please see the references provided.

Modern Heterocyclic Chemistry, First Edition.
Edited by Julio Alvarez-Builla, Juan Jose Vaquero, and José Barluenga.

Figure 7.1 Structure numbering and frontier electron populations of benzo[*b*]furan.

7.2
General Structure and Reactivity

7.2.1
Relevant Physicochemical Data, Computational Chemistry and NMR Data

Benzo[*b*]furan is an aromatic compound and is usually recognized as a heterocyclic analog of naphthalene. Each of the ring atoms in the benzo[*b*]furan rings is in the same plane and has a p-orbital perpendicular to the ring plane. Additionally, (4*n* + 2)p electrons are associated with each ring. The oxygen in benzo[*b*]furan acquires considerable positive charge since it provides two p-electrons for the aromatic sextet. Accordingly, the same amount of negative charge is displayed in the all ring carbon atoms.

Figure 7.1 illustrates the frontier electron populations of the parent benzo[*b*]furan according to frontier orbital theory [11], and Table 7.1 shows the UV absorption bands and NMR signals of benzo[*b*]furan.

In benzo[*b*]furan, the π-electron excess of C2 is lower than that of C3. Consequently,. ^{13}C NMR chemical shifts show that C2 (141.5 ppm) is deshielded in comparison with C3 (106.9 ppm). It is noticeable that C7 is in the upfield position compared with the other benzenoid carbons at positions 4, 5, and 6. The benzenoid protons H4 and H7 appear downfield from H5 and H6. Also noteworthy is that the long-range coupling between H3 and H7 is of considerable diagnostic value in establishing the orientation of a substituent on the benzo[*b*] furan ring [12b].

Table 7.1 UV and NMR data of benzo[*b*]furan [12].

UV (ethanol) λ (nm) (ε, mol^{-1} dm^3 cm^{-1})	^{1}H NMR (acetone-d_6) [δ (ppm)]		^{13}C NMR [δ (ppm)]	
244 (4.03)	H2: 7.79	H6: 7.30	C2: 141.5	C6: 124.6
274 (3.39)	H3: 6.77	H7: 7.52	C3: 106.9	C7: 111.8
281 (3.42)	H4: 7.64		C4: 121.6	C3a: 127.9
	H5: 7.23		C5: 123.2	C7a: 155.5

7.3 Isolation of Naturally Occurring Benzofurans

Numerous biologically active benzofurans have been isolated from natural sources [1]. Table 7.2 shows some examples of naturally occurring compounds containing the benzo[*b*]furan skeleton.

For instance, a new benzofuran dimer, 5,6,5′,6′-tetrahydroxy[3,3′]bibenzofuranyl-2,2′-dicarboxylic acid dimethyl ester (kynapcin-24, entry 5) has been isolated from *Polyozellus multiflex* and shown to noncompetitively inhibit prolyl endopeptidase (PEP), with an IC_{50} of 1.14 μM. Kynapcin-24 is less inhibitory to other serine proteases such as chymotrypsin, trypsin and elastase [1c].

Recently, naturally occurring furocarbazole alkaloids were also identified [13]. These molecules have a broad range of useful pharmacological activities [14]. Their useful bioactivities and their interesting structural features attracted the attention of synthetic chemists and has led, over the last decade, to the development of many different synthetic strategies [13a, 15].

Table 7.2 Naturally occurring compounds containing benzo[*b*]furan skeleton.

Entry	Compound name	Source	Biological activity	Reference
1	Ulexins C, A and D	*Ulex europaeus* ssp. *europaeus*	No inhibition of the growth of *Cladosporium cucumerinum*	[1a]
2	Paradisin C	Grapefruit juice	Inhibition of cytochrome P450 (CYP) 3A4 ($IC_{50} = 1.0\,\mu M$)	[1b]
3	Skimmianine and dictamnine	*Teclea trichocarpa* Enge. (Rutaceae)	N/A	[1g]
4	Millettocalyxin C and pongol methyl ether	Stem bark of *Millettia erythrocalyx*	N/A	[1d]
5	Kynapcin-24	Fruiting bodies of *Polyozellus multiflex*	Non-competitively inhibited prolyl endopeptidase (PEP) ($IC_{50} = 1.14\,\mu M$)	[1c]
6	Stemofurans A–K	Roots of *Stemona collinsae*	Antifungal activity against *Cladosporium herbarum*	[1f]
7	Tournefolal and tournefolic acid B	Stems of *Tournefortia sarmentosa*	Anti-LDL-peroxidative activity	[1e]

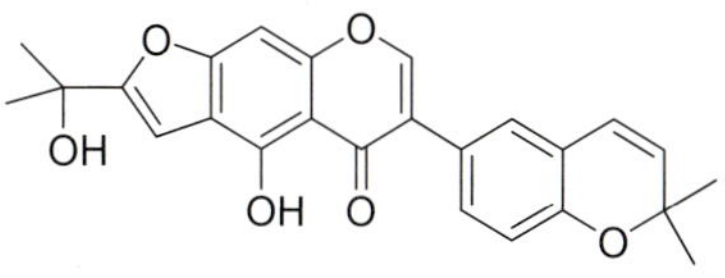

Ulexin A

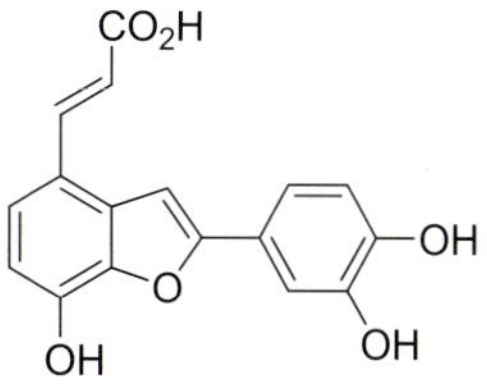

Tournefolic acid B

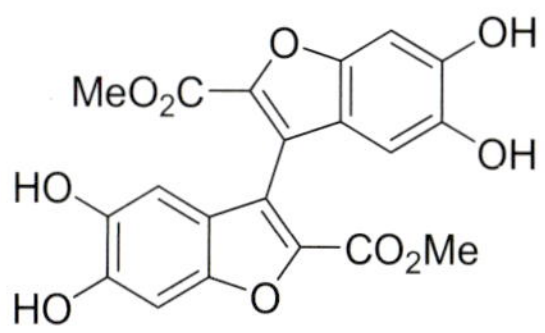

Kynapcin-24

furocarbazole

7.4 Synthesis of Benzofuran

As described above, major synthetic strategies for benzofurans include (Scheme 7.1): (i) dehydrative cyclization of α-(phenoxy)alkyl ketones [6a–6e]; (ii) dehydration of *o*-hydroxybenzyl ketones under acidic conditions [7a, 7b]; (iii) decarboxylation of *o*-acylphenoxyacetic acids or esters on treatment with a base [8a–8d]; (iv) cyclofragmentation of oxiranes, prepared in three or four steps from the corresponding *o*-hydroxybenzophenones [9]; and (v) palladium(II)-catalyzed cyclization of arylacetylenes [10a–10e].

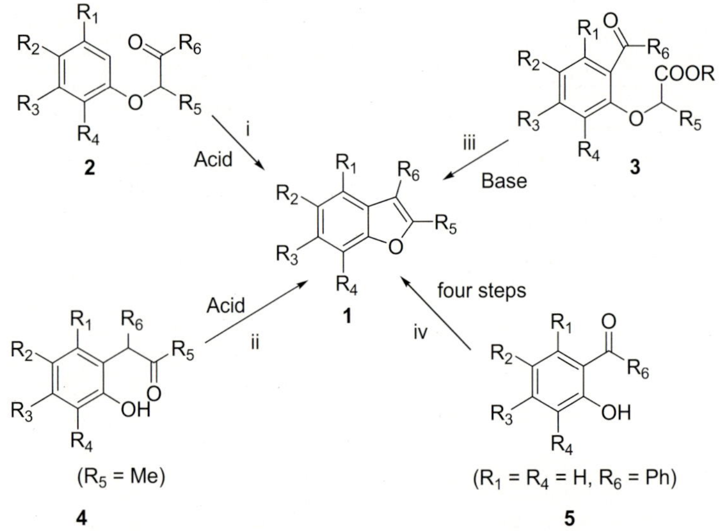

Scheme 7.1

7.4.1 Transition Metal Catalyzed Benzofuran Synthesis

7.4.1.1 Synthesis of 2,3-Disubstituted Benzo[*b*]furans

The metal-catalyzed intramolecular cyclization of aryl-substituted alkynes possessing a nucleophile in proximity to the triple bond has proven effective for the synthesis of five-membered heterocycles (Scheme 7.2) [16]; the first report describing synthetic

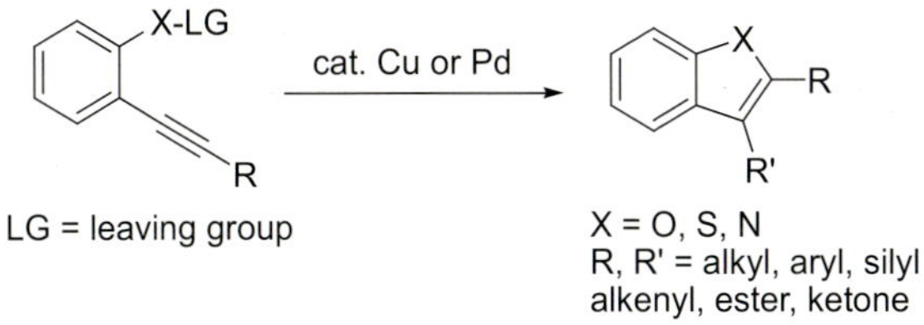

LG = leaving group

X = O, S, N
R, R' = alkyl, aryl, silyl
alkenyl, ester, ketone

Scheme 7.2

efforts relevant to 2,3-diarylbenzo[*b*]furan with this approach was given by Arcadi in 1996 [17a].

The reaction of *o*-ethynylphenols with a wide variety of unsaturated halides or triflates RX (R=vinyl, aryl; X=Br, I, OTf) in the presence of a palladium catalyst gives 2-vinyl- and 2-arylbenzo[*b*]furans **10** in good to high yield, through an intramolecular cyclization step (Scheme 7.3). Small amounts of 2,3-disubstituted benzo[*b*]furans **9** are usually isolated as side products. In some cases, however, benzofurans **9** are generated in significant yield or even as the main products. The formation of **10** can be prevented by employing alternative procedures that use *o*-[(trimethylsilyl)ethynyl] phenyl acetates as starting building blocks. One procedure is based on the palladium-catalyzed reaction of *o*-[(trimethylsilyl)ethynyl]phenyl acetates with RX in the presence of $Pd(PPh_3)_4$, Et_3N and n-Bu_4NF, followed by the hydrolysis of the resultant coupling derivative under basic conditions. The other procedure affords **10** through an *in situ* coupling/cyclization of *o*-[(trimethylsilyl)ethynyl]phenyl acetates with RX in the presence of $Pd(PPh_3)_4$ and KOBut. The utilization of *o*-alkynylphenols as the starting alkynes in the palladium-catalyzed reaction with RX leads to the formation of 2,3-disubstituted benzo[*b*]furans through an annulation process promoted by

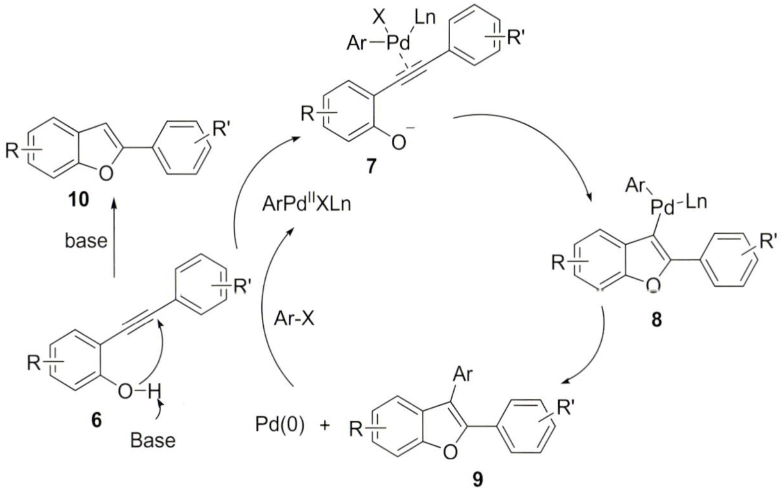

Scheme 7.3

σ-vinyl- and σ-arylpalladium complexes generated *in situ*. The best results in this case are obtained by using KOAc and $Pd(PPh_3)_4$. In the presence of KOAc and $Pd(PPh_3)_4$, and under an atmosphere of carbon monoxide, the reaction of *o*-alkynylphenols with RX provides 2-vinyl- and 2-aryl-3-acylbenzo[b]furans.

Later, Arcadi reported that the 5-*endo-dig*-iodocyclization of 2-alkynylphenols with I_2 in the presence of $NaHCO_3$ at room temperature produces 2-substituted 3-iodobenzo[b]furans, which are useful synthetic intermediates for the preparation of 2,3-disubstituted benzo[*b*]furans via Pd-catalyzed cross-coupling reactions [17b].

More recently, Flynn and colleagues [18] have disclosed an efficient approach to synthesize 2-substituted-3-arylbenzo[*b*]furan by a palladium-catalyzed multicomponent sequential coupling strategy, starting from iodophenol and terminal phenyl acetylene. In this reaction, MeMgBr was used as an essential base to form the corresponding magnesium salts of phenolate. Although the method was applied successfully to one substituted iodobenzene, utilization of MeMgBr to form the magnesium salts could hamper application of this method to other substrates having functional groups such as ketone, ester or amide, thus limiting the universality required in diversity oriented synthesis (Scheme 7.4).

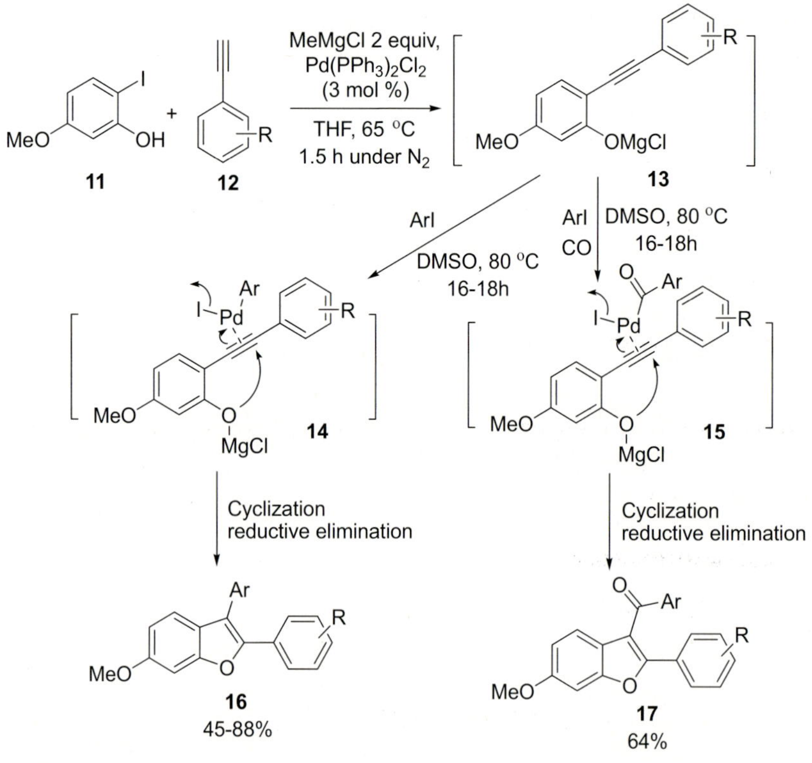

Scheme 7.4

To realize a strategy of diversity oriented synthesis and branching reaction pathways Yang [19] has described the palladium/bpy-catalyzed annulation of *o*-alkynylphenol with various aryl halides to generate diversified 2,3-diarylbenzo[*b*]furans (**19**). In the reaction process, the presence of bpy ligand is essential for the successful transformation. This method provides an efficient synthetic pathway for the combinatorial synthesis of conformationally restricted 2,3-diarylbenzo[*b*]furan (Scheme 7.5).

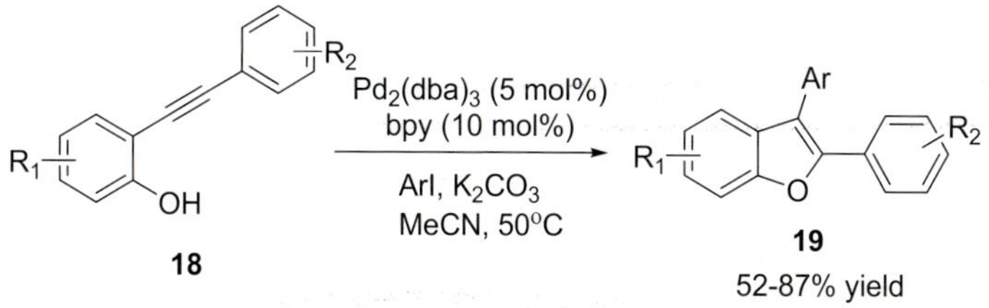

Scheme 7.5

Yang [20a] has also developed a highly effective co-catalysis system (PdI_2-thiourea and CBr_4) for carbonylative cyclization of both electron-rich and electron-deficient *o*-hydroxylarylacetylenes to the corresponding methyl benzo[*b*]furan-3-carboxylates. This was the first time using carbon tetrabromide (CBr_4) as a superior oxidative agent for the turnover of palladium(0) to palladium(II) (Scheme 7.6).

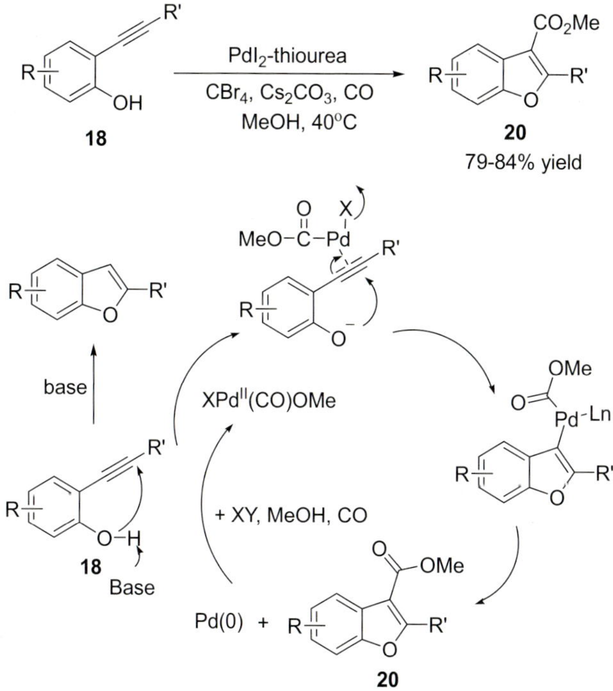

Scheme 7.6

The overall process may involve attack of a carboalkoxypalladium(II) intermediate on the alkyne **18** to generate a complex, followed by nucleophilic addition of the phenolic oxide to the $XPd^{II}(CO)OR$-activated alkyne to give intermediate, which went through reductive elimination to produce ester **20** and palladium(0). The palladium(0) is then oxidized to palladium(II), which re-enter the catalytic cycle. The method has been successfully applied in the solid-phase synthesis of benzo[*b*]furan-3-carboxylates [20e].

Further studies shows the carbonylative annulation of *o*-alkynylphenols mediated by $PdCl_2(PPh_3)_2$ and dppp in the presence of CsOAc at 55 °C in acetonitrile under a balloon pressure of CO generates functionalized benzo[*b*]furo[3,4-*d*]furan-1-ones **25** in good yields [20b]. This novel synthetic approach provides a highly efficient method for diversification of the benzofuran scaffold for combinatorial synthesis. The authors speculated that the overall process may involve attack of alcohol **21** on the $Pd^{II}XYLn$ to generate complex **22**, followed by insertion of CO to give intermediate **23**. Intramolecular nucleophilic addition of the phenolic oxide to the resulting acyl-palladium complex **23** leads to formation of intermediate **24**, which might undergo reductive elimination to produce the five-membered lactone **25** and palladium(0). The palladium(0) is then oxidized to palladium(II), thereby completing the cycle (Scheme 7.7).

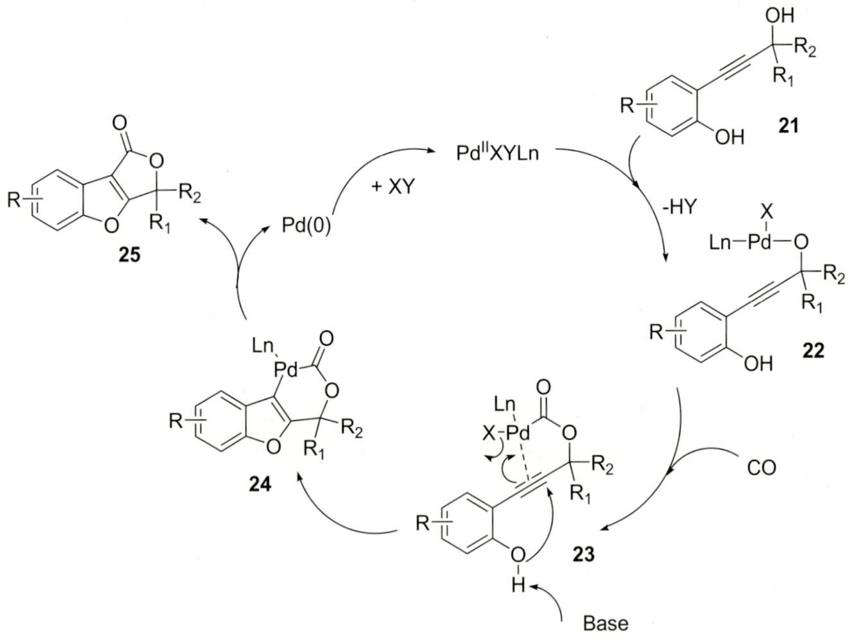

Scheme 7.7

Recently, Yang [20h] further described a novel, mild method for the rapid synthesis of benzo[*b*]furan-3-carboxylic acids **26** directly from the substituted *o*-alkynyl- phenols

18 in good yields by utilizing a Pd^{II}-mediated carboxylative annulation since, from a drug discovery perspective, synthesis of benzofuran-based carboxylic acids could be more interesting because of their increased solubility in aqueous media and the potential enhancement of ionic interactions with basic residues in their association with biological receptors (Scheme 7.8).

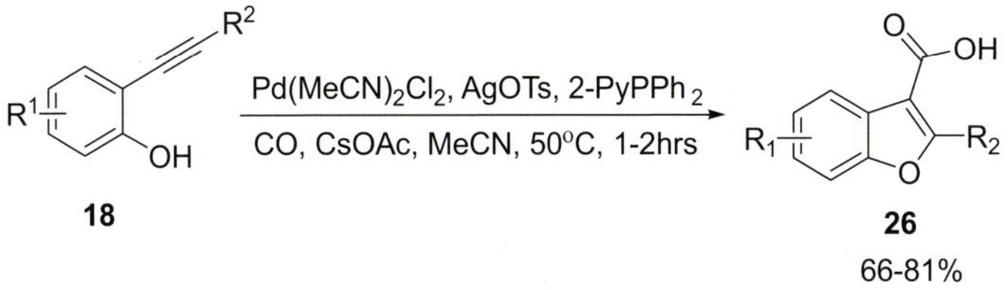

$Pd(MeCN)_2Cl_2$, AgOTs, 2-$PyPPh_2$
CO, CsOAc, MeCN, 50°C, 1-2hrs

18 → **26** 66-81%

Scheme 7.8

Cacchi has also described palladium catalysis in the construction of the benzo[*b*]furan ring from alkynes and organic halides or triflates [16m,16n].

3-Spiro-fused benzofuran-2(3*H*)-ones can be conveniently synthesized starting from vinyl triflates and o-iodophenols through palladium-catalyzed chemoselective carbonylation and subsequent regioselective intramolecular Heck reaction. For example, 2-iodo-4-methylphenol reacted with trifluoromethanesulfonic acid 4-(1,1-dimethylethyl)-1-cyclohexen-1-yl ester leading to 4-(1,1-dimethylethyl)-1-cyclohexene-1-carboxylic acid 2-iodo-4-methylphenyl ester. Subsequent ring closure of this intermediate furnishes spirobenzofuranone C (85% yield) [17c].

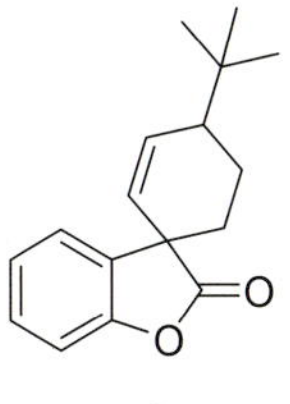

Spirobenzofuranone C

Arcadi [17d] has reported the synthesis of 2,3-disubstituted furo[3,2-*b*]pyridines, 2,3-disubstituted furo[2,3-*b*]pyridines and 2,3-disubstituted furo[2,3-*c*]pyridines under mild conditions via the palladium-catalyzed cross-coupling of 1-alkynes with *o*-iodoacetoxy- or *o*-iodobenzyloxypyridines, followed by electrophilic cyclization by I_2 or by $PdCl_2$ under a balloon of carbon monoxide (Scheme 7.9).

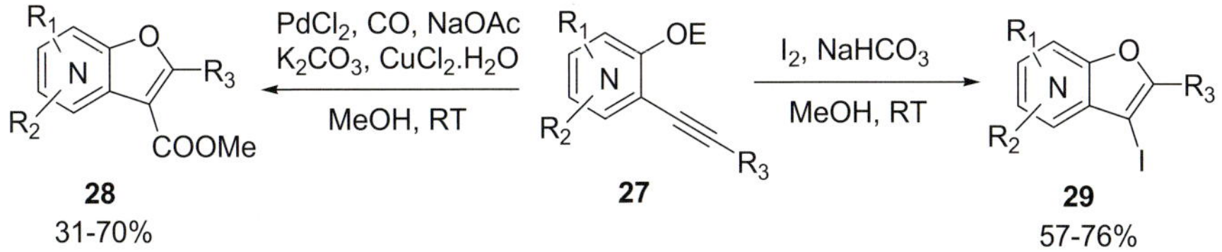

$PdCl_2$, CO, NaOAc, K_2CO_3, $CuCl_2.H_2O$, MeOH, RT
I_2, $NaHCO_3$, MeOH, RT

28 31-70% ← **27** → **29** 57-76%

Scheme 7.9

Dibenzofuran-type molecules can be formed from the reaction of *o*-iodophenols with silylaryl triflate in the presence of a palladium catalyst. Nucleophilic addition of *o*-iodophenols to benzyne (generated by treatment of silylaryl triflate with CsF) and subsequent Pd-catalyzed intramolecular arylation are involved in the reaction [21] (Scheme 7.10).

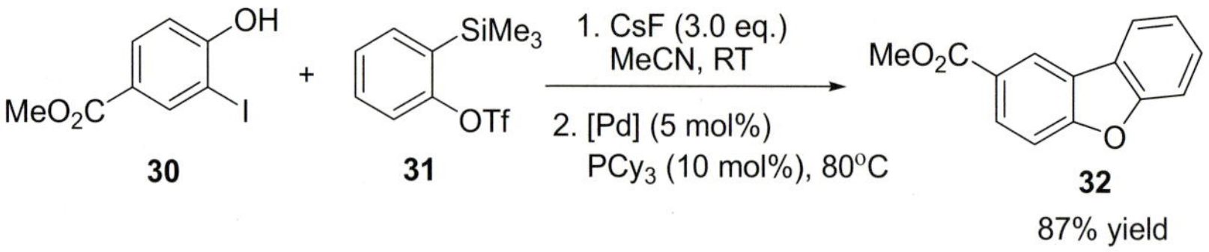

Scheme 7.10

Recently, Stoltz [22] has developed a method for the synthesis of electron-rich, highly substituted benzofuran and dihydrobenzofuran derivatives (50–80% yields) via an intramolecular Fujiwara-Moritani/oxidative Heck reaction (Scheme 7.11); 15 examples are demonstrated in the article. No extra functionalization step was required for palladium(II)-catalyzed oxidative carbocyclizations, which provided highly substituted benzofurans and dihydrobenzofurans by net dehydrogenation. The direct C–H bond functionalization of the aromatic ring and cyclization with unactivated olefins is involved in the oxidation process. Furthermore, the products contain quaternary carbon stereocenters that can be obtained in diastereomerically pure form.

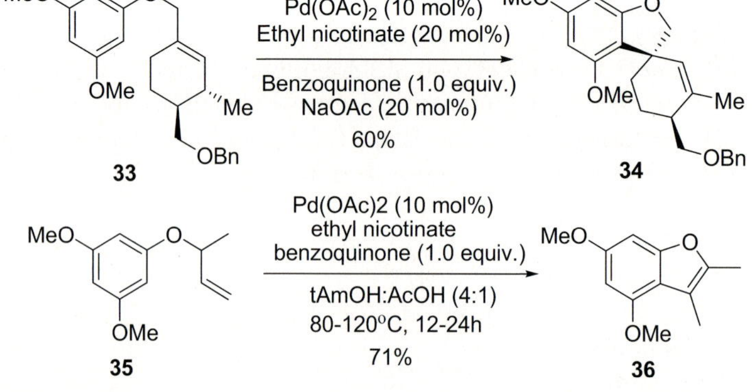

Scheme 7.11

Ionic liquid ([BMIm]BF_4) has been utilized as an effective solvent for the $PdCl_2$-catalyzed benzofuran formation via an intramolecular Heck reaction [23]. This strategy has been applied to construct the seven-membered ring based (−)-frondosin B [24].

The palladium-catalyzed annulation of 1,3-dienes with *o*-iodoacetoxyflavonoids for the generation of biologically interesting dihydrofuroflavonoids was realized (Scheme 7.12). This reaction is quite general and regioselective, and a wide variety of terminal, cyclic and internal 1,3-dienes are applicable [25].

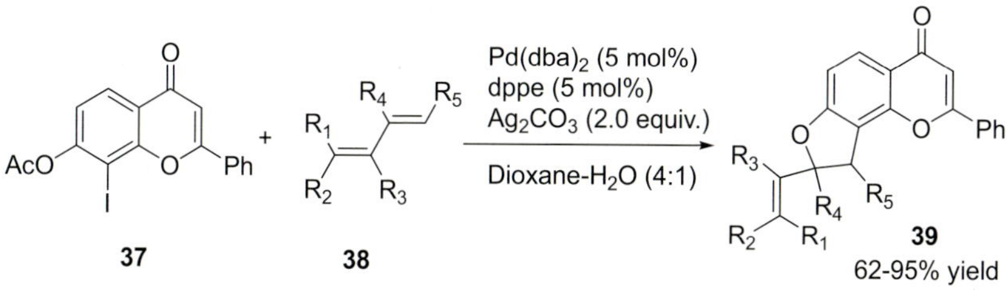

AcO, I, O, O, Ph — **37** + R_1, R_2, R_3, R_4, R_5 — **38**; $Pd(dba)_2$ (5 mol%), dppe (5 mol%), Ag_2CO_3 (2.0 equiv.), Dioxane-H_2O (4:1) → **39** 62-95% yield

Scheme 7.12

One of the key intermediates (**43**) towards the total synthesis of frondosin B has been synthesized by a sequential reaction from phenol **40**, enyne **41**, and bromide **42** in a one-pot operation (Scheme 7.13) [26]. Palladium-catalyzed intramolecular C−O bond formation between aryl halides and enolates has been employed to form 2,3-disubstituted benzo[*b*]furans [27].

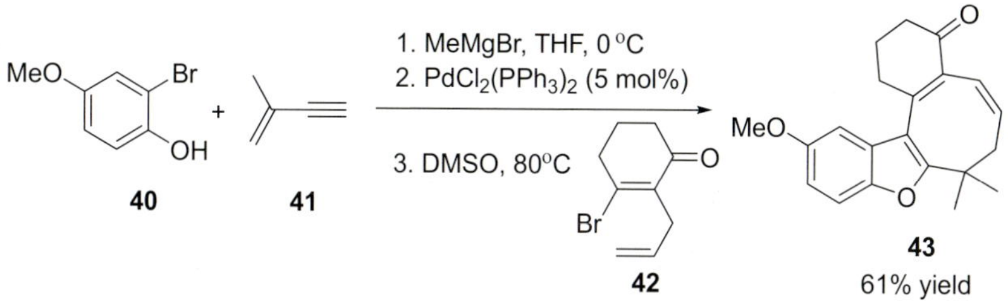

MeO, Br, OH — **40** + **41**; 1. MeMgBr, THF, 0 °C; 2. $PdCl_2(PPh_3)_2$ (5 mol%); 3. DMSO, 80°C, **42** (Br, =O) → MeO, O, =O — **43** 61% yield

Scheme 7.13

Synthesis of 3-fluoroalkylated benzo[*b*]furans was achieved via a palladium-catalyzed reaction of fluorine-containing internal alkynes with various 2-iodophenols in the presence of $P(^tBu)_3$ as an essential ligand [28] (Scheme 7.14).

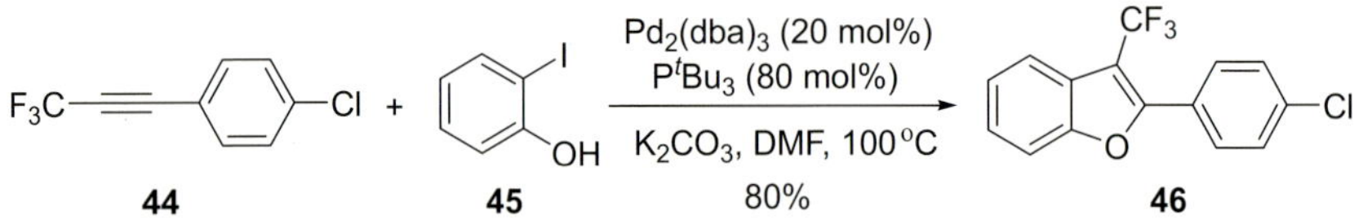

Scheme 7.14

As shown in Scheme 7.15, the optically active dihydrobenzo[*b*]furan-ring **48** has been constructed efficiently via a C–H insertion reaction, leading to the total synthesis of (−)-ephedradine A [29].

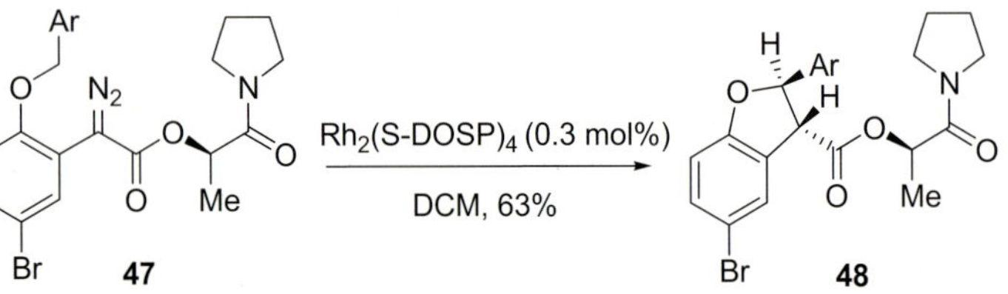

Scheme 7.15

Herndon has reported benzannulation of heterocyclic ring systems through coupling of Fischer carbene complexes and heterocycle-bridged enynes [30]. The benzofuran rings are easily annulated onto furan, thiophene and imidazole ring systems in a reaction process involving the coupling of Fischer carbene complexes with either 2-alkenyl-3-alkynyl- or 3-alkenyl-2-alkynyl-heteroaromatic systems (Scheme 7.16).

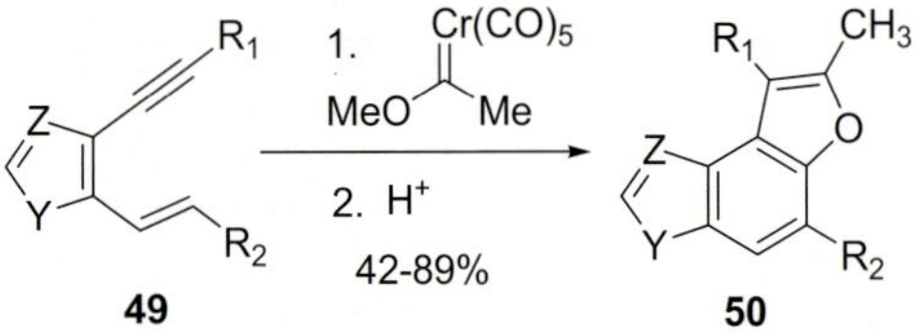

Scheme 7.16

Arylboronic acids reacted with nitriles catalyzed by a cationic palladium complex leading to aryl ketones in moderate to good yields [31]. Based on this result, a one-step synthesis of benzofurans from phenoxyacetonitriles under the catalysis of $[(bpy)Pd^{+}(OH)]_2(^{-}OTf)_2$ or $[(bpy)Pd^{2+}(H_2O)_2](^{-}OTf)_2$ has been developed that shows that the cationic palladium catalyst is highly active for these addition reactions (Scheme 7.17).

2-Chloroaryl alkynes, which are generated from 1,2-dihaloarenes using known methodology, undergo the reaction conditions shown in Scheme 7.18 successfully providing benzofurans **54** in good yields [32]. While the cyclization of 2-hydroxyalk-

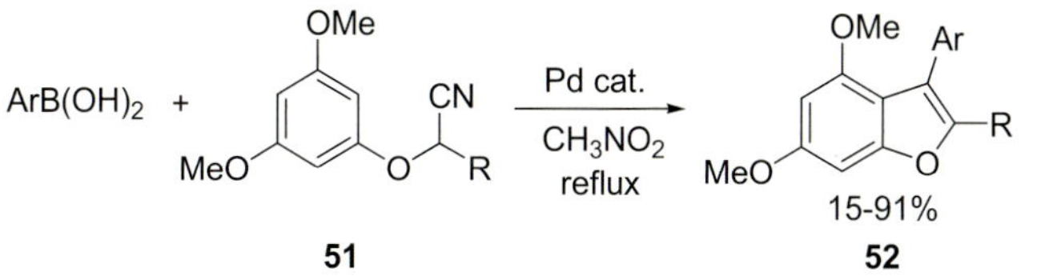

Scheme 7.17

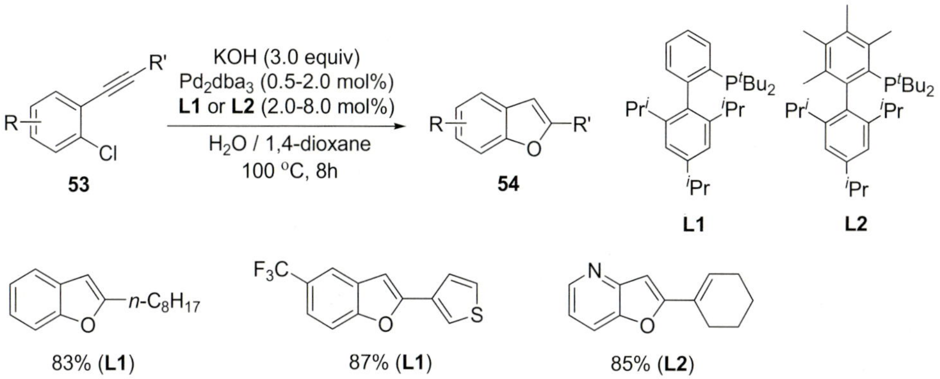

Scheme 7.18

ynyl arenes is known, this is the first time this strategy has been employed starting with a 2-haloaryl alkyne.

The formation of 2-alkynyl benzofurans is described via a new tandem coupling approach [33]. This reaction utilizes easily accessible *gem*-dibromovinyl substrates and terminal alkynes and proceeds via Pd/C- and CuI-catalyzed tandem Ullman/Sonogashira couplings (Scheme 7.19).

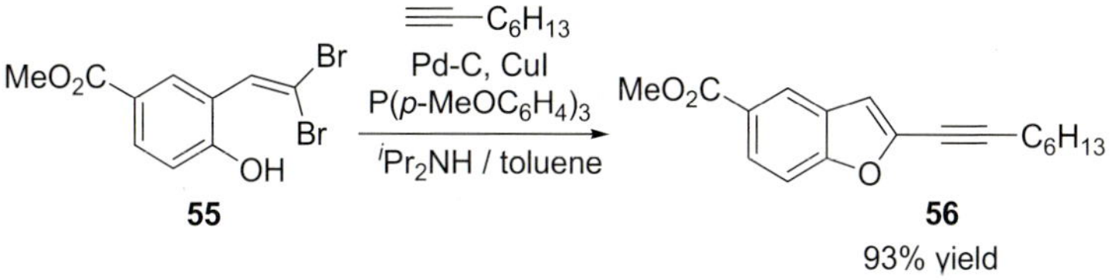

Scheme 7.19

An efficient synthesis of benzofurans from *o*-anisole-substituted ynamides has been reported by Hsung and co-workers via an unexpected Rh(I)-catalyzed demethylation-cyclization sequence (Scheme 7.20) [34]. The silver salt is critical for the successful transformation in the reaction process.

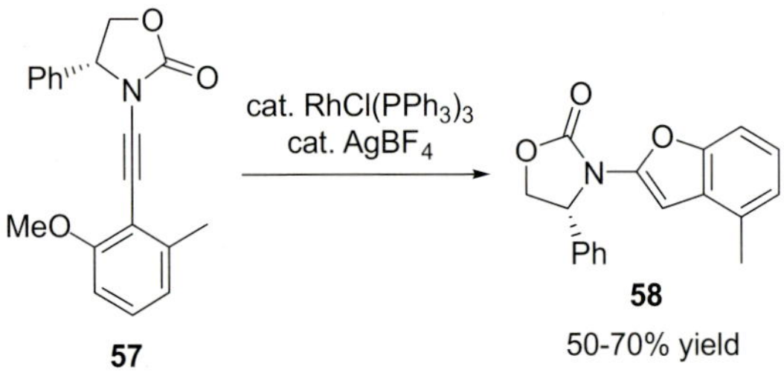

Ph O N O MeO

cat. $RhCl(PPh_3)_3$
cat. $AgBF_4$

O O N O Ph

57

58
50-70% yield

Scheme 7.20

Trost has reported that benzofurans can be formed chemoselectively from the Rh-catalyzed cyclo-isomerization reaction of easily prepared 2-alkynylphenol substrates (e.g., 2,4-dichloro-6-ethynylphenol) [35]. The reaction may proceed by nucleophilic capture of a vinylidene intermediate (Scheme 7.21).

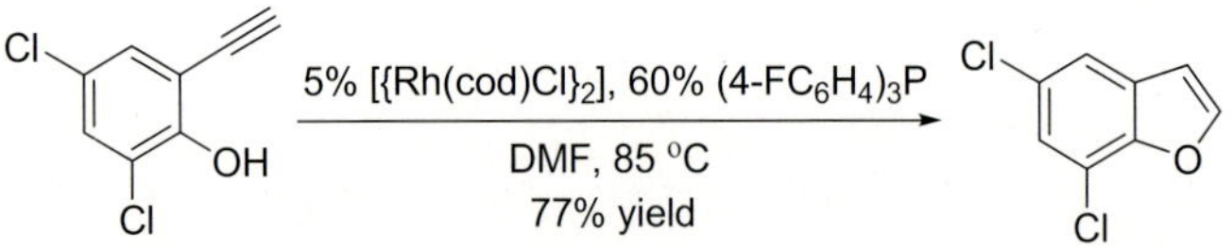

Cl OH Cl

5% $[\{Rh(cod)Cl\}_2]$, 60% $(4\text{-}FC_6H_4)_3P$
DMF, 85 °C
77% yield

Cl O Cl

59

60

Scheme 7.21

Benzofuran products can be delivered in good yields through palladium-catalyzed intramolecular C–O bond formation of enolates derived from α-(ortho-haloaryl)-substituted ketones (Scheme 7.22) [36]. A catalyst generated from $Pd_2(dba)_3$ and the ligand DPEphos effects the key bond formation to produce various substituted products from both cyclic and acyclic precursors. In the meantime, a cascade sequence that produces the required α-aryl ketones *in situ* has been developed. However, the substrate scope is more restricted.

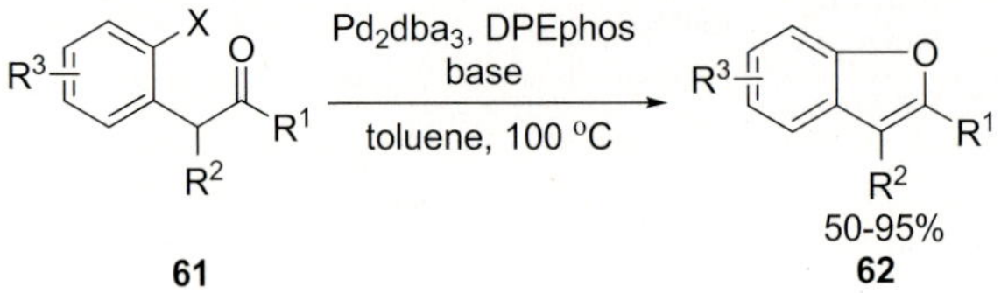

R^3 X O R^1 R^2

Pd_2dba_3, DPEphos
base
toluene, 100 °C

R^3 O R^1 R^2
50-95%

61

62

Scheme 7.22

Li has presented a novel and selective palladium-catalyzed annulation of 2-alkynylphenols method for the synthesis of 2-substituted 3-halobenzo[*b*]furans [37]. In the presence of PdX_2, 2-substituted benzo[*b*]furans were afforded in good yields, whereas in the presence of 5–10 mol% of PdX_2, 3.0 equiv of CuX_2 and 0.2 equiv of

HEt_3NI, 2-disubstituted 3-halobenzo[*b*]furans were selectively produced as the major products (Scheme 7.23). A possible mechanism was also proposed.

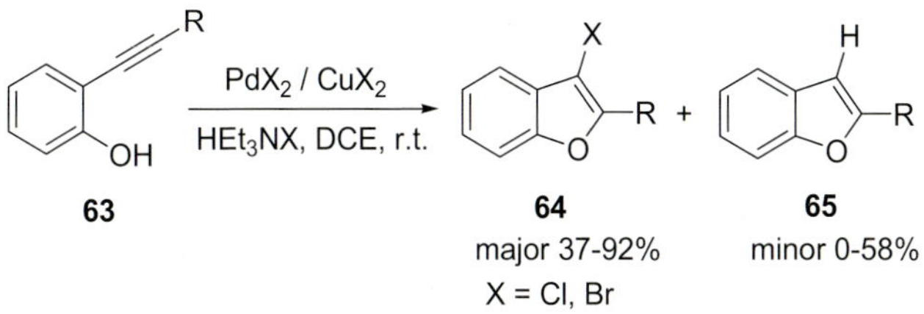

R
PdX_2 / CuX_2
HEt_3NX, DCE, r.t.
OH
63
X
R +
O
64
major 37-92%
X = Cl, Br
H
R
O
65
minor 0-58%

Scheme 7.23

3-Zinciobenzofurans **66**, generated in excellent yield through a metallative 5-*endo-dig* cyclization reaction of 2-alkynylphenoles in the presence of BuLi and ZnCl2, can be further transmetallated to the corresponding cuprates **67**, which then react with electrophiles to produce various 2,3-disubstituted benzofurans **68** [38] (Scheme 7.24).

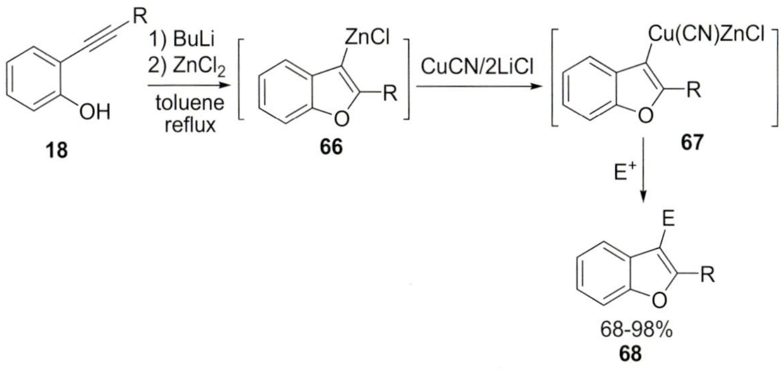

R
1) BuLi
2) $ZnCl_2$
toluene
reflux
OH
18
ZnCl
R
O
66
CuCN/2LiCl
Cu(CN)ZnCl
R
O
67
E^+
E
R
O
68-98%
68

Scheme 7.24

7.4.2
Oxidative Cyclization

Recently, Bellur [39] has reported an efficient synthesis of functionalized benzofurans based on a [3 + 2] cyclization/oxidation strategy. The functionalized benzofurans were prepared by DDQ oxidation of 2-alkylidenetetrahydrofurans, which are readily available by one-pot cyclizations of 1,3-dicarbonyl dianions or 1,3-bis-silyl enol ethers (Scheme 7.25).

The first total and biomimetic synthesis of violet-quinone has been accomplished by utilizing an oxidative dimerization of a substituted 4-methoxy-1-naphthol with a ZrO_2/O_2 system, the so-formed dimer eventually led to the target molecule (Scheme 7.26) [40]. The same research group later published the $SnCl_4$-mediated oxidative biaryl coupling reaction to build up the dinaphthanofuran framework [41]. Silver(I) acetate is an efficient agent with which to obtain the dimer of resveratrol in

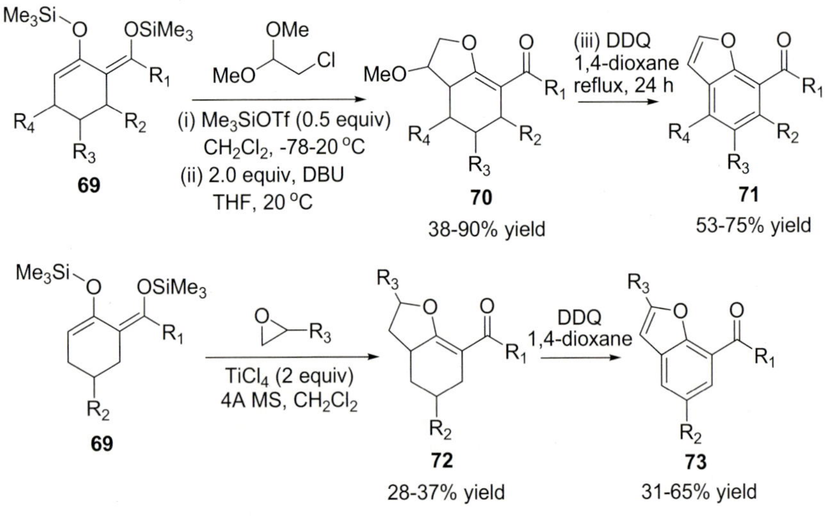

Scheme 7.25

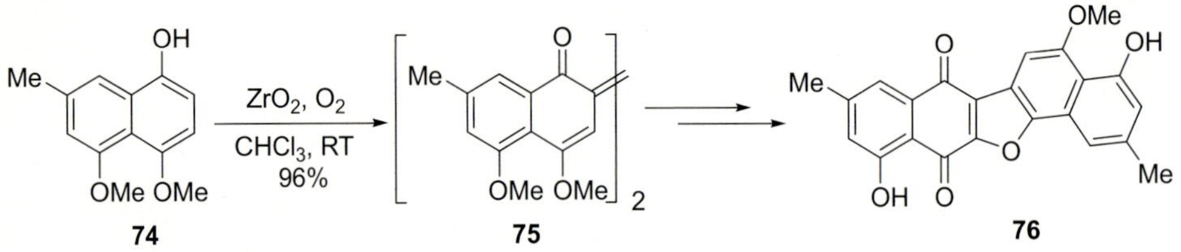

Scheme 7.26

high yield [42]. Oxidation of phenol with PIFA has also been applied to construct the framework of (−)-galanthamine [43].

A new family of benzo[*b*]furans has been synthesized by an anodic oxidation of an aqueous solution of 3-substituted catechols followed by coupling with dimedone (**78**) (Scheme 7.27) [44].

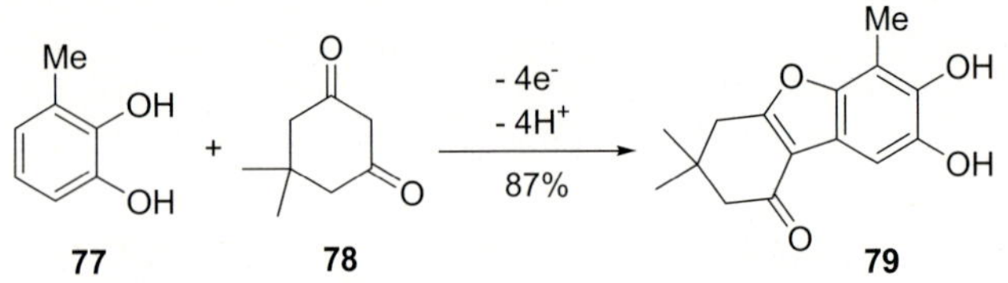

Scheme 7.27

Compound **80** can be cyclized in the presence of either mercury acetate in acetic acid or bromine in chloroform to give 3-chloromercurio- or 3-bromobenzofuran, respectively. The 3-chloromercurio intermediates could be reduced to proton or derivatized to ester or bromide, leading to the synthesis of ailanthoidol (30% yield), XH-14 (15% yield) and obovaten (11% yield), respectively [45] (Scheme 7.28).

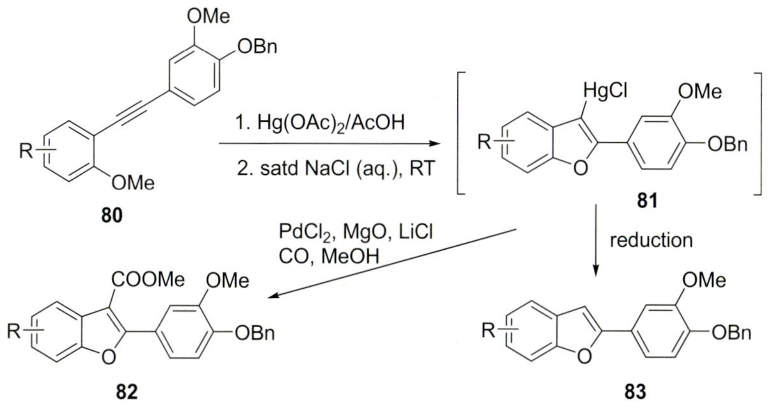

Scheme 7.28

5-[2-(4-Hydroxyphenyl)vinyl]benzene-1,3-diol **84** (resveratrol) was treated with an equimolar amount of silver(I) acetate in dry MeOH to afford its (*E*)-dehydrodimer, 5-{5-[2-(3,5-dihydroxyphenyl)vinyl]-2-(4-hydroxyphenyl)-2,3-dihydrobenzofuran-3-yl} benzene-1,3-diol **85**, as a racemic mixture in high yield [46] (Scheme 7.29). This method is applicable to the oxidative dimerization of 4-hydroxystilbenes such as *trans*-styrylphenol and 5-{6-hydroxy-2-(4-hydroxyphenyl)-4-[2-(4-hydroxyphenyl)vinyl]-2,3-dihydrobenzofuran-3-yl}benzene-1,3-diol (viniferin), giving rise to the corresponding 2-(4-hydroxyphenyl)-2,3-dihydrobenzofurans.

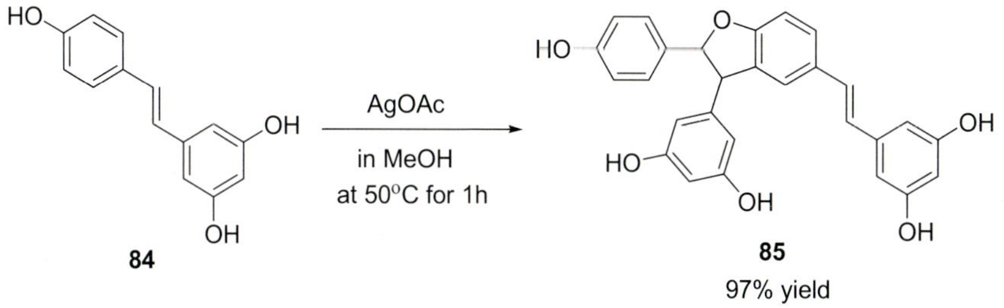

Scheme 7.29

7.4.3
Radical Cyclization

A radical initiated benzo[*b*]furan formation has been applied to the synthesis of spiro [chroman-3,3′-(2′*H*)-benzofurans] in the presence of *n*-Bu_3SnCl and $Na(CN)BH_3$ [47] (Scheme 7.30).

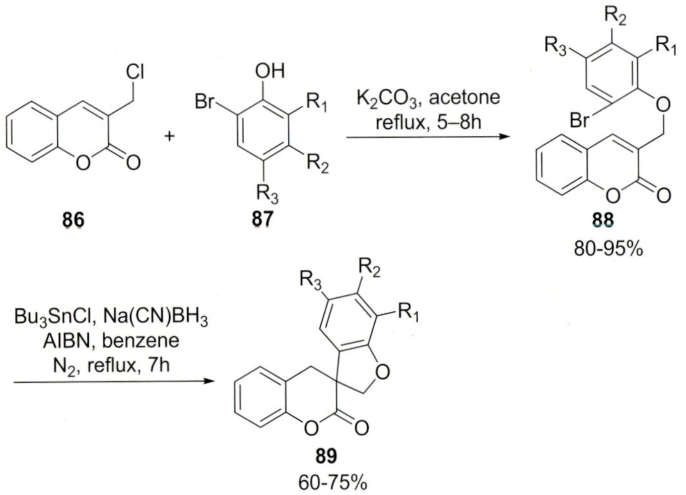

Cl
OH
Br
R_1
R_2
R_3
O O
+
K_2CO_3, acetone
reflux, 5–8h
86
87
88
80-95%
Bu_3SnCl, $Na(CN)BH_3$
AIBN, benzene
N_2, reflux, 7h
89
60-75%

Scheme 7.30

A similar approach has also been demonstrated by the same group to obtain spiro [pyrimidine-6,3′-2′,3′-tetrahydrobenzofuran]-2,4-diones [48]. On the other hand, *n*-Bu_3GeH is reported to be an effective alternative compared with *n*-Bu_3SnH in the synthesis of 3-substituted-2,3-dihydrobenzo[*b*]furans [49]. Moreover, a photo-induced fast tin-free reductive radical dehalogenation has found use for the synthesis of 2,3-dihydrobenzo[*b*]furans [50].

2,3-Disubstituted benzofuran derivatives have been synthesized from *o*-acylphenols in two steps. The β-aryloxyacrylates prepared from the o-acylphenols react with n-Bu_3SnH/AIBN and then with 5% HCl-EtOH leading to 2,3-disubstituted benzofurans [51].

A radical [3 + 2] annulation reaction with an N-centered radical has been developed. The reaction of alkenes with *N*-allyl-*N*-chlorotosylamide yields the corresponding pyrrolidine derivatives **90** in good yields, in the presence of Et_3B as a radical initiator, via an atom-transfer process [52] (Scheme 7.31).

Grimshaw [53] has reported that the cyclization of 2′-bromodeoxybenzoin with Cu powder in refluxing $AcNMe_2$ gives 2-phenylbenzofurans in 65–70% yield (Scheme 7.32).

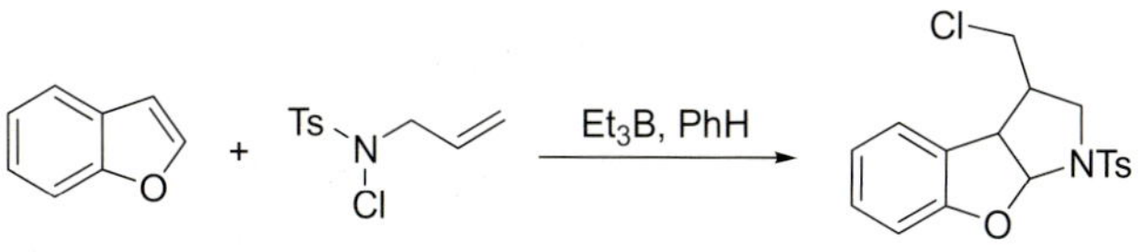

Et$_3$B, PhH

90
82% yield

Scheme 7.31

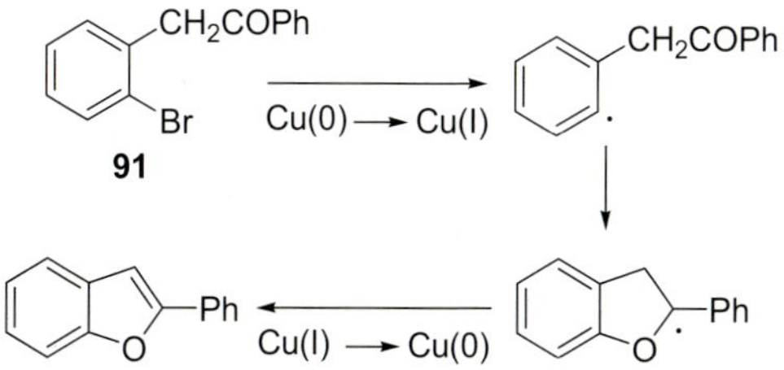

91

Cu(0) → Cu(I)

Cu(I) → Cu(0)

92
65-70% yield

Scheme 7.32

7.4.4
Acid- and Base-Mediated Cyclization

Katritzky [54] has disclosed the preparation of 2,3-disubstituted benzofurans by reactions of *o*-hydroxyphenyl ketones or *o*-(1-hydroxy-2,2-dimethylpropyl)phenol with 1-benzotriazol-1-ylalkyl chlorides in two or three steps (Scheme 7.33). These

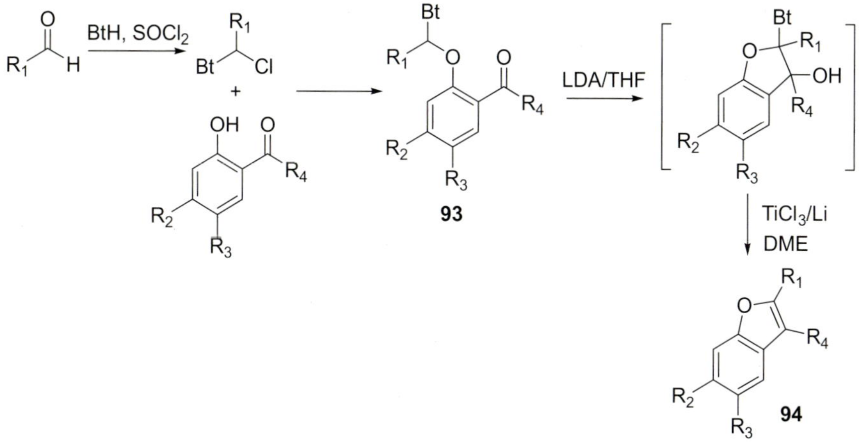

BtH, SOCl$_2$

LDA/THF

TiCl$_3$/Li
DME

93

94

Scheme 7.33

approaches provide a facile route to a variety of benzofurans in good overall yields and complements a previous benzotriazole-mediated preparation of benzofurans.

2-Arylbenzo[*b*]furan can be cyclized in a modest yield from a non -symmetrical diarylethyne, which was generated via palladium -catalyzed Sonogashira reaction of an aryl bromide and an aryl acetylene (Scheme 7.34) [55]. Other types of 2-alkyl/aryl substituted benzo[*b*]furans have also been obtained by the palladium-catalyzed coupling reaction of *o*-iodophenols (even *o*-iodophenols with a base-labile nitro group) with various alkynes in the presence of prolinol as base in water. This environmental friendly procedure does not need the phase transfer or water-soluble phosphine ligands and is free from the use of any organic co-solvent [56]. A similar process has also been reported with an amphiphilic polystyrene–polyethylether (PS-PEG) resin-supported palladium-phosphine complex as a catalyst in water to give the corresponding aryl-substituted alkynes in high yield under copper-free conditions [57]. In the total synthesis of pterulinic acid, the core structure of 2-substituted benzofuran was generated by the palladium-catalyzed heteroannulation of an *o*-iodophenol derivative with methyl 3-butynoate [58].

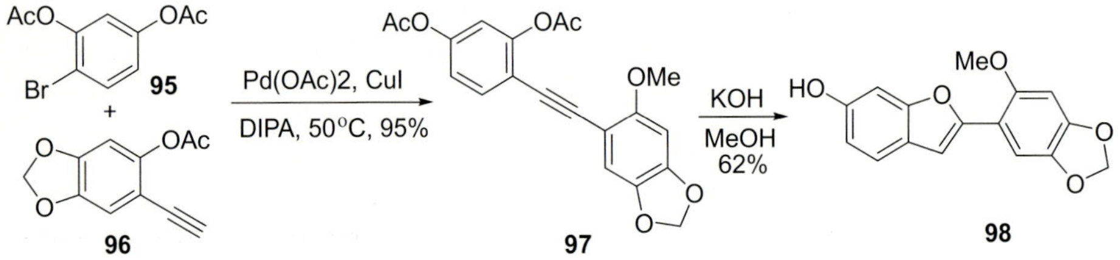

Scheme 7.34

Base-mediated conditions have also been applied to afford 2-arylbenzo[*b*]furan, employing the coupling product generated from the ultra-fine nickel-catalyzed Sonogashira reaction of iodophenols with phenylacetylenes [59]. Four 2,6-linked and 2,5-linked benzo[*b*]furan trimers as organic electroluminescent materials have been prepared by the base-mediated cyclization of 2-alkynylphenols [60].

In the synthesis of furoclausine A, acid-catalyzed conditions were used to produce the furo[3,2-*a*]carbazole framework from a ketal as depicted (Scheme 7.35) [61]. An acid-catalyzed intramolecular cyclization to form the scaffold of furoquinoline

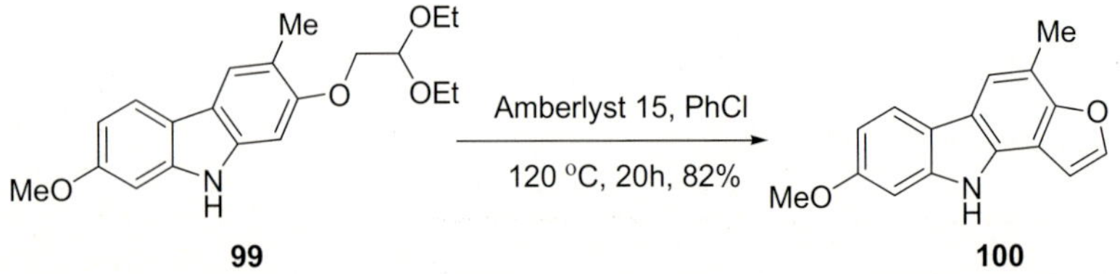

Scheme 7.35

alkaloids has also been achieved from 3-oxiranylquinolines [62]. Furanoeremophilane sesquiterpenes have been synthesized by acid-mediated furan ring formation from their corresponding phenolic-ketone ethers [63]. 3-Aryl-2,2-dialkyl-2,3-dihydrobenzo[*b*]furans have been delivered from phenols and 2-aryl-2,2-dialkylacetaldehydes in the presence of a catalytic amount of CF_3SO_3H [64]. A $ZnCl_2$-mediated benzo[*b*]furan formation has been utilized to produce 2-carboxylate benzo[*b*]furans from 3-dimethylaminopropenoates [65].

Synthesis of different types of substituted benzoheterocycles under metal-free protocols have been developed. The combination of *o*-alkynylbenzaldehyde derivatives, iodonium ions and alkenes was demonstrated to effectively produce the corresponding benzofurans [66]. The rapid access to benzofurans with interesting di- and tri-substitution patterns and featuring the 2,3-unsubstituted ring motif has been established (Scheme 7.36).

1. IPy_2BF_4 / HBF_4 (1.1 equiv), CH_2Cl_2, 0 °C to r.t.
2. Ph-CH=CH$_2$ (1.2 equiv), r.t.

101 → **102** (36% yield)

Scheme 7.36

A simple procedure for the construction of the *trans*-5,6-ring system existing in phenylmorphans was developed by the displacement of nitro-activated aromatic fluorine with a hydroxyl group [67] (Scheme 7.37).

103 → (NaH, THF, reflux, 97%) → **104** → (Pd-C, $HCOONH_4$, 90%) → **105**

Scheme 7.37

Synthesis of coumestrol (**108**) has been achieved by sequential condensation between phenyl acetate and benzoyl chloride, followed by demethylation and cyclization (Scheme 7.38) [68].

Hellwinkel has reported 2-arylbenzofurans formation using MF/Al_2O_3 base-systems [69]. Saito has described a novel method for the generation of 4-acetoxy-2-amino-3-arylbenzofurans from 1-aryl-2-nitroethylenes and cyclohexane-1,3-diones [70]. This one-pot operation shows high efficiency (Scheme 7.39).

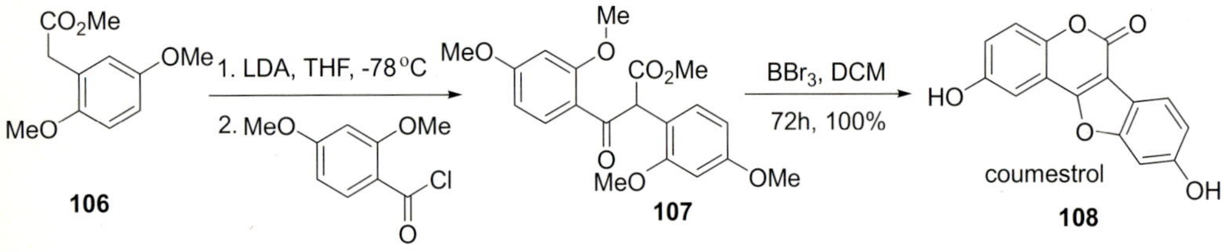

Scheme 7.38

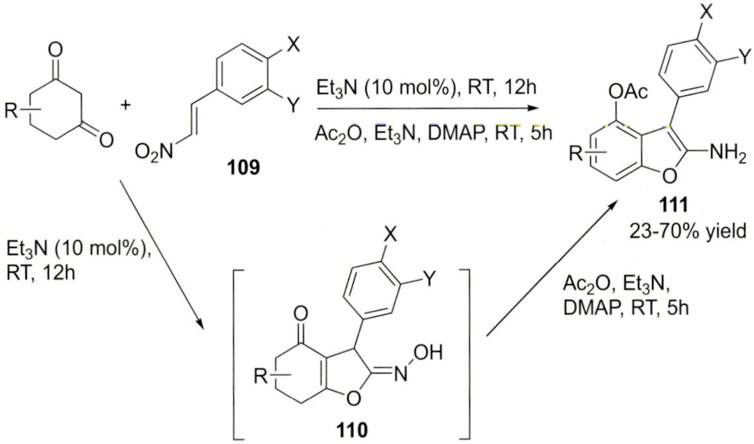

Scheme 7.39

Wagner [71] has described the synthesis of 2-(6-hydroxy-2-methoxy-3,4-methylenedioxyphenyl)benzofuran (**113**, Scheme 7.40) via an acid mediated cyclization.

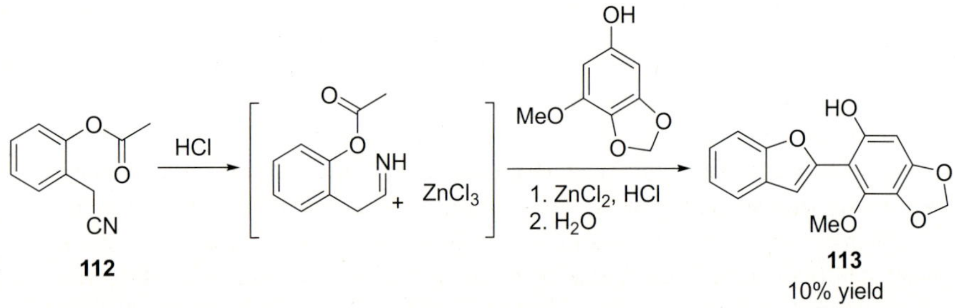

Scheme 7.40

Johnson [72] has reported benzofuran formation via a organolithium-induced cyclization. Treatment of 2,2,2-trifluoroethyl phenyl ethers with 4 equivalents of an aryl or alkyl lithium reagent (R^3Li) causes in almost all cases complete dehalogenation of the trifluoroethyl side chain with the concomitant introduction of an alkyl or aryl group (R^3) at the acetylenic 2-position (Scheme 7.41). This is followed apparently by ortholithiation to give the lithio intermediates; the latter then spontaneously cyclize to the 2-lithioheterocyles. Subsequent electrophilic quenching leads to the corresponding benzofurans depending whether the electrophile is a proton or another electronegative species.

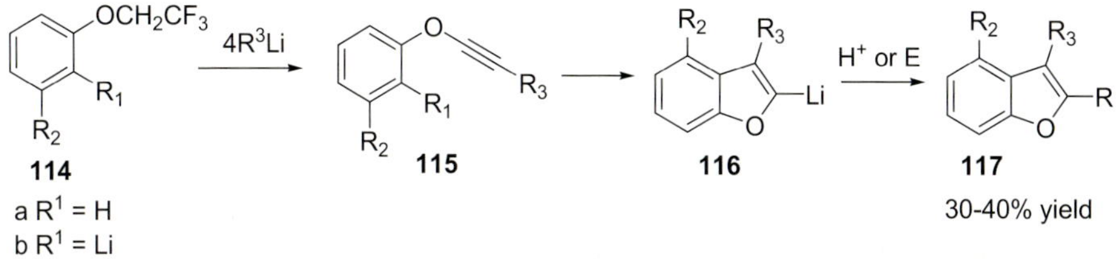

OCH_2CF_3 $4R^3Li$ R_1 R_2 **114** a R^1 = H b R^1 = Li → **115** → **116** → H^+ or E → **117** 30-40% yield

Scheme 7.41

Banerji [73] [Scheme 7.42 (1)] and Clerici [74] [Scheme 7.42 (2)] have reported titanium-mediated benzofuran synthesis, respectively.

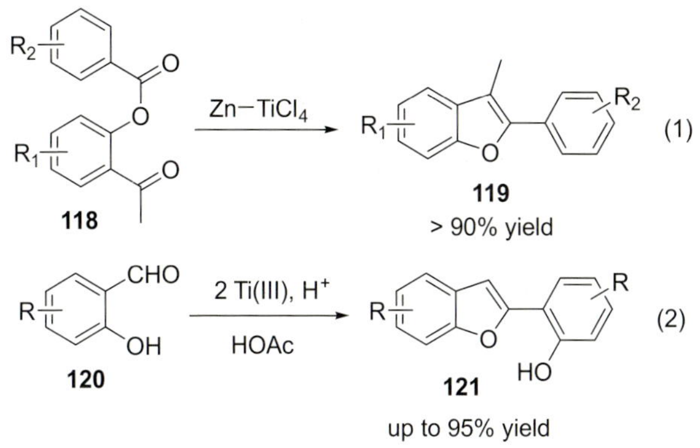

118 → $Zn-TiCl_4$ → **119** > 90% yield (1)

120 → 2 Ti(III), H^+ / HOAc → **121** up to 95% yield (2)

Scheme 7.42

Jha has also reported a facile synthesis of 2-arylbenzo[*b*]furans through an unusual acid-catalyzed 1,2-elimination [75]. 2-Phenoxyalkanals react with methanol at room temperature under homo- or heterogeneous acid-catalysis conditions leading to the formation of diacetal as well as some quantities of appropriate 2-alkylbenzofurans. 2-Alkylbenzofurans can be obtained in high yields via cyclization of the 1,1-dimethoxy-2-phenoxyalkanes under mild conditions over Amberlyst 15 [76].

An effective route to chiral optically active 2-substituted benzofurans directly from carboxylic acids has been reported (Scheme 7.43) [77]. This procedure, which allows

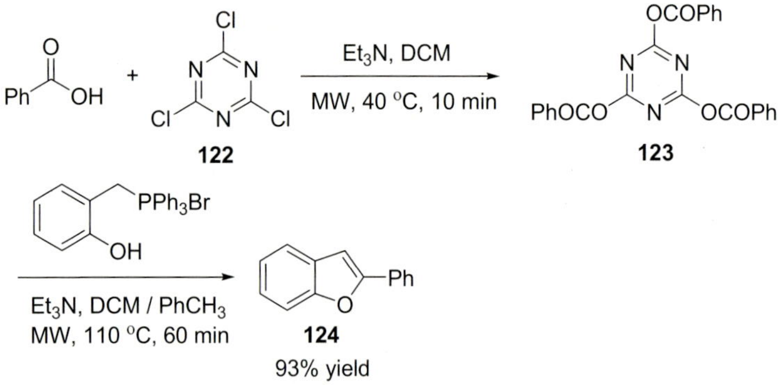

Scheme 7.43

the preparation of R-alkyl-2-benzofuranmethanamines from N-protected R-amino acids without sensible racemization phenomena, proceeds in good yields under mild conditions with the help of microwave irradiation.

Treatment of benzyl 2-halophenyl ethers with 3 equivalents of t-BuLi results in Li–halogen exchange and lithiation at benzylic methylene simultaneously. These dianions can be trapped with electrophiles. Their reactions with carboxylic esters afford the corresponding 2-aryl-3-hydroxy-2,3-dihydrobenzo[*b*]furans, which subsequently undergo acid-catalyzed or mediated dehydration to give moderate to good overall yield of various 2-aryl-3-substituted benzo[*b*]furans (**127**) (Scheme 7.44) [78].

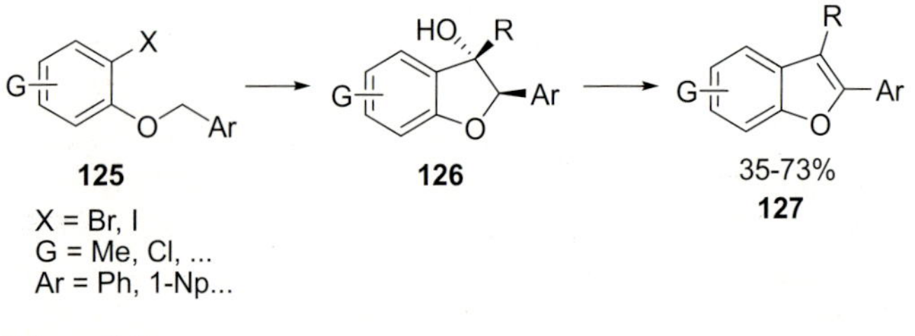

Scheme 7.44

$Zn(OTf)_2$ (10 mol%) catalyzes the cyclization of propargyl alcohols with PhOH in hot toluene (100 °C) without additive to give benzofuran products in good yields. Its mechanism has been elucidated. This catalytic cyclization is also applicable to the synthesis of oxazoles through the cyclization of propargyl alcohols and amides without a 1,2-nitrogen shift (Scheme 7.45) [79].

Asao has described an efficient synthesis of functionalized aromatic compounds from enynals and carbonyl compounds [80]. The reaction most probably proceeds through the reverse electron demand-type Diels–Alder reaction between the pyrylium intermediate and enol 2π-system, derived from carbonyl compounds. The scope of the reaction was extended to the synthesis of benzofused heteroaromatic compounds. For instance, benzofuran synthesis using furan derivatives **130** has been

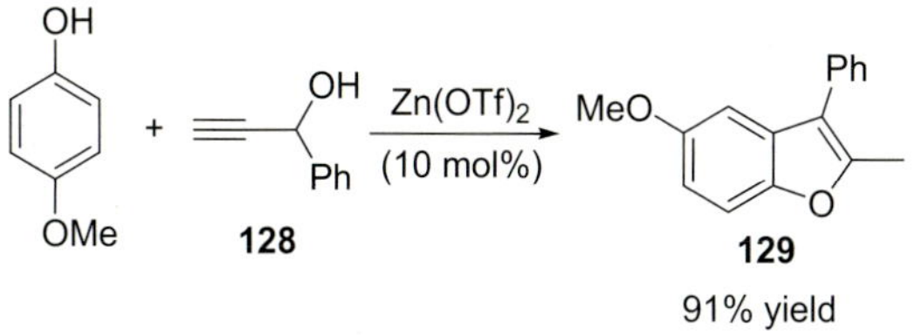

OH
OMe
+
OH
Ph
128
$Zn(OTf)_2$
(10 mol%)
MeO
Ph
O
129
91% yield

Scheme 7.45

examined. As expected, the reaction of **130** with propionaldehyde proceeded in the presence of $AuBr_3$ or $Cu(OTf)_2$ catalyst to give the corresponding product **131** in 58 or 67% yields, respectively. The $AuBr_3$-catalyzed reaction of **130** with β-methoxystyrene gave **132** in 50% yield (Scheme 7.46).

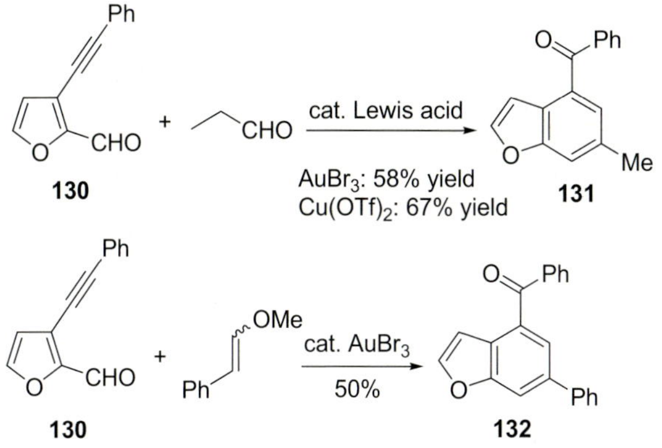

Ph
O
CHO
130
+
CHO
cat. Lewis acid
$AuBr_3$: 58% yield
$Cu(OTf)_2$: 67% yield
O
Ph
O
Me
131

Ph
O
CHO
130
+
OMe
Ph
cat. $AuBr_3$
50%
O
Ph
O
Ph
132

Scheme 7.46

SanMartin has reported copper-catalyzed straightforward synthesis of benzo[*b*]furan derivatives in neat water [81]. This on-water methodology delivers a range of benzo[*b*]furans (**134**) in good to excellent yields starting from readily available substrates **133** (Scheme 7.47).

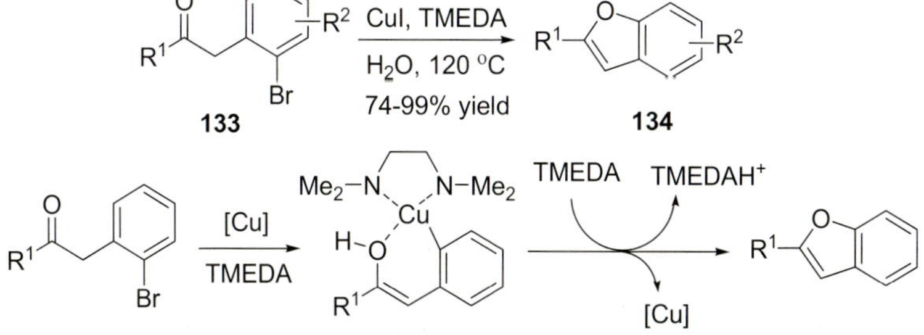

O
R^1
R^2
Br
133
CuI, TMEDA
H_2O, 120 °C
74-99% yield
R^1
O
R^2
134

O
R^1
Br
[Cu]
TMEDA
Me_2–N
N–Me_2
Cu
H–O
R^1
TMEDA
$TMEDAH^+$
[Cu]
R^1
O

Scheme 7.47

A microwave-mediated solvent-free Rap–Stoermer reaction has been reported for the synthesis of benzofurans from various salicylaldehydes and phenacyl halides (Scheme 7.48) [82]. Some of the advantages and highlights of this microwave protocol include, solvent-free clean reaction conditions and high yields of benzofurans obtained in short reaction times. In addition, the 2-aroyl benzofurans formed using this method are also important as the corresponding carbinols (reduction products) are known to have hypolipidemic activity.

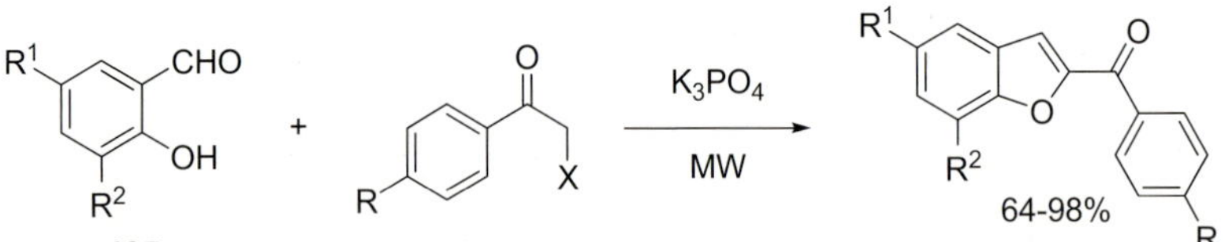

Scheme 7.48

DiMauro has reported a rapid, efficient synthesis of various substituted fused benzofurans using a microwave-assisted one-pot cyclization-Suzuki coupling approach [83]. The benzofuran scaffold was formed under base conditions. Further elaboration via Suzuki cross-coupling reactions afforded the desired products **140** in moderate yields (Scheme 7.49).

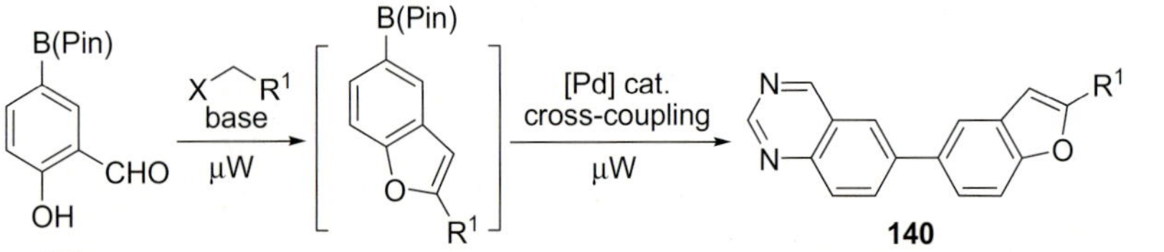

a: R^1 = CO_2Et, X = Br, 33% yield
b: R^1 = CONHPh, X = Cl, 40% yield
c: R^1 = COPh, X = Br, 74% yield
d: R^1 = CO^tBu, X = Br, 72% yield

Scheme 7.49

7.4.5
Olefin-Metathesis Approach

Sequential isomerization and ring-closing metathesis for the synthesis of benzo-fused heterocycles has been reported [84]. 2-Allylphenol is converted into allyl-2-(allyloxy)benzene (**141**) under suitable conditions. $[RuClH(CO)(PPh_3)_3]$ is then added to a solution of **141** in benzene-d_6 or toluene-d_8 and the reaction mixture heated at 60–80 °C for 18 h (Scheme 7.50). Analysis by ^1H NMR spectroscopy confirmed that the isomerization of both allyl substituents had occurred to afford

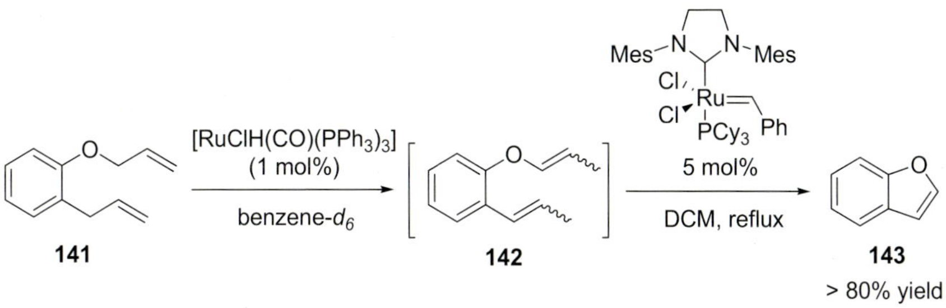

Scheme 7.50

the acid-labile compound, which was not isolated. The introduction of catalyst (Grubbs Generation I) then readily affords the benzo[*b*]furan in excellent conversion, as determined by further ^{1}H NMR spectroscopy. This result constitutes a novel approach to the ubiquitous benzo[*b*]furan skeleton.

Grubbs *et al.* have described a method for the synthesis of cyclic enol ethers (including benzofuran) via molybdenum alkylidene-catalyzed ring-closing metathesis (Scheme 7.51) [85]. To demonstrate an application of this process, the authors chose the naturally occurring benzofuran 2,4,2′,4′-dihydroxyphenyl-5,6-(methylenedioxy)benzofuran (*Sophora* compound I), the antifungal phytoalexin isolated from aerial part of *Sophora tomentosa* L [86], as a synthetic target.

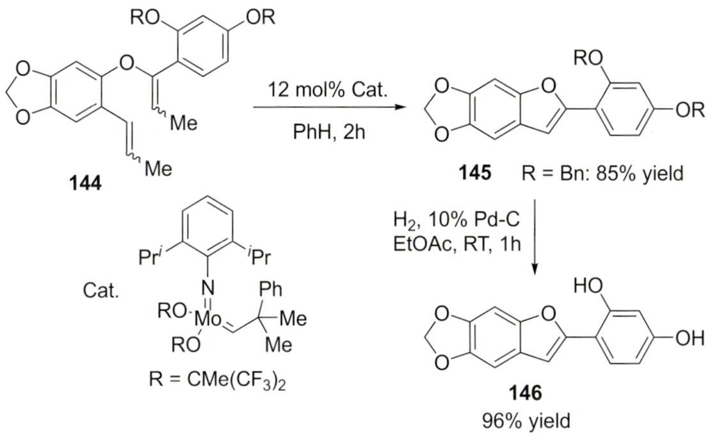

Scheme 7.51

Several 2,3-dihydrobenzo[*b*]furans can be made by the Ru-catalyzed olefin metathesis approach in the presence of trimethylsilyl vinyl ether (Scheme 7.52) [87]. The isovanillin derived benzo[*b*]furan has also been synthesized by the C-propenylation-O-vinylation and olefin metathesis approach [88].

A strategy employing a second generation Grubbs catalyst to facilitate the production of various cyclic enol phosphates, including benzofuran-2-yl enol phosphate scaffolds, has been described. This work represents the first case of an olefin

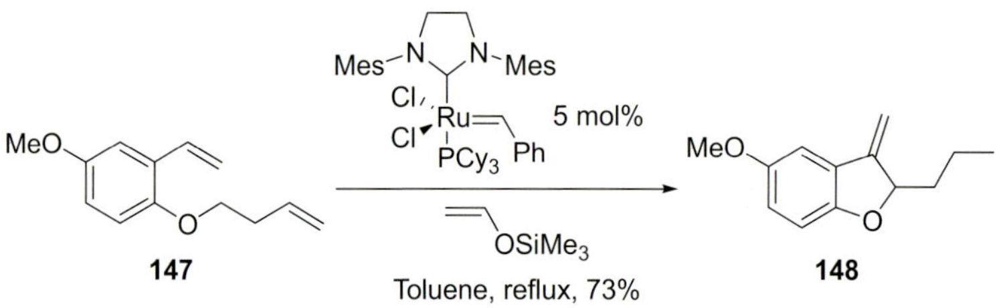

Scheme 7.52

metathesis reaction in which one of the groups participating in the metathesis event is an enol phosphate moiety [89].

7.4.6 Miscellaneous

3-Cyano or ethoxycarbonyl-2-methyl-benzo[*b*]furans have been prepared in a one-step synthesis by the microwave induced Claisen rearrangement under solvent-free conditions (Scheme 7.53) [90]. The Fries rearrangement has been employed in the synthesis of benzo[*b*]naphtha[2,3-*d*]furan-6,11-dione [91]. A [2,3]-Stille–Wittig rearrangement has also been utilized to make 2,3-disubstituted benzo[*b*]furans from 2-stannane substituted benzo[*b*]furans [92].

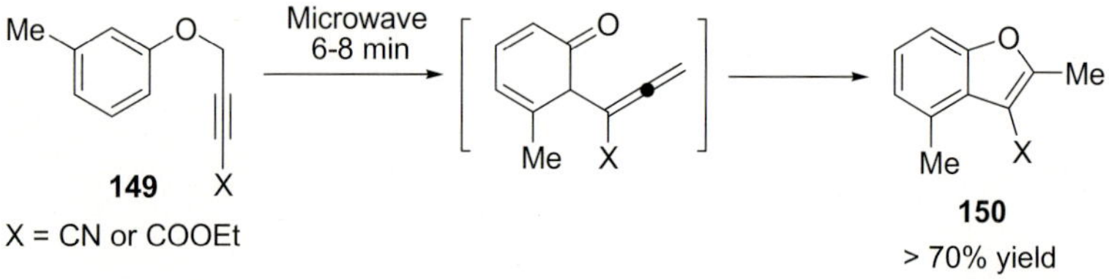

Scheme 7.53

An efficient generation of 2-arylbenzofurans proceeds via a route involving acylation and subsequent [3,3]-sigmatropic rearrangement of oxime ethers [93]. Its synthetic utility is demonstrated by a short synthesis of stemofuran A (**153**) and eupomatenoid (**154**) in which no procedure for protection of the phenolic hydroxyl groups is needed (Scheme 7.54).

The synthetic strategy involving an intramolecular hydroxyl epoxide opening has been applied to build up the cyclopenta[*b*]benzofuran ring for the total synthesis of naturally occurring rocaglaol [94] (Scheme 7.55).

Horaguchi has reported the synthesis of benzofurans using photocyclization reactions of aromatic carbonyl compounds [95]. Benzofurans functionalized with hydroxy and acetyl functionalities are not only the core structures found in numerous biological important natural products but are also the vital precursors for several

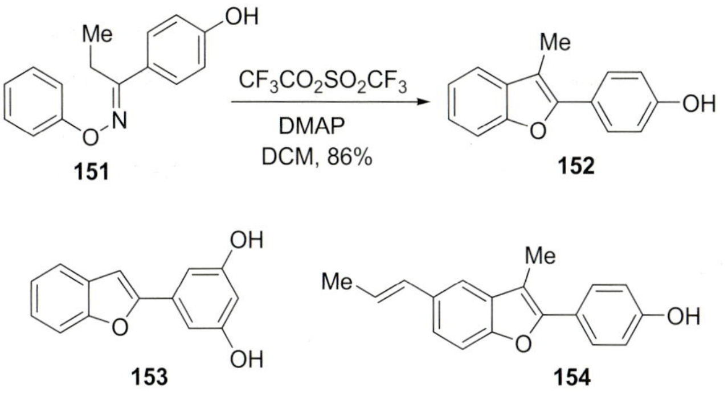

Scheme 7.54

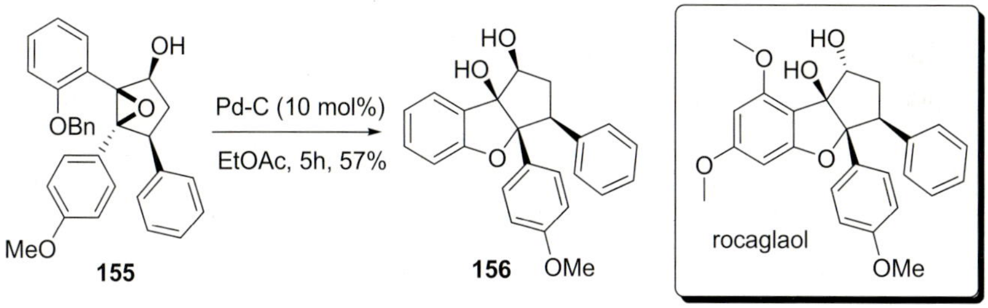

Scheme 7.55

naturally occurring furanoflavonoids. Numerous synthetic methodologies are available in the literature for the synthesis of functionalized benzofurans, but few references appear on the access of benzofurans with adjacent hydroxy and acetyl functionalities. Dixit [96] has reported a highly convenient synthesis of nature-mimicking benzofurans and their dimers from easily accessible precursors. The crystal structure of 5,5′-diacetyl-2′,3′-dihydro-2,3′-bibenzofuran-6,6′-diol is reported.

7.4.7 Progress in Solid-Phase Synthesis

Novel titanium benzylidenes (Schrock carbenes) bearing an arylboronate group are generated from thioacetals with low-valent titanium species, $Cp_2Ti[P(OEt)_3]_2$, and alkylidenate Merrifield resin-bound esters to give enol ethers. Treatment with 1% TFA gives 2-substituted (benzo[*b*]furan-5-yl)boronates, and solid-phase Suzuki cross-coupling gives 2,5-disubstituted benzofurans (Scheme 7.56) [97].

Yang *et al.* have reported a combinatorial synthesis of a 2,3-disubstituted benzo[*b*]furan library via palladium(II)-mediated cascade carbonylative annulation of *o*-alkynylphenols on silyl linker-based macrobeads (Scheme 7.57) [98].

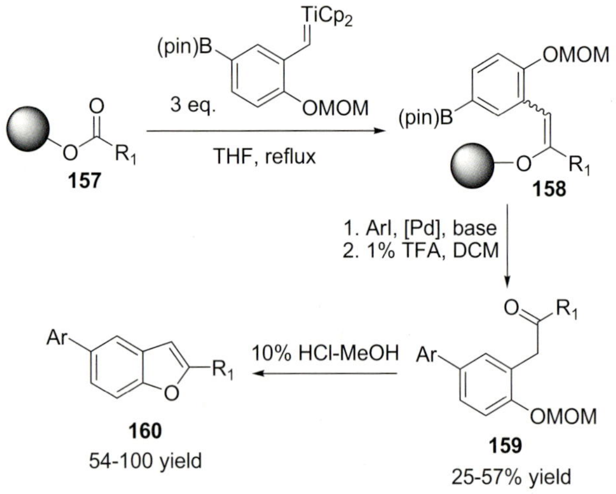

Scheme 7.56

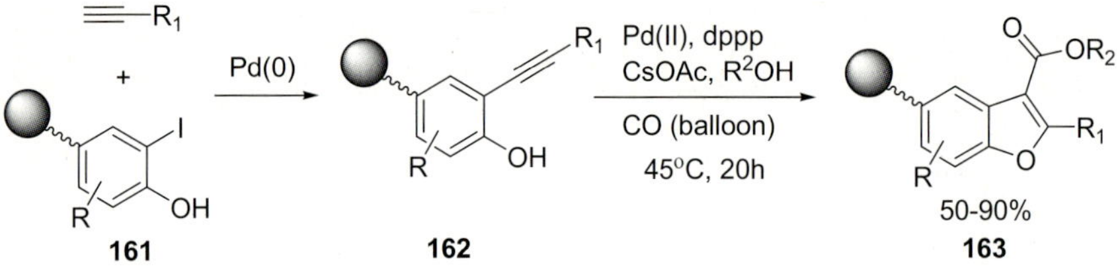

Scheme 7.57

The same authors have developed a novel catalytic system of AgOTs-$CuCl_2$-TMEDA for the homocoupling of aliphatic acetylenes on solid support. It is the first observation that an Ag(I)-activated triple bond can facilitate Cu(II)-mediated oxidative acetylenic homocoupling. This procedure provides an efficient way to synthesize a diversified symmetric 1,3-alkadiynediol bis(benzo[*b*]furan-carboxylate) library on solid support [99].

Furthermore, a split-pool synthesis of dimeric benzo[*b*]furans has been developed employing the Sonogashira reaction, palladium-mediated carbonylative annulation and olefin cross-metathesis as the key steps on high-capacity, lightly cross-linked, silyl-linker-based polystyrene macrobeads. This protocol provides direct access to a range of dimeric molecules that are ideal for high-throughput screening of protein–protein interactions in a cell-based assay system [100].

Subsequently, the authors described a conformationally restricted 2,3-diarylbenzo [*b*]furan library built up on a solid-phase by the palladium/bipyridyl-catalyzed annulation of *o*-alkynyl phenols with aryl halides (Scheme 7.58) [101].

A 2-substituted furo[3,2-*b*]quinolines library has been made on solid support by K_2CO_3-mediated sequential deprotection and cyclization [102] (Scheme 7.59).

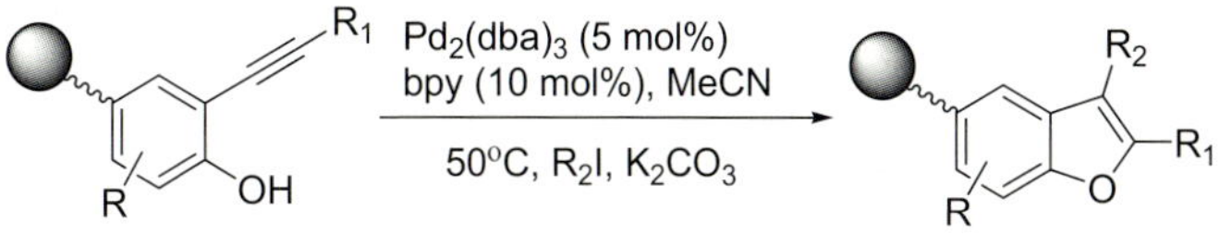

Scheme 7.58

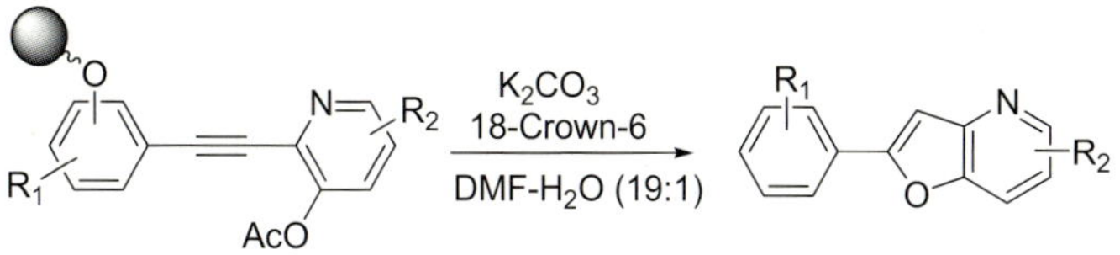

Scheme 7.59

7.5
Uses of Benzofuran

7.5.1
Uses of Benzofuran in Drug Discovery

Jones has synthesized benzbromarone analogs (**168**) screened for inhibitory potency against 2C19, or used as substrates or metabolite standards [103]. The findings illustrate the increased utility of benzbromarone analogues since they have now been adapted to act as 2C19 inhibitors. According to this study with benzbromarone and previous works on phenobarbital analogues and proton pump inhibitors, it is demonstrated that for 2C19, ligands with two hydrophobic regions separated by a polar group are the most complementary. Interestingly, high-affinity binding of benzbromarone ligands to 2C19 appears to be achieved without constraining substrate mobility within the enzyme.

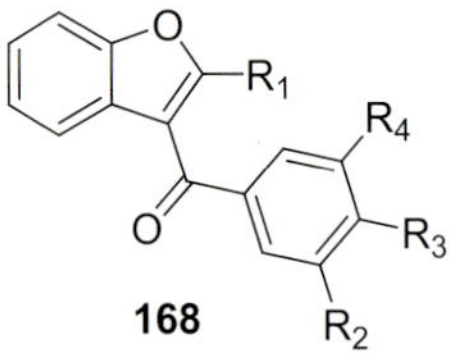

168

A series of benzofuran-2-yl-(phenyl)-3-pyridylmethanol derivatives have been prepared in good yields using an efficient one-step procedure. Additionally, to determine the effect of the benzene ring in benzofuran with respect to inhibitory activity, furan-2-yl-(phenyl)-3-pyridylmethanol derivatives have been synthesized in the meantime. The pyridylmethanol derivatives were all evaluated *in vitro* for inhibitory activity against aromatase (P 450AROM, CYP19), using human placental microsomes. The benzofuran-2-yl-(phenyl)-3-pyridylmethanol derivatives displayed good to moderate activity ($IC_{50} = 1.3–25.1\,\mu M$), which was either better than or comparable with aminoglutethimide ($IC_{50} = 18.5\,\mu M$) but lower than arimidex ($IC_{50} = 0.6\,\mu M$), with the 4-methoxyphenyl substituted derivative displaying optimum activity. Moreover, it shows the activity to reside with the (S)-enantiomer based on molecular modeling of the benzofuran-2-yl-(4-fluorophenyl)-3-pyridylmethanol derivatives. The essential role of the benzene ring of the benzofuran component for enzyme binding is demonstrated since the furan-2-yl-(phenyl)-3-pyridylmethanol derivatives were devoid of activity [104].

Histamine H_3 receptor antagonists are being developed to treat various neurological and cognitive disorders that may be ameliorated by enhancement of central neurotransmitter release. A nonimidazole, benzofuran ligand ABT-239 [4-(2-{2-[(2*R*)-2-methylpyrrolidinyl] ethyl}-benzofuran-5-yl)benzonitrile] (169) has been utilized in the *in vitro* pharmacological and *in vivo* pharmacokinetic profiles and compared with several previously described imidazole and nonimidazole H_3 receptor antagonists. The assay results demonstrate that ABT-239 is a selective, nonimidazole H_3 receptor antagonist/inverse agonist with similar high potency in both human and rat and favorable drug-like properties [105]. The potency and selectivity of this compound and of analogs from this class support the potential of H_3 receptor antagonists for the treatment of cognitive dysfunction [106].

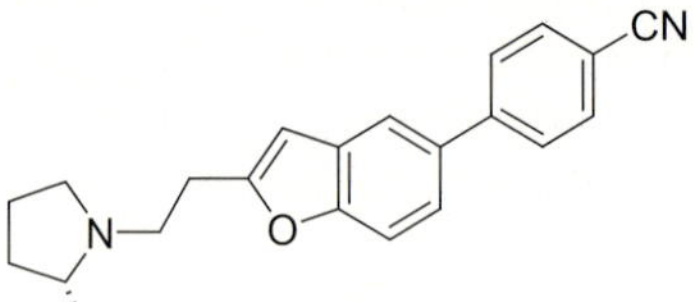

169 (ABT-239)

A series of 2-(4-hydroxyphenyl)benzofuran-5-ols with relatively lipophilic groups in the 7-position of the benzofuran has been synthesized for measurement of the affinity and selectivity for ERβ. Some analogs are active as potent and selective ERβ ligands. The structural modifications at the benzofuran 4-position as well as at the 3′-position of the 2-Ph group (e.g., **170**) further increase selectivity. Such modifications have lead to compounds with <10 nM potency and >100-fold selectivity for ERβ [107].

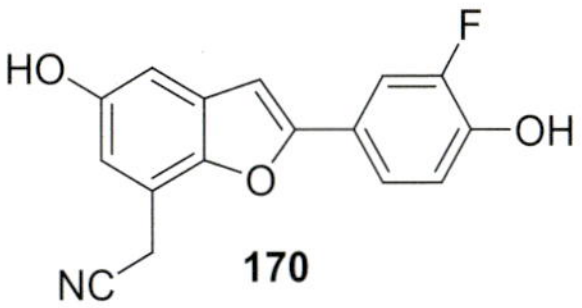

170

5-Chloro-3-methyl-2-acetylbenzofuran reacts with bromine in acetic acid leading to 5-chloro-3-methyl-2-bromoacetylbenzofuran, which then undergoes condensation with various substituted aromatic amines to afford 2-*N*-arylaminoacetyl-5-chloro-3-methylbenzofurans. These compounds have been screened for their antibacterial, antifungal, analgesic, antiinflammatory and diuretic activities and some hits were discovered [108].

A series of 1-(1-benzofuran-2-yl-ethylidene)-4-substituted thiosemicarbazides along with some derived ring systems, substituted-2,3-dihydro-thiazoles and thiazolidin-4-ones, have been synthesized and evaluated for their *in vitro* anti-HIV, anticancer, antibacterial and antifungal activities. Among the tested compounds, two produced a significant reduction the viral cytopathic effect (93.19% and 59.55%) at concentrations of $>2.0 \times 10^{-4}$ M and 2.5×10^{-5} M, respectively. One compound displayed moderate anti-HIV activity. Several compounds showed mild antifungal activity. However, no significant anticancer activity was discovered for the tested compounds [109].

7.5.2
Uses of Benzofuran in Material Science

New functionalized mono- and bis-benzo[*b*]furan derivatives which possess a CN, CHO, CH=CHPh, $CH{=}CPh_2$, or CH=CHCOOH group at C4 have been synthesized and developed as blue-light emitting materials [110]. Two benzo[*b*]furan nuclei in bis-benzo[*b*]furan derivatives have been connected by a divinylbenzene bridge. Bis-benzo[*b*]furan **171** was fabricated as a device with good volatility and thermal stability. It emitted blue light with brightness 53 430 cd m^{-2} (at 15.5 V) and a high maximum external quantum efficiency 3.75% (at 11 V) (Scheme 7.60).

The direct anodic oxidation of 2,3-benzofuran on stainless steel sheet in boron trifluoride di-Et etherate (BFEE) contained 10% poly(ethylene glycol) (PEG) with a molar mass of 400 (by vol.) affords a visible-light transparent high-quality substrate-supported poly(2,3-benzofuran) (PBF) film [111]. The oxidation potential of 2,3-benzofuran in this medium was measured to be only 1.0 V vs SCE, which is lower than that determined in acetonitrile + 0.1 M Bu_4NBF_4 (1.2 V vs SCE). Good electrochemical behavior and good thermal stability with a condense of 10^{-2} S cm^{-1}, are displayed for these PBF films, and the doping level of as-prepared PBF films was determined to be only 8.9%. The structure of the polymer has been studied by UV/Vis, IR spectroscopy and SEM.

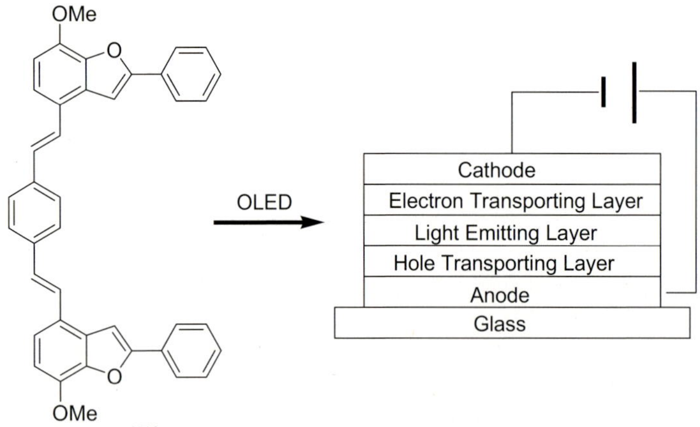

OMe
OMe
171

OLED

Cathode
Electron Transporting Layer
Light Emitting Layer
Hole Transporting Layer
Anode
Glass

Scheme 7.60

Yang has reported the synthesis and spectral properties of novel 4-benzofuranyl-1,8-naphthalimide derivatives [112]. A series of 4-benzofuranyl-*N*-alkyl-1,8-naphthalimides has been prepared from 4-ethynyl-*N*-alkyl-1,8-naphthalimides and substituted *o*-iodophenols catalyzed by a $Pd(PPh_3)_2Cl_2/CuI$ system under mild conditions. The absorption and fluorescence spectra of these benzofuran-1,8-naphthalimides have been recorded and the quantum yields are measured using quinine sulfate as the standard. The UV/Vis absorption spectra were in the range of 380–400 nm and the emission spectra were in the range of 500–520 nm.

A novel fluorescence active 12*H*-benzo[*e*]indolo[3,2-*b*]benzofuran and its derivatives (**172**) has been prepared from 6-R^1-2-naphthols in good yields. The synthesized compounds with planar geometry and extended conjugation exhibit excellent fluorescence properties [113].

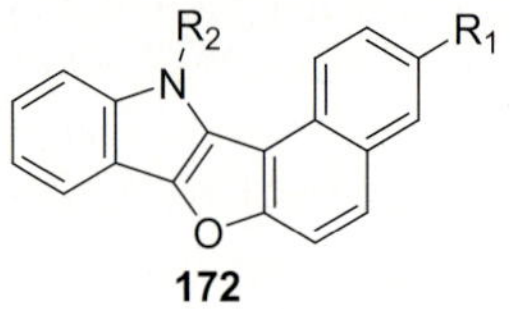

R_2 R_1
N
O
172

Four linear benzofuran trimers have been prepared and tested as materials for organic electroluminescence (OEL). The solubility, aggregation, and film-forming properties were modulated by *tert*-butyl and *n*-hexyl substituents on the benzofurans. Additionally, two *tert*-butyl groups prevented aggregation in the solid state, thus maintaining emission in the blue region of the visible spectrum. The OEL characteristics of the *tert*-butyl-substituted benzofuran trimer have been explored, and blue emission observed. The two-stage synthetic procedure employed for the preparation of these benzofuran trimers may be applied to a wide variety of benzofuran oligomer and polymer targets [114].

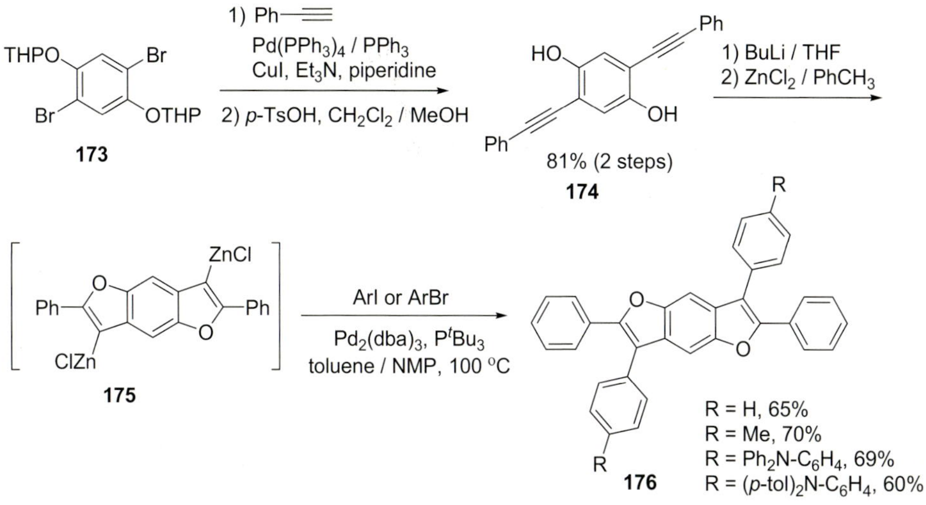

Scheme 7.61

Recently, Nakamura has described a facile route to 2,3,6,7-tetraarylbenzo[1,2-*b*:4,5-*b*]difurans (BDFs) (Scheme 7.61), which could be functioned as the hole-transporting material (HTM) in layered organic light-emitting diodes (OLEDs) [115]. The high performance of the compounds is primarily due to the BDF core itself, which is in sharp contrast to the fact that the biphenyl scaffold in R-NPD alone does not function as a HTM. They also found that there is a synergetic effect of the BDF core and the substituents. The physical properties can be improved by suitable functionalization. It can be expected that the BDF molecule will serve as a useful new molecular scaffold on which multiple functional groups can be attached to obtain new properties.

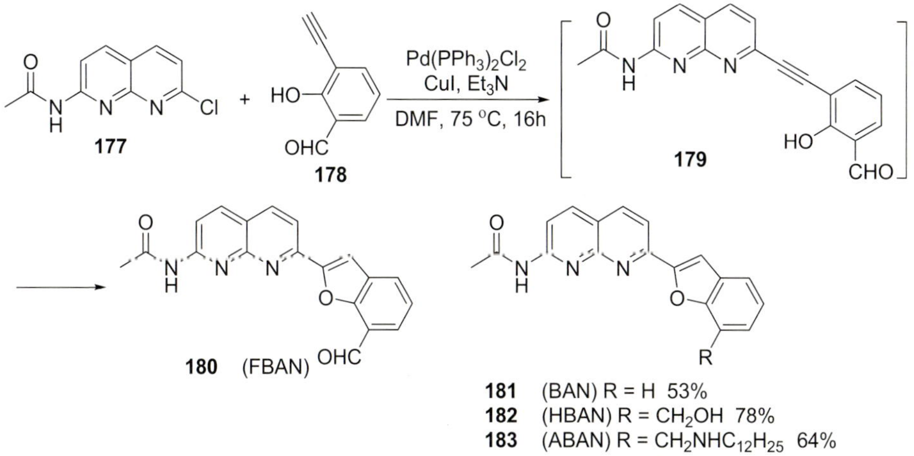

Scheme 7.62

Benzofuran-naphthyridine links show high-yield fluorescence with solvatochromic properties. For instance, after formation of fluorescent organic nanoparticles (FONs1.) of **ABAN** (**183**, Scheme 7.62), the photophysical properties (such as the spectral features and intensity) are remarkably different from those at the molecular level (solution) and in bulk material [116].

References

1 (a) For selected examples, see: Máximo, P., Lourenço, A., Feio, S.S., and Roseiro, J.C. (2002) *Journal of Natural Products*, **65**, 175; (b) Ohta, T., Maruyama, T., Nagahashi, M., Miyamoto, Y., Hosoi, S., Kiuchi, F., Yamazoe, Y., and Tsukamoto, S. (2002) *Tetrahedron*, **58**, 6631; (c) Song, K.-S. and Raskin, I. (2002) *Journal of Natural Products*, **65**, 76; (d) Sritularak, B., Likhitwitayawuid, K., Conrad, J., Vogler, B., Reeb, S., Klaiber, I., and Kraus, W. (2002) *Journal of Natural Products*, **65**, 589; (e) Lin, Y.-L., Chang, Y.-Y., Kuo, Y.-H., and Shiao, M.-S. (2002) *Journal of Natural Products*, **65**, 745; (f) Pacher, T., Seger, C., Engelmeier, D., Vajrodaya, S., Hofer, O., and Greger, H. (2002) *Journal of Natural Products*, **65**, 820; (g) Muriithi, M.W., Abraham, W.-R., Addae-Kyereme, J., Scowen, I., Croft, S.L., Gitu, P.M., Kendrick, H., Njagi, E.N.M., and Wright, C.W. (2002) *Journal of Natural Products*, **65**, 956; (h) Weng, J.-R., Yen, M.-H., and Lin, C.-N. (2002) *Helvetica Chimica Acta*, **85**, 847; (i) Iliya, I., Tanaka, T., Iinuma, M., Furusawa, M., Ali, Z., Nakaya, K., Murata, J., and Darnaedi, D. (2002) *Helvetica Chimica Acta*, **85**, 2394; (j) Yan, K.-X., Terashima, K., Takaya, K.-I., and Niwa, Y.M. (2002) *Tetrahedron*, **58**, 6931; (k) Cagniant, P. and Carniant, D. (1975) *Advances in Heterocyclic Chemistry*, **18**, 337; (l) Donnelly, D.M.X. and Meegan, M.J. (1984) in *Comprehensive Heterocyclic Chemistry, Vol. 4* (eds A.R. Katritzky and C.W. Rees), Pergamon Press, Oxford, pp. 657–712.

2 (a) Felder, C.C., Joyce, K.E., Briley, E.M., *et al.* (1998) *The Journal of Pharmacology and Experimental Therapeutics*, **284**, 291; (b) Yang, Z., Hon, P.M., Chui, K.Y., Chang, H.M., Lee, C.M., Cui, Y.X., Wong, H.N.C., Poon, C.D., and Fung, B.M. (1991) *Tetrahedron Letters*, **32**, 2061; (c) Carter, G.A., Chamberlain, K., and Wain, R.L. (1978) *The Annals of Applied Biology*, **88**, 57; (d) Ingham, J.L. and Dewick, P.M. (1978) *Phytochemistry*, **17**, 535; (e) Takasugi, M., Nagao, S., and Masamune, T. (1979) *Tetrahedron Letters*, **20**, 4675; (f) Davies, W. and Middleton, S. (1957) *Chemistry & Industry (London)*, 599; (g) McGarry, D.G., Regan, J.R., Volz, F.A., Hulme, C., Moriarty, K.J., Djuric, S.W., Souness, J.E., Miller, B.E., Travis, J.J., and Sweeney, D.M. (1999) *Bioorganic and Medicinal Chemistry*, **7**, 1131; (h) Hayakawa, I., Shioya, R., Agatsuma, T., Furukawa, H., Naruto, S., and Sugano, Y. (2004) *Bioorganic and Medicinal Chemistry Letters*, **14**, 455; (i) Fukai, T., Oku, Y., Hano, Y., and Terada, S. (2004) *Planta Medica*, **70**, 685; (j) Tsuji, E., Ando, K., Kunitomo, J.-i., Yamashita, M., Ohta, S., Kohno, S., and Ohishi, Y. (2003) *Organic Biomolecular Chemistry*, **1**, 3139.

3 (a) Example in material science: Hwu, J.R., Chuang, K.-S., Chuang, S.H., and Tsay, S.-C. (2005) *Organic Letters*, **7**, 1545, In Organic Synthesis: (b) Cagniant, P. and Cagniant, D., (1975) in *Advances in Heterocyclic Chemistry, Vol. 18* (eds A.R. Katritzky and A.J. Boulton), Academic Press New York, p. 337; (c) Katritzky, A.R. and Rees, C.W. (1984) *Comprehensive Heterocyclic Chemistry, Vol. 4*, Pergamon Press, Oxford, p. 531; (d) Katritzky, A.R. Rees, C.W., and Scriven, E.F.V. (1996) *Comprehensive Heterocyclic Chemistry II, Vol. 2*, Pergamon Press, Oxford, p. 259.

4 (a) McCallion, G.D. (1999) *Current Organic Chemistry*, **3**, 67; (b) Hou, X.-L., Yang, Z., and Wong, H.N.C. (2003)

Progress in Heterocyclic Chemistry, **15**, 167; (c) Dell, C.P. (2001) *Science of Synthesis*, **10**, 11; (d) Kadieva, M.G. and Oganesyan, E.T. (1997) *Chemistry of Heterocyclic Compounds*, **33**, 1245; (e) Reck, S. and Friedrichsen, W. (1997) *Progress in Heterocyclic Chemistry*, **9**, 117; (f) Friedrichsen, W. (1996) *Comprehensive Heterocyclic Chemistry II*, **2**, 351; (g) Hurst, D.T. (1997) *Rodd's Chemistry of Carbon Compounds*, 2nd edn, 4(Pt. A), Elsevier, Amsterdam, 283–335; (h) Reck, S. and Friedrichsen, W. (1996) *Progress in Heterocyclic Chemistry*, **8**, 121; (i) Friedrichsen, W. and Pagel, K. (1995) *Progress in Heterocyclic Chemistry*, **7**, 130; (j) Bird, C.W. (1994) *Progress in Heterocyclic Chemistry*, **6**, 129; (k) Boswell, D.E., Landis, P.S., Givens, E.N., and Venuto, P.B. (1968) *Industrial & Engineering Chemistry Product Research and Development*, **7**, 215; (l) Hou, X.-L., Yang, Z., and Wong, H.N.C. (2002) *Progress in Heterocyclic Chemistry*, **14**, 139; (m) Hou, X.-L., Yang, Z., and Wong, H.N.C. (2001) *Progress in Heterocyclic Chemistry*, **13**, 130; (n) Pang, J.-Y. and Xu, Z.-L. (2005) *Chinese Journal of Organic Chemistry*, **25**, 25.

5 (a) Schroeter, S., Stock, C., and Bach, T. (2005) *Tetrahedron*, **61**, 2245, and references cited therein; (b) Kao, C.-L. and Chern, J.-W. (2002) *The Journal of Organic Chemistry*, **67**, 6772. (c) Huang, N. T., Hussain, M., Malik, I., Villinger, A. and Langer, P. (2010) *Tetrahedron Letters*, 51, 2420.

6 (a) Examples of dehydrative cyclization of α-(phenoxy)alkyl ketones: Wright, J.B. (1960) *The Journal of Organic Chemistry*, **25**, 1867; (b) Royer, R., Bisagni, E., Hudry, C., Cheutin, A., and Desvoye, M.-L. (1963) *Bulletin de la Societe Chimique de France*, 1003; (c) Pene, C., Demerseman, P., Cheutin, A., and Royer, R. (1966) *Bulletin de la Societe Chimique de France*, 586; (d) Kawase, Y., Takata, S., and Hikishima, E. (1971) *Bulletin of the Chemical Society of Japan*, **44**, 749; (e) Horaguchi, T., Iwanami, H., Tanaka, T., Hasegawa, E., and Shimizu, T. (1991) *Journal of the Chemical Society. Chemical Communications*, 44.

7 (a) Examples of dehydration of *o*-hydroxybenzyl ketones under acidic conditions: Dams, R. and Whitaker, L. (1956) *Journal of the American Chemical Society*, **78**, 8; (b) Kalyanasundaram, M., Rajagopalan, K., and Swaminathan, S. (1980) *Tetrahedron Letters*, **21**, 4391.

8 (a) Examples of decarboxylation of *o*-acylphenoxyacetic acids or esters on treatment with a base: Muller, A., Meszaros, M., and Kormendy, K. (1954) *The Journal of Organic Chemistry*, **19**, 472; (b) Horaguchi, T., Tanemura, K., and Suzuki, T. (1988) *Journal of Heterocyclic Chemistry*, **25**, 39; (c) Horaguchi, T., Kobayashi, H., Miyazawa, K., Hasegawa, E., and Shimizu, T. (1990) *Journal of Heterocyclic Chemistry*, **27**, 935; (d) Boehm, T.L. and Showalter, H.D.H. (1996) *The Journal of Organic Chemistry*, **61**, 6498.

9 Examples of cyclofragmentation of oxiranes, prepared in three or four steps from the corresponding *o*-hydroxybenzophenones: Nicolaou, K.C., Snyder, S.A., Bigot, A., and Pfefferkorn, J.A. (2000) *Angewandte Chemie, International Edition*, **39**, 1093. 10.

10 (a) Examples of palladium(II)-catalyzed cyclization of arylacetylenes: Kondo, Y., Shiga, F., Murata, N., Sakamoto, T., and Yamanaka, H. (1994) *Tetrahedron*, **50**, 11803; (b) Arcadi, A., Cacchi, S., Rosario, M.D., Fabrizi, G., and Marinelli, F. (1996) *The Journal of Organic Chemistry*, **61**, 9280; (c) Cacchi, S., Fabrizi, G., and Moro, L. (1998) *Tetrahedron Letters*, **39**, 5101; (d) Monteiro, N., Arnold, A., and Blame, G. (1998) *Synlett*, 1111; (e) Cacchi, S., Fabrizi, G., and Goggiomani, A. (2002) *Heterocycles*, **56**, 613.

11 Fleming, I. (1976) *Frontier Orbitals and Organic Chemical Reactions*, John Wiley & Sons Inc., New York, p. 58.

12 (a) Abraham, R.J. and Reid, M. (2002) *Journal of the Chemical Society, Perkin Transactions 2*, 1081; (b) Black, P.J. and Heffernan, M.L. (1965) *Australian Journal of Chemistry*, **18**, 353.

13 (a) Knoelker, H.-J. and Reddy, K.R. (2002) *Chemical Reviews*, **102**, 4303; (b) Wu, T.-S., Huang, S.-C., and Wu, P.-L. (1997) *Heterocycles*, **45**, 969.

14 (a) Balasubramanian, B.N., St. Laurent, D.R., Saulnier, M.G., Long, B.H., Bachand, C., Beaulieu, F., Clarke, W., Deshpande, M., Eummer, J., Fairchild, C.R., Frennesson, D.B., Kramer, R., Lee, F.Y., Mahler, M., Martel, A., Naidu, B.N., Rose, W.C., Russell, J., Ruediger, E., Solomon, C., Stoffan, K.M., Wong, H., Zimmermann, K., and Vyas, D.M. (2004) *Journal of Medicinal Chemistry*, **47**, 1609; (b) Hudkins, R.L., Diebold, J.L., Angeles, T.S., and Knight, E. (1997) *Journal of Medicinal Chemistry*, **40**, 2994.

15 (a) Froehner, W., Krahl, M.P., Reddy, K.R., and Knoelker, H.-J. (2004) *Heterocycles*, **63**, 2393; (b) Knoelker, H.-J. and Krahl, M.P. (2004) *Synlett*, 528.

16 (a) Arcadi, A., Cacchi, S., and Marinelli, F. (1992) *Tetrahedron Letters*, **333**, 3915; (b) Cacchi, S., Carnicelli, V., and Marinelli, F. (1994) *Journal of Organometallic Chemistry*, **475**, 289; (c) Cacchi, S., Fabrizi, G., and Moro, L. (1998) *Tetrahedron Letters*, **63**, 5306; (d) Larock, R.C., Yum, E.K., and Refvik, M.D. (1998) *The Journal of Organic Chemistry*, **63**, 7652; (e) Cacchi, S. (1999) *Journal of Organometallic Chemistry*, **576**, 42; (f) Nan, Y., Miao, H., and Yang, Z. (2000) *Organic Letters*, **2**, 297; (g) Bellina, F., Biagetti, M., Carpita, A., and Rossi, R. (2001) *Tetrahedron*, **57**, 2857; (h) Flynn, B.L., Verdier-Pinard, P., and Hamel, E. (2001) *Organic Letters*, **3**, 2973; (i) Roesch, K.R. and Larock, R.C. (2001) *The Journal of Organic Chemistry*, **66**, 412; (j) Arcadi, A., Cacchi, S., Giuseppe, S.D., Fabrizi, G., and Marinelli, F. (2002) *Synlett*, 453; (k) Yue, D. and Larock, R.C. (2002) *The Journal of Organic Chemistry*, **67**, 1905; (l) Hu, Y.-H., Zhang, Y., Yang, Z., and Fathi, R. (2002) *The Journal of Organic Chemistry*, **67**, 2365; (m) Cacchi, S., Fabrizi, G., and Goggiomani, A. (2002) *Heterocycles*, **56**, 613; (n) Cacchi, S. (1996) *Pure and Applied Chemistry*, **68**, 45; (o) Sakamoto, T., Kondo, Y., and Yamanaka, H. (1988) *Heterocycles*, **27**, 2225.

17 (a) Arcadi, A., Cacchi, S., Rosario, M.D., Fabrizi, G., and Marinelli, F. (1996) *The Journal of Organic Chemistry*, **61**, 9280; (b) Arcadi, A., Cacchi, S., Fabrizi, G., Marinelli, F., and Moro, L. (1999) *Synlett*, 1432; (c) Anacardio, R., Arcadi, A., D'Anniballe, G., and Marinelli, F. (1995) *Synthesis*, 831; (d) Arcadi, A., Cacchi, S., Giuseppe, S.D., Fabrizi, G., and Marinelli, F. (2002) *Organic Letters*, **4**, 2409.

18 (a) Chaplin, J.H. and Flynn, B.L. (2001) *Chemical Communications*, 1594; (b) Flynn, B.L., Hamel, E., and Jung, M.K. (2002) *Journal of Medicinal Chemistry*, **45**, 2670.

19 Hu, Y., Nawoschik, K.J., Liao, Y., Ma, J., Fathi, R., and Yang, Z. (2004) *The Journal of Organic Chemistry*, **69**, 2235.

20 (a) Nan, Y., Miao, H., and Yang, Z. (2000) *Organic Letters*, **2**, 297; (b) Hu, Y.-H. and Yang, Z. (2001) *Organic Letters*, **3**, 1387; (c) Liao, Y., Fathi, R., Reitman, M., Zhang, Y., and Yang, Z. (2001) *Tetrahedron Letters*, **42**, 1815; (d) Hu, Y.-H., Zhang, Y., Yang, Z., and Fathi, R. (2002) *The Journal of Organic Chemistry*, **67**, 2365; (e) Liao, Y., Reitman, M., Zhang, Y., Fathi, R., and Yang, Z. (2002) *Organic Letters*, **4**, 2607; (f) Liao, Y., Fathi, R., and Yang, Z. (2003) *Organic Letters*, **5**, 909; (g) Liao, Y., Fathi, R., and Yang, Z. (2003) *Journal of Combinatorial Chemistry*, **5**, 79; (h) Liao, Y., Hu, Y.-H., Wu, J., Zhu, Q., Donovan, M., Fathi, R., and Yang, Z. (2003) *Current Medicinal Chemistry*, **10**, 2285; (i) Liao, Y., Smith, J., Fathi, R., and Yang, Z. (2005) *Organic Letters*, **7**, 2707.

21 Liu, Z. and Larock, R.C. (2004) *Organic Letters*, **6**, 3739.

22 Zhang, H., Ferreira, E.M., and Stoltz, B.M. (2004) *Angewandte Chemie, International Edition*, **43**, 6144.

23 Xie, X., Chen, B., Lu, J., Han, J., She, X., and Pan, X. (2004) *Tetrahedron Letters*, **45**, 6235.

24 Hughes, C.C. and Trauner, D. (2004) *Tetrahedron*, **60**, 9657.

25 Rozhkov, R.V. and Larock, R.C. (2004) *Tetrahedron Letters*, **45**, 911.

26 Kerr, D.J., Willis, A.C., and Flynn, B.L. (2004) *Organic Letters*, **6**, 457.

27 Willis, M.C., Taylor, D., and Gillmore, A.T. (2004) *Organic Letters*, **6**, 4755.

28 Konno, T., Chae, J., Ishihara, T., and Yamanaka, H. (2004) *Tetrahedron*, **60**, 11695.

29 Kurosawa, W., Kobayashi, H., Kan, T., and Fukuyama, T. (2004) *Tetrahedron*, **60**, 9615.

30 Zhang, Y., Candelaria, D., and Herndon, J.W. (2005) *Tetrahedron Letters*, **46**, 2211.

31 Zhao, B. and Lu, X. (2006) *Organic Letters*, **8**, 5987.

32 Anderson, K.W., Ikawa, T., Tundel, R.E., and Buchwald, S.L. (2006) *Journal of the American Chemical Society*, **128**, 10694.

33 Nagamochi, M., Fang, Y.-Q., and Lautens, M. (2007) *Organic Letters*, **9**, 2955.

34 Oppenheimer, J., Johnson, W.L., Tracey, M.R., Hsung, R.P., Yao, P.-Y., Liu, R., and Zhao, K. (2007) *Organic Letters*, **9**, 2361.

35 Trost, B.M. and McClory, A. (2007) *Angewandte Chemie, International Edition*, **46**, 2074.

36 Willis, M.C., Taylora, D., and Gillmore, A.T. (2006) *Tetrahedron*, **62**, 11513.

37 Liang, Y., Tang, S., Zhang, X.-D., Mao, L.-Q., Xie, Y.-X., and Li, J.-H. (2006) *Organic Letters*, **8**, 3017.

38 Nakamura, M., Ilies, L., Otsubo, S., and Nakamura, E. (2006) *Organic Letters*, **8**, 2803.

39 Bellur, E., Freifeld, I., and Langer, P. (2005) *Tetrahedron Letters*, **46**, 2185.

40 Ogata, T., Okamoto, I., Kotani, E., and Takeya, T. (2004) *Tetrahedron*, **60**, 3941.

41 Takeya, T., Doi, H., Ogata, T., Otsuka, T., Okamoto, I., and Kotani, E. (2004) *Tetrahedron*, **60**, 6295.

42 Bowman, W.R., Krintel, S.L., and Schilling, M.B. (2004) *Organic and Biomolecular Chemistry*, **2**, 585.

43 Vaillard, S.E., Postigo, A., and Rossi, R.A. (2004) *The Journal of Organic Chemistry*, **69**, 2037.

44 Nematollahi, D., Habibi, D., Rahmati, M., and Fafiee, M. (2004) *The Journal of Organic Chemistry*, **69**, 2637.

45 Kao, C.-L. and Chern, J.-W. (2002) *The Journal of Organic Chemistry*, **67**, 6772.

46 Sako, M., Hosokawa, H., Ito, T., and Iinuma, M. (2004) *The Journal of Organic Chemistry*, **69**, 2598.

47 Majumdar, K.C. and Chattopadhyay, S.K. (2004) *Tetrahedron Letters*, **45**, 6871.

48 Majumdar, K.C. and Mukhopadhyaya, P.P. (2004) *Synthesis*, 1864.

49 Bowman, W.R., Krintel, S.L., and Schilling, M.B. (2004) *Organic and Biomolecular Chemistry*, **2**, 585.

50 Vaillard, S.E., Postigo, A., and Rossi, R.A. (2004) *The Journal of Organic Chemistry*, **69**, 2037.

51 Kim, K.-O. and Tae, J. (2005) *Synthesis*, 387.

52 Tsuritani, T., Shinokubo, H., and Oshima, K. (2001) *Organic Letters*, **3**, 2709.

53 Grimshaw, J. and Thompson, N. (1987) *Journal of the Chemical Society, Chemical Communications*, 240.

54 Katritzky, A.R., Ji, Y., Fang, Y., and Prakash, I. (2001) *The Journal of Organic Chemistry*, **6**, 5613.

55 Novák, Z., Timári, G., and Kotschy, A. (2003) *Tetrahedron*, **59**, 7509.

56 Pal, M., Subramanian, V., and Yeleswarapu, K.R. (2003) *Tetrahedron Letters*, **4**, 8221.

57 Uozumi, Y. and Kobayashi, Y. (2003) *Heterocycles*, **59**, 71.

58 Lin, Y.-L., Kuo, H.-S., Wang, Y.-W., and Huang, S.-T. (2003) *Tetrahedron*, **59**, 1277.

59 Wang, L., Li, P., and Zhang, Y. (2004) *Chemical Communications*, 514.

60 Anderson, S., Taylor, P.N., and Verschoor, G.L.B. (2004) *Chemistry - A European Journal*, **10**, 518.

61 Knölker, H.J. and Krahl, M.P. (2004) *Synlett*, 528.

62 Bhoga, U., Mali, R.S., and Adapa, S.R. (2004) *Tetrahedron Letters*, **45**, 9483.

63 Hirai, Y., Doe, M., Kinoshita, T., and Morimoto, Y. (2004) *Chemistry Letters*, 136.

64 Yamashita, M., Ono, Y., and Tawada, H. (2004) *Tetrahedron*, **60**, 2843.

65 del Carmen Cruz, M. and Tamariz, J. (2004) *Tetrahedron Letters*, **45**, 2377.

66 Barluenga, J., V1zquez-Villa, H.,Merino, I., Ballesteros, A., and Gonz1lez, J.M. (2006) *Chemistry - A European Journal*, **12**, 5790.

67 Hashimoto, A., Przybyl, A.K., Linders, J.T.M., Kodato, S., Tian, X., Deschamps, J.R., George, C., Flippen-Anderson, J.L., Jacobson, A.E., and Rice, K.C. (2004) *The Journal of Organic Chemistry*, **69**, 5322.

68 Al-Maharik, N. and Botting, N.P. (2004) *Tetrahedron*, **60**, 1637.
69 Hellwinkel, D. and Goeke, K. (1995) *Synthesis*, 1135.
70 Ishikawa, T., Miyahara, T., Asakura, M., Higuchi, S., Miyauchi, Y., and Saito, S. (2005) *Organic Letters*, **7**, 1211.
71 (a) Wagner, A.F. (1962) US 3068265 [*Chem. Abstr.* (1963) 58, 12514f]. (b) Wagner, A.F., Wilson, A.N., and Folkers, K. (1959) *Journal of the American Chemical Society*, **81**, 5441.
72 Johnson, F. and Subramanian, R. (1986) *The Journal of Organic Chemistry*, **51**, 5040.
73 Banerji, A. and Nayak, S.K. (1990) *Journal of the Chemical Society, Chemical Communications*, 150.
74 Clerici, A. and Porta, O. (1990) *The Journal of Organic Chemistry*, **55**, 1240.
75 Jha, A.K., Sharma, P.C., Maulik, P.R., Yadav, U., and Hajela, K. (2004) *Indian Journal of Chemistry, Section B: Organic Chemistry Including Medicinal Chemistry*, **43B**, 1341.
76 Kwiecien, H., Witczak, M., and Rosiak, A. (2004) *Polish Journal of Chemistry*, **78**, 249.
77 De, L., Lidia Giacomelli, G., and Nieddu, G. (2007) *The Journal of Organic Chemistry*, **72**, 3955.
78 Sanz, R., Miguel, D., Martinez, A., and Perez, A. (2006) *The Journal of Organic Chemistry*, **71**, 4024.
79 Kumar, M.P. and Liu, R.-S. (2006) *The Journal of Organic Chemistry*, **71**, 4951.
80 Asao, N. and Aikawa, H. (2006) *The Journal of Organic Chemistry*, **71**, 5249.
81 Carril, M., SanMartin, R., Tellitu, I., and Dominguez, E. (2006) *Organic Letters*, **8**, 1467.
82 Rao, M.L.N., Awasthi, D.K., and Banerjee, D. (2007) *Tetrahedron Letters*, **48**, 431.
83 DiMauro, E.F. and Vitullo, J.R. (2006) *The Journal of Organic Chemistry*, **71**, 3959.
84 Otterlo, V., Willem, A.L., Ngidi, E.L., and de Koning, C.B. (2003) *Tetrahedron Letters*, **44**, 6483.
85 Fujimura, O., Fu, G.C., and Grubbs, R.H. (1994) *The Journal of Organic Chemistry*, **59**, 4029.
86 (a) McKittrick, B.A., Scannell, R.T., and Stevenson, R. (1982) *Journal of the Chemical Society-Perkin Transactions*, **1**, 3017; (b) Komatsu, M., Yokoe, I., and Shirataki, Y. (1978) *Chemical & Pharmaceutical Bulletin*, **26**, 1274.
87 Terada, Y., Arisawa, M., and Nishida, A. (2004) *Angewandte Chemie, International Edition*, **43**, 4063.
88 Tsai, T.-W., Wang, E.-C., Huang, K.-S., Li, S.-R., Wang, Y.-F., Lin, Y.-L., and Chen, Y.-H. (2004) *Heterocycles*, **63**, 1771.
89 Whitehead, A., Moore, J.D., and Hanson, P.R. (2003) *Tetrahedron Letters*, **44**, 4275.
90 RamaRao, V.N.S., Venkat Reddy, G., Maitraie, D., Ravikanth, S., Yadla, R., Narsaiah, B., and Shanthan Rao, P. (2004) *Tetrahedron*, **60**, 12231.
91 Azevedo, M.S., Alves, G.B.C., Cardoso, J.N., Lopes, R.S.C., and Lopes, C.C. (2004) *Synthesis*, 1262.
92 Caruana, P.A. and Frontier, A.J. (2004) *Tetrahedron*, **60**, 10921.
93 Miyata, O., Takeda, N., and Naito, T. (2004) *Organic Letters*, **6**, 1761.
94 Thede, K., Diedrichs, N., and Ragot, J.P. (2004) *Organic Letters*, **6**, 4595.
95 Horaguchi, T. (1999) *Trends in Heterocyclic Chemistry*, **6**, 1.
96 Dixit, M., Sharon, A., Maulik, P.R., and Goel, A. (2006) *Synlett*, 1497.
97 McKiernan, G.J. and Hartley, R.C. (2003) *Organic Letters*, **5**, 4389.
98 Liao, Y., Reitman, M., Zhang, Y., Fathi, R., and Yang, Z. (2002) *Organic Letters*, **4**, 2607.
99 Liao, Y., Fathi, R., and Yang, Z. (2003) *Organic Letters*, **5**, 909.
100 Liao, Y., Fathi, R., and Yang, Z. (2003) *Journal of Combinatorial Chemistry*, **5**, 79.
101 Hu, Y., Nawoschik, K.J., Liao, Y., Ma, J., Fathi, R., and Yang, Z. (2004) *The Journal of Organic Chemistry*, **69**, 2235.
102 Cironi, P., Tulla-Puche, J., Barany, G., Albericio, F., and Alvarez, M. (2004) *Organic Letters*, **6**, 1405.
103 Locuson, C.W., II, Suzuki, H., Rettie, A.E., and Jones, J.P. (2004) *Journal of Medicinal Chemistry*, **47**, 6768.
104 Saberi, M.R., Shah, K., and Simons, C. (2005) *Journal of Enzyme Inhibition and Medicinal Chemistry*, **20**, 135.

105 Esbenshade, T.A., Fox, G.B., Krueger, K.M., Miller, T.R., Kang, C.H., Denny, L.I., Witte, D.G., Yao, B.B., Pan, L., Wetter, J., Marsh, K., Bennani, Y.L., Cowart, M.D., Sullivan, J.P., and Hancock, A.A. (2005) *Journal of Pharmacology and Experimental Therapeutics*, **313**, 165.

106 Cowart, M., Faghih, R., Curtis, M.P., Gfesser, G.A., Bennani, Y.L., Black, L.A., Pan, L., Marsh, K.C., Sullivan, J.P., Esbenshade, T.A., Fox, G.B., and Hancock, A.A. (2005) *Journal of Medicinal Chemistry*, **48**, 38.

107 Collini, M.D., Kaufman, D.H., Manas, E.S., Harris, H.A., Henderson, R.A., Xu, Z.B., Unwalla, R.J., and Miller, C.P. (2004) *Bioorganic & Medicinal Chemistry Letters*, **14**, 4925.

108 Basawaraj, R., Parameshwarappa, G., and Sangapure, S.S. (2006) *Indian Journal of Heterocyclic Chemistry*, **16**, 75.

109 Rida, S.M., El-Hawash, S.A.M., Fahmy, H.T.Y., Hazza, A.A., and El-Meligy, M.M.M. (2006) *Archives of Pharmacal Research*, **29**, 16.

110 Hwu, J.R., Chuang, K.-S., Chuang, S.H., and Tsay, S.-C. (2005) *Organic Letters*, **7**, 1545.

111 Xu, J., Nie, G., Zhang, S., Han, X., Pu, S., Shen, L., and Xiao, Q. (2005) *European Polymer Journal*, **41**, 1654.

112 Yang, J.-X., Wang, X.-L., Tu, So., and Xu, L.-H. (2005) *Dyes and Pigments*, **67**, 27.

113 Karnik, A.V. and Upadhyay, S.P. (2004) *Indian Journal of Chemistry, Section B: Organic Chemistry Including Medicinal Chemistry*, **43B**, 1345.

114 Anderson, S., Taylor, P.N., and Verschoor, G.L.B. (2004) *Chemistry - A European Journal*, **10**, 518.

115 Tsuji, H., Mitsui, C., Ilies, L., Sato, Y., and Nakamura, E. (2007) *Journal of the American Chemical Society*, **129**, 11902.

116 Sun, Y.-Y., Liao, J.-H., Fang, J.-M., Chou, P.-T., Shen, C.-H., Hsu, C.W., and Chen, L.-C. (2006) *Organic Letters*, **8**, 3713.